2020 22nd European Conference on Power Electronics and Applications (EPE'20 ECCE Europe)

Lyon, France
7-11 September 2020

Pages 1373-2058

IEEE Catalog Number: CFP20850-POD
ISBN: 978-1-7281-9807-1

Copyright © 2020, EPE Association
All Rights Reserved

*** *This is a print representation of what appears in the IEEE Digital*
Library. Some format issues inherent in the e-media version may also
appear in this print version.

IEEE Catalog Number: CFP20850-POD
ISBN (Print-On-Demand):   978-1-7281-9807-1
ISBN (Online):   978-9-0758-1536-8

Additional Copies of This Publication Are Available From:

Curran Associates, Inc
57 Morehouse Lane
Red Hook, NY 12571 USA
Phone: (845) 758-0400
Fax: (845) 758-2633
E-mail: curran@proceedings.com
Web: www.proceedings.com

2020 22nd European Conference on Power Electronics and Applications (EPE'20 ECCE Europe)

Lyon, France
7-11 September 2020

Pages 1373-2058

IEEE Catalog Number: CFP20850-POD
ISBN: 978-1-7281-9807-1

TABLE OF CONTENTS

VALIDATION OF THERMAL STRESS MODELING IN PV INVERTERS UNDER MISSION PROFILE OPERATION 1

    Ariya Sangwongwanich, Huai Wang, Frede Blaabjerg

ON THE LIMITATIONS OF USING A LTI MODELLING APPROACH FOR CONTROL TUNING OF VSC-HVDC SYSTEMS 9

    Pablo Briff, Julián Freytes, Guillaume De-Preville, Jiaqi Li, Omar Jasim

A VOLTAGE CONTROL METHOD FOR POWER DISTRIBUTION LINES UTILIZING DISPERSED CUSTOMER RESOURCES 19

    Hiroki Ishihara, Kaho Nada, Miwako Tanaka, Sadayuki Inoue, Akiko Kuwata, Tomihiro Takano

PERFORMANCE COMPARISON BETWEEN SIC AND SI INVERTER MODULES IN AN ELECTRICAL VARIABLE TRANSMISSION APPLICATION 27

    Mauricio Dalla Vecchia, Simon Ravyts, Florian Verbelen, Jeroen Tant, Peter Sergeant, Johan Driesen

SEAMLESS INTEGRATION OF FEEDFORWARD AND FEEDBACK CONTROL OF BALANCE OF ARM CAPACITOR VOLTAGES IN STATCOMS BASED ON CHAIN LINKS OF H BRIDGE MODULES 37

    D. Basic, N. Lapassat

ASYNCHRONIZED ELECTROMECHANICAL CONVERTER IN THE ELECTRICAL SUPPLY SYSTEM OF POWERFUL ENERGY CONSUMERS... 47

    Aleksey G. Vorontsov, Mikhail V. Pronin, Anastasiia D. Stotckaia, Vasiliy V. Glushakov, Pavel V. Sokur

SYMMETRIC AND ASYMMETRIC OPERATING MODES OF HYBRID CASCADE FREQUENCY CONVERTERS 56

    Aleksey G. Vorontsov, Vasiliy V. Glushakov, Mikhail V. Pronin, Anastasiia D. Stotckaia

SYSTEM FREQUENCY DYNAMIC RESPONSE OF A NOVEL, SELF-SYNCHRONIZING INVERTER IN A HIGH RENEWABLE PENETRATION GRID 65

    Christian Perenyi, Moath Alqatamin, Thibaut Harzig, Michael McIntyre, Brandon M. Grainger

ROTOR POSITION ESTIMATION WITH HALL-EFFECT SENSORS IN BEARINGLESS DRIVES 75

    Patricio Peralta, Jacopo Leo, Yves Perriard

NON-UNIT ROCOV SCHEME FOR PROTECTION OF MULTI-TERMINAL HVDC SYSTEMS.............. 85

    María José Pérez-Molina, Pablo Eguia, Marene Larruskain, Garikoitz Buigues, Esther Torres

MODELLING OF CONVERTER SYSTEMS PARALLELED VIA INTERPHASE TRANSFORMERS IN CYCLIC CASCADE TOPOLOGY AND OPTIMIZATION OF PWM CARRIER SHIFTS 95

    D. Basic, H. Baërd, S. Siala

MEASUREMENT AND CALCULATION METHOD OF WIRELESS POWER TRANSFER COIL EQUIVALENT SERIES RESISTANCE UNDER THE VEHICLE....................................... 105

    Norihito Kimura, Hiroaki Yuasa

DESIGN OF A CIRCUMSCRIBING POLYGON WIDE BANDGAP BASED INTEGRATED MODULAR MOTOR DRIVE TOPOLOGY WITH THERMALLY DECOUPLED WINDINGS AND POWER CONVERTERS 115

Abdalla Hussein Mohamed, Hendrik Vansompel, Peter Sergeant

LIMITS OF ENHANCED DESATURATION DETECTION METHOD WITH ADAPTIVE BLANKING FOR GAN HEMTS 124

Jan Schmitz, Markus Meißner, Steffen Bernet

CURRENT CONTROL OF A GRID-CONNECTED SINGLE-PHASE VOLTAGE-SOURCE INVERTER WITH LCL FILTER 134

Alfonso Parreño Torres, Fco. Javier López-Alcolea, Pedro Roncero-Sánchez, Javier Vázquez, Emilio J. Molina-Martínez, Felix García-Torres

FOUR SWITCH BUCK/BOOST CONVERTER FOR DC MICROGRID APPLICATIONS 143

Matthias Schulz, Nico Schleippmann, Kilian Gosses, Bernd Wunder, Martin März

STABILITY INVESTIGATION OF THREE-PHASE GRID-TIED PV INVERTERS WITH IMPEDANCE-BASED METHOD 153

Zhiqing Yang, Wanchao Gou, Xian Luo, Chirag Shah, Nurhan Rizqy Averous, Rik W. De Doncker

STABILITY INVESTIGATION OF LARGE-SCALE PV PARKS WITH EIGENVALUE-BASED METHOD 163

Zhiqing Yang, Christian Bendfeld, Jin Qiang, Benedict Mortimer, Rik W. De Doncker

COMPACT CORE LOSS MODEL BASED ON AN EFFECTIVE FREQUENCY FOR ARBITRARY CORE EXCITATIONS INCLUDING DC-BIAS 173

Erika Stenglein, Manfred Albach, Thomas Dürbaum

ASSESSMENT OF AGING AND PERFORMANCE DEGRADATION OF SUPERCAPACITORS INTEGRATED INTO A MODULAR MULTILEVEL CONVERTER 183

F. Errigo, L. Chédot, F. Morel, P. Venet, A. Sari, A. Hijazi, R. A. Peña

SEPARATION OF MAGNETIC FLUX DENSITY TRAJECTORIES INTO SUBLOOPS FOR THE PREDICTION OF HYSTERESIS LOSS 193

Erika Stenglein, Manfred Albach, Thomas Dürbaum

INFLUENCE OF GENERALIZED DISCONTINUOUS PULSE WIDTH MODULATION (GDPWM) ON THE DC-LINK CURRENT AND VOLTAGE RIPPLE IN BATTERY-FED PWM INVERTER SYSTEMS 203

Panagiotis Mantzanas, Alexander Bucher, Daniel Kuebrich, Alexander Pawellek, Christian Hasenohr, Harald Hofmann, Thomas Duerbaum

AUTOMATED DESIGN METHOD FOR SINE WAVE FILTERS IN MOTOR DRIVE APPLICATIONS WITH SIC-INVERTERS 213

Thorben Schobre, Regine Mallwitz

A SYMMETRICAL BOOST CONVERTER WITH REDUCED COMMON-MODE LEAKAGE CURRENTS FOR EV APPLICATIONS 223

Caniggia Viana, Netan Yakop, Damien Frost, Peter Lehn

MODELING AND ANALYSIS OF CONDUCTED EMI ON FLYBACK CONVERTER USING POWER MANAGEMENT IC WITH CHAOTIC SUPPRESSION EMI 231

Diao Jiaqi, Yang Ru, Liu Zuolian, Yang Hong, Jie Hai

HIGH PERFORMANCE DRIVE INVERTER FOR AN ELECTRIC TURBO COMPRESSOR IN FUEL CELL APPLICATIONS 241
  N. Langmaack, G. Tareilus, R. Mallwitz

DEVELOPMENT OF AN ALGORITHM FOR THE AUTOMATION OF THE MODELLING PROCESS OF POWER CONVERTERS 251
  Jon Anzola, Iosu Aizpuru, Asier Arruti

A NOVEL FULLY DISTRIBUTED COST OPTIMAL CONTROL METHOD FOR DC MICROGRID 260
  Qingping Xia, Hua Han, Yao Liu, Zhangjie Liu, Yao Sun, Mei Su

MEASUREMENT OF DYNAMIC ON-STATE RESISTANCE OF HIGH-VOLTAGE GAN-HEMTS UNDER REAL APPLICATION CONDITIONS 266
  Benedikt Kohlhepp, Carsten Kuring, Stefan Peller, Daniel Kübrich

ANALYSIS OF DC-SIDE FAULT RESPONSE OF MMCS WITH CONTROLLED FAULT BLOCKING CAPABILITY FOR DIFFERENT TRANSMISSION LINE TYPES 276
  Willem Leterme, Paul D. Judge, Tim C. Green

A HYBRID SERIES-PARALLEL MICROGRID AND ITS LOW-DEPENDENT COMMUNICATION CONTROL 285
  Lang Li, Yao Sun, Hua Han, Mei Su

ADAPTIVE VOLTAGE CONTROL OF ISLANDED RES-BASED RESIDENTIAL MICROGRID WITH INTEGRATED FLYWHEEL/BATTERY HYBRID ENERGY STORAGE SYSTEM 292
  Linda Barelli, Gianni Bidini, Ermanno Cardelli, Dana-Alexandra Ciupageanu, Andrea Ottaviano, Dario Pelosi, Simone Castellini, Gheorghe Lazaroiu

AN IMPROVED λ-CONSENSUS CONTROL METHOD FOR DC MICROGRIDS 302
  Siqi Fu, Yao Sun, Zhangjie Liu, Hua Han, Mei Su

DECREASE OF POWER ELECTRONIC SWITCHING LOSSES USING VARIABLE SWITCHING EVENTS 307
  Hannes Ramm, Michael Homann, Torben A. Schulze, Faical Turki, Heiko Rabba

OPTIMIZATION OF MEDIUM-FREQUENCY TRANSFORMERS WITH LARGE CAPACITY AND HIGH INSULATION REQUIREMENT 317
  Xuan Guo, Chi Li, Zedong Zheng, Yongdong Li

IMPROVED SOC BALANCING AND ACTIVE POWER SHARING CONTROL METHOD IN HIGHLY RESISTIVE LINE MICROGRID 326
  Yuanhao Zhu, Hua Han, Guangze Shi, Zhangjie Liu, Yao Sun, Mei Su

TECHNO-ECONOMIC ANALYSIS OF SECOND-LIFE LITHIUM-ION BATTERIES INTEGRATION IN MICROGRIDS 332
  Camille Birou, Xavier Roboam, Hugo Radet, Fabien Lacressonnière

DESIGN, MODELLING, AND TEST OF A SOLID-STATE MAIN BREAKER FOR HYBRID DC CIRCUIT BREAKER 342
  Jiawen Xi, Xiaoze Pei, Xianwu Zeng, Liyong Niu

MODEL PREDICTIVE CONTROL FOR THREE-PHASE SPLIT-SOURCE INVERTER 352
  Youssuf Elthokaby, Islam Mohamed, Naser Abdel-Rahim

HARDWARE IMPLEMENTATION STUDY OF VARIABLE SPEED WIND-TURBINE-DFIG IN STAND-ALONE MODE 362
Fayssal Amrane, Bruno Francois, Azeddine Chaiba

INFLUENCE OF WIRE-BONDING LAYOUT ON RELIABILITY IN IGBT MODULE 370
Lubin Han, Lin Liang, Wei Xin, Fang Luo

RAIL POTENTIAL CALCULATION MODEL FOR DC RAILWAY POWER SUPPLY EQUIPPED WITH VOLTAGE LIMITING DEVICE 377
Shota Kimura, Tsutomu Miyauchi, Kenji Oguma, Hirotaka Takahashi, Keiko Teramura

HOMOGENIZATION OF CURRENT DISTRIBUTION IN PARALLEL CONNECTION OF INTERLEAVED WINDING LAYERS OF HIGH-FREQUENCY TRANSFORMERS BY OPTIMIZING DISTANCE BETWEEN WINDING LAYERS 386
Ryo Murata, Tomohide Shirakawa, Kazuhiro Umetani, Eiji Hiraki, Hiroto Mizutani, Takaaki Takahara, Osamu Mori

REAL-TIME PARAMETERS IDENTIFICATION OF LITHIUM-ION BATTERIES MODEL TO IMPROVE THE HIERARCHICAL MODEL PREDICTIVE CONTROL OF BUILDING MICROGRIDS 396
Daniela Yassuda Yamashita, Ionel Vechiu, Jean-Paul Gaubert

IMPACT OF DC FAULT BLOCKING CAPABILITY ON THE SIZING OF THE DC-DC MODULAR MULTILEVEL CONVERTER 406
J. D. Paez, F. Morel, S. Bacha, Piotr Dworakowski, D. Frey

OPTIMIZATION OF HIGH FREQUENCY MAGNETIC DEVICES WITH CONSIDERATION OF THE EFFECTS OF THE MAGNETIC MATERIAL, THE CORE GEOMETRY AND THE SWITCHING FREQUENCY 416
Sobhi Barg, Muhammad Farhan Alam, Kent Bertilsson

REAL TIME CONTROL HARDWARE IN THE LOOP TEST OF A NOVEL MVDC SOLID-STATE BREAKER 424
Alessio Clerici, Riccardo Chiumeo, Chiara Gandolfi

IGBT LIFETIME ESTIMATION IN A MODULAR MULTILEVEL CONVERTER FOR BIDIRECTIONAL POINT-TO-POINT HVDC APPLICATION 433
Diego Velazco, Guy Clerc, Emmanuel Boutleux, François Wallart, Laurent Chédot

OPTIMIZATION DESIGN FOR SIC DRIFT STEP RECOVERY DIODE (DSRD) 443
Xiaoxue Yan, Lin Liang, Ziyue Wang, Guoqiang Tan

DISCRETE SUPER-TWISTING SLIDING MODE CURRENT CONTROLLER FOR INDUCTION MOTOR DRIVES 450
Tianqing Wang, Bo Wang, Yong Yu, Yangming Zhu, Dianguo Xu

NEW GRID-CONNECTED MULTILEVEL BOOST CONVERTER TOPOLOGY WITH INHERENT CAPACITORS VOLTAGE BALANCING USING MODEL PREDICTIVE CONTROLLER 460
Rasoul Shalchi Alishah, Kent Bertilsson, Frede Blaabjerg, Mohd. Ali Jagabar Sathik, Ali Yahya Rezaee

DCM OPERATION OF SINGLE-SWITCH HIGH STEP-UP DC-DC CONVERTER WITH THREE-WINDING COUPLED INDUCTOR 467
Masataka Minami, Genki Hase

POWER LOSSES CALCULATION FOR MEDIUM VOLTAGE DC/DC CURRENT-FED SOLID STATE TRANSFORMER FOR BATTERY GRID-CONNECTED 471
E. K. Hussain, Mohammad Abusara, S. M. Sharkh

MODELLING AND EXPERIMENTAL VALIDATION OF A POLE-TO-GROUND PROTECTION DEVICE IN LOW VOLTAGE DC MICROGRIDS 480
L. Hallemans, G. Govaerts, G. Van Den Broeck, S. Ravyts, M. M. Alam, P. Van Tichelen, J. Driesen

DESIGN OF A DUAL ACTIVE BRIDGE CONVERTER FOR ON-BOARD VEHICLE CHARGERS USING GAN AND INTO TRANSFORMER INTEGRATED SERIES INDUCTANCE 490
K. Siebke, M. Giacomazzo, R. Mallwitz

AN EXPERIMENTAL ANALYSIS OF CIRCULATING CURRENT CONTROL CIRCUIT FOR OUTPUT POWER FROM VIBRATION GENERATOR FOR VIBRATION INCLUDING THE THIRD HARMONICS 498
Masataka Minami, Akito Nakagaki, Genki Hase

IMPLEMENTATION OF CONTROL STRATEGY FOR STEP-DOWN DC-DC CONVERTER BASED ON PIEZOELECTRIC RESONATOR 503
Mustapha Touhami, Ghislain Despesse, François Costa, Benjamin Pollet

THERMAL IMPEDANCES AND TEMPERATURE SENSORS: A COMBINED APPROACH FOR A NOVEL THERMAL MODEL OF POWER SEMICONDUCTORS 512
Maria De Lauretis, Jonas Millinger, Erik Baker, Martin Karlsson, Diane -Perle Sandik

A 3A LOW VOLTAGE LASER DIODE DRIVER IC IN A CMOS TECHNOLOGY FOR AN ITOF-BASED 3D IMAGE SENSOR 522
Romain David, Bruno Allard, Xavier Branca, Charles Joubert

COMPARISON OF DECOUPLING TECHNIQUES VIA DISCRETE LUENBERGER STYLE OBSERVER FOR VOLTAGE ORIENTED CONTROL 532
Gyanendra Kumar Sah, Michael Schütt, Hans-Günter Eckel

VARIABLE SWITCHING POINT PARALLEL PREDICTIVE CURRENT CONTROL (VSP3CC) FOR INDUCTION MOTOR 542
Qing Chen, Ralph Kennel

OPERATION OF AN EXTERNALLY EXCITED SYNCHRONOUS MACHINE WITH A HYBRID MULTILEVEL INVERTER 551
C. Terbrack, J. Stöttner, C. Endisch

A FACILITY FOR MIXED FLOWING GAS TESTING OF AND EXPERIMENTATION WITH POWER ELECTRONIC COMPONENTS AND SYSTEMS 563
Juuso Rautio, Janne Jäppinen, Tommi J. Kärkkäinen, Markku Niemelä, Pertti Silventoinen, Mika Kiviniemi, Joonas Leppänen, Jonny Ingman

IMPACT OF IMPLEMENTATION OF AUXILIARY BIAS-WINDINGS ON CONTROLLABLE INDUCTORS FOR POWER ELECTRONIC CONVERTERS 571
Jonas Pfeiffer, Pierre Küster, Yeliz Erenler, Ziyad H. S. Qashlan, Peter Zacharias

APPROXIMATED SLIDING-MODE CONTROL OF PARALLEL-CONNECTED GRID INVERTERS 581
Albrecht Gensior

EQUIVALENT MODEL AND CONTROL OF A NEUTRAL POINT SUPPLY SYNRM DRIVE.............. 590
Xiaokang Zhang, Jean-Yves Gauthier, Xuefang Lin-Shi

IMPROVEMENTS ON SIGNAL-TO-NOISE RATIO IN FEEDBACK MEASUREMENT IN
DC/DC CONVERTERS... 598
Fernando Davalos Hernandez, Federico Ibanez, Sebastian Gutierrez, Wilmar Martinez

APPROACH OF AN ACTIVE DEVICE PROTECTION FOR DRIVE INVERTERS AGAINST
SHORT CIRCUIT FAULTS IN AN OPEN INDUSTRIAL DC GRID.. 608
Simon Puls, Urs Obernolte, Martin Ehlich, Holger Borcherding

A NEW DESIGN OF AN AIR CORE TRANSFORMER FOR ELECTRIC VEHICLE ON-
BOARD CHARGER.. 618
Valentin Rigot, Tanguy Phulpin, Daniel Sadarnac, Jihen Sakly

ENABLING FOIL WINDINGS OF MEDIUM-FREQUENCY TRANSFORMERS FOR HIGH
CURRENTS 627
Thomas B. Gradinger, Uwe Drofenik, Filip Grecki

A HIGH-EFFICIENCY WIRELESS POWER TRANSFER SYSTEM FOR UNMANNED
AERIAL VEHICLE CONSIDERING CARBON FIBER BODY 637
Kai Song, Peng Zhang, Zhengxin Chen, Guang Yang, Jinhai Jiang, Chunbo Zhu

ANALYTICAL COMPUTATION OF NORMAL AND FAULT-TOLERANT ACTIVE SHORT
CIRCUIT OPERATION OF ANISOTROPIC SYNCHRONOUS DOUBLE STAR MACHINES 644
Michael Gleissner, Johannes Häring, Wolfgang Wondrak, Mark-M. Bakran

FULL-SILICON 98.7% EFFICIENT THREE-PHASE FIVE-LEVEL 3-PORT UPS
ARCHITECTURE WITH WIDE VOLTAGE RANGE BATTERY BASED ON MULTIPLEXED
TOPOLOGY 654
Kepa Odriozola, Thierry A. Meynard, Alain Lacarnoy

ON-GRID/OFF-GRID DC MICROGRID OPTIMIZATION AND DEMAND RESPONSE
MANAGEMENT 667
Wenshuai Bai, Manuela Sechilariu, Fabrice Locment

SHEDDING AND RESTORATION ALGORITHMS FOR AN EV CHARGING STATION TO
MAXIMIZE AVAILABLE POWER.. 677
Dian Wang, Fabrice Locment, Manuela Sechilariu

EFFICIENCY AND COST COMPARISON OF B6 AND HYBRID ANPC CONVERTERS FOR
TRACTION DRIVES 686
Johannes Häring, Michael Gleissner, Wolfgang Wondrak, Mark-M. Bakran

DESIGN AND CONTROL OF A KE (KINETIC ENERGY) - COMPENSATED
GRAVITATIONAL ENERGY STORAGE SYSTEM 696
Alfred Rufer

A NOVEL POWER FLOW CONTROL STRATEGY FOR HETEROGENEOUS BATTERY
ENERGY STORAGE SYSTEMS BASED ON PROGNOSTIC ALGORITHMS FOR
BATTERIES 707
Markus Muehlbauer, Samantha Klier, Herbert Palm, Oliver Bohlen, Michael A. Danzer

AN IGCT-BASED MULTI-FUNCTIONAL MMC SYSTEM WITH COMMUTATION AND
SWITCHING... 718
Chaoqun Xu, Mingzhu Guo, Biao Zhao, Bojin Tang, Zhanqing Yu, Dongling Zhai, Chunpin
Ren

COMMON-MODE NOISE MODELLING AND RESONANT ESTIMATION IN A THREE-PHASE MOTOR DRIVE SYSTEM: 9-150 KHZ FREQUENCY RANGE 726
Hansika Rathnayake, Amir Ganjavi, Firuz Zare, Dinesh Kumar, Pooya Davari

POLYNOMIAL MULTI-VARIABLE CONTROL STRATEGY FOR FLUX BALANCING IN DUAL ACTIVE BRIDGE CONVERTER........ 736
Pierre-Baptiste Steckler, Jean-Yves Gauthier, Xuefang Lin-Shi, François Wallart

ENHANCED POWER SYSTEM DAMPING ESTIMATION VIA OPTIMAL PROBING SIGNAL DESIGN........ 745
S. Boersma, X. Bombois, L. Vanfretti, V. Peric, J-C. Gonzalez-Torres, R. Segur, A. Benchaib

IMPROVED HIGH STEP-UP BOOST-BASED DC/DC CONVERTER WITH BUILT-IN TRANSFORMER AND ACTIVE CLAMP FOR DC MICROGRIDS........ 755
Konstantinos Zaoskoufis, Emmanuel C. Tatakis

ELIMINATION/MITIGATION OF OUTPUT VOLTAGE HARMONICS FOR MULTILEVEL CONVERTERS OPERATED AT FUNDAMENTAL SWITCHING FREQUENCY USING MATLAB'S GENETIC ALGORITHM OPTIMIZATION 765
Anton Kersten, Manuel Kuder, Arthur Singer, Weiji Han, Torbjörn Thiringer, Thomas Weyh, Richard Eckerle

EVALUATION OF DRIVE TOPOLOGIES FOR MACRO SCALE SYNCHRONOUS ELECTROSTATIC MACHINES 777
Peter Killeen, Daniel C. Ludois

DECENTRALIZED VOLTAGE REGULATION IN ISLANDED DC MICROGRIDS IN THE PRESENCE OF DISPATCHABLE AND NON-DISPATCHABLE DC SOURCES........ 787
Mohammadreza Nabatirad, Reza Razzaghi, Behrooz Bahrani

AN ULTRA-FAST GATE DRIVER WITH OVER CURRENT PROTECTION FOR GAN POWER TRANSISTORS 797
Qingqing Nie, Han Peng, Yong Kang

A NEW GAN HYBRID RESONANT-CLAMPING GATE DRIVER FOR HIGH FREQUENCY SIC MOSFETS........ 804
Ziyue Dang, Han Peng, Hao Peng, Yong Kang, Yu Chen, Xudan Liu, Maojun He

MAINTENANCE SCHEDULING IN POWER ELECTRONIC CONVERTERS CONSIDERING WEAR-OUT FAILURES........ 810
Saeed Peyghami, Frede Blaabjerg, Jose Rueda Torres, Peter Palensky

AC/DC DYNAMIC INTERACTIONS OF MMC-HVDC IN GRID-FORMING FOR WIND-FARM INTEGRATION IN AC SYSTEMS 820
Rayane Mourouvin, Kosei Shinoda, Jing Dai, Abdelkrim Benchaib, Seddik Bacha, Didier Georges

A DESIGN OF SOLID STATE POWER CONTROLLER FOR A BIDIRECTIONAL DC-DC CONVERTER IN AN AERONAUTIC CONTEXT........ 829
Hassan Cheaito, Bruno Allard, Guy Clerc, Joris Pallier, Pascal Pommier-Petit

A NEW APPROACH OF RESONANT CONVERTER USING LARGE AIR GAP TRANSFORMER 835
Michael Finkenzeller, Monika Poebl, Thomas Komma

REDUCED CAPACITOR SIZE AND ON-STATE LOSSES IN ADVANCED MMC SUBMODULE TOPOLOGIES 843

    Christopher Dahmen, Rainer Marquardt

STABILITY AND ROBUSTNESS ANALYSIS OF FRACTIONAL PROPORTIONAL RESONANT CONTROLLERS IN CURRENT-CONTROLLED VOLTAGE-SOURCE-INVERTERS 853

    Daniel Heredero-Peris, Cristian Chillón-Antón, Daniel Montesinos-Miracle

EMPLOYING VIRTUAL SYNCHRONOUS GENERATOR WITH A NEW CONTROL TECHNIQUE FOR GRID FREQUENCY STABILIZATION 863

    Meysam Saeedian, Bahman Eskandari, Kumars Rouzbehi, Shamsodin Taheri, Edris Pouresmaeil

A HYBRID PULSE WIDTH MODULATION TECHNIQUE WITH TEMPERATURE CONTROL FOR MODULAR MULTILEVEL CONVERTERS 871

    Ara Bissal, Waqas Ali, Rob Leedham, Mark Snook, Ibrahim Elsabrouty, Ilknur Colak

DESIGN FLOW OF A COMPACT HIGH-FREQUENCY DC/DC CONVERTER WITH OPTIMUM AVERAGE EFFICIENCY IN A WIDE OPERATION RANGE 880

    Maximilian Nitzsche, Matthias Zehelein, Julian Weimer, Dominik Koch, Jörg Roth-Stielow

ANALYSIS OF THE TRANSFORMER MODULARIZATION FOR HIGH FREQUENCY ISOLATED HIGH VOLTAGE GENERATOR WITH THE SILICON CARBIDE DEVICES 892

    Saijun Mao, Popovic Jelena, Jan Abraham Ferreira

IMPROVED DIRECT-MODEL PREDICTIVE CONTROL WITH A SIMPLE DISTURBANCE OBSERVER FOR DFIGS 900

    Mohamed Abdelrahem, Christoph Hackl, José Rodríguez, Ralph Kennel

MODELING OF SIC-MOSFET CONVERTER LEG INCLUDING PARASITICS OF PRINTED CIRCUIT BOARD LAYOUT AND DEVICE PACKAGING 909

    M. Pulvirenti, L. Salvo, A. G. Sciacca, G. Scelba, M. Cacciato

PERFORMANCE ANALYSIS OF RL DAMPER IN GAN-BASED HIGH-FREQUENCY BOOST CONVERTER 919

    A. Gutierrez, E. Marcault, C. Alonso, D. Tremouilles

RAPID IMPEDANCE ESTIMATION ALGORITHM FOR MITIGATION OF SYNCHRONIZATION INSTABILITY OF PARALLELED CONVERTERS UNDER GRID FAULTS 927

    Mads Graungaard Taul, Robert Eric Betz, Frede Blaabjerg

ADAPTIVE THERMAL CONTROL FOR MOSFET-BASED MODULAR MULTILEVEL CONVERTER 937

    Tianxiang Yin, Lei Lin, Chen Xu

ELECTRIC IMPULSE TECHNOLOGY – BREAKING ROCK 944

    Matthias Voigt, Erik Anders, Franziska Lehmann, Margarita Mezzetti, Frank Will

IMPACT OF COMBINED THERMO-MECHANICAL AND ELECTRO-CHEMICAL STRESS ON THE LIFETIME OF POWER ELECTRONIC DEVICES 954

    Felix Hoffmann, Stefan Schmitt, Nando Kaminski

CURRENT CONTROL AND FPGA–BASED REAL–TIME SIMULATION OF GRID–TIED INVERTERS 962

    Sabin Carpiuc, Matthias Schiesser, Carlos Villegas

IMPACT OF CONTROL LOOPS ON THE LOW-FREQUENCY PASSIVITY PROPERTIES OF GRID-FORMING CONVERTERS 969

Mebtu Beza, Massimo Bongiorno, Anant Narula

GRID IMPEDANCE ESTIMATION WITH OVERSAMPLING FOR GRID-CONNECTED CONVERTERS 979

Niklas Himker, Robin Strunk, Axel Mertens

LOW SPEED SENSORLESS CURRENT CONTROL FOR PMSM WITH SEARCH-BASED OBSERVER (SBO) 989

K. Scicluna, C. Spiteri Staines, R. Raute

INSIGHT INTO THE PECULIARITIES OF OPTIMIZED PULSE PATTERNS FOR PERMANENT-MAGNET SYNCHRONOUS MACHINES 998

Georgios Darivianakis, Ioannis Tsoumas

INVESTIGATING THE EFFECT OF DIFFERENT PARAMETERS ON HARMONICS AND EMI EMISSIONS AT THE FREQUENCY RANGE OF 0–9 KHZ 1006

Amir Ganjavi, Hansika Rathnayake, Firuz Zare, Dinesh Kumar, Amin Abbosh, Pooya Davari

FIVE-LEVEL NESTED INVERTER WITH NEUTRAL POINT CONNECTION 1016

Juhamatti Korhonen, Aleksi Mattsson, Heikki Järvisalo, Pertti Silventoinen, William Giewont, Dan Isaksson

ELECTRIC SPRING-BASED SMART WATER HEATER FOR LOW VOLTAGE MICROGRIDS 1025

Alexander Micallef, Racquel Ellul, John Licari

ENERGY-BALANCING OF A MODULAR MULTILEVEL CONVERTER USING AN ONLINE TRAJECTORY PLANNING ALGORITHM 1030

Qiuye Gui, Jan Lasse Gnärig, Hendrik Fehr, Albrecht Gensior

CAPACITOR SIZE COMPARISON ON HIGH-POWER DC-DC CONVERTERS WITH DIFFERENT TRANSFORMER WINDING CONFIGURATIONS ON THE AC-LINK 1040

Babak Khanzadeh, Torbjörn Thiringer, Yuhei Okazaki

DYNAMIC CHARACTERISTICS VERIFICATION OF LINEAR INDUCTION MOTOR BY SIMULTANEOUS PROPULSION AND LEVITATION CONTROL 1047

Shota Nakatani, Daichi Okamori, Toshimitsu Morizane, Hideki Omori

'IG,VGS' MONITORING FOR FAST AND ROBUST SIC MOSFET SHORT-CIRCUIT PROTECTION WITH HIGH INTEGRATION CAPABILITY 1057

Yazan Barazi, François Boige, Nicolas Rouger, Jean-Marc Blaquiere, Frédéric Richardeau

FAULT-TOLERANT CONTROL OF SERIES CONNECTABLE MODULAR FULL-BRIDGE INVERTER MITIGATING OPEN SWITCH FAULTS 1067

Juris Arrozy, Darian V. Retianza, Jorge L. Duarte, Henk Huisman

DESIGN AND CONTROL OF A MODULAR POWER ELECTRONIC BACK-TO-BACK CONVERTER FOR WAVE ENERGY HARVESTING APPLICATIONS 1076

Mattia Mantellini, Riccardo Morici, Marcos Blanco, Marcos Lafoz, Gustavo Navarro, Jorge Torres, Jorge Najera, Miguel Santos

INTELLIGENT HIGH CURRENT SENSOR FOR VARIOUS FREQUENCY 1086

Bohumil Skala, Vladimir Kindl, Pavel Turjanica, Ales Voborník, Libor Polacek, Josef Stengl, Vladimir Pavlicek, Jiri Fort

FAIL-SAFE SWITCHING-CELLS ARCHITECTURES BASED ON MONOLITHIC ON-CHIP FUSE 1096
 Amirouche Oumaziz, Emmanuel Sarraute, Frédéric Richardeau, Abdelhakim Bourennane

HOW GOOD ARE THE DESIGN TOOLS IN POWER ELECTRONICS? 1106
 Thomas Lagier, Piotr Dworakowski, Laurent Chédot, François Wallart, Bruno Lefebvre, Jose Maneiro, Juan Páez, Philippe Ladoux, Cyril Buttay

ANALYSIS OF THE IMPACT OF MANUFACTURING DISSYMMETRY ON CURRENT DISTRIBUTION FOR MAGNETICALLY COUPLED INTERLEAVED INVERTERS 1118
 Rita Mattar, Mickael Petit, Eric Monmasson, Stéphane Lefebvre, Christelle Saber, Cyrille Gautier, Marwan Ali

POWER FLOW CONTROL USING A BIDIRECTIONAL Z-SOURCE INVERTER–BASED STATIC SYNCHRONOUS SERIES COMPENSATOR 1128
 Xuejiao Pan, Han Huang, Li Zhang

INVESTIGATION OF HARMONICS CONTENT IN PWM NATURAL AND REGULAR SAMPLING INCLUDING DEAD TIME AND LOAD CURRENT PHASE 1138
 Tonny Wederberg Rasmussen, Anushruti Vashishtha, Ankit Jotwani

USING A WEB SCRAPING ALGORITHM FOR COMPONENT MODEL GENERATION IN MULTIOBJECTIVE OPTIMIZATION OF POWER ELECTRONIC APPLICATIONS 1148
 Marcel Gladen, Volker Staudt

IMPACT ON THE ELECTRICAL CHARACTERISTICS, WAVEFORMS AND LOSSES OF THE ZERO-SEQUENCE INJECTION ON THE MODULAR MULTILEVEL CONVERTER 1158
 Francois Gruson, Pierre Vermeersch, Philippe Delarue, Philippe Le Moigne, Frédéric Colas, Haibo Zhang, Moez Belhaouane, Xavier Guillaud

WIDE BANDWIDTH CURRENT SENSOR FOR COMMUTATION CURRENT MEASUREMENT IN FAST SWITCHING POWER ELECTRONICS 1168
 Philipp Ziegler, Nathan Tröster, Dimitri Schmidt, Johannes Ruthardt, Manuel Fischer, Jörg Roth-Stielow

A SERIES–PARALLEL-TYPE RESONANT CIRCUIT WIRELESS POWER TRANSFER SYSTEM WITH A DUAL ACTIVE BRIDGE DC–DC CONVERTER 1177
 Kohei Sugiyama, Taishi Kitamura, Shuto Uwai, Takahiro Yano, Yoshitaka Kawabata

STRAY VOLTAGE CAPTURE FOR ROBUST AND ULTRA-FAST SHORT CIRCUIT DETECTION IN POWER ELECTRONICS WITH HALF-BRIDGE STRUCTURE: THE LIMITATION AND IMPLEMENTATION 1186
 Darian Verdy Retianza, Jeroen Van Duivenbode, Henk Huisman

ON THE INFLUENCE OF THE STATOR WINDING TOPOLOGY ON THE ELECTROMAGNETIC EMISSIONS OF FRACTIONAL HORSEPOWER BLDC MOTORS 1196
 Felix Krall, Annette Muetze

IMPACT OF SILICON CARBIDE DEVICES IN 2 MW DFIG BASED WIND ENERGY SYSTEM 1205
 Antxon Arrizabalaga, Aitor Idarreta, Mikel Mazuela, Iosu Aizpuru, Unai Iraola, José Luis Rodriguez, Daniel Labiano, Ibrahim Alisar

SMALL-SIGNAL STABILITY OF HVDC SYSTEM COMPRISING DC REACTORS 1215
 Kosei Shinoda, Abdelkrim Benchaib, Jing Dai

MODEL PREDICTIVE CONTROL FOR THE REDUCTION OF DC-LINK CURRENT RIPPLE IN TWO-LEVEL THREE-PHASE VOLTAGE SOURCE INVERTERS 1224
Junzhong Xu, Fei Gao, Thiago Batista Soeiro, Linglin Chen, Luca Tarisciotti, Houjun Tang, Pavol Bauer

CARRIER-BASED MODULATED MODEL PREDICTIVE CONTROL FOR VIENNA RECTIFIERS 1233
Junzhong Xu, Fei Gao, Thiago Batista Soeiro, Linglin Chen, Luca Tarisciotti, Houjun Tang, Pavol Bauer

NEW HIGH-EFFICIENCY POWER GENERATION USING POSITION SENSOR-LESS PERMANENT MAGNET SYNCHRONOUS GENERATOR 1243
Somi Takeuchi, Hiroyuki Takahashi, Shota Yamada, Yoshitaka Kawabata

ACTIVE CLAMPING METHOD FOR SIC MOSFET HIGH POWER MODULES - BENEFITS AND LIMITS 1252
Robert W. Maier, Mark-M. Bakran

PREDICTIVE TORQUE CONTROL OF INDUCTION MACHINE WITH AN ADAPTIVE OBSERVER FOR TRAJECTORY PLANNING OF SERVO PRESS 1262
Qi Li, Jianbo Gao, Qiwu Wang, Ralph Kennel

FUTURE GRID STABILITY, A COST COMPARISON OF GRID-FORMING AND SYNCHRONOUS CONDENSER BASED SOLUTIONS 1270
Thibault Prevost, Guillaume Denis, Clementine Coujard

DEMONSTRATION OF THE SHORT-CIRCUIT RUGGEDNESS OF A 10 KV SILICON CARBIDE BIPOLAR JUNCTION TRANSISTOR 1279
Besar Asllani, Hervé Morel, Pascal Bevilacqua, Dominique Planson

LOSS MINIMIZATION OF TRACTION SYSTEMS IN BATTERY ELECTRIC VEHICLES USING VARIABLE DC-LINK VOLTAGE TECHNIQUE — EXPERIMENTAL STUDY 1289
Libo Liu, Boyang Li, Gunther Götting, Yusheng Xiang, Qusay Salem, Muhammad Hamid, Jian Xie

DIRECT MULTIVARIABLE CONTROL FOR MMC: DIGITAL SIGNAL PROCESSING AND EXPERIMENTAL RESULTS 1297
Daniel Dinkel, Claus Hillermeier, Rainer Marquardt

STATE OF CHARGE CONTROL FOR A FREQUENCY-SUPPORTING STORAGE SYSTEM BASED ON AN AUTO-REGRESSIVE FREQUENCY FORECAST 1306
A. Bolzoni, R. Todd, Q. Zhu, A. J. Forsyth

DESIGN OF A WIDE INPUT VOLTAGE RANGE CURRENT-FED DC/DC CONVERTER WITHIN A REDUCED DUTY-CYCLE RANGE 1316
Michael Gerstner, Martin Maerz, Armin Dietz

AN IMPROVED CONTROL STRATEGY FOR RENEWABLE ENERGY SOURCES (RES) BASED DC MICROGRID WITH ENHANCED SYSTEM STABILITY AND CONTROL PERFORMANCE 1326
Muhammad Adnan Mumtaz, Zheng Yan

TRANSIENT VOLTAGE DIP MITIGATION SYSTEM BASED ON HYBRID MODULAR MULTILEVEL CONVERTERS 1336
Manuel Colmenero, Francisco R. Blanquez, Karsten Kahle

A LOSS-COMPENSATED CONTROL SCHEME FOR SIC-BASED DUAL ACTIVE BRIDGE CONVERTER 1346

Ishan Pendharkar, Tobias Strittmatter, Paula Diaz Reigosa, Nicola Schulz

EXPERIMENTAL HYBRID AC/DC-MICROGRID PROTOTYPE FOR LABORATORY RESEARCH 1354

Enrique Espina, Claudio Burgos-Mellado, Juan S. Gomez, Jacqueline Llanos, Erwin Rute, Alex Navas F., Manuel Martínez-Gómez, Roberto Cárdenas, Doris Sáez

EXPERIMENTAL AND NUMERICAL CHARACTERIZATION OF PCB-EMBEDDED POWER DIES USING SOLDERLESS PRESSED METAL FOAM 1363

S. Bensebaa, M. Berkani, S. Lefebvre, M. Petit, N. Schmitt

FEASIBILITY STUDY OF A SUPERCONDUCTING POWER FILTER FOR HVDC GRIDS 1373

Loïc Quéval, Olivier Despouys, Frédéric Trillaud, Bruno Douine

POWER DECOUPLING METHOD OF DC TO SINGLE-PHASE AC CONVERTER USING FLYING CAPACITOR DC/DC CONVERTER WITH BOUNDARY CURRENT MODE 1380

Hiroki Watanabe, Keisuke Kusaka, Jun-Ichi Itoh

AN ARCHITECTURE FOR LEVEL-3 EV BATTERY CHARGER STATIONS USING INTEGRATED SOLID STATE TRANSFORMER (I-SST) 1390

Erick I. Pool-Mazun, Prasad Enjeti, Gerardo Escobar, Ira Pitel

LQR AND H-INFINITY CONTROL OF VOLTAGE SOURCE INVERTERS FOR AC MICROGRIDS 1400

Tenorio Jorge, Jose Miguel Ramirez Scarpetta, Fabio Andrade

FAMILY OF SPLITTING CURRENT SINGLE-LOOP CONTROL FOR LCL- TYPE GRID-CONNECTED INVERTER 1410

Yuying He, Xuehua Wang, Xinbo Ruan, Guoxing Su, Fuxin Liu

ANALYSIS AND DESIGN OF HIGH-POWER SINGLE-STAGE THREE-PHASE DIFFERENTIAL-BASED FLYBACK INVERTER FOR PHOTOVOLTAIC APPLICATIONS 1417

Ahmed Ismail M. Ali, Mahmoud A. Sayed, Takaharu Takeshita

INVESTIGATION OF IMPROVEMENT OF MODELING PRECISION FOR CONDUCTED NOISE ON ISOLATED AC/DC CONVERTER USING SIC DEVICES 1425

Kazuki Kuwana, Kohei Mitani, Wataru Kitagawa, Takaharu Takeshita

PASSIVITY-BASED DESIGN FOR THE PLUG-AND-PLAY SINGLE-LOOP CONTROLLED LCL-FILTERED INVERTER 1435

Yuying He, Xuehua Wang, Xinbo Ruan, Yixiao Ma, Fuxin Liu

CHARACTERISTICS OF AN INTEGRATED MOTOR CONTROLLED INDEPENDENTLY BY MULTI-INVERTERS TO ACHIEVE HIGH EFFICIENCY AND A WIDE SPEED RANGE 1442

Kazuto Sakai, Yano Hideaki

AN ISOLATED MEDIUM-VOLTAGE AC-DC CONVERTER USING LEVEL-SHIFTED PWM CONTROL OF A MODULAR MATRIX CONVERTER 1450

Kohei Budo, Takaharu Takeshita

DETAILED SIMULATION MODEL OF AN ASYMMETRICAL HALF-BRIDGE PWM CONVERTER WITH SYNCHRONOUS RECTIFICATION INCLUDING PARASITIC ELEMENTS 1460

Benedikt Kohlhepp, Valentin Zeller, Markus Barwig, Thomas Dürbaum

ELECTRICAL PROPERTY VARIABILITY OF GAN TRANSISTORS IN PARALLEL AND THEIR IMPACT ON FAST SWITCHING OPERATIONS 1470

Thilini Wickramasinghc, Bruno Allard, Réne Escofficr, Marc Plissonnicr

A COMPARISON BETWEEN DIFFERENT MODELS OF THE MODULAR MULTILEVEL CONVERTER 1479

Rafael Coelho-Medeiros, Bogdan Džonlaga, Jean-Claude Vannier, Jing Dai, Loic Queval, Philippe Egrot

PACKAGING TECHNOLOGY FOR THE IMPROVEMENT OF POWER CYCLING CAPABILITY OF HVIGBTS 1489

Kenji Hatori, Keiichi Nakamura, Nobuhiko Tanaka, Yasuhiro Sakai, Norikazu Sakai, Kenji Ota, Takeshi Higashihata, Eckhard Thal, Nils Soltau

A BIDIRECTIONAL DAB-LLC DCX TO ACHIEVE VOLTAGE REGULATION AND WIDE ZVS RANGE CAPABILITY 1498

Yuefeng Liao, Tao Peng, Mei Su, Yao Sun, Weijing Xiong, Guo Xu

SALIENCY SELECTION FOR SEARCH-BASED AC MACHINE LOW AND ZERO SPEED ESTIMATION METHODS 1506

K. Scicluna, C. Spiteri Staines, R. Raute

GENETIC ALGORITHM BASED MULTI OBJECTIVE OPTIMIZATION FOR INDUCTOR DESIGN 1515

Thorben Schobre, Raquel González Aríztegui, Regine Mallwitz

DIGITAL SMART DRIVER FOR SIC MOSFETS 1524

Nerea Arandia, José Ignacio Garate, Jon Mabe, Ander Ordoño

FASTER SWITCHING WITH LESS OVERVOLTAGE - OPERATING A SIC-MOSFET AT ITS SPEED LIMIT 1533

Pablo Rodriguez De Mora, Mark-M. Bakran

THE ENERGY RING TO SUPPLY THE EXPOELECTRIC'18 SHOW WITH RENEWABLE ENERGY SOURCES AND ELECTRIC VEHICLES 1542

Cristian Chillón-Antón, Daniel Heredero-Peris, Francesc Girbau-Llistuella, Paula González-Fontderubinat, Marc Llonch-Masachs, Daniel Montesinos-Miracle, Oriol Gomis-Bellmunt

IMPEDANCE-BASED MODELING OF A THREE-LEVEL CONVERTER UNDER BALANCED AND UNBALANCED CONDITION FOR THE STABILITY ANALYSIS OF BIPOLAR LVDC GRIDS 1551

T. Roose, G. Van Den Broeck, M. M. Alam, J. Beerten

LCL FILTER DESIGN FOR THREE PHASE AC-DC CONVERTERS CONSIDERING SEMICONDUCTOR MODULES AND MAGNETICS COMPONENTS PERFORMANCE 1561

Marco Stecca, Thiago Batista Soeiro, Laura Ramirez Elizondo, Pavol Bauer, Peter Palensky

SWITCHING BEHAVIOR AND COMPARISON OF 600V SMD WIDE BANDGAP POWER DEVICES 1569

Markus Meißner, Jan Schmitz, Steffen Bernet

ANALYSIS OF THE COUPLING BETWEEN THE OUTER AND INNER CONTROL LOOPS OF A GRID-FORMING VOLTAGE SOURCE CONVERTER 1579

T. Qoria, F. Gruson, F. Colas, X. Kestelyn, X. Guillaud

INFLUENCE OF DIFFERENT PULSE-WIDTH MODULATION METHODS ON MAGNET LOSSES IN PERMANENT MAGNET SYNCHRONOUS MACHINES 1589
Narciso G. Marmolejo, Xiaohu Tang, Martin Doppelbauer

RESONANT DC/DC CONVERTER WITH CLASS ϕ_2 INVERTER AND CLASS DE RECTIFIER BASED ON GAN HEMT 1599
Cai Si-Yuan, He Jun-Ping, Li Zi-Fan

FOUR-LEVEL INVERTER WITH VARIABLE VOLTAGE LEVELS FOR HARDWARE-IN-THE-LOOP EMULATION OF THREE-PHASE MACHINES 1605
Manuel Fischer, Johannes Ruthardt, Vasken Ketchedjian, Philipp Ziegler, Maximilian Nitzsche, Jörg Roth-Stielow

POWDER INJECTION MOLDING IN THE FABRICATION OF SOFT FERRITE MATERIAL FOR POWER ELECTRONICS 1613
J-S Ngoua-Teu, U. Soupremanien, P. Sallot, G. Delette, M. Bohnke

MODULATION SCHEME WITH COMMON MODE AND DIFFERENTIAL MODE VOLTAGE ELIMINATION FOR A FIVE LEVEL INVERTER FED OPEN END WINDING INDUCTION MOTOR DRIVE 1619
Greeshma Nadh, Durga Nair S., Arun Rahul S.

A FAST AND ROBUST MODEL OF DUAL-ACTIVE BRIDGE CONVERTERS IN REAL-TIME SIMULATION 1627
Ming Jia, Philipp Joebges, Rik W. De Doncker

DUAL INTERLEAVED 3.6 KW LLC CONVERTER OPERATING IN HALF-BRIDGE, FULL-BRIDGE AND PHASE-SHIFT MODE AS A SINGLE-STAGE ARCHITECTURE OF AN AUTOMOTIVE ON-BOARD DC-DC CONVERTER 1638
Philipp Rehlaender, Sergey Tikhonov, Frank Schafmeister, Joachim Bocker

SWITCHING LOSS ESTIMATION USING A VALIDATED MODEL OF 650 V GAN HEMTS 1648
Joao Oliveira, Florent Loiselay, Hervè Morel, Dominique Planson

REDUCTION OF CONDUCTION LOSSES IN RESONANT CONVERTERS BY CONNECTING THREE SINGLE-PHASE INVERTERS TO A COMMON GENERATOR 1658
Sergio Tárraga, John Paul Mayorga, Esther De Jódar, José Villarejo

COMPARISON OF DIFFERENT LOW VOLTAGE MULTILEVEL CONVERTER TOPOLOGIES FOR DISTRIBUTED POWER GENERATION 1666
Ingmar Kaiser, Hans-Günter Eckel

LOSS DISTRIBUTION COMPARISON OF VARIABLE AND FIXED INDUCTOR DAB CONVERTERS 1675
Erik Smailus, Gerd Griepentrog, Markus Pfeifer, Marcel Lutze

DESIGN BY OPTIMIZATION OF MULTIPHASE INVERTER FOR ELECTRIC VEHICLE DRIVE 1685
Nasreddine Kesbia, Jean-Luc Schanen, Hadi Alawieh, Lauric Garbuio, Yvan Avenas

OPTIMAL TORQUE/SPEED CHARACTERISTICS OF A FIVE-PHASE SYNCHRONOUS MACHINE UNDER PEAK OR RMS CURRENT CONTROL STRATEGIES 1693
Tiago José Dos Santos Moraes, Hailong Wu, Eric Semail, Ngac Ky Nguyen, Duc Tan Vu

COMPARATIVE STUDY OF TWO CONTROL TECHNIQUES OF REGENERATIVE
BRAKING POWER RECOVERING INVERTER BASED DC RAILWAY SUBSTATION 1700
  Youssef Krim, Khaled Almaksour, Hervé Caron, Tony Letrouvé, Christophe Saudemont,
  Bruno Francois, Benoit Robyns

JUNCTION TEMPERATURE CONTROL STRATEGY FOR LIFETIME EXTENSION OF
POWER SEMICONDUCTOR DEVICES 1709
  Johannes Ruthardt, Hendrik Schulte, Philipp Ziegler, Manuel Fischer, Maximilian Nitzsche,
  Jörg Roth-Stielow

HIGH DYNAMIC POWER BALANCING FOR DUAL TWO-LEVEL INVERTERS DURING
HIGH-SPEED MACHINE OPERATION 1718
  Johannes Büdel, Johannes Teigelkötter, Alexander Stock, Christian Herkommer, Kai
  Kuhlmann

CHARGING HIGH VOLTAGE CAPACITORS IN PULSED POWER APPLICATIONS WITH A
CAPACITOR DIODE VOLTAGE MULTIPLIER OF REDUCED SIZE AND LOWER RIPPLE
CURRENTS 1727
  Tristan Weinert, Wolfgang Oberschelp, Günter Schröder

REVIEW OF OPTIMIZATION METHODS FOR THE DESIGN OF POWER ELECTRONICS
SYSTEMS 1737
  Mylène Delhommais

A FLEXIBLE POWER CROSSBAR-BASED ARCHITECTURE FOR SOFTWARE-DEFINED
POWER DOMAINS 1747
  Francesco Di Gregorio, Gilles Sassatelli, Abdoulaye Gamatié, Arnaud Castelltort

IMPACT OF GRID-FORMING CONTROL ON THE INTERNAL ENERGY OF A MODULAR
MULTILEVEL CONVERTER 1756
  Ebrahim Rokrok, Taoufik Qoria, Antoine Bruyere, Bruno Francois, Haibo Zhang, Moez
  Belhaouane, Xavier Guillaud

COMBINING MULTIPLE TEMPERATURE-SENSITIVE ELECTRICAL PARAMETERS
USING ARTIFICIAL NEURAL NETWORKS 1766
  Daniel Herwig, Torben Brockhage, Axel Mertens

SINGLE-PHASE MEASUREMENT OF THE OUTPUT IMPEDANCE OF THE FOUR-
QUADRANT CASCADED H-BRIDGE CONVERTER CELL USING WIDEBAND SIGNALS 1776
  Marko Petkovic, Dražen Dujic

A NOVEL THREE-PHASE PFC DIODE RECTIFIER BY LC NETWORK CIRCUITS FOR
HIGH FREQUENCY GENERATOR 1786
  Shin-Ichi Motegi, Yasuyuki Nishida

FREQUENCY-DOMAIN SIMULATION OF POWER ELECTRONIC SYSTEMS BASED ON
MULTI-TOPOLOGY EQUIVALENT SOURCES MODELLING METHOD 1793
  Stephane Vienot, Arnaud Videt, Nadir Idir, Lamine Kone, Sébastien Weiss, Frederic Lafon

MODULAR MULTILEVEL CONVERTER WITH DISTRIBUTED GALVANIC ISOLATION: A
DECENTRALIZED VOLTAGE BALANCING ALGORITHM WITH SMART GATE DRIVERS 1803
  Darbas Corentin, Ginot Nicolas, Olivier Jean-Christophe, Poitiers Frédèric

COMPARISON AND OPTIMIZATION OF MAGNETICALLY COUPLED AND NON-
COUPLED MAGNETIC DEVICES IN INTERLEAVED OPERATION 1813
  Peter Zacharias, Alejandro Aganza-Torres

EXPERIMENTAL TUNING AND DESIGN GUIDELINES OF A DYNAMICALLY
RECONFIGURED WEIGHTING FACTOR FOR THE PREDICTIVE TORQUE CONTROL OF
AN INDUCTION MOTOR.. 1823
  Ilker Sahin, Ozan Keysan, Eric Monmasson

COMPENSATION OF TEMPERATURE DEPENDENCE IN A MODULE PARASITIC BASED
CURRENT MEASUREMENT SYSTEM 1831
  Frank Lautner, Mark-M. Bakran

DEVELOPMENT AND IMPLEMENTATION OF A LOW-COST RESEARCH PLATFORM
FOR CONTROL APPLICATIONS FOR INVERTER-BASED GENERATORS 1841
  Jesus D. Vasquez Plaza, Juan F. Patarroyo-Montenegro, Fabio Andrade

CONTROL OF PARALLEL CONNECTED VOLTAGE SOURCE INVERTERS IN A
MICROGRID FOR EXPERIMENTAL TESTING 1850
  Jesus D. Vasquez-Plaza, Jorge Tenorio, J. M. Ramírez-Scarpetta, Jose Alex Restrepo, Fabio
  Andrade

OPTIMIZATION STRATEGY FOR THE SIZING OF PASSIVE MAGNETIC COMPONENTS 1858
  Guillaume Devos, Maya Hage-Hassan, Philippe Dessante, Cyrille Gautier, Adrien Mercier,
  Eric Labouré

EXPLOITING A MULTI-PORT TRANSFORMER FOR MINIMAL DC-LINK CAPACITANCE
FOR AN AUTOMOTIVE ONBOARD CHARGER 1866
  Franz Vollmaier, Alexander Connaughton, Thomas Langbauer, Klaus Krischan

DESIGN AND OPTIMIZATION OF HIGH-EFFICIENCY 1W 500V-12V ISOLATED LOW-
COST DC/DC CONVERTER... 1874
  Etienne Foray, Christian Martin, Bruno Allard

CHALLENGES IN CALIBRATING AN UNCONVENTIONAL PARTIAL DISCHARGE
MEASUREMENT SYSTEM FOR PULSED VOLTAGES 1885
  Markus Fürst, Mark-M. Bakran

ELECTROTHERMAL MODELING OF GAN POWER TRANSISTOR FOR HIGH
FREQUENCY POWER CONVERTER DESIGN... 1895
  Loris Pace, Florian Chevalier, Arnaud Videt, Nicolas Defrance, Nadir Idir, Jean-Claude De
  Jaeger

MODELING AND FAULT DETECTION IN PHOTOVOLTAIC SYSTEMS USING THE I-V
SIGNATURE 1905
  Abdelhadi Benzagmout, Thierry Talbert, Olivier Fruchier, Thierry Martire, Philippe
  Alexandre, Carolina Penin

EFFICIENCY REQUIREMENTS FOR PASSIVELY COOLED CONVERTERS WITH
THERMAL MEASUREMENT BASED 3D-FEM SIMULATION 1915
  Julian Weimer, Dominik Koch, Maximilian Nitzsche, Matthias Zehelein, Ingmar Kallfass

GENERIC CONTROL LAW FOR DC AND AC MACHINES... 1923
  Pierre-Philippe Robet, Maxime Gautier, Yannick Aoustin

A HIGH PERFORMANCE 48-TO-8 V MULTI-RESONANT SWITCHED-CAPACITOR
CONVERTER FOR DATA CENTER APPLICATIONS.. 1934
  Rose A. Abramson, Zichao Ye, Robert C. N. Pilawa-Podgurski

SISO CONTROL STRATEGY OF RESONANT DUAL ACTIVE BRIDGE WITH A TUNED CLC NETWORK 1944
Meiqi Wang, Bo Yang, Lie Xu, Jing Li, David Gerada, Chunyang Gu, He Zhang, Chris Gerada, Yongdong Li

IMPACT OF STEADY-STATE GRID-FREQUENCY DEVIATIONS ON THE PERFORMANCE OF GRID-FORMING CONVERTER CONTROL STRATEGIES 1952
Anant Narula, Massimo Bongiorno, Mebtu Beza, Jan R Svensson, Xavier Guillaud, Lennart Harnefors

A GENERAL METHOD TO DAMP WIND TURBINE SSR WITH DIFFERENT TRANSMISSION SYSTEMS 1962
Ignacio Vieto, Jian Sun

A TEST SCHEME FOR THE COMPREHENSIVE QUALIFICATION OF MMC SUBMODULE BASED ON 10 KV SIC MOSFETS UNDER HIGH DV/DT 1972
Xingxuan Huang, Shiqi Ji, Dingrui Li, Cheng Nie, William Giewont, Leon M. Tolbert, Fred Wang

PWM GAIN LINEARIZATION ALGORITHM FOR MEDIUM VOLTAGE SOURCE INVERTER 1982
Hamza El Jihad, Sami Siala, Elise Savarit

AUTO-COMMISSIONING OF ACOUSTIC CONTROL OF IM DRIVE USING BAYESIAN OPTIMIZATION 1992
Michal Kroneisl, Václav Šmídl

EXPERIMENTAL EMI STUDY OF A 3-PHASE 100KW 1200V DUAL ACTIVE BRIDGE CONVERTER USING SIC MOSFETS 2000
Hadiseh Geramirad, Florent Morel, Piotr Dworakowski, Philippe Camail, Bruno Lefebvre, Thomas Lagier, Christian Vollaire

MODELING OF A DAB UNDER PHASE-SHIFT MODULATION FOR DESIGN AND DM INPUT CURRENT FILTER OPTIMIZATION 2010
Glauber De Freitas Lima, Yves Lembeye, Fabien Ndagijimana, Jean-Christophe Crebier

ACTIVE CURRENT AND ENERGY CONTROL FOR THE QUASI-THREE-LEVEL OPERATION MODE OF AN EXTENDED MODULAR MULTILEVEL CONVERTER TOPOLOGY 2020
Malte Lorenz, Jakub Kucka, Axel Mertens

TORQUE RIPPLE REDUCTION TECHNIQUE FOR A SWITCHED RELUCTANCE MOTOR 2029
Krzysztof Jackiewicz, Arkadiusz Kaszewski, Andrzej Stras, Bartlomiej Ufnalski, Tomasz Balkowiec

EXPERIMENTAL VALIDATION OF THE PERFORMANCES OF AN INVERTER SIZED WITH OPTIMIZATION METHODS 2039
Adrien Voldoire, Jean-Luc Schanen, Jean-Paul Ferrieux, Alexis Derbey, Cyrille Gautier, Marwan Ali

INFLUENCE OF SYSTEM PARAMETERS IN VARIABLE SPEED AC-INDUCTION MOTOR DRIVES ON PARASITIC ELECTRIC BEARING CURRENTS 2049
Martin Weicker, Guilherme Bello, Dennis Kampen, Andreas Binder

PLASMA IMPACT ON OVERVOLTAGE SHORT-CIRCUIT FAILURES IN ANPC CONVERTERS 2059
David Hammes, Sidney Gierschner, Dietmar Krug, Hans-Günter Eckel

NOVEL SOFT-SWITCHING INTERLEAVED BOOST CONVERTERS FOR RENEWABLE ENERGY CONVERSION SYSTEMS 2068
Madhuchandra Popuri, V. V. Subrahmanya Kumar Bhajana, Pavel Drabek, Manoj Kumar Maharana

POWER DENSITY OF PLANAR TRANSFORMERS DESIGNED WITH COMMERCIAL STANDARD CORES 2078
Reda Bakri, Xavier Margueron, Jean Sylvio Ngoua Teu Magambo, Philippe Le Moigne, Nadir Idir

EFFECTS OF PV PANEL AND BATTERY DEGRADATION ON PV-BATTERY SYSTEM PERFORMANCE AND ECONOMIC PROFITABILITY 2088
Monika Sandelic, Ariya Sangwongwanich, Frede Blaabjerg

FULL SENSORLESS OPERATION OF INDUCTION MACHINES BASED ON ONLINE IDENTIFICATION OF SALIENCIES USING HARMONIC COMPENSATION LUTS IN TRACTION APPLICATIONS 2098
E. Rodriguez Montero, M. Vogelsberger, T. Wolbank

MITIGATING DRAIN SOURCE VOLTAGE OSCILLATION WITH LOW SWITCHING LOSSES FOR SIC POWER MOSFETS USING FPGA-CONTROLLED ACTIVE GATE DRIVER 2106
Zheming Li, Robert W. Maier, Mark-M. Bakran

ONLINE TRAJECTORY PLANNING DURING LOW-VOLTAGE FRT OF A MODULAR MULTILEVEL CONVERTER 2116
Hendrik Fehr, Albrecht Gensior

EVALUATING FREQUENCY STABILITY WITH CONSIDERATION OF LOAD TYPE IN DIFFERENT SHARE OF RENEWABLES AND EMULATED INERTIA IN CASE OF SYSTEM SPLIT 2126
Nastaran Fazli, Sidney Gierschner, Hans-Günter Eckel

DISCRETE-TIME DIRECT POLE PLACEMENT FOR STABILITY ENHANCEMENT OF LCL-FILTERED INVERTERS IN THE SYNCHRONOUS-REFERENCE FRAME 2135
Pei Cai, Xiaohua Wu, Yongheng Yang, Wenli Yao, Weilin Li, Frede Blaabjerg

ON THE SWITCHING-INDUCED DC-LINK VOLTAGE RIPPLE IN THREE-LEVEL CONVERTERS WITH A NEUTRAL POINT 2145
Ioannis Tsoumas, Tobias Geyer

EFFECT OF PASSIVE INVERTER OUTPUT MOTOR FILTERS ON DRIVE SYSTEMS 2153
Dennis Kampen, Martin Weicker

IMPACT OF THE NEUTRAL POINT POTENTIAL RIPPLE ON THE GRID SIDE HARMONICS OF A 3LNPC BACK-TO-BACK CONVERTER EMPLOYED IN A MEDIUM VOLTAGE WECS 2163
Ioannis Tsoumas

TWO-LAYER GENETIC ALGORITHM FOR THE CHARGE SCHEDULING OF ELECTRIC VEHICLES 2172
Nikolaos T. Milas, Dimitris A. Mourtzis, Panagiotis I. Giotakos, Emmanuel C. Tatakis

SIX-PHASE PMSM DRIVE INVERTER TESTING ON A HIGH PERFORMANCE POWER HARDWARE-IN-THE-LOOP TESTBED 2182
Yasser Rahmoun, Patrick Winzer, Alexander Schmitt, Horst Hammerer

AN IMPROVED BIDIRECTIONAL HYBRID SWITCHED INDUCTOR CONVERTER.........................2192

Dan Hulea, Mihaita Gireada, Danut Vitan, Octavian Cornea, Nicolae Muntean

HYBRID MULTIPLE CHOPPER CELLS OF PWM AND SQUARE-WAVE OPERATION FOR
SOLID-STATE TRANSFORMER 2200

Naoto Kikuchi, Jun-Ichi Itoh, Keisuke Kusaka, Hoai Nam Le

A NEW ZVS ZONE IDENTIFICATION FOR DUAL ACTIVE BRIDGE WITH A GENERAL
MODULATION OBJECTIVE.........................2210

Suman Maharana, Dipankar De, Alberto Castellazzi

SINGLE-STAGE BOOST MODULAR MULTILEVEL CONVERTER (BMMC) FOR ENERGY
STORAGE INTERFACE.........................2220

Ahmed Abdelhakim, Frede Blaabjerg, Hans-Peter Nee

LOW VOLTAGE GAN-BASED GATE DRIVER TO INCREASE SWITCHING SPEED OF
PARALLELED 650 V E-MODE GAN HEMTS 2230

Raffael Risch, Jürgen Biela

GATE STRESSES AND THRESHOLD VOLTAGE INSTABILITY IN NORMALLY-OFF GAN
HEMTS 2241

Jose Ortiz Gonzalez, Burhan Etoz, Olayiwola Alatise

NEW ENERGY MANAGEMENT ALGORITHM BASED ON FILTERING FOR ELECTRICAL
LOSSES MINIMIZATION IN BATTERY-ULTRACAPACITOR ELECTRIC VEHICLES 2251

*Bakou Traoré, Moustapha Doumiati, Cristina Morel, Jean-Christophe Olivier, Ousmane
Soumaoro*

MECHANISTIC POWER MODULE DEGRADATION MODELLING CONCEPT WITH
FEEDBACK.........................2258

Martin Bendix Fogsgaard, Paula Diaz Reigosa, Francesco Iannuzzo, Michael Hartmann

EXPERIMENTAL VALIDATION AND COMPARISON OF A SIC MOSFET BASED 100 KW
1.2 KV 20 KHZ THREE-PHASE DUAL ACTIVE BRIDGE CONVERTER USING TWO
VECTOR GROUPS 2265

*Thomas Lagier, Piotr Dworakowski, Cyril Buttay, Philippe Ladoux, Andrzej Wilk, Philippe
Camail, Elissa Cresenta Anak Justin*

IMPEDANCE ANALYSIS OF AN AUTOMOTIVE DC BUS.........................2274

Michael Schlüter, Marius Gentejohann, Sibylle Dieckerhoff

A NEW DUAL-MODE MPPT ALGORITHM APPLIED TO A QUADRATIC CONVERTER IN
A SOLAR ENERGY SYSTEM 2284

Ahmad Ghamrawi, Jean-Paul Gaubert, Driss Mehdi

THERMAL MODEL DEVELOPMENT FOR SIC MOSFETS ROBUSTNESS ANALYSIS
UNDER REPETITIVE SHORT CIRCUIT TESTS 2293

M. Pulvirenti, D. Cavallaro, N. Bentivegna, S. Cascino, E. Zanetti, M. Saggio

COMPENSATION OF THE RADIAL AND CIRCUMFERENTIAL MODE 0 VIBRATION OF A
PERMANENT MAGNET ELECTRIC MACHINE BASED ON AN EXPERIMENTAL
CHARACTERISATION.........................2303

Jan Andresen, Stephan Vip, Axel Mertens, Sebastian Paulus

MEASUREMENT BASED MODEL FOR THE CALCULATION OF CURRENT DISTRIBUTIONS BETWEEN PARALLELED POWER SEMICONDUCTORS DURING HIGH CURRENT OPERATION 2312
Julian Da Cunha

DUAL-LOOP CONTROL SCHEME WITH OPTIMIZED TYPE-III CONTROLLER BASED ON GENETIC ALGORITHM FOR 6-PHASE INTERLEAVED CONVERTER IN ELECTRIC VEHICLE DRIVETRAINS 2320
Dai-Duong Tran, Sajib Chakraborty, Thomas Geury, Joeri Van Mierlo, Mohamed El Baghdadi, Omar Hegazy

HIGH SENSITIVITY CURRENT TRANSFORMER WITH LOW SETTLING TIME, FOR MAGNIFIED AC CURRENT MEASUREMENTS IN PULSED APPLICATIONS 2331
Georgios Tsolaridis, Pascal Seiler, Juergen Biela

LOSS SEPARATION IN HARD- AND SOFT-SWITCHING GAN HEMTS OPERATED IN A 10 KW ISOLATED DC/DC CONVERTER 2341
Jan Böcker, Sören Heucke, Sibylle Dieckerhoff

A SWITCHED-MODE POWER AMPLIFIER FOR ION ENERGY CONTROL IN PLASMA ETCHING 2350
Qihao Yu, Erik Lemmen, Korneel Wijnands, Bas Vermulst

EXPLORING THE BOUNDARIES AND EFFECTS OF THE DISCONTINUOUS CONDUCTION MODE IN H-BRIDGE INVERTER WITH DEAD-TIME 2358
Qihao Yu, Erik Lemmen, Korneel Wijnands, Bas Vermulst

FIGURES-OF-MERIT AND CURRENT METRIC FOR THE COMPARISON OF IGCTS AND IGBTS IN MODULAR MULTILEVEL CONVERTERS 2366
Arthur Boutry, Cyril Buttay, Dong Dong, Rolando Burgos, Bruno Lefebvre, Florent Morel, Colin Davidson

ZERO-CURRENT SWITCHING WITH LC RESONANT TANK CIRCUIT AND CAPACITOR ISOLATION DC-DC CONVERTER 2376
Hideki Jonokuchi, Osamu Nakashima, Daichi Hiwatari, Hiroshi Hirayama

A FULL STATE-VARIABLE PREDICTIVE CONTROL OF BI-DIRECTIONAL BOOST CONVERTERS WITH GUARANTEED STABILITY 2386
Yu Li, Zhenbin Zhang, Ralph Kennel

SYSTEM-LEVEL RELIABILITY ANALYSIS OF A REPAIRABLE POWER ELECTRONIC-BASED POWER SYSTEM CONSIDERING NON-CONSTANT FAILURE RATES 2393
Amirali Davoodi, Yongheng Yang, Tomislav Dragicevic, Frede Blaabjerg

AN EFFICIENCY ANALYSIS OF A FERRITE MAGNET ASSISTED SYNCHRONOUS RELUCTANCE MACHINE FOR LOW POWER DRIVES INCLUDING FLUX WEAKENING 2403
Matthias Hofer, Mario Nikowitz, Thomas Kirowitz, Manfred Schrödl

HIGH PERFORMANCE LQR CONTROL OF MODULAR MULTILEVEL CONVERTERS WITH SIMPLE CONTROL STRUCTURE AND IMPLEMENTATION 2409
Min Jeong, Simon Fuchs, Jürgen Biela

FAULT DETECTION AND CLASSIFICATION BASED ON DEEP LEARNING IN LVDC OFF-GRID SYSTEM 2419
Iurii Demidov, Antti Pinomaa, Andrey Lana, Olli Pyrhönen

AN INPUT-SERIES OUTPUT-INDEPENDENT FULL-BRIDGE DUAL ACTIVE BRIDGE CONVERTER WITH SOFT-SWITCHING CHARACTERISTICS FOR CHARGING AND BALANCING ELECTRIC VEHICLE BATTERY STACKS 2429
Alex V. Mirtchev, Emmanuel C. Tatakis

A METHOD TO SEARCH GLOBAL MAXIMA BY PERMANENT MONITORING OF VOLTAGE AND CURRENT OF EACH PV PANEL 2439
Shailendra Rajput, Moshe Averbukh

SURVEY AND COMPARISON OF 1D/2D ANALYTICAL MODELS OF HF LOSSES IN LITZ WIRE.. 2446
Qingchao Meng, Jürgen Biela

HIGH-FREQUENCY SIC-BASED MEDIUM VOLTAGE QUASI-2-LEVEL FLYING CAPACITOR DC/DC CONVERTER WITH ZERO VOLTAGE SWITCHING.................... 2457
Rafal Kopacz, Przemyslaw Trochimiuk, Grzegorz Wrona, Jacek Rabkowski

SMART FUEL CELL MODULE (6.5 KW) FOR A RANGE EXTENDER APPLICATION 2467
Pascal Bazin, Bruno Beranger, Jacques Ecrabey, Laurent Garnier, Sylvain Mercier

IMPACT OF THE INITIAL TRANSIENT INTERRUPTION VOLTAGE (ITIV) ON THE DESIGN AND OPERATION OF HYBRID CURRENT-INJECTION DC CIRCUIT BREAKERS............ 2475
Andreas Jehle, Jürgen Biela

FOUR QUADRANT BUS-TIE SWITCH FOR PROTECTION OF SHIPBOARD POWER SYSTEMS 2486
Gabriele Ulissi, Seong-Yong Lee, Drazen Dujic

ESTIMATION OF AN UNBALANCED GRID IMPEDANCE USING A THREE-PHASE POWER CONVERTER 2495
Jarno Kukkola, Ville Pirsto, Mikko Routimo, Marko Hinkkanen

FAULT DIAGNOSIS OF HVDC TRANSMISSION SYSTEM USING WAVELET ENERGY ENTROPY AND THE WAVELET NEURAL NETWORK 2505
Cuicui Liu, Feng Wang, Fang Zhuo, Ziqian Zhang

REDUCING THE ENERGY STORAGE REQUIREMENTS OF MODULAR MULTILEVEL CONVERTERS WITH OPTIMAL CAPACITOR VOLTAGE TRAJECTORY SHAPING 2513
Simon Fuchs, Min Jeong, Jürgen Biela

LEAKAGE INDUCTANCE MODELLING OF TRANSFORMERS: ACCURATE AND FAST MODELS TO SCALE THE LEAKAGE INDUCTANCE PER UNIT LENGTH.................... 2524
Richard Schlesinger, Jürgen Biela

A GAN-BASED DC/DC CONVERTER FOR E-VEHICLES APPLICATIONS 2535
Eduardo F. De Oliveira, Sebastian Sprunck, Jonas Pfeiffer, Peter Zacharias

THEORY OF INFLUENCING THE BREATHING MODE AND TORQUE PULSATIONS OF PERMANENT MAGNET ELECTRIC MACHINES WITH HARMONIC CURRENTS 2545
Jan Andresen, Stephan Vip, Axel Mertens, Sebastian Paulus

POWER HARDWARE IN THE LOOP SYSTEM BASED ON INTERLEAVED CONVERTER AND FPGA - APPLICATION TO DC AND AC SIDE EMULATION FOR PHOTOVOLTAIC INVERTER TESTING.. 2554
R. Kadri, R. Bakri, A. Omrane, F. Colas, F. Delpech

IMPLEMENTATION OF TAPIR SWITCHING CELLS WITH INTEGRATED DIRECT AIR-COOLING FOR SIC POWER DEVICES 2564
Wendpanga Fadel Bikinga, Kouceila Alkama, Bachir Mezrag, Jean Michel Guichon, Yvan Avenas

EFFECT OF UNIPOLAR AND BIPOLAR SPWM ON THE LIFETIME OF DC-LINK CAPACITORS IN SINGLE-PHASE VOLTAGE SOURCE INVERTERS 2573
Silpa Baburajan, Saeed Peyghami, Dinesh Kumar, Frede Blaabjerg, Pooya Davari

TRANSIENT THERMAL MODELS OF CAPACITORS AND INDUCTORS FOR SYSTEM OPTIMIZATION 2583
Vasilios Karaventzas, Juergen Biela, Felix Rodriguez Mateos

ENERGY MANAGEMENT FOR ISOLATED RENEWABLE-POWERED MICROGRIDS USING REINFORCEMENT LEARNING AND GAME THEORY 2594
Rui Hu, Alexis Kwasinski

ALL-GAN BIDIRECTIONAL ANPC-BASED RESONANT DC-DC CONVERTER 2603
Tino Kahl, Laurenz Wernicke, Sibylle Dieckerhoff, Christopher Fromme, Marvin Tannhäuser, Ag Siemens

LIFETIME ESTIMATION AND DIMENSIONING OF THE MACHINE-SIDE CONVERTER FOR PUMPING-CYCLE AIRBORNE WIND ENERGY SYSTEM 2613
Bakr Bagaber, Patrick Junge, Axel Mertens

A DESIGN OF HIGH-POWER INVERTER CIRCUIT INCLUDING GAN POWER DEVICES 2623
Takashi Sawada, Hiroshi Tadano, Koji Shiozaki

SPEED SENSORLESS COMMISSIONING OF RESONATING MECHANICAL SYSTEM IN ELECTRIC DRIVES 2630
A. Putkonen, N. Nevaranta, O. Liukkonen, M. Niemelä, O. Pyrhönen

CONTROL OF A TWO-STAGE, SINGLE-PHASE GRID-TIED, GAN BASED SOLAR MICRO-INVERTER 2638
Anthony Bier, Van Sang Nguyen, Stéphane Catellani, Jérémy Martin

A DC/DC BUCK-BOOST CONVERTER CONTROL USING SLIDING SURFACE MODE CONTROLLER AND ADAPTIVE PID CONTROLLER 2648
Bassem Saleh, Ahmed Teirelbar, Amr Wasfi

SENSORLESS NEUTRAL POINT VOLTAGE STABILIZATION IN THREE-PHASE FOUR-WIRE CONVERTERS 2656
Xinwei Xu, Gabriel Tibola, Jorge L. Duarte

BIDIRECTIONAL ISOLATED RIPPLE CANCEL TRIPLE ACTIVE BRIDGE DC-DC CONVERTER 2666
Takahiro Ohta, Pin-Yu Huang, Yuichi Kado

DESIGN OF THE SPEED SENSORLESS FIELD ORIENTED CONTROL SYSTEM FOR INDUCTION MOTORS CONSIDERING SUDDEN CHANGE OF THE ROTOR SPEED 2675
Yoshiki Sakurazawa, Osamu Yamazaki, Kazuaki Yuki, Yosuke Nakazawa, Kenji Natori, Keiichiro Kondo

EFFICIENCY POTENTIAL OF SOLID-STATE PULSE MODULATORS USING SIC DEVICES 2684
Spyridon Stathis, Michael Jaritz, Sebastian Blume, Jürgen Biela

EFFICIENT AND SCALABLE POWER CONTROL IN MULTI-PORT ACTIVE-BRIDGE CONVERTERS 2695
Soleiman Galeshi, David Frey, Yves Lembeye

COMPARISON OF PRESS-PACK AND WIRE-BONDING TECHNOLOGIES FOR SIC MOSFETS UNDER SHORT-CIRCUIT CONDITIONS 2704
Ran Yao, Francesco Iannuzzo, Amir Sajjad Bahman, Hui Li

ERROR INDUCED BY THE OPTICAL PATH OF A HIGH ACCURACY AND HIGH BANDWIDTH OPTICAL CURRENT MEASUREMENT SYSTEM 2712
Stefan Rietmann, Jürgen Biela

ANALYSIS OF THE RMS CURRENT STRESS ON THE DC LINK CAPACITORS OF THE FOUR PHASE 3-LEVEL T-TYPE VOLTAGE SOURCE CONVERTER 2723
Zoran Miletic, Werner Tremmel, Roland Bründlinger, Johannes Stöckl, Petar J. Grbovic

AN ADAPTIVE DROOP CONTROL METHOD FOR INTERLINK CONVERTER IN HYBRID AC/DC MICROGRIDS 2733
Mohammad S. Golsorkhi, Rasool Heydari, Mehdi Savaghebi

SIMPLIFIED CALCULATION OF PARASITIC ELEMENTS AND MUTUAL COUPLINGS OF WIDE-BANDGAP POWER SEMICONDUCTOR MODULES 2743
Mohammad Ali, Jens Friebe, Axel Mertens

VARIABLE-SPEED-DRIVE-BASED SENSORLESS ESTIMATION OF PUMP SYSTEM RESERVOIR FLUID LEVEL 2753
Santeri Pöyhönen, Aleksi Simola, Jero Ahola

ANALYSIS OF SWITCHING PERFORMANCE AND EMI EMISSION OF SIC INVERTERS UNDER THE INFLUENCE OF PARASITIC ELEMENTS AND MUTUAL COUPLINGS OF THE POWER MODULES 2763
Mohammad Ali, Jan-Kaspar Müller, Jens Friebe, Axel Mertens

WIRE-WOUND MULTI-PHASE STATOR BASED EMEH WITH MPPT SELF-POWERED ENERGY MANAGEMENT SYSTEM 2773
Mahmoud Shousha, Dragan Dinulovic, Talha Zafar, Michael Brooks, Martin Haug

COMPARISON OF OPTIMIZED MOTOR-INVERTER SYSTEMS USING A STACKED POLYPHASE BRIDGE CONVERTER COMBINED WITH A 3-, 6-, 9-, OR 12-PHASE PMSM 2780
Thilo Bringezu, Jürgen Biela

DESIGN OF A PULSE MODULATOR BASED ON TRANSMISSION LINES FOR GENERATING FAST CURRENT PULSES FOR PLASMA DRILLING 2791
Oliver Keel, Melissa Artiglia, Juergen Biela

ANALYSIS OF CURRENT IN PULSATING DC LINK CONVERTER WITH ZERO VOLTAGE TRANSITION 2802
Daniele Marciano, Giovanni Busatto, Carmine Abbate, Annunziata Sanseverino, Davide Tedesco, Francesco Velardi

SIGNAL INJECTION FOR SENSORLESS CURRENT SHARING WITH EXPERIMENTAL VERIFICATION ON 1 MHZ GAN PROTOTYPE 2812
N. Boškovic, J. Duarte, E. A. Lomonova

MODELLING AND ANALYSIS OF SENSORLESS CURRENT SHARING APPROACH 2820
N. Boškovic, J. Duarte

PWM-INDUCED HARMONIC POWER IN 75 KW IM DRIVE SYSTEM 2829
Lassi Aarniovuori, Hannu Kärkkäinen, Markku Niemelä, Juha Pyrhönen

PROPOSAL OF BOOST CONVERTER WITHOUT REACTOR USING OPEN-ENDED
WINDING PMSM FOR PHOTOVOLTAIC PUMP SYSTEM.. 2838
Akihiro Okazaki, Sari Maekawa

THE PROPOSAL OF DISCRIMINATING STABLE CONTROL BANDWIDTH USING ANN IN
SENSORLESS SPEED CONTROL SYSTEM FOR PMSM.. 2844
Ami Tanaka, Sari Maekawa

COST FUNCTION DESIGN FOR STABILITY ASSESSMENT OF MODULATED MODEL
PREDICTIVE CONTROL.. 2851
Jordan P. Zucuni, Fernanda Carnielutti, Humberto Pinheiro, Margarita Norambuena, Jose
Rodriguez

A ROBUST FUZZY-BASED CONTROL TECHNIQUE FOR WIND FARM TRANSIENT
VOLTAGE STABILITY USING SVC AND STATCOM: COMPARISON STUDY 2860
Reza Ebrahimi, Vahid Eslampanah, Hossein Madadi Kojabadi, Mohammadreza Azizian,
Naser Nourani Esfetanaj, Dao Zhou

TEMPERATURE EVOLUTION AS AN EFFECT OF WIRE-BOND FAILURES IN A MULTI-
CHIP IGBT POWER MODULE... 2865
N. Degrenne, R. Delamea, S. Mollov

COST OF ENERGY ASSESSMENT OF WIND TURBINE CONFIGURATIONS 2873
Catalin Dincan, Philip Kjær, Lars Helle

ENERGY MANAGEMENT IN A MULTI-SOURCE SYSTEM USING ISOLATED DC-DC
RESONANT CONVERTERS... 2881
M. Arazi, A. Payman, M. B. Camara, B. Dakyo

LONG-TERM CLIMATE IMPACT ON IGBT LIFETIME.. 2888
Martin Vang Kjaer, Yongheng Yang, Huai Wang, Frede Blaabjerg

COMMUNICATION-FREE SECONDARY FREQUENCY AND VOLTAGE CONTROL OF
VSC-BASED MICROGRIDS: A HIGH-BANDWIDTH APPROACH 2898
Rasool Heydari, Mohammad S. Golsorkhi, Mehdi Savaghebi, Tomislav Dragicevic, Frede
Blaabjerg

OFFSHORE WIND FARM LAYOUT OPTIMIZATION CONSIDERING WAKE EFFECTS 2907
Asma Dabbabi, Salvy Bourguet, Rodica Loisel, Mohamed Machmoum

SMALL-SIGNAL STABILITY ANALYSIS OF SMART GRIDS CONSIDERING HIGH
PENETRATION OF POWER ELECTRONICS CONVERTERS AND ENERGY MARKETS 2917
Javiera Meneses, Patricio Mendoza-Araya

COMPONENT-LEVEL RELIABILITY ASSESSMENT OF A DIRECT-DRIVE PMSG WIND
POWER CONVERTER CONSIDERING LONG-TERM AND SHORT-TERM THERMAL
CYCLES.. 2928
Shuaichen Ye, Dao Zhou, Frede Blaabjerg

A SUBMODULE IMPLEMENTATION FOR PARALLEL CONDUCTION OF DIODES IN
MODULAR MULTILEVEL CONVERTERS... 2938
Martin Geske, Duro Basic, Christian Keller, Thomas Brückner

EVALUATION OF THE I_{MAX}-F_{SW}-DV/DT TRADE-OFF OF HIGH VOLTAGE SIC MOSFETS BASED ON AN ANALYTICAL SWITCHING LOSS MODEL 2946
Anliang Hu, Jürgen Biela

PROTECTION MEASURES FOR MODULAR MULTILEVEL CONVERTERS IN CASE OF DC SHORT-CIRCUIT FAULTS 2957
Martin Geske, Duro Basic, Roland Jakob, Christian Keller, Thomas Brückner

INVESTIGATION ON PARALLEL OPERATION OF TWO MMC-HVDC LINKS IN GRID FORMING CONNECTED TO AN EXISTING NETWORK 2967
H. Saad, P. Rault, S. Dennetière

MODELLING AND EXPERIMENTAL VALIDATION OF A LABORATORY-SCALED HVDC CABLE EMULATOR TESTED IN AN MMC-BASED PLATFORM 2977
Enric Sánchez-Sánchez, Adrià Junyent-Ferré, Eduardo Prieto-Araujo, Oriol Gomis-Bellmunt, Tim Green

DAISY CHAIN PN CELL FOR MULTILEVEL CONVERTER USING GAN FOR HIGH POWER DENSITY 2987
Faheem Ahmad, Asger Bjørn Jørgensen, Szymon Michal Beczkowski, Stig Munk-Nielsen

GRID-FREQUENCY VIENNA RECTIFIER AND ISOLATED CURRENT-SOURCE DC-DC CONVERTERS FOR EFFICIENT OFF-BOARD CHARGING OF ELECTRIC VEHICLES 2996
Jacek Rabkowski, Andrei Blinov, Denys Zinchenko, Grzegorz Wrona, Mariusz Zdanowski

UNIDIRECTIONAL THYRISTOR-BASED DC-DC CONVERTER FOR HVDC CONNECTION OF OFFSHORE WIND FARMS 3006
Pierre Le Métayer, Piotr Dworakowski, Jose Maneiro

INDUCTOR SIZE EVALUATION OF AN ELECTROMAGNETIC INTERFERENCE FILTER FOR A TWO-LEVEL POWER FACTOR CORRECTION RECTIFIER USING DIFFERENT MODULATION TECHNIQUES 3015
Mohammad Najjar, Alireza Kouchaki, Morten Nymand

EVALUATION OF MMCS FOR HIGH-POWER LOW-VOLTAGE DC-APPLICATIONS IN COMBINATION WITH THE MODULE LLC-DESIGN 3024
Roland Unruh, Frank Schafmeister, Joachim Böcker

IRON LOSS CHARACTERISTICS OF MNZN FERRITES UNDER GAN INVERTER EXCITATION IN THE MHZ ORDER 3034
Wilmar Martinez, Camilo Suarez, Federico Ibanez

VIBRATION SUPPRESSION AND CONTROL PARAMETER DESIGN OF A SENSORLESS PMSM ROTARY COMPRESSOR DRIVE 3044
Tao Li, Chaohui Liang

3D PCB PACKAGE FOR GAN INVERTER LEG WITH LOW EMC FEATURE 3054
Pawel B. Derkacz, Jean-Luc Schanen, Pierre-Olivier Jeannin, Piotr Musznicki, Piotr J. Chrzan, Mickael Petit

ESTIMATION OF THE WINDING LOSSES OF MEDIUM FREQUENCY TRANSFORMERS WITH LITZ WIRE USING AN EQUIVALENT PERMEABILITY AND CONDUCTIVITY METHOD 3064
Mohammad Kharezy, Morteza Eslamian, Torbjörn Thiringer

IMPROVEMENT OF DRIVING EFFICIENCY OF PMSM BY USING MODIFIED TRAPEZOIDAL MODULATING SIGNAL 3071
Kento Betto, Satoshi Joryo, Toshimitsu Morizane

DESIGN AND CONTROL OF A VIRTUAL DC-LINK FOR A FULL GAN-BASED SINGLE PHASE CONVERTER WITH HIGH POWER DENSITY 3081
Yugandhara H. Wankhede, Leon Fauth, Jens Friebe

USING BOTH THE CIRCULATING CURRENTS AND THE COMMON-MODE VOLTAGE FOR THE BRANCH ENERGY CONTROL OF MODULAR MULTILEVEL CONVERTERS 3091
Rebecca Dierks, Jakub Kucka, Axel Mertens

ANALYTICAL HARMONIC CURRENT MODEL FOR A PERMANENT MAGNET ASSISTED SYNCHRONOUS RELUCTANCE MOTOR (PMA-SYNRM) FED BY PWM INVERTER 3101
Jessica Neumann, Carole Hénaux, Maurice Fadel, Etienne Founier, Dany Prieto, Mathias Tientcheu Yamdeu

GENERALIZED SMALL-SIGNAL AVERAGED SWITCH MODEL ANALYSIS OF A WBG-BASED INTERLEAVED DC/DC BUCK CONVERTER FOR ELECTRIC VEHICLE DRIVETRAINS 3111
Sajib Chakraborty, Dai-Duong Tran, Joeri Van Mierlo, Omar Hegazy

ADAPTIVE PREDICTIVE-DPC FOR LCL-FILTERED GRID CONNECTED VSC WITH REDUCED NUMBER OF SENSORS 3119
Hosein Gholami-Khesht, Pooya Davari, Frede Blaabjerg

FPGA IMPLEMENTATION OF MODIFIED SPACE VECTOR MODULATION (SVM) FOR HIGH-FREQUENCY HYBRID ACTIVE NEUTRAL-POINT-CLAMPED (NPC) POWER FACTOR CORRECTION RECTIFIER 3129
Mohammad Najjar, Alireza Kouchaki, Morten Nymand

ENHANCED FLUX CONTROL INCLUDING A CLOSED LOOP VOLTAGE CONTROLLER TO OPTIMIZE THE VOLTAGE USAGE AND THE TORQUE COMPUTATION FOR A 48V IPMSM 3137
Felix Bertele, Ulrich Ammann, Christoph Cheshire, Tobias Röser

EXTENDED BOOST PV INVERTER TOPOLOGY FOR THE REDUCTION OF COMMON-MODE LEAKAGE CURRENT IN THREE-PHASE APPLICATIONS 3146
Georgios I. Orfanoudakis, Eftychios Koutroulis, Michael A. Yuratich, Suleiman M. Sharkh

A ROBUST CONTROL DESIGN TO REAL-TIME CONDITIONS AND MODELLING OF A MICROGRID 3156
Iréna Horvatic, Delphine Riu, Moataz Elsied, Sébastien Benjamin

DESIGN OF MODULAR LOW-PROFILE FREQUENCY CONVERTER FOR MULTI-MOTOR MANIPULATORS 3166
Tomas Glasberger, Zdenek Kehl, Tomas Kosan, Jan Molnar

STUDY OF THE CONTROL OF A NEW AC VOLTAGE STABILIZER USING LINEAR CONTROLLER WITH REFERENCE FRAME TRANSFORMATION 3172
Bunthern Kim, Etienne Boulaud, Emile Boisaubert, Sokchea Am, Phok Chrin

HYBRID ENERGY STORAGE SYSTEM FOR MVDC-GRIDS 3179
Florian Mahr, Johann Jaeger, Stefan Henninger, Hubert Rubenbauer

A COMBINED MODEL FOR OPTIMAL POWER FLOW APPLIED TO MT-HVDC SYSTEMS 3189
Fernando Torres, Javier Muñoz, Fredy Muñoz, Claudio Roa

CHARACTERIZATION OF LITHIUM ION SUPERCAPACITORS 3198
Zeyang Geng, Felix Mannerhagen, Torbjöm Thiringer

GREY WOLF OPTIMIZER BASED PREDICTIVE TORQUE CONTROL FOR ELECTRIC
VEHICLE APPLICATIONS 3205
Ali Djerioui, Azeddine Houari, Mohamed Machmoum, Malek Ghanes, Tedjani Mesbahi,
Mohamed Fouad Benkhoris

OPERATION PRINCIPLE AND PERSPECTIVE PERFORMANCES OF METAL OXIDE
VACUUM FIELD EFFECT TRANSISTOR - MOVFET 3210
Davide Patti, G. Busatto, G. Golluccio, D. Marciano, A. Sanseverino, F. Velardi

IMPROVED METHODOLOGY FOR PREDICTING CORRELATED COLOR TEMPERATURE
IN MIXED LED LIGHTING SOURCES 3217
Thais E. Bolzan, Bruno F. Almeida, Renan R. Duarte, Vitor C. Bender, Rafael A. Pinto

DC MICROGRID CONCEPT FOR MINE ENVIRONMENT 3227
Jooa Pursiainen, Jenni Rekola, Raimo Juntunen, Mikko Valtee, Pasi Peltoniemi

A COMPARISON OF TWO-STAGE INVERTER AND QUASI-Z-SOURCE INVERTER FOR
HYBRID ENERGY STORAGE APPLICATIONS 3237
V. Castiglia, R. Miceli, F. Blaabjerg, Y. Yang

STATE ESTIMATION FOR MEDIUM AND LOW VOLTAGE DISTRIBUTION GRIDS
BASED ON NEAR REAL-TIME GRID MEASUREMENTS AND DELAYED SMART
METERS DATA 3247
Mohammad Rayati, Thomas Pidancier, Mauro Carpita, Mokhtar Bozorg

GROUND FAULT ACTIVE COMPENSATION IN EMULATED DISTRIBUTION GRID OF 10
KV 3257
Tomáš Komrska, Antonín Glac, Jakub Talla, Bohumil Skala, Jan Štepánek, Lubeš Streit,
Zdenek Peroutka

MODELING OF A POWER TRANSFORMER INCLUDING HIGHER ORDER RESONANCES 3263
Lukas Reißenweber, Alexander Stadler

A COMPARISON OF TWO STATE-SPACE MODELS OF AN INDUCTION MACHINE
CONSIDERING DIFFERENT SETS OF WINDING DISTRIBUTION HARMONICS 3272
Julien Cordier, Stefan Klass, Ralph Kennel

PERFORMANCE IMPROVEMENT FOR PLUG-IN REVERSE CONDUCTING IGBTS
THROUGH GATE-VOLTAGE OBSERVATION 3282
Daniel Lexow, Hans-Günter Eckel

DIFFERENTIAL FLATNESS FOR SMOOTH TRANSITION BETWEEN GRID-CONNECTED
AND STANDALONE MODE OF THREE-PHASE INVERTER 3289
Abdelhakim Saim, Azeddine Houari, Mourad Ait-Ahmed, Mohamed Machmoum, Josep. M
Guerrero

DIFFERENTIAL MODEL EMI FILTER ANALYSIS FOR INTERLEAVED BOOST PFC
CONVERTERS CONSIDERING OPTIMAL PHASE SHIFTING 3295
Naser Nourani Esfetanaj, Yamen Saad, Omar Ahmed Sakaria, Huai Wang, Pooya Davari

MODULAR HYBRID DC BREAKER-BASED ADAPTIVE AUTO-RECLOSING METHOD
FOR MMC-HVDC SYSTEMS 3305
Hossein Iman-Eini, M. Langwasser, L. Camurca, Marco Liserre

MULTISTEP MPC OF DUAL INVERTER FOR SWITCHING LOSSES OPTIMIZATION 3314
Martin Votava, Tomas Glasberger, Zdenek Peroutka

A HIGH-EFFICIENCY CONTROL OF A DOUBLE-INPUT CONVERTER FOR RENEWABLE
ENERGIES AND HYBRID VEHICLES 3321
Mario Marchesoni, Massimiliano Passalacqua, Luis Vaccaro

DEAD-TIME INFLUENCE ON FAST SWITCHING PULSED POWER CONVERTERS
DESIGN - A HIGH CURRENT APPLICATION FOR ACCELERATOR'S MAGNETS 3330
Ludovic Horrein, Jean-Marc Cravero, Philippe Delarue, Alain Bouscayrol, Davide Aguglia,
Carmen Ortega-Perez

DYNAMIC CHARACTERIZATION OF A SIC-MOSFET HALF BRIDGE IN HARD- AND
SOFT-SWITCHING AND INVESTIGATION OF CURRENT SENSING TECHNOLOGIES 3340
Janine Ebersberger, Jan-Kaspar Müller, Axel Mertens

POWER SUPPLY DESIGN CONSIDERATIONS FOR 400HZ AIRCRAFT APPLICATIONS 3348
Bilal Ahmad, Jorma Kyyrä, Juha Mäkelä

DC CAPACITOR VOLTAGE FEEDBACK METHOD FOR A PEAK VOLTAGE
SUPPRESSION CONTROL WITH MULTIPLE LEG-SHORT-CIRCUITS USING SIC-
MOSFETS EMPLOYED IN POWER CONVERTERS 3358
Tomoyuki Mannen, Takanori Isobe, Keiji Wada

INVESTIGATION OF BOND WIRE LIFT-OFF BY ANALYZING THE CONTROLLER
OUTPUT VOLTAGE HARMONICS FOR THE PURPOSE OF CONDITION MONITORING 3366
Firat Yüce, Marc Hiller

FRUGAL INNOVATION FOR SUSTAINABLE RURAL ELECTRIFICATION 3376
Bunthern Kim, Phok Chrin, Maria Pietrzak-David, Pascal Maussion

A CURRENT-MODULUS DERIVATIVE-BASED PROTECTION METHOD IN A FLEXIBLE
DC GRID 3385
Jianquan Liao, Niancheng Zhou, Qianggang Wang

COMPARATIVE ASSESSMENT OF VOLTAGE MODULATION METHODS FOR
ASYMMETRIC SIX-PHASE MACHINES 3393
R. S. Kanchan, Omer Ikram Ul Haq, Luca Peretti

SIMULATION AND MEASUREMENT-BASED ANALYSIS OF EFFICIENCY
IMPROVEMENT OF SIC MOSFETS IN A SERIES-PRODUCTION READY 300 KW / 400 V
AUTOMOTIVE TRACTION INVERTER 3403
A. Nisch, M. Heller, W. Wondrak, A. Bucher, C. Hasenohr, K. Kefer, B. Lunz, A. Pawellek, A.
Smit, M. Gärtner, N. Twardon, U. Kirchenberger

VALIDITY OF POWER CYCLING LIFETIME MODELS FOR MODULES AND EXTENSION
TO LOW TEMPERATURE SWINGS 3413
Josef Lutz, Christian Schwabe, Guang Zeng, Lukas Hein

ROADMAP FOR DC 3422
Pavol Bauer

THE ROLE OF COLLABORATIVE RESEARCH TO SUPPORT INNOVATION FOR CLEAN
ENERGY TRANSITION 3424
Hubert De La Grandiere

THOMAS EDISON VINDICATED — THE RESURGENCE OF DC IN MV AND HV POWER GRIDS 3425

Colin Davidson

INTEGRATION OF ELECTRIC MOBILITY IN THE FRENCH PUBLIC ELECTRICITY DISTRIBUTION NETWORK 3426

Anne-Sophie Cochelin

A CRITICAL ROLE FOR R&I FOR CLEAN ENERGY FOR THE EU GREEN AND DIGITAL RECOVERY 3427

Hélène Chraye

Author Index

Feasibility Study of a Superconducting Power Filter for HVDC grids

Loïc Quéval[1,2], Olivier Despouys[3], Frédéric Trillaud[4], Bruno Douine[5]

[1] Université Paris-Saclay, CentraleSupélec, CNRS, Laboratoire de Génie Electrique et Electronique de Paris, 91192, Gif-sur-Yvette, France. [2] Sorbonne Université, CNRS, Laboratoire de Génie Electrique et Electronique de Paris, 75252, Paris, France. [3] Reseau de Transport d'Electricité, Paris la Defense, France. [4] Instituto de Ingeniería, National Autonomous University of Mexico (UNAM), Mexico city, Mexico. [5] GREEN, Université de Lorraine, Nancy, France. Email: loic.queval@geeps.centralsupelec.fr

Keywords

≪HVDC≫, ≪Passive filter≫, ≪Device≫, ≪Superconductors≫, ≪Emerging technology≫.

Abstract

A new application of high temperature superconducting technology for HVDC grids is introduced. The device referred to as "superconducting power filter" (ScPF) aims at increasing the stability of HVDC grids by adding a current-dependent resistance to the grid. In comparison with other stabilization techniques, a ScPF achieves a fully passive stabilization with virtually no losses in nominal operation. To clarify its feasibility for HVDC grids, the CIGRE B4 DSC1 benchmark is considered. Using an electrical-thermal model of the device, the stability of the DC grid is numerically assessed. The first steps towards an experimental proof of concept are presented.

Introduction

With the rapid advance of power electronics technologies, DC electrical grids are expanding at every scale and various high temperature superconducting devices are being developed to operate in such grids. Up to now applications of superconductivity with a clear impact in the electrical sector have been high temperature superconducting (HTS) power cables and fault current limiters. Various prototypes have been built and tested around the world both in the industry and in research institutions. These technologies have demonstrated the potential to address increasing power transmission demand and protection against faults in electrical grids. In the present work, a new application of HTS technology for HVDC grids is introduced. The device referred to as "superconducting power filter" aims at increasing the stability of HVDC grids by adding a current-dependent resistance to the grid. The operation of the device is closely related to the one of a resistive fault current limiter (r-ScFCL), but here the superconductor is expected to remain in the superconducting state during transients, with the option of providing timely over-current protection if required. In comparison with other stabilization techniques, a superconducting power filter achieves a fully passive stabilization with virtually no losses in nominal operation. To clarify its feasibility for HVDC grids, the DCS1 HVDC benchmark proposed by CIGRE B4 [1] is considered for the sake of simplicity; yet, it should be emphasized that the same principle could be observed on a more complex DC grid. Using an electrical-thermal model of the device the stability of the DC grid is assessed depending on the design parameters of the device. The hereafter reported results confirm that a ScPF can indeed provide a stabilizing functionality in HVDC grids.

Stability of a 2-terminal HVDC link

The CIGRE B4 DCS1 benchmark is a 2-terminal symmetric monopole HVDC link (±200 kV). Since were are interested in the behavior of the DC grid, we make the following simplifications: (i) the length

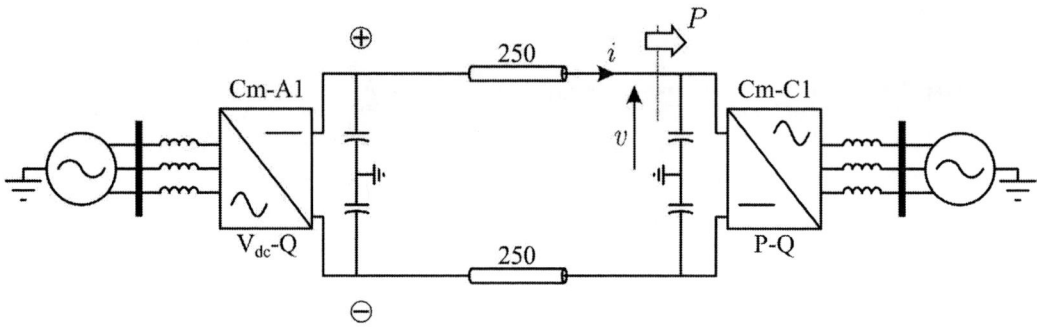

AC/DC converter station	Power rating [MVA]	Operation mode setpoints
Cm-A1	800	Vdc = 1 pu and Q = 0 pu
Cm-C1	800	P variable and Q = 0 pu

Fig. 1: CIGRE B4 DC grid test system DCS1 with simplifications (i)-(iv).

of the AC lines is neglected, (ii) the controller of the converter Cm-C1 is modified to P-Q mode, (iii) 2-level VSC converters are considered instead of modular-multilevel converters. In addition, (iv) we used a 250 km long line instead of the 200 km long one to decrease the system stability. The resulting circuit is shown in Fig. 1 with the parameters summarized in the appendix. Fig. 2 shows the equivalent circuit of the positive pole of the HVDC grid DCS1 (assuming balanced operation and ideal controllers). The constant voltage source models the DC side of the HVDC converter Cm-A1 operating in Vdc-Q mode. The RLC elements models the HVDC cable (and the DC capacitor of the 2-level VSC). The power-controlled current source (power load) models the DC side of the HVDC converter Cm-C1 operating in P-Q mode.

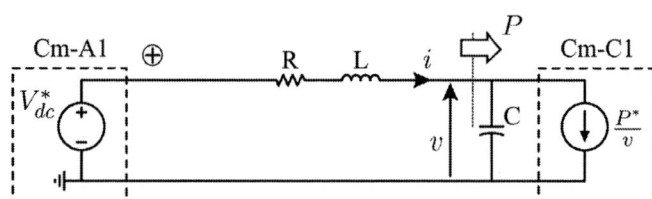

Fig. 2: CIGRE B4 DC grid test system DCS1 : equivalent circuit of the positive pole.

Because the power load behaves like a negative resistance, it reduces system damping and can lead to instability. The topic has been extensively addressed in the literature (see [3] for example). The approximated expression for the stability condition of the DC grid of Fig. 2 is recalled here,

$$P^* \leq \frac{RC}{L} V_{dc}^{*2} \tag{1}$$

In practice, the control loops modify the dynamics of the equivalent circuit, making it difficult to calculate precisely the stability limit. But this expression is useful to illustrate that above a given power reference P^*, the system becomes unstable.

To increase the stability limit, one could, for example, increase the DC grid equivalent resistance R [4]. But in nominal operation, the power losses would increase. This issue can be solved by using a superconducting power filter (ScPF). This is a non-inductive superconducting coil that behaves as a current-dependent resistance. In nominal operation for which the critical current I_c of the device is much larger than the operating current i, there is no dissipation introduced in the system. However during unstable operation, the operating current oscillates approaching I_c, the device resistance increases and damps the fluctuations to reach a new stable point of operation as shown in the next sections.

Simulation

Modeling

The grids 1 and 2 are modeled by ideal 3-phase voltage sources. The VSC Cm-A1 and Cm-B1 are modeled with an averaged VSC model. The PI controllers of the VSCs are tuned using the symmetrical optimum method [2]. The cable is modeled with a π model. The ScPF is modeled using a thermo-electric lumped-parameter model [5] that allows us to realistically represent the dynamic response of the superconductor. The parameters of the simulation model are summarized in the appendix. We consider a sequence where the active power reference for Cm-C1 P^* varies.

Results without ScPF

The results of the simulation without ScPF are shown in Fig.3. We observe that the DC voltage starts oscillating when the active power transfer P increases. The stability limit was found to be around 350 MW per pole for this test case. Note that for a line length of 200 km, corresponding to the original DCS1 benchmark, the DC voltage is stable for an active power transfer of 400 MW ie. 1 pu (not shown here because of space limitation). By increasing the line length to 250 km, the system becomes unstable for a lower power transfer and allows us to illustrate the behavior of the ScPF.

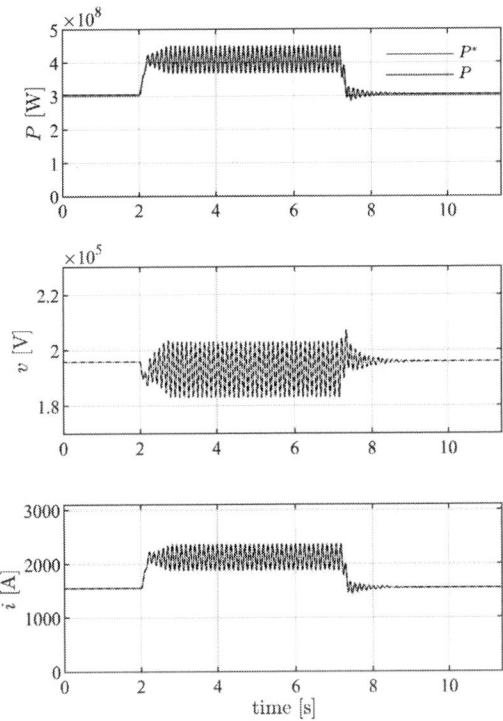

Fig. 3: Simulation results without ScPF. The continuous lines mark the signals of the positive pole. The dashed lines mark the signals of the negative pole. The active power reference ramp rate is 500 MVA/s per pole.

Results with ScPF

In this section, we insert one ScPF in series with each cables. The parameters of the ScPF are summarized in the appendix. Note that those parameters correspond to commercially available ReBCO HTS coated conductors. The device has been designed to have a critical current $n_t I_c$ of 1600 A, corresponding to the DC current around the stability limit.

Feasibility Study of a Superconducting Power Filter for HVDC grids   QUEVAL Loic

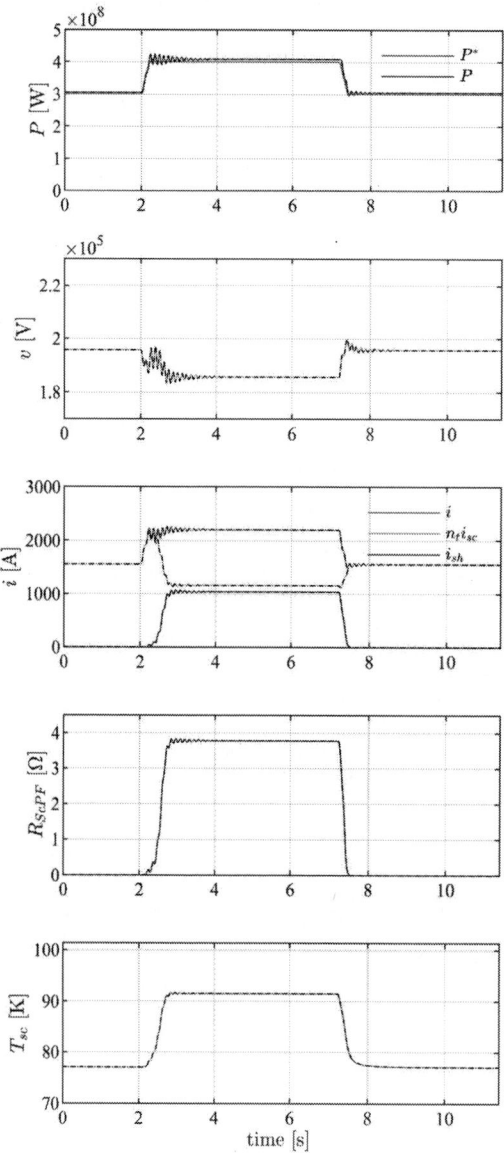

Fig. 4: Simulation results with ScPF. The continuous lines mark the signals of the positive pole. The dashed lines mark the signals of the negative pole. The active power reference ramp rate is 500 MVA/s per pole.

The results of the simulation with ScPF are shown in Fig. 4. Initially, the current going through the ScPF is lower than the device critical current $i < n_t I_c$, its resistance R_{ScPF} is zero and its temperature T_{sc} stays constant. All the current flows through the superconducting layer $i \approx n_t i_{sc}$ and no current flows through the parallel shunt $i_{sh} \approx 0$. When the power reference increases above the stability limit, the current going through the ScPF becomes larger than the device critical current $i > n_t I_c$, both its resistance R_{ScPF} and its temperature T_{sc} increase. This is because the superconductor starts to enter its resistive state thereby dissipating active power. The current now flows simultaneously through the superconductor and through the shunt. Following Equation (1), the stability limit is higher thus explaining the stabilization of the DC grid. When the power reference goes back bellow the stability limit, the temperature decreases and the superconductor resistivity falls back to zero.

This demonstrates that the ScPFs are providing the stabilization function. In addition, this shows that the ScPFs can recover their superconducting state in few tens of milliseconds after the transient, since they operated around their critical current without thermal runaway (quench).

Towards an experimental proof of concept

A laboratory scale experimental setup is being assembled to back up the numerical results. The DC test grid is constituted of an ideal constant DC voltage source and a controlled power load connected through a RLC filter in series with a superconducting power filter (Figure 5). The ideal DC voltage source emulates the HVDC converter operating in Vdc-Q mode. The controlled power load emulates the HVDC converter operating in P-Q mode.

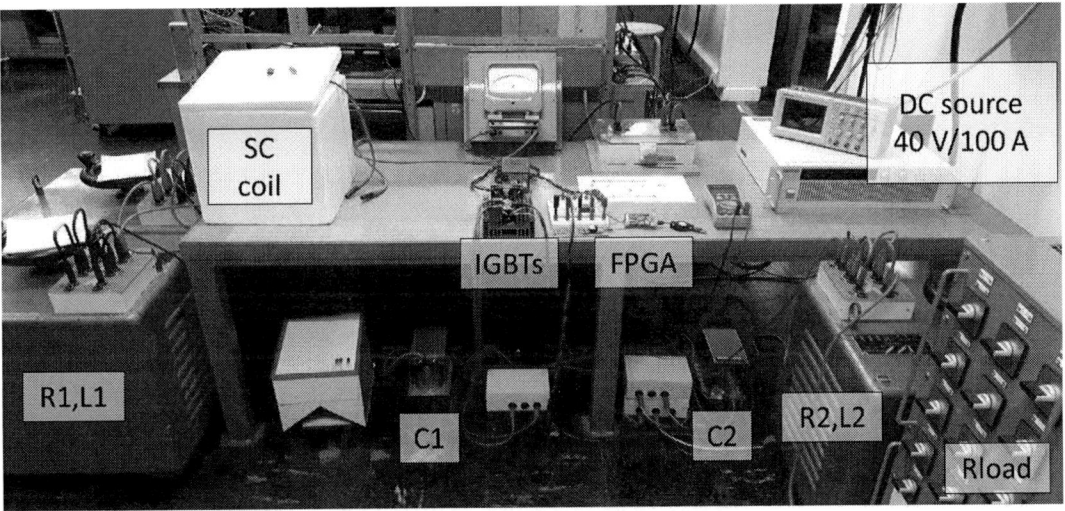

Fig. 5: DC test grid for the proof of concept.

Two ScPF prototypes have been realized. The first one, shown in Fig. Fig. 6(a), has been wound with 2 m of ReBCO HTS coated conductor manufactured by reactive co-evaporation by deposition & reaction (SuNAM SCN04150-151201-01). The second one, shown in Fig. 6(b), has been wound with 5 m of BSCCO HTS powder in tube conductor manufactured by the controlled over pressure sintering technique (Sumitomo DI-BSCCO type H). The IV curve of the BSCCO prototype at 77 K in liquid nitrogen is plotted in Fig. 7. The measured critical current obtained for the usual 1 µV/cm criteria is 175 A, which is similar to the datasheet value. This demonstrates that the superconductor has not been damaged during winding. Note that at low current, the voltage is zero (zero resistance) but when the current approaches the critical current, the voltage increases exponentially (positive resistance); it is expected that the ScPF operates in this region when providing the stabilization function, without quenching. The next step is to connect the ScPF prototype to the DC test grid and to confirm experimentally the functionality.

(a) ReBCO prototype   (b) BSCCO prototype

Fig. 6: Superconducting power filter prototypes for the proof of concept.

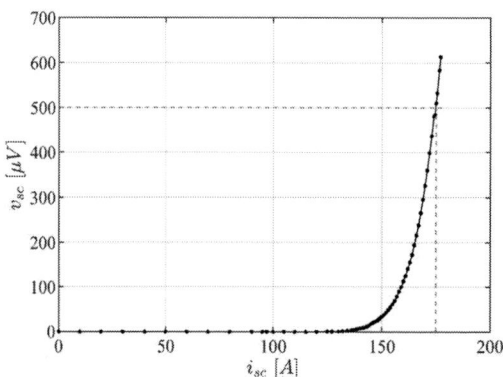

Fig. 7: IV curve of the BSCCO prototype.

Conclusion

In the present work, we proposed a "superconducting power filter" (ScPF) to increase the stability of HVDC grids. The device is fully passive and behaves like a current-dependent resistance, with zero voltage drop in nominal operation. To study the feasibility of the device for HVDC grids, we considered a test HVDC grid proposed by the Cigre Study Committee B4, and used an electrical-thermal lumped-parameter model of the device. It was demonstrated that the device operates as expected. Finally two ScPF prototypes have been built and preliminary tests have been carried out. Proof-of-concept studies are still being worked on today.

References

[1] CIGRE B4, "Guide for the Development of Models for HVDC Converters in a HVDC Grid," Technical brochure, no. 604, 2014.

[2] L. Quéval, H. Ohsaki, "Back-to-back converter design and control for synchronous generator-based wind turbines," *International Conference on Renewable Energy Research and Applications (ICRERA 2012)*, Nagasaki, Japan, Nov. 2012.

[3] M. Cupelli, F. Ponci, G. Sulligoi, A. Vicenzutti, CS. Edrington, T. El-Mezyani, A. Monti, "Power flow control and network stability in an all-electric ship," *Proceedings of the IEEE*, vol. 103, no. 12, pp. 2355-2380, Dec. 2015.

[4] R.W. Erickson, "Optimal single resistor damping of input filters," *Applied Power Electronics Conf. (APEC)*, pp. 1073-1079, Dallas, Texas, USA, March 1999.

[5] J.J. Perez-Chavez, F. Trillaud, L.M. Castro, L. Quéval, A. Polasek, R. de Andrade Jr, "Generic model of three-phase (RE)BCO resistive superconducting fault current limiters for transient analysis of power systems," *IEEE Transactions on Applied Superconductivity*, vol. 29, no. 6, pp. 1-11, Sept. 2019.

Appendix

Table I: DCS1 parameters

Parameter	Symbol	Value	Notes
Nominal 3-ph power [VA]	S_{nom}	800e6	[1, p.134]
Nominal voltage [V L-L rms]	V_{nom}	220e3	[1, p.134]
Nominal frequency [Hz]	F_{nom}	50	[1, p.128]
Nominal DC bus voltage [V]	$V_{dc,nom}$	$\pm$200e3	[1, p.128]
Tie reactor resistance [Ω]	R_r	1.21	
Tie reactor inductance [H]	L_r	0.0039	
DC bus capacitor [F]	C_{dc}	3.9789e-04	[2]
PWM generator carrier frequency [Hz]	f_{PWM}	1e3	
modulation index saturation [pu]	m_{max}	1	
PI inner loop (current) saturation [pu]	$i_{dq,lim}$	1.1	[1, p.145]
PI outer loop (DC voltage) saturation [pu]	$V_{dc,lim}$	1.2	[1, p.145]
PI inner loop (current) P gain [si]	Kp_{idq}	1.21	[2]
PI inner loop (current) I gain [si]	Ki_{idq}	119.4222	[2]
PI outer loop (DC voltage) P gain [si]	Kp_{vdc}	-0.0704	[2]
PI outer loop (DC voltage) I gain [si]	Ki_{vdc}	-0.6234	[2]
DC cable resistance [Ω/km]	r_{cable}	0.011	[1, p.145]
DC cable inductance [H/km]	l_{cable}	2.615e-3	[1, p.145]
DC cable capacitance [F/km]	c_{cable}	0.2185e-6	[1, p.145]

Table II: ScPF parameters

Variable	Value	Unit	Description
Tc	92	[K]	critical temperature at 77 K, SF
Ic0	200	[A]	tape critical current at 77 K, SF
n0	21	[-]	n-value at 77 K, SF
th_m	0.145	[mm]	thickness of metallic layer
th_sc	1	[um]	thickness of superconducting layer
tw	4	[mm]	tape width
tl	3500	[m]	length of tapes
nt	8	[-]	number of tapes in parallel
RSh	8	[Ω]	shunt resistance
Tref	77	[K]	temperature of reference

Power Decoupling Method of DC to Single-phase AC Converter using Flying Capacitor DC/DC Converter with Boundary Current Mode

Hiroki Watanabe, Keisuke Kusaka, and Jun-ichi Itoh
Nagaoka University of Technology
1603-1 Kamitomioka-machi
Nagaoka, Niigata, Japan
Tel.: +81/(258)-47.9533
E-Mail: hwatanabe@vos.nagaokaut.ac.jp, kusaka@vos.nagaokaut.ac.jp,
itoh@vos.nagaokaut.ac.jp
URL: http://itohserver01.nagaokaut.ac.jp/itohlab/index.html

Keywords

«Single phase inverter», «Power decoupling», «Boundary current mode», «Flying capacitor DC/DC converter»

Abstract

In this paper, a flying capacitor DC/DC converter with the active power decoupling capability utilizing small inductor and capacitor is introduced. The single-phase AC power converters are widely employed such as the Photovoltaic (PV) generation systems. Typically, a bulky electrolytic capacitor is required as the energy buffer for the compensation of the double-line frequency power ripple in these applications. However, the drawback of the electrolytic capacitor is often mentioned such as the large volume, and short life-time due to the high environmental temperature. On the other hand, an active power decoupling method achieves the power ripple compensation with small firm or ceramic capacitor. As the one of the active power decoupling approach, the power decoupling control method for the flying capacitor DC/DC converter has been proposed. However, this control method is mentioned under the Continuous Current Mode (CCM) condition. Therefore, large inductor is necessary to limit the current ripple within CCM.

The active power decoupling control method of the flying capacitor DC/DC converter with the Boundary Current Mode (BCM) is proposed in order to minimize the inductor in this paper. The validity of the proposed power decoupling control is confirmed by the experiment with 1-kW prototype circuit. As the experimental result, the DC-link voltage fluctuation due to the double-line frequency power ripple is reduced by 85.6% owing to the proposed control.

Introduction

The renewable energy sources as the PVs and the wind power systems are the key of the environmental problem solution, and this trend will be continue in the future power generation systems [1]-[2]. The grid-tied inverters have been widely employed for PV systems owing to market growth of PVs.

Various grid-tied inverter topology has been proposed for PV inverters. The typical inverter consists of the boost converter, Voltage Source Inverter (VSI), and the power decoupling capacitor which compensates the double-line frequency power ripple. However, the typical configuration requires the large capacitance for the power decoupling capacitor, and an electrolytic capacitor is usually accepted for the power decoupling capacitor owing to high energy density. On the other hand, it is mentioned that the electrolytic capacitor may limit the life-time of the power converter because the life-time of the capacitor depends on the environmental temperature.

The active power decoupling circuit, and it control method have been actively researched to remove the bulky electrolytic capacitor [3]-[10]. This method reduces the capacitance requirement for the power decoupling capacitor. Therefore, a small firm or ceramic capacitor with long life-time is accepted for energy buffer of the power decoupling. The basic solution is that the additional boost or buck type

DC/DC converter connect on DC-link part for the active power decoupling. However, additional component including inductor and heat sink may increases the circuit volume. On the other hand, the active power decoupling techniques without the additional components have been also proposed [11]. This method utilizes the energy buffer of the basic circuit component such as the filter capacitor or the snubber capacitor for the active power decoupling. Therefore, the additional components become minimum.

The active power decoupling circuit based flying capacitor converter (FCC) is one of this method. In this method, a flying capacitor is utilized for the energy buffer of the active power decoupling. Note that the conventional control method is considered only the condition of CCM, which means the large inductance is required for the filter inductor.

In this paper, the active power decoupling techniques with FCC under BCM is proposed in order to improve the power density of FCC. The originality of this paper is that the inductor current is regulated as the trapezoidal waveforms in order to reduce the RMS current in comparison with when this current becomes typical triangular waveforms. As the result, the conduction losses on each switching devices decreases from the typical BCM operation. Finally, the experimental results are demonstrated in order to confirm the validity of the proposed control.

Circuit configuration

Fig.1 shows the typical Active Power Decoupling (APD) circuit using boost or buck type DC/DC converter. The APD circuit regulates the capacitor voltage of C_{buf}, and the double-line frequency power ripple is compensated. The capacitor energy w_c is expressed as

$$w_c = \frac{1}{2} C_{buf} \left\{ \left(V_{ave} + \Delta \frac{V_c}{2} \right)^2 - \left(V_{ave} - \Delta \frac{V_c}{2} \right)^2 \right\} \tag{1}$$

where V_{ave} is the average voltage of the power decoupling capacitor, ΔV_c is the peak to peak voltage of the voltage fluctuation. The active power decoupling obtains the buffer energy from ΔV_c, which means the capacitance of C_{buf} becomes small. The boost type APD circuit reduces the C_{buf} more than the buck type APD circuit owing to boost-up operation of v_{cbuf}. On the other hand, the buck type APD circuit reduces the voltage rating for each device in comparison with the boost type. These methods achieve the active power decoupling operation with simple configuration and easy control. However, additional component may increase the circuit volume and cost. Especially, two inductors of L_{boost} and L_{buf} are necessary on DC side.

Fig.2 shows the circuit configuration which consists of FCC and VSI. FCC operates as the boost converter as shown in fig. 1 (a) and (b). In addition, the flying capacitor C_{fc} is utilized for the power decoupling capacitor when the active power decoupling control is applied. Therefore, the additional circuit does not necessary for the active power decoupling capability in this system.

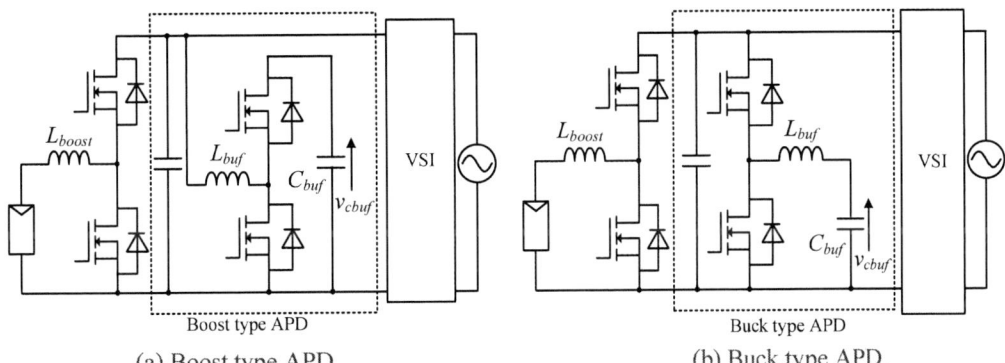

(a) Boost type APD.　　　　　　　　　　(b) Buck type APD.

Fig.1: Typical active power decoupling circuit using boost or buck type DC/DC converter.

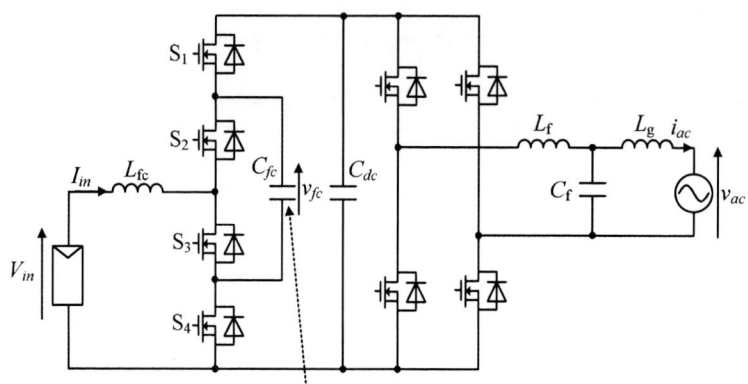

Small capacitor for power decoupling

Fig.2: Active power decoupling circuit using flying capacitor DC/DC converter.

Principle of power decoupling

Fig 3 shows the waveforms of input and output power between the DC and the single-phase side. Note that the grid voltage and the inverter output current waveforms are the unity sinusoidal waveform, the instantaneous power of single-phase AC grid pout is obtained as

$$p_{out} = \frac{V_{acp}I_{acp}}{2}(1 - \cos 2\omega t) \qquad (2)$$

where V_{acp} is the peak grid voltage, I_{acp} is the peak inverter output current. According to (2), the double-line frequency occurs at DC-link voltage. In order to suppress the power ripple, the instantaneous buffer power p_{buf} should be controlled as

$$p_{buf} = \frac{1}{2}V_{acp}I_{acp}\cos 2\omega t \qquad (3)$$

where the polarity of the p_{buf} is defined as positive when the buffer energy discharges. As the result, the input power is matched to the output power of AC side, which means power decoupling is achieved.

$$p_{in} = \frac{1}{2}V_{acp}I_{acp} = V_{IN}I_{IN} \qquad (4)$$

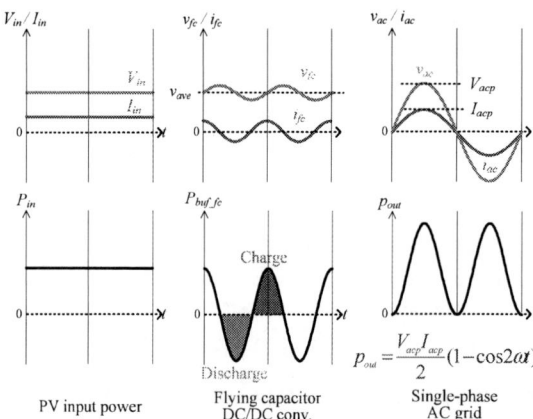

Fig. 3: Compensation method of power ripple due to double-line frequency. Buffer power is obtained from small flying capacitor by active power decoupling control.

Control Strategy of proposed circuit

Duty calculation of boundary current mode

Fig.4 shows the operation mode of the FCC, and fig.5 shows the inductor current of i_{Lfc} on the switching state at each operation modes. In mode I, the inductor L_{fc} is charged for boost-up of v_{in}, and i_{in} linearly increases. In mode II, C_{fc} is charged by i_{in}, and the inductor voltage v_{Lfc} is given by

$$v_{Lfc} = v_{fc} - v_{in} \tag{5}$$

where v_{fc} is the flying capacitor voltage. In this mode, i_{in} slowly decreases as shown in fig. (4). Then L_{fc} and C_{fc} are discharged to the load in mode III. In this period, v_{Lfc} is expressed as

$$v_{Lfc} = v_{dc} - v_{fc} - v_{in} \tag{6}$$

where v_{dc} is the DC-link voltage. i_{in} is also slowly decreases as same as mode II. Finally, L_{fc} is discharged until i_{in} reaches zero. Note that BCM operation achieves when the mode I is started after mode IV. The proposed BCM actively utilizes mode II and III in order to generate trapezoidal current waveforms for reduction of the RMS current of i_{in}. Note that, each peak current of I_{pk1} and I_{pk2} are given by

$$\begin{cases} I_{pk1\_charge} = \dfrac{v_{in}}{L} D_1 T_{sw} \\ I_{pk2\_charge} = \dfrac{v_{in} - v_{fc}}{L} D_2 T_{sw} + \dfrac{v_{in}}{L} D_1 T_{sw} \\ \qquad = -\dfrac{v_{in} - v_{dc}}{L} D_4 T_{sw} \end{cases} \tag{7}$$

$$\begin{cases} I_{pk1\_discharge} = \dfrac{v_{in}}{L} D_1 T_{sw} \\ I_{pk2\_discharge} = \dfrac{v_{in} - \left(v_{dc} - v_{fc}\right)}{L} D_3 T_{sw} + \dfrac{v_{in}}{L} D_1 T_{sw} \\ \qquad = -\dfrac{v_{in} - v_{dc}}{L} D_4 T_{sw} \end{cases} \tag{8}$$

where D_1, D_2, D_3, and D_4 are the duty of each mode. The inductor average current is calculated from duty and peak current in each modes, and it is expressed as

$$I_{ave\_charge} = \frac{1}{2} I_{pk1\_charge} D_1 + \frac{1}{2} \left(I_{pk1\_charge} + I_{pk2\_charge} \right) D_2 + \frac{1}{2} I_{pk2\_charge} D_4 \tag{9}$$

$$I_{ave\_discharge} = \frac{1}{2} I_{pk1\_discharge} D_1 + \frac{1}{2} \left(I_{pk1\_discharge} + I_{pk2\_discharge} \right) D_3 + \frac{1}{2} I_{pk2\_discharge} D_4 \tag{10}$$

where I_{ave_charge} is the average input current in charge mode of fig. 4 (a), $I_{ave\_discharge}$ is the average input current in discharge mode of fig. 4 (b).

Note that, the sum of each duty in BCM is expressed as

$$\begin{cases} D_1 + D_2 + D_4 = 1 \\ D_1 + D_3 + D_4 = 1 \end{cases} \tag{11}.$$

In addition, the inductor average energy on the switching period is zero, and it is expressed as

$$v_{in} D_1 + \left(v_{in} - v_{fc} \right) D_2 + v_{dc} D_4 = 0 \tag{12}$$

$$v_{in}D_1 + \left(v_{in} - v_{dc} - v_{fc}\right)D_3 + v_{dc}D_4 = 0$$

(13).

According to these equations, the relationship between each duty and the average input current is expressed as

$$D_{1\_charge\_BCM} = \cfrac{\left\{\cfrac{\left(v_{dc}v_{fc} - v_{in}v_{fc}\right)}{-\sqrt{\left(v_{dc}v_{fc} - v_{in}v_{fc}\right)^2 - v_{dc}v_{fc}\left\{\begin{array}{l}2LI_{ave}f_{sw}\left(v_{dc} - v_{fc}\right)\\ + \left(v_{dc} - v_{in}\right)\left(v_{fc} - v_{in}\right)\end{array}\right\}}}\right\}}{v_{dc}v_{fc}}$$

$$D_{2\_charge\_BCM} = \frac{v_{dc} - v_{in} - v_{dc}D_1}{v_{dc} - v_{fc}}$$

$$D_{4\_charge\_BCM} = 1 - D_1 - D_2$$

(14)

$$D_{1\_discharge\_BCM} = \cfrac{\left\{\cfrac{\left(v_{dc}^{\ 2} + v_{in}v_{fc} - v_{dc}v_{fc} - v_{in}v_{dc}\right)}{-\sqrt{\left(v_{dc}^{\ 2} + v_{in}v_{fc} - v_{dc}v_{fc} - v_{in}v_{dc}\right)^2 \\ -v_{dc}\left(v_{dc} - v_{fc}\right)\left\{\begin{array}{l}2LI_{ave}f_{sw}v_{fc}\\ +\left(v_{dc} - v_{in}\right)\left(v_{dc} - v_{fc} - v_{in}\right)\end{array}\right\}}}\right\}}{v_{dc}\left(v_{dc} - v_{fc}\right)}$$

$$D_{3\_discharge\_BCM} = \frac{v_{dc} - v_{in} - v_{dc}D_1}{v_{fc}}$$

$$D_{4\_discharge\_BCM} = 1 - D_1 - D_3$$

(15).

Note that the BCM condition depends on the output power condition because the zero current period occurs at light load. Therefore, the proposed operation mode is changed to DCM or BCM by the load condition.

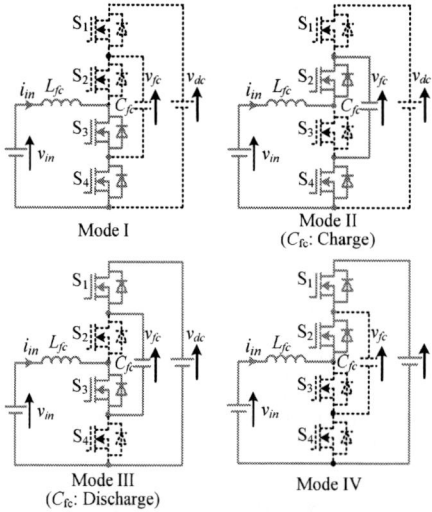

Fig. 4: Operation mode of FCC

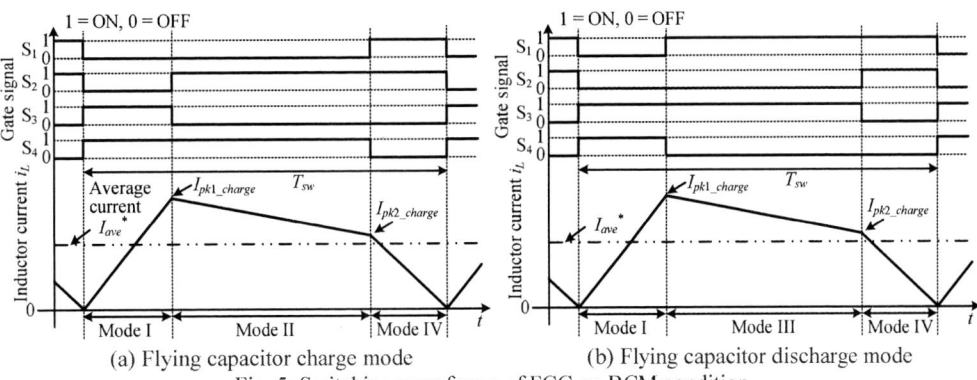

(a) Flying capacitor charge mode   (b) Flying capacitor discharge mode

Fig. 5: Switching waveforms of FCC on BCM condition.

Duty calculation of discontinuous current mode for right load condition

Fig.6 shows the switching waveforms under DCM condition on right load condition. di/dt in mode 2 and 3 is decided by the inductor voltage in each mode, and the output power depends on the period of D_1. In this case, D_1 of right load condition becomes short in comparison with fig. 5. As the result, inductor current reaches zero at mode II and III. In this case, the relationship between each duty and the average input current under DCM is expressed as

$$D_{1\_charge\_DCM} = \sqrt{2LI_{ave}f_{sw}\frac{v_{fc}-v_{in}}{v_{in}v_{fc}}}$$

$$D_{2\_charge\_DCM} = \frac{v_{in}}{v_{fc}-v_{in}}D_1$$

(16)

$$D_{1\_discharge\_DCM} = \sqrt{2LI_{ave}f_{sw}\frac{v_{dc}-v_{fc}-v_{in}}{v_{in}\left(v_{dc}-v_{fc}\right)}}$$

$$D_{3\_discharge\_DCM} = \frac{v_{in}}{v_{dc}-v_{fc}-v_{in}}D_1$$

(17).

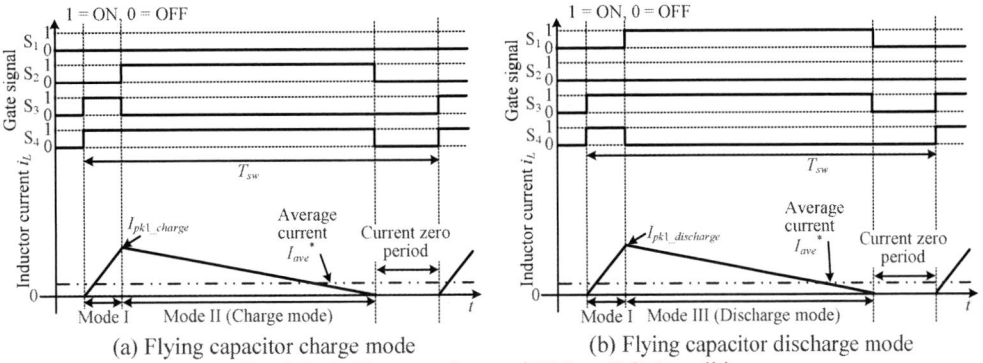

(a) Flying capacitor charge mode   (b) Flying capacitor discharge mode

Fig. 6: Switching waveforms of FCC on DCM condition.

Control block diagram of FCC

Fig.7 shows the control block diagram of FCC. This block consists of the duty calculation, operation mode selector, and carrier comparison to generate the gate pulse. The mode selector decides the operation mode based on the duty of D_4 and output of the hysteresis control for the flying capacitor voltage. The flying capacitor voltage is regulated in order to achieve the active power decoupling, and the flying capacitor voltage command $v_{fc\_ref}{}^*$ is expressed as

$$v_{fc}{}^*(t) = \sqrt{\left(\frac{v_{dc}}{2} - \frac{p_{out}}{\omega_{out}C_{fc}V_{dc}}\right)^2 - \frac{p_{out}}{\omega_{out}C_{fc}}\left\{\sin(2\omega t) - 1\right\}} \tag{17}$$

where p_{out} is the output power, ω_{out} is the angular frequency of the single-phase AC grid, C_{fc} is the capacitance of the flying capacitor.

Fig.8 shows the control method of the flying capacitor voltage using hysteresis control. The two threshold voltage of V_{fc_H} and V_{fc_L} are applied based on the flying capacitor voltage command of (17). The capacitor charge mode and discharge mode are actively changed as shown in fig.8. Note that the hysteresis width is set to narrow as soon as possible in order to reduce the voltage ripple due to the hysteresis control.

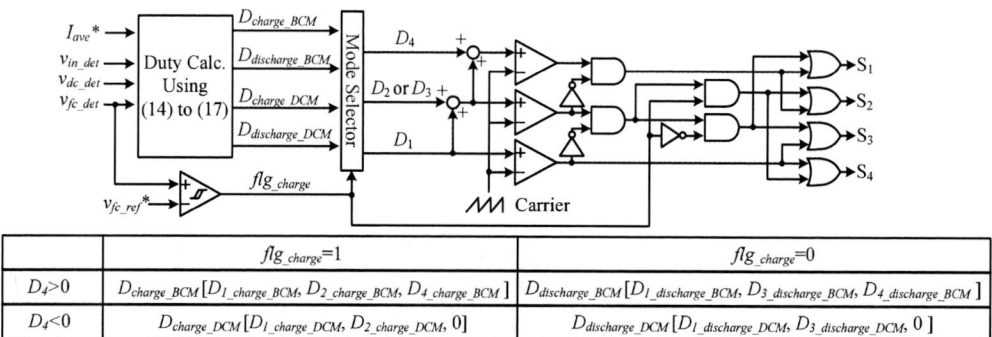

	$flg_{charge}=1$		$flg_{charge}=0$	
$D_4>0$	$D_{charge\_BCM}[D_{1\_charge\_BCM}, D_{2\_charge\_BCM}, D_{4\_charge\_BCM}]$		$D_{discharge\_BCM}[D_{1\_discharge\_BCM}, D_{3\_discharge\_BCM}, D_{4\_discharge\_BCM}]$	
$D_4<0$	$D_{charge\_DCM}[D_{1\_charge\_DCM}, D_{2\_charge\_DCM}, 0]$		$D_{discharge\_DCM}[D_{1\_discharge\_DCM}, D_{3\_discharge\_DCM}, 0]$	

Fig.7 Control block diagram of FCC.

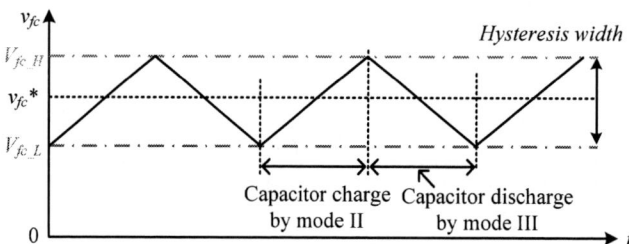

Fig.8 Control method of flying capacitor voltage using hysteresis control.

Simulation and Experimental result

Simulation result of power decoupling control

Table I shows the simulation conditions, and fig.10 shows the simulation results of the active power decoupling control. According to fig. 10 (a), the DC-link voltage fluctuates at the double-line frequency due to the power ripple. In particular, the inverter output current is distorted due to the fluctuation of the DC-link voltage. On the other hand, according to fig.10 (b), the DC-link voltage becomes constant owing to the active power decoupling control, and the inverter output current distortion is improved.

Fig.11 shows the harmonic analysis result of the DC-link voltage. Note that each harmonic components with the proposed control is normalized by the harmonic components without proposed control. According to fig.11, the second-order harmonics of the DC-link voltage is reduced by 94.4% owing to the proposed active power decoupling control. From these simulation result, the fundamental operation of the proposed control is confirmed.

Table I: Simulation parameter.

Rated power	P	1 kW
Input voltage	v_{in}	100 V
DC voltage	v_{dc}	300 V
Output voltage	v_{out}	200 V_{rms}
Grid frequency	f_{out}	50 Hz
Switching frequency	f_{sw}	20 kHz
Boost inductor	L	124 μH
Flying Capacitor	C_{fc}	240 μF

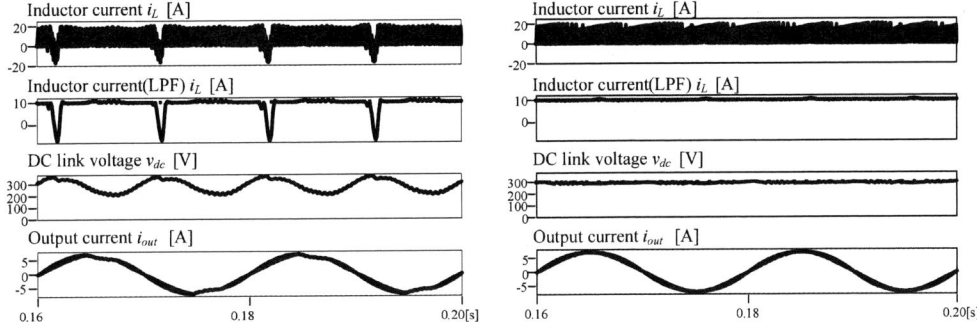

(a) Without power decoupling   (b) With active power decoupling

Fig.9: Simulation results of active power decoupling control.

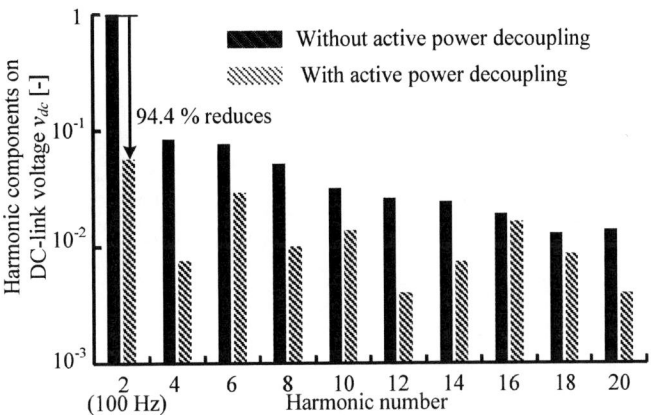

Experimental result

The validity of the proposed method was demonstrated by the prototype circuit. Table II shows the experimental parameter, and fig.11 shows the experimental results of the active power decoupling control. Note that the output power is 1-kW, input voltage is 100 V, DC-link capacitor is 60 µF, the flying capacitor is 180 µF, and the boost inductor is 124 µF. In this experiment, VSI is operated under the open loop control, and R-L load is connected on the AC output.

According to fig.11 (a), the DC-link voltage is fluctuated at the double-line frequency of the inverter output frequency. As the result, the output current is distorted due to this frequency component. On the other hand, according to fig. 11(b), the DC-link voltage fluctuation is reduced by the power decoupling control. Accordingly, the flying capacitor voltage is regulated to the sinusoidal waveforms with double-line frequency.

Fig. 12 shows the switching waveforms with charge mode. According to Fig.12, it is confirmed that the inductor current becomes trapezoidal waveforms as fig.5 (a). In particular, the inductor current is kept the BCM condition without zero current period.

Fig.13 shows the harmonic analysis result of the DC-link voltage. According to Fig.13, the second-order harmonics is reduced by 85.6% by the proposed power decoupling control. From these experimental results, the validity of the power decoupling control is confirmed.

Table II: Experimental parameter.

Output power	P	1 kW
Input voltage	v_{in}	100 V
DC voltage	v_{dc}	300 V
Output voltage	v_{out}	200 V$_{rms}$
Switching frequency	f_{sw}	20 kHz
Boost inductor	L	124 µH
Flying Capacitor	C_{fc}	180 µF
DC-link Capacitor	C_{dc}	60 µF

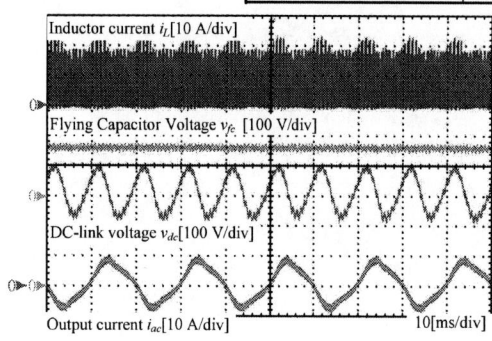 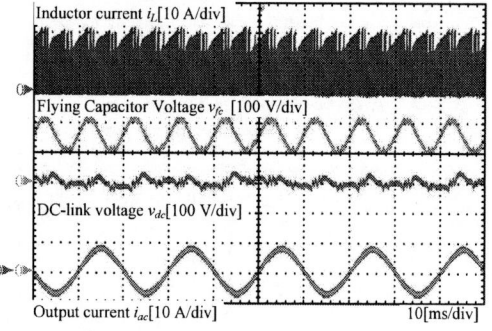

(a) Without power decoupling   (b) With active power decoupling

Fig.11: Experimental results of active power decoupling control.

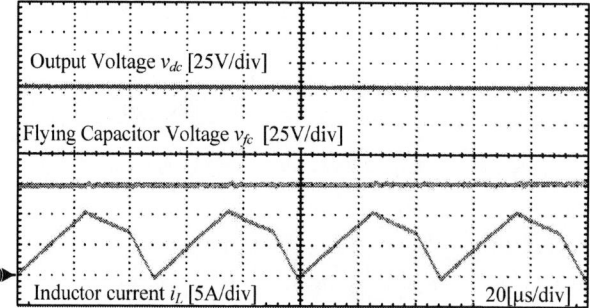

Fig. 12: Switching waveforms of FCC.

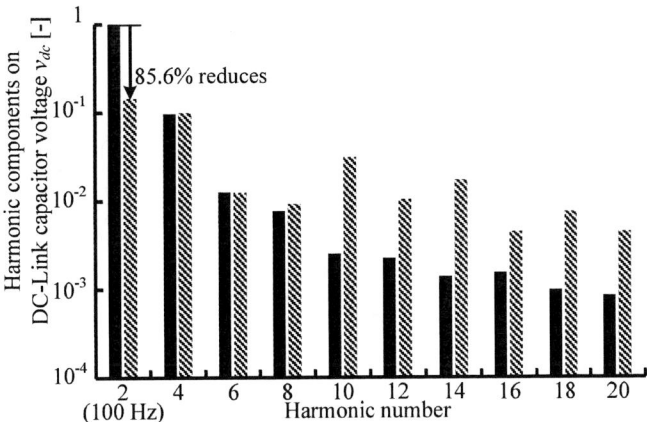

Fig.13: Harmonic analysis result of DC-link voltage.

Conclusion

This paper proposed the power decoupling method using flying capacitor DC/DC converter with the BCM operation. The proposed BCM generates the inductor current of the trapezoidal waveforms in order to reduce the RMS current for reduction of the conduction losses. The validity of the proposed power decoupling control was demonstrated by the experiment with 1-kW prototype circuit. As the experimental result, the DC-link voltage fluctuation due to the double-line frequency power ripple was reduced by 85.6% owing to the proposed control.

References

[1] H. Radwan, M. A. Sayed, T. Takeshita, A. A. Elbaset, G. Shabib "Boost Inverter Topology with High-Frequency Link Transformer for PV Grid-Tied Applications", *IEEE Journal of Industry Applications*, May 2019, vol.8, No.5, pp.849-856.

[2] M. Tanemo, K. Matsudate, S. Nomura "Change in Circuit Configuration of Photovoltaic Modules Using Series/Parallel Switching Circuits Composed of Power MOSFETs", *IEEE Journal of Industry Applications*, Sep. 2019, vol.9, No.9, pp.73-81.

[3] B, Liu.; Z, Liu.; J, Liu.; T, Wu.; S, Wang. A feedforward control based power decoupling scheme for voltage-controlled grid-tied inverters. *In Proc. IEEE Appl. Power Electron. Conf. Expo.*, Mar. 2016; pp. 3328-3332.

[4] X, Liu.; D, Dong.; M, Harfman-Todorovic.; L, Garaces. A PV Micro-inverter With PV Current Decoupling Strategy. *In Proc. IEEE Appl. Power Electron. Conf. Expo.*, Mar. 2016; pp. 3403-3408.

[5] T. Onodera, T. Shimizu "Fault-Ride-Through (FRT) Characteristics of a Power-Decoupling-Type Photovoltaic System", *In Proc. IEEE Appl. Power Electron. Conf. Expo.*, Mar. 2019; pp. 3207-3212.

[6] Haibing Hu, Souhib Harb, Nasser Kutkut, Issa, Batarseh, Z. John Shen: "Power Decoupling Techniques for Micro-inverters in PV Systems — a Review", *In proc. IEEE Energy Conver. Congr. and Expo.*, pp.3235-3240.

[7] F. Shinjo, K. Wada, T. Shimizu: "A Single-Phase Grid-Connected Inverter with a Power Decoupling Function", *2007 IEEE Power Electronics Specialists Conference*, pp.1245-1249.

[8] F. Schimpf, L. Norum: "Effective Use of Film Capacitors in Single-Phase PV-inverters by Active Power Decoupling", *IECON 2010 – 36th Annual Conference on IEEE Industrial Electronics Society*, pp. 2784-2789.

[9] K.-H. Chao, P.-T. Cheng : "Power decoupling methods for single-phase three-poles AC/DC converters", *2009 IEEE Energy Conversion Congress and Exposition*, pp.3742-3747.

[10] R. Wang, F. Wang, R. Lai, P. Ning, R. Burgos, D. Boroyevich : "Study of Energy Storage Capacitor Reduction for Single Phase PWM Rectifier" *2009 Twenty-Fourth Annual IEEE Applied Power Electronics Conference and exposition*, pp1177-1183.

[11] Y. Tang, W. Yao, P. C. Loh, F. Blaabjerg: "Highly Reliable Transformerless Photovoltaic Inverters With Leakage Current and Pulsating Power Elimination", IEEE Trans. on Ind. Electon., Vol. 63, No. 2, pp. 1016-1026.

An Architecture for Level-3 EV Battery Charger Stations Using Integrated Solid State Transformer (I-SST)

Erick I. Pool-Mazun*, Prasad Enjeti*, Gerardo Escobar** and Ira Pitel†

*Texas A&M University, Texas, USA.

** ITESM Campus Monterrey, Nuevo León, México.

†Magna-Power Electronics Inc. New Jersey, USA

Tel.:+1 (979) 703 0499

E-Mail: epooltamu@tamu.edu , enjeti@tamu.edu

Keywords

<<Battery charger>>,<< soft switching>>, <<Switched-mode power supply>>, <<Electric vehicle>>, <<Transformer>>.

Abstract

This paper proposes a new battery charger station architecture based on Integrated Solid-State Transformer (I-SST). The system consists of modular AC-AC converters with 20 kHz isolation and direct three-phase AC-DC Medium Voltage (MV) conversion. Output DC voltage regulation is achieved with a simple control scheme. Zero Voltage Switching (ZVS) is achievable for a wide output voltage range using phase-shift modulation. To comply with IEEE-519 standard, an Active Power Filter circuit block is employed. Simulation and experimental results are discussed.

I.   Introduction

The two main state of the art architectures used for Level 3 battery chargers are shown in Fig. 1. Because of the high power requirements for Level 3 chargers (>50kW) [1-2], battery charger stations are built by rectifying a 3 phase Medium Voltage (MV) from the utility. In both cases, the MV is stepped down by a line frequency transformer [1-5] and then followed by power electronics circuitry. Fig. 1 (a) shows an AC bus architecture. This approach makes use of individual AC-DC modules to rectify the three-phase MV into a fixed DC output and to generate a clean sinusoidal input current with close to unity power factor. Some of the most popular circuits employed for this are described in [6-8]. After the rectification stage, a DC-DC converter is connected in cascade through DC link capacitor. It can be noticed that this fist architecture has the advantage of being modular. However, the main disadvantage is the unreliable DC link capacitor which interconnects each one of the modular converters. Fig. 1 (b) shows a common DC bus configuration. In this approach, a single AC-DC converter is used to rectify the MV utility input. Once again, a bulky line frequency step down transformer is used to isolate the overall system and also to bring down the voltage to a proper level for the rectification stage. The AC-DC rectifier used for this stage is similar to those used in the common AC-bus case [5-8]; however, in this case a single converter must handle all the power of the station. The DC bus generated is then connected to multiple DC-DC converters to regulate the output voltage and current for battery charging purposes. Since the overall system is already isolated by the MV transformer, the DC-DC converter can be non-isolated circuits as the ones described in [9-12]. This architecture is currently the most popular due to its high efficiency. However, the system still consists of a 2-stage rectification system interconnected with a DC link capacitor. Furthermore, the step-down transformer used consists of a bulky line frequency transformer, which affects the overall size of the station. This paper proposes a new architecture for level-3 battery chargers based on an integrated solid-state transformer (I-SST) approach as shown in Fig. 2. The proposed topology consists of AC-DC converters with direct three-phase rectification and 20 kHz high frequency isolation. The AC-DC converters consist of modular front-end AC-AC converters as shown in Fig. 2 (b). With the modulation scheme discussed in section II, output voltage is shown to be regulated along with simultaneous soft-switching of AC-AC modules over a wide load range. Fig. 3 shows series stacking of modular AC-AC cells for higher utility input voltages.

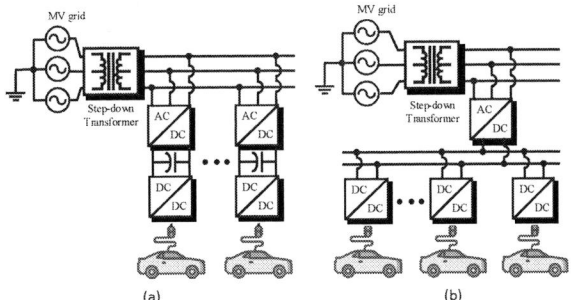

Fig. 1 : System Level Architecture for Level 3 battery charging station (a) AC bus approach (b) DC bus approach

To achieve high quality input current and to comply with IEEE 519 standard [13], an Active Power Filter (APF) is connected (Fig. 2 (a)) to compensate the harmonics generated by the three-phase diode rectifier at the DC side. The proposed architecture has the following advantages:

- Direct AC-DC power conversion (Fig. 2, no DC link stage).
- No bulky MV line frequency step down transformer.
- Easily scalable to MV using stacked modules (Fig. 3).
- Turn-On Zero Voltage Switching (ZVS) of the 20 kHz AC-AC converter in the main power conversion stages.
- Simple control scheme of the AC-DC converter, equivalent to a buck type converter.

II. Proposed Architecture

The proposed architecture is based in a common AC-bus approach using an I-SST. In this case, a group of High-Frequency (HF) isolated AC-DC converters are connected in parallel to the three-phase MV utility as illustrated in Fig. 2 (a). Therefore, a line-frequency step down transformer is avoided. The AC-DC converters in the proposed approach are capable of direct three-phase rectification. This means that a bulky DC-link capacitor is not needed. Each AC-DC converter with I-SST consists of three AC-AC converters as depicted in Fig. 2 (b). Moreover, each AC-AC module can be extended to "n" number of modules in series as depicted in Fig. 3 to reduce voltage and current stress in the switching devices. In order to simplify the analysis, from now on this paper will only address the case shown in Fig. 2 (b) when the AC-AC converter consists of a single AC-AC cell.

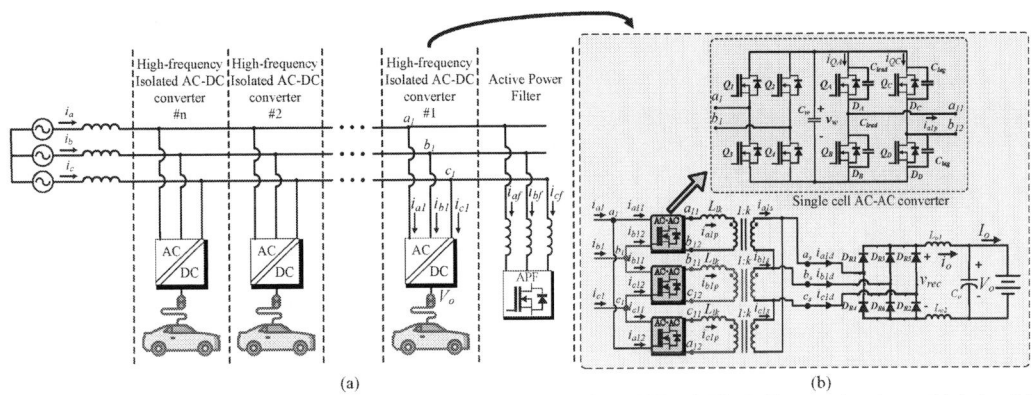

Fig. 2:Proposed architecture for Level 3 EV battery charger station using I-SST. (a) Block diagram showing multiple level-3 charge ports (b) circuit topology for the AC-DC high-frequency isolated converters with single-stage rectification and ZVS capability.

Each AC-AC converter cell consists of two full bridge converters. The first full bridge converter, which is connected to the line-to-line voltage $v_{a1,b1}$, commutates at line frequency (60 Hz) with a switching function s_{rec} as depicted in Fig. 4 (b). The rectifier section is then followed by a small film capacitor c_w, which provides a weak DC-link connection. On the other hand, the second full-bridge converter commutates at high frequency (20 kHz) with a switching function s_{inv} depicted in Fig. 4 (c). The overall switching function of the AC-AC block s_{sw} is a high-frequency square-wave as illustrated in Fig. 4 (d). The voltage $v_{a11,b12}$ is a high-frequency square-wave with a sine envelope as depicted in Fig. 4 (e) which results from the multiplication of the input line-to-line voltage $v_{a1,b1}$ and the AC-AC converter switching function s_{sw}. Therefore, the three transformers depicted in Fig. 2 (b) are high-frequency transformers with reduced size and volume compared to a conventional line-frequency transformer. A plot of the secondary side voltage $v_{as,bs}$ is depicted in Fig. 5 (a). The three single-phase transformers are connected in delta configuration at the secondary side. Then, a three-phase diode rectifier is connected to the transformer. Since the three-phase diode rectifier is connected to HF voltages (as depicted in Fig. 5 (a)), currents i_{a1d}, i_{b1d} and i_{c1d} have a HF component as depicted in Fig. 5 (b). Therefore, in order to implement this topology in practice, Silicon Carbide (SiC) diodes are required. The resulting voltage v_{rec}, depicted in Fig.5 (c), is passed through an LC filter to eliminate the high frequency components resulting in a DC output voltage V_o. On the other hand, from Fig.5 (a), it can be noticed that the pulse-width of the HF square-wave can be adjusted to generate zero voltage states in $v_{as,bs}$ when duty cycle $D < 1$. This consequently affects the average value of v_{rec} at the DC output and allows output voltage regulation as shown in Fig. 5 (c).

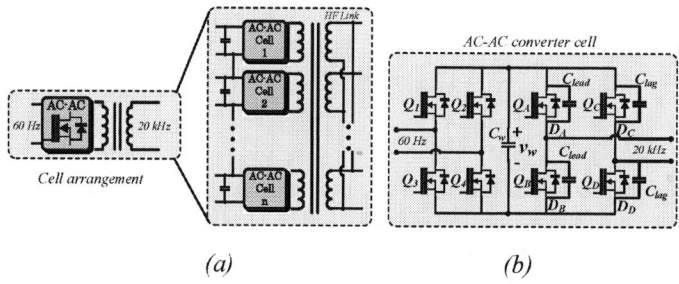

(a)

(b)

Fig. 3: Modular extension of the AC-AC converter stage. (a) Series arrangement of AC-AC converter cells, and (b) AC-AC back-to-back converter cell

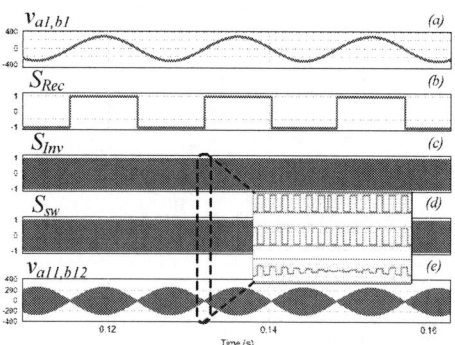

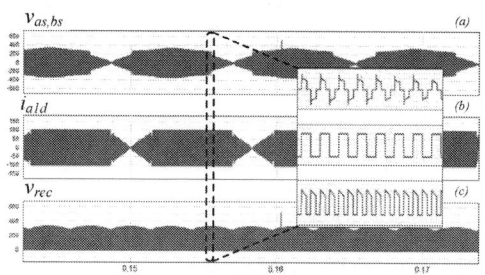

Fig. 4: Switching function of AC-AC converter for D=1 (a) Input line to line voltage $v_{a1,b1}$,(b) Rectifier switching function s_{rec}, (c) Inverter switching function s_{inv}, (d) overall AC-AC converter switching function s_{sw}. Notice that $s_{sw} = s_{rec} \cdot s_{inv}$, and (e) High frequency link primary side voltage $v_{a11,b12}$. Notice that $v_{a11,b12} = v_{a1,b1} \cdot s_{sw}$

Fig. 5: Secondary side waveforms for $D<1$ (a) secondary winding voltage $v_{as,bs}$. Notice that a change of duty cycle introduces zero states which modify the average value of the voltage and therefore the value of V_o at the DC side, (b) diode current i_{a1d} with high frequency component, and (c) rectifier's output voltage v_{rec}.

EPE'20 ECCE Europe

Assigned jointly to the European Power Electronics and Drives Association & the Institute of Electrical and Electronics Engineers (IEEE)

II.A. Three-phase diode rectifier operation at high frequency

The proposed converter behaves as a push-pull three-phase rectifier. Therefore, the characteristics of both a full-bridge push pull converter and a three-phase diode rectifier are present in this circuit. To get an expression of the DC output voltage V_o as a function of a duty cycle D, the diode switching functions are analyzed. The switching function of each diode rectifier leg is divided into two main components: a low frequency switching function coming from the line frequency voltage, and a HF component from the AC-AC converter at the primary side as shown in Fig. 6.

The rectified voltage v_{rec} is equivalent to a the line frequency three-phase diode rectifier voltage but modulated by a high frequency signal $s_{sw} \cdot s_{sw}$ as expressed in (1):

$$v_{rec} = s_{sw} \cdot s_{sw} \cdot \frac{3\sqrt{2}}{\pi} V_{LL} \left[1 - \sum_{n=1}^{\infty} \frac{2}{36^2 - 1} \cos(6n\omega t) \right] \tag{1}$$

Where V_{LL} is the RMS line-to-line voltage a from the utility, and signal $s_{sw} \cdot s_{sw}$ is the switching function obtained when s_{sw} (see Fig. 6) is rectified and it has an average value equal to D. If the turn ratio k of the transformer is taken into account the DC component of v_{rec} is given by:

$$V_{oi} = \frac{3\sqrt{2}}{\pi} k D V_{LL} \tag{2}$$

Where V_{oi} is the ideal DC output voltage of the converter as a function of the duty cycle D.

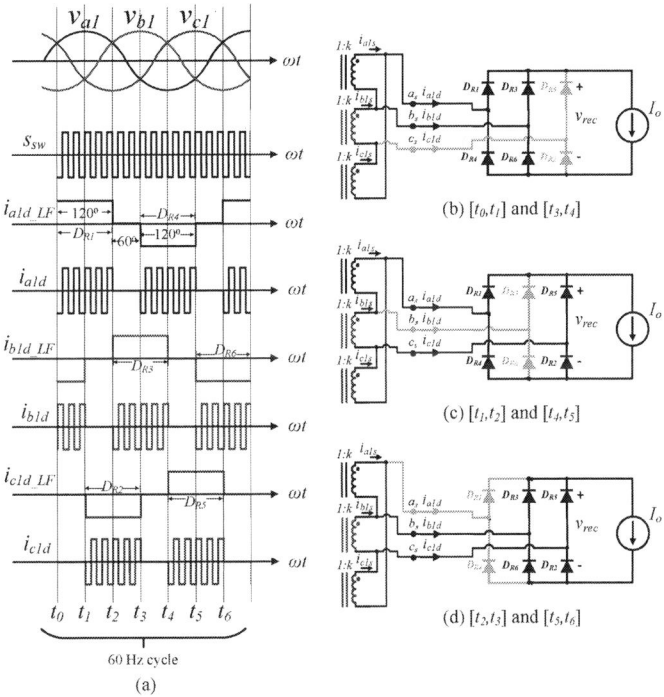

Fig. 6: Secondary side rectification stage of the I-SST (Fig. 2) for full duty cycle range $0 < D < 1$. (a) Switching pattern of the diode rectifier currents i_{a1d}, i_{b1d}, and i_{c1d} for the proposed modulation (b) active diode legs during intervals $[t_0, t_1]$ and $[t_3, t_4]$ (c) active diode legs during intervals $[t_1, t_2]$ and $[t_4, t_5]$ and (d) active diode legs during intervals $[t_2, t_3]$ and $[t_5, t_6]$

The proposed I-SST is intended to be isolated by a three-phase delta-delta transformer. The secondary side of the system is depicted in Fig. 2 (b). Due to the primary side HF switching, voltages $v_{as.bs}$, $v_{bs,cs}$, $v_{cs,as}$ consist of HF signals and therefore currents i_{a1d}, i_{b1d} and i_{c1d} are also HF currents as depicted in Fig.6 (a). From Fig. 6 (a) it can be notice that each line frequency cycle can be divided into six different intervals of 60 degrees each one. During any of the 60 degrees intervals, only two legs from the three-phase diode rectifiers remain active as shown in Fig. 6 (b) , (c) , and (d). Because of the delta connection the following relationships can be obtained.

During intervals $[t_0 - t_1]$ and $[t_3 - t_4]$:

$$i_{a1s} = \frac{2}{3}I_o \tag{3}$$

$$i_{b1s} = i_{c1s} = \frac{1}{3}I_o \tag{4}$$

During intervals $[t_1 - t_2]$ and $[t_4 - t_5]$:

$$i_{c1s} = \frac{2}{3}I_o \tag{5}$$

$$i_{a1s} = i_{b1s} = \frac{1}{3}I_o \tag{6}$$

during intervals $[t_2 - t_3]$ and $[t_5 - t_6]$:

$$i_{c1s} = \frac{2}{3}I_o \tag{7}$$

$$i_{a1s} = i_{b1s} = \frac{1}{3}I_o \tag{8}$$

This behavior is exemplified in Fig. 7 for currents i_{a1d} and i_{a1s}. This particular behavior of the currents is used for the ZVS constrain equations discussed in section II.B.

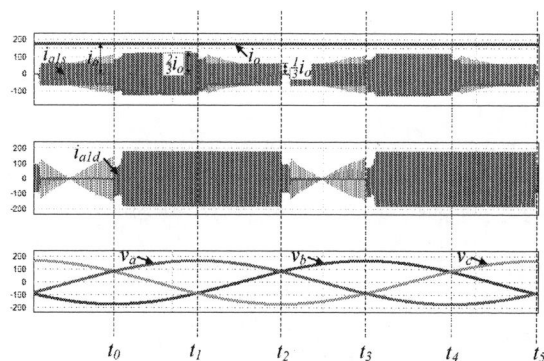

Fig. 7: Secondary side current waveforms of line "a". Time interva $[t_0 - t_5]$ is equivalent to the same time interval form Fig. 6. Since at every moment only two legs of the diode rectifier are active, the current seen by the secondary has two current levels that is , $i_{a1s} = \frac{2}{3}i_o$ during intervals $[t_0 - t_1]$ and $[t_3 - t_4]$, and $i_{a1s} = \frac{1}{3}i_0$ during interval $[t_1 - t_2]$, $[t_2 - t_3]$, $[t_4 - t_5]$

II.B. Zero Voltage Switching

As in a single-phase full bridge converter, phase shift modulation can be employed in order to achieve ZVS and output voltage regulation. The switching pattern of a HF cycle for the switches of the AC-AC converter results to be the same as explained in previous literature [14-17] and depicted in Fig. 8. As in the case of a DC-DC full-bridge push pull converter, each one of the converter legs have diferent requierement to achieve ZVS.By considering the analysis of the primary-secondary side currents, the worst-case current seen by the high frequency converter occurs for one third of the output current, then the minimum current to achieve ZVS for the leading leg of the primary side HF converter is:

$$I_{o\ min,lead} > 6C_{lead}\frac{V_{in}}{t_{d(lead)}} \tag{9}$$

Since C_{lead} is the output capacitance of the switching device and cannot be modified, the parameter which allows to achieve ZVS for the leading leg is $t_{d(lead)}$.

The ZVS routine for the lagging leg depends on the primary side leakage inductance L_{lk} and the value of the device's output capacitance C_{lag}. To achieve ZVS the following equation most be satisfied [14-17]:

$$\frac{1}{2}L_{lk}I_2^2 > C_{lag}V_{in}^2 + \frac{1}{2}C_{TR}V_{in}^2 \tag{10}$$

where C_{TR} is the transformer's leakage capacitance.

During interval $[t_2', t_3']$: $I_2 = kI_o$.

Since again due to delta-delta connection of the transformer, the worst case for the current seen at the primary side is for one third of the output current I_o. By taking this consideration and neglecting C_{TR}.

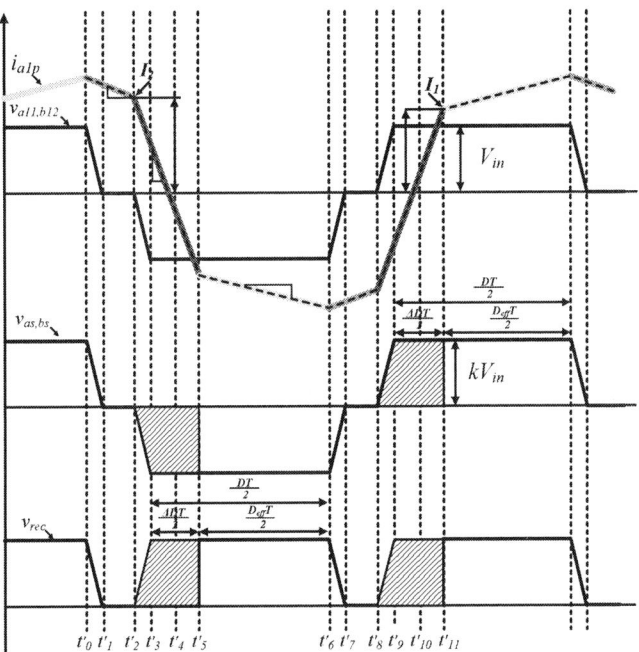

Fig.8: Phase shift modulation for ZVS as described in [14-17].

$$I_{o\,min,lag} \geq \frac{3V_{in}}{k}\sqrt{\frac{2C_{lag}}{L_{lk}}} \tag{11}$$

Due to the three-phase diode rectifier operation, V_{in} can be approximated as:

$$V_{in} \approx \frac{3\sqrt{2}}{\pi}V_{LL} \tag{12}$$

Substituting (12) into (11)

$$I_{o\,min,lag} \geq \frac{9\sqrt{2}}{\pi k}V_{LL}\sqrt{\frac{2C_{lag}}{L_{lk}}} \tag{13}$$

II.C. Duty cycle loss

Alternatively, as in the converter presented in [14-17] and depicted in Fig. 8, a duty cycle loss can be seen in the converter's output voltage v_{rec}. For this specific case with a HF three-phase diode rectifier duty cycle lost is given by equation (14):

$$\Delta D = \frac{8\pi L_{lk}I_o f_s k}{9\sqrt{2}V_{LL}} \tag{14}$$

Defining duty cycle D of the converter as in (15):

$$D = D_{eff} + \Delta D \tag{15}$$

By replacing D for D_{eff} in the ideal output voltage equation (2), an expression considering the duty cycle loss is obtained:

$$V_o = \frac{3\sqrt{2}}{\pi} k V_{LL} (D - \Delta D) \tag{16}$$

Where V_o represents the effective DC output voltage and ΔV_o is the voltage drop due to duty cycle loss.

II.D. Active Power Filter

The proposed isolated AC-DC converter has the capability to achieve output voltage regulation with ZVS. However, the harmonic content of the AC-DC modules is the same as in a regular line frequency three-phase diode rectifier. Because of this, an active power filter can be employed to inject current to cancel out low frequency harmonics at the utility side. By adding this block to the system, a sinusoidal current i_a with THD < 5% is obtained as shown in Fig. 11 (b) which comply with IEEE 519 standard [13]. The same control scheme for a conventional APF can be used for this case. This is because the load currents i_{a1}, i_{b1}, and i_{c1} are still the same line frequency signals as in a conventional three-phase diode rectifier load as depicted in Fig. 11 (c) for i_{a1}. APFs have been widely studied and multiple approaches have been described in the literature. For this case, an APF based on Selective Harmonic compensation as described in [18-20] was used.

III. Design example for ZVS ISST

A 50 kW design example high frequency three-phase rectifier which achieves voltage regulation in a range from 500 V to 250 V is designed in this section. The system will be designed to operate at CCM for up to 10% of full load, and to achieve ZVS in the primary side converters from 100% to 50% load. A three-phase line to line input voltage of 480V RMS is considered. The main parameters of the design example are listed in TABLE I.

TABLE I
PARAMETERS FOR SIMULATION DESIGN EXAMPLE

Operating conditions used for simulation in PLECS	
Grid voltage (line-to-line rms V_{LL})	480 V rms (1 p.u)
Grid frequency (ω)	60Hz
Output dc voltage ($V_{out,max}$)	500 V (1.042 p.u)
Rated power	50 kW
Switching Frequency (ω_s)	19.98 kHz
Leakage inductance (L_{lk})	10 µH (0.0008 p.u)
Output Inductor ($L_o = L_{o1} + L_{o2}$)	0.651 mH (0.05 p.u)
Output Capacitor (C_o)	3.5 mF
Transformer's turn ratio (k)	5/6
Output capacitance of devices ($c_{oss@600V}$)	131 pF

Equation (2) can be used to calculate the ideal output DC voltage, that is without considering duty cycle loss ΔD. To do so, a value of k has to be selected in such a way that the initial guess for V_{oi} exceeds $V_{o,max}$ by an amount ΔV_o. Considering $k = \frac{5}{6}$.

$$V_{oi} = \frac{3\sqrt{2}}{\pi} k V_{LL} = 540.1897 \ V \tag{17}$$

Since the desired output voltage is $V_o = 500 \ V$ then $\Delta V_o = 40.1897$. Therefore the given value of k can be used to calculate the desired parasitic inductance L_{lk}. The duty cycle loss can be calculated from (16)

$$\Delta D = \frac{\Delta V_o \pi}{3\sqrt{2} V_{LL} k} = 0.073399 \tag{18}$$

From (14)

$$L_{lk} = \frac{9\sqrt{2}V_{LL}\Delta D}{8\pi I_o f_s k} = 10.71\ \mu H \approx 10\ \mu H\ (0.0008\ p.u.) \tag{19}$$

The output filter is designed using the same procedure as in a three-phase diode rectifier and is not discussed in this paper [21]. The values of L_o and C_o are listed in Table I. The value of the parasitic inductance L_{lk} can be substituted in equation (13) to calculate the minimum current for ZVS in the lagging leg of the HF full bridge converter. Using an approximation model for the output capacitance of the devices as in [17] the plots for the minimum currents for the leading and lagging legs are obtained as shown in Fig. 9.

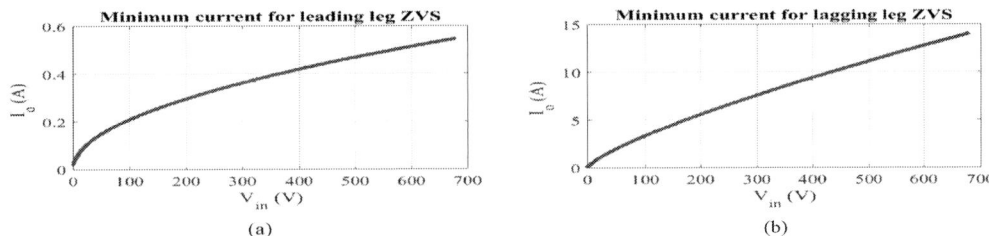

Fig. 9: Minimum output current to achieve ZVS. (a) Minimum output current to achieve lagging leg ZVS, (b) Minimum output current to achieve leading leg ZVS for 50 kW design example.

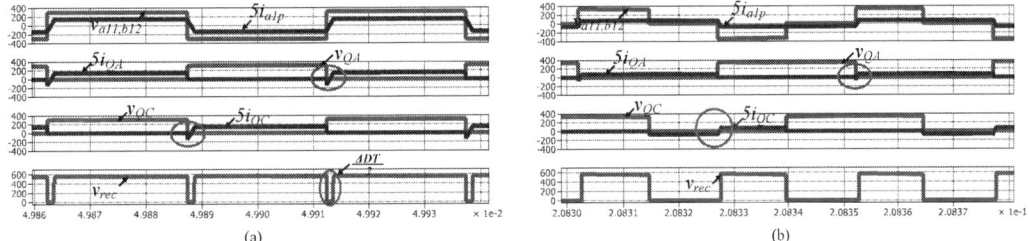

Fig. 10: Primary side voltage and current (a) primary side voltage $v_{a11,b12}$ and primary side current i_{a1p} (b) ZVS for high frequency converter at full load and (c) ZVS for high frequency converter at 50% output power from 50 KW design example

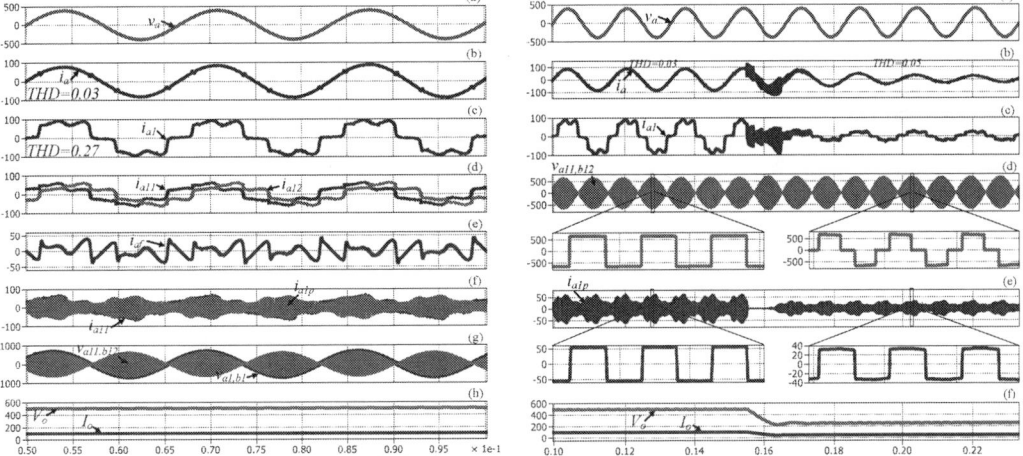

Fig. 11: Current and voltage waveforms with APF added to the system (a) Utility input current i_a and voltage v_a ,(b) direct AC-DC converter load current i_{a1} ,(c) load currents i_{a11} and i_{a12} where $i_{a1} = i_{a11} + i_{a12}$ (d) APF current i_{af} ,(e) Primary side current i_{a1p} ,(f) line-to-line voltage input of AC-AC converter $v_{a1,b1}$ and primary side voltage $v_{a11,b12}$, and (g) Output DC voltage V_o and current I_o

Fig. 12 Output voltage step by change of duty cycle. (a) line-to neutral voltage v_a , (b) utility input current i_a , (c) HF rectifier current i_{a1} , (d) primary HF voltage $v_{a11,b12}$, (e) primary HF current i_{a1p} , and (f) DC output voltage V_o and current I_o

Fig. 10 illustrates the voltage and current seen by switches Q_A and Q_C from the leading and lagging legs respectively (refer to Fig,2). As indicated by the circles in Fig. 10 Turn-On ZVS is achieved at 100% and 50% output power load. Alternatively, Fig. 11 shows the waveforms of the system including the APF circuit. Since the current generated by the rectifier is equal to the conventional line frequency three-phase diode rectifier, the same APF control schemes described in literature can be used. Fig. 12 depicts the waveforms of the overall system for a duty cycle change from $D = 1$ to $D = 0.5$ (50% of rated load).

IV. Preliminary experimental results

Preliminary experimental results from a scaled down laboratory prototype using three ferrite core transformers rated 5 kW at 20 kHz are discussed. Fig.11 depicts the secondary side waveforms for one AC-DC converter of the three-phase I-SST module. Similar to the simulation results, from Fig. 13 it can be notice that with the proposed modulation scheme a high-frequency link voltage $v_{a11,b12}$ results at the primary side. Moreover, from Fig.13 it can be seen that the pulse width of v_{rec} is controlled by the duty cycle at the primary side. Alternatively, Fig. 14 shows the diode currents at the secondary side. It can be seen that the diodes operate at high frequency and the utility input current i_{a1} remains the same as in a conventional three-phase diode rectifier.

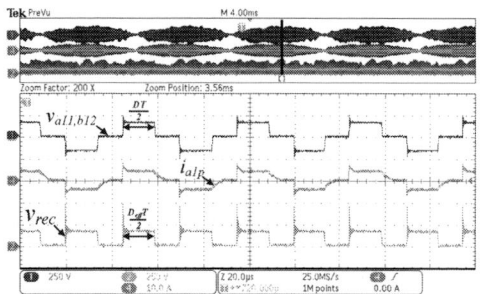
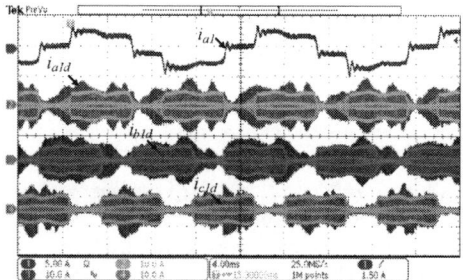

Fig. 13: Output voltage regulation by changing duty cycle at the primary side voltage $v_{a11,b12}$ from scaled down experimental prototype

Fig. 14: High-frequency diode currents i_{a1d}, i_{b1d}, i_{c1d} and rectifier's current i_{a1} from scaled down experimental prototype

V. Conclusion

In this paper, a new architecture for level-3 battery charger station based on an Integrated Solid State Transformer (I-SST) was porposed. The proposed architecture allows to replace the bulky line frequency transformer with a 20 kHz high frequency ferrite core type to achieve higher power dessity. Series stacking the AC-AC converter modules allows for direct medium voltage (MV) ac to dc conversion with 20 kHz isolation. ZVS of the AC-AC converters improves the conversion efficiency and duty cycle control has been shown to regulate the output DC voltage/power. An active filter block is shown to comply with IEEE 519 harmonic current limits. Simulation results of the system were discussed and preliminary experimental results from an scaled down experimental prototype were shown.

References

[1] M. Yilmaz and P. T. Krein, "Review of Battery Charger Topologies, Charging Power Levels, and Infrastructure for Plug-In Electric and Hybrid Vehicles," in *IEEE Transactions on Power Electronics*, vol. 28, no. 5, pp. 2151-2169, May 2013.

[2] A. Khaligh and S. Dusmez, "Comprehensive Topological Analysis of Conductive and Inductive Charging Solutions for Plug-In Electric Vehicles," in *IEEE Transactions on Vehicular Technology*, vol. 61, no. 8, pp. 3475-3489, Oct. 2012.

[3] V. M. Iyer, S. Gulur, G. Gohil and S. Bhattacharya, "An Approach Towards Extreme Fast Charging Station Power Delivery for Electric Vehicles with Partial Power Processing," in IEEE Transactions on Industrial Electronics, vol. 67, no. 10, pp. 8076-8087, Oct. 2020, doi: 10.1109/TIE.2019.2945264.

[4] S. Srdic and S. Lukic, "Toward Extreme Fast Charging: Challenges and Opportunities in Directly Connecting to Medium-Voltage Line," in *IEEE Electrification Magazine*, vol. 7, no. 1, pp. 22-31, March 2019.

[5] S. Bai and S. M. Lukic, "Unified Active Filter and Energy Storage System for an MW Electric Vehicle Charging Station," in IEEE Transactions on Power Electronics, vol. 28, no. 12, pp. 5793-5803, Dec. 2013.

[6] P. Verdelho and G. D. Marques, "Four-wire current-regulated PWM voltage converter," in *IEEE Transactions on Industrial Electronics*, vol. 45, no. 5, pp. 761-770, Oct. 1998.

[7] R. Zhang, F. C. Lee and D. Boroyevich, "Four-legged three-phase PFC rectifier with fault tolerant capability," *2000 IEEE 31st Annual Power Electronics Specialists Conference. Conference Proceedings (Cat. No.00CH37018)*, Galway, Ireland, 2000, pp. 359-364 vol.1.

[8] J. W. Kolar and F. C. Zach, "A novel three-phase utility interface minimizing line current harmonics of high-power telecommunications rectifier modules," in *IEEE Transactions on Industrial Electronics*, vol. 44, no. 4, pp. 456-467, Aug. 1997.

[9] O. Garcia, P. Zumel, A. de Castro and A. Cobos, "Automotive DC-DC bidirectional converter made with many interleaved buck stages," in IEEE Transactions on Power Electronics, vol. 21, no. 3, pp. 578-586, May 2006.

[10] B. Stevanović, D. Serrano, M. Vasić, P. Alou, J. A. Oliver and J. A. Cobos, "Highly Efficient, Full ZVS, Hybrid, Multilevel DC/DC Topology for Two-Stage Grid-Connected 1500-V PV System With Employed 900-V SiC Devices," in IEEE Journal of Emerging and Selected Topics in Power Electronics, vol. 7, no. 2, pp. 811-832, June 2019.

[11] F. Alhuwaishel, A. Allehyani, S. Al-Obaidi and P. Enjeti, "A New Medium Voltage DC Collection Grid for Large Scale PV Power Plants with SiC Devices," 2018 IEEE 19th Workshop on Control and Modeling for Power Electronics (COMPEL), Padua, 2018, pp. 1-8.

[12] C. -. Lin, L. -. Yang and G. W. Wu, "Study of a non-isolated bidirectional DC-DC converter," in IET Power Electronics, vol. 6, no. 1, pp. 30-37, Jan. 2013.

[13] IEEE, "IEEE Recommende Practice and Requirements for Harmonic Control in Electric Power Systems," in IEEE Std 519-2014 (Revision of IEEE Std 519-1992), ed, pp. 1-29, 2014.

[14] L. H. Mweene, C. A. Wright and M. F. Schlecht, "A 1 kW 500 kHz front-end converter for a distributed power supply system," in IEEE Transactions on Power Electronics, vol. 6, no. 3, pp. 398-407, July 1991.

[15] J. A. Sabate, V. Vlatkovic, R. B. Ridley, F. C. Lee and B. H. Cho, "Design considerations for high-voltage high-power full-bridge zero-voltage-switched PWM converter," Fifth Annual Proceedings on Applied Power Electronics Conference and Exposition, Los Angeles, CA, USA, 1990, pp. 275-284.

[16] O. D. Patterson and D. M. Divan, "Pseudo-resonant full bridge DC/DC converter," in IEEE Transactions on Power Electronics, vol. 6, no. 4, pp. 671-678, Oct. 1991.

[17] X. Ruan, Soft-Switching PWM Full-Bridge Converters: Topologies, Control, and Design, Singapore Wiley Science press, 2014.

[18] P. Mattavelli, "A closed-loop selective harmonic compensation for active filters," in IEEE Transactions on Industry Applications, vol. 37, no. 1, pp. 81-89, Jan.-Feb. 2001.

[19] G. Escobar, A. M. Stankovic and P. Mattavelli, "An adaptive controller in stationary reference frame for D-statcom in unbalanced operation," in IEEE Transactions on Industrial Electronics, vol. 51, no. 2, pp. 401-409, April 2004.

[20] A. A. Valdez-Fernandez, G. Escobar, P. R. Martinez-Rodriguez, J. M. Sosa, D. U. Campos-Delgado and M. J. Lopez-Sanchez, "Modelling and control of a hybrid power filter to compensate harmonic distortion under unbalanced operation," in IET Power Electronics, vol. 10, no. 7, pp. 782-791, 10 6 2017.

[21] M.H. Rashid, Power electronics: circuits, devices, and applications, second edition, New Jersey Prentice-Hall,inc, 1993.

LQR and H-infinity Control of Voltage Source Inverters for AC Microgrids

Tenorio Jorge, Jose Miguel Ramirez Scarpetta, Member, IEEE, Fabio Andrade, Member, IEEE.
Dept. of Electrical Engineering, Universidad del Valle, Cali, Colombia
Dept. of Electrical Engineering, Universidad de Puerto Rico, Mayaguez, Puerto Rico
jorge.tenorio@correounivalle.edu.co
jose.ramirez@correounivalle.edu.co
fabio.andrade@upr.edu

Keywords

≪Grid-isolated≫, ≪LC-filter≫, ≪Active-damper≫, ≪Voltage-source≫, ≪Robust-control≫.

Abstract

This paper presents a voltage controller design of three-phase voltage source inverters using *LQR*-based active damping and H∞ control strategy. Active damping consists of an *LQR* voltage and current feedback of *LC* filter in the αβ frame. In addition to the damping obtained in the *LC* filter, the proposed *LQR* strategy allows increasing the bandwidth of the system to reject high frequency disturbances. Three-phase voltage regulation is achieved in the αβ frame using the H∞ control strategy. In this strategy is used a new four order sensitivity functions to obtain very low tracking error and rejection of harmonic disturbances generated by nonlinear loads. The controller has been validated in laboratory tests using 2kW-220V three-phase inverters and non-linear loads. This controller has been compared to resonant proportional control and better performance has been verified. A $THD_v = 1.3\%$ has been experimentally obtained when a nonlinear load is connected.

Introduction

In a microgrid, the efficiency and performance of the load sharing control depends on the dynamics of the voltage control of the generators. Also, the microgrid voltage must meet some standards. The standards IEC 62257 and IEEE 1547.4 define the requirements and specifications for the design and commissioning of isolated microgrids. Moreover, tracking references for inverter voltages are generated by the power control law that governs them. Achieving voltage tracking with zero error is very important for the strategy. Chen in 1970 [1] has established that for a control system to track a reference it is necessary that the forward path gain must be infinite at the frequencies that set up the reference signal. The voltage control of a three-phase inverter can be done by transforming the coordinate system *abc* to αβ or *dq* coordinates using the Clark and Park transformations respectively. The voltage control in *dq* coordinates, traditionally, consists of two cascade controllers for current and voltage of the inverter [2]. This is possible by decoupling, canceling interactions of variables. In αβ coordinates, the cascade control is traditionally implemented without decoupling variables, using the resonant controllers for voltages and currents. See Fig. 1. The Voltage Source Inverters (VSI) are formed generally by a processing unit, an inverter and a *LC* filter. The *LC* filter has poor damping. A simple, but not efficient strategy to damp *LC* filter is to place a small resistance in series with the capacitors, [3]. Active damping makes it possible to stabilize the *LC* filter efficiently and at low cost. Authors in [4, 5] use a notch filter; in [6, 7] use virtual resistance and capacitance with a multi-loop feedback voltage and current; [8] uses poles assignment by state feedback.

Regarding the regulation and tracking of the *VSI* voltage, several control techniques are found in the literature. Repetitive control [9, 10] is not robust to non-periodic disturbances; proportional resonant

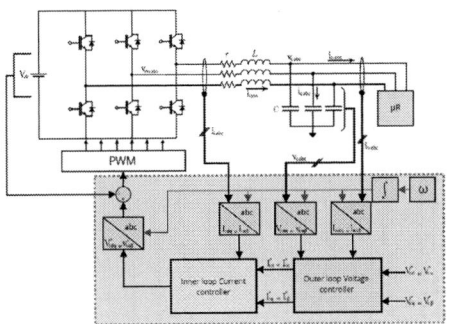

Fig. 1: Typical scheme for the control of a *VSI* connected to the microgrid. V_c are capacitor voltages, I_i are inverter currents an Io are output currents. All signals are in *abc*, αβ or *dq* frames

control [11, 12] uses band-pass filters to reject harmonic disturbances, but are difficult to determine how stable is the system. Proportional integral control [14, 13] requires decoupling of variables, which in isolated microgrids is not accurate because the frequency is not constant. In addition, non-linear loads generates harmonics in currents and voltages. Robust control techniques have also been proposed to control the voltage of a *VSI*. In [15] a H∞ and Lineal Matrix Inequality (LMI) control is stated for single-phase UPSs. Also, the design of a H∞ control with *LMI* is presented in [16]. This control reaches a low tracking error for the inner current of an inverter with *LCL* output filter. In [17, 18] the voltage controller of three-phase *VSI* is designed in an isolated microgrid using robust control.

The paper is developed as described below. The section "Active damping *LC* filter by *LQR* contol" presents a mathematical model of a *VSI* in αβ in the reference frame and implements an active damping function by *LQR* theory. Section "Voltage regulation by H∞ control" presents new weight functions that allow to synthesize an optimal controller that meets the infinite norm applied to the sensitivity function in proving the voltage error until 17th harmonic. Section "H∞ control synthesis" presents an application of criterion design for this H∞ control.

Finally, section "Experiment results", validates the proposed controllers through experimental tests, comparing the performance with the resonant proportional controller.

Active Damping *LC* filter by *LQR* control

The mathematical model of a *VSI* with and *RLC* output filter is obtained using the three-phase equations from the circuit presented in Fig. 1. This model is shown below. In Fig. 1, v_{invabc} is voltage switched on inverter output in *abc* frame, v_{cabc} represents a capacitor voltage in *abc* frame, i_{iabc} is an inductance current in *abc* frame, i_{oabc} is output current in *abc* frame. The biphasic representation of these voltages and currents is obtained with the Clark (αβ) transformations. $v_{c\alpha\beta}$, $i_{i\alpha\beta}$, $i_{o\alpha\beta}$ are the capacitor voltage, input *LC* filter current and output *LC* filter current respectively. Using Kirchhoff laws in an *LC* filter, the model in state space is presented, equation 1 (where $x(t)$ is state variable, $u(t)$ is input variable, $d(t)$ is disturbance input, $y_v(t)$ is output and f_x, g_u, h_d and c_x are the respective vector coefficients for an assumed system within the time domain). It is set as shown in 1, 2 and 3. I_2 is Identity matrix.

$$\begin{cases} \frac{d}{dt}x = f_x \cdot x(t) + g_u \cdot u(t) + h_d \cdot d(t) \\ \qquad y_v(t) = c_x \cdot x(t) \end{cases} \tag{1}$$

$$\frac{d}{dt}\begin{bmatrix} v_{c\alpha\beta} \\ i_{i\alpha\beta} \end{bmatrix} = \begin{bmatrix} 0 & (1/C)I_2 \\ (-1/L)I_2 & (-r/L)I_2 \end{bmatrix} \begin{bmatrix} v_{c\alpha\beta} \\ i_{i\alpha\beta} \end{bmatrix} + \begin{bmatrix} 0 \\ (1/L)I_2 \end{bmatrix} v_{inv\alpha\beta} - \begin{bmatrix} (1/C)I_2 \\ 0 \end{bmatrix} i_{o\alpha\beta} \tag{2}$$

$$\begin{bmatrix} v_{\alpha\beta} \\ i_{\alpha\beta} \end{bmatrix} = \begin{bmatrix} I_2 & 0 \end{bmatrix} \begin{bmatrix} v_{c\alpha\beta} \\ i_{i\alpha\beta} \end{bmatrix} \tag{3}$$

Control law is $u(t) = -[k_i I_2, k_v I_2]x(t)$ as is show in figure 2. $x(t)$ is states vector composed by capacitors voltage $(v_{C\alpha\beta})$ and inductance currents $(i_{L\alpha\beta})$. k_i and k_v are the gain of feedback states. At low

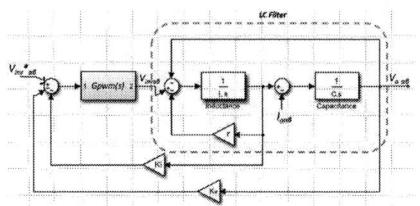

Fig. 2: Model LC with feedback state.

frequencies, delays caused by sampling and PWM may not be considered, and the VSI model with active damping can be established as show in equations 4 and 5 by $\alpha\beta$ reference frame.

$$\frac{d}{dt}\begin{bmatrix} v_{c\alpha\beta} \\ i_{i\alpha\beta} \end{bmatrix} = \begin{bmatrix} 0 & \left(^1\!/_C\right)I_2 \\ \left(^{-1}\!/_{L_n}\right)I_2 & \left(^{-r_n}\!/_{L_n}\right)I_2 \end{bmatrix}\begin{bmatrix} v_{c\alpha\beta} \\ i_{i\alpha\beta} \end{bmatrix} + \begin{bmatrix} 0 \\ \left(^1\!/_{L_n}\right)I_2 \end{bmatrix}v_{inv\alpha\beta} - \begin{bmatrix} \left(^1\!/_C\right)I_2 \\ 0 \end{bmatrix}i_{o\alpha\beta} \tag{4}$$

$$\begin{bmatrix} v_{\alpha\beta} \\ i_{\alpha\beta} \end{bmatrix} = \begin{bmatrix} I_2 & 0 \end{bmatrix}\begin{bmatrix} v_{c\alpha\beta} \\ i_{i\alpha\beta} \end{bmatrix} \tag{5}$$

where, $r_n = \left(^{(r+k_i)}\!/_{(1+k_v)}\right)$, $L_n = \left(^{L}\!/_{(1+k_v)}\right)$ and $u_v = (1+k_v)$. From equation 4 is obtain that natural frequency is $\omega_n = 1/\sqrt{L_nC}$, and the damping is $\zeta = \left(r_n/2\right)\sqrt{C/L_n}$. Then, state feedback does change the damping and the natural frequency of the system. k_i virtually increases the resistance of the inductance, increasing the damping of the system, while k_v virtually permits to decrease the inductance value of the LC filter, thus increasing the bandwidth of the system, which allows rejection of high frequency disturbances. A k_v less or equal to -1 destabilizes the system.

The continuous model has allowed observing the effect of the state feedback. For controller design purpose, the model is discretized as show in equation 6.

$$x_{k+1} = G \cdot x_k + H \cdot u_k$$
$$G = \begin{bmatrix} e^{-\alpha T}\left(\cos(\omega T) + \frac{\alpha}{\omega}\sin(\omega T)\right) & \frac{e^{-\alpha T}}{\omega C}\sin(\omega T) \\ -\frac{e^{-\alpha T}}{\omega L}\sin(\omega T) & \cos(\omega T) \end{bmatrix}, \; H = \begin{bmatrix} \frac{1}{LC(\omega^2+\alpha^2)}\left(e^{-\alpha T}\left(\frac{\alpha}{\omega}\sin(\omega T) - \cos(\omega T)\right) + 1\right) \\ \frac{\sin(\omega T)}{\omega L} \end{bmatrix} \tag{6}$$
$$\alpha = \frac{r}{2L}, \quad \omega^2 = \frac{1}{LC} - \left(\frac{r}{2L}\right)^2$$

The LQR control problem defines a quadratic objective function that depends on the input $u(t)$ and output $y(t)$ as defined in equation 7 [23], considering the model presented in equation 6. textit ρ represent the optimal state feedback vector, which maintains a balance between control effort and performance as it is demonstrated by the Pareto curve [19].

$$J = \lim_{\tau \to \infty} \frac{1}{\tau}\sum_{t=0}^{\tau-1} y_v^2(t) + \rho \cdot u^2(t) \tag{7}$$

As a case study, next, it is proceeding to the design of the control of an inverter whose parameters are shown in table I. The control is implemented by a digital unit. The sampling frequency, f_s, must be at least six times greater than the resonant frequency, f_r, [25]. The inverter meets this criterion because the resonance frequency is equal to 7860 rad/s and the sampling frequency is 10k samples per second.

Figure 3 shows the design of the controller. The poles map shown in figure 3a, is obtained from equations 6 and 7 and the parameters for LC filter shown in table I. This map is obtained by iteratively varying the parameter ρ in equation 7. In this map, it can be seen that when $\rho = 0.05$, it corresponds to the state feedback $k_v = 1.22$ and $k_i = 36.14$. With this feedback, the closed loop poles are located at $p = 0.276 \pm j0.396$. Without state feedback, $\omega_n = 5618.3 rad/s$ and $\zeta = 2.47 \times 10^{-3}$; and with state feedback $\omega_n = 8375.3 rad/s$ and $\zeta = 0.6$.

Figure 3b shows the Bode diagram with LQR control (red line) and without LQR control (blue line). This figure compares the frequency response of the LC filter without and with LQR control. According

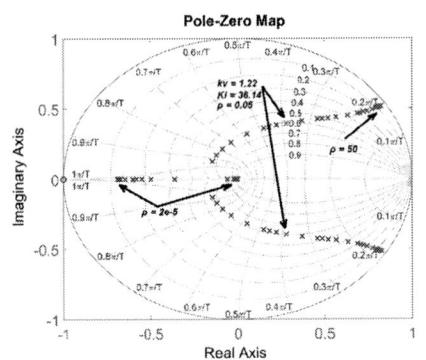

(a) Geometric place of the roots to select the optimal poles.

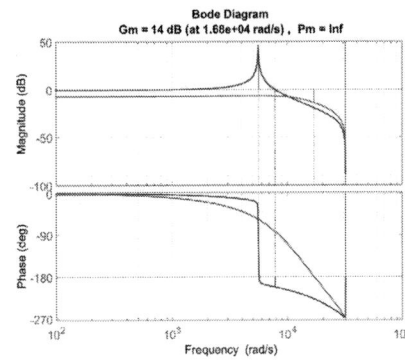

(b) Frequency for LC filter voltage: the blue line is the response frequency of LC filter without LQR control; the red line is the Bode response with LQR control.

Fig. 3: Geometric place of the roots and Bode diagram for LC filter without and with LQR control.

to [20], a delay of less than 34 degrees is ideal for rejecting disturbances. This means (as obtained from the phase diagram in figure 3b) that this design allows rejection of disturbances caused up to the 17th harmonic.

The design of the LQR control has been validated through Matlab® simulation. Figure 4 validates the design of the LQR controller. Figure 4a shows a step test of the VSI control with LQR. An overshoot of less than 10% and a stabilization time of 600 μs validate the design. On this platform, a single inverter has been used, isolating from the microgrid, to which a linear load has been connected. Figure 4b shows three-phase voltage without LQR control. A lineal load is switched in $t = 0.3\ s$. In this figure, the oscillatory mode, at the frequency ω_n, is observed in the three-phase voltage. As show in figure

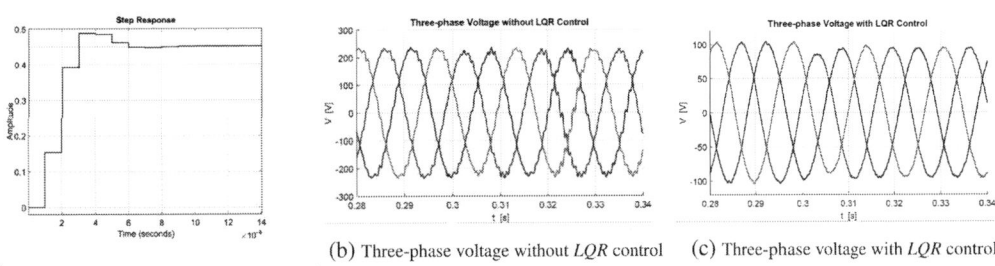

(a) Step test for the LC filter with LQR control

(b) Three-phase voltage without LQR control

(c) Three-phase voltage with LQR control

Fig. 4: Time response of the LC filter controlled with LQR control.

4b, the oscillations at frequency ω_n have been reduced by the LQR control, but a loss of gain and a dampened response is obtained. The loss of gain is due to the feedback of states. These effects are not an inconvenience because they will be compensated by the voltage control proposed below.

Voltage regulation by H∞ control

The H∞ control is a strategy control introduced by Doyle (1983). This control can be applied for both SISO systems and MIMO systems. Here, control H∞ is adopted to regulate the voltage in VSI capacitors by to meet the following control objectives: Tracking for minimal error to ω_g; disturbance rejection from the DC bus and load currents; and moderated control effort. The formulation makes use of the control scheme shown in figure 5a.

Here, P represents the generalized plant and K is the H∞ controller. In this notation, the subscripts denote the input and output ports of P in this order. P includes the processing delay and PWM (Transfer function

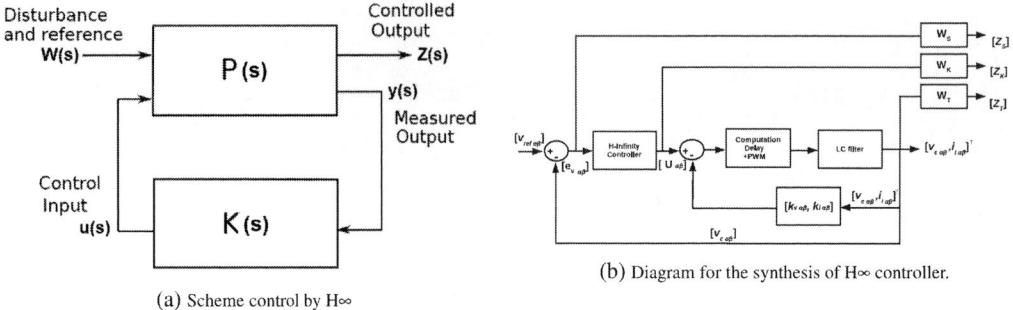

(a) Scheme control by H∞

(b) Diagram for the synthesis of H∞ controller.

Fig. 5: H∞ control design.

delay, $Gpwm(s)$, due to computation delay, sampler, ZOH, and PWM delay is modeled as in [26].), plus the LC filter damped by the LQR control. Closed loop transfer function from Z to W, T_{ZW}, is shown in equation 8, [21]. In this equation, P_{Wy}, P_{WZ}, P_{uy} and P_{uZ} represent the transfer functions obtained in the diagram shown in figure 5a.

$$T_{ZW}(S) = P_{WZ} + P_{Wy}K(I - P_{uy}K)^{-1}P_{uZ} \tag{8}$$

The optimal control problem lies in finding a K controller that, based on the information of the y output (capacitors voltage), generates a u control signal, which counteracts the influence of W on Z, thus minimizing the norm of the closed loop from W to Z. A sub-optimal control problem consists in, given a $\gamma \in \mathbb{R}$ pre-set, to design a stabilizing controller which ensures equation 9.

$$\|T_{ZW}(s)\|_\infty = \max_W \overline{\sigma}(T_{ZW}(j\omega)) \leq \gamma \tag{9}$$

The scheme shown in figure 5a is extended in figure 5b. $v_{ref\,\alpha\beta}$ is the reference W; Z_S, Z_K and Z_T are the Z output; $v_{c\,\alpha\beta}$ and $i_{i\,\alpha\beta}$ are the y measured output (capacitor voltage and inductance current respectively); $e_{v\,\alpha\beta}$ is error voltage and $u_{\alpha\beta}$ is control input in $\alpha\beta$ frame.

As is shown in figure 5b, W_S, W_K and W_T are transfer functions, defined such that the inverse of these functions delimit the frequency response of the sensitivity function, the control signal and the response of the closed loop of the system respectively. Using W_S, W_K and W_T, H∞ control is optimally synthesized to satisfy the requirements that singular values of the transfer function between Z and $v_{ref\,\alpha\beta}$ are less than γ, as indicated in equation 9, and that the infinite norm of the composite matrix by $W_S S$ and $W_K KS$ be less than unity, as shown in equation 10. In this equation, K represents the H∞ control and S is the sensibility function transfer.

$$\left\| \begin{array}{c} W_S S \\ W_K KS \end{array} \right\|_\infty < 1 \tag{10}$$

In principle, a tracking error is null at frequencies where the direct path gain is very large or infinite [1]. Consequently, having high gain in the direct path $|L(j\omega)|$, in the frequencies of interest ω, to obtain a voltage tracking error equal to or near zero, is similar to making the magnitude of the sensitivity function zero at such frequencies [22].

Some researchers have proposed functions to obtain W_s [24, 15]. In [15], it is presented a weight function type second-order band-pass filter for W_s in VSI control. As the transfer function is not strictly proper, the authors add to W_s a low pass filter with a cutoff frequency greater than the VSI voltage frequency.

In this research, the structure that shapes the sensitivity function is defined by a sum of pass-band filters tuned to the fundamental frequency, $\omega_1 = \omega_g$, and some harmonics, $h = 2, 3, 4, ...$, as is shown in equation 11. In comparison, [24] use a second order low pass filter, and [15] use a second order band pass filter. The transfer function proposed here is stable and strictly proper. These functions seek a sensitivity function magnitude of zero at the fundamental frequency and its harmonics, to obtain a minimum tracking error and as well as the ability to reject disturbances in the main harmonic frequencies that typically occur in three-phase power networks when non-linear loads are connected to the power grid (such as rectifiers, for example).

$$W_s(s) = \sum_{h=1}^{\infty} k_{sh} \left(\frac{\left(\frac{\omega_h}{Q_h}\right)s}{s^2 + \left(\frac{\omega_h}{Q_h}\right)s + \omega_h^2} \right)^2 \tag{11}$$

W_k function must be chosen so that $\|KS\|_\infty$ is less than 6 dB in working frequencies to prevent actuator saturation. Equation 12 shows the function selected to fulfill this purpose.

$$W_k = k_k \frac{(s + a\omega_{nk})^2}{s^2 + 1.44\omega_{nk} + \omega_{nk}^2} \tag{12}$$

where, a, k_k are adjusted parameters and ω_k is grid frequency. Finally, W_T is a constant value equal a less than $0\ dB$.

H∞ control synthesis

In this section the H∞ control for VSI is designed. Control H∞ is synthesized using the diagram in figure 5b. In this diagram, W_S and W_K are obtained from equations 11 and 12 respectively. Controller has been synthesized using the mixsyn() Matlab® function. Using this function, the parameter γ of the equation 9 has been set to a value of 0.997, taking into account the weight functions W_S, W_K and W_T whose settings are explained below.

Regarding W_s proposed in this investigation, equation 11, the gains K_{s1} and K_{sh} are adjusted to satisfy the conditions shown in equations 9 and 10. These gains, which have been found iteratively, are shown in table I. The inverse function of the W_s gains in the respective frequencies, fundamental and harmonic, delimit the depth that the frequency response gain curve will have on such frequencies. Increasing K_h reduces the voltage tracking error but compromises the stability margin of the system. Harmonics of higher order are not taken into account due to the bandwidth of the LC filter and the waterbed effect that occurs when trying to shape the magnitude of the sensitivity function in the frequency domain [22].

Figure 7a (dashed line) shows the Bode diagram of W_s corresponding to equation 11, which has been adjusted with the parameters of table I and the other parameters mentioned. As shown in the figure, this function, $1/W_s$, allows sensitivity to be minimized only at the frequencies of interest. This is favorable for the optimization algorithm to find a solution that satisfies the infinity norm of sensitivity function shown in equation 10.

The sensitivity function obtained with the H∞ controller synthesis is examined in figure 7 (continuous line). According to Nyquist criteria, the module margin is the minimum distance between the Nyquist curve and the critical point $(-1, j0)$. The inverse of this minimum distance is the infinity norm of the sensitivity function, $\|S\|_\infty$. From the $SISO$ approach, $\|S\|_\infty$ is the bode diagram corresponding to the magnitude of S. A good margin of stability establishes that the maximum magnitude obtained in the Bode diagram of S is less than 6 dB [22]. Compliance with this margin, can be seen in figures 7b, 7c the sensitivity function gain not exceed 6 dB. The waterbed effect is observed by setting $|S| > 1$ at the harmonic frequencies of the fundamental frequency, as is shown in figure 7c. This means that signals with inter-harmonic frequencies could be amplified or, which is the equally, no rejection of inter-harmonic frequency disturbances would occur. $|S| \approx 1$ and $|T| \to 0$ implies rejection of high frequency measurement noise. Compliance with this feature is observed by approaching the Bode diagrams for S and T, figures 7c and 7d, at $\omega > 11000 rad/s$. As shown in figure 7e, the closed loop gain is 0 dB at the frequency ω_g and its harmonics, as a necessary condition for signal tracking and disturbance rejection to occur.

The purpose of W_k is to ensure that the actuator does not operate under saturation and stress conditions.

[15] propose a W_k scalar. The W_k proposed in equation 12 has been configured with the following parameters: $k_k = 5 \times 10^3$, $a = 11$, $\omega_{nk} = 0.01\omega_g$. The W_k function must ensure that $|KS| < 6\,dB$ in the disturbance rejection frequencies. With these aforementioned parameters, this purpose has been met, as shown in figure 7f.

W_T used in this investigation is a scalar equal to 0.15. The open-loop transfer function, L, is the series between H∞ control (K) and the plant (damped LC filter with state feedback). Figure 7h shows how the open loop gain, $|L|$, is sufficiently high in the frequency ω_g and its harmonics. This result guarantees voltage monitoring and rejection of disturbances in these frequencies.

Figure 7j shows how it holds that $|W_kKS| < 1$ and also shows that $|W_kKS| \rightarrow 0$ at high frequencies, which indicates noise immunity.

Another condition that indicates zero tracking error is that the sum between the closed loop function, T, and the sensitivity transfer function, S, is equal to 0 dB. Compliance with this condition in this design is shown in figure 7k.

Experimental Results

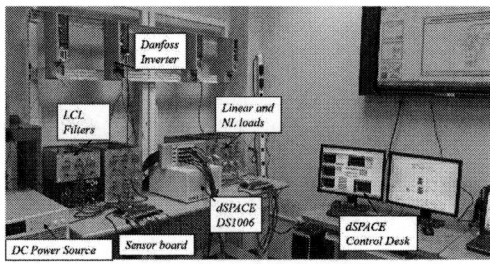

Fig. 6: The Universidad de Puerto Rico Mayaguez Microgrid Laboratory.

To evaluate the performance of the proposed controllers, the experiment is carried out in a microgrid system, figure 6, is composed of four Danfoss inverters ($2.2\,kW$), LCL filters and measurement sensors. The inverters are supplied by a DC voltage of $400V$ which emulates the dc-link between the grid-side inverter and the energy source. A real-time platform, dSPACE 1006, is used to test both proposed controllers (Proportional Resonant (PR) and H∞-controller) in different scenarios and evaluate their performance.

The electrical setup and control system parameters are detailed in table I. All the experimental results were extracted from dSPACE control desk and plotted using MATLAB®.

The voltage regulation controls have been tested on an inverter without connection to the microgrid. The experimental setup considers a single inverter with a LCL filter and linear and non-linear loads. An inverter has supplied voltage the non-linear loads using PR and H∞ control. Figures 8a and 8b show the performance of both controllers, PR and H∞, before a non-linear load and steady-state regime. These results are obtained through the data acquisition system of the dSPACE. It has been possible to configure the controller H∞ and PR (using the parameters in table I) to obtain a THD_v equal to 1.3%, lower than the THD_v obtained with the resonant controller (equal to 4.6%). It is also observed how a better result is obtained with the H∞ controller as it can significantly attenuate the 13th and 17th harmonics. Figures 8a and 8b also reveal the voltages in the capacitors when both controls are used (resonant and H∞ respectively). The error obtained with the H∞ control is 4 Volts while the error of the PR control is 10 Volts. It is then appreciated that using a H∞ controller it is possible to obtain a lower voltage error, which is desirable in terms of efficiency and precision in the load sharing between generators of the microgrid. Figure 9 presents the results obtained using a digital oscilloscope, Fluke 43. The THD_v, using the H∞ control, is 1%, compared to the voltage THD_v voltage, using PR control, which is 2.8%. Figure 10 presents evidence of the performance of the controls, resonant and H∞ controllers, before the step of a non-linear load. This test was carried out using the Tektronix MSO 4034 oscilloscope. The voltage and

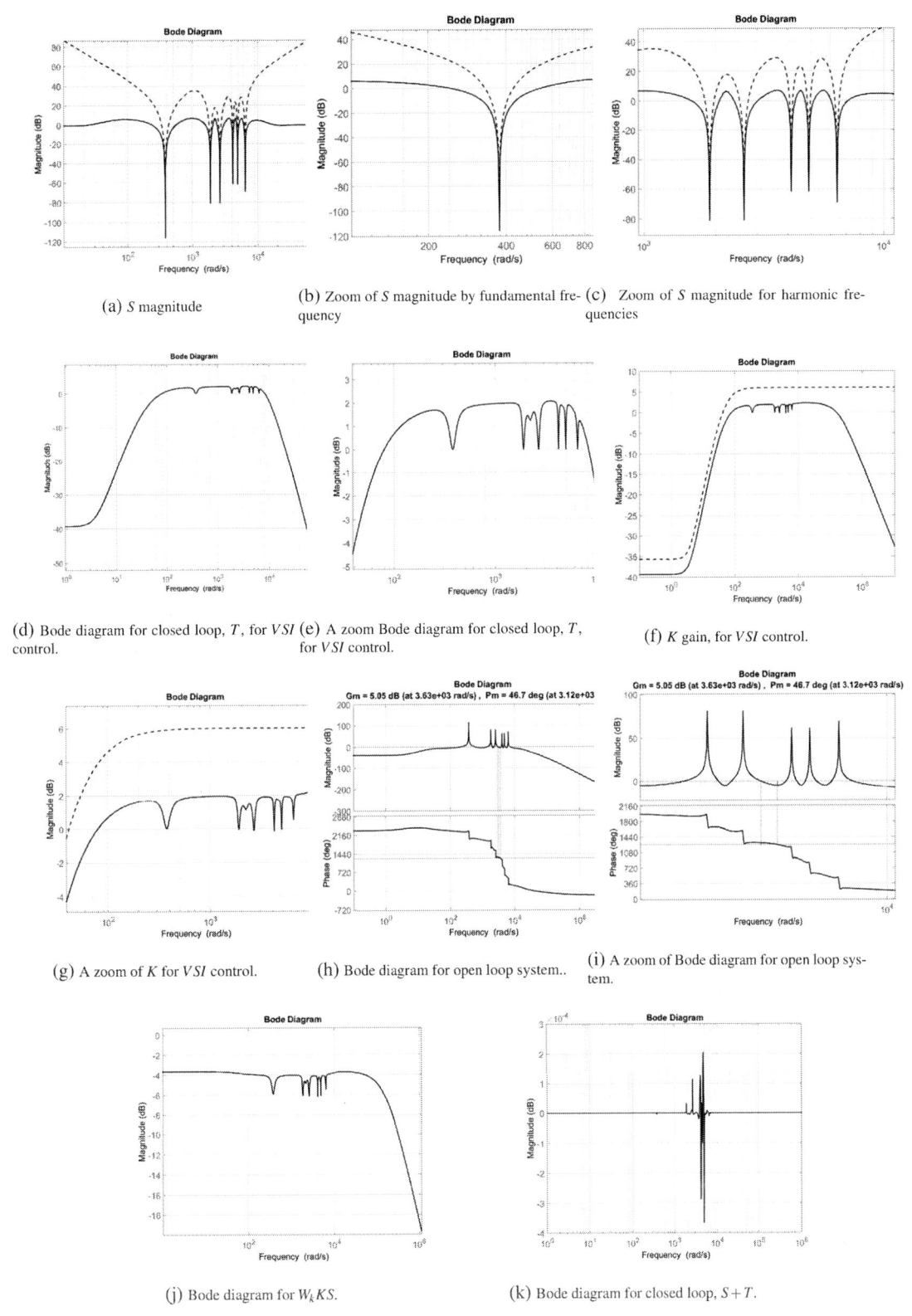

(a) S magnitude

(b) Zoom of S magnitude by fundamental frequency

(c) Zoom of S magnitude for harmonic frequencies

(d) Bode diagram for closed loop, T, for VSI control.

(e) A zoom Bode diagram for closed loop, T, for VSI control.

(f) K gain, for VSI control.

(g) A zoom of K for VSI control.

(h) Bode diagram for open loop system..

(i) A zoom of Bode diagram for open loop system.

(j) Bode diagram for $W_k KS$.

(k) Bode diagram for closed loop, $S + T$.

Fig. 7: H∞ control design.

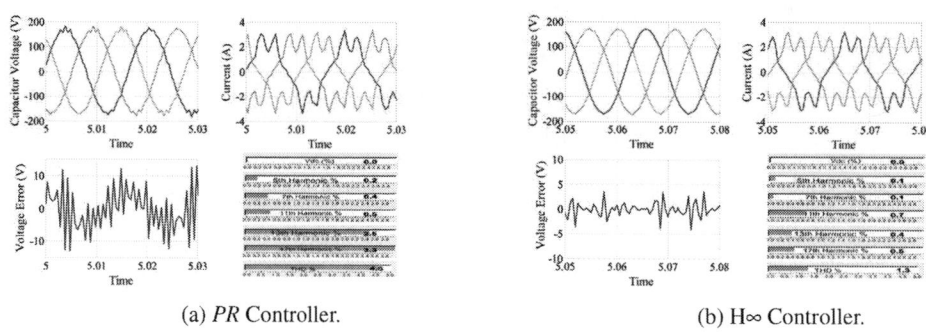

(a) *PR* Controller.   (b) H∞ Controller.

Fig. 8: Results obtained in hardware in the loop.

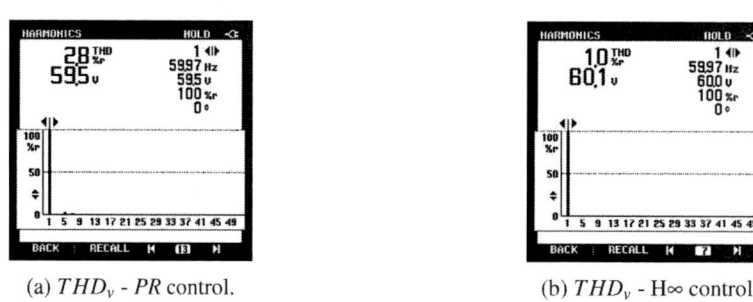

(a) *THD$_v$* - *PR* control.   (b) *THD$_v$* - H∞ control.

Fig. 9: *THD$_v$* obtained with H∞ and *PR* control.

current delivered by the inverter have been sampled for a long period of time. This can be seen in the upper part of figure 10. In these samples, in a short period of time, the images have been enlarged to observe the voltage and current transient response to the step-type input of a non-linear load. Figure 10 shows that the voltage controlled with H∞ stabilizes in less time than that a period of the frequency of the network, but at the cost of a significant overshoot.

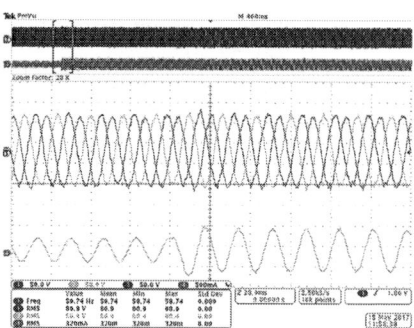

Fig. 10: Voltage obtained with H∞ control.

Table I: PARAMETERS

PARAMETER	SYMBOL	VALUE	UNITS
Grid Voltage	V_g	120	V_{rms}
Grid Frequency	ω_g	376.99	rad/s
Output Impedance	L_o	3.6	mH
Inner Inductance	L_i	3.6	mH
Capacitance	C_f	8.8	μF
DC Voltage	V_{dc}	400	V

PARAMETER	$h=1$	$h=5$	$h=7$	$h=11$	$h=13$	$h=17$
k_h	94 dB	73 dB	73 dB	54 dB	54 dB	60 dB
Q_h	2000	800	800	800	800	800

Voltage PR	$K_{pV1}, K_{iV1}, K_{V5,7,11,13}$	0.3 200 75 75 75
Current PR	K_{pI1}, K_{iI1}	0.3 100

Conclusion

This paper shows the design of a voltage control for VSI using the LQR and H∞ control techniques. Active damping is achieved by feedback of the voltage signals in capacitors and currents signals in the inductances which are available more easily than the currents signals in the capacitors required by other active damping techniques. Additionally, a design of a robust controller using the H∞ norm with a new sensitivity function W_s is developed and a comparison with classical proportional resonant controller shows that H∞ controller has better performance when nonlinear loads are connected to the VSI.

References

[1] Chi-Tsong Chen. (1970). Linear system theory and design. In Oxford University Press (p. 679).
[2] Pogaku, N., Prodanovi, M., Green, T. C. (2007). Modeling, analysis and testing of autonomous operation of an inverter-based microgrid. IEEE Transactions on Power Electronics, 22(2), 613625. https://doi.org/10.1109/TPEL.2006.890003
[3] Wu, W., Sun, Y., Huang, M., Wang, X., Wang, H., Blaabjerg, F., Liserre, M., Chung, H. S. H. (2014). A robust passive damping method for LLCL-filter-based grid-tied inverters to minimize the effect of grid harmonic voltages. IEEE Transactions on Power Electronics, 29(7), 32793289. https://doi.org/10.1109/TPEL.2013.2279191
[4] Bosch, S., Lebsanft, D., Steinhart, H. (2017). Self-adaptive resonance frequency tracking for digital notch-filter-based active damping in LCL-filter-based active power filters. 2017 19th European Conference on Power Electronics and Applications (EPE17 ECCE Europe), 2017-Janua, P.1-P.10. https://doi.org/10.23919/EPE17ECCEEurope.2017.8099104
[5] Byk, M., Tan, A., Inci, M., Tmay, M. (2017). A notch filter based active damping of llcl filter in shunt active power filter. 19th International Symposium on Power Electronics, Ee 2017, 2017-Decem, 15. https://doi.org/10.1109/PEE.2017.8171701
[6] Wang, X., Blaabjerg, F., Loh, P. C. (2015). Virtual RC Damping of LCL -Filtered Voltage Source Harmonic Compensation. 30(9), 47264737. https://doi.org/10.1109/TPEL.2014.2361853
[7] Xu, J., Member, S., Xie, S., Tang, T. (2014). Active Damping-Based Control for Grid-Connected LCL -Filtered Inverter With Injected Grid Current Feedback Only. 61(9), 47464758. https://doi.org/10.1109/TIE.2013.2290771
[8] Chowdhury, V. R., Mukherjee, S., Shamsi, P., Ferdowsi, M. (2017). State Feedback Control to Damp Output LC Filter Resonance for Field Oriented Control of VSI Fed Induction Motor Drives. IEEE Green Technologies Conference, 4, 370375. https://doi.org/10.1109/GreenTech.2017.60
[9] Kim, E. K., Mwasilu, F., Choi, H. H., Jung, J. W. (2015). An Observer-Based Optimal Voltage Control Scheme for Three-Phase UPS Systems. IEEE Transactions on Industrial Electronics, 62(4), 20732081. https://doi.org/10.1109/TIE.2014.2351777
[10] Escobar, G., Valdez, A. A., Leyva-Ramos, J., Mattavelli, P. (2007). Repetitive-based controller for a UPS inverter to compensate unbalance and harmonic distortion. IEEE Transactions on Industrial Electronics, 54(1), 504510. https://doi.org/10.1109/TIE.2006.888803
[11] Ayad, A., Hashem, M., Hackl, C., Kennel, R. (2016). Proportional-resonant controller design for quasi-Z-source inverters with LC filters. IECON Proceedings (Industrial Electronics Conference), 35583563. https://doi.org/10.1109/IECON.2016.7793282
[12] Monfared, M., Golestan, S., Guerrero, J. M. (2014). Analysis, design, and experimental verification of a synchronous reference frame voltage control for single-phase inverters. IEEE Transactions on Industrial Electronics, 61(1), 258269. https://doi.org/10.1109/TIE.2013.2238878
[13] Li, B., Yao, W., Hang, L., Tolbert, L. M. (2012). Robust proportional resonant regulator for grid-connected voltage source inverter (VSI) using direct pole placement design method. IET Power Electronics, 5(8), 1367. https://doi.org/10.1049/iet-pel.2012.0102
[14] Richter, S. a., Doncker, R. W. De. (2011). Digital proportional-resonant (PR) control with anti-windup applied to a voltage-source inverter. Proceedings of the 2011 14th European Conference on Power Electronics and Applications, 110.
[15] Sari, B., Mohamed Fouad Benkhoris, Le Claire, J. C., Rabhi, B. (2012). Robust H_{∞} output feedback control design applied to Uninterruptible Power Supplies. 2012 IEEE International Symposium on Industrial Electronics, 490495. https://doi.org/10.1109/ISIE.2012.6237136
[16] Maccari, L. A., Massing, J. R., Schuch, L., Rech, C., Pinheiro, H., Montagner, V. F., Oliveira, R. C. L. F. (2012). Robust H_{∞} control for grid connected PWM inverters with LCL filters. 2012 10th IEEE/IAS International Conference on Industry Applications, 16. https://doi.org/10.1109/INDUSCON.2012.6451389
[17] Sedghi, L., Fakharian, A. (2016). Robust voltage regulation in islanded microgrids: A LMI based mixed H2/H_{∞} control approach. 2016 24th Mediterranean Conference on Control and Automation (MED), 431436. https://doi.org/10.1109/MED.2016.7535926
[18] Wang, Y., Jiang, H., Zhou, L., Xing, P. (2016). Inverter seamless mode tranfer strategy based on H_{∞} robust control in micro-grid. 2016 IEEE International Conference on Aircraft Utility Systems (AUS), 10941098. https://doi.org/10.1109/AUS.2016.7748222
[19] Hindi, H. A., Hassibi, B., Boyd, S. P. (1998). Multiobjective H2/H-optimal control via finite dimensional Q-parametrization and linear matrix inequalities. Proceedings of the American Control Conference, 5(June), 32443249. https://doi.org/10.1109/ACC.1998.688463
[20] Liu, Y., Wu, W., He, Y., Lin, Z., Blaabjerg, F., Chung, H. S. (2016). An Efficient and Robust Hybrid Damper for LCL- or $LLCL$-Based Grid-Tied Inverter with String Grid-Side Harmonic Voltage Effect Rejection. IEEE Transactions on Industrial Electronics, 63(2), 926936. https://doi.org/10.1109/TIE.2015.2478738
[21] Kemin Zhou, John C. Doyle. (1997). Essentials of Robust Control. Pearson. ISBN=0135258332.
[22] Werner, Herbert. (2013). Optimal and robust control. Technische Universitat Hamburg-Harburg.
[23] F. L. Lewis, D. Vrabie, and V. L. Symons. Optimal control. John Wiley Sons, 2012.
[24] Yang, S., Lei, Q., Peng, F. Z., Qian, Z. (2011). A Robust Control Scheme for Grid-Connected Voltage-Source Inverters. IEEE Transactions on Industrial Electronics, 58(1), 202212. https://doi.org/10.1109/TIE.2010.2045998
[25] Pan, D., Ruan, X., Bao, C., Li, W., Wang, X. (2014). Capacitor-Current-Feedback Active Damping With Reduced Computation Delay for Improving Robustness of LCL-Type Grid-Connected Inverter. IEEE Transactions on Power Electronics, 29(7), 34143427. https://doi.org/10.1109/TPEL.2013.2279206
[26] Agorreta, J. L., Borrega, M., Lpez, J., Marroyo, L. (2011). Modeling and control of N-paralleled grid-connected inverters with LCL filter coupled due to grid impedance in PV plants. IEEE Transactions on Power Electronics, 26(3), 770785. https://doi.org/10.1109/TPEL.2010.2095429

Family of Splitting Current Single-Loop Control for *LCL*-Type Grid-Connected Inverter

Yuying He[1], Xuehua Wang[1], Xinbo Ruan[1,2], Guoxing Su[1], Fuxin Liu[2]

1. Huazhong University of Science and Technology
Wuhan, China
2. Nanjing University of Aeronautics and Astronautics
Nanjing, China
Tel.: +86 / 18627864801.
E-Mail: heyuying@hust.edu.cn
URL: http://ceee.hust.edu.cn

Acknowledgements

This work was supported by the National Key Research and Development Program of China under Award 2016YFB0900100.

Keywords

«Single-loop», «weighted average current control », «stabilization », «robustness», «*LCL* filter».

Abstract

The single-loop weighted average current (WAC) control is attractive for the *LCL*-type grid-connected inverter owing to its inherent active damping. High robustness can be naturally harvested without any additional damping methods, but it requires two current sensors. Depending on the different weight values, a family of the splitting current single-loop control schemes with the associated filter configurations is developed in this paper, where only one current sensor is needed. As a result, high robustness, high-quality grid current and cost-efficiency can be concurrently achieved. Finally, experimental results from a 6-kW setup confirm the effectiveness and practicality of the splitting current single-loop control.

Introduction

The energy sector is moving to the era of smart grids, and the grid-connected voltage-source inverters (VSIs) have been widely installed as the effective interfaces in the renewable power plants and distribution networks [1]. The *LCL* filter is commonly equipped at the output of grid-connected inverters, due to its better harmonic attenuation compared to L and LC filters. However, the inherent resonance of the *LCL* filter poses a challenge to the system stability.

The stabilization of the *LCL*-type grid-connected inverters has been extensively studied. It is revealed that, with the single-loop inverter-current-feedback (ICF) and grid-current-feedback (GCF) control, the grid-connected inverters can be conditionally stabilized thanks to the inherent digital control delays T_d [2]. Typically, under $T_d = 1.5\ T_s$ (T_s is the sampling period), the stability is subject to the relation of resonance frequency (f_r) and one-sixth of sampling frequency ($f_s/6$), where $f_r < f_s/6$ and $f_r > f_s/6$ are suitable for the ICF and GCF, respectively. Nevertheless, whether ICF or GCF is used, f_r may shift across $f_s/6$ due to the grid impedance variation, and the stable operation would be easily broken, resulting in a poor robustness.

Besides the single-loop ICF and GCF control, the weighted average current (WAC) control gains much attention [3], [4]. Instead of the inverter current or the grid current, their average value serves as the control variable. It is revealed that, the WAC control is actually equivalent to a dual-loop current control, consisting of an inner capacitor-current loop and an outer grid-current loop, where the inner loop provides inherent active damping. The damping performance is determined by the weight value [3]. If properly designed, sufficient gain margin and phase margin can be reserved, even with the wide range of grid impedance variation. Thus, compared with the single-loop ICF or GCF control, the single-loop WAC control can easily harvest robustness against grid impedance variation.

Although the conventional single-loop WAC control is attractive, it requires two current sensors to sense both the inverter and grid currents [4]. Alternatively, it is realized by splitting the filter capacitor into two ones, i.e., the *LCCL* configuration, and using the current flowing between the two equipotential junctions connected by the two split capacitors as the target control variable [3]. As a result, a current sensor is saved. However, the *LCCL* configuration limits the reachable range of the proportion of the two split capacitors, i.e., the weight value. In fact, it is also worth concerning those weight values beyond the reachable range relevant to the *LCCL* configuration and the corresponding single-loop control. Their potential filter configurations like *LCCL* are necessary to be explored. These, to the best knowledge of the authors, have hitherto remained unveiled. This paper attempts to address the missing pieces and complete this work with a family of splitting current single-loop control.

This paper is organized as follows. In Section II, the WAC control is reviewed. Thereafter, a family of splitting current single-loop control schemes together with the associated filter configurations is explored in Section III, and their general model is given in Section IV. To verify the theoretical expectations, experimental results are presented in Section V. Finally, Section VI concludes this paper.

Weighted Average Current Control with Unit PCC Voltage Feedforward

Fig. 1 shows the configuration of a single-phase grid- connected inverter feeding the power grid through an *LCL* filter, consisting of the inverter-side inductor L_1, grid-side inductor L_2 and filter capacitor C. The power grid is modeled as an ideal voltage source v_g in series with a grid inductance L_g. Depending on the grid configuration, L_g may vary in a wide range, and thus imposes challenge to the inverter stability. To achieve robust operation, the WAC control gains much attention, where the control variable is the weighted average current, expressed as

$$i_{WA} = \beta i_{L1} + (1-\beta) i_{L2} \tag{1}$$

where, β and $1-\beta$ are the weight values of the inverter current i_{L1} and the grid current i_{L2}, respectively. The primary control objective is to regulate i_{WA} to track its reference i^*, constituted by the demanded amplitude I^* and the PCC phase θ extracted by a phase-locked loop (PLL). The error signal of i^* and i_{WA} is processed by a current regulator G_i. Meanwhile, in order to be immune from the grid voltage distortion, v_{PCC} is usually fed forward with the unit feedforward gain [5]. Summing up the feedforward signal and the output of G_i, the modulation reference v_M is obtained and sent to a digital PWM modulator, generating the control signals of the power switches.

Considering $i_{L1} = i_{L2} + i_C$, the feedback of i_{L1} can be converted into the ones of i_{L2} and i_C, so the part in the dotted box in Fig. 1 can be equivalently transformed into Fig.2. It tells, the WAC control is actually equivalent to the grid current control with the well-known capacitor-current-feedback active damping. As reported in [6], both the positive and negative coefficients of capacitor current are possible to retain a stable operation. So, from the perspective of active damping, the weight value β should not be subconsciously limited between 0 and 1, while $\beta < 0$ and $\beta > 1$ are also necessary to be considered. From the perspective of implementation, they are all feasible, but note that two current sensors are required. To reduce the cost, the single-loop control with only one current sensor is necessary to be studied.

Family of Splitting Current Single-Loop Control for LCL-Type Grid-Connected Inverter   HE Yuying

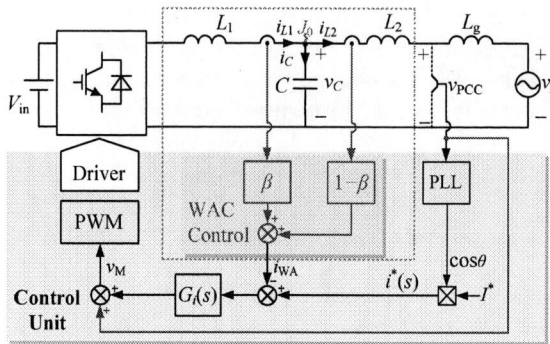

Fig. 1. Configuration of a single-phase *LCL*-type grid-connected inverter with WAC control and PCC voltage feedforward.

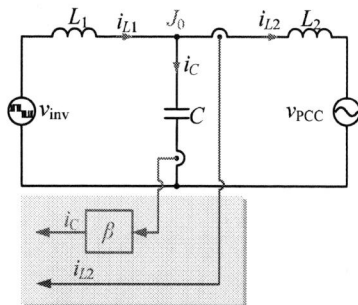

Fig. 2. The equivalent representation of WAC control.

Family of Splitting Current Single-Loop Control Schemes

According to Fig. 1, we have $i_{L1} = i_{L2} + i_C$. Substitution of this expression into (1), yields

$$i_{WA} = i_{L2} + \beta i_C .$$

(2)

As an alternative, i_{WA} can also be expressed as

$$i_{WA} = i_{L1} - (1-\beta) i_C.$$

(3)

■ **Splitting Capacitor-Current Single-Loop Control Scheme ($0 \leq \beta \leq 1$)**

It can be observed from (2) and (3) that, if the capacitor current i_C is split into two parts, i.e., $(1-\beta)i_C$ and βi_C, as shown in Fig. 3(a), the current flowing between the two equipotential junctions J_{01} and J_{02} is exactly the expected weighted average current i_{WA}. To implement it in real circuit, the filter capacitor C should be split into C_1 and C_2, as shown in Fig. 4(a), and $C_1 = (1-\beta)C$ and $C_2 = \beta C$. Obviously, only one current sensor is needed by this termed *splitting capacitor-current single-loop control*. Nonetheless, it should be noted that, the split capacitances of C_1 and C_2 cannot be negative, i.e., $(1-\beta)C \geq 0$ and $\beta C \geq 0$, resulting in $0 \leq \beta \leq 1$. It means this control scheme is suitable for replacing the conventional WAC control only when $0 \leq \beta \leq 1$.

■ **Splitting Grid-Current Single-Loop Control Scheme ($\beta \geq 1$)**

Eq. (2) can be rewritten as

$$i_{WA} = \beta \left(\frac{1}{\beta} i_{L2} + i_C \right).$$

(4)

EPE'20 ECCE Europe

Assigned jointly to the European Power Electronics and Drives Association & the Institute of Electrical and Electronics Engineers (IEEE)

Taking $0 < 1/\beta < 1$ into account, the grid current i_{L2} can be split into two parts, i.e., $(1/\beta)i_{L2}$ and $(1-1/\beta)i_{L2}$, as shown in Fig. 3(b). As a result, the current flowing between J_{01} and J_{02} is i_{WA}/β. Likewise, in real circuit, the grid inductor L_2 can be split into paralleled L_{21} and L_{22}, as shown in Fig. 4(b), where $L_{21}/L_{22} = 1/(\beta-1)$, and i_{WA}/β would be obtained. Multiplying the sensed current by β, yields the expected weighted average current i_{WA}. Likewise, only one current sensor is needed by this termed *splitting grid-current single-loop control*.

In fact, $\beta = 1$ can be treated as a special case of this control scheme by letting $L_{21} = \infty$, which implies the branch of L_{21} shown in Fig. 4(b) is open. In this manner, the control scheme naturally evolves into the ICF single-loop control. To sum up, this single-loop control scheme is suitable for replacing the conventional WAC control only when $\beta \geq 1$.

■ Splitting Grid-Current Single-Loop Control Scheme ($\beta \leq 0$)

When $\beta < 0$, Eq. (3) can be rewritten as

$$i_{WA} = i_{L1} - \left(1+|\beta|\right)i_C = \left(1+|\beta|\right)\left(\frac{1}{1+|\beta|}i_{L1} - i_C\right). \tag{5}$$

Due to $0 < 1/(1+|\beta|) < 1$, the inverter current i_{L1} can be split into two parts, i.e., $(1/(1+|\beta|))i_{L1}$ and $(|\beta|/(1+|\beta|))i_{L1}$, as shown in Fig. 3(c). As a result, the current flowing between J_{01} and J_{02} is $i_{WA}/(1+|\beta|)$. Similarly, in real circuit, splitting the inverter inductor L_1 into two paralleled ones L_{11} and L_{12}, as shown in Fig. 4(c), where $L_{11}/L_{12} = 1/|\beta|$, $i_{WA}/(1+|\beta|)$ would be obtained, which flows between the two equipotential junctions connected by the two split inductors. Multiplying the sensed result by $(1+|\beta|)$, yields the expected weighted average current i_{WA}. Likewise, only one current sensor is needed to implement this termed *splitting inverter-current single-loop control*.

Similarly, $\beta = 0$ can also be treated as a special case of this control scheme by letting $L_{11} = \infty$, which implies the branch of L_{11} shown in Fig. 4(c) is open, and the control scheme naturally evolves into the GCF single-loop control. Likewise, this control scheme is suitable for replacing the conventional WAC control only when $\beta \leq 0$.

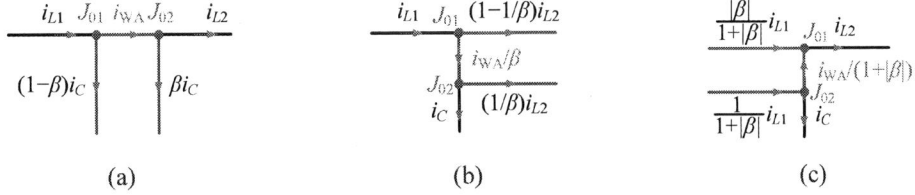

(a) (b) (c)

Fig. 3. Representations of (a) splitting capacitor current, (b) splitting grid current, and (c) splitting inverter current.

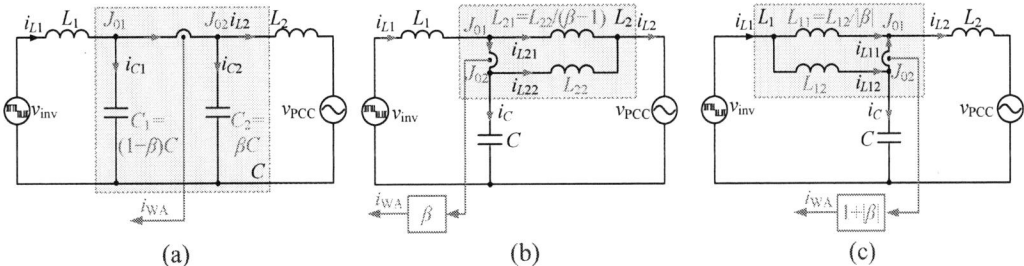

(a) (b) (c)

Fig. 4. The splitting current single-loop control schemes with (a) splitting capacitor-current, (b) splitting grid-current, and (c) splitting inverter-current.

In conclusion, by splitting the capacitor current (see Fig. 4(a)), the grid current (see Fig. 4(b)), or the inverter current (see Fig. 4(c)), i_{WA} in the cases of $0 \leq \beta \leq 1$, $\beta \geq 1$ and $\beta \leq 0$ can be directly sensed, respectively. Accordingly, the single-loop control schemes can be applied, saving a current sensor.

General Model of Splitting Current Single-Loop Controlled Inverter with PCC Voltage Feedforward

Recall that all the three splitting current single-loop control schemes origin from Fig. 1, which means their models can be reunified into that of the conventional WAC control, as shown in Fig. 5. Here, $G_d(s)$ represents the s-domain model of the control delays, commonly approximated as $G_d(s) \approx e^{-1.5\,sT_s}$. Thanks to the PCC voltage feedforward, the disturbing error caused by the grid voltage is almost eliminated, so a Proportional-Integral (PI) current regulator is enough to sufficiently reduce the tracking error between i_{WA} and i^*, yielding $G_i(s) = K_p + K_i/s$. According to Fig. 5, the loop gain $T(s)$ can be derived as

$$T(s) = \frac{G_i(s)}{sL_1(L_2+L_g)C} \cdot \frac{K_{PWM}G_d(s)}{s^2 + s \cdot \left[\beta G_i(s) - L_g/\left[sC(L_2+L_g)\right]\right]G_d(s)/L_1 + \omega_r^2} \tag{6}$$

where

$$\omega_r = 2\pi f_r = \sqrt{\frac{L_1 + L_2 + L_g}{L_1(L_2+L_g)C}}. \tag{7}$$

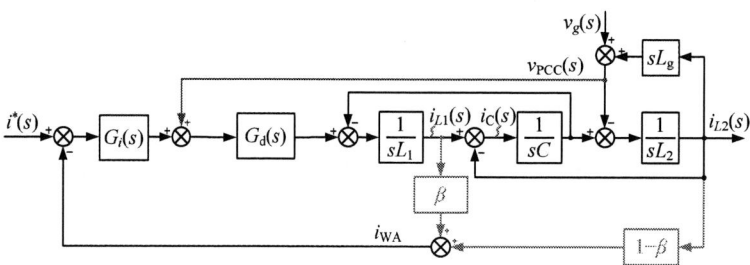

Fig. 5. Block diagram of the splitting current single-loop controlled inverter with the unit PCC voltage feedforward.

Experimental Verification

In order to verify the practicality of the splitting current single-loop control schemes, a 6-kW setup with main parameters listed in Table I was built and tested, whose parameters are listed in Table I. According to root locus method, $\beta = 1.2$ or $\beta = 0.9$ can be selected for Filter I, corresponding to the splitting grid- and splitting capacitor- current ones, respectively; $\beta = -1$ can be selected for Filter II, corresponding to the splitting inverter-current one.

Table I: Main Parameters of the *LCL*-Type Grid-Connected Inverter

Parameter		Value	Parameter		Filter I	Filter II
Grid voltage	V_{in}	220 V	Inverter-side inductor	L_1	600 μH	
Input voltage	V_g	360 V	Grid-side inductor	L_2	150 μH	
Output power	P_o	6 kW	Filter capacitor	C	30 μF	3 μF
Switching frequency	f_{sw}	10 kHz	Resonance frequency	f_{r0}	2.7 kHz	8.4 kHz

With the selected parameters, Fig. 6 depicts the closed-loop root locus maps of the three splitting current single-loop control schemes. Here, since the pair of the closed-loop poles introduced by the PI regulator suffer little from the grid impedance variation, the corresponding root locus are not given in these maps. As seen, with the selected β in the robust region, the root loci always stay inside the unit circle with the increase of L_g. It confirms that the selected parameters can harvest the strong robustness for inverter.

Next, experiments were exerted under the real grid. Taking the Filter II with $\beta = -1$ as an example, splitting inverter-current single-loop control is applied. The steady-state waveforms and the dynamic responses with a step change of the current reference between half- and full-load and vice versa, are shown in Fig. 7(a) and (b) respectively. As seen from Fig. 7(a), the grid current is stable, the total harmonic distortion (THD) of the grid current is low, and the measured power factors almost reach 1. Thus, high-quality injected grid currents are harvested. As seen from Fig. 7(b), satisfactory dynamic responses are also achieved. They confirm that the splitting inverter-current single-loop control is effective. Likewise, the effectiveness of other splitting current control schemes can also be verified, which due to the space limitation, is not presented here.

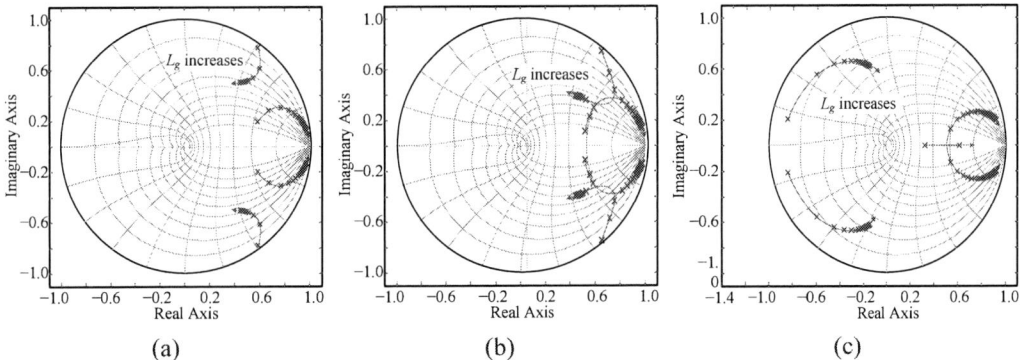

Fig. 6. The root locus of the inverter with: (a). splitting grid-current single-loop control. (b). splitting capacitor-current single-loop control. (c). splitting inverter-current single-loop control.

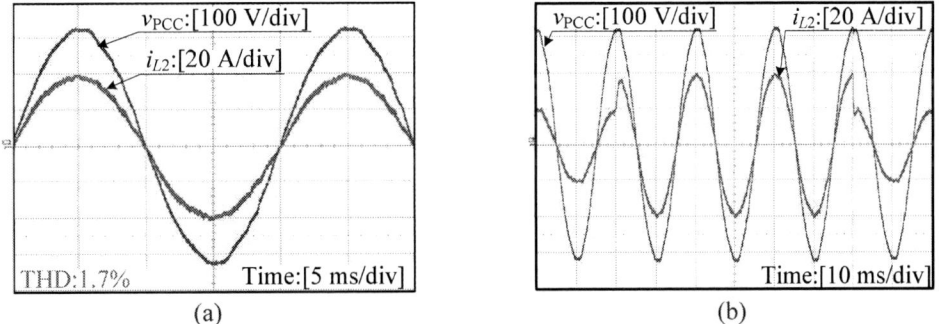

Fig. 7. Experimental results at full load under the real grid. (a). Steady-state performance. (b). Dynamic response with a step change of the current reference between half- and full-load and vice versa.

Conclusion

In this paper, a family of splitting current single-loop control schemes is proposed for LCL-type grid-connected inverter. These schemes origin from the conventional weighted average current (WAC) control, but address the problems of cost and robustness. Finally, the experimental results verified that, low cost, high robustness and low distortion can be juggled by the proposed control schemes.

References

[1] F. Blaabjerg, R. Teodorescu, M. Liserre, and A. V. Timbus, "Overview of control and grid synchronization for distributed power generation systems," *IEEE Trans. Ind. Electron.*, vol. 53, no. 5, pp. 1398–1409, Oct. 2006.

[2] S. G. Parker, B. P. McGrath, and D. G. Holmes, "Regions of active damping control for *LCL* filters," *IEEE Trans. Ind. Appl.*, vol. 50, no. 1, pp. 424–432, Jan. 2014.

[3] D. Pan, X. Ruan, X. Wang, F. Blaabjerg, X. Wang and Q. Zhou, "A highly robust single-loop current control scheme for grid-connected inverter with an improved LCCL filter configuration," *IEEE Trans. Power Electron.*, vol. 33, no. 10, pp. 8474–8487, Oct. 2018.

[4] G. Shen, X. Zhu, J. Zhang, and D. Xu, "A new feedback method for PR current control of *LCL*-filter-based grid-connected inverter," *IEEE Trans. Ind. Electron.*, vol. 57, no. 6, pp. 2033-2041, Jun. 2010.

[5] M. Lu, A. Al-Durra, S. M. Muyeen, S. Leng, P. C. Loh, and F. Blaabjerg, "Benchmarking of stability and robustness against grid impedance variation for *LCL*-filtered grid-interfacing inverters," *IEEE Trans. Power Electron.*, vol. 33, no. 10, pp. 9033–9046, Oct. 2018.

[6] D. Pan, X. Ruan, C. Bao, W. Li, and X. Wang, "Optimized controller design for *LCL*-type grid-connected inverter to achieve high robustness against grid-impedance variation," *IEEE Trans. Ind. Electron.*, vol. 62, no. 3, pp. 1537–1547, Mar. 2015.

Analysis and Design of High-Power Single-Stage Three-Phase Differential-Based Flyback Inverter for Photovoltaic Applications

Ahmed Ismail M. Ali[1,2]   Mahmoud A. Sayed[2]   Takaharu Takeshita[1]

[1]Nagoya Institute of Technology, Dept. of Electrical and Mechanical Engineering, Graduate School of
Engineering, 466-8555, Gokiso, Showa, Nagoya, JAPAN
[2]South Valley University, Dept. of Electrical Engineering, 83523, Qena, Egypt
Email: a.ali.404@stn.nitech.ac.jp

Keywords

<<Differential-based flyback inverter (DBFI)>>, <<Continuous Modulation Scheme (CMS)>>, <<Static Linear Method (SLM)>>, <<Flyback high-frequency transformer (FB-HFT)>>, <<Negative-sequence second-order harmonic component (NS-SOHC)>>.

Abstract

This paper offers mathematical analysis, design, and experimental verification of three-phase differential-based flyback inverter (DBFI) based-on SiC-MOSFET for photovoltaic (PV) applications. The DBFI utilizes continuous modulation-scheme (CMS) combined with the static linearization method (SLM), for low-order harmonics mitigation, based-on a simple and low-cost flyback converter. In addition, design investigations are detailed for the high-power isolated single-stage inverter. Therefore, the paper reveals the design practical issues of the DBFI such as; power-stage design, HFT-based Nano-Crystalline Core design, passive elements selection, harmonics compensation strategy. The system is simulated by PSIM computer-aided simulator. Consequently, the 1.6 kW, 200 V, 50 kHz DBFI prototype has been experimentally validated with and without negative-sequence second-order harmonic component (NS-SOHC).

Introduction

Photovoltaic (PV) energy is one of the most promising energy resources for required energy generation [1]. Therefore, many PV-inverters are introduced recently, which satisfies the inverter reliability, size compactness, low-cost, and inverter reduced footprint [2-5]. Transformer-less inverters are used in many PV applications [4, 6-9]. However, these topologies require a separate DC-DC converter-stage to attain the requires voltage gain in the DC-AC conversion, which increase the system cost at low power density. Moreover, it suffers from the leakage of grid isolation and requires a line-frequency transformer for grid-integration [10, 11]. Therefore, the transformer-based inverters are introduced for isolation requirements. The isolated/non-isolated single-stage differential inverters are proposed in the recent few years for single and three-phase operation, which suppress the requirement of the unfolding H-bridge circuit [12-14]. Compared with the different isolated buck-boost converter-based inverters types, differential-based flyback inverter (DBFI) offers number of features such as; low circuit construction complexity, low-cost, voltage boosting/bucking property, and reduced number of required passive elements as the flyback high-frequency transformer (FB-HFT) executes two operational functions; it stores the energy in the magnetic core as an inductor, as well as the ideal transformer operation. However, other converters have separate storage element and transformer that increase the inverter size and footprint [3, 15, 16]. The practical implementation of high-power DBFI forms the major challenge of the inverter design due to the large storage energy capability. Consequently, the inverter design success is mainly relying on the robustness of the design of the FB-HFT as well as the different DBFI loss parts such as; input LC low-pass filter power losses, FB-HFT losses, SIC-MOSFET losses, SIC-diode losses, and the grid-side filter loss. That was the power loss studying motivation for each component of the DBFI to propose an enhanced efficiency design. In addition, the implementation and practical design

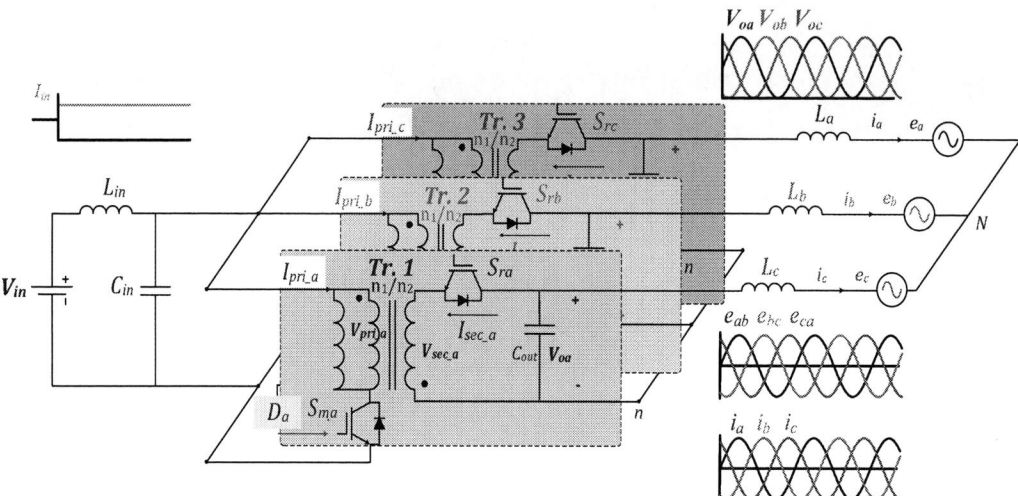

Fig. 1: Circuit configuration of the three-phase isolated DBFI.

issues of the DBFI parts are quite important for the inverter operation such as; input LC filter design, FB-HFT design, PCB power-stage design, and the decent control system design. Consequently, this paper introduces a loss study based on the derivation of the mathematical loss model for each component in terms of the time and the system parameters. Also, the different component loss participation can be reduced in a well revised inverter design. Ultimately, the single-stage three-phase isolated DBFI is deeply analyzed for efficiency boosting purpose based-on its unique features listed in [17]. The validity of the system has been investigated using experimental system prototype, which is controlled by MWPE3C6713A-Expert III digital controller.

DBFI Circuit Configuration and Operation Principals

The proposed isolated single-stage three-phase high-power DBFI consists of three separate flyback DC-DC converters, which are connected in parallel at the DC-side and differentially at the load/grid-side as depicted Fig. 1. Therefore, the three-phase DBFI consists of an input LC low-pass filter, three FB-HFT, three stringent designed snubber-circuits, six SiC-MOSFET power switches, three output film-capacitors, and three grid-interfaced chock coils. The DBFI operation in the three-phase applications is mainly depends on the modulation of three DC-DC modules with variable duty cycles and 120° phase shift. Consequently, the full-bridge unfolding circuit is suppressed in the DBFI inverter, which reduced the required elements and the system footprint. The operation of each single-module is divided into two modes of operation as illustrated in [18], which reveals that the three-phase DBFI operates like a voltage-controlled current-source inverter (VC-CSI). In addition, the continuous modulation-scheme (CMS) with the static linearization-method (SLM) is used to control the DBFI to mitigate the low-order harmonics. Moreover, the input LC filter and the proposed control strategy decrease the input current ripples for PV applications suitability. For the modulation of the proposed DBFI, sinusoidal pulse-width modulation (SPWM) is applied, see Fig. 2. Fig. 2 illustrate the three-phase duty cycles generation (D_a, D_b, D_c) based on the VC-CSI operation concept of the proposed DBFI. Hence, the duty cycles are compared with the high-frequency carrier waveform (V_{Car}) to synthesize the required control pulses for the inverter switches. Therefore, Fig. 2 show all current components for single-module of the DBFI; the high-frequency component (i_{pri_aH}), the line-frequency component (i_{pri_aL}) that illustrates the instantaneous average of (i_{pri_aH}) over one switching period, the average DC component (i_{pri_aDC}), and the grid sinusoidal current, i_a, of phase (a). Obviously, the switched-waveforms of the high-frequency primary and secondary currents have sinusoidal line-frequency envelope, which is distorted with a negative-sequence second-order harmonic component (NS-SOHC) that results from the mismatch between the three flyback DC-DC modules.

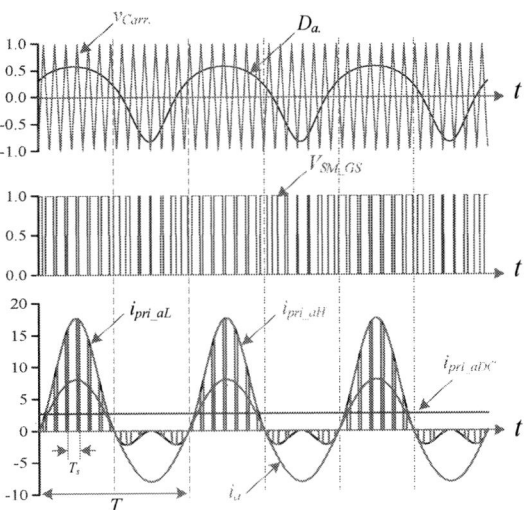

Fig. 2: Flyback input and output current components.

Therefore, with the application of DC input voltage to the flyback modules paralleled input side, the output-to-input voltage conversion-ratio can be expressed as follows;

$$M(D) = \frac{V_o}{V_{in}} = \frac{nD}{D'} \cdot \left(1 - \frac{D' \cdot V_D}{n \cdot D \cdot V_{in}}\right) * \left(\frac{1}{1 + \frac{R_L \cdot D \cdot n^2}{R_{eq} \cdot D'^2} + \frac{R_{on} \cdot D \cdot n^2}{R_{eq} \cdot D'^2} + \frac{R_D}{R_{eq} \cdot D'}}\right) \tag{1}$$

where; V_o is the flyback module output voltage,
V_{in} is the input DC-voltage,
n is the transformer turns-ratio, $n=n2/n1$
D is the main-switch duty cycle, $(D'=1-D)$
V_D is the voltage drop over the flyback diode
R_L is the primary inductor resistance
R_{eq} is the grid equivalent resistance
R_{on} is the MOSFET on-resistance
R_D is the diode on-resistance

Also, the grid three-phase voltages can be formulated as follows;

$$\begin{bmatrix} V_{aN}(t) \\ V_{bN}(t) \\ V_{cN}(t) \end{bmatrix} = \sqrt{\frac{2}{3}} \cdot E \cdot \begin{bmatrix} \sin(\omega t + \alpha) \\ \sin\left(\omega t + \alpha - \frac{2\pi}{3}\right) \\ \sin\left(\omega t + \alpha + \frac{2\pi}{3}\right) \end{bmatrix} \tag{2}$$

where; E is the RMS line voltage, ω is the grid angular frequency, α is an arbitrary angle

The grid three-phase injected currents can be formulated as follows;

$$\begin{bmatrix} I_a(t) \\ I_b(t) \\ I_c(t) \end{bmatrix} = \sqrt{2} I * \begin{bmatrix} \sin(\omega t + \alpha) \\ \sin\left(\omega t + \alpha - \frac{2\pi}{3}\right) \\ \sin\left(\omega t + \alpha + \frac{2\pi}{3}\right) \end{bmatrix} \tag{3}$$

where; I is the RMS injected grid current.

Three-Phase DBFI Design

Generally, most of the flyback PV-inverters are designed as small PV micro-inverters for stand-alone and grid-tied PV applications. In the proposed three-phase DBFI, high-power flyback-based inverter is considered as a central-type PV-inverter. Hence, the three-phase DBFI is used at power rating of 1.6 kW in the experimental prototype verification, where the experimental system-parameters are listed in Table I. As well, the inverter switching frequency is used as 50 kHz in order to decrease the system size by reducing the size of the required magnetics in the proposed inverter and to improve the output voltage and current THD. However, increasing the switching frequency increases the switching loss in the inverter, which adversely affects the DBFI operation efficiency. Therefore, fast rising/falling response SiC MOSFETs are utilized to minimize the switching losses of the DBFI. Moreover, the flyback switches undergo high turn-off voltage-stress due to the leakage inductance of the HFT. Therefore, a rigid snubber-circuit is required to maintain the transient voltage in safe operation region. Another important aspect of the DBFI operation is the optimal design of the different parts of DBFI;

Table I: Proposed DBFI experimental parameters

Design parameters	Specifications
Rated inverter power, P	1.6 kW
Input DC voltage, V_{in}	100 V
Input filter, L_{in}, C_{in}	150 μH, 10 μF
Input filter resistance, r_{in}	4 Ω
Grid voltage (L.L), E, ω	200 V, 2×π×60 rad/s
HFT magnetizing inductance, L_m	100 μH
HFT primary resistance, r_m	2 mΩ
Output capacitor, C_o	10 μF
HFT leakage inductance, $L_{Leakage}$	2.25 μH
HFT turns ratio, n	1:1
Grid inductance, L_g	4 mH
Grid inductor resistance, r_g,	5 mΩ
Switching Frequency, Fsw	50 kHz
PI controller gains, K_P, K_I	0.097 A/V, 280 rad.sec^{-1}

- Input LC filter design
- Flyback high-frequency transformer design
- PCB power stage design
- Control system design

The design of different parts are as follows;

Input LC Filter Design

The input LC low-pass filter is used to deliver a continuous input DC-current, which emulates the PV central-type inverter continuous current. The DBFI utilizes a single small-size LC input filter to enhances the power density, size, cost, stability, and footprint of the inverter. Also, using the single LC input low-pass filter, at the common-coupling of three flyback modules, reduces the number and size of required passive elements. Moreover, it uses a small LC filter due to the phase-shift between the three-modules that minimizes the ripple magnitude and triples their frequency. Hence, as the number of modules/phases increase, the magnitude of required passive elements decease. However, it increases the number of components in the multi-phase operation. Therefore, the three flyback-modules with a single LC-filter structure forms the optimal choice for three-phase DBFI operation. Another important aspect of the LC input filter is the effect of filter dynamics and parasitic on the system stability, which depend on the inductance and capacitance of the input filter. Therefore, the utilized optimal values are listed in Table I.

PCB Design

The design of the PCB board is an important issue for the optimal design of the proposed DBFI. The PCB board is considered as the main cause of the high-frequency noise in the inverter circuit due to the presence of the board high parasitic-inductance. Hence, the power-circuit and gate-drivers loop inductances need to be shortened as possible for industrial applications suitability to mitigate the high-frequency noise main causes. Therefore, a short parasitic inductances loops are used in the power-stage and gate-driver circuits as well as optimized blocking capacitors are utilized. As well, high technology PCB design is utilized for the power-stage and gate-driver circuits as depicted in Fig. 3. The PCB boards design are performed using a thin trace thickness with a wide width for higher RMS current operation capability with reduced parasitic inductances. Furthermore, this DBFI design is not the final design for practical implementation. another adjustment is now under study for efficiency enhancement, size-compactness, loss-minimization, and footprint issues.

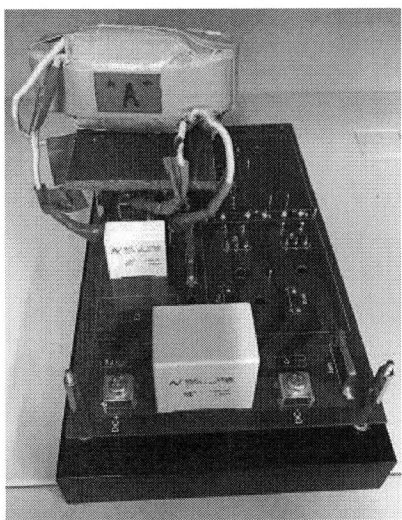

Fig. 3: A single module of the DBFI and its gate-driver circuit.

High-frequency Transformer Design

Actually, the flyback converters are available in the industrial markets with lower power ratings; i.e.; 200 Watts or lower, due to the required storage tank for the temporarily operation. Therefore, the success operation of the proposed DBFI is mainly related to the flyback high-frequency transformer design. In the proposed DBFI, each flyback converter is designed to handle around three-times of the available converters, i.e.; 550 Watts. As well, the major challenge of the DBFI is to maintain the inverter size as compact as possible, which directly refers to the FB-HFT design. Therefore, the selection of the inverter parameters is made by an optimization study and thus, the FB-HFT magnetizing inductance is considered as (L_M=115 µH) at switching-frequency of 50 kHz, which is responsible for energy storing and release during the periods (DT) and ($D'T$), respectively. Also, for DBFI power density, small size Nano-crystalline Core is utilized for FB-HFT design having cross-sectional area of 312 mm^2 and air-gap length of 1.45 mm, which can be premeditated by the formula;

$$l_e = \frac{N^2 \cdot \mu_o \cdot A_{Core}}{L_m} \tag{3}$$

where; l_e is the length of required air-gap.

In addition, the FB-HFT leakage-inductance increases the DBFI power loss. Therefore, keeping lower leakage-inductance is considered as an important design issue. Hence, the transformer design can be concluded as follows [19];

- Lowering leakage-inductance for FB-HFT efficient operation.
- Litz-wire is utilized in the design of FB-HFT windings to decrease the power loss.
- Air-gap distribution strategy is used by the magnetic cores. However, it is not applicable for the Nano-crystalline C-Core. The final design of the HFT core is now within optimization strategy for better footprint.

Control System Design

This section discusses the single-stage three-phase DBFI closed-loop control system. As previously mentioned, the operation of the proposed inverter utilizes a wide-range of duty cycle variation, which disturbs the system dynamic poles and zeros location and then the inverter stability. Therefore, the proposed inverter stability is a critical design issue. Therefore, a robust compensator design is required to maintain the inverter stability with improved phase-margin (PM) and band-width (BW) that confirms the inverter accurate operation over a wide-range of duty cycles and frequency variations. Fig. 4 portrays the closed-loop control scheme of the proposed DBFI. It consists of the two control-loops; the main control-loop that responsible for the output voltages and grid currents regulation to follow the reference values, however, the second control-loop is used for NS-SOHC compensation. It is worth to mention that NS-SOHC compensation loop uses the single-pole integrator, which furnishes the main control-loop with a pole at the origin that reduces the compensator order in the main control-loop. Hence, the control system complexity is mitigated by reducing the required computational time. The bode-plot of the overall control system of the proposed DBFI is depicted in Fig. 5. The system phase-margin is 20° and the inverter band-width is 3 kHz, which affirms the stability of the DBFI over wide-range of frequency variations.

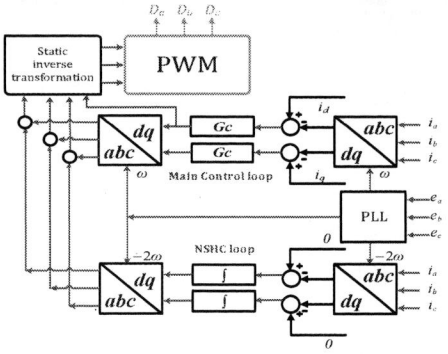

Fig. 4: Closed-loop control scheme of the DBFI.

Fig. 5: Closed-loop control scheme bode plot of the DBFI with the two control-loops

System Verification

The DBFI results based on 200 V, 1.6 kW, and 50 kHz switching frequency is verified in simulation software as well as experimentally to prove its efficacy for high-power with design practical issues. Fig. 6 depicts the DBFI simulation results. Fig. 6(a and b) portrays the DC input voltage and current before and after NS-SOHC compensation, respectively. Obviously, the harmonic compensation techniques minimize the input current ripples (from 5.2% to 1.4%), which is a vital issue in the PV applications. Also, Fig. 6 (c and d) depicts the AC-side results (three-phase grid voltages, grid currents, and module terminal voltages) before and after the NS-SOHC compensation, respectively. Before the NS-SOHC compensation, the three-phase grid-injected current are distorted by a high second-order second order harmonic that distorts the module's terminal voltages. However, the compensator strategy compensates the harmonic in the grid currents for sinusoidal grid current operation with THD that follows the IEEE harmonic standard limits.

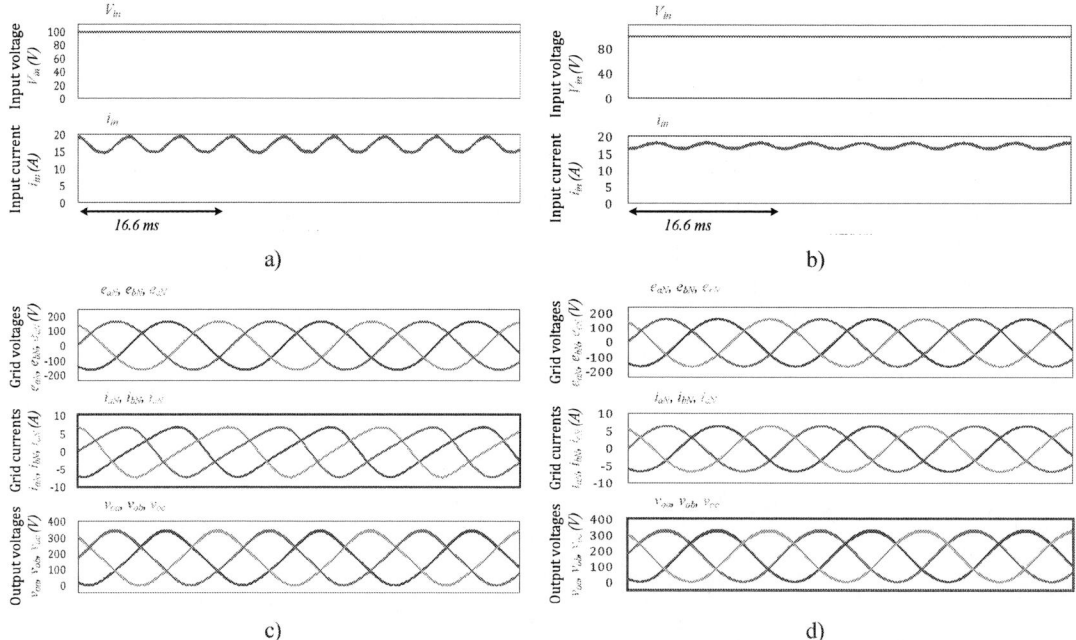

Fig. 6: Simulation results of the proposed isolated DBFI; (a) DC-side waveform before harmonic compensation, (b) DC-side waveform after harmonic compensation, (c) AC-side waveform before harmonic compensation, (d) AC-side waveform after harmonic compensation.

Fig. 7: The photograph of the proposed three-phase isolated DBFI.

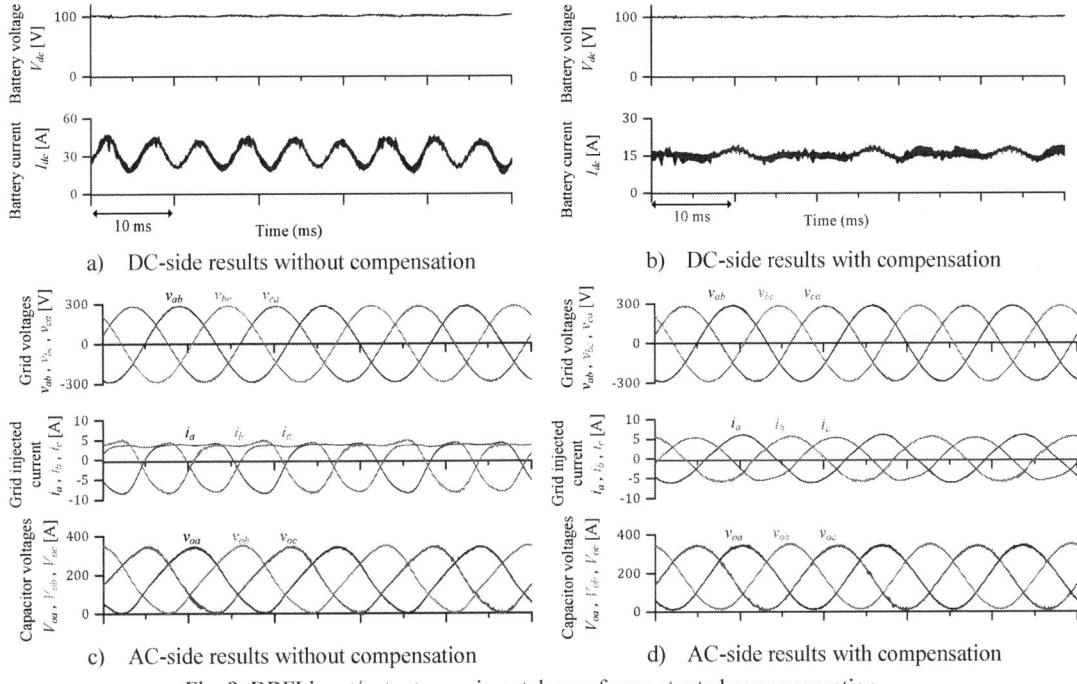

a) DC-side results without compensation

b) DC-side results with compensation

c) AC-side results without compensation

d) AC-side results with compensation

Fig. 8: DBFI input/output experimental waveforms at rated power operation.

The three-phase DBFI system photograph is shown in Fig. 7. Obviously, Fig. 8 portrays the DBFI stable operation at high-power rating. The steady state input and output waveforms, before and after NS-SOHC compensation, are depicted is Fig. 8 (a, b, c, and d). Also, Fig. 8(a and b) shows the DC-side waveforms before and after NS-SOHC compensation strategy, however, Fig. 8(a and d) shows the AC-side results before and after NS-SOHC compensation, respectively. The harmonic compensation technique using a single-pole integrator reduced the grid currents THD from 39.3% to 3.95%, which follows the harmonic standard limitations. Also, it reduced the input current ripples from 5.2% to 1.4%, which is an important issue for PV applications.

Conclusion

A single-stage three-phase isolated differential-based flyback inverter design, and experimentation rated at 1.6 kW is presented as a PV central-type inverter for small power-system applications. Each flyback module transfers about 550W, which is higher than the common flyback converter ratings in the markets. The NS-SOHC strategy decreased the input current ripples from 5.2% to 1.4%, which is an important issue in the PV inverters. In addition, the second compensation-loop reduced the grid-currents THD from 39% to 3.95% at 1.6kW that follows the IEEE harmonic standard limits.

References

[1] A. I. Ali, M. A. Sayed, and E. E. Mohamed, "Modified efficient perturb and observe maximum power point tracking technique for grid-tied PV system," International Journal of Electrical Power & Energy Systems, vol. 99, pp. 192-202, 2018.

[2] S. B. Kjaer, J. K. Pedersen, and F. Blaabjerg, "A review of single-phase grid-connected inverters for photovoltaic modules," IEEE Transactions on industry applications, vol. 41, pp. 1292-1306, 2005.

[3] T. Lodh, N. Pragallapati, and V. Agarwal, "Novel control scheme for an interleaved flyback converter based solar PV microinverter to achieve high efficiency," IEEE Transactions on industry applications, vol. 54, pp. 3473-3482, 2018.

[4] A. I. Ali, M. A. Sayed, E. E. Mohamed, and A. M. Azmy, "Advanced Single-Phase Nine-Level Converter for the Integration of Multiterminal DC Supplies," IEEE Journal of Emerging and Selected Topics in Power Electronics, vol. 7, pp. 1949-1958, 2018.

[5] A. M. Hassan, X. Yang, A. I. Ali, T. A. Ahmed, and A. M. Azmy, "A Study of Level-Shifted PWM Single-phase 11-Level Multilevel Inverter," in 2019 21st International Middle East Power Systems Conference (MEPCON), 2019, pp. 170-176.

[6] M. A. Sayed, E. Mohamed, and A. Ali, "Maximum power point tracking technique for grid tie PV system," in 7th International Middle-East Power System Conference,(MEPCON'15), Mansoura University, Egypt, 2015.

[7] L. Zhang, M. J. Waite, and B. Chong, "Three-phase four-leg flying-capacitor multi-level inverter-based active power filter for unbalanced current operation," IET Power Electronics, vol. 6, pp. 153-163, 2013.

[8] A. I. Ali, M. A. Sayed, and E. E. Mohamed, "Maximum Power Point Tracking technique applied on partial shaded grid connected PV system," in 2016 Eighteenth International Middle East Power Systems Conference (MEPCON), 2016, pp. 656-663.

[9] C. E. Feloups, A. I. Ali, and E. E. Mohamed, "Single-phase seven-level pwm inverter for pv systems employing multi-level boost converter," in 2018 International Conference on Innovative Trends in Computer Engineering (ITCE), 2018, pp. 403-409.

[10] M. A. Rezaei, K.-J. Lee, and A. Q. Huang, "A high-efficiency flyback micro-inverter with a new adaptive snubber for photovoltaic applications," IEEE transactions on Power Electronics, vol. 31, pp. 318-327, 2015.

[11] T. A. Ahmed, A. I. Ali, A. Youssef, and E. E. Mohamed, "Study of Multilevel Inverter Fed Single Phase Induction Motor," in 2018 Twentieth International Middle East Power Systems Conference (MEPCON), 2018, pp. 603-607.

[12] F. Gao, P. C. Loh, R. Teodorescu, F. Blaabjerg, and D. M. Vilathgamuwa, "Topological design and modulation strategy for buck–boost three-level inverters," IEEE transactions on Power Electronics, vol. 24, pp. 1722-1732, 2009.

[13] A. Darwish, A. M. Massoud, D. Holliday, S. Ahmed, and B. Williams, "Single-stage three-phase differential-mode buck-boost inverters with continuous input current for PV applications," IEEE transactions on Power Electronics, vol. 31, pp. 8218-8236, 2016.

[14] S. Mehrnami, S. K. Mazumder, and H. Soni, "Modulation scheme for three-phase differential-mode Ćuk inverter," IEEE transactions on Power Electronics, vol. 31, pp. 2654-2668, 2015.

[15] Z. Zhang, X.-F. He, and Y.-F. Liu, "An optimal control method for photovoltaic grid-tied-interleaved flyback microinverters to achieve high efficiency in wide load range," IEEE transactions on Power Electronics, vol. 28, pp. 5074-5087, 2013.

[16] B. Tamyurek and B. Kirimer, "An interleaved high-power flyback inverter for photovoltaic applications," IEEE transactions on Power Electronics, vol. 30, pp. 3228-3241, 2014.

[17] F. Zhang, Y. Xie, Y. Hu, G. Chen, and X. Wang, "A Hybrid Boost–Flyback/Flyback Microinverter for Photovoltaic Applications," IEEE Transactions on Industrial Electronics, vol. 67, pp. 308-318, 2019.

[18] Y. Li and R. Oruganti, "A low cost flyback CCM inverter for AC module application," IEEE Transactions on Power Electronics, vol. 27, pp. 1295-1303, 2011.

[19] B. Tamyurek and D. A. Torrey, "A three-phase unity power factor single-stage AC–DC converter based on an interleaved flyback topology," IEEE transactions on Power Electronics, vol. 26, pp. 308-318, 2010.

Investigation of Improvement of Modeling precision for Conducted Noise on Isolated AC/DC Converter using SiC Devices

Kazuki Kuwana, Kohei Mitani, Wataru Kitagawa and Takaharu Takeshita
Dept. of Electrical and Mechanical Engineering, Graduate School of Engineering,
Nagoya Institute of Technology
Gokiso, Showa
Nagoya, 466-8555 Japan
Phone: +81 (52) 735-5441
Fax: +81 (52) 735-5432
Email: kitagawa.wataru@nitech.ac.jp
URL: http://motion.web.nitech.ac.jp

Acknowledgments

A part of this work was supported by JSPS KAKENHI Grant Number JP19K04326

Keywords

≪EMC/EMI≫, ≪Noise≫, ≪Silicon Carbide (SiC)≫, ≪Modeling≫, ≪Simulation≫

Abstract

This paper presents the simulation method of conducted noise on isolated AC/DC converter using SiC devices. The leakage current is measured to derive the equivalent circuit. In addition, this paper proposes the simulation method with EMI filters. The usefulness of the proposed simulation method is confirmed by comparing the simulation result and the experimental result.

Introduction

The development and spread of power generation using renewable energy such as photovoltaic and wind turbine generation system are increasing. In order to keep the stable power supply system, it is necessary to level the power by using battery systems. Because the battery systems are connected to the utility grid, the power conversion by AC/DC converter is necessary. In addition, AC/DC converter for EVs or PHVs are required to be isolated for safety [1].

In recent year, high speed switching modules such as SiC devices will be widely spread. SiC devices have the advantage of high voltage, high temperature operation and high speed switching. However, when high voltage in SiC devices is achieved, switching noise increase for the reason of structure. In addition, there is parasitic capacitance in transformer for isolation, and the high frequency leakage current which is the cause of noise flow through the transformer. Therefore, it is necessary to consider the problem of the high frequency leakage current and electromagnetic interference (EMI) [2]-[7].

Moreover, in noise countermeasure process, optimal filters are selected to repeat filter design and experiment and evaluation. Thus, noise countermeasure need cost and time. Simulation technique which can predict the reduction of noise by filter is necessary to reduce cost and time.[8]-[13].

This paper presents the simulation method of conducted noise on the isolated AC/DC converter by using the leakage current waveforms, and proposes the simulation method with EMI filters. Moreover, this paper presents the equivalent circuit of EMI filter on the modeling circuit. The usefulness of the proposed the simulation method with EMI filter is confirmed by comparing the simulation result and the experimental result.

Evaluation Method

Fig.1 shows the experimental system for measurement of conducted noise on the isolated AC/DC converter. In addition, Tab.I shows the experimental conditions. As shown in Fig.1, the high frequency transformer is used for isolation, and SiC devices are used for each switches. Line Impedance Stabilization Network (LISN), the primary side heat sink and the secondary side heat sink are connected to the copper plate. By connecting in this way, the leakage current occurred by switching flow through the copper plate to LISN. The leakage current i_{le1}, i_{le2} were measurred by current prove. In addition, Fig.2 shows internal circuit of LISN. LISN has the same three circuit in each u,v and w phase. Conducted noise is evaluated by measuring the voltage across $50\,\Omega$ in LISN with spectrum analyzer (N9010A,keysight) as a disturbance voltage.

Experimental Result

A. Experimental result

Fig.3 shows the measurement results of voltage S_{ug}, S_{uh} and the leakage current i_{le1}. S_{ug} is the voltage between u phase and g phase. S_{uh} is the voltage between u pahase and h phase. i_{le1} is the leakage current flowing from the primary side heat sink to the copper plate. Fig.4 shows the measurement results of voltage S_{jn}, S_{kn} and the leakage current i_{le1}. S_{jn} is the voltage between j phase and n phase. S_{kn} is the voltage between k pahase and n phase. i_{le2} is the leakage current flowing from the secondary side heat sink to the copper plate. As shown in Fig.3 and Fig.4, the leakage current are occured by AC/DC

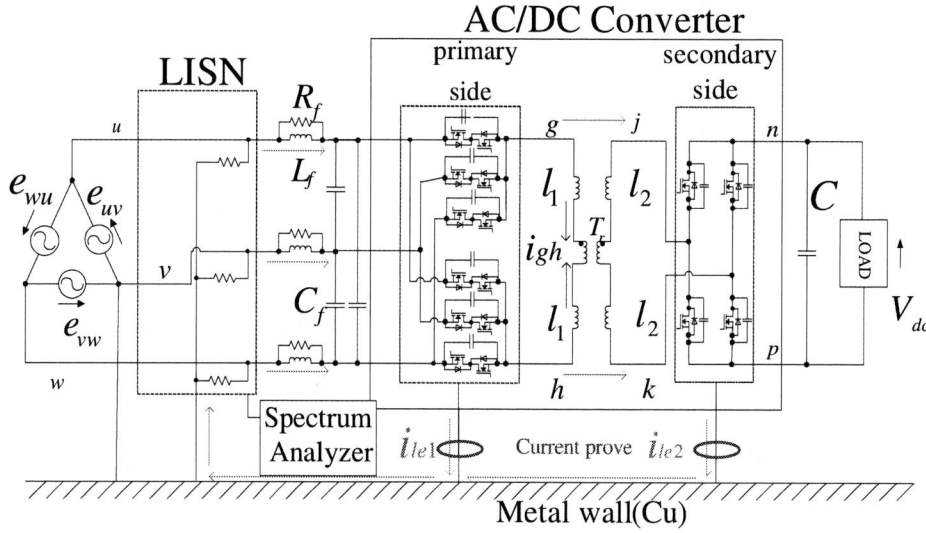

Fig. 1: Experimental system

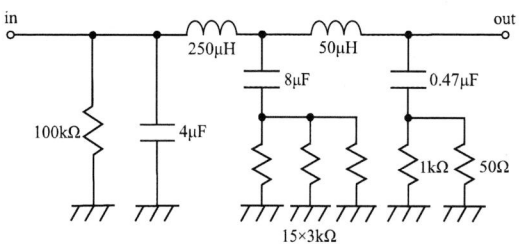

Fig. 2: Internal circuit of LISN

Table I: Experimental conditions

Source voltage E, ω	$200\,\text{V}, 2\pi \times 60\,\text{rad/s}$
DC voltage reference v_{dc}	$280\,\text{V}$
Switching frequency f_s	$10\,\text{kHz}$
Inductors l_1, l_2	$0.1\,\text{mH}, 0.1\,\text{mH}$
Capacitor C	$300\,\mu\text{F}$
Input filter L_f, C_f, R_f	$0.6\,\text{mH}, 5.6\,\mu\text{F}, 27\,\mu\Omega$
Damping resistor R_f	$27\,\Omega$
Load R	$52.3\,\Omega$
Output power P_{out}	$1500\,\text{W}$

converter switching bacause they occure at the same time. i_{le1} flow to heatsink from switch through the parastic capacitance and make the route that is heatsink to metal wall to LISN to primary side. i_{le2} flow to heatsink from switch through the parastic capacitance and make the route that is heatsink to metal wall to LISN to primary side to secondary side. In addition, Fig.10 shows the enlarged wave form of i_{le1}. Fig.11 shows the enlaged wave form of i_{le2}. Fig.5 shows the measurement result of disturbance voltage. There are the resonance points at the frequency of 700 kHz, 5.0 MHz and 6.0 MHz. In addition, Because the noise level exceed international standard (CISPR11), noise countermeasures is necessary.

B. Experimental result with filter

To confirm reduction of noise by filters, and to simulate a disturbance voltage with filters, the experiment with filter is performed. In this paper, the experiment with two type of CMCC (Common Mode Choke Coil) was used for the filters. these are inserted one by one, and experiment was performed. As shown in Fig.6, CMCC of 3 mH (744839003460, wurth elektronik) and CMCC of 47 mH(74483904160, wurth elektronik) are inserted between R_f and C_f. Fig.7 shows the measurement result of disturbance voltage with filter. As shown in Fig.7, CMCC reduce the noise level of disturbance voltage at 700 kHz. particularly, CMCC of 47 mH reduces more noise level of disturbence voltage than CMCC of 3 mH in all frequency range. This is because that impedance characteristics of CMCC of 47 mH is more noise level than CMCC of 3 mH.

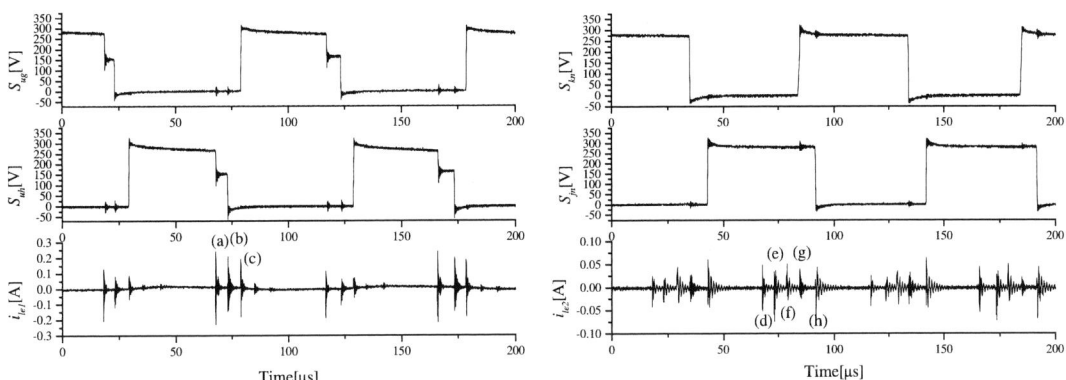

Fig. 3: Experimental result of voltage and leakage current i_{le1}

Fig. 4: Experimental result of voltage and leakage current i_{le2}

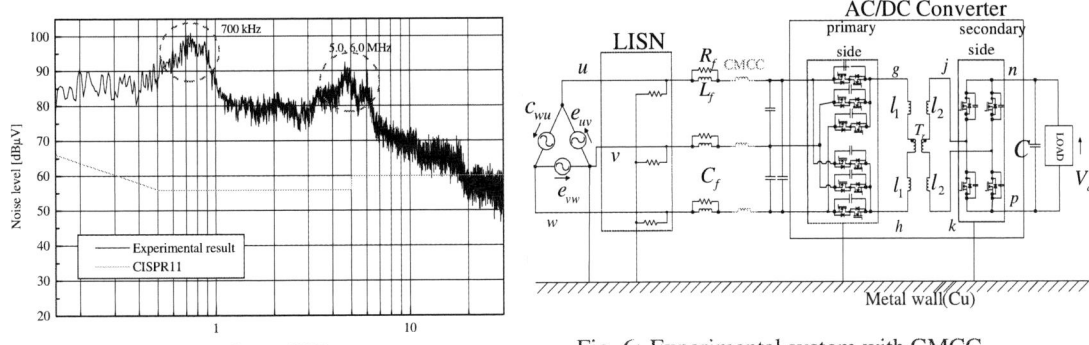

Fig. 5: Experimental result of disturbance voltage

Fig. 6: Experimental system with CMCC

Simulation

This section proposes the simulation method. The common mode modeling circuit is made baced on the leakage current to simulate a disturbance voltage.

A. Common mode modeling circuit

The common mode modeling circuit is derived by using experimental results of the leakage current. In Fig.10 and Fig.11, the leakage current waveforms are damped vibration. In addition, damped vibration is modeled by the *RLC* series circuit and step voltage. Therefore, the leakage current is modeled using the modeling circuit which is approximated by *RLC* series circuit and step voltage. The equation of the damped vibration current after applying the step voltage E_c to the *RLC* series circuit is shown by the following equation.

$$i(t) \cong \frac{E_c}{Z_0} e^{-\zeta \omega_n t} \sin \omega_n t \tag{1}$$

$$\omega_n = 2\pi f, \quad Z_0 = \frac{E_c}{i_{peak}}, \quad \zeta = \frac{R}{2Z_0} \tag{2}$$

where the parameter ω_n is the eigen frequency, f is the resonance frequency, Z_0 is the characteristic impedance, i_{peak} is the maximum value of the leakage current and ζ is the damping coefficient. The measured voltage of switch $S_{ug}, S_{uh}, S_{kn}, S_{jn}$ are used for E_c. The maximum value of leakage current is used for i_{peak}. FFT analysis result of leakage current is used for f. The eigen frequency ω_n and the characteristic impedance Z_0 are decided by the following equation.

$$\omega_n = \frac{1}{\sqrt{LC}}, \quad Z_0 = \sqrt{\frac{L}{C}} \tag{3}$$

where the parameters L and C are inductance and capacitance of the *RLC* series circuit. the parameters L and C are decided by the following equation.

$$L = \frac{Z_0}{\omega_n}, \quad C = \frac{1}{\omega_n Z_0} \tag{4}$$

where the parameter R is resistance of the *RLC* series circuit. The parameter R is decided from the attenuation situation. The parameters of the *RLC* are calculated by using the equations (2), (4).

For example, the common mode modeling circuit of i_{le1} (a) is shown in Fig8. In addition, Tab.II shows

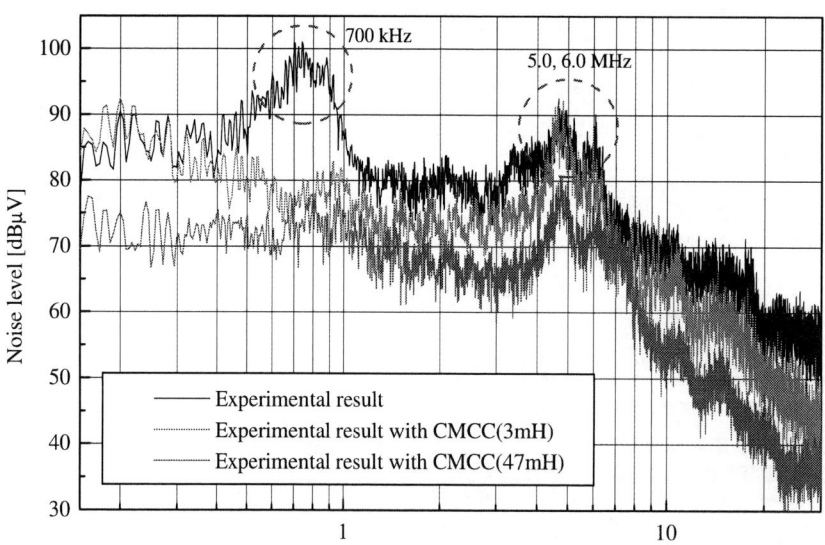

Fig. 7: Experimental result of disturbance voltage with CMCC

the parameter of the *RLC*. The parameters are determined by the above method. In this case, the leakage current has two frequency, 5.1 MHz and 5.9 MHz . Thus, the *RLC* series circuits are connected in parallel. In addition, the modeling circuits are made for each leakage current (a)-(h). Fig.10 and Fig.11 show the simulation result of the leakage current. In Fig.10 and Fig.11, it is confirmed that the simulation result agree with the experimental result from view points of the amplitude, the attenuation and the frequency.

B. Simulation of disturbance voltage

Fig.9 shows the modeling circuit of the all leakage currents (a)-(h). In Fig.9, Each modeling circuit of leakage curent (a)-(h) are connected in parallel for simulation of plural leakage currents. The equivalent circuit of LISN is connected to modeling circuit to simulate the disturbance voltage. The equivalent circuit of LISN is from datasheet.

1/3 of the leakage current i_{le1} and i_{le2} flows in the resistance of LISN. Therefore, in the simulation, the voltage across the resistance $(50/3\,\Omega)$ is equivalent to conducted noise. The value obtained by FFT analysis of the voltage is converted to the disturbance voltage by the following equation.

$$\text{Level (dB}\mu\text{V)} = 20\log_{10}(\frac{V}{\sqrt{2}} \times 10^6) \qquad (5)$$

where V is the voltage across the resistance $(50/3\,\Omega)$ in Fig9.

Fig.12 shows the simulation result of disturbance voltage. The simulation result agree with the experimental result. The resonance points and the noise voltage level are well simulated

Simulation with CMCC

This section proposes the simulation method with CMCC. In the simulation, CMCC is modeled to the equivalent circuit from impedance and phase characteristics. The modeling circuit of CMCC is added to the common mode modeling circuit.

A. Conventional modeling method

The impedance characteristics of the CMCC is measured with the impedance analyzer. Fig.13 and Fig.14 show the impedance and phase characteristics. The impedance and phase characteristics are modeled by RLC parallel circuit. In addition, Fig.13 and Fig.14 show the modeling result of the CMCC. In Fig.13 and Fig.14, it is confirmed that the modeling impedance and phase characteristics agree with the measured impedance and phase characteristics from 150 kHz to 30 MHz.

Table II: *RLC* series circuit parameters

f	5.1 MHz	5.9 MHz
E_c	100 V	100 V
i_{peak}	0.107 A	0.103 A
ω_n	32.0 Mrad/s	37.1 Mrad/s
Z_0	934 Ω	972 Ω
L	29.1 μH	27.85 μH
C	33.4 pF	27.8 pF
R	40 Ω	40 Ω

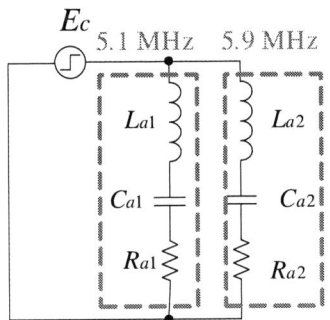

Fig. 8: Modeling circuit of i_{le1}(a)

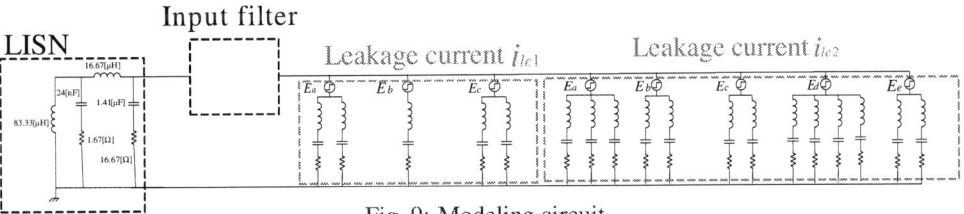

Fig. 9: Modeling circuit

Fig.9 shows the common mode modeling circuit with CMCC. In Fig.9, It is shown the position where the model of CMCC is inserted. this position is correspond to the position where CMCC is inserted in the experiment. The simulation inputed the model of CMCC is performed.

Fig.15 and Fig.16 show the simulation result of disturbance voltage with CMCC. In Fig.16, the simulation result with CMCC 47 mH agree with the experimental result at view point of reasonance point. In Fig.15, the simulation result don't agree with the experimental result at view point of resonance point. This is because the modeling impedance and phase characteristics of CMCC at 100 Hz to 150kHz in Fig.13 does't agree with measured impedance and phase characteristics. Thus, it is need that the equivalent circuit modeled impedance and phase characteristics in wide frequency band.

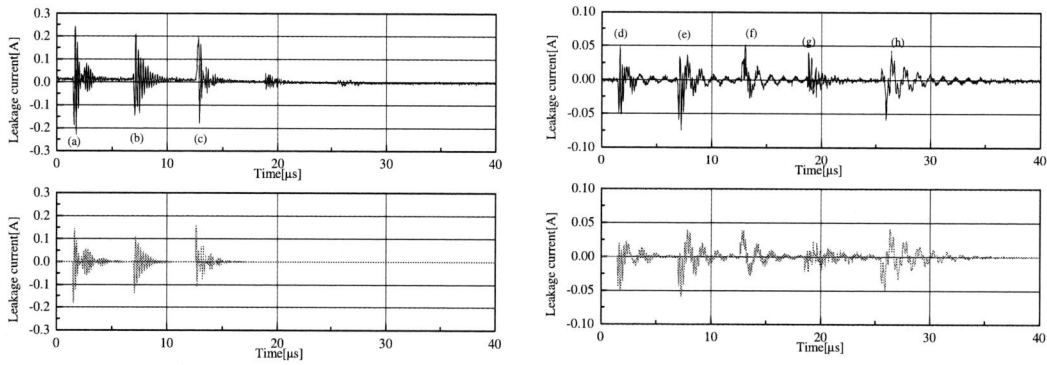

Fig. 10: Simulation result of leakage current i_{le1} Fig. 11: Simulation result of leakage current i_{le2}

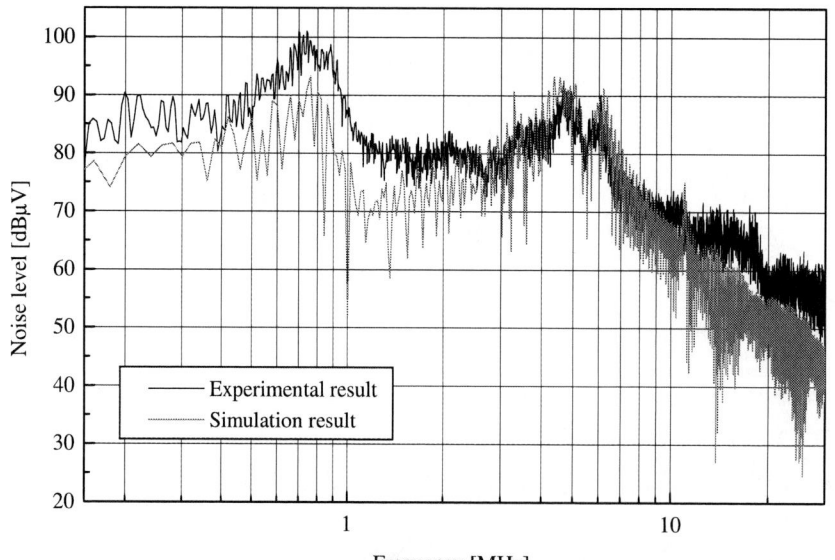

Fig. 12: Simulation result of disturbance voltage

B.Proposal modeling method

The measured impedance and phase characteristics at 100 Hz to 150 kHz can't be modeled by RLC parallel circuit because the measured impedance and phase characteristics is complex. Thus, For modeling

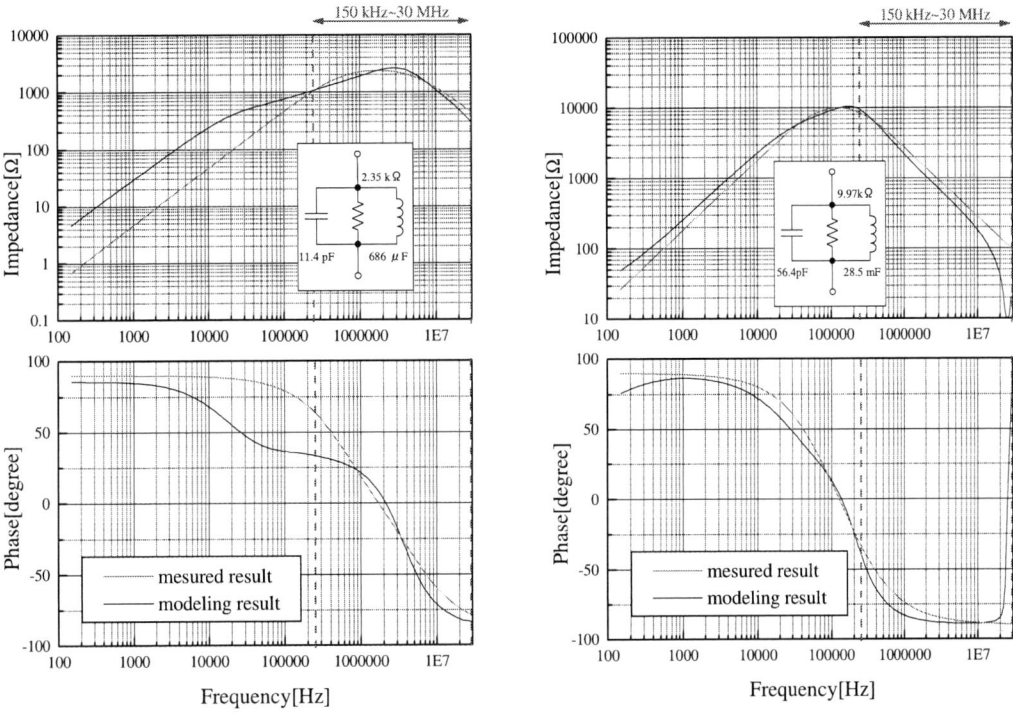

Fig. 13: Impedance and phase characteristics of CMCC with convetional model (3 mH)

Fig. 14: Impedance and phase characteristics of CMCC with conventional model (47 mH)

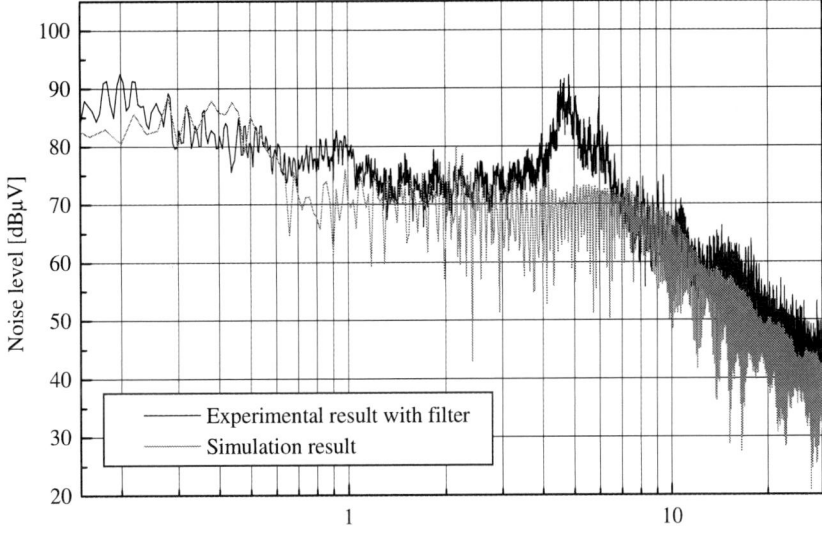

Fig. 15: Simulation result of disturbance voltage with conventional CMCC modeling circuit (3 mH)

Investigation of Improvement of Modeling precision for Conducted Noise on an Isolated AC/DC Converter using SiC Devices KUWANA Kazuki

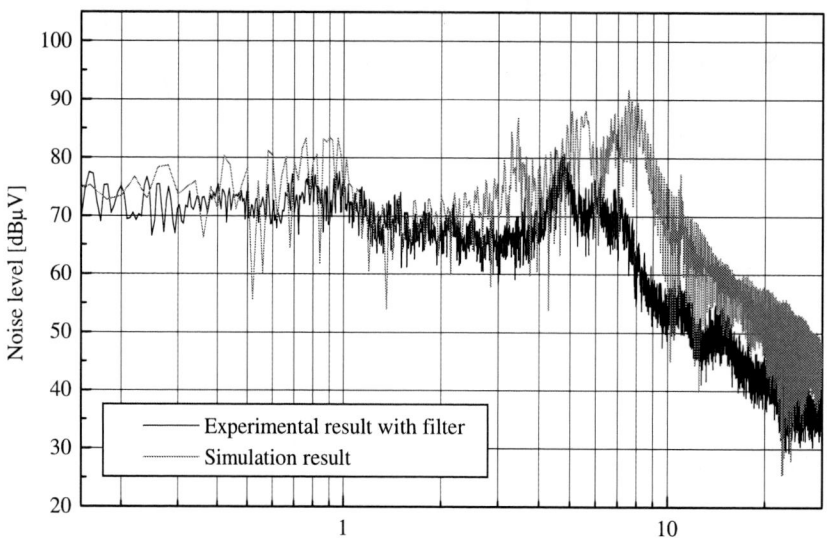

Fig. 16: Simulation result of disturbance voltage with conventional CMCC modeling circuit (47 mH)

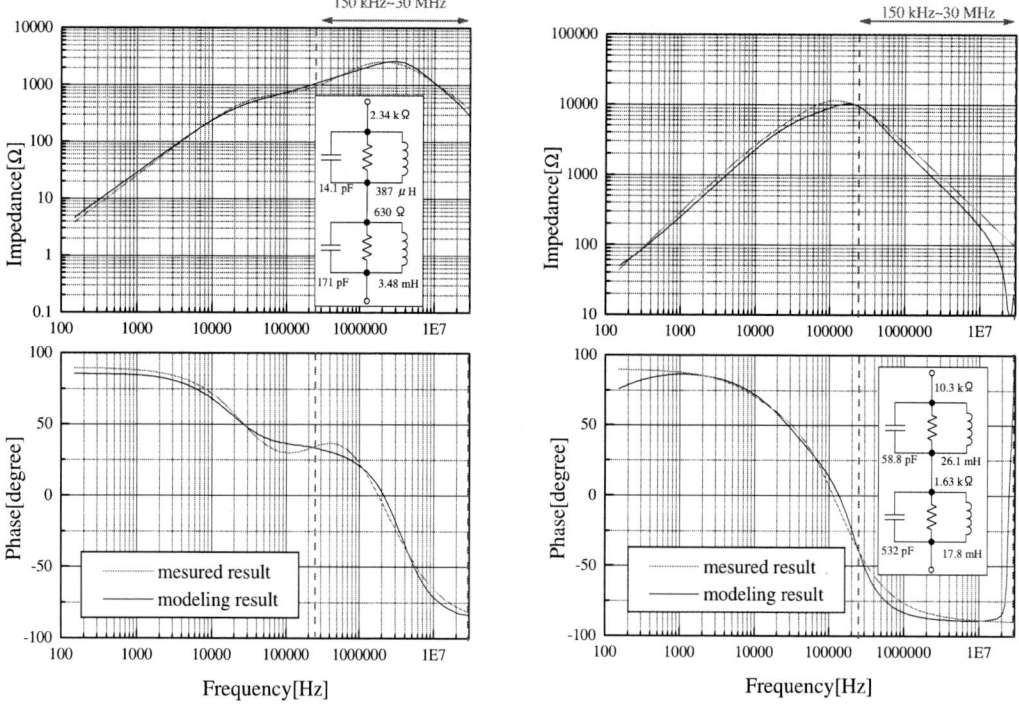

Fig. 17: Impedance and phase characteristics of CMCC with proposal model (3 mH)

Fig. 18: Impedance and phase characteristics of CMCC with proposal model (47 mH)

the impedance and phase characteristics of CMCC at 100 Hz to 150 kHz, the modeling circuit is approximated two RLC parallel circuit connected in series. Fig.17 and Fig.18 show the impedance and phase characteristics. In Fig.17 and Fig.18, it is confirmed that impedance and phase characteristics agree with the measured impedance characteristics from 150 Hz to 30 Hz.

The simulation inserted the model of CMCC to Fig.9 is performed. Fig19 and Fig20 show the simulation result of disturbance voltage with this circuit. In Fig19 and Fig20, it is confirmed that the simulation result agree with the experimental result at view point of resonance point. Therefor, the propose modeling circuit of CMCC is valid , and the propose simulation method is useful. In addition, in Fig20 simulation result does't agree with experimental result at view point of view noise level. The reason is that only three paths are cosidered. It is thoght that there are more noise paths. Therefore, it is necessary to clarify the noise paths and to model it in order to perform more accurate simulation.

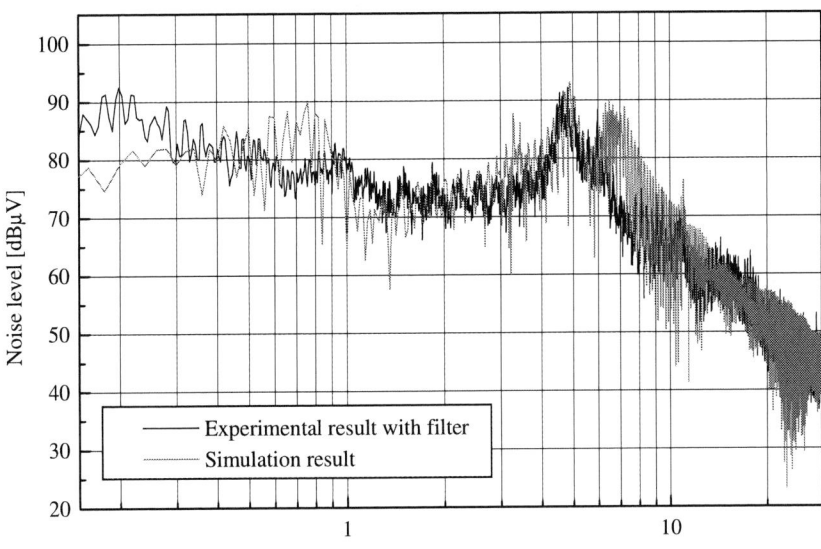

Fig. 19: Simulation result of disturbance voltage with proposal CMCC modeling circuit (3 mH)

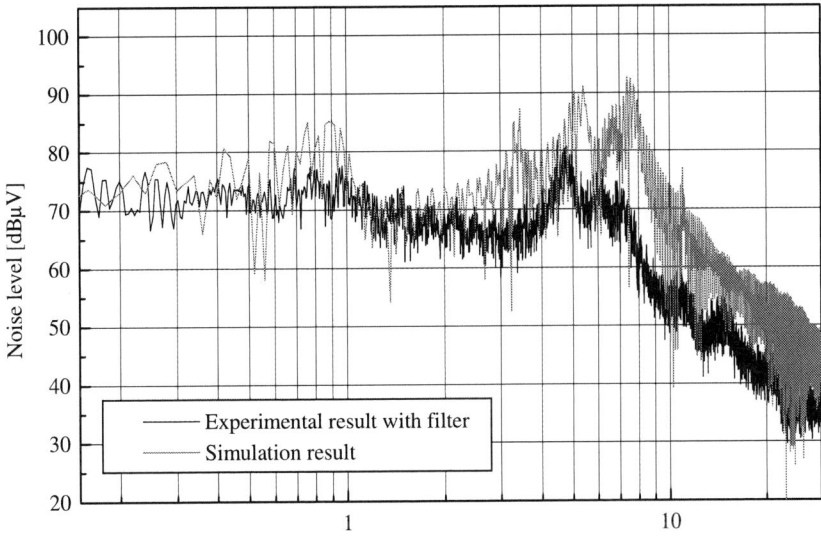

Fig. 20: Simulation result of disturbance voltage with proposal CMCC modeling circuit (47 mH)

Conclusions

This paper presents the modeling method of conducted noise on the AC/DC converter using SiC devices, and proposes the simulation method with filters.The modeling circuit of the conducted noise is derived by approximated the leakage current to the *RLC* series circuit. The simulation result using the modeling circuit agree with the experimental result.

In addition, when the simulation with CMCC is performed, the modeling circuit of CMCC is inserted to the common mode modeling circuit. The modeling circuit constituted the *RLC* parallel circuit well model impedance and phase characteristics of CMCC from 150 kHz to 30 MHz. However, the measured impedance and phase characteristics can't be approximated by the *RLC* parallel circuit at under 150 kHz because the measured impedance and phase characteristics is complex. Thus, this paper propose two *RLC* parallel circuits connected in series to model the phase characteristics and impedance characteristics in wide frequency band. The modeling circuit constituted the two RLC parallel circuit connected in series well model impedance and phase characteristics of CMCC from 100 Hz to 30 MHz. In addition, the simulation result of the disturbance voltage is improved by using the proposed modeling circuit of CMCC at view point of resonance point. It is confirmed that modeling both the impedance characteristics and the phase characteristics in wide frequency band is necessary.

References

[1] Takumi Hamaguchi, Kazuma Suzuki, Wataru Kitagawa, and Takaharu Takeshita: "Variable Output Voltage Control of an Isolated Bi-directional AC/DC Converter with a Soft-Switching Technique" The 2018 International Power Electronics Conference -ECCE Asia- (IPEC-Niigata 2018)

[2] Taekyun Kim, Dong Feng, Minsoo Jang and Vassilios Agelidis : "Common Mode Noise Analysis for Cascaded Boost Converter with Silicon Carbide Devices", *IEEE Transactions on Power Electronics, vol.PP, issue.99 (2016)*

[3] Hui-Chen Yang and Rejeki Simanjorang and Kye Yak See : "A Method of Junction Temperature Estimation for SiC Power MOSFETs via Turn-on Saturation Current Measurement" *IEEJ Journal of Industry Applications, Vol.8, No.2, pp.306-313(2019)*

[4] Paco Bogónez-Franco and Josep Balcells Sendra : "EMI comparison between Si and SiC technology in a boost converter", *Electromagnetic Compatibility (EMC EUROPE), 2012 International Symposium on (2012-9)*

[5] Hiroyoshi Komatsu, Toshihiko Noguchi : "Development of Drive Circuits for Super High-Speed Switching Devices to Reduce High-Frequency EMI Noise", The papers of Technical Meeting on Semiconductor Power Converter, IEE Japan, vol.SPC-07, no.11-20, pp.43-48 (2007)

[6] Mitio Tamate, Tamiko Sasaki, Akio Toba, Yasushi Matsumoto, Keiji Wada and Toshihisa Shimizu : "Method for Evaluating Insertion Loss of EMI Filter Connected to Semiconductor Power Converters", *The transactions of the Institute of Electrical Engineers of Japan*, D, A publication of Industry Applications Society 132(7), 727-735, 2012-07-01

[7] Kazuhiro Umetani and Takahiro Tera and Kazuhrio Shirakawa : "A Magnetic Structure Integrating Differential Mode and Common-Mode Inductors with Improved Tolerance to DC Saturation", *IEEJ Journal of Industry Applications, Vol.4, No.3, pp.166-173(2015)*

[8] Tadashi Aoki, Satoshi Ogasawara, Hirohito Funato and Yasuyuki Kobayashi : "Frequency Analysis of Conducted EMI Produced a PWM Inverter", *The papers of Technical Meeting on Semiconductor Power Converter, IEE Japan*, vol.SPC-07, no.11-20, pp.37-42 (2007-1)

[9] Satoshi Ogasawara : "Reduction of Leakage Current, Surge Voltage and Shaft Voltage in Variable-Speed AC Drives", *The transactions of the Institute of Electrical Engineers of Japan*, D, A publication of Industry Applications Society 118(9), 975-980, 1998-09

[10] Kohei Mitani, Yuki Kawamura, Wataru Kitagawa and Takaharu Takeshita : "Circuit Modeling fo Common Mode Noise on AC/DC Converter Using SiC Device", *2019 21th European Conference on Power Electronics and Applications (EPE'18 ECCE Europe)* (2019)

[11] Hidetoshi Tanaka, Kazuma Suzuki, Wataru Kitagawa, Takaharu Takeshita : "Design for Conducted Noise Reduction on AC/DC Converter using SiC-MOSFET", *Electrical Machines and Systems (ICEMS), 2016 19th International Conference on* (2016-11)

[12] Sari Maekawa, Junichi Tsuda, Atsuhiko Kuzumaki, Shuhei Matsumoto, Hiroshi Mochikawa and Hisao Kubota : "EMI Prediction Method for SiC Inverter by Developing an Accurate Model of Power Device", *IEEJ trans.IA, Vol.134, No.4, (2014)*

[13] Yuki Kawamura, Hidetoshi Tanaka, Wataru Kitagawa and Takaharu Takeshita : "Investigation of Modeling for Conducted Noise Reduction on Isolated AC/DC Converter using SiC Devices", *2018 20th European Conference on Power Electronics and Applications (EPE'18 ECCE Europe)* (2018)

Passivity-Based Design for the Plug-and-Play Single-Loop Controlled *LCL*-Filtered Inverter

Yuying He[1], Xuehua Wang[1], Xinbo Ruan[1,2], Yixiao Ma[1], Fuxin Liu[2]

1. Huazhong University of Science and Technology
Wuhan, China
2. Nanjing University of Aeronautics and Astronautics
Nanjing, China
Tel.: +86 / 18627864801.
E-Mail: heyuying@hust.edu.cn
URL: http://ceee.hust.edu.cn

Acknowledgements

This work was supported by the National Key Research and Development Program of China under Award 2016YFB0900100.

Keywords

«Passivity», «stabilization», «single-loop», «grid impedance», «Feedforward».

Abstract

The grid impedance variation poses a challenge to the stability of *LCL*-filtered inverter. Critical stable ranges have been revealed for the single-loop inverter-current-feedback (ICF) and grid-current-feedback (GCF) controlled ones. However, they are all derived under the assumption that the grid impedance is inductive, whereas the presence of parasitic grid capacitances will lead to different results. In this paper, considering the capacitive condition and based on *passivity* concept, it is found that the stability is hardly secured for either the ICF or GCF controlled ones. From perspective of passivity, the phase of inverter output impedance would better be shaped between −90° and 90°. In view of this, the feedforward of the voltage of the point of common coupling (PCC) is applied in this paper to achieve this target, and the proper feedforward solutions are given. As a result, the inverter could harvest the robust plug-and-play functionality regardless of grid impedance. The correctness of the theoretical analysis and effectiveness of the proposed methods are verified by simulation results.

Introduction

As an efficient power conversion interface, the *LCL*-filtered grid-connected inverter has been widely applied in the distributed power generation systems (DPGS) to harvest the sustainable energy. The *LCL* filter contributes to desirable high-frequency attenuation, but on the other hand, its resonance peak will challenge the system stability, depending on the controlled current (the inverter current or the grid current), the control delays and grid condition.

In practice, both inverter current feedback (ICF) and grid current feedback (GCF) control schemes are available and have their own merits. The former is commonly selected in industry from the cost perspective since current sensors are integrated in inverter side for overcurrent protection [1]. The latter directly controls the injected current and thus its stability can be intuitively read [2]. Under the typical control delays $T_d = 1.5 \ T_s$ (T_s is the sampling period), it was revealed that the stability is subject to the relation of resonance frequency (f_r) and one-sixth of sampling frequency ($f_s/6$), where $f_r < f_s/6$ and $f_r > f_s/6$ are suitable for the ICF and GCF control, respectively [3]. Considering that the grid impedance Z_g may vary in a wide range, letting $f_{r0} < f_s/6$ would benefit a stable operation for the ICF-controlled system, and $f_{r0} > f_s/6$ and $f_{r\infty} \ge f_s/6$ would be desirable for the GCF-controlled one, where f_{r0} and $f_{r\infty}$ represent the *LCL* resonance frequency under $Z_g = 0$ and $Z_g = \infty$, respectively. However, it is worth noting that, the above damping regions are all obtained in the presence of an inductive grid impedance. If a capacitive grid impedance is connected, the damping regions may be far different. This issue has not been paid much attention and studied.

In fact, the passivity concept provides an attractive solution, which guides the design to secure a stable operation regardless of the inductive or the capacitive grid impedance [4], [5]. According to the passivity concept, as long as the real part of the inverter output impedance is positive, the system will stay stable irrespective of the changeable grid impedance and the number of parallel inverters. In other words, if the system is designed to have a *passive* behavior, the plug-and-play functionality could be achieved.

In this paper, based on the passivity concept, the single-loop ICF and GCF control are investigated. It is revealed that, the passivity can hardly be ensured for the single-loop either ICF or GCF controlled inverter. To address this issue, the proper feedforward of PCC voltage is proposed to adjust the output impedance phases. The rest of this paper is organized as follows. In Section II, an overall description of the system is briefly presented. In Section III, the passivity-based stability analysis is provided for the single-loop ICF and GCF control. In Section IV, proper feedforward scheme is given to achieve the plug-and-play functionality. In Section V, the effectiveness of the proposed methods is validated through simulation results. Finally, Section VI concludes this paper.

System Description

Fig. 1(a) illustrates the configuration of a single-phase grid-connected inverter feeding into power grid through an *LCL* filter, where V_{in} and v_{PCC} represent the input dc voltage and the PCC voltage, respectively. The control objective is to regulate grid current to track its reference i^*, whose amplitude is directly given as I^* and phase is extracted by a phase-locked loop (PLL) to synchronize i_g with v_{PCC}. Either inverter-current i_{L1} or grid-current i_{L2} can be sensed to achieve this objective, called inverter-current feedback (ICF) and grid-current feedback (GCF) control respectively. H_i is the sense gain and G_i is the grid current regulator. To reduce the steady-state error, the proportional-resonant (PR) regulator is employed here. Z_g represents the grid impedance, which varies in an uncertain and wide range in practice, posing a great challenge on the stability of grid-connected inverter.

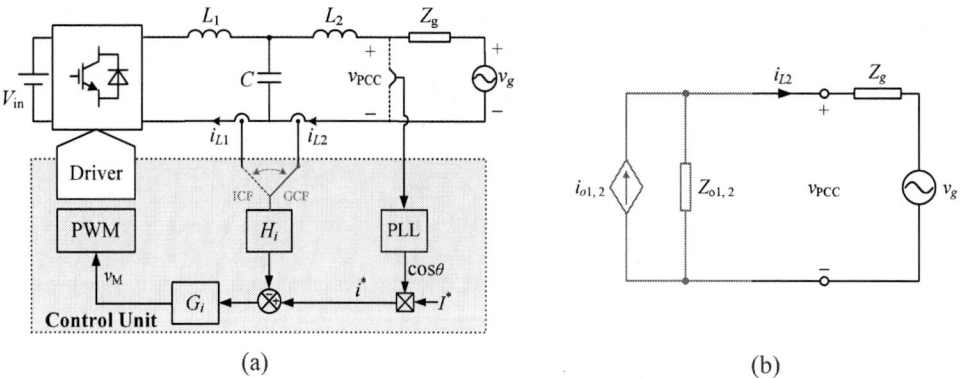

(a)　　　　　　　　　　　　　　　　　(b)

Fig.1. The single-loop controlled *LCL*-filtered inverter. (a). Configuration. (b) Equivalent impedance model.

Passivity-Based Study on the Single-Loop Controlled Inverter without PCC Voltage Feedforward

According to the *passivity* concept [4], [5], the system can be stable regardless of the grid impedance variation if:

1) The system is stable when $Z_g = 0$.

2) The inverter has a passive (dissipative) behavior, i.e., the phase of the output impedance is between $-90°$ and $90°$ for all frequencies (or the real part of output impedance is nonnegative for all frequencies).

As known, the first requirement, i.e., stable operation under $Z_g = 0$, can be easily met by setting $f_{r0} < f_s/6$ and $f_{r0} > f_s/6$ for the single-loop ICF and GCF controlled inverter respectively. Therefore, the output impedance is the key to secure the system robustness. According to Fig. 1(a), the equivalent impedance model can be obtained, as depicted in Fig. 1(b), where the inverter is modeled as the parallel connection of ideal current source $i_{o1,\,2}$ and original output impedance $Z_{o1,\,2}$. Here, i_{o1}, Z_{o1} correspond to the ICF control, and i_{o2}, Z_{o2} correspond to the GCF control. The output impedances are expressed as

$$Z_{o1}(s) = sL_2 + \cfrac{1}{1/sL_1 + sC + \underbrace{CH_iK_{\mathrm{PWM}}G_d(s)/L_1}_{1/Z_d}} + \underbrace{\frac{H_iG_i(s)K_{\mathrm{PWM}}G_d(s)}{s^2L_1C_f + sCH_iG_i(s)K_{\mathrm{PWM}}G_d(s)+1}}_{Z_{\mathrm{reg1}}} \tag{1}$$

$$Z_{o2}(s) = sL_2 + \frac{1}{1/sL_1 + sC} + \underbrace{\frac{H_iG_i(s)K_{\mathrm{PWM}}G_d(s)}{s^2L_1C_f + 1}}_{Z_{\mathrm{reg2}}} \tag{2}$$

In (1) and (2), $G_d(s) = e^{-1.5sTs}$ represents the digital control delays, $K_{\mathrm{PWM}} = V_{in}/V_{tri}$ is the gain of the PWM inverter and V_{tri} is the amplitude of the triangular carrier. As shown in (1) and (2), both of Z_{o1} and Z_{o2} consist of three parts: sL_2, the paralleled impedance of sL_2 and $1/sC$ (and the inherent damping Z_d in Z_{o1}), Z_{reg} introduced by current regulator, which are illustrated in circuit notation (see Fig. 2).

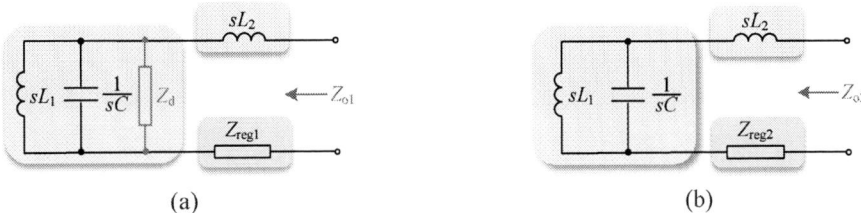

(a)   (b)

Fig. 2. Circuitry representation of the output impedances with: (a) single-loop ICF control. (b) single-loop GCF control.

For the ease of illustration, two sets of filter parameters, i.e., filter 1 and filter 2 which correspond to $f_{r0} < f_s/6$ and $f_{r0} > f_s/6$ respectively, are given, as listed in Table I. Correspondingly, single-loop ICF control is applied for filter I and single-loop GCF control is applied for filter II. According to the main parameters listed in Table I, the Bode diagrams of Z_{o1} and Z_{o2} are plotted in Fig. 3(a) and (b) respectively.

Table I: Main Parameters of the *LCL*-filtered Grid-Connected Inverter

Parameter	Symbol	Value	Parameter	Symbol	**Filter I**	**Filter II**
Grid voltage	V_{in}	220 V	Inverter-side inductor	L_1	600 μH	600 μH
Input voltage	V_g	360 V	Grid-side inductor	L_2	150 μH	150 μH
Output power	P_o	6 kW	Filter capacitor	C	30 μF	10 μF
Switching frequency	f_{sw}	10 kHz	Resonance frequency	f_{r0}	2.65 kHz	4.59 kHz

As for single-loop ICF controlled inverter, as seen from Fig. 3(a), $\angle Z_{o1}$ will exceed $90°$ in the frequency band $(f_x, f_s/2)$ where f_x is approximately $f_s/6$. The reason is simple. In high frequency band, Z_{reg1} is small and thus the real part of Z_{o1} is almost decided by the inherent damping resistance Z_d. Due to control delays, the real part of Z_d is definitely negative in the frequency band $(f_s/6, f_s/2)$, thus resulting in $\angle Z_{o1} > 90°$. Accordingly, the single-loop ICF controlled inverter is not passive.

As for single-loop GCF controlled inverter, as seen from Fig. 3(b), an infinite resonance peak arises in the magnitude-frequency curve and the corresponding phase frequency curve drops 180°. Recalling (2), it can be easily found that the resonance happens at the resonance frequency of L_1 and C, denoted as f_{r_L1C}. From the physical insight, due to no inherent damping Z_d, parallel resonance between L_1 and C could happen (see Fig. 2(b)). As depicted in Fig. 3(b), if $f_{r\_L1C} = f_s/6$, $\angle Z_o$ can remain between $-90°$ and 90°. Yet, if $f_{r\_L1C} > f_s/6$, $\angle Z_o$ would exceed 90° in the frequency band ($f_s/6$, f_{r_L1C}), and if $f_{r\_L1C} < f_s/6$, $\angle Z_{o2}$ would below $-90°$ in the frequency band (f_{r_L1C}, $f_s/6$). Thus, the passivity is only achieved under $f_{r\_L1C} = f_s/6$. This condition is tough to be met in practice, since f_{r_L1C} will shift with the filter parameter fluctuation.

Therefore, the passivity can hardly be satisfied for the single-loop either ICF or GCF controlled inverter. Therefore, a simple and effective approach is urgently demanded to make the system behave passive.

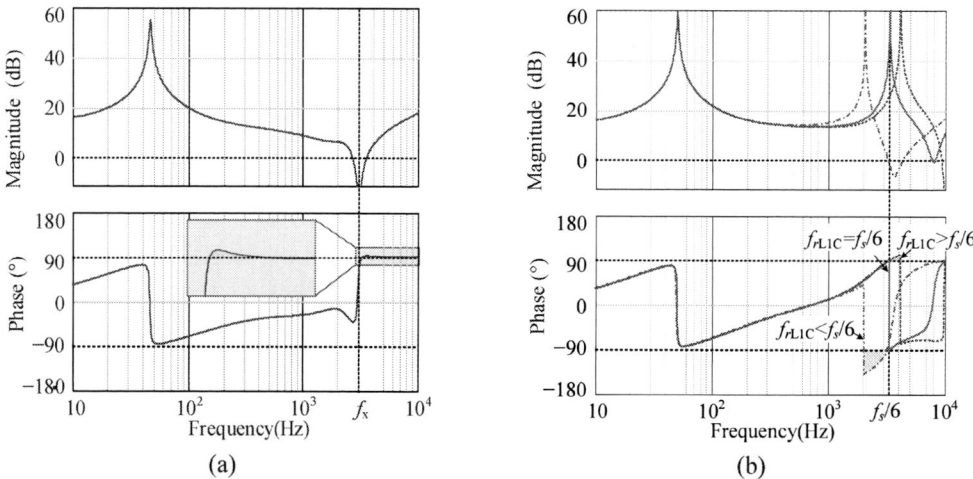

(a)   (b)

Fig. 3. Bode diagram of the output impedances with: (a) single-loop ICF control. (b) single-loop GCF control.

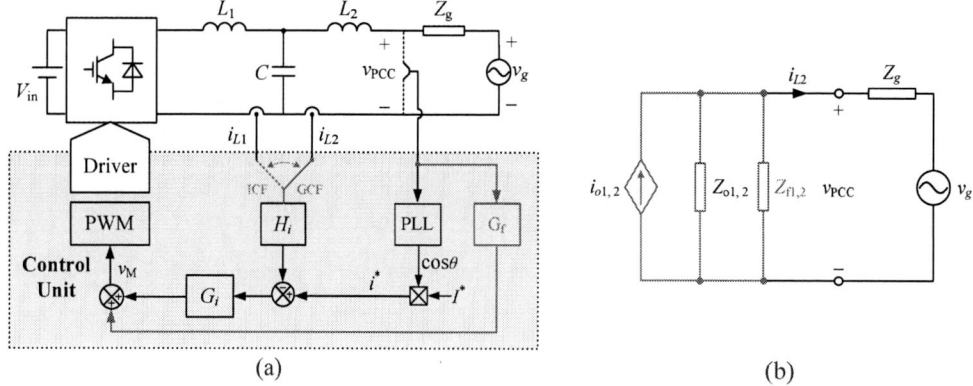

(a)   (b)

Fig. 4. The single-loop controlled *LCL*-filtered inverter with the PCC voltage feedforward. (a). Configuration. (b) Equivalent impedance model.

Passivity Enhancement Using the Feedforward of PCC Voltage

To achieve our objective, the output impedance should be reshaped between $-90°$ and 90°. The PCC voltage is commonly sensed for synchronization. If feeding the sensed PCC voltage forward with the gain G_f to the output of current regulator, as shown in Fig. 4(a), virtual impedances $Z_{f1,2}$ would accordingly be introduced in paralleled with $Z_{o1,2}$ (see Fig. 4(b)). Thus, it would be promising to achieve $\angle(Z_o//Z_f)\in(-90°, 90°)$ in the entire frequency band. Hereinafter, proper PCC voltage feedforward schemes will be explored.

A. Passivity Enhancement for the Single-Loop ICF Controlled Inverter

According to Fig. 4(a), the expression of Z_{f1} can be obtained, i.e.,

$$Z_{f1}(s) = -\frac{s^3 L_1 L_2 C + s(L_1 + L_2) + sCH_i G_i(s) K_{PWM} G_d(s) + H_i G_i(s) K_{PWM} G_d(s)}{K_{PWM} G_d(s) G_f(s)} \tag{3}$$

The common-used proportional feedforward scheme, denoted as $G_f = a$, firstly comes into mind. The Bode diagram of reshaped output impedance $Z_o//Z_{f1}$ with $G_f = a$ ($a < 0$) is plot in Fig. 5. As depicted, compared with the range without feedforward, the non-passive frequency range shrinks. But as for the frequencies close to the Nyquist frequency, the passivity is still not met. Naturally, an idea is adding a differential term into the feedforward function, i.e., $G_f = a+b\cdot s$, so as to act at the high frequencies. Accordingly, as shown in Fig. 5, the phase of reshaped output impedance with $G_f = a+b\cdot s$ ($a < 0$, $b < 0$) can be successfully amended between $-90°$ and $90°$ in the entire frequency band, meeting our expectation.

It is worth noting that, although the passivity is met by $G_f = a+b\cdot s$ ($a < 0$, $b < 0$), the low-frequency magnitude is unfortunately driven down, compromising the harmonic attenuate ability. Further, a high-pass-filter can be inserted to reduce such adverse effect, yielding

$$G_{f\_ICF} = (a+bs)\cdot G_{HPF} = (a+bs)\cdot\frac{s}{s+\omega_c}, \text{where } a < 0 \text{ and } b < 0 \tag{4}$$

As a result, it can be observed from Fig. 5 that the passivity would be ensured and meanwhile the 3th, 5th, 7th harmonic attenuate ability nearly remain unchanged.

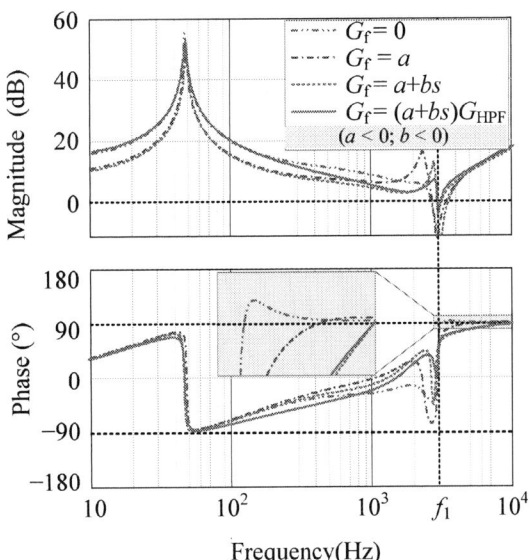

Fig. 5. Bode diagram of the paralleled impedance of Z_{o1} and Z_{f1} with different G_f.

B. Passivity Enhancement for the Single-Loop GCF Controlled Inverter

According to Fig. 4(a), the expression of Z_{f2} can also be obtained, i.e.,

$$Z_{f2}(s) = -\frac{s^3 L_1 L_2 C + s(L_1 + L_2) + H_i G_i(s) K_{\text{PWM}} G_d(s)}{K_{\text{PWM}} G_d(s) G_f(s)}. \tag{5}$$

Likewise, the Bode diagram of reshaped output impedance $Z_o//Z_{f2}$ with simple proportional feedforward scheme is plotted, as shown in Fig. 6. Fortunately, the positive proportional feedforward is found effective, expressed as

$$G_{f\_GCF} = a, \text{where } a > 0 \tag{6}$$

As shown in Fig. 6, thanks to Z_{f2}, the −180 degrees step of the phase frequency curve is avoided. By selecting a proper proportional coefficient, the phase can be amended into (−90°, 90°) in the entire frequency band, meeting our expectation. Besides, the positive feedforward of PCC voltage also helps improve the harmonic attenuate ability.

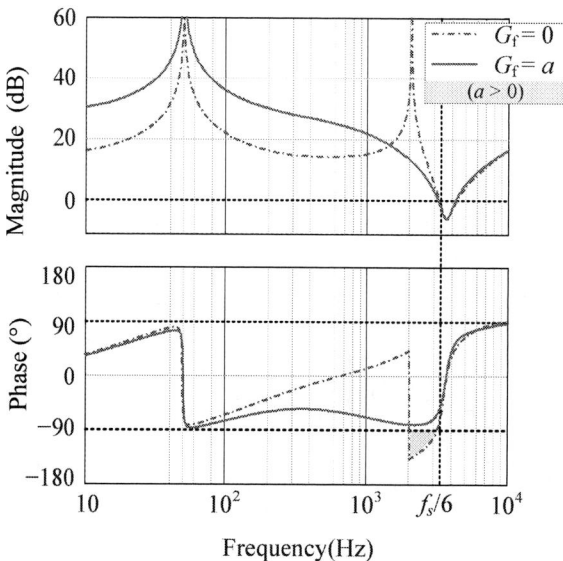

Fig. 6. Bode diagram of the paralleled impedance of Z_{o2} and Z_{f2} with the proportional PCC voltage feedforward.

Simulation Verification

Finally, in order to verify the effectiveness of the proposed feedforward schemes, simulation with main parameters listed in Table I is performed. Referring to the above analysis, as for filter I, the feedforward function expressed in (4) is applied, where $a = -0.0026$, $b = -2.75 \times 10^{-7}$ and $\omega_c = 2\pi \cdot 50$ rad/s are designed; as for filter II, the feedforward function expressed in (6) is applied, where $a = 0.01$ is designed.

Fig. 7 presents the simulation results of the single-loop ICF control under the grid condition of $C_g = 15$ uF and $L_g = 100$ uH. As shown in Fig. 7, the current is oscillate without the PCC voltage feedforward, in agreement with the theoretical analysis in Section III; after enabling the proposed feedforward scheme, the current becomes stable, in agreement with Section IV. A. The results confirm the validation of the proposed feedforward scheme for single-loop ICF control.

Fig. 8 presents the waveforms of the single-loop GCF control under $C_g = 5$ uF and $L_g = 500$ uH. As shown in Fig. 8, the current is initially oscillate, and stabilized after enabling the PCC voltage feedforward scheme. It agrees with our expectation and clarifies that the proposed feedforward scheme for single-loop GCF control is effective.

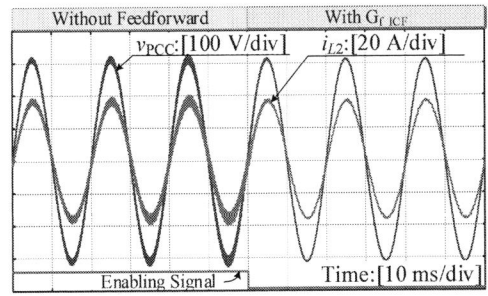

Fig. 7. Simulation results of inverter with single-loop ICF control.

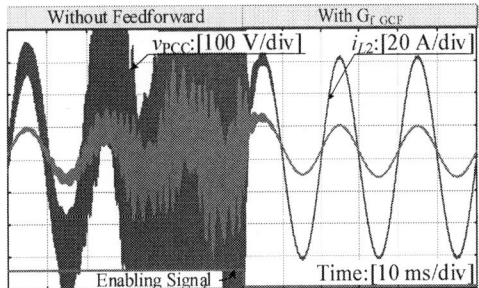

Fig. 8. Simulation results of inverter with single-loop GCF control.

Conclusion

This paper reveals that the passivity can hardly be satisfied for the single-loop either ICF or GCF controlled inverter. Driven by this, the proper PCC voltage feedforward schemes are proposed, which can push the phases of inverter output impedance into the expected range, i.e., $(-90°, 90°)$. As a result, the system passive behavior can be remained. Accordingly, the high robustness against grid impedance and the plug-and-play functionality are harvested. Simulation results have verified the effectiveness of the proposed methods.

References

[1] B. Liu, Q. Wei, C. Zou, and S. Duan, "Stability analysis of *LCL*-type grid-connected inverter under single-loop inverter-side current control with capacitor voltage feedforward," *IEEE Trans. Ind. Inform.*, vol. 14, no. 2, pp. 691–702, Feb. 2018.

[2] D. Pan, X. Ruan, C. Bao, W. Li, and X. Wang, "Capacitor-current-feedback active damping with reduced computation delay for improving robustness of *LCL*-type grid-connected inverter," *IEEE Trans. Power Electron.*, vol. 29, no. 7, pp. 3414–3427, Jul. 2014.

[3] J. Dannehl, C. Wessels, and F. W. Fuchs, "Filter-based active damping of voltage source converters With *LCL* filter," *IEEE Trans. Ind. Electron.*, vol. 58, no. 8, pp. 3623–3633, Aug. 2011.

[4] L. Harnefors, A. G. Yepes, A.Vidal, and J. Doval-Gandoy, "Passivity-based controller design of grid-connected VSCs for prevention of electrical resonance instability," *IEEE Trans. Ind. Electron.*, vol. 62, no. 2, pp. 702–710, Feb. 2015.

[5] L. Harnefors, X. Wang, A. G. Yepes, and F. Blaabjerg, "Passivity-based stability assessment of grid-connected VSCs — An overview," *IEEE J. Emerg. Sel. Topics Power Electron.*, vol. 4, no. 1, pp. 116–125, Mar. 2016.

Characteristics of an integrated motor controlled independently by multi–inverters to achieve high efficiency and a wide speed range

Kazuto Sakai, Yano Hideaki
TOYO UNIVERSITY
2100 Kujirai
Kawagoe, Saitama, Japan
Tel.: +81 / (49) – 239–1353.
Fax: +81 / (49) – 231–1400.
E-Mail: k.sakai@toyo.jp
URL: http://www.toyo.ac.jp/en/

Acknowledgements

This work was supported by JSPS KAKENHI, Grant Number JP17K06313.

Keywords

«Adjustable speed drive», «Efficiency», «Electric vehicle», «Electrical Machines», «Variable speed drive»

Abstract

For electric vehicles, motor operation is achieved by using a highly efficient variable-speed drive. In this paper, we describe the characteristics of a novel motor capable of pole and phase changes; it is a high-performance motor because it has multiple windings combined with multi-inverters. The pole changes will allow the motor to operate with a large torque at a low speed by using a small number of poles; the motor can operate at high efficiency at high speeds by using a large number of poles. The motor had three groups of three-phase windings consisting of 18 coils. One group of the three-phase windings was consisted of 6 coils. Three three-phase inverters that connected the three groups of three-phase windings directly controlled the current phase of each winding, which helped change the number of poles and phases. The results of our analyses confirmed that the motor produced a large torque at a low speed of 3000 rpm for the eight-pole mode, and operated with a high efficiency of approximately 96% at a high speed of 6000 rpm for the four-pole mode.

Introduction

For electric vehicles, it is vital to have a highly efficient motor over a wide speed range. Fig. 1 shows the relationship between the motor torque and the speed for vehicles and trains. Here, the motor performance can decline because of the impedance factors, back electromotive forces, and the associated frequency characteristics, which affect the upper limits of the voltage source for systems with variable-speed drives. For a high-power permanent-magnet motor, it is possible to vary the magnetic force of a permanent magnet, which would increase the speed range and efficiency [1–4]. Conventional induction motors can operate at variable speeds by controlling the magnetization current, which regulates the motor voltage; however, these motors operate at a low power factor and low efficiency. In this paper, we propose a machine technique combined with power electronics to

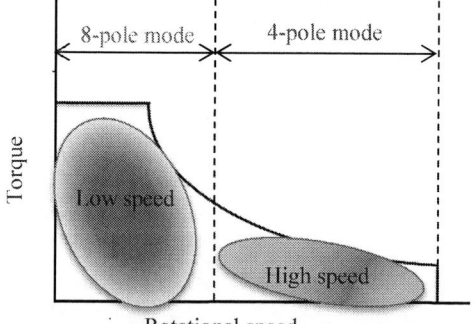

Fig. 1: Pole change for high efficiency in variable speed system.

achieve a wider speed range and higher efficiency than those of conventional motors. In particular, pole changes were used to achieve a wide speed range (see Fig. 1); the motor operates in the eight-pole mode in a low-speed range and in the four-pole mode in a high-speed range. The number of poles vary according to the driving conditions, which helps maintain high efficiency. Induction motors can change poles by changing the winding connections [5–8]. Permanent magnet motors can change the pole by changing the magnetization of permanent magnets [9–15]. The motor-drive systems that comprise a motor with multiple coils or multiple groups of windings and multi-inverters allow the pole and phase to change [13, 15–17]. Furthermore, we proposed a motor-drive system that consisted of a motor with a multiple three-phase winding and a three-phase multi-inverter. Also, we assume that the power factor increases with a decrease in the motor reactance at low frequencies because of the pole change. Accordingly, we have examined the characteristics of a novel motor-drive system with variable poles and phases. We discuss the motor characteristics with respect to the electrical pole changes and the transient characteristics of the pole-change intervals using magnetic analysis. The motor produced a high torque at low speeds and operated with high efficiency at low and high rotational speeds.

Multi-inverter pole changing and phase changing

We propose a motor-drive system using electrical pole changing to ensure high efficiency over a wide speed range. Integrated motors capable of controlling current (IMCC) have multiple three-phase windings connected to three-phase multi-inverters. Poles and phases in the motor can be changed using a motor-controlled coil current in combination with multiple units of three-phase inverter circuits that control the current vector of each motor coil. Each inverter connects each group of the three-phase motor windings and controls the three-phase current for each winding group. The pole change caused the coil pitch to assume a different value for each pole mode. Therefore, we determined the number of slots by considering the coil pitch in each pole mode. In this study, we selected a concentrated winding by considering a reduction in the copper loss because of the short coil end and the excellent manufacturability. Fig. 2 illustrates an induction-type IMCC with a three three-phase inverter configuration and a conventional squirrel-cage rotor, which is a system that consists of a motor and three three-phase inverter circuits. Here, 18 coils of the stator winding are subdivided into three groups; each group is connected to a corresponding individual three-phase inverter circuit. Table I illustrates the 18-coil stator arrangement of a three-phase motor for eight- and four-pole operations. The set coils in the first winding group of the motor are labeled a1, b1, and c1, which are connected to the output terminals of a1, b1, and c1 of the three-phase Inverter 1, respectively. The second winding

Table I: Current phase for pole change.

		Pole changing	
Phase		9	9
Pole		8	4
Current phase in inverter 1 (degree)	a1	0	0
	b1	120	240
	c1	240	120
Current phase in inverter 2 (degree)	a2	40	200
	b2	160	80
	c2	280	320
Current phase in inverter 3 (degree)	a3	80	40
	b3	200	280
	c3	320	160

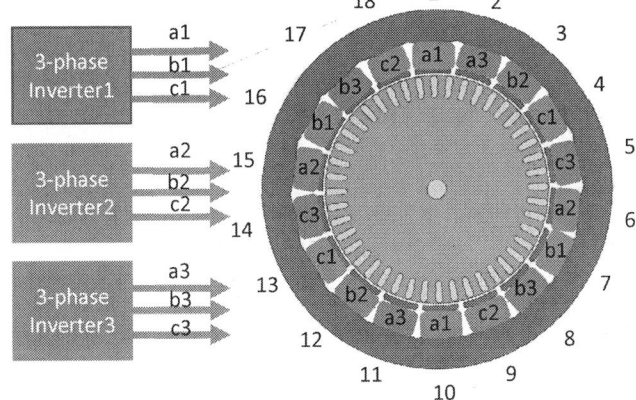

Fig. 2: Arrangement of the coils connecting the three three-phase inverters.

group consists of the coils a2, b2, and c2, which are connected to the output terminals a2, b2, and c2 of the three-phase Inverter 2, respectively.

The third winding group consists of the coils a3, b3, and c3, which are connected to the output terminals a3, b3, and c3, respectively, of the three-phase Inverter 3. The current in each of the three inverters drives the 18 coils in the proposed IMCC, which shifts the phase of the current in the Inverters 1, 2, and 3 according to the number of poles. The coils form a rotating field generating a specified number of poles, which generates a multi-pole rotating field.

Magnetic field analysis and the motor model

We performed magnetic field analysis using the JMAG finite element modeling (FEM) software to verify and understand the mechanisms behind pole changes and to obtain the essential physical characteristics and efficiencies of the IMCC for each pole state. Thereafter, we configured the analytical IMCC models to examine and discuss the effects of multi-inverters on the torque and operational characteristics. The analytical model of the proposed motor is shown in Fig. 3; their corresponding specifications are listed in Table II. The rated current in each winding (or each inverter) is 6.26 Arms.

Verification of electrical pole-change analysis and the motor model

For each inverter, the current phase was set to form eight or four poles, as shown in Table I. Fig. 4 shows the distribution of the magnetic-flux density when the IMCC was operating at a slip of 0.05 and at a synchronous speed of 3,000 rpm in the set current phase for the eight- and four-pole modes. Fig. 4(a) shows that the magnetic-flux density distributes to the eight poles when the current phases are set based on the data given in Table I. With respect to the four-pole mode, if the current phases change, then the magnetic-flux density distributes to the four poles, as shown in Fig. 4 (b).

Table II: Specifications of IMCC.

Items	Motor controlled by three 3-phase inverters
Phase	9
Number of poles in pole change	8/4
Number of slots	18
Stator outer diameter (mm)	123
Rotor outer diameter (mm)	78.3
Number of rotor bars	44
Air gap length (mm)	0.3
Core length (mm)	47.5
Number of turns per coil	49
Rated current per winding (A)	6.26
Rated current density (A/mm^2)	5.00
Frequency at 3000 rpm (Hz)	200/100

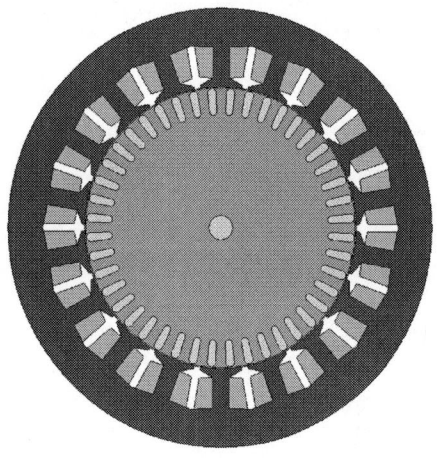

Fig. 3: Analytical model of the motor with three three-phase inverters.

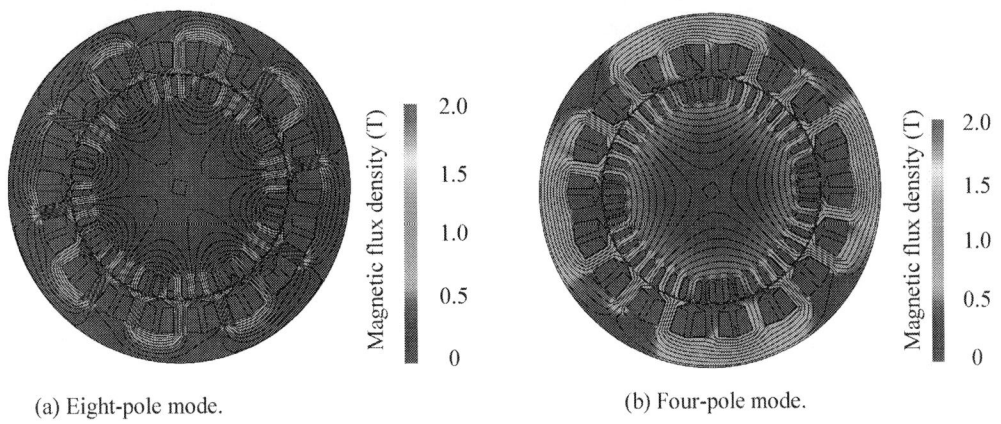

(a) Eight-pole mode.　　　　　　　　(b) Four-pole mode.

Fig. 4: Distribution of magnetic-flux density at a slip of 0.05 and a synchronous speed of 3,000 rpm.

Motor characteristics

In this section, we discuss the motor characteristics of each pole mode. Torque analysis was performed using FEM with respect to the slip parameter to obtain the motor performance for each pole mode. The conditions were set as follows: a rated current of 6.26 A and synchronous speeds of 3,000 and 6,000 rpm. Figs. 5(a) and (b) show the torque characteristics at different slips for the four- and eight-pole modes at the synchronous speeds of 3,000 and 6,000 rpm. For the eight-pole mode, the maximum torques were 1.72 and 1.75 times larger than those of the four-pole mode at the synchronous speeds of 3,000 and 6,000 rpm, respectively. The output power, losses, power factor, and efficiency were calculated using the magnetic-analysis results for evaluating the output performance. Table III and IV illustrate the motor characteristics at different slips and at the synchronous speed of 3,000 rpm for both modes. Here, the efficiencies of both modes were high: 90.0% for the eight-pole mode and 93.9% for the four-pole mode. In other words, the IMCC was highly efficient at producing large torques at the synchronous speed of 3,000 rpm. Similarly, Tables V and VI show the motor characteristics at different slips and at the synchronous speed of 6,000 rpm for both the modes. Here, the efficiencies for both the modes were again high: 94.8% for the eight-pole mode and 95.9% for the four-pole mode. In other words, the IMCC was highly efficient at producing high powers at the synchronous speed of 6,000 rpm. The efficiencies of the four-pole mode were higher at different slips because the iron losses were minimized at low frequencies. Also, the results imply that the iron loss varied according to the

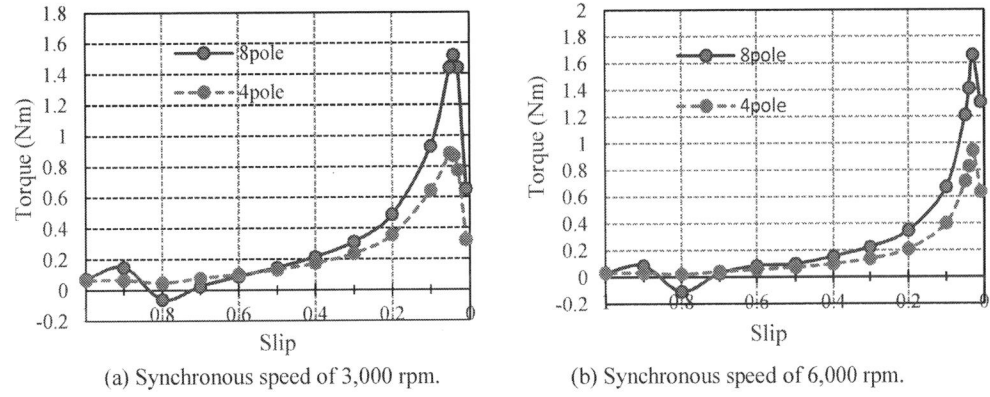

(a) Synchronous speed of 3,000 rpm.　　　　(b) Synchronous speed of 6,000 rpm.

Fig. 5: Torque characteristics of IMCC motor for eight- and four-pole modes at different slips.

torque. If the torque increases, the secondary current increases, and the magnetization current decreases simultaneously because a constant primary current is being used. For a synchronous speed of 6000 rpm, the primary voltage at the slip of 0.03 produced the maximum torque for the eight-pole mode; the torque for the eight-pole mode was 1.69 times higher than that for the four-pole mode. Therefore, if the motor operates in the four-pole mode under the condition of the same primary voltage for the eight-pole mode, it can operate at a speed 1.69 times higher than that for the eight-pole mode. Furthermore, in case of the four-pole mode, the motor for the four-pole mode operating at the same voltage of 14.38 V at the slip of 0.04 produced the maximum torque at 3000 rpm for the eight-pole mode the motor operating at 6875 rpm at a slip of 0.03 produced the maximum torque for 6000 rpm. Also, in this operating point, the power for the eight-pole mode was 458 W at the synchronous speed

Table III: Motor characteristics for eight-pole mode at synchronous speed of 3,000 rpm for the rated current.

Slip	Torque (Nm)	Primary voltage(V)	Primary current(A)	Secondary current(A)	Input (W)	Output (W)	Iron loss(W)	Copper loss(W)	Efficiency (%)	Power factor(%)
0.01	0.65	17.72	6.26	17.00	246.18	202.47	36.90	6.81	82.24	24.65
0.03	1.44	15.92	6.26	45.47	489.52	438.82	31.80	18.90	89.64	54.59
0.04	1.52	14.38	6.26	54.24	509.57	458.42	27.35	23.80	89.96	62.92
0.05	1.44	12.50	6.26	59.52	479.37	429.77	22.80	26.80	89.65	68.07
0.1	0.93	8.20	6.26	68.28	304.62	262.95	9.57	32.10	86.32	65.91
0.2	0.49	5.77	6.26	70.53	160.13	122.65	3.68	33.80	76.59	49.26

Table IV: Motor characteristics for four-pole mode at synchronous speed of 3,000 rpm for the rated current.

Slip	Torque (Nm)	Primary voltage(V)	Primary current(A)	Secondary current(A)	Input (W)	Output (W)	Iron loss(W)	Copper loss(W)	Efficiency (%)	Power factor(%)
0.01	0.322	9.06	6.26	9.62	119.08	100.15	17.70	1.23	84.10	23.34
0.03	0.773	8.46	6.26	21.40	254.04	235.56	13.40	5.08	92.73	53.29
0.04	0.869	7.69	6.26	27.50	280.18	262.08	11.44	6.66	93.54	64.64
0.05	0.885	7.02	6.26	31.19	281.39	264.13	9.60	7.66	93.87	71.17
0.1	0.643	4.39	6.26	37.68	196.16	181.80	4.94	9.42	92.68	79.31
0.2	0.356	2.20	6.26	39.96	102.87	89.47	3.45	9.95	86.97	83.00

Table V: Motor characteristics for eight-pole mode at synchronous speed of 6,000 rpm for the rated current.

Slip	Torque (Nm)	Primary voltage(V)	Primary current(A)	Secondary current(A)	Input (W)	Output (W)	Iron loss(W)	Copper loss(W)	Efficiency (%)	Power factor(%)
0.01	1.31	31.03	6.26	33.73	888.92	814.87	68.4	5.65	91.67	50.84
0.03	1.66	21.27	6.26	64.68	1067.72	1011.72	35.2	20.8	94.76	89.10
0.04	1.41	18.14	6.26	71.86	897.89	850.49	23.7	23.7	94.72	87.83
0.05	1.21	15.39	6.26	73.66	764.95	722.25	17.3	25.4	94.42	88.22
0.1	0.67	11.34	6.26	79.69	414.04	378.88	6.46	28.7	91.51	64.80
0.2	0.347	9.46	6.26	79.50	209.17	174.42	3.15	31.6	83.39	39.23

of 3000 rpm, and for the four-pole mode, the power was estimated to be 660 W at 6875 rpm. This shows that the pole-change techique was effective for achieving high power for operating at a wide speed range. In addition, the switching loss of the inverter decreased by approximately half of the switching loss of the eight-pole mode when the motor changed from the eight-pole mode to four-pole mode. Thus, the proposed system was capable of changing the pole, which led to an increase in the total efficiency of the motor drive system.

Table VI: Motor characteristics for four-pole mode at the synchronous speed of 6,000 rpm for the rated current.

Slip	Torque (Nm)	Primary voltage(V)	Primary current(A)	Secondary current(A)	Input (W)	Output (W)	Iron loss(W)	Copper loss(W)	Efficiency (%)	Power factor(%)
0.01	0.634	17.34	6.26	15.51	440.11	394.37	44	1.74	89.61	45.04
0.03	0.946	12.55	6.26	33.47	602.20	576.56	19.4	6.24	95.74	85.15
0.04	0.827	10.16	6.26	36.40	520.21	498.83	14.2	7.18	95.89	90.87
0.05	0.714	8.47	6.26	37.49	445.58	426.19	11.7	7.69	95.65	93.42
0.1	0.402	4.93	6.26	39.92	243.57	227.33	7.54	8.7	93.33	87.61
0.2	0.208	2.59	6.26	40.55	119.67	104.55	5.68	9.44	87.37	81.95

Transient characteristics of changing poles

Here, we discuss the transient motor characteristics during pole changing. After operating in the eight-pole mode for 0.5 s, the motor changed to the four-pole mode. Following this pattern, we performed magnetic field analysis with the rated current of 6.26 A and a slip of 0.05. Figs. 6, 7, and 8 show the torque, secondary current, and voltage waveform, respectively, of the motor at the synchronous speed of 3000 rpm. Clearly, the motor generates a torque when the current phase of the inverters changes from the eight-pole mode to the four-pole mode. The torque ripple was 48% for the eight-pole mode, which was the result of secondary-current harmonics (there were 8th harmonics of the secondary current). The torque ripple was less for the four-pole mode. Figs. 9. 10, and 11 show the torque, secondary current, and voltage waveform of the motor, respectively, at the rotational speed of 6000 rpm. The rise time to reach the torque of the four poles after changing from the eight-pole mode to four-pole mode at the synchronous speed of 6000 rpm was approximately half the rise time at 3000 rpm.

Conclusion

In this paper, a novel motor system was proposed with three winding groups controlled by three three-phase inverters that were capable of changing the poles and phases for variable-speed drives at high efficiencies. The FEM was performed using the JMAG software to predict the pole changes under specific operating conditions; the results confirmed the proposed motor's ability to change poles and phases. In particular, for the eight-pole modes at 3,000 and 6,000 rpm, the IMCC produced a large torque at the efficiencies of 90.0% and 94.8%, respectively. For the four-pole modes at 3,000 and 6,000 rpm, the efficiencies were 93.9% and 95.9%, respectively, and the primary voltage was approximately half that for the eight-pole mode. Therefore, the IMCC produced a high torque with high efficiency for the eight-pole mode in the low-speed region and produced high power with high efficiency for the four-pole mode in the high-speed region. This means that the pole-change technique was effective for achieving high power with high efficiency in a wide speed range.

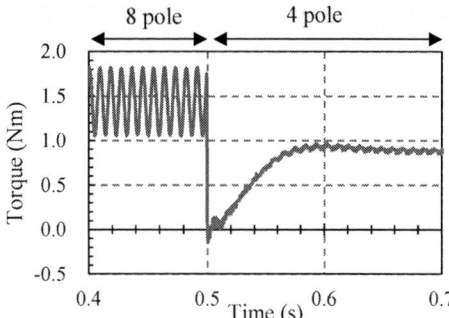

Fig. 6: Transient characteristics of the torque during pole changing at synchronous speed of 3000 rpm.

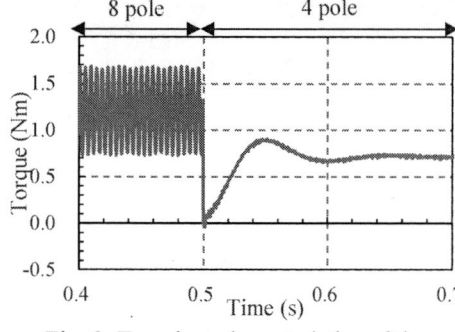

Fig. 9: Transient characteristics of the torque during pole changing at synchronous speed of 6000 rpm.

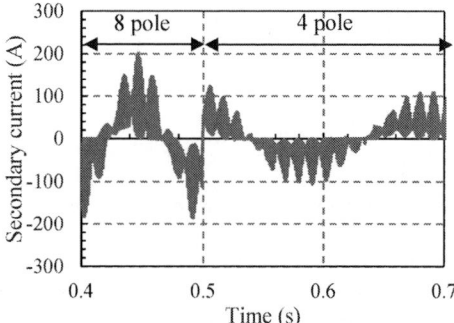

Fig. 7: Transient characteristics of the secondary current during pole changing at synchronous speed of 3000 rpm.

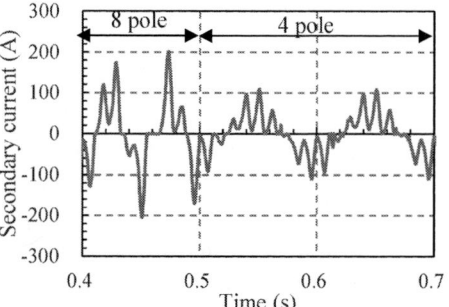

Fig. 10: Transient characteristics of the secondary current during pole changing at synchronous speed of 6000 rpm.

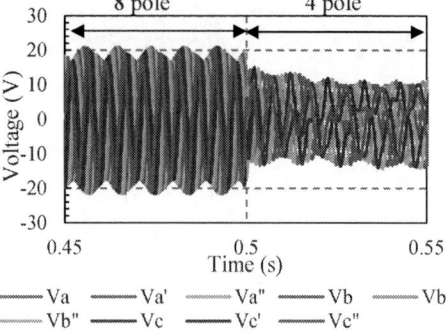

Fig. 8: Transient characteristics of the primary voltage during pole changing at synchronous speed of 3000 rpm.

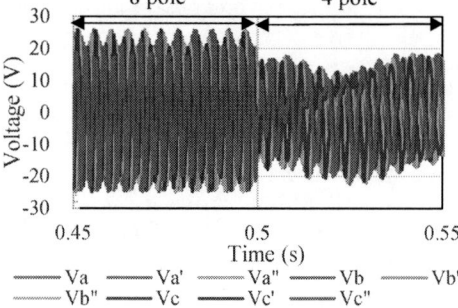

Fig. 11: Transient characteristics of the primary voltage during pole changing at synchronous speed of 6000 rpm.

References

[1] Ostovic V.: Memory Motors, IEEE Industry Applications Magazine, Vol. 9, pp. 52-61

[2] Sakai K., Yuki K., Hashiba Y., Takahashi N., Yasui K.: Principle of the variable-magnetic-force memory motor," in ICEMS 2009, paper LS6A-1

[3] Gagas B.S., Sasaki K., Fukushige T., Athavale A., Kato T., Lorenz R. D.: Analysis of magnetizing trajectories for variable flux pm synchronous machines considering voltage, high-speed capability, torque ripple, and time duration, IEEE Transactions Industry Applications, Vol. 52 no 5, pp. 4029-4038

[5] Eastham J. F., Laithwaite E. R.: Pole-change motors using phase-mixing techniques, Proceedings IEE power Engineering Vol. 109 no. 47, pp. 397-409

[6] Osama M., Lipo T. A.: Modeling and analysis of a wide-speed-range induction motor drive based on electronic pole changing, IEEE Transactions on Industry Applications, Vol. 33 no. 5, pp. 1177-1184.

[7] Mizuno T., Tsuboi K., Hirotsuka I., Suzuki S., Matsuda I., Kobayashi T.: Basic principle and maximum torque characteristics of a six-phase pole change induction motor for electric vehicle, Electrical Engineering in Japan, Vol. 118 no 3, pp. 78-81

[8] Osama M., Lipo T. A.: Experimental and finite-element analysis of an electronic pole-change drive ole-changing permanent-magnet machines, IEEE Transactions on Industry Applications, Vol. 36 no 6, pp. 1637-1644

[9] Ostovic, V.: Pole-changing permanent-magnet machines, IEEE Transactions on Industry Applications, Vol. 38 no 6, pp. 1493-1499

[10] Sakai K., Hashimoto H.: Permanent magnet motors to change poles and machine constants, in Proc. 20th International Conference on Electrical Machines 2012, pp. 433-438

[11] Sakai K., Yuzawa N.: Permanent magnet motors capable of pole changing and three-torque-production mode using magnetization, IEEJ Journal of Industry Applications 2013, Vol. 2 no. 6, pp. 269-275V

[12] Sakai K., Yuzawa N.: Permanent magnet motor capable of pole changing for high efficiency, in Proc. IEEE Energy Conversion Congress and Exposition 2013, paper 5311

[13] Okayasu M., Sakai K.: Novel integrated motor design that supports phase and pole changes using multiphase or single phase inverters, in Proc. 18th European Conf. on Power Electronics and Applications, EPE'16-ECCE Europe 2016, paper 135

[14] Li F., Chau K. T., Liu. C: Pole-changing flux-weakening dc-excited dual-memory machines for electric vehicles, IEEE Transactions on Energy Conversion, Vol. 31 no.1, pp. 27-35

[15] Sakai K., Okayasu M.: Integrated motor controlled by multi-inverter with pole-changing functionality and fault tolerance, in Proc. 19th European Conf. on Power Electronics and Applications, EPE'17-ECCE Europe 2017, paper 0057

[16] Hidaka Y., Komatsu T., Arita H.: A novel pole-changing method with a multiple three-phase inverter, in Proc. IPEC2018-ECCE-Asia 2018, pp. 2820-2825

[17] Yano H., Sakai K.: Integrated Motor-Controlled Independently by Multi-Inverters with Pole and Phase Changes, in Proc. 20th European Conf. on Power Electronics and Applications, EPE'19-ECCE Europe 2019, paper 0245

An Isolated Medium-Voltage AC-DC Converter Using Level-Shifted PWM Control of a Modular Matrix Converter

Kohei Budo and Takaharu Takeshita
Dept. of Electrical and Mechanical Engineering, Graduate School of Engineering,
Nagoya Institute of Technology
Gokiso, Showa, Nagoya, 466-8555 Japan
Phone: +81 (52) 735-5441
Fax: +81 (52) 735-5432
Email: take@nitech.ac.jp
URL: http://motion.elcom.nitech.ac.jp

Keywords

≪Battery charger≫, ≪Power converters for EV≫, ≪Multilevel converters≫, ≪High frequency power converter≫, ≪Insulation≫

Abstract

This paper presents level-shifted PWM control in a Modular Matrix Converter (MMxC) for an isolated medium-input AC-DC converter. The proposed control method can generate the multi-level output voltage in the MMxC even when the MMxC generates the high-frequency AC voltage. The effectiveness of this control method is verified by experiments.

Introduction

Electric vehicles (EVs) are attracting public attentions to solve environmental problems. The one of the problems for spread of EVs is the long charging time. Because the demand of the quick battery chargers for EVs is increasing, the battery charging time must be shortened. The high electric power must be charged to the battery to realize the quick charge. However, the conventional circuit of the isolated AC-DC converters [1]-[4] for the battery charging system converts a low input AC voltage to the DC voltage. When the conventional circuits convert the high electric power, a large current flows in the converters due to the low input voltage. In recent year, a Modular Matrix Converter (MMxC) [5]-[8] is attracting attentions, to be directly connected with the power distribution system. The MMxC can convert the medium input three-phase AC voltage to the three-phase or the single-phase AC voltage without the commercial transformers which convert the medium voltage to the low voltage. Because the MMxC does not need the commercial transformer, the overall system from the input medium voltage to the load is downsized.

The authors have proposed an isolated AC-DC converter for a quick battery charger using the MMxC which converts the three-phase AC voltage to the high-frequency single-phase AC voltage [9]. This proposed circuit can directly charge the electric power to the battery from the input medium voltage. The proposed circuit consists of the MMxC, a high-frequency transformer, and an H-bridge circuit. In the proposed circuit, because the MMxC output voltage is the high-frequency AC voltage, it is difficult to apply the phase-shifted PWM control [5]-[8] to the proposed circuit.

The authors propose level-shifted PWM control [10] in the MMxC which converts the three-phase AC voltage to the high-frequency single-phase AC voltage to generate the multi-level output voltage. The proposed control method can reduce the harmonics in the MMxC output voltage and the switching loss. The effectiveness is verified by the experiments using the 5-kW small power system.

EPE'20 ECCE Europe

Assigned jointly to the European Power Electronics and Drives Association & the Institute of Electrical and Electronics Engineers (IEEE)

Main Circuit Configuration and Theoretical Waveforms

Main Circuit Configuration

Fig. 1 shows a main circuit configuration of an isolated medium-voltage AC-DC converter using the MMxC. The source voltages e_{su}, e_{sv}, and e_{sw} are connected to the MMxC through inductors L_f. The inductors L_f suppress the harmonics in the source current i_{su}, i_{sv}, and i_{sw}. The MMxC consists of six arms, and the one arm has three modules in series. Each module is composed of an H-bridge and a capacitor C_m. Because each arm of the MMxC has the three modules in series, the MMxC has the capability of the high breakdown voltage. Therefore, the MMxC can be directly connected to the power distribution system. The primary side of the high-frequency transformer T_r is connected between the g-phase and the h-phase of the MMxC. The three arms connected to the terminal g are defined as sub-converter g and three arms connected to the terminal h are defined as sub-converter h.

The secondary side of the high-frequency transformer T_r is connected to the H-bridge circuit with four switches $S_{jp} - S_{kn}$. The H-bridge circuit with switches of $S_{jp} - S_{kn}$ is connected to the battery V_{dc}. The secondary voltage v_2 is generated by the switching of the H-bridge circuit with four switches $S_{jp} - S_{kn}$. The relationship between the voltage v_1 and v_2 is $v_1 = v_2$, because the turn ratio of the high-frequency transformer T_r is 1:1.

Theoretical Waveforms

From Fig. 1, the voltage equation of the sub-converter g in the MMxC is obtained as the following equation.

$$\begin{bmatrix} e_{su} \\ e_{sv} \\ e_{sw} \end{bmatrix} = L_f \frac{d}{dt} \begin{bmatrix} i_{su} \\ i_{sv} \\ i_{sw} \end{bmatrix} + L_b \frac{d}{dt} \begin{bmatrix} i_{ug} \\ i_{vg} \\ i_{wg} \end{bmatrix} + \begin{bmatrix} v_{ug} \\ v_{vg} \\ v_{wg} \end{bmatrix} + \frac{1}{2} v_2 \begin{bmatrix} 1 \\ 1 \\ 1 \end{bmatrix} . \tag{1}$$

Similarly, the voltage equation of the sub-converter h in the MMxC is obtained as the following equation.

$$\begin{bmatrix} e_{su} \\ e_{sv} \\ e_{sw} \end{bmatrix} = L_f \frac{d}{dt} \begin{bmatrix} i_{su} \\ i_{sv} \\ i_{sw} \end{bmatrix} + L_b \frac{d}{dt} \begin{bmatrix} i_{uh} \\ i_{vh} \\ i_{wh} \end{bmatrix} + \begin{bmatrix} v_{uh} \\ v_{vh} \\ v_{wh} \end{bmatrix} - \frac{1}{2} v_2 \begin{bmatrix} 1 \\ 1 \\ 1 \end{bmatrix} . \tag{2}$$

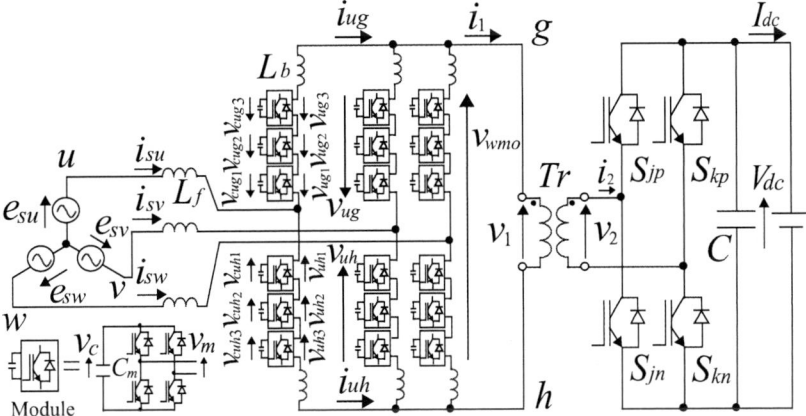

Fig. 1: Proposed circuit configuration.

The arm voltages $v_{ug} - v_{uh}$ and the arm currents $i_{ug} - i_{wh}$ are defined as the following equations.

$$\begin{bmatrix} v_{ug} & v_{uh} \\ v_{vg} & v_{vh} \\ v_{wg} & v_{wh} \end{bmatrix} = \begin{bmatrix} v_{su} & v_{su} \\ v_{sv} & v_{sv} \\ v_{sw} & v_{sw} \end{bmatrix} + \frac{1}{2} v_{mo} \begin{bmatrix} -1 & 1 \\ -1 & 1 \\ -1 & 1 \end{bmatrix}. \tag{3}$$

$$\begin{bmatrix} i_{ug} & i_{uh} \\ i_{vg} & i_{vh} \\ i_{wg} & i_{wh} \end{bmatrix} = \frac{1}{2} \begin{bmatrix} i_{su} & i_{su} \\ i_{sv} & i_{sv} \\ i_{sw} & i_{sw} \end{bmatrix} + \frac{1}{3} i_1 \begin{bmatrix} -1 & 1 \\ -1 & 1 \\ -1 & 1 \end{bmatrix}. \tag{4}$$

The voltages v_{su}, v_{sv} and v_{sw} in (3) are the commercial-frequency three-phase AC voltages and the positive-sequence voltages. The voltage v_{mo} in (3) is the MMxC output voltage which is the single-phase high-frequency AC voltage. The arm current $i_{ug} - i_{wh}$ consist of the half of the source current and the one third of the transformer current i_1. The equation of the source current is obtained by the following equations, adding the voltage equation of the sub-converter g in (1) to the sub-converter h in (2).

$$\begin{bmatrix} e_{su} \\ e_{sv} \\ e_{sw} \end{bmatrix} = \left(L_f + \frac{1}{2} L_b \right) \frac{d}{dt} \begin{bmatrix} i_{su} \\ i_{sv} \\ i_{sw} \end{bmatrix} + \begin{bmatrix} v_{su} \\ v_{sv} \\ v_{sw} \end{bmatrix}. \tag{5}$$

The source currents i_{su}, i_{sv}, and i_{sw} are controlled by the voltages v_{su}, v_{sv}, and v_{sw} in (5).

The equation of the transformer current i_1 is obtained by the following equation, subtracting the voltage equation of the sub-converter g in (1) from the sub-converter h in (2).

$$v_{mo} - v_2 = \frac{2}{3} L_b \frac{di_1}{dt}, \tag{6}$$

where the voltage v_{mo} and v_2 are the square-waveforms. The peak value of the voltage v_{mo} and v_2 is V_{dc} of the output DC voltage in Fig. 1. The phase-difference between v_{mo} and v_2 controls transformer current i_1. This control method is derived by the dual-active-bridge converter [11].

Fig. 2 shows the theoretical waveforms in the proposed circuit. The waveforms are the source voltage e_{su}, the arm voltages v_{ug} and v_{uh}, the commercial-frequency voltage v_{su}, the source current i_{su}, the MMxC output voltage v_{mo}, the secondary voltage v_2, the transformer current i_1. Because the commercial-frequency sinusoidal waveform AC voltage v_{su} in the arm voltages v_{ug} and v_{uh} controls the source current i_{su} in (5), the source current i_{su} can be controlled to the sinusoidal waveform. The transformer current i_1 is controlled by the phase-difference θ_d between the MMxC output voltage v_{mo} and the secondary voltage v_2 in (6). The transformer current i_1 of the trapezoidal waveform is obtained.

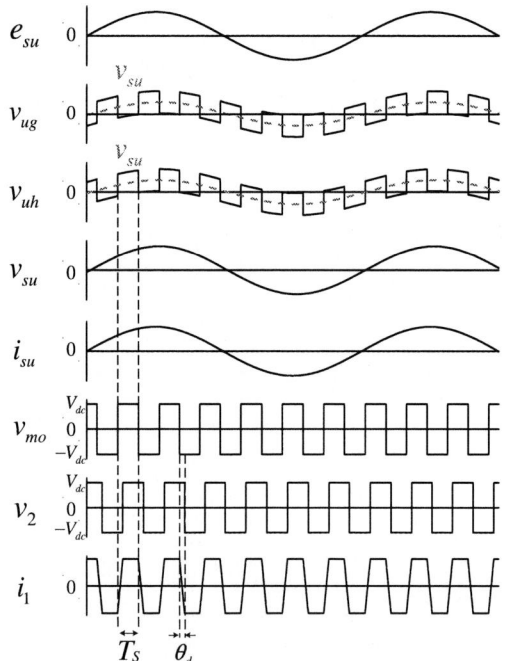

Fig. 2: Theoretical waveforms in proposed circuit.

Control Method

Fig. 3 shows an overall control block diagram of the proposed circuit. The control system consists of the source current control, the output DC current control, the capacitor voltage average control, and the level-shifted PWM control. The source currents i_{sd} and i_{sq} on the $d-q$ rotating coordinate are controlled by the arm voltage $v_{su} - v_{sw}$ in (3). The output DC current I_{dc} is controlled by the phase-difference θ_d between the MMxC output voltage v_{mo} and the secondary voltage v_2. The capacitor voltage v_c are controlled by the active power in the power source.

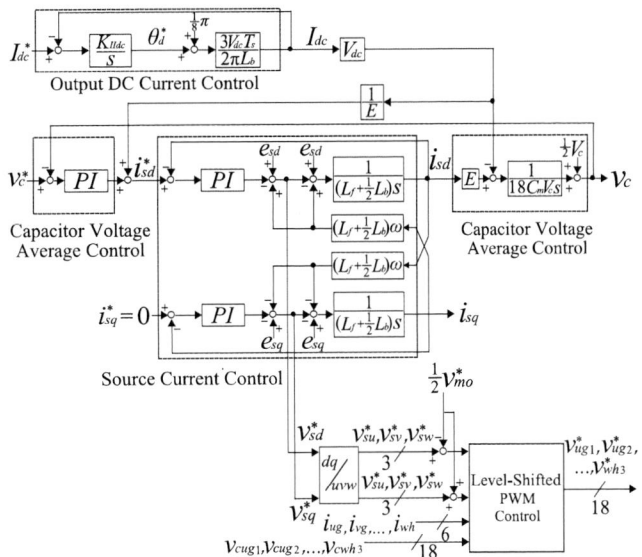

Fig. 3: Overall block diagram.

The source voltage e_{su}, e_{sv}, and e_{sw} are obtained by

$$\begin{bmatrix} e_{su} \\ e_{sv} \\ e_{sw} \end{bmatrix} = \sqrt{\frac{2}{3}} E \begin{bmatrix} \cos \theta_s \\ \cos (\theta_s - 2\pi/3) \\ \cos (\theta_s - 4\pi/3) \end{bmatrix} \qquad \theta_s = \omega t, \tag{7}$$

where E is the effective value of the source line voltage and ω is the source angular frequency. The source current i_{su}, i_{sv}, and i_{sw} are obtained by

$$\begin{bmatrix} i_{su} \\ i_{sv} \\ i_{sw} \end{bmatrix} = \sqrt{2} I_s \begin{bmatrix} \cos (\theta_s + \varphi_s) \\ \cos (\theta_s - 2\pi/3 + \varphi_s) \\ \cos (\theta_s - 4\pi/3 + \varphi_s) \end{bmatrix}, \tag{8}$$

where I_s is the effective value of the source current and φ_s is the power factor angle of the power source. The transformation matrix $\mathbf{C_{dq}}$ of the $d-q$ rotating coordinate is defined by

$$\mathbf{C_{dq}} = \sqrt{\frac{2}{3}} \begin{bmatrix} \cos \theta_s & \cos (\theta_s - 2\pi/3) & \cos (\theta_s - 4\pi/3) \\ -\sin \theta_s & -\sin (\theta_s - 2\pi/3) & -\sin (\theta_s - 4\pi/3) \end{bmatrix}. \tag{9}$$

The source voltage e_{sd} and e_{sq}, and the source current i_{sd} and i_{sq} on the $d-q$ rotating coordinate are obtained by the following equation.

$$\begin{bmatrix} e_{sd} \\ e_{sq} \end{bmatrix} = \mathbf{C_{dq}} \begin{bmatrix} e_{su} \\ e_{sv} \\ e_{sw} \end{bmatrix} = \begin{bmatrix} E \\ 0 \end{bmatrix}, \qquad \begin{bmatrix} i_{sd} \\ i_{sq} \end{bmatrix} = \mathbf{C_{dq}} \begin{bmatrix} i_{su} \\ i_{sv} \\ i_{sw} \end{bmatrix} = \sqrt{3} \mathbf{I_s} \begin{bmatrix} \cos \varphi_s \\ \sin \varphi_s \end{bmatrix}. \tag{10}$$

Capacitor Voltage Average Control

The capacitor voltage average control can control the average of the capacitor voltage v_c. The average of the capacitor voltage v_c in the MMxC is obtained by the following equation:

$$v_c = \frac{1}{18} \sum_{z=1}^{3} \left(v_{cugz} + v_{cuhz} + v_{cvgz} + v_{cvhz} + v_{cwgz} + v_{cwhz} \right). \tag{11}$$

The references i_{sd}^* and i_{sq}^* of the source current on the $d-q$ rotating coordinate are given as

$$i_{sd}^* = \left(K_{Pvc} + \frac{K_{Ivc}}{s} \right) (v_c^* - v_c) + \frac{V_{dc}I_{dc}}{E}, \qquad\qquad i_{sq}^* = 0, \qquad\qquad (12)$$

where K_{Pvc} is the proportional gain and K_{Ivc} is the integral gain. The first term on the right side in (12) is the PI control calculation for controlling the capacitor voltage v_c. The capacitor voltage average control uses the d-axis-current i_{sd}, because the average of the capacitor voltage v_c is controlled by the active power of the power source. The second term on the right side in (12) is the compensation of the output power $P_{out}(= V_{dc}I_{dc})$. The current reference of the q-axis-current i_{sq}^* is zero for the source power factor of 1.

Source Current Control

The voltage equation concerning the source current on the $d-q$ rotating coordinate are obtained the following equation by applying the d-q transformation to (5).

$$\begin{bmatrix} e_{sd} \\ 0 \end{bmatrix} = \begin{bmatrix} v_{sd} \\ v_{sq} \end{bmatrix} + \left(L_f + \frac{1}{2}L_b \right) \frac{d}{dt} \begin{bmatrix} i_{sd} \\ i_{sq} \end{bmatrix} + \omega \left(L_f + \frac{1}{2}L_b \right) \begin{bmatrix} 0 & -1 \\ 1 & 0 \end{bmatrix} \begin{bmatrix} i_{sq} \\ i_{sd} \end{bmatrix}, \qquad (13)$$

where

$$\begin{bmatrix} v_{sd} \\ v_{sq} \end{bmatrix} = \mathbf{C_{dq}} \begin{bmatrix} v_{su} \\ v_{sv} \\ v_{sw} \end{bmatrix}. \qquad\qquad (14)$$

The references v_{sd}^* and v_{sq}^* of the arm voltage on the $d-q$ rotating coordinate are given as

$$\begin{bmatrix} v_{sd}^* \\ v_{sq}^* \end{bmatrix} = - \left(K_{Pis} + \frac{K_{Iis}}{s} \right) \begin{bmatrix} i_{sd}^* - i_{sd} \\ i_{sq}^* - i_{sq} \end{bmatrix} + \begin{bmatrix} e_{sd} \\ 0 \end{bmatrix} + \omega \left(L_f + \frac{1}{2}L_b \right) \begin{bmatrix} 0 & -1 \\ 1 & 0 \end{bmatrix} \begin{bmatrix} i_{sq} \\ i_{sd} \end{bmatrix}, \qquad (15)$$

where K_{Pis} is the proportional gain and K_{Iis} is the integral gain. The reference of the arm voltage v_{sd}^* and v_{sq}^* in (15) consist of the PI control and the decoupling control between the d-axis and the q-axis. The first term on the right side in (15) shows the PI control calculation for controlling the source current i_{sd} and i_{sq} in (12). The third term on the right side in (15) works as the decoupling control between i_{sd} and i_{sq}.

Output DC Current Control

The output DC current I_{dc} can be controlled by the phase-difference θ_d between the voltage v_{mo} and the secondary voltage v_2 in (6). This output DC current control is derived by the operating principle of the dual active bridge converter [11]. The output DC current I_{dc} is obtained by the following equation [9].

$$I_{dc} = \frac{3V_{dc}T_s}{L_b} \left\{ \frac{1}{4} - \left(\frac{\theta_d}{\pi} - \frac{1}{2} \right)^2 \right\}, \qquad\qquad (16)$$

where T_s is the half period of the MMxC output voltage v_{mo} and the secondary voltage v_2. Because the phase-difference θ_d is controlled by the range of $0 \leq \theta_d \leq \pi/2$, the output DC current I_{dc} performs the approximation by performing Taylor expansion around $\theta_d = \frac{\pi}{4}$. The output DC current I_{dc} is obtained by the following equation.

$$I_{dc} = \frac{3V_{dc}T_s}{2\pi L_b} \left(\theta_d + \frac{1}{8}\pi \right). \qquad\qquad (17)$$

The output DC current I_{dc} in (17) can be controlled by the phase-difference θ_d between the voltage v_{mo} and the secondary voltage v_2. The reference θ_d^* of the phase-difference is given as

$$\theta_d^* = \frac{K_{IIdc}}{s}\left(I_{dc}^* - I_{dc}\right), \tag{18}$$

where K_{IIdc} is the integral gain. The reference θ_d^* of the phase-difference is generated by the I control of the output DC current I_{dc}.

Level-Shifted PWM Control

Phase-shifted PWM control [5]-[8] in the MMxC is widely utilized to generate the multi-level output voltage in the arm of the MMxC. When the frequency of the MMxC output voltage is lower than the switching frequency, phase-shifted PWM control can be applied to the MMxC. However, the frequency of the MMxC output AC voltage in the proposed circuit is the same as the switching-frequency f_s. Therefore, phase-shifted PWM control can not be applied to the proposed circuit. Fig. 4 shows the difference between phase-shifted PWM control and level-shifted PWM control [10] in the proposed circuit. Fig. 4(a) shows phase-shifted PWM control, and Fig. 4(b) shows level-shifted PWM control. The red line is the voltage reference, and the black line is the output voltage. The theory of phase-shifted PWM control gives the phase-difference of $2\pi/3$ among the carrier signals for the three modules in the arm. However, the phase-difference is smaller than $2\pi/3$ in the Fig. 4(a), because the arm voltage v_{ug} must generate the voltage reference v_{ug}^*. Therefore, phase-shifted PWM control has more harmonics in the arm output voltage v_{ug} than level-shifted PWM control in Fig. 4.

The authors derive the theory of level-shifted PWM control [10] in the MMxC which converts the three-phase AC voltage to the high-frequency single-phase AC voltage to generate the multi-level output voltage in the arm of the MMxC. The control method for the arm of the MMxC between the u-phase and the g-phase is derived. For simplicity of explanation, the module capacitor voltages v_{cug1}, v_{cug2}, and v_{cug3} are balanced in Fig. 1, and then the module capacitor voltages v_{cug1}, v_{cug2}, and v_{cug3} are given as

$$v_{cug1} = v_{cug2} = v_{cug3} = v_c, \tag{19}$$

where the voltage v_c is the average of all module capacitor voltage in the MMxC. Each module in the MMxC can generate the three-level output voltage of $\pm v_c$ and 0 using the module capacitor voltage v_c. Level-shifted PWM control adjusts the parameter k. The parameter k is defined as the number of the modules which generate the voltage $+v_c$ or $-v_c$. For example, when the arm voltage reference v_{ug}^* is the positive value in Fig. 4(b), the module output voltage reference $v_{ug1}^* - v_{ug3}^*$ and the parameter k are obtained by the following equations.

$$\begin{cases} v_{ug1}^* = v_c \\ v_{ug2}^* = v_{ug}^* - v_c \\ v_{ug3}^* = 0 \\ k = 2. \end{cases} \tag{20}$$

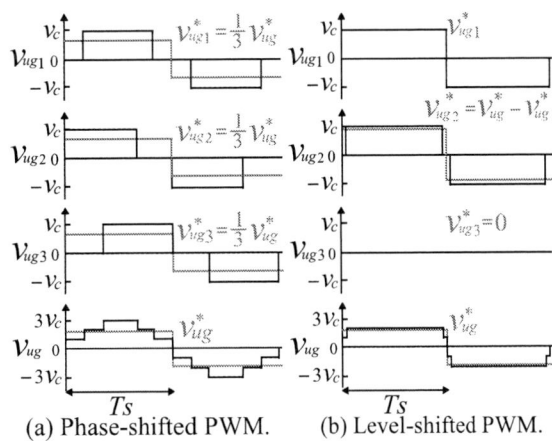

(a) Phase-shifted PWM. (b) Level-shifted PWM.

Fig. 4: The diffrence between phase-shifted PWM control and level-shifted PWM control in the proposed circuit.

Because the number of the modules which generate the voltage v_c is two, the parameter k is two. The voltage level of the arm voltage v_{ug} is v_c and $2v_c$ in the case of (20).

Under the conditions of the arm voltage reference v_{ug}^* range of $(k-1)v_c \leq v_{ug}^* \leq kv_c$, the module output voltage references v_{ug1}^*, v_{ug2}^*, and v_{ug3}^* are given as

$$
\begin{cases}
v_{ug1}^* = \cdots = v_{ug(k-1)}^* = v_c \\
v_{ugk}^* = v_{ug}^* - (k-1)v_c \\
v_{ug(k+1)}^* = \cdots = v_{ug3}^* = 0.
\end{cases}
\tag{21}
$$

In (21), $(k-1)$ modules generate the positive voltage v_c, one module generates the voltage v_{ugk}^*, and $(3-k)$ modules generate the zero voltage. In this case, the arm output voltage v_{ug} can generate the voltage-levels of $(k-1)v_c$ and kv_c. Under the condition of the arm voltage reference v_{ug}^* range of $-kv_c \leq v_{ug}^* \leq -(k-1)v_c$, the module output voltage references v_{ug1}^*, v_{ug2}^*, and v_{ug3}^* are given as

$$
\begin{cases}
v_{ug1}^* = \cdots = v_{ug(k-1)}^* = -v_c \\
v_{ugk}^* = v_{ug}^* + (k-1)v_c \\
v_{ug(k+1)}^* = \cdots = v_{ug3}^* = 0.
\end{cases}
\tag{22}
$$

In (22), $(k-1)$ modules generate the negative voltage $-v_c$, one module generates the voltage v_{ugk}^*, and $(3-k)$ modules generate the zero voltage. In this case, the arm output voltage v_{ug} can generate the voltage-levels of $-(k-1)v_c$ or $-kv_c$. In the proposed control method, the number k of the modules which generate the voltages $+v_c$ and $-v_c$ are controlled because of generating the seven voltage-level of $0, \pm v_c, \pm 2v_c$, and $\pm 3v_c$ in the arm of the MMxC. In addition, when the module voltage reference is zero voltage, these modules have no switching. Therefore, the switching loss can be reduced in the MMxC.

Fig. 5 shows the theoretical voltage waveforms of the arm voltage v_{ug} and the arm voltage reference v_{ug}^*, and the module output voltages $v_{ug1} - v_{ug3}$ and the module voltage references $v_{ug1}^* - v_{ug3}^*$ in Fig. 1. The arm output voltage reference v_{ug}^* is obtained by adding the commercial-frequency AC voltage v_{su}^* and the MMxC output voltage v_{mo}. The arm voltage v_{ug} level shifts because of controlling the number k of the modules which generate the voltages $+v_c$ or $-v_c$, according to the arm voltage reference v_{ug}^*. Therefore, the arm output voltage has the seven levels of $0, \pm v_c, \pm 2v_c$, and $\pm 3v_c$. In conventional control method[9], all of the switches of the three modules in the arm switch in every output voltage period. However, in the proposed control method, when the parameter k is not three, the no-switching modules exist. Therefore, the switching loss can be reduced compared to the conventional control method [9].

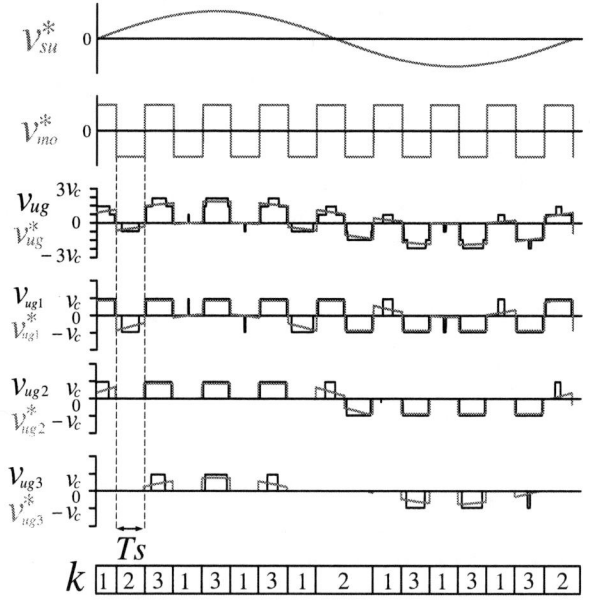

Fig. 5: Theoretical waveforms in level-shifted PWM control.

Table I: The module output voltage references in the condition of $v_{cug3} \leq v_{cug2} \leq v_{cug1}$.

v_{ug}^* state	v_{ug1}^*	v_{ug2}^*	v_{ug3}^*
$v_{ug}^* \leq -2v_c$	$-v_c$	$-v_c$	$v_{ug}^* + 2v_c$
$-2v_c \leq v_{ug}^* \leq -v_c$	$-v_c$	$v_{ug}^* + v_c$	0
$-v_c \leq v_{ug}^* \leq 0$	v_{ug}^*	0	0
$0 \leq v_{ug}^* \leq v_c$	0	0	v_{ug}^*
$v_c \leq v_{ug}^* \leq 2v_c$	0	$v_{ug}^* - v_c$	v_c
$2v_c \leq v_{ug}^*$	$v_{ug}^* - 2v_c$	v_c	v_c

Capacitor Voltage Balancing Control in The Arm

When the module capacitor voltages v_{cug1}, v_{cug2} and v_{cug3} are unbalanced in Fig. 1, the imbalance of the capacitor voltages must be compensated. The module capacitor voltage v_{cugz} is given as

$$
\begin{cases}
v_{cugz} = v_c + \dfrac{1}{3C_m v_c} \displaystyle\int p_{ugz} dt \\
z = \{1, 2, 3\},
\end{cases}
\tag{23}
$$

where p_{ugz} is the instantaneous electric power of the modules in the arm $u - g$ in Fig. 1. The instantaneous electric power p_{ugz} are given using the module output voltage v_{ugz} and the arm current i_{ug} in Fig. 1 as follows:

$$
\begin{cases}
p_{ugz} = v_{ugz}^* i_{ug} \\
z = \{1, 2, 3\}.
\end{cases}
\tag{24}
$$

The module capacitor voltages v_{cug1}, v_{cug2}, and v_{cug3} are controlled by the instantaneous electric power p_{ugz} in (23). Because the three modules in the one arm are given to the difference voltage reference $v_{ug1}^* - v_{ug3}^*$ using the level-shifted PWM control in (21) and (22), the instantaneous electric power p_{ugk} is imbalance. Therefore, the capacitor voltages v_{cug1}, v_{cug2}, and v_{cug3} are unbalanced in (23). The positive instantaneous electric power p_{ugz} charges the module capacitors. On the other hand, the negative instantaneous electric power p_{ugz} discharge the module capacitors. The module which generates the voltage of $+v_c$ or $-v_c$ must be decided on the basis of the module capacitor voltage conditions and the sign of the instantaneous electric power p_{ugz} to achieve the capacitor voltage balance in the arm.

For example, Table I shows the module voltage references v_{ug1}^*, v_{ug2}^*, and v_{ug3}^* when the state of the module capacitor voltage is $v_{cug3} \leq v_{cug2} \leq v_{cug1}$ and the arm current i_{ug} is positive. When the arm voltage reference v_{ug}^* is positive, the module capacitors are charged, and when the arm voltage reference v_{ug}^* is negative, the module capacitors are discharged in Table I. For instance, when the state of the arm voltage reference v_{ug}^* is $v_c \leq v_{ug}^* \leq 2v_c$, the module capacitors are charged. Because the minimum capacitor voltage v_{cug3} should be charged the highest electric power of the three modules, the maximum voltage reference $v_{ug3}^* (= v_c)$ is given to the module 3. The module voltage reference $v_{ug2}^* (= v_{ug}^* - v_c \leq v_{ug3}^*)$ is given to the module 2 because the module capacitor v_{cug2} is charged less than the module 3. The module voltage reference $v_{ug1}^* (= 0)$ is given to the module 1 because the maximum module capacitor voltage v_{cug1} doesn't have to be charged.

Experimental Results

Table II and Fig. 6 shows the experimental conditions and the experimental systems in the small power system. The effective value of the source line voltage E is 200 V and the frequency of the source voltage is 60 Hz. The output power P_{out} is 5 kW. Each arm of the MMxC consists of the three modules in series and the module capacitor voltage reference v_c^* is 125 V. The frequency of the MMxC output voltage, the voltage and the current of the high-frequency transformer is 7.5 kHz.

Table II: Experimental conditions in the small power systems.

Source voltage E, ω	200V , $2\,\pi\times$ 60 rad/s
Power factor angle φ^*	0 rad
Inductors L_f, L_b	1.0 mH , 0.4 mH
Number of series modules n	3
Module capacitors C_m	1200 μF
Module capacitor voltages v_c^*	125 V
Turn ratio of transformer a	1
Frequency of transformer f_s	7.5 kHz
Output capacitor C	1200 μF
Output DC voltage V_{dc}	300 V
Output power P_{out}	5 kW

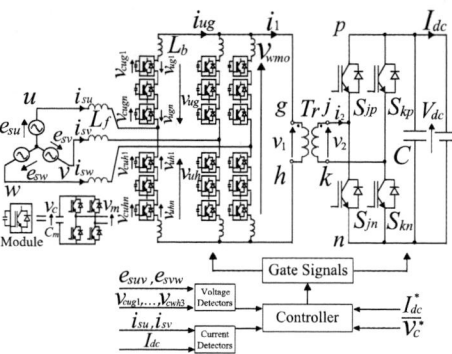

Fig. 6: Experimental system configuration.

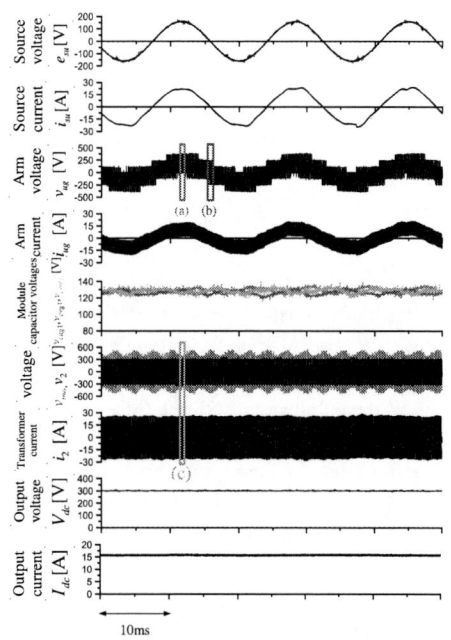

Fig. 7: Experimental waveforms.

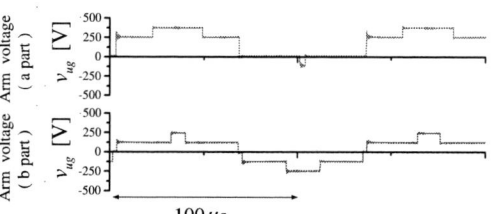

Fig. 8: Magnification waveforms of the arm voltae v_{ug}.

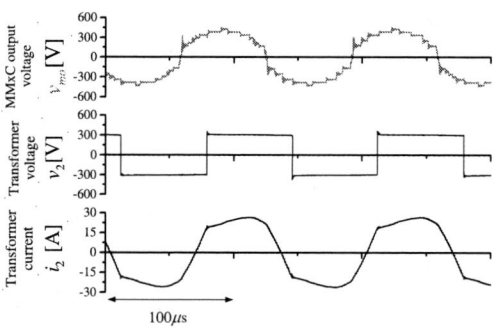

Fig. 9: Magnification waveforms.

Fig. 7 shows the experimental results under the conditions of Table II. The sinusoidal source current i_{su} is obtained by the source current control. The waveforms of the module capacitor voltages v_{cug1}, v_{cvg1}, and v_{cwg1} are kept to the module capacitor voltage reference v_c^*=125 V by the capacitor voltage average control. The output power P_{out} of 5 kW is generated because of the waveforms of $V_{dc} = 300$ V and $I_{dc} = 16.7$ A by the output current control.

Fig. 8 shows the magnification waveforms v_{ug} of (a) and (b) parts in Fig. 7. The waveform of the (a) part in Fig. 7 is obtained when the voltage v_{su}^* is the neighborhood of a maximum value in Fig. 5. The arm voltage v_{ug} can generate the voltage level of the $-v_c$, $+2v_c$, and $+3v_c$ at (a) part. The waveform of the (b) part is obtained when the voltage v_{su}^* is the zero voltage in Fig. 5. The arm voltage v_{ug} can generate the voltage level of the $\pm v_c$ and $\pm 2v_c$ at (b) part. Therefore the arm voltage v_{ug} can generate the seven voltage-level of 0, $\pm v_c$, $\pm 2v_c$, and $\pm 3v_c$ by the level-shifted PWM control.

EPE'20 ECCE Europe

Assigned jointly to the European Power Electronics and Drives Association & the Institute of Electrical and Electronics Engineers (IEEE)

Fig. 9 shows the magnification waveforms of the (c) part in Fig. 7. The waveforms show the MMxC output voltage v_{mo}, the high-frequency transformer voltage v_2, and the high-frequency transformer current i_2. The frequency of these waveforms is 7.5 kHz. The transformer current i_2 of the trapezoidal waveforms is obtained by the control of the phase-difference between the MMxC output voltage v_{mo} and the secondary voltage v_2.

Conclusions

This paper has presented level-shifted PWM control in the MMxC which converts the three-phase AC voltage to the high-frequency single-phase AC voltage to generate the multi-level output voltage. The effectiveness of the proposed control method has been verified by the experiments using the 5-kW small power system.

References

[1] K. Suzuki, W. Kitagawa, and T. Takeshita, " Soft-Switching Three-Phase AC to DC Converter Isolated By High-Frequency Transformer, " *18th European Conference on Power Electronics and Applications (EPE) 2016*, pp. 1-10, Sept. 2016.

[2] Chushan Li, Yulin Zhong, and David Xu, "Soft-switching three-phase matrix based isolated AC-DC converter for DC distribution system," *2015 IEEE Energy Conversion Congress and Exposition (ECCE)*, pp. 6755-6761, Sep. 2015.

[3] J. J. Sandoval, S. Essakiappan, and P. Enjeti, " A bidirectional series resonant matrix converter topology for electric vehicle dc fast charging, " *in 2015 IEEE Applied Power Electronics Conference and Exposition (APEC)*, pp. 3109-3116, March 2015.

[4] Fanxiu Fang, and Yun Wei Li, "Modulation and control method for bidirectional isolated AC/DC matrix based converter in hybrid AC/DC microgrid," *2017 IEEE Energy Conversion Congress and Exposition (ECCE)*, pp. 37-43, Oct. 2017.

[5] Y. Yamada and T. Takeshita, "Distribution and Balancing Control of Capacitor Voltages among Arms of a Modular Matrix Converter, " *in Con,Rec. International Power Electronics and Motion Control Conference-ECCE Asia (IPEMC 2016-ECCE Asia)*, pp.1028-1035 (2016)

[6] W. Kawamura and H. Akagi, "Control of the Modular Multilevel Cascade Converter Based on Triple-Star Bridge-Cells (MMCC-TSBC) for Motor Drives, " *2012 IEEE Energy Conversion Congress and Exposition (ECCE)*, pp. 3506-3513 (2012)

[7] Boran Fan, Kui Wang, Pat Wheeler, Chunyang Gu, Yongdong Li, "An Optimal Full Frequency Control Strategy for the Modular Multilevel Matrix Converter Based on Predictive Control, " *IEEE Transactions on Power Electronics, Volume: 33, Issue: 8*, pp. 6608-6621, Aug. 2018

[8] N. Niimura and H. Akagi, "Decoupled Control of a Three-Phase Modular Multilevel Cascade Converter Based on Double-Star Chopper-Cells , " *IEEJ Trans.IA*, Vol. 132, No11, pp. 1050-1064 (2012)

[9] K. Suzuki and T. Takeshita, "AC/DC Converter Using Modular Matrix Converter for Quick Battery Charger, " *10th International Conference on Power Electronics-ECCE Asia (ICPE 2019)*, pp. 1623-1630 (2019)

[10] Paul Sochor, H. Akagi, "Theoretical and Experimental Comparison Between Phase-Shifted PWM and Level-Shifted PWM in a Modular Multilevel SDBC Inverter for Utility-Scale Photovoltaic Applications, " *IEEE Transactions on Industry Applications*, Vol. 53, No. 5, pp. 4695-4707 (2017)

[11] S. Inoue and H. Akagi: "A Bi-Directional DC/DC Converter for an Energy Storage System", *IEEE Trans. Power Electron.*, Vol. 22, No. 6, pp. 2299-2306 (2007)

Detailed Simulation Model of an Asymmetrical Half-Bridge PWM Converter with Synchronous Rectification including Parasitic Elements

Benedikt Kohlhepp, Valentin Zeller, Markus Barwig, Thomas Dürbaum
Electromagnetic Fields, Friedrich-Alexander University Erlangen-Nürnberg (FAU)
Cauerstraße 7
91058, Erlangen
Tel.: +49 (0)9131 85 28951
Fax: +49 (0)9131 85 27787
E-Mail: benedikt.kohlhepp@fau.de
URL: http://emf.eei.uni-erlangen.de/

Keywords

«ZVS converters», «Gallium Nitride (GaN)», «Simulation», «Soft switching», «Switched-mode power supply».

Abstract

The asymmetrical half-bridge converter implemented with GaN-switches represents a great candidate for achieving high power density and high efficiency. These requirements call for the implementation of a synchronous rectifier on the secondary side to push the efficiency. Furthermore, higher switching frequencies lead to smaller passive components needed for compact power supplies. In order to gain accurate and reliable simulation results for these high switching frequencies, this paper presents a detailed simulation model featuring parasitics of all semiconductors and transformer capacitances. Besides current and voltage waveforms of one switching cycle, this model allows for studying ZVS transitions as well. Measurements on a practical test setup deliver waveforms very close to the simulation results and thus prove the validity of the simulation model.

Introduction

In order to achieve high efficiency and power density, which are typical customer demands in case of power electronic converters, an optimization procedure is needed. Besides this, high cost pressure linked to consumer electronics also calls for a reliable optimization routine. To be able to carry out an optimization, a suitable simulation model is necessary. Furthermore, soft switching techniques like zero voltage switching (ZVS) are indispensable for high efficiency. Additionally, wide bandgap semiconductors like GaN featuring very low parasitic properties and low on-state resistance must be used to achieve high efficiency. This paper analyzes the asymmetrical half-bridge PWM converter as it is a promising candidate for this need [1] [2] [3] [4] [5]. Besides ZVS, synchronous rectification is mandatory for high efficient operation for low output voltage applications [6] [7] [8]. [9] introduces a simple simulation model for the asymmetrical half-bridge converter. This model allows to gain basic waveforms for loss predictions within the converter. However, [10] shows that this relatively simple simulation model is not suitable for accurate loss predictions when implementing synchronous rectification. Parasitic oscillations within practical test setups demonstrate that besides inaccurate loss predictions also ZVS estimations can be hampered when using the simple simulation model. Furthermore, the simple simulation model is no longer suitable for high switching frequencies. As the ZVS transitions take a considerable amount of the switching period's time, the assumptions for the simple model are no longer valid. Therefore, an appropriate simulation model for gaining reliable results during optimization is needed. The proposed simulation model features, besides inductances, also parasitic capacitances and resistances of the transformer. Furthermore, the synchronous rectifier model is more detailed in order to get accurate loss predictions. The semiconductor's parasitics are considered as well. This allows for gaining reliable ZVS predictions. Moreover, the reverse conduction characteristic of the GaN-high electron mobility transistors (HEMTs) must be modeled as even short reverse conduction phases impact efficiency.

EPE'20 ECCE Europe

Assigned jointly to the European Power Electronics and Drives Association & the Institute of Electrical and Electronics Engineers (IEEE)

Converter model and assumptions

The most important step for simulating power electronic circuits is the creation of an appropriate converter model. Fig. 1 (left) depicts the circuit diagram of the asymmetrical half-bridge converter with synchronous rectification including the most important components. The half-bridge on the input side consists of the two GaN-HEMTs S_1 and S_2. Additionally, a series capacitance C_s and a transformer T are provided. Since the transformer's leakage inductance L_s participates in the functionality of the asymmetrical half-bridge converter, it is additionally placed in the circuit diagram besides the magnetizing inductance L_m and the ideal transformer with the effective turns ratio n. The secondary side switch S_3 is also a GaN-based transistor and used for synchronous rectification. At the output there is a capacitor C_{out} and the resistor R_L representing the load. Based on this, a simplified simulation model is developed in [9] which is utilized to get the converter's characteristic waveforms. Fig. 1 (right) shows this simplified converter model. It consists of the series capacitor C_s and a square wave voltage source which models the half-bridge. The model neglects the output capacitor and its parasitic effects. As many applications require a constant output voltage, a sufficiently large capacitor at the output must be used. This ensures a negligibly small voltage ripple and therefore the output capacitor can be modeled as a constant voltage source U_{out}. The transformer is modeled by a magnetizing inductance L_m and a leakage inductance L_s. The circuit diagram includes a resistor on the primary side, which represents the on-resistance of the half-bridge switches and the transformer primary ohmic losses. The resistor R_s' considers the ohmic losses on the secondary side. The apostrophe within the symbols of the simulation model indicates that the secondary side components are already transformed to the primary side, according to the effective turns ratio of the transformer n. With that simulation model, it is not possible to study the waveforms during the dead times and thus the ZVS transitions as the half-bridge is modeled as a rectangular voltage source neglecting the half-bridge transitions. The ZVS prediction is only based on the charge carried by the current i_{Lm}. Because of parasitic oscillations in a practical test setup, this can result in wrong loss estimations as the waveforms within the circuit differ from those gained by simulation using the simple model [10]. Additionally, with increasing switching frequency the dead times take a larger share of the total switching period which results in deviations between the simplified model and reality. For that reason, dead times can no longer be neglected in the simulation model.

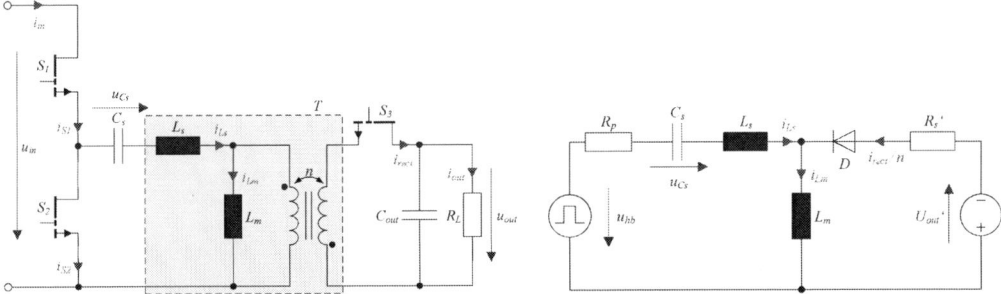

Fig. 1: Circuit diagram of the proposed asymmetrical half-bridge PWM converter (left) and the simplified simulation model [9] (right).

Thus, a new simulation model is needed which considers ZVS transitions and therefore the device's parasitics. In the following sections, the GaN-switches, the synchronous rectifier and the transformer will be examined in detail to expand the existing simulation model in order to gain realistic waveforms by simulation. In the last section, all parts are combined to form the entire simulation model.

GaN-Switch equivalent circuit diagram

According to the usual procedure, the GaN-HEMTs are modeled by an ideal switch and a series resistor $R_{ds,on}$ (Fig. 2, path in the middle). Considering the conduction losses of the GaN-HEMTs by simply inserting a resistor is valid, as [11] shows that the dynamic on-state resistance is negligible for the applied devices. Taking the waveforms during dead times of the half-bridge into account also requires considering the reverse conduction characteristic of the GaN-HEMTs and the output capacitance. During reverse conduction, these devices show a behavior similar to a diode. Therefore, reverse conduction is represented by a piecewise linear approximation. The corresponding simulation model consists of an ideal Diode D, a voltage source $U_{D,0}$ and a series resistor R_D representing the forward characteristics of the diode like behavior (Fig. 2, path on the left). In contrast to body-diodes, the reverse conduction of GaN-HEMTs shows a notably higher forward voltage drop $U_{D,0}$. This leads to higher power dissipation during reverse conduction in contrast to e. g. conventional silicon MOSFETs. The aforementioned issue expresses the importance of extending the switches equivalent circuit diagram. Fortunately, GaN-HEMTs do not exhibit reverse recovery, which simplifies reverse conduction modeling. The third path consists of the output capacitance C_{oss}. Normally, this capacitance shows a nonlinear behavior depending on its voltage. The voltage-dependent characteristic of a 100 V GaN-HEMT's output capacitance C_{oss} is the blue line shown in Fig. 2 (right).

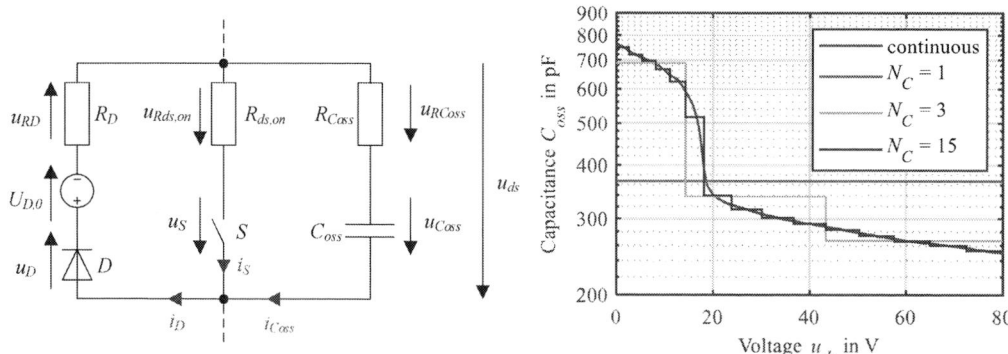

Fig. 2: Equivalent circuit diagram of the switches (left) and the continuous [12] and stepped curve of the 100 V GaN-HEMT's output capacitance (right).

This nonlinear behavior significantly impacts current and voltage waveforms during dead time and thus ZVS transitions. Especially, the waveforms during dead times may differ considerably from the assumption of linear capacitances. Nevertheless, simulation models often neglect the nonlinearity because of the huge effort for the implementation. Conversion to a step like function enables circuit simulation based on time discrete simulation to consider the nonlinear output capacitances. Therefore, the voltage-dependent curve of the output capacitance is approximated by piecewise constant capacitances. The determination of the stepped approximation should lead to the same amount of charge Q_{tot} compared to the exact solution [12]:

$$Q_{tot} = \int_0^{U_{ds,max}} C_{oss}(u) \, \mathrm{d}u \tag{1}$$

This charge is split into several equal partial charges Q_{st} according to the given number of steps N_{st}:

$$Q_{st} = \frac{Q_{tot}}{N_{st}} \tag{2}$$

In order to calculate the capacitance valid in each interval i, the partial charge Q_{st} must be divided by difference of the voltage boundaries of the corresponding interval:

$$C_{st,i} = \frac{Q_{st}}{U_{st,i} - U_{st,i-1}} \tag{3}$$

As an example the following equation shows the capacitance values for the GaN-HEMT GS61008P [13] used as synchronous rectifier switch for $N_{st} = 3$:

$$C_{oss}(u) = \begin{cases} C_{st,1} = 688 \, \mathrm{pF} & u < 14.3 \, \mathrm{V} \\ C_{st,2} = 338 \, \mathrm{pF} & \text{for} \quad 14.3 \, \mathrm{V} \leq u < 43.4 \, \mathrm{V} \\ C_{st,3} = 267 \, \mathrm{pF} & 43.4 \, \mathrm{V} \leq u \end{cases} \tag{4}$$

The number of steps represents a tradeoff between required accuracy and simulation time. A larger number of steps leads to higher accuracy but concurrently to longer simulation time, and vice versa.

Fig. 2 (right) shows the voltage-dependent stepped curve of the GaN-HEMT's output capacitance for one, three and fifteen steps. When using one step, the resulting capacitance is the charge equivalent capacitance of the nonlinear output capacitance of the GaN-HEMT. In order to achieve safe operation, the voltage across the 100 V GaN-HEMT should not exceed 80 V. Additionally, a series resistor R_{Coss} in series to the output capacitance is provided. This is required because of the so called C_{oss}-losses, which can emerge during the charging and discharging process of the output capacitance C_{oss}. Various publications confirm this phenomenon especially for super junction MOSFETs but recently also for GaN-HEMTs [14] [15]. This loss can impact the performance of soft-switching converters drastically. The presented aspects result in the equivalent circuit diagram of the switches which is shown in Fig. 2 (left).

Synchronous Rectifier model

To increase efficiency, a synchronous rectifier instead of a rectifier diode is used. The GaN-HEMT, used as synchronous rectifier switch, is modeled by the circuit in Fig. 2 (left). Usually, a synchronous rectifier is modeled as a diode with no forward voltage drop $U_{D,0}$. This means it switches on and off like a conventional diode. However, in reality there is an integrated circuit (IC) which drives the transistor used for synchronous rectification. Typically, these ICs observe the drain-source voltage to drive the power switch accordingly to achieve diode like behavior. With increasing switching frequencies, the internal delay times of the IC can no longer be neglected. In order to achieve simulation results close to reality, the whole functionality of the synchronous rectifier control IC must be modeled. The behavior emulates a commercially available synchronous rectifier IC [16]. It includes propagation delays during turn-on and turn-off transitions. Also, the threshold levels for the detection of turn-on and turn-off events are included. Additionally, on and off blanking times are implemented which prevent falsely detected switching transitions due to ringing induced by the PCB layout or other parasitic elements. Fig. 3 shows the principle waveforms of the implemented synchronous rectifier.

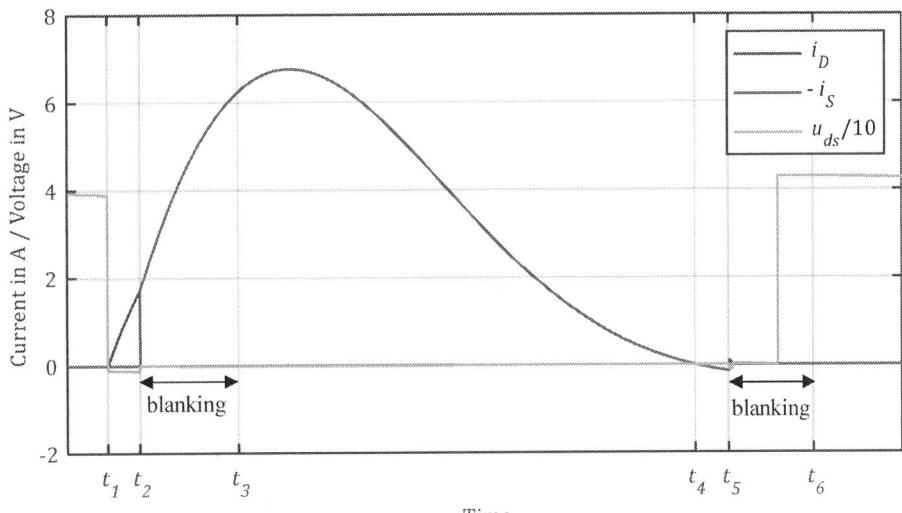

Fig. 3: Principle waveforms of synchronous rectifier operation.

For the detection of the switching-on and switching-off actions, the control IC observes the GaN-HEMTs drain-source voltage u_{ds}. Starting from the blocking state with positive u_{ds} the GaN-HEMT changes to reverse conduction with a low negative voltage u_{ds} and rising current i_D. The crossing of the on-level threshold is detected at the point t_1. Because of the internal propagation delay times, the GaN-HEMT is not turned on immediately but after some ns at t_2. The current commutates to the channel of the switch and the voltage u_{ds} decreases as the on-state voltage is significantly smaller than the voltage drop during reverse conduction. At this point the on-blanking time starts, which prevents a premature turn-off due to parasitic oscillations or noise coupled into the detection circuitry. The blanking time ends at t_3. From that point on, a turn-off action can be detected through a polarity change of u_{ds}, crossing the threshold off-level. This happens at t_4 and after the internal propagation delay time the GaN-HEMT turns off at the time instant t_5. During this interval, the current can rise with changed polarity, which eventually leads to ringing when the switch turns off. The off-blanking time takes from t_5 to t_6 and disables the turn-on detection. When the input half-bridge of the converter switches (high side switch turns on), the synchronous rectifier's drain-source voltage u_{ds} rises to its blocking voltage. After that, a new cycle can start. For GaN-HEMTs, especially the consideration of the turn-on propagation delay is crucial. During this time, the GaN-HEMT is reverse conducting which leads to higher losses caused by the higher forward voltage drop [8]. Also, premature turn-off of the synchronous rectifier significantly impacts the occurring losses and therefore should be avoided [8]. Turning off the synchronous rectifier GaN-HEMT too late can damage it due to overvoltage at drain-source during turn-off.

Transformer model

A transformer provides galvanic isolation between input and output. Typically, the transformer model only describes the magnetic behavior and comprises of a magnetizing inductance L_m, a leakage inductance L_s and an effective turns ratio n. Additionally, winding losses can be represented through a primary resistance R_p and secondary resistance R_s. With increasing switching frequencies and the use of switches with very low output capacitances, like the mentioned GaN-HEMTs, the impact of parasitic capacitances of the transformer is not negligible anymore. For that reason, a simple transformer model only considering inductances is no longer sufficient. In reality, there are many small capacitances between all turns distributed over the whole winding of the transformer. Normally, the capacitances between the winding and the transformer core can be neglected because of relatively huge distances between core and windings. For black box modelling with concentrated elements the transformer with two windings is considered as a system with three independent voltages. As a consequence, an equivalent circuit with six discrete capacitances can be used. [17]. The winding capacitances C_1 to C_6 can be determined by measurement [18]. Alternatively, they can be estimated through an analytic calculation method [19]. In switched mode power supplies electromagnetic noise is produced due to rapid current and voltage changes. Typically, in order to comply with EMC standards, a Y-capacitor C_Y with a few nF (its value is limited by safety standards in case of power supplies connected to the mains) is used between the primary and the secondary side of the transformer to supply a return path of common-mode currents. This helps to reduce common-mode noise. Fig. 3 (left) shows the resulting circuit diagram of the transformer including the Y-capacitor C_Y.

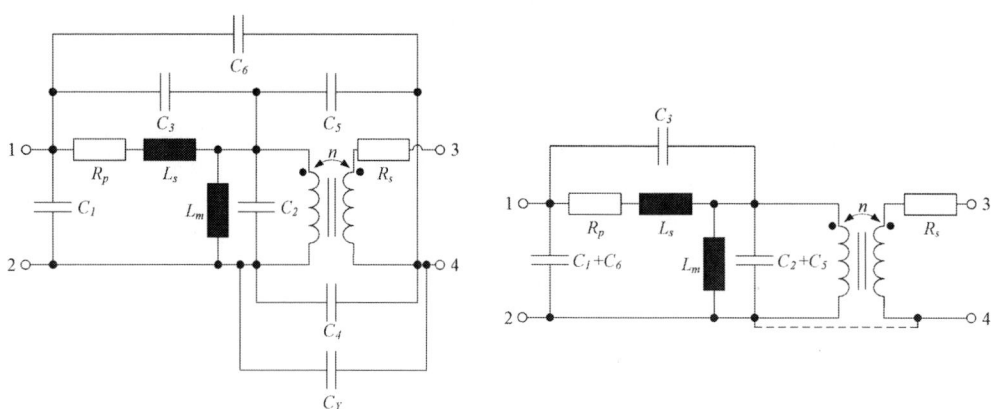

Fig. 4: Transformer equivalent circuit with six capacitors and Y-capacitor (left) and the reduced equivalent circuit through the high-frequency effective short circuit in consequence of the Y capacitor (right).

Simulation results and impedance measurements within the relevant frequency range reveal that the Y - capacitor C_Y (in nF-range) can be modeled as a short circuit across the transformer capacitance C_4 which typically is in pF-range. In Fig. 4, C_4 and C_Y are parallel connected and therefore, the larger capacitor dominates the resulting capacitance. In the relevant frequency range, the impedance of the parallel connection of both capacitors is sufficiently small that it can be replaced by a short circuit. As a consequence, the latter circuit can be converted into a simplified one. Due to the short circuit, the capacitance C_4 is eliminated. Furthermore, again using the short circuit, the capacitances C_1 and C_6 are now connected in parallel and can be consolidated to one capacitance on the input side. The same applies for the capacitances C_2 and C_5 on the output side of the transformer. The transformer capacitance C_3 remains unchanged. These assumptions result in the circuit diagram shown in Fig. 3 (right). The mentioned short circuit is indicated by the dashed line. For the verification of the aforementioned simplifications Fig. 5 shows the small signal measurement of the transformer's impedance magnitude $\left|\underline{Z}_0\right|$ measured from the primary side with open secondary side winding and a short circuit instead of the Y-capacitor. Additionally, various simulated waveforms are depicted. The first simulation belongs to the transformer model with six capacitances and a short circuit over the capacitance C_4. For the second one a Y-capacitor with 1 nF is placed instead of the short circuit. The last one results from the simplified transformer model with three capacitances. For the three simulation configurations, all small signal simulations deliver the same behavior within the relevant frequency range. It can be recognized that the measurement and the three simulation results show a very good correlation up to ca. 10 MHz. Above that, the impedance of the transformer exhibits additional resonances which cannot be modeled with the limited number of passive components used in the simulation models. Furthermore, that frequency range does not influence the converter behavior from power electronics side of view.

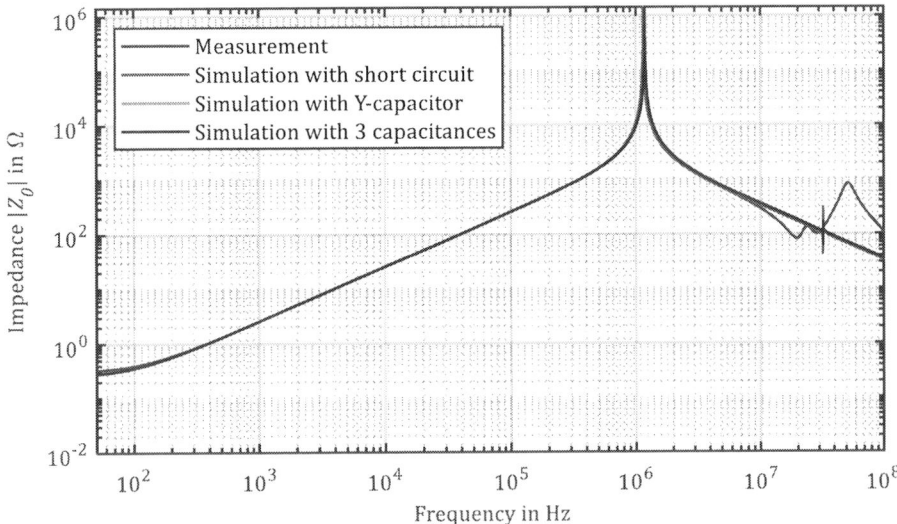

Fig. 5: Comparison between measured und simulated transformer impedance magnitude $\left|\underline{Z}_0\right|$.

Final simulation model

After describing the important parts of the simulation model separately, they are combined to the circuit diagram of the asymmetrical half-bridge converter. This results in the new circuit diagram. However, it consists of a large number of components. To reduce the equivalent circuit diagram component number and thereby decrease simulation time and complexity, additional simplifications in the complete simulation model are carried out. In good approximation, the capacitances on the transformers input side $C_1 + C_6$ (see Fig. 3(right)) are connected in parallel to the output capacitances of the upper and lower half-bridge switch C_{oss1} and C_{oss2}. As a consequence, they are merged into two capacitances belonging to the half-bridge switches C_{S1} and C_{S2}. The linear (no voltage dependency) transformer capacitance is thereby added to the voltage-dependent stepped capacitance characteristic of the output capacitances. For the stepped capacitance curve, the transformer capacitance part is implemented as a constant offset to the stepped output capacitance. The same approach is carried out at the output side of the transformer. Transforming the capacitance $C_2 + C_5$ (see Fig. 3 (right)) to the secondary side, it is approximately connected in parallel to the output capacitance of the synchronous rectifier GaN-HEMT C_{oss3} as the output is modeled as a constant voltage source and the resistances are relatively small. Consequently, they are analogously merged in the capacitance C_{S3}. Simulation results prove that the two simplifications show negligible small impact on the voltage and current waveforms. Fig. 4 shows the final equivalent circuit diagram. Finally, the simulation model based on the detailed circuit diagram is implemented using the state space representation. In this routine, the synchronous rectifier's functionality is realized as well. With this model, it is now possible to simulate detailed converter waveforms including the ZVS transitions during the dead times.

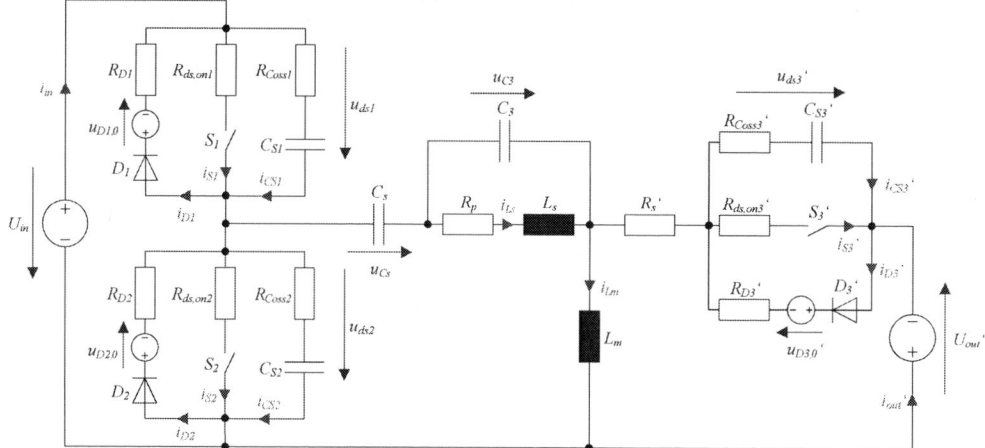

Fig. 6: Equivalent circuit diagram of the final simulation model.

Verification by measurement

For the verification of the proposed simulation model, a test setup is realized. The accuracy of the simulation model needs to be validated before it can be used for intensive optimization. The prototype of the converter bases on a specification of a 90 W laptop power supply with a DC input voltage of 400 V and an output voltage of 19.5 V. A power factor correction (PFC) circuit (e.g. a boost converter) supplies the DC input voltage of the converter studied here. Therefore, the asymmetrical half-bridge serves as converter for galvanic isolation of the output voltage from the mains. Furthermore, it converts the high voltage (400 V) to a level suitable for laptops. Transistors from GaN Systems are used in the built prototype. The 650 V GaN-HEMTs GS-065-011-1-L realize the half-bridge switches [20]. For the synchronous rectifier, the 100 V GaN-HEMT GS61008P is used [13]. The transformer is realized based on a RM 6 core with two primary layers and the secondary layer in between. Because of electrical safety regulations triple insulated litz wire is used. One film capacitor with 24 nF is placed as the series capacitance C_s. For the comparison between the simulation model and the test setup, the voltage waveform across the lower half-bridge GaN-HEMT u_{ds2} is studied. Fig. 7 shows a comparison of the simulated and measured waveforms. The small deviation between simulation and measurement during the rising edge of the half-bridge voltage results from a slightly different half-bridge current. As the half-bridge capacitances are very small, small deviations of the half-bridge current between simulation and measurement results in different voltage slew rates during the charging process of the half-bridge capacitances.

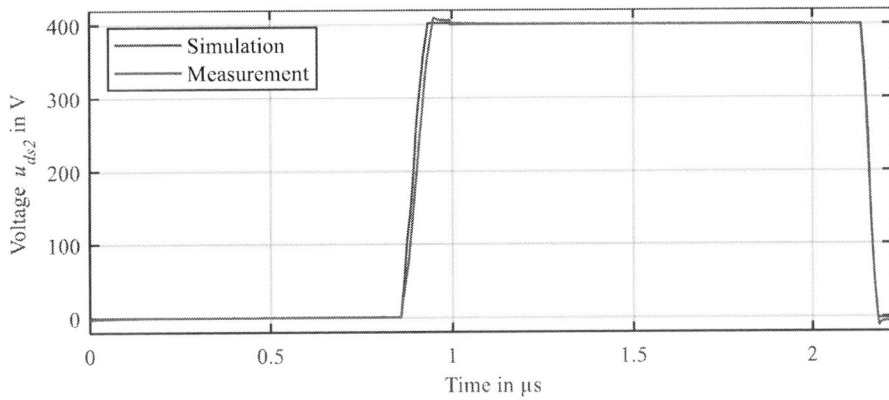

Fig. 7: Comparison between simulated and measured voltage across the lower half-bridge switch u_{ds2}.

Besides the half-bridge voltage, shown in Fig. 7, also the series capacitor voltage serves as verification of the simulation model. Fig. 8 depicts a comparison of the voltage of the series capacitor and shows that the simulated waveform agrees with that gained by measurements in the test setup. In contrast to the well-known resonant LLC converter, the asymmetrical half-bridge converter exhibits a relatively small voltage ripple of the series capacitor, as the resonant tank of the converter operates well above its resonant frequency [4].

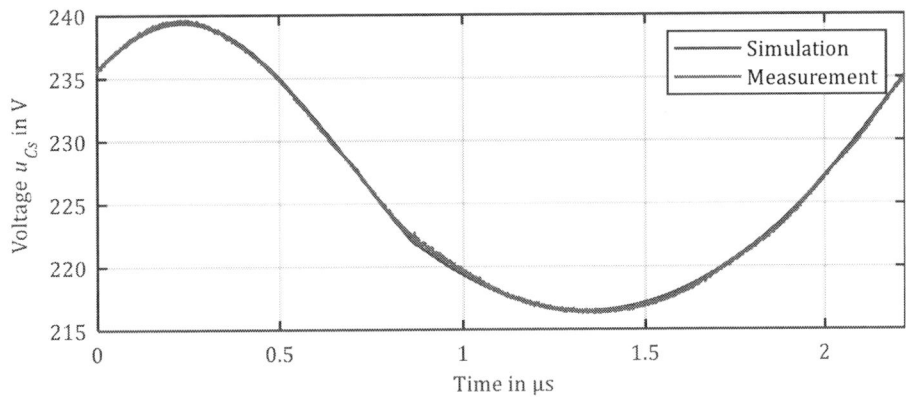

Fig. 8: Comparison between simulated and measured voltage across the series capacitance u_{Cs}.

The comparison between simulation and measurement shows only small deviations. This proves the assumptions made and, thus, shows the accuracy of the proposed simulation model. As the GaN-devices for building the half-bridge feature very low device inductance and an optimized PCB layout is used, the half-bridge commutation inductance of the test setup is very small and thus only negligible ringing occurs during switching transitions [20]. Besides this, soft switching reduces ringing due to the lower voltage slew rate compared to hard switching.

Conclusion

In this paper, based on a simplified simulation model, a new, more detailed simulation model of the asymmetrical half-bridge PWM converter with the use of GaN-HEMTs and a synchronous rectifier is developed. This includes a detailed equivalent circuit of the GaN-HEMTs, a comprehensive model of the synchronous rectifier and a transformer model including parasitic capacitances. Comparing the half-bridge voltage waveform of the model with a test setup verifies the presented assumptions and the new simulation model. With this, it is now actually possible to study the ZVS transitions during the dead times of the half-bridge and the impact of parasitic oscillations. In addition to that, it allows a realistic optimization of the asymmetrical half-bridge converter using wide bandgap semiconductors, like GaN-HEMTs, and also enables to gain reliable voltage and current waveforms with high switching frequencies. This offers the opportunity to achieve compact and efficient power supplies based on the asymmetrical half-bridge converter.

References

[1] P. C. Heng and R. Oruganti, *Family of two-switch soft-switched asymmetrical PWM DC/DC converters,* Proceedings of 25th Annual IEEE PESC 1994; Power Electronics Specialists Conference, pp. 85-94, 1994.

[2] J. H. Jung and J. G. Kwon, *Soft Switching and Optimal Resonance Conditions of APWM HB Flyback Converter for High Efficiency under High Output Current,* Proceedings of 39th Annual IEEE PESC 2008, pp. 2994-3000, 2008.

[3] B. Kohlhepp, M. Barwig and T. Dürbaum, *Efficiency Comparison of an Asymmetrical Half-bridge PWM Converter with Schottky and Synchronous Rectification,* Proceedings of UPEC 2018; 53th International Universities Power Engineering Conference, 2018.

[4] B. Kohlhepp, M. Barwig and T. Dürbaum, *Improving Efficiency of an Asymmetrical Half-Bridge PWM Converter with Synchronous Rectification by a Modified Design Process,* Proceedings of EPE 2018; International Conference and Exposition on Electrical and Power Engineering, 2018.

[5] B. Kohlhepp, J. Göttle, E. Schmidt and T. Dürbaum, *A Novel Combination of Algorithms for Accelerated Convergence to Steady-State,* Proceedings of PCIM Europe 2018; International Exhibition and Conference for Power Electronics, Intelligent Motion Renewable Energy and Energy Management, 2018.

[6] J. Cho, J. Kwon and S. Han, *Asymmetrical ZVS PWM flyback converter with synchronous rectification for ink-jet printer,* Proceedings of 37th Annual IEEE PESC 2006; Power Electronics Specialists Conference, pp. 1-7, 2006.

[7] B. R. Lin, *Implementation of the ZVS converter with synchronous rectifier,* IEEE Proceedings of Electric Power Applications, pp. 361-368, 2006.

[8] B. Kohlhepp, M. Barwig and T. Dürbaum, *GaN Improves Efficiency of an Asymmetrical Half-Bridge PWM Converter with Synchronous Rectification,* Proceedings of PCIM Europe 2019; International Exhibition and Conference for Power Electronics, Intelligent Motion Renewable Energy and Energy Management, 2019.

[9] J. Goettle, T. Hieke and T. Duerbaum, *Detailed Analysis and Optimization of the Asymmetrical Half-Bridge PWM Converter including Parasitics,* Proceedings of PCIM Europe 2014; International Exhibition and Conference, pp. 728-735, 2014.

[10] B. Kohlhepp, M. Barwig and T. Duerbaum, *Influence of the Rectifier's Output Capacitance on the ZVS Transition in Asymmetrical Half-Bridge PWM Converters,* Proceedings of ICPE 2019; 10th International Conference on Power Electronics and ECCE Asia (ICPE 2019 - ECCE Asia), 2019.

[11] B. Kohlhepp, D. Kübrich, C. Kuring and S. Peller, *Measurement of Dynamic On-State Resistance of High Voltage GaN-HEMTs under Real Application Conditions,* Proceedings of 22th European Conference on Power Electronics and Applications (EPE'20 ECCE Europe), to be published 2020.

[12] E. Schmidt and T. Dürbaum, *Fast converter simulation method including parasitic nonlinear capacitances,* Proceedings of 15th International Conference on Synthesis, Modeling, Analysis and Simulation Methods and Applications to Circuit Designs, 2019.

[13] Datasheet: GaN Systems, *GS61008,* 2018.

[14] M. Guacci, M. Heller, D. Neumayr, D. Bortis, J. W. Kolar, G. Deboy, C. Ostermaier and O. Häberlein, *On the Origin of the Coss-Losses in Soft-Switching GaN-on-Si Power HEMTs,* IEEE Journal of Emerging and Selected Topics in Power Electronics, 2018.

[15] B. Kohlhepp, D. Kübrich and T. Dürbaum, *Experimental Study of the Coss-Losses Occurring During ZVS Transitions – Emphasis on Low and High Voltage GaN-HEMTs,* Proceedings of CIPS 2020, 11th International Conference on Integrated Power Electronics Systems, 2020 to be published.

[16] Datasheet: ON Semiconductor LLC, *NCP4305*, 2019.

[17] M. Albach, *Induktivitäten in der Leistungselektronik*, Springer Vieweg, 2017.

[18] B. Cotigore, J. P. Keradec and J. Barbaroux, *The two-winding transformer: an experimental method to obtain a wide frequency range equivalent circuit*, IEEE Transactions on Instrumentation and Measurement, 1994.

[19] T. Duerbaum and G. Sauerlaender, *Energy based capacitance model for magnetic devices*, Conference Proceedings of Applied Power Electronics Conference and Exposition, 2001.

[20] Datasheet: GaN Systems, *GS-065-011-1-L*, 2019.

Electrical property variability of GaN transistors in parallel and their impact on fast switching operations

Thilini Wickramasinghe[1], Bruno Allard[1], Réne Escoffier[2], Marc Plissonnier[2]

[1]Univ Lyon, INSA Lyon, CNRS, Ampere UMR 5005, F-69621 Villeurbanne, France.
[2]Dpartement DCOS, Laboratoire LC2E, CEA-Leti, Grenoble, France.
thilini.wickramasinghe@insa-lyon.fr
http://www.ampere-lab.fr

Keywords

≪GaN HEMTs≫, ≪nonidentical devices≫, ≪paralleling≫, ≪high-speed switching≫, ≪parasitics≫

Abstract

In this paper, the impact of paralleling nonidentical Gallium Nitride High Electron Mobility transistors (GaN HEMTs) in a high-speed switch is investigated. GaN HEMTS with a packaged p-gate and a die of MIS-gate types are being evaluated theoretically. The p-type simulation results are verified with experimental prototype of a fast switching cell.

Introduction

The figure of merit of the GaN HEMTs is allowing the devices to apply in very high-speeds switches over 100 kHz. Due to high frequencies, the size of passive components in a switching power supply can be reduced to make compact devices; therefore, the reductions in copper losses and the material cost. To improve high current capabilities, the GaN HEMTs can easily configure to devices in parallel as they have a relatively stable threshold voltages over the range of 25° to 150° C and a positive temperature coefficient in the on-resistance.

However, high frequencies lead to many challenges in circuit designs due to rigorous constraints in the gate operation of the GaN HEMTs. Compared to the Si counter parts, the absolute maximum limits of the gate voltage are low—generally in the range of −10 to +10V, and the gate threshold voltage is around 1 V. The voltage transients caused by fast current transitions or the high frequency-dependent parasitic components in both the package and the circuit design can exceed the gate operation limits and permanently damage a GaN device [1]. However, the packaging and circuit design solutions with low parasitic effect for GaN-based high-speed switching cells are still unrealized tasks.

A possible solution to reduce parasitics and to establish a reliable control of the component is to integrate required functions into the transistor. However, the technology is yet to be matured enough to make complex integrations of GaN-based power circuits with a rich set of functions and further to generalize them for many applications. Moreover, the switching behavior and the operation of the device have to be thoroughly investigated prior to the decisions of integration. Most commonly used strategy to address the issues related to parasitic components is decelerating the gate slew rates using passive gate control. However, this method under-utilizes the full operation capabilities of the GaN transistors and increases the losses [2]. Different techniques have been used to address this issue [3, 4]. But, their applications for different gate types are yet to be investigated.

The discrepancies in the device parameters between the transistors in parallel can affect the switching behavior. Although, many studies have been carried out to investigation of the board level parasitic effect, there are very few studies for the device parameter discrepancies. Especially, the non-identical devices at a current imbalance situation in a high-speed power cell have not been investigated. The information is useful to identify the factors that are essential in fast switching characterization.

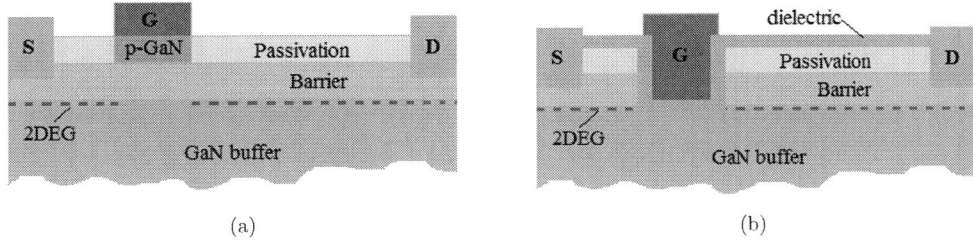

(a)   (b)

Figure 1: The normally-off GaN HEMT gates: (a) p-type (GIT), (b) MIS recess gate.

Table 1: The basic properties of the two normally-off GaN HEMTs

Gate type	Drain voltage V_{gs} (V)	Gate voltage V_{ds}(V)	Drain current I_{ds} (A)	Package type	Threshold voltage V_{th} (V)	On resistance R_{on} (Ω)
GIT p-gate	650	$-10/+7$	30	GaN$_{PX}$	1.1 –2.6	$<50-63$
MIS recess gate	650	$-10/+10$	30	die	0.7–1.8	$<50-100$

The main objective of this study is to identify the acceptable dispersion range of the electrical properties of GaN transistors in parallel to obtain a stress-free switch. The gate threshold voltage (V_{th}) and the on-resistance (R_{on}) were selected as the variable properties. Further, circuit parasitic influence on the switching behavior is also evaluated.

The current trends are in normally-off GaN gates than the normally-on devices due to their less complexity of the gate control. Compare to the various gate structures of the normally-off devices, the gate injection (GIT) and the metal insulated semiconductor (MIS) devices have similar gate operation requirements. In this study, two normally-off gate-types of GaN HEMTs (from GaN Systems Inc.—GS66508T and a GaN transistor developed by STMicroelectronics and CEA, Grenoble under the *POWERGAN* project) are investigated to identify the impact of paralleling nonidentical components. The GS66508T has a GIT type gate structure while the *POWERGAN* GaN transistor has a MIS recess gate (MISR). The basic gate structures of the two gates are illustrated in Fig. 1 and their basic properties of interest are in Table 1. These normally-off GaN gates can be driven by Si-based gate drivers provided that the gate constraints are satisfied.

The board level parasitic effect

The board level parasitic effects and their significance on the switching operations are examined in [1,5,6]. They have been found that a high impact on switching behavior is caused by (i) the stray inductance of main power loop, (ii) the mutual inductance between power loop and gate control loop, and (iii) the quasi-common source inductance [1]. Multi-layered circuits with symmetrical designs were proposed in [1] to reduce length of routing and therefore minimize parasitic inductance losses. However, to identify the impact of the layout and to evaluate the current distribution among transistors, single layered PCBs were sufficient [5,6].

In [5], the parasitic inductance mismatch in parallel branches of a micro-inverter switch has been evaluated. An asymmetrical layout was built by making unequal lengths between the input and drains of two low voltage GaN HEMTs (200 V) in parallel. The DC-DC stage of the micro-inverter was 25–380 V and the grid voltage was at 220 V. It was confirmed that the inductance in the power and control paths in the asymmetrical layout can lead to driver over voltage. However, neither mention of the device parameter variations nor information about how the current was measured in each branch in these experiments.

In [6], the current distribution among four parallel GaN HEMTs (650 V) in two symmetrical circuit layouts of a power cell was investigated. In this study, the parameters variations of the devices in parallel were less than 10%. Current through each GaN HEMT was measured using stud-type, high bandwidth, coaxial shunt current sensors (current viewing resistors—CVRs). The same circuit was further investigated by removing the current sensors. Due to the elimination of the CVRs a high current was achieved. In the circuit without CVRs, only the voltage measurements and thermal images were obtained. As consequences of adding inline CVRs in the power path: the amplitude of oscillations during the switching transition have been increased. The issue of adding inline CVRs was further discussed in [7].

The accurate current measurements are difficult to obtain with inline sensing devices. In [6,7], this issues was addressed by building a fairly precise SPICE simulation model. The ANSYS Q3D software was used to extract the parasitic elements of the layout. For the circuit without the CVRs, the model was used to predict the current through the devices. Thermal images were used to evaluate the power dissipation between the four HEMTs and to correlate them with the current measurements.

The device discrepancies

In [8], the switching behavior of two parallel Silicon Carbide (SiC) transistors with dissimilar R_{on} and threshold voltages V_{th} in parallel has been investigated. The dissimilarities in R_{on} and V_{th} of the two parallel devices were approximately 10% and 25% respectively. The experiments were conducted from 30 to 100 kHz switching range. The study showed that the R_{on} influences the static current sharing while V_{th} influences the dynamic. Further, less switching losses can be expected in devices with low gate charges and low gate resistances. Similar to [5], here the effect of utilizing inline current sensors has not been considered for the measurements. The GaN-based switches operate faster than SiC. The impact of GaN device parameter variations (R_{on} and V_{th}) on high-speed switching can be different.

Methodology

This paper presents a theoretical analysis of two nonidentical GaN HEMTs in parallel. The simulation results of a GIT and a MIS recess gate-based devices are compared. The impact of implementing nonidentical components in parallel in an asymmetrical circuit layout of the GIT devices is presented with an experimental analysis. The significance of device parameter mismatches in an asymmetrical layout was investigated using GS66508T based switching cell.

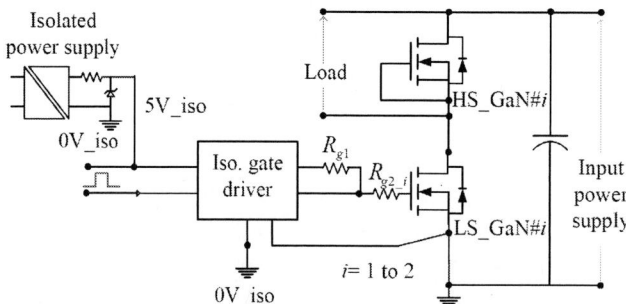

Figure 2: The schematic diagram of the preliminary test prototype. GaN HEMTs in both high and low sides.

A low-side switch of a simple h-bridge converter in a double pulse test environment was used as the device under test. Figure 2 illustrates a schematic diagram of the test prototype. In this circuit, two GaN HEMTs in the low-side are driven by an isolated gate driver providing uniform signals to the gates.

Initially, parametric tests were performed to identify the dispersion and differentiate devices (i.e. identical/ nonidentical). Figure 3(a) and 3(b) illustrate the GS66508T basic characteristics of a sample of five devices while Fig. 3(c) shows the dispersion of typical output characteristics of the MISR transistor. The Power Device Analyzer (Keysight B1505) was used to obtain the device parameters. As seen in the results of GS66508T in Fig. 3, the device GaN#2 and GaN#3 were considered as the nonidentical components in parallel. The MISR transistors were categorized with respect to the theoretical characteristics curve. The devices over 5% variation with the theoretical value were considered for the nonidentical component-based circuit.

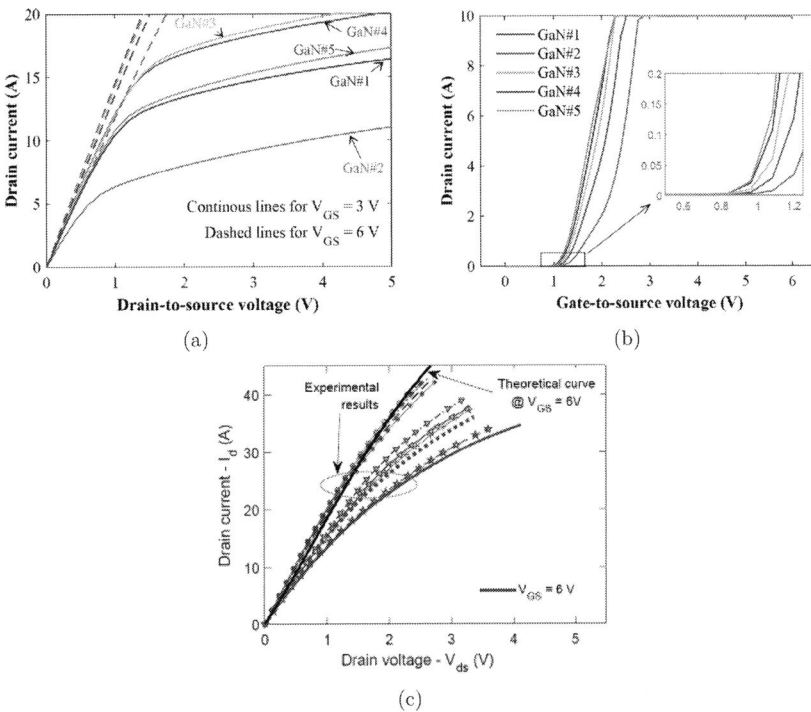

Figure 3: Parametric test results of the samples: (a) typical output characteristics of GS66508T, (b) transfer characteristics of GS66508T (c) a comparison of the typical output characteristics of the MISR transistor with its theoretical model.

Two experimental prototypes were built with (i) identical and (ii) nonidentical devices to compare the impact of nonidentical devices in parallel. The voltage waveform of the circuits (gate–source, drain–source, etc.) of the two circuits were compared. To estimate the current through each devices, a simulation model was used.

Theoretical Analysis

Two simulation models (i.e. large signal modeling) of the power cells of the GIT p-gate and the MIS recess gate were used to analyze the switching behavior devices in parallel. The p-gate model has included the package parasitic (i.e. gate connection is 0.4 nH, source is 0.04 nH and gate is 1 nH in series with 0.72 Ω) which are very low compared to the wire bonded packages. Figure. 4 shows the effect on threshold voltage variation in two parallel devices (the dispersion of V_{th} in the range of 1 to 10%). As depict in the Fig. 4, the change in the peak current due to the change in threshold voltage is approximately 28 A/V for GS66508T transistors in parallel.

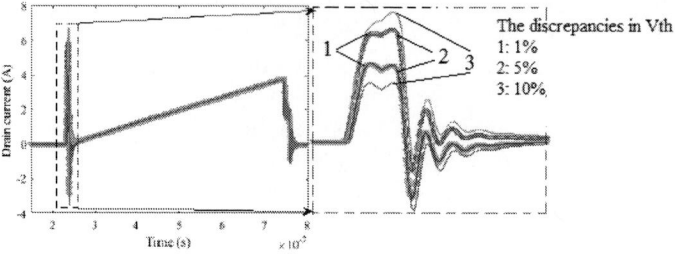

Figure 4: Current through two parallel GS66508T devices for 3 different cases of V_{th} discrepancies.

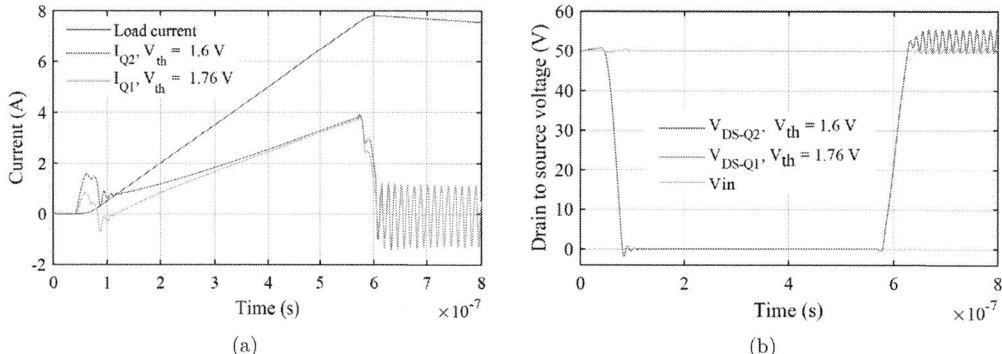

(a)   (b)

Figure 5: Theoretical analysis: (a) current through two parallel devices of the MISR transistor with different V_{th}, (b) drain voltages of the MISR transistors.

The MIS recess gate device is a die and wire bonding technique was considered to attached it to the circuit. Hence, the simulation results in the Fig. 5 were obtained with the consideration of the parasitics in the wire bonds and the PCB connections (the connections of the gate is 3.6 nH, source is 3.9 nH, drain is 2.3 nH and 8 mΩ in series with each connection). The oscillations in off-state are due to those parasitic components while the variation in drain currents are due to the discrepancies in the V_{th} (Fig. 5(a)). As seen in Fig. 5(b), there was no deviation in the drain-source voltages (V_{ds}) of the two nonidentical components in parallel due to the symmetrical circuit.

Experimental verification

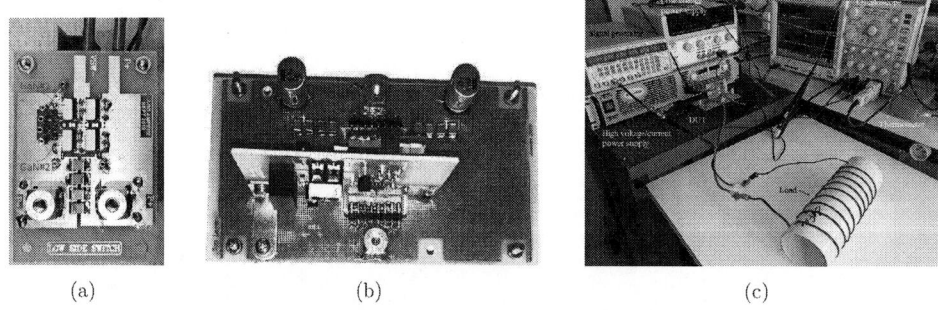

(a)   (b)   (c)

Figure 6: Experimental prototypes and the setup: (a) GS66508T-based prototype, (b) MISR transistor-based prototype (c) test setup.

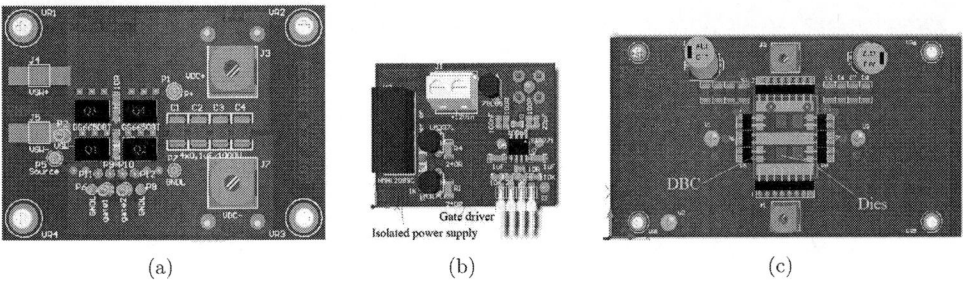

Figure 7: Layouts of the experimental prototypes: (a) GS66508T-based power loop circuit PCB layout and (b) gate control loop circuit (c) MISR transistor-based power circuit PCB layout

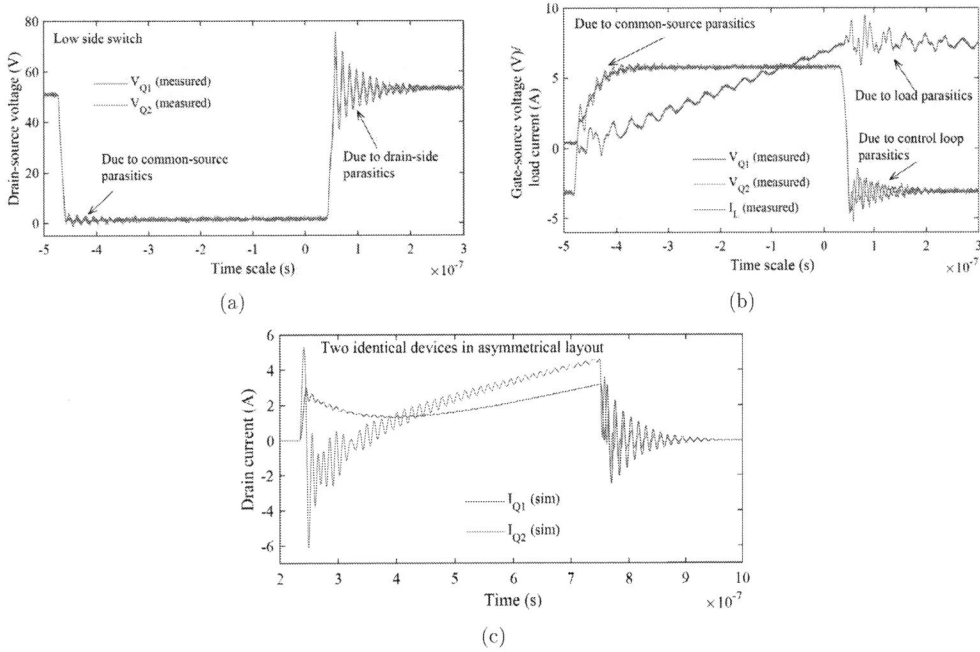

Figure 8: The experimental results of the asymmetrical circuit with identical components and the prediction of the drain current using the simulation model.

The prototype in Fig. 6(a) was built to identify the impact on nonidentical GS66508T components in an asymmetrical circuit. A symmetrical layout in Fig. 6(b) was used to identify the impact of the nonidentical MISR transistors in parallel. The main test apparatus used in the setup were digital oscilloscope (Tektronix DPO 4034B, 350 MHz) combines with passive voltage probes (Tektronix P6139B, 500 Mhz), a current probe (TCP0030A , DC to >120MHz, 30 A—maximum range), and an arbitrary signal generator (DG645 digital delay/pulse generator for PWM generation).

The first experiment was to investigate the significance of device parameter mismatches when unbalanced layout parasitics are presented. Figure 7(a) shows the PCB design. In the layout, the input to the high-side drains and the ground to the low-side sources of the switching cell were unequal. The experiment results were obtained for identical and non-identical components. The Vth and Ron of the identical components were in the ranges of 1.17–1.3 V and 50–70 mΩ respectively while non-identical components were in 1.17–1.44 V and 50–87 mΩ. No current sensors utilized in the circuit.

Figure 8(a) shows a comparison of the two identical GS66508T in parallel in the asymmetrical circuit. As seen in Fig. 8(b), the circuit parasitics caused the oscillations in voltages and currents.

Due to more parasitic inductance in the drain-side of Q1, the amplitude of the V_{ds1} oscillations during off-transition is higher than V_{ds2}.

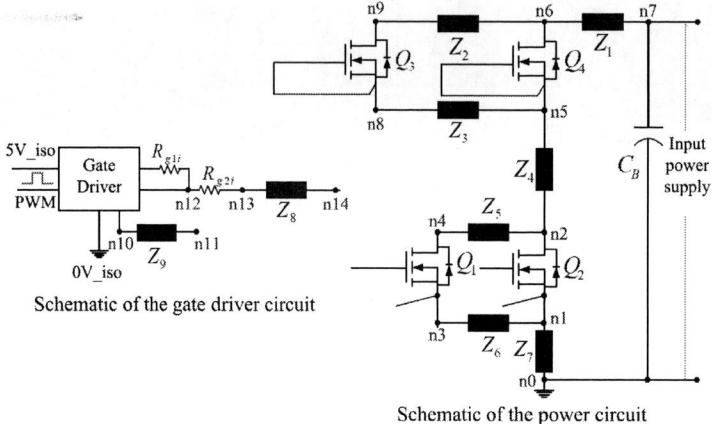

Figure 9: A schematic of the PCB parasitic model.

To predict the current via the parallel devices, the same technique in study [6] was applied (i.e. a simulation model). The PCB parasitic components indicated in Fig. 9 were extracted using ANSYS Q3D software. These models remained the same for both identical and nonidentical prototypes as only the GaN component being replaced for experiments. In the simulation models, the GaN devices were modified to adjust the parameter variations as in the test results.

As depict in Fig. 8(c), the device close to the power supply draw more current at steady state but high oscillations can be seen during on-state due to the parasitic inductance in the control loop.

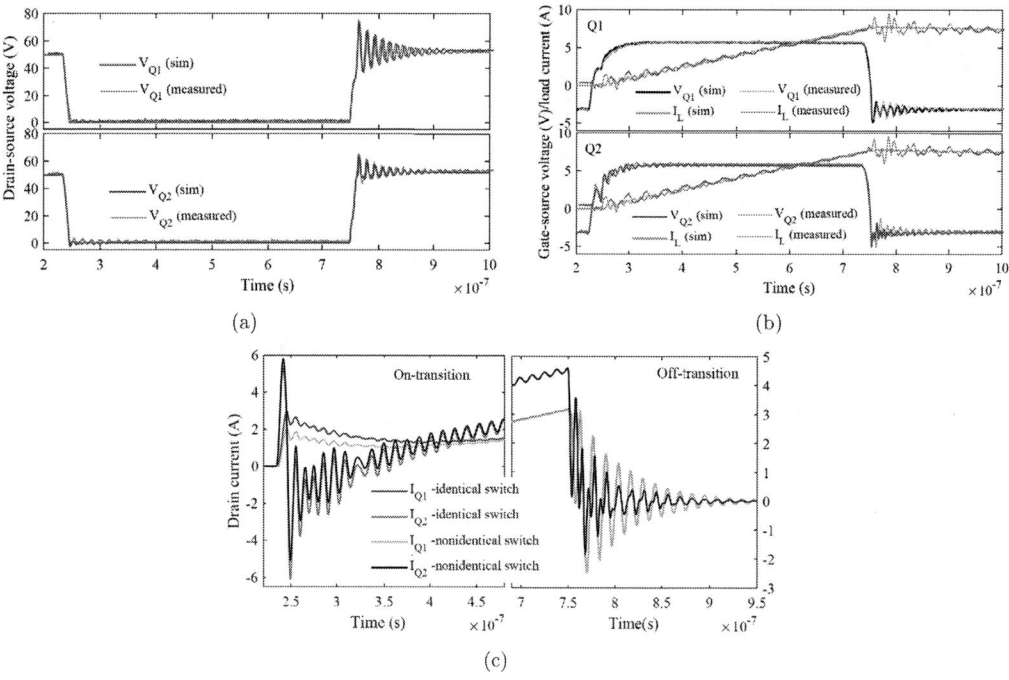

Figure 10: A comparison: (a) and (b) experimental results of the asymmetrical circuit with non-identical components and (c) on and off transitions of the predicted drain currents with identical and non-identical components in parallel.

A device in the switch (Q1 in Fig. 7(a)) was replaced by another device with 1.44 V of V_{th} and 87 mΩ of R_{on}. The comparable results of simulation and experiments depict in Fig.s 10(a) and 10(b) indicate an achievement of a reasonable simulation model. As seen in Fig. 10(c), there is less impact on the drain currents at steady state and off transient. The high V_{th} device take longer time to switch-on hence more current flow through the low V_{th} device during on-transition. Although, the discrepancies in threshold voltages of parallel GaN HEMTs lead to unbalanced current sharing among devices during on-transients, the on-resistance has shown a less significance for this case.

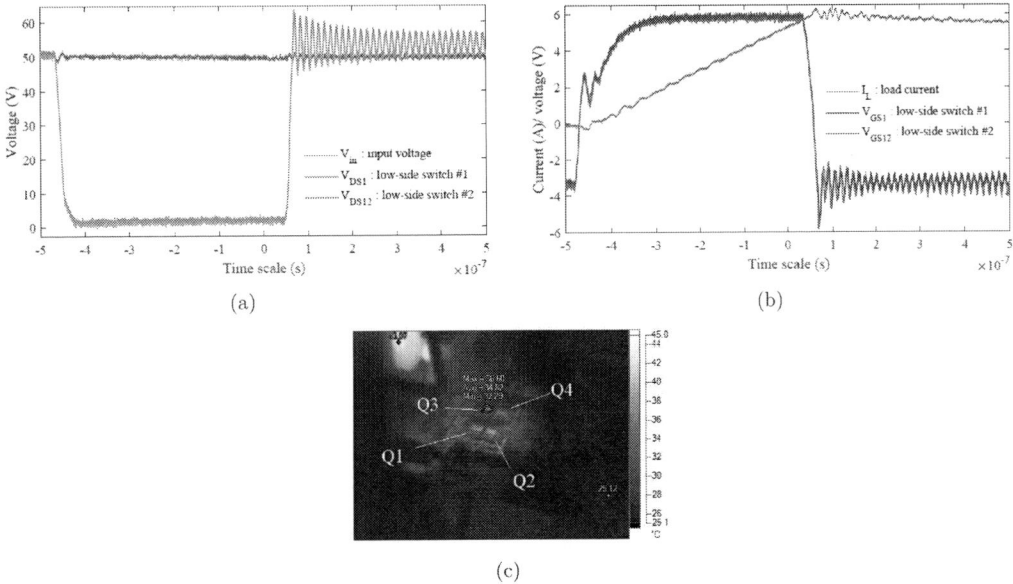

Figure 11: A comparison of nonidentical components in parallel in a symmetrical layout: (a) drain–source voltages and (b) gate–source voltages and (c) a thermal image of the components.

The prototype in Fig. 6(b) was built to identify the impact on nonidentical MISR transistors in parallel. The results in Fig. 11 is a comparison of the two components. The layout of this circuit is symmetrical. As predicted in the simulation results, the drain–source voltages of the two devices were similar. Further, the gate–source voltages are also comparable as seen in Fig. 11(b). However, the thermal imbalance in devices of the Fig. 11(c) depicts an unbalance of current through the devices.

Conclusion

Both experiment results and the simulations verified that the PCB parasitics affect the performance of power distribution among transistors. Higher oscillations can be expected with increase of parasitic inductance. In die based designs, it would be more appropriated to apply bonding techniques with low parasitics than the wire bonding. The board level parasitics in these circuits are more prominent than that of the device discrepancies (V_{th} within 10% and R_{on} within 25%) on the switching characteristics in very fast transitions. Although, threshold voltage mismatch is less significant for drain and gate voltage oscillations in a symmetrical circuit layout, it creates more unbalanced current distribution among transistors during on-transition. The effect of on-resistance on the switching transitions compared to threshold voltage is insignificant.

Acknowledgements

This work was funded by the French national program "programme d'Investissements d'Avenir IRTNanoelec" ANR-10-AIRT-05.

References

[1] Lu J, Bai H, Brown A, McAmmond M, et al. Design consideration of gate driver circuits and PCB parasitic parameters of paralleled E-mode GaN HEMTs in zero-voltage-switching applications. In 2016 IEEE Applied Power Electronics Conference and Exposition (APEC) 2016 Mar 20 (pp. 529-535). IEEE.

[2] GaN Systems Inc.: "Design with GaN Enhancement mode HEMT", Feb. 2018, [available online] Design-with-GaN-EHEMT.pdf.

[3] Beye,M. L., Mogniotte, J. F., Phung, L. V., et al. An application of open-loop active gate voltage control for GaN transistors. In 21st European Conference on Power Electronics and Applications. pp. P-1, IEEE, 2019.

[4] Dymond HC, Wang J, Liu D, et al. A 6.7 GHz active driver for GaN FETs to combat overshoot, ringing, and EMI. IEEE Trans. on Power Electronics. 33, no.1 (2017):581594.

[5] Y. Zhang, J.Li, and J.Wang, "Investigations on Driver and Layout for Paralleled GaN HEMTs in Low Voltage Application" IEEE IEEE Access, pp. 179134–179142 , Dec. 2019.

[6] T. Wickramasinghe, C. Buttay, C.Martin, et al., "An investigation of current distribution over four GaN HEMTs in parallel configurations," IEEE The 7th Workshop on Wide Bandgap Power Devices and Applications (WiPDA 2019) , Oct. 2019.

[7] T. Wickramasinghe, B. Allard, C. Buttay, et al., "A Study on Shunt Resistor-based Current Measurements for Fast Switching GaN Devices" in Proc. IEEE Industrial Electronics Society vol. 1, pp. 1573–1578, Oct. 2019.

[8] G. Wang, J. Mookken, J. Rice, and M. Schupbach, "Dynamic and static behavior of packaged silicon carbide MOSFETs in paralleled applications" in Proc. Appl. Power Electron. Conf. Expo. (APEC), pp. 1478–1483, Mar 2014.

A Comparison between Different Models of the Modular Multilevel Converter

Rafael Coelho-Medeiros[1,2,3], Bogdan Džonlaga[3], Jean-Claude Vannier[1,2], Jing Dai[1,2], Loic Queval[1,2] and Philippe Egrot[3]

[1] Université Paris-Saclay, CentraleSupélec, CNRS, Laboratoire de Génie Electrique et Electronique de Paris, 91192, Gif-sur-Yvette, France.

[2]Sorbonne Université, CNRS, Laboratoire de Génie Electrique et Electronique de Paris, 75252, Paris, France

[3]EDF R&D - Electrical Equipment Laboratory EDF Lab Les Renardières, 77250, Moret-sur-Loing, France

Email: rafael.coelho-medeiros@edf.fr

URL: [1,2] http://www.lgep.supelec.fr, [3]https://www.edf.fr

Keywords

≪HVDC≫, ≪Multilevel converters≫, ≪Modeling≫, ≪Converter circuit≫, ≪Voltage Source Converter≫.

Abstract

In this article, we compare the detailed switching (DS) model of the modular multilevel converter (MMC) with the arm average model (AA), the state-space time-invariant (SSTI) model, and the harmonic state-space (HSS) model. We consider a five sub-modules (SM) and a 50 SM application. We evaluate their performance in terms of accuracy on the representation of dynamics and simulation speed. The results show that the AA can be 700 times faster than the DS model, while the SSTI and the HSS can be more than 7000 times faster than the DS model. The highest relative deviation between models is kept under 8%.

1 Introduction

The modular multilevel converter (MMC) is widely used in high-voltage direct current (HVDC) power transmission links. Compared to the 2-level VSC, this topology presents advantages such as modular design, voltage-level scalability, high efficiency, and low distortion of the AC-side voltage [1]. The MMC presents complex internal dynamics related to the circulating current and to the capacitor voltage ripple. These dynamics influence the stability of the converter operation, and therefore modeling their behavior is critical in the early stages of design.

When modeling the MMC, one can rely on time-dependent models such as the detailed switching (DS) model [2] and the arm average (AA) model [3]. The time-dependent differential equations of the converter are then solved using fixed or variable time-step numerical methods; the simulation of systems with large numbers of state variables can be a major challenge in terms of time and memory requirement. Alternatively, the state-space time-invariant (SSTI) model [4] and the harmonic state-space (HSS) model [5] eliminate the time dependence of the converter equations; therefore they can be solved through sparse-matrix inversion leading to short simulation time. Until this date, works on time-independent models have mainly focused on stability analysis [6, 7], but they may also be useful to the design problem.

The DS model [2] reproduces the sub-module (SM) topology with power switches (transistors and diodes), where one must provide the turn ON/OFF gate signals for each SM. The reference modulated voltage is converted into the individual switching patterns of SMs by a modulation technique associated with a capacitor voltage balancing algorithm (VBA).

The AA model [3] assumes equal voltages of all the SMs capacitors in each arm, which makes it possible to replace all of them with a pair of controlled voltage and current sources. Thus, the power switches are not explicitly represented. The equivalent SM of each arm is controlled by the arm switching function corresponding to the sum of the individual switching functions of each arm. Additionally, the arm switching function is averaged [3].

The SSTI model [4] adopts the same assumptions as the AA. This time-independent model is obtained by truncating the converter equation at the second harmonic and writing them in two rotating reference frames at the fundamental frequency f_0 and at the second harmonic $2f_0$ respectively.

The HSS model [5] adopts also the same assumptions as the AA and is truncated at second harmonic where the time dependence is eliminated by transposing the converter equations from the time domain to the frequency domain using the HSS decomposition of linear time-periodic systems.

The rest of this paper is organized as follows. In Section 2, we recall the MMC topology and the converter equations that are common for all the models. In Section 3, we give the simulation results of two study cases (five SM and 50 SM) for all the models and discuss their behavior in terms of accuracy and computation time, and we highlight their purpose and applicability to the converter analysis.

2 MMC converter modeling

We consider an MMC with N half-bridge (HB) SMs connected, as represented in Fig.1. Each arm coil is modeled by a resistance R in series with an inductance L. The MMC is connected to an AC grid modeled by a resistance R_g and an inductance L_g in series with a voltage source. The phase index is represented by j for clarity.

In Fig. 1, $v_{nj}(t)$ is the grid voltage, $i_{uj}(t)$ and $i_{lj}(t)$ are the upper and lower-arm currents, $u_{dc}(t)$ is the DC-bus voltage and $i_j^{\Delta}(t)$ is the AC-side current.

For the ith SM of phase j, the modulation switching pattern is given by $u_{uj}^{(i)}(t)$ and $u_{lj}^{(i)}(t)$ for the upper and lower SMs respectively. The ith modulated voltage is

$$v_{m,uj}^{(i)}(t) = u_{uj}^{(i)}(t)\, v_{cuj}^{(i)}(t) \tag{1a}$$

$$v_{m,lj}^{(i)}(t) = u_{lj}^{(i)}(t)\, v_{clj}^{(i)}(t) \tag{1b}$$

where $i \in \{1, 2, ..., N\}$, $v_{cuj}^{(i)}(t)$ and $v_{clj}^{(i)}(t)$ are the ith capacitor voltage of phase j for the upper and lower arm respectively.

The modulated voltages of arm j are

$$v_{m,uj}(t) = \sum_{i=1}^{N} v_{m,uj}^{(i)}(t) \tag{2a}$$

$$v_{m,lj}(t) = \sum_{i=1}^{N} v_{m,lj}^{(i)}(t) \tag{2b}$$

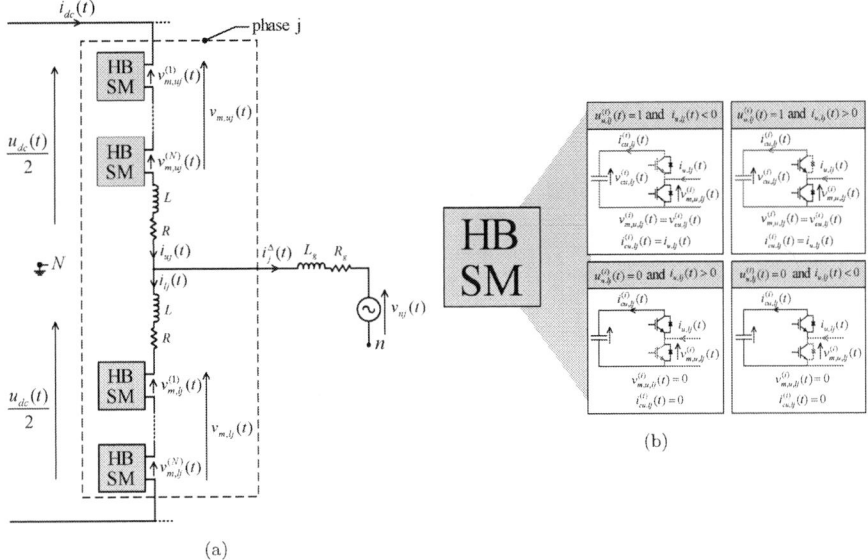

Fig. 1: (a) Circuit of the three-phase MMC with HB SM. (b) HB SM operation.

The current dynamics of the upper and lower arms of the MMC are coupled. An approach to decouple them is to use the UL-$\Delta\Sigma$ transformation [4, 7].

$$i_j^\Delta(t) \triangleq i_{uj}(t) - i_{lj}(t) \tag{3a}$$

$$i_j^\Sigma(t) \triangleq \frac{i_{uj}(t) + i_{lj}(t)}{2} \tag{3b}$$

$$v_{cj}^\Delta \triangleq \frac{-v_{cuj}(t) + v_{clj}(t)}{2} \tag{3c}$$

$$v_{cj}^\Sigma \triangleq \frac{v_{cuj}(t) + v_{clj}(t)}{2} \tag{3d}$$

$$v_{mj}^\Delta \triangleq \frac{v_{m,uj}(t) - v_{m,lj}(t)}{2} \tag{3e}$$

$$v_{mj}^\Sigma \triangleq \frac{v_{m,uj}(t) + v_{m,lj}(t)}{2} \tag{3f}$$

$$u_j^\Delta(t) \triangleq u_{uj}(t) - u_{lj}(t) \tag{3g}$$

$$u_j^\Sigma(t) \triangleq u_{uj}(t) + u_{lj}(t) \tag{3h}$$

where $v_{cuj}(t)$ and $v_{clj}(t)$ are the sum of all individual capacitor's voltage, and $u_{uj}(t)$ and $u_{lj}(t)$ are the sum of all individual modulation switching pattern for the upper and lower arm respectively.

The external dynamic refers to the AC-side dynamic, while the internal dynamic corresponds to the capacitor voltage fluctuations and the circulating current.

The external current dynamic of the MMC is driven by [7]

$$L_{ac} \frac{\mathrm{d}}{\mathrm{d}t} i_j^\Delta(t) + R_{ac}\, i_j^\Delta(t) + v_{nj}(t) - v_{mj}^\Delta(t) + v_{nN}(t) = 0 \tag{4}$$

with

$$L_{ac} \triangleq \frac{L + 2L_g}{2} \tag{5a}$$

$$R_{ac} \triangleq \frac{R + 2R_g}{2} \tag{5b}$$

and $v_{nN}(t)$ is the differential voltage between the two reference points which is adopted as zero. The internal current dynamic of the MMC is driven by [7]

$$L \frac{\mathrm{d}}{\mathrm{d}t} i_j^\Sigma(t) + R \, i_j^\Sigma(t) + v_{mj}^\Sigma(t) - \frac{u_{dc}(t)}{2} = 0 \tag{6}$$

The internal capacitor voltage dynamics are given by [7]

$$C \frac{\mathrm{d}}{\mathrm{d}t} v_{cj}^\Delta(t) = -\frac{u_j^\Sigma(t)}{4} \, i_j^\Delta(t) + \frac{u_j^\Delta(t)}{2} \, i_j^\Sigma(t) \tag{7a}$$

$$C \frac{\mathrm{d}}{\mathrm{d}t} v_{cj}^\Sigma(t) = -\frac{u_j^\Delta(t)}{4} \, i_j^\Delta(t) + \frac{u_j^\Sigma(t)}{2} \, i_j^\Sigma(t) \tag{7b}$$

The reference modulation indices for the upper and lower arm are adopted as [8]

$$m_{uj}^*(t) = \frac{N}{2} \left(1 - M \cos\left(\omega_0 t + \frac{2\pi k}{3} - \theta_m \right) \right) \tag{8a}$$

$$m_{lj}^*(t) = \frac{N}{2} \left(1 - M \cos\left(\omega_0 t + \frac{2\pi k}{3} - \theta_m \right) \right) \tag{8b}$$

where $k \in \{-1, 0, 1\}$ for phase $\{a, b, c\}$ respectively and M and θ_m are the amplitude and the phase of the modulation index.

The power and current control blocs are not included in any of the models. We set the operation point by imposing of M et θ_m. The grid voltage and the DC-bus voltage are assumed as

$$v_{nj}(t) = V_{nj} \, \cos\left(\omega_0 t + \frac{2\pi k}{3} \right) \tag{9a}$$

$$u_{dc}(t) = U_{dc} \tag{9b}$$

MMC detailled switching model

For the DS model, $u_{uj}^{(i)}(t)$ and $u_{lj}^{(i)}(t)$ are obtained from the upper and lower arm voltage references, $v_{m,uj}^*(t)$ and $v_{m,lj}^*(t)$, as illustrated in Fig. 2.

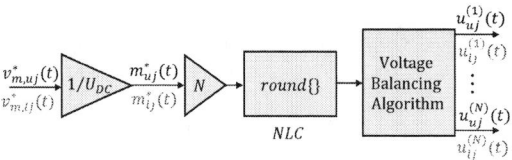

Fig. 2: NLC modulation scheme and capacitor voltage balancing of the DS.

MMC arm average model

For the AA, we assume that all SMs are identical and that the capacitor voltage balancing is ideal, thus assuring that all the SM have the same capacitor voltage,

$$v_{cuj}^{(i)}(t) = \frac{v_{cuj}(t)}{N} \; \forall i \in \{1, N\} \tag{10a}$$

$$v_{clj}^{(i)}(t) = \frac{v_{clj}(t)}{N} \; \forall i \in \{1, N\} \tag{10b}$$

Eq. (2) becomes

$$v_{m,uj}(t) = u_{uj}(t)\frac{v_{cuj}(t)}{N} \tag{11a}$$

$$v_{m,lj}(t) = u_{lj}(t)\frac{v_{clj}(t)}{N} \tag{11b}$$

In order to analyze the converter behavior in low frequency, we assume the functions $u_{uj}(t)$ and $u_{lj}(t)$ to be [3]

$$u_{uj}(t) = m_{uj}^*(t) \tag{12a}$$

$$u_{lj}(t) = m_{uj}^*(t) \tag{12b}$$

MMC $\Delta\Sigma$ SSTI model

The $\Delta\Sigma$ SSTI model in dq0 frame with notations adapted from [4] is

$$v_{m,dq0}^\Delta = M_{A1a}^{\Delta*}\, v_{c,dq0}^\Sigma + M_{A1b}^{\Sigma*}\, v_{c,dq0}^\Delta \tag{13a}$$

$$v_{m,dq0}^\Sigma = M_{A2a}^{\Sigma*}\, v_{c,dq0}^\Sigma + M_{A2b}^{\Delta*}\, v_{c,dq0}^\Delta \tag{13b}$$

$$L_{ac}\, J_\omega i_{dq0}^\Delta = v_{m,dq0}^\Delta - v_{n,dq0} - R_{ac}i_{dq0}^\Delta \tag{13c}$$

$$L\, J_{-2\omega}i_{dq0}^\Sigma = \frac{1}{2}\begin{bmatrix}0 & 0 & U_{dc}\end{bmatrix}^T - v_{m,dq0}^\Sigma - R\, i_{dq0}^\Sigma \tag{13d}$$

$$J_\omega\, v_{c,dq0}^\Delta = \frac{N}{2C}\left(M_{3a}^{\Sigma*}\, i_{dq0}^\Delta + M_{3b}^{\Delta*}\, i_{dq0}^\Sigma\right) \tag{13e}$$

$$J_\omega\, v_{c,dq0}^\Delta = \frac{N}{2C}\left(M_{4a}^{\Delta*}\, i_{dq0}^\Delta + M_{4b}^{\Sigma*}\, i_{dq0}^\Sigma\right) \tag{13f}$$

This system is rewritten as,

$$A\, x_{dq0}^{\Delta\Sigma} = B \tag{14}$$

where A is a 18×18 matrix composed by the control variables $m^{\Delta*}$ and $m^{\Sigma*}$, the passive elements L, C, R, L_g, R_g, and N. B is a 18×1 vector of AC and DC grid voltages. The solution of the SSTI is obtained by inverting Eq. (14).

MMC $\Delta\Sigma$ HSS model

Equations (4),(6) and (7) can be rewritten as the linear time-periodic state-space representation,

$$\frac{\mathrm{d}}{\mathrm{d}t}x_j(t) = A_j(t)x_j(t) + B_j(t)u_j(t) \tag{15a}$$

$$y_j(t) = C_j(t)x_j(t) + D_j(t)u_j(t) \tag{15b}$$

with

$$x_j(t) = \begin{bmatrix}i_j^\Delta(t) & i_j^\Sigma(t) & v_{cj}^\Delta(t) & v_{cj}^\Sigma(t)\end{bmatrix}^T \tag{16a}$$

$$u_j(t) = \begin{bmatrix}v_{nj}(t)\\ u_{dc}(t)\end{bmatrix} \tag{16b}$$

$$A_j(t) = \begin{bmatrix}-\frac{R_{ac}}{L_{ac}} & 0 & \frac{u_j^\Sigma(t)}{2NL_{ac}} & \frac{u_j^\Delta(t)}{2NL_{ac}}\\ 0 & -\frac{R}{L} & -\frac{u_j^\Delta(t)}{2NL} & -\frac{u_j^\Sigma(t)}{2NL}\\ -\frac{u_j^\Sigma(t)}{4C} & \frac{u_j^\Delta(t)}{2C} & 0 & 0\\ -\frac{u_j^\Delta(t)}{4C} & \frac{u_j^\Sigma(t)}{2C} & 0 & 0\end{bmatrix} \tag{16c}$$

$$B_j(t) = \begin{bmatrix} -\frac{1}{L_{ac}} & 0 \\ 0 & \frac{1}{2L} \\ 0 & 0 \\ 0 & 0 \end{bmatrix} \tag{16d}$$

$$C_j(t) = \mathbb{1}_4 \tag{16e}$$

$$D_j(t) = \mathbb{0}_{4 \times 2} \tag{16f}$$

where $\mathbb{0}$ and $\mathbb{1}$ are the zero and the identity matrix.

The HSS model is

$$sX = (\mathcal{A} - N_t)X + \mathcal{B}U \tag{17a}$$

$$Y = \mathcal{C}X + \mathcal{D}U \tag{17b}$$

where $\mathcal{A}$, $\mathcal{B}$, $\mathcal{C}$ and $\mathcal{D}$ are Toeplitz matrix and N_t is a block diagonal matrix.

The steady-state solution $(s \to 0)$ of the HSS is obtained from Eq. (17)

$$Y = \left(-\mathcal{C}(\mathcal{A} - N)^{-1}B + \mathcal{D} \right)U \tag{18}$$

3 Comparison of the models

The DS is built with PLECS® using the implicit model vectorization. Ideal switching devices model the HB SMs, and the Nearest Level Control (NLC) [2] and the VBA are implemented with a C-Script bloc.

The AA is built with PLECS®, with each arm assimilated by a pair of controlled voltage/current sources associated with the equivalent SM capacitor.

Both the SSTI and the HSS are implemented with MATLAB® scripts.

We consider two applications, denoted as Cases 1 and 2. Their parameters are given in Table I. While Case 1 reproduces a lab-scale prototype [9], Case 2 reproduces an industrial-scale prototype [2].

Table I: Simulation parameters

Parameter	Variable	Case 1	Case 2
Nominal active power	P_{nom}	140 W	1000 MW
Nominal reactive power	Q_{nom}	30 VAr	300 MVAr
DC-bus voltage	U_{dc}	140 V	640 kV
Number of SMs per arm	N	5	50
Arm inductance	L	10 mH	50 mH
Arm resistance	R	0.1 Ω	0.9 Ω
AC-side inductance	L_g	20 mH	32.6 mH
AC-side resistance	R_g	1.5 Ω	2.28 Ω
AC grid line to ground voltage amplitude	V_{nj}	$50\sqrt{2}$ V	$226\sqrt{2/3}$ kV
SM capacitor	C	3.3 mF	10 mF

The simulation results for Case 1 are illustrated in Fig. 3 and 4. The simulation results for Case 2 are illustrated in Fig. 5 and 6. Note that for conciseness, only the quantities of the upper arm are shown since both arms display the same behavior. The simulation time per period in steady state for all the models are summarized in Table. II.

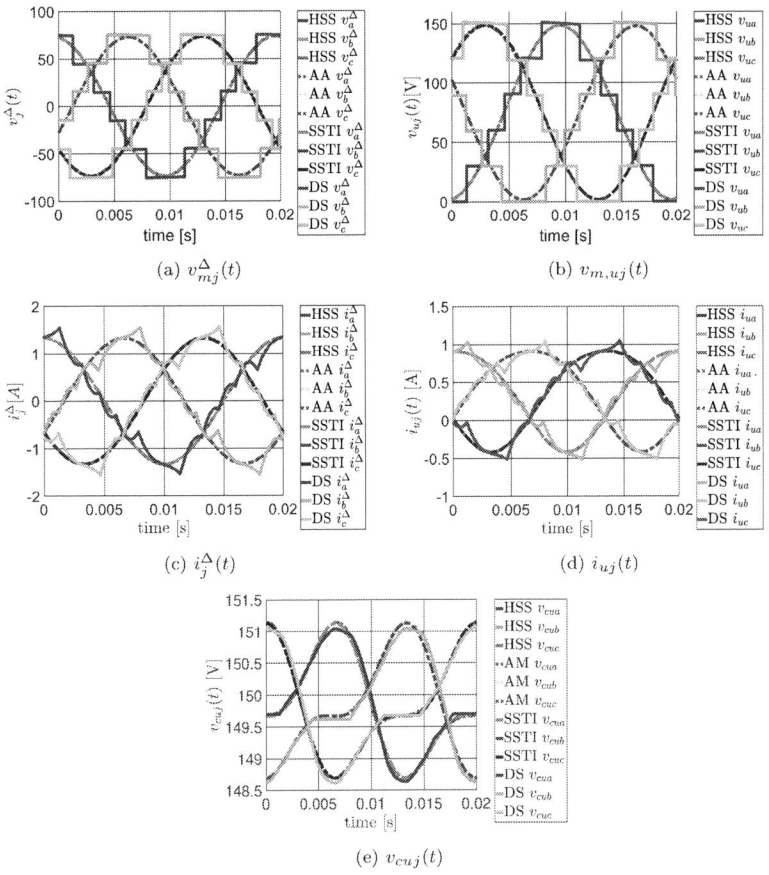

Fig. 3: Waveforms for Case 1 for $P = 140$ W and $Q = 0$ VAr ($M = 0.9789$ and $\theta_m = 0.1345$ rad for the AA, the HSS and the SSTI models, and $M = 0.9374$ and $\theta_m = 0.1335$ rad for the DS model)

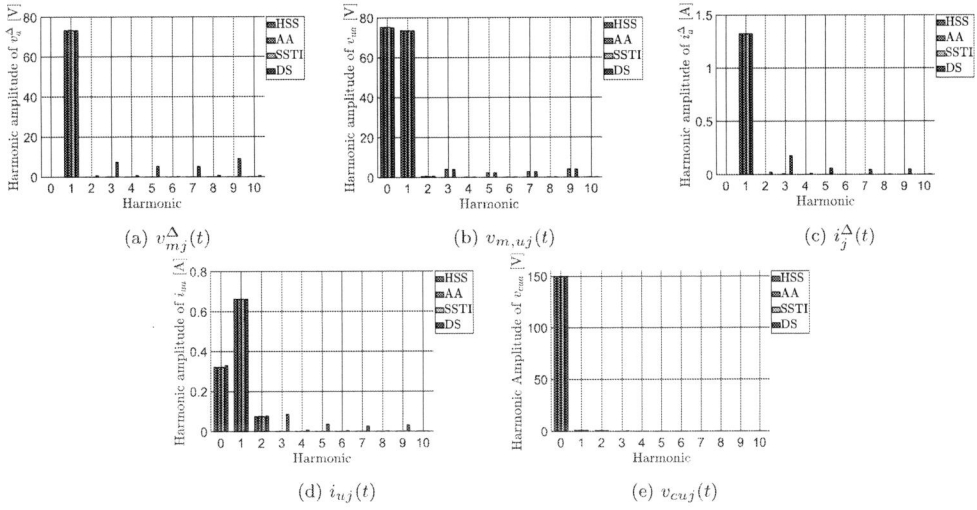

Fig. 4: Harmonic decomposition of waveforms for Case 1

A Comparison between Different Models of the Modular Multilevel Converter   MEDEIROS Rafael

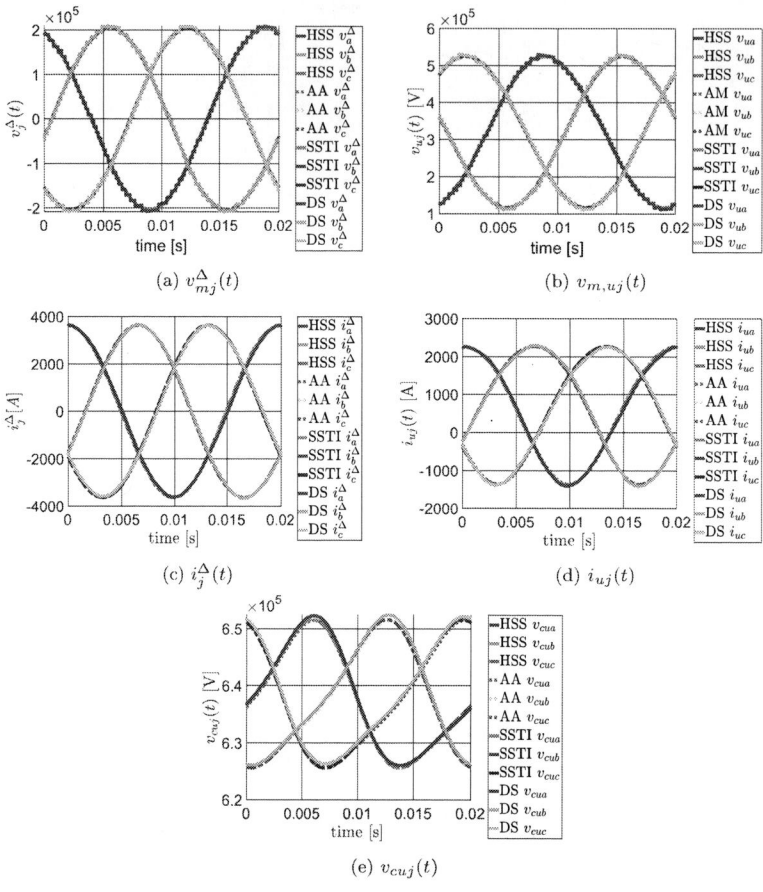

Fig. 5: Waveforms for Case 2 for $P = 1000$ MW and $Q = 0$ VAr ($M = 0.6368$ and $\theta_m = 0.3093$ rad for the AA, the HSS and the SSTI models, and $M = 0.6402$ and $\theta_m = 0.3008$ rad for the DS model)

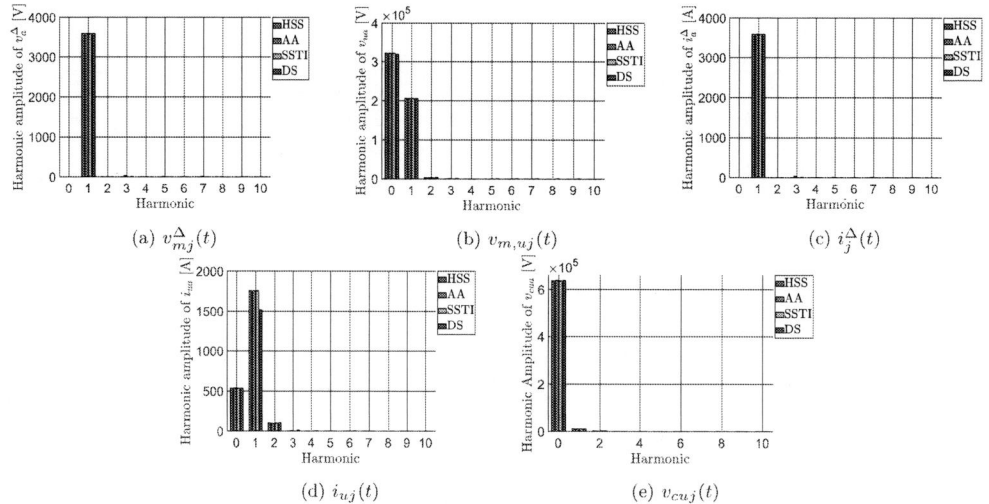

Fig. 6: Harmonic decomposition of waveforms for Case 2

EPE'20 ECCE Europe

Assigned jointly to the European Power Electronics and Drives Association & the Institute of Electrical and Electronics Engineers (IEEE)

Table II: simulation time per fundamental period in steady state

Model	DS	AA	SSTI	HSS
Case 1	17.4400 s	0.3596 s	0.0082 s	0.0129 s
Case 2	578.4000 s	0.8230 s	0.0625 s	0.0755 s

Intel Core i3-8130U CPU @ 2.20 GHz 8.00 Go RAM, Windows 10 Professional operating system.

For Case 1, the AA is 48 times faster than the DS model, while the SSTI and the HHS are more than 1000 times faster than the DS model. For Case 2, the AA is 700 times faster than the DS model, while the SSTI and the HSS are more than 7000 times faster than the DS model.

For both study cases, the DC component of the capacitor voltage (harmonic of order zero) is equal for the AA, SSTI, and the HSS model, and it slightly differs from the DS model. As a consequence, the modulation index to impose the same power flow is somewhat different. This difference is likely due to the voltage balancing control and the modulation technique since the DS is the only model that reproduces them.

While the nearest level control is modeled only by the DS model, all the other models represent the internal dynamics with good accuracy. By comparing all the results for the harmonics of order 0, 1, and 2, the highest relative error is 8% corresponding to the second harmonic of $i_{uj}(t)$ for Case 2.

Both the HSS and the SSTI used in this article are truncated at second harmonic. This truncation may lead to a loss of accuracy if the order of the main harmonics of internal dynamics are higher than two. To increase the harmonic order in the SSTI model, one needs to implement as many Park transforms as harmonic orders. For the HSS model, the extension of the harmonic order could be easily handled by the harmonic decomposition of the LTP system.

When the number of SM is high (Case 2), the external dynamics are reproduced with good precision by all the models. On the contrary, when the number of SMs is small (Case 1), the AA, HHS, and SSTI models lack the precise description of the external dynamics since they do not account for modulation harmonics.

In conclusion, the DS model is suitable for studying capacitor voltage balancing control, modulation technique, and the accurate description of semiconductors' losses. The AA can overcome the computational burden of the DS, and is suitable for power flow, control, and stability analysis. The SSTI and HSS models have the same application range as the AA model, as long as the harmonic truncation requirement is fulfilled. Furthermore, due to their good performance in computation time, they are suitable for studies of parametric variation/design of MMC converters for HVDC applications.

4 Conclusions

In this article, we compared the performance of four models for the MMC, for a five SMs application and a 50 SMs application. Overall, the time-independent models (SSTI, HSS) are much lighter in terms of the computational burden.

The internal dynamics are not affected by NLC for either of the two cases, and, as a consequence, the AA, SSTI, and HSS models provide a very good estimation.

For the application with a high number of levels, all the dynamics are reproduced with good precision by all the models. This last statement cannot be generalized to the low number of SMs, where the DS is more precise when reproducing the external dynamics.

Each model is suitable for a range of application. In particular, the DS model is suitable for studying capacitor voltage balancing control, modulation technique, and the accurate description of semiconductors' losses. The AA is suitable for power flow, control, and stability analysis.

The SSTI and HSS models have the same range as the AA and, and due to their good performance in computation time, are suitable for studies of parametric variation/design of MMC converters.

References

[1] R. Marquardt, "Modular Multilevel Converters: State of the Art and Future Progress," in IEEE Power Electronics Magazine, vol. 5, no. 4, pp. 24-31, Dec. 2018.

[2] H. Saad, S. Dennetière, J. Mahseredjian, P. Delarue, X. Guillaud, J. Peralta and S. Nguefeu, "Modular Multilevel Converter Models for Electromagnetic Transients," in IEEE Transactions on Power Delivery, vol. 29, no. 3, pp. 1481-1489, June 2014.

[3] K. Shinoda, J. Freytes, A. Benchaib, J. Dai, H. Saad, X. Guillaud (2016). "Energy Difference Controllers for MMC without DC Current Perturbations" in eProceedings of HVDC, 2016.

[4] B. Džonlaga, L. Quéval and J. Vannier, "Impact of the arm resistance and inductance on the PQ diagram of a modular multilevel converter," 2019 20th International Symposium on Power Electronics (Ee), Novi Sad, Serbia, 2019, pp. 1-6, doi: 10.1109/PEE.2019.8923290.

[5] Z. Xu, B. Li, S. Wang, S. Zhang and D. Xu, "Generalized Single-Phase Harmonic State Space Modeling of the Modular Multilevel Converter With Zero-Sequence Voltage Compensation," in IEEE Transactions on Industrial Electronics, vol. 66, no. 8, pp. 6416-6426, Aug. 2019, doi: 10.1109/TIE.2018.2885730.

[6] J. Lyu, X. Cai and M. Molinas, "Frequency Domain Stability Analysis of MMC-Based HVdc for Wind Farm Integration," in IEEE Journal of Emerging and Selected Topics in Power Electronics, vol. 4, no. 1, pp. 141-151, March 2016, doi: 10.1109/JESTPE.2015.2498182.

[7] G. Bergna-Diaz, J. Freytes, X. Guillaud, S. D'Arco and J. A. Suul, "Generalized Voltage-Based State-Space Modeling of Modular Multilevel Converters With Constant Equilibrium in Steady State," in IEEE Journal of Emerging and Selected Topics in Power Electronics, vol. 6, no. 2, pp. 707-725, June 2018, doi: 10.1109/JESTPE.2018.2793159.

[8] D. Jovcic and A. A. Jamshidifar, "MMC converter detailed phasor model including second harmonic," IEEE PES Innovative Smart Grid Technologies, Europe, Istanbul, 2014, pp. 1-5, doi: 10.1109/ISGTEurope.2014.7028946.

[9] N. Stanković, G. Bergna, A. Arzandé, E. Berne, P. Egrot and J. -. Vannier, "A digital control algorithm for modular multilevel converters," 2015 17th European Conference on Power Electronics and Applications (EPE'15 ECCE-Europe), Geneva, 2015, pp. 1-10, doi: 10.1109/EPE.2015.7309242.

Packaging Technology for the Improvement of Power Cycling Capability of HVIGBTs

Kenji Hatori*, Keiichi Nakamura*, Nobuhiko Tanaka*, Yasuhiro Sakai*,
Norikazu Sakai*, Kenji Ota*, Takeshi Higashihata*, Eckhard Thal **, Nils Soltau**
*Mitsubishi Electric Corporation, 1-1-1 Imajukuhigashi, Nishi-Ku, Fukuoka, Japan
**Mitsubishi Electric Europe B.V., Germany
Tel.: +81-92-805-3406
Fax: +81-92-805-3676
E-Mail: Hatori.Kenji@dx.MitsubishiElectric.co.jp
URL: https://www.mitsubishielectric.com/semiconductors/products/powermod/index.html

Keywords

«IGBT», «Power cycling», «Power semiconductor device», «Packaging»

Abstract

Power cycling capability is one of the major reliability topics of IGBT modules. Therefore, the packaging technology for the improvement of power cycling capability is described in the paper. Bond-wire lift off, due to the mismatch between silicon (Si) and aluminum (Al) bond wire, is well-known failure root cause of power-cycling fatigue [1]. Therefore, the improved bond wire technology was introduced [2]. However, it is not enough just to improve wire bonding technology because increased operating temperature brings additional stress on other materials. In addition to bond wire technology, three components and their impact are described in this paper: silicone gel material, solder material and metallization material of the ceramic. Finally, the improvement with the combination of all established technologies is confirmed.

Introduction

High-Voltage IGBTs (HVIGBTs) are mainly installed in the railway application and therefore high reliability is required for them. Especially power-cycling capability is one of the most important criteria because rolling stock's repetitive acceleration and braking phases which cause temperature cycles.
Bond-wire lift off, due to the mismatch of thermal-expansion coefficients between silicon (Si) and aluminum (Al) bond wire, is well-known failure root cause of power-cycling fatigue. Therefore, the improved bond wire technology was introduced. However, it is not enough just to improve wire bonding technology because increased operating temperature brings additional stress on other materials.

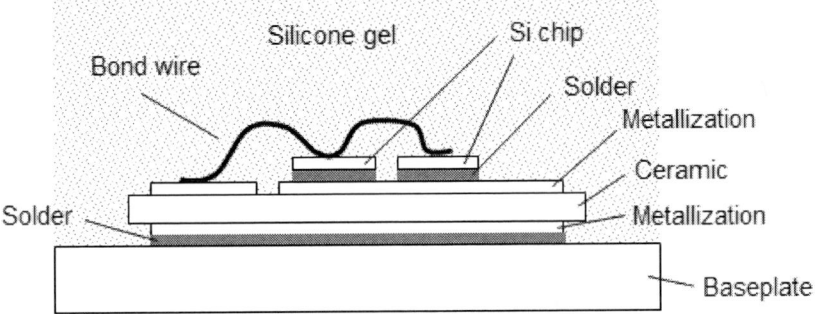

Fig. 1: A general structure of an IGBT module

As shown in Fig. 1, an IGBT module consists of many components and according interfaces between them. After the optimization of the bonding technology and with increasing operating temperatures, also other components and interfaces (beside the bond wires themselves) are worth optimizing in regards of power-cycling lifetime.

The silicone gel, for example, with which the module is filled, is in permanent contact with the bond wires. It is assumed therefore, that the silicone gel directly affects the bond wires and with it the power-cycling capability of the module.

Moreover, the Si chips are attached to the ceramic via the die-bonding solder. A degradation of this solder, and the worsening of the thermal resistance, will increase the local chip temperature and will accelerate the bond-wire fatigue. Consequently, the die-bonding solder as well is important for the power-cycling capability of the module.

Finally, proper metallization of the ceramic substrate and the solder connection with the baseplate is required to ensure sufficient cooling. Again, a degradation of substrate metallization or the substrate solder increases the thermal resistance accelerating the bond-wire fatigue.

These examples show that, in addition to the bond-wire technology itself power cycling capability is decided by other additional components also. Therefore, this paper will analyze and quantify the silicone gel material, the solder material and the metallization material of the ceramic, in this regard.

Impact of Silicone Gel

In general, IGBT modules are filled with silicone gel. Hence, the silicone gel fills also the space around the Al bond wires. When modules are heated up, silicone gel is expanded. Despite of the gel's softness, its expansion brings tension to Al bond wires. The tension to bond wires depends on the softness of silicone gel. The gel softness is usually described by its penetration. The penetration is defined with the depth where the defined cone penetrate in the gel. The measurement method of penetration is in accordance with JIS K 2220, the Japanese Industrial Standard regarding lubricating grease [3] and described in Fig. 2. The bigger number of penetration means softer. Two modules filled with different softness gel were evaluated by power cycling test. The softness comparison of tested silicone gel is shown in Table I. Gel B is 1.5 times softer than Gel A.

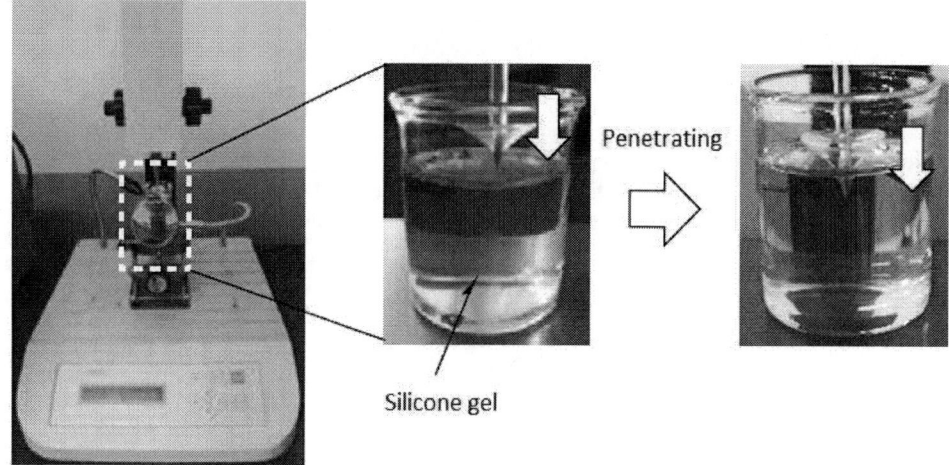

Fig. 2: Penetration measurement method

The test conditions are shown in Table II regarding testing current, maximum junction temperature T_{jmax} and junction temperature swing ΔT_j. The tested devices are 3.3kV-IGBTs rated 1800A. The modules are filled with different gel material, Gel A and Gel B. Despite of that, the modules are identical. As shown in Table II, the test conditions for both gel types are almost same. The test result is shown in Fig. 3. Accordingly, the power cycling test result of Gel B shows 1.5 times better capability than Gel A. Failure appearance after power cycling test is shown in Fig. 4. Al wires are lifted off. The test results indicate that softer gel can relax stress on bond wires and results in better power cycling capability.

Table I: Softness of Tested Silicone gel

	Penetration (Softness)
Gel A	100%
Gel B	150%

Table II: Power Cycling Test Condition

	Current	T_{jmax}	ΔT_j
Gel A	1800A	150°C	68.0 K
Gel B	1800A	150°C	67.3 K

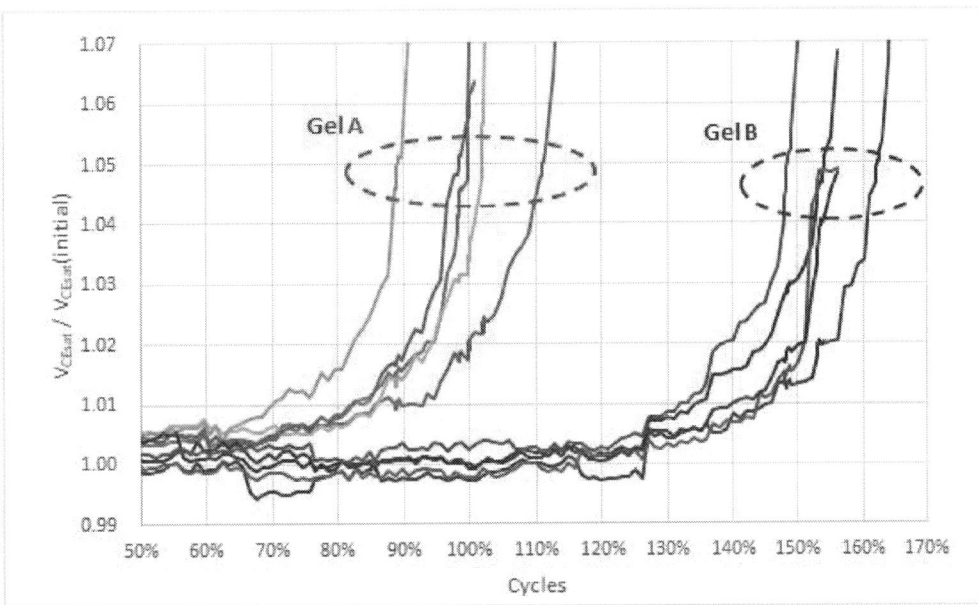

Fig. 3. The power cycling test results of different gel material

Fig. 4. Appearance of bond wires after power cycling test
(Left: Sample of Gel A, Right: Sample of Gel B)

Impact of substrate bonding material and metallization of ceramic

Crack in substrate bonding solder layer is caused by temperature swing due to power cycling. That solder crack causes increase of local junction temperature during power cycling. As a result, bond wires are finally lifted off earlier than the case of no degradation in solder layer. Therefore, reliability of substrate bonding solder is important to improve power cycling capability.

Solder degradation is affected by metallization of ceramic. Especially, the metallization stability at high temperature is important [4]. The effects of creep deformation generally become larger as the material is approaching to its melting point. The melting point of copper is 1358K, which is almost 1.5 times of aluminum melting point being 933K. Consequently, creep deformation of copper is more relaxed compared to aluminum. As a result, stress on solder is more relaxed with copper metallization. Moreover, the solder material itself is important for reliable connection.

Thus, two modules with different combination of metallization and solder are evaluated by power cycling tests. The structures of both modules are shown in Table III and the property of solder is described in Table IV. Sample A has Aluminum metallization of ceramic. In contrast, sample B has copper metallization of ceramic. Solder B, used for sample B, has a higher Young Modulus and tensile strength compared to solder A. The test conditions are shown in Table V. The tested devices are 3.3kV-IGBTs rated 1800A.

Table III: Samples structures of tested modules

	Ceramic metallization	Substrate bonding solder
Sample A	Aluminum (Melting point: 933K)	Solder A
Sample B	Copper (Melting point: 1358K)	Solder B

Table IV: Properties of tested solder

	Young Modulus	Tensile strength
Solder A	100%	100%,
Solder B	180%	116%,

Table V: Power cycling Test condition

	Current	T_{jmax}	ΔT_j
Sample A	1800A	141°C	73.1 K
Sample B	1800A	145°C	78.3 K

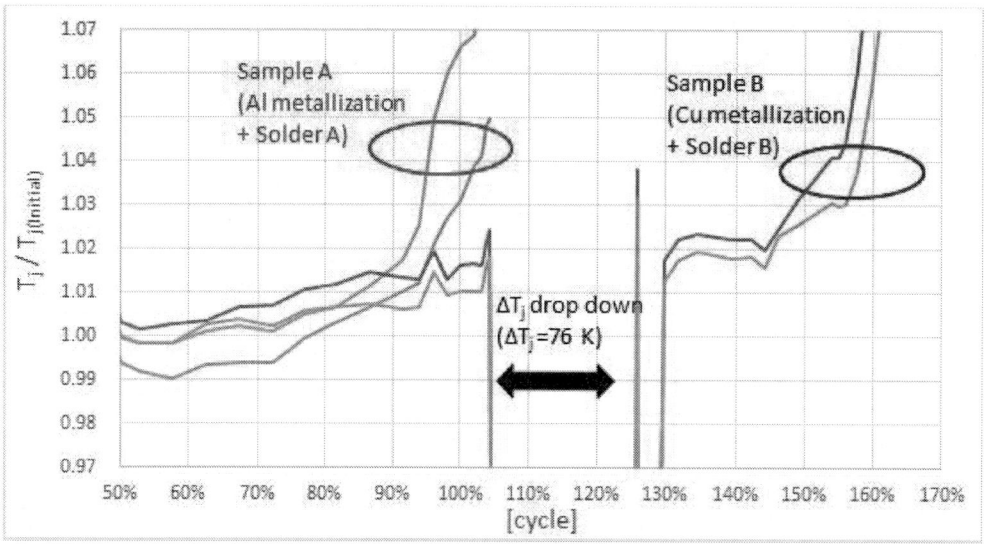

Fig. 5: Test results of different metallization and substrate bonding solder

The test results are shown in Fig. 5. It should be noted that, between 105% and 130% of cycles, ΔT_j of sample B has been unintentionally reduced by 2 K due to external reasons. Regardless of that, sample B is constantly applied with a larger ΔT_j compared to sample A, at any time. And yet sample B is tested with the severer stress, Fig. 5 shows a 1.6-times better power-cycling capability compared to sample A. Solder condition after the test is analyzed by SAT (Scanning Acoustic Tomography) as shown in Fig. 6. Crack in solder layer is observed after power cycling test in the case of solder A. On the other hand, solder B is still stable after the power cycling test even with 1.6-times larger number of cycles. The test results indicate that reliability of substrate solder is important to improve power cycling capability.

	Sample A (Solder A)	Sample B (Solder B)
Initial		
End of the test		

Fig. 6: Substrate Solder condition analyzed by SAT (Scanning Acoustic Tomography)

Impact of die bonding material

Crack in die bonding solder layer accelerates crack of Al wire bonding and causes wire lift-off as well as substrate solder. Therefore, reliability of die bonding solder is important to improve power cycling capability. Three different solder are evaluated by power cycling test; solder C, D and E. Solid phase points of solder C and solder D are below 250˚C . In contrast to that, the solid phase point of solder E is higher than 280˚C as shown in Table VI. In general, with increasing solid phase point temperature, a solder is more stable at high temperatures because creep deformation effect is more relaxed. The test conditions are shown in Table VII. The tested devices are 3.3kV-IGBTs rated 450A.

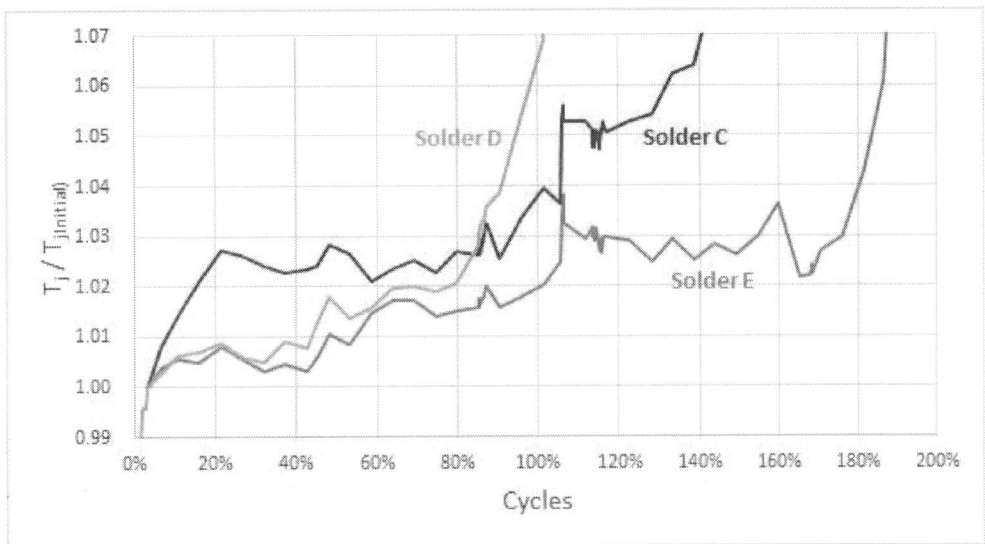

Fig. 7: Test results of different die bonding solder

The test result is shown in Fig. 7. Power cycling capability of solder E is 1.8 times better than solder C. Solder condition after the test is analyzed by SAT (Scanning Acoustic Tomography) as shown in Fig. 8. Solder degradation of solder C and solder D is observed after power cycling test. On the contrary, solder E is still stable after the power cycling test even with 1.8 times larger number of cycles. The test results indicate that reliability of die bonding solder is important to improve power cycling capability.

Table VI: Properties of tested solder

	Solid phase point
Solder C	< 250°C
Solder D	< 250°C
Solder E	> 280°C

Table VII: Power Cycling Test Condition

	Current	T_{jmax}	ΔT_j
Solder C	500A	150°C	79.5 K
Solder D	500A	150°C	81.0 K
Solder E	500A	150°C	80.0 K

	Sample of solder C (Solid phase point < 250°C)	Sample of solder D (Solid phase point < 250°C)	Sample of solder E (Solid phase point > 280°C)
Initial	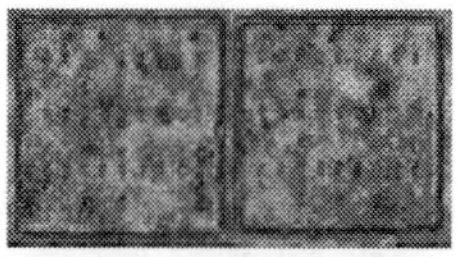	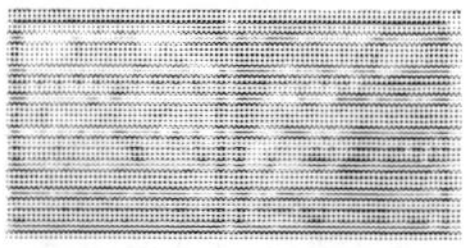	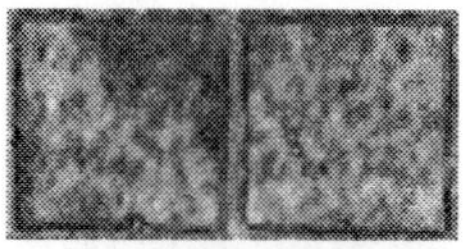
End of the test		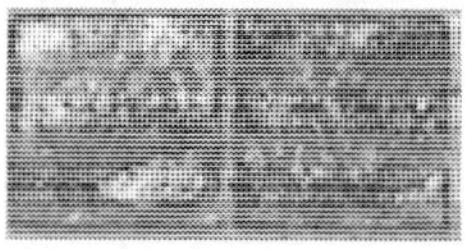	

Fig. 8: Die bonding solder condition analyzed by SAT (Scanning Acoustic Tomography)

Power cycling improvement with combination of technologies

As already discussed, it is confirmed that it is possible to improve power cycling capability by silicone gel, metallization of ceramic, die bonding solder and substrate solder. The established technologies are implemented into the new generation X-Series HVIGBTs [5] [6].
The comparison between X-series HVIGBT modules and conventional module is described in Table VIII. Conventional HVIGBT of CM1500HC-66R implements gel A, aluminum for ceramic metallization, solder A for substrate bonding and solder D for die bonding. X-series HVIGBT of CM1800HC-66X implements all established technologies: gel B, copper metallization for ceramic, solder B for substrate bonding and solder E for die bonding. CM1500HC-66R is rated for 1500A/3300V and CM1800HC-66X is rated for 1800A/3300V.

Table VIII: Samples structures of tested modules

	Silicone Gel	Ceramic metallization	Substrate bonding solder	Die bonding solder
CM1500HC-66R (Conventional HVIGBT)	Gel A	Aluminum	Solder A	Solder D
CM1800HC-66X (X-series HVIGBT)	Gel B	Copper	Solder B	Solder E

Table IX: Power Cycling Test Condition

	Current	T_{jmax}	ΔT_j
CM1500HC-66R (Conventional HVIGBT)	1500A	142°C	76.8 K
CM1800HC-66X (X-series HVIGBT)	1800A	150°C	78.5 K

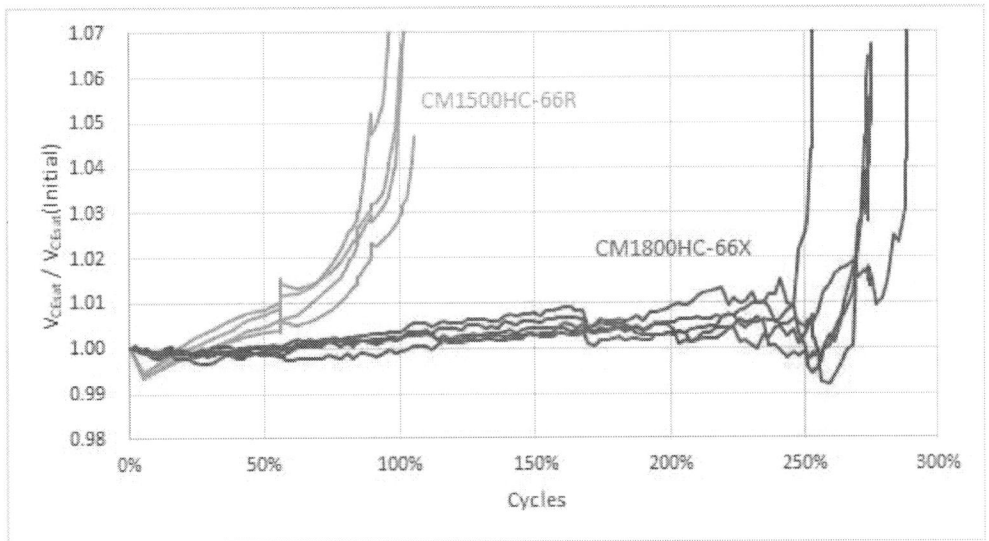

Fig. 9: Test results comparison between conventional module and X-series HVIGBT

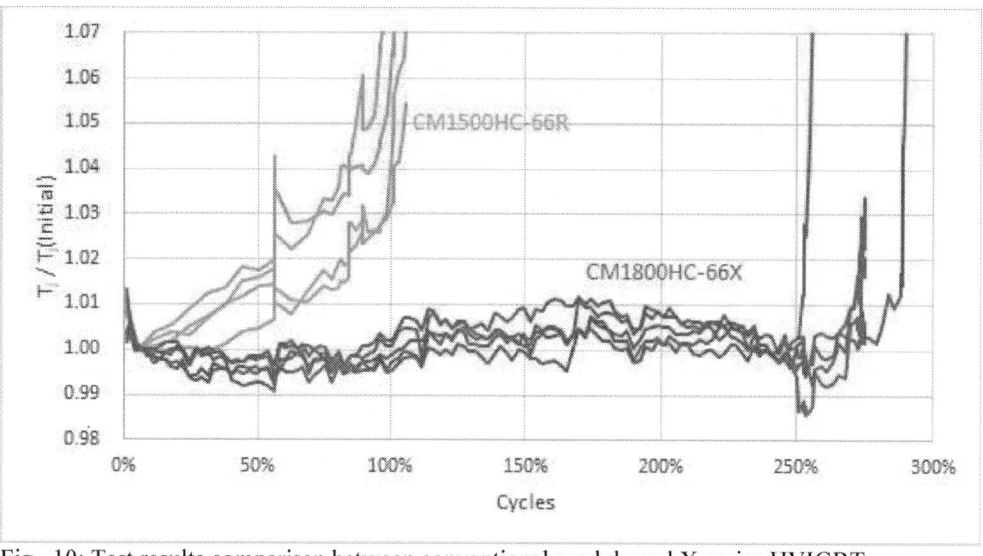

Fig. 10: Test results comparison between conventional module and X-series HVIGBT

The test condition of both modules is described in Table IX and test result is shown in Fig. 9 and Fig. 10. As shown in Table IX, CM1800HC-66X is tested with 8 °C higher T_{jmax} than CM1500HC-66R. Even with such severe condition, CM1800 HC-66X shows 2.7 times power cycling capability as CM1500HC-66R.

Fig. 11. Appearance of bond wires after power cycling test
(Left: Sample of CM1500HC-66R, Right: Sample of CM1800HC-66X)

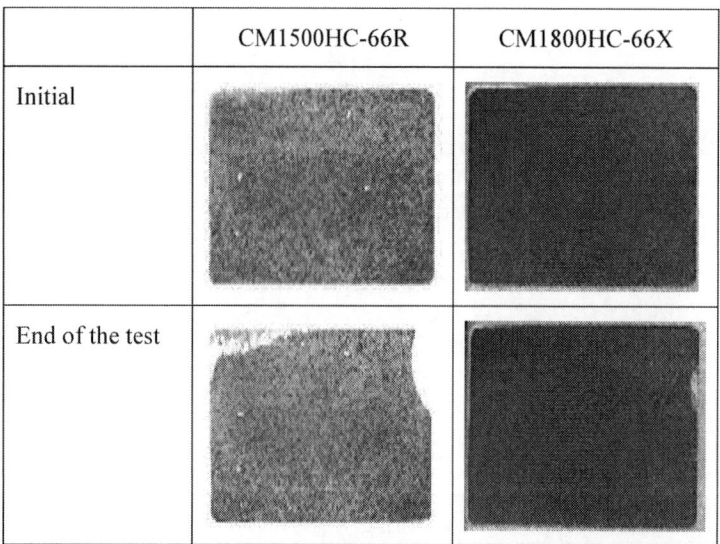

Fig. 12: Substrate Solder condition analyzed by SAT (Scanning Acoustic Tomography)

	CM1500HC-66R	CM1800HC-66X
Initial	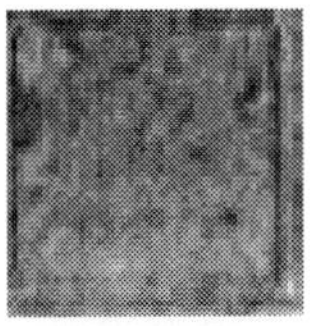	
End of the test	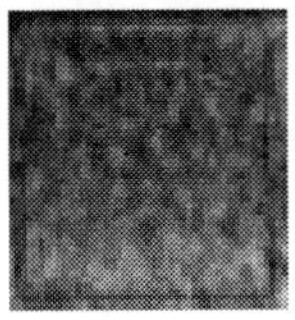	

Fig. 13: Die bonding solder condition analyzed by SAT (Scanning Acoustic Tomography)

The appearance of bond wires after power cycling test is shown in Fig. 11. Both modules show failures of bond wires lift-off. Also, Solder condition after the test is analyzed by SAT (Scanning Acoustic Tomography) as shown as shown in Fig. 12 and Fig. 13. Die bonding solder degradation and substrate solder degradation of CM1500HC-66R is observed after power cycling test. On the other hand, both die solder and substrate solder of CM1800HC-66X is still stable after the power cycling test even with 2.7 times larger number of cycles. This analysis also confirms the benefit of solder B and solder E as regards to power cycling capability.

Conclusion

This paper has demonstrated an improvement of power cycling capability by changing the silicone gel, the metallization of ceramic, the die bonding solder and the substrate solder. It is confirmed that 1.5-times higher power cycling capability is achieved with 1.5-times softer gel. 1.6-times higher power cycling capability is achieved by more reliable substrate bonding solder with copper metallization. The later in particular benefits from higher Young Modulus and higher tensile strength. It is also confirmed that the die bonding solder with higher solid phase point can achieve 1.8 times better power cycling capability.
Finally, a power cycling test has been performed with the combination of all established technologies. It has been confirmed that power cycling capability increases by factor of 2.7 compared to the combination of conventional technologies. All established technologies are implemented to the new generation HVIGBTs of X-series.

References

[1] J. Lutz, "Packaging and Reliability of Power Modules", 8th International Conference on Integrated Power Electronics Systems (CIPS 2014), Nuremberg, Germany, 2014, pp. 1-8.

[2] T. Yanagimoto, et al, "Bonding Technology for High Operation Temperature Power Semiconductor Module", International Exhibition and Conference for Power Electronics, Intelligent Motion, Renewable Energy and Energy Management (PCIM Europe 2015), Nuremberg, Germany, 2015, pp. 430-433

[3] https://www.jisc.go.jp/eng/index.html

[4] T. Poller, J. Lutz, B. Böttge and H. Knoll, "Analysis of the plastic deformation in aluminium metallizations of Al2O3-based DAB substrates," *15th European Conference on Power Electronics and Applications (EPE 2013)*, Lille, 2013, pp. 1-10

[5] E. Wiesner et al., "High Power Density, High Performance X-Series 4500V IGBT Power Modules", Bodo's Power Systems, December 2017, p34-37

[6] Y. Sakai, et al., "Power cycle lifetime improvement by reducing thermal stress of a new dual HVIGBT module," *18th European Conference on Power Electronics and Applications (EPE'16 ECCE Europe)*, Karlsruhe, 2016, pp. 1-7

A Bidirectional DAB-LLC DCX to Achieve Voltage Regulation and Wide ZVS Range Capability

Yuefeng Liao[1,2], Tao Peng[1,2], Mei Su[1,2], Yao Sun[1,2], Weijing Xiong[1,2], Guo Xu[1,2]
1.School of Automation, CENTRAL SOUTH UNIVERSITY
2.Hunan Provincial Key Laboratory of Power Electronics Equipment and Grid
Changsha Hunan P.R. China
E-Mail: liaoyuefeng_uav@126.com

Acknowledgements

This work was supported in part by the National Natural Science Foundation of China under Grant 51907206, in part by the Major Project of Changzhutan Self-Dependent Innovation Demonstration Area under Grant 2018XK2002, in part by the Hunan Provincial Key Laboratory of Power Electronics Equipment and Grid under Grant 2018TP1001, and in part by the Development of Advanced Power Conversion Technology and Equipment for Smart Microgrid under Grant 2018SK2140.

Keywords

«DAB-LLC Converter », «Tight voltage regulation », «Zero voltage switching», « Bidirectional power flow ».

Abstract

To achieve both voltage regulation and high efficiency in a DCX, a bidirectional DAB-LLC converter is studied. In this structure, most of power flows through LLC, and DAB is used to regulate the output voltage, which can achieve wide ZVS range capability in both forward and backward direction.

Introduction

In recent years, LLC resonant converters are widely used, due to its high efficiency, wide soft switching range, electrical isolation in medium-power transmission occasions [1]-[3]. Among those applications, using LLC converter as a DC-DC transformer (DCX)][4]-[5], which operates at the resonant frequency, is a suitable solution to achieve maximum conversion efficiency. However, because the frequency is not regulated, when the load fluctuates, the output gain will be affected [6]-[7]. In order to achieve voltage regulation capability, the cascade structure is adopted which uses a DC/DC converter such as buck or boost circuit as voltage regulating conversion stage, but two-stage conversion may reduce the conversion efficiency [8]. CPES proposed a sigma converter structure in [9]. It is a kind of quasi-parallel converter, which has two converters in series from the input and parallel from the output. This architecture can achieve higher conversion efficiency and also has voltage regulation ability. However, it is a non-isolated converter and the soft switching for buck converter is not easy to realize. In addition, when LLC converter is applied to bidirectional power transmission, its forward and backward operating modes are different, so for realize ZVS in the backward process, it need add an auxiliary inductor or modify the modulation strategy.

Dual active bridge (DAB) converter is an isolated bidirectional topology with Zero Voltage Switching (ZVS) capability, and can realize the regulation of output power through controlling of the phase shift angle between primary bridge and secondary bridge [10] - [11]. Referring to the sigma converter structure in this paper, the non-isolated DC-DC converter is replaced by DAB to form a DAB-LLC converter for DCX application in this paper. Besides, they share the same output bridges. Through proper design, most of the power can flow through LLC converter, and DAB converter regulates the output voltage. In this way, the advantages of the two converters can be fully utilized to realize output voltage regulation when the load fluctuates. At the same time, DAB converter can provide current to help the proposed converter achieve wide ZVS range capability in both forward and backward direction.

Thereby improving system efficiency. Finally, the experiment of a 750W prototype are shown to verify the effectiveness of the proposed converter.

Circuit topology and operation principle

A. Circuit Topology

The structure diagram of the proposed bidirectional DAB-LLC DCX is shown in Fig. 1, which is composed of an LLC converter and a DAB converter. Different from the traditional series input and parallel output (ISOP) structure, its inputs are in series and the outputs of the secondary side of transformers are in parallel, and the secondary side full bridge structure is shared, as shown in Fig.2. The input voltages of the two modules determine their power distribution, which is related to the transformer turns ratios. In this structure, DAB converter can be equivalent to a bidirectional current source.

The proposed circuit topology is shown in Fig. 2, where V_1, V_2, V_{C1} and V_{C2} represent the input voltage, the output voltage, the input voltages of LLC and DAB, respectively. C_1, C_2 and C_o are LLC input capacitor, DAB input capacitor and output capacitor, respectively. R is the load resistance. K_1 and K_2 are transformers turns ratio. L_r, L_m and C_r are the resonant inductor, the magnetizing inductor and resonant capacitor of LLC, respectively. L_k is the leakage inductor of DAB, Q_1-Q_4 are the primary side switches of LLC converter, S_1-S_4 are the secondary side switches. Q_5-Q_8 are the primary side switches of DAB converter.

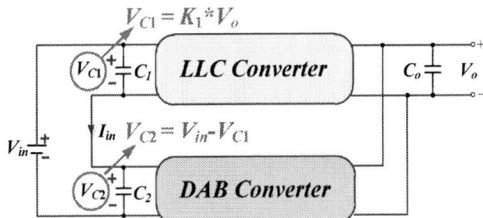

Fig. 1: The proposed structure of the bidirectional DAB-LLC DCX.

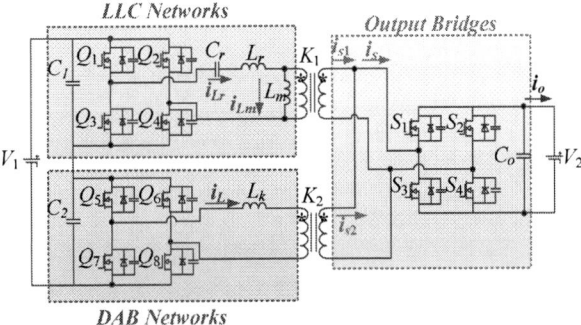

Fig. 2: The proposed circuit topology of the bidirectional DAB-LLC DCX.

B. Operation Principle

During the operating process, LLC converter is open-loop controlled and the switch driving signals of Q_i at the primary side are the same with driving signals of S_i at the secondary side (i = 1, 2, 3 and 4), which are operate at fixed switching frequency of 50% duty cycle. The key waveforms of the proposed converter under the forward power transmission and backward power transmission are shown in Fig. 3. The key equivalent circuits path during each time interval are shown in Fig. 4.

Under this modulation strategy, LLC converter can be regarded as a non-regulated DC transformer (DCX). Due to that DAB converter shares the secondary side bridge with LLC converter, it regulates the output power through adjusting the phase shift angle between the primary side switches and the secondary side switches.

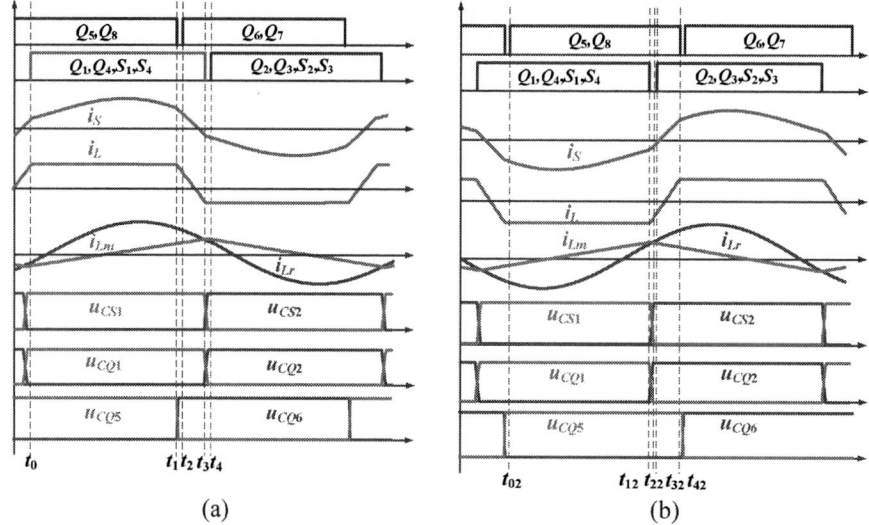

(a)   (b)

Fig. 3: The key waveforms of proposed converter. (a) Forward power transmission. (b) Backward power transmission.

At t_0, Q_1, Q_4, Q_5, Q_8, S_1 and S_4 are turned on, as shown in Fig. 4(a). The primary side of LLC converter begins the traditional resonant process. The secondary current i_s is shown as follow, where i_{s1} and i_{s2} are the secondary current of the transformer of LLC converter and DAB converter, respectively.

$$i_s = i_{s1} + i_{s2} = K_1 \left(i_{Lr} - i_{Lm} \right) + K_2 i_L \tag{1}$$

At t_1, Q_5 and Q_8 are turned off, as shown in Fig. 4(b). The leakage current $i_L(t_1)$ is bigger than zero, which can discharge/charge the junction capacitor of the primary side of DAB converter. Then, Q_6 and Q_7 are turned on with ZVS. The leakage current i_L drops to negative.

Q_1, Q_4, S_1 and S_4 are turned off when the resonant current is equal to the magnetizing current, as shown in Fig. 4(c). Then, the secondary current $i_s(t_3)$ equals $i_{s2}(t_3)$, which is reach the minimum value and help to achieve ZVS, as (2) shown. The magnetizing inductor current i_{Lm} can discharge/charge the junction capacitor of the primary side of LLC converter. At the same time, i_{s2} can discharge/charge the junction capacitor of the secondary side. At the end of this mode, the diodes D_2 D_3, D_{s2} and D_{s3} are turned on. Then, Q_2, Q_3, S_2 and S_3 are turned on with ZVS, LLC converter enters the resonant process of the negative half cycle.

$$\begin{cases} i_s(t_3) = i_{s2}(t_3) = K_2 i_L < 0 \\ i_{Lr}(t_3) > 0 \end{cases} \tag{2}$$

Since the power transfer direction is negative, the phase shift angle of DAB converter is also negative. According to the above analysis, equation (2) is still satisfied. Then, all the switches can still realize ZVS, as shown in Fig. 4(e) – Fig. 4(f).

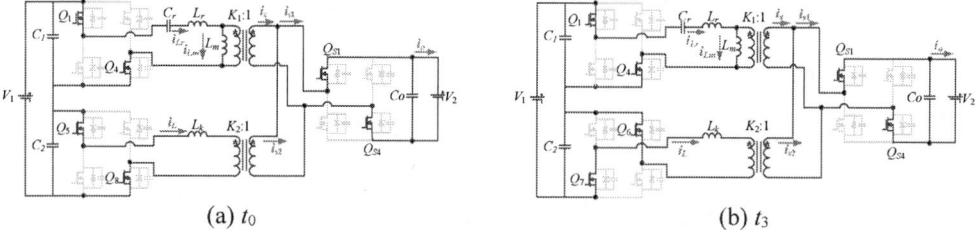

(a) t_0 (b) t_3

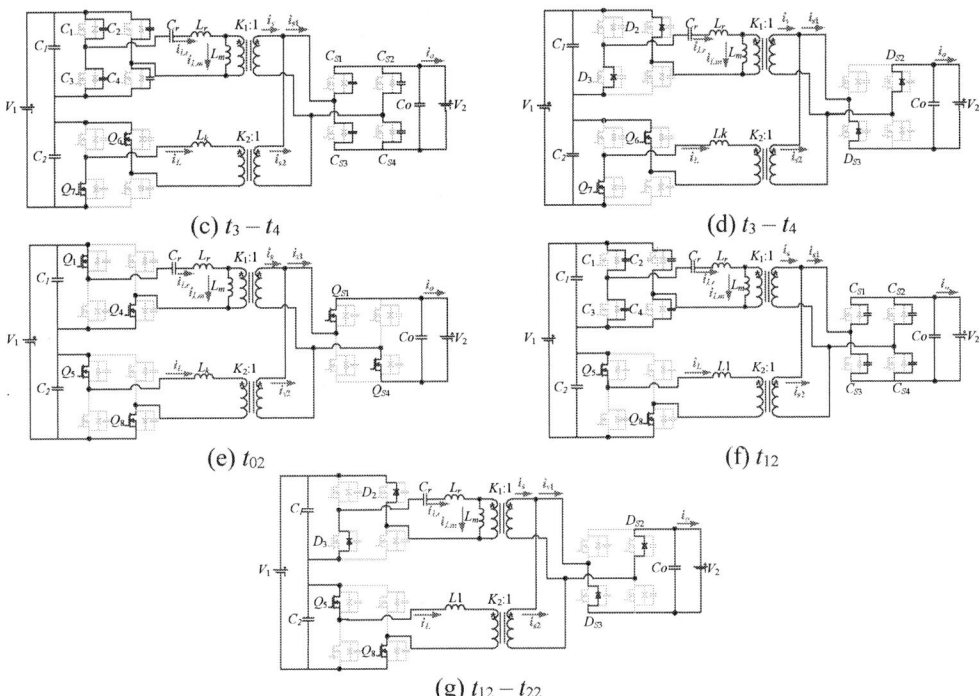

(c) $t_3 - t_4$ (d) $t_3 - t_4$

(e) t_{02} (f) t_{12}

(g) $t_{12} - t_{22}$

Fig. 4: Operation modes of the proposed converter in the forward and backward power transmission.

ZVS analysis and parameters design

A. The ZVS analysis of primary side of DAB converter

Considering to achieve ZVS of the primary side of DAB converter under light load (10% rated power), the energy stored in the leakage inductor needs to be higher than the energy stored in the junction capacitor of the switches. The leakage inductor L_k also needs to satisfy the condition under the rated power. The steady phase angle is defined as 0.25 for the need for sufficient margin when considering dynamics. The analysis is shown in Fig.5(a).

$$L_k \leq \min \left[\frac{K_2 R V_{in} D_\phi \left(1 - D_\phi \right)}{2 V_o f_s}, \frac{128 C_p f_s^2}{\left(\dfrac{4 V_o f_s}{5 K_2 V_{in} R} + 32 C_p f_s^2 \right)^2} \right] \qquad (3)$$

B. ZVS analysis of primary side of LLC converter and secondary side

As above analyzed, ZVS of the secondary side in the forward power transmission and the ZVS of primary side of LLC in the backward power transmission can always be guaranteed. However, the ZVS of primary side of LLC converter in the forward power transmission and secondary side in the backward power transmission may be lost during the dead time. It can be deduced that the ZVS range is more narrower than that in the forward power transmission. Therefore, the ZVS implementation in backward power transmission should be guaranteed first.

Due to the proposed structure, in the backward power transmission, the voltage crossing the resonant inductor L_r will be abruptly changed if the charge and discharge times of the primary and secondary junction capacitors are inconsistent, as (4) shown, where t_p and t_s are the primary and secondary side discharging time, respectively. C_p, C_s, f_s and D_ϕ are the primary junction capacitor of LLC converter, the secondary junction capacitor, switching frequency and steady state phase shift angle, respectively.

$$
\begin{cases}
V_{Lr} = V_{C1}(t_s) + K_1 V_o(t_s) + V_{Cr} & , t_p = 8C_p L_m f_s > t_s = \dfrac{4C_s L_k f_s}{K_2{}^2 D_\phi} \\
V_{Lr} = -V_{C1}(t_p) - K_1 V_o(t_p) + V_{Cr} & , t_p < t_s
\end{cases}
\tag{4}
$$

Then, the secondary side current in the dead time during the backward power transmission is shown as

$$
i_s(t) = K_1\left(i_{Lr}(t) - i_{Lm}(t)\right) + K_2 i_L(t) = K_1 \Delta i_{Lr} + K_2 i_{L\_off} + K_2 \frac{V_{C2} + K_2 V_o}{L_k}(t - t_{dis})
\tag{5}
$$

where i_{L_off} is the leakage current when Q_1-Q_4, S_1-S_4 are turned off, and t_{dis} is the total discharge time of the junction capacitors. If i_s can remain negative, the ZVS of the secondary side will not be lost. And then the ZVS range can be extend if Δi_{Lr} is negative, where Δi_{Lr} is the changes in resonant current caused by V_{Lr}. Therefore, the primary junction capacitor discharging time is designed to be smaller than the secondary side in the proposed converter. Then, the ratio of DAB converter and L_m can be designed, where the dead time is set to be 300 ns, as shown in Fig. 5(b). To reduce the number of variables, the power distribution of the proposed converter is characterized as

$$
\frac{P_{DAB}}{P_o} = \frac{P_{DAB}}{P_{LLC} + P_{DAB}} = \frac{V_{C2} i_1}{V_{C1} i_1 + V_{C2} i_1} = \frac{K_2}{K_1 + K_2} = \lambda
\tag{6}
$$

where λ is power distribution coefficient ignoring the conversion loss and i_1 is the input current.

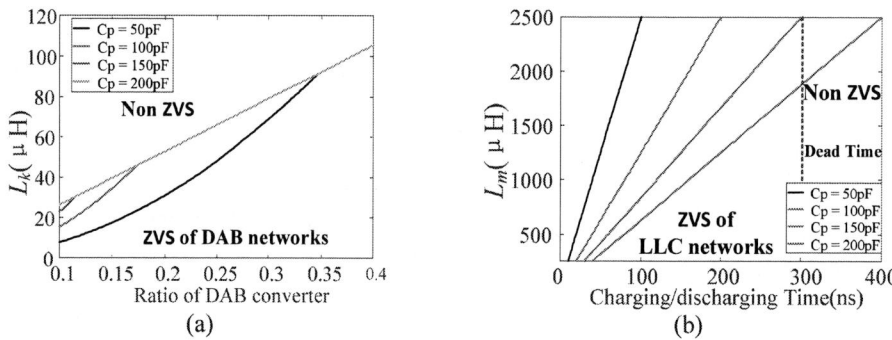

Fig. 5: Parameters range analysis. (a) L_k range analysis. (b) L_m range analysis

Experiments results

A 750W bidirectional DAB-LLC DCX prototype is built up with 350 V input and 48 V output voltage, as shown in Fig. 6. The system circuit parameters are shown in table I. It can seen that the transformer turns ratio of LLC is bigger than DAB, which means the main power still flows through LLC resonant converter. The output voltage is used as feedback and the phase shift angle of DAB is obtained by PI controller.

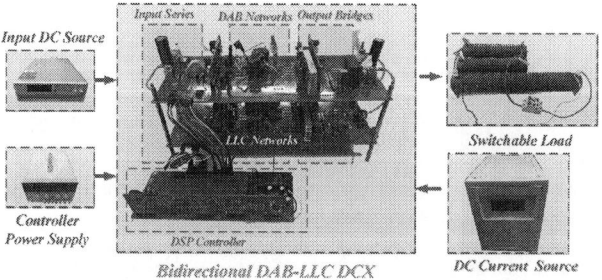

Fig. 6: The prototype of the bidirectional DAB-LLC DCX.

TABLE II: CIRCUIT PARAMETERS

Symbol	Quantity	Value
L_r	LLC Resonant Inductor	25µH
C_r	LLC Resonant Capacitor	0.1µF
L_k	DAB Leakage Inductor	25µH
R	Load Resistance	3Ω-6Ω
f_s	Switching Frequency	100KHz
K_1	Transformer Turns ratio of LLC	35/6
K_2	Transformer Turns ratio of DAB	35/24
λ	Power Distribution Coefficient	0.2

The steady state waveforms of proposed converter under half load and full load condition are shown in Fig. 7, which include output voltage, output current, LLC resonant current and DAB inductor current. It can be seen from Fig. 7(a) and Fig. 7(b) that when the load is different, the steady state output voltage can remains 48V. Fig. 7(c) shows that the proposed converter has voltage regulation capability.

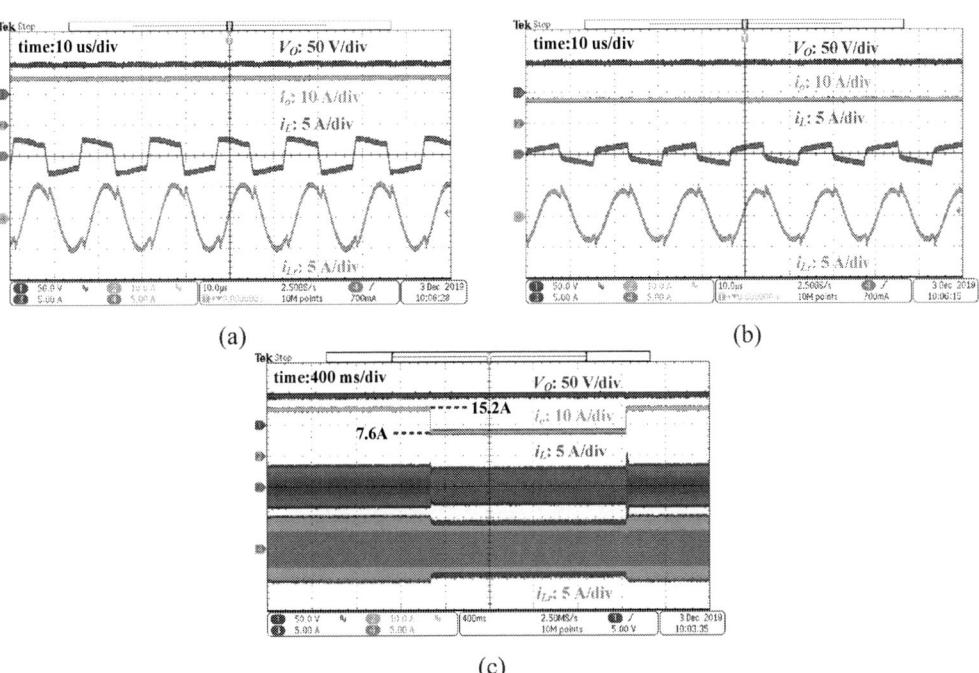

(a)   (b)

(c)

Fig. 7: The key waveforms under different load. (a) Full load. (b) Half load. (c) Load is step changing between half load and full load.

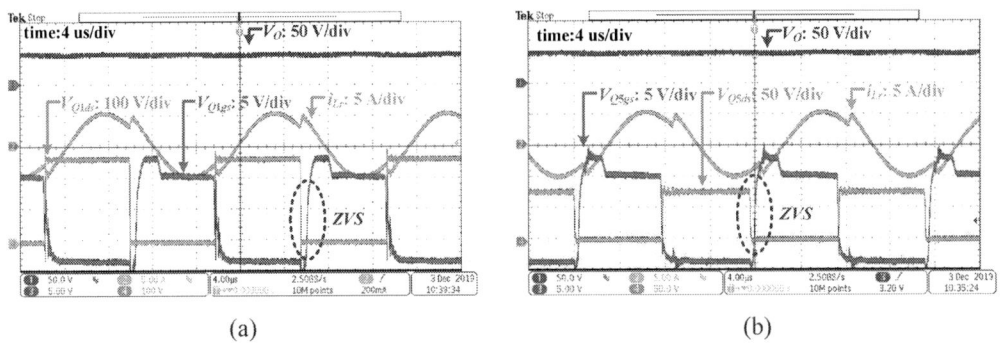

(a)   (b)

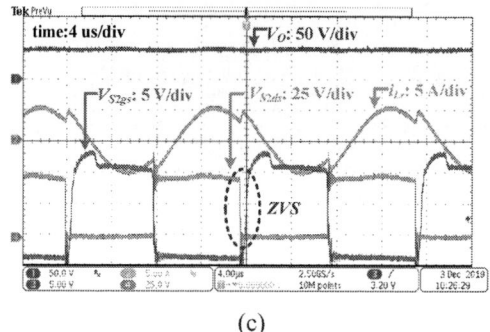

(c)

Fig. 8: Soft switching waveforms. (a) Switches of LLC networks. (b) Switches of DAB networks. (c) Switches of output bridges.

Fig. 8 shows the soft switching waveforms for forward and backward power transmission under full load. which include the output voltage, LLC resonant current, gate signal and drain-source voltage of switches of LLC networks, DAB networks and output bridges, respectively. It can be seen that all switches can achieve ZVS. Bidirectional power switching experiment waveforms are shown in Fig.9. As seen, the resonant current is smoothly changed. Fig. 10 shows the measured efficiency under full load. It can be seen that the efficiency under full load of the proposed converter is 96.58%.

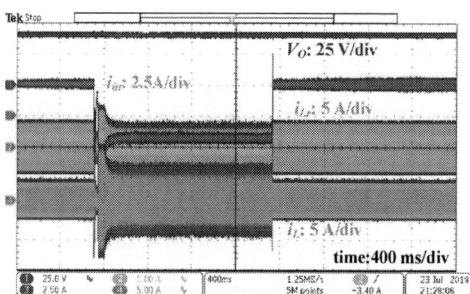

Fig.9: The key waveforms with power direction switching.

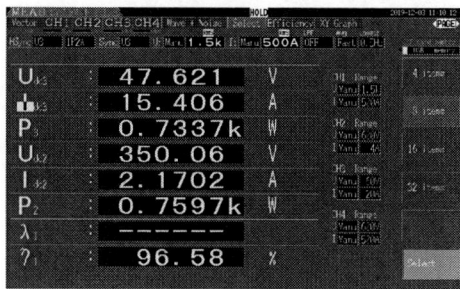

Figure 10: Measured efficiencies under full load.

Conclusion

In this paper, a bidirectional DAB-LLC DCX is proposed for achieving both voltage regulation and high efficiency, whose inputs are in series and outputs are in parallel. Through proper design, the main power flows through LLC resonant converter, which can achieve high-efficiency. While DAB converter is used for voltage regulation and help to achieve wide ZVS range capability in both forward and backward power transmission. The ZVS range is widened by designing inconsistently of the charging and discharging time of the primary and secondary side junction capacitor in the dead time. Finally, the experimental results show that the proposed topology is effective.

References

[1] R. W. A. A. DeDoncker, D. M. Divan, and M. H. Kheraluwala, "A three-phase soft-switched high-power-density dc/dc converter for high power applications," IEEE Trans. Ind. Appl, vol. 27, no. 1, pp. 63–73

[2] M. N. Kheraluwala, R. W. Gascoigne, D. M. Divan, and E. D. Baumann,"Performance characterization of a high-power dual active bridge dc-to dc converter," IEEE Trans. Ind. Appl, vol. 28, no. 6, pp. 1294–1301

[3] D. Doncker, D. Divan et al. "A three-phase soft-switched high power density DC/DC converter for high power applications," in Conference Record of the 1988 IEEE Industry Applications Society Annual Meeting, 1988, pp. 796–805.

[4] J. Xu, J. Yang, G. Xu, et al, "PWM Modulation and Control Strategy for LLC-DCX Converter to Achieve Bidirectional Power Flow in Facing With Resonant Parameters Variation," IEEE Access, vol. 7, pp. 54693–54704

[5] X. Wu , H. Chen and Z. Qian, "1-MHz LLC Resonant DC Transformer (DCX) With Regulating Capability," IEEE Transactions on Industrial Electronics, vol. 63, no. 5, pp. 2904–2912

[6] Q. Ting, and C. Qian, "An Adaptive Frequency Optimization Scheme for LLC Converter With Adjustable Energy Transferring Time," IEEE Transactions on Power Electronics, vol. 34, no. 3, pp. 2018–2024

[7] Shen Y, Zhao W, Chen Z, et al, "Full-bridge LLC resonant converter with series-parallel connected transformers for electric vehicle on-board charger," IEEE Access, vol. 6, pp. 13490–13500

[8] M. Fariborz et al., "An LLC Resonant DC–DC Converter for Wide Output Voltage Range Battery Charging Applications," IEEE Transactions on Power Electronics, vol. 28, no. 12, pp. 5437–5445

J. Sun, M. Xu, D. Reusch, et al, "High Efficiency Quasi-Parallel Voltage Regulators," in Proc. IEEE Appl. Power Electron. Conf. Expo. (APEC'08), 2008, pp. 811–817.

[9] M. Ahmed, C. Fei, Lee F, et al, "Single-Stage High-Efficiency 48/1V Sigma Converter with Integrated Magnetics," IEEE Transactions on Industrial Electronics, to be published. DOI: 10.1109/TIE.2019.2896082.

[10] X. Guo et al., "Leakage current suppression of three-phase flying capacitor pv inverter with new carrier modulation and logic function," IEEE Transactions on Power Electronics, vol. 33, no. 3, pp. 2127–2135

[11] H. Qin, J. Kimball, "Generalized Average Modeling of Dual Active Bridge DC–DC Converter," IEEE Transactions on Power Electronics, vol. 27, no. 4, pp. 2078–2084

Saliency Selection for Search-based AC Machine Low and Zero Speed Estimation Methods

K. Scicluna[1][2], C. Spiteri Staines[1], R. Raute[1]

[1]UNIVERSITY OF MALTA
Msida, MSD2080, Malta

[2]MALTA COLLEGE OF ARTS, SCIENCE & TECHNOLOGY
Paola, PLA9032, Malta
E-Mail: Kris.Scicluna@mcast.edu.mt

Keywords

«Sensorless control», «Permanent magnet motor», «Estimation technique», «Variable speed drive», «Vector control»

Abstract

This paper presents the analysis and selection of a suitable injection angle to be used with continuous pulsating injection and search-based sensorless control. Pulsating injection offers the possibility of different injection orientations in the stationary reference frame. The injection results in High-Frequency current signatures, which vary in shape and magnitude due to the multiple saliency harmonics on the experimental Permanent Magnet Synchronous Machine. The analysis and selection process presented in this paper aims to reduce the implementation time for robust sensorless control. Experimental results are presented for different injection angles and i_q-current operating points.

Introduction

Historically, basic zero to low-speed sensorless estimators assumed a single saliency [1, 2]. However, this is a significant practical limitation, as identified in [3, 4]. Experimental AC machines, including Permanent Magnet Synchronous Machines (PMSMs), typically exhibit additional harmonic saliencies due to deviations from the ideal model. When the harmonic component is significant trigonometric based estimators fail. Various techniques have been used to overcome the multiple saliency problem while still using a trigonometric observer; one of these is Space Harmonic Profiling [5, 6] which is an offline compensation technique. In order to overcome the shortcomings of offline compensation, search-based sensorless estimation methods have been proposed in [7, 8] which rely on a search and comparison method of real-time saliency with a previously offline/online calibrated Look-Up Table (LUT).

Search-based sensorless algorithms are significantly different from other methods as they rely on the previously estimated position and the integrity of the measured saliency with respect to the commissioned LUT. This paper presents a numerical method for differentiating between different magnetic signatures in order to select the one which is most likely to result in the optimum sensorless control. A pulsating HF injection is used as this has been reported to offer advantages over other continuous injection methods [2, 9, 10]. A numerical analysis of deviation (based on standard deviation) and gradients of the saliency in the stationary $\alpha\beta$ frame is discussed with an analysis carried out for an experimental 400 W PMSM at different injection angles.

Pulsating High-Frequency Injection

In this paper, a continuous HF pulsating injection at a frequency ω_i ($2\pi f_i$) was superimposed on the fundamental stator voltage signals in the stationary $\alpha\beta$ frame of the form (1). In this research, it was observed that the injection of HF voltages at different angles γ in the stationary frame results in different magnetic signatures. Saliencies are noted to vary in deviation (difference of instantaneous saliency from the mean) and gradients (change in saliency per unit change in rotor position).

The pulsating vector (1) is injected at an angle γ to the stationary α axis and has a maximum magnitude of V_i at the pulsating vector peak. The coefficients A_1 and A_2 are defined in (2-3).

$$v_{i\alpha\beta} = \begin{bmatrix} A_1 \cos \omega_i t \\ A_2 \cos \omega_i t \end{bmatrix} \tag{1}$$

$$A_1 = V_i \cos \gamma \tag{2}$$

$$A_2 = V_i \sin \gamma \tag{3}$$

For an ideal PMSM with a single saliency, the injection in (1) results in the HF currents $i_{i\alpha}$ and $i_{i\beta}$ in (4) which can be simplified as an amplitude modulated function in (5). The saliency in the single saliency model is a result of the difference in the synchronous frame inductances (L_d/L_q).

$$\begin{bmatrix} i_{i\alpha} \\ i_{i\beta} \end{bmatrix} = \begin{bmatrix} I_1 A_1 \sin(\omega_i t) \\ +I_2 A_1 \cos(2\theta_e) \sin(\omega_i t) \\ +I_2 A_2 \sin(2\theta_e) \sin(\omega_i t) \\ \\ I_1 A_2 \sin(\omega_i t) \\ -I_2 A_2 \cos(2\theta_e) \sin(\omega_i t) \\ +I_2 A_1 \sin(2\theta_e) \sin(\omega_i t) \end{bmatrix} \tag{4}$$

$$\begin{bmatrix} i_{i\alpha} \\ i_{i\beta} \end{bmatrix} = \begin{bmatrix} A_{i\alpha} \sin(\omega_i t) \\ A_{i\beta} \sin(\omega_i t) \end{bmatrix} \tag{5}$$

Where:

θ_e is the electrical rotor position, and

I_1, I_2 are amplitude coefficients resulting from the single saliency model and are defined in (6-7)

$$I_1 = \frac{V_i L}{\omega_i (L^2 - \Delta L^2)} \tag{6}$$

$$I_2 = \frac{V_i \Delta L}{\omega_i (L^2 - \Delta L^2)} \tag{7}$$

Where $\quad L = \frac{L_q + L_d}{2}$ $\qquad\qquad\qquad\qquad\qquad\qquad \Delta L = \frac{L_q - L_d}{2}$

The amplitudes ($A_{i\alpha}$ and $A_{i\beta}$) of the HF currents are shown to be saliency dependent in (8) and can be obtained by demodulation.

$$\begin{bmatrix} A_{i\alpha} \\ A_{i\beta} \end{bmatrix} = \begin{bmatrix} I_1 A_1 + I_2 A_1 \cos(2\theta_e) + I_2 A_2 \sin(2\theta_e) \\ I_1 A_2 - I_2 A_2 \cos(2\theta_e) + I_2 A_1 \sin(2\theta_e) \end{bmatrix} \tag{8}$$

Saliency Requirements for Search-Based Methods

In general, saliencies in practical machines are observed to have harmonics besides the fundamental saliency such that the accuracy of traditional trigonometric observers is unsatisfactory for sensorless estimation at low to zero speed. Search-based methods [7, 8] overcome the multiple saliency problem as they depend on the comparison of $A_{i\alpha}$, $A_{i\beta}$ components measured in real-time with a previously commissioned LUT which can be generated both online and offline. In order to obtain accurate sensorless position estimation, the saliency measurements on the machine should have the following properties:

1. Low Deviation (High Signal-to-Noise (SNR))
2. High Gradient (A/°) in the combined αβ saliency components

The deviation in the measured saliency affects the instantaneous error in the estimated sensorless position. A high deviation from the mean (commissioned LUT) of the saliency will result in a high deviation from the actual position being estimated.

As regards to the gradient, the ideal $A_{i\alpha}$, $A_{i\beta}$ saliency of the SM-PMSM should consist of two sinusoidal components phase shifted by 90°. This can be easily demonstrated by setting one of the injected components such as $A_2 = 0$. The resultant saliency components are shown in (9).

$$\begin{bmatrix} A_{i\alpha} \\ A_{i\beta} \end{bmatrix} = \begin{bmatrix} I_1 A_1 + I_2 A_1 \cos(2\theta_e) \\ I_2 A_1 \sin(2\theta_e) \end{bmatrix} \tag{9}$$

The complex plot of (9) results in a perfect circle which has a constant gradient (A/°) at all positions of θ_e. In this research, it was observed that machines of similar rating and construction to the one described in Table IV tend to have a saliency which is more oval/triangular rather than circular. This results in specific rotor positions where the gradient of $A_{i\alpha}$, $A_{i\beta}$ is low. This results in low sensitivity of the search-based observer and should be minimized.

Magnetic Signature Analysis for Search-Based Methods

Definition of Statistical Parameters for Analysis

The numerical analysis presented in [11] assumes that a saliency sample for $A_{i\alpha}$, $A_{i\beta}$ has been collected for a period T s. All samples are sorted with respect to an integer element number N (Fig. 1) which is a scaled quantity of the mechanical rotor position θ_m. The sorted samples are defined as $A_{i\alpha\_sort}$, $A_{i\beta\_sort}$. The mean of the sorted saliency measurements for each value of N is averaged out to obtain the mean saliency for the α and β stationary frame of references $\overline{A_{i\alpha\_sort}}$, $\overline{A_{i\beta\_sort}}$. The deviation components are defined in (10-11). $\overline{A_{i\alpha\_sort}}$, $\overline{A_{i\beta\_sort}}$ can also used for the LUTs in the search-based sensorless control algorithm.

$$D_\alpha = \frac{\sigma A_{i\alpha\_sort}}{\max|\overline{A_{i\alpha\_sort}}|} \tag{10}$$

$$D_\beta = \frac{\sigma A_{i\beta\_sort}}{\max|\overline{A_{i\beta\_sort}}|} \tag{11}$$

Where: $\sigma A_{i\alpha\_sort}$, $\sigma A_{i\beta\_sort}$ is the standard deviation of the measured saliency $A_{i\alpha}$, $A_{i\beta}$ from the mean $\overline{A_{i\alpha\_sort}}$, $\overline{A_{i\beta\_sort}}$. The deviation components are normalized with respect to the maximum value of the sorted measurements. The average of the deviation signals (12-13) and the maxima (14-15) are used for saliency analysis.

$$\overline{D_\alpha} = \overline{\left|\frac{\sigma A_{i\alpha\_sort}}{\max|\overline{A_{i\alpha}}|}\right|} \tag{12}$$

$$\overline{D_\beta} = \overline{\left|\frac{\sigma A_{i\beta\_sort}}{\max|\overline{A_{i\beta}}|}\right|} \tag{13}$$

$$D_{\alpha max} = \max\left(\left|\frac{\sigma A_{i\alpha\_sort}}{\max|\overline{A_{i\alpha}}|}\right|\right) \tag{14}$$

$$D_{\beta max} = \max\left(\left|\frac{\sigma A_{i\beta\_sort}}{\max|\overline{A_{i\beta}}|}\right|\right) \tag{15}$$

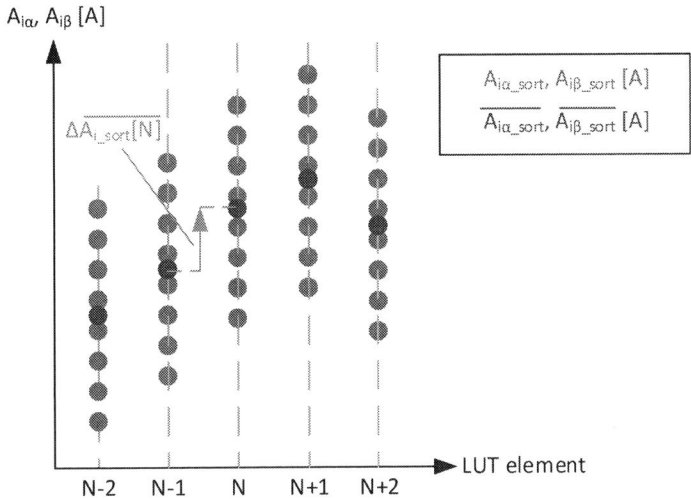

Fig. 1: Sorting and Averaging of $A_{i\alpha}$, $A_{i\beta}$ Measurements

The gradient which is the change in the mean saliency between one element value and another is defined as $\Delta\overline{A_{i\alpha\_sort}}$, $\Delta\overline{A_{i\beta\_sort}}$. Two parameters dependent on the gradient of the saliency are defined in (16-17).

$$\min\_\Delta\alpha\beta = \min(|\Delta\overline{A_{i\alpha\_sort}}| + |\Delta\overline{A_{i\beta\_sort}}|) \tag{16}$$

$$\overline{\Delta\alpha\beta} = \overline{|\Delta\overline{A_{i\alpha\_sort}}| + |\Delta\overline{A_{i\beta\_sort}}|} \tag{17}$$

The expression in (16) results in the minimum change in the saliency of the sum of the gradients of the absolute of $\Delta\overline{A_{i\alpha\_sort}}$ and $\Delta\overline{A_{i\beta\_sort}}$ while (17) is the mean of the sum of the same absolute gradients. The parameters defined in (12-17) are used to differentiate between the saliency at different injection angles γ and different values of the i_q-current. Lower values of $\overline{D_\alpha}$, $\overline{D_\beta}$ and higher values of min_$\Delta\alpha\beta$ and $\overline{\Delta\alpha\beta}$ will tend to improve the sensitivity of the observer and reduce the mean error in the estimation.

Experimental Saliency Determination with Pulsating Injection at No Load

The saliency measurements on the PMSM were carried out with $i_q^* = 0$ A with injection in the stationary frame of reference angle γ varied between $0 \leq \gamma \leq 165°$ in steps of $15°$. The rotor speed of the machine is set by another coupled machine at $\omega_m = 1$ rad/s. The injection was carried out at a frequency $f_i = 2.5$ kHz with a 20 kHz PWM and a maximum pulsating voltage vector of $V_i = 3$ V. The parameters defined in (12-17) were calculated for all of the operating points. $A_{i\alpha}$, $A_{i\beta}$ were measured over a period $T = 60$ s are shown in Figs 2 – 5 for the unloaded machine with $i_q^* = 0$ A. The saliency is shown as a function of the mechanical rotor position (with one mechanical rotor revolution mapped to 4095 elements in the search-based algorithm). As predicted from the ideal model in (8) the saliency is a function of $2\theta_e$ since it repeats itself 12 times in one revolution of θ_m for a 6-pole machine. The shape of the saliency, however, was observed to vary significantly in the complex plane shown in Fig. 5.

The results for the numerical analysis of the saliency at $i_q^* = 0$ A are shown in Table I. The numerical results show that the deviations vary with the injection angle γ. The $\overline{D_\alpha}$ varies from a minimum of 8.98E-04 at $\gamma = 0°$ to a maximum of 0.0056 at $\gamma = 45°$. Similarly, for the β axis $\overline{D_\beta}$ varies from a minimum of 8.47E-04 at $\gamma = 135°$ to a maximum of 0.0028 at $\gamma = 0°$. The minimum combined gradient min_$\Delta\alpha\beta$ was observed to vary from a minimum of 22.7 µA/element at $\gamma = 30°$ to a maximum of 371 µA/element at $\gamma = 120°$. The mean combined gradient $\overline{\Delta\alpha\beta}$ was observed to have a maximum of 1.6 mA/element at $\gamma = 75/165°$. From Table I the component $\overline{\Delta\alpha\beta}$ for all injection values is of the same order of magnitude. Hence $\gamma = 120°$ was selected as a suitable injection angle due to the maximum min_$\Delta\alpha\beta$ occurring at this point. From Fig. 5 it can be observed that the saliency at $\gamma = 120°$ is closer to the ideal circular saliency found in single saliency machines compared to that at $\gamma = 30°$, which is more triangular in shape. These saliencies are a result of the same injection frequency, injection amplitude and load with the only difference being the injection angle in the stationary $\alpha\beta$ frame.

The deviation components in Table I were only considered to discriminate between saliencies when the selection based of the gradient components does not result in satisfactory sensorless control. This may occur due to the high noise component in the measured saliency, albeit of the high gradient in the saliency. If such a case arises a saliency with lower $\overline{\Delta\alpha\beta}$ would have to be chosen in order to have lower deviation values.

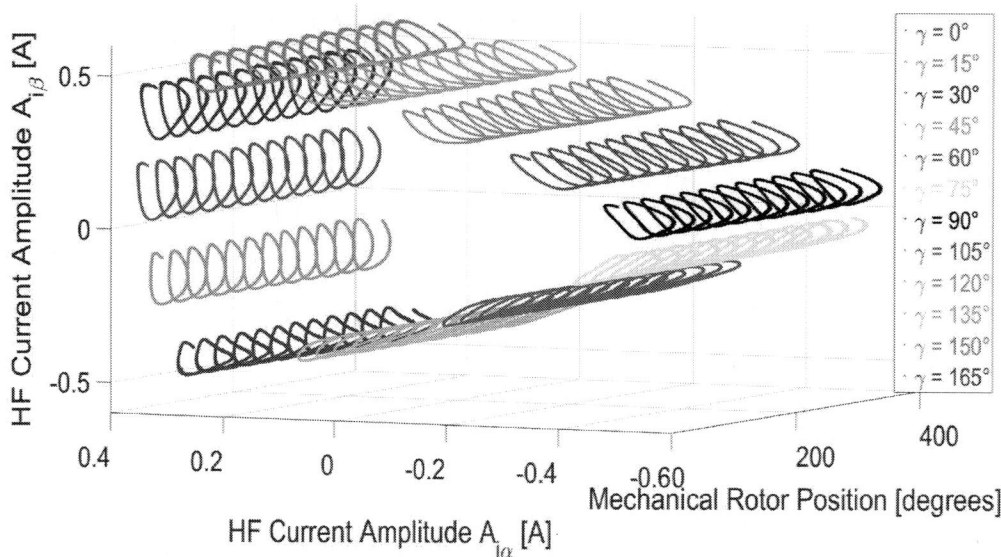

Fig. 2: 3D Plot of $A_{i\alpha}$ [A], $A_{i\beta}$ [A] and Mechanical Rotor Position [°], $f_i = 2.5$ kHz and $V_i = 3$ V.

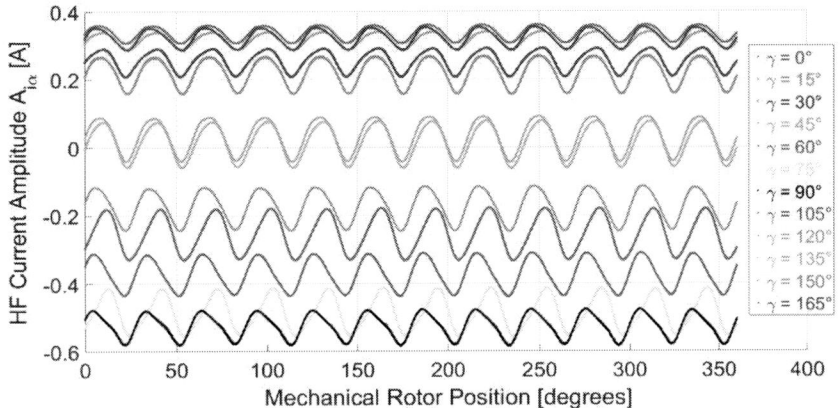

Fig. 3: Plot of $A_{i\alpha}$ [A] vs. Mechanical Rotor Position [°], f_i = 2.5 kHz and V_i = 3 V.

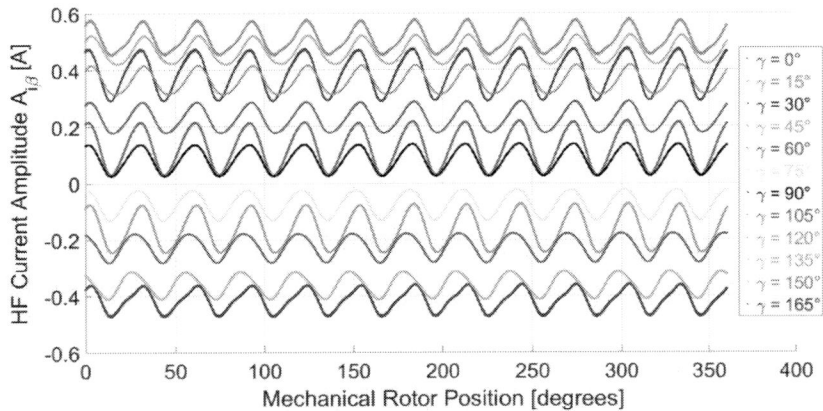

Fig. 4: Plot of $A_{i\beta}$ [A] vs. Mechanical Rotor Position [°], f_i = 2.5 kHz and V_i = 3 V.

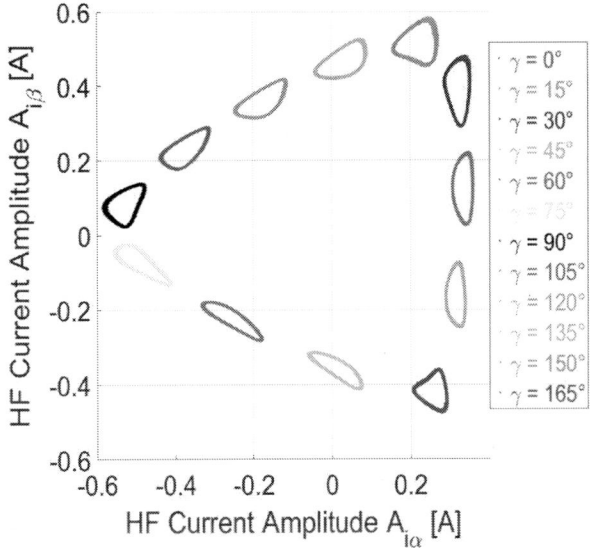

Fig. 5: Plot of $A_{i\alpha}$ [A] vs. $A_{i\beta}$ [A] f_i = 2.5 kHz and V_i = 3 V.

Table I: Saliency Numerical Analysis for $0° \leq \gamma \leq 165°$, $f_i = 2.5$ kHz, $V_i = 3$ V, $i_q^* = 0$ A, $\omega_m = 1$ rad/s.

γ	$\overline{D_\alpha}$	$D_{\alpha max}$	$\overline{D_\beta}$	$D_{\beta max}$	min_$\Delta\alpha\beta$ (A/element)	$\overline{\Delta\alpha\beta}$ (A/element)
0	8.98E-04	0.0019	0.0028	0.0063	3.31E-05	0.0014
15	9.28E-04	0.0018	0.0022	0.0049	9.88E-05	0.0013
30	0.0016	0.0041	0.0015	0.0038	2.27E-05	0.0013
45	0.0056	0.012	9.13E-04	0.0019	2.47E-04	0.0014
60	0.0015	0.0035	0.0013	0.0025	4.33E-05	0.0015
75	0.0014	0.0038	0.0027	0.0051	4.88E-05	0.0016
90	0.0012	0.0033	0.0027	0.0059	7.08E-05	0.0014
105	0.0011	0.0024	0.0013	0.0027	6.62E-05	0.0014
120	0.0018	0.0037	9.00E-04	0.002	3.71E-04	0.0014
135	0.005	0.0114	8.47E-04	0.0021	2.70E-04	0.0014
150	0.0022	0.0048	0.0013	0.0038	8.07E-05	0.0015
165	0.0014	0.0036	0.0017	0.004	7.01E-05	0.0016

Experimental Saliency Determination with Pulsating Injection in Loaded Condition

In the previous section, the numerical analysis for selecting a suitable HF current signature with a pulsating injection angle at different injection angles γ was presented for an unloaded machine. This signature is expected to change significantly as a function of the machine load [12]. This was shown experimentally for the same experimental setup in this paper in [13]. Since the signatures are subject to signification variation, the numerical analysis was repeated for different values of the i_q-current. The numerical analyses for $i_q^*=5$A and $i_q^*=-5$A are shown in Tables II and III, with all other injection parameters unchanged.

The numerical parameters for both deviation and gradient change significantly at different values of the of the i_q-current. If the selection is still carried out based on the minimum gradient min_$\Delta\alpha\beta$ the best injection angle changes from $\gamma=120°$ at no load, to $\gamma=165°$ (5 A) and $\gamma=75°$ (-5 A). The mean deviation parameters are shown for -10 A $\leq i_q^* \leq 10$ A in steps of 2.5 A in Fig 6. Both the average deviation and the maximum deviation increase with additional loading on the machine. This is attributed to various factors including higher variation in the injection and possibly a less pronounced saliency. From these results, a higher Root Mean Square Error (RMSE) in the estimated rotor position is expected for higher values of the i_q-current.

Table II: Saliency Numerical Analysis for $0° \leq \gamma \leq 165°$, $f_i = 2.5$ kHz, $V_i = 3$ V, $i_q^* = 5$ A, $\omega_m = 1$ rad/s

γ	$\overline{D_\alpha}$	$D_{\alpha max}$	$\overline{D_\beta}$	$D_{\beta max}$	min_$\Delta\alpha\beta$ (A/element)	$\overline{\Delta\alpha\beta}$ (A/element)
0	0.0022	0.0053	0.006	0.0145	2.68E-05	0.0013
15	0.0022	0.0054	0.005	0.0152	2.66E-05	0.0012
30	0.0054	0.0176	0.0054	0.0196	5.84E-06	0.0019
45	0.1240	0.032	0.0022	0.0074	1.05E-06	0.0013
60	0.0033	0.0086	0.0022	0.0076	4.81E-06	0.0012
75	0.0043	0.013	0.0042	0.0088	2.42E-06	0.0016
90	0.0041	0.014	0.0045	0.0101	1.81E-06	0.0017
105	0.0025	0.0087	0.0024	0.0061	2.44E-06	0.0012
120	0.0037	0.0096	0.0020	0.0041	2.94E-06	0.0012
135	0.0112	0.0275	0.0021	0.0053	3.51 E-06	0.0013
150	0.0062	0.0185	0.0042	0.0123	1.08 E-06	0.0020
165	0.0040	0.0124	0.0049	0.0146	5.06 E-06	0.0018

Table III: Saliency Numerical Analysis for $0° \leq \gamma \leq 165°$, $f_i = 2.5$ kHz, $V_i = 3$ V, $i_q^* = -5$ A, $\omega_m = 1$ rad/s.

γ	$\overline{D_\alpha}$	$D_{\alpha max}$	$\overline{D_\beta}$	$D_{\beta max}$	min_$\Delta\alpha\beta$ (A/element)	$\overline{\Delta\alpha\beta}$ (A/element)
0	0.0027	0.0074	0.0076	0.0206	2.65 E-05	0.0014
15	0.0031	0.0065	0.0067	0.0197	4.99E-06	0.0013
30	0.0080	0.302	0.0077	0.0303	1.57 E-05	0.0020
45	0.0174	0.0466	0.0331	0.0128	1.07 E-05	0.0013
60	0.0046	0.0134	0.0029	0.0108	7.74E-06	0.0019
75	0.0061	0.0214	0.0054	0.0134	4.05 E-05	0.0018
90	0.0056	0.0177	0.0060	0.0144	3.78 E-05	0.0017
105	0.0034	0.0098	0.0032	0.0081	3.42 E-05	0.0012
120	0.0052	0.0135	0.0027	0.0066	3.21 E-05	0.0012
135	0.0150	0.0451	0.0028	0.0092	1.70 E-05	0.0013
150	0.0088	0.0267	0.0059	0.0186	1.19 E-05	0.0020
165	0.0053	0.0171	0.0062	0.0199	6.30E-06	0.0019

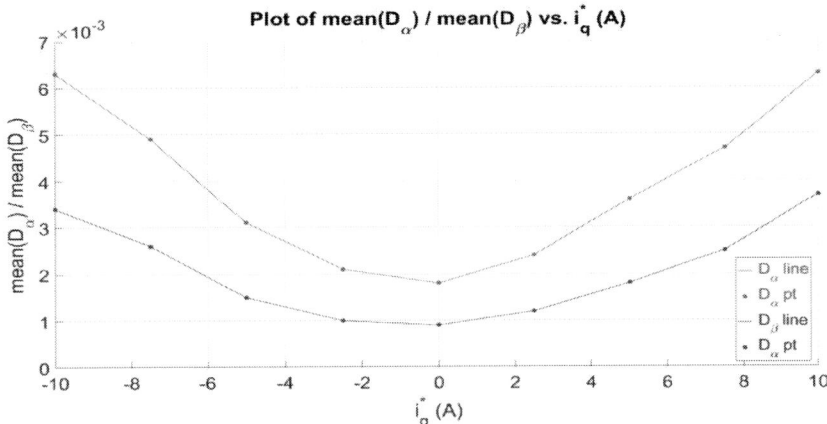

Fig. 6: Plot of $\overline{D_\alpha}$, $\overline{D_\beta}$ with $\omega_m^* = 1$ rad/s on M4, $i_d^* = 0$ A and -10 A $\leq i_q^* \leq 10$ A, $\gamma = 120°$, $f_i = 2.5$ kHz and $V_i = 3$ V.

Conclusions

This paper presented a numerical analysis of magnetic signatures at different injection angles γ in the stationary $\alpha\beta$ frame. The signatures of the PMSM machine under test were shown not to follow the single saliency model due to the presence of multiple saliency harmonic components. The different signatures obtained at the same injection frequency (f_i=2.5 kHz) and pulsating injection amplitude ($V_i = 3$ V) were analyzed in terms of both deviation and gradient parameters. The best signature at no load condition following the analysis was manually chosen to be at $\gamma = 120°$.

The numerical analysis was extended to different i_q-current values as the saliency was found to vary the load of the machine. It was shown that both deviation and gradient parameters change significantly at different load operating points. The suitable injection angle γ was shown to change at each operating point tested. This indicates that for optimum sensorless control, the injection angle γ should be also varied. This is currently being investigated further in order to provide a mechanism for switching between different angles in real-time.

While the selection of a suitable injection angle in this paper was carried out manually, it provides a more structured way based on statistical analysis rather than a trial-and-error approach. Currently, further research is being carried out in order to improve upon the proposed saliency selection process and possibly automate the process entirely.

Table IV: PMSM Machine Parameters

Symbol	Description	Value	Units
P	Rated Power	400	W
N_{rated}	Rated Speed	600	RPM
V_{rms}	Rated RMS Voltage	24	V
I_{rms}	Rated RMS Current	7	A
L_d	d-axis stator inductance	807.9	μH
L_q	q-axis stator inductance	641.1	μH
R	Stator resistance	262	mΩ
p	Number of Pole Pairs	6	-

References

[1] P. L. Jansen and R. D. Lorenz, "Transducerless position and velocity estimation in induction and salient AC machines," *Industry Applications, IEEE Transactions on,* vol. 31, pp. 240-247, 1995.

[2] D. Raca, P. Garcia, D. D. Reigosa, F. Briz, and R. D. Lorenz, "Carrier-Signal Selection for Sensorless Control of PM Synchronous Machines at Zero and Very Low Speeds," *IEEE Transactions on Industry Applications,* vol. 46, pp. 167-178, 2010.

[3] M. W. Degner and R. D. Lorenz, "Using multiple saliencies for the estimation of flux, position, and velocity in AC machines," *Industry Applications, IEEE Transactions on,* vol. 34, pp. 1097-1104, 1998.

[4] D. Paulus, P. Landsmann, S. Kuehl, and R. Kennel, "Arbitrary injection for permanent magnet synchronous machines with multiple saliencies," in *Energy Conversion Congress and Exposition (ECCE), 2013 IEEE,* 2013, pp. 511-517.

[5] N. Teske, G. M. Asher, M. Sumner, and K. J. Bradley, "Analysis and suppression of high-frequency inverter modulation in sensorless position-controlled induction machine drives," *IEEE Transactions on Industry Applications,* vol. 39, pp. 10-18, 2003.

[6] N. Teske, G. M. Asher, K. J. Bradley, and M. Summer, "Analysis and suppression of inverter clamping saliency in sensorless position controlled induction machine drives," in *Conference Record of the 2001 IEEE Industry Applications Conference. 36th IAS Annual Meeting (Cat. No.01CH37248),* 2001, pp. 2629-2636 vol.4.

[7] K. Scicluna, C. S. Staines, and R. Raute, "Sensorless Position Estimation using a Search based Online Commissionable Method (SONIC)," presented at the 20th European Conference on Power Electronics and Applications (EPE '18), Riga, Lativa, 2018.

[8] K. Scicluna, C. S. Staines, and R. Raute, "Sensorless Position Tracking in Steer-by-Wire Using the SONIC Method," in *2018 New Generation of CAS (NGCAS),* 2018, pp. 122-125.

[9] X. Luo, Q. Tang, A. Shen, and Q. Zhang, "PMSM sensorless control by injecting HF pulsating carrier signal into estimated fixed-frequency rotating reference frame," *IEEE Transactions on Industrial Electronics,* vol. 63, pp. 2294-2303, 2016.

[10] J. M. Liu and Z. Q. Zhu, "A new sensorless control strategy by high-frequency pulsating signal injection into stationary reference frame," in *2013 International Electric Machines & Drives Conference,* 2013, pp. 505-512.

[11] K. Scicluna, C. S. Staines, and R. Raute, "High frequency injection-based sensorless position estimation in permanent magnet synchronous machines," *Mathematics and Computers in Simulation,* 2020.

[12] C. Zhe, C. Xinbo, R. Kennel, and W. Fengxiang, "Enhanced sensorless control of SPMSM based on Stationary Reference Frame High-Frequency Pulsating Signal injection," in *2016 IEEE 8th International Power Electronics and Motion Control Conference (IPEMC-ECCE Asia),* 2016, pp. 885-890.

[13] K. Scicluna, C. Spiteri-Staines, and R. Raute, "Sensorless low/zero speed Estimation for PMSM using a Search-based Real-Time Commissioning Method," *IEEE Transactions on Industrial Electronics,* pp. 1-1, 2020.

Genetic Algorithm Based Multi Objective Optimization for Inductor Design

Thorben Schobre, Raquel González Aríztegui, Regine Mallwitz
TU Braunschweig, Institute for Electrical Machines, Traction and Drives
Hans-Sommer-Straße 66
Braunschweig, Germany
Phone: +49 (531) 391-3910
Email: t.schobre@tu-braunschweig.de
URL: https://www.tu-braunschweig.de/imab

Keywords

≪Converter circuit≫, ≪Modeling≫, ≪Design≫, ≪Software≫, ≪Artificial Intelligence≫

Abstract

In this work an optimal design approach for inductors for buck or boost converters is presented. The optimization routine is based on Non-Dominated Sorting Genetic Algorithm (NSGA2) which is well suited for such multi objective optimization applications with discrete design parameters. The optimization objectives are size and loss. Therefore an adequate loss model is established. The whole algorithm and the results are presented. To verify the models a test inductor is realized and characterized.

Introduction

Power electronics design is usually dominated by finding the best trade-off of efficiency, power density and costs with the goal to find the optimal solution. The goals usually oppose each other which result in a multi objective optimization problem. One approach to solve these kind of problems are strategies based on parameter variations. Also a lot of power electronics design rules are based on best practice [1] or experience of the power electronic design engineer which often will find a good solution, but it is not ensured that it is the optimal solution.

The technique of parameter variations reaches the limit of complexity very quickly. A power electronic converter can have easily more than 10 different design parameters. Each parameter can then have a large variation range of maybe more than 100 variations per parameter. This calculation example results in a total number of $1 \cdot 10^{20}$ possible variations resulting in a high demand of computation time.

To minimize the computation complexity one beneficial approach is the use of multi objective optimization algorithms. Several different algorithms have been developed in the past. Some are inspired by natural processes like the Multi Objective Particle Swarm Optimization [2] or the genetic algorithms [3]. In this work a Non-Dominated Sorting Genetic Algorithm (NSGA2) algorithm is applied to optimize the inductor of a conventional buck converter. This algorithm is suited because it can deal with a discrete design space which is present in most power electronic applications.

NSGA2 Algorithm

The Elitist Non-Dominated Sorting Genetic Algorithm 2 [3] is well suited to solve problems where the decision variables are discontinues like the design of power electronics. It is based on the genetic principle. This algorithm has already been applied to the design of power electronic components [4]. The goal of the algorithm to find a solution which is not dominated by others.

Genetic algorithms are based on generations of individuals with a a fixed size N_{gen}. In this work an individual is one possible design solution for an inductor. Each individual consists its design variables.

Based on the design variables the objective functions and the constraints can be calculated. The result of the objective functions like total loss or volume define the fitness of each individual.

The general principle of NSGA2 is explained in the following. At the beginning of the algorithm a first parent generation is generated by random with the number of N_{gen} individuals, also shown in Fig. 1. This parent generation P_t is combined with an offspring generation Q_t. This offspring generation of length N_{gen} is created from the parent generation by binary tournament selection, crossover and mutation. These genetic operations occur with different probabilities which are control parameters of the algorithm. The binary tournament selection takes two random individuals and compares them. The one with the higher fitness can reproduce. When the fitness is equal, the individual with the higher crowding distance can reproduce. From 2 tournaments two parents are generated which then create two offsprings by crossover. A mutation is a second genetic operator which is applied to the population after the crossover step. Randomly a mutation changes a parameter of a individual.

The combined population is called R_t. R_t is sorted by non-dominated sorting which means that all individuals are assigned to fronts F_i. The fronts are created by searching for individuals which dominate others. The individuals of the highest ranking front F_1 are not dominated by any other individuals. The second front is only dominated by individuals of F_1 but the individuals of F_2 dominate all other remaining individuals. Whenever the count of individuals of all fronts is $\geq N_{gen}$ the non-dominated sorting can be stopped. Elitism is ensured, because the non-dominated sorting algorithm works on the whole set of parent and offspring generation and high ranking solutions are preserved from one to the next parent generation. To generate the next parent generation P_{t+1} all highest ranking fronts are directly taken to the next generation as long as the count of N_{gen} is exceeded. In the example Fig. 1 this is the case for front F_1 and F_2. Adding front F_3 would exceed the number of individuals per generation N_{gen} therefore the last front need to be truncated. Before the truncation the last front (F_3 in the example) is sorted by a crowding distance factor. The individuals with the higher crowding distance are ranked higher than with the lower one. This crowding distance sorting ensures diversity of the solution set and prevent to fall into local minima. All individuals with a higher crowding distance value will go into the next parent generation P_{t+1}.

This procedure is repeated until a stop criterion is met, which for example can be a maximum number of populations or a minimal change in the results of the objective functions between one to the next generation.

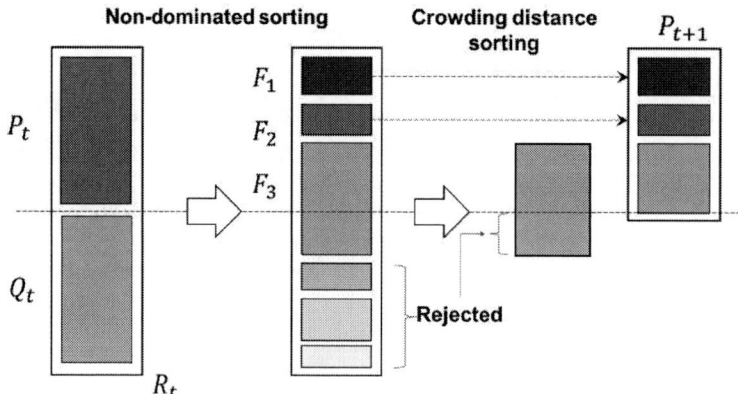

Fig. 1: Principle of generating a new Population with NSGA2 [3].

Objective Functions and Constrains

To apply the NSGA2 the optimization objectives need to be modeled. In this example the loss and the geometrical size of the inductor are to be minimized. These objectives are partly contrary so a trade-off between both will give the optimal solution. Constraints are introduced to fulfill certain requirements. If a constraint is not met by an individual, it will be deleted and replaced.

Objective Function: Losses of the Inductor

The inductor loss can be separated into core losses and winding losses. Which both need to be modeled sufficiently exact but also computable with minimal complexity. For this purpose analytical models are suited well. The core loss can be modeled by the Improved Generalized Steinmetz Equation (IGSE) [5], which gives a good approximation of the losses for non-sinusoidal flux density waveforms based on Steinmetz coefficients provided by the manufacturer. However the influence of a magnetic bias field is not modeled. But to find the optimal solution out of thousands of possible solutions this already is a good measure. Especially because this simplification is present in the whole design process and therefore the prediction error is systematic. This optimization routine give a good idea how the final inductor will look like. But for the final design of the actual inductor 3D FEM modeling with a more sophisticated representation of core losses should be applied. The calculation with IGSE is well suited whenever the current can be described with a piece-wise linear (PWL) waveform. The following equation (1) shows the IGSE for PWL waveforms [5] which is adapted to a buck converter's current respectively flux density waveform giving the result of (3).

$$P_{\text{core}} = \frac{1}{T} V_{\text{Core}} k_i \Delta B^{\beta-\alpha} \sum_m \left| \frac{B_{m+1} - B_m}{t_{m+1} - t_m} \right|^\alpha (t_{m+1} - t_m) \tag{1}$$

Where m is the number of piece wise linear elements, T is the switching period, V_{Core} is the core volume, B_m and t_m are the flux density and the time values of the PWL waveform, α and β are the Steinmetz' coefficients from the datasheet and k_i is calculated in (4). With $m = 2$ for the triangular waveform of a buck converter the following expression follows.

$$P_{\text{core}} = \frac{1}{T} V_{\text{Core}} k_i \Delta B^{\beta-\alpha} \left[\left| \frac{\delta B}{DT} \right|^\alpha DT + \left| \frac{\delta B}{(1-D)T} \right|^\alpha (1-D)T \right] \tag{2}$$

Where D is the duty cycle of the buck converter. After some computation and simplification the following expression based on the IGSE is obtained which is part of the objective function for the losses.

$$P_{\text{core}} = V_{\text{Core}} k_i \mid \Delta B \mid^\beta \frac{1}{T^\alpha} [D^{1-\alpha} + (1-D)^{1-\alpha}] \tag{3}$$

For k_i the following approximation is given in [5] with the Steinmetz datasheet coefficient k and is applied to the core loss calculation.

$$k_i = \frac{k}{2^{\beta+1} \pi^{\alpha-1} (0.2761 + \frac{1.7061}{\alpha+1.354})} \tag{4}$$

The winding loss can be split into low frequency losses and high frequency losses caused by skin and proximity effect. In case of the buck converter the low frequency losses are caused by the DC current through the DC resistance of the inductor. The high frequency winding losses are difficult to approximate. In this work the simple Dowell's method is applied. However this method has some drawbacks, particularly it is not considering the air gap's fringing field. Therefore in future work it is preferable to apply a more complex approximation method like [6]. With Dowell's method [7] the eddy current factor F_R is calculated according to (5) in [8], which describes the ratio of the ac resistance over the DC resistance.

$$F_R = \frac{R_{\text{AC}}}{R_{\text{DC}}} = X \left[\frac{\sinh(2X)\sin(2X)}{\cosh(2X) - cos(2X)} + \frac{2(M_L^2 - 1)}{3} \frac{\sinh(X) - \sin(X)}{\cosh(X) + \sin(X)} \right] \tag{5}$$

M_L is the number of winding layers and X is described in (6).

$$X = \frac{h}{\delta} \sqrt{\frac{N_L b}{b_w}} \tag{6}$$

h is the layer height, N_L is the number of turns per layer, H_w is the width of the winding window, b is the conductor width and δ the skin depth. With this resistance ratio the total winding losses can then be approximated according to equation (7).

$$P_{\text{wind}} = R_{\text{DC}} F_R I_{rms}^2 \tag{7}$$

The sum of winding losses and core losses are the total losses P_{loss} and therefore one of the objective functions of this optimization routine.

$$P_{\text{loss}} = P_{\text{core}} + P_{\text{wind}} \tag{8}$$

Objective Function: Size of the Inductor

The second objective function in the optimization procedure is the boxed volume around the core, based on its geometric parameters. The calculation is based on the longest dimensions in each x,y,z-plane. This approach simplifies the additional volume caused by the winding, however it can be easily calculated based on datasheet parameters of the different cores. To estimate the volume caused by the winding, the winding window outer dimension D is assumed in the y-axis. In the z-axis the core height H is assumed and in the x-axis the core width W is assumed. An example core with its dimensions is shown in Fig. 2. Based on these dimensions the boxed volume can be calculated according to (9). This calculation is working for all cores with a center post which can either be round, square or any other shape.

$$V_{\text{boxed}} = H \cdot W \cdot D \tag{9}$$

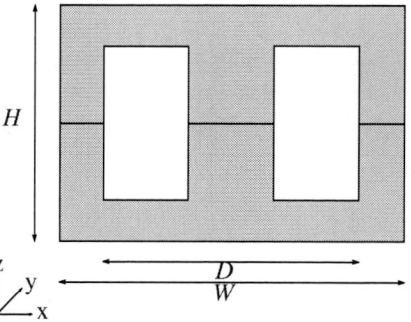

Fig. 2: Example representation of core dimensions for the volume calculation.

Constraint: Maximum Flux Density

In order to have a working inductor the saturation limit of the magnetic core should never been reached. Therefore the safety margin a is introduced. The constraint for the flux density B is given in equation (10).

$$B \leq B_{\text{max}} = aB_{\text{sat}} \tag{10}$$

The flux density is dependent of the air gap in a ferrite core and can be calculated according to equation (11) found in [9].

$$B = I \frac{\mu_0 N}{\frac{1}{\mu_a}(l_e - l_g \frac{A_e}{A_c}) + l_g \frac{A_e}{A_g}} \tag{11}$$

Where N is the number of turns, l_e in the effective length of the magnetic path, l_g is the gap length, A_e is the effective cross sectional area of the core, A_c is the cross sectional area of the core and A_g is the cross sectional area of the air gap. The permeability of the core material is highly nonlinear and dependent of temperature, frequency and field amplitude. For the calculation of the constraint the worst case amplitude permeability μ_a of the materials is assumed to ensure the saturation stability under any temperature and B-field condition.

Constraint: Thermal Limit

The thermal constraint which is expressed in equation (12) ensures safe operation in terms of not reaching critical temperatures like the curie temperatures or the limiting temperature of insulation materials. The thermal behavior of the inductor is modeled with a simplified thermal equivalent circuit model. Consisting of a loss source P_{loss} and a thermal resistance R_{th}.

$$R_{\text{th}}P_{\text{loss}} \leq T_{\max} - T_{\text{amb,max}} = \Delta T_{\max} \tag{12}$$

The maximum allowed temperature rise of the component is defined as $\Delta T_{\max}$. Which is the difference between the maximum component temperature $T_{\max}$ and the maximum allowed ambient temperature $T_{\text{amb,max}}$. The component temperature rise caused by the loss is calculated in (12). The loss P_{loss} is provided by the objective functions. The thermal resistance is approximated according to [10] and shown in (13).

$$R_{\text{th}} = \frac{0.0457}{\sqrt{V_{\text{core}}}} \tag{13}$$

For a fast calculation, which is required for the optimization routine, this approach is good enough, however in future developments of the optimization process a more accurate model is desirable. The thermal resistance R_{th} can also be calculated from more complex thermal models which will be done in future work.

Optimization Process

All constraints and objective functions are included in the optimization process. The purpose of the optimization process of this work is to design an optimal inductor for a buck converter with a fixed duty cycle operation. The entire optimization process is shown in Fig. 3. Based on the desired ripple the required inductance is calculated. Three design parameters are used, which are the air gap length, the ferrite material of the core and the core geometry (core type). These design variables are varied by the NSGA2 algorithm. The NSGA2 algorithm is real coded in this work. Which means that the decision variables are real valued. The core sizes are usually discrete in this approach a core database is used, where the cores types/sizes are coded as integer numbers. The termination criterion of this implementation is the number of generations N_{gen}. The parameters of the algorithm are the size of the population, and the probabilities of mutation p_m and of crossover p_c. The buck converter is defined with its input voltage V_{in}, its duty cycle D, its continuous dc current I_{DC} and its desired current ripple ΔI. The limit value of the maximum ambient temperature $T_{\text{amb,max}}$ and the maximum temperature rise $\Delta T_{\max}$ are also predefined. The number of turns N is calculated based on the required inductance L with the lowest amplitude permeability $\mu_{\text{a,min}}$. The calculation is shown in equation (14) according to [9].

$$N = \sqrt{\frac{L}{\mu_0}\left[\frac{1}{\mu_{\text{a,min}}}\left(C_1 - \frac{l_g}{A_c}\right) + \frac{l_g}{A_g}\right]} \tag{14}$$

Where C_1 is the geometrical core factor given in the manufacturers datasheet and $\mu_{\text{a,min}}$ is the lowest amplitude permeability in the desired operating range. Based on the number of turns, which are calculated in the algorithm, and the core geometry the possible cross sectional area of the copper wire $A_{mathrmwire}$ is calculated in (15).

$$A_{mathrmwire} = \frac{K_u A_{\text{W}}}{N} \tag{15}$$

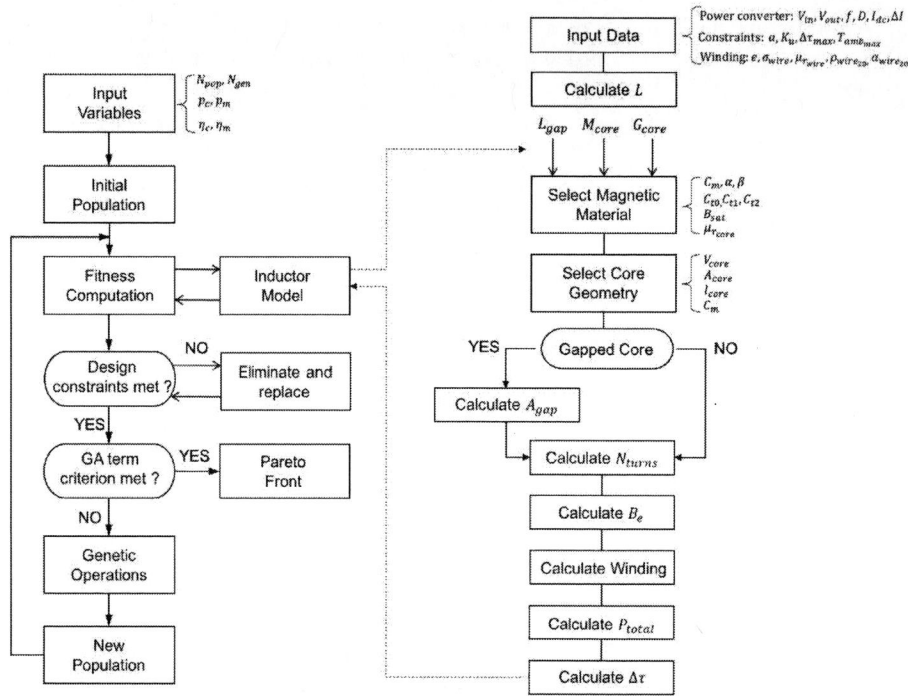

Fig. 3: Flow chart of entire optimization process.

Where A_W is the area of the winding area and K_u the packing factor of the round copper wire, which includes the insulation distances, the bobbin, and the arrangement of the conductors. The design optimization process is implemented in *Python* with the DEAP toolbox [11].

Results of the Optimization Process

To apply the optimization routine a test case is defined in the following Table I. The gap length is limited to 10 mm. The database of core materials offers 9 different Materials of the manufacturer *Ferroxcube*.

Table I: Parameters of optimization experiment.

NSGA2 Parameters	Number of generations N_{gen}	250
	Number of individuals N_{pop}	300
	Cross over probability p_c	0.90
	Mutation probability p_m	0.33
Constraints	Maximum ambient temperature $T_{amb,max}$	70°C
	Maximum temperature rise ΔT_{max}	70°C
	Saturation limit B_{max}	350 μT
	Design interval of air gap length	[0; 10 mm]
	Database ferrite materials	9 materials
	Database core geometries	17 types of PQ cores
Buck converter parameters	Input voltage V_{in}	12 V
	Output voltage V_{out}	6 V
	Duty cycle D	0.5
	Desired inductance L	12 μH
	DC current I_{DC}	25 A
	Maximum current ripple ΔI	$0.1 I_{DC}$
	Switching frequency f_{sw}	100 kHz

The core geometry database includes the whole range of PQ cores. Both databases of course can be extended to any core shape or material by any manufacturer.

The result of the optimization routine shows the pareto front of optimal solutions in Fig. 4 The designer can now choose, which objective is most important and select one individual. Although in the design example there were 300 individuals only 11 are shown in the plot. This is because the solutions are repeated and therefore discarded. Table II shows the values of the individuals in detail. All contraints are met, the temperature rise is less than 70°C and the maximum flux density is less than $B_{\max}$. For a further design verification the inductor number #6 is chosen because it is close to the knee of the pareto front.

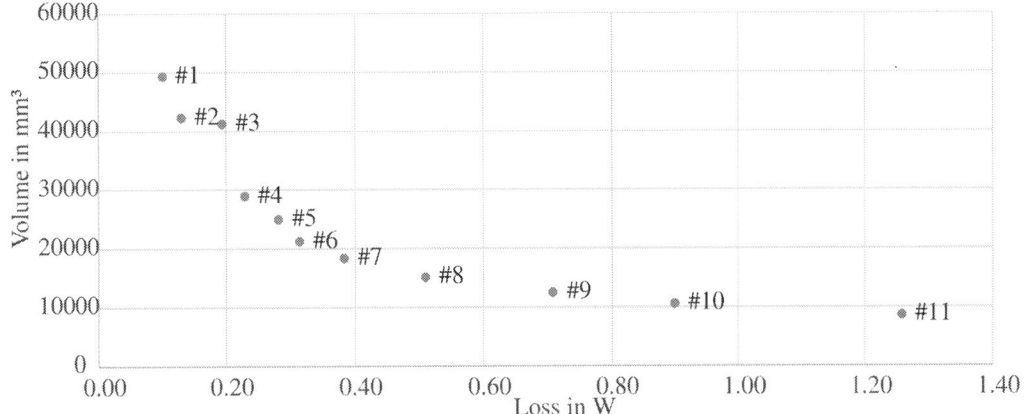

Fig. 4: Result of optimization process with NSGA2.

Table II: Parameters of optimization experiment.

	l_g (mm)	Core material	N	Core geometry	V_{core} (mm^3)	P_{loss} (W)	B (mT)	ΔT (°C)
#1	0.28	3C98	3	PQ50/35	49280	0.101	349	0.66
#2	0.28	3C98	3	PQ50/30	42240	0.130	350	0.92
#3	0.48	3F36	5	PQ40/40	41232	0.194	332	1.38
#4	0.48	3C98	5	PQ35/35	28870	0.229	345	1.95
#5	0.48	3C98	5	PQ35/30	24960	0.282	340	2.58
#6	0.63	3F36	6	PQ32/35	21175	0.314	329	3.12
#7	0.62	3F36	6	PQ32/30	18331	0.383	330	4.09
#8	0.61	3F36	6	PQ32/25	15125	0.509	330	5.99
#9	0.61	3F36	6	PQ32/20	12463	0.708	325	9.17
#10	1.03	3F36	8	PQ26/25	10559	0.898	334	12.64
#11	0.99	3F36	8	PQ26/20	8635	1.258	333	19.56

Experimental Validation

The inductor with the number #6 is built up according to the results in Table II with an all legs gapped configuration, shown in Fig. 6. A small signal impedance analysis with the *Keysight* E4990A is done to prove that the modeling of the inductor is correct. The result of the small signal measurement is shown in Fig. 5. It can be observed, that the target on 12 µH is met almost perfectly. The increase in inductance for higher frequency can be explained by the frequency dependency of the core material which is dominant due to the small air gap.

Besides the small signal analysis also a measurement of saturation current and the determination of the inductance under load current is done to verify the design process. In Fig. 7 the results of a saturation

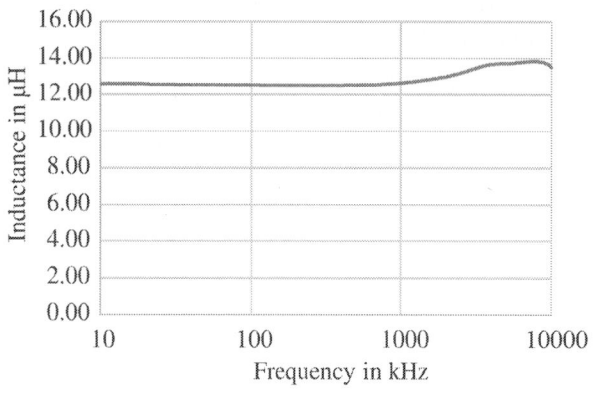

Fig. 5: Small signal measurement of inductance.

Fig. 6: Inductor #6 with PQ 32/35

measurement at 30°C are plotted. The measurement to ensure that no saturation is occurring is done by applying a constant voltage to the inductor and switching off the current, when the desired test current is exceeded. In this example the applied voltage is switched off when the inductor current of $I_L = 27.5\,\text{A}$ is reached. In Fig. 7 it can be observed, that the current rises linear. Therefore no saturation event occurs and the inductor is suited for the designed application in terms of saturation.

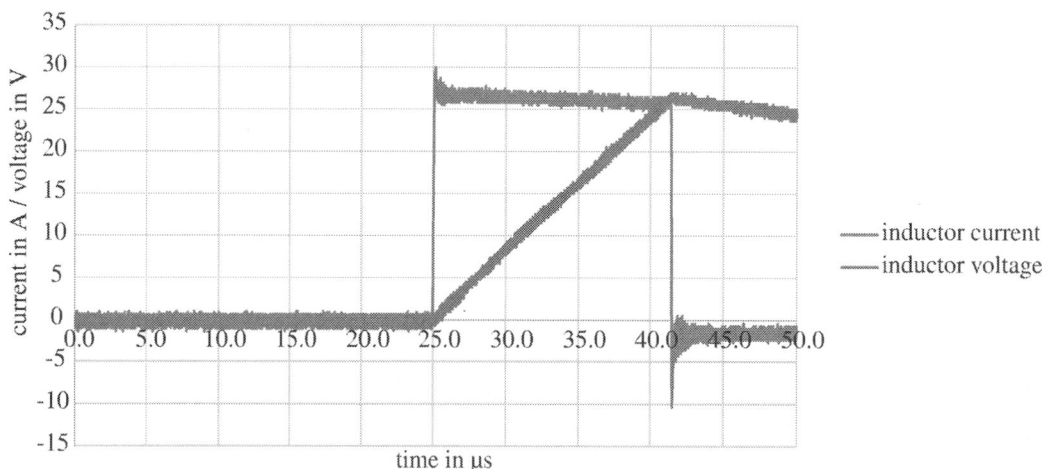

Fig. 7: Saturation measurement under load current conditions.

From the current and the voltage waveform of Fig. 7 also the inductance can be calculated. The inductance of the inductor under the load current conditions is approximately $L_{\text{load}} = 16.7\,\mu\text{H}$. This result is slightly higher than the small signal frequency measurement. It can be explained by the nonlinear behavior of the core material *3F36*. According to the datasheet the amplitude permeability rises from $\mu_a = 1800$ for small signal to $\mu_a = 2700$ for load conditions resulting in the higher inductance. But since the inductor is designed to the worst case conditions the minimum required inductance can be ensured over the whole operating current area without saturation.

Conclusion

To find the optimal trade-off in power electronics design a NSGA2 based multi objective optimization routine is developed. The Algorithm is used to design an inductor for DC to DC converters. Therefore a loss model and a size model is introduced as objective functions. The solutions of the algorithm are limited by saturation and temperature constraints. The algorithm is applied for an experimental converter and a Pareto-Front of optimal solutions is obtained. One solution from the optimal front is built up and a small signal measurement is carried out, which proves that the required inductance is obtained. Also measurements under load current conditions are carried out to prove the constraint modeling of maximum flux density and the minimum required inductance.

In future work the models of the inductor can be refined. Also a whole converter optimization with consideration of the semiconductor and capacitor losses can be implemented. Applying genetic algorithms for optimization in power electronics helps to find most suitable design solutions from wide parameter sets.

References

[1] Y. Liu et al, "LCL Filter Design of a 50-kW 60-kHz SiC Inverter with Size and Thermal Considerations for Aerospace Applications," in IEEE Transactions on Industrial Electronics, vol. 64, no. 10, pp. 8321 - 8333, October, 2017

[2] C.A. Coello, M.S. Lechuga, "MOPSO: a proposal for multiple objective particle swarm optimization," in Proceedings of the 2002 Congress on Evolutionary Computation. CEC'02, pp. 1051-1056, vol.2, May 2002

[3] K. Deb, A. Pratap, S. Agarwal, T. Meyarivan, "A fast and elitist multiobjective genetic algorithm: NSGA-II," in IEEE Transactions on Evolutionary Computation, vol. 6, no. 2, pp. 182-197, April 2002

[4] C. Versele, O. Deblecker, J. Lobry, "Multiobjective optimal design of high frequency transformers using genetic algorithm," in 13th European Conference on Power Electronics and Applications 2009, pp. 1-10, September 2009

[5] K. Venkatachalam, C.R. Sullivan, T. Abdallah, H. Tacca, "Accurate prediction of ferrite core loss with non-sinusoidal waveforms using only Steinmetz parameters," in IEEE Workshop on Computers in Power Electronics, IEEE Power Electronics Society 2002 – COMPEL 2002, June 2002

[6] Jiankun Hu, C.R. Sullivan, "Optimization of shapes for round-wire high-frequency gapped-inductor windings," in IEEE Industry Applications Conference. Thirty-Third IAS Annual Meeting 1998, pp. 907-912, vol. 2, October 1998

[7] P.L. Dowell, "Effects of eddy currents in transformer windings," in Proceedings of the Institution of Electrical Engineers, vol. 113, no. 8, pp. 1387-1394, August 1966

[8] W.-J. Gu, R. Liu, "A study of volume and weight vs. frequency for high-frequency transformers," in Proceedings of IEEE Power Electronics Specialist Conference - PESC '93, pp. 1123-1129, June 1993

[9] M. Allbach, "Induktivitäten in der Leistungselektronik," Erlangen, Springer Vieweg, 2017, pp. 170-174

[10] EPCOS AG, "Application notes. Ferrites and accessories," 2006

[11] F.-M. De Rainville, F.-A. Fortin, M.-A. Gardner, M. Parizeau, C. Gagné, "DEAP: Evolutionary Algorithms Made Easy," in Machine Learning Research, vol. 13, pp. 2171-2175, July 2012

Digital smart driver for SiC MOSFETs

Nerea Arandia[1], José Ignacio Garate[2], Jon Mabe[1], Ander Ordoño[1]
[1] Tekniker, Eibar, Spain
[2] University of the Basque Country (UPV/EHU), Bilbao, Spain
Tel.: +34 / 943.20.67.44.
E-Mail: nerea.arandia@tekniker.es
URL: http://www.tekniker.es

Keywords

«Intelligent drive», «Silicon Carbide (SiC)», «Efficiency», «Reliability», «MOSFET».

Abstract

This paper presents a new concept of an isolated smart gate driver platform for SiC (Silicon Carbide) MOSFETs. It describes the required hardware to implement advanced functions to improve converter performance, switching behavior and reliability. It also includes GHz bandwidth voltage monitorization and low speed current and temperature monitorization.

Introduction

The massive use of power converters in all kind of applications and the markets' increasing demand of more power density and efficiency have led to the development of new solutions to design power converters. The use of Wide-Bandgap (WBG) semiconductors and the optimization of the gate drivers constitute two research lines to improve the efficiency and reliability of power systems.

Traditional Silicon (Si) power devices are reaching their theorical maximum performance ratings. Hence, they have become a limiting technology due to their losses, relatively low switching frequency and reduced temperature dissipation capabilities. New opportunities have arisen with the development of SiC semiconductors which are able to improve the standard Si characteristics. SiC devices have been known to possess significantly lower losses compared with silicon devices. However, in order to attain such low losses, the devices must be switched quickly, and this presents technical challenges such as the increase of electromagnetic interference (EMI), and voltage or current overshoots which can reduce the life span of the device and eventually destroy it. Moreover, it is not yet a mature technology, and its reliability is a critical factor compared to conventional technologies.

A power converter has a power unit and a control unit which are interconnected through gate drivers. These devices are responsible of the charge and discharge of the power semiconductor's gate, making it switch on and switch off according to control signals. As the element in charge of the interconnection of power semiconductors and control algorithms, the driver is paramount to optimize the performance of the converter. In addition, some drivers have advanced functionalities to optimize switching behavior and guarantee a safe commutation using sophisticated monitorization and protection mechanisms.

The improvement in gate drivers can positively impact power converters performance in any type of applications. There are three key aspects of power conversion stages that gate drivers can improve:
- Efficiency: so far switching losses are the main contributor to the overall system losses in high voltage switching, thus reducing switching losses is critical.
- Power density: high switching frequency converters with small passive components are being developed to achieve high power densities. High current gate drivers and the development of new gate drivers compatible with WBG technologies such as Gallium nitride (GaN) and SiC are required.

- Reliability: robust protections are required to guarantee safe shut-down or recovery under critical faults. Consequently, WBG components require more complex protections to ensure its integrity.

The combination of SiC semiconductors' advantages and the use of smart gate drivers to optimize its operation is one method to improve future power converters performance. In this paper, existing smart gate drivers and its hardware and control requirements are presented. Due to the characteristics of SiC MOSFET, it is necessary to propose a new concept of intelligent driver. For this reason, the requirements and difficulties are analyzed and a hardware platform that serves as a smart gate driver for SiC power modules is presented.

Smart Gate Driver concept

Traditional gate driver's main function is to amplify control signals applied to the semiconductor, providing signal isolation, and protecting the semiconductor under fault conditions. Smart gate drivers go further, they can improve switching performance and protection functionalities. To achieve this, they have an advanced monitorization stage and a FPGA or microprocessor based digital control unit to execute advanced gate control algorithms.

To comply with its functionality, the smart gate driver requires information about the semiconductors' parameters and its real-time performance. This is achieved by monitoring devices' main voltages and currents. Both gate side and power side voltage and currents used can be measured with the proper circuit. Temperature can also provide relevant information for the control algorithm, as the performance of the semiconductor and its parameters are temperature dependent. Current Intelligent Power Modules (IPM) usually include an on-chip temperature sensor to measure semiconductors' junction temperature. However, for traditional discrete and power modules obtaining a valid junction temperature is difficult. The junction temperature is usually estimated by using case temperature and electrical measurements.

Conventional drivers apply an ON/OFF voltage to the semiconductor through a current limiting resistor. A smart driver can use the measured variables and the dependence between all of them to improve the switching performance. A gate control algorithm can be designed and executed in the control unit, using as control variable a current controlled gate source circuit or a gate resistor array. The measurements can also be used to obtain relevant information about the status of the semiconductor.

In figure 1 the concept and functionalities of the smart gate drivers are presented.

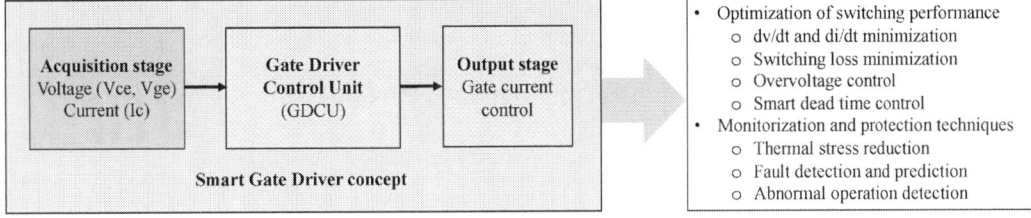

Fig. 1: smart gate driver concept and its functionalities

Smart driver functionalities

In the recent years, due to the benefits of these devices, several smart gate drivers have been designed for Si-based semiconductors. This developments has two main functionalities: optimization of switching waveform characteristics [1]–[3] and improvements in monitorization and protection [4]–[6] techniques.

Optimization of switching performance

The optimization of switching behavior allows any power converter to work with higher efficiency and to reduce undesirable effects generated during commutation. Using smart gate drivers, critical parameters such as dead time, di/dt, overvoltage, dv/dt, switching power loss and EMI can be optimized online.

In this line, adaptive dv/dt and di/dt control is widely implemented for Si IGBT, as the rate of change of voltage generated by power semiconductor has a significant impact on any component connected to its output. In [7], a FPGA based driver is developed to reduce and control offline voltage derivatives that suffer the connected loads. Thus, current measurements are used, and gate resistor is modified. In [2] highly dynamic di/dt and dv/dt control is performed just with passive components. Authors in [8] achieve a trade-off between switching losses and EMI using a closed loop control for IGBTs gate circuit.

Monitorization and protection

Traditional driver's protection mechanisms detect certain operating anomalies and, if necessary, they launch the corresponding protections. Typical ones are short circuit protection (desaturation detection) and active, gate or Miller clamping circuits. These protections are used to guarantee that the semiconductor is not destroyed due to a malfunction of the circuit, they are not used with the aim of improving system performance.

The use of smart drivers makes easier the development of functionalities to extend the life of the converter using self-diagnosis techniques. With a self-diagnosis algorithm, the performance of the semiconductors can be optimized for different criteria:
- Dynamic thermal stress
- Maximum junction temperature
- Power sharing in parallelized/serialized components

Many studies have focused on how semiconductors' life can be extended by monitoring and controlling its junction temperature, as it is directly related with the degradation of the components. In [6], a control algorithm is designed for real time monitorization of the semiconductor. Power losses and critical temperatures are estimated. The driver can adjust power loss and temperature modifying gate resistance and switching frequency. In [10] a temperature monitorization gate driver is also implemented. The development and the results were validated using an IR camera.

The self-diagnosis of power semiconductor can also be used to estimate the aging of the components and prevent unwanted faults before they happen. When the fault has already occurred, they can also provide a fault diagnosis, reducing maintenance costs. A fault detection methodology is implemented for IGBTs in [9]. It is based on online monitoring of the collector current slope signal during the turn on transient. In [6], a health monitoring framework for IGBT modules is implemented using online measurements of the voltage drop between collector-emitter (V_{CE}) and the device's threshold voltage (Vth).

Smart gate drivers for SiC MOSFET

Si semiconductors based smart gate drivers have not yet been transferred to SiC MOSFETs. The characteristics of these semiconductors make traditional solutions not directly transferable: an adaption to the high-speed switching characteristics of WBG devices is necessary in most of the cases.

In the literature, there are some developments regarding to the monitorization of degradation of SiC MOSFETs. In [11] a real time aging detection methodology is defined. This tool can be integrated inside a smart gate driver or a converter control unit and it is based on the monitorization voltage drop across the gate turn-on resistance.

In [12] a high-speed gate driver development with an embedded Rogowski current sensor is presented. However, the bandwidth of the current is not defined. In [13], a high voltage intelligent gate driver for SiC is presented, this driver is based on electrical measurements (drain-source voltage drop V_{DS} and drain current I_D) and module temperature. This development assumes the use of inbuilt shunt resistor and a thermistor temperature sensor, which is not available for all commercial modules.

The development of gate drivers for SiC require some design considerations which are not crucial in Si: voltage and current ranges, parasitic inductances in the gate loop or short circuit protection are critical [14]. Besides, the last remarks, combined with the fact that the high switching frequency imply high speed acquisition and control requirements to the sensing and control stages, makes traditional smart gate driver solutions not suitable for new semiconductors.

Challenges

Switching behavior optimization implies monitorization of switching waves: current and voltage evolution during the commutation. Silicon based semiconductors switch in the range of tens of microseconds, whereas SiC MOSFET switch in the range of nanoseconds. Sensing and driving requirements raise considerably for sensing circuitry and should be adapted to capture signals with bandwidth in the range of GHz and minimum delays.

Concerning current monitoring, some authors have analyzed the different sensor technologies available in the market and its integrability in the power stage [15]. The only sensor that provides a bandwidth in the range of GHz is the shunt coaxial resistors. These sensors are non-efficient resistors and it is not feasible to integrate them into commercial power converters due to their large size. All in all, it is not possible to use current measurement to drive SiC MOSFET with a smart gate driver so voltage measurement must be the base of the driving algorithm.

Voltage measurement must be also done with a bandwidth of GHz. Commercial voltage transductors are in the range of kHz, so methods such as voltage resistors with high speed ADCs must be considered. The measuring circuit and the ADC should have small parasitic components to maximize the bandwidth and the PCB design becomes complex.

In addition, control algorithms must be adapted to work without current monitoring in the range of GHz: optimization of switching behavior must be only based in voltage measurements. Moreover, the increase of switching speed increases control processor requirements and high-speed microprocessors or FPGA should be used. In this line, close loop-based controls are not a suitable solution, as several GHz bandwidth is required, and new control techniques should be developed.

As in Si based smart drivers, the variable to control the switching behavior is the gate resistor. To fully optimize the switching performance of the SiC devices, the gate resistor must be modified in the range of ns. This is not feasible and efficient with current technologies. For these reasons, the control should anticipate to the semiconductor requirements and modify the gate resistor during the conduction and blocking time of the device.

Proposed Smart Gate Driver

A digital gate driver platform has been developed for the execution of control algorithms that adds smart functionality to traditional drivers. For that, an innovative architecture is proposed. The core of the gate driver is the Gate Driver Control Unit (GDCU) which will use voltage, current and temperature variables to optimize switching performance and lifetime of the power module.

As input signals, the following are provided to the GDCU:
- High frequency drain-source-voltage (V_{DS}) and gate-source-voltage (V_{GS}) acquisitions at a 1 GHz. This measurement is made using a high bandwidth resistive sensor.
- Low frequency drain-current (I_D) acquisition at 20 kHz to measure output current.

- Temperature measurement to allow gate driver control unit the estimation of the junction temperature. For that, a temperature sensor must be placed in the case of the power module.

Once the GDCU studies all the input signals, it will modify the semiconductor's switching behavior. This will be possible through the modification of the gate resistor value. This modification will be done using a low latency gate resistor modification circuitry.

The hardware is designed for a half-bridge SiC MOSFET power module. Specifically, this design fits CREE's CAS120M12BM2 power module, which is a 1200V, 13 mΩ all-SiC half-bridge module. For SiC MOSFET driving purpose the Infineon 1ED020I12-F2 EiceDRIVER 1200V high-side gate driver IC is used. This driver includes galvanic isolation, active Miller clamp, desaturation and short circuit clamping.

The digital gate driver board is mechanically designed to fit 62 mm module's terminals and it has three stages: low voltage stage, isolation stage and low voltage stage (see Fig. 2).

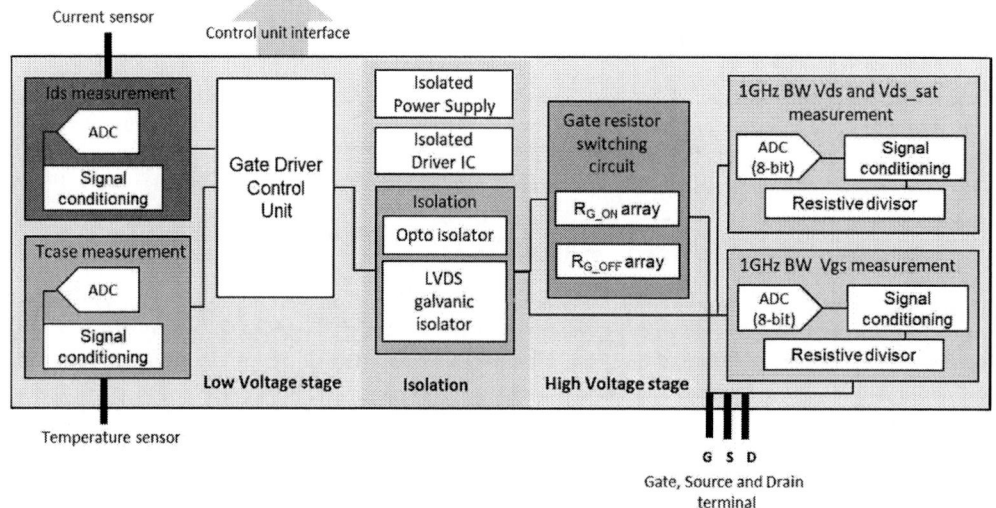

Fig. 2: developed smart gate driver block diagram.

High voltage stage

High voltage side includes voltage measurement circuit and R_G switching circuit which integrates the control signal boosting stage.

V_{DS} and V_{GS} voltage measurement circuit is made using HMCAD1511 ADC. Voltage sensing circuit can measure drain-source voltages up to 1200V and gate-source voltages between -5V and 20V. MOSFET voltages are adapted using a filtering stage and a resistive divider. The selected ADC has a bandwidth of 1 GHz and a resolution of 8 bits. In addition, this ADC is connected to the control unit through LVDS interface.

The use of a 1 GHz 8-bit ADC converter provides a good trade-off between costs and voltage measurement resolution. For a 1000 V application, the voltage resolution is around 4 V, which is enough to identify dv/dt. In addition, when on-state voltage drop is required, the range of voltage measurement is adjusted, and a resolution of 35 mV is achieved.

R_G switching circuit purpose is to change R_{G_ON} and R_{G_OFF} value to modify switching ON and OFF wave characteristics. For that, a controlled R_G array is designed for each semiconductor. This array is controlled using several optocoupled outputs of the FPGA. The design of this stage is made guarantying that array commutation time is smaller than the 50% of the PWM cycle because the commutation of the array must be performed while the MOSFET is in the permanent saturation or cut-off region.

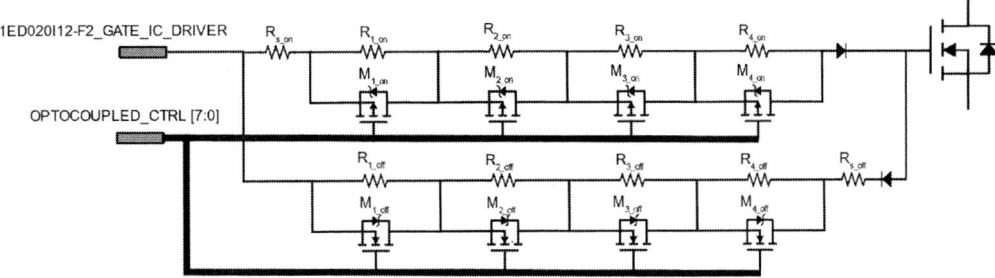

Fig. 3: R_G switching circuit

In figure 4 the simulation of the commutation of M1 MOSFET is presented. V(n004) represent the input voltage at the OPTOCOUPLED_CTRL signal, it presents a delay of about 1us in its activation. Once the activation is performed, a current peak (I_G) discharges M_1 gate and the MOSFET M_1 becomes an open circuit. The value of R_{G_ON} increases from R_{S_ON} to $R_{S\_ON} + R_{1\_ON}$. As is shown the commutation takes 2μs.

The design considers the use of 5 resistors connected in series that allow to cover the range of 1 Ω - 16 Ω with a resolution of 1 Ω. For that, the following resistor values are used: 1 Ω, 1 Ω, 2 Ω, 4 Ω and 8 Ω.

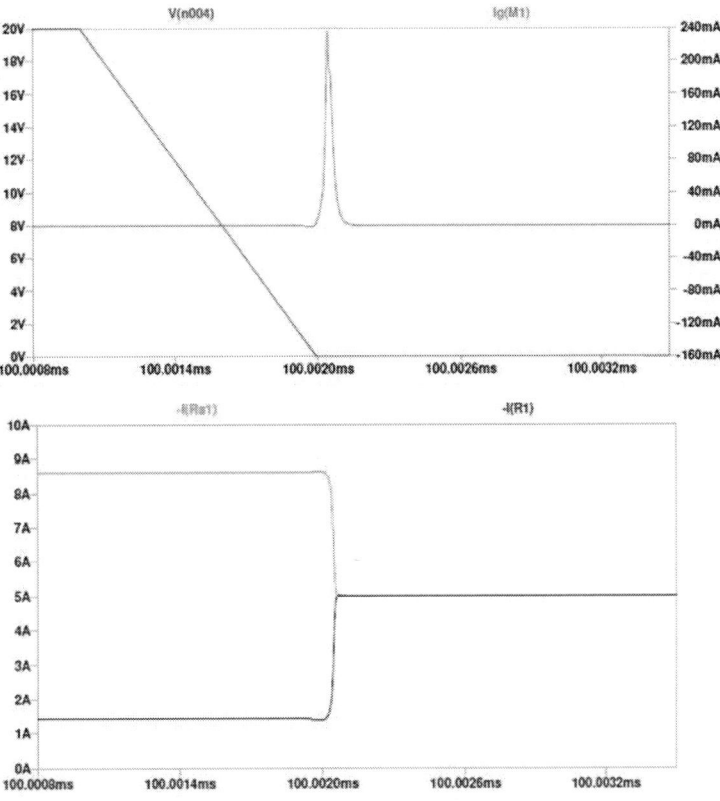

Fig. 4: R_{G_ON} switching simulation

Low voltage stage

Low voltage side integrates an Altera SoC (FPGA and a microprocessor) to run control algorithms. This control unit includes a high-speed communication interface to receive PWM signals and to send information about the status of the power module.

Due to the limitations of current sensors' bandwidth, it is not feasible to characterize I_D during commutation. Moreover, I_D measurement must be done without increasing parasitic inductance of the converter. For that, load current is measured, as it is equivalent to the current through the semiconductor in conduction state. This current is measured using a developed external current sensor with a bandwidth of 20 kHz.

In addition, a temperature sensor is placed in the module's case to estimate junction temperature. Temperature measurement is made using PT100 sensors which can be placed in the surface of the 62 mm power module and provide galvanic isolation. For sensing, digitization and communication of temperature measurements, Texas Instruments ADS1248 integrated circuit is used.

Isolation stage

Low power stage and high-power stage are divided by an isolation stage. An isolated DC-DC is used to power the voltage measurement and R_G switching circuitry. The transmission of voltage measurement from the ADC is made by LVDS interface, to isolate this high-speed signal a galvanic LVDS isolator is used. For signals that not required GHz bandwidth such as the control of gate resistor commutation, an optocoupled isolator is used. In addition, the selected SiC MOSFETs driving IC includes galvanic isolation.

Switching losses optimization

When a power converter is scaled for a certain application, the selection of the R_G is made to satisfy some requirements, such as the maximum allowable dv/dt, di/dt or device overvoltage. Meeting these requirements involves selecting a gate resistor high enough so that the requirements are met in the whole operation range; however, the power converter will work in non-optimum conditions in a wide range.

The developed smart driver platform can be used to optimize several parameters of the switching waveform. The switching behavior can be controlled by modifying the gate resistor according to the measurements. In this way, an advanced algorithm has been developed to minimize the switching losses during turn-on. This approach increases the efficiency of the power conversion and ensures that the dv/dt of the semiconductors does not exceed a configurable value.

In figure 5, the comparison between the non-optimized switching losses and the optimized ones is presented. The comparison is made for a current load between 0 A and 200 A. In a conventional driver, a R_G value of 7 Ω is required to fulfil all the requirements of the application. However, for the smart gate driver, the value of the gate resistor can be reduced for low load currents (figure 5.b), as the dv/dt reduce when the switched current does. The minimum R_G value for each current is presented in Table 1. For example, when the load current is between 40 A and 80 A, the gate resistor will be 4 Ω.

Table I: minimum R_G value for each load current

Load current (A)	40	80	120	160	200
Gate resistor (Ω)	3	4	5	6	7

In figure 5.c, the evolution of switching losses over time are presented. Both non-optimized and optimized switching losses are depicted. The conventional driver with a fixed gate resistor and the smart driver will have equal switching losses for high currents. However, when the load current is reduced (for example, in a sine shaped current), the gate smart driver can considerably reduce the switching losses compared to a fixed gate resistor.

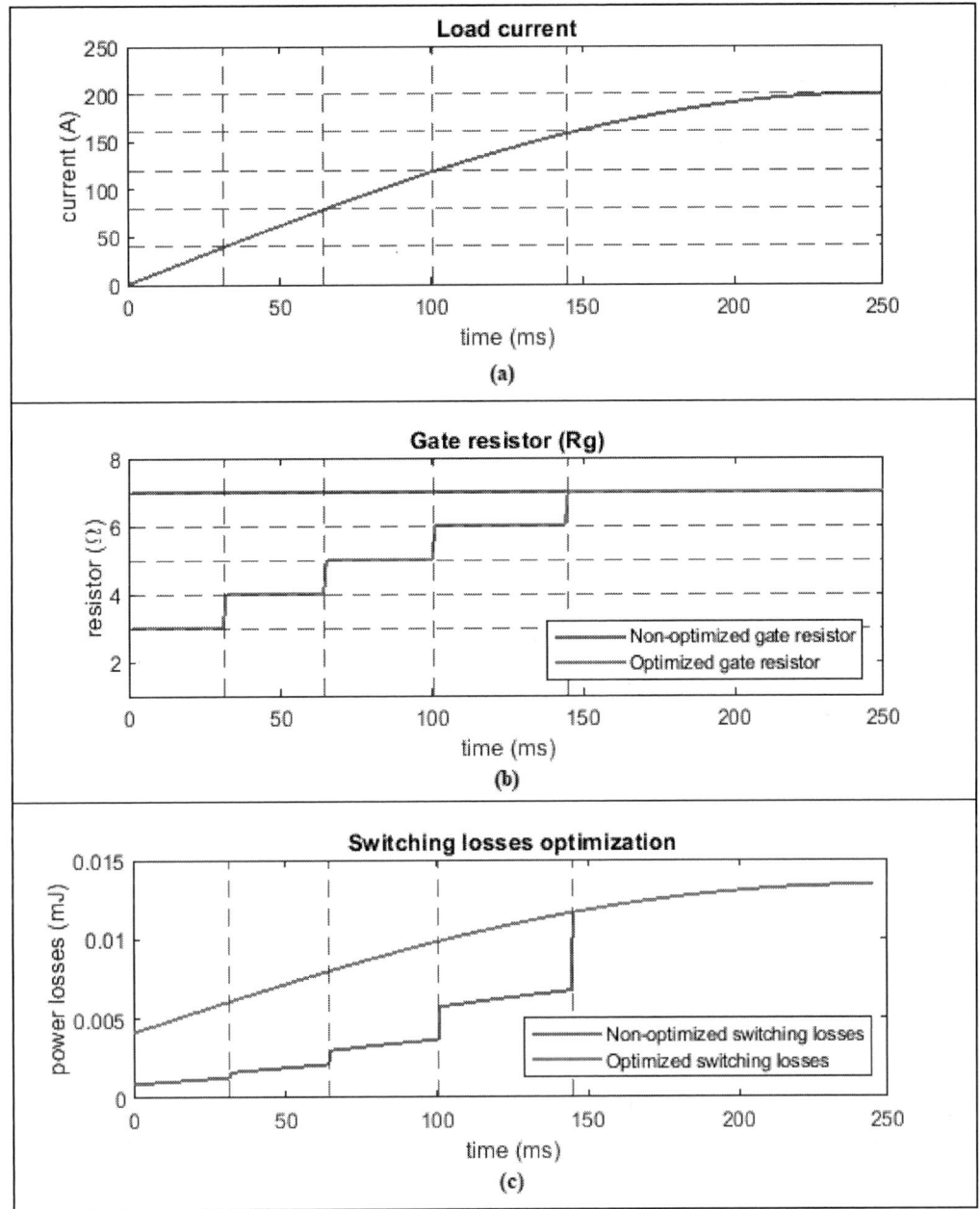

Fig. 5: optimization of switching losses using the smart gate driver

Conclusion

Smart drivers that increase the monitorization capabilities and functionalities of drivers are an alternative approach to increase power conversion efficiency and improve SiC MOSFET protection mechanism, and consequently their lifetime. However, smart gate drivers developed for Si are not suitable for SiC semiconductors. Thus, current hardware platforms must be adapted to high frequency commutation rates and the control strategies must be redesigned.

This paper introduces a new concept of smart gate driver for SiC MOSFETs. The proposed design can monitor the most relevant voltages of the semiconductor with a GHz bandwidth and the ON current at kHz. The smart driver can use these variables to optimize the switching and conduction performance of the semiconductor by changing the gate resistor value. The measurements can also be used to improve faulte detection and monitor the ageing of the SiC devices.

Likewise, this platform can ease the transition from Si semiconductors to SiC since it offers mechanisms for controlling and monitoring the unwanted effects that these new semiconductors can generate.

References

[1]   L. Michel, X. Boucher, A. Cheriti, P. Sicard, and F. Sirois, "FPGA implementation of an optimal IGBT gate driver based on posicast control," IEEE Trans. Power Electron., vol. 28, no. 5, pp. 2569–2575, 2013.

[2]   L. Shu, J. Zhang, F. Peng, and Z. Chen, "Active Current Source IGBT Gate Drive with Closed-Loop di/dt and dv/dt Control," IEEE Trans. Power Electron., vol. 32, no. 5, pp. 3787–3796, 2017.

[3]   Z. Wang, X. Shi, L. M. Tolbert, F. Wang, and B. J. Blalock, "A di/dt feedback-based active gate driver for smart switching and fast overcurrent protection of IGBT modules," IEEE Trans. Power Electron., vol. 29, no. 7, pp. 3720–3732, 2014.

[4]   J. Chen et al., "A Smart IGBT Gate Driver IC with Temperature Compensated Collector Current Sensing," IEEE Trans. Power Electron., vol. 34, no. 5, pp. 4613–4627, 2019.

[5]   F. Di Napoli et al., "On-line junction temperature monitoring of switching devices with dynamic compact thermal models extracted with model order reduction," Energies, vol. 10, no. 2, 2017.

[6]   M. A. (University of N. Eleffendi and C. M. (University of N. Johnson, "A Health Monitoring Framework for IGBT Power Modules," in 9th International Conference on Integrated Power Electronics Systems, 2016.

[7]   A. (Amantys) Bryant, "Optimisation of dV/dt - Losses Trade Off Using Switchable Gate Resistance," Bodo's Power Syst., 2018.

[8]   H. Ghorbani, V. Sala, A. P. Camacho, and J. L. R. Martinez, "A simple closed-loop active gate voltage driver for controlling di C /dt and dv CE /dt in IGBTs," Electron., vol. 8, no. 2, 2019.

[9]   M. A. Rodriguez-Blanco, M. Cervera-Cevallos, J. L. Vazquez-Avila, and M. S. Islas-Chuc, "Fault detection methodology for the IGBT based on measurement of collector transient current," Int. Power Electron. Congr. - CIEP, vol. 2018-Octob, pp. 44–48, 2018.

[10]   E. U. Benefits, "Junction Temperature Estimation Technology Demonstrator Junction Temperature Estimation Technology Demonstrator."

[11]   F. Erturk, E. Ugur, J. Olson, and B. Akin, "Real-time aging detection of SiC MOSFETs," IEEE Trans. Ind. Appl., vol. 55, no. 1, pp. 600–609, 2019.

[12]   J. Wang, S. Mocevic, Y. Xu, C. Dimarino, R. Burgos, and D. Boroyevich, "A High-Speed Gate Driver with PCB-Embedded Rogowski Switch-Current Sensor for a 10 kV, 240 A, SiC MOSFET Module," 2018 IEEE Energy Convers. Congr. Expo. ECCE 2018, pp. 5489–5494, 2018.

[13]   A. Tripathi, K. Mainali, S. Madhusoodhanan, A. Yadav, K. Vechalapu, and S. Bhattacharya, "A MV intelligent gate driver for 15kV SiC IGBT and 10kV SiC MOSFET," Conf. Proc. - IEEE Appl. Power Electron. Conf. Expo. - APEC, vol. 2016-May, pp. 2076–2082, 2016.

[14]   J. Rice and J. Mookken, "SiC MOSFET gate drive design considerations," IEEE Int. Work. Integr. Power Packag. IWIPP 2015, pp. 24–27, 2015.

[15]   G. Laimer and J. W. Kolar, "Accurate Measurement of the Switching Losses of Ultra High Switching Speed CoolMOS Power Transistor / SiC Diode Combination Employed in Unity Power Factor PWM Rectifier Systems," Proc. 8th Eur. Power Qual. Conf., pp. 71–78, 2001.

Faster switching with less overvoltage - operating a SiC-MOSFET at its speed limit

Pablo Rodriguez de Mora, Mark-M. Bakran
UNIVERSITY of BAYREUTH
Department of Mechatronics
Centre for Energy Technology - ZET
Universitätsstraße 30
95447 Bayreuth, Germany
+49 (0) 921 55-7808
Pablo.Rodriguez-de-Mora@uni-bayreuth.de
https://www.mechatronik.uni-bayreuth.de

Acknowledgments

This project was supported by the ZF Friedrichshafen AG. Special thanks go to the Business Unit Electronic Systems, ZF Bayreuth.

Keywords

≪Silicon Carbide (SiC) ≫, ≪MOSFET≫, ≪Switching losses≫, ≪Wide bandgap devices≫, ≪Power semiconductor device≫.

Abstract

This paper explores the turn-off switching behavior of a third-generation SiC MOSFET encapsulated in a TO-247-4 package. The work is focused on the selection of the optimal gate resistance to reduce the turn-off switching losses. The selection criteria is based on the limitation of the inductive over-voltage peak (OVPK) for the worst-case scenario i.e. maximum DC-link voltage and switched current. The gate resistance is tuned to induce the allowed OVPK, gradually, with decreasing gate resistance the switching losses would be reduced but the OVPK would increase. Contrary to the expected behavior, it is observed that there exists a threshold value from which, the decrease of the gate resistor reduces the OVPK and, moreover, the turn-off losses do also decrease.

Introduction

The pursuit of lower converter losses is leading to the adoption of SiC MOSFET technology. For the successful implementation of the technology, it is well documented the importance of reducing the DC-link parasitic inductance as much as possible to limit the inductive over-voltage ($V_{DC} + L_\delta \cdot di/dt$) and to reduce switching oscillations [1]. Once the minimum parasitic inductance is achieved in a design, and only if there is no limit for the dV/dt rate, the gate resistance can be tuned by looking at the over-voltage. This component modulates the di/dt which induces the over-voltage peak (OVPK) thus, reducing the gate resistance will gradually increase the peak, but also reduce the switching losses.

This paper shows a turn-off mechanism by which the over-voltage peak decreases (i.e. limiting behavior) when the gate resistance decreases under a given value. This effect contrasts with the typical expected behavior for turn-off switching, by which a lower turn-off gate resistance leads to a higher OVPK.

In [2] the concept of "self turn-on" is introduced which may aid in the understanding of the phenomena. However, the results of the studied case cannot be fully exported to the observations. In this work, the kelvin source pin used for turn-off and, therefore the characteristic behavior cannot be explained through partial turn-on during turn-off induced by the source inductive coupling. In [3], the theoretical switching waveforms for a GaN device is shown. A voltage knee effect in the drain-source voltage is shown, which has a certain resemblance to the observed behavior. Similarly to [2], it is associated with the source common coupling inductance. In [4] a similar effect could experimentally be observed, but it is not studied in detail. Thus, it is considered necessary to perform a detailed study of the phenomena to understand the working mechanism and implications.

Test Setup

The test setup is based on the well-known Double Pulse Test (DPT) circuit [5]. Typically for SiC MOSFET switching, the parasitic inductance is minimized; however, in this setup, it has been scaled up to represent the switching of a high current module. An equivalent $L_\delta \approx 350\,\text{nH}$ was determined, which would represent the switching behavior of a high current module of 600 A [6]. The Device Under Test (DUT) is a third-generation 1.2kV/ 75 mΩ SiC MOSFET encapsulated in a TO-247-4 package with an internal gate resistance of $R_{G\,\text{int}} = 9\,\Omega$. The schematic of the setup and the measuring positions is shown in Fig. 1a.

In accordance with the semiconductor manufacturer, the gate driver has a voltage swing of $V_{GS} = -4/+15$ V. Galvanic isolation is provided with a fibre optic cable and isolated power supplies with common mode chokes in the supply lead. The gate driver PCB is placed as near as possible to the DUT, Fig. 1b.

Testing at high speed requires measurement equipment with sufficient bandwidth. The used oscilloscope has a bandwidth of 1 GHz and the voltage probes have a bandwidth of at least 400 MHz; care has been taken to reduce the size of the ground lead to avoid introducing unnecessary inductance in the voltage probe loop [7]. For the drain-source current measurements, a custom-designed and manufactured current probe [8] is employed, and the gate current is measured with a wideband current monitor with a bandwidth of 200 MHz.

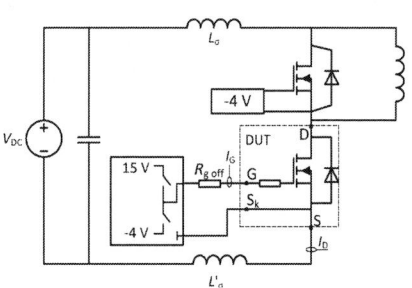

(a) Schematic diagram of the test circuit. (b) Detail of the gate connection.

Fig. 1: Test setup.

Measured switching waveforms

The main topic of this work is the influence of the gate resistance value on the switching behavior when the gate resistor is drastically reduced. This behavior can be observed in Fig. 2, where switching is performed with $R_{G\,\text{off}} = 0$, 5.6, 12 and 20 Ω for a test current of $I_{\text{test}} = 14$ A at a DC-link voltage of V_{DC}=830 V. The Gate-Source voltage (V_{GS}), Drain-Source voltage (V_{DS}) and Source current (I_S) are shown.

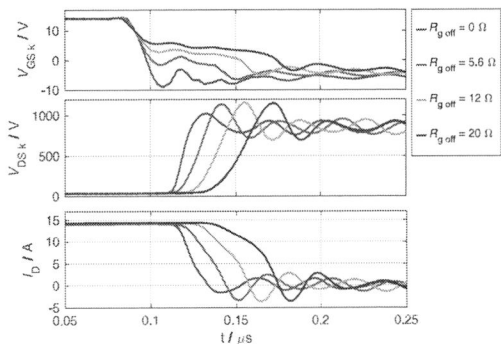

Fig. 2: Turn off measurement at DC-link voltage of 830 V and $I_{\text{test}} = 14\,A$.

As can be observed in Fig. 2, for the V_{DS}, when the gate resistor is set at 12 and 20 Ω the OVPK decreases with increasing $R_{\text{G off}}$, which is the usual behavior. By contrast, when the gate resistance is lower than $R_{\text{G off}} < 12\,\Omega$, a reduction of gate resistance does not lead to higher OVPK or oscillations, and actually, the lowest over-voltage peak is obtained when $R_{\text{G off}} = 0\,\Omega$. This behavior is in contrast with the typical switching process, by which the di/dt rate is inversely proportional to the gate resistor and, the over-voltage rises together with lower $R_{\text{G off}}$.

As for I_{S} waveform shown in Fig. 2, one can observe that in all cases the waveform presents two gradients. The first part, is due to the capacitive drop during dV/dt, by which, part of the switched current flows through drain-source capacitor (C_{DS}) with ($I_{\text{C}} = C_{\text{DS}} * dV_{\text{DS}}/dt$). In none of the tested cases I_{D} reaches null amperes during V_{DS} rise time. In the second part, the current derivative which induces the over-voltage occurs; it can be noticed that the di/dt with $R_{\text{G off}} = 0\,\Omega$ is not the highest.

The results show that there exists a mechanism that becomes activated with lower gate resistances. This process results in the MOSFET becoming restrained at the end of the V_{DS} rise time, i.e. limiting the OVPK. The behavior can be observed in Fig. 3, where the OVPK is is analysed over a range of gate resistances, from 0 to 47 Ω. Additionally, it is shown that the turn off losses decrease gradually with lower gate resistance.

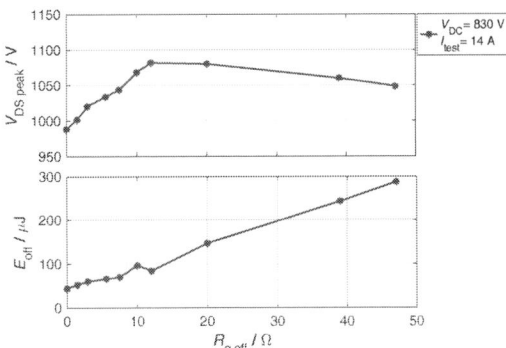

Fig. 3: Overvoltage peak and turn off losses as a function of the gate resistance: $V_{\text{DC}} = 830$ V and $I_{\text{test}} = 14$ A.

The key parameter that activates the limitation appears to be the gate resistance, but one should not consider this behavior as a clamping phenomenon. As can be observed in Fig. 4a, the over-voltage increases at higher test currents. Therefore it may be expected that, although this mechanism allows for very fast switching without excessive OVPK, a current level may be reached that produces an over-voltage that can reach the limit.

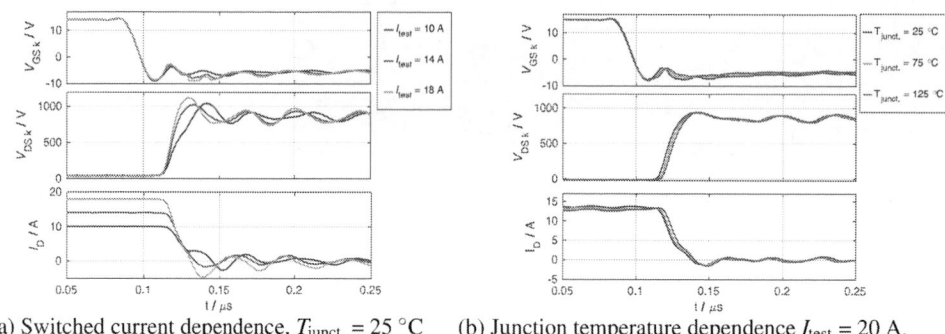

(a) Switched current dependence, $T_{\text{junct.}} = 25\,°C$ (b) Junction temperature dependence $I_{\text{test}} = 20$ A.

Fig. 4: Switching measurements at 830 V and $R_{\text{g off}} = 0\,\Omega$..

The influence of the junction temperature ($T_{\text{junct.}}$) in Fig. 4b, shows that the phenomena also occurs at higher temperatures.

In order to measure the gate current, due to the form factor of the current monitor, the gate loop had to be modified. A higher inductive gate loop is implemented and the gate driver is no longer connected near to the DUT. Nonetheless, as shown in Fig. 5a, this added inductance has a very limited impact on the switching behavior. In Fig. 5b the measurement of the gate current (I_G) is introduced in the comparison between $R_{\text{G off}} = 0$ and 20 Ω. During fast switching, observing I_G, it is possible to observe that de device does not suffer a sudden/partial turn-on during turn-off because it does not show a change in the current direction.

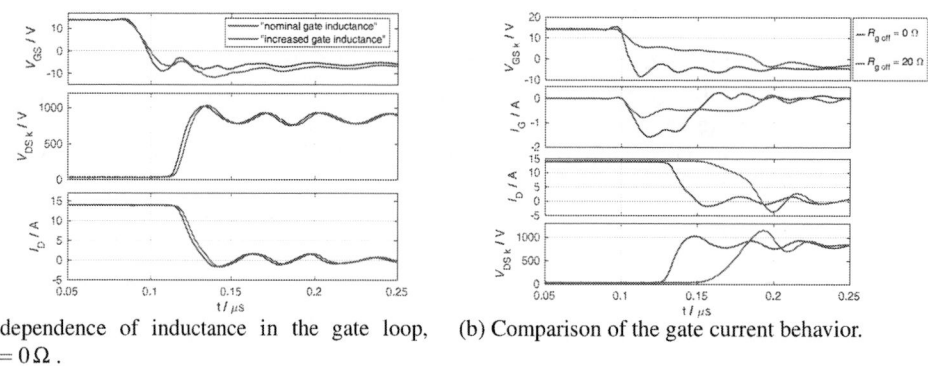

(a) Independence of inductance in the gate loop, $R_{\text{G off}} = 0\,\Omega$.   (b) Comparison of the gate current behavior.

Fig. 5: Behavior examining the gate loop, $V_{\text{DC}} = 830$ V and $I_{\text{test}} = 14$ A.

Finally, one could point out that the parasitic inductance is excessively high for the tested device and current, and thus distorting the switching procedure. Fig. 6 shows that even if the stray inductance is significantly reduced ($L_\delta \approx 80\,\text{nH}$), the voltage limitation does also occurs.

EPE'20 ECCE Europe

Assigned jointly to the European Power Electronics and Drives Association & the Institute of Electrical and Electronics Engineers (IEEE)

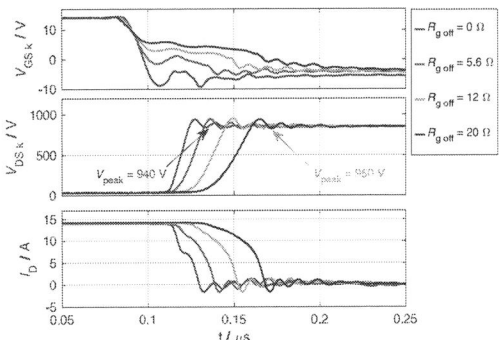

Fig. 6: Turn off measurement at 830 V and 14 A.

Simulation and behavior description

To understand the mechanism, the devices are simulated and the internal behavior is examined. The objective is to observe if the characteristic behavior also occurs during simulation, even if the waveforms do not completely comply with the measurements, and then propose an explanation.

The simulated device has an internal structure as shown in Fig. 7, where the current source is controlled by V_{GS}.

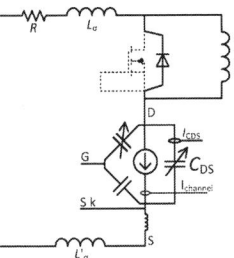

Fig. 7: Internal structure of the simulated device.

Due to several uncertainties between the test circuit and simulation model, it is accepted that the simulation should only show a similar behavior as a function of the gate resistance. In Fig. 8 the comparison between the measured and simulated waveforms is shown.

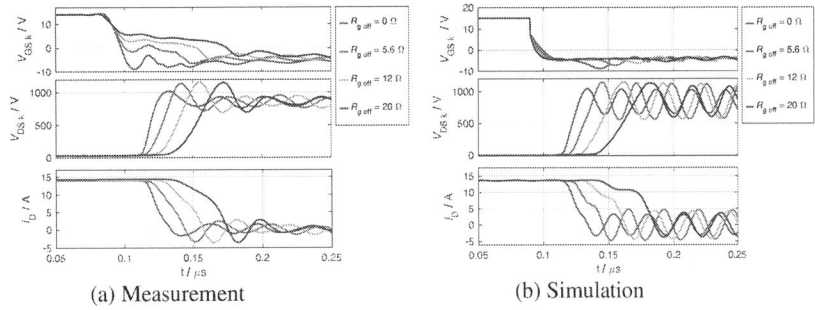

(a) Measurement   (b) Simulation

Fig. 8: Comparison between simulated and measured waveforms at $V_{DC} = 830$ V and $I_{test} = 14$ A.

In the simulations in Fig. 8b, the V_{DS} waveform shows a similar behavior as in the test, Fig. 8a. When the gate resistance is switched from 20 to 12 Ω, the over-voltage increases slightly. If a lower gate resistance is used, a gradual reduction of the OVPK is observed. Nonetheless, the simulation shows significant oscillation after turn-off which the measurement does not show; this is attributed to uncertainties of the simulation model. The drain current (I_D) presents similarly a good agreement with the aforementioned uncertainties being also present.

The comparison in simulation, between the two gate resistors, $R_{g\,off} = 0$ and $20\ \Omega$, including the internal variables is shown in Fig. 9a. And Fig. 9b shows the corresponding transfer characteristics, with the variables taken from the simulation.

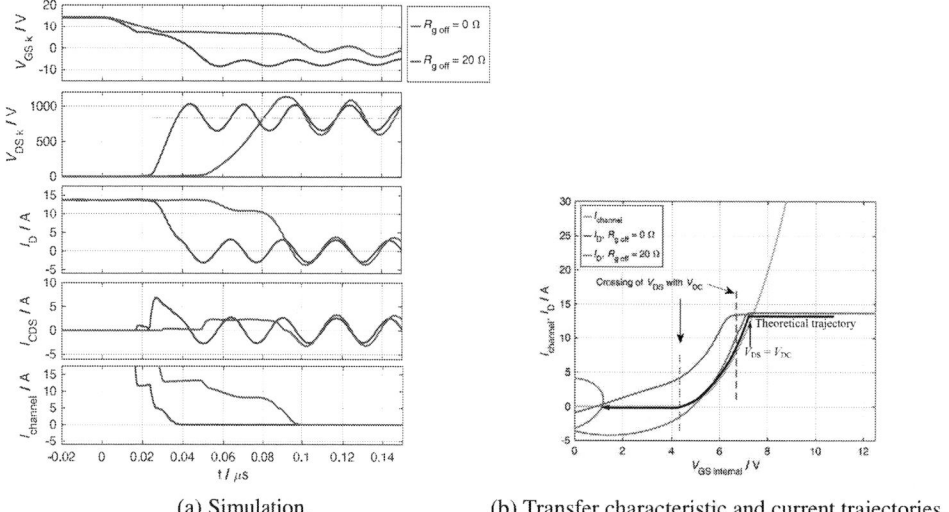

| (a) Simulation. | (b) Transfer characteristic and current trajectories. |

Fig. 9: Simulation of the device behavior at $V_{DC} = 830$ V and $I_{test} = 14$ A.

In Fig. 9a, if the current waveforms are observed, it is possible to notice a number of characteristics that differentiate the slow and fast switching. On the one hand, observing the waveform corresponding to the capacitor current (I_{CDS}), a significantly higher maximum value peak can be observed with $R_{g\,off} = 0\ \Omega$. In the case of the transfer current, it is possible to observe that with $R_{g\,off} = 0\ \Omega$, it reaches zero before the complete drop of I_D.

This difference can be observed in Fig. 9b, where the transfer characteristic curve and current trajectories are shown. The usual behavior requires that I_D follows $I_{channel}$. Additionally, the instants in each case at which V_{DSk} reaches V_{DC} are marked with vertical lines. The results show that switching with $0\ \Omega$, in contrast to $20\ \Omega$, produces a decoupling between the channel and drain currents. This indicates that during the current fall, the device current is no longer being controlled by the gate. Moreover, one should observe that in contrast to the usual switching behavior, where the transfer current does not begin to fall until the drain-source voltage reaches the DC-link; with $0\ \Omega$ the V_{DSk} reaches V_{DC}, and at that instant the transfer current has already reached 0. This means that the channel is closed prematurely.

One could represent the resulting equivalent circuit between the controlled switching procedure in Fig. 10a, becoming Fig.10b with $R_{g\,off} = 0\ \Omega$.

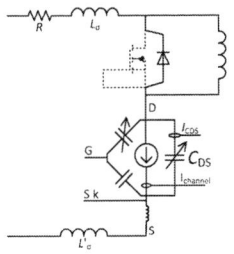

(a) Usual equivalent circuit.

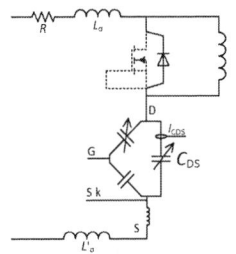

(b) Proposed equivalent circuit when switching with very low or zero gate resistance.

Fig. 10: Proposed equivalent circuits.

The proposed equivalent circuit in Fig.10b can actually be recognized as a resonant circuit between the different lump model components. The transition to a resonant occurs at the instant in which the channel current reaches 0 A, when the channel is closed.

Therefore if the initial conditions and temporal position of the resonance are correctly selected, the waveforms should overlie correctly, as shown in Fig. 11.

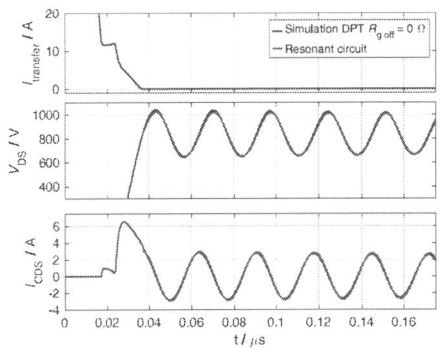

Fig. 11: Overlaying switching and resonant simulations.

Mathematical analysis

In order to obtain an expression that explains the over-voltage, the resonant circuit is mathematically modelled, where the expression (1) represents the voltage and (2) the current oscillation. The damping is disregarded due to its limited impact on the first maximum. $V_{DS}(0)$ and $I_{CDS}(0)$ are the voltage and capacitor current at the instant when the channel closes and thus the initial conditions for the oscillation, these two terms are closely related to the switched current.

$$V_{DS}(t) = V_{DC} - \left[(V_{DC} - V_{DS}(0)) \cdot \cos(\omega_0 \cdot t) - I_{CDS}(0) \cdot \sqrt{\frac{L_{\delta total}}{C_{DS}(V_{DS})}} \cdot \sin(\omega_0 \cdot t) \right] \tag{1}$$

$$I_D(t) = I_{CDS}(0) \cdot \cos(\omega_0 \cdot t) - \sqrt{\frac{C_{DS}(V_{DS})}{L_{\delta total}}} \cdot (V_{DS}(0) - V_{DC}) \sin(\omega_0 \cdot t) \tag{2}$$

$$\omega_0 = \frac{1}{\sqrt{(L_{\delta total} \cdot C_{DS}(V_{DS}))}} \tag{3}$$

The agreement between the simulations and the proposed expression is shown in Fig. 12, where a good agreement is seen.

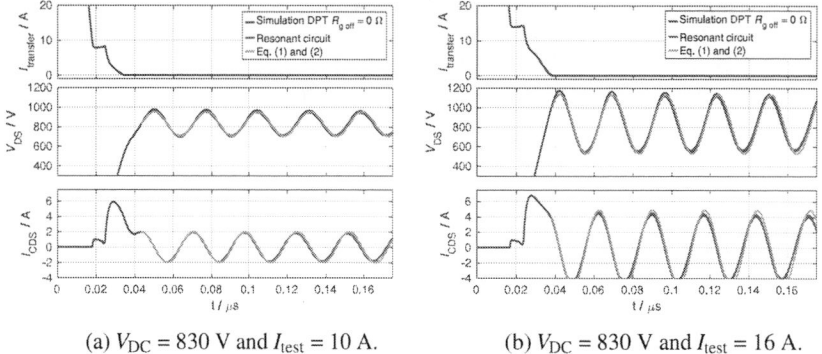

(a) $V_{DC} = 830$ V and $I_{test} = 10$ A.

(b) $V_{DC} = 830$ V and $I_{test} = 16$ A.

Fig. 12: Comparison between simulated switching, simulated resonant circuit and modelled oscillation.

The two expressions (1) and (2) result in an over-voltage peak (V_{DSpk}) that can be expressed as in (4).

$$V_{DSpk} = V_{DC} + \sqrt{(V_{DC} - V_{DS}(0))^2 + \left(I_{CDS}(0)\sqrt{\frac{L_{\delta total}}{C_{DS}}}\right)^2} \tag{4}$$

The expression in (4) may be evaluated as a function of the initial conditions for a given circuit (constant $L_{\delta total}$ and C_{DS}), creating a 3D representation of the expected OVPK as shown in Fig. 13. The maximum over-voltage can be inserted in the graph (blue plane) to know if a certain initial condition can be allowed.

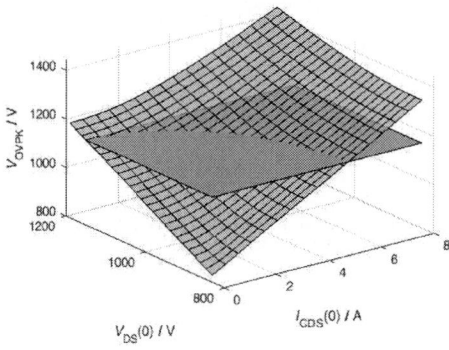

Fig. 13: Expected area for semi-intrinsic turn off, initial conditions and expected over-voltage peak.

Conclusion

In this paper the switching procedure of a SiC MOSFET operating at is maximum turn-off speed is examined. Usual turn-off behavior dictates that with lower gate resistance, lower switching losses are obtained at the cost of a higher inductive over-voltage caused by an increasing di/dt. Thus a minimum gate resistance is required to operate the device safely. However, in this work it is shown that the lower gate resistance higher over-voltage relationship can be partially broken. It is observed that when the gate resistor is reduced under a certain threshold, the inductive over voltage decreases.

The exceptional behavior allows for the further reduction of the turn-off losses. It is observed that the behavior is sensitive to the DC-link parasitic inductance and switched current. Nonetheless, it does not show dependency of the junction temperature, which is an interesting point if the switching procedure is to be applied in a practical setup.

The switching process can be simulated and an explanation of the switching process is found. The device closes its conducting channel prematurely; then, the current falls following a damped resonant circuit between the device capacitance and circuit parasitic inductance. Even though the resonant circuit shows a lower over-voltage in comparison with the usual switching procedure, an increased switched current would push the peak voltage up to the limit.

References

[1] J. Lutz, et al. , Semiconductor Power Devices: Physics, Characteristics, Reliability. Springer-Verlag Berlin Heidelberg, 2011.
[2] W. Zhang, et al., "Common source inductance introduced self-turn-on in MOSFET turn-off transient," in Conference Proceedings - IEEE - APEC, 2017.
[3] A. Lidow and J. Strydom, "eGaN® FET Drivers and Layout Considerations," White Pap., no. WP008, pp. 1–7, 2016.
[4] A. P. Arribas, et al., "Accurate characterization of switching losses in high-speed, high-voltage Power MOS-FETs," IEEE. IWIPP 2015, pp. 95–98, 2015.
[5] Z. Zhang, et al. , "Methodology for Wide Band-Gap Device Dynamic Characterization," IEEE Trans. Power Electron., vol. 32, no. 12, pp. 9307–9318, 2017.
[6] R. W. Maier and M. Bakran, "Switching SiC MOSFETs Under Conditions of a High Power Module," 2018 20th European Conference on Power Electronics and Applications (EPE'18 ECCE Europe), Riga, 2018, pp. P.1-P.9.
[7] GaN Systems Inc., "GN003 Application Note, Measurement Techniques for High-Speed GaN E-HEMTs," no. August, pp. 1–26, 2018.
[8] S. Hain and M. M. Bakran, "New Rogowski Coil Design with a High dV/dt Immunity and High Bandwidth," EPE J., vol. 25, no. 1, pp. 22–27, 2015.

The Energy Ring to supply the Expoelectric'18 show with Renewable Energy Sources and Electric Vehicles

Chillón-Antón, Cristian; Heredero-Peris, Daniel; Girbau-Llistuella, Francesc; González-Fontderubinat, Paula; Llonch-Masachs, Marc, Montesinos-Miracle, Daniel; Gomis-Bellmunt, Oriol
Centre d'Innovació Tecnològica en Convertidors Estàtics i Accionaments (CITCEA-UPC)
Departament d'Enginyeria Elèctrica, Universitat Politècnica de Catalunya
ETS d'Enginyeria Industrial de Barcelona, Av. Diagonal 647
08028 Barcelona, Spain
Phone: +34 934016727
Email: daniel.heredero@citcea.upc.edu
URL: http://citcea.upc.edu

Acknowledgements

This work has been supported by Ministerio de Ciencia, Innovación y Universidades under the project RTI2018-099540 "Flexibilidad de los recursos energéticos distribuidos parar optimizar la operación de las redes de distribución".

Keywords

«Microgrid », «Renewable energy systems», «Voltage Source Inverters», «Three-phase system».

Abstract

The Expoelectric fair takes place every year in Barcelona (Spain) and is devoted to Renewable energy and Electric Mobility. Street trade shows and fairs usually use Diesel generators to provide energy to fair booths because the street infrastructure is not designed to supply these events. In the case of Expoelectric and its objectives, the organization has proposed to supply the fair booths with on-site renewable energy and the use of exposed electrical vehicles to provide storage, which makes excellent sense in this kind of events.

This paper aims to show this experience and detail the system developed for this objective. The result is the development of a power electronics-based system able to generate the grid where other resources and storage elements are connected. This paper details the system but also the control and operation algorithms to operate the microgrid without any type of communication between all the elements connected to them.

During the two days of the fair, the Energy Ring microgrid delivered energy to the booths, assuring the supply and power quality needed for all the equipment connected (lighting, freezers, EV chargers, and computers).

Introduction

The society is increasingly aware of renewable technologies and clean mobility. Proof of this is the wide variety of fairs aligned with the electric mobility sector that nowadays takes place around Europe. Some examples are Expoelectric, Electric & Hybrid Vehicle Technology Expo, Mobility Electronics Suppliers (MES) Expo, Electric Energy Storage fair, AEDIVE Electric vehicle's fair, eMove360° or Autonomy & the Urban Mobility Summit. In all of them, the leading-edge topics are autonomy, new flexibility services thanks to the hyper-digitalization, quick-charging, new storage technologies or new mobility devices such as scooters, skateboards, cars or motorbikes.

Nonetheless, energetically speaking, all these shows require not only from a grid-connection but also from an energy contingency alternative plan. The usual solution is to consider gensets sized for feeding the exhibitors' consumptions. The use of fossil fuels as a backup plan results in a great contradiction considering nowadays distributed resources potential, even more, if the show context involves electric mobility and renewables. Although many efforts on Vehicle to Grid (V2G) [1], operation or other capabilities like Vehicle to Home (V2H) [2], have been included as extra functionalities in commercial electric vehicles, EVs, is not common to find real examples of use. One promising benefit is to use an EV's storage system to create a digital electric Genset. Thus, the V2G technology plus the inclusion of other renewable sources and the use of power electronics can become a new entity understood as a gamechanger for e-mobility acting as a feeder used in places without connection to a power grid, or as an emergency power supply.

Last October 6th-7th 2018, the eighth edition of the Expoelectric show was held at the Arc de Triomf in Barcelona, Spain. The Expoelectric show promotes electrical mobility and renewables, the deployment of electric vehicles and has had an increasing relevance and participation throughout the last editions. Nowadays, it can be considered a key event for the proliferation of electric mobility in Barcelona, Catalonia and Spain. In the last edition, the organization was ready to tackle a new challenge: to deploy an energetic electrical ring to supply the electrical loads of the fair, using electric vehicles, optimally hybridized with renewable energy source and electrical energy storage systems. In previous editions, it had been necessary to use diesel generators to ensure the electric supply of the show. This fact continues to be necessary for many outdoor events since the distribution network has limitations in many locations. As Expoelectric deals with electrical mobility, it was possible a transition to sustainable electrical systems for shows, without relying on diesel generators. This transition ensures a coherent message without any sort of contradictions, in alignment with a new emerging energy trend, where the integration of several types of actives agents such as Distributed Energy Resources (DERs), and prosumers are pushing for an evolved electric system. Concepts like microgrids, smart grids, or local markets are paving the way for this paradigm [3]. For instance, microgrids provide potential economic, environmental and flexibility benefits. Still, their implementation is hand in hand with technical challenges as high dynamics control needs, optimal energy management and protective coordination [4]. An electrical energy ring was the proposed solution for the Expoelectric show. The energy ring concept is an autonomous electrical system isolated from the distribution network, which supplies loads of the ring employing renewable energies and batteries of electric vehicles operating in V2G mode. The system is an integrated solution with multiple different devices from different involved manufacturers and partners.

System overview

The energetic ring is a proposal derived from the concept of isolated microgrids that requests for a grid forming equipment, energy resources, storage and loads. The system developed is depicted in Figure 1, and includes a grid forming converter, supplied by a battery, a grid feeding storage, PV generation and loads.

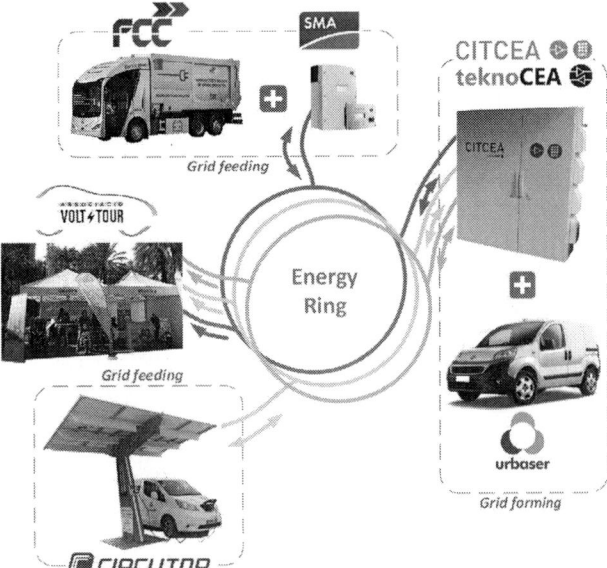

Figure 1. General scheme of the Energy Ring

The main component is the focus of this paper and is the Distributed Energy Resource (DER) responsible for forming the grid, fixing the voltage and frequency. The Energy Ring Generator (ERG) was developed for this purpose by the authors of the paper at CITCEA-UPC. The ERG is a SiC-based power electronics system able to synthesize the requested voltage for the three-phase, four-wire system, 400 V, 50 Hz. The ERG can provide up to 40 kVA (50 A maximum current). It also includes some functionalities needed for the correct operation of the complete energy ring as frequency control to manage energy flows in the microgrid, short circuit capability to trip circuit breakers and advance voltage control for non-linear loads. The energy to this equipment is provided by a battery of an electric van from URBASER. The second life Li-ion battery provides up to 30 kWh at 295 V (rated values). The ERG includes a DC/DC converter to interface and manage the battery correctly.

As a DER, the energy ring has a PV system that acts as a standard PV generator, injecting a specific power depending on solar irradiation provided by CIRCUTOR. The PV inverter includes anti-islanding and acts as a current source to the system, in a grid feeding scheme.

Also, a backup battery base system is connected as a storage system and controlled as a second current source managed externally, in a grid feeding scheme. An electric track provides the battery from FCC, and the battery inverter is from SMA.

For proper operation and protection of the ERG, a TT neutral scheme is considered. A YNyn transformer is used at the output of the ERG from where the microgrid is created and where all the loads, DERs and storage is connected.

Loads are connected at different single-phase and three-phase lines protected by circuit-breakers and RCD.

Operation, protection and control

The key point on the operation of the energy ring is the correct management of the grid-feeding equipment according to the real-time energy balance of the system. In the Energy Ring developed, any communication protocol between all the equipment has been avoided for different reasons. The first and most pragmatic is the difficulty in some cases to have access to the protocol of the equipment, and the second one is willing to do that. The solution adopted is based on the standard VDE-AR-N-4105 [5]. In this standard, the power control is done by the frequency of the grid. In that sense, the grid-feeding systems adapt the injected power to the energy ring according to the profile depicted in Figure 2.

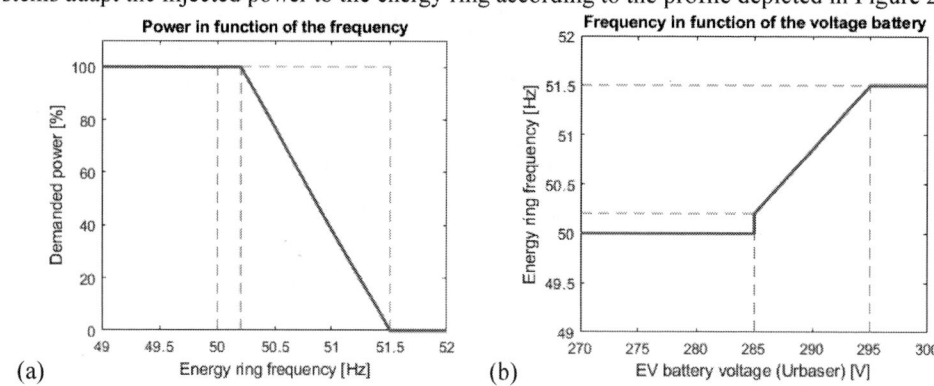

Figure 2. (a) Demand power management of the grid-feeding DERs depending on the frequency in the Energy Ring. (b) Frequency profile imposed by the ERG depending on the SoC of the battery

In this sense, the ERG can implement a kind of energy management system and use the surplus of energy to charge the primary energy source (directly related with its State of Charge of the battery) connected to the ERG.

Battery

One of the challenges of the project was to estimate the requested capacity on the battery of the ERG. The rated capacity of a new battery was 30 kWh. Usually, second-life batteries are supposed to be at 80% of State of Health, so the estimated capacity of the battery is supposed to be 24 kWh.

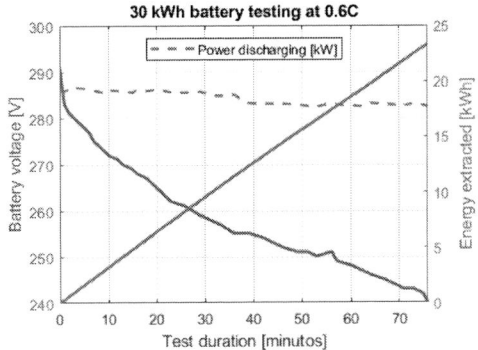

Figure 3. Second-life battery of the ERG energy testing results

Supposing a power consumption of 50 W to 200 W per booth, the total estimated power consumption in the Energy Ring was about 6 kW. Figure 3 shows the result of the capacity tests done with the battery. The battery was discharged at 18 kW of power (about 0.6C discharge rate). The energy capacity of the battery was estimated to be 24 kWh.

Voltage control and non-linear loads

The second challenge of the project was to generate a pure sinusoidal voltage in the presence of heavy non-linear loads and grid-feeding DERs connected to the Energy Ring. The quality of the output voltage can compromise the proper operation of the connected systems.

Before installing the equipment, some tests were conducted to properly tune the voltage controlled with a heavy non-linear load as depicted in Figure 4. As can be seen, at the time instant that the load is connected a large inrush current (about 30 A peak) distorts the voltage, but the voltage controller can restore the value in few ms. Once the non-linear load reaches the steady-state, the voltage distortion is lower than 2 %.

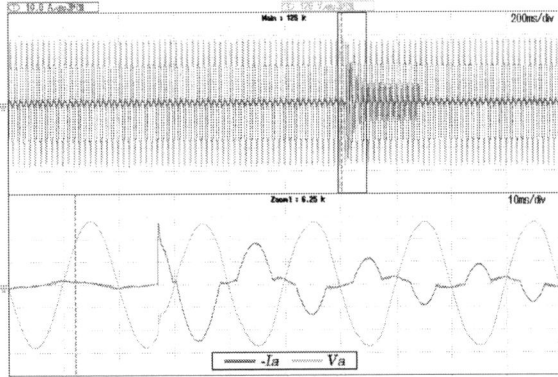

Figure 4. Energy Ring voltage supplying a non-linear load

The output voltage controller has to address the fact that the system has three-phases plus neutral, so single-phase loads are accepted. Independent control per phase has to be used, and different current consumption (and harmonic content) per phase can be present. The controller is a double high-bandwidth control loop based in Fractional proportional-resonant controllers [6].

Short-circuit and selective fault

Short circuits and faults are not conventional in electrical systems but have to be addressed for proper system protection. Usually, circuit breakers are used to protect the installation against short circuits and overcurrents. These devices have a trip curve where the trip time depends on the severity of the short circuit or over current, as can be seen in Figure 5. The generator in the system is responsible for providing short-circuit current to trip the protection device, protecting the installation from the short-circuit. Rotating machines have a substantial overload capability to provide these current levels, to the short-circuit, but this is not the case when using power electronics-based systems as the ERG. In that case, because of cost constraints, these systems can not be designed to have such overload capacities to trip the protection devices properly. So, adequate protection and current generation system for ERG was developed.

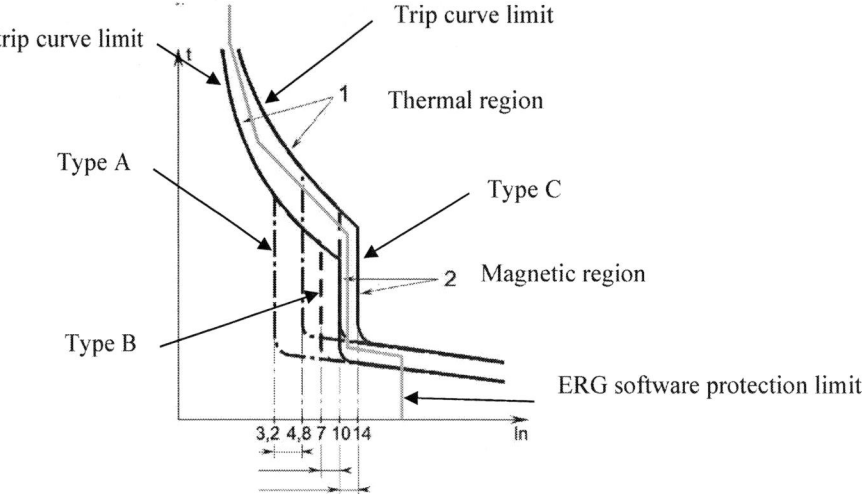

Figure 5. Circuit breaker trip curves and the emulated curve in the ERG

Because the Energy Ring is a three-phase four-wire system, the faults can be of different type and consequently, the fault current will be different, so the current limit has to be different [7]. Figure 6 shows the test results for different types of short-circuits. After the short-circuit, the voltage is re-established to the rated value.

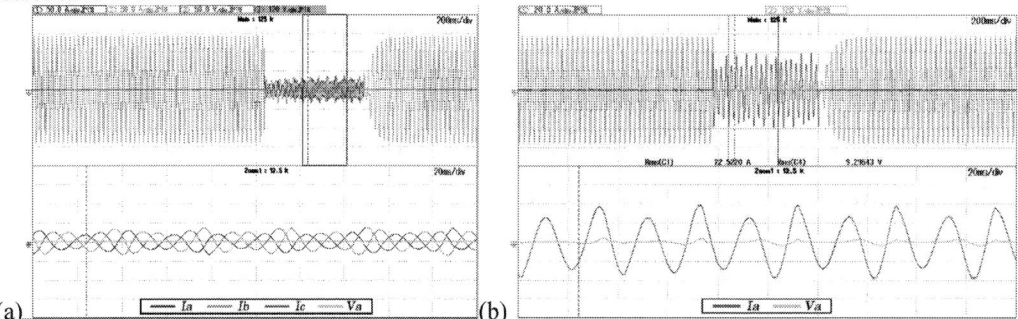

Figure 6. (a) Energy ring voltage under a low impedance three-phase short-circuit. (b) Energy ring voltage under a low impedance single-phase short-circuit

In the energy ring, different lines exist to provide energy to different areas of the trade show. Selectivity for different circuits has to be assured in this type of installation. In the case of a fault in one line, that has not to affect other lines. Due to limits on the current capability of the ERG in the case of overload or short-circuit, the Energy Ring adopts different mechanism for proper selectivity.

Three different components have been implemented in the Energy Ring. A main three-phase four-pole 16 A Type C circuit breaker is placed at the output of the ERG. For each individual single-phase line, single phase 6 A Type B circuit breakers are adopted. An extra virtual (software) circuit breaker tripping curve was implemented in the ERG, as shown in Figure 5.

Different tests were conducted to test the protection capabilities of this protection and selectivity scheme. Figure 7a, at the output of the ERG, a 16 A Type C circuit breaker is placed. The lines are protected with 6 A Type B circuit breakers, and the ERG virtual breaker is set to 10 A Type C curve. The load is set at 16 A. In that case, at 2.5 s, the 6 A Type B breaker opens the fault with selectivity.

The second test is shown in Figure 7b. In that case, at the output of the ERG, a 16 A Type C circuit breaker is placed. The lines are protected with 16 A Type C circuit breakers, and the ERG virtual breaker is set to 10 A Type C curve. The load is set at 16 A. In that case, no selectivity is expected, and at 3 s, the ERG software breaker trips and stops the operation.

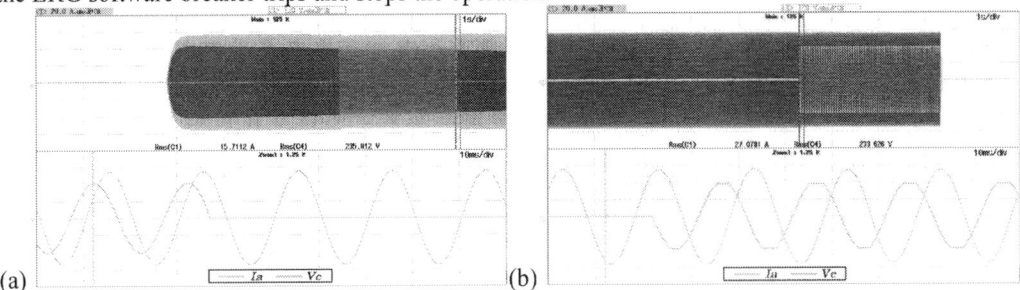

(a)   (b)

Figure 7. (a) Selectivity test between the software protection of 10 A and the 6 A (type C) circuit breaker. (b) Selectivity test between the software protection of 10 A and the 16 A (type C) circuit breaker

Field experience

The Expoelectric trade show took place in Barcelona (Spain) on October 6 and 7 of 2018. Figure 8 shows some pictures of the booth where the ERG was installed. For the two days, the Energy Ring was operated satisfactory but in completely different conditions. The first day was sunny, and the loads were lower than the second day. The second day was rainy, with no PV production and the load were expanded following the good results of the first day. In the following lines, the results obtained are explained in detail.

Figure 8. Pictures of the setup and the EV used to supply the ERG

First day – 6th October 2018

The ring operation began at 10:15 and continued uninterrupted until 18:00. It can be divided into three periods, clearly defined in Figure 9. From 10:15 to 12:00 the ring was supplied by the ERG and the PV system. At 12:00, the ERG battery has supplied a total amount of 0.7 kWh. That can also be seen in Figure 10, were from 10:15 to 12:00 the frequency decreases as the battery is being discharged and indicating to other DERs to provide more power to the ring, if possible.

From 12:00 to 15:00, the grid-feeding battery-based system connected to phase A was started and fixed a set-point of 1.5 kW. This amount of power and energy was used to recharge the ERG battery, as can be seen in Figure 9, were in this period the power is negative in phase A. Also, the frequency increases because of the SoC of the ERG battery increases.

At 15:00, the grid-feeding battery-based system was stopped down to the end of the day at 18:00, as can be seen in Figure 9. The battery of the ERG finishes with a surplus of 3 kWh of energy, both, from the PV system and the other storage system in the ring.

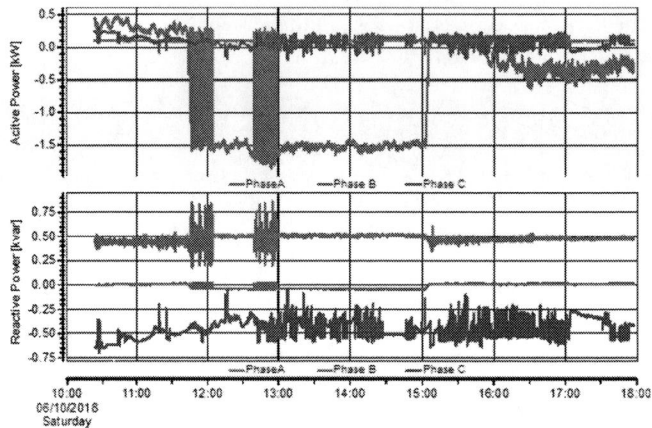

Figure 9. First day active and reactive power delivered by the ERG

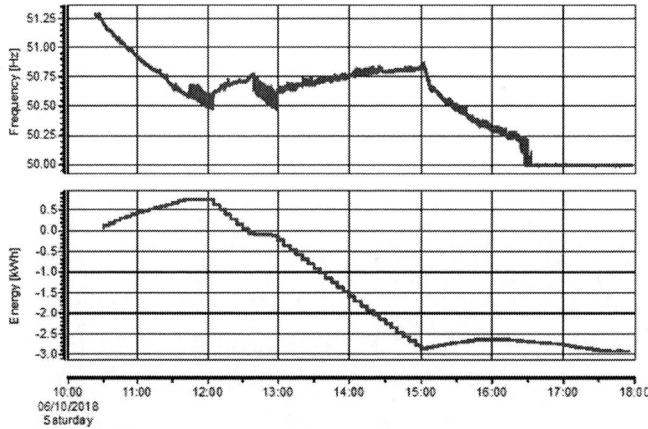

Figure 10. First day: frequency and energy provided by the ERG

Second day – 7th October 2018

In the second day, the operation started at 9:30 and continued to 18:00 with two interruptions. As the results of the first day were satisfactory, the Energy Ring was expanded to cover more booths. But the weather conditions were not good for the PV system. It was a cloudy and rainy day.

This second day can also be divided into three parts. From 9:30 to 12:30, the system was operating as expected, as can be seen in Figure 11. Figure 12 shows a deeper slope on the frequency as the battery of the ERG is discharged faster because it was supplying higher power.

The second part starts at 12:30. At that point, the ERG battery has discharged at about half of the SoC, and the system was stopped.

The third period starts at 15:30, and the Energy Ring was started again, with a reduced load, and operated satisfactory up to 18:00, when the system was definitively stopped.

As can be seen in Figure 12, the ERG battery has supplied a total amount of 16 kWh, from the 24 kWh estimated energy capacity.

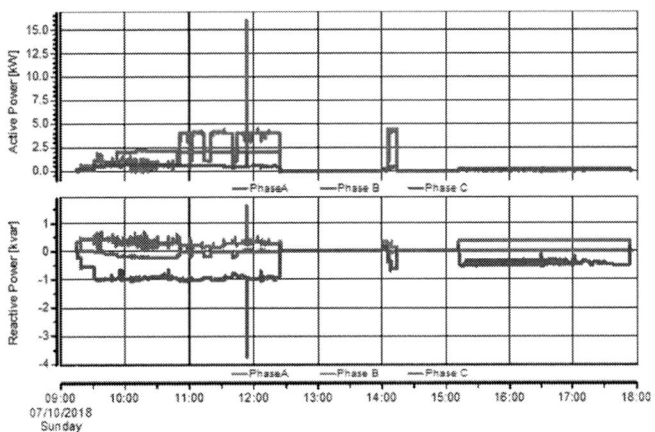

Figure 11. Second day active and reactive power delivered by the ERG

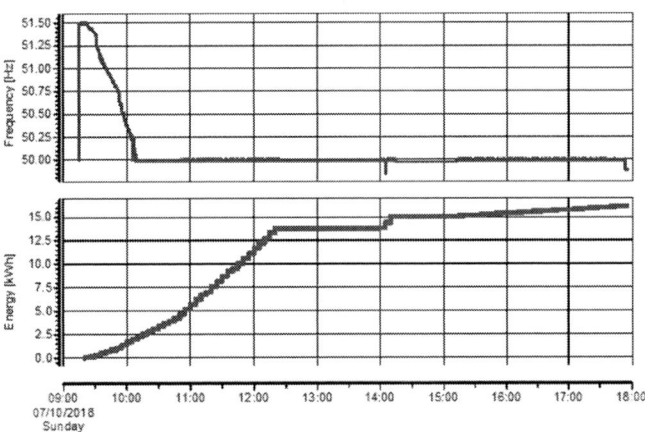

Figure 12. Second day: frequency and energy provided by the ERG

Conclusions

Distributed energy resources and power electronics have transformed the energy paradigm. The new model of the electric sector is changing to become bidirectional, smart, auto-configurable and auto-adaptable thanks to the inclusion of digital technology and power electronics.

In specific environments, due to the constraints of the electric system, power electronics-based solutions in combination with non-stationary batteries and DER can create a microgrid to provide energy, as it has been demonstrated in this paper, tested in real conditions.

The proposed Energy Ring is an AC microgrid where the primary source of energy is based on V2G capabilities and where the AC output voltage is controlled per phase using a highly flexible power electronics-based device. Other active agents based on renewables and huge e-mobility devices as electric trucks complement the energy ring allowing to create an "electronic Genset". The autonomy range of the primary energy source can be extended using these active agents.

The functioning of the ring concept has been previously evaluated in the laboratory and then tested in a real environment. The operation of this energy ring creates not only a high performance independent three-phase output voltage but also implements an adaptive fault current limiting algorithm, operative in the case of a short-circuit. However, a selective based protective device is proposed to isolate any possible fault due to the intrinsic limitation of the short-circuit current on power electronics-based devices.

As the energy ring integrates other active sources, it is also able to control the exchanged power of those external sources by variating the AC frequency of the generated voltage set according to VDE-AR-N-4105. In this sense, it is avoided the use of dedicated communication for this purpose. Through the ERG, the energy ring concept can reduce the dependence of conventional gensets which contaminates and generates noise, stepping forward into the sustainable supply.

References

[1] J. Wang, Z. Wang, P. Zeng, X. Jin, D. Li, M. Wan and F. Kong, «Software-Defined Wi-V2G: A V2G Network Architecture,» *IEEE Intelligent Transportation Systems Magazine,* vol. 10, n° 2, pp. 167-179, 2018.

[2] C. Liu, K. T. Chau, D. Wu and S. Gao, «Opportunities and challenges of vehicle-to-home, vehicle-to-vehicle, and vehicle-to-grid technologies,» *Proceedings of the IEEE,* vol. 101, n° 11, pp. 2409-2427, 2013.

[3] B. Hartono, Budiyanto and R. Setiabudy, «Review of microgrid technology,» de *2013 International Conference on Quality in Research, QiR 2013 - In Conjunction with ICCS 2013: The 2nd International Conference on Civic Space,* 2013.

[4] A. Hooshyar and R. Iravani, «Microgrid Protection,» *Proceedings of the IEEE,* vol. 105, n° 7, pp. 1332-1353, 2017.

[5] VDE, «Power Generating Plants in the Low Voltage Grid (VDE-AR-N 4105),» VDE, 2019.

[6] D. Heredero-Peris, C. Chillón-Antón, E. Sánchez-Sánchez and D. Montesinos-Miracle, «Fractional proportional-ressonant current controllers for voltage source converters,» *Electric Power Research,* vol. 168, pp. 20-45, March 2019.

[7] A. Junyent-Ferré, O. Gomis-Bellmunt, T. Green and D. Soto-Sanchez, «Current control reference calculation issues for the operation of renewable source grid interface VSCs under balanced voltage sags,» *IEEE Trans. On Power Electronics,* vol. 26, n° 12, pp. 3744-3753, 2011.

[8] D. Heredero-Peris, C. Chillón-Antón, M. Pagès-Giménez, D. Montesinos-Miracle, M. Santamaría, D. Rivas and M. Aguado, «An enhancing fault current limitation hybrid droop/V-f control for grid-tied four-wire inverters in AC microgrids,» *Applied Sciences,* vol. 8, n° 10, p. 1725, 2018.

[9] J. Rocabert, Á. Luna, F. Blaabjerg and P. Rodríguez, «Control of power converters in AC microgrids,» *IEEE Trans. On Power Electronics,* vol. 27, n° 11, pp. 4734-4749, 2012.

Impedance-Based Modeling of a Three-level Converter Under Balanced and Unbalanced Condition for the Stability Analysis of Bipolar LVDC Grids

T. Roose[*], G. Van den Broeck[*†], M. M. Alam[‡] and J. Beerten[*]

[*]KU LEUVEN, ELECTA, Kasteelpark Arenberg 10, 3001 Leuven, Belgium
[†]DCINERGY, Langdorpsesteenweg 106, 3200 Aarschot, Belgium
[‡]VITO, Boeretang 200, 2400 Mol, Belgium
Email: thomas.roose@kuleuven.be

Acknowledgments

The work of T. Roose is funded by a research grant from the Research Foundation - Flanders (FWO) and the Flemish Institute for Technological Research (VITO), Grant no. 1182519N.

Keywords

≪Converter control≫, ≪DC power supply≫, ≪Modeling≫, ≪Power quality≫.

Abstract

This paper presents an impedance-based assessment to identify potential instabilities in bipolar LVDC grids where the voltage is controlled by a three-level DC-DC converter. As an essential step to perform the assessment, the small-signal admittance matrix of the three-level DC-DC converter is analytically derived and verified by a non-linear switched model implemented in the EMT-type software PSCAD. It is shown that the converter admittance matrix exhibits non-passive behavior at frequencies above 500 Hz. Furthermore, the non-passivity is found to be dependent on the operating point of the converter. This behavior can negatively interact with resonances of bipolar LVDC systems, e.g. LC-filter resonances, leading to unstable high-frequency oscillations and system instability.

Introduction

The increasing share of renewable energy sources, power electronic loads and energy storage systems installed at the distribution level revive the interest in low-voltage direct current (LVDC) grids [1, 2]. Interconnecting these converter-interfaced applications, which are mainly DC-based, through a LVDC grid significantly improves the power transfer capability and reduces the amount of conversion steps, increasing the energy efficiency [3].

More particularly, bipolar LVDC networks allow connecting high-power loads to the pole-to-pole voltage, while low-power loads are connected to the pole-to-neutral voltages [4]. However, an asymmetric power distribution in the two poles leads to voltage unbalance as the neutral terminal voltage shifts [5]. Consequently, converters with voltage balancing capability (VBC) are required. According to [4], it is possible to efficiently integrate the VBC in a non-isolated three-level DC-DC converter (TLC), capable of directing current asymmetrically towards the positive and negative pole in order to equalize the pole-to-neutral voltages.

Unfortunately, converter dynamics can exert pressure on the stability of LVDC systems. It has been shown that the constant power load (CPL) behavior of converters, i.e. consuming a fixed amount of power independent of the supply voltage, causes non-passivity of the converter impedance [6]. This non-passivity due to a negative incremental resistance can adversely interact with the remaining elements of the system, leading to unstable oscillations and system instability [7,8]. CPL behavior can be invoked by several types of feedback control, e.g. active power control [9], voltage control [10–12] and torque/speed-control in drives [12–15]. As the CPL behavior is only exhibited within the bandwidth of the control loop, the unstable oscillations caused by these slower outer control loops primarily occurred at lower frequencies [16]. Hence, the effect of the faster inner control is typically neglected in these stability studies as the focus is on the lower frequency range. It is thus unclear to what extent LVDC systems can also exhibit unstable behavior due to faster converter dynamics.

This paper seeks to address this research question by investigating the risk of high-frequency unstable oscillations in bipolar LVDC systems as a result of converter non-passivity associated with the faster inner control. In particular, the effect of the three-level DC-DC converter dynamics on the system stability is analyzed by the use of an impedance-based approach [17]. As an essential step in the impedance-based stability analysis, the admittance matrix of the three-level DC-DC converter is derived based on the equations describing its dynamics and control structure. The small-signal impedance-based model of the TLC is then verified by a non-linear switched model implemented in the EMT-type software PSCAD and the passivity of the matrix elements is investigated. Finally, the risk of instability for a bipolar LVDC test system including a three-level DC-DC converter is assessed and the results are compared to time-domain simulations of the non-linear switched model.

Dynamics and control structure of three-level DC-DC converter

Fig. 1a depicts the non-isolated step-down three-level DC-DC converter as described in [4, 18]. The converter consists out of four power semi-conductor devices. The positive (d_p, d'_p) and negative semi-conductors (d_n, d'_n) are switched in a complementary manner by using pulse-width modulation. Consequently, the duty cycles d_p and d_n are two controllable input variables. If the discrete effects of switching are disregarded, the following equations govern the dynamics of the averaged model of the TLC,

$$L\frac{di_L}{dt} = d_p v_p + d_n v_n - v_2, \quad C\frac{dv_p}{dt} = i_p - d_p i_L \quad \text{and} \quad C\frac{dv_n}{dt} = i_n - d_n i_L \tag{1}$$

where i_L represents the inductor current, L is the inductance, v_p is the positive pole-to-neutral voltage, v_n is the neutral-to-negative pole voltage, v_2 is the back-end voltage and C is the pole-to-neutral capacitance. The voltages and currents can be decomposed in balanced (b) and unbalanced (u) components according to the transformation matrix T [5],

$$\begin{bmatrix} x_b \\ x_u \end{bmatrix} = T \begin{bmatrix} x_p \\ x_n \end{bmatrix} \quad \text{where} \quad T = \begin{bmatrix} 1/2 & 1/2 \\ 1/2 & -1/2 \end{bmatrix} \quad \text{and} \quad T^{-1} = \begin{bmatrix} 1 & 1 \\ 1 & -1 \end{bmatrix} \tag{2}$$

which is similar to the symmetrical component decomposition in three-phase AC systems. Applying the decomposition to (1) results in

$$L\frac{di_L}{dt} = 2d_b v_b + 2d_u v_u - v_2, \quad C\frac{dv_b}{dt} = i_b - d_b i_L \quad \text{and} \quad C\frac{dv_u}{dt} = i_u - d_u i_L \tag{3}$$

which gives the dynamics of the TLC in the balanced and unbalanced (BU) frame. The balanced duty cycle d_b primarily controls i_L, while the unbalanced duty cycle d_u controls the unbalanced voltage v_u [4]. The control structure of the three-level converter is depicted in Fig. 1b [19]. Three control loops can be distinguished, i.e. inductor current control, balanced voltage control and unbalance voltage control. The inductor current control, which is considered to be the inner control, is ten times faster than the outer controllers, which are the balanced and unbalanced voltage control. According to Fig. 1b, the following equations summarize the control dynamics of the TLC,

$$i_L^* = C_{v_b}(v_b^* - F_{v_b} v_b), \quad u_i = C_{i_L}(i_L^* - F_{i_L} i_L) \quad \text{and} \quad d_b = \frac{u_i + v_2}{2v_b}, \tag{4}$$

$$u_u = C_{v_u}(v_u^* - F_{v_u} v_u) \quad \text{and} \quad d_u = -\frac{u_u}{i_L}. \tag{5}$$

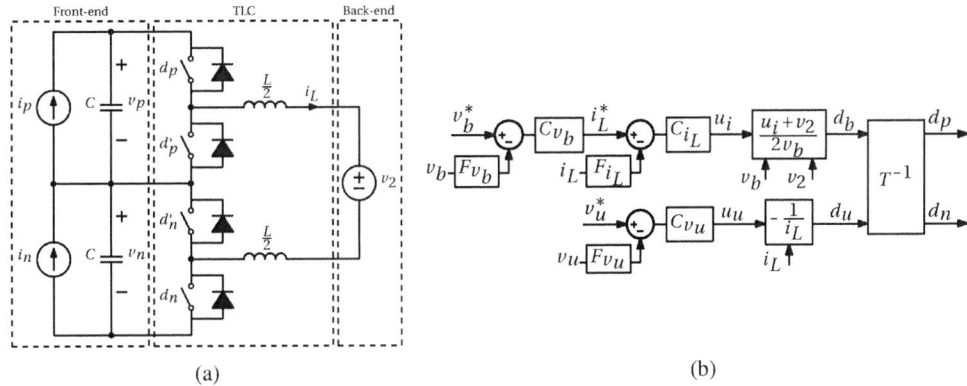

Fig. 1: (a) Step-down three-level DC-DC converter   (b) TLC control structure in BU-frame

The proportional-integral (PI) controller of each loop is indicated with a transfer function $C(s)$ and the first-order measurement filters by $F(s)$,

$$C_{i_L}(s) = K_{p,i_L} + \frac{K_{i,i_L}}{s}, \quad C_{v_b}(s) = K_{p,v_b} + \frac{K_{i,v_b}}{s} \quad \text{and} \quad C_{v_u}(s) = K_{p,v_u} + \frac{K_{i,v_u}}{s}, \tag{6}$$

$$F_{i_L}(s) = \frac{\alpha_{f,i_L}}{s + \alpha_{f,i_L}}, \quad F_{v_b}(s) = \frac{\alpha_{f,v_b}}{s + \alpha_{f,v_b}} \quad \text{and} \quad F_{v_u}(s) = \frac{\alpha_{f,v_u}}{s + \alpha_{f,v_u}} \tag{7}$$

where s is the Laplace variable.

Derivation of admittance matrix of three-level DC-DC converter

In order to perform an impedance-based stability analysis, the small-signal admittance matrix of the three-level converter is defined in the BU-frame as

$$\begin{bmatrix} \Delta i_b \\ \Delta i_u \end{bmatrix} = \begin{bmatrix} y_{bb} & y_{bu} \\ y_{ub} & y_{uu} \end{bmatrix} \begin{bmatrix} \Delta v_b \\ \Delta v_u \end{bmatrix}. \tag{8}$$

Linearizing the equations of (3) and (4) which determine the balanced dynamics of the converter and converting to the Laplace-domain gives

$$\Delta i_b = Cs\Delta v_b + i_{L,0}\Delta d_b + d_{b,0}\Delta i_L, \tag{9}$$

$$\Delta i_L = \frac{2v_{b,0}}{Ls}\Delta d_b + \frac{2d_{b,0}}{Ls}\Delta v_b + \frac{2d_{u,0}}{Ls}\Delta v_u \quad \text{and} \tag{10}$$

$$\Delta d_b = -\frac{C_{i_L}C_{v_b}F_{v_b}}{2v_{b,0}}\Delta v_b - \frac{C_{i_L}F_{i_L}}{2v_{b,0}}\Delta i_L - \frac{v_2}{2v_{b,0}^2}\Delta v_b \tag{11}$$

where the unbalanced voltage is controlled to zero in steady-state, $v_{u,0} = 0$, and the back-end voltage v_2 is assumed to be a constant voltage source. If (10) is inserted in (11), then

$$\Delta d_b = \left(-\frac{C_{i_L}C_{v_b}F_{v_b}}{2v_{b,0}} - \frac{v_2}{2v_{b,0}^2}\right)\Delta v_b - \frac{C_{i_L}F_{i_L}}{2v_{b,0}}\left(\frac{2v_{b,0}}{Ls}\Delta d_b + \frac{2d_{b,0}}{Ls}\Delta v_b + \frac{2d_{u,0}}{Ls}\Delta v_u\right). \tag{12}$$

Subsequently, the small-signal balanced duty cycle Δd_b is written as a function of Δv_b and Δv_u,

$$\Delta d_b = \left(\frac{Ls}{Ls + C_{i_L}F_{i_L}}\right)\left[\left(-\frac{C_{i_L}C_{v_b}F_{v_b}}{2v_{b,0}} - \frac{v_2}{2v_{b,0}^2} - \frac{C_{i_L}F_{i_L}d_{b,0}}{v_{b,0}Ls}\right)\Delta v_b - \left(\frac{C_{i_L}F_{i_L}d_{u,0}}{v_{b,0}Ls}\right)\Delta v_u\right]. \tag{13}$$

Inserting (11) in (10) yields

$$\Delta i_L = -\frac{C_{i_L} C_{v_b} F_{v_b}}{Ls}\Delta v_b - \frac{v_2}{v_{b,0} Ls}\Delta v_b + \frac{2d_{b,0}}{Ls}\Delta v_b + \frac{2d_{u,0}}{Ls}\Delta v_u - \frac{C_{i_L} F_{i_L}}{Ls}\Delta i_L. \tag{14}$$

Writing Δi_L as a function of Δv_B and Δv_u results in

$$\Delta i_L = \left(\frac{1}{Ls + C_{i_L} F_{i_L}}\right)\left(-C_{i_L} C_{v_b} F_{v_b} - \frac{v_2}{v_{b,0}} + 2d_{b,0}\right)\Delta v_b + \left(\frac{2d_{u,0}}{Ls + C_{i_L} F_{i_L}}\right)\Delta v_u. \tag{15}$$

If (13) and (15) are inserted in (9), Δi_b is exclusively dependent on Δv_b and Δv_u,

$$\begin{aligned}
\Delta i_b &= \left[Cs + \left(\frac{i_{L,0} Ls}{Ls + C_{i_L} F_{i_L}}\right)\left(-\frac{C_{i_L} C_{v_b} F_{v_b}}{2v_{b,0}} - \frac{v_2}{2v_{b,0}{}^2} - \frac{C_{i_L} F_{i_L} d_{b,0}}{v_{b,0} Ls}\right) - \left(\frac{d_{b,0} C_{i_L} C_{v_b} F_{v_b}}{Ls + C_{i_L} F_{i_L}}\right)\right]\Delta v_b \\
&\quad + \left[\left(-\frac{i_{L,0} Ls}{Ls + C_{i_L} F_{i_L}}\right)\left(\frac{C_{i_L} F_{i_L} d_{u,0}}{v_{b,0} Ls}\right) + \left(\frac{2d_{u,0} d_{b,0}}{Ls + C_{i_L} F_{i_L}}\right)\right]\Delta v_u \\
&= y_{bb}\Delta v_b + y_{bu}\Delta v_u
\end{aligned} \tag{16}$$

taking into account that $v_2 = 2d_{b,0} v_{b,0}$. The expression for y_{bb} and y_{bu} is now determined. Linearizing and combining the equations of (3) and (5) which determine the unbalanced dynamics of the converter and converting to the Laplace-domain gives the linearized unbalanced admittance which is independent of the balanced voltage or current,

$$y_{uu} = Cs + C_{v_u} F_{v_u} \quad \text{and} \quad y_{ub} = 0. \tag{17}$$

Consequently, the transfer functions of the four elements of the admittance matrix are derived. According to (3), the steady-state duty cycles and inductor current in balanced voltage condition are

$$d_{b,0} = \frac{v_2}{2v_{b,0}}, \quad i_{L,0} = \frac{i_{b,0}}{d_{b,0}} \quad \text{and} \quad d_{u,0} = \frac{i_{u,0}}{i_{L,0}}. \tag{18}$$

The balanced and unbalanced active power are defined as

$$P_b = \frac{P_p + P_n}{2} = \frac{v_p i_p + v_n i_n}{2} = v_b i_b + v_u i_u \quad \text{and} \quad P_u = \frac{P_p - P_n}{2} = \frac{v_p i_p - v_n i_n}{2} = v_b i_u + v_u i_b \tag{19}$$

where the power is positive in the direction towards the converter. Based on (19), the balanced and unbalanced currents are defined as

$$i_b = \frac{v_b P_b - v_u P_u}{v_b{}^2 - v_u{}^2} \quad \text{and} \quad i_u = \frac{v_b P_u - v_u P_b}{v_b{}^2 - v_u{}^2}. \tag{20}$$

Subsequently, the steady-state balanced and unbalanced currents in case the pole-to-neutral voltages are balanced are

$$i_{b,0} = \frac{P_{b,0}}{v_{b,0}} \quad \text{and} \quad i_{u,0} = \frac{P_{u,0}}{v_{b,0}}. \tag{21}$$

If the active power in the positive and negative pole is the same ($P_u = 0$), the steady-state unbalanced duty cycle $d_{u,0}$ and consequently the element y_{bu} are equal to zero. The admittance matrix is then diagonal.

Verification of TLC admittance matrix by non-linear model

The small-signal admittance matrix derived in the previous section is now verified against a non-linear switched model of the three-level DC-DC converter implemented in the EMT-type software PSCAD. The parameters of the converter are summarized in Table I and the verification is performed at the operating point $P_p = -2\,\text{kW}$ and $P_n = -2\,\text{kW}$. Consequently, the element y_{bu} is equal to zero and the admittance

Table I: Converter and control parameters

Par.	Value	Par.	Value	Par.	Value	Par.	Value	Par.	Value	Par.	Value
C	$220\,\mu\text{F}$	v_2	$200\,\text{V}$	K_{p,i_L}	5.72	K_{i,i_L}	11679	BW_{i_L}	$650\,\text{Hz}$	α_{f,i_L}	$40841\,\text{rad/s}$
L	$1.4\,\text{mH}$	i_L	$20\,\text{A}$	K_{p,v_b}	-0.31	K_{i,v_b}	-64.24	BW_{v_b}	$65\,\text{Hz}$	α_{f,v_b}	$4084\,\text{rad/s}$
v_p, v_n	$350\,\text{V}$	P_2	$4\,\text{kW}$	K_{p,v_u}	0.09	K_{i,v_u}	18.35	BW_{v_u}	$65\,\text{Hz}$	α_{f,v_u}	$4084\,\text{rad/s}$

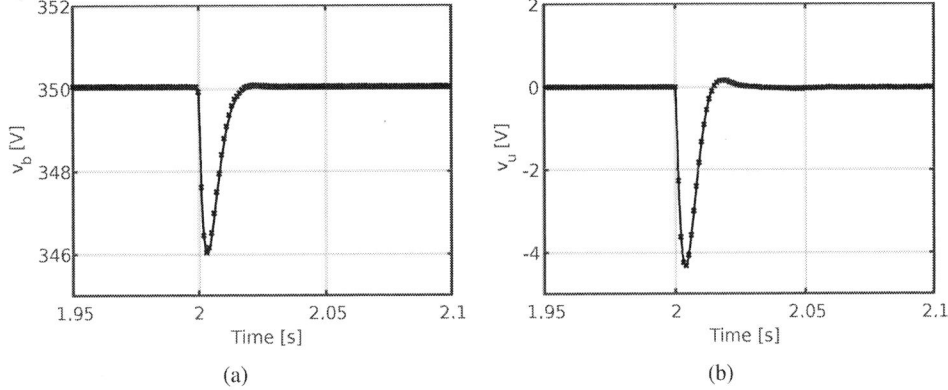

(a) (b)

Fig. 2: Step response of (a) v_b and (b) v_u, for linearized model (solid black line) and non-linear model (black crosses) at operating point $P_p = -2\,\text{kW}$, $P_n = -2\,\text{kW}$.

matrix reduces to

$$\begin{bmatrix} \Delta i_b \\ \Delta i_u \end{bmatrix} = \begin{bmatrix} y_{bb} & 0 \\ 0 & y_{uu} \end{bmatrix} \begin{bmatrix} \Delta v_b \\ \Delta v_u \end{bmatrix}. \tag{22}$$

The verification of the admittances y_{bb} and y_{uu} is thereafter performed by the use of step response analysis and a frequency sweep method.

Verification via step response

For analyzing the step response of the small-signal converter model, a step in the balanced current i_b and the unbalanced current i_u is applied and the resulting response of the balanced voltage v_b and unbalanced voltage v_u is obtained. In case of a step in the current, the response of the voltage is determined by the impedance matrix which is the inverse of the admittance matrix,

$$\begin{bmatrix} \Delta v_b \\ \Delta v_u \end{bmatrix} = \begin{bmatrix} y_{bb} & 0 \\ 0 & y_{uu} \end{bmatrix}^{-1} \begin{bmatrix} \Delta i_b \\ \Delta i_u \end{bmatrix} \quad \text{where} \quad \begin{bmatrix} y_{bb} & 0 \\ 0 & y_{uu} \end{bmatrix}^{-1} = \frac{1}{y_{bb}y_{uu}} \begin{bmatrix} y_{uu} & 0 \\ 0 & y_{bb} \end{bmatrix} = \begin{bmatrix} z_{bb} & 0 \\ 0 & z_{uu} \end{bmatrix}. \tag{23}$$

In Fig. 2a, the balanced voltage v_b is depicted for a step of 10% in the balanced current i_b. The solid black line represents the step response of the small-signal impedance z_{bb}. The data points obtained from the time-domain simulations of the non-linear switched model are indicated with black crosses. As both results closely coincide, it is demonstrated that the element y_{bb} of the admittance matrix is indeed capable of accurately representing the converter dynamics for small-signal stability studies. A similar analysis is performed in Fig. 2b for a step in the unbalanced current i_u, which also confirms the validity of the element y_{uu} of the admittance matrix.

Verification via frequency sweep

As a second verification method, a frequency sweep is applied to the non-linear switched model. The frequency sweep is based on the injection of known current harmonics at the converter terminals and measuring the amplitude and phase of the resulting voltage harmonics. The converter admittance as a function of the frequency is then obtained by performing a Fast Fourier Transform (FFT) on the voltage

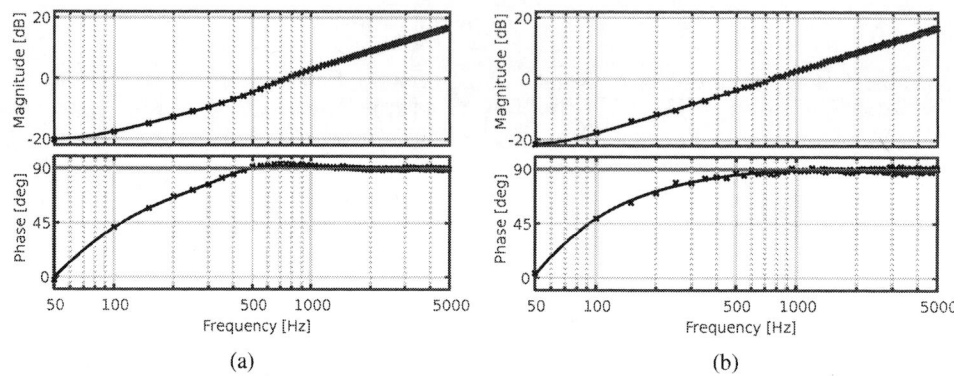

Fig. 3: Frequency sweep of (a) y_{bb} and (b) y_{uu}, for the linearized model (solid black line) and the non-linear model (black crosses) at the operating point $P_p = $ -2 kW, $P_n = $ -2 kW.

and current signals. According to this procedure, the admittances y_{bb} and y_{uu} are determined and depicted in Fig. 3 where harmonics are added to either the balanced or the unbalance current. The injected harmonics range from 50 Hz to 5000 Hz with a frequency increment of 50 Hz. Again it shows that the data points obtained from the non-linear switched model closely coincide with the linearized elements y_{bb} and y_{uu}.

Analysis of passivity of TLC admittance matrix elements

In this section, the passivity of the admittance matrix elements is analyzed for different operating points of the TLC. The converter can be in balanced condition when the consumed or produced power is equal in both poles or in unbalanced condition if the these powers are not equal. According to [20], the maximum value of the unbalanced duty cycle is

$$| d_{u,0} | \leq \begin{cases} d_{b,0} & \text{if} \quad d_{b,0} \leq 0.5 \\ 1 - d_{b,0} & \text{if} \quad d_{b,0} > 0.5 \end{cases} \tag{24}$$

For a converter with operating condition according to Table I, the absolute value of $d_{u,0}$ is limited to the value of $d_{b,0}$. Hence, it is possible to transfer the full back-end power P_2 to one pole. This results in four different operating points of the TLC for which the total consumed or produced power is at its limit: $P_p = 2$ kW, $P_n = 2$ kW; $P_p = $ -2 kW, $P_n = $ -2 kW; $P_p = 4$ kW, $P_n = 0$ kW and $P_p = $ -4 kW, $P_n = 0$ kW. In Fig. 4, the Bode plots of y_{bb}, y_{bu} and y_{uu} are shown for the four operating points and with converter parameters according to Table I. As previously discussed, $y_{bu} = 0$ for the balanced condition in Fig. 4a and Fig. 4b.

The Bode plots of Fig. 4 indicate that element y_{bb} of the three-level converter admittance matrix has an area of non-passivity above 500 Hz. However, the element y_{uu} remains passive over the entire frequency range. Due to the limited bandwidth of the voltage controllers, the non-passivity at the higher frequency range is mainly provoked by the dynamics of the inductor current control, which are included in y_{bb}. The non-passive behavior of the converter potentially amplifies resonances located in this frequency region if the system damping of the resonance is low. Important to notice is that the non-passivity is higher when the converter transfers power from the back-end to the front-end as depicted in Fig. 4b and Fig. 4d. For these two cases, the phase angle reaches a maximum value of 92.26° at 683 Hz. According to Fig. 4c and Fig. 4d, the non-diagonal element y_{bu} also shows non-passivity in particular frequency ranges. As the magnitude of y_{bu} is low compared to y_{bb} and y_{uu}, the impact of the non-diagonal element on the stability is expected to be limited.

From Fig. 4, it can be seen that the admittances y_{bb} and y_{uu} are the same for operating points $P_p = 2$ kW, $P_n = 2$ kW and $P_p = 4$ kW, $P_n = 0$ kW. The back-end power P_2 is the equal in both cases, which results in the same steady-state inductor current $i_{L,0}$. As the other parameters of y_{bb} and y_{uu} remain the same according to (16) and (17), the admittance are independent from unbalanced power distribution. This is

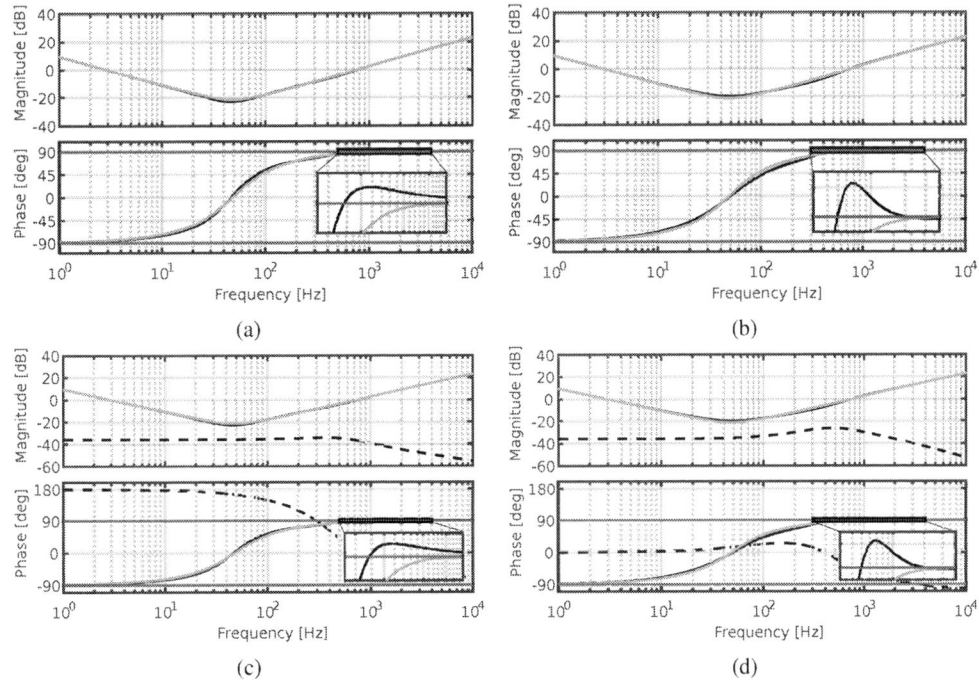

(a)

(b)

(c)

(d)

Fig. 4: Bode plot of $y_{bb}(s)$ (solid black line), $y_{bu}(s)$ (dashed black line) and $y_{uu}(s)$ (solid grey line) for (a) $P_p = 2\,\text{kW}$, $P_n = 2\,\text{kW}$; (b) $P_p = -2\,\text{kW}$, $P_n = -2\,\text{kW}$; (c) $P_p = 4\,\text{kW}$, $P_n = 0\,\text{kW}$ and (d) $P_p = -4\,\text{kW}$, $P_n = 0\,\text{kW}$. The border of passivity is indicated with a red line.

also applicable for y_{bb} and y_{uu} at operating points $P_p = -2\,\text{kW}$, $P_n = -2\,\text{kW}$ and $P_p = -4\,\text{kW}$, $P_n = 0\,\text{kW}$.

Switching the unbalanced power from the positive pole, $P_p = \pm 4\,\text{kW}$, $P_n = 0\,\text{kW}$, to the negative pole, $P_p = 0\,\text{kW}$, $P_n = \pm 4\,\text{kW}$, changes the sign of $d_{u,0}$ and influences the passivity of y_{bu} according to (16) and (18). However, the other steady-state values and the elements y_{bb} and y_{uu} remain the same. Therefore, the cases of $P_p = 0\,\text{kW}$, $P_n = 4\,\text{kW}$ and $P_p = 0\,\text{kW}$, $P_n = -4\,\text{kW}$ are not added to Fig. 4.

Impedance-based stability analysis of bipolar LVDC system with TLC

As a test case, the stability of the bipolar LVDC system depicted in Fig. 5 is analyzed. The system consists of the three-level DC-DC converter and two unipolar converters, which are represented by a constant current source with input LC-filter. The LC-filter of the unipolar converter is designed to sufficiently attenuate the switching harmonics of a 4 kW converter with a switching frequency of 65 kHz [21].

According to [22], the closed-loop stability of a multiple input multiple output system can be assessed by applying the generalized Nyquist stability criterion to the loop-gain matrix. The generalized Nyquist stability criterion determines the system stability based on the encirclements of the -1 point by the eigenloci of the loop-gain matrix [23]. If the loop-gain matrix has no unstable poles, the closed-loop system is unstable if the eigenloci encircle the -1 point. The loop-gain matrix in the BU-frame $L_{bu}(s)$ for the bipolar LVDC system depicted on Fig. 5 is equal to

$$L_{bu}(s) = Y_{TLC}(s)Z_{bus}(s). \tag{25}$$

The impedance matrix $Z_{bus}(s)$ represents the impedances of the currents sources with LC-filters in the BU-frame. The impedances of the LC-filters in the PN-frame are

$$z_p(s) = L_p s + \frac{1}{C_p s} \quad \text{and} \quad z_n(s) = L_n s + \frac{1}{C_n s}. \tag{26}$$

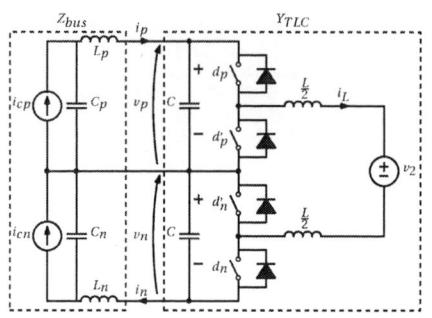

Fig. 5: Bipolar LVDC test system

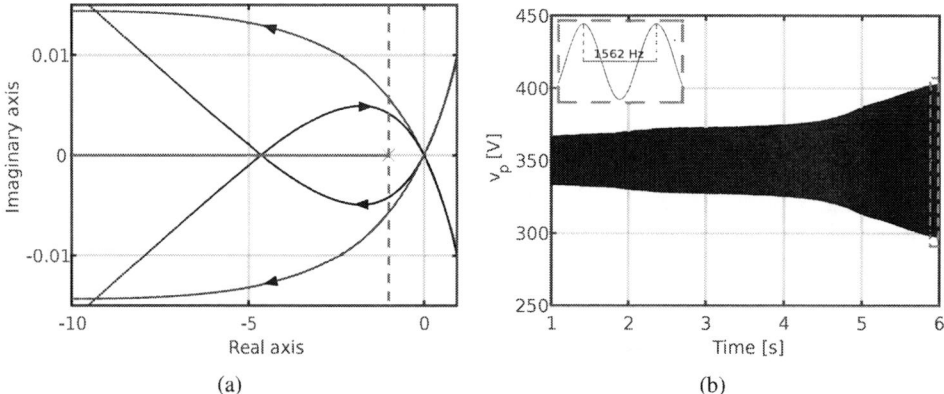

(a)

(b)

Fig. 6: Case 1 (a) Nyquist curve of first eigenlocus (solid black line) and second eigenlocus (solid blue line) (b) Time-domain signal of v_p

Converting these impedances to the BU-frame according to (2) gives $Z_{bus}(s)$,

$$Z_{bus}(s) = \begin{bmatrix} 1/2 & 1/2 \\ 1/2 & -1/2 \end{bmatrix} \begin{bmatrix} z_p & 0 \\ 0 & z_n \end{bmatrix} \begin{bmatrix} 1 & 1 \\ 1 & -1 \end{bmatrix} = \begin{bmatrix} \frac{z_p(s)+z_n(s)}{2} & \frac{z_p(s)-z_n(s)}{2} \\ \frac{z_p(s)-z_n(s)}{2} & \frac{z_p(s)+z_n(s)}{2} \end{bmatrix}. \tag{27}$$

The generalized Nyquist stability criterion can then be applied to the loop-gain matrix defined as

$$L_{bu}(s) = \begin{bmatrix} y_{bb}(s) & y_{bu}(s) \\ 0 & y_{uu}(s) \end{bmatrix} \begin{bmatrix} \frac{z_p(s)+z_n(s)}{2} & \frac{z_p(s)-z_n(s)}{2} \\ \frac{z_p(s)-z_n(s)}{2} & \frac{z_p(s)+z_n(s)}{2} \end{bmatrix}. \tag{28}$$

Case 1: TLC in balanced condition and equal LC-filters

For equal LC-filters where $z_p(s) = z_n(s) = z(s)$ and if the TLC is in balanced condition at the operating point $P_p = -2\,\text{kW}$, $P_n = -2\,\text{kW}$, the loop-gain matrix simplifies to a diagonal matrix

$$L_{bu}(s) = \begin{bmatrix} y_{bb}(s) & 0 \\ 0 & y_{uu}(s) \end{bmatrix} \begin{bmatrix} z(s) & 0 \\ 0 & z(s) \end{bmatrix}. \tag{29}$$

If the LC-filters have values $L_p = L_n = 250\,\mu\text{H}$ and $C_p = C_n = 50\,\mu\text{F}$, the system is unstable according to the eigenloci in Fig. 6. The solid black line represents the Nyquist curve of the first eigenlocus and the solid blue line of the second eigenlocus. The instability is confirmed by the time-domain simulation in Fig. 6b. A time equals 1 s, the TLC is at the desired operating point and an oscillation in the voltage is amplified. The unstable oscillation has a frequency of 1562 Hz.

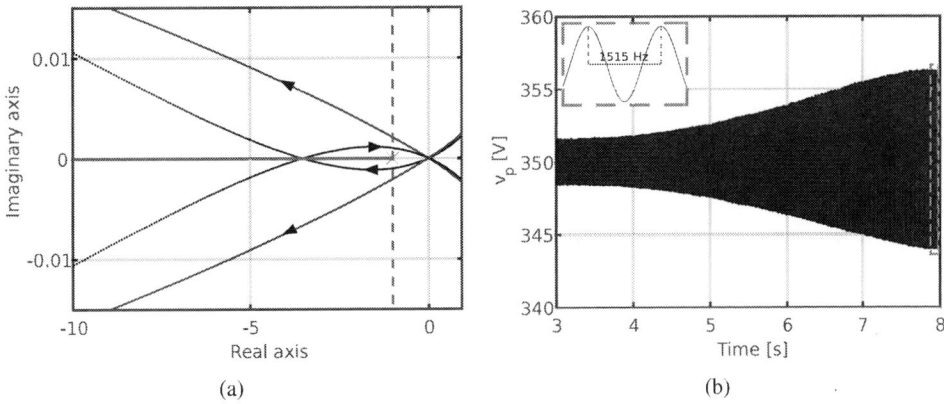

(a)   (b)

Fig. 7: Case 2 (a) Nyquist curve of first eigenlocus (solid black line) and second eigenlocus (solid blue line) (b) Time-domain signal of v_p

Case 2: TLC in unbalanced condition and unequal LC-filters

In case of unequal LC-filters and if the converter is in unbalanced condition, the loop-gain matrix contains non-diagonal elements as shown in (28). For LC-filters with $L_p = 250\,\mu H$, $L_n = 500\,\mu H$, $C_p = 50\,\mu F$ and $C_n = 25\,\mu F$ and an active power demand in the poles of $P_p = -4\,kW$, $P_n = 0\,kW$, the LVDC system is unstable according to Fig. 7. This is again confirmed by Fig. 7b, where an unstable oscillation with frequency 1515 Hz is depicted.

Conclusion

The derivation of the small-signal admittance matrix of the three-level DC-DC converter with voltage balancing capability revealed that the TLC shows non-passive behavior beyond 500 Hz, where the non-passivity is dependent on the operating point. This indicates that the TLC can negatively interact with resonances in the LVDC grid, causing unstable oscillations in the grid voltages and system instability. The analytical admittance matrix is verified by the use of a non-linear switched model of the converter implemented in the EMT-type software PSCAD, showing that it is capable of accurately representing the converter dynamics for small-signal stability studies. Furthermore, a test system is modeled and analyzed to demonstrate the risk of instability in bipolar LVDC grids including a three-level DC-DC converter.

References

[1] J. J. Justo, F. Mwasilu, J. Lee, and J.-W. Jung, "AC-microgrids versus DC-microgrids with distributed energy resources: A review," *Renew. Sustain. Energy Rev.*, vol. 24, pp. 387–405, Mar. 2013.

[2] T. Dragičević, J. C. Vasquez, J. M. Guerrero, and D. Škrlec, "Advanced LVDC electrical power architectures and microgrids: A step toward a new generation of power distribution networks," *IEEE Electrif. Mag.*, vol. 2, no. 1, pp. 54–65, Mar. 2014.

[3] A. Agustoni, E. Borioli, M. Brenna, G. Simioli, E. Tironi, and G. Ubezio, "LVDC distribution network with distributed energy resources: analysis of possible structures," in *Proc. 18th Int. Conf. Electr. Distrib.*, Turin, Italy, Jun. 2005, 5 pages.

[4] G. Van den Broeck, S. De Breucker, J. Beerten, J. Zwysen, M. Dalla Vecchia, and J. Driesen, "Analysis of three-level converters with voltage balancing capability in bipolar DC distribution networks," in *Proc. ICDCM*, Nürnberg, Germany, Jun. 2017, pp. 248–255.

[5] Y. Gu, W. Li, and X. He, "Analysis and control of bipolar LVDC grid with DC symmetrical component method," *IEEE Trans. Power Syst.*, vol. 31, no. 1, pp. 685–694, Jan. 2016.

[6] A. Emadi, A. Khaligh, C. H. Rivetta, and G. A. Williamson, "Constant power loads and negative impedance instability in automotive systems: Definition, modeling, stability, and control of power

electronic converters and motor drives," *IEEE Trans. Veh. Technol.*, vol. 55, no. 4, pp. 1112–1125, Jul. 2006.

[7] G. Pinares and M. Bongiorno, "Modeling and analysis of VSC-based HVDC systems for DC network stability studies," *IEEE Trans. Power Deliv.*, vol. 31, no. 2, pp. 848–856, Apr. 2016.

[8] T. Roose, A. Bayo-Salas, and J. Beerten, "Impedance-based DC side stability assessment of VSC-HVDC systems with control time delay," in *Proc. EPE'18 ECCE*, Riga, Latvia, Sep. 2018, 10 pages.

[9] A. Riccobono and E. Santi, "Stability analysis of an all-electric ship MVDC power distribution system using a novel passivity-based stability criterion," in *Proc. ESTS 2013*, Arlington, USA, Apr. 2013, pp. 411–419.

[10] ——, "Comprehensive review of stability criteria for DC power distribution systems," *IEEE Trans. Ind. Appl.*, vol. 50, no. 5, pp. 3525–3535, Oct. 2014.

[11] T. Dragičević, "Dynamic stabilization of DC microgrids with predictive control of point-of-load converters," *IEEE Trans. Power Electron.*, vol. 33, no. 12, pp. 10 872–10 884, Dec. 2018.

[12] F. Gao, S. Bozhko, A. Costabeber, C. Patel, P. Wheeler, C. I. Hill, and G. Asher, "Comparative stability analysis of droop control approaches in voltage-source-converter-based DC microgrids," *IEEE Trans. Power Electron.*, vol. 32, no. 3, pp. 2395–2415, Mar. 2017.

[13] P. Liutanakul, A. B. Awan, S. Pierfederici, B. Nahid-Mobarakeh, and F. Meibody-Tabar, "Linear stabilization of a dc bus supplying a constant power load: A general design approach," *IEEE Trans. Power Electron.*, vol. 25, no. 2, pp. 475–488, Feb. 2010.

[14] K. Pietiläinen, L. Harnefors, A. Petersson, and H. P. Nee, "DC-link stabilization and voltage sag ride-through of inverter drives," *IEEE Trans. Ind. Electron.*, vol. 53, no. 4, pp. 1261–1268, Aug. 2006.

[15] S. D. Sudhoff and S. F. Glover, "Three-dimensional stability analysis of dc power electronics based systems," in *Proc. PESC*, Galway, Ireland, Jun. 2000, pp. 101–106.

[16] M. Cespedes, L. Xing, and J. Sun, "Constant-power load system stabilization by passive damping," *IEEE Trans. Power Electron.*, vol. 26, no. 7, pp. 1832–1836, Jul. 2011.

[17] R. Middlebrook, "Input filter considerations in design and application of switching regulators," in *Proc. IEEE IAS Annual Meeting*, Chicago, USA, Oct. 1976, pp. 366–382.

[18] P. J. Grbović, P. Delarue, P. Le Moigne, and P. Bartholomeus, "A bidirectional three-level DC-DC converter for the ultracapacitor applications," *IEEE Trans. Ind. Electron.*, vol. 57, no. 10, pp. 3415–3430, Oct. 2010.

[19] G. Van den Broeck, "Voltage control of bipolar DC distribution systems," Ph.D. dissertation, KU Leuven, Nov. 2019.

[20] L. Tan, B. Wu, V. Yaramasu, S. Rivera, and X. Guo, "Effective voltage balance control for bipolar-DC-bus-fed EV charging station with three-level DC-DC fast charger," *IEEE Trans. Ind. Electron.*, vol. 63, no. 7, pp. 4031–4041, Jul. 2016.

[21] R. W. Erickson and D. Maksimovic, *Fundamentals of power electronics*, 2nd ed. New York, US: Kluwer Academic Publishers, 2004.

[22] M. Amin and M. Molinas, "Small-signal stability assessment of power electronics based power systems: A discussion of impedance- and eigenvalue-based methods," *IEEE Trans. Ind. Appl.*, vol. 53, no. 5, pp. 5014–5030, Sep. 2017.

[23] C. A. Desoer and Y.-T. Wang, "On the generalized Nyquist stability criterion," *IEEE Trans. Automat. Contr.*, vol. 25, no. 2, pp. 187–196, Apr. 1980.

LCL Filter Design for Three Phase AC-DC Converters Considering Semiconductor Modules and Magnetics Components Performance

Marco Stecca, Thiago Batista Soeiro, Laura Ramirez Elizondo, Pavol Bauer, and Peter Palensky
DELFT UNIVERSITY OF TECHNOLOGY
Delft, The Netherlandsy
Phone: +31 (0) 15-278-9042
Email: m.stecca@tudelft.nl

Keywords

≪Voltage Source Converter (VSC)≫, ≪Passive filter≫.

Abstract

LCL filters are commonly adopted to attenuate the current harmonics produced by Pulse Width Modulation (PWM) Voltage Source Converters (VSC). Due to the nature of LCL filters, several combinations of L and C can deliver the attenuation required by the standards. The optimal configuration is generally evaluated, considering power density, costs, and filter efficiency. This paper shows that semiconductor efficiency should also be considered as an important design variable. It is shown that the AC ripple across the converter side inductor can reduce, to a certain extent, the overall semiconductor losses, when commercial IGBTs and the respective anti-parallel diodes are used. Reduced losses have benefits in terms of semiconductor module lifetime, chip area and cost reduction, and simplification of cooling requirements. Higher AC ripple, however, negatively affect the filter losses. Nonetheless, inductive components are typically much less critical in terms of losses dissipation and lifetime than semiconductors.

Introduction

Voltage Source Converters (VSCs) are used to interface, among others, renewable energy-based generators, battery energy storage systems, and electric motors with the electrical network [1]. Pulse Width Modulation (PWM) techniques for the control of VSCs, intrinsically generate harmonics in the AC output terminal. However, the connection to the main network requires compliance to several standards that regulate, i.e., the current harmonic limits [2]. In this context, LCL filters are widely adopted for the reduction of the high order harmonics. The design of LCL filters has already been widely treated in the literature. Methods for defining the boundary values of the filter components and their design have been proposed in [3, 4]. Due to the nature of LCL filters, a specific harmonic attenuation can be obtained with several values of the inductive and capacitive components; therefore other variables, such as cost, weight, volume and power losses, can play a significant role in the selection of the LCL filter optimal parameters [4]. Furthermore, also the amplitude of the ripple current flowing in the converter is defined by the filter parameters. In medium-high power systems, the LCL filter assumes relevant weight and size, becoming a key design variable; hence, various studies include efficiency and power density as optimization criterion [5, 6]. On the other hand, in previous studies, the direct influence of the AC ripple amplitude, driven by the selection of the LCL parameters, in the power losses of the semiconductor modules, is often neglected. Therefore, this paper will address the influence of the AC current ripple on the semiconductor modules efficiency of VSCs when designing its LCL filter.

In this paper, a three phase three-wire 100 kW DC-AC converter, as shown in Fig. 1, is taken as a case study. The filter parameters are analytically calculated considering the relevant standards, such as the IEEE 519-2014 [2], and the losses in the passive components are evaluated through well-established

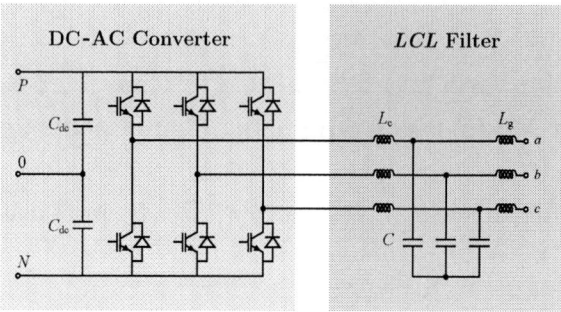

Fig. 1: Three-wire Voltage Source Converter with LCL output filter

methods in literature [5, 7, 8]. The resulting possible filter designs are benchmarked in terms of weight, size, and power losses. In this respect, a particular focus on the impact of the AC ripple on the semi-conductors' losses is given. It is shown that, for the selected commercial semiconductors IGBT-Diode modules, the AC ripple can reduce switching losses, increasing the lifetime of the modules and decreasing the required cooling resources. By contrast, the AC ripple has a negative impact on the filter losses. Thereby a trade-off between semiconductor and filter efficiency has to be made. In this respect, it has to be considered that inductive components are not as demanding as the semiconductor modules in terms of cooling, thermal management, and losses dissipation. Consequently, selecting the LCL filter parameters to have the minimum semiconductor losses can lead to lower cooling efforts and lower semiconductor stress at the expense of a slight reduction of the LCL filter overall efficiency.

Filter parameter selection

In grid-connected VSCs, LCL filters are widely used to suppress the injected current harmonics generated by the converter AC output voltages. According to the IEEE 519-2014 standard, high frequency odd current harmonics ($h > 35$) need to be contained to less than 0.3% of the nominal line frequency current, I_n, and even harmonics to 0.075% of it [2]. The amplitude of the hth current harmonic produced by the converters can be calculated as the ratio between the peak harmonic voltage and the filter impedance at that specific frequency. The voltage harmonic spectrum for the three-wire two-level VSC operated with PWM modulation can be analytically calculated through double Fourier integration [9]. The first relevant harmonic is the sideband $f_s - 2f_g$, thus, starting from this harmonic, the filter needs to be able to provide the minimum required attenuation that guarantees the compliance to the standards. The -60 dB slope of the filter transfer function will effectively further attenuate the higher-order harmonics. Once obtained the harmonic voltage amplitude, if $f_s - 2f_g > 35f_g$, the required attenuation, according to the IEEE 519 standard, can be found through equation (1):

$$Att_{\text{IEEE}-519} = 0.003 \frac{P_{\text{nom}}}{\sqrt{3}V_{\text{ll}}V_{1,-2}}. \tag{1}$$

where P_{nom} is the nominal power of the converter, V_{ll} is the line-to-line AC voltage, and $V_{1,-2}$ is the $f_s - 2f_g$ sideband harmonic voltage. The transfer function of a LCL filter is given by equation (2):

$$H(s) = \frac{\omega_{\text{res}}^2}{s\left(L_g + L_c\right)\left(s^2 + \omega_{\text{res}}^2\right)}, \tag{2}$$

where:

$$\omega_{\text{res}} = k_{\text{res}}\omega_s = \sqrt{\frac{L_c + L_g}{L_c L_g C}} \tag{3}$$

Table I: Specifications for the LCL filter design

Parameter	P_{nom}	V_{dc}	$V_{\text{ac,ll}}$	f_s	k_{res}	f_g
Value	100 kW	900 V	400 V	8 kHz	0.35	50 Hz

and L_c and L_g are respectively the converter side and grid side inductances and C the capacitance of the filter, as indicated in Fig. 1. The resonance frequency of the filter needs to be carefully evaluated to avoid the amplification of the sideband harmonics and to guarantee control stability [10]. Knowing the attenuation required, as defined in Equation (1), and given the LCL filter transfer function, Equation (2), the minimum total inductance that guarantees standards' compliance can be found as:

$$L_{\text{tot,min}} = \frac{k_{\text{res}}^2 \omega_s^2}{\omega_h \left(\omega_h^2 - k_{\text{res}}^2 \omega_s^2 \right) Att_{\text{IEEE}-519}}.$$ (4)

The maximum value of L_{tot}, instead, is limited by the voltage drop across the filter. For S-PWM, Equation (5) gives the upper boundary [4]:

$$L_{\text{tot,max}} = \frac{\sqrt{\frac{V_{\text{dc}}^2}{8} - \frac{V_{\text{ll}}^2}{3}}}{2\pi f_c I_n}$$ (5)

As shown in Equation (2), the harmonic attenuation in the grid side current for a fixed filter resonance frequency is given by the sum of the converter side L_c and grid side L_g inductances. Additionally, the converter side inductance, for a fixed filter resonance frequency, defines the amplitude of the AC current ripple flowing through the semiconductors, since the transconductance $Y_{1,1} = i_1/v_1$ that defines the convert side current harmonics is expressed as:

$$Y_{11}(s) = \frac{s^2 + \omega_{\text{lc}}^2}{sL_c \left(s^2 + \omega_{\text{res}}^2 \right)},$$ (6)

where:

$$\omega_{\text{lc}} = \frac{1}{CL_c}.$$ (7)

The grid side inductance, L_g, can be found subtracting L_c from L_{tot}. Having the values of the inductances L_c and L_g and fixing the resonance frequency, the capacitance value is derived rearranging Equation (2). The maximum reactive power absorption at grid frequency gives the upper boundary for the capacitance value, usually limited to 5% the nominal power. The reactive power injected by the capacitor needs to be compensated by the converter, lowering its efficiency, especially at low partial loads.

Once defined the parameter boundaries, several possible designs for the LCL filter can be found by varying L_c. The feasible LCL parametric combinations, derived according to the parameters listed in Table I are plotted in Fig. 2. Increasing the value of L_c the peak current ripple decreases and as well L_g, since the L_{tot} is kept constant and equal to the minimum required value. To evaluate the impact of the AC ripple in the efficiency of the semiconductor and the LCL filter, several combinations of LCL parameters are selected and further designed. These are chosen to have linear variations of the AC ripple, and they are indicated in Fig. 2 by the vertical brown dashed lines. To compare the L_c, L_g and C combinations, it is necessary to analyze more in-depth the inductor design, and the power losses on the semiconductor modules, as presented in the following Sections.

Inductors design

As mentioned in the previous Section, to evaluate the performance of the LCL filter, it is necessary to design its components. The bulk of the losses in the LCL filter are usually found in the converter side

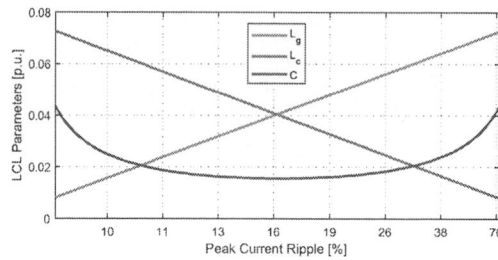

Fig. 2: LCL filter possible design for a grid connected VSC with the specifications listed in Table I.

inductor since the high frequency ripple will flow through it, affecting both high frequency winding losses and core losses. In this paper, the inductors are designed considering commercially available toroidal-shape cores from ChangSung [11]. Starting from the core material permeability μ, the mean magnetic path length $l[m]$ and the effective cross area section $A[m^2]$, the number of turns that give the required inductance $L[H]$ can be found as:

$$N = \sqrt{\frac{Ll}{0.4\mu\pi A 10^{-2}}}. \tag{8}$$

Then, the total winding length l_w can be found according to the number of turns N and the core material dimensions. The power losses in the inductors can be separated in winding losses, due to the skin and proximity effect, and core losses:

$$P_{ind} = P_{skin} + P_{prox} + P_{core}. \tag{9}$$

The skin effect losses are evaluated through:

$$P_{skin} = l_w R_{ac} \hat{I}_{hf}^2 + \frac{1}{2} l_w R_{dc} \hat{I}_{ac}^2, \tag{10}$$

where $\hat{I}_{hf}$ is the high frequency peak current, $\hat{I}_{ac}$ the peak sinusoidal AC current, R_{ac} and R_{dc} are the AC and DC resistance of the windings, and δ the skin depth which are found according to the wire diameter d and its conductance σ:

$$R_{dc} = \frac{4}{\sigma d^2 \pi}, \tag{11}$$

$$\delta = \frac{1}{\sqrt{\pi \mu_0 \sigma f}}. \tag{12}$$

The AC resistance of the inductor winding, instead, is calculated through the analytical approximation given by [7]:

$$R_{ac} = R_{dc} \frac{\gamma}{4\sqrt{2}} \left(\frac{\mathrm{ber}_0\gamma\mathrm{bei}_1\gamma - \mathrm{bei}_0\gamma\mathrm{ber}_1\gamma}{\mathrm{ber}_1^2\gamma + \mathrm{bei}_1^2\gamma} - \frac{\mathrm{bei}_0\gamma\mathrm{ber}_1\gamma - \mathrm{bei}_0\gamma\mathrm{bei}_1\gamma}{\mathrm{ber}_1^2\gamma + \mathrm{bei}_1^2\gamma} \right) \tag{13}$$

Table II: Specifications of the selected LCL filter designs and calculated power losses in the inductors at rated power.

Design #	L_g [μH]	L_c [μH]	C [μF]	Ripple [%]	Losses in L_c[W]	Losses in L_g[W]
1	370	41	86	9	96	69
2	74	337	53	42	80	50
3	41	370	86	76	122	65

with ber_r and bei_i the real and imaginary parts of the Kelvin function of the ith order, and γ defined as:

$$\gamma = \frac{d}{\sqrt{2}\delta}. \tag{14}$$

The proximity effect losses depend on the external magnetic field $\hat{H}_e$, derived following the approach of [12], and can be calculated as [5, 7]:

$$P_{\text{prox}} = R_{\text{dc}} G_R \hat{H}_e^2 \tag{15}$$

$$G_R = \frac{\gamma \pi^2 d^2}{2\sqrt{2}} \left(\frac{ber_2 \gamma ber_1 \gamma - ber_2 \gamma bei_1 \gamma}{ber_0^2 \gamma + bei_0^2 \gamma} + \frac{bei_2 \gamma bei_1 \gamma - bei_2 \gamma ber_1 \gamma}{ber_0^2 \gamma + bei_0^2 \gamma} \right) \tag{16}$$

Equations (10)-(16) are then applied for evaluating the winding losses caused by each current harmonic in which the current flowing through the inductor can be decomposed. Finally, the total winding losses are found through summing the contribution of each harmonics.

The core losses are calculated through the *improved Generalized Steinmetz Equation* (iGSE). For a triangular waveform the iGSE takes the following formulation [13]:

$$P_{\text{core}} = k f_s \left(\frac{2}{\pi^2} 4 f_s \right)^{\alpha - 1} \hat{B}^{\beta} V_{\text{core}} \tag{17}$$

$$k = \frac{k}{(2\pi)^{\alpha - 1} \int_0^{2\pi} |\cos\theta|^\alpha 2^{\beta - \alpha} d\theta} \tag{18}$$

where V_{core} is the core volume, k, α and β are the Steinmetz parameters derived from the core material, and $\hat{B}$ is the peak to peak flux density given in Equation (19) as a function of the peak to peak current ripple $I_{\text{r,pp}}$ and the inductor geometry:

$$\hat{B} = \frac{L I_{\text{r,pp}}}{2NA} \tag{19}$$

As previously mentioned, the *iGSE* estimates the core losses derived by a triangular shaped current. The total inductor core losses are then calculated summing the contribution of each triangular minor loop in which the switched current can be divided and averaging the total in the 50 Hz period.

Applying the method described through Equations (8) - (19) the power losses on the LCL filter inductors can be analytically calculated. This procedure is repeated varying the core diameter, the number of stacked elements, and the core material, according to the specifications of the commercially available toroidal Powder Core [11] to design the inductors of three possible LCL configurations, whose parame-

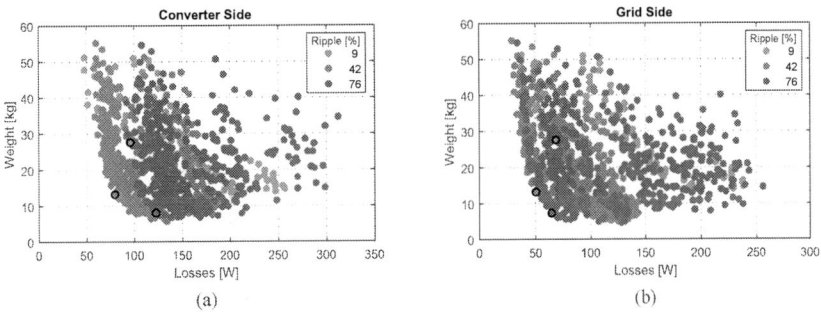

Fig. 3: L_c and L_g power losses and weight design space for different ripple amplitudes and so inductance value, according to the parameters of Table II. The selected designs are circled by a black line.

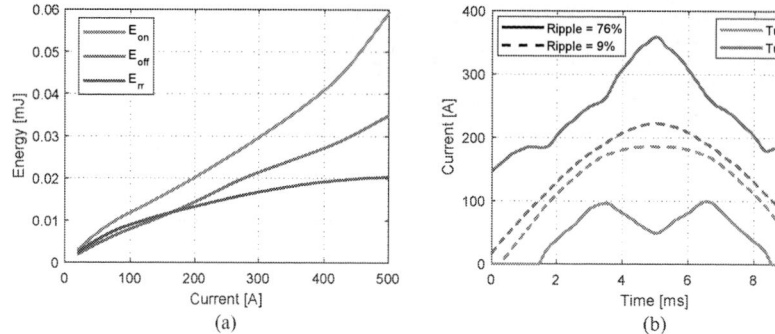

Fig. 4: (a) Semikron SKM400GB125D module switching characteristics from datasheet; and (b) turn on and turn off currents of the upper IGBT with different ripple amplitudes at rated power.

ters are specified in Table II. These three configurations lead to different AC current ripple values so that its influence on the system performance can be later estimated.

In Fig. 3 the losses and weight design space of L_c and L_g in these three configurations are plotted. The designs are checked for thermal compliance so that their maximum temperature does not exceed 150°C. Between all the feasible designs, the selected designs of L_c and L_g in the three configurations are highlighted by a black circle in Fig. 3. These are selected to be in the knee of the weight-losses Pareto front so that they exhibit the optimal trade-off between the two parameters.

Full system efficiency

The current flowing through the semiconductor modules consists of the sum of the fundamental 50 Hz current and the high frequency AC ripple. Given a specific current waveform, it is possible to derive conduction, P_c, and switching, P_s, power losses in the semiconductor, through equations (20)-(22).

$$P_{c,i} = v_i I_{avg,i} + r_i I_{rms,i}^2 \tag{20}$$

$$P_{s,IGBT} = \frac{f_s V_{dc}}{4\pi V_b} \int_{-\varphi}^{2\pi-\varphi} \left[E_{on}\left(I_g\right) + E_{off}\left(I_g\right) \right] dt \tag{21}$$

$$P_{s,Diode} = \frac{f_s V_{dc}}{4\pi V_b} \int_{-\varphi}^{2\pi-\varphi} E_{rr}\left(I_g\right) dt \tag{22}$$

v_i and r_i are the on state characteristics of the component i, from the semiconductor datasheet, φ the phase shift between the fundamental AC output voltage and current, V_{dc} the switching voltage, V_b the datasheet measured switching voltage, and $E_{on}(I_g)$, $E_{off}(I_g)$, and $E_{rr}(I_g)$ the switching energy functions, extracted from the semiconductor datasheet and plotted in Fig. 4(a). The commercially available Semikron IGBT-Diode half-bridge module SKM400GB125D, rated 1200V-300A, has been considered [14]. The conduction losses are marginally affected by the ripple since the average, and RMS value of the current through the semiconductor do not see significant variations. However, the IGBT turn-off current significantly increases. At the same time, the turn-on decreases, leading to soft switching at the beginning and the end of the half period, as indicated in Fig. 4(b). It is then expected an increase in the turn off losses and a decrease in the turn-on losses. These variations are also linked to the switching energy of the semiconductors modules. If the IGBTs are selected to have turn-off losses comparable lower to the sum of the turn-on and the reverse recovery of the diode, then it is expected an overall decrease in the switching losses when the AC ripple increases. The IGBT module SKM400GB125D has turn-on energy higher than turn off and reverse recovery energy; consequently, its performance will benefit from the AC ripple superimposition, as detailed in Fig. 5(a).

Fig. 5(b) displays the semiconductor module efficiency varying the converter output power for different peak values of the high frequency ripple, according to the LCL parameters described in Table II. What

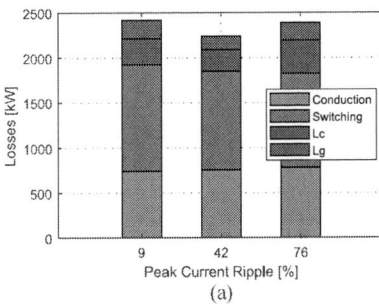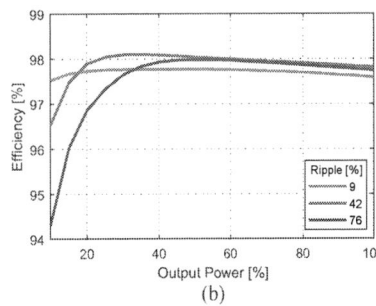

(a)   (b)

Fig. 5: (a) Semiconductor module and LCL filter losses breakdown at rated power; and (b) efficiency curves at partial load for the VSC employing the three selected LCL filter designs.

stands out from Fig. 5(b) is the fact that, for the selected IGBT module, a low ripple amplitude shows a relatively flat efficiency curve at partial loads with considerable benefits at low partial loads. However, for a high partial load operation, the high AC current ripple's superposition improves the overall system performances and reduces the semiconductor power losses.

In this context, the LCL filter parameters can also be designed according to the VSC's application requirements and mission profile, seeking to maximize the semiconductor module efficiency with the appropriate amplitude of the high frequency AC current ripple. The semiconductor modules can benefit in terms of lower losses from the AC ripple current. Lower losses translate in lower cooling requirements and less degradation of the switches due to junction temperature variations. However, the impact of the AC ripple is strongly related to the semiconductor switching characteristics. The LCL configuration that assures the best semiconductor performance has to be evaluated case by case.

Conclusions and future work

An analytical procedure for the LCL filter design and the evaluation of the filter and AC-DC converter efficiency based on well-established methods in literature have been presented. Following the proposed approach, it has been shown how the AC ripple affects not only the design and efficiency of the output filter but also the performances of the semiconductor modules. More in detail it has been shown that the AC ripple, until a certain extent, and depending on the output power of the AC-DC converter, can have a beneficial impact on the IGBT-Diode losses, leading to lower thermal stress of the semiconductors and reducing the requirements of the thermal management system. This, however, comes at the price of lower efficiency of the LCL filter. Future work will focus on extending the analysis, considering more broadly the converter operating conditions, i.e., reactive power generation. Experimental verification of the results will be performed to verify the models adopted.

References

[1] M. Stecca, L. Ramirez Elizondo, T. Batista Soeiro, P. Bauer, and P. Palensky, "A Comprehensive Review of the Integration of Battery Energy Storage Systems into Distribution Networks," *IEEE Open J. Ind. Electron. Soc.*, vol. 1, pp. 46–65, 2020.

[2] IEEE Standards Association, "IEEE Std. 519-2014. IEEE Recommended Practice and Requirements for Harmonic Control in Electric Power Systems," pp. 1–29, 2014.

[3] M. Liserre, F. Blaabjerg, and S. Hansen, "Design and control of an LCL-filter-based three-phase active rectifier," *IEEE Trans. Ind. Appl.*, vol. 41, no. 5, pp. 1281–1291, 2005.

[4] K. Jalili and S. Bernet, "Design of LCL filters of active-front-end two-level voltage-source converters," *IEEE Trans. Ind. Electron.*, vol. 56, no. 5, pp. 1674–1689, 2009.

[5] J. Mühlethaler, M. Schweizer, R. Blattmann, J. W. Kolar, and A. Ecklebe, "Optimal design of lcl harmonic filters for three-phase pfc rectifiers," *IEEE Trans. Power Electron.*, vol. 28, no. 7, pp. 3114–3125, 2013.

[6] K. B. Park, F. D. Kieferndorf, U. Drofenik, S. Pettersson, and F. Canales, "Weight Minimization of LCL Filters for High-Power Converters: Impact of PWM Method on Power Loss and Power Density," *IEEE Trans. Ind. Appl.*, vol. 53, no. 3, pp. 2282–2296, 2017.

[7] J. Mühlethaler, "Modeling and multi-objective optimization of inductive power components," no. 20217, 2012. [Online]. Available: http://e-collection.library.ethz.ch/view/eth:5781

[8] J. A. Ferreira, "Improved Analytical Modeling of Conductive Losses in Magnetic Components," *IEEE Trans. Power Electron.*, vol. 9, no. 1, pp. 127–131, 1994.

[9] D. G. Holmes and T. A. Lipo, "Pulse Width Modulation for Power Converters: Principles and Practice," *Pulse Width Modul. Power Convert.*, pp. 531–554.

[10] Y. Wu, A. Shekhar, T. B. Soeiro, and P. Bauer, "Voltage Source Converter Control under Distorted Grid Voltage for Hybrid AC-DC Distribution Links," *IECON 2019 - 45th Annu. Conf. IEEE Ind. Electron. Soc.*, vol. 1, pp. 5694–5699, 2019.

[11] Chang Sung Powder Core Material. [Online]. Available: http://www.changsung.com/_eng/index.php

[12] P. Dowell, "Effects of eddy currents in transformer windings," *Proc. Inst. Electr. Eng.*, vol. 113, no. 8, p. 1387, 1966.

[13] J. Reinert, A. Brockmeyer, and R. W. De Doncker, "Calculation of losses in ferro- and ferrimagnetic materials based on the modified Steinmetz equation," *IEEE Trans. Ind. Appl.*, vol. 37, no. 4, pp. 1055–1061, 2001.

[14] Semikron, "Semikron IGBT Modules." [Online]. Available: https://www.semikron.com/products/product-classes/igbt-modules.html

Switching Behavior and Comparison of 600V SMD Wide Bandgap Power Devices

Markus Meißner, Jan Schmitz, Steffen Bernet
CHAIR OF POWER ELECTRONICS, DRESDEN UNIVERSITY OF TECHNOLOGY
Helmholtzstr. 9
01069 Dresden
Tel.: +49 / (351) – 463 – 39212.
Fax: +49 / (351) – 463 – 42138.
E-Mail: markus.meissner@tu-dresden.de
URL: https://tu-dresden.de/ing/elektrotechnik/eti/le

Keywords

«Gallium Nitride (GaN)», «Silicon Carbide (SiC)», «Switching losses», «Measurement», «Transistor»

Abstract

This Paper compares the switching characteristics of silicon, silicon carbide and gallium nitride based semiconductor power devices over a wide operating range. Several 600V SMD packaged devices are measured in a similar test setup. To consider different R_{DSon} values a switching figure of merit (FOM) is used. The investigations focus on the switching losses, the presented FOM and the dv/dt during the switching process.

Introduction

Since wide bandgap semiconductor devices are increasing its market share, it is important to evaluate the performance and compare the characteristics of different devices and technologies. Few works presented the switching characteristics of wide bandgap power devices [3], [4], [5], [6] in different setups. This paper presents the characteristics of 600V - 650V GaN-HEMTs, SiC and Si Mosfets in similar low inductive hard switching setups. A new SMD packaged 650V - SiC Mosfet is used which enables a lower parasitic inductance. Due to different on resistance values a switching figure of merit is used. This takes R_{DSon} and switching losses into account.

Measurement Setup

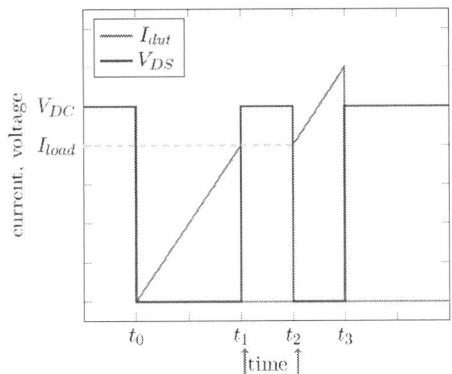

Fig. 1 Double pulse current and voltage transients

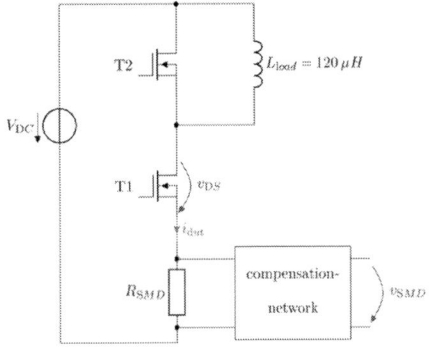

Fig. 2 Double pulse test setup

A double pulse test setup is used for the characterization of the switching behavior. Fig. 2 presents the double pulse test setup. As freewheeling diode the intrinsic body diode and for GaN the intrinsic reverse conduction behavior of each power switch is used. Switching T1 in a double pulse pattern, results in the in Fig. 2 shown current and voltage waveforms. The switching transients at t_1 and t_2 are used for turn-on and turn-off characterization. For current measurement the in [1] shown SMD shunt method, utilizing four 100 mΩ shunts with compensation network, is used. The resulting voltage V_{SMD} is measured differential with two oscilloscope channels set to 50 Ω. The drain source voltage is measured by the passive 800 MHz probe TPP 0850 from Tektronix. An oscilloscope MSO5204B and a DPO 5204B are used. The case temperature is adjusted with a heat plate connected at the top side of the DUT.

Tab. I presents the chosen DUTs with nominal voltages from 600V up to 650V. All are packaged in SMD cases. This is important to achieve similar stray inductance values. The R_{DSon} ranges from 25 mΩ to 60 mΩ. To achieve better comparability, the switching figure of merit shown on equation (1) is used. This takes the on state drain source resistance R_{DSon} and the switching energy E_{SW} into account.

$$FOM = R_{DSon} \cdot E_{SW} \tag{1}$$

All devices are measured using a separate, adapted PCB using the gate drivers and gate voltages shown in Tab. I. The power loop design is mostly equivalent with small differences due to the different packages. The total commutation loop inductance differs from 2 nH to 6 nH. The gate resistances are set to datasheet values and are depicted in Tab. I as well. The turn-on resistance of the silicon device is significantly larger since the turn-on di/dt must be limited to protect the body diode of the high side switch. The silicon Mosfet is designed for resonant switching. To achieve full hard switching performance a SiC diode as commutation partner should be used. Nevertheless, the internal body diode of silicon device is used, to achieve similar conditions in the used setups.

Tab. I: Selected properties of tested power semiconductor devices. Typical values.

Component	Material	Type	V_{nom} [V]	I_{nom} [A]	R_{dson} [mΩ]	package	Driver	Ron, ext [Ω]	Roff, ext [Ω]
GS66516T	GaN	eHEMT	650	60	25	GaNPX TOP	UCC27511 @ +/-6V	10	1
PGA26E07BA	GaN	GiT	600	26	56	DFN-8	AN34092B @ 0/12V	6.2	4.7
SCT3030AW7	SiC	Mosfet	650	70	30	TO-263-7L	ADuM4135 @ 0/18V	0	0
SCT3060AW7	SiC	Mosfet	650	38	60	TO-263-7L	ADuM4135 @ 0/18V	0	0
IPL60R060CFD7	Si	Mosfet	650	40	48	PG-VSON-4	UCC27511 @ 0/10V	33	3.3

Data evaluation

The raw measurement data need to be adjusted to compensate offsets and time delays. Especially for GaN devices small time delays can lead to high differences in loss calculation as shown in [2]. Therefore, turn-on voltage drop and di/dt during current rise were adjusted. The turn-on losses are determined by equation (2) and turn off losses by equation (3). The mean dv/dt's are calculated were v_{DS} changes between 10% - 90% V_{DC}.

$$E_{on} = \int_{t1}^{t2} v_{DS} \cdot i_S dt \quad with \qquad t_1 = t(i_S = 10\% \cdot i_{load}) \quad and \tag{2}$$

$$E_{off} = \int_{t1}^{t2} v_{DS} \cdot i_S dt \quad with \qquad t_1 = t(v_{DS} = 10\% \cdot v_{DC}) \quad and \quad t_2 = t(i_S = 2\% \cdot i_{load}) \tag{3}$$

Measurement Results

Fig. 3 presents the turn-off and Fig. 4 shows the turn-on behavior of all tested DUTs at V_{DC}=400V, T_c=25°C and I_{load}=I_{nom} as given in Tab. I. The turn-off transients are faster the turn-on transients. All SiC transients contain more ringing. This is caused by a larger stray inductance of the package and a larger common source coupling as well. The GaN GiT device shows few oscillations which is caused by a higher slew rate and slighty larger parasitic inductance than the GaN eHEMT. At turn-on the GaN GiT shows a high dv/dt slope while the GaN eHEMT is showing a very low dv/dt at lower voltages. The measurements show clearly the absence of a reverse recovery charge in the freewheeling path for the GaN devices, because no pn junction is forward biased during reverse conduction of the highside switch. The SiC Mosfets have a high current peak during turn-on. This is caused by the reverse recovery charge and parasitic turn-on due to the high internal gate resistance. The Si Mosfet has a very large reverse recovery current peak. This is caused by the body diode of the high side switch, which is turned off with a very high di/dt. However, the di/dt is kept below the maximum di/dt of 1300 A/µs which is depicted in the datasheet. Therefore a large turn-on gate resistance must be used. This leads to a slow turn-on di/dt in comparison with the other switches.

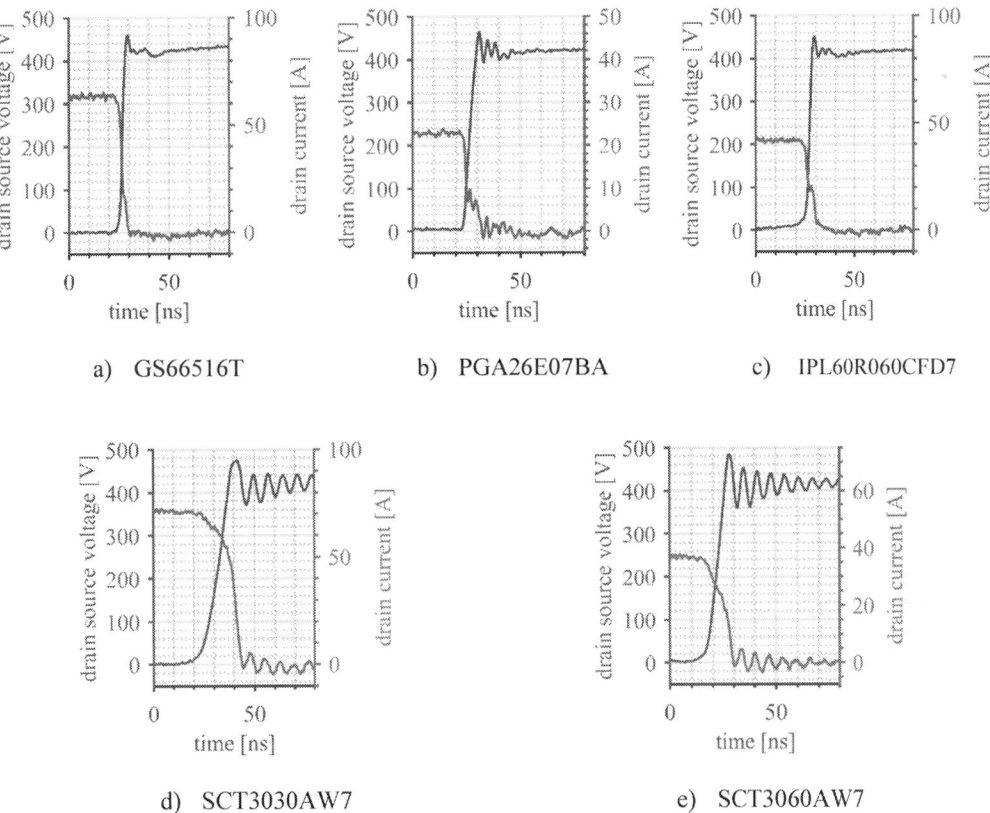

Fig. 3 Turn-off transients of measured power devices at V_{DC}=400V, T_c=25°C and I_{load}=I_n

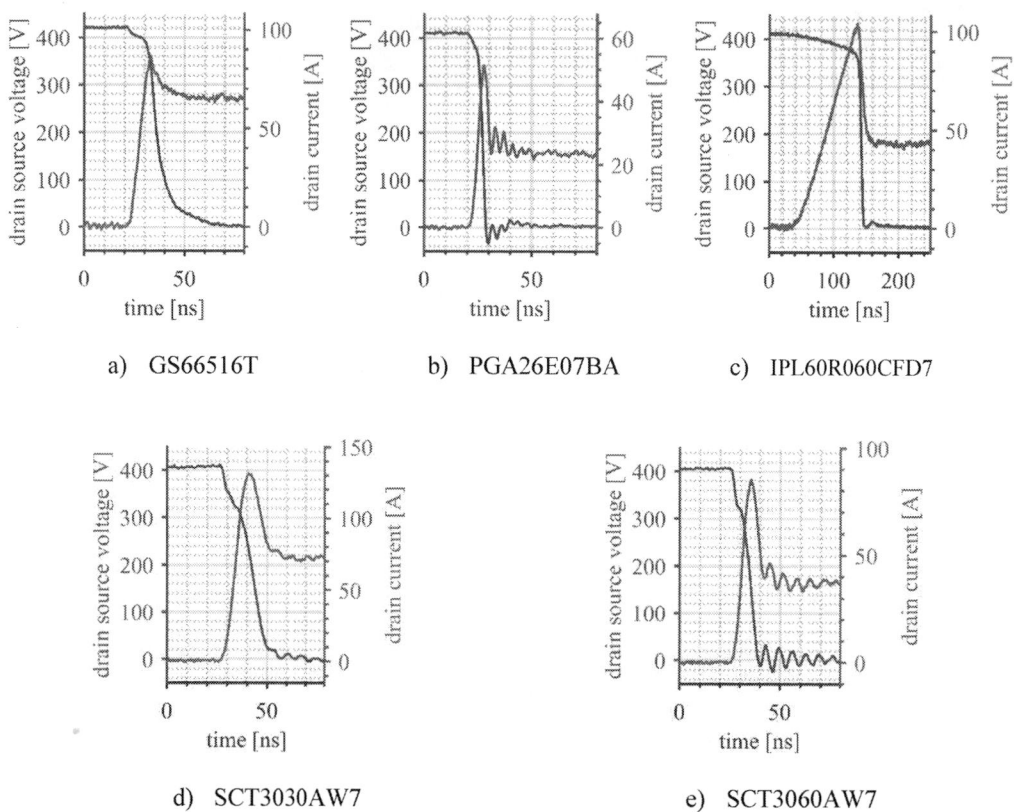

a) GS66516T   b) PGA26E07BA   c) IPL60R060CFD7

d) SCT3030AW7   e) SCT3060AW7

Fig. 4 Turn-on transients of measured power devices at V_{DC}=400V, Tc=25°C and I_{load}=I_n

Switching Characteristic Comparision

This paper focus on the switching loss characterisitcs and the voltage transient speed. Losses are a crucial property of the most power electronic circuits. Regarding fast switching devices as shown in this paper the dv/dt gets more and more attention since it causes EMI, common mode and isolations issues in power electronics systems.

Fig. 5 shows the turn-on and turn-off losses of all tested devices at dc link voltage V_{dc}= 400V and case temperature T_c = 25°C. The turn-on and turn-of losses beetween the different power devices vary strongly. Especially the Si device turn-on performance is limited due to the body diode of the high side switch. To achieve a better comparability, the in equation (1) given FOM shall be compared instead of switching losses. By using this, the R_{DSon} value is considered as well, because conduction losses must be minimized in many applications as well. Additionally all measurement values were displayed related to their nominal current at 25°C to achieve comparable workload for each chip. The turn-on FOM of the Si device must be displayed scaled by 10, because it is much higher than the other values. The results for different DC voltages are shown in Fig. 6 and Fig. 7. As expected, the losses increase with higher dc voltages and higher drain currents. The turn-off losses are much smaller than the turn-on losses. The turn-off losses of the SiC devices are much higher than the ones of the GaN devices. This is caused by the reverse recovery charge of the high side SiC body diode. Additonally the SiC power switches have a large internal gate resistances which cause parasitic turn-on of the high side mosfet at the high switching speed. This leads to a higher turn-on current peak of the low side switch.

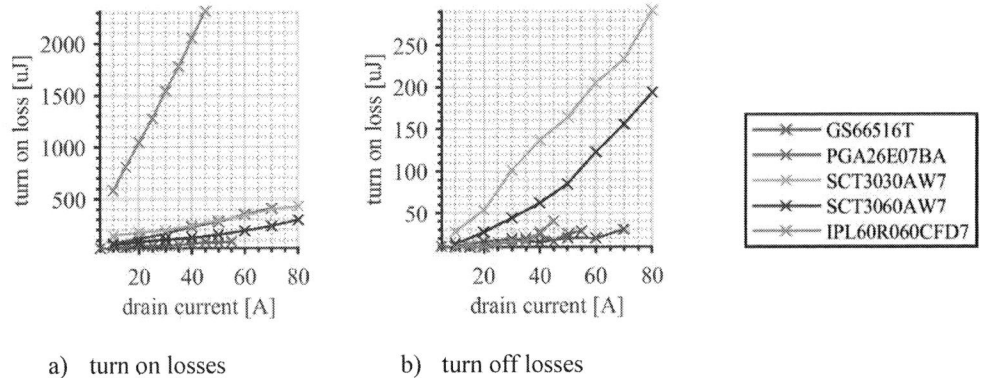

a) turn on losses   b) turn off losses

Fig. 5 Turn-on and turn-off switching loss of measured power devices at V_{DC}=400 V, T_c=25 °C at different load currents.

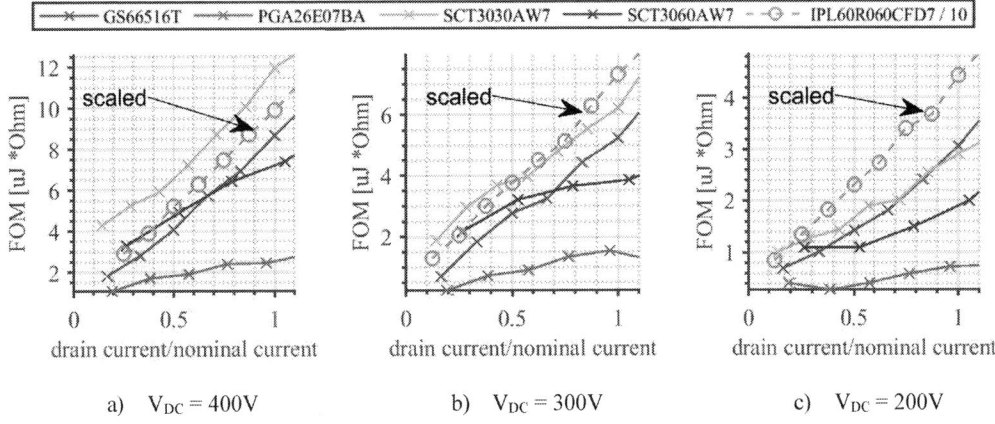

a) $V_{DC} = 400V$ b) $V_{DC} = 300V$ c) $V_{DC} = 200V$

Fig. 6 FOM during turn-on transient at different dc link voltages at $T_c = 25°C$. The Silicon device FOM must be divided by 10 to be displayable.

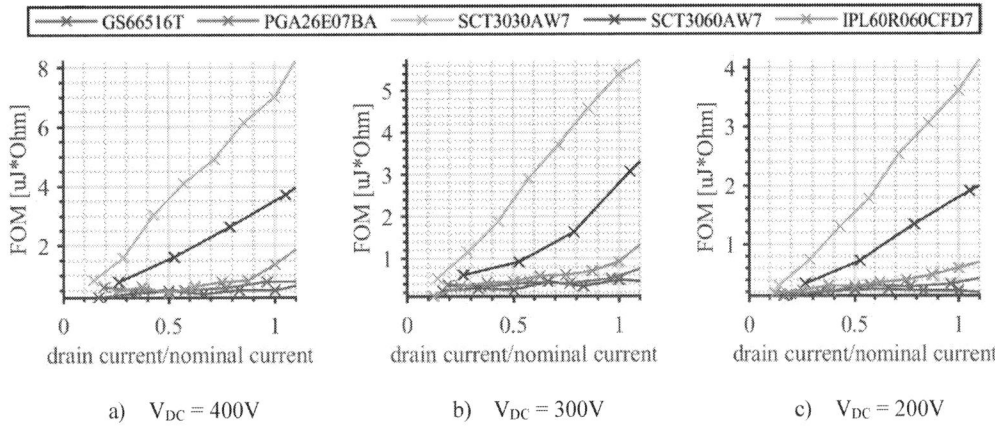

a) $V_{DC} = 400V$ b) $V_{DC} = 300V$ c) $V_{DC} = 200V$

Fig. 7 FOM during turn-off transient at different dc link voltages at $T_c = 25°C$.

The bahavior regarding dv/dt during turn-off is displayed in Fig. 8 and Fig. 9. The measured values are displayed standardized to each device nominal current as well. Since the load current charges the output capacitance during turn off process, the dv/dt increases according to the load current. The maximimum dv/dt during turn-off is visbile for the GaN eHEMT and the Si Mosfet. Both have a very small output capacitance in relation to their nominal current. The mean dv/dt is smaller for all chips since the voltage slope is not linear during the switching process. The SiC devices have the lowest dv/dt during turn off. For all DUTs the dv/dt increases according to the DC link voltage.

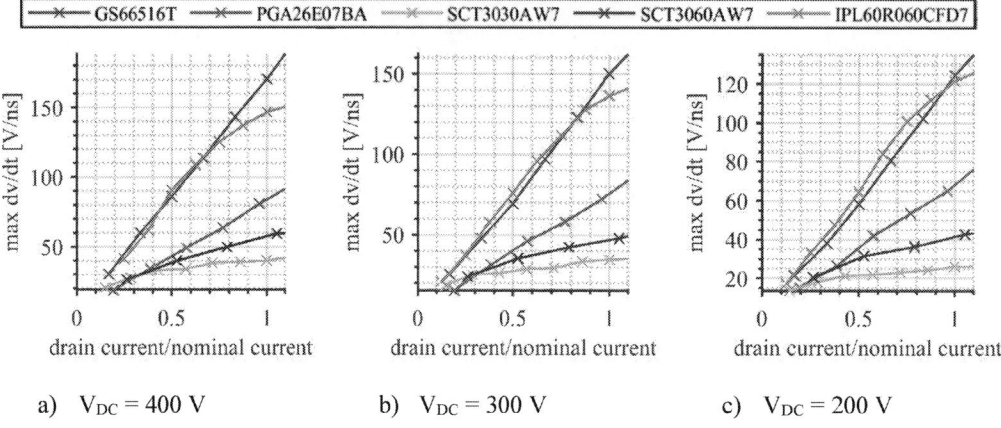

a) $V_{DC} = 400$ V   b) $V_{DC} = 300$ V   c) $V_{DC} = 200$ V

Fig. 8 Maximum dv/dt during turn-off at different dc link voltages at $T_c = 25°C$

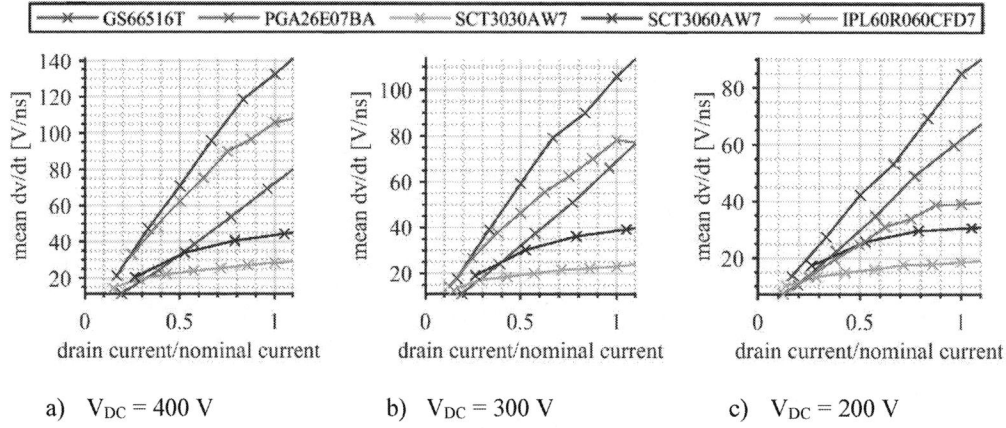

a) $V_{DC} = 400$ V   b) $V_{DC} = 300$ V   c) $V_{DC} = 200$ V

Fig. 9 Mean dv/dt during turn-off at different dc link voltages at $T_c = 25°C$

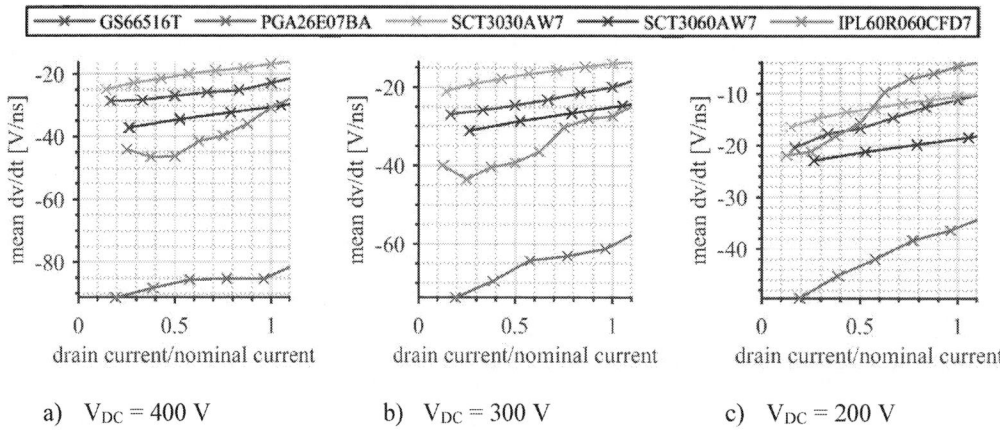

a) $V_{DC} = 400$ V   b) $V_{DC} = 300$ V   c) $V_{DC} = 200$ V

Fig. 10 Maximum dv/dt during turn-on at different dc link voltages at $T_c = 25°C$

The bahavior regarding dv/dt during turn-on is displayed in Fig. 10 and Fig. 11. The dv/dt's are lower than during turn-off except for the GaN GiT. The GaN eHEMT has a much lower dv/dt than during turn-off. Additonally there is a large difference between mean and maximum dv/dt which is caused by the slow voltage slope at the end of the switching process. The dv/dt is less current dependend than during the device turn-off process.

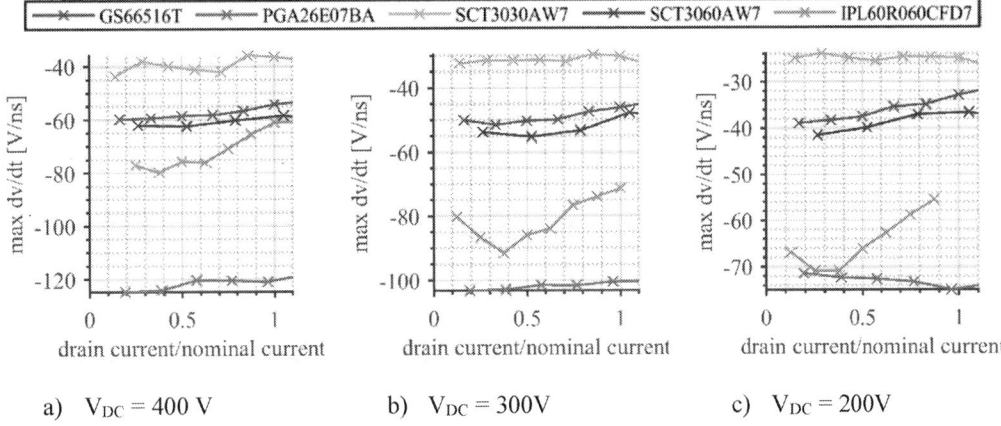

a) $V_{DC} = 400$ V   b) $V_{DC} = 300$V   c) $V_{DC} = 200$V

Fig. 11 Mean dv/dt during turn-on at different dc link voltages at $T_c = 25°C$

Thermal behavior

To study the thermal behavior, the FOM given in equation (1) is used. The R_{DSon} for the FOM calculation is interpolated from datasheet values for 25 °C and 150 °C, to consider the on resistance temperature dependency as well. The turn-on losses of the Si Mosfets needs to be scaled again to achieve a suitable presentation in a single diagram. All power devices showing higher losses at higher temperatures. This is caused by lower switching speed. The turn-on behavior of the GaN eHEMT changes strongly. Looking at the temperature dependent dv/dt, the GaN eHEMT showing a significant decrease in switching speed. The turn off losses are less temperature dependent and the FOM is influenced significantly by the temperature dependency of the R_{DSon}. The GaN eHEMT showing the worst behavior of all wide band gap devices at high temperatures, because the turn on losses dominating the total losses. Only the Si Mosfet has higher losses due to the worse behavior of the high side body diode.

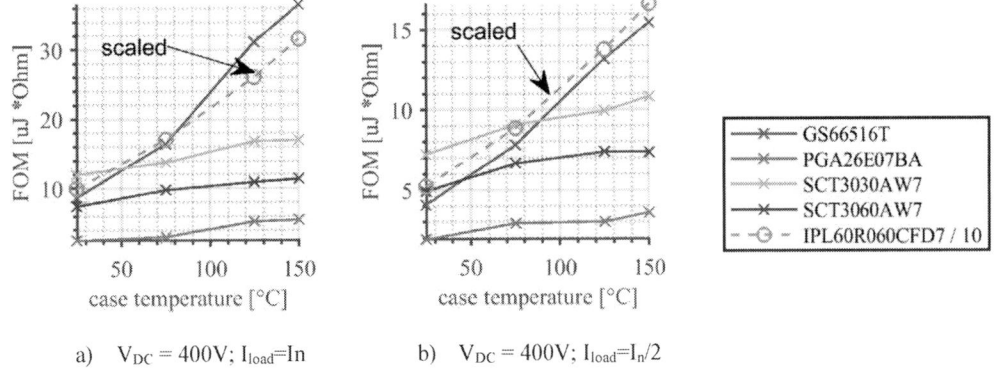

a) $V_{DC} = 400$V; $I_{load}=In$ b) $V_{DC} = 400$V; $I_{load}=I_n/2$

Fig. 12 FOM temperature dependency during turn-on at $V_{DC} = 400$V

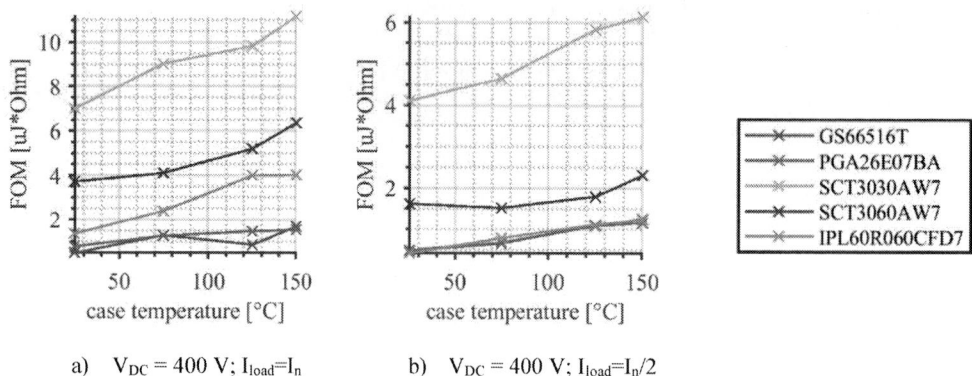

a) $V_{DC} = 400$ V; $I_{load}=I_n$ b) $V_{DC} = 400$ V; $I_{load}=I_n/2$

Fig. 13 FOM temperature dependency during turn-off at $V_{DC} = 400$V. At half load, the GaN eHEMT line is hidden behind GaN GiT and Si Mosfet line.

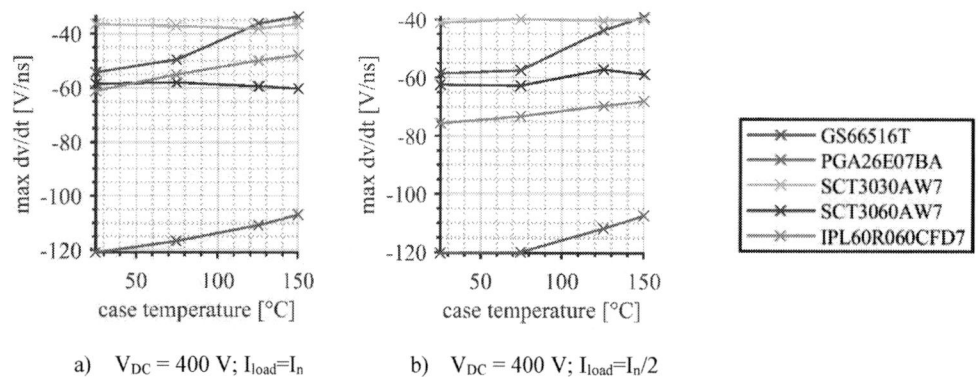

a) $V_{DC} = 400$ V; $I_{load}=I_n$ b) $V_{DC} = 400$ V; $I_{load}=I_n/2$

Fig. 14 Temperature dependency of max dv/dt during turn-on at $V_{DC} = 400$V

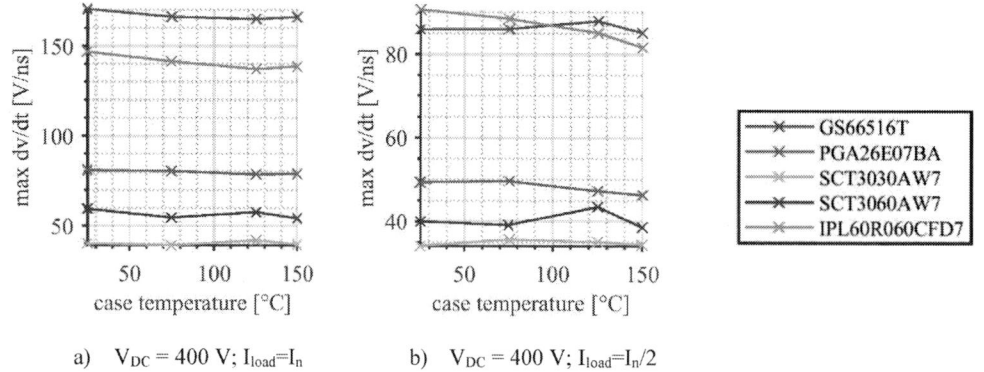

a) $V_{DC} = 400$ V; $I_{load}=I_n$ b) $V_{DC} = 400$ V; $I_{load}=I_n/2$

Fig. 15 Temperature dependency of max dv/dt during turn-off at $V_{DC} = 400$V

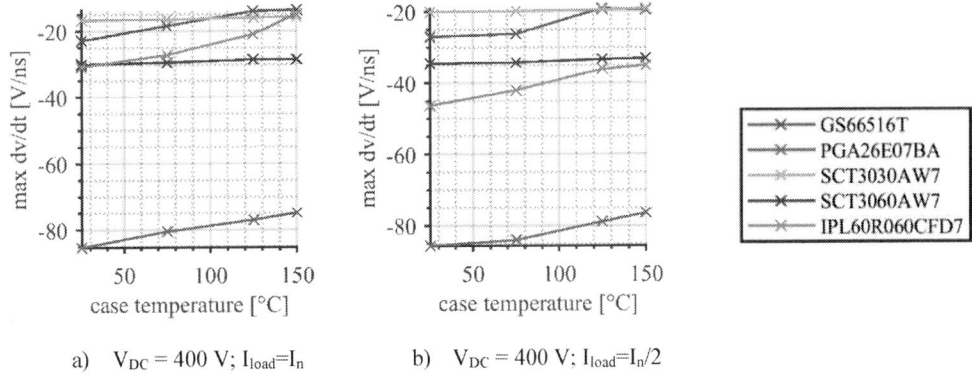

a) $V_{DC} = 400$ V; $I_{load} = I_n$ b) $V_{DC} = 400$ V; $I_{load} = I_n/2$

Fig. 16 Temperature dependency of mean dv/dt during turn-on at $V_{DC} = 400$V

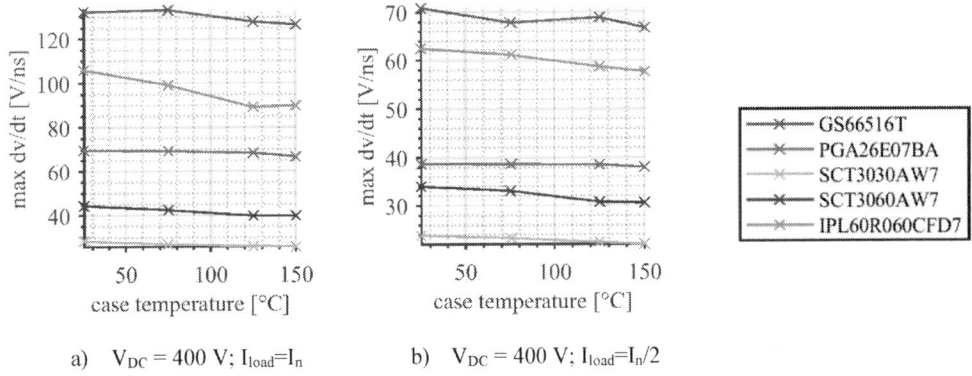

a) $V_{DC} = 400$ V; $I_{load} = I_n$ b) $V_{DC} = 400$ V; $I_{load} = I_n/2$

Fig. 17 Temperature dependency of mean dv/dt during turn-off at $V_{DC} = 400$V

Conclusion

The comparison of GaN, SiC and Si power switches showing at turn-off and nominal current the highest dv/dt for the Si Mosfet and the GaN eHEMT device. This must be considered by designers, since high dv/dt can cause EMI, common mode and isolation difficulties.

Additionally, the measurements have shown, the Si Mosfet is limited by its commutation partner. In the in this paper used test setup, the internal body diode or intrinsic reverse conduction ability were used. Hence the Si device cannot compete in this hard switching test setup. To achieve the full performance of the Si Mosfet a SiC diode is recommended. Otherwise it should be used in a soft or resonant switching application.

The GaN GiT has the best turn-on and the GaN eHEMT the best turn-off behavior. The turn-on losses are dominating the total losses. Therefore, the results regarding FOM comparison of the GaN devices indicates, if ideal parallelization is assumed, that it may be better to use multiple GaN GiTs at a lower load than one eHEMT at full load. Often the GaN devices have advantages regarding losses in comparison to the SiC devices. Nevertheless, the conditions can vary at lower DC voltages or higher temperatures.

Regarding thermal behavior, all devices showing lower switching speed at high temperatures. The turn-on process is influenced more than the turn-off process. Especially the GaN eHEMT has a strong temperature dependency which must be considered at high case temperatures.

References

[1] M. Meissner, J. Schmitz, F. Weiss and S. Bernet, "Current measurement of GaN power devices using a frequency compensated SMD shunt," *PCIM Europe 2019*; Nuremberg, Germany, 2019, pp. 1-8.

[2] Z. Zhang et al., "Methodology for switching characterization evaluation of wide band-gap devices in a phase-leg configuration," 2014 IEEE Applied Power Electronics Conference and Exposition - APEC 2014, Fort Worth, TX, 2014, pp. 2534-2541.

[3] M. Danilovic, Z. Chen, R. Wang, F. Luo, D. Boroyevich and P. Mattavelli, "Evaluation of the switching characteristics of a gallium-nitride transistor," 2011 IEEE Energy Conversion Congress and Exposition, Phoenix, AZ, 2011, pp. 2681-2688.

[4] G. Sorrentino and A. Gaito, "Advantages of using 650V SiC MOSFETs in High-Frequency DC-DC Converters," PCIM Europe 2017; International Exhibition and Conference for Power Electronics, Intelligent Motion, Renewable Energy and Energy Management, Nuremberg, Germany, 2017, pp. 1-4.

[5] T. Bertelshofer, R. Horff, A. März and M. Bakran, "Comparing 650V and 900V SiC MOSFETs for the application in an automotive inverter," 2016 18th European Conference on Power Electronics and Applications (EPE'16 ECCE Europe), Karlsruhe, 2016, pp. 1-10.

[6] S. Buetow and R. Herzer, "Characterization of GaN-HEMT in cascode topology and comparison with state of the art-power devices," 2018 IEEE 30th International Symposium on Power Semiconductor Devices and ICs (ISPSD), Chicago, IL, 2018, pp. 196-199.

Analysis of the coupling between the outer and inner control loops of a Grid-forming Voltage Source Converter

T. QORIA[*], F. GRUSON[*], F. COLAS[*], X. KESTELYN[*], X. GUILLAUD[*]

[*]L2EP, Univ. Lille, Arts et Metiers Institute of Technology, Centrale Lille, Yncrea Hauts-de-France,
ULR 2697 - L2EP - Laboratoire d'Electrotechnique de Puissance, F-59000 Lille, France.
Taoufik.qoria@ensam.eu

Acknowledgements

This project has received funding from the European Union's Horizon 2020 research and innovation program under grant agreement No 691800. This paper reflects only the author's views and the European Commission is not responsible for any use that may be made of the information it contains.

Keywords

«Grid-forming based voltage source converter», «State-Space modeling», «Small-signal analysis», «Participation factors», «Parametric sensitivities»

Abstract

The question of grid forming control is very different depending on the connection to a low voltage or high voltage grid. In case of higher power application, the low switching frequency may induce some stability issues. This question has been studied and some solutions have been proposed through new inner current and voltage control tuning methods. However, the possible interactions between the inner and the outer controls have not been discussed yet. Actually, in large power system, the phasor modeling approximation is used in order to ease the analysis and reduce the time computations. It assumes a good decoupling between the control loops, which allows neglecting the inner loop dynamics. This paper investigates the effectiveness of this assumption by taking some examples of tuning methods proposed in the literature and showing the ability of each method to guarantee the decoupling between controllers. In this paper, small-signal analysis tool, participation factors and parametric sensitivities are used.

Introduction

Conventionally, the VSC based on grid-forming control is always connected to the AC system through an LC filter. It is used to pass the fundamental frequency and attenuate the rest of undesired high order harmonics [1] e.g. pulse width modulation (PWM) switching effect, which appears in the current and voltage profiles. The presence of the LC filter requires a voltage control feedback across the filter capacitor. The aim is to guarantee AC voltage stiffness when the power converter is connected to the load and to avoid exciting the resonance of the LCL filter in a grid-connected mode [2], [3].

When the studies focus on the active power exchange, inertial effect and the transient stability, the LC filter and its inner control dynamics are often neglected for simplicity purpose [4]–[7]. This assumption considers a sufficient time-decoupling between control loops i.e.; the dynamic of the voltage and current loops should be faster than the outer loop. This assumption hold true for power converters with high bandwidth, which are mainly used for low-power applications (Microgrid and UPS). However, in high power applications such as transmission systems, the control bandwidth is limited by the switching frequency, which can result in a slower response, narrow stability regions and instability issues in the grid-connected mode following the analysis in [8], [9]. To deal with this issue, some tuning methods have been proposed in the literature [8], [10], [11]. However, the question that arises is: Does the proposed tuning methods ensure a decoupling between inner and outer controls?

This paper aims to respond this question. The studies focuses first on the inner controllers design proposed in [8], [10]. To analyze possible interactions between the control loops, participation factors [12] are then used. Besides highlighting the coupling issues between control loops, this paper proposes simplified models and a linearized full dynamic state-space models that can serves readers in other applications.

Grid-Forming Based on the Cascaded Control Structure

The system illustrated in the lower part of Fig. 1 consists of a three phase 2-Level voltage source converter represented by a switching model. It is supplied by a DC voltage source that is assumed to be a DC storage and/or primary source i.e.; PV, wind turbine, etc., and connected to an AC system through an $L_fC_fL_c$ filter. The AC system is modeled by an equivalent AC voltage source in series with its equivalent impedance $Z_g = R_g + jX_j$. The AC system is assumed to be a stiff symmetrical AC system.

Following the notations on the AC side, the state variables are the VSC output current i_s, the AC voltage e_g and the grid current i_g. For the analysis, only the average modeling of the system is used.

The upper part of Fig. 1 presents the grid-forming cascaded control structure. It consists of an inner cascaded AC voltage and a VSC output current control represented in the synchronous reference frame (SRF). The inner voltage control is ensured by two proportional-integral (PI) controllers considering the feedforward decoupling terms $C_f\omega e_{g_{dq}}$ and compensations $kFFi \cdot i_{g_{dq}}$. The inner current control is also ensured by two proportional-integral (PI) controllers considering the feedforward decoupling terms $L_f\omega i_{s_{dq}}$ and compensations $kFFv \cdot e_{g_{dq}}$. k_{pv} and k_{iv} are the proportional gain and integral gain for the AC voltage control, respectively. Meanwhile, k_{pc} and k_{ic} are the proportional gain and integral gain for the VSC current control, respectively. The current control loop generates the modulated voltage to the linearization stage that delivers the modulation signals to the switching stage of the power converter.

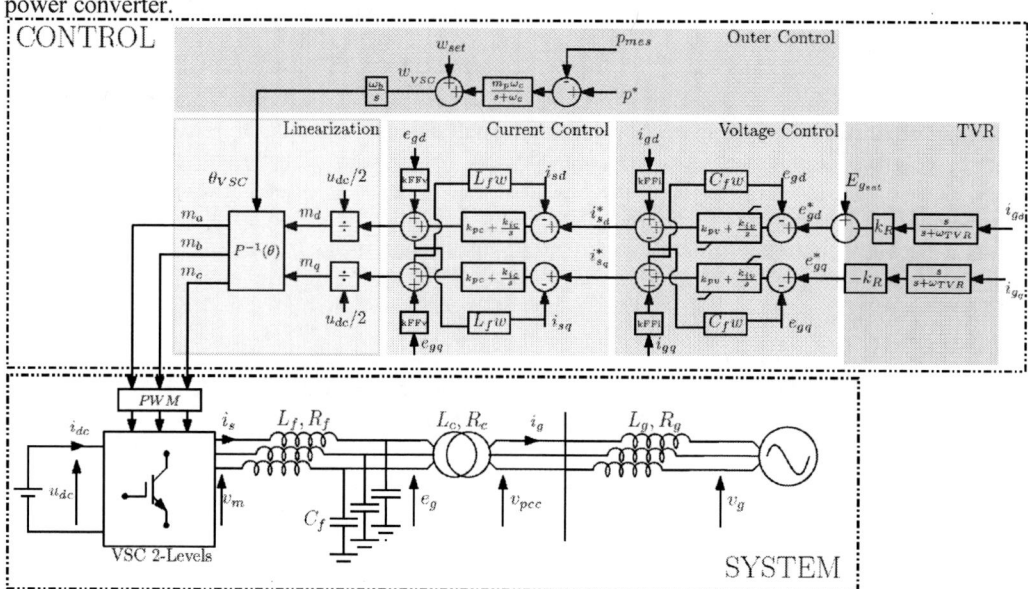

Fig.1. Voltage source converter based on grid forming control

The control angle θ_{VSC} and AC voltage magnitude E_g are provided by the outer droop control. To avoid particular resonance around the grid frequency a transient virtual resistor (TVR) used.
The active power droop control is expressed by the following equations:

$$\omega_{VSC} - \omega_{set} = \frac{m_p\omega_c}{\omega_c + s}(p_{mes} - p^*) \tag{1}$$

where ω_{VSC} and m_p are the VSC output frequency and active power droop gain, respectively. The low-pass filter used in the active power droop aims to simultaneously filter the measurement noises and emulate the inertial effect of a synchronous machine [4], [6].

Inner and Outer Controls Coupling

Simplified modeling of a VSC based on the grid-forming control

In the simplified model, e_g is assimilated to v_m i.e., the voltage and current loops are not considered. Only the outer loop is implemented. This leads to a 6th order linear system (Fig.2a):

$$\Delta \dot{x}_i = A_S\,\Delta x_i + B_S\,\Delta u_j \tag{2}$$

"Δ" refers to a small variation of the state variable around an operating point. $\Delta x_i = [\Delta i_{g_d}\; \Delta i_{g_q}\; \zeta_{TVR_d}\; \zeta_{TVR_q}\; \Delta\delta_{VSC}\; \Delta\omega_{VSC}]$ represents the state variables, where $\Delta\delta_{VSC}$ and $\Delta\omega_{VSC}$ are the phasor angle and the converter frequency, respectively. $\zeta_{TVR_{dq}}$ are the state variables of the TVR. $u_j = [\Delta p^*]$ is the control inputs. The index "S" in the control matrices A_S and B_S refers to the simplified model.

$$
A_S = \begin{bmatrix}
-\frac{R_T}{L_T}\omega_b & \omega_0\omega_b & \frac{\omega_b}{L_T} & 0 & \frac{V_{gd}sin\delta_0 - V_{gq}cos\delta_0}{L_T}\omega_b & 0 \\
-\omega_0\omega_b & -\frac{R_T}{L_T}\omega_b & 0 & \frac{\omega_b}{L_T} & \frac{V_{gd}cos\delta_0 + V_{gq}sin\delta_0}{L_T}\omega_b & 0 \\
\frac{R_v R_T}{L_T}\omega_b & -R_v\omega_0\omega_b & -(\omega_{LF} + \frac{R_v\omega_b}{L_c}) & 0 & -\frac{V_{gd}sin\delta_0 - V_{gq}cos\delta_0}{L_T}\omega_b & 0 \\
R_v\omega_0\omega_b & \frac{R_v R_T}{L_T}\omega_b & 0 & -(\omega_{LF} + \frac{R_v\omega_b}{L_c}) & -\frac{V_{gd}cos\delta_0 + V_{gq}sin\delta_0}{L_T}\omega_b & 0 \\
0 & 0 & 0 & 0 & 0 & \omega_b \\
-m_p\omega_c E_{gd_0} & 0 & -m_p\omega_c l_{gd_0} & 0 & 0 & -\omega_c
\end{bmatrix},\;
B_S = \begin{bmatrix} 0 \\ 0 \\ 0 \\ 0 \\ 0 \\ m_p\omega_c \end{bmatrix}
$$

In the present power system dominated by synchronous machines, the grid current dynamics are often neglected in the analysis against the active power dynamics i.e., $\frac{d}{dt}i_{g_{dq}} = 0$. This consideration is also effective for dynamic model of power converters when a transient virtual resistor is introduced [13], [14]. The modeling simplifications are illustrated in Fig.2b.

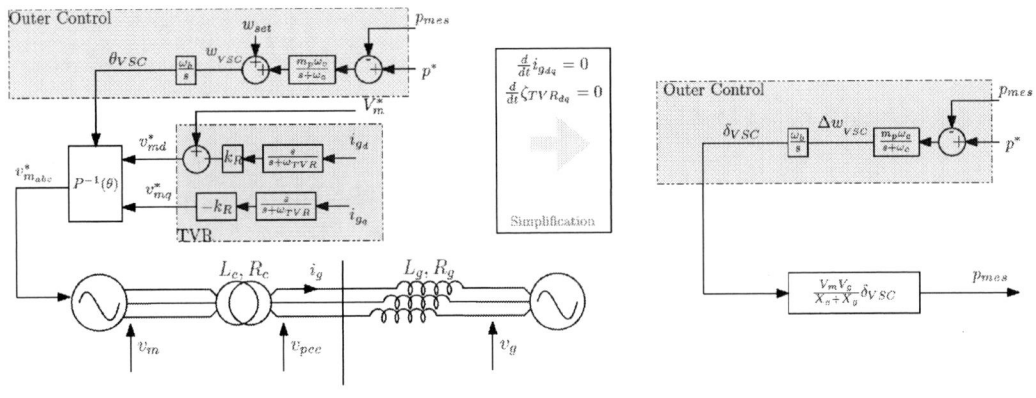

(a) 6th order model (b) 2nd order quasi-static model

Fig.2. Simplified models

Based on the quasi-static model in Fig.2.b, the equation of the active power can be expressed by a simplified 2nd order transfer function:

$$\Delta p_{mes} = \frac{1}{1 + \frac{X_c + X_g}{m_p\omega_b E_g V_g}s + \frac{X_c + X_g}{m_p\omega_b\omega_c E_g V_g}s}\,\Delta p^* \tag{3}$$

From (3), the active power dynamics are mainly imposed by the droop gain, and the low-pass filter cut-off frequency and the grid impedance.

Considering the system and control parameters listed in Table I. The theoretical response time is equal to $T_{r5\%}=185$ ms.

TABLE I
SYSTEM AND CONTROL PARAMETERS

Symbol	Value	Symbol	Value
P_n	1 GW	$Cos\phi$	0.95
U_{ac}	320 kV ph-ph	f_n	50 Hz
E_{gset}	1 p.u	m_p	0.02 p.u.
X_c	0.15 p.u	X_g	0.1 p.u
ω_c	31.4 rad/s	R_c	0.005 p.u.
R_g	0.01 p.u	ω_b	100π
k_R	0.02 p.u	ω_{TVR}	60 rad/s

The check the accuracy and the rightness of the 2nd order compared to the 6th order system in terms of active power dynamics. Pole map and time-domain simulations are presented in Fig. 3.

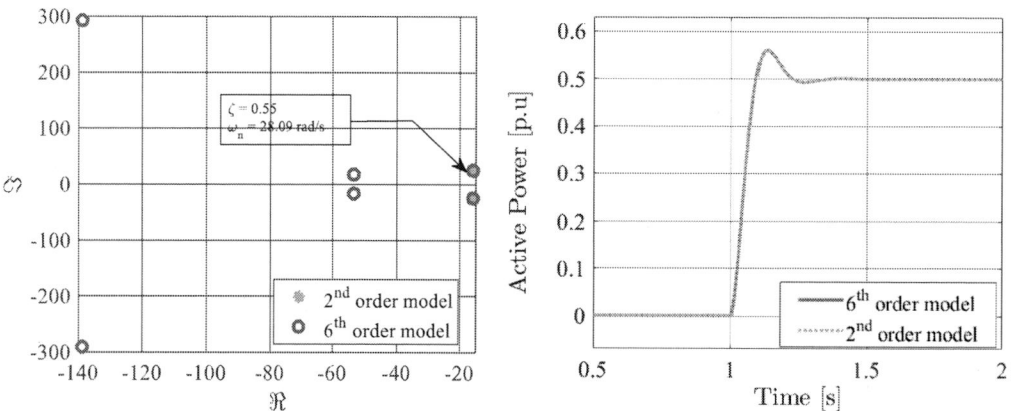

Fig.3. Comparison between 2nd order and 6th order system in case of active power change

From the obtained results, it can be noticed that the power control loop has two dominant eigenvalues $\lambda^S_{1-2} = -15.67 \pm 23.38i$ that impose the active power dynamics, and thereby, the current dynamics has no effect on the active power. Moreover, the performances obtained theoretically based on the 2nd system are still effective for the 6th order system ($T_{r5\%}=185$ ms).

Full order model of a VSC based on the grid-forming control

In this subsection, all the system dynamics are taken into account (LC filter, voltage and current control loops). Consequently, the system moves form a 6th order to 14th order linear system, where:

$$\Delta \dot{x}_k = \mathbf{A_F} \, \Delta x_k + \mathbf{B_F} \, \Delta u_j \tag{4}$$

$$\Delta x_k = [\Delta i_{s_d} \; \Delta i_{s_q} \; \Delta e_{g_d} \; \Delta e_{g_q} \; \Delta i_{g_d} \; \Delta i_{g_q} \; \Delta \zeta_{TVR_d} \; \Delta \zeta_{TVR_q} \; \Delta C_{c_d} \Delta C_{c_q} \; \; \Delta C_{v_d} \; \Delta C_{v_q} \; \Delta \delta_{VSC} \; \Delta \omega_{VSC}] \, , \, u_j = [\Delta p^*]$$

with,

$$A_{F21} = \begin{bmatrix} 1 & 0 & -k_{pv} & C_f\omega_0 & -1 & 0 & k_{pv} & 0 \\ 0 & 1 & -C_f\omega_0 & -k_{pv} & 0 & -1 & 0 & k_{pv} \\ 0 & 0 & 1 & 0 & 0 & 0 & -1 & 0 \\ 0 & 0 & 0 & 1 & 0 & 0 & 0 & -1 \\ 0 & 0 & 0 & 0 & 0 & 0 & 0 & 0 \\ 0 & 0 & -m_p\omega_c I_{gd_0} & 0 & -m_p\omega_c E_{gd_0} & 0 & 0 & 0 \end{bmatrix} , A_{F22} = \begin{bmatrix} 0 & 0 & -k_{iv} & 0 & 0 & 0 \\ 0 & 0 & 0 & -k_{iv} & 0 & 0 \\ 0 & 0 & 0 & 0 & 0 & 0 \\ 0 & 0 & 0 & 0 & 0 & 0 \\ 0 & 0 & 0 & 0 & 0 & -\omega_c \\ 0 & 0 & 0 & 0 & 0 & \omega_b \end{bmatrix}$$

$$A_{F11}=\begin{bmatrix}
\frac{\omega_b}{L_f}(k_{pc}-R_f) & 0 & \frac{\omega_b}{L_f}k_{pv}k_{pc} & \frac{\omega_b}{L_f}C_f\omega_0 k_{pc} & -\frac{\omega_b}{L_f}k_{pc} & 0 & \frac{\omega_b}{L_f}k_{pv}k_{pi} & 0 \\[4pt]
0 & \frac{\omega_b}{L_f}(k_{pc}-R_f) & -\frac{\omega_b}{L_f}k_{pv}k_{pc} & \frac{\omega_b}{L_f}k_{pv}k_{pc} & 0 & -\frac{\omega_b}{L_f}k_{pc} & 0 & \frac{\omega_b}{L_f}k_{pv}k_{pi} \\[4pt]
\frac{\omega_b}{c_f} & 0 & 0 & \omega_b\omega_0 & -\frac{\omega_b}{c_f} & 0 & 0 & 0 \\[4pt]
0 & \frac{\omega_b}{c_f} & -\omega_b\omega_0 & -\left(\omega_{LF}+\frac{R_v\omega_b}{L_c}\right) & 0 & -\frac{\omega_b}{c_f} & 0 & 0 \\[4pt]
0 & 0 & \frac{\omega_b}{L_T} & 0 & -\frac{R_T\omega_b}{L_T} & \omega_b\omega_0 & 0 & 0 \\[4pt]
0 & 0 & 0 & \frac{\omega_b}{L_T} & -\omega_b\omega_0 & -\frac{R_T\omega_b}{L_T} & 0 & 0 \\[4pt]
0 & 0 & -\frac{R_v\omega_b}{L_T} & 0 & R_v\frac{R_T\omega_b}{L_T} & -R_v\omega_b\omega_0 & \omega_{TVR} & 0 \\[4pt]
0 & 0 & 0 & -\frac{R_v\omega_b}{L_T} & R_v\omega_b\omega_0 & R_v\frac{R_T\omega_b}{L_T} & 0 & \omega_{TVR}
\end{bmatrix}$$

$$A_{F12}=\begin{bmatrix}
\frac{\omega_b}{L_f}k_{ic} & 0 & -\frac{\omega_b}{L_f}k_{iv}k_{pc} & 0 & 0 & 0 \\[4pt]
0 & \frac{\omega_b}{L_f}k_{ic} & 0 & -\frac{\omega_b}{L_f}k_{iv}k_{pc} & 0 & 0 \\[4pt]
0 & 0 & 0 & 0 & 0 & 0 \\[4pt]
0 & 0 & 0 & 0 & 0 & 0 \\[4pt]
0 & 0 & 0 & 0 & \frac{V_{gd}\sin\delta_0-V_{gq}\cos\delta_0}{L_T}\omega_b & 0 \\[4pt]
0 & 0 & 0 & 0 & \frac{V_{gd}\cos\delta_0+V_{gq}\sin\delta_0}{L_T}\omega_b & 0 \\[4pt]
0 & 0 & 0 & 0 & -R_v\frac{V_{gd}\sin\delta_0-V_{gq}\cos\delta_0}{L_T}\omega_b & 0 \\[4pt]
0 & 0 & 0 & 0 & -R_v\frac{V_{gd}\cos\delta_0+V_{gq}\sin\delta_0}{L_T}\omega_b & 0
\end{bmatrix}$$

$$B=\begin{bmatrix}0 & 0 & 0 & 0 & 0 & 0 & 0 & 0 & 0 & 0 & 0 & 0 & 0 & m_p\omega_c\end{bmatrix}^T$$

The goal of this section is to verify if the accuracy of simplified model compared to the full dynamic model. The studies focus first on the control design made by [8]. The parameters of the inner control loops and the LC filter are listed in Table II.

TABLE II
SYSTEM AND CONTROL PARAMETERS

Symbol	Value	Symbol	Value
E_{set}	1 p.u	L_f	0.15 p.u.
R_f	0.005 p.u.	C_f	0.066 p.u.
$kFFi$	1	$kFFv$	1
k_{pv}	0.52 p.u	k_{iv}	1.16 p.u.
k_{ic}	1.19 p.u.	k_{pc}	0.73 p.u.

Based on the parameters of table I, table II and the control matrix A_F, the eigenvalues of the full order model are given in table III. The eigenvalues index "F" refers to the full dynamic model.

TABLE III
FULL ORDER MODEL EIGENVALUES

Eigenvalue	Location	Eigenvalue	Location
λ^F_{1-2}	$-780.52 \pm 3286i$	λ^F_{11}	$-2.38 + 0i$
λ^F_{3-4}	$-748.97 \pm 2779.2i$	λ^F_{12}	$-2.18.49 + 0i$
λ^F_{5-6}	$-23.51 \pm 112.94i$	λ^F_{13}	$-1.64 + 0i$
λ^F_{7-8}	$-60.986 \pm 3.9938i$	λ^F_{14}	$-1.54 + 0i$
λ^F_{9-10}	$\mathbf{-16.061 \pm 24.39i}$		

One can remark that the $\Re$ and $\Im$ part of λ^S_{1-2} are approximatively equal to those of λ^F_{9-10}. For the simplified 2nd order model, it is clear that λ^S_{1-2} are linked to the state variables $\Delta\omega_{VSC}$ and $\Delta\delta_{VSC}$ that impose the active power dynamic. However, it is not certain that λ^F_{9-10} are also linked to the same state variables. The answer to this question requests establishing a link between the state variables and the eigenvalues. Hence, the participation factors tool is used for this aim.

Based on the control matrix A_F the participation factors are given in table IV.

One can notice from the results that $\Delta\omega_{VSC}$ and $\Delta\delta_{VSC}$ are mostly participating in λ^F_{9-10}. They are also participating in λ^F_{5-6}, λ^F_{7-8}, , λ^F_{11}, λ^F_{12}, λ^F_{13} and λ^F_{14}. Despite that the participation rate of $\Delta\omega_{VSC}$ and $\Delta\delta_{VSC}$ in these latter is negligible, it creates a small coupling between them and λ^F_{9-10}, which results in the error between λ^F_{9-10} and λ^S_{1-2}.

TABLE IV
PARTICIPATION FACTORS (NORMALIZED)

	λ^F_{1-2}	λ^F_{3-4}	λ^F_{5-6}	λ^F_{7-8}	λ^F_{9-10}	λ^F_{11}	λ^F_{12}	λ^F_{13}	λ^F_{14}
$\Delta\omega_{VSC}$	0	0	0.1	0.02	1	0.003	0.003	0	0
$\Delta\delta_{VSC}$	0	0	0.1	0	1	0.04	0.04	0.02	0.02

To demonstrate the correctness of this analysis, the simplified and the full dynamic models are compared through time-domain simulations. The results are gathered in Fig. 4.

The result shows that the error in between the simplified model and the full dynamic model could be considered as negligible. In this case, the simplified model is enough for the active power analysis.

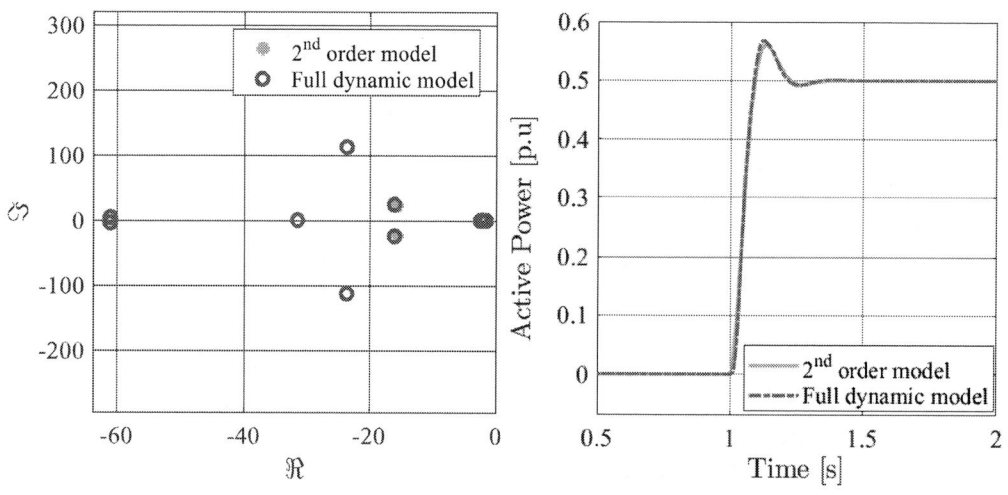

Fig.4. Comparison between 2nd order and full order system in case of active power change

Parametric sensitivities

The goal of this section is to study the impact of some parametric variations on the coupling between the inner and outer control loops. The focus is made on two parameters:
- The short circuit ratio (SCR) that defines the grid stiffness.
- The cut-off frequency ω_c of the low-pass filter used in the outer droop control that imposes the inertia constant ($H = \frac{1}{2\omega_c m_p}$) [6].

It is briefly recalled that the SCR $= 1/L_g$. The impact of the droop gain is not studied since it is a fixed value linked with the load sharing functionally.

The first test case consists in varying the SCR from 3 (Strong grid) to SCR = 1.5 (Very weak grid), while, the second test case consists in varying ω_c from 31.4 rad/s (equivalent to H = 0.79 s) to 12.5 rad/s (equivalent to H = 2 s). In both cases, a comparison between the eigenvalues evolution of the 2nd order and full dynamic model is performed. The results are gathered in Fig.5.

The change of the SCR results in a decrease of the imaginary part of λ_{1-2}^{S}, which yields an improved active power damping and slower dynamics. Compared to the full dynamic model, the evolution of λ_{9-10}^{S} is a bit different, however, looking to the real part, this difference is negligible. This result is confirmed through time-domain simulations, where both model are very close for three values of SCR. A decrease of ω_c yields a low damping factor and slow response of the active power. This behavior is verified through the evolution of the eigenvalues and confirmed via time-domain. From these results, the 2nd order model response replicates exactly the response of the full dynamic model.

From the performed test cases, it can be concluded that : The controllers tuning proposed in [8], guaranties a well time-decoupling between inner and outer droop control loops.

The impact of the inner control on the choice of the outer control structure

Beside the droop control, many outer control structures have been proposed in the literature to ensure a self-synchronization, power control and inertia emulation such as Virtual Synchronous Machine (VSM) depicted in Fig.6. This control concept is basically equivalent to the droop control since it fulfils the same features. However, the VSM contains additional dynamics such as a phase-locked loop (PLL), which induces more dynamics, a thereby, the probability of a coupling between controller is higher. The effectiveness of an inner control tuning should not be limited to one outer control loop. Thus, the effectiveness of [8] is demonstrated, when the outer control is a VSM.

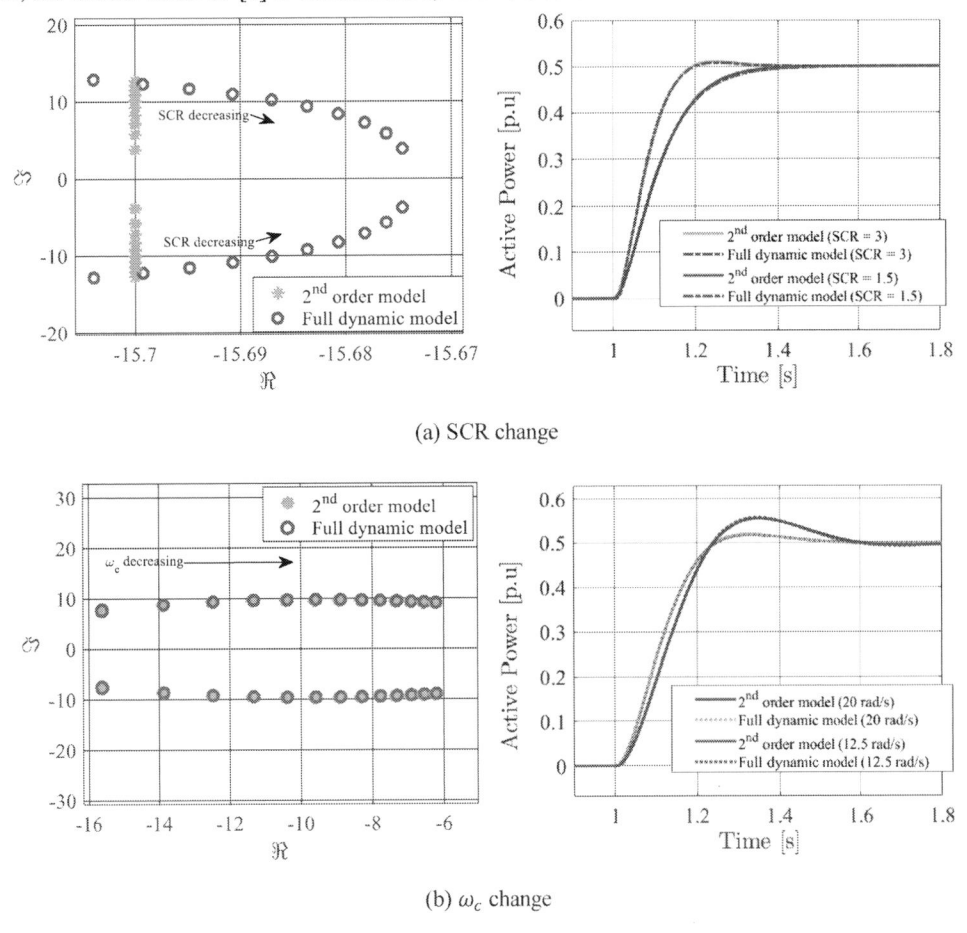

(a) SCR change

(b) ω_c change

Fig.5. Impact of SCR and cut-off frequency of LPF on the inner and outer controls coupling

When the VSM is used, the full dynamic model is a 16th order one. The quasi-static model of the system based on the VSM is given by a simplified 2nd order transfer function in (5). This equation is effective assuming a fast PLL dynamics.

$$\Delta p_{mes} = \frac{1}{1 + \frac{X_c K}{\omega_b E_g V_g} s + \frac{2H(X_c + X_g)}{\omega_b E_g V_g} s} \Delta p^* \tag{5}$$

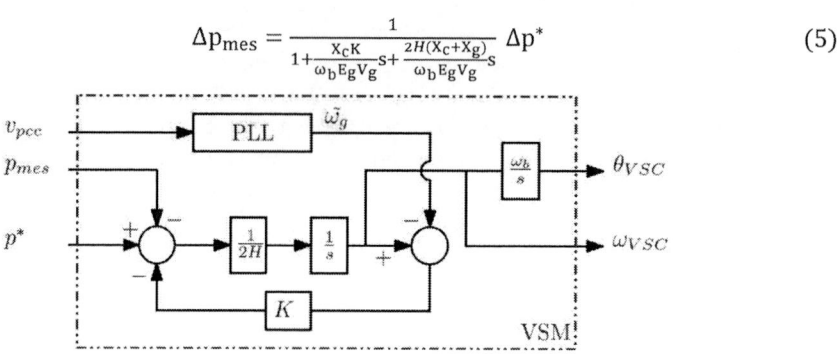

Fig.6. Virtual Synchronous Machine

The VSM parameters are listed in table IV.

TABLE IV
SYSTEM AND CONTROL PARAMETERS

Symbol	Value	Symbol	Value
$T_{R_{PLL}}$	10 ms	$K_{p_{PLL}}$	2.2282 p.u
$K_{i_{PLL}}$	795.7 p.u.	H	5 s
K	333 p.u.		

To assess the decoupling between the outer and the inner loops, the same test cases in Fig.5 are performed in Fig.7.

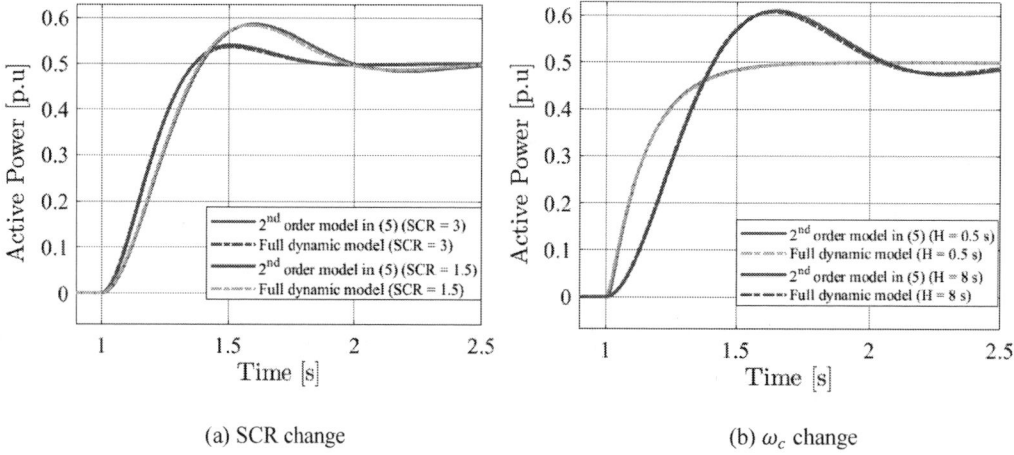

(a) SCR change   (b) ω_c change

Fig.7. Impact of SCR and the inertia on the inner control and VSM based-outer controls coupling

The results show that the response of the full dynamic model is close to the response of the simplified model in (5). Thus, it can be concluded that the system analysis based on the phasor approximation are of high accuracy.

Let's new consider another outer control tuning method proposed in [10]. This method is based on a simple algorithm that deduces the cascaded PI controller gains based on eigenvalues location. In [10], a VSM outer loop is used. The system and control parameters taken in this paper are listed in table V.

To assess the ability of the method proposed in [10] to guarantee a decoupling between controller loops, different parameters of the VSM damping are taken. Simulation results are gathered in Fig. 8, where, a power step of 50% is applied as previously.

Contrary to the proposed method in [8], where the system remains stable for different values of inertia constants and a damping of K =333 p.u, the results in Fig. 8 demonstrate the high sensitivity of the system to the outer control parameters. Besides the inability of the control to guarantee a decoupling between inner and outer control loops, the dynamic model may lead to instability in particular conditions (e.g. K = 900 p.u), whereas the phasor model is still given a stable results, which yield to wrong analysis.

TABLE V
SYSTEM AND CONTROL PARAMETERS [10]

Symbol	Value	Symbol	Value
E_{set}	1 p.u	$L_c = L_f$	0.1 p.u.
$R_c = R_f$	0.003 p.u.	C_f	0.2 p.u.
$kFFi$	0	$kFFv$	1
k_{pv}	1.795 p.u	k_{iv}	80 p.u.
k_{ic}	20 p.u.	k_{pc}	0.6366 p.u.

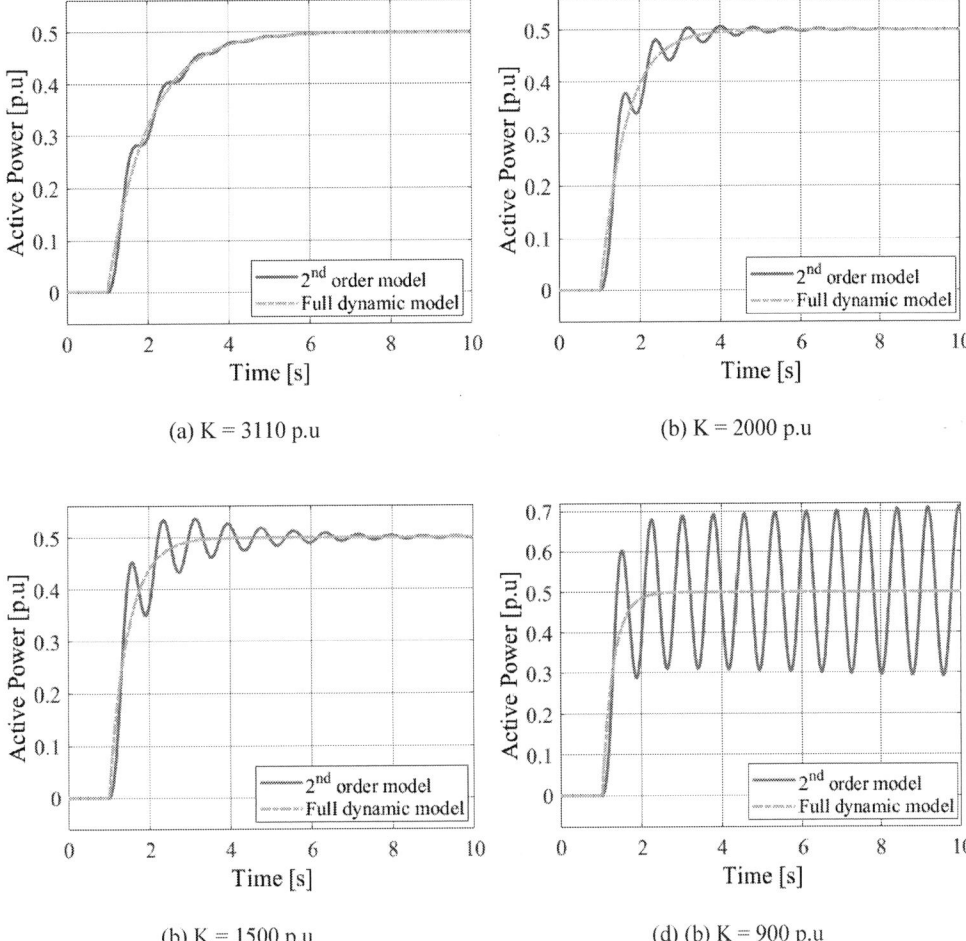

(a) K = 3110 p.u

(b) K = 2000 p.u

(b) K = 1500 p.u

(d) (b) K = 900 p.u

Fig.8. Impact of VSM damping on the inner control and outer controls coupling

Conclusion

The paper addresses the subject of possible interactions between inner and outer control loops due to the limited bandwidth of the high-power voltage source converters. This paper takes first a case study of one controller tuning proposed in the literature. Based on the eigenvalues analysis and time-domain simulation, the obtained results lead to the main conclusion: The first tuning method studied allows a good decoupling between control loops, which results in a high consistency between the phasor model and the full dynamic one, while, the second tuning method does not give the same results. This clearly highlights the importance of the choice of the inner control in order to be able to use some simplified power electronic converters modelling for large power system studies.

References

[1] B. Hoseinzadeh and C. L. Bak, "Impact of Grid Impedance Variations on Harmonic Emission of Grid-Connected Inverters," *ArXiv161202045 Cs*, Apr. 2017, Accessed: Nov. 17, 2019. [Online]. Available: http://arxiv.org/abs/1612.02045.

[2] T. Qoria, C. Li, K. Oue, F. Gruson, F. Colas, and X. Guillaud, "Direct AC voltage control for grid-forming inverters," *J. Power Electron.*, vol. 20, no. 1, pp. 198–211, Jan. 2020, doi: 10.1007/s43236-019-00015-4.

[3] E. Rokrok, T. Qoria, A. Bruyere, B. Francois, and X. Guillaud, "Effect of Using PLL-Based Grid-Forming Control on Active Power Dynamics Under Various SCR," in *IECON 2019 - 45th Annual Conference of the IEEE Industrial Electronics Society*, Oct. 2019, vol. 1, pp. 4799–4804, doi: 10.1109/IECON.2019.8927648.

[4] J. Liu, Y. Miura, and T. Ise, "Comparison of Dynamic Characteristics Between Virtual Synchronous Generator and Droop Control in Inverter-Based Distributed Generators," *IEEE Trans. Power Electron.*, vol. 31, no. 5, pp. 3600–3611, May 2016, doi: 10.1109/TPEL.2015.2465852.

[5] L. Huang, H. Xin, Z. Wang, L. Zhang, K. Wu, and J. Hu, "Transient Stability Analysis and Control Design of Droop-Controlled Voltage Source Converters Considering Current Limitation," *IEEE Trans. Smart Grid*, vol. 10, no. 1, pp. 578–591, Jan. 2019, doi: 10.1109/TSG.2017.2749259.

[6] T. Qoria, F. Gruson, F. Colas, G. Denis, T. Prevost, and X. Guillaud, "Inertia effect and load sharing capability of grid forming converters connected to a transmission grid," in *15th IET International Conference on AC and DC Power Transmission (ACDC 2019)*, Feb. 2019, pp. 1–6, doi: 10.1049/cp.2019.0079.

[7] T. QORIA, F. Gruson, F. COLAS, X. Kestelyn, and X. GUILLAUD, "Current Limiting Algorithms and Transient Stability Analysis of Grid-Forming VSCs," *PSCC 2020*, p. 9, Jun. 2020.

[8] T. Qoria, F. Gruson, F. Colas, X. Guillaud, M. Debry, and T. Prevost, "Tuning of Cascaded Controllers for Robust Grid-Forming Voltage Source Converter," in *2018 Power Systems Computation Conference (PSCC)*, Jun. 2018, pp. 1–7, doi: 10.23919/PSCC.2018.8443018.

[9] Z. Li, C. Zang, P. Zeng, H. Yu, S. Li, and J. Bian, "Control of a Grid-Forming Inverter Based on Sliding-Mode and Mixed H_2/H_∞ Control," *IEEE Trans. Ind. Electron.*, vol. 64, no. 5, pp. 3862–3872, May 2017, doi: 10.1109/TIE.2016.2636798.

[10] S. D'Arco, J. A. Suul, and O. B. Fosso, "Automatic Tuning of Cascaded Controllers for Power Converters Using Eigenvalue Parametric Sensitivities," *IEEE Trans. Ind. Appl.*, vol. 51, no. 2, pp. 1743–1753, Mar. 2015, doi: 10.1109/TIA.2014.2354732.

[11] G. Denis, T. Prevost, P. Panciatici, X. Kestelyn, F. Colas, and X. Guillaud, "Improving robustness against grid stiffness, with internal control of an AC voltage-controlled VSC.," in *2016 IEEE Power and Energy Society General Meeting (PESGM)*, Jul. 2016, pp. 1–5, doi: 10.1109/PESGM.2016.7741341.

[12] S. Danielsen, O. B. Fosso, and T. Toftevaag, "Use of participation factors and parameter sensitivities in study and improvement of low-frequency stability between electrical rail vehicle and power supply," in *2009 13th European Conference on Power Electronics and Applications*, Sep. 2009, pp. 1–10.

[13] L. Zhang, L. Harnefors, and H. Nee, "Power-Synchronization Control of Grid-Connected Voltage-Source Converters," *IEEE Trans. Power Syst.*, vol. 25, no. 2, pp. 809–820, May 2010, doi: 10.1109/TPWRS.2009.2032231.

[14] T. Qoria, Q. Cossart, C. Li, X. Guillaud, F. Gruson, and X. Kestelyn, "Deliverable 3.2: Local control and simulation tools for large transmission systems," p. 89.

Influence of Different Pulse-Width Modulation Methods on Magnet Losses in Permanent Magnet Synchronous Machines

Narciso G. Marmolejo[1], Xiaohu Tang[1*], Martin Doppelbauer[2]
[1] MERCEDES-BENZ AG; [2] KARLSRUHE INSTITUTE OF TECHNOLOGY
[1] Mercedes Str. 130/6; [2] Engelbert-Arnold-Str. 5
[1] Stuttgart, Germany; [2] Karlsruhe, Germany
[1] +49 176-309-67150, [*] +49 176-257-43027; [2] +49 721-608-42473
[1] narciso_genovese.marmolejo@daimler.com, [*] st158318@stud.uni-stuttgart.de;
[2] martin.doppelbauer@kit.edu

Keywords

«Permanent magnet motor», «Conduction losses», «Pulse Width Modulation (PWM)», «Automotive application», «Converter machine interactions», «Variable speed drive».

Abstract

This paper compares the effects of different PWM methods and different switching frequencies on the eddy current loss in the magnets of a dual three-phase PMSM. Results of the 3D-FEM simulation display a significant difference in eddy current loss caused by different PWM methods for high-torque and high-speed operating points under the continuous torque boundary.

Introduction

Permanent magnet synchronous machines (PMSMs) have received much attention in the last decades in academia and in industry, especially in the automotive sector, for their high power density, wide operating range, and comparatively less rotor losses [1]. Although PMSM operating principles result in low rotor losses, the rotor is the thermally limiting component [2] in low-salience interior permanent magnet and surface permanent magnet machines [1]. Authors in [3] found that magnet losses increase substantially in machines with distributed windings upon exciting the machine with an inverter, so that they become a considerable portion of the rotor losses. The main reason for said increase in losses is the additional speed-dependent harmonics centred about integer multiples of the switching frequency. Said harmonics are generated by voltage-sourced pulse-width modulation (PWM) machine excitation, and are typically referred as called carrier groups [4]. Because magnet losses also heavily depend on high-frequency rotor magnetic flux components [3], it is shown that the different PWM exciting spectra induce different magnet eddy current losses at different operating points. This paper also clarifies the causal chain from exciting spectra to magnet eddy current losses. Lastly, the best PWM method is selected for each operating point according to the induced magnet losses.

Simulation Method

For this study, a commercially available finite element analysis (FEA) software is used. The switching action of the inverter is coupled to the machine for a wide operating range. This, in turn, necessitates defining the operating points in terms of current combinations. Maximum torque per Ampère (MTPA) is used to generate said current combinations for each desired torque-speed operating point. Figure 1 shows the overall simulation workflow. In order to reduce the influence of the controller on the results, torque and current values were verified with open-loop voltage sources. The FEM model is then fed with those d- and q-voltages after transforming them into the stator frame, applying the PWM method, and generating the duty cycles. Only operating points at the lowest speeds use feedback for stability; all other investigated points use open-loop control to avoid that the controller damp current harmonics due to 30-degree offset dual three-phase machines and bias the results. Although space-vector decomposition is not used for that purpose nor to control the machine, a controller-plant system with high bandwidth may also damp those current harmonics.

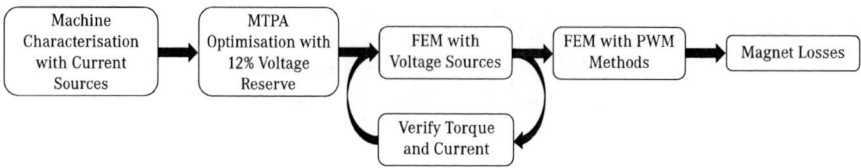

Fig. 1: Simulation workflow

In order to model magnet losses accurately, the machine is simulated in three dimensions. Simulating otherwise would risk over-estimating the magnet losses, as seen in [5] for concentrated windings. Furthermore, the time-step size and mesh size have an influence on the result, as shown in Figure 2. Figure 2 shows the magnet loss convergence as a function of mesh elements and step size. The step size is represented as the number of steps per carrier signal period. Due to the results in Figure 2, 40 steps per carrier signal period and 40 thousand elements are chosen for the simulation.

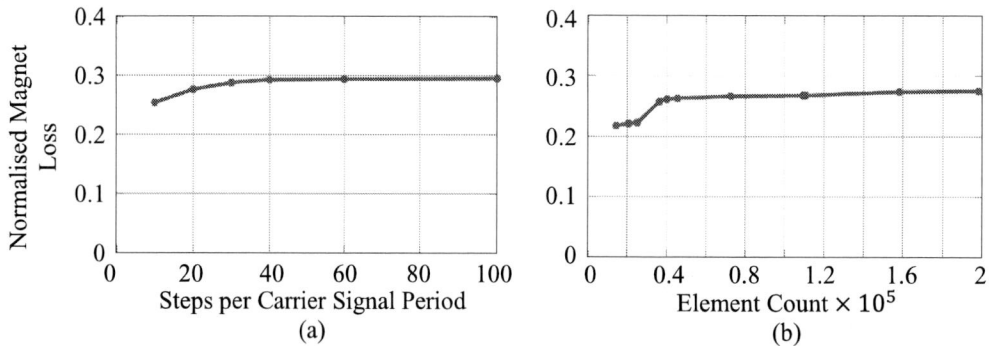

Fig. 2: Magnet loss convergence as a function of simulation step-size (a) and mesh element count at 20 steps per carrier signal period (b).

The utilised software calculates the mean magnet losses as eddy current losses according to equation (1), where T is one electrical period, σ is the permanent magnet conductivity, n is the number of symmetries modelled, t is time, $\boldsymbol{J}$ is the 3D current density vector, and dv is a volume infinitesimal. The machine is segmented and only one-half segment of one pole is simulated with appropriate boundary conditions to spare computation time [3] [5]. There are eight segments and four pole-pairs in total. We assume throughout the paper that one may neglect hysteresis losses in the magnet and that eddy currents in the steel laminations are negligible. As an additional note, all values are normalised.

$$P_{\text{pm}} = \frac{n}{T} \int_0^T \left(\frac{1}{\sigma} \int \|\boldsymbol{J}(t)\|^2 dv \right) dt \tag{1}$$

Dual Three-Phase Machine

A 30-electrical degree dual three-phase machine with distributed windings is simulated in this study. In order to reduce the amount of operating points simulated due to the small time-steps and 3D FEA calculation required, some representative operating points (OP's) are selected both on the constant torque region, i.e. below the base speed, and on the field-weakening region and close to the points most used in a worldwide harmonized light-duty vehicles test cycles (WLTC). Figure 3 (a) shows the peak torque line, continuous torque line, field weakening bound, and the investigated operating points. Figure 3 (b) depicts a cross-section of the motor with labelled windings and permanent magnet magnetisation directions. The dotted lines bound one pole. The tangential symmetry is anti-symmetric, meaning that the phase designation after $V2+$ in the counter-clockwise direction is $U1-$ and the magnetisation points toward the centre of the rotor.

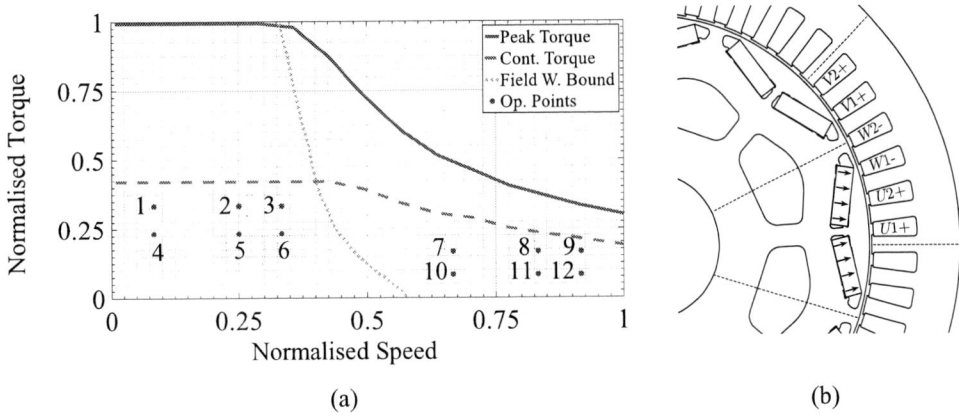

Fig. 3: (a) Simulated operating points (OP's) based on the WLTC cycle and continuous operation, the peak torque line, and field weakening bound. Values are arbitrarily normalised. Operating points are numbered for later reference. (b) Motor cross-section with dotted boundaries for one pole.

Review of PWM Methods

Five pulse-width modulation (PWM) methods are investigated, each distinguished by a different reference signal: sinusoidal (SPWM), space-vector (SVPWM), discontinuous PWM 1 (DPWM1), DPWM2, and DPWM3. The procedure to generate the discontinuous PWM methods uses one parameter, δ, as described in [6] [7] and repeated in equations (2) and (3). The index, i, is used as a subscript to the phase voltage, v, to indicate the three phases, meaning that $i = \{U, V, W\}$. The variables with a star superscript, v_i^*, indicate the voltages commanded from a controller. Since we analysed a permanent magnet machine, θ signifies the synchronous electrical rotor angle with the corresponding voltage vector offset such that $v_U^*(t = 0) = \|v_U^*\| \cos\theta = \|v_U^*\|$, where $\|v_U^*\|$ is the voltage amplitude of phase U. Lastly, v_{DC} represents the DC-bus voltage.

Table I summarizes the correspondence between discontinuous PWM methods and the δ-parameter. Furthermore, since DPWM methods switch, on average, less per electrical period than continuous methods, the effect of PWM methods on magnet losses under two conditions is investigated first: same carrier signal frequency and same electrical-period average switching rate. In the latter case, DPWM methods must have a carrier frequency 1.5 times higher than that of continuous methods, i.e. SPWM and SVPWM [8].

$$v_i = v_i^* + \frac{1}{2}(1 - 2\alpha)v_{DC} - \alpha \min(v_U^*, v_V^*, v_W^*) + (\alpha - 1)\max(v_U^*, v_V^*, v_W^*) \tag{2}$$

$$\alpha = \frac{1}{2}(1 - \text{sgn}(\cos(3(\theta + \delta)))) \tag{3}$$

Table I: Different PWM Methods and Corresponding δ

	DPWM1	DPWM2	DPWM3	SVPWM
δ	0	$-\pi/6$	$-\pi/3$	$\pi/6 - \theta$

Figure 4 shows the fundamental commanded voltage, the reference voltage, and the phase voltage fed into the machine for OP 10 at 9 kHz switching frequency. The duty cycles not shown would confirm module state 'clamping' when the reference voltage has a flat-top and make the 2/3 switching factor more visible.

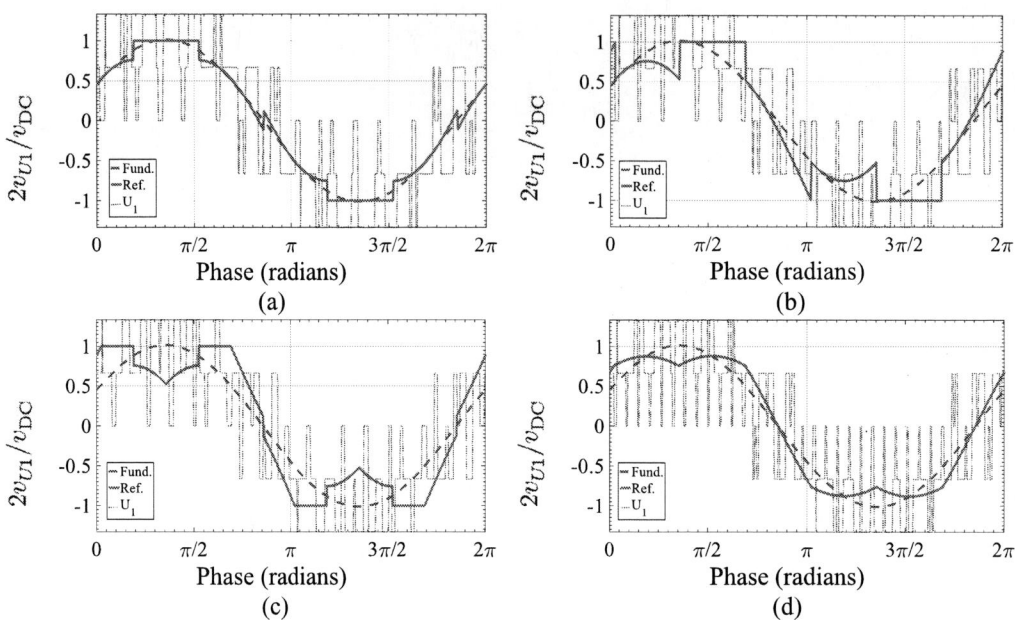

Fig. 4: PWM fundamental, reference, and phase voltage waveforms for phase U_1 (M=1.016, 9 kHz, OP10). (a) DPWM1. (b) DPWM2. (c) DPWM3. (d) SVPWM.

Results and Discussion

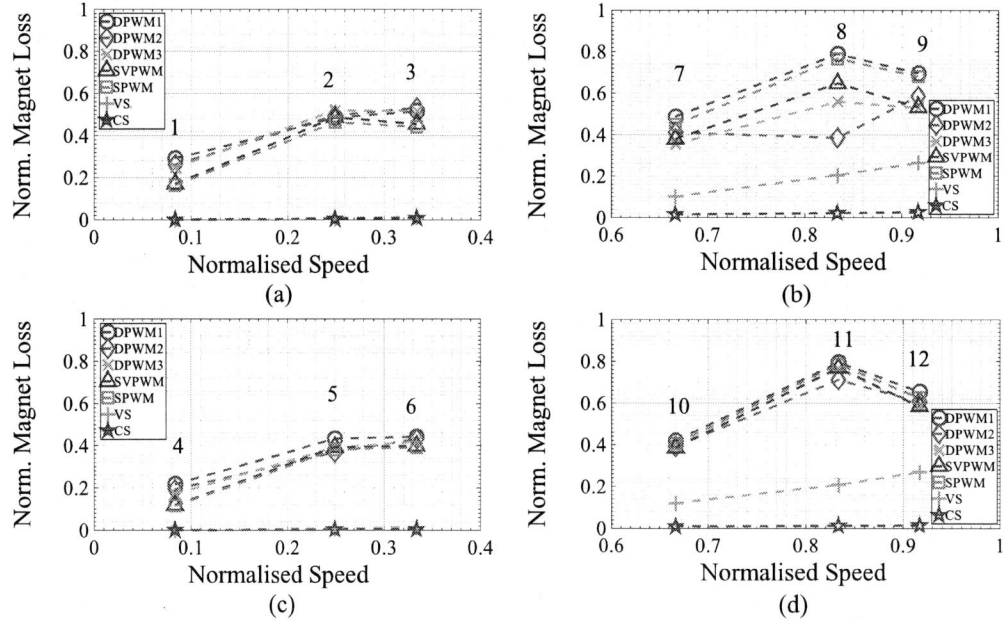

Fig. 5: Magnet losses with the same carrier frequency across all PWM methods and sinusoidal voltage and current sources. (a) OP 1, 2, and 3; (b) OP 7, 8, and 9; (c) OP 4, 5, and 6; and (d) OP 10, 11, and 12.

Figure 5 shows the magnet losses for the same carrier frequency and Figure 6 shows the magnet losses for the same average switching rate over one electrical period. Both plots also include the magnet losses with sinusoidal current and voltage source excitation as a baseline. A slight upward trend with speed is also recognisable, especially with the voltage source. This is mainly due to Faraday's law, which establishes the proportionality between frequency and the magnetic field component at said frequency to the curl of the current density field. The stator teeth harmonics as seen from the rotor contribute the most. From the voltage baseline, Figure 5 suggest that at least 40% of the losses occur due to inverter excitation. As seen in Figure 5 (a) and (c), magnet losses at operating points below the base speed hardly differ. When they do, however, SVPWM and SPWM yield the lowest losses.

The case for operating points in the field-weakening region is mixed. Figure 5 (b) shows that SPWM and DPWM1 excite the most losses, and DPWM2 excites the least losses for OP 8, and does not excite many more losses compared to SVPWM at other speeds. Figure 5 (d) shows a similar situation, albeit with less variance between the losses caused by different PWM methods and more losses caused by DPWM2, DPWM3, and SVPWM.

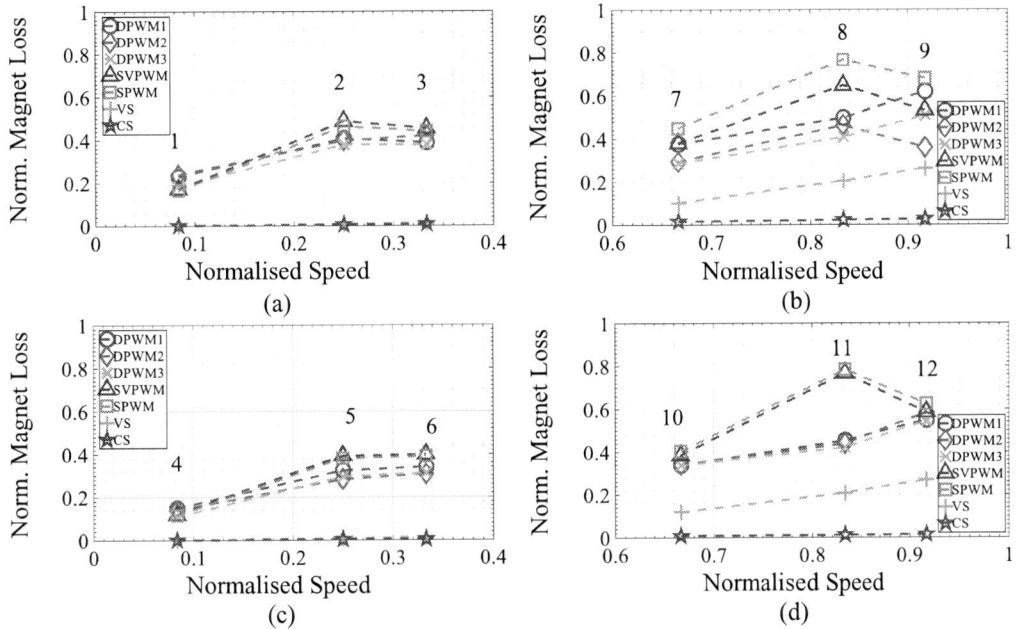

Fig. 6: Magnet losses with the same average switching frequency over one electrical period across all PWM methods and sinusoidal voltage and current sources. (a) OP 1, 2, and 3; (b) OP 7, 8, and 9; (c) OP 4, 5, and 6; and (d) OP 10, 11, and 12.

Like in Figure 5, the losses shown in Figure 6 (a) and (c) for the same average switching rate below the base speed hardly differ; however, when they do, DPWM3 yields the lowest magnet losses for operating points 2, 3, and 4. In the field-weakening region, the results again are more mixed, but only when comparing different DPWM methods. Figure 6 shows that, except for OP 9, all DPWM methods yield less than or equal to the same losses as continuous methods. The overall fewer losses induced by DPWM methods with the same average switching frequency may be attributed to the fact that these methods were driven with a higher carrier frequency. It is known that a higher carrier frequency results in less magnetic losses [9]. This is because, whereas the power loss increases proportionally to the square-root of the carrier frequency for constant space-vector currents [10], their carrier component magnitudes are proportional to ω^{-1}, where ω is the frequency of the current. Given that the power loss is proportional to the square of the exciting currents, then the power loss is approximately proportional to $\omega^{-3/2}$. To summarise, comparing Figure 5 with Figure 6, one may posit that the magnet losses tend to decrease when the carrier frequency increases, as expected, except for one operating point at OP 8 in Figure 5.

The anomaly at OP 8, Figure 5 motivates researching the effects of the PWM methods in the field-weakening region for large torques. Figure 7 shows the results of the investigation in said region –OP 7, 8, and 9 –and points at the same speeds with lower torque –OP 10, 11, and 12. The losses for the low torque points are displayed on the left column and those for larger torques on the right column. The points in the low torque column behave like their counterparts below the base speed: the PWM method does not influence magnet losses. Nevertheless, those low-torque operating points show a travelling loss peak with increasing speed and switching frequency. When the switching frequency creates a first carrier group 3rd subharmonic that coincides with the 12th harmonic, the loss peak occurs at that switching frequency. For higher torques, however, PWM methods seem to amplify, flatten, or invert the loss peak. OP 8, Figure 7 (d), for example, shows a travelling power loss trough corresponding to OP 8 in Figure 5 (b) for DWPM2.

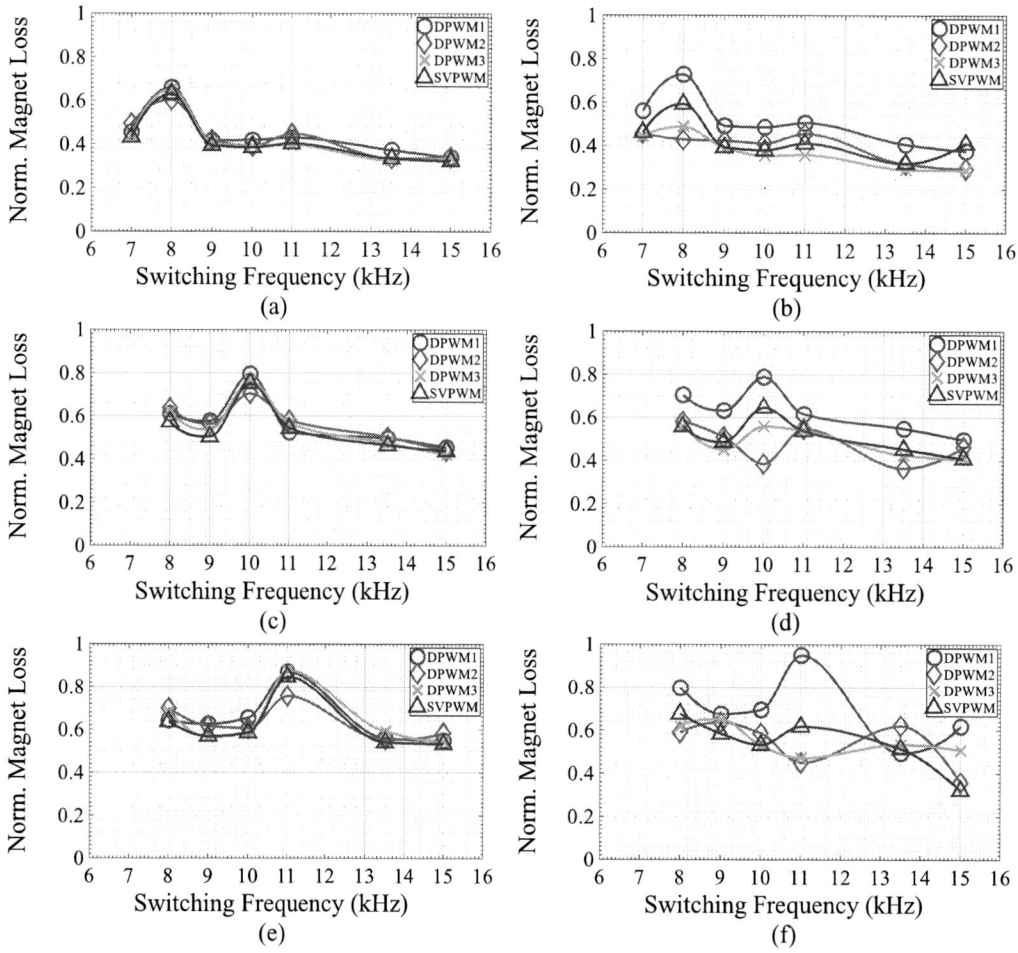

Fig. 7: Normalised magnet loss as a function of switching frequency for different PWM. (a) OP 10. (b) OP 7. (c) OP 11. (d) OP 8. (e) OP 12. (f) OP 9.

At said operating point 12th harmonic eddy current density component is more significant in the field-weakening region. Because inverter harmonics approach the vicinity of the stator teeth harmonic in this region, clarifying the interaction between the harmonics from the two sources could explain why some PWM methods result in different magnet eddy current losses. Proceeding with the clarification, it is important to note that in the stator frame the harmonics from the first carrier group are $f_s \pm 2nf_e$, according to [4], where n is a positive integer, f_s is the switching frequency, and f_e is the synchronous electrical frequency. In Figure 9 (a), $f_s/f_e = 15$. In the rotor frame, the frequencies from the first carrier group shift to $f_s \pm (2n-1)f_e$, which includes the 3rd harmonic from the first carrier group as seen in Figures 9 (c), 9 (e) and 9 (g). For example, in Figure 9 (c), the 11th, 13th, 17th, and 19th resolve to the 12th and 18th harmonics in Figure 9 (e) and 9 (g). The 5th and 7th harmonics, labelled as 'PM Flux' in Figure 9 (c) and 9 (d), are resolved to the 6th harmonic in Figures 9 (e) and 9 (f). The fact that the 'PM Flux' component only appears in d-axis current plots suggests that said current influences the permanent magnet flux the most, as is expected from field weakening control, for example.

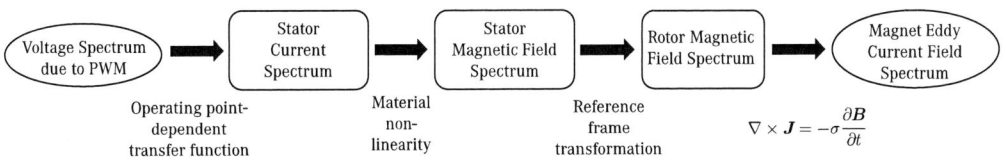

Fig. 8: Simplified process chain of magnet eddy current creation due to inverter feeding.

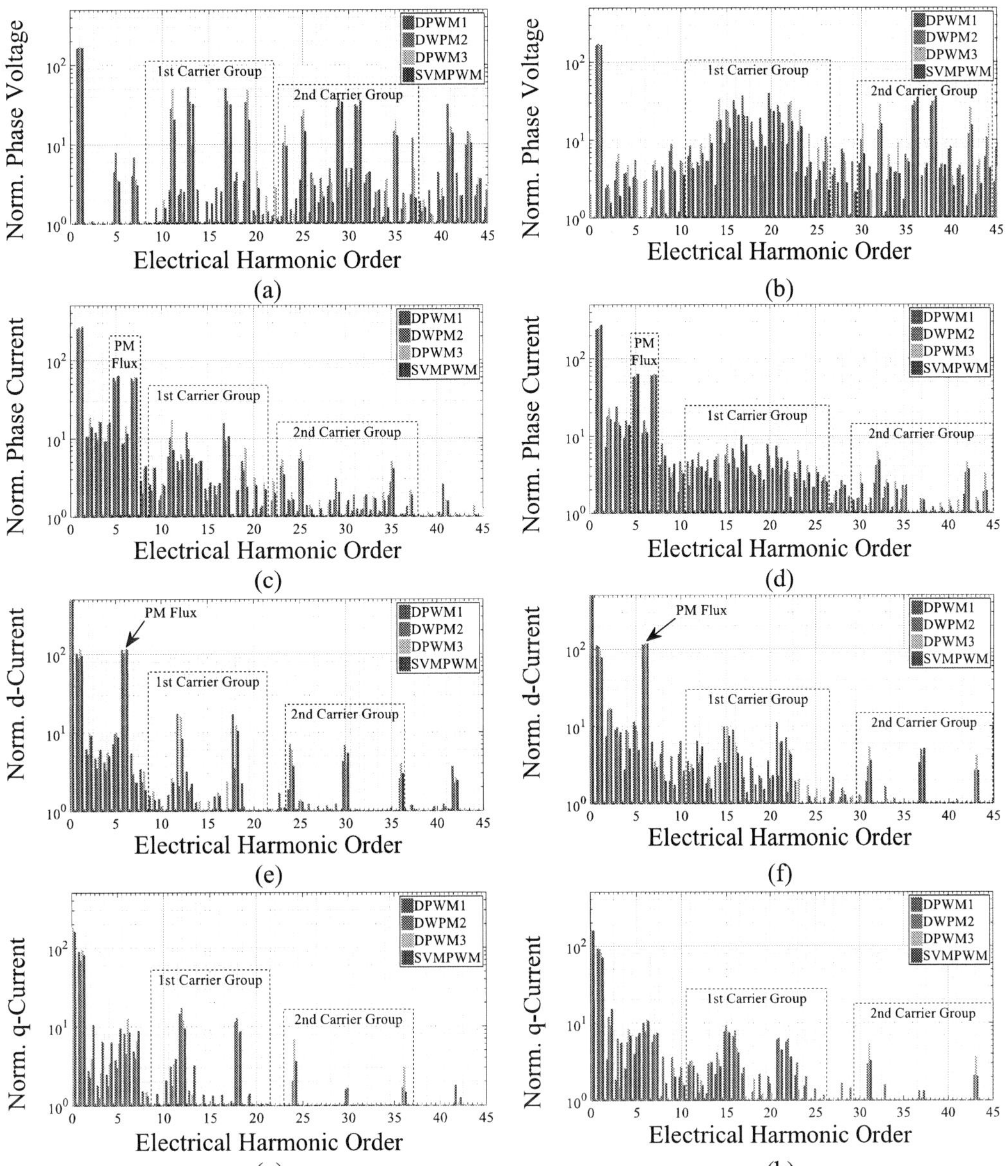

(a)

(b)

(c)

(d)

(e)

(f)

(g)

(h)

Fig. 9: Frequency response of phase U_1 voltage, phase U_1 current, i_{d1} and i_{q1} for various PWM methods at OP 9. The first column, (a), (c), (e), and (g), has 11 kHz excitation and the second column, (b), (d), (f), and (h), has 13.5 kHz excitation.

What is important to note from Figures 9 (e), (f), (g), and (h) is that the 12th harmonic in the d-axis current differs the most among PWM methods when the 3rd subharmonic from the first carrier group coincides with the 12th harmonic from the stator teeth. The d-axis current correlates the most with the axial current density and joule loss density at the point of maximum joule loss density in the magnet, shown in Figures 10 (c), (d), (e), and (f). The magnitude of the magnetic field density at the point of maximum joule loss density, shown in Figures 10 (a) and 10 (b), displays no significant difference between the methods at the 12th and 24th harmonic, suggesting that they represent the stator teeth harmonics only –furthermore, the DC component is not shown. Combining the observations from the d-axis current plots and the magnetic field density plots, one may state the following: the eddy current response in the magnet due to the 3rd subharmonic from the first carrier group overlaps with the eddy current response in the magnet due to the 12th harmonic from the stator teeth.

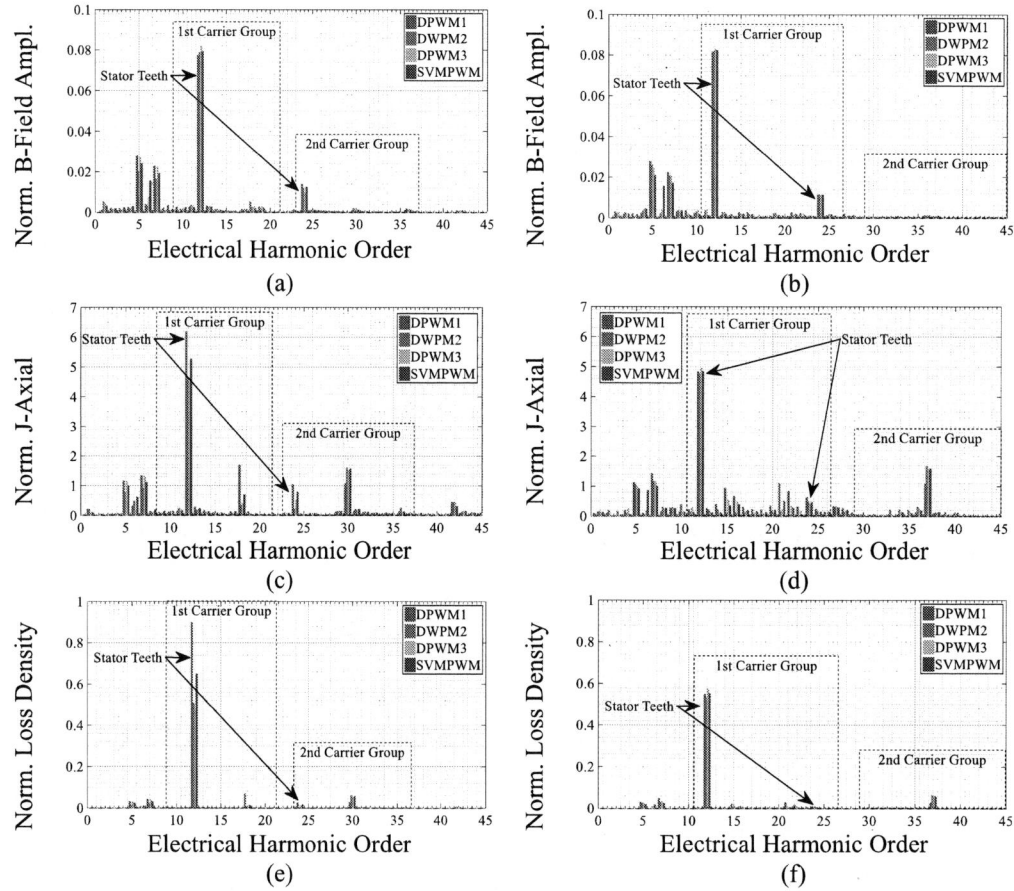

Fig. 10: Frequency response of the magnetic flux density amplitude, axial current density, and power loss density at the maximum joule loss density point for various PWM methods at OP 9. The first column, (a), (c), and (e) has 11 kHz excitation and the second column, (b), (d), and (f) has 13.5 kHz excitation.

As seen in Figure 10, no significant difference between PWM methods appear on the magnetic field density, although they do differ in the axial current density. The reason is that current values take into account the gradient of the scalar potential, which, in the windings, actually 'drives' the field equations, whereas the magnetic field density is the curl of the vector potential only. Equation (4) shows the Maxwell field equations that the FEM software solves [3], equation (5) is the circuit-coupling equation [3], equation (6) states the definition of the magnetic field density, B, as a function of the vector potential, A, and equation (7) states the definition of the current density, J, in in terms of the vector and scalar, ϕ, potentials. Due to the superposition of derivatives of those two quantities in equation (7), the interaction between them –and their corresponding harmonics –becomes more transparent. M represents the permanent magnet magnetisation, μ the magnetic permeability, σ the conductivity of the permanent magnet, Φ the stator winding flux linkage, i the stator winding current, R the stator winding resistance, and v the stator winding voltage. The subscript w denotes a winding quantity or a restriction thereto.

$$\nabla \times \left(\frac{1}{\mu}\nabla \times A\right) = J_{\mathrm{w}} - \sigma\left(\frac{\partial A}{\partial t} + \nabla\phi\right) + \frac{1}{\mu_0}\nabla \times M \tag{4}$$

$$v_{\mathrm{w}} = \nabla\phi|_{\mathrm{w}} = \frac{d\Phi_{\mathrm{w}}}{dt} + R_{\mathrm{w}}i_{\mathrm{w}} \tag{5}$$

$$B = \nabla \times A \tag{6}$$

$$J = -\sigma\left(\frac{\partial A}{\partial t} + \nabla\phi\right) \tag{7}$$

To summarise, the interaction of the 3rd subharmonic from the first carrier group with the 12th harmonic from the stator teeth can reduce the net amplitude of the rotor field at the 12th harmonic, reducing overall magnetic losses. This is independent of the overall magnetic loss reduction due to increasing the switching frequency. As shown in Figure 7, the resonances may also increase their intensity at higher speeds, which, depending on the choice of switching frequency, may unknowingly increase the magnet losses, thereby increasing the temperature and likelihood of demagnetisation. Such a situation may be especially plausible in a traction application when the vehicle accelerates on a highway to pass another vehicle, since such a manoeuvre requires high speed and torque. Figure 8 displays a simplified process chain from the exciting voltage to the magnet eddy current spectrum as detailed in Figures 9 and 10. As an aside, although Figures 9 and 10 show the detailed process chain for two configurations of one operating point, the analysis illustrates what happens at other points as well. Lastly, Figure 11 shows the simulated half-segment with the resulting magnet eddy currents for the 1st stator tooth harmonic and the joule loss contours. Figure 11 shows that the maximum current density and joule loss density occurs at the corner nearest to the air-gap in the direction of rotation and near the centre of the magnet segment. Figure 11 (a) also indicates where the tangential and segment symmetries are located, along with the insulation condition. Figure 11 (b) contains two views of the magnet: a top view above the thick arrow and a cross-section view below it.

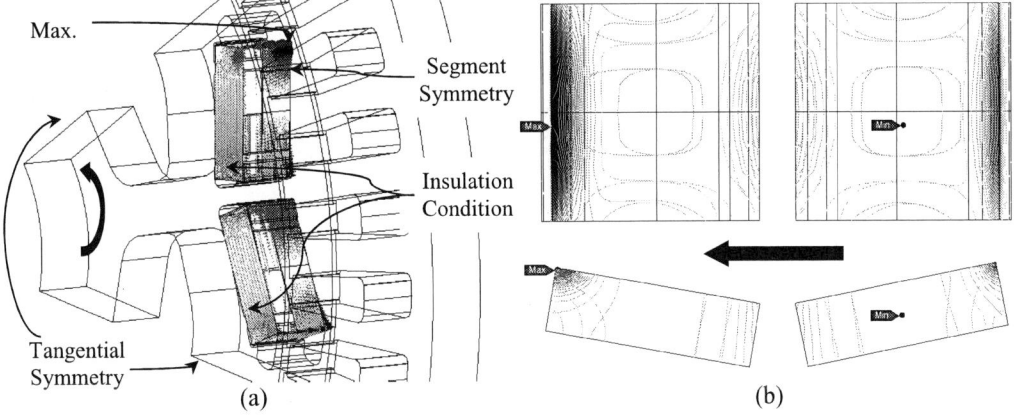

Fig. 11: 3D FEM field results for SVPWM, 11 kHz excitation at OP 9, 1st stator tooth harmonic. The thick arrows indicate the direction of rotation. (a) FEM model of a half-segment with eddy currents. (b) Joule loss density contours with maximum and minimum joule loss density points.

Conclusion

In this paper, the reduction of magnet eddy current losses with increasing carrier frequency is confirmed. Under the requirement of equivalent average switching frequency, DPWM3 is recommended for the reduction of magnet losses except at extremely high and low speeds. For the same carrier frequency, the conventional SVPWM is recommended except at the anomalous point. Furthermore, it is found that for an operating point at high speed and torque, the inverter harmonics interact differently with different PWM methods with the 12th harmonic from the stator teeth. Combined with the knowledge that, below the base speed, carrier harmonics dominate the magnet losses and above the base speed a combination of carrier harmonics and stator teeth harmonics dominate, one may design a scheme – design parameters and other inverter parameters aside –that minimizes magnet losses over the entire operating range, or at least avoids the loss peaks. Such a scheme would, for example, increase the carrier frequency below the base speed and vary the carrier frequency above the base speed to avoid magnet loss peaks. Increasing the carrier frequency to a point where stator teeth harmonics at maximum speed do not interact with inverter subharmonics is also a strategy to avoid loss peaks. Lastly, additional segmentation could further damp the loss peaks.

References

[1] G. Pellegrino, A. Vagati, B. B. and P. Guglielmi, "Comparison of Induction and PM Synchronous Motor Drives for EV Application Including Design Examples," *IEEE Transactions on Industry Applications,* vol. 48, no. 6, pp. 2322-2332, November/December 2012.

[2] J. Lange, S. Wachter, T. Engelhardt, S. Oechslen and A. Heitmann, "Highly Integrated Electric Axle," in *E-MOTIVE Expert Forum Electric Vehicle Drives,* Schweinfurt, Germany, 2019.

[3] K. Yamazaki and A. Abe, "Loss Investigation of Interior Permanent-Magnet Motors Considering Carrier Harmonics and Magnet Eddy Currents," *IEEE Transactions on Industry Applications,* vol. 45, no. 2, pp. 659-665, March/April 2009.

[4] D. Holmes and T. Lipo, Pulse Width Modulation for Power Converters: Principles and Practice, Piscataway, NJ: IEEE Press, 2003.

[5] K. Yamazaki, Y. Kanou, Y. Fukushima, S. Ohki, A. Nezu, T. Ikemi and R. Mizokami, "Reduction of Magnet Eddy-Current Loss in Interior Permanent-Magnet Motors With Concentrated Windings," *IEEE Transactions on Industry Applications,* vol. 46, no. 6, pp. 2434-2441, November/December 2010.

[6] O. Ojo, "The Generalized Discontinuous PWM Scheme for Three-Phase Voltage Source Inverters," *IEEE Transactions on Industrial Electronics,* vol. 51, no. 6, pp. 1280-1289, December 2004.

[7] V. Blasko, "Analysis of a Hybrid PWM Based on Modified Space-Vector and Triangle-Comparison Methods," *IEEE Transactions on Industry Applications,* vol. 33, no. 3, pp. 756-764, 1997.

[8] J. Kolar, H. Ertl and F. Zach, "Influence of the Modulation Method on the Conduction and Switching Losses of a PWM Converter System," *IEEE Transactions on Industry Applications,* vol. 27, no. 6, pp. 1063-1075, 1991.

[9] M. van der Geest, H. Plinder and J. Ferreira, "Influence of PWM switching frequency on the losses in PM machines," in *2014 International Conference on Electrical Machines (ICEM),* Berlin, Germany, 2014.

[10] S. Mukerji, M. George, M. Ramamurthy and K. Asaduzzaman, "Eddy Currents in Solid Rectangular Cores," *Progress In Electromagnetics Research B,* vol. 7, pp. 117-131, 2008.

Resonant DC/DC converter with Class Φ_2 inverter and Class DE rectifier based on GaN HEMT

Cai Si-yuan, HE Jun-ping, Li Zi-fan
Harbin Institute of Technology (Shenzhen)
Shenzhen, 518055, P.R. China
E-Mail: hejunping@hit.edu.cn

Acknowledge

This work is supported by Basic and Application Basic Research Fund of Guangdong Prov. P. R. China under Grant No. 2020A1515010913.

Keywords

«GaN switch», «Very high frequency», «DC/DC converter», «ON/OFF control»

Abstract

A resonant DC/DC converter with Class Φ_2 inverter and Class DE rectifier based on GaN HEMT is designed and validated. A PCB plane transformer is designed to realize the isolation of the primary side and the secondary side of the converter. The output voltage is regulated by hysteresis ON/OFF control. The operation principle and design procedure are introduced in detail.

Introduction

With the rapidly development of information and semiconductor technology, higher performance requirements are been proposed for energy management todays. High efficiency, high switching frequency and high power density have become the development trend of isolated DC/DC converter [1, 2]. Therefore, the high-efficiency converter is no doubt a promising and challenging research field that fits for GaN. With the increase of frequency, the switching frequency reaches above MHz, the loss of the power switch will be larger. Resonant topology is usually adapted for a very high frequency (VHF) converter when considering VHF working conditions [3-5]. Class DE resonant inverter topology has a simple structure, but it requires a compatible drive circuit design [6]. The voltage stress of traditional Class E resonant VHF converter is larger and its closed-loop regulation of output voltage is complex to design [7,8]. The Class Φ_2 resonant converter is proposed to decrease the voltage stress [9-11]. Galvanic isolation is generally requested for many power supplies. Although an air-core planar transformer can meet the demand of VHF converter, its efficiency is low and area is large [12,13]. Selecting a suitable magnetic core, such as Mn-Zn or Ni-Zn materials, can also meet the requirements of high frequency converter [14].

The main objective of this paper is to design a new type isolated DC/DC converter. This converter mainly consists of Class Φ_2 inverter stage and Class DE rectifier stage. The principles of Class Φ_2 converter and Class DE converter are introduced firstly. Then, the PCB planar transformer is designed. In addition, the hysteresis ON/OFF control is adopted to regulate the output voltage of this converter. Finally, a 4-MHz converter experiment prototype with 24V input, 5V/2A output is built up. The characteristics of ZVS of GaN HEMT, topological structure of Class Φ_2-DE and ON/OFF control are verified functionally by experimental results.

Design of power-stage of VHF DC/DC converter

The main power topology of the novel VHF DC-DC converter proposed in this paper is shown in Fig. 1, it is made up of a Class Φ_2 inverter and a Class DE rectifier. The design procedure of this converter can be described in two phases: The first stage is to complete the Class DE rectifier with an ideal sinusoidal input voltage source to take the place of its input source, the load of Class DE rectifier is simplified as a constant voltage source to get the same circuit characteristic of the steady state of this VHF converter; the second stage is to design the Class Φ_2 inverter with the assumptions that the resonant current of the output to the inverter is in sinusoidal waveforms without any other harmonic component, the parameter of this current source is set according to the input voltage source of Class DE rectifier mentioned above.

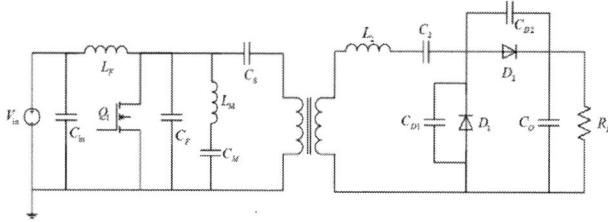

Fig.1 Topology of converter

A) Design of Class Φ2 inverter-stage

Fig. 2 shows the topology of Class Φ_2 inverter, which consist of the main power switch Q_1, input resonant inductor L_F and drain-source resonant capacitor C_F. C_s is DC isolation capacitance. The equivalent load of secondary is simplified as a resistor R_L, the single tuned filter network is composed by L_M and C_M, which filters out the second harmonics of the drain-source voltage of the switch Q_1. The working principle of Class Φ_2 inverter is similar to Class E inverter analyzed [9], the parameter design considerations are derived in the way as follows. The design considerations of Class Φ_2 inverter must meet the following three design criteria: (1) The main power switch Q_1 turns on with soft-switching condition. (2) The drain-source voltage of Q_1 is shaped into a quasi-trapezoidal wave to restrain the voltage stress across switches. (3) The output of the inverter meeting the output requirements of the desired power rating.

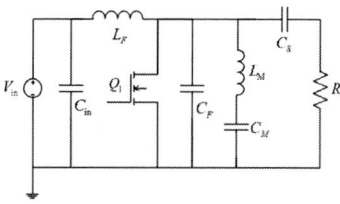
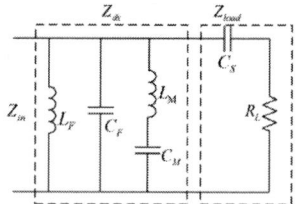

Fig.2 Topology of Class Φ_2 inverter-stage   Fig.3 Drain-source impendence of the switch

The obvious advantage of the Class Φ_2 inverter topology is the restrained voltage stress across the main power switch, its main contribution is to board the application of the high-frequency resonant converter in high input voltage fields, the single tuned filtering network resonant at the double switching frequency, so the second harmonic of the drain to source voltage V_{ds} can be filtered, the drain-source impendence is shown as Fig.3. The design procedure of initial value of the components in Class Φ_2 is follow the steps derived in [9], the impedance of the drain-source is focused and limited to the desired shaping so that the propose to shaping the voltage can be reached, according to the given impedance in equation (1), the resonant peak and valley could be determined, the initial parameter except C_F could be determined in this way. The result is shown in equation (2) ~ (4). The value of C_F depends on engineering experience.   When the V_{ds} is shaped with single-tuned filtering network, it can be seen as the sum of fundamental and third harmonics, which reduce the voltage stress of V_{ds} diminutions from 4 times V_{in} in Class E inverter to 2.5 times V_{in} in Class Φ_2 inverter, V_{ds} in Class Φ_2 inverter is shown as Fig.4.

$$Z_{ds} = \frac{j\omega\left(1-\omega_S^2 L_M C_M\right)}{\omega_S^4 L_F C_F L_M C_M - \omega^2\left(L_F C_F + L_M C_M + L_F C_M\right)+1} \tag{1}$$

$$\omega_S^4 L_F C_F L_M C_M - \omega^2 \left(L_F C_F + L_M C_M + L_F C_M \right) + 1 = 0 \tag{2}$$

$$(3\omega_S)^4 L_F C_F L_M C_M - (3\omega_S)^2 \left(L_F C_F + L_M C_M + L_F C_M \right) + 1 = 0 \tag{3}$$

$$(2\omega_S)^2 L_M C_M - 1 = 0 \tag{4}$$

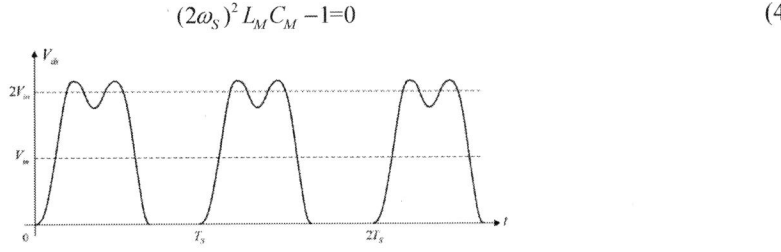

Fig.4 Drain-source waveforms of the main power switch in Class Φ_2 inverter

B) Design of Class DE rectifier-stage

The Class DE rectifier consists of diodes D_1 and D_2, shut capacitors C_{D1} and C_{D2}, a filter capacitance C_f, and load resistance R_L. The diodes work as half-wave voltage rectifiers and the rectified voltages ate converted into a dc voltage through the filter capacitance. At the turn-off transition of the diodes, both the diode voltages v_{D1}, v_{D2} and their slopes of $dv_{D1}/d\theta$, $dv_{D2}/d\theta$ are zero. Therefore, the class DE rectifier can also achieve the high power conversion efficiency at high frequency. The rectifier stage is designed using the model shown in Fig.5. The design detail of the Class DE rectifier was shown in [15,16]. The maximum power occurs at 0.25 of duty ratio, and the diode shunt capacitances C_{D1}, C_{D2} are expressed as:

$$C_{D1} = C_{D2} = \frac{2\pi}{\omega R_L} \tag{5}$$

The Class DE rectifier can be approximately equivalent to the input capacitance C_{in} and the input resistance R_{in} in series. The input capacitance C_{in} and the input resistance R_{in} are expressed as.

$$C_{in} = 2 \left(C_{D1} + C_{D2} \right) \tag{6}$$

$$R_{in} = \frac{V_{Rin}}{I_m} = \frac{1}{\pi \omega \left(C_{D1} + C_{D2} \right)} \tag{7}$$

For L_2 resonates with C_i, C_2 can be attained.

$$C_2 = \frac{C_i}{\omega^2 L_2 C_i - 1} \tag{8}$$

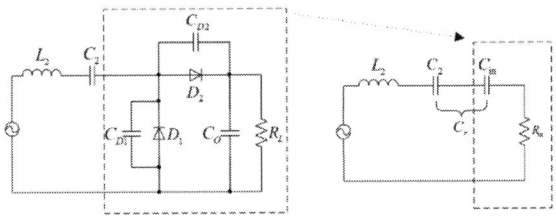

Fig.5 Class DE rectifier equivalent circuit model

C) Design of PCB planar transformer

Fig.6 shows the geometrical figure of a PCB planar transformer. For the PCB planar transformer, the design criteria include: (1) Suitable core material is used to meet the requirement of VHF converter. (2) The number of turns and winding parameters are reasonably designed for the purpose of miniaturization. (3) The leakage inductance should be as small as possible, for the purpose that it has little effect on the resonant network of the rectifier topology.

Appropriate magnetic core materials should be selected by the working frequency of the transformer, then the selection of magnetic materials mainly refers to its saturation magnetic flux density, core loss and other parameters. Soft ferrite cores are widely used in switching power supply due to its low saturation flux density and high resistivity. Mn-Zn ferrite is the most common soft magnet material. The 3F46 of FERROXCUBE has good permeability at 4MHz, so it is used as the core of PCB plane transformer in this paper.

According to the parameters of transformer core and topological design requirements, the parameters of transformer winding can be calculated. The geometric parameters of the transformer include the primary winding turns n_1 and the secondary winding turns n_2, PCB copper width w, copper thickness h, distance t between primary and secondary winding, inner radius R_i of winding.

$$n_1 = 4, n_2 = 2, w = 1.8mm, h = 0.07mm, t = 0.15mm, R_i = 3.4mm$$

Finite Element Method (FEM) is used to simulate the transformer, the leakage inductance parameter of transformer is obtained.

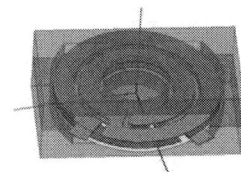

Fig.6 HFSS simulation model of PCB transformer

Hysteresis ON/OFF control

The closed-loop regulation for output voltage of DC/DC converter is achieved by hysteresis ON/OFF control, its some primary waveforms and block diagram are shown in Fig. 7 and Fig. 8. Control voltage represents the low frequency ON/OFF signal generated by comparator, and voltage V_{gs} represents the driving signal of the power switch. The ON/OFF signal holds high level when the output voltage of the DC/DC converter is lower than V_H. During this time, the converter is at the state of ON and the output voltage will increase to V_H. The ON/OFF signal will turn to low level when the output voltage of the converter rises up to V_H. After that, output capacitance C_o discharges by resistance R_L, as a result, the output voltage decreases and the stage ends until it reaches to lowest level V_L. The ON/OFF signal will turn to high level again when the output voltage lower than V_L. Not only can the output voltage of converter be controlled, but the ZVS of power semiconductor device can achieve by this control method.

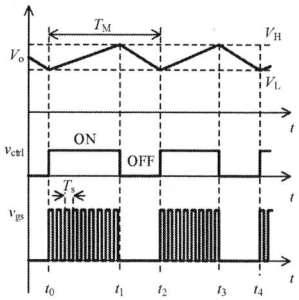

Fig.7 key waveforms of ON/OFF control

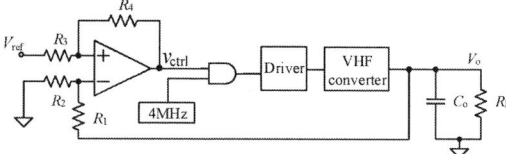

Fig.8 structure diagram of hysteresis ON/OFF control

Simulation Results and Analysis

According to the design method of Class Φ_2 inverter-stage, Class DE rectifier-stage and PCB planar transformer, the components value of inverter and rectifier topology are given as shown in equation (2)-(8). The simulation model of 4-MHz isolated resonant DC/DC converter is built up, and the simulation waveforms are shown in Fig. 9-Fig. 10.

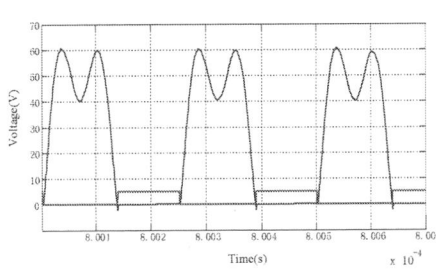

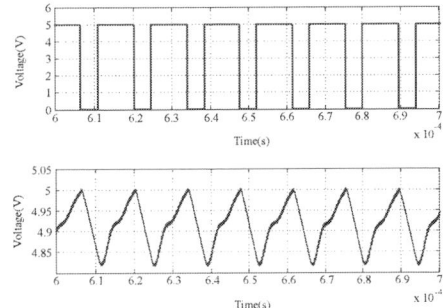

Fig.9 waveforms of drain-source and driving signal

Fig.10 the ON/OFF signal and voltage ripple

Fig.9 shows the waveforms of the drain-source of power switch and driving signal. Before the driving signal turns high level, the drain-source voltage has been zero. The peak voltage of drain-source is about 2.5 times as much as Vin. Fig.10 shows the waveforms of ON/OFF signal and output voltage ripple. The output voltage of DC/DC converter increases when ON/OFF signal is high level, and the ripple of output voltage is 200 mV which satisfy the design indices.

Experiment Result and Analysis

Based on the analytical study and simulation results, a 4-MHz isolated DC/DC converter experiment prototype with 24V input, 5V/2A output is built up to verify the simulation results as show in Fig.11. The GS61004B GaN transistors of GaN Systems is adopted in this design according to the requirements of the design specifications and topology on the pressure resistance, current resistance, switching frequency and heat dissipation condition.

(a) Front side of prototype

(b) Reverse side of prototype

Fig.11 Experimental prototype of the converter

The waveforms of drain-source voltage are shown in Fig.12, the results are in accordance with simulation waveforms. Fig.13 shows the waveform of ON/OFF signal and driving signal. The experiment results are in keeping with the principle of ON/OFF control analysis.

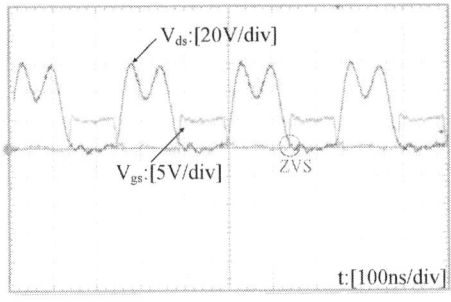
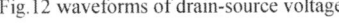

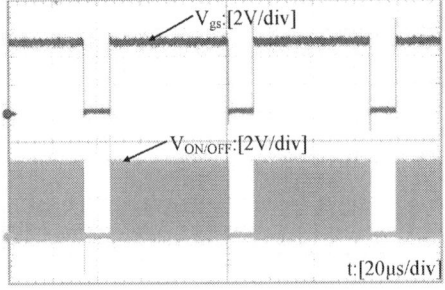

Fig.12 waveforms of drain-source voltage

Fig.13 waveforms of ON/OFF and driving signal

Fig.13 shows the experiment results of ON/OFF signal and output voltage ripple. Output voltage takes on an upward trend when the converter in the state of ON. Output voltage decreases when the converter in the state of OFF. Fig.14 is efficiency curve of the converter, the converter achieved 76.9% overall efficiency at 10W (2.5Ω) output power.

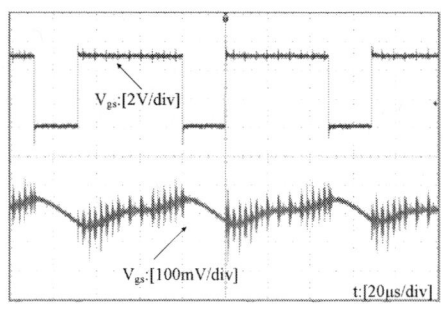
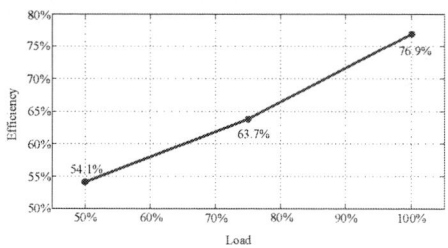

Fig.13 waveforms of ON/OFF and output voltage ripple

Fig.14 efficiency curve of the converter

Conclusion

A 4 MHz isolated resonant DC/DC converter composed of Class Φ_2 inverter and Class DE rectifier is designed and built up in this paper. This converter achieves electrical isolation through PCB planar transformer and its output voltage of converter is regulated by hysteresis ON/OFF control. Both the experiment results and simulated results matched with theoretical analyses well and the validated the feasibility of the VHF converter design. Improvements in efficiency and power density are the goals of the next phase of this project.

References

[1] Reusch D, Strydom J. Evaluation of Gallium Nitride Transistors in High Frequency Resonant and Soft-Switching DC–DC Converters[J]. IEEE Transactions on Power Electronics, 2015, 30(9): 5151-5158.

[2] Jones E A, Wang F, Costinett D. Review of Commercial GaN Power Devices and GaN-Based Converter Design Challenges [J]. IEEE Journal of Emerging & Selected Topics in Power Electronics, 2016, 4(3):707-719.

[3] Zhang Y, Rodriguez M, Maksimovic D. Very High Frequency PWM Buck Converters Using Monolithic GaN Half-Bridge Power Stages with Integrated Gate Drivers[J]. IEEE Transactions on Power Electronics, 2016, 31(11): 7926-7942.

[4] Huang X, Du W, Lee F C, et al. A novel driving scheme for synchronous rectifier in MHz CRM flyback converter with GaN devices[C]// Energy Conversion Congress & Exposition. IEEE, 2015.

[5] Cai W, Zhang Z, Ren X, et al. A 30-MHz isolated push-pull VHF resonant converter[C]// Applied Power Electronics Conference & Exposition. IEEE, 2014.

[6] Ezawa T , Sekiya H , Yahagi T . Design of class DE amplifier with nonlinear shunt capacitances for any output Q[C]// IEEE International Symposium on Circuits & Systems. IEEE, 2008.

[7] Andersen T M, Søren K. Christensen, Knott A, et al. A VHF Class E DC-DC Converter with Self-Oscillating Gate Driver[C]// Applied Power Electronics Conference & Exposition. IEEE Xplore,2011.

[8] Madsen M, Knott A, Andersen M A E . Low power very high frequency resonant converter with high   step   down ratio[C]// Africon. IEEE, 2013.

[9] J. M. Rivas, Y. Han, O. Leitermann, A. D. Sagneri and D. J. Perreault. A High-Frequency Resonant Inverter Topology With Low-Voltage Stress[J]. IEEE Transactions on Power Electronics, 2008, 23(4): 1759-1771.

[10] J. M. Rivas, O. Leitermann, Y. Han and D. J. Perreault. A Very High Frequency DC–DC Converter Based on a Class Φ2 Resonant Inverter[J]. IEEE Transactions on Power Electronics, 2011, 26(10): 2980-2992.

[11] Zou X, Zhang Z, Dong Z, et al. A 10-MHz eGaN FETs based isolated class-Φ2 DCX[C]// Proceedings of the IEEE Applied Power Electronics Conference and Exposition, 2016: 2518-2524.

[12] Sepahvand A, Zhang Y, Maksimovic D. 100 MHz isolated DC-DC resonant converter using spiral planar PCB transformer[C]// Control & Modeling for Power Electronics. IEEE, 2015.

[13] Mao S, Li C, Song T, et al. High frequency high voltage generation with air-core transformer[C]// IEEE International Workshop on Integrated Power Packaging. 2017.

[14] C. M. Arturi and A. Gandelli. High frequency models of PCB-based transformers. IEEE 2001 Midwest Symposium on Circuits and Systems. MWSCAS 2001, Dayton, OH, USA, 2001, pp.797-801 vol.2.

[15] K. Fukui and H. Koizumi. Analysis of Half-Wave Class DE Low dv/dt Rectifier at Any Duty Ratio[J]. IEEE Transactions on Power Electronics, 2014: 29(1), 234-245.

[16] K. Fukui and H. Koizumi. Half-wave Class DE Low dv/dt rectifier[C]// Proceedings of the IEEE Asia Pacific Conference on Circuits and Systems, Kaohsiung, 2012: 69-72.

Four-Level Inverter with Variable Voltage Levels for Hardware-in-the-Loop Emulation of Three-Phase Machines

Manuel Fischer, Johannes Ruthardt, Vasken Ketchedjian, Philipp Ziegler, Maximilian
Nitzsche, Jörg Roth-Stielow
INSTITUTE FOR POWER ELECTRONICS AND ELECTRICAL DRIVES
University of Stuttgart
Pfaffenwaldring 47
70569 Stuttgart, Germany
Tel.: +49 711 – 685.67388
E-Mail: manuel.fischer@ilea.uni-stuttgart.de
URL: http://www.ilea.uni-stuttgart.de

Keywords

«Converter Control», «Multilevel Converters», «Machine Emulation», «Electric Machine», «Voltage
Source Inverters (VSI)».

Abstract

In order to emulate the behavior of three-phase machines in a hardware-in-the-loop test bench, an
applicable power electronic device is required. For this purpose, an inverter concept with four variable
voltage levels per phase is presented. This concept precisely fulfills the requirements of three-phase
machine emulation. The presented inverter is theoretically able to adjust the ideal countervoltage course
which is needed to emulate three-phase machines' electrical behavior. In contrast, the countervoltages
of commonly used power electronic devices suffer from deviation caused by switched-mode operation
or limited bandwidth.
Within this paper the setup of the four-level inverter is presented. Besides, the paper discusses a control
strategy for operating the four-level inverter as a three-phase machine emulator.

Introduction

In general, drive trains consist of an inverter and an electric machine. In the development process of
drive train systems, testing is an important element to prove functionality, safety and reliability of the
system's components. For this purpose, a machine test bench is set up. The investigated machine is
coupled to a load machine to set several mechanical operating points. The device under test (DUT)
inverter feeds the investigated machine, as shown in Fig. 1. However, machine test benches are very
expensive and complex to set up. Moreover, they are not flexible in varying machine settings. An
alternative approach for testing the DUT inverter at variable machine settings is the use of power-
hardware-in-the-loop (PHIL) machine emulation. In this process, the DUT inverter is connected to a
machine emulator, which has the same electrical
characteristic at the output terminals as the
investigated machine [1-3], see Fig. 1. Two main
components are essential for a PHIL machine
emulator: a real-time machine model, which
calculates the machine's behavior as exactly as
possible, and a suitable power electronic device.
This power electronic device has to adjust
countervoltages in such a way, that the occurring
phase currents follow the same course as the
machine currents would. Especially, emulating the
slew rate of the current ripple is a challenging task.

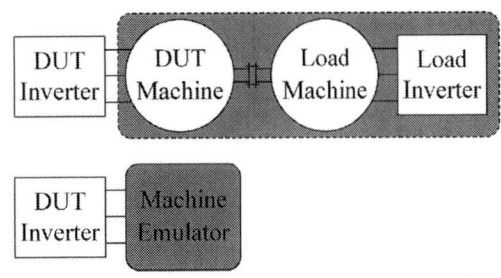

Fig. 1: Schematic diagram of a common machine
test bench in comparison to a PHIL test bench

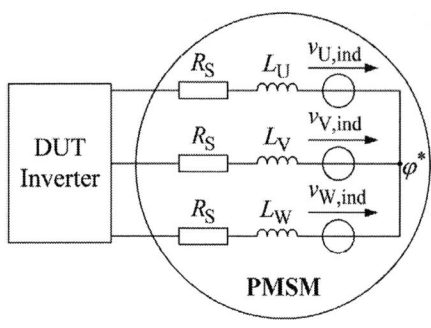

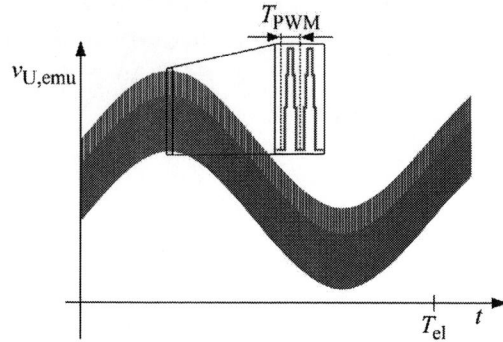

Fig. 2: Simplified equivalent circuit diagram of a three-phase PMSM

Fig. 3: Exemplary course of the desired counter-voltage for a three-phase two-level DUT inverter

In Fig. 2, a simplified equivalent circuit diagram of a three-phase permanent-magnet synchronous machine (PMSM) is depicted. The circuit diagram considers the leg resistances R_S, its variable inductances L_ν with $\nu = \{U, V, W\}$, as well as the back EMF voltage $v_{\nu,ind}$ in each leg. Due to the DUT inverter's switched-mode operation, the input voltages of the machine change stepwise. Thus, the machine's neutral point potential φ^* jumps with every switching transition. Consequently, the voltage across the leg inductances L_ν changes stepwise, too, although the back EMF voltage varies continuously. In order to emulate this behavior, the PHIL power electronic device has to be capable to adjust discontinuous countervoltages. The number of the adjustable countervoltage levels depends on the number of different machine's neutral point potentials φ^* occurring within one switching period. With an increasing number of DUT inverter's phases n_{phase} and DUT inverter's voltage levels $n_{DUT\text{-}level}$ also the number of different machine's neutral point potentials φ^* increases. Thereby, the number of the adjustable countervoltage levels $n_{Emu\text{-}level}$ increases, too. For a joint neutral point the number of required countervoltage levels is calculated by eq. 1 with respect to the number of the DUT inverter's phases n_{phase} and the DUT inverter's voltage levels $n_{DUT\text{-}level}$.

$$n_{Emu\text{-}level} = \left(n_{DUT\text{-}level} - 1 \right) \cdot n_{phase} + 1 \tag{1}$$

The course of one phase's reference countervoltage for a three-phase two-level DUT inverter ($n_{phase} = 3$, $n_{DUT\text{-}level} = 2$) is exemplarily shown in Fig. 3. Within one switching period T_{PWM} of the DUT inverter the countervoltage has six discontinuous steps between four voltage levels. These levels are dedicated to the machine's neutral point potentials φ^*, respectively to the switching state of the DUT inverter. The steps represent the jumpwise voltage change across the machine leg inductances L_ν. The voltage of each of these four voltage level changes sinusoidally and relatively slow within one electrical period T_{el} of the machine. This course represents the back EMF voltage and the variable machine inductances.

An already reviewed power electronic concept, that fulfils these requirements, is a linear amplifier [4]. Its bandwidth must be high enough to approximate the discontinuous countervoltages adequately. A disadvantage is its low efficiency. Further analyzed concepts are a modular multilevel converter [5] or an inverter with multiple branches per phase [6-9]. Another requirement is minimum ripple content inserted additionally into the output currents. Therefore, a high switching frequency is chosen or a special interleaved modulation technique is used for these topologies [10,11]. However, more switching transitions can cause more disturbances. A further approach to reduce the additional ripple content is adding a passive LC filter either in the coupling path [12] or at the output terminals of the emulator's inverter [13,14]. These topologies improve the electromagnetic compatibility but have to deal with the challenge of lower dynamic behavior.

In order to emulate a machine driven by a three-phase two-level DUT inverter, this paper presents a four-level inverter with variable voltage levels, which combines the advantages of the previous topologies: smoothed voltage levels and stepwise changes between these levels at the output terminals. Therefore, the unwanted additional current ripple is nearly zero and the dynamic is sufficient to emulate the current ripple's slew rate. The use of four voltage levels is especially tailored to the emulation of three-phase machines fed by a three-phase two-level inverter. The four levels represent the countervoltages during the four DUT inverter's switching states: zero, one, two or all three high-side MOSFETs in on-state.

Four-Level Inverter with Variable Voltage Levels

Fig. 4 depicts the circuit diagram of the considered inverter. All three inverter phases are supplied by a constant voltage source V_{DC}. Each phase is divided into four legs consisting of a buck converter equipped with MOSFETs, a passive low pass filter (filter inductor L_F and filter capacitor C_F) and two antiparallel MOSFETs for voltage level selection.

The four buck converters adjust the four variable countervoltage levels $v_{v,Cn}$, with $v = \{U, V, W\}$ and $n = 1..4$. For that matter, the passive filters' cutoff frequency can be chosen much lower than the switching frequency of the MOSFETs S_{v1} to S_{v8}. The reason for this is the course of the desired reference countervoltage within one leg. By neglecting the switched-mode operation of the DUT inverter, the emulator's countervoltages only have to represent the machines back EMF voltage, the voltage drops across machine's leg resistances and its variable leg inductances (compare to Fig. 2 and Fig. 3). Approximately, these voltage components change their value periodically with the machine's rotational speed multiplied by its number of pole pairs, hereinafter called "machine's electrical frequency". Thus, the variable voltage levels $v_{v,Cn}$ have to change its value with machine's electrical frequency, too. This can be considered by using the machine's maximum rotational speed and its pole pair number to dimension the filters' cutoff frequency. Hence, the filters' cutoff frequency is much lower than the buck converters' switching frequency. A lower cutoff frequency means less additional ripple content caused by switched-mode operation of the half bridges.

For each leg, the desired countervoltage is adjusted by a cascaded closed-loop voltage control system. Besides the voltage $v_{v,Cn}$ across the filter capacitors, also the capacitor current $i_{v,Cn}$ is measured and fed back to the control system. A subordinated current controller is able to compensate the disturbance values very fast. The whole control system is presented in the following section "Voltage Controller".

In consideration of the DUT inverter's switched-mode operation, the voltages across the machine's leg inductances L_v step discontinuously. These stepwise changes are realized by the antiparallel arranged MOSFETs S_{v9} to S_{v16}, which allow an abrupt switchover between the four voltage levels. So at each time, only one leg is active and carries the whole phase's output current. To avoid short circuits between two voltage levels during switchover, a turn-on delay (dead time) has to be implemented. During this dead time, the output current can commutate to either the freewheeling diode D_{v1} or D_{v2} depending on its direction. To keep the error in the output voltage as low as possible, the dead time has to be very small.

When dimensioning the filter elements one has to bear in mind that the output current acts as disturbance value for the capacitor voltage in the appropriate leg. After the output current commutates abruptly to a different leg, the current through the filter inductor is not able to decrease fast enough due to the inductance's limited rate of current change. The surplus charge carriers flow into the filter capacitor and

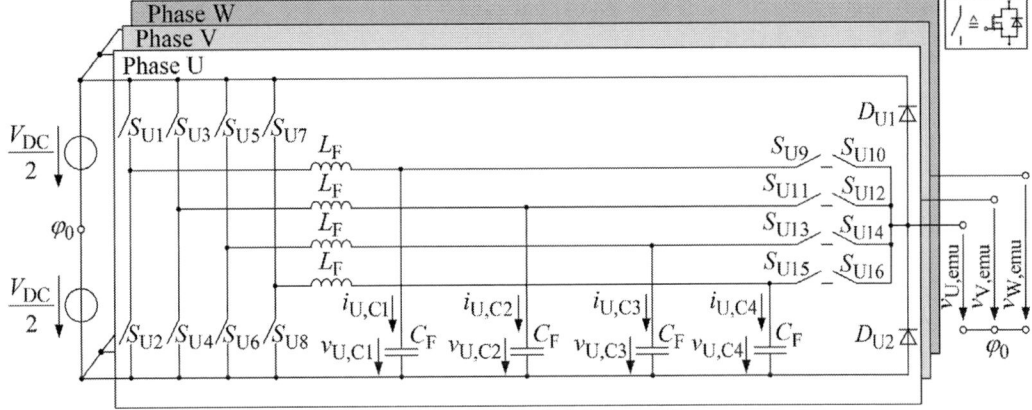

Fig. 4: Schematic of four-level inverter with variable voltage levels

effect a short-term voltage deviation. In order to reduce this deviation one should prefer a smaller absolute value of the filter inductance and a bigger absolute value of the filter capacitance instead of vice versa.

Machine Emulator Operation

The four-level inverter with variable voltage levels is coupled to the DUT inverter by using coupling inductors L_C to actuate as a machine emulator. The overview of the setup is depicted in Fig. 5.
A signal processing system calculates the electric state variables. This signal chain consists of a machine model, a reference voltage calculation, a voltage controller and a leg selector, which are explained in the following sections.

Machine Model

The DUT inverter's output voltages $v_{v,DUT}$ are measured and fed into a machine model. This model calculates the electrical and mechanical behavior of the investigated machine in real time. In general, the implementation of an accurate and non-linear machine model such as a flux model [15,16] is reasonable. But within this paper – only showing the functionality of the presented power electronic device – a simple fundamental component model for a permanent-magnet synchronous machine (PMSM) is used [17,18], compare Fig. 2. This model takes constant leg resistances, variable leg inductances and a sinusoidal back EMF voltage into account.
Irrespective of its type, the machine model calculates the occurrent currents in the investigated machine. They act as reference currents $i_{v,ref}$ for the whole emulating system.

Reference Voltage Calculation

The reference value $v_{v,emu,ref}$ of the countervoltage $v_{v,emu}$, that the emulator's power electronic has to adjust evoking the reference currents $i_{v,ref}$, is calculated for each phase. Hence, eq. 2 for the voltage drop over the coupling inductor L_C is transformed to get the reference value $v_{v,emu,ref}$ (see eq. 3).

$$v_{v,DUT} - v_{v,emu} = L_C \cdot \frac{d i_v}{d t} \qquad \text{with } i_v = i_{v,ref} \text{ and } v_{v,emu} = v_{v,emu,ref} \tag{2}$$

$$v_{v,emu,ref} = v_{v,DUT} - L_C \cdot \frac{d i_{v,ref}}{d t} \tag{3}$$

Additionally, a proportional current controller with the parameter K_{corr} compensates differences between the reference currents $i_{v,ref}$ and the measured values of the currents i_v, see Fig. 6. The controller does not need an integral element because the controlled system has an integral behavior.
The reference voltages $v_{v,emu,ref}$ show a discontinuous course, caused by DUT inverter's switching behavior and the stepwise change of the virtual machine's neutral point potential φ^*. In order to get continuous courses, the reference voltage of one phase is divided into four reference voltages $v_{v,Cn,ref}$, which are dedicated to the four legs of the four-level inverter. The selection depends on the switching

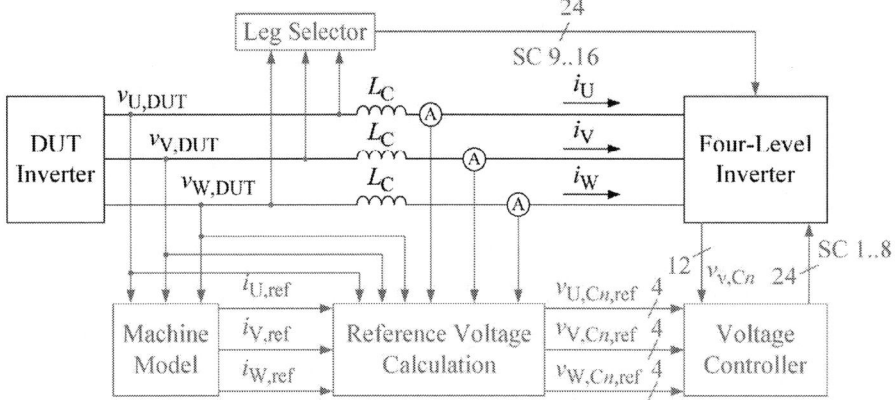

Fig. 5: Overview of the machine emulator setup

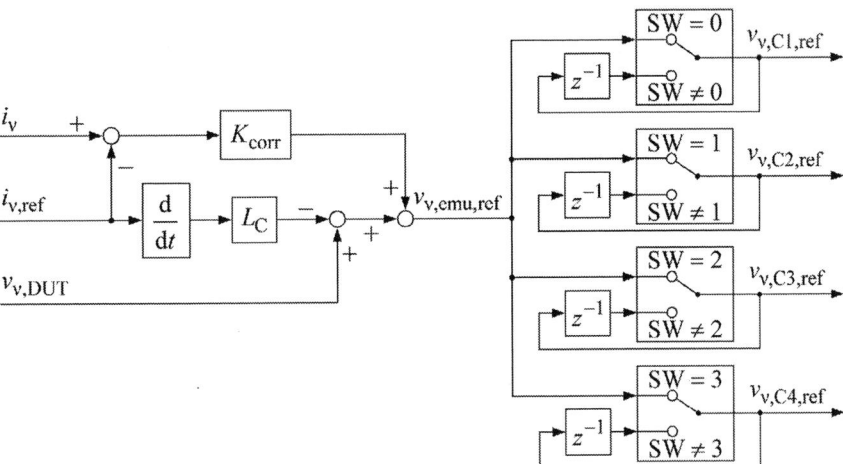

Fig. 6: Block diagram of reference voltage calculation

state of the DUT inverter. For that matter, the variable $SW = 0..3$ represents the number of DUT inverter's high-side MOSFETs in on-state. Consequently, each leg receives only reference voltages when the same switching state of the DUT inverter is active. As long as the switching state is different, the leg's reference voltages $v_{v,Cn,ref}$ stay unmodified. Although the courses of the reference voltages $v_{v,emu,ref}$ step discontinuously with six steps per DUT switching period, the courses of leg's reference voltages $v_{v,Cn,ref}$ change relatively slow with the machine's electrical frequency. These reference voltages can also be adjusted with a lower filter cutoff frequency in the range of machine's maximum electrical frequency. If each leg had to adjust discontinuous voltages, this cutoff frequency would be infinite in theory.

Voltage Controller

The voltage control system determines the switching commands for the MOSFETs S_{v1} to S_{v8}. Each leg has a dedicated voltage controller which controls the output voltage $v_{v,Cn}$ at the capacitor C_F. For this purpose, a cascaded control system is implemented, see Fig. 7. The superimposed control system uses the reference value $v_{v,Cn,ref}$ and the measured value of the leg's output voltage $v_{v,Cn}$ as input variables to calculate the desired current $i_{v,Cn,ref}$ flowing into the filter capacitor C_F. For the calculation, two setting parameters K and K_1 have to be chosen to tune the transient response of the controller. Within this paper, the controller has a transient response of a critically damped low pass filter. Additionally, a bypassed integrator with time constant T_I is added to avoid steady-state deviation. In this paper the time constants of proportional part of the controller and its bypassed integrator are chosen equal.

The subordinate control loop has to control the current $i_{v,Cn}$ flowing into the filter capacitor. Especially after a switch process between two legs, the filter inductor in the turned-off leg keeps on driving a current into or from the filter capacitor. This disturbance current has to be compensated as fast as possible. On this account, a time-triggered bang-bang controller is implemented for the capacitor current control.

The reference current $i_{v,Cn,ref}$ is compared to the actual value of the current $i_{v,Cn}$. If the reference value is bigger than the actual value, the high-side MOSFET of the corresponding half bridge has to be in on-

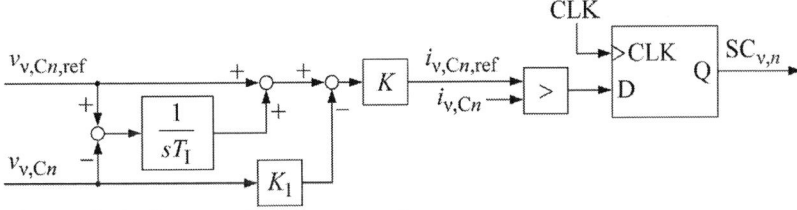

Fig. 7: Block diagram of voltage control system

state. Otherwise, the low-side MOSFET has to be turned on. By feeding back the current into the filter capacitor instead of the current through the filter inductor, the influence of the output current as a disturbance value is eliminated.

In order to limit the MOSFETs' switching frequency, the desired switching command is sent to a clock triggered D flip-flop. Its trigger frequency should be chosen as high as possible but is limited by the switching speed of the chosen MOSFETs and their gate drivers. The output of the flip-flop is the actual switching command $SC_{v,n}$ of the dedicated half bridge.

Leg Selector

The leg selector system determines the switching commands for the MOSFETs S_{v9} to S_{v16}. It counts the number of the DUT inverter's high-side MOSFETs in on-state, formerly introduced as variable SW, and generates the switching commands in accordance with Table I. Thereby, this leg is connected with the output, which supplies the currently desired countervoltage. A dead time is implemented to avoid compensating currents between two voltage levels during switchover.

Table I: Leg selector map

SW	Active MOSFETS
0	S_{v9} and S_{v10}
1	S_{v11} and S_{v12}
2	S_{v13} and S_{v14}
3	S_{v15} and S_{v16}

Simulation Results

The functionality of the four-level inverter is proven by simulations. Therefore, the presented inverter is dimensioned to emulate machines fed by automotive on-board supply systems from 12 V to 24 V. The DUT inverter is a common three-phase two-level inverter equipped with MOSFET half bridges, which are driven by a PWM unit. Table II lists the parameters used for the following simulation results. For this purpose a numerical optimization method has calculated the values of the passive filter components and the inductance of the coupling inductor with the aim of getting the least deviation between reference and actual current. The filter cutoff frequency is dimensioned for a machine with maximum rotational speed of $3000\,\mathrm{min}^{-1}$. In addition, the power electronic system should be dimensioned to be able to emulate currents with harmonic content up to the seventh harmonic. In doing so, the machine model has to be changed (not relevant in this paper).

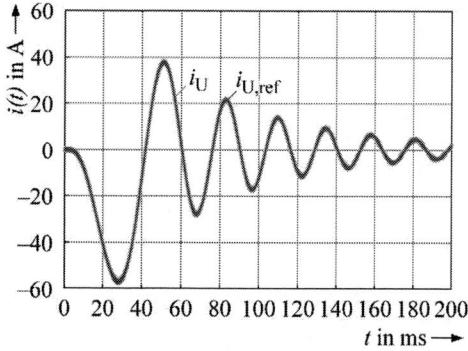

Fig. 8: Simulation results of step response

In the simulation, the machine starts at standstill. At $t = 0$ the DUT inverter's output voltage steps from $0\,\mathrm{V}$ to $3\,\mathrm{V}$ in q-direction and arouses a torque-forming current. The torque leads to an acceleration of the virtual machine. Hence, the machine currents decrease caused by rising back EMF voltage.

Fig. 8 shows the reference value $i_{U,ref}$ and measured value i_U of the phase U machine current. Fig. 9 presents the difference between these two values. Although the current reaches an absolute maximum value of about 58 A, the difference between reference and actual current is less than 0.2 A.

In Fig. 10, a zoomed view of these courses is depicted. It can be seen, that the emulation system is also able to emulate the current's slew rate.

The reason for this is the fact, that the four-level inverter with variable voltage levels is able to follow the course of the discontinuous reference countervoltage nearly perfectly, see Fig. 10.

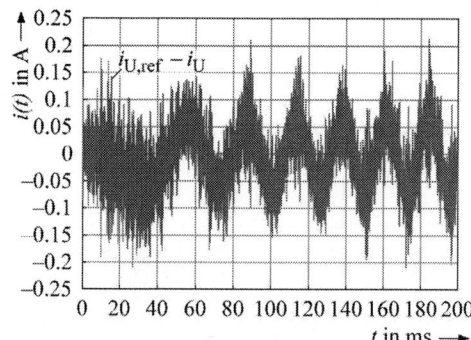

Fig. 9: Difference between reference and actual current

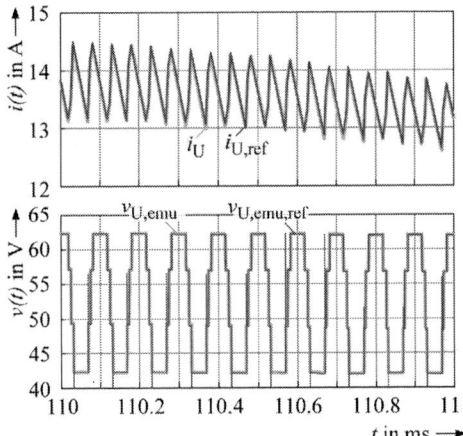

Fig. 10: Detailed view on machine currents and emulator's countervoltages

Especially the four voltage levels can be seen in this plot.

By calculating the reference voltage corresponding to eq. 3, the simulation tool has a serial processing. The reference current $i_{U,ref}$ cannot be calculated before the voltage $v_{U,emu,ref}$ is identified to determine the switching commands. For that matter, the reference current for eq. 3 is delayed by one sample period of the simulation tool. This leads to the small spikes which occur in the course of the reference voltage $v_{U,emu,ref}$.

Table II: Simulation parameters

DUT inverter	
DC link voltage	$V_{DC,DUT} = 20\,V$
Switching frequency	$f_{PWM} = 10\,kHz$
Four-level inverter (variable voltage levels)	
DC link voltage	$V_{DC} = 80\,V$
Clock frequency (D flip-flop)	$f_{PWM} = 400\,kHz$
Coupling inductance	$L_C = 65\,\mu H$
Filter cutoff frequency	$f_F = 2.6\,kHz$
Filter inductance	$L_F = 5.5\,\mu H$
Filter capacitance	$C_F = 680\,\mu F$
Voltage controller	
Setting parameters	$K = 11.1\,\dfrac{A}{V}$ $K_1 = 2$
Controller time constant	$T = 61.2\,\mu s$
Time constant of bypassed integrator	$T_I = 61.2\,\mu s$
Virtual PMSM	
Pole pair number	$z_p = 4$
Leg resistance	$R_S = 32.5\,m\Omega$
Leg inductances	$L_d = 56\,\mu H$ $L_q = 72\,\mu H$
Magnet's flux	$\Psi_{PM} = 10.2\,mVs$
Max. rotational speed	$n_{max} = 3000\,min^{-1}$

Conclusion

In this paper, the concept of a four-level inverter with four variable voltage levels per phase in a power-hardware-in-the-loop machine emulator is shown.

Due to the fact that the desired countervoltage in emulating three-phase machines, which are driven by a two-level DUT inverter, has a discontinuous course with four stepwise levels, the presented topology fulfils the requirements for the output voltage perfectly. For that reason, the system is able to emulate the slew rate of the machine's current ripple accurately.

Moreover, unwanted additional current ripple is reduced by using low-pass filter with lower cutoff frequency.

The hardware setup of the four-level inverter, a control strategy to use this device as a machine emulator and simulation results, which prove the functionality of this concept, are explained.

All in all, the additional expenses in controllability and hardware components lead to better results than using other power electronic devices like a two-level inverter with an output filter.

Future Work

With the aim of verifying the presented system, a hardware setup of the three-phase inverter with LC output filter will be built up. The half bridges are set up in a modular way. The machine model, the reference voltage calculation and the leg selector run on an FPGA which monitors the DUT inverter's output voltages. Furthermore, one FPGA per phase is responsible for the voltage control.

Concluding, the simulated results will be compared to appropriate measurement results.

References

[1] D. J. Atkinson, A. G. Jack, and H. J. Slater, "The virtual machine," IEE Colloquium on Vector Control Revisited, 1998, pp. 7/1–7/6.

[2] A. G. Jack, D. J. Atkinson, and H. J. Slater, "Real-time emulation for power equipment development. I. Realtime simulation," IEE Proceedings – Electric Power Applications, vol. 145, issue 2, pp. 92-97, 1998.

[3] H. J. Slater, D. J. Atkinson, and A. G. Jack, "Real-time emulation for power equipment development. II. The virtual machine," IEE Proceedings – Electric Power Applications, vol. 145, issue 3, pp. 153-158, 1998.

[4] M. Fischer, R. Malic, N. Tröster, J. Ruthardt, and J. Roth-Stielow, "Design of a Three-Phase 100 Ampere Linear Amplifier for Power-Hardware-in-the-Loop Machine-Emulation," IET The Journal of Engineering, no.17, pp. 4041-4044, 2019.

[5] M. Schnarrenberger, L. Stefanski, C. Rollbühler, D. Bräckle, and M. Braun, "A 50 kW Power Hardware-in-the-Loop Test Bench for Permanent Magnet Synchronous Machines based on a Modular Multilevel Converter," 20th European Conference on Power Electronics and Applications, 2018.

[6] C. Nemec, O. Lehmann, M. Heintze, and J. Roth-Stielow, "Optimal inductor setup for a power-hardware-in the-loop machine emulator," IEEE 10th International Conference on Power Electronics and Drive Systems, pp. 364-369. 2013.

[7] A. Schmitt, J. Richter, M. Braun, and M. Doppelbauer, "Power Hardware-in-the-Loop Emulation of Permanent Magnet Synchronous Machines with Nonlinear Magnetics – Concept & Verification," PCIM Europe, pp. 393-400, 2016.

[8] A. Schmitt, J. Richter, M. Gommeringer, T. Wersal, and M. Braun, "A Novel 100 kW Power Hardware-in-the-Loop Emulation Test Bench for Permanent Magnet Synchronous Machines with Nonlinear Magnetics," 8th IET International Conference on Power Electronics, Machines and Drives, 2016.

[9] M. Fischer, S. Petzner, J. Ruthardt, J. Schuster, S. Bintz, and J. Roth-Stielow, "Current Control for a Multiphase Interleaved-Switched Inverter Using Field Oriented Coordinates," 20th European Conference on Power Electronics and Applications, 2018.

[10] C. Nemec and J. Roth-Stielow, "Ripple current minimization of an interleaved-switched multi-phase PWM inverter for three-phase machine-emulation," 14th European Conference on Power Electronics and Applications, 2011.

[11] M. Fischer, J. Ruthardt, M. Nitzsche, P. Ziegler, and J. Roth-Stielow, "Investigation on Carrier Signals to Minimize the Overall Current Ripple of an Interleaved-Switched Inverter," PCIM Europe, pp. 1644-1649, 2019.

[12] R. Sudharshan Kaarthik, and P. Pillay, "Emulation of a Permanent Magnet Synchronous Generator in Real-time Using Power Hardware-in-the-loop," International Conference on Power Electronics, Drives and Energy Systems, 2016.

[13] M. Fischer, F. Gliese, J. Ruthardt, M. Zehelein, and J. Roth-Stielow, "Investigation on a Three-Phase Inverter with LC Output Filter for Machine Emulation," 21st European Conference on Power Electronics and Applications, 2019.

[14] M. Fischer, D. Erthle, P. Ziegler, J. Ruthardt, and J. Roth-Stielow, "Comparison of Two Power Electronic Topologies for Power Hardware in the Loop Machine Emulator", IEEE Applied Power Electronics Conference and Exposition (APEC), 2020.

[15] M. Boesing, M. Niessen, T. Lange, and R. De Doncker, "Modeling spatial harmonics and switching frequencies in PM synchronous machines and their electromagnetic forces," 20th International Conference on Electrical Machines, 2012.

[16] T. Lange, M. Boesing, and R. W. De Doncker, "Measurement parameterized synchronous machine model with spatial-harmonics," 17th International Conference on Electrical Machines and Systems (ICEMS), 2014.

[17] T. Sebastian and G.R. Slemon, "Transient modeling and performance of variable-speed permanent-magnetmotors," IEEE Transactions on Industry Applications, vol. 25, issue 1, pp. 101-106, 1989.

[18] O. Lehmann, M. Heintze, C. Nemec, and J. Roth-Stielow, "Using a multiphase interleaved-switched inverter as power-hardware-in-the-loop machine emulator to test sensorless control techniques," PCIM Europe, 2015.

Powder Injection Molding in the Fabrication of Soft Ferrite material for Power Electronics

J-S Ngoua-Teu[1], U. Soupremanien[2], P. Sallot[1], G. Delette[2] and M. Bohnke[2]

[1]SAFRAN TECH,
Magny-Les-Hameaux, 78114, FRANCE

[2]Université Grenoble Alpes, CEA Grenoble, CEA/LITEN/DTNM/SA3D/LMCM,
38054 Cedex 9, FRANCE

jean-sylvio.ngoua-teu@safrangroup.com

https://www.safran-group.com/

Keywords

«PIM», «Uniaxial pressing», «Ferrite», «Permeability», «Core loss».

Abstract

This work presents the Powder Injection Molding (PIM) process for the fabrication of soft ferrite material. The traditional method commonly used by manufacturer is pressing which was compared to PIM focusing on core losses and permeability.

Introduction

In power electronics applications, magnetic components are essential in power converters where energy management and electrical insulation are required. In embedded systems, thermal management is also an important aspect that has to be taken into account. However, in some cases, magnetic core sold in the market [1-4] can be limited in their use due to geometry restriction related to the fabrication process. It is the case when considering particular cores, like matrix transformer [5-7] that may require specific drilling techniques to be machined in order to fulfill integration constraints. The matrix core Fig. 1(b) is an interesting architecture in application regarding the reduction of magnetic induction and core losses. A relevant way to fabricate that kind of geometry could be eased using PIM process.

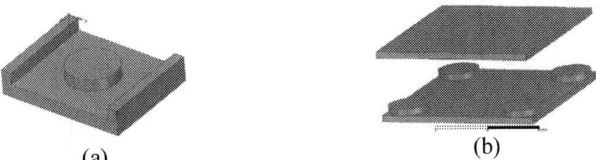

(a)   (b)

Fig. 1: Ferrite core geometry (a) Typical ER64 core (b) Matrix core with circular legs

Fabrication techniques

• Pressing

Pressing is the common technique used by magnetic component manufacturers to mass-produce most of existing parts. This technique requires a mold (metallic matrix) to shape the part. Usually, powders are supplied in a "ready to press" state, which corresponds to a mix of powder with a very small amount of binder. For the specific case of the ferrite used is this study, the amount and type of binder was of 1% of poly vinyl acetate (PVA). After pressing the powder particles remained attached to form a ring-shape core (toroid), which is resistant enough to be handle for the debinding and sintering stages (Fig. 2).

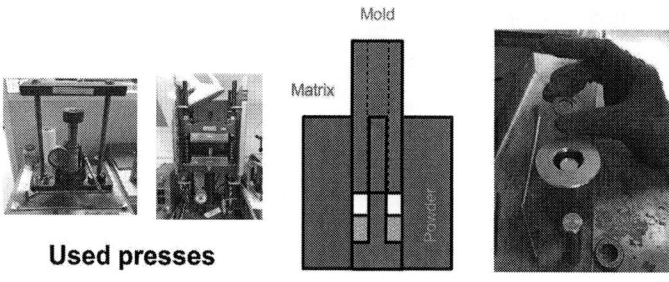

Used presses

Fig. 2: Pressing technique

- **Powder Injection Molding Process**

For fabrication of ferrite parts, powder injection molding is an interesting process that allows the manufacturing of complex and net-shape components [8-10]. Such a technology is very attractive in industry because massive production of complex parts could be achieved. This technique is nowadays largely used in ceramics production. The PIM process is separated in several steps: feedstocks' formulation and preparation, injection molding, chemical and thermal debinding, and sintering (Fig. *3*). The feedstock preparation consists in mixing the charge (ferrite powder) with several polymers.

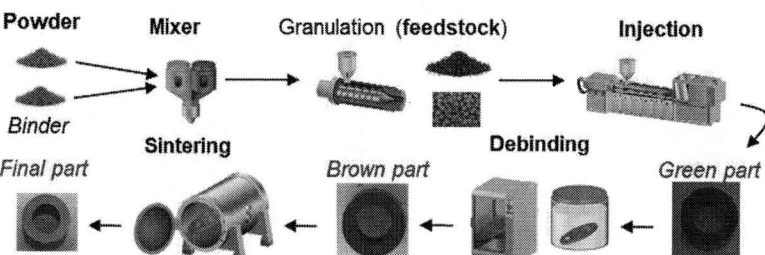

Fig. 3: Powder Injection Molding's flow chart (adapted from [11])

The ferrites' mass fraction (powder) in the feedstock should be defined as high as possible to facilitate the debinding steps but not too high (>65% vol.) regarding injection molding aspect. The targeted magnetic material was selected to operate at high frequency (HF: from several 100 kHz to few MHz). Two main class of soft magnetic materials can be used in HF applications: Nickel-Zinc (Ni-Zn) and Manganese-Zinc (Mn-Zn) oxides. In power electronics, Mn-Zn is more attracting because of its high permeability and wide working frequency range (up to few MHz). However, the fabrication of Mn-Zn ferrites is more complex than Ni-Zn due to the fact that the sintering stage requires controlled level of oxygen content.

The ferrite powder (Z70-L2G) contains a theoretical mixture of $(Mn_{0.6}Zn_{0.1})$ Fe_2O_4 (~97-99%), PVA (1%), SiO_2 (<0.2%) and $Ca(OH)_2$ (<0.1%)[1] and was supplied from Japan Metals and Chemicals (JMC) [12]. The recipe for sintering the Mn-Zn powder and the summary of the main ferrite's characteristics (reported in the JMC datasheets) are shown in Fig. *4*.

---

[1]Theoretical composition specified by JMC

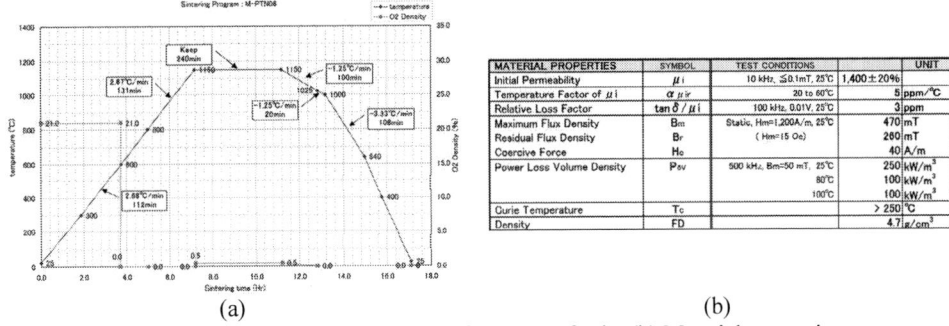

MATERIAL PROPERTIES	SYMBOL	TEST CONDITIONS		UNIT
Initial Permeability	μ_i	10 kHz, ≦0.1mT, 25℃	1,400±20%	
Temperature Factor of μ_i	$\alpha\,\mu_{ir}$	20 to 60℃	5	ppm/℃
Relative Loss Factor	tan δ / μ_i	100 kHz, 0.01V, 25℃	3	ppm
Maximum Flux Density	Bm	Static, Hm=1,200A/m, 25℃	470	mT
Residual Flux Density	Br	(Hm=15 Oe)	280	mT
Coercive Force	Hc		40	A/m
Power Loss Volume Density	Psv	500 kHz, Bm=50 mT, 25℃	250	kW/m³
		60℃	100	kW/m³
		100℃	100	kW/m³
Curie Temperature	Tc		> 250	℃
Density	FD		4.7	g/cm³

(a) (b)

Fig. 4: JMC data (a) Sintering cycle for Mn-Zn ferrite (b) Material properties

For the analysis, ring-shape geometry was selected as core-losses were easier to determine using such shapes. A picture showing the parts through the fabrication steps is presented in Fig. 5. As it can be seen, the general dimensions did not evolved too much between injection and debinding steps. However, after sintering step the parts shrinkage is noticeable (-15%).

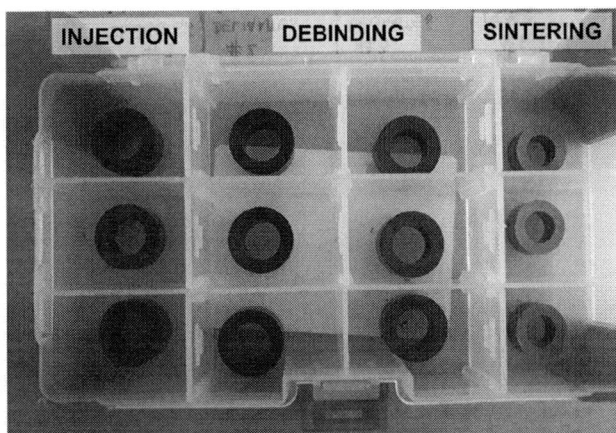

Fig. 5: Aspect evolution of the fabricated toroids during the process' steps

- **Bench description and power losses calculation method**

To determine the behavior of the fabricated cores, a dedicated bench (Fig. 6) has been set-up for core losses calculations *(and initial permeability estimation)* at four temperatures (25, 50, 85 and 110°C). Core losses evolutions were analyzed with the variation of the sintering temperature. Five fabrications' batch were applied to the toroids (sintering temperatures: 1140, 1145, 1150, 1155 and 1160°C). Each batch consists in 4 parts made using #1 pressing and #2 PIM processes. Core losses were determined using a direct method by measuring the voltage and the current through the toroids [13, 14]. These measurements are relevant only if the extra phase shift due to the component characteristics could be evaluated correctly. To do so, a decade capacitance box (for achieving resonance in the circuit and reduced the phase shift between voltage and current) was added.

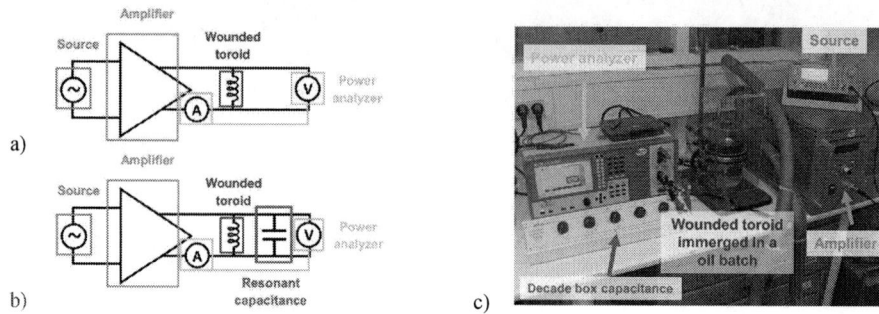

Fig. 6: Schematic of the test bench for core losses determination

The toroids core were coiled using a Litz wire of 1.2 mm diameter (320 strands 40 µm) .The generated induction inside the toroids was calculated using the effective area (A_{eff}), the RMS excitation voltage (U), the frequency (f) and the number of winding turns (N). This first measurement was made without the capacitance decade box (Fig. 6a).

$$B = \frac{U_{rms} \times \sqrt{2}}{N \times 2\pi f \times A_{eff}} \qquad (1)$$

This first measurement allow for the calculation of the needed capacitance value (which should be mounted in parallel with the ring-shape core in the electrical circuit) to achieve resonance and having the phase angle (φ) between voltage and current to be close to zero.

$$C_{th} = \frac{1}{(2\pi f)^2 \times L_s} \qquad (2)$$

The inductance value (L_s) can be calculated using (without capacitance decade box):

$$L_s = \frac{1}{2\pi f} \frac{U_{rms}}{I_{rms}} \sqrt{1 - \cos(\varphi)^2} \qquad (3)$$

By adding the calculated resonant capacitance to the electrical circuit, the phase shift angle (φ) was close to 0 and voltage and current became in phase. The power losses were thus determined as follow:

$$P_{losses} = \int_o^t U(t) \times I(t) \times \cos(\varphi)dt = \int_o^t U(t) \times I(t)dt \qquad (4)$$

- **Power losses measured on commercial ring-shape core (TDK, N49, R22) to validate the bench working**

To validate the bench for core losses and permeability estimation, 8 different commercial ring-shape cores from TDK (N49) [2] were tested and compared to the datasheet given by the supplier. The measurements showed that the relative permeability equaled 1358.8 with a standard deviation of 36.6. The permeability of these cores were all in the tolerances' range given by TDK (1500±25% at 25°C).

We measure magnetics losses for the ring-shape cores between 100 kHz and 1 MHz and for three magnetic inductions (12.5, 25 and 50 mT). Between 100 kHz and 700 kHz, the core losses measurements were close to supplier data (less than 20% error) and the difference was more important at 1 MHz (*Fig. 7*). This difference was attributed to the cross section change between used cores (R22: 32.5mm²) compared to reference datasheet cores (R34: 82.6 mm²).

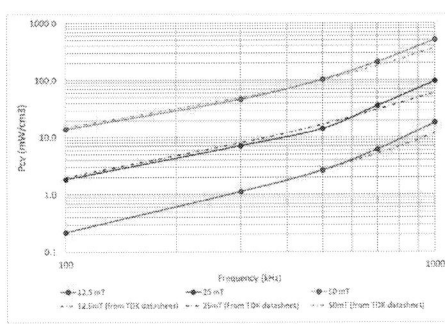

f kHz	B (mT)	Pcv (mW/cm3)	Pv datasheet (mW/cm3)	diff. %
100	12.5	0.2	N/A	N/A
300	12.5	1.1	N/A	N/A
500	12.5	2.7	2.6	2.2%
700	12.5	6.1	5.2	17.4%
1000	12.5	18.2	12	51.8%
100	25	1.9	2	-7.3%
300	25	7.1	8	-11.0%
500	25	14.2	17	-16.2%
700	25	35.8	31	15.5%
1000	25	96.7	60	61.2%
100	50	14.0	15	-6.8%
300	50	46.9	50	-6.2%
500	50	104.5	100	4.5%
700	50	210.9	179	17.8%
1000	50	515.3	366	40.8%

Fig. 7: Core losses measurement vs datasheet

- **Power losses measured on fabricated toroid to compare pressing and PIM processes**

The power losses were compared regarding the fabrication process (pressing versus PIM) on frequency and induction values, where no significant difference was noticed between our laboratory measurements and the data from TDK. The selected induction and frequency were 50 mT at 500 kHz.

For both processes, initial permeability and core losses have been plotted for different operating (oil batch temperature) and sintering temperatures. The analysis of the data from Fig. 8 shows that the ring-shape cores processed by the PIM technique at a sintering temperature of 1160°C had magnetic loss characteristics closer to the JMC datasheet (black dots). In addition, for similar thermal treatments conditions PIM process lead to a higher permeability and to lower losses at 500 kHz and 50 mT than pressing.

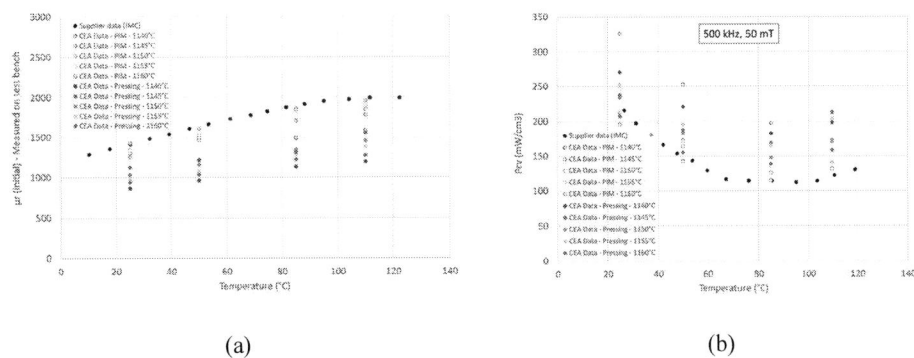

(a)   (b)

Fig. 8: a) Initial permeability and b) magnetic losses in function of oil temperature

Conclusion

This work presents two different techniques to fabricate soft magnetic components from a commercial ferrite powder. When comparing permeability and core losses, PIM process seemed to achieve better performance than pressing for same sintering conditions (same batch for toroid fabricated by pressing and PIM). The sintering temperature effect on performances has been shown. The results obtained with the PIM fabrication technique are very interesting and allow the ferrite properties to be very close compared to the magnetic characteristics given by the powder manufacturer. This technique should allow designing complex ferrite geometry with good magnetic performance as long as the sintering cycle is well controlled.

References

1. *Global Leader in High Performance Ferrite-Ferroxcube.* Available from: https://ferroxcube.com/en-global.
2. *TDK Corporation.* Available from: https://www.tdk.com/corp/en/index.htm.
3. *Hitachi Global.* Available from: http://www.hitachi.com/.
4. *Micrometals Powder Core Solutions.* Available from: https://www.micrometals.com/.
5. Fei, C., F.C. Lee, and Q. Li. *High-efficiency high-power-density 380V/12V DC/DC converter with a novel matrix transformer.* in *2017 IEEE Applied Power Electronics Conference and Exposition (APEC).* 2017.
6. Knabben, G.C., et al. *New PCB Winding "Snake-Core" Matrix Transformer for Ultra-Compact Wide DC Input Voltage Range Hybrid B+DCM Resonant Server Power Supply.* in *2018 IEEE International Power Electronics and Application Conference and Exposition (PEAC).* 2018.
7. Wang, S., et al. *Integrated Matrix Transformer with Optimized PCB Winding for High-Efficiency High-Power-Density LLC Resonant Converter.* in *2019 IEEE Energy Conversion Congress and Exposition (ECCE).* 2019.
8. Stanimirović, Z. and I. Stanimirović. *Injection molded Mn-Zn ferrite ceramics.* in *2010 27th International Conference on Microelectronics Proceedings.* 2010.
9. Zlatkov, B.S., et al., *Properties of MnZn ferrites prepared by powder injection molding technology.* Materials Science and Engineering: B, 2010. **175**(3): p. 217-222.
10. Lukovic, M.D., et al., *Mn-Zn Ferrite Round Cable EMI Suppressor With Deep Grooves and a Secondary Short Circuit for Different Frequency Ranges.* IEEE Transactions on Magnetics, 2013. **49**(3): p. 1172-1177.
11. *Metal injection molding.* Available from: https://en.wikipedia.org/wiki/Metal_injection_molding.
12. *Japan Metals & Chemicals Co., Ltd. | 日本重化学工業株式会社.* Available from: http://www.jmc.co.jp/en.html.
13. Tan, F.D., J.L. Vollin, and S.M. Cuk, *A practical approach for magnetic core-loss characterization.* IEEE Transactions on Power Electronics, 1995. **10**(2): p. 124-130.
14. Foo, C.F., D.M. Zhang, and X. Li, *A Simple Approach for Determining Core-loss of Magnetic Materials.* Journal of the Magnetics Society of Japan, 1998. **22**(S_1_ISFA_97): p. S1_277-279.

Modulation scheme with common mode and differential mode voltage elimination for a five level inverter fed open end winding induction motor drive

Greeshma Nadh, Durga Nair S, Arun Rahul S
Indian Institute of Technology Palakkad
Palakkad, Kerala, India
Email: 121704002@smail.iitpkd.ac.in

Keywords

≪Pulse Width Modulation (PWM)≫, ≪Multilevel converters≫, ≪Variable speed drive≫, ≪Induction motor≫, ≪Harmonics≫

Abstract

A five level inverter topology with single DC supply is analysed for an open end winding induction motor (OEWIM) drive. A carrier based modulation scheme is proposed to eliminate the common mode (CMV) and differential mode voltages (DMV). The modulation scheme utilises the voltage vector and switching state redundancy of the inverter topology. The voltage vector redundancy is used to eliminate the DMV and switching state redundancy is used to eliminate the CMV. Compared to the conventional level shifted PWM method, the proposed modulation scheme offers improved harmonic performance and also ensures self-balancing of capacitor voltages. The simulation studies are carried out in MATLAB for an OEWIM drive operating with open-loop constant Volts-per-Hertz control. The modulation scheme is implemented using TMS320F28379D controller board and an FPGA module.

Introduction

Multi level inverters (MLI) constitute an integral part of medium to high voltage motor drive and power system applications. In addition to the three classical multi level inverters, numerous new and hybrid topologies have been proposed for applications in single phase as well as multi phase systems [1, 2, 3]. Emergence of new topologies are aimed at reducing the i) total component count, ii) voltage stress across the devices, iii) DC link voltage requirement and iv) complexity of voltage balancing circuit. Among the new topologies, stacked multi level inverter (SMI) [4, 5] is a promising configuration which exhibits higher modularity and employs fewer components. SMI topology is explored for star connected motor drive systems in [5]. However, motors and transformers, which forms the key components in electric drives and power systems, can be connected in open end winding fashion. Similar to dual inverter topology, SMI can be used to supply open end winding motor. In this paper, a five level (5L) SMI topology is explored for three phase open end winding motor drive application.

In an open end winding motor drive system, the reliability of the motor is affected by both the common mode and differential mode voltages [6][7]. Common mode voltage causes voltage build up in shaft and leads to flow of current which slowly damages the motor bearing. On the other hand, differential mode voltage causes circulating current in the motor phase winding for topologies fed with single DC link supply. Another key concern in multi level inverter topologies, especially derived from standard neutral point clamped topology with single DC link, is the balancing of the neutral point voltages. In addition to causing unequal switching stress in the device, the neutral point imbalance also leads to over sizing of capacitors and switching devices and causes an increase in harmonic distortion of the output current. Selection of proper pulse width modulation (PWM) scheme for the inverter is the most cost effective way to eliminate these voltages and to ascertain the neutral point voltage balance. The modulation scheme utilises one of the key features of MLI topologies, i.e., the presence of numerous redundant voltage vectors. PWM technique used in [8] exploits the switching state redundancy to balance all the capacitor voltages in the topology. The PWM methods developed in [9, 10, 11], also uses the vector redundancy of topology but eliminates only the differential mode voltages. In [7], computationally intense space vector based modulation scheme is analysed for eliminating the common mode voltages.

In this work, a carrier based modulation scheme is proposed to eliminate both the common mode and the differential mode voltage of the five level inverter. The modulation method exploits the voltage vector and switching state redundancy of the inverter topology. The proposed modulation scheme also ensures neutral point balance and has better harmonic performance compared to conventional level shifted PWM.

Five level inverter topology

The topology of five level inverter fed induction motor is shown in Fig. 1a. Each phase of the motor winding is connected across an H-bridge. The two ends of the H-bridge can be connected to the positive, negative or the DC link midpoint using the switches S_{1X} and S_{2X}, where X denotes the phase R, Y and B. Thus, the two complementary switch pairs gated by $S_{1X} : S'_{1X}$ and $S_{2X} : S'_{2X}$ decide the level of output voltage. The H-bridge switches, operated in bipolar mode by signal S_X and S'_X, decide the polarity of the output. Each phase has 8 possible combinations and results in five distinct voltage levels: $\pm 0.5 V_{dc}$, $\pm 0.25 V_{dc}$ and 0. The gating signals and corresponding states of the inverter for different voltage levels are listed in Table I. The topology has, in total, 512 (8^3) switching states and 125 (5^3) voltage vectors. However, these vectors occupy 61 locations as shown in the space vector structure of Fig. 1b. This implies that the topology has numerous redundancy in the voltage vectors which can be employed to design new PWM methods.

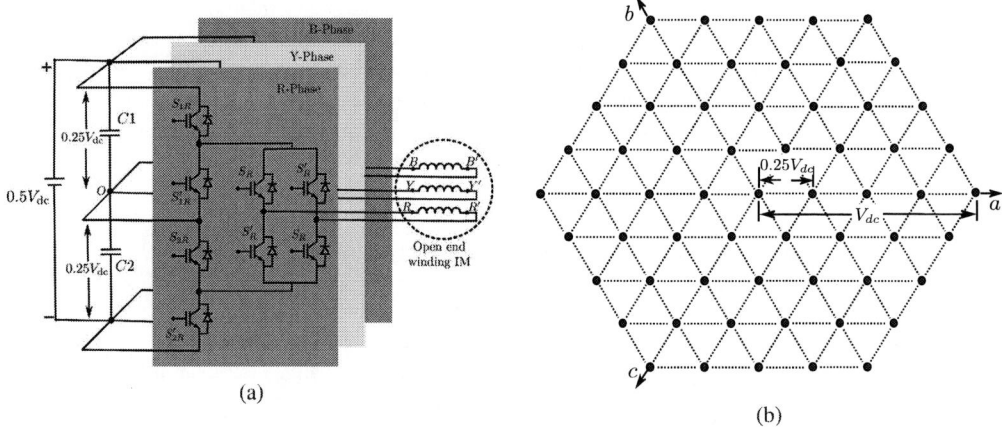

(a)

(b)

Fig. 1: Five level inverter fed OEWIM drive: (a) circuit topology and (b) space vector structure

Table I: Various voltage levels of proposed 5L inverter

S_{1X}	S_{2X}	S_X	Voltage level	State	S_{1X}	S_{2X}	S_X	Voltage level	State
1	0	0	$-0.5V_{dc}$	-2	1	0	1	$+0.5V_{dc}$	2
1	1	0	$-0.25V_{dc}$	-1	1	1	1	$+0.25V_{dc}$	1
0	0	0	$-0.25V_{dc}$	-1	0	0	1	$+0.25V_{dc}$	1
0	1	0	0	0	0	1	1	0	0

In addition to the redundancy in voltage vectors, the inverter possesses a simple and highly modular structure. A comparison of proposed topology with the classical 5L inverter topologies is given in Table II. With single DC link configuration, the topology consists of two DC link capacitors and 24 (8×3)

switches. The proposed topology has less components due to the absence of clamping capacitors, diodes and fewer independent power sources. The reduced number of electrolytic capacitors, (here only 2 compared to 4 in NPC, 10 in FC and 6 in CHB) increases the reliability of the topology. The DC link voltage requirement is also half as compared to the classical NPC and FC inverters. Although the total number of switches is same as that of the classical inverters, the rating of switches are higher in the proposed topology. The H-bridge switches should be rated for the DC link voltage ($0.5V_{dc}$). However, these switches can be made to commute at the fundamental frequency thereby minimising their loss contribution.

Table II: Comparison with classical 5L inverter topologies

Topology	Switch		Capacitors		Diodes	DC sources
	$0.25V_{dc}$	$0.5V_{dc}$	DC link	Clamping		
Diode Clamped (NPC)	24	0	4	0	18	1
Capacitor Clamped (FC)	24	0	1	9	0	1
Cascaded H bridge (CHB)	24	0	6	0	0	6
Proposed topology	12	12	2	0	0	1

Carrier Based Modulation Technique

One of the simplest and most common modulation method for MLI topology is the level shifted carrier PWM. The gating pulse generation using level shifted conventional space vector (LS-CSV) PWM for the proposed topology is shown in Fig. 2. The modulating reference signals (v_x^*) of the three phase is compared with two level shifted carriers (v_l and v_u) to generate the gating pulse for the low voltage switches, S_{1X} and S_{2X}. Note that the logic comparison for generating S_{2X} pulse is opposite to that of S_{1X}. The gating pulse of H-bridge switch, S_X is high for the positive half and low for the negative half cycle of the reference signal.

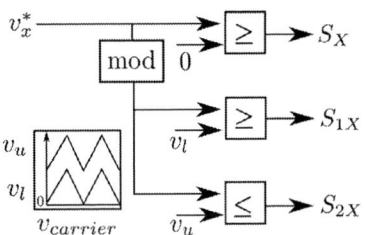

Fig. 2: Gating pulse generation using LS-CSV PWM

The output voltage of the proposed topology along with the gating signal is shown in Fig. 3. The inverter is fed with single DC link voltage ($0.5V_{dc}$) of 150 V. Induction motor is operated with constant Volts-per-Hertz control. Triangular carrier with synchronous sampling of 24 samples per cycle is considered for PWM generation. From Fig. 3a, we can see that the high voltage switch (S_R) commutes at the fundamental frequency. This reduces the switching loss. Also, at lower modulation index (m), the upper switches (S_{2R}) remains clamped to logic high and do not contribute to switching loss.

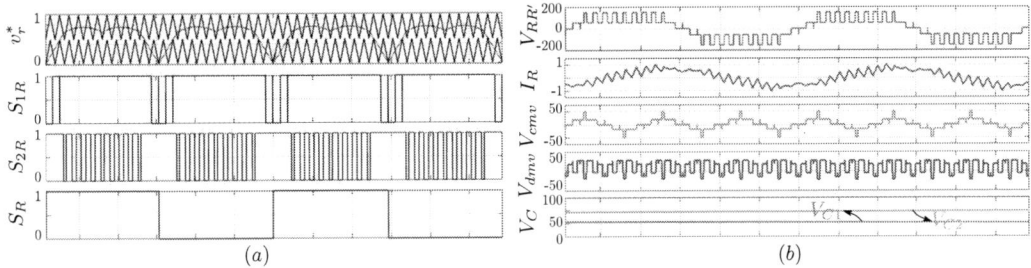

Fig. 3: Waveforms showing (a) R-phase modulating signal (v_r^*) and gating pulses (S_{1R}, S_{2R}, S_R); (b) phase voltage ($V_{RR'}$), phase current (I_R), common mode voltage (V_{cmv}), differential mode voltage (V_{dmv}) and capacitor voltages (V_{C1}, V_{C2}) at $m = 0.89$ (35 Hz) and 0.7 load power factor (pf) for two fundamental cycles using LS-CSV PWM

One of the major drawbacks of the conventional modulation scheme is the presence of common mode (V_{cmv}) and differential mode voltages (V_{dmv}) as observed in Fig. 3b. The differential mode voltage (DMV) is expressed as the average of the phase voltages, $\sum V_{XX'}/3$ and common mode voltage (CMV) is expressed as the average of all the pole voltages, $\sum(V_{XO} + V_{X'O})/6$ [7]. Due to single DC supply, presence of DMV results in circulating current through the motor phase winding and the presence of CMV leads to shaft voltage build up due to the parasitic capacitance. Using the conventional LS-CSV PWM, the DMV reaches peak magnitude of $V_{dc}/10$ and CMV is around $V_{dc}/6$ at higher modulation indicies. Also, from the plot of capacitor voltage (V_C in Fig. 3b), we observe that the conventional PWM fails to balance the DC link capacitor voltages (V_{C1} and V_{C2}) leading to imbalance in neutral point potential. Hence, for the conventional PWM method, additional control is required to maintain the capacitor balance and to eliminate the harmful effects of common mode voltage.

Proposed Modulation Technique

The voltage vector redundancy of the topology can be utilised to eliminate the common mode and differential mode voltages. In the 5L inverter, out of the 61 locations, there are 19 locations which has zero DMV vectors. The voltage vectors which gives zero DMV are highlighted in the space vector structure of Fig. 4a. The zero DMV vectors resembles a $90°$ shifted three level inverter space vector structure with 6 sectors (1 to 6) as shown in Fig. 4b. However, note that the magnitude of large voltage vectors of conventional 3L inverter is V_{dc}, while that constituting zero DMV vectors are $0.866V_{dc}$ ($V_{dc}\cos 30°$). In general, the magnitude of zero DMV vectors of 5L inverter are 0.866 times the magnitude of active vectors of conventional 3L inverter.

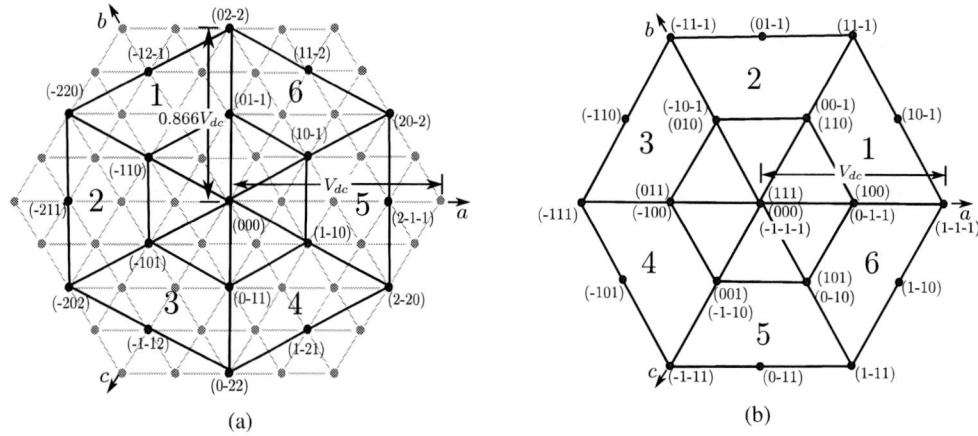

Fig. 4: Space vector structure depicting (a) zero DMV vectors of the 5L inverter and (b) voltage vectors of 3L inverter

In the zero DMV vectors, the states -1, 1 and 0 have switching state redundancy i.e., they can be obtained by two possible switching combinations (refer Table I). The switching state for the zero DMV vectors are selected such that the resulting CMV is zero. One of the possible switching state selection is given in Table III. For each of the zero DMV vectors, we can find corresponding switching state (S_{1X}, S_{2X} and S_X) such that the CMV is zero. From the table, for the zero DMV vector combination of (1,1,-2) given in column-1, the switching state for R phase is 001, Y phase is 111 and B phase is 110 as given in column-3. The corresponding pole voltage at each phase is given in column-4. The average of all pole voltages gives CMV (column-5). Here we can see that the redundancy of state 1 (takes 001 for R phase while 111 for Y phase) is used to make the average of pole voltages (CMV) as zero. Similarly for the vector combination of (2,-1,-1) the redundancy of state -1 (bold) is utilised to eliminate the CMV.

The switching state redundancy of zero vector can be utilised to reduce the switching transition in the H-bridge switches. The redundancy in the zero vectors, 010 and 011 (refer Table I) is in the H-bridge state

Table III: Selection of switching state for zero CMV

1. Zero DMV vectors of 5L inverter	2. Voltage state	3. Switching state			4. Pole voltage		5. CMV
		S_{1X}	S_{2X}	S_X	V_{X0}	$V_{X'0}$	$\sum(V_{XO}+V_{X'O})/6$
(000)	0	0	1	x	0	0	0
$f(1,0,-1)$	1	0	0	1	0	$-0.25V_{dc}$	$\left.\begin{array}{l} \\ \\ \end{array}\right\}$ 0
	0	0	1	x	0	0	
	-1	1	1	0	0	$+0.25V_{dc}$	
$f(1,1,-2)$	**1**	**0**	**0**	**1**	0	$-0.25V_{dc}$	$\left.\begin{array}{l} \\ \\ \end{array}\right\}$ 0
	1	**1**	**1**	**1**	$+0.25V_{dc}$	0	
	-2	1	0	0	$-0.25V_{dc}$	$+0.25V_{dc}$	
$f(2,0,-2)$	2	1	0	1	$+0.25V_{dc}$	$-0.25V_{dc}$	$\left.\begin{array}{l} \\ \\ \end{array}\right\}$ 0
	0	0	1	x	0	0	
	-2	1	0	0	$-0.25V_{dc}$	$+0.25V_{dc}$	
$f(2,-1,-1)$	2	1	0	1	$+0.25V_{dc}$	$-0.25V_{dc}$	$\left.\begin{array}{l} \\ \\ \end{array}\right\}$ 0
	-1	**1**	**1**	**0**	0	$+0.25V_{dc}$	
	-1	**0**	**0**	**0**	$-0.25V_{dc}$	0	

(can take 0 or 1). The zero vector for each phase is chosen to obtain H-bridge switching at fundamental frequency similar to LS-CSV PWM. As the H-bridge switches commute at higher voltage (twice the other switches), this is beneficial in reducing switching loss. Referring to the space vector structure of Fig. 4a, vectors constituting Sector-5 are 000, 1-10, 2-20, 2-1-1, 20-2, 10-1. Here, note that, R-phase takes zero or positive (1 or 2) state only, which implies that the H bridge switch, (S_R) is gated high always (on similar notes S_Y and S_B are gated low). Hence in Sector-5, to avoid unnecessary commutation, the zero vector for R phase is chosen such that S_R is high i.e, 011 and for Y and B phase, the zero vector is 010. The zero vector for each phase in other sectors can be derived in similar manner and are listed in Table IV.

Table IV: Selection of zero vector

Phase	Sector-1	Sector-2	Sector-3	Sector-4	Sector-5	Sector-6
R- phase	010	010	010	011	011	011
Y- phase	011	011	010	010	010	011
B- phase	010	011	011	011	010	010

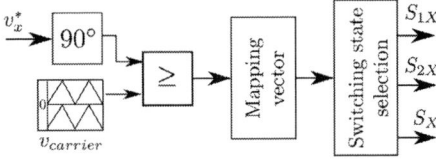

Fig. 5: Gating pulse generation using proposed modulation scheme

The gating pulse generation of proposed modulation scheme is depicted as a block diagram in Fig. 5. The reference modulating signal (v_x^*) is shifted by $90°$ and compared with level shifted carriers to obtain the space vector structure of Fig. 4b. Then, each possible vector combination of three level inverter is mapped to the corresponding zero DMV vectors of the 5L inverter. Vectors in Sector-1 of 3L (say 1-1-1 in Fig. 4b) is mapped to Sector-1 of 5L inverter (02-2 in Fig. 4a) and so on. Then, the information of selected switching state for redundant vectors are encoded to obtain the gating pulse. Due to switching of only zero DMV vectors, the modulation range using the proposed PWM is limited to $m = 1$ ($0.5V_{dc}$), as opposed to other third harmonic injection PWM methods which gives a maximum output phase voltage of $0.577V_{dc}$ ($m = 1.15$). This is because the modulation limit of proposed PWM is decided by the circle inscribed within the zero DMV vector hexagon and not by the outer hexagon of 5L inverter [12]. Further, for the proposed PWM method, because of the reduced active vector magnitude of the inscribed zero DMV vectors (as indicated in Fig. 4) the magnitude of instantaneous reference modulating signal is 1.15 ($1/\cos 30°$) times that of LS-CSV PWM to ensure same modulation index (fundamental voltage).

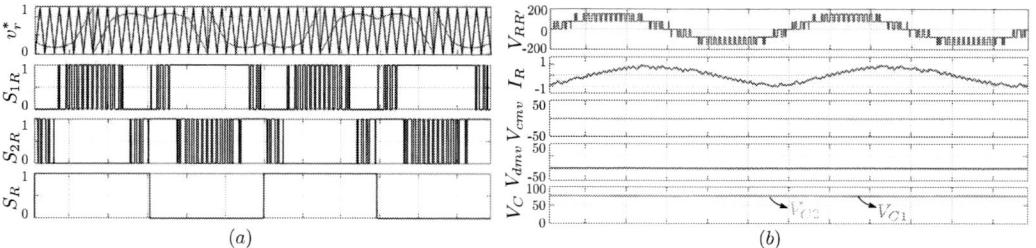

Fig. 6: Waveforms showing (a) R-phase modulating signal (v_r^*) and gating pulses (S_{1R}, S_{2R}, S_R) and (b) phase voltage ($V_{RR'}$), phase current (I_R), common mode voltage (V_{cmv}), differential mode voltage (V_{dmv}) and capacitor voltages (V_{C1}, V_{C2}) at $m = 0.89$ (35 Hz) and 0.7 pf for two fundamental cycles using proposed PWM

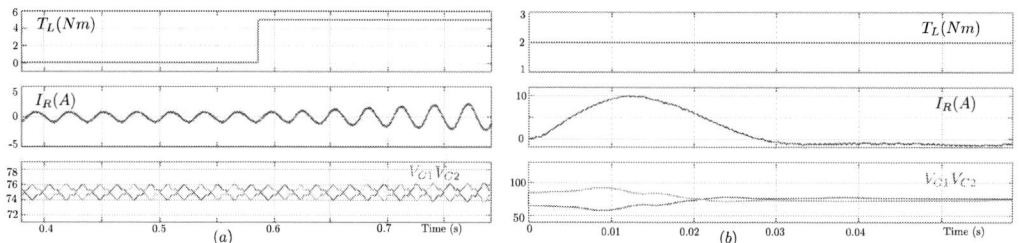

Fig. 7: Analysis of self balancing of capacitor voltage (V_{C1}, V_{C2}) under (a) sudden load change from T_L =0.1 to T_L =5 Nm and (b) different initial voltage (V_{C0}) in capacitors, $V_{C10} = 30V$ and $V_{C20} = 50V$.

The voltage waveforms and gating pulse using the proposed PWM are shown in Fig. 6. For ease of implementation, the R-phase modulating signal (v_r^*) is level shifted and compared with a single carrier. The mapping and switching state selection result in gating pulse S_{1X}, S_{2X} and S_X. Using the proposed PWM, the CMV and DMV are zero as observed from Fig. 6b (V_{cmv} and V_{dmv}). The switching in the H-bridge switches (S_X) is at fundamental frequency. From the phase current profile (I_R), we observe that the current waveform of proposed modulation scheme is better than that of the conventional PWM (Fig. 3b). Also, the PWM scheme is able to balance the capacitor voltages (V_{C1} and V_{C2}) to half the DC link voltage, without any additional controller. The capacitor voltage balancing is analysed under sudden load change and under different capacitor initial voltage condition. In both the cases, the capacitor voltages converge to 75 V ($V_{dc}/4$) as shown in Fig. 7.

The quality of current waveform is analysed based on stator current ripple for the entire modulation range of the proposed PWM. Stator current ripple is obtained as the integral of voltage error between the reference voltage and the actual output voltage of the inverter [13]. From the plot of Fig. 8, it is observed that the proposed PWM results in reduced ripple (distortion) in current for the entire operating condition as compared to LS-CSV PWM. The higher distortion in LS-CSV PWM is due to capacitor unbalance and the switched voltage vector. The voltage vectors switched for LS-CSV PWM and proposed PWM over a fundamental cycle of 35 Hz ($m = 0.89$) is shown in Fig. 9. In LS-CSV PWM, the reference voltage is obtained as switched average of inner medium vectors and outermost large vectors (highlighted in red in

Fig. 9a). In the proposed PWM method, the reference voltage is obtained as switched average only zero DMV vectors (highlighted in red in Fig. 9b). The outermost vectors increases the error voltage, this may be the reason for higher distortion in LS-CSV PWM as compared to proposed PWM which has fewer outermost vectors.

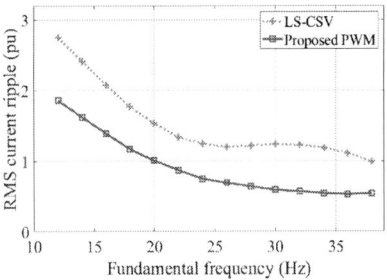

Fig. 8: Stator current ripple of LS-CSV and proposed PWM with respect to fundamental frequency (modulation index).

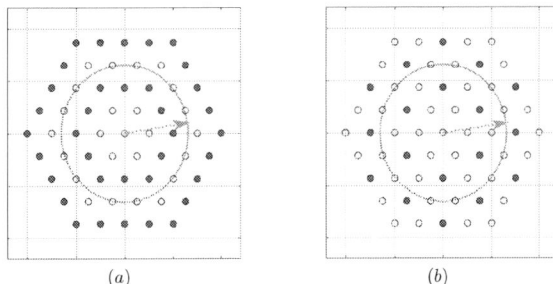

Fig. 9: Vectors switched for (a) LS-CSV PWM and (b) Proposed PWM in a fundamental cycle of 35 Hz. Blue trace depicts the reference voltage signal. Black dots represents the available vectors of 5L inverter. Red dots represents the vectors switched in a fundamental cycle.

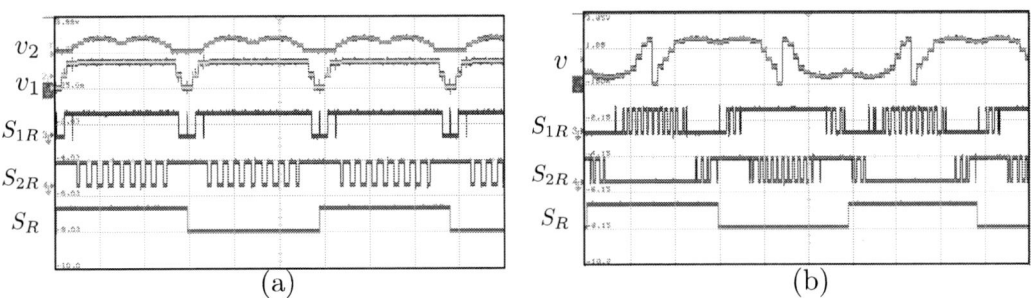

Fig. 10: Modulating wave and gating pulse of R-phase with (a) LS-CSV PWM and (b) proposed PWM at 35 Hz. X-axis:5 ms/div

The implementation of modulation scheme is done using TMS320F28379D DSP board and Xilinx Spartan-6 XC6SLX4 FPGA module. The triangular carrier comparison, vector mapping and switching state selection is done in DSP board. A dead time of 2 μsec to avoid shoot-through among complementary switches is given using FPGA module. The modulating wave and gating pulse obtained using both the PWM methods are shown in Fig. 10. For the LS-CSV PWM, the modulating signal v_1 generates the gating pulse S_{1R} and v_2 generates the gating pulse S_{2R}. In the proposed PWM, the modulating wave is level shifted and compared with a single carrier. We observe that for both the PWM methods, the H-bridge switch (S_R) commutes at fundamental frequency. However the switching in low voltage switches, S_{1R} and S_{2R} of proposed PWM are higher than LS-CSV PWM. To reduce the switching transitions in the low voltage switches of the proposed modulation scheme, discontinuous PWM can be applied instead of CSV PWM. In discontinuous PWM (bus clamping PWM), as the modulating wave of each phase remains clamped for certain duration [13], the switching in low voltage switch can be reduced.

Conclusion

The paper proposes a carrier based modulation scheme for a five level inverter topology to eliminate common mode and differential mode voltages. The five level topology has a simple and modular structure with reduced component count as compared to the classical five level inverter topologies. The modulation scheme uses voltage vector and switching state redundancy of the topology to eliminate the DMV and CMV in the motor. Zero DMV avoids circulating current in the motor winding and zero CMV avoids the bearing current. Thus, the proposed modulation scheme plays a key role in applications where reliability is of utmost importance such as deep excavation, mining etc. The modulation scheme also ensures balancing of DC link capacitor voltage and has superior harmonic performance as compared to the conventional PWM method.

References

[1] S. Kouro, M. Malinowski, K. Gopakumar, J. Pou, L. G. Franquelo, B. Wu, J. Rodriguez, M. A. Perez and J. I. Leon:Recent Advances and Industrial Applications of Multilevel Converters, IEEE Transactions on Industrial Electronics, vol. 57, pp. 2553-2580, Aug 2010

[2] A. Akbari, F. Poloei and A. Bakhshai.:A Minimum Number of Switch Five-Level Inverter for Renewable Energy Sources Applications,2019 1st Global Power, Energy and Communication Conference (GPECOM), pp. 134-139, June 2019

[3] G. Buticchi and E. Lorenzani and G. Franceschini.:A Five-Level Single-Phase Grid-Connected Converter for Renewable Distributed Systems, IEEE Transactions on Industrial Electronics, vol. 60, pp. 906-918, March 2013

[4] K. Wang, Z. Zheng, D. Wei, B. Fan and Y. Li.:Topology and Capacitor Voltage Balancing Control of a Symmetrical Hybrid Nine-Level Inverter for High-Speed Motor Drives, IEEE Transactions on Industry Applications, vol. 53, pp. 5563-5572, Nov 2017

[5] V. Nair R, K. Gopakumar and L. G. Franquelo.: A Very High Resolution Stacked Multilevel Inverter Topology for Adjustable Speed Drives, IEEE Transactions on Industrial Electronics, vol. 65, pp. 2049-2056, March 2018

[6] A. Willwerth and M. Roman.: Electrical bearing damage a lurking problem in inverter-driven traction motors, in 2013 IEEE Transportation Electrification Conference and Expo (ITEC), pp. 14, June 2013

[7] J. Kalaiselvi and S. Srinivas.:Bearing currents and shaft voltage reduction in dual-inverter-fed open-end winding induction motor with reduced cmv pwm methods, IEEE Transactions on Industrial Electronics, vol. 62, pp. 144152, Jan 2015

[8] M. G. Majumder, A. K. Yadav, K. Gopakumar, K. R. R, U. Loganathan and L. G. Franquelo.:A 5-Level Inverter Scheme Using Single DC Link With Reduced Number of Floating Capacitors and Switches for Open-End IM Drives, IEEE Transactions on Industrial Electronics, vol. 67, pp. 960-968, Feb 2020

[9] P. N. Tekwani, R. S. Kanchan, K. Gopakumar, and A. Vezzini.: A five-level inverter topology with common mode voltage elimination for induction motor drives, in EPE 2005, pp. 10, Sept 2005

[10] N. Bodo, M. Jones and E. Levi.:A Space Vector PWM with Common-Mode Voltage Elimination for Open-End Winding Five-Phase Drives With a Single DC Supply, IEEE Transactions on Industrial Electronics, vol. 61, pp. 2197-2207, May 2014

[11] P. P. Rajeevan and K. Gopakumar.:A Hybrid Five-Level Inverter With Common-Mode Voltage Elimination Having Single Voltage Source for IM Drive Applications, IEEE Transactions on Industry Applications, vol. 48,pp. 2037-2047, Nov 2012

[12] M. R. Baiju, K. K. Mohapatra and R. S. Kanchan and K. Gopakumar.:A dual two-level inverter scheme with common mode voltage elimination for an induction motor drive, IEEE Transactions on Power Electronics, vol. 19, pp. 794-805, May 2004

[13] G. Nadh, S. Durga Nair and S. Arun Rahul.:Evaluation of modulation methods on switching loss, common mode voltage and current ripple for an open end winding induction motor drive, IEEE international conference on sustainable energy technologies and systems, pp. 126-131, Feb 2019

A Fast and Robust Model of Dual-Active Bridge Converters in Real-Time Simulation

Ming Jia, Philipp Joebges, Rik W. De Doncker
Institute for Power Generation and Storage Systems
E.ON Energy Research Center
RWTH Aachen University
Mathieustrasse 10, 52074
Aachen, Germany
Phone: +49 24180-49940
Email: post_pgs@eonerc.rwth-aachen.de

Acknowledgments

This work is supported by the European Regional Development Fund (EFRE-0500029).

Keywords

≪Modelling≫, ≪Real-time simulation≫, ≪Converter circuit≫, ≪DC grid≫, ≪Stability analysis≫

Abstract

Real-time simulation is becoming an emerging and powerful tool for the development of power electronics conversion systems. One widely employed methods used in real-time (RT) simulations is the time average method. In this paper, a C-code average model is proposed for a two-level three-phase dual-active bridge dc-dc converter (2L-DAB3) to perform an efficient RT simulation. The calculation speed and accuracy as well as its stability with different integration methods are investigated and compared with state-of-the-art auto-generated models.

Introduction

In recent years, control hardware-in-the-loop (CHiL) tests are getting more and more popular in industry. It enables the test of a controller under thousands of valid scenarios without the risk of damage in hardware components. In CHiL scenarios, a physical controller is connected to a virtual plant in an RT simulator environment instead of a real plant [1]. An RT simulator executes each time-step with a fixed time-step, synchronizes with the PWM signals and therefore synchronizes with the real-world time. However, an offline simulation with variable time-step can be more accurate, but it can run much faster or slower than the real plant. The execution time of the offline simulation depends not only on the time-step but also on the calculation capability and the system complexity [2]. A digital RT simulator is only capable of operating a fixed-step model. For a fixed-step simulation, an average model is more accurate than a switched model. It will be explained in the next section. It is a very promising and attractive approach to verify controller before performing experiments with hardware components. The average model includes the averaged switching states during a switch change in the calculation, while the switched model doesn't. Different fixed-step solvers, forward Euler (FE), backward Euler (BE) and trapezoidal (TR), require also different time-steps to calculate accurate results from the model.

Averaging techniques and the corresponding equivalent model are widely employed to model power electronics in RT simulations, as they are capable of handling decimal gate signals. Previously published studies of the time average model perform the averaging over a switching period [3]. More recently, a sub-cycle average model for half-bridge power modules is proposed in [4]. Two integrated diodes are utilized in the model to realize the bidirectional current flow. Besides, an averaging of the PWM signals is performed during one sampling period, which improves the precision and suppressing the switching harmonics. It is implemented as a sub-cycle averaged mode of a half-bridge in the library of the simulation software PLECS. The corresponding C-code can be generated through the code generation function and employed to the RT simulator directly. However, auto-generated code can be substantially longer with increased complexity, which results in larger binary files and execution time for the RT simulator. In this work, a self-developed C-code based average model is investigated as an alternative of an auto-generated average model taking the example of a 2L-DAB3 converter. It minimizes the size of the model description and accelerates the calculation and execution process of the RT simulation. The efficiency of the RT simulation is analyzed and the numerical stability is investigated. The proposed models are validated in simulation and experiments.

C-Code Average Model of 2L-DAB3

The average gate signals can be calculated as a duty cycle over one sub-cycle time-step T, as demonstrated in [5]. Hence, an average model can process the decimal gate signals between zero and one. In this work, a 2L-DAB3 converter, which is a promising topology for application in future dc grids, is investigated as an example. As depicted in Fig. 1, it consists of three half-bridges in both primary side and secondary side connected by a transformer in between. A straight forward average model of the 2L-DAB3 is implemented with the half-bridge component in the library of PLECS [6] for comparison.

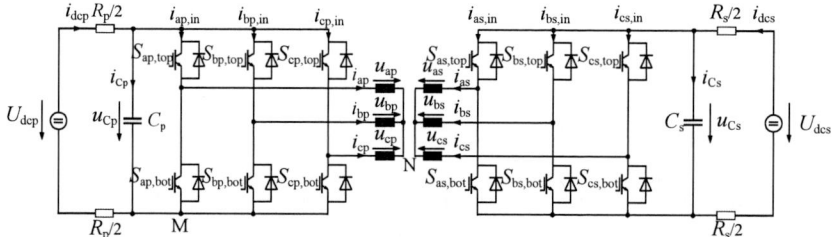

Fig. 1: Circuit diagram of a 2L-DAB3 converter

Modelling Methodology

A PLECS average model of a half-bridge consists of controlled currents, voltage sources, and a diode pair [4]. It is capable of processing decimal switching gate signals. A further explanation can be found in [7]. To improve its efficiency, a mathematical model of a 2L-DAB3 is derived and then converted into a C-code description. To visualize the sub-cycle averaging, a commutation from the bottom switch to the top switch of one phase-leg is illustrated in Fig. 2. The red lines S_{top} and S_{bot} indicate the real top and bottom gate signals, as utilized in the switched model. The blue lines with shadowed areas $\overline{S}_{top}$ and $\overline{S}_{bot}$ represent the averaged gate signals over one time-step with consideration of dead time. In a fixed time-step simulation, it is impossible to capture the exact time when the current changes its direction in one sampling period. During the dead time both the upper and lower switches are switched off, as shown in Fig. 2. This cannot be precisely included in the switched model. Therefore, the average model is more precise than the switched model for a fixed time-step simulation, as the precise start and end time of the dead time are included. The averaged switching signals are calculated based on (1).

$$\overline{S}_{bot(top)} \text{ of one switching state} = \frac{\text{number of sampling points with value 1 of } S_{bot(top)}}{\text{number of total sampling points of } S_{bot(top)}} \tag{1}$$

In the RT simulator, the PWM signals are sampled every 20 ns by a field-programmable gate array (FPGA). Then the average PWM values are calculated for every time-step, and used for the further calculation in digital signal processor (DSP), as shown in Fig. 2 with blue lines.

The developed C-code average model can also deal with decimal input signals. Taking the modelling of

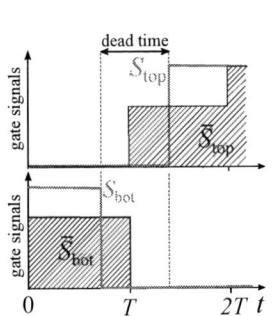

Fig. 2: A switching state of one phase-leg

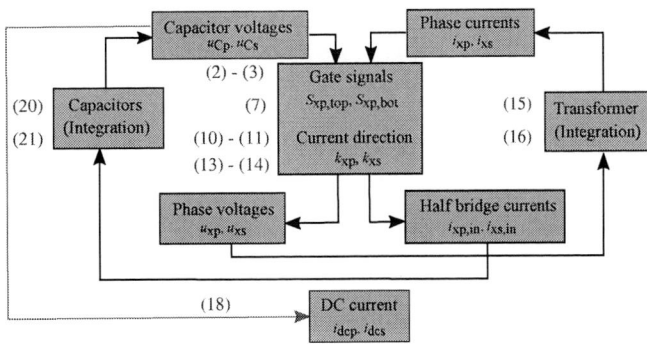

Fig. 3: Flowchart of the C-code realization for a 2L-DAB3 converter

the primary side in the 2L-DAB3 as an example, k_{xp} is introduced in (2) to represent the direction of the current in the average model. Hence, k_{xp} is either one for a positive ac current or zero for a negative ac current as shown in (3). $\overline{S}_{xp,top}$ and $\overline{S}_{xp,bot}$ represent the averaged switching signals for the top switch and the bottom switch in one phase leg of the primary side. The secondary side can be calculated similarly. $x \in \{a, b, c\}$ represents the phase leg. The representative parameters are corresponding to the variables in Fig. 1. The phase voltage in respect to the negative dc rail M of the primary side, is indicated as $\overline{u}_{xp,M}$. The current flowing through each half-bridge is denoted as $\overline{i}_{xp,in}$.

$$
\begin{cases}
\overline{u}_{xp,M} = k_{xp} \cdot \overline{S}_{xp,top} \cdot u_{Cp} + (1 - k_{xp}) \cdot (1 - \overline{S}_{xp,bot}) \cdot u_{Cp} \\
\qquad = \overline{S}_{uxp,M} \cdot u_{Cp} \\
\overline{i}_{xp,in} = k_{xp} \cdot \overline{S}_{xp,top} \cdot i_{xp} + (1 - k_{xp}) \cdot (1 - \overline{S}_{xp,bot}) \cdot i_{xp} \\
\qquad = \overline{S}_{ixp} \cdot i_{xp}
\end{cases}
\quad (2) \qquad
k_{xp} = \begin{cases} 1 & i_{xp} \geq 0 \\ 0 & i_{xp} < 0 \end{cases}
\quad (3)
$$

$\overline{S}_{uxp,M}$ and $\overline{S}_{ixp}$ are the equivalent switching gate signals for the phase voltage with negative dc rail M of primary side as reference and the equivalent switching gate signals for the ac current.

$$
\begin{cases}
k_{xp} \cdot \overline{S}_{xp,top} + (1 - k_{xp}) \cdot (1 - \overline{S}_{xp,bot}) = \overline{S}_{uxp,M} \\
k_{px} \cdot \overline{S}_{xp,top} + (1 - k_{xp}) \cdot (1 - \overline{S}_{xp,bot}) = \overline{S}_{ixp}
\end{cases}
\tag{4}
$$

As the topology is assumed to be symmetrical, the voltage of the transformer midpoint O is obtained as

$$
\overline{u}_{Op} = \frac{1}{3} \cdot (\overline{u}_{ap,M} + \overline{u}_{bp,M} + \overline{u}_{cp,M}),
\tag{5}
$$

which results in the transformer phase-to-midpoint voltage

$$
\overline{u}_{ap} = \overline{u}_{ap,M} - \overline{u}_{Op} = \frac{2}{3} \cdot \overline{u}_{ap,M} - \frac{1}{3} \cdot \overline{u}_{bp,M} - \frac{1}{3} \cdot \overline{u}_{cp,M}.
\tag{6}
$$

And therefore, taking phase a as example, the equivalent switching state of the phase voltage is derived as

$$
\begin{aligned}
\overline{S}_{uap} &= \frac{2}{3} \cdot \overline{S}_{uap,M} - \frac{1}{3} \cdot \overline{S}_{ubp,M} - \frac{1}{3} \cdot \overline{S}_{ucp,M} \\
&= \frac{2}{3} \cdot k_{ap} \cdot \overline{S}_{ap,top} + \frac{2}{3} \cdot (1 - k_{ap}) \cdot (1 - \overline{S}_{bp,bot}) - \frac{1}{3} \cdot k_{bp} \cdot \overline{S}_{bp,top} - \frac{1}{3} \cdot (1 - k_{bp}) \cdot (1 - \overline{S}_{bp,bot}) \quad (7) \\
&\quad - \frac{1}{3} \cdot k_{cp} \cdot \overline{S}_{cp,top} - \frac{1}{3} \cdot (1 - k_{cp}) \cdot (1 - \overline{S}_{cp,bot}).
\end{aligned}
$$

Similarly, for the other two phases, $\overline{S}_{ubp}$ and $\overline{S}_{ucp}$ can be derived. The equation set is based on the primary side parameters, but valid for both the primary side and secondary side. The self-defined code model complies with the power conservation rules, that the input power is equal to the output power for each half-bridge. The C-code calculation procedure is roughly presented in Fig. 3. First, the dc-link capacitor voltages (u_{Cp}, u_{Cs}) and ac currents (i_{xp}, i_{xs}) are initialized. The phase voltages (u_{xp}, u_{xs}) and currents flowing through each half-bridge ($i_{xp,in}$, $i_{xs,in}$) can then be determined by the given switching signals and current directions with (2). Through the integration (15) and (16), the capacitor voltages and ac currents are updated. The calculation can thereafter be iteratively executed once for each time-step.

The switching functions can be derived according to the current relationship. The derived equations should be transferred from abc-frame to $\alpha\beta$-frame for the stability analysis in the following section. Only in the $\alpha\beta$-frame, the characteristic state-space matrix is with full-rank and is thus capable of the matrix inversion calculation. It will be discussed later. According to [8], the inverse Park transformation of the current in a balanced state can be expressed as

$$
\begin{bmatrix} i_{ap} \\ i_{bp} \\ i_{cp} \end{bmatrix} = \begin{bmatrix} 1 & 0 \\ -\frac{1}{2} & \frac{\sqrt{3}}{2} \\ \frac{1}{2} & -\frac{\sqrt{3}}{2} \end{bmatrix} \cdot \begin{bmatrix} i_{\alpha p} \\ i_{\beta p} \end{bmatrix}.
\tag{8}
$$

The current i_p flowing through the primary side of the dc link can be derived in a stationary $\alpha\beta$ frame

$$
i_p = \begin{bmatrix} \overline{S}_{iap} & \overline{S}_{ibp} & \overline{S}_{icp} \end{bmatrix} \cdot \begin{bmatrix} i_{ap} \\ i_{bp} \\ i_{cp} \end{bmatrix} = \begin{bmatrix} \overline{S}_{iap} & \overline{S}_{ibp} & \overline{S}_{icp} \end{bmatrix} \cdot \begin{bmatrix} 1 & 0 \\ -\frac{1}{2} & \frac{\sqrt{3}}{2} \\ -\frac{1}{2} & -\frac{\sqrt{3}}{2} \end{bmatrix} \cdot \begin{bmatrix} i_{\alpha p} \\ i_{\beta p} \end{bmatrix}
$$

$$
= \begin{bmatrix} \overline{S}_{iap} - \frac{1}{2} \cdot \overline{S}_{ibp} - \frac{1}{2} \cdot \overline{S}_{icp} & \frac{\sqrt{3}}{2} \cdot \overline{S}_{ibp} - \frac{\sqrt{3}}{2} \cdot \overline{S}_{icp} \end{bmatrix} \cdot \begin{bmatrix} i_{\alpha p} \\ i_{\beta p} \end{bmatrix}.
\tag{9}
$$

The equivalent switching states of the current calculation in the $\alpha\beta$-frame can be concluded as

$$
\overline{S}_{i\alpha p} = \overline{S}_{iap} - \frac{1}{2} \cdot \overline{S}_{ibp} - \frac{1}{2} \cdot \overline{S}_{icp}
\tag{10}
$$

$$
\overline{S}_{i\beta p} = \frac{\sqrt{3}}{2}(\overline{S}_{ibp} - \overline{S}_{icp}).
\tag{11}
$$

In the similar way, the transformation of the voltages are

$$
\begin{bmatrix} u_{\alpha p} \\ u_{\beta p} \end{bmatrix} = \frac{2}{3} \begin{bmatrix} 1 & -\frac{1}{2} & -\frac{1}{2} \\ 0 & \frac{\sqrt{3}}{2} & -\frac{\sqrt{3}}{2} \end{bmatrix} \begin{bmatrix} u_{ap} \\ u_{bp} \\ u_{cp} \end{bmatrix}
$$

$$
\begin{bmatrix} \overline{S}_{u\alpha p} \\ \overline{S}_{u\beta p} \end{bmatrix} \cdot u_{Cp} = \begin{bmatrix} \frac{2}{3} & -\frac{1}{3} & -\frac{1}{3} \\ 0 & \frac{1}{\sqrt{3}} & -\frac{1}{\sqrt{3}} \end{bmatrix} \begin{bmatrix} \overline{S}_{uap} \\ \overline{S}_{ubp} \\ \overline{S}_{ucp} \end{bmatrix} \cdot u_{Cp}.
\tag{12}
$$

The equivalent switching states for the voltage calculation in the $\alpha\beta$-frame can be concluded.

$$
\overline{S}_{u\alpha p} = \frac{2}{3} \cdot \overline{S}_{uap} - \frac{1}{3} \cdot \overline{S}_{ubp} - \frac{1}{3} \cdot \overline{S}_{ucp}
\tag{13}
$$

$$
\overline{S}_{u\beta p} = \frac{1}{\sqrt{3}} \cdot (\overline{S}_{ubp} - \overline{S}_{ucp})
\tag{14}
$$

To summarize, applying (4) the equivalent switching state in abc-frame can be calculated refer to the negative dc rail M of the primary side. Thereafter applying (7) to the switching states, they can be transferred into a system with the transformer midpoint O as reference point. The switching states of the current and voltage in the $\alpha\beta$-frame can be derived with (10), (11), (13) and (14). The concept of the equivalent switching state is introduced for the calculation of the corresponding voltages and currents.

Only the switching states $\overline{S}_{\text{xp,top}}$ and $\overline{S}_{\text{xp,bot}}$ represent the states of the real semiconductor switches. The further mentioned equivalent switching states $\overline{S}_{\text{uxp}}$, $\overline{S}_{\text{ixp}}$, $\overline{S}_{\text{u}\alpha(\beta)\text{p}}$ and $\overline{S}_{\text{i}\alpha(\beta)\text{p}}$ are derived to establish the mathematical model of 2L-DAB3, and have no physical meanings.

Simplified Transformer Model

As the power transfer in the 2L-DAB3 converter is determined by the voltage across the transformer's leakage inductance, the transformer is simplified as a series-connected inductor between the primary side and the secondary side. The mutual inductance is omitted, as shown in Fig. 4. The transformer is related to the primary side, with the total leakage inductance $L_\sigma = L_\text{p} + L'_\text{s}$, where L_p is the leakage inductance of the primary side and L'_s is the primary referred secondary side leakage inductance. This applies analog for the copper resistance $R_\text{w} = R_\text{p} + R'_\text{s}$. The current flowing through the transformer can be calculated according to the primary side. The primary side ac currents in $\alpha\beta$-frame $i_{\alpha\text{p}}$ and $i_{\beta\text{p}}$ are defined as i_α and i_β. Therefore, the secondary side ac currents $i_{\alpha\text{s}}$ and $i_{\beta\text{s}}$ are equal to $-N \cdot i_\alpha$ and $-N \cdot i_\beta$.

$$
\begin{aligned}
L_\sigma \cdot \frac{di_\alpha}{dt} &= u_{\alpha\text{p}} - N \cdot u_{\alpha\text{s}} - R_\text{w} \cdot i_\alpha \\
\leftrightarrow L_\sigma \cdot \frac{di_\alpha}{dt} &= \overline{S}_{\text{u}\alpha\text{p}} \cdot u_{\text{Cp}} - \overline{S}_{\text{u}\alpha\text{s}} \cdot N \cdot u_{\text{Cs}} - R_\text{w} \cdot i_\alpha \\
\leftrightarrow \frac{di_\alpha}{dt} &= \frac{\overline{S}_{\text{u}\alpha\text{p}}}{L_\sigma} \cdot u_{\text{Cp}} - \frac{\overline{S}_{\text{u}\alpha\text{s}}}{L_\sigma} \cdot N \cdot u_{\text{Cs}} - \frac{R_\text{w}}{L_\sigma} \cdot i_\alpha
\end{aligned}
\tag{15}
$$

It can be derived similarly for the β component of the current.

$$
\frac{di_\beta}{dt} = \frac{\overline{S}_{\text{u}\beta\text{p}}}{L_\sigma} \cdot u_{\text{Cp}} - \frac{\overline{S}_{\text{u}\beta\text{s}}}{L_\sigma} \cdot N \cdot u_{\text{Cs}} - \frac{R_\text{w}}{L_\sigma} \cdot i_\beta
\tag{16}
$$

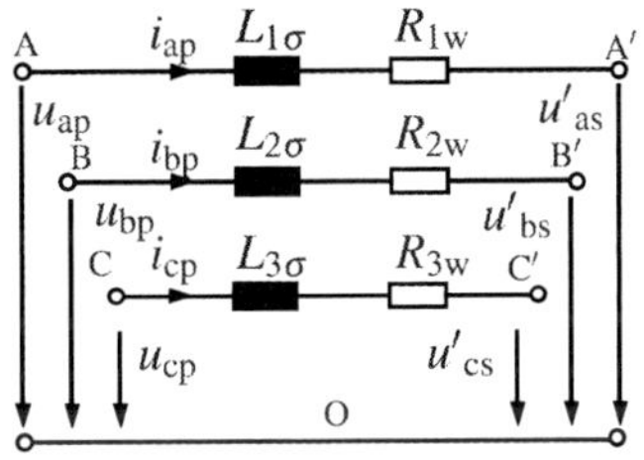

Fig. 4: Three-phase transformer with infinite mutual inductance, with $L_{1\sigma} = L_{2\sigma} = L_{3\sigma} = L_\sigma$ and $R_{1\text{w}} = R_{2\text{w}} = R_{3\text{w}} = R_\text{w}$

DC Link

According to the Kirchhoff's voltage law (KVL), the dc-link capacitor voltage in Fig. 1 can be obtained from

$$
\begin{aligned}
i_{\text{Cp}} = C \cdot \frac{du_{\text{Cp}}}{dt} &= i_{\text{dcp}} - i_{\text{ap}} - i_{\text{bp}} - i_{\text{cp}} \\
\leftrightarrow \frac{du_{\text{Cp}}}{dt} &= \frac{1}{C_\text{p}} \cdot (i_{\text{dcp}} - i_\text{p}).
\end{aligned}
\tag{17}
$$

The dc-link current i_{dcp} and the sum of the half-bridge currents i_p in (17) can be expressed as

$$
i_{\text{dcp}} = \frac{1}{R_\text{p}} \cdot (U_{\text{dcp}} - u_{\text{Cp}})
\tag{18}
$$

$$
i_\text{p} = i_{\alpha\text{p}} \cdot \overline{S}_{\text{i}\alpha\text{p}} + i_{\beta\text{p}} \cdot \overline{S}_{\text{i}\beta\text{p}} = i_\alpha \cdot \overline{S}_{\text{i}\alpha\text{p}} + i_\beta \cdot \overline{S}_{\text{i}\beta\text{p}}.
\tag{19}
$$

Substituting (18) and (19) into (17) results in

$$\leftrightarrow \frac{du_{C_p}}{dt} = \frac{1}{C_p \cdot R_p} \cdot (U_{dcp} - u_{Cp}) - \frac{1}{C_p} \cdot (i_{\alpha p} \cdot \overline{S}_{i\alpha p} + i_{\beta p} \cdot \overline{S}_{i\beta p})$$

$$\leftrightarrow \frac{du_{C_p}}{dt} = -\frac{1}{C_p \cdot R_p} \cdot u_{Cp} - \frac{1}{C_p} \cdot \overline{S}_{i\alpha p} \cdot i_{\alpha} - \frac{1}{C_p} \cdot \overline{S}_{i\beta p} \cdot i_{\beta} + \frac{1}{C_p \cdot R_p} \cdot U_{dcp}. \tag{20}$$

The i_{ap}, i_{bp} and i_{cp} are respectively the phase current of phase a, b and c in primary side. Similarly, The secondary side equations can be derived.

$$\frac{du_{C_s}}{dt} = -\frac{1}{C_s \cdot R_s} \cdot u_{Cs} + \frac{1}{C_s} \cdot \overline{S}_{i\alpha s} \cdot N \cdot i_{\alpha} + \frac{1}{C_s} \cdot \overline{S}_{i\beta s} \cdot N \cdot i_{\beta} + \frac{1}{C_s \cdot R_s} \cdot U_{dcs} \tag{21}$$

A 2L-DAB3 model with simplified transformer can be implemented. Compared to the PLECS auto-generated state-space code, where the state matrix has to be constant during the run time, the self-defined state-space description is changed during the run time [9].

Numeric Solvers

It is indispensable to evaluate the precision and numerical stability of different integration methods to calculate the capacitor voltage and the current flowing through the transformer [10]. In this paper, the precision achieved by three different integration methods is discussed. A higher-order integration method is more precise than the single-step integration. Nevertheless, to achieve higher precision is calculation-power-demanding and therefore time-demanding, which could cause an overrun and fail to fulfill the RT requirements [2]. Three single-step integration methods, FE, BE and TR are compared in this paper. A further explanation has been investigated in [11] and the comparison of these different integration methods in power systems is illustrated in [12]. The integration method is illustrated based on the generic state vector x as follows. The derivative of the state variable vector x can be expressed as,

$$\dot{x}(t) = A \cdot x(t) \tag{22}$$

where A is the characteristic state-space matrix of vector x. The relationship between the adjacent state variables in the discrete system is defined as

$$x[n+1] = F \cdot x[n]. \tag{23}$$

Especially for the stability calculation, the stable region for each integration method is transferred to the unit circle around the origin point with the discrete state matrix F. The relationship between the current variables and variables in the next step using FE, BE and TR integration are respectively derived in (24), (25) and (26). The matrix F for each integration method can be concluded from the state-space matrix A and the time-step T.

$$\begin{aligned} x(t+T) &= x(t) + \dot{x}(t) \cdot T \\ \leftrightarrow x[n+1] &= (1+T \cdot A) \cdot x[n] \Rightarrow F_{FE} = 1 + T \cdot A \end{aligned} \tag{24}$$

$$\begin{aligned} x(t+T) &= x(t) + \dot{x}(t+T) \cdot T \\ \leftrightarrow x[n+1] &= (1-T \cdot A)^{-1} \cdot x[n] \Rightarrow F_{BE} = (1-T \cdot A)^{-1} \end{aligned} \tag{25}$$

$$\begin{aligned} x(t+T) - x(t) &= \frac{1}{2} \cdot (\dot{x}(t) + \dot{x}(t+T)) \cdot T \\ \leftrightarrow x[n+1] &= \frac{1+\frac{1}{2} \cdot T \cdot A}{1-\frac{1}{2} \cdot T \cdot A} \cdot x[n] \Rightarrow F_{TR} = \frac{1+\frac{1}{2} \cdot T \cdot A}{1-\frac{1}{2} \cdot T \cdot A} \end{aligned} \tag{26}$$

In PLECS, two solvers, Tustin and BI45 integration methods, are provided for the code generation. Tustin approach is based on the trapezoidal method while BI45 is based on a 5th order backward interpolation method, which is a high order Taylor extension [13]. In the following simulations, the two methods are used for the PLECS auto-generated code model to compare with the investigated implementation.

Simulation Result

Simulations have been conducted to verify the optimal integration method for the modelling. In Fig. 5(a) and (b), the PLECS average model developed from library defined blocks is compared with the self-defined code average model with respect to line-to-line voltage u_{L-L} and ac current i_{ph}. For the simulation, both primary- and secondary-side dc-link voltages of 5000 V, dc-link capacitors of 5 mF, dc-link resistances of 0.1 mΩ and leakage inductances of 135 μH are assumed. It can be seen from Fig. 5(a), that the errors between two models for the dc-link capacitor voltages are rather small with TR. The other two integration methods show in this scenario similar results. As depicted in Fig. 5(b), the errors of ac current are smallest with TR integration method. It will be shown later that, the model with FE integration becomes unstable with a time-step larger than 1 μs. Remarkable current errors happen only at the switching events because the ac currents integration serves as an accumulation of the voltage errors.

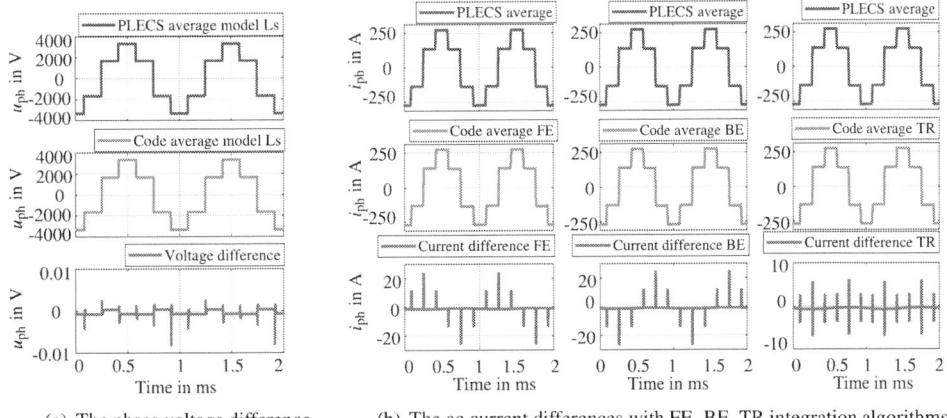

(a) The phase voltage difference (b) The ac current differences with FE, BE, TR integration algorithms

Fig. 5: Simulation results of the implemented models with simplified transformer with quasi infinite mutual inductance: PLECS average model compared with self-defined C-code average model with duty cycle $D = 0.5$, phase shift $\phi = 30°$, switching frequency $f_{sw} = 1\,\text{kHz}$, dead time $T_{dead} = 1\,\mu\text{s}$ and sub-cycle sampling (averaging) time-step $T_{ave} = T = 1\,\mu\text{s}$

HiL Verification

In this work, the RT simulator XRS7070 with an FPGA and a DSP is utilized [14]. The setup of the HiL test is shown in Fig. 6. The average model of 2L-DAB3 can be implemented either directly based on a PLECS library average model or by a self-defined C-code using the aforementioned modelling methodology. The C-code is generated automatically from the PLECS model with the "Code Generation" function in PLECS, which is subsequently downloaded into the RT simulator.

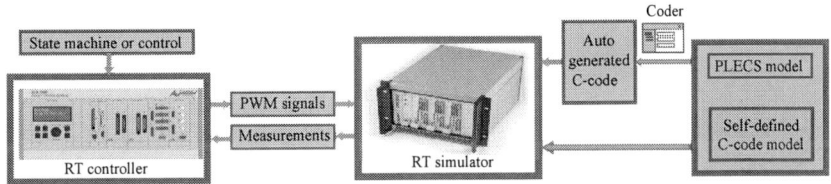

Fig. 6: Setup of the HIL test for a 2L-DAB3 converter

In this work, a comparison is conducted to evaluate the two methods. Fig. 7 depicts the voltages and currents of both the PLECS model and the C-code model tested with an RT simulator. The phase voltages and ac currents of both models are coincident with each other. The self-defined C-code average model reduces the number of the auto-generated C-code lines from PLECS average model by 99.4%, from 68,000 lines to 400 lines, thus it is much more compact. Consequently, the compiling and the downloading process of the code average model is faster.

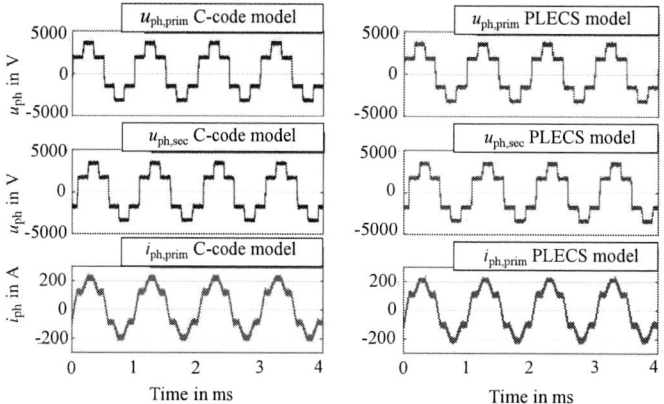

Fig. 7: Results of RT simulator XRS7070 using TR with duty cycle $D = 0.5$, phase shift $\phi = 30°$, switching frequency $f_{sw} = 1$ kHz, dead time $T_{dead} = 5\,\mu s$ and time-step $T = 10\,\mu s$

Table I: Comparison of the test results with the auto-generated and self-defined code in the RT simulator

	Total (μs)	Model (μs)	Input (μs)	Output (μs)	T setting (μs)
BE self-defined code	7.18	**2.525**	1.91	2.165	10
TR self-defined code	7.192	**2.578**	1.92	2.148	10
Tustin auto-generated code	8.248	**3.537**	1.925	2.197	10
BI45 auto-generated code	8.57	**3.883**	1.923	2.187	10

The run time of all compared models is shown in Table I. The total run time includes the input and output communication time as well as the model calculation time. The execution period T should be set longer than the total utilized time to ensure that the complete procedure is finished within one time-step. This avoids the overrun of the RT simulation. The input and output communication time makes no large difference because it depends almost only on the number of used ports and slots of the RT simulator. According to the integration complexity, the corresponding run time is also longer. And for the same integration method (TR and Tustin), the self-defined model runs approximately $1\,\mu s$ faster than the auto-generated C-code model. Total execution time is reduced from $8.248\,\mu s$ to $7.192\,\mu s$. Therefore, a smaller time-step can be applied to the RT simulator with a self-defined C-code average model. Thus, compared with using a auto-generated model, more precise results using the self-defined C-code average model can be obtained from the simulator by reducing the integration step size.

Numerical Stability Analysis of C-Code Average Model

In this section, the numerical stability of the C-code average models for 2L-DAB3 using different integration algorithms is analyzed. Based on the stability criterion, a discrete-time system is analytically stable if and only if all of its eigenvalues are located inside an unit circle.

State Space Representation of C-Code Average Model

The numerical stability of continuous systems is equivalent to the analytical stability for the discrete counterpart [11]. An analogous stability analysis is adopted in [15] to verify the stability of a nodal-reduced modelled 2L-DAB3. The code average model is firstly formulated in state-space form,

$$\dot{x} = A \cdot x + B \cdot u \tag{27}$$

$$y = C \cdot x + D \cdot u \tag{28}$$

with the state variables $x = \begin{bmatrix} u_{Cp} \\ u_{Cs} \\ i_{\alpha} \\ i_{\beta} \end{bmatrix}$, the inputs $u = \begin{bmatrix} U_{dcp} \\ U_{dcs} \end{bmatrix}$ and the outputs $y = \begin{bmatrix} u_{\alpha p} \\ u_{\beta p} \\ u_{\alpha s} \\ u_{\beta s} \end{bmatrix}$.

The eigenvalues can be obtained from the matrix A of the state-space description and utilized to analyze the stability. There are four eigenvalues for each set of specifications. the state variables x of the system include the dc-link capacitor voltages u_{Cp} and u_{Cs}, as well as the ac current in $\alpha\beta$-frame i_{α} and i_{β} as shown in Fig. 1. The dc source voltages U_{dcp} and U_{dcs} have no impact on the system stability. The characteristic state matrix A, the input matrix B, the output matrix C and the input-output matrix D for the self-defined code model can be then derived with (15), (16), (20) and (21)

$$A = \begin{bmatrix} -\frac{1}{C_p \cdot R_p} & 0 & -\frac{1}{C_p} \cdot \overline{S}_{i\alpha p} & -\frac{1}{C_p} \cdot \overline{S}_{i\beta p} \\ 0 & -\frac{1}{C_s \cdot R_s} & \frac{N}{C_s} \cdot \overline{S}_{i\alpha s} & \frac{N}{C_s} \cdot \overline{S}_{i\beta s} \\ \frac{1}{L_\sigma} \cdot \overline{S}_{u\alpha p} & -\frac{N}{L_\sigma} \cdot \overline{S}_{u\alpha s} & -\frac{1}{L_\sigma} \cdot R_w & 0 \\ \frac{1}{L_\sigma} \cdot \overline{S}_{u\beta p} & -\frac{N}{L_\sigma} \cdot \overline{S}_{u\beta s} & 0 & -\frac{1}{L_\sigma} \cdot R_w \end{bmatrix}, B = \begin{bmatrix} \frac{1}{C_p \cdot R_p} & 0 \\ 0 & \frac{1}{C_s \cdot R_s} \\ 0 & 0 \\ 0 & 0 \end{bmatrix},$$

$$C = \begin{bmatrix} \overline{S}_{u\alpha p} & 0 & 0 & 0 \\ \overline{S}_{u\beta p} & 0 & 0 & 0 \\ 0 & \overline{S}_{u\alpha s} & 0 & 0 \\ 0 & \overline{S}_{u\beta s} & 0 & 0 \end{bmatrix} \text{ and } D = \begin{bmatrix} 0 & 0 \\ 0 & 0 \\ 0 & 0 \\ 0 & 0 \end{bmatrix}.$$

The specifications for the influencing factors in matrix A are listed in Table II. The three integration methods have the same matrix A, but they have different stability regions according to A. Therefore, the continuous state-space matrix A is transferred to the discrete state-space matrix F with (24) - (26) to normalize the stability regions to the unit circle with the center point at the origin (0, 0). The discrete state-space of the self-defined code average model is expressed as

$$x[n+1] = F \cdot x[n] + G \cdot u[n] \tag{29}$$

with F denoting the discrete characteristic state matrix and G denoting the discrete input matrix. The innovative part of the paper is that the voltages and currents are calculated using the equations of the average model. There are unlimited sets of decimal gate signals, with the only constraint that the sum of the top and the bottom switching gate signals should not exceed one. An exemplary array of switching signals [0 0.25 0.5 0.75 1] is chosen for validation performed here. In total, 184,320 poles from the combinations of the specifications and switch states are considered in the test.

Results of Stability Analysis

Table II shows both the specified values, which are used in the simulation, and the specifications of extended ranges, which serve to get an overview of the stability for the three integration methods. Fig. 9 depicts the poles of 2L-DAB3 converter using FE, BE and TR integration method as an example of the extended range specifications from Table II.

Table II: Extended range and specified values for stability analysis

Parameter	Extended Range	Specified Values
C_p in F	5×10^{-3}	5×10^{-3}
C_s in F	5×10^{-3}	5×10^{-3}
L_σ in H	1.34×10^{-6} to 134×10^{-6}	134×10^{-6}
R_w in Ω	1×10^{-2}	1×10^{-2}
T in s	1×10^{-7} to 5×10^{-5}	1×10^{-7} to 5×10^{-5}
R_p in Ω	1×10^{-4} to 1×10^{-2}	1×10^{-4}
R_s in Ω	1×10^{-4} to 1×10^{-2}	1×10^{-4}
N	1 to 3	1

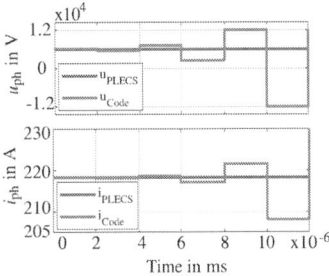

Fig. 8: Unstable simulation results with FE

The natural frequency (around the boundary) and the damping ratio (along the radius) are depicted in Fig. 9. There are more unstable poles with a larger time-step (light colour in Fig. 9(a) and (b)) than with a

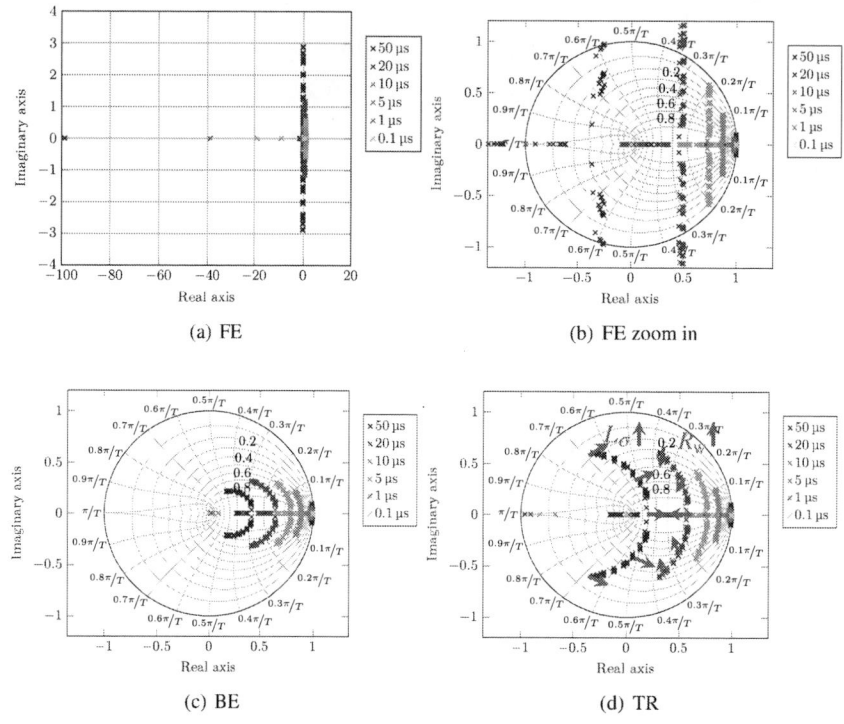

Fig. 9: Stability test of code average model of 2L-DAB3 with an extended range of specifications

smaller time-step with FE (dark blue in Fig. 9(a) and (b)). There are 39,792 out of 184,320 poles unstable with FE integration. The unstable poles are far away from the unit circle with FE as shown in Fig. 9(a) and are zoomed-in in Fig. 9(b). The stability margin for FE with specified values is 1 μs, which is consistent with the simulation results for FE with time-step 2 μs, as shown in Fig. 8. The results from the self-defined code model are divergent. No unstable poles exist for BE and TR methods in Fig. 9(c) and (d). It can be seen that the poles of smaller time-steps tend to the right side boundary of the unit circle than the ones of larger time-steps. The reason is that the positions of the poles represent the relative natural frequency of the system. For a system with different time-steps, the absolute natural frequency, which is multiplied by the time-step, should be compared instead of the relative natural frequency. Although the poles of a smaller time-step are intuitively closer to the unit circle boundary, they have better dynamic performances than the poles of larger time-steps. As shown in Fig. 9(d), the poles move to the right side when the L_σ is increasing. It concludes that the L_σ impacts mainly the natural frequency. With a larger R_w, the poles move to the left side approximately in the direction of a larger damping factor in the unit circle. The similar tendency can also be proven with FE and BE. Both of BE and TR are suitable for the self-defined code average model of 2L-DAB3 converter with the specified parameters.

Conclusion

The proposed C-code average model of 2L-DAB3 converter improves the calculation speed compared to the one of auto-generated average code model. It has less lines of code, is valid with smaller time-step and therefore needs less calculation and execution time for the RT simulation. The C-code average models using different single-step integration methods are evaluated and compared in respect of accuracy and stability. Above the three algorithms, TR counterpart owns the best precision. The C-code average models with the specified parameters using BE and TR are always stable. Despite the self-defined C-code average model requires more efforts than the auto-generated model, it serves as an more execution-efficient alternative of the auto-generated average model of 2L-DAB3 converter for RT simulation.

References

[1] Seung Tae, Cha, Qiuwei Wu, Arne Hejde Nielsen, Jacob: Real-Time Hardware-in-the-Loop (HIL) Testing for Power Electronics Controllers

[2] J. Belanger and P. Venne and J.-N.Paguin: The What, Where and Why of Real-Time Simulation, OPAL-RTs 10th International Conference on Real-Time Simulation, 2010

[3] G.W. Wester, R.D. Middlebrook, Low-Frequency Characterization of Switched dc-dc Converters, 1973

[4] Jost Allmeling and Niklaus Felderer: Sub-Cycle Average Models with Integrated Diodes for Real-Time Simulation of Power Converters, 2017 IEEE Southern Power Electronis Conference, 2017

[5] K. L. Lian and P. W. Lehn: Real-Time Simulation of Voltage Source Converters Based on Time Average Method, IEEE Transactions on Power Systems, 2005

[6] Jost Allmeling, Niklaus Felderer and Min Luo: High Fidelity Real-Time Simulation of Multi-Level Converters, The 2018 International Power Electronics Conference, 2018

[7] Plecs: Plecs User Manual, the Simulation Platform for Power Electronic System

[8] Edith Clark: Circuit Analysis of AC Power Systems, Vol. I. J. Wiley & Sons, New York 1943

[9] Jost Allmeling, Niklaus Felderer: Sub-Cycle Average Models with Integrated Diodes for Real-Time Simulation of Power Converters, 2017

[10] Christian Graf and Jurgen Maas and Thomas Schulte and Johannes Weise-Emden: Real-Time HIL-Simulation of Power Electronics, 2008 34th Annual Conference of IEEE Industrial Electronics, 2008

[11] Francois E. Cellier and Ernesto Kofman: Continuous System Simulation, Springer, 2005

[12] J. Marti and J. Lin: Suppression of Numerical Oscillations in the EMTP, IEEE trans. Power System, Vol 4, No. 2, 1989

[13] Wei Xie, Backinterpolation Methids for the Numerical Solution of Ordinary Differential Equations and Applications, Master's thesis, Dept. of Elelctrical & Computer Engineering, University of Arizona, Tucson, Ariz., 1995

[14] XRS7070 Getting Start, 2015

[15] Robert Uhl and Amir Arasteh and Antonello Monti and Arne Hinz and Rik W. De Doncker: Nodal-Reduced Modeling of Three-Phase Dual-Active Converters for EMTP-type Simulations, 2017 IEEE 26th International Symposium on Industrial Electronics (ISIE), 2017

Dual Interleaved 3.6 kW LLC Converter Operating in Half-Bridge, Full-Bridge and Phase-Shift Mode as a Single-Stage Architecture of an Automotive On-Board DC-DC Converter

Philipp Rehlaender[1], Sergey Tikhonov[2], Frank Schafmeister[1], Joachim Böcker[1]

[1] Paderborn University, Power Electronics and Electrical Drives
[2] Delta Energy Systems (Germany) GmbH

Warburger Str. 100
33098 Paderborn Germany
Tel.: +49 (0)5251 60-2159
E-Mail: rehlaender@lea.upb.de
URL: http://wwwlea.upb.de

Keywords

»Resonant Converter«, »LLC Converter«, »Phase-Shift Modulation«, »Asymmetrical Duty Cycle« »LLC Balancing«, »Current Sharing«, »Onboard DC-DC Converter«

Abstract

On-board DC-DC converters are required to operate over a wide input and output voltage range depending on the state-of-charge of the input and output battery. Conventionally, the power transfer between these batteries is enabled by a two-stage converter concept where a galvanically-coupled DC-DC converter regulates the input voltage of the second-stage galvanically-isolated DC-DC converter. This paper presents a single-stage interleaved 3.6 kW LLC converter for this purpose. While LLC converters are usually not suitable for such a wide voltage range, this LLC converter is operated in full-bridge mode for large gains and in half-bridge mode for low gains. For intermediate gains and loads, the LLC makes use of phase-shift mode. To operate the interleaved LLCs at an equal switching frequency enabling output current ripple cancellation, again phase-shift mode is utilized to balance the output currents during full-bridge mode while asymmetrical duty-cycle mode is proposed for current balancing during half-bridge mode. This paper analyzes the converter design for these modes of operation. A 3.6 kW-prototype employing Si-superjunction MOSFETs achieves a power density of 2.1 kW/l. The maximum efficiency reaches 96.5 % while for most operating points it is kept well above 90 %.

1 Introduction

While the driving motor of electric vehicles is usually connected via an inverter to the high-voltage traction battery, other on-board consumers such as lights, servomotors, etc. are powered by a conventional low-voltage auxiliary battery. Both battery voltages vary largely with their respective state-of-charge (SOC). The traction battery voltage may vary in a range of 240 to 420 V while the auxiliary battery voltage may cover a range between 8 and 16 V. To transfer power from the traction battery to the auxiliary battery, an on-board DC-DC converter is needed. Due to both distinctively varying voltages, the voltage-transfer ratio can get large and so the normalized gain G_{norm} also results much larger as compared to many other applications.

$$G_{norm} = \frac{V_{in,max}V_{out,max}}{V_{in,min}V_{out,min}} = 3.5 \tag{1}$$

The topology of the LLC resonant converter is typically not suitable for a voltage-transfer ratio being that large. Generally, the LLC-voltage-transfer ratio is adjusted by the switching frequency, here resulting in a tremendously varying switching-frequency range or in a design with very large resonant currents. Common approaches rely, therefore, on a two-stage concept [1]. A first-stage, galvanically-coupled DC-DC converter supplies the optimum input voltage for a second-stage isolated LLC converter. This allows the LLC to run in its optimum operating point close to the resonant frequency.

To increase the power density, this paper presents an interleaved single-stage LLC converter. It is operated in full-bridge mode for large voltage-transfer ratios. For low voltage-transfer ratios S_{i4} is permanently turned-on and S_{i2} is kept off (see Figure 1). This reduces the converter gain by a factor of two. Mode changes during operation can be performed through the on-the-fly topology morphing presented in [2]. To cover intermediate voltage-transfer ratios at a low output load, the converter can be operated with freewheeling intervals in a phase-shift operation to avoid very high switching frequencies (c.f. Figure 2a). The overall concept was compared to other topology candidates qualifying for an automotive on-board DC-DC application in [3]. There it was identified to be one of the best solutions to cover the wide input and output voltage range. In [4] this concept was analyzed for a single-rail design and an appropriate dimensioning methodology has been presented as well. This paper deals with an interleaved single-stage LLC converter. Both parallel LLC rails are operated with an equal switching frequency, but 90°-phase-shifted ("interleaved"), to significantly reduce the current ripple at the input and – more importantly – at the output capacitor ("ripple cancellation"). Typically, paralleled LLC rails operate with slightly different switching frequencies resulting from component deviations in the resonant tank and, as a consequence, preventing ripple cancellation. For balancing the two rails, the transferred power is adjusted dynamically - in this paper by a phase-shift modulation in full-bridge mode and by an asymmetrical duty-cycle operation in the half-bridge mode.

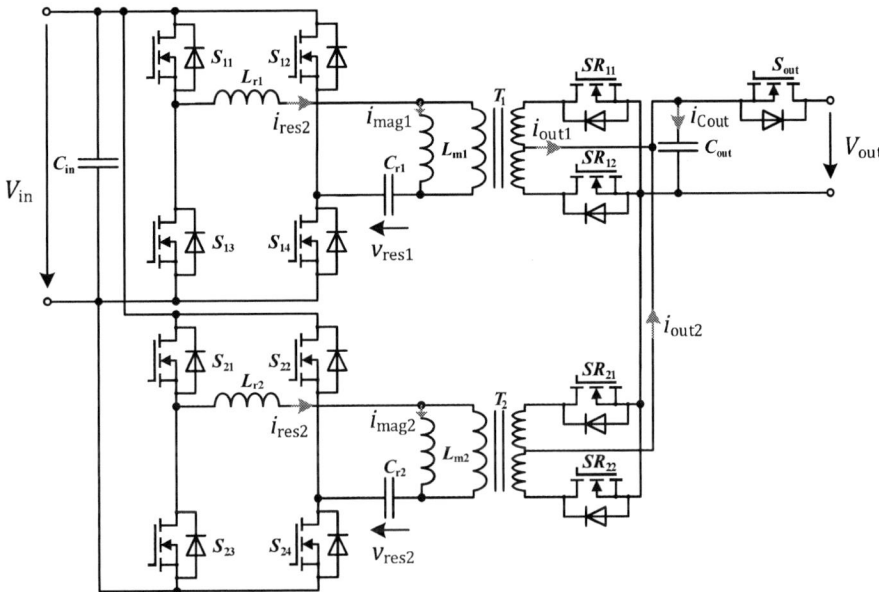

Figure 1: Interleaved single-stage LLC resonant converter with output reverse-polarity protection S_{out}

The paper is structured as follows: chapter 2 reviews the current balancing concepts for interleaved LLC converters and presents the balancing concepts used for the proposed converter. Chapter 3 analyzes the converter operation and defines the operating regions of the different operating modes. A design methodology is presented in chapter 4. Finally, experimental results taken on a compact 3,6 kW-prototype verify the concept in chapter 5.

2 Interleaving of LLC resonant converters

When paralleling converters, it is not possible to operate them completely equally since component deviances cause a slightly different system behavior. Typical component deviations can be as high as 10 % and more. Therefore, it is practically necessary to control the transferred power of each paralleled rail. For PWM-based converters operating at a constant switching frequency, the transferred power is adjusted through the duty cycle or phase shift of the respective rails. In case of paralleled converters, this translates into an identical (constant) switching frequency allowing synchronizing and interleaving the switching pulses of the individual converter rails. As an advantage, the current ripple, especially of the output capacitor, can be largely reduced yielding a significantly smaller output capacitor component size. For the LLC converter, however, the transferred power is adjusted by the switching frequency such that in conventional operation, paralleled LLC rails are operated at slightly different switching frequencies preventing the advantageous ripple cancellation (see Figure 2b). The asymmetrical transfer behavior is largely affected by the deviations of the resonant components. Especially for large output currents this is a significant drawback since the output capacitor needs to be designed much larger. If a C-L-C-filter is used, same applies for the inductor (L), since the effective switching frequency is halved compared to the 90 °-interleaved variant. However, since balancing LLC converter rails via switching frequency is an easy and robust method, it is applied in [5, 6].

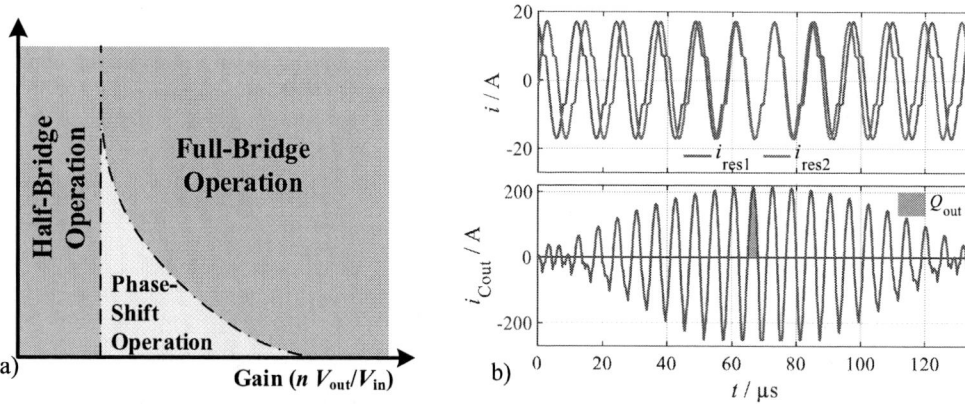

Figure 2: a) operating principle of the LLC converter, b) LLC interleaving through adjustment of the individual switching frequencies: A high current ripple with a low-frequency beat arises at the output capacitor. The frequency of this beat depends on the difference of the switching frequencies.

To operate paralleled LLC converters with an equal switching frequency, many methods have been presented to achieve current cancellation: In [7] an additional switched capacitor was described, which balances the resonant tank actively depending on the required operating point. This method requires at least two additional power MOSFETs. In [8, 9] the balancing of the converters was achieved by connecting them in series on the input and in parallel on the output. The transferred power is hereby inherently adjusted by their respective input voltage. However, this method relies on the series connection. Phase shedding (i.e. the selective shutdown of one leg/rail at low power transfer) is not an option and the converters need to be designed for a lower input voltage such that conduction losses increase. In [10] the balancing was achieved by adjusting the delay time of the secondary low-side synchronous rectifiers of a full-bridge rectification circuit. With this control, a freewheeling interval was introduced on the secondary side, allowing the balancing of the two rails while the two converters can be operated with an equal switching frequency. As a drawback however, this method can only be applied to full-bridge synchronous rectifier circuits, but not to center-tapped rectifiers. Furthermore, self-controlled synchronous rectification ICs, driving the rectifier MOSFETs locally, cannot be used since latter have to be synchronized to the central LLC-controller. Finally, in [11, 12] a two-stage balancing concept was described. The system consists of two-stage rails. A first stage, galvanically-coupled DC-DC converter generates the individual input voltage level for the second-stage LLC converter. By adjusting this input voltage level depending on the transferred power, the balancing of the two rails is achieved while operating them with an equal switching frequency allowing synchronized interleaving and thus the current ripple cancellation at the output. However, in principle this method can only be applied to two-stage concepts but not to a single-stage design.

In [9, 12] it was shown that power-balancing is possible in full-bridge mode by applying a phase shift to that converter rail transferring more power. An exemplary phase-shift modulation is depicted in Figure 3a. The effect on the output capacitor current is visualized in Figure 4a. For half-bridge operation, however, it is not possible to apply a phase shift since one leg is generally turned off. To ensure equal switching frequencies in half-bridge mode, this paper proposes to apply an asymmetrical duty-cycle modulation for balancing: The switches S_{i1} and S_{i3} of a same bridge-leg are not operated with equal, but with individual duty cycles [13, 14]. Asymmetrical duty-cycle operation has been introduced as a low-load modulation method in [13] since it leads to an imbalance in the resonance- and switch currents for the primary and secondary semiconductors. In the half-bridge mode, the resonant capacitor is charged to a DC offset of $V_{in}/2$. Depending on the choice of duty cycle, the average resonant capacitor voltage either increases or decreases. Therefore, the maximum resonant capacitor voltage appears in half-bridge mode and the resonant tank needs to be designed accordingly. In the following the larger duty cycle is consequently applied to the low-side switch S_{i3} whereas the smaller one is assigned to the high-side switch S_{i1}. This paper analyzes the effect of these balancing methods in chapter 3 to propose a design methodology in chapter 4. Figure 3b visualizes the modulation, Figure 4b shows the effect on the output capacitor.

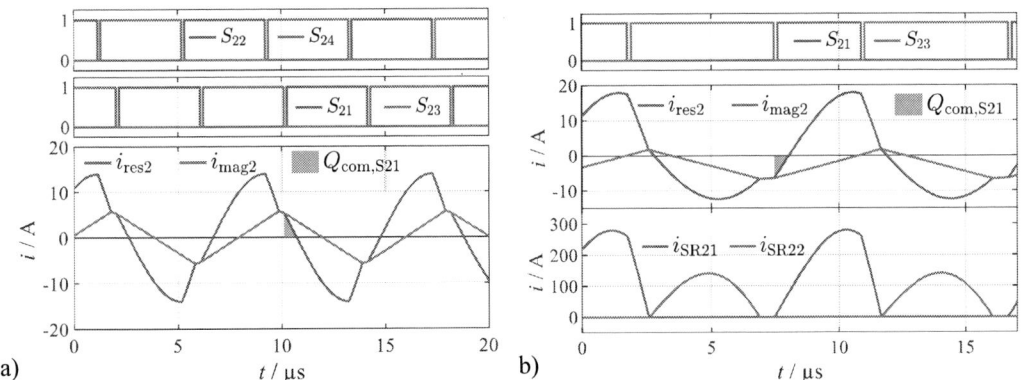

Figure 3: a) phase-shift modulation of the LLC converter, b) asymmetrical duty-cycle modulation

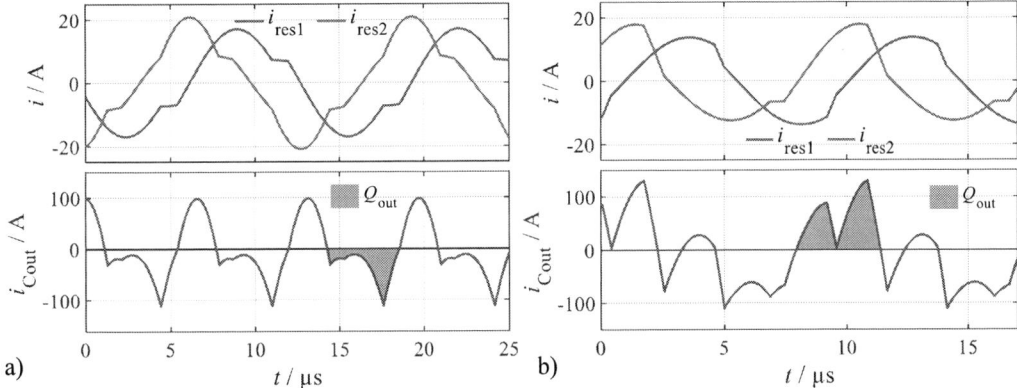

Figure 4: output capacitor current for a) phase-shift modulation, b) asymmetrical duty-cycle modulation.

3 System Analysis

The on-board DC-DC converter is required to run at an input voltage range between 240 V and 420 V. While the output voltage varies between 8 V and 16 V for a maximum output power of 3.64 kW and a maximum output current of 260 A. For output voltages above 11 V, the converter is required to run at any output power while for voltages below 11 V, only the maximum output current of 260 A is required – the output current is always 260 A.

The balancing operation is highly critical to the component deviations of the two LLC resonant tank. While capacitors can be relatively well controlled by selecting components with a tolerance of 1 %, controlling the deviations of the inductive components is much more difficult such that these components are subject to a wider variation in their respective inductance. In the analysis, it is assumed that the inductances can be matched by a variation of ± 5 % while the resonant capacitor's capacitance varies by only $\pm 1\,\%$. In [12] the different possibilities of imbalances were analyzed and it was found that the worst-case imbalance are apparent when one of two parallel LLC converters has a mismatch of the resonant inductance and magnetizing inductance as well as the resonant capacitance in the positive direction while the other LLC has a mismatch in the negative direction. This leads to two different resonant frequencies. For these reasons, the analysis assumes the following conditions: $L_{r1} = 1.05\,L_r$, $C_{r1} = 1.01\,C_r$, $L_{m1} = 1.05\,L_m$ and $L_{r2} = 0.95\,L_r$, $C_{r2} = 0.99\,C_r$, $L_{m2} = 0.95\,L_m$ where L_r, L_m and C_r are the nominal parameters being calculated. The first step of analysis is the designation of the operating regions for full-bridge, half-bridge and phase-shift operation. In [4] a detailed analysis on how the different operating regions can be defined for a non-interleaved operation was presented. The half-bridge operation was limited to the maximum resonant capacitor voltage, which was set to 500 V for the low-tolerance resonant capacitors with a maximum blocking voltage of 630 V. This region definition shall be the basis of the following system analysis of the interleaved version. Vital for the LLC operation with Si-MOSFETs in the inverter stage is that zero-voltage switching (ZVS) is ensured. To achieve ZVS, the turn-off current of the adjacent half-bridge switch needs to be sufficiently high in order to discharge the output capacitance when commutating to the respective switch intended to turn-on at zero-voltage. Looking at Figure 3b, it is evident that the asymmetrical balancing increases the turn-off current of S_{21} such that ZVS is basically ensured for turn-on of S_{23}. However, the turn-off current of S_{23} is reduced such that ZVS for S_{21} may be lost. To ensure ZVS, it is critical that the commutation charge $Q_{\text{com},S21}$ is larger than the output charge at the respective input voltage [15].

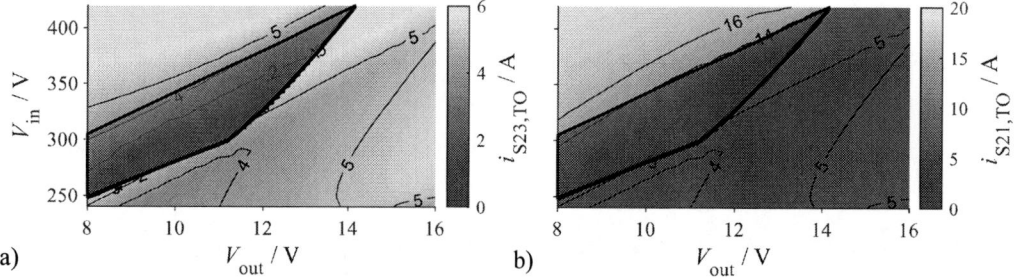

Figure 5: a) turn-off current $i_{S23,TO}$ for switch S_{23}, and b) turn-off current $i_{S21,TO}$ for switch S_{21} for a 5 % resonant tank imbalance over the required operating region

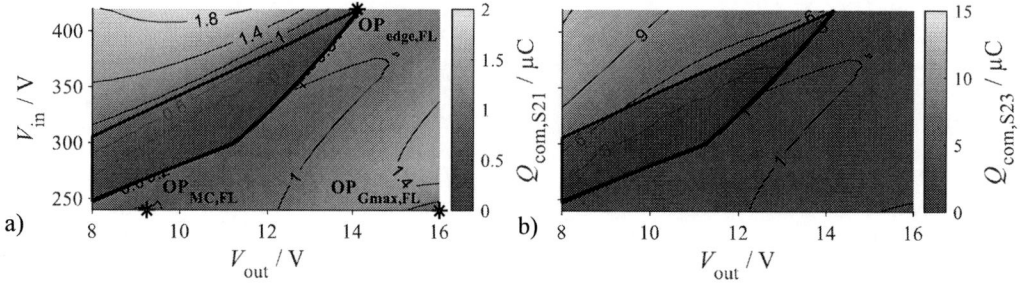

Figure 6: a) commutation charge $Q_{\text{com},S21}$ for switch S_{21}, and b) commutation charge $Q_{\text{com},S23}$ for switch S_{23} over the required operating region. For full-bridge operation, the commutation charge is equal for switch S_{21} and S_{23}. Operating point $OP_{\text{MC,FL}}$ and $OP_{\text{edge,FL}}$ are iteratively defined.

Figure 5 shows the turn-off current $i_{S23,TO}$ of switch S_{23} and the turn-off current $i_{S21,TO}$ for the switch S_{21} assuming a $\pm 5\,\%$ imbalance for the inductances and a $\pm 1\,\%$ imbalance for the resonant capacitance. The commutation charge is depicted in Figure 6. Notice that a small turn-off current of S_{23} leads to a small commutation charge for S_{21} and reverse. It is evident that in certain operation regions the turn-off current of S_{23} and the commutation charge of S_{21} reduces largely such that ZVS may be lost. The IPW65R080CFDA-MOSFETs used in [4] have an output charge $Q_{\text{oss}}(V_{\text{in}}) = \int_0^{V_{\text{in}}} C_{\text{oss}}(v)\mathrm{d}v$ of approximately 0.3 µC at the maximum input voltage of 420 V. To ensure a sufficient commutation charge, the half-bridge operating region is limited to the non-greyed-out area. With this choice, the commutation charge is well above the necessary value such that ZVS should be ensured. In phase-shift balancing, there are three operating points where ZVS may be lost. At $OP_{MC,FL}$, the operating mode in phase-shift mode changes from the ACD-mode to the CDE-mode (modes named according to [16]). For operating points with a lower output voltage, the resonant current is at the switching instant still higher than the magnetizing current. For $OP_{MC,FL}$ and lower output voltages, the resonant current drops during the freewheeling interval onto the magnetizing currents. However, as switching frequencies reduce with increasing output voltages, the turn-off current and commutation charge become larger with increasing output voltages. Therefore, $OP_{MC,FL}$ is a critical operating point to ensure ZVS. For large gains, the phase-shift modulation may enter the CDF-mode where ZVS may also be lost. As a consequence, it is also crucial to ensure ZVS in $OP_{Gmax,FL}$. The respective commutation charge is depicted in Figure 6.

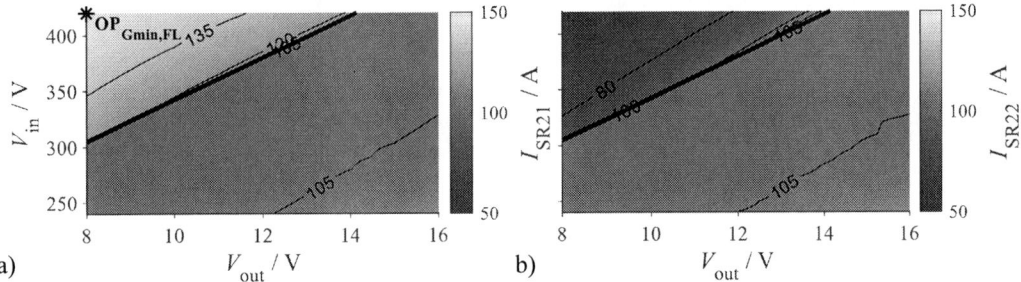

Figure 7: a) RMS current I_{SR21}, and b) RMS current for switch I_{SR22} over the required operating region

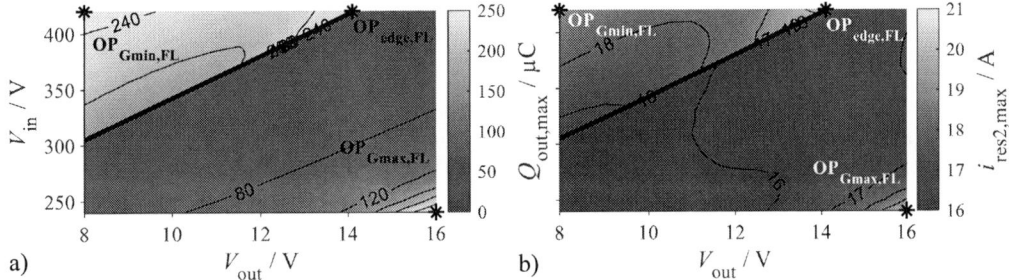

Figure 8: a) output capacitor charge $Q_{out,max}$ and b) maximum resonant current $i_{res2,max}$ for variable input and output voltages

Figure 3 showed that the asymmetrical LLC balancing method leads to an imbalance in the synchronous rectifier RMS current. Figure 7 shows the synchronous rectifier current I_{SR21} and I_{SR22}. It is evident that in half-bridge mode, a smaller gain leads to an increasing imbalance in the rectifier currents while for the phase-shift balancing in full-bridge mode, the rectifier current is relatively constant. For the depicted parameters, the rectifier current increases to about 150 A compared to a maximum current of 107 A in full-bridge mode. While the balancing method has a large influence on the commutation charge and the rectifier current, the influence on the primary RMS currents is comparably small. For neither switch, the current increases significantly compared to their respective maximum values for the normal operation. However, the maximum resonant current increases compared to normal operation. Figure 8 b) shows the maximum resonant current. Furthermore, the output charge is depicted in Figure 8 a). Both figures show that their respective maximum may appear in $OP_{Gmin,FL}$, $OP_{edge,FL}$ or $OP_{Gmax,FL}$.

4 System design

4.1 Design methodology

To keep the design process as general as possible, the resonant tank parameters shall be independent of a pre-defined resonant frequency. For that purpose, the resonant tank parameters are normalized with respect to the resonant frequency as follows:

$$f_r = \frac{1}{2\pi}\sqrt{\frac{1}{L_r C_r}}, \quad Z = \sqrt{\frac{L_r}{C_r}}, \quad \lambda = \frac{L_r}{L_m} \tag{2}$$

The normalization has the advantage that the parameters of the resonant tank are reduced from L_r, C_r and L_m to λ and Z, which reduces the number of design parameters from 3 to 2. By calculating the current waveform for a set of Z and λ, relevant stress values (RMS current, turn-off current, resonant capacitor voltage, normalized switching frequency etc.) are valid for any choice of f_r. This makes it possible to define a suitable set of resonant tank parameters in terms of Z and λ and define the resonant frequency in a second design step. The effect of λ and Z on the stress values can hereby be analyzed through an intuitive representation of the λ-Z-plane in a 3D-plot such as a contour figure.

4.2 Stress value definition and calculation

To calculate the stress values of the different converter designs, the LLC converter has been simulated (in PLECS) for different choices of λ, Z and the winding ratio n. Although for the simulation a certain "dummy" resonant frequency was selected, due to the normalization (2), the results are valid for any choice of f_r. While currents and voltages are directly valid, frequency-independent stress values have to be defined for those parameters that depend on the switching frequency. An example of these parameters are the output and commutation charge. They directly depend on the choice of resonant frequency. To define a resonant-frequency-independent value, the charge can be multiplied with the "dummy" frequency of the time-series solution creating a resonant-frequency-independent variable for the charge, which is of the unit Ampere:

$$Q_n = Q\,f_r \tag{3}$$

If the output charge Q_{out} is normalized according to (3), the output capacitance C_{out} can be calculated upon selection of the resonant frequency f_r and the maximum allowed output voltage ripple ΔV_{out} as

$$C_{out} = \frac{Q_{n,out}}{f_r\,\Delta V_{out}}. \tag{4}$$

Considering the normalized commutation charge, it is possible to calculate the resonant frequency, at which ZVS may be lost with the normalized commutation charge $Q_{n,com} = Q_{com}f_r$ as

$$f_{r,ZVS} = \frac{Q_{n,com}}{Q_{out,MOS}} \tag{5}$$

4.3 Converter design

The resonant tank has been designed by analyzing the frequency-independent parameters defined in 4.2 for λ, Z and n in the important operating points depicted in Table 1. For reasons of limited space, the following figures show the stress values for $n = 15$ only while other winding ratios have also been analyzed. Figure 9 shows the stress values of the normalized output charge $Q_{n,out}$, the ZVS resonant frequency $f_{r,ZVS}$ and the RMS current for the synchronous rectifier SR_{21}. Figure 9 shows the stress values for the thermal worst-case operating point $OP_{FB,edge,FL}$ including the normalized switching frequency f_{sw}/f_r, the turn-off current $i_{S22,TO}$ and the RMS current I_{S22}.

Table 1: Operating points for the resonant tank design. Values with a star were iteratively defined.

OP	V_{in} / V	V_{out} / V	I_{out} / A	Mode
$OP_{Gmin,FL}$ (OP_1)	420	8	260	HB
$OP_{edge,FL}$ (OP_2)	420	*	260	HB
$OP_{Gmax,FL}$ (OP_3)	240	16	228	FB
$OP_{MC,FL}$ (OP_4)	240	*	260	FB
$OP_{FB,edge,FL}$ (OP_5)	420	*	*	FB

5 Experimental Results

According to the stress value analysis of section 4.3, the parameter pair of $\lambda = 0.33$ and $Z = 19.3\,\Omega$ was selected. After performing a loss analysis of the primary MOSFETs and comparing different types of MOSFETs, the resonance frequency was designated to be 100 kHz such that all resonant tank parameters can be calculated with

$$L_{\mathrm{r}} = \frac{Z}{2\pi f_{\mathrm{r}}}, \quad L_{\mathrm{m}} = \frac{L_{\mathrm{r}}}{\lambda}, \quad C_{\mathrm{r}} = \frac{1}{2\pi f_{\mathrm{r}} Z} .\tag{6}$$

to $L_{\mathrm{r}} = 31\ \mu\mathrm{H}$, $C_{\mathrm{r}} = 82\ \mathrm{nF}$ and $L_{\mathrm{m}} = 93\ \mu\mathrm{H}$.

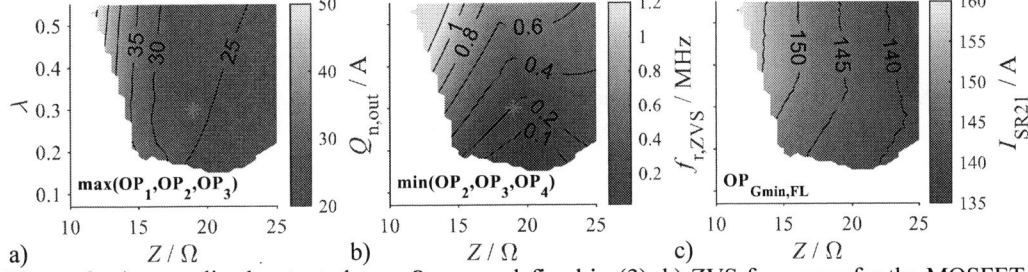

a) Z / Ω b) Z / Ω c) Z / Ω

Figure 9: a) normalized output charge $Q_{\mathrm{n,out}}$ as defined in (3), b) ZVS frequency for the MOSFETs IPW65R080CFDA as defined in (5), and c) maximum synchronous rectifier current I_{SR21} for various design configurations in terms of λ and Z. The selected parameters are depicted with a red star.

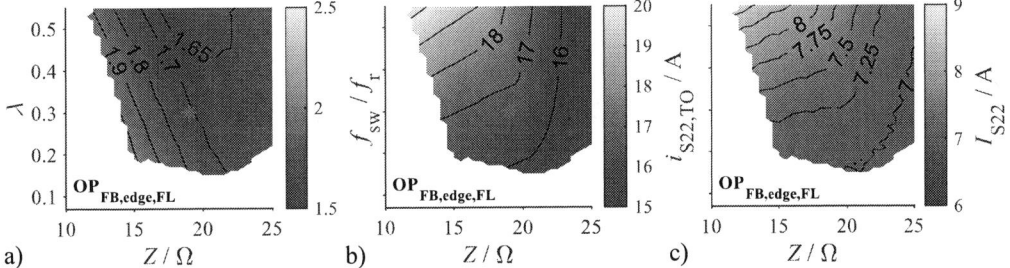

a) Z / Ω b) Z / Ω c) Z / Ω

Figure 10: a) normalized switching frequency $f_{\mathrm{sw}}/f_{\mathrm{r}}$, b) turn-off current $i_{\mathrm{S22,TO}}$, and c) RMS current I_{S22} for the thermal worst-case operating point $\mathrm{OP_{FB,edge,FL}}$ for various design configurations in terms of λ and Z. The selected parameters are depicted with a red star.

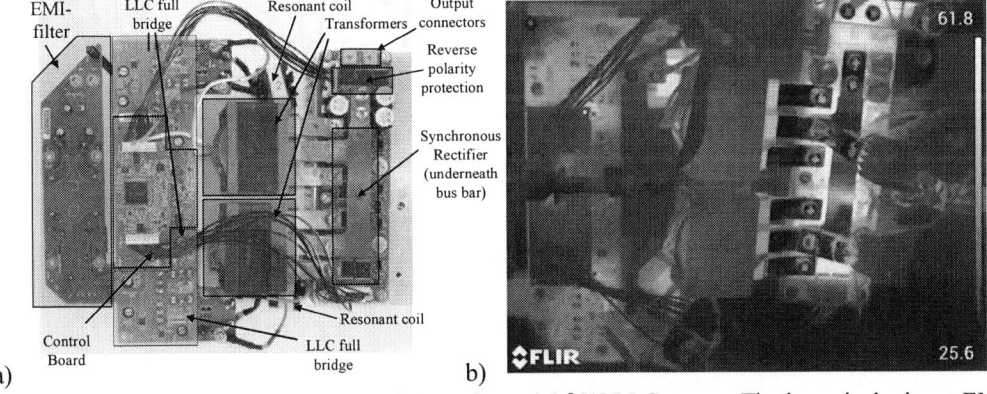

a)   b)

Figure 11: a) developed prototype consisting of two 1.8 kW LLC stages. The input is the input EMI-filter on the very left. In the center is the control board with the two full-bridge inverter stages. Between the primary and secondary side are the two transformers with the resonant coils. The output rectifier is on the right side with a reverse-polarity protection on the very right. Disregarding the input EMI-filter the achieved power density is 2.1 kW/l, b) the thermal image of the converter operating at $V_{\mathrm{in}} = 360\ \mathrm{V}$, $V_{\mathrm{out}} = 14\ \mathrm{V}$, $I_{\mathrm{out}} = 260\ \mathrm{A}$ shows a moderate temperature increase of approx. 35 K versus ambient.

The prototype is depicted in Figure 11. On the left side is the input EMI-filter connected to the two full-bridge LLC inverters with the control board placed by board-to-board connectors onto the main board. Between the main board and the rectifier boards are the two transformers and the resonant coils. The secondary side of the transformers is directly screwed onto the high-current terminals of the rectifier board. A busbar connects the center-tapped 12 V-potential directly to the low-voltage reverse-polarity protection MOSFETs S_{out}. For the inverter automotive-qualified 80 mΩ-COOLMOS-MOSFETs of type IPW65R080CFDA have been selected. The synchronous rectifier MOSFETs are two parallel Opti-MOS switches of type IAUT300N08S5N012, which are bottom cooled through the PCB. To limit the computational complexity of driving the synchronous rectifiers in every of the four modes of operation, they are driven by a self-controlled synchronous driver (IR11688STRPBF). The transformer has a 15-turn primary and two 1-turn secondary windings on an ER54/28/38 ferrite core. While the primary winding is made of litz wire having a strand diameter of 0.1 mm, the secondary side windings are made of two parallel 0.5 mm copper sheets. The resonant inductor utilizes a 17-turn winding on a PQ-32-25 core. The output MOSFETs required for the reverse-polarity protection are three parallel MOSFETs of type IPLU300N04S4 – R8.

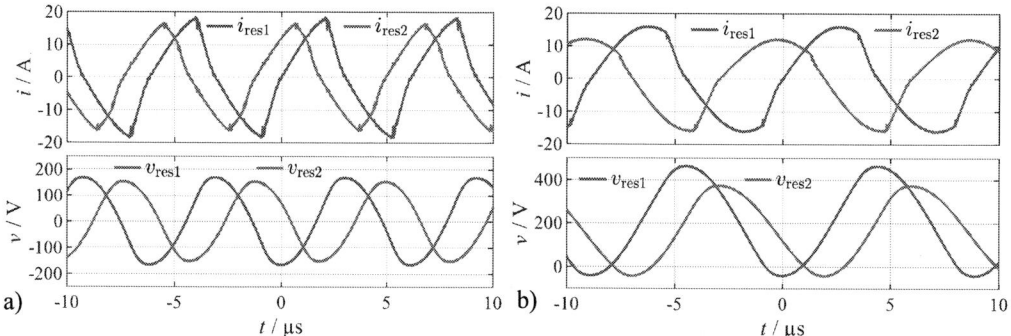

Figure 12: Measurement results for the interleaved operation with balancing control enabled: a) interleaved full-bridge operation (V_{in} = 240 V, V_{out} = 14 V, 280 A, 164 kHz). b) interleaved half-bridge operation (V_{in} = 420 V, V_{out} = 10 V, 260 A, 112 kHz).

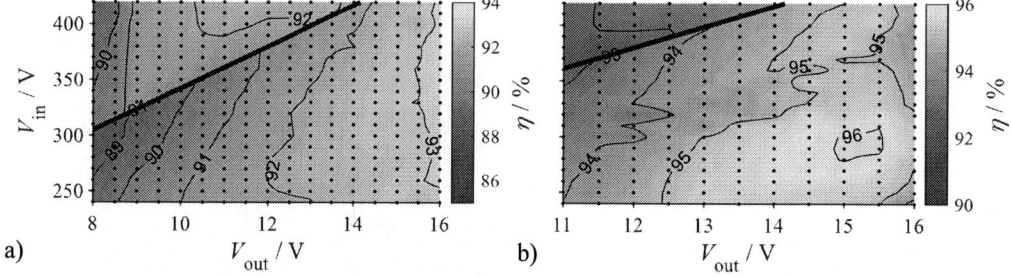

Figure 13: Measured converter efficiency for a) maximum output current (260 A) and b) an output power of 2200 W. Notice that for an output power of 2200 W, operation below an output voltage of 11 V is not required. Measurement points are marked with black stars.

Figure 12 shows some measurement results for the primary resonant currents and the resonant capacitor voltages. a) depicts the balancing in full-bridge operation. LLC_1 is operated with a phase shift for balancing the power equally to both rails. Due to the full-bridge operation, there is no DC-component in the resonant capacitor voltage. The phase shift of the balancing control is 55,4°. Figure 12b) shows an exemplary operation in the half-bridge mode. LLC_1 is operated with an asymmetrical duty cycle. Due to the half-bridge operation, a DC-component offset of approximately $V_{in}/2$ in the resonant capacitor voltage is present, which is reduced for v_{res1} by the asymmetrical duty-cycle mode balancing control resulting in a reduced peak voltage by 45 V. The duty cycle is 0.6 for the low-side switch and 0.4 for the high-side switch. The converter achieves a peak efficiency of approximately 96.5 % while it reaches an efficiency above 90 % for almost all operating points. Only for output voltages below 10 V at full load, the efficiency is below 90 %. The efficiencies were measured with a HIOKI PW6001. Measurement of the input current was hereby done with a HIOKI CT6841 while the output current was measured with a HIOKI CT6844 current probe. Figure 13 shows the efficiency results over the entire operating region for a) an output current of 260 A and b) an output power of 2200 W. For better compatibility, in a) the output current was held to 260 A also for output voltages above 14 V, even though the converter is required to operate at a constant maximum output power of 3640 W at these voltages. The thermal limit of 90°C was kept at all times.

6 Conclusion

This paper proved that LLC converters can be used for a wide input and output voltage range by using several modes of operation. In half bridge mode, low gains can be achieved while for large voltage-transfer ratios, the full-bridge mode can be applied. For intermediate loads and gains, phase-shift mode is applied to limit the maximum switching frequencies. It is proven that output current cancellation is possible by operating the interleaved LLC converters with the equal switching frequency and balancing the two synchronized converters by using the phase-shift mode for full-bridge operation and asymmetrical duty-cycle mode for half-bridge operation. However, it is vital to consider the balancing methods in the converter design as ZVS must be ensured and the increased rectifier current in asymmetrical duty-cycle mode must be considered. A 3,6 kW-prototype design achieved a maximum efficiency of 96.5 % while an efficiency of well over 90 % was ensured for most operating points. A power density of 2.1 kW/l was achieved through the proposed single-stage concept.

References

[1] T. Rüschenbaum, P. Rehlaender, P. Ha, T. Grote, F. Schafmeister, and J. Böcker, "Two-stage automotive DC-DC converter design with wide voltage-transfer range utilizing asymmetric LLC operation," in *PCIM Europe 2020: International Exhibition and Conference for Power Electronics, Intelligent Motion, Renewable Energy and Energy Management*, Nuremberg, Jul. 2020.

[2] M. M. Jovanovic and B. T. Irving, "On-the-Fly Topology-Morphing Control—Efficiency Optimization Method for LLC Resonant Converters Operating in Wide Input- and/or Output-Voltage Range," *IEEE Transactions on Power Electronics*, vol. 31, no. 3, pp. 2596–2608, 2016, doi: 10.1109/TPEL.2015.2440099.

[3] P. Rehlaender, F. Schafmeister, J. Bocker, and T. Grote, "Analytical Topology Comparison for a Single Stage On-Board EV-Battery Converter," in *2019 IEEE 28th International Symposium on Industrial Electronics: Proceedings : Pinnacle Hotel Harbourfront, Vancouver, BC, Canada, 12-14 June, 2019*, Vancouver, BC, Canada, 2019, pp. 2477–2482.

[4] P. Rehlaender, T. Grote, S. Tikhonov, M. Schröder, F. Schafmeister, and J. Böcker, "A 3,6 kW Single-Stage LLC Converter Operating in Half-Bridge, Full-Bridge and Phase-Shift Mode for Automotive Onboard DC-DC Conversion," in *PCIM Europe 2020: International Exhibition and Conference for Power Electronics, Intelligent Motion, Renewable Energy and Energy Management*, Nuremberg, Jul. 2020.

[5] G. Yang, "Design of a High Efficiency High Power Density DC/DC Converter for Low Voltage Power Supply in Electric and Hybrid Vehicles," PhD thesis, École supérieure d'électricité, Gif-sur-Yvette, France, 2014.

[6] G. Yang, P. Dubus, and D. Sadarnac, "Double-Phase High-Efficiency, Wide Load Range High- Voltage/Low-Voltage LLC DC/DC Converter for Electric/Hybrid Vehicles," *IEEE Trans. Power Electron.*, vol. 30, no. 4, pp. 1876–1886, 2015, doi: 10.1109/TPEL.2014.2328554.

[7] Y.-F. Liu and Z. Hu, "Interleaved Resonant Converter," WO/2014/040170, Canada, 20.03.

[8] J. Xu, J. Xiong, Z. Feng, L. Xu, H. Dong, and X. Xu, "A kind of parallel LLC resonant DC/DC power converters," CN106230268A, China.

[9] H. Figge, F. Schafmeister, and T. Grote, "LLC balancing," US9263951B2, USA, 09.01.

[10] Y. Jang, M. M. Jovanovic, J. M. Ruiz, M. Kumar, and G. Liu, "A novel active-current-sharing method for interleaved resonant converters," in *IEEE Applied Power Electronics Conference and Exposition (APEC), 2015: 15 - 19 March 2015, Charlotte Convention Center, Charlotte, North Carolina*, Charlotte, NC, USA, 2015, pp. 1461–1466.

[11] H. Figge, T. Grote, N. Froehleke, J. Boecker, and P. Ide, "Paralleling of LLC resonant converters using frequency controlled current balancing," in *IEEE Power Electronics Specialists Conference, 2008: PESC 2008 ; 15 - 19 June 2008, Capsis Hotel and Convention Center, Rhodes, Greece ; proceedings*, Rhodes, Greece, 2008, pp. 1080–1085.

[12] H. Figge, "High Power LLC Resonant Converter Optimized for High Efficiency and Industrial Use," PhD thesis, Faculty of Computer Science, Electrical Engineering and Mathematics, Paderborn University, Paderborn, 2016.

[13] P. Imbertson and N. Mohan, "Asymmetrical duty cycle permits zero switching loss in PWM circuits with no conduction loss penalty," *IEEE Trans. on Ind. Applicat.*, vol. 29, no. 1, pp. 121–125, 1993, doi: 10.1109/28.195897.

[14] Z. Shang, Y. Zhao, and Y. Lian, "An APWM controlled LLC resonant converter for a wide input range and different load conditions," in *Proceedings 2017 12th IEEE International Conference on ASIC: Oct. 25-28, 2017, Guiyang, China*, Guiyang, 2017, pp. 608–611.

[15] L. Keuck, P. Hosemann, B. Strothmann, and J. Böcker, "A Comparative Study on Si-SJ-MOSFETs vs. GaN-HEMTs Used for LLC-Single-Stage Battery Charger," in *PCIM Europe 2017: International Exhibition and Conference for Power Electronics, Intelligent Motion, Renewable Energy and Energy Management*, Nuremberg, Jul. 2020.

[16] W. Liu, B. Wang, W. Yao, Z. Lu, and X. Xu, "Steady-state analysis of the phase shift modulated LLC resonant converter," in *ECCE 2016: IEEE Energy Conversion Congress & Expo : Sept. 18-22, Milwaukee, WI : proceedings*, Milwaukee, WI, USA, 2016, pp. 1–5.

Switching Loss Estimation Using a Validated Model of 650 V GaN HEMTs

Joao Oliveira[1], Florent Loiselay[4]
VEDECOM ITE
23 bis Alle des Marronniers, 78000, Versailles, France
Phone: [1]+33 (0) 6 25 71 40 30 and [4]+33 (0) 6 51 49 81 14
Email: [1]joao-andre.soares-de-oliveira@vedecom.fr and [2]florent.loiselay@vedecom.fr
Hervé Morel[2], Dominique Planson[3]
Univ Lyon, INSA Lyon, Université Claude Bernard Lyon 1, Ecole Centrale de Lyon, CNRS, AMPERE
F-69621, Lyon, France
Phone: [2]+33 (0) 4 72 43 82 38 and [3]+33 (0) 4 72 43 87 24
Email: [2]herve.morel@insa-lyon.fr and [3]dominique.planson@insa-lyon.fr

Keywords

≪Gallium Nitride (GaN)≫, ≪Switching losses≫, ≪Device characterisation≫, ≪Automotive application≫.

Abstract

GaN power devices allow building more compact power converters. In order to study these new devices, it is important to measure and estimate switching losses. Therefore, an instrumented PCB is developed including the measurement points needed for this purpose. The parasitic elements of the PCB layout extracted by ANSYS Q3D and the models of the measurement instruments are also included in the simulation model. In this way, by means of a validated model, it will be possible to evaluate the losses in an optimized circuit. Simulation and experimental results are presented to validate the simulation approach.

Introduction

The high power density capability of Wide Band Gap (WBG) power devices with the potentially fast switching and low losses make them suitable for automotive applications where performance and low weight are keys for the development of more compact power converters. GaN power devices exhibit a better performance when compared with Si power devices thanks to their superior critical electric field and their high electron mobility in the 2DEG. This allows having lower parasitic capacitance values and therefore, high-speed commutation, in addition to having a lower on-resistance. However, these improvements bring new challenges when it comes to the loss measurements of the converters based on GaN material. A half bridge topology is thus developed including the measurement probes needed to evaluate the switching losses in a double pulse test. Furthermore, an evaluation of the parasitic elements of the PCB is considered.

The experimental results are used to validate the SPICE model and also for the loss estimation. Moreover, with the validated model, it will be possible to evaluate by simulation the losses in an optimized circuit without measurement probes [1,2].

The development of a method that allows the estimation of switching losses is important for the design of converters. Also due to the fast switching, the effect of circuit parasitic elements is strongly important. For this reason, a good method used by converter designers is to analyse, by using simulations, the device switching behaviour under the influence of circuit parasitic elements before performing the physical circuit. This is especially helpful because it allows adjusting circuit layouts or comparing different layouts based on the simulation results [3–5].

GaN HEMTs 650 V Static Characteristics

The GaN device presents some particularities due to the fast switching capability. The transistor characteristics were measured with a Curve Tracer Keysight B1506a. In regard to the loss estimation and efficiency calculation, the I-V characteristics of the device need to be accurately measured. The linear mode for I-V graph is represented by the region when drain-source voltage is below the limit $V_{gs} - V_{th}$. We have to regard this region for a good estimation of the on-losses. At the moment that drain-source voltage is above $V_{gs} - V_{th}$, the curve represents the saturation region, being important for a good estimation of the switching losses [6–9].

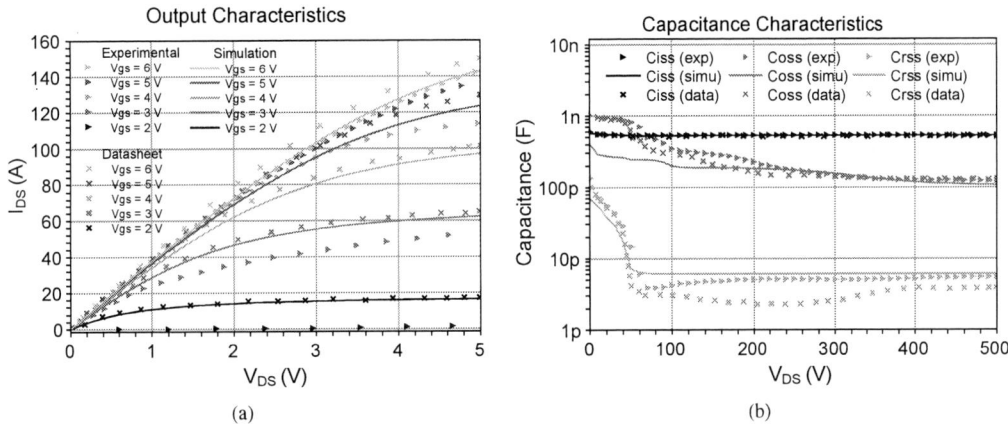

Fig. 1: Output characteristics for the GS66516B (a) and (b) input, output and reverse transfer capacitances.

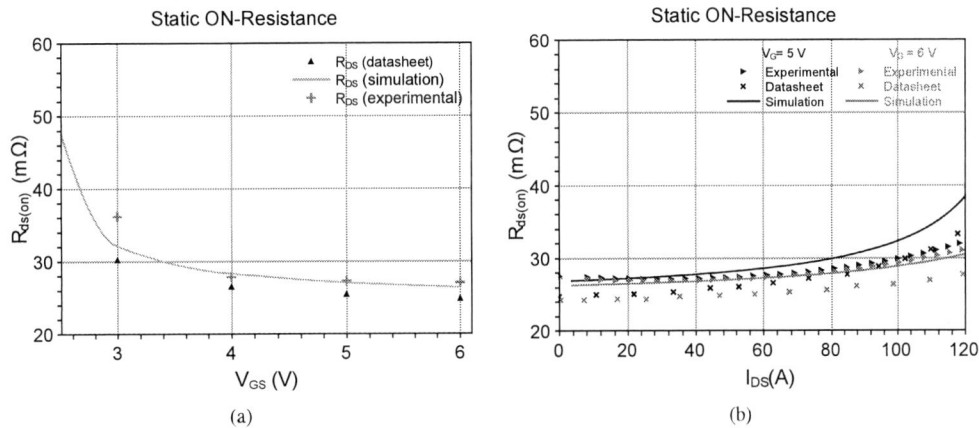

Fig. 2: (a) Static On-resistance and (b) its variation versus drain-source current.

Fig. 1(a) shows a comparison between measurements, datasheet values and the simulations obtained using the manufacturer model (LTspice). Some difference between experimental and simulation results can be noticed, which can be explained by components with different threshold voltages. These results show that some improvements in the default model are needed. The modelling of the input capacitance must be also precisely performed, since it determines the switching speed, influencing consequently the switching losses. The miller capacitance (C_{rss}) is the main component for cross conduction effect, as demonstrated in dynamic characterisation tests [1]. Fig. 1(b) shows a good correlation between simulation and measurement results. This implies that for switching loss estimation the SPICE model presents a behaviour close to the real component. The on-state resistance can be read directly from the output characteristic curves and are presented in Fig. 2. Dispersion of V_{th} values could mainly explain the waveform variations found for the on-resistance.

Protocol Test

During measurements, it is noticed that the output characteristic for the GaN component was modified if a drain leakage current characteristic was measured just before the output characteristics. The leakage current test consists of a gate-source voltage applied to hold the switch turned-off (0 V, in this study). At the same time, a drain-source voltage is applied to slowly increase its value (up to 400 V). For example, Fig. 3 shows that the curve for the gate voltage equal to 2 V was significantly modified. This effect is due to the current collapse associated with the GaN components [10–13].

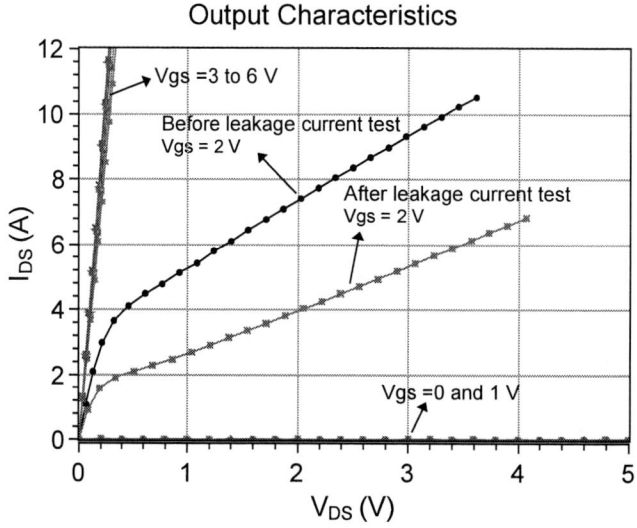

Fig. 3: Output Characteristics for GS66516B GaN Systems.

This is an important point that will be considered for the measurements. To avoid this phenomenon, our protocol includes a measurement of the transfer characteristics from −8 V to 4 V (gate-source voltage) just before performing the output characteristic test. The drain-source on-resistance is measured from the Fig. 3. Hence, at ambient temperature and before the drain leakage test it is found 22.3 mΩ. Thereafter, the on-resistance increases up to 23.5 mΩ.

GaN HEMT 650 V Dynamic Characteristic

Instrumented PCB Analyses

The developed PCB is instrumented to measure voltages (drain-source and gate-source) and currents (gate and drain-source) on the low side switch, as seen in Fig. 4. The command circuit is connected from another PCB, positioned ninety degrees in order to obtain the lowest coupling effect between the power and drive circuits. Furthermore, it is used a ground plane connected to the earth, in order to minimize the mutual inductance values. In the Table I, the measurement instruments are described.

The estimation of switching losses is performed considering three main elements on the spice simulation: the GaN component spice model, measurement instruments modelling and the parasitic elements from 3D wiring model. In Fig. 5(a) the circuit for the double-pulse test (DPT) and its measurement points. The 3D model seen in Fig. 5(b) is also developed in SpaceClaim and simulated on ANSYS Q3D to extract the RLC parasitic matrix.

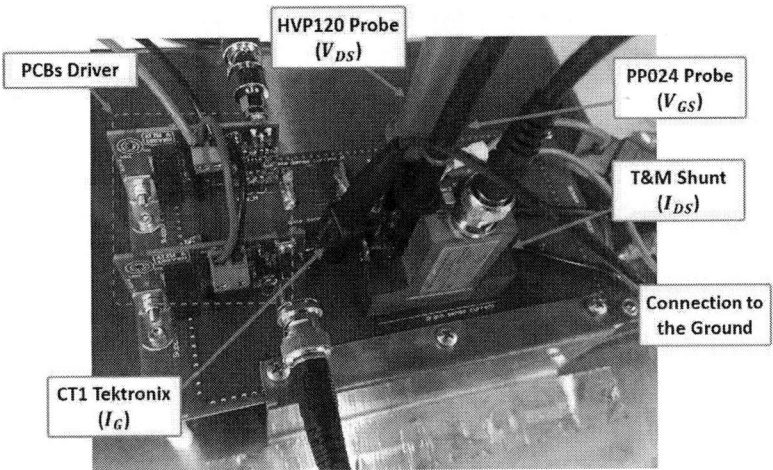

Fig. 4: PCB developed to the double pulse test with the GaN components behind the voltage probes.

Table I: List of measurement instruments used on this study.

Probe Reference	Manufacturer	Usage	Symbol
PP024	Lecroy	Gate-source voltage	V_{GS}
HVP120	Lecroy	Drain-source voltage	V_{DS}
W-1-01C-1FC	TandM Reasearch	Drain-source Current	I_{DS}
CT1	Tektronix	Gate current	I_G

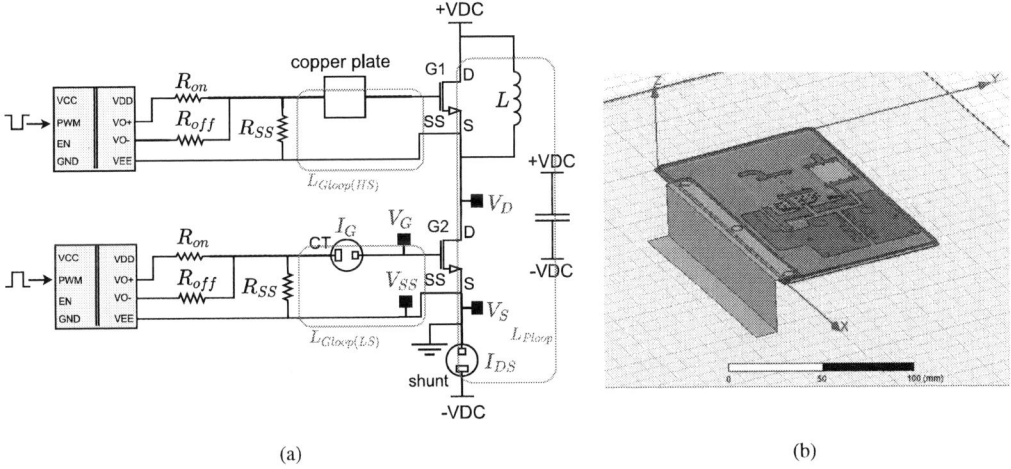

(a)   (b)

Fig. 5: (a) Schematic showing all measurement points used to the tests. (b) 3D Model simulated on ANSYS Q3D.

The modelling steps consist of characterizing the PCB without components, thus being able to analyse the parasitic elements associated to the switch, and then perform switching tests for different charge levels. In Fig. 6, the voltage waveforms of gate-source and drain-source on low side switch for different operation points can be seen. In Fig. 7, the current waveforms are shown. The correlation between the spike current and the drain-source voltage applied is clearly verified. The maximum steady drain current reached is about 19 A rated at 400 V. By means of the Fig. 7(a), it is possible to observe that the drain-source current shapes under different operation points are about the same, therefore it indicates that the influencing of parasitic capacitors is not dependent on the operating point [3].

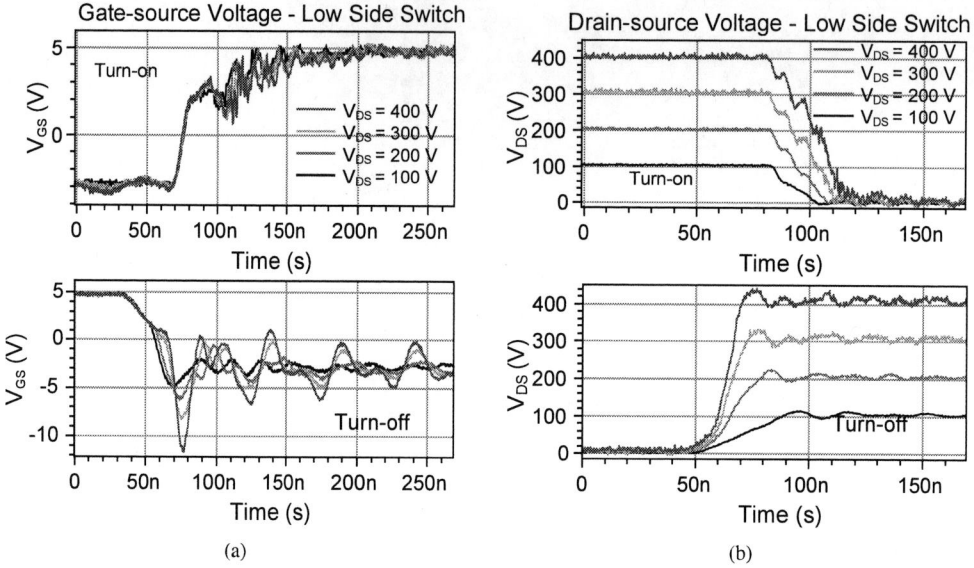

Fig. 6: Experimental results: (a) V_{GS} low side switch and (b) V_{DS} low side switch.

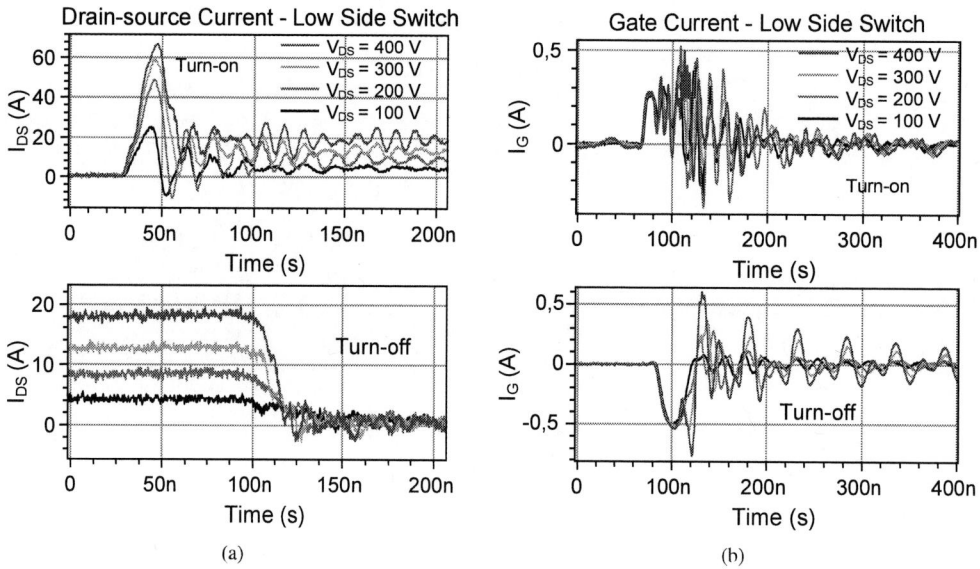

Fig. 7: Experimental results: (a) I_{DS} low side switch and (b) I_G low side switch.

In Fig. 8(a) the effect of measuring the gate current can be seen. Some disturbances are detected, but the waveform remains nearly the same. The main oscillation frequencies obtained by FFT (Fast Fourier Transform) applied on measured V_{GS} indicates a resonance effect between the input capacitance of the switches and the gate loop inductance. In Fig. 8(b), the analytical calculus is performed to confirm this proposition. Therefore, by adding a probe current for I_G measurement the impact on signals is limited.

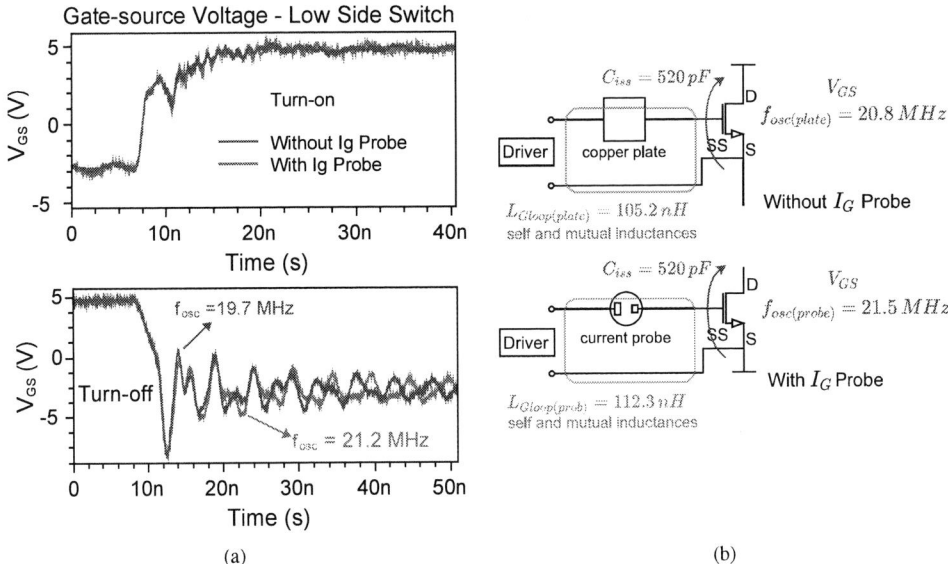

Fig. 8: Experimental result: (a) Comparison of measured gate voltage V_{GS} considering the impact of using a current probe for I_G measurement. (b) Main oscillation frequency from V_{GS} for both cases: without I_G probe and with I_G probe. Test condition: 300 V and 12.6 A.

Analyses of Switching Losses on GaN HEMTS 650 V

In order to evaluate the switching losses in hard switching mode, the high side GaN is driven by a negative voltage (-3 V), thus behaving like a diode. The turning-on period starts with the charging time of the input capacitance (C_{iss}). When the V_{GS} reaches the threshold voltage, the low side switch can begin to conduct current. This phase is terminated when the drain current reaches the steady value for the inductor current. Right after, the drain-source voltage starts decreasing (period of voltage plateau), as long as the Miller capacitance (C_{rss}) is discharged. In addition, it is important to point out that there are parasitic effects which directly influence the oscillations detected during the switching [1, 14].

During the turning-on of a hard-switching, the effective current through the low side switch is a result of the sum of parasitic currents and the charge inductor current, as seen in Fig. 9(a), causing a spike current effect. The energy associated with the output capacitance of both switches (C_{oss}) and (C_{qoss}) is discharged towards the 2DEG channel of the low side switch. Furthermore, we should regard the PCB parasitic capacitance (C_{pcb}) and the intrinsic capacitance associated with the inductor (C_{pl}). Both elements generate high-frequency currents towards the low side switch and are not voltage-dependent. For the turning-off, the drain current is equal to the load current minus the currents related to the parasitic capacitance values, as seen in Fig. 9(b) [3].

The loss estimation is performed by using four different simulations. The first one is a simple simulation which is considered the switch spice model (GS66516B), driver circuit, power loop inductance (L_{Ploop}), and the equivalent circuit (RLC) for the load inductor. For a second simulation, the measurement instrument models (voltage probe, shunt and current transformer) are considered in order to add the delay effect coming from the cable length and the disturbances associated to the impedance insertion. Afterwards, the gate loop parasitic inductance (L_{Gloop}, extracted by ANSYS Q3D) is added to the simulation. Finally, a simulation is performed by adding the full matrix Q3D (RLC) coupled with the electrical circuit of the instrumented PCB, including also the probe models.

In Fig. 10 and Fig. 11, the comparison between the performed simulations and the experimental data can be seen. Based on the results, it is reasonable considering the simulation which includes the power and

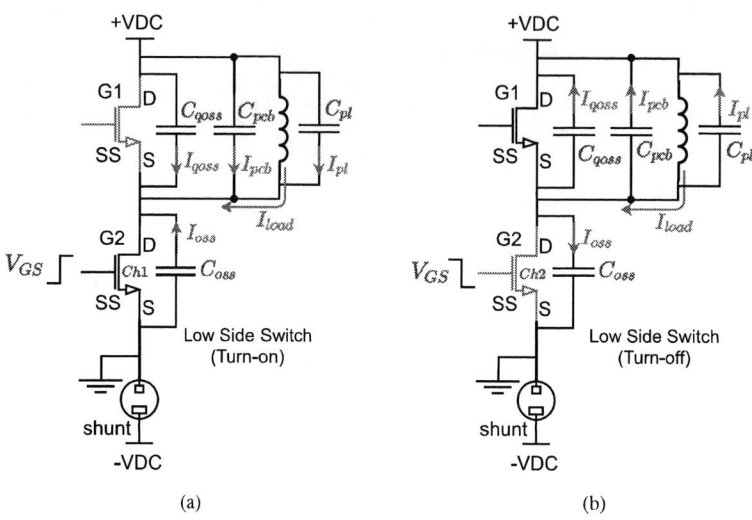

(a)   (b)

Fig. 9: (a) Current distribution at the moment of turning-on (low side switch). (b) Current on low side switch at turning-off.

gate loop inductances extracted by Q3D to calculate the switching losses. The modelling of parasitic elements and measurement instruments allows us to achieve a good correlation between simulation and experimental results, as seen in Fig. 11 by means of the curves overlay. In fact, the width of the power loss triangle increases with the power loop inductance. Moreover, the width and height of the power loss triangle also increase with the gate loop inductance of the high side switch. The delay added by the cable length of the measurement instruments and the disturbances found on the shape of the drain-source voltage are also strong elements that contribute to a good estimation of switching losses.

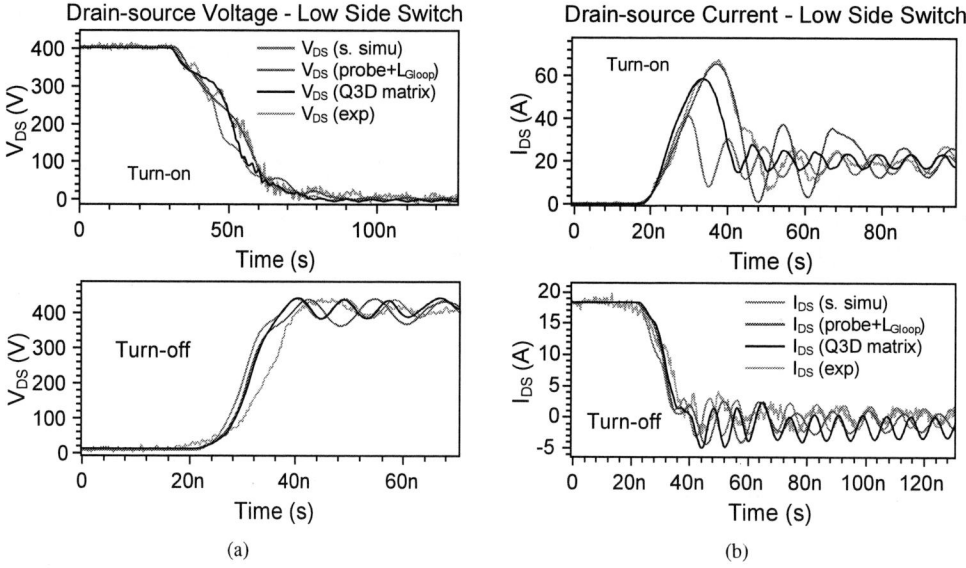

(a)   (b)

Fig. 10: (a) Drain-source voltage comparison for simulation and experimental results. (b) Drain-source current comparison for simulation and experimental results.

The loss calculation is estimated using the signals from the switch terminal and the scope circuit, in

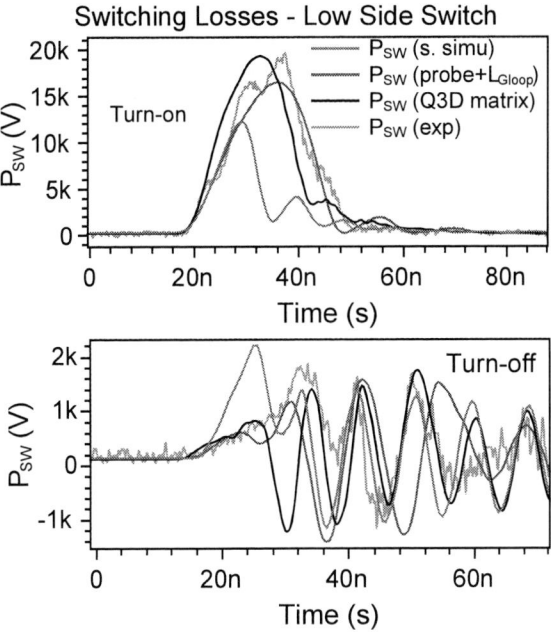

Fig. 11: Power switching loss for GS66516B GaN Systems on PCB instrumented.

order to compare the effect caused by the measurement instruments. For the experimental results, a new column (Delay Comp.) is calculated shifting the voltage signal at 4 ns towards the left of the time axis, implementing an advance on the signal. This value corresponds to the delay added by the cable models from the shunt and voltage probe. Some values of losses are shown in Table II and Table III for different operation points. It is possible to verify that by adding parasitic elements on the simple simulation, the difference between simulation and experimental values becomes smaller. By performing several measurements an uncertainty of 10% was detected on the values of switching losses. This is related to different integration points chosen for loss calculation, different voltages probes used, and the effect of measuring the gate current.

Table II: Switching loss energy of GS66516B for turning-on.

Operation Points (V | A)	Turn-on (μJ)								
	Simple Simulation	Probe Models		Adding L_{Gloop}		With Full Q3D		Experimental	
		Switch	Scope	Switch	Scope	Switch	Scope	Delay Comp.	Scope
100 | 4.3	7.2	6.5	10.3	7.8	13.2	9.4	14.4	10.1	15.2
200 | 8.5	29.1	26.4	39.5	42.2	62.4	49.9	68.7	51.6	71.6
300 | 12.6	65.8	65.1	89.5	113.2	158.6	128.2	160.2	136.5	171.8
400 | 18.3	126.3	127.5	170.5	228.2	285.2	232.7	284.2	261.8	314.7

Table III: Switching loss energy of GS66516B for turning-off.

Operation Points (V | A)	Turn-off (μJ)								
	Simple Simulation	Probe Models		Adding L_{Gloop}		With Full Q3D		Experimental	
		Switch	Scope	Switch	Scope	Switch	Scope	Delay Comp.	Scope
100 | 4.3	2.5	3.6	1.8	1.9	1.2	2.7	1.5	7.47	4.5
200 | 8.5	6.7	6.5	2.5	6.9	2.8	6.2	3.2	13.3	9.5
300 | 12.6	10.8	11.0	2.9	12.3	4.4	11.5	3.7	20.2	11.8
400 | 18.3	14.9	15.7	4.1	20.1	6.3	16.5	5.2	36.8	19.9

It is important to highlight that the oscillations on current waveforms are due to the parasitic elements and does not affect the integral loss calculation, once the average value of the high frequency of oscillation is almost zero after integration. It can be also noticed that only with the power and gate loop inductances (L_{PLoop} and L_{GLoop}) extracted by ANSYS Q3D and with the instrument measurement models, the divergence between the theoretical values and measured values are relatively small.

Conclusion

The comprehensive analysis of the impact of the wiring parasitic elements can improve the accuracy of switching loss estimation. Furthermore, by means of experimental results the need of a good modelling of instrument measurements and a 3D Model for parasitic extraction is verified. A simulation considering the power and gate loop inductances can yield a good match with experimental data. This approach allows us to validate all models used in spice simulations, and thus to be able to evaluate more accurately the overall efficiency of converters based on GaN components, even those that will be not instrumented.

References

[1] Zheng Chen. *Characterization and modeling of high-switching-speed behavior of SiC active devices*. PhD thesis, Virginia Tech, 2009.

[2] Xiao Shan Liu, Bertrand Revol, and François Costa. Parasitic elements modeling and experimental identification in a GaN HEMT based power module. In *19 ème Colloque International et Exposition sur la Compatibilité ÉlectroMagnétique (CEM 2018)*, 2018.

[3] Ruoyu Hou, Juncheng Lu, and Di Chen. Parasitic capacitance Eqoss loss mechanism, calculation, and measurement in hard-switching for GaN-HEMTs. In *2018 IEEE Applied Power Electronics Conference and Exposition (APEC)*, pages 919–924. IEEE, 2018.

[4] Olivier Goualard. *Utilisation de semi-conducteurs GaN basse tension pour l'intégration des convertisseurs d'énergie électrique dans le domaine aéronautique*. PhD thesis, 2016.

[5] Fadi Nader Fouad Zaki. *Characterization, modeling and aging behavior of GaN power transistors*. PhD thesis, Université Paris-Saclay, 2018.

[6] Perrin Rémi. *Characterization and design of high-switching speed capability of GaN power devices in a 3-phase inverter*. PhD thesis, Lyon, 2018.

[7] Juncheng Lucas Lu and Di Chen. Paralleling GaN E-HEMTs in 10kW-100kW systems. *Conference Proceedings - IEEE Applied Power Electronics Conference and Exposition - APEC*, pages 3049–3056, 2017.

[8] Ke LI. *Wide Bandgap (SiC/GaN) Power Devices Characterization and Modeling: Application to HF Power Converters*. PhD thesis, Lille, 2014.

[9] Guillaume Regnat. *Onduleur à forte intégration utilisant des semi-conducteurs à grand gap*. PhD thesis, Grenoble, 2016.

[10] Gaudenzio Meneghesso, Fabiana Rampazzo, Peter Kordos, Giovanni Verzellesi, and Enrico Zanoni. Current collapse and high-electric-field reliability of unpassivated GaN/AlGaN/GaN-HEMTs. *IEEE Transactions on Electron Devices*, 53(12):2932–2941, 2006.

[11] Zhikai Tang, Qimeng Jiang, Yunyou Lu, Sen Huang, Shu Yang, Xi Tang, and Kevin J Chen. 600-v normally off SiNx/AlGaN/GaN MIS-HEMT with large gate swing and low current collapse. *IEEE Electron Device Letters*, 34(11):1373–1375, 2013.

[12] Donghyun Jin. *Dynamic ON-resistance in high voltage GaN field-effect-transistors*. PhD thesis, Massachusetts Institute of Technology, 2014.

[13] Ke Li, Paul Evans, and Mark Johnson. GaN-HEMT dynamic ON-state resistance characterisation and modelling. In *2016 IEEE 17th Workshop on Control and Modeling for Power Electronics (COMPEL)*, pages 1–7. IEEE, 2016.

[14] Zheyu Zhang, Ben Guo, Fei Fred Wang, Edward A. Jones, Leon M. Tolbert, and Benjamin J. Blalock. Methodology for Wide Band-Gap Device Dynamic Characterization. *IEEE Transactions on Power Electronics*, 32(12):9307–9318, 2017.

Reduction of conduction losses in resonant converters by connecting three single-phase inverters to a common generator

Sergio Tárraga, John Paul Mayorga, Esther de Jódar, José Villarejo
UNIVERSIDAD POLITÉCNICA DE CARTAGENA
Dpto. de Automática, Ingeniería Eléctrica y Tecnología Electrónica
Cartagena, España
Tel.: +34/ 968 325461
Fax.: +34/ 968 325345
E-Mail: sergio.tarraga@upct.es
URL: http://www.upct.es

Acknowledgements

This work was supported by the Ministry of Economy and Competitiveness, Spain, under Research Project TEC2016-80136-P.

Keywords

«Interleaved converters», «Resonant Converter», «Conduction losses», «Photovoltaic».

Abstract

This paper proposes the use of a resonant converter consisting of a high frequency bridge and three resonant circuits with their corresponding transformer and rectifier. This configuration eliminates the low frequency pulsating power in the oscillator and reduces the conduction losses compared to the use of three independent DC/DC converters considering unitary power factor. For situations when this conditions can´t be fulfilled a method is proposed in order to obtain and calculate the filter capacitor, as well as a dynamic model that works for both situations. Finally, a scale prototype has been built, with experimental results.

Introduction

Utility-scale photovoltaic (PV) inverters are predominantly built with single-stage topologies connected to a low-voltage to medium-voltage line frequency transformer. Given the costs, and power losses associated with line-frequency transformers, manufacturers are investigating architectures that produce medium-voltage ac (MVAC) directly. Multilevel inverters are the natural choice because the large number of series-connected devices not only allows for increased voltage blocking but also enables the synthesis of high-quality waveforms [1]–[5]. The multilevel inverters bulky passive components and the phase power imbalance are the main challenges of these converters. Several CHB (Cascaded H-Bridge) based topologies have been proposed to deal with the aforementioned issue. In [6] a CHB-based PV system is proposed, where three independent dc-dc converters with isolation are in parallel in each full-row. However, it has the disadvantages of complex control system and difficult transformation towards higher powers due to the DC/DC used topology. [7] presents a multilevel medium-frequency link inverter where the PV power can be distributed among three phases by a common magnetic link. However, the complex magnetic design poses serious challenges for its application. Different QAB (Quadruple Active Bridge)-based CHB converters are presented in [8-9] where both improve the ability of phase power balancing and improve the ability of phase power balancing and bypasses the need for bulk energy storage. [8] focuses on the modeling and control of the converter. On the other hand, [9] places more emphasis on the modularity of the system and uses the QAB with a constant transformation ratio in order to ensure ZVS (Zero Voltage Switching) during the grid pulsating power conversion.

Of the four bridges of the QAB one works with constant power and the rest with a power that varies twice line frequency. This means that whatever power the CHB processes, the power of three of the bridges goes through very low values. So, there is only a voltage ratio that ensures ZVS. This operation can be achieved with SRC (Series Resonant Converter) or LLC resonant converters switching at the resonant frequency. These converters are easily scalable and include the parasites of the transformers in their operation taking advantage of them to obtain soft switching. The use of a resonant converter consisting of a high frequency bridge and three resonant circuits with their corresponding transformer and rectifier is proposed. The converter works at the resonant frequency and ensures isolation and constant transformation relationship without a control loop. PV power is transferred as constant balanced three-phase ac power, instantaneous input–output power balance bypasses the need for bulk energy storage, also eliminating the low frequency pulsating power in the oscillator and reducing the conduction losses with respect to the use of three independent DC/DC converters. All this, simplifies the design of the CHB converter and reduce its cost. The insulation voltage that this DC/DC converter must provide is very high and it is not cheap to pass signals through this barrier. However, since it is a unidirectional converter, it will not be able to regulate its output voltage, HB input voltage, when the CHB does not work with unity power factor. In order to study the converter behavior during "no output voltage regulation" a dynamic model of the converter has been developed. A calculation method for the filter capacitors of the converter is proposed when the inverter works with low power factor. Finally, a scale prototype has been built.

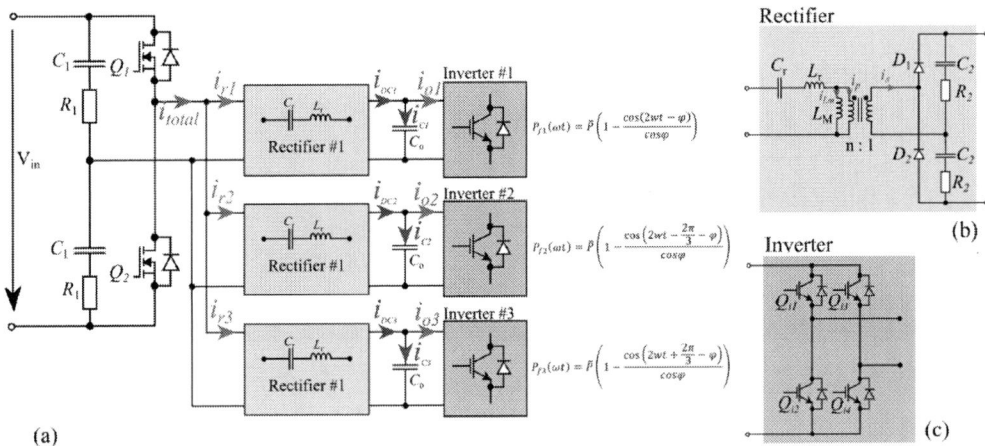

Fig. 1: Proposed approach of the resonant converter (a). Resonant rectifier (b). Full-Bridge inverter (c).

Analysis of the Proposed Approach

Fig. 1 shows the resonant converter proposed in this paper. Based on the operation of the converter at the resonance frequency, the output current of each converter can be calculated as (1)

$$\langle \bar{i}_o \rangle_{Ts} = \langle \bar{i}_{D1} \rangle_{Ts} = \frac{1}{2\pi} \int_0^\pi i_{D1}(\omega t) d\omega t = \frac{1}{2\pi} \int_0^\pi \hat{i}_{2p} \cdot sin(\omega t) d\omega t = \frac{\hat{i}_{2p}}{2\pi} \tag{1}$$

Each inverter connected to the output of the rectifiers can be modeled as a power source using (2), (3) and (4), with $\omega = 2\pi f_{grid}$

$$P_{p1}(\omega t) = \widehat{V_g} \cdot sin(\omega t) \cdot \widehat{I_g} \cdot sin(\omega t \text{-} \varphi) = \overline{P} \cdot \left(1 - \frac{cos(2\omega t \text{-} \varphi)}{cos(\varphi)} \right) \tag{2}$$

$$P_{p2}(\omega t)=\widehat{V_g}\cdot sin(\omega t)\cdot \widehat{I_g}\cdot sin\left(\omega t-\frac{4\pi}{3}-\varphi\right)=\overline{P}\cdot\left(1-\frac{cos\left(2\omega t-\frac{4\pi}{3}-\varphi\right)}{cos(\varphi)}\right) \quad (3)$$

$$P_{p3}(\omega t)=\widehat{V_g}\cdot sin(\omega t)\cdot \widehat{I_g}\cdot sin\left(\omega t+\frac{4\pi}{3}-\varphi\right)=\overline{P}\cdot\left(1-\frac{cos\left(2\omega t+\frac{4\pi}{3}-\varphi\right)}{cos(\varphi)}\right) \quad (4)$$

Where

$$\overline{P}=\frac{\widehat{V_g}\cdot\widehat{I_g}}{2}\cdot cos(\varphi) \quad (5)$$

Thus, the current flowing through the secondary winding, is, can be obtained by (6)

$$i_s(\omega t)=i_o(\omega t)\cdot\pi\cdot sin(\omega_s t)=\frac{\overline{P}}{V_o}\left(1-\frac{cos(2\omega t-\varphi)}{cos(\varphi)}\right)\pi\cdot sin(\omega_s t) \quad (6)$$

Being $\omega_S = 2\pi f_s$. The current flowing through each resonant tank would be the sum of the magnetizing current, i_{Lm}, and the current flowing through the primary transformer winding. Since the magnetizing current is normally much lower than the current flowing through the primary winding, the current of each resonant tank may approximate to that of the primary, neglecting the effects of the magnetizing current (7).

$$i_{r1}\approx i_s(\omega t)\cdot\frac{N_2}{N_1} \quad (7)$$

Therefore, the total current that would circulate through the Half Bridge would be (8)

$$i_T(\omega t)=\frac{N_2}{N_1}\cdot\frac{P\pi}{V_o}sin(\omega_s t)\left[3-\left(\frac{cos(2\omega t-\varphi)+cos\left(2\omega t-\frac{4\pi}{3}-\varphi\right)+cos\left(2\omega t+\frac{4\pi}{3}-\varphi\right)}{cos(\varphi)}\right)\right]=$$
$$(...)=3\cdot\frac{N_2}{N_1}\cdot\frac{P\pi}{V_o}sin(\omega_S t) \quad (8)$$

In (8) it is checked how for the input current the effect of the low frequency pulsation f_{grid} has disappeared, leaving only the effect of the switching frequency.

Semiconductor Losses

As previously mentioned, one of the main advantages of resonant converters is their wide range soft switching, so the losses will be mostly due to conduction. For the conventional approach where there are three individual converters, the conduction losses in the half bridge MOSFETs, when $\varphi = 0$ would be (9):

$$P_{1conv}=I_{1rms}^2 R_{DS} \quad (9)$$

Where $k = \frac{\pi P N_2}{V_o N_1}$, the current can be described as (10):

$$I_1(t) = k \cdot (1 - cos(2\omega t)) \cdot sin(\omega_s t) = k \cdot sin(\omega_s t) - \frac{k}{2} sin(\omega_s t - 2\omega t) - \frac{k}{2} sin(\omega_s t + 2\omega t) \qquad (10)$$

On the other hand, the rms value of (10) would be (11):

$$I_{1rms}^2 = \frac{k^2}{2} + 2\left(\frac{k}{2\sqrt{2}}\right)^2 = \frac{3}{4}k^2 \qquad (11)$$

Therefore, power (9) can be reformulated as (12):

$$P_{1conv} = \frac{3}{4}k^2 R_{DS} \qquad (12)$$

So the power for all three converters would be (13):

$$P_{3conv} = \frac{9}{4}k^2 R_{DS} \qquad (13)$$

The conduction losses for the system in Fig. 1(a), considering the same amount of Si as for the calculation of the individual converters, would be (14)

$$P_{cond1x3} = I_{Trms}^2 \frac{R_{DS}}{3} \qquad (14)$$

Where:

$$I_{Trms}^2 = \left(\frac{3}{\sqrt{2}} \frac{\pi P N_2}{V_o N_1}\right)^2 \qquad (15)$$

If it is redefined (15) as a function of k, the conduction losses of the system under consideration (14) remain as (16):

$$P_{cond1x3} = \frac{3}{2} k^2 R_{DS} \qquad (16)$$

Thus, the losses of the system proposed in Fig. 1(a) compared to the traditional mode of three individual converters are reduced by 33.33% (17)

$$\frac{P_{cond1x3}}{P_{3conv}} = \frac{\frac{3}{2} k^2 R_{DS}}{\frac{9}{4} k^2 R_{DS}} = \frac{2}{3} \qquad (17)$$

The above development is only true if $\varphi = 0$. In the case of unidirectional converters when $\varphi \neq 0$, the inverters will deliver power to the DC/DC, which, being unidirectional, cannot process it. Under these conditions the capacitor, C_o, stores the reactive energy by increasing its voltage above the programmed one (Fig. 2(a. θ_1)). During the second section (Fig. 2(a. θ_2)), the inverter injects energy back to the grid, but the DC/DC converter remains inactive, as the voltage of the capacitor has to drop to the voltage at which the control will operate again. The same energy that was stored during θ_1 is returned during θ_2, which will be used to calculate the duration of the sections. During the final segment (Fig. 2(a. θ_3)), the DC/DC converter keeps the voltage constant over C_o, so that no low-frequency current will circulate through it and the power variation will be reflected towards the DC/DC converter input.

Calculation method for the filter capacitor Co

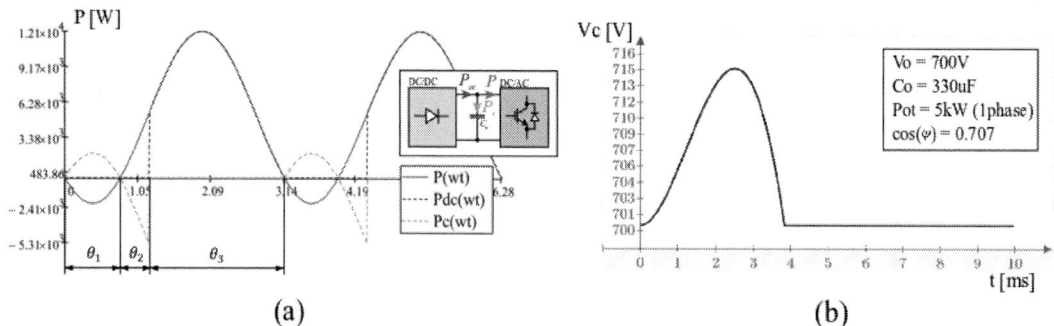

Fig. 2: Power flowing through the converters and C_o (a). In green appear the power processed by C_o, in blue the one processed by the DC/DC and in red by the DC/AC. (b) Example of voltage calculation in C_o. Variation of the voltage at the input of the Full Bridge.

For a Full Bridge inverter that injects power $\bar{P}$ with a power factor, $\cos(\varphi)$, the instantaneous power processed by the DC/AC converter can be written as (2). The energy stored by the capacitor at θ_1 is returned at θ_2, and can be mathematically expressed as (18).

$$\int_{0}^{\theta_1+\theta_2} 1-\frac{\cos(2\omega t-\varphi)}{\cos(\varphi)}\,d\omega t=0;\ \ \theta_1+\theta_2<\pi \qquad (18)$$

Using (2) and solving numerically (18) the value $\theta_1+\theta_2$ can be obtained. According to the diagram shown in Fig. 2(a), the equations for permanent regime power can be found in (19) and (20).

$$P_{DC}(\omega t)=\begin{cases}0, & 0<\omega t<\theta_1+\theta_2\\ P_{p1}(\omega t), & \theta_1+\theta_2<\omega t<\pi\end{cases} \qquad (19)$$

$$P_C(\omega t)=P_{DC}(\omega t)-P_{p1}(\omega t) \qquad (20)$$

Once the power equations have been obtained, the differential equation (21) can be proposed, which models the operation of the capacitor in the section $\theta_1+\theta_2$. Fig. 2(b) shows an example of how the voltage in the capacitor would look like.

$$C_o\frac{dv_c}{dt}=i_c=\frac{P_C(\omega t)}{v_c} \qquad (21)$$

Dynamic model of the converter

An averaged model is developed to evaluate the transient operation of the converter. This model helps to better understand the behaviour of the resonant converter, as well as simplifying the simulation of complex switched systems. The dynamic equivalent model (Fig. 3(b)) can be obtained in a generic way, based on a power/loss and stored energy approach. This approach is based on [11] and [12].

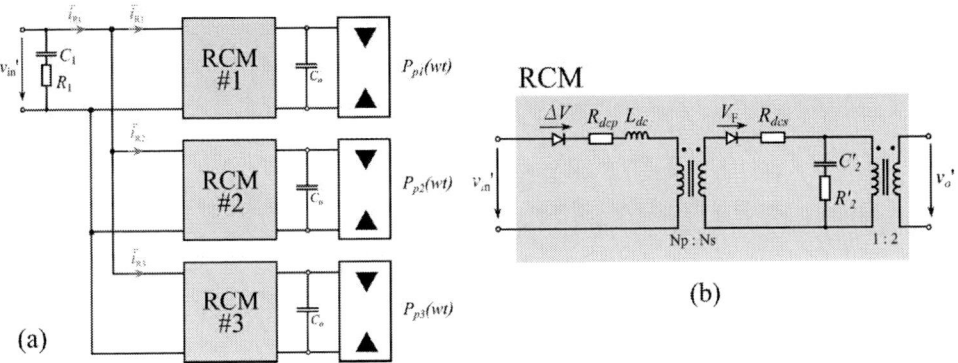

Fig. 3: Equivalent circuit for Fig. 1(a), where the switching system has been replaced by the dynamic equivalent model of the rectifier/Half Bridge and power sources instead of Full Bridge inverters. (b) Dynamic Equivalent Model of the converter.

The results obtained after applying the dynamic equivalent model to the configuration studied (Fig. 3(a)) are shown in Fig. 4. The model has been simulated considering: $\varphi = \pi/4$ (Fig. 4 (a)) and $\varphi = 0$ (Fig. 4(b)). In both situations the model mimics properly its equivalent switched model.

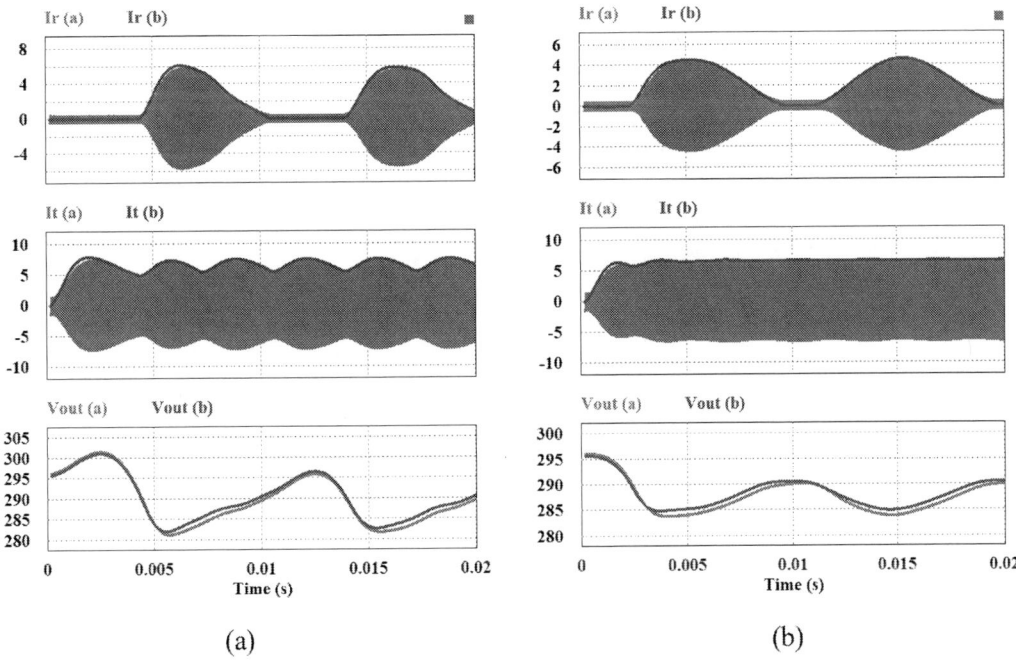

Fig. 4: Simulated results of the dynamic equivalent model considering $\varphi = \pi/4$ (a), $\varphi = 0$ (b). The results show the current in a branch (top), total input current (middle) and output voltage (bottom).

Experimental Verification

Table I: Design specifications

Parameter	Unit	Value
V_{in}	V	200
V_o	V	257
P_o	W	150
f_s	kHz	57
L_r	µH	165.5
L_m	µH	1000
C_r	nF	47
$N_p{:}N_s$		42:54
C_o	µF	69
C_1	µF	1
C_2	µF	1
R_{onDS}	Ω	0.28
deadtime	ns	400

In order to validate the previous results, a scale prototype has been built. This prototype has been designed and built so it can be easily scaled to higher potencies, respecting insulation properties, modularity… The design specifications can be found in Table I. The primary side Half Bridge MOSFETs are IPW50R280CE, whereas the secondary diodes are STPSC406D. The measured results (Fig. 5) show that the prototype behaves as expected on the simulations and theoretical analysis, reducing the low frequency pulsating power in the oscillator. Concerning the magnetizing current it can be seen that, whereas the peak and frequency values fully agree, the waveform is slightly different than the theoretical. This might be due to the transformer real parameters of construction and the fact that in the simulation the transformer model is a simplification that not necessarily fully agrees with reality.

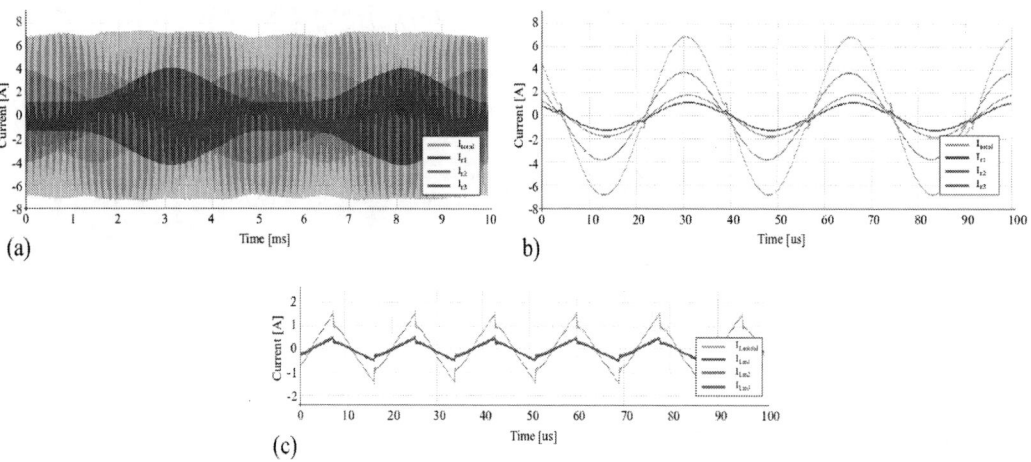

Fig. 5: Experimental results obtained for the specifications of Fig. 4(c), where it can be seen that the input current is almost flat after eliminating the effects of the low frequency pulsating power (a), detailed view of the currents (b); magnetizing currents (c).

Conclusion

In this paper a resonant converter consisting of a high frequency bridge and three resonant circuits with their corresponding transformer and rectifier working at resonant frequency has been presented. An analytical model has been developed from which it has been possible to demonstrate how this association allows to eliminate the low frequency pulsating power when $\varphi=0$. In these situations of zero power factor it has also been possible to verify that the conduction losses are reduced by up to 33.33% with respect to those that would be obtained for the association of three single-phase inverters individually. For situations in which reactive power flowing from the DC/AC to DC/DC is considered ($\varphi\neq0$), a calculation method has been proposed for the output capacitor and a dynamic model that has proved to be valid to mimic the behaviour of the converter in both situations. Finally, a scale prototype has been built where the theoretical and simulated results have been experimentally tested.

References

[1] R. Naderi and A. Rahmati, "Phase-shifted carrier PMW technique for general cascaded inverters," IEEE Trans. Power Electron., vol. 23, no. 3, pp. 1257–1269, May 2008.

[2] G. S. Konstantinou and V. G. Agelidis, "Performance evaluation of halfbridge cascaded multilevel converters operated with multicarrier sinusoidal PWM techniques," in Proc. IEEE Conf. Ind. Electron. Appl., May 2009, pp. 3399–3404.

[3] B. Johnson, P. Krein, and P. Chapman, "Photovoltaic ac module composed of a very large number of interleaved inverters," in Proc. Appl. Power Electron. Conf. Expo., Mar. 2011, pp. 976–981.

[4] C. D. Fuentes, C. A. Rojas, H. Renaudineau, S. Kouro, M. A. Perez and T. Meynard, "Experimental Validation of a Single DC Bus Cascaded H-Bridge Multilevel Inverter for Multistring Photovoltaic Systems," in *IEEE Transactions on Industrial Electronics*, vol. 64, no. 2, pp. 930-934, Feb. 2017.

[5] Y. Yu, G. Konstantinou, B. Hredzak and V. G. Agelidis, "Power Balance Optimization of Cascaded H-Bridge Multilevel Converters for Large-Scale Photovoltaic Integration," IEEE Trans. Power Electron, vol. 31, no. 2, pp. 1108-1120, Feb. 2016.

[6] C. D. Townsend, Y. Yu, G. Konstantinou, and V. G. Agelidis,"Cascaded H-bridge multilevel PV topology for alleviation of per-phase power imbalances and reduction of second harmonic voltage ripple," IEEE Trans. Power Electron., vol. 31, no. 8, pp. 5574–5586, Aug. 2016.

[7] H. S. Krishnamoorthy, S. Essakiappan, P. N. Enjeti, R. S. Balog and S. Ahmed, "A new multilevel converter for Megawatt scale solar photovoltaic utility integration," *2012 Twenty-Seventh Annual IEEE Applied Power Electronics Conference and Exposition (APEC)*, Orlando, FL, 2012, pp. 1431-1438.

[8] K. Wang, M. Andresen, S. Pugliese and M. Liserre, "Phase Power Balancing of Interphase Grid-Connected CHB-QAB PV Systems," *IECON 2018 - 44th Annual Conference of the IEEE Industrial Electronics Society*, Washington, DC, 2018, pp. 3363-3368.

[9] P. K. Achanta, B. B. Johnson, G. Seo and D. Maksimovic, "A Multilevel DC to Three-Phase AC Architecture for Photovoltaic Power Plants," in *IEEE Transactions on Energy Conversion*, vol. 34, no. 1, pp. 181-190, March 2019.

[10] H. Choi, W. Zhao, M. Ciobotaru and V. G. Agelidis, "Large-scale PV system based on the multiphase isolated DC/DC converter," *2012 3rd IEEE International Symposium on Power Electronics for Distributed Generation Systems (PEDG)*, Aalborg, 2012, pp. 801-807.

[11] Jonas E. Huber, Johann Miniböck, and Johann W. Kolar. "Generic Derivation of Dynamic Model for Half-Cycle DCM Series Resonant Converters", in *IEEE Transactions on Power Electronics, vol. 33, no. 1, January 2018*.

[12] Jonas E. Huber and Johann W. Kolar. "Analysis and Design of Fixed Voltage Transfer Ratio DC/DC Converter Cells for Phase-Modular Solid-State Transformers", in *Proc. IEEE Energy Convers. Congr. Expo., Montreal, QC, Canada. Sep. 2015, pp. 5021-5029*.

Comparison of different Low Voltage Multilevel Converter Topologies for Distributed Power Generation

Ingmar Kaiser, Hans-Günter Eckel
UNIVERSITY OF ROSTOCK
Albert-Einstein-Str. 2
18059 Rostock, Germany
Phone: +49 (0) 381 498 7135
Fax: +49 (0) 381 498 7102
Email: ingmar.kaiser@uni-rostock.de
URL: http://www.iee.uni-rostock.de

Acknowledgments

This work is supported by the Federal Ministry for Economic Affairs and Energy on basis of a decision by the German Bundestag (Grant Number: 03EE2005A).

Keywords

≪Converter circuit≫, ≪Design≫, ≪Distributed power≫, ≪MOSFET≫, ≪Multilevel converters≫, ≪Voltage Source Converter (VSC)≫, ≪Wind energy≫

Abstract

Typically converters for distributed power generation built with IGBTs have a poor partial load efficiency. Multilevel converters with low voltage SI MOSFETs can be an attractive alternative regarding partial load efficiency and semiconductor cost. In this paper, different multilevel topologies with low voltage SI MOSFETs are compared.

Introduction

Multilevel converters play an important role in many high power applications, such as MVDC and HVDC networks, machine drives, and the integration of renewable energy resources to the grid [8, 9]. Due to fluctuating supply to the grid, especially of wind and solar energies, converters spend a large part of their operating time in partial load range. In medium and high power applications, typically IGBTs are used. Due to their diode-like forward voltage drop and higher switching losses IGBTs suffer from underproportional loss reduction at lower currents. MOSFETs feature a resistive forward characteristic and have lower switching losses compared to IGBTs. At lower current, the total losses are reduced proportionally which makes them superior in partial load operation. Because of lower blocking voltages, the use of MOSFETs in medium and high power applications is only possible with multilevel converters. The demanded efficiency and power level can be reached by parallelising multiple MOSFETs per switch. For this reason, a low power loss density is obtained, which will simplify the cooling concept [11]. The idea of using low voltage SI MOSFETs will be investigated by designing and comparing different multilevel converters for use in distributed power generation feeding-in into the medium voltage grid. This includes the following topologies:

- Modular Multilevel Converter (MMC)

- Flying Capacitor Multilevel Converter (FCM)

- Active Neutral-Point-Clamped Multilevel Converter (ANPCM)

Boundary Conditions for Designing the Converters

For the reason of comparison, all converters are designed with the same electrical parameters, which are listed in Tab. I. The values, except voltages, are normalised to nominal AC voltage and nominal AC current. The design calculations are done analytically without simulation models. Only stationary operation is considered.

Table I: Electrical parameters for the design of the multilevel converters

Parameter	Value
Nominal AC voltage	630 V
Nominal frequency	50 Hz
Efficiency	99 %
Normalised nominal active power	1,5
DC voltage	1235 V
MOSFET blocking voltage	150 V
On-state resistance @ 75 °C	4,5 mΩ

Furthermore, several assumptions are made which are listed and explained in the following:

- The MOSFET on-state losses are about 80 % of the total MOSFET losses. It is assumed that this ratio can be reached by an appropriate switching frequency and control concept.

- The requirements of the German grid codes for the medium voltage grid are considered. This includes unbalanced grid conditions, mains overvoltage, as well as the reactive power control range [5].

- Apart from the Q2L-controlled converters, all multilevel converters are assumed to have a sufficient output voltage quality. Therefore output filter components are not investigated.

Description of the Detailed Converter Design

In the following, the converter design is described for every topology. Concerning future industrial applications, three criteria are of main interest for the following converter designs - *efficiency, cost,* and *volume*. Due to a specified efficiency, the remaining design variables are *cost* and *volume*. For a practical design, the total number of MOSFETs can be compared to the total amount of energy stored on the capacitors (denoted as capacitor-energy), as both are reliable indicators for the cost and volume of a converter.

The MMC is designed first. Therefore the design of the FCM and ANPCM is partly based on the results of the MMC. For reasons of clarification, only one MOSFET is drawn per switch in every depicted topology (Fig. 1, 4 and 6). To achieve the specified efficiency, it will be necessary to connect multiple MOSFETs in parallel per switch as it will decrease the conduction losses.

Modular Multilevel Converter

The basic structure of an MMC, depicted in Fig. 1, consists of three phases, each having a positive and negative arm. Each arm is built by *m* half-bridge submodules (also denoted as cells).

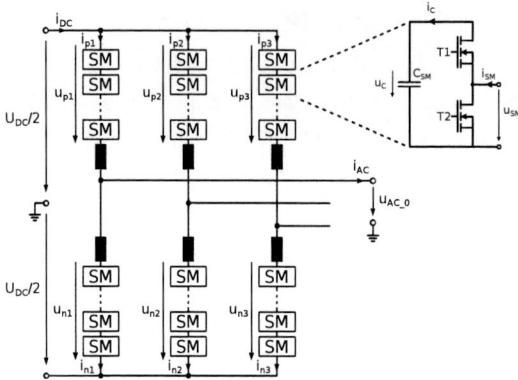

Fig. 1: Modular Multilevel Converter with a separate half-bridge submodule

The fundamentals of the MMC are described in detail in the literature [6, 10]. Assuming predetermined AC and DC voltage and power levels as well as a phase-shifted pulse-width modulation for all submodules, the submodule voltage ($a \in \{p,n\}, b \in \{1\cdots3\}$)

$$u_{SMab} = \frac{u_{ab}}{m} \tag{1}$$

and submodule current i_{SMab} (which corresponds to the arm current) can be easily derived. For a simplified design process, sinusoidal short-term mean values are assumed for currents and voltages, as currents with mainly fundamental frequency will flow through the submodules. The submodule capacitor voltage u_{Cab} has to be equal or higher than u_{SMab} and lower than the blocking voltage of the MOSFET. These boundaries limit the possible submodule capacitor voltage ripple, and as a result, the required submodule capacitance C_{SM} and the capacitor voltage u_{Cab} can be derived. The current stress of the submodule capacitor can be reduced by injecting circulating currents [4]. This method will reduce the required submodule capacitance and therefore is used for the converter design.

For the same blocking voltage, a lower number of submodules per arm leads to a higher submodule voltage u_{SMab}, which limits the possible voltage ripple of u_{Cab}. This consequence can be seen in Fig. 2.

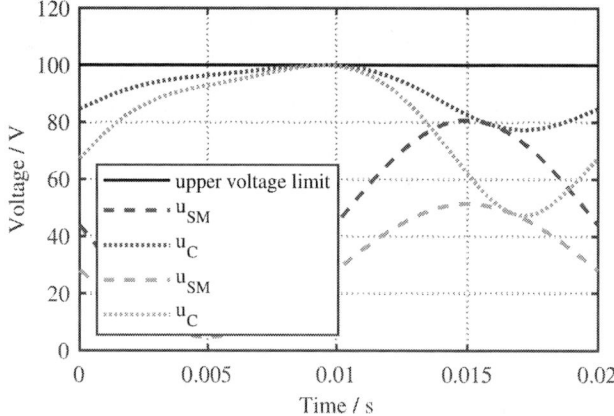

Fig. 2: Voltages in a submodule for 16 (dark blue) and 22 (light blue) submodules per arm. The upper voltage limit is defined by the MOSFET blocking voltage.

A lower number of submodules per arm leads to a disproportionate higher demand of capacitor-energy per submodule, which leads to a more significant total converter volume. On the other hand, a higher number of submodules per arm leads to an increased number of parallel-connected MOSFETs m_{MOSFET} to achieve the specified efficiency, which again will lead to a more significant total converter volume. In the following, the calculation of m_{MOSFET} is described. The drain currents which are flowing through the upper and lower switch of the submodule (see Fig. 1) can be deduced by the relative duty cycle of the switches

$$\tau_{T1} = \frac{u_{SMab}}{u_{Cab}} \tag{2}$$

$$\tau_{T2} = 1 - \tau_{T1} \tag{3}$$

and the submodule current as

$$i_{T1} = \tau_{T1} \cdot i_{SMab} \tag{4}$$

$$i_{T2} = \tau_{T2} \cdot i_{SMab} \tag{5}$$

The specified efficiency leads, with the assumption of equally spread losses, to the maximum conduction losses per submodule P_{Vd-SM}. The maximum total on-state resistance $R_{DSon\Sigma}$ (which corresponds to the combined resistance of all parallel-connected MOSFETs per switch) can be calculated by

$$R_{DSon\Sigma} = \frac{P_{Vd-SM}}{I_{T1}^2 + I_{T2}^2} \tag{6}$$

Rounding up

$$m_{MOSFET} = \frac{R_{DSon}}{R_{DSon\Sigma}} \tag{7}$$

will lead to the minimum number of parallel connect MOSFETs. The value of R_{DSon} can be extracted from the MOSFETs datasheet. The result can be further optimized by distributing the total amount of MOSFETs in accordance with the current load of the respective switch.

The increased number of parallel-connected MOSFETs will also result in higher semiconductor costs. To quantify the described effects, several infineon SI MOSFETs in the low voltage range of 150 V to 300 V are compared in Fig. 3 by estimating the influence on costs and volume of an MMC.

The *total component surface* expresses the area which results when summing the footprint of all MOS-FETs and capacitors of an MMC together. The footprint of the MOSFET is given in the datasheet . The footprint of the capacitor is calculated by creating a virtual capacitor which is available in all voltage levels but with a constant height and a constant energy density. The *total semiconductor costs* are determined by online price information, which is valid for small quantities.

The comparison shows the superior characteristics of the infineon OptiMOS 5 MOSFET, mainly due to its low on-state resistance. Therefore the IPB044N15N5 with 150 V blocking voltage and a number of submodules per arm $m = 18$ is chosen for further design. For the reason of comparison, the IPB044N15N5 is considered for all other converters.

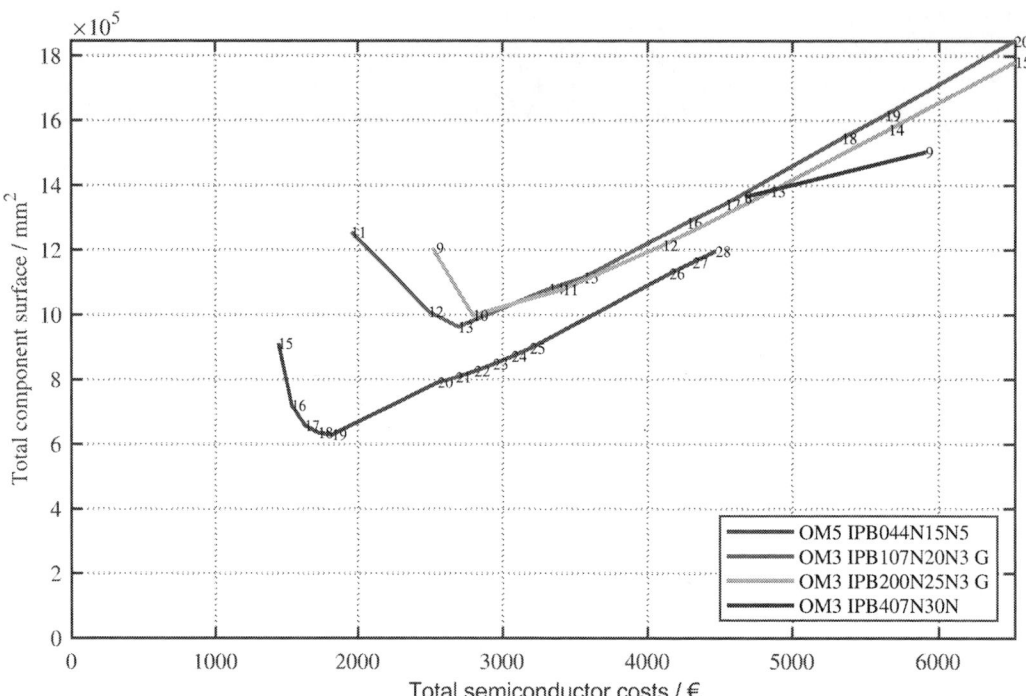

Fig. 3: Comparison of the total semiconductor costs and the total component surface (MOSFETs + capacitors) for a three-phase MMC. The numbers on the lines specify the number of submodules per arm.

Flying Capacitor Multilevel Converter

The FCM consists of multiple cells n which are connected in series. Each cell comprises two complementary working switches and a capacitor. The cell capacitor voltages are stepped as follows ($x \in \{1 \cdots n\}$):

$$U_{Cx} = U_{DC} - \frac{U_{DC} \cdot (x-1)}{n} \tag{8}$$

By applying a $\frac{360°}{n}$ phase-shifted pulse width modulation (PS-PWM) at each cell, a $n+1$ level stepped sinusoidal AC voltage from $\frac{-U_{DC}}{2}$ to $\frac{U_{DC}}{2}$ will be generated [12]. The FCM topology is depicted in Fig. 4.

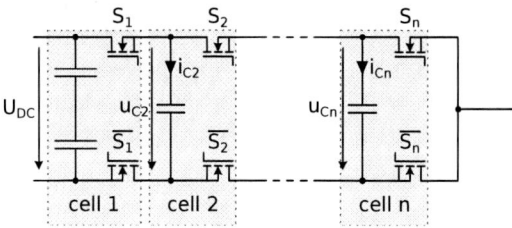

Fig. 4: One phase of a Flying Capacitor Multilevel Converter

The design of the FCM is done similarly to the MMC. The cell capacitance, which leads to the total amount of capacitor-energy, results from the limited voltage ripple of the cell capacitor voltages. The average cell voltage difference between two cells is

$$\Delta U_C = \frac{U_{DC}}{n} \tag{9}$$

The maximum cell voltage difference must not exceed a defined maximum voltage U_{Cmax} which yields from the MOSFET blocking voltage. So the allowable cell voltage ripple is

$$\hat{U}_{C-ripple} = U_{Cmax} - \Delta U_C \tag{10}$$

Subsequently, the cell capacitance of all cells can be calculated by

$$C_x = \frac{\int i_{Cx} dt}{\hat{U}_{C-ripple}} \tag{11}$$

In Fig. 5, the total amount of semiconductors is compared to the total amount of installed capacitor-energy for different numbers of cells.

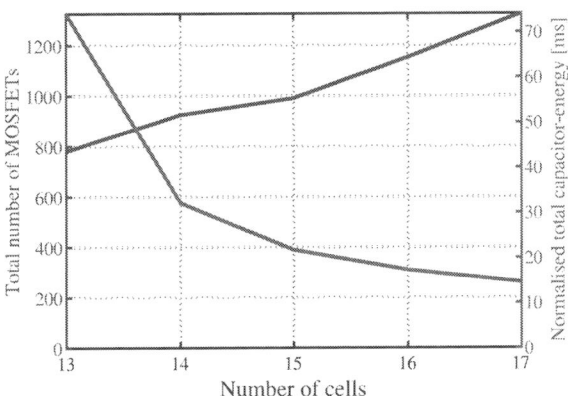

Fig. 5: Comparison of the total number of MOSFETs and the total amount of capacitor-energy for different numbers of cells of a three-phase FCM

As the number of cells increases, the number of MOSFETs increases to meet the specified efficiency. On the other hand, the amount of installed capacitor-energy is decreasing due to the acceptable capacitor voltage ripple, which increases with the number of cells. As a compromise, a number of $n = 15$ cells per phase is chosen.

Active Neutral-Point-Clamped Multilevel Converter

The ANPCM is a combination of an Active Neutral-Point-Clamped Converter (ANPC, DC side) and a FCM (AC side) [2]. In Fig. 6, one phase of an ANPCM is depicted.

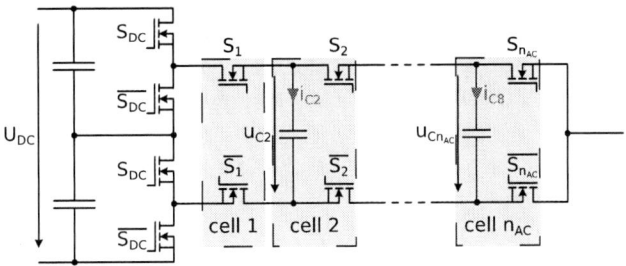

Fig. 6: Active neutral-point-clamped multilevel converter (one-phase)

The FCM part is built similarly as described in the previous section. The cell capacitor voltages are stepped as follows ($y \in \{1 \cdots n_{AC}\}$):

$$U_{Cy} = \frac{U_{DC}}{2} - \frac{U_{DC} \cdot (y-1)}{2 \cdot n_{AC}} \tag{12}$$

Due to the high DC voltage and the low blocking voltage of the semiconductor, a series connection of multiple switches n_{DC} at DC side is necessary. These switches are switched at a fundamental frequency depending on the polarity of the AC voltage, generating the major voltage-levels [7]. By applying a $\frac{360°}{n_{AC}}$ phase-shifted pulse width modulation (PS-PWM) at each AC side cell, a $2 \cdot n_{AC} + 1$ level stepped sinusoidal AC voltage from $\frac{-U_{DC}}{2}$ to $\frac{U_{DC}}{2}$ will be produced. For the reason of comparison, the sum of in series-connected switches n_{DC} and in series-connected cells n_{AC} is equal to n (defined for the FCM). The calculation of the cell capacitance is done as described for the FCM.

Quasi 2-Level Control (Q2L)

The Q2L control was investigated first for the Diode-Clamped Converter [1] and later for the MMC [3]. By switching all levels of one arm/phase at the same time, a two-level output voltage is generated. The resulting du/dt can be adjusted by switching the levels with a small time delay (within nano- or microseconds). This time delay is mentioned as dwell time T_d. The difference between the output voltages of an FCM and a Q2L-controlled FCM (FCM-Q2L) is depicted in Fig. 7.

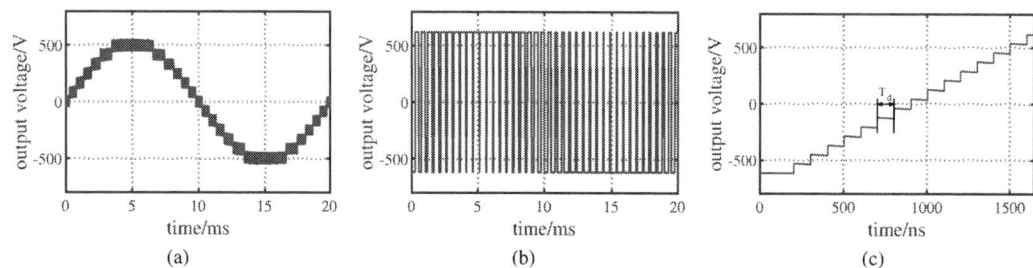

Fig. 7: Output voltage of an FCM (a) and an FCM-Q2L (b). Example of a rising edge of the output voltage of an FCM-Q2L (c).

With the application of the Q2L control, some advantages of multilevel converters are lost. The effort for the output filter will increase. But the cell capacitances can be reduced considerably as the intermediate voltage levels are only used for a very short time. The Q2L control is considered and applied to the FCM and ANPCM (denoted as ANPCM-Q2L). The dwell time is calculated to achieve a $du/dt = 800V/\mu s$. With this information, the current which flows through the cell capacitors can be calculated. The calculation of the cell capacitance is done as described for the FCM.

Design Results

Tab. II shows the design results.

The MMC features a total number of MOSFETs, which is higher than the FCM but lower than the ANPCM. The total number of MOSFETs of the MMC are distributed in accordance with the current load, which leads to the uneven number of parallel-connected MOSFETs for the switches T1 and T2. The maximum number of parallel-connected switches is eight (MMC) compared to eleven (FCM and ANPCM), which is a slight advantage for the MMC as a high number of parallel-connected switches can complicate the drive electronics. The maximum losses per MOSFET are on a similar level for all converters. The MMC has the highest cell capacitance but which will make the DC-link capacitance negligible. The cell capacitances of the FCM and ANPCM are in a similar low range but with a higher DC-link capacitance. In the case of the ANPCM this value refers to one DC-link capacitor. The current stress of the cell capacitor is at the lowest for the MMC and at the highest for the ANPCM. When comparing the total energy stored in a three-phase converter (denoted as converter-energy) and the total number of MOSFETs, the FCM is in a leading position. The MMC and ANPCM are on a similar level with a slight advantage of the MMC in case of the total number of MOSFETs. When comparing the Q2L-controlled converters, the FCM-Q2L is in a leading position. The application of the Q2L control on

Table II: Converter design results (the results for the Q2L-controlled converters are given in the brackets)

	MMC	FCM (-Q2L)	ANPCM (-Q2L)
Number of cells	18	15	N_AC = 8
In series connected switches	-	-	N_DC = 7
Switching frequency [kHz]	5	5	5
In parallel connected MOSFETs per switch	T1=4, T2=8	11	11
Normalised max. effective current per MOSFET	T2=0,069	0,064	0,064
Normalised max. losses per MOSFET	T2=10,9E-6	9,9E-6	9,9E-6
Total number of MOSFETs	1296	990	1452
Normalised effective current (cell capacitor)	0,12	0,36 (0,03)	0,49 (0,03)
Normalised cell capacitance [s]	22,6	0,83 (0,007)	1,41 (0,006)
Normalised effective current (DC capacitor)	0	0,57 (0,57)	0,57 (0,57)
Normalised DC-link capacitance [s]	negligible	1,9 (1,9)	25,4 (25,4)
Normalised total capacitor-energy [ms]	30,70	22,1 (0,18)	4,79 (0,02)
Normalised DC-link energy [ms]	0	3,61 (3,61)	24,39 (24,39)
Normalised total converter-energy [ms]	30,70	25,70 (3,79)	29,18 (24,41)

the ANPCM is impractical as the value of the DC-link capacitance is very high. This is due to the third harmonics of the current flowing through the DC-link capacitor.

Conclusion

In this paper, three multilevel converter topologies are designed with low voltage SI MOSFETs for use in distributed power generation. For a demonstrative design, the focus is placed on comparing the total number of MOSFETs, needed to gain the specified converter efficiency, to the total amount of installed capacitor-energy for a three-phase converter. The design results of the multilevel converters are compared. The FCM features the smallest number of MOSFETs combined with the lowest amount of total converter-energy. Furthermore, a quasi two-level control method is mentioned and applied to the FCM and ANPCM. The control considerably reduces the amount of installed capacitor-energy, which can be seen especially for the FCM-Q2L. This can be very attractive for several industrial applications. The comparison of Q2L-controlled converters will be extended to the MMC in the future.

As only stationary operation is considered in this work, the design will be validated by simulating dynamical operations. For a more comprehensive comparison, appropriate capacitors have to be chosen which meet the calculated cell capacitance and the cell capacitor current stress. Thereby the lifetime of the capacitors should be considered. The use of a printed circuit board with a high number of MOSFETs and capacitors will lead to high mechanical and electrical stress, which should be investigated. A final comparison should be made after a complete hardware design and the estimation of the cost for the components and the production. This final design will reflect the complexity of production and the effort for control engineering.

References

[1] Grain P. Adam et al. "Capacitor Balance Issues of the Diode-Clamped Multilevel Inverter Operated in a Quasi Two-State Mode". In: *IEEE Transactions on Industrial Electronics* 55.8 (Aug. 2008), pp. 3088–3099. ISSN: 0278-0046, 1557-9948. DOI: 10.1109/TIE.2008.922607.

[2] P. Barbosa et al. "Active-neutral-point-clamped (ANPC) multilevel converter technology". In: *2005 European Conference on Power Electronics and Applications*. 2005 European Conference on Power Electronics and Applications. Sept. 2005, 10 pp.–P.10. DOI: 10.1109/EPE.2005.219713.

[3] I. A. Gowaid et al. "Quasi Two-Level Operation of Modular Multilevel Converter for Use in a High-Power DC Transformer With DC Fault Isolation Capability". In: *IEEE Transactions on Power Electronics* 30.1 (Jan. 2015), pp. 108–123. DOI: 10.1109/TPEL.2014.2306453.

[4] K. Ilves et al. "Steady-State Analysis of Interaction Between Harmonic Components of Arm and Line Quantities of Modular Multilevel Converters". In: *IEEE Transactions on Power Electronics* 27.1 (Jan. 2012), pp. 57–68. ISSN: 0885-8993, 1941-0107. DOI: 10.1109/TPEL.2011.2159809. URL: http://ieeexplore.ieee.org/document/5887423/ (visited on 08/06/2019).

[5] Ingmar Kaiser and Hans-Gunter Eckel. "Investigation of Submodule Capacitor Voltage Fluctuation of a Modular Multilevel Converter Under Unbalanced Grid Conditions". In: *2018 20th European Conference on Power Electronics and Applications (EPE'18 ECCE Europe)*. 2018 20th European Conference on Power Electronics and Applications (EPE'18 ECCE Europe). Sept. 2018, P.1–P.9.

[6] Johannes Kolb. *Optimale Betriebsführung des modularen Multilevel-Umrichters als Antriebsumrichter für Drehstrommaschinen*. OCLC: 878977974. Karlsruhe: KIT Scientific Publ, 2014. 310 pp. ISBN: 978-3-7315-0183-1.

[7] Georgios Konstantinou et al. "The seven-level flying capacitor based ANPC converter for grid intergration of utility-scale PV systems". In: (2012), p. 6.

[8] Samir Kouro et al. "Recent Advances and Industrial Applications of Multilevel Converters". In: *IEEE Transactions on Industrial Electronics* 57.8 (Aug. 2010), pp. 2553–2580. ISSN: 0278-0046, 1557-9948. DOI: 10.1109/TIE.2010.2049719.

[9] Rainer Marquardt. "Modular Multilevel Converters: State of the Art and Future Progress". In: *IEEE Power Electronics Magazine* 5.4 (Dec. 2018), pp. 24–31. ISSN: 2329-9207, 2329-9215. DOI: 10.1109/MPEL.2018.2873496.

[10] André Schön. *Gleichspannungswandler für die Hochspannungsgleichstromübertragung*. 1. Auflage. Elektrotechnik. OCLC: 932022807. München: Verlag Dr. Hut, 2015. 240 pp. ISBN: 978-3-8439-2358-3.

[11] Mario Schweizer and Thiago B. Soeiro. "Heatsink-less Quasi 3-level flying capacitor inverter based on low voltage SMD MOSFETs". In: *2017 19th European Conference on Power Electronics and Applications (EPE'17 ECCE Europe)*. 2017 19th European Conference on Power Electronics and Applications (EPE'17 ECCE Europe). Sept. 2017, P1–P.10. DOI: 10.23919/EPE17ECCEEurope.2017.8098916.

[12] Richardt H. Wilkinson, Thierry A. Meynard, and Hendrik du Toit Mouton. "Natural Balance of Multicell Converters: The General Case". In: *IEEE Transactions on Power Electronics* 21.6 (Nov. 2006), pp. 1658–1666. ISSN: 0885-8993, 1941-0107. DOI: 10.1109/TPEL.2006.882951.

Loss distribution comparison of variable and fixed inductor DAB converters

Erik Smailus and Gerd Griepentrog
Institute for Power Electronics and Control of Drives
Technical University of Darmstadt
Fraunhoferstr. 4
64283 Darmstadt, Germany
URL: http://www.lea.tu-darmstadt.de

Markus Pfeifer and Marcel Lutze
Siemens AG
Gleiwitzer Str. 555
90475 Nürnberg, Germany

Keywords

≪Magnetic device≫, ≪Efficiency≫, ≪Passive component≫, ≪ZVS converters≫

Abstract

This paper deals with the aspects of loss distribution and magnitude under heavy step-down conditions comparing low load and full load operating points of a DAB converter with fixed and variable inductor. A simulation model concerning the losses in semiconductors and magnetic components is developed and verified by the overall losses using accurate DC measurements. Changes in loss distribution are discussed comparing both inductor variants in the DAB setup.

Introduction

Variable inductors have already been used in various power electronic converters. Besides quasi-active power factor correction [1] and keeping converters in continuous conduction mode for better MPPT in PV applications [2] a variable magnetic device can also be used for the improvement of efficiency of Dual Active Bridge (DAB) converters at low load condition [3]. The latter approach is using an additional winding on the magnetic core of the series inductor which injects a DC current that shifts the operational point from high to low inductance. Having to control the current injection is the major drawback of this solution and can be avoided by taking a partly saturable magnetic core as introduced in [4]. This technique ensures a high DAB efficiency improvement in the low load operational area especially for output to input voltage ratios $M = U_{out}/U_{in}$ other than the transformer ratio $n = N_s/N_p$ which will be discussed in the results section. As power measurements at $100\,\text{kHz}$ fundamental frequency have no suitable accuracy to identify the losses in the semiconductors a loss simulation is done using the simulation software PLECS. The further content is structured with the following sections:

- DAB with variable inductor: An explanation of the basic idea is given and the difference in operation of fixed and variable inductor DAB is clarified.
- Loss simulation: Several adaptions have to be made for suitable loss calculation with PLECS.
- Measurement setup: The loss calculation is verified by DC power measurement results on a DAB test setup, which has certain limitations discussed in this section.
- Results: Verification of simulation results is made and differences of loss distribution in variable and fixed inductor DAB are developed.

The findings are summarized in the conclusion section.

DAB with variable inductor

Dual Active Bridge converters offer the ability of isolated bidirectional power flow and consists of two full bridge converters connected by a medium frequency transformer and an inductor as depicted in Fig 1. The single phase shift modulation technique is very simple: A fixed frequency bipolar voltage with 50 %

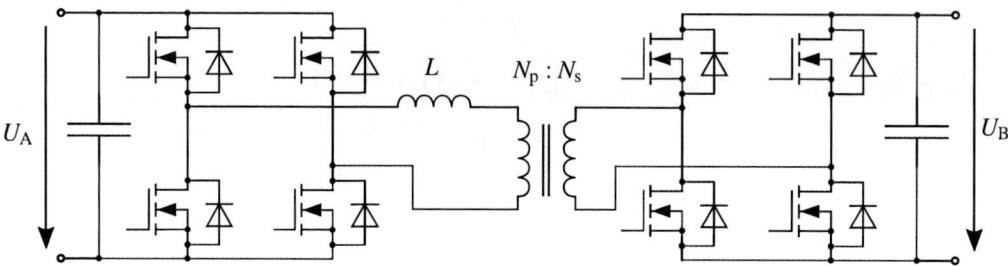

Fig. 1: DAB setup with series inductor

duty cycle is applied to the AC-Link from each full bridge. The transferred power can be controlled by applying a phase shift φ between the two rectangular AC voltages. The power flowing from bridge A to bridge B can be calculated as

$$P_{\mathrm{AB}} = \frac{U_{\mathrm{A}}U_{\mathrm{B}}}{2\pi n f_{\mathrm{sw}}L}\varphi\left(1 - \frac{\varphi}{\pi}\right), \tag{1}$$

with φ in radians. From (1) it is clear that the maximum power is limited by the value of L. If possible the inductor as a component is omitted and only the stray inductance of the transformer is used as reactive series element, needed to shape and also limit the current waveform. Higher stray flux produces additional losses in the transformer windings, so only in very high power applications or applications with very high operational frequency (increasing the impedance of low inductance elements) an additional inductive component can be omitted. Applications with wide operating ranges require an additional inductor due to the big voltage-time-area in non-unity voltage translation ratio $\frac{U_{\mathrm{out}}}{nU_{\mathrm{in}}}$. To achieve zero-voltage-switching (ZVS) the stored magnetic energy

$$W_{\mathrm{m}} = 0.5Li_{\mathrm{L}}^{2} \tag{2}$$

has to be sufficiently high to charge and discharge the parasitic drain source capacitance of each MOS-FET during the interlock interval of a half bridge. Otherwise the commutation is hard and EMI as well as higher losses occur. Therefore in low load operation, a DAB converter designed with small inductor for high power transfer is lacking ZVS operation and has therefor poor efficiency. The inductor introduced in [4] with partly saturating core material leads to both, a high inductance value for low power operating points as well as low inductance at high power range. It is compared to an inductor of the same core without the additional I-cores (red in Fig. 2) which leads to a fixed air gap of 4 mm as can be seen from the core parameters in Tab. I. The coils are identical in both inductors. Because of high magnetic field strength in the area of the air gap which leads to additional copper loss the windings are seperated as can be observed in Fig. 2. This measure is common in power electronic inductive elements with air-gap as described in detail in [5].

Table I: used ferrite cores in the lab setup

component	cores	details
fixed inductor	2x E55/28/21 N87	2 mm air-gap per core
variable inductor	2x E55/28/21 N87 + 2x I14/1.5/5 N87	2 mm air-gap per core (non-saturating) saturating cores
transformer	UI93/104/30 + UI93/104/20 parallel	Mf102 material (comparable to N87)

For both inductors the converter is controlled by an output voltage PI-control using single phase shift modulation. The waveforms in Fig. 3 show the controlled state of $U_{\mathrm{in}} = 500\,\mathrm{V}$, $U_{\mathrm{out}} = 350\,\mathrm{V}$ at a load

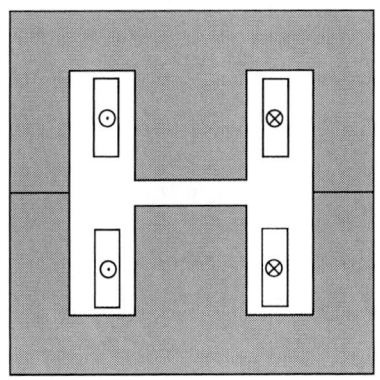 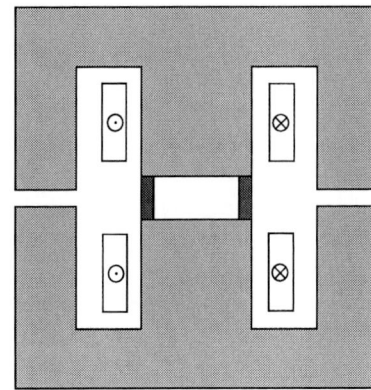

(a) fixed inductor with centered air gap

(b) variable inductor with introduced I-type cores for nonlinear behavior colored in red

Fig. 2: profiles of fixed and nonlinear inductors with seperated windings

$P_{DAB} = 2.5\,\mathrm{kW}$. As can be observed from the waveforms the phase shift is bigger for the variable inductor, but the current zero crossing is during the phase shift time interval which is a requirement for ZVS on both sides. The edges in the current waveform of the nonlinear inductor show the ending and beginning saturation of the introduced I-cores. For this operating point the characteristic nonlinear waveform is beneficial compared to the fixed inductor. The additional effort of introducing these highly nonlinear

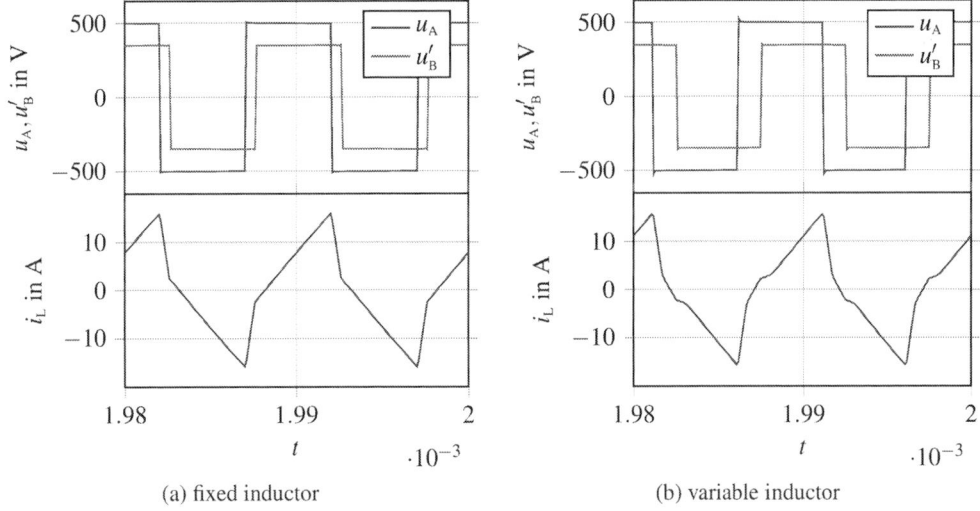

(a) fixed inductor

(b) variable inductor

Fig. 3: simulated waveforms of fixed and nonlinear inductors in DAB circuit in the same operating point

waveforms into traditional linear loss calculation methods is huge which is the reason for the PLECS loss simulation model developed in the following section.

Loss simulation

Semiconductor Losses

PLECS is a fast simulation platform for power electronic converters as it models all switches ideally. Unfortunately this is not suitable when soft switching boundaries come into focus. To find the ZVS boundaries one can insert MOSFET drain to source capacities parallel to the switches and introduce a turn-on delay for each gate signal. It should be mentioned that all following remarks regarding the PLECS loss calculation algorithms assume version 4.3.4 of the software. The following issues affect the calculated losses when leaving the ideal ZVS region:

    a. High peaks in the conduction losses occur during switching instants because the not completely discharged capacities C_{DS} are short circuited and produce extremely high currents.

    b. The PLECS loss generation algorithm assumes an ideally fast commutation and calculates the turn on loss using the current right after the switching instant, which is the aforementioned high peak from short circuiting the capacities.

    c. While the turn on losses are too high, the turn off losses appear to be calculated too small as the PLECS algorithm uses the switch voltage right after the switching instant for the turn off loss calculation. The latter is still very small after the turn off, because the parallel capacitor has to be charged by an externally driven current first.

A switching cell consists of a half bridge with two MOSFET and two antiparallel diodes, which can be the body diodes intrinsic to MOSFETs or an externally added commutation diode. Additionally the parasitic MOSFET drain to source capacity can be added for a better behavioral description regarding soft switching. C_{DS} is a nonlinear capacity depending on the drain to source voltage but can be assumed to be constant using the equivalent charge when charged to the reference voltage $Q_{DS,ref} = C_{DS,Q,eq}(U_{DS,ref}) \cdot U_{DS,ref}$ [6]. Together with the probing, PCB parasitic capacities and transformer winding capacity of overall approximately $250\,\mathrm{pF}$ this forms the full bridge output capacity $C_{FB,out}$ that has to be charged/discharged during the interlock interval of the half bridges. This is considered in the lumped full bridge output capacitor in the simulation model. PLECS offers the ability to add externally measured variables and custom tables into the loss calculation using the formula option in the thermal description dialog menu. Thereby the output current of a half bridge can be used and the conduction losses can be calculated according to a custom table which uses the given forward voltage data from the original conduction loss table. Additionally the factor `IL/i` is multiplied to the forward voltage so the original FET current is

Table II: customized formulas for accurate ZVS boundary loss calculation with drain to source capacities

loss type	formula
conduction loss (forward voltage)	`lookup('conduction custom', IL, T)*IL/i`
turn on loss (energy)	`lookup('turn-on custom 3D', abs(IL), v, T)`
turn off loss (energy)	`lookup('turn-off custom 3D', i, UDC, T)`

canceled. For accurate turn on loss calculation the original table is copied and the absolute value of the output current is used. Turn off loss calculation can be adapted by introducing the half bridge DC voltage into the thermal description dialog and using this variable instead of the original MOSFET voltage. All customized formulas are summarized in table II. Having no hysteric block suitable for ferrite material available in the magnetic model, the influence on the semiconductor losses is modeled using an additional controlled current source injecting a constant current depending on the sign of the rate of change of the flux in the core. This is explained in further detail in the following section.

Magnetic Loss

The losses of the magnetic cores are calculated by applying the improved generalized Steinmetz equation (iGSE) introduced in [7]:

$$q_{iGSE} = \frac{1}{T} \int_0^T k_i \left| \frac{dB}{dt} \right|^\alpha (\Delta B)^{\beta-\alpha} dt, \tag{3}$$

with

$$\Delta B = B_{\text{max}} - B_{\text{min}} \tag{4}$$

and

$$k_i = \frac{k}{(2\pi)^{\alpha-1} \int_0^{2\pi} |\cos\theta|^{\alpha} 2^{\beta-\alpha} d\theta}, \tag{5}$$

using the original Steinmetz parameters α, β and k. For accurate loss calculation according to iGSE the rate of change of the flux density needs to be evaluated. Therefore all magnetic components of the test setup are modeled in the magnetic domain of PLECS. The magnetic PLECS models of fixed and variable inductor are given in fig. 4. Due to the permeance-capacity approach used in PLECS the rate of change

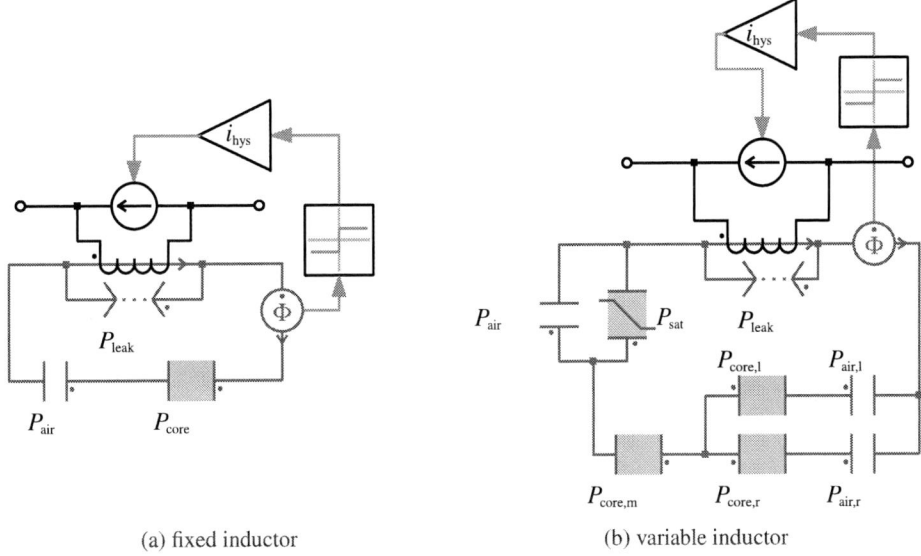

(a) fixed inductor   (b) variable inductor

Fig. 4: models of fixed and variable inductor in PLECS with simplified hysteresis model

of flux is directly available. Simply divided by the cross-sectional area, the squared rate of change in flux density can be integrated for each period. The terms $\left|\frac{dB}{dt}\right|^{\alpha}$ and $(\Delta B)^{\beta-\alpha}$ cannot be calculated during runtime and are substituted by precalculated lookup tables. The losses $Q_{\text{L,core}}$ are then provided by multiplication of q_{iGSE} of the core with the respective volume of the core V_c which assumes that the flux density is homogeneous in the whole core. For better specification of the loss components the inductor and transformer are handled seperately. As the core losses also depend on the core temperature, an additional term

$$C(\tau) = c_0 - c_1\tau + c_2\tau^2 \qquad \text{with} \quad \tau = \frac{T}{100\,^\circ\text{C}} \tag{6}$$

is multiplied to apply for the quadratic thermal behavior as described in detail in [5]. Please note that the aforementioned Steinmetz parameters have to be taken from loss measurements or datasheet values at $100\,^\circ\text{C}$. Aiming for the core temperature the empirical formula

$$R_\theta = 0.06 / \sqrt{V_c} \tag{7}$$

taken from [8] for the thermal resistance leads to the temperature difference

$$\Delta T = R_\theta Q_{\text{L,core}} \tag{8}$$

relative to the ambient. The absolute temperature is fed back into (6) for the next simulation time step to avoid algebraic loops. Modeling of the variable inductor can be done with a saturatable core in parallel to an air gap of the same length. Additionally there are the two 1 mm air gaps of the outer paths of the E-cores due to the 5 mm saturated ferrite blocks in the centered air-gap. This model is given in fig. 4b. As can be seen from fig. 4 the sign of the rate of change of the magnetic flux inside the inductors is measured and a fixed additional current

$$i_{hys} = H_c \cdot l_e / N_{ind} \tag{9}$$

with the coercitive field strength of the core material H_c, the effective core length l_e and the number of windings in the inductor N_{ind} is injected. This measure aims to model the hysteresis effect of the magnetic core on the current for more accurate conduction and switching loss computation as figured in the semiconductor loss section. In addition, the ohmic resistance of transformer and inductor is estimated and modeled as a single resistor $R_{s,20} = 250\,\mathrm{m\Omega}$ in series with the inductor. This resistor is connected to a thermal model, that allows for resistance changes due to temperature variation according to

$$R(T) = (1 + \alpha_T(T - 20\,^\circ\mathrm{C}))R_{20\,^\circ\mathrm{C}}. \tag{10}$$

The model is depicted in Fig. 5. It should be noted that PLECS models a heat sink with a thermal capacity that collects all losses inside the blue frame. As seen in Fig. 5 PLECS uses a capacitor parallel to the variable resistor to damp oscillating reaction of the model due to minor time step changes in resistance.

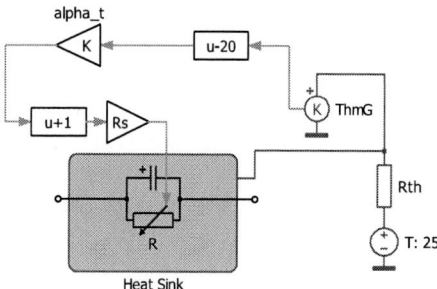

Fig. 5: Series resistor with thermal dependency

Measurement setup

The lab setup consists of two seperate full bridge converters, an UI93/104/50 transformer with winding ratio $N_p = N_s = 12$, two DC blocking series capacitors $C_{block} = 10\,\mu\mathrm{F}$ and the inductors to be tested with. The blocking capacitors are installed to prevent saturation of the transformer due to non-ideal 50% duty cycle of the setup. These have no further function and will be neglected regarding the loss calculation. All component details are shown in table III. As mentioned before the measurement of real power at 100 kHz fundamental frequency is a big challenge for digital power meters which is visible in the increased specified maximum measuring error of

$$\Delta P_{err,max,100kHz} = \pm(1\,\%\ \mathrm{of\ reading} + 1.1\,\%\ \mathrm{of\ the\ upper\ range\ value} + 30\,\mu\mathrm{A\,A}^{-2} \cdot I_{trms}^2 \cdot U_{trms}). \tag{11}$$

This is an absolute uncertainty of the measured value i.e. at 3.5 kW of $\Delta P_{err,max} = \pm 1091.735\,\mathrm{W}$ and is bigger than the expected losses. The power analyzer is therefor used at the DC terminals only with a much better accuracy specified in the ZES LMG671 manual (version 02/2019).

$$\Delta P_{err,max,DC} = \pm(0.032\,\%\ \mathrm{of\ reading} + 0.09\,\%\ \mathrm{of\ the\ upper\ range\ value} + 50\,\mu\mathrm{A\,A}^{-2} \cdot I_{trms}^2 \cdot U_{trms}). \tag{12}$$

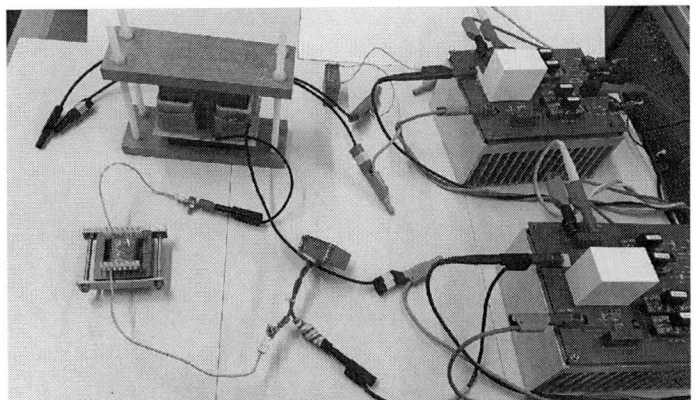

Fig. 6: Laboratory setup

A measurement time of 1 s is used and an averaging over 10 measurment cycles leads to the results displayed in fig. 7. The uncertainty is still relatively high because the ADC range is designed for AC measurements with high peaks, but low rms value. This leaves less ADC bit resolution to measure DC-values without exceeding the rated values for each measurement range.

Table III: component details (besides cores) of the lab setup

component	name	details
MOSFET	C3M0075120K	1200 V, 0.75 mΩ
litz wire	245x0.1 mm	
precision power analyzer	LMG671	A-Channel specification

Results

As can be seen from fig. 7a the introduced loss simulations match the measurement results with their specified uncertainty band (see (12)) for the fixed inductor very well except the very low load range. Fig. 7b shows the comparison of the variable inductor measurements and loss simulation results which correlate very well within the measurement error range. With the simulation results one can differentiate

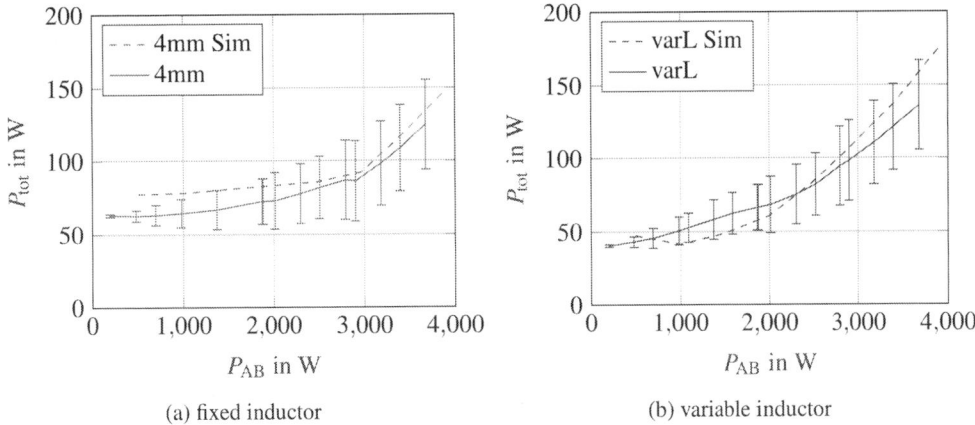

(a) fixed inductor

(b) variable inductor

Fig. 7: comparison of overall losses measured at DC terminals with the specified uncertainty area of the power meter at $U_A = 500\,\text{V}$ and $U_B = 350\,\text{V}$ with a summarized ohmic resistance of 250 mΩ for all passives and connections between the loss components as outlined in fig. 8. It is obvious, that the variable inductor leads to fewer

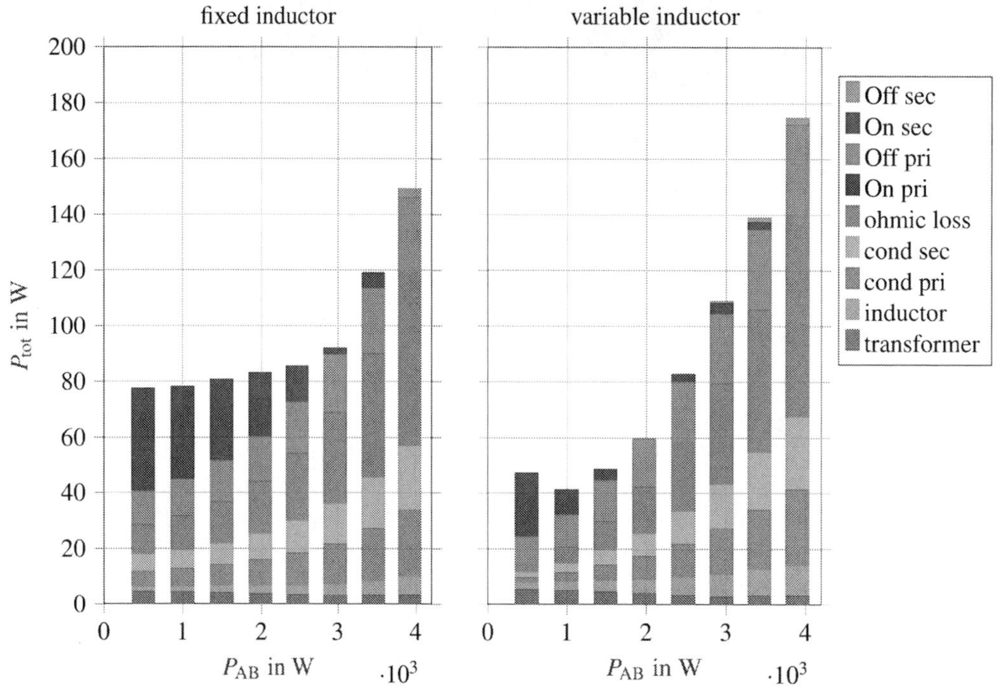

Fig. 8: Loss distribution from simulation: left: fixed inductor, right: variable inductor

losses in the low load region and more losses at full power. The main findings are as follows:
- the benefit in low load region can be mostly traced back to the smaller turn on losses.
- The use of the variable inductor design leads to greater losses in the inductor itself as can be observed by the green bars at the bottom of the chart. This increase of loss due to core losses is neglectable in terms of the overall efficiency but has to be taken into account concerning the thermal design and packaging of the inductor.
- In case of the ohmic losses one has to differ:
  - in the low load region the higher inductance value of the variable inductor leads to smaller current amplitudes, which contributes to smaller ohmic losses as well.
  - the smaller inductance value in the high load range leads to greater currents amplitudes and higher conduction, turn off and ohmic losses.

Comparing the simulated overall losses of variable and fixed inductor DAB the low load loss advantage of the variable inductor based design is up to 40 %. In high load range the fixed inductor design is up to 20 % better in terms of losses. Comparing these results to the interpolated measurements (without uncertainty intervals) the tendency can be verified (see fig. 9). The low load benefit is up to 30 % and the full load disadvantage is around 10 %. Being able to differentiate between the loss components, one can compare the semiconductor losses and inductor core losses in fig. 10. The reduction in semiconductor stress during low load operation is up to 50 % whereas the additional loss at high load is only between 15 and 20 %, which makes the variable inductor design a valid option for mainly low load operation. When comparing the inductor core losses, the partly saturating design is utterly worse, but as shown by fig. 8 the absolute core losses are only few compared to the semiconductor losses and fig. 11 shows that the temperature rise is rather low compared to the benefit. Concerning the semiconductor losses fig. 10 is assuming the same chips, which means a higher resulting maximum junction temperature in case of the variable inductor. When adjusting the chip size for this case (assuming $T_{j,max}, P \propto \frac{1}{A}$), one would have to increase the variable inductor converter chip size by 7.7 %, which makes it more expensive. Therefor the variable inductor is obviously no good choice for full load applications.

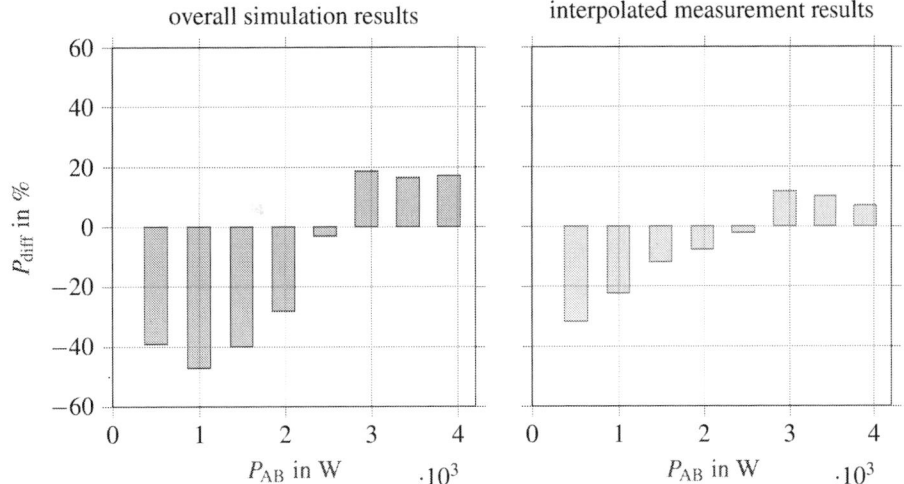

Fig. 9: verification of overall loss advantages and disadvantages of variable inductor setup compared to fixed inductor setup

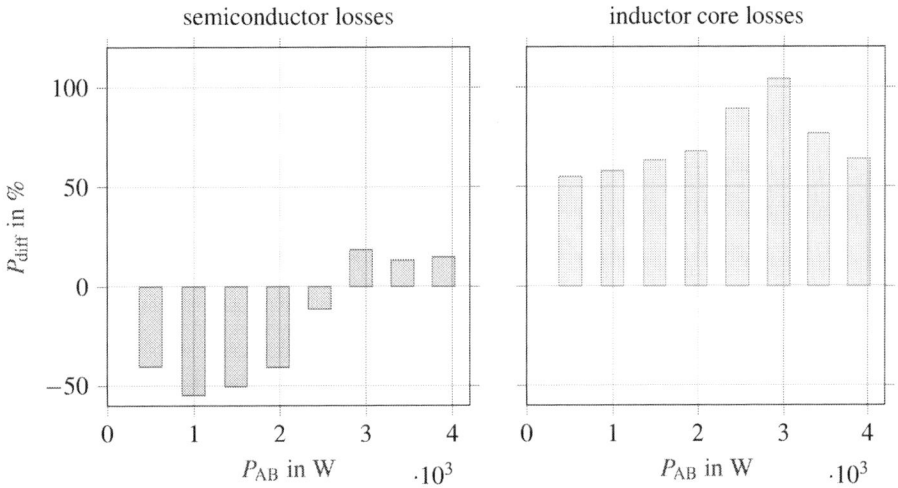

Fig. 10: comparison of semiconductor and inductor losses of variable inductor setup relative to fixed inductor setup

Though each inductor has its benefits and the choice should be done according to the expected use case. One can see that the use of a variable inductor has benefits if the normal use case is a partial load application and the higher power ability is only needed as a short time regulation buffer. For mainly high load use cases the fixed inductor is the better solution but lacks the good efficiency at low load condition. Figure 11 shows the temperature rise in inductor and transformer in the DAB with the two inductors. One can clearly see that the bottleneck of this setup is the inductor, as the temperature rises up to 116.1 °C at the variable inductor. The transformer temperature never reaches 100 °C. Within the inductor the hotspot is at the coil, as far as one can observe. The real hotspot should be at the saturating I-cores inside the coil bobbin as this spot is heated up from the saturating material itself and by the surrounding windings. Even though the temperature is above the measured 116.1 °C at the outside of the winding, the core losses remain stable. This is exceptional because of the typical positive temperature coefficient of ferrite material above 100 °C, but the cooling of the bigger E-cores' outer legs suffice. Not only the variable inductor itself but also the transformer winding is experiencing higher temperatures due to the greater

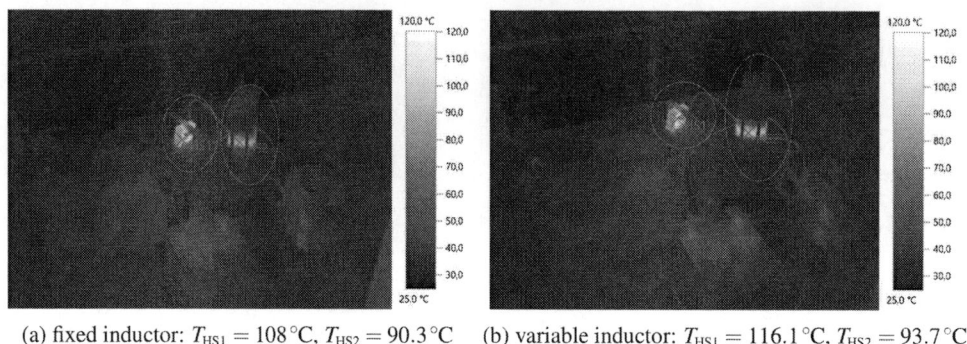

(a) fixed inductor: $T_{\text{HS1}} = 108\,°\text{C}$, $T_{\text{HS2}} = 90.3\,°\text{C}$ (b) variable inductor: $T_{\text{HS1}} = 116.1\,°\text{C}$, $T_{\text{HS2}} = 93.7\,°\text{C}$

Fig. 11: infrared hotspot measurements at $U_{\text{in}} = 500\,\text{V}$, $U_{\text{out}} = 350\,\text{V}$ and $P_{\text{out}} = 3500\,\text{W}$

current magnitude, when using the variable inductor at this high load operating point.

Conclusion

A loss model taking into account parasitic capacities and a variable inductor design for DAB applications has been developed and verified by DC efficiency measurements. Furthermore the variable inductor DAB was compared to a fixed inductor based DAB in terms of component losses. It was shown how the losses shift from semiconductors to inductor in low load application, which shows the main benefit of the variable inductor DAB. Behavior in different operating points in terms of voltage translation will be subject to further publications.

References

[1] Wölfle W. H. and Hurley W. G.: Quasi-active power factor correction with a variable inductive filter: theory, design and practice, IEEE Transactions on Power Electronics, vol. 18, no. 1, pp. 248-255, 2003.
[2] Zhang L., Wölfle W. H. and Hurley W. G.: A New Approach to Achieve Maximum Power Point Tracking for PV System With a Variable Inductor, IEEE Transactions on Power Electronics, vol. 26, no. 4, pp. 1031-1037, 2011.
[3] Saeed S. and Garcia J.: Extended Operational Range of Dual-Active-Bridge Converters by using Variable Magnetic Devices, 2019 IEEE Applied Power Electronics Conference and Exposition (APEC), IEEE, 3/17/2019 - 3/21/2019, pp. 1629-1634.
[4] Smailus E., Griepentrog G., Lutze M. and Pfeifer M.: Improvement of ZVS range in Dual Active Bridge Converters using nonlinear inductors by ferrite block insertion, PCIM Europe 2020, 2020.
[5] Albach M.: Induktivitäten in der Leistungselektronik, Springer Vieweg, Wiesbaden, 2017
[6] Kasper M., Burkhart R.M., Deboy G., Kolar J.W.: ZVS of Power MOSFETs Revisited, IEEE Transactions on Power Electronics, vol. 31, no. 12, pp. 8063-8067, 2016.
[7] Venkatachalam K., Sullivan C.R., Abdallah T., Tacca H.: Accurate prediction of ferrite core loss with nonsinusoidal waveforms using only Steinmetz parameters, 2002 IEEE Workshop on Computers in Power Electronics, Mayaguez, Puerto Rico, USA, 2002.
[8] Hurley W.G. and Wölfle W.H.: Transformers and inductors for power electronics - Theory, design and applications, Wiley-Blackwell, Chichester, West Sussex, 2013.

Design by Optimization of multiphase inverter for electric vehicle drive

Nasreddine KESBIA[1,2], Jean-Luc Schanen[2], Hadi Alawieh[1], Lauric Garbuio[2], Yvan Avenas[2]
[1] VEDECOM Institute, 78000 Versailles, France
[2] Univ. Grenoble Alpes, CNRS, Grenoble INP,G2Elab, 38000 Grenoble, France
Tel.:+33 (0) 6 40 86 30 94
E-Mail: nasreddine.kesbia@vedecom.fr, jean-luc.schanen@g2elab.grenobe-inp.fr,
hadi.alawieh@vedecom.fr, lauric.garbuio@g2elab.grenoble-inp.fr
URL: http://www.vedecom.fr/ http://www.g2elab.grenoble-inp.fr

Keywords

«Multiphase drive», «Voltage Source Inverters (VSI)» ,«Design», «Modelling», «Optimization».

Abstract

This paper proposes a gradient-based design optimization of multiphase inverter for electric vehicle application. The aim is to optimize the total volume of the converter taking account the number of phases, the sizing of dc-link capacitor and total surface of semi-conductors. The developed method shows that the optimum design is given for a number of phases more than three.

Introduction

The increasing awareness on climate change due to the pollutant emissions and the price of oil are rapidly impacting the sustainability of the transportation sector. For these reasons, the transports industry is interested in the electrification of vehicles [1]. However, this application involves specific requirements in terms of performances, volume and cost, especially concerning the design of the electric drive. To satisfy the requirements and achieve the optimal point system volume and efficiency, the design of main components of the drive system have to be optimized together. This is challenging owing to the tradeoff between design variables, as the switching frequency and the power density [2].

Embedded systems as electrical vehicles offer degree of freedom of choosing the number of phases of the drive system. Using multiphase drive systems offers several advantages over conventional three phase drives for both inverters and AC motors [3]. Increasing the number of phases reduces the torque ripple caused by the 2N±1 harmonics of the supply. Furthermore, increasing the number of phases offers the possibility of harmonic current injection for torque enhancement. Concerning inverter, increasing the number of phases reduces the current rating of individual devices in each leg and offers the possibility to work in fault tolerant mode.

The optimization of individual machine and power converter using stochastics algorithms has been studied in literature in case of three-phase drives [2,4,5]. These methods are time consuming and cannot easily handle many constraints. Design using deterministic algorithms can be used in power electronics [6], providing less computation efforts and exhibiting very quick optimization time, what is mandatory for considering a full mission profile for instance. However, they require continuous and derivable models.

This paper focuses on volume optimization of multiphase inverter using gradient-base algorithm software. Therefore, a derivable and continuous model of each part of the converter is necessary. In the first part, an analytical model of semiconductor losses is presented and validated with experiment measurements. Then, an accurate analytical model of DC link capacitors rms current is developed in case of multiphase inverter. In the last part, an example of optimization of the converter under several constraints is presented.

Multiphase inverter continuous losses model

An accurate losses model is critical to reliably perform an accurate design of the multiphase inverter. To this end, a comprehensive model of losses of IGBT semiconductors of multiphase inverter is developed based on analytical equations. In PWM VSI, based on Si IGBT and diodes, the conduction losses can be expressed as function of the IGBT and diode characteristics Eq.(1). The switching losses depend strongly on the value of switched current. The usual approximation is a quadratic interpolation of switching energy as a function of switched current (I_{sw}). by averaging that switching losses in fundamental period Eq.(2) [7]:

$$P_{cond-IGBT/diode} = V_{ce0/d0}I_{av-IGBT/diode} + R_{ce/d}I^2_{rms-IGBT/diode} \qquad (1)$$

$$P_{sw-IGBT/diode} = f_{sw}\frac{V_{dc}}{V_{rated}}\left(\frac{a_T}{\frac{\overline{D}}{4}}I^2_{sw} + \frac{b_T}{\frac{\overline{D}}{\pi}}I_{sw} + \frac{c_T}{\frac{\overline{D}}{2}}\right) \qquad (2)$$

where V_{ce}, V_{d0} are the threshold voltage of the IGBT and diode respectively. R_{ce}, R_d are the on-state dynamic resistance of the IGBT and diode respectively. I_{av}, I_{rms} represent the device average and RMS currents respectively.

By replacing expressions of average and RMS currents through the IGBT and diode in Eq.(1), the conductions losses become:

$$P_{cond-IGBT} = \frac{V_{ce0}I_{max}}{2\pi}\left[1 + r\frac{\pi\cos(\phi)}{4}\right] + R_{ce}I^2_{max}[\frac{1}{8} + \frac{r\cos(\phi)}{3\pi}] \qquad (3)$$

$$P_{cond-Diode} = \frac{V_{d0}I_{max}}{2\pi}\left[1 - r\frac{\pi\cos(\phi)}{4}\right] + R_dI^2_{max}[\frac{1}{8} - \frac{r\cos(\phi)}{3\pi}] \qquad (4)$$

where r is the modulation index, $\cos(\phi)$ the power factor, I_{max} is the amplitude of the load current.

The operating junction temperature of the device is estimated using a thermal resistance model considering that the junction temperature ripple is negligible on a fundamental period. The maximum temperature is calculated with the total losses for each considered device and ambient temperature.

In multiphase VSI, increasing the number of phases results in a reduction of the phase leg current. Therefore, the current rating of the semi-conductors can be reduced if the number of phases is increased. For this reason, using multiphase inverter can limit the number of semi-conductors associated in parallel [4]. Since a continuous and derivable model is required for gradient-based algorithm, the concept of imaginary die of IGBT and diode is adopted. This die area represents a continuous evolution of the rated current, and clearly changes the on-state losses and switching energies.

To obtain the characteristics of this imaginary device, an interpolation of the on-state dynamic resistance, threshold voltage and switching energies as function of the die area is needed. Therefore, a full characterization of several half-bridge arrangements based on discrete IGBT transistors has been carried out. To measure the losses of these components arrangements, calorimetric measurements method is used.

Table 1: Prototypes characteristics

Prototypes	P1	P2	P3	P4	P5	P6	P7
Rated current(A)	1*20	1*40	2*20	3*20	1*75	2*40	4*20
IGBT_area (mm²)	10	20	20	30	40	40	40
Diode_area (mm²)	5	10	10	15	20	20	20

For the characterization and the validation of the continuous losses model, seven prototypes have been manufactured with different area of IGBT and Diode up to 40mm² for the IGBT and 20mm² for the diode using discrete TO-247 fig.1.c. considering that the anti-parallel diode present in the discrete package of TO-247 is assumed to have the half area of the IGBT die. Each prototype is built by using an IMS card to fix the discrete IGBT and ensure a better heat transfer to the cold plate fig.1.a. And a PCB to make the interconnections between power discrete components, DC link and driver parts fig.1.b.

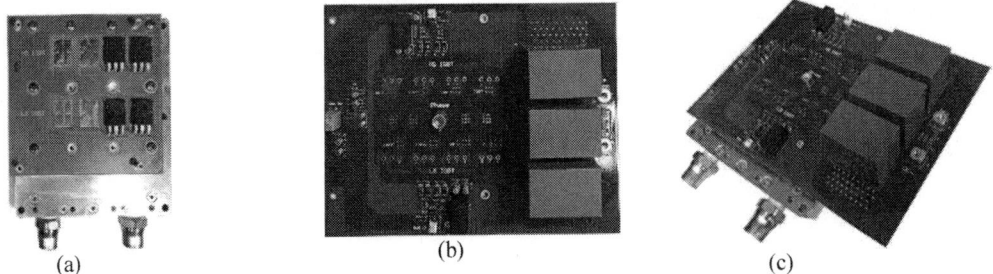

(a) (b) (c)

Figure 1: buck converter prototype (a) IMS Board (b) Power Board (c) full converter

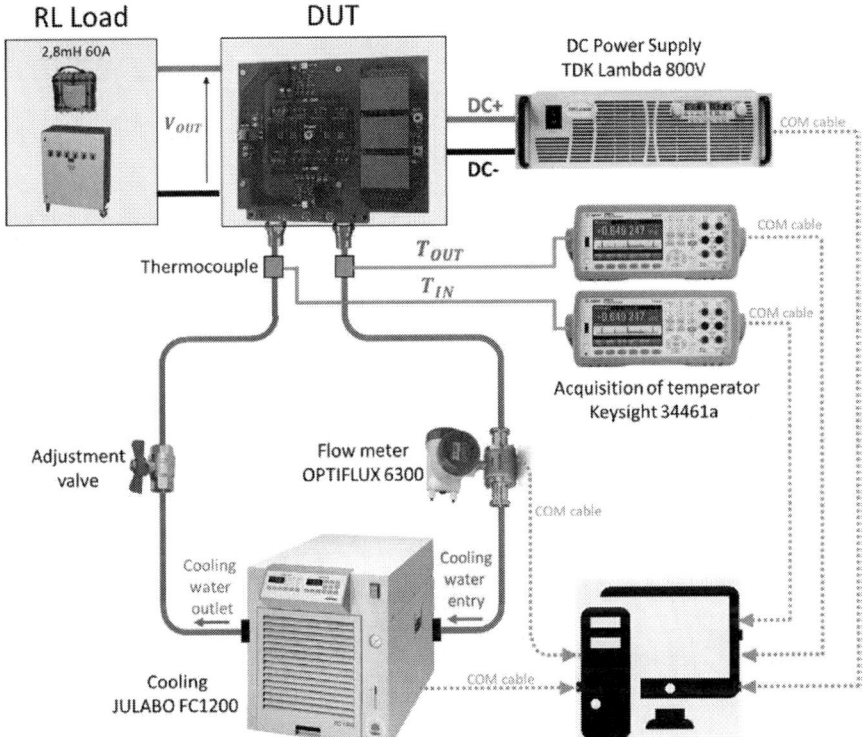

Figure 2: Losses measurement test bench

The prototypes P1,P2 and P5 are tested to create the model of the imaginary die using the calorimetric losses measurement test bench presented in fig.2 [12]. The conductions parameters V_{ce0}/V_d and R_{ce}/R_d are identified with the model in static using a duty cycle equal to one for two different current levels. Furthermore, the curve of switching energy are determinate by increasing the frequency from 10kHz to 40kHz for different current levels. The slope of the curve of losses as function the frequency represents the switching energy.

Table 2: IGBTs characterization results

	P1		P2		P5	
	IGBT	Diode	IGBT	Diode	IGBT	Diode
Threshold voltage (V)	1.12	1.2	1.08	1.14	1.1	0.9
R_c, R_d (Ohm)	0.04	0.034	0.023	0.017	0.0128	0.008
E(J)	$E(I) = 4\ 10^{-6}I^2 + 10^{-6}I + 2\ 10^{-5}$		$E(I) = 4\ 10^{-6}I^2 + 1\ 10^{-5}I + 2\ 10^{-5}$		$E(I) = 9\ 10^{-7}I^2 + 5.4\ 10^{-5}I + 5.3\ 10^{-6}$	
Imaginary die parameters(V)	$V_{ce0}(IGBT_{area}) = -0.0041 * IGBT_{area} + 1.1644$; $R_c(IGBT_{area}) = 0.2794 * IGBT_{area}^{-0.837}$ $V_d(Diode_{area}) = -0.0213 * Diode_{area} + 1.3434$; $R_d(Diode_{area}) = 0.25 * Diode_{area}^{-1.134}$					

The results of characterization are presented in Table.2 and can be interpolated as function of the die area. The parameters of the imaginary die are used in the analytic model presented in eq (3, 4, 5, and 6) in the case of buck converter with 400 dc-bus voltage and and 2.8mH and 18.6 A with 3.5A minimum current ripple. The conductions losses decrease by increasing the die area fig.3.a due to the resistance and EMF drop voltage reduction. Contrarily to the switching losses which have an optimal value for minimum losses fig.3.B. The total losses are given in fig.3.c by summing the conduction and switching losses and we can notice that there is an optimal IGBT die area minimizing the losses. The total losses increase by increasing the switching frequency due to the improvement of the switching losses.

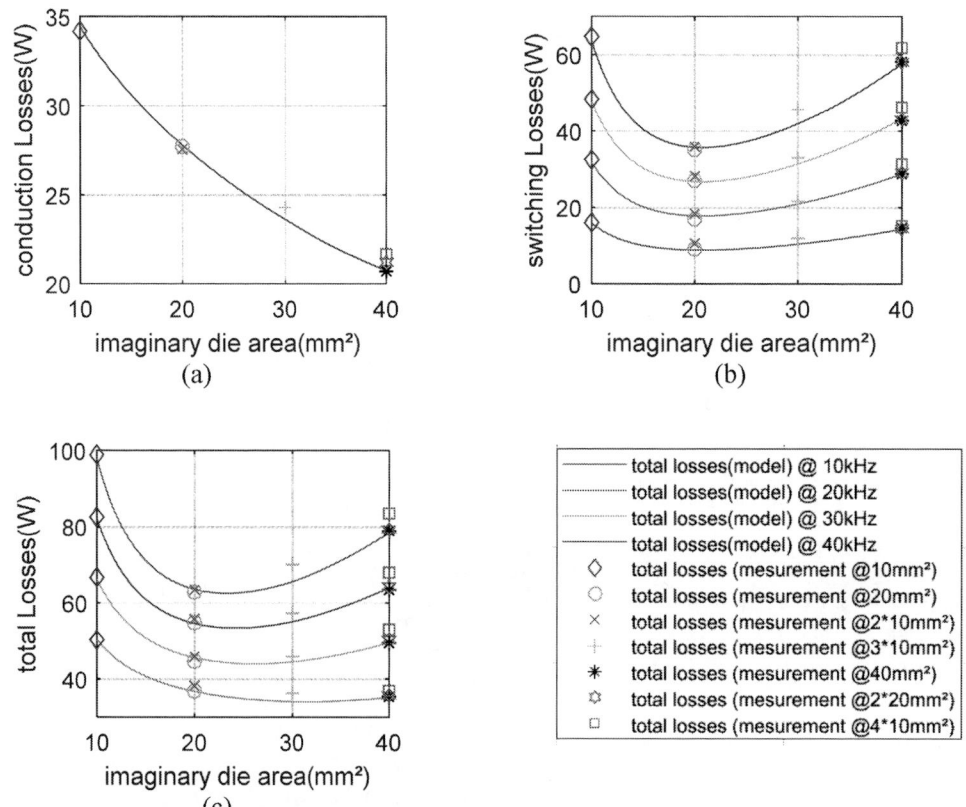

Figure 3: Analytical model and experiment validation (a) conduction losses (b) switching losses (c) total losses @400VDC , 10kHz to 40kHz Switching frequency, 18.6A output current

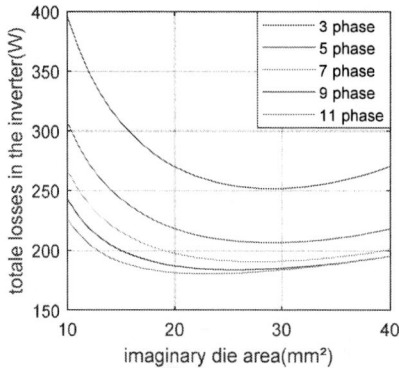

Figure 4: simulation for 48kW inverter for several number of phases

The prototypes losses measurements using calorimetric test bench matches with the analytical model, the prototype using single device gives more accurate results than the prototypes using several devices in parallel. This difference is due to current unbalance in the case of paralleling device. However, this is not significant in a predesign process, and can be solved with more symmetrical layout.

Fig.4 presents the total losses in inverter from 3 to 11 phases using the analytical model of imaginary die presented in equ (3, 4, 5 and 6) as function of die area. We can notice that there is an optimal IGBT die area for each number of phases, minimizing the total losses. The optimal point of losses is due to the reduction of conduction and switching by increasing the die area until we achieve the optimal die area minimizing losses, for larger die area total losses increases due to the switching losses. It is also worth noting that the losses are reduced with the increase of phase number, but the improvement seems not that much important from 9 to 11.

Finally, another effect has been considered: with larger die area, the thermal resistance of the device is decreased, what leads to better cooling capability. The thermal resistance variation as a function of the device area for the imaginary has been obtained from actual datasheets and interpolated with a mathematical function. The same kind of results has been obtained for the diode.

Analytical model of DC link for multiphase inverter

In electric vehicle application, the input capacitor is sized to ensure a low voltage ripple in dc link side. The dc link capacitor represents an important part of the inverter volume. In addition, the dc link capacitor volume is linked to its capacitance, as well as its allowed RMS current. The sizing of this capacitance is linked to the voltage and current ripple requirement [8]. Therefore, a generic and analytical model of the current and voltage ripple on the Dc link capacitor in case of multiphase inverter is needed.

The first step in the development of the analytical approach for the computation of the dc-link current is to define the instantaneous equation of different currents in each leg. Fig.5.

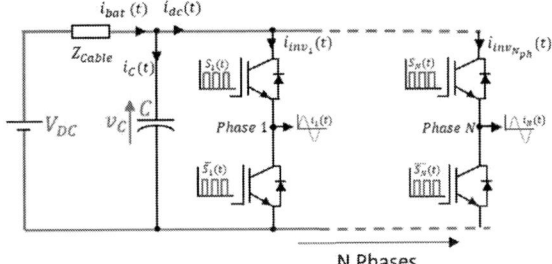

Figure 5: Multiphase Voltage source inverter

The multiphase VSI input current can be expressed as a function of switching states S_x and the load current as follows:

$$I_{dc}(t) = \sum_{x=1}^{Nph} I_{inv_x}(t) = \sum_{x=1}^{Nph} S_x(t) i_x(t) \tag{7}$$

where S_x and i_x are the switching function of phase x and the load current of the multiphase inverter, respectively. The switching function depends on the used PWM strategy [8]. The symmetrical sine-triangle modulation is used in our study, which compares sinusoidal fundamental with a triangular carrier wave form. The expression of the switching function in the frequency domain uses a double Fourier transformer and Bessel function

The inverter current is given by time domain multiplication (equ.9) and its equivalent to convolution operation in frequency domain. The inverter current expression for each leg considering a sine load current with an amplitude of I_0 and ϕ phase angle is as fellow:

$$I_{inv_x}(t) = \frac{r}{4} I_0 \cos(\phi) + \frac{1}{2} I_0 \cos(\phi) \cos(\phi_x) \cos(\omega_0 t) - \frac{1}{2} I_0 \cos(\phi) \cos(\phi_x) \sin(\omega_0 t) + \frac{r}{4} I_0 \cos(\phi) \cos(\phi_x) \cos(\omega_0 t) - \frac{r}{4} I_0 \cos(\phi) \cos(\phi_x) \sin(\omega_0 t) + \sum_{m=1}^{+\infty} \sum_{n=-\infty}^{+\infty} A_{mn} \cos(m\omega_c t + n\omega_0 t) + B_{mn} \sin(m\omega_c t + n\omega_0 t) \tag{9}$$

with:
$$\begin{cases} A_{mn} = \frac{I_0}{m\pi} \sin\left([m+n]\frac{\pi}{2}\right) \left[J_{n+1}\left(m\frac{\pi}{2}r\right) - J_{n-1}\left(m\frac{\pi}{2}r\right) \right] \cos(\phi) \cos(n\,\phi_x) \\ B_{mn} = \frac{I_0}{m\pi} \sin\left([m+n]\frac{\pi}{2}\right) \left[J_{n-1}\left(m\frac{\pi}{2}r\right) + J_{n+1}\left(m\frac{\pi}{2}r\right) \right] \sin(\phi) \cos(n\,\phi_x) \end{cases}$$

Dc-link current harmonics are calculated from the inverter current using current divider for each harmonic and considering the cable impedance between the battery and the DC-link capacitor [11]. The current spectrum consists of the multiple of the switching frequency and its side-band frequency components. The analytical dc-link RMS current is calculated from the analytical spectrum of the dc-link current and gives a good accuracy with an error less than 1% fig.6.

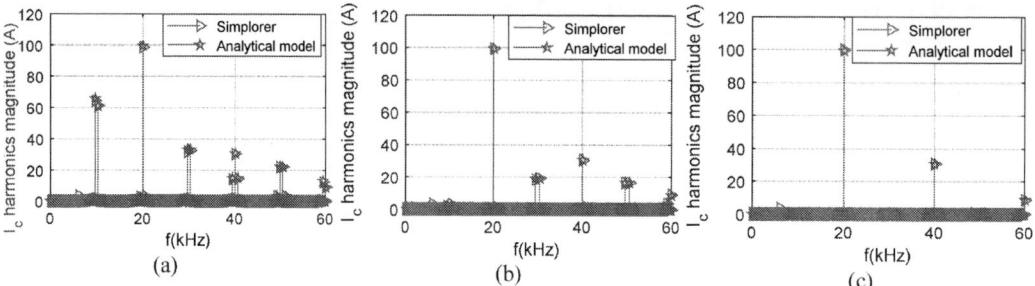

(a)　　　　　　　　　(b)　　　　　　　　　(c)

Figure 6:comparison between analytical model spectrum and simulation spectrum(b) for 3-phase inverter with m=0.8 r=0.8 fsw=10kHz (a) three-phase inverter (b) 5 phase inverter (c) 11 phase inverter

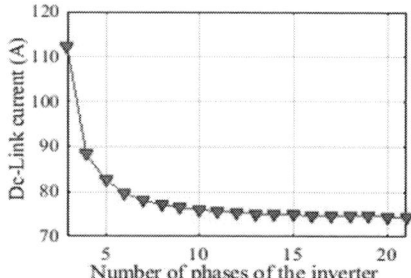

Figure 7:dc-link RMS current as function the number of phases of the inverter

It is noticed that the RMS current decreases by increasing the number of inverter's phases fig.7. The value of this current is reduced by 21% from 3 to 4 phases, and by 27% from 3 to 5 phases, in addition, when the number of phases exceeds 11 the reduction of dc-link RMS current is no more noticeable due to the elimination of all sidebands harmonics [10].

The reduction of the current ripple in the Dc link capacitor implies a reduction of the voltage ripple on the capacitance. These results are promising for the reduction of the Dc link volume in case of multiphase inverter.

Optimization of multiphase inverter

The technology used to build the multiphase inverter is presented in this section. Three boards are composing this inverter: IMS board (fig.8.a), used to fix the different IGBTs and play only a thermal interface role. A Power PCB (fig.8.b) is used as a low-cost DC link. This PCB is connecting the different inverter legs, DC bus. A driver PCB. is used to generates the drives voltages of different IGBTs while ensuring the galvanic isolation. Fig.8.c illustrates an example of 48kW,12 phases inverter using this technology. The power card is made up by several cells of legs of the inverter, and a capacitor between two cells. the presented technology of realization is optimized to enhance the power density of the multiphase inverter while ensuring a moderate cost of fabrication. Noting that technology of realization is used below to evaluate the overall volume of multiphase.

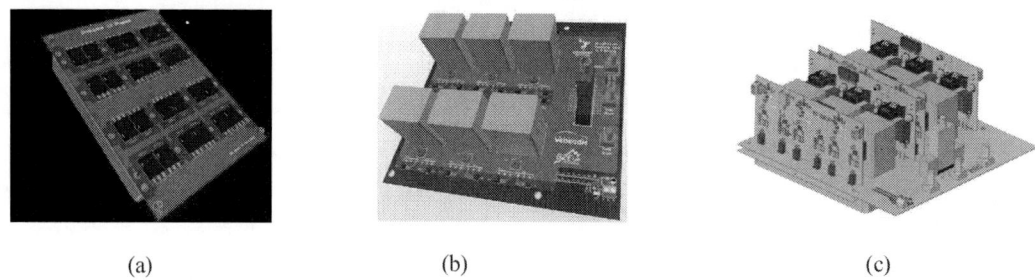

(a)　　　　　　　　　(b)　　　　　　　　　(c)

Figure 8: 12 phases 48kW inverter
a) IMS board, b: power PCB board, c) full converter with driver PCBs

To optimize the inverter a continuous model of the dc link capacitors volume and rated rms current of film capacitors have been use (fig.9). Volume and rated RMS current are from the datasheets of KEMET film capacitors.

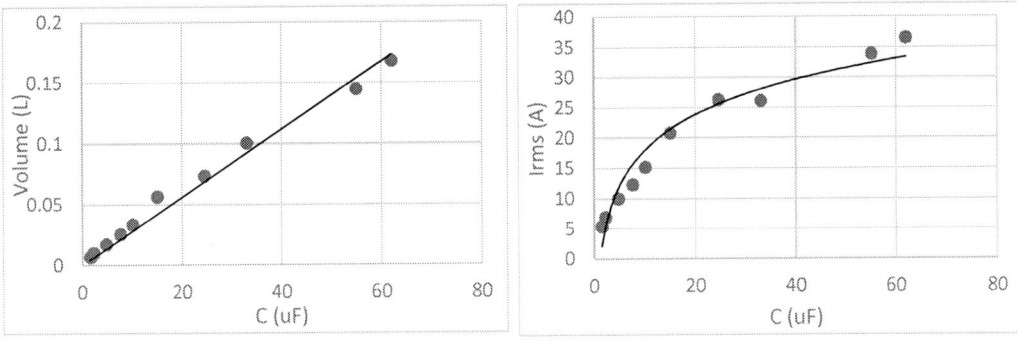

(a)   (b)

Figure 9: KEMET Film capacitors interpolation @650VDC
(a) Volume of the capacitor as function of capacitance (b) Rated RMS current as function of capacitance

A parametric design optimization of a 48kW inverter is made in this part, the specifications and constraints are summarized in table.3. The calculated volume of the inverter taking into account the dc link capacitor, drivers, and the number of inverter legs.

Table 3: optimization specifications

Objective function : minimizing Volume	
Operating point	Power=48kW
	V_{dc}=420V
	$\cos(\phi)$=0.8, r=0.8
Optimization variables	f_{sw}
	$IGBT_{area}$; $Diode_{area}$
	C
Constraints	$Efficiency \geq 95\%$
	$T_{j-IGBT/DIODE} \leq 130°C$
	$I_{c-max} < I_{crms-rated}$ $\Delta V < 8\%$

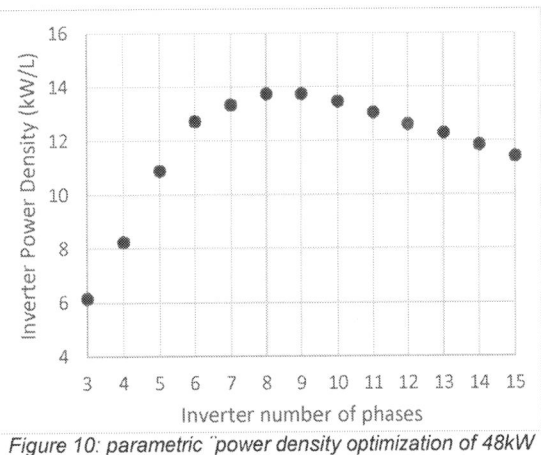

Figure 10: parametric power density optimization of 48kW inverter

The results of the parametric optimization are shown in fig.10. The power density increases with number of phases and achieve an optimal point at 8 phases. By increasing the number of phases, the RMS current passing through the dc-link capacitor decrease which reduce the volume of the capacitor. Noting that the overall volume of inverter increases linearly with the number of phases after reaching optimum point. In this case, the volume added by additional inverter's phase leg can't compensate the reduction in the volume of the DC link capacitor (which is no more changing, according to Figure 7).

Conclusion

This paper focused on the optimization of a multiphase inverter used in a drive for electric vehicle. Analytical model of semi-conductors devices and dc-link capacitor design are developed in case of multiphase inverter, and validated using time-domain simulation and experiments. The volume of the converter is computed according to a specific integration technology using IMS and PCBs. The optimization shows the enhancement of the power density by increasing the number of phases up to 8 in this case of study but will depend on the specifications of the converter.

References

[1] A. Emadi, "Transportation 2.0," in *IEEE Power and Energy Magazine*, vol. 9, no. 4, pp. 18-29, July-Aug. 2011.

[2] B. Cheong, P. Giangrande, X. Zhang, M. Galea, P. Zanchetta and P. Wheeler, "System-Level Motor Drive Modelling for Optimization-based Designs," *2019 21st European Conference on Power Electronics and Applications (EPE '19 ECCE Europe)*, Genova, Italy, 2019, pp. P.1-P.9.

[3] Emil Levi, "Multiphase electric machines for variable-speed applications ", IEEE Trans. Ind. Electron., vol. 55, no. 5, pp. 1893– 1909, May 2008.

[4] I. Laird, X. Yuan, J. Scoltock and A. J. Forsyth, "A Design Optimization Tool for Maximizing the Power Density of 3-Phase DC–AC Converters Using Silicon Carbide (SiC) Devices," in *IEEE Transactions on Power Electronics*, vol. 33, no. 4, pp. 2913-2932, April 2018.

[5] Y. Chen, Z. Yuan and F. Luo, "A Model-Based Multi-Objective Optimization for High Efficiency and High Power Density Motor Drive Inverters for Aircraft Applications," *NAECON 2018 - IEEE National Aerospace and Electronics Conference*, Dayton, OH, 2018, pp. 36-42.

[6] M. Delhommais & al., "First order design by optimization method: Application to an interleaved buck converter and validation," *2018 IEEE Applied Power Electronics Conference and Exposition (APEC)*, San Antonio, TX, 2018, pp. 944-951.

[7] N. Kesbia, J. Schanen, L. Garbuio and M. Ameziani, "Impact of the number of phases on losses of a multiphase inverter for electric vehicle drive," *2019 10th International Power Electronics, Drive Systems and Technologies Conference (PEDSTC)*, Shiraz, Iran, 2019, pp. 589-59

[8] D.G Holmes, T.A Lipo, "Pulse width modulation for power converters", IEEE press series on power engineering,2003.

[9] J. W. Kolar and S. D. Round, "Analytical calculation of the RMS current stress on the DC-link capacitor of voltage-PWM converter systems," IEE Proceedings - Electric Power Applications, vol. 153, no. 4, pp. 535-543, July 2006.

[10] N. Kesbia, JL. Schanen, H. Alawieh, L. Garbuio, " An Analytical model of the Dc-Link Current Ripple in multiphase PWM inverter," *PCIM Europe 2020; International Exhibition and Conference for Power Electronics, Intelligent Motion, Renewable Energy and Energy Management*, Nuremberg, Germany, 2020.

[11] McGrath, Brendan Peter, and Donald Grahame Holmes. "A general analytical method for calculating inverter DC-link current harmonics." *IEEE Transactions on Industry Applications* 45.5 (2009): 1851-1859.

[12] JL. Schanen *et al.*, "Teaching how to characterize and implement high speed power devices for tomorrow's engineers," *2019 IEEE Energy Conversion Congress and Exposition (ECCE)*, Baltimore, MD, USA, 2019, pp. 404-411, doi: 10.1109/ECCE.2019.8912181.

Optimal torque/speed characteristics of a Five-Phase Synchronous Machine under Peak or RMS current control strategies

Tiago José dos Santos Moraes, Hailong Wu, Eric Semail, Ngac Ky Nguyen, Duc Tan Vu

Univ. Lille, Arts et Metiers Institute of Technology, Centrale Lille, Yncrea Hauts-de-France, ULR 2697
L2EP, F-59000 Lille, France
E-mail. {tiago.dossantosmoraes ; hailong.wu ; eric.semail ; ngacky.nguyen ;ductan.vu}@ensam.eu

Keywords

«Multiphase drive», «Harmonic injection», «Torque optimization», «Control under constraints», «Traction drive».

Abstract

Torque density is usually improved by injecting the third current harmonic for five-phase permanent magnet synchronous machine (PMSM). It increases the degrees of freedom of a multiphase drive. However, it also separates the current limitations of the motor and the transistors, respectively related to the RMS and peak values of the currents. These two constraints are represented by Maximum Torque Per Ampere (MTPA) strategy and Maximum Torque Per Peak Current (MTPPC) strategy. In this paper, these two strategies are studied and analyzed in order to optimize the generated torque with injection of the third current harmonic. Torque improvement principle and the optimizing algorithm considering two constraints are illustrated. Then, the analytical results of these two strategies are compared and discussed. It is shown that injecting the third current harmonic can improve the torque especially at flux-weakening region. Besides, compared with MTPA, MTPPC could produce higher torque for the same inverter current limit.

1. Introduction

Multiphase PMSMs have been widely studied for transport applications because of their fault-tolerant ability [1]. One particular feature of the PMSM with (2k+1) phases is that vector control with high torque quality can be achieved even with non-sinusoidal back electromotive forces (back-EMF) in transient operation. In [1]-[5] and [8], the first and third harmonics of currents contribute to the torque of a five-phase machine and of a nine-phase machine [6]. It has been shown that the third harmonic current injection may lead to high torque density of a multiphase machine, depending on the harmonic distribution of the back-EMF obtained by machine design.

Besides, in order to estimate the maximum capabilities of a drive in steady states and transient operations, voltage and constraints must be considered during the control of multiphase PMSM. We will consider two current constraints for the optimization and control strategy: MTPA [2][7] and MTPPC [9][10]. The first one maximizes the torque for a given RMS current, directly related to the copper losses, while the second strategy maximizes the torque for a given peak current, much important for the Voltage Source inverter sizing. When only the first harmonic of stator currents is used, the mathematical relationship between its RMS value and its peak value is fixed. Hence, both MTPA and MTPPC strategies result in the same torque and current values.

However, if the first and third harmonics of stator currents are considered, the mathematical relationship between the RMS value and the peak value can be changed by supplying different amplitudes and phases for each harmonic. Consequently, each strategy will result in different drive's behaviors.

These different strategies could have various influences on the control performance of multiphase PMSMs. However, the impact on the frontiers of the torque speed characteristics has not been studied. This paper aims to explain the impact of the MTPA and the MTPPC strategies on the maximum torque and copper losses of a 5-phase PMSM when the third current harmonic is injected.

In this paper, the improvement of torque density by adding the third current harmonics of a five-phase PMSM is introduced in the second section. Then the optimizing algorithms considering two constraints are illustrated in the third section. Section four gives and discusses the results. Section five summarizes the analysis of this paper.

2. Studied five-phase machine and inverter

An open-end winding (OEW) five-phase PMSM is designed and studied in this paper. This prototype has 20 stator slots and 14 poles as presented in Fig. 1(a). The rare-earth magnets are buried in the rotor in radial disposition called spoke. Besides, the inverter structure of this open-end winding machine is illustrated in Fig. 1(b). The inductances in natural stator frame are obtained in stator natural frame by a finite element software:

$$\boldsymbol{L}_{abcde} = \begin{bmatrix} L_p & M_{ab} & M_{ac} & M_{ad} & M_{ae} \\ M_{ba} & L_p & M_{bc} & M_{bd} & M_{be} \\ M_{ca} & M_{cb} & L_p & M_{cd} & M_{ce} \\ M_{da} & M_{db} & M_{dc} & L_p & M_{de} \\ M_{ea} & M_{eb} & M_{ec} & M_{ed} & L_p \end{bmatrix} \tag{1}$$

The cyclic inductances in dq rotating frame can be calculated by

$$\boldsymbol{L}_{dq} = PC\boldsymbol{L}_{abcde}C^{-1}P^{-1} \tag{2}$$

where the Clark transformation and Park transformation are introduced respectively

$$C = \sqrt{\frac{2}{5}} \begin{bmatrix} 1 & \cos\alpha & \cos 2\alpha & \cos 3\alpha & \cos 4\alpha \\ 0 & \sin\alpha & \sin 2\alpha & \sin 3\alpha & \sin 4\alpha \\ 1 & \cos 2\alpha & \cos 4\alpha & \cos 6\alpha & \cos 8\alpha \\ 0 & \sin 2\alpha & \sin 4\alpha & \sin 6\alpha & \sin 8\alpha \\ \sqrt{\frac{1}{2}} & \sqrt{\frac{1}{2}} & \sqrt{\frac{1}{2}} & \sqrt{\frac{1}{2}} & \sqrt{\frac{1}{2}} \end{bmatrix} \tag{3} \qquad P = \begin{bmatrix} \cos\theta_e & \sin\theta_e & 0 & 0 & 0 \\ -\sin\theta_e & \cos\theta_e & 0 & 0 & 0 \\ 0 & 0 & \cos 3\theta_e & -\sin 3\theta_e & 0 \\ 0 & 0 & \sin 3\theta_e & \cos 3\theta_e & 0 \\ 0 & 0 & 0 & 0 & 1 \end{bmatrix} \tag{4}$$

$\alpha = 2\pi/5$

According to (2), the characteristics of the windings in dq rotating frame, obtained from inductances determined in stator natural frame by a finite element software, are listed in **Table I**.

Table I: Winding characteristics.

Parameter	Value	Parameter	Value
L_{d1}	154.8 µH	L_{q1}	199 µH
L_{d3}	96.5 µH	L_{q3}	97.1 µH
L_h	159.2 µH	R	0,016 Ω

After analysis of the results of table I, it is will considered only three cyclic inductances $L_1 = (L_{d1}+L_{q1})/2 = 177$ µH, $L_3 = (L_{d3}+L_{q3})/2 = 97$ µH, $L_h = 159$µH. It can be remarked that the value of the minimum time constant among L_1/R, L_3/R, L_h/R is equal to about 6.1ms. A PWM frequency of 10 kHz will ensure a good control of the average currents of the machine even for the zero-sequence current which is present in open-winding configuration.

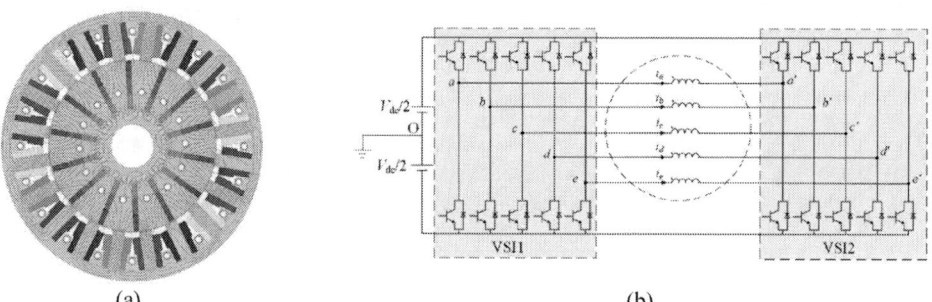

(a)　　　　　　　　　　　　　　　　　　　(b)

Fig. 1: Finite element model of the machine (a) and drive scheme (b).

In [3], it has been shown that with MTPA strategy the optimal injection of a third harmonic of a machine leads to a torque.

$$T_e = \frac{5}{2} p \left(I_1 \lambda_{m1} \cos\theta_1 + I_3 \lambda_{m3} \cos\theta_3 \right) = \frac{5}{2} \sqrt{2} p \left(\lambda_{m1} r I_{RMS} \cos\theta_1 + 3\lambda_{m3} \sqrt{1-r^2} I_{RMS} \cos\theta_3 \right) \tag{5}$$

Where p is the number of pole pairs, I_1 and I_3 are the peak amplitudes of the fundamental and the third stator current harmonics respectively; λ_{m1} and λ_{m3} are the peak amplitudes of the first and the third flux-linkage produced by permanent magnet respectively; θ_1 and θ_3 are the angles between the back-EMF harmonic and current harmonic respectively; r is the ratio between I_1 and I_{RMS} ($r = I_1/\sqrt{I_1^2 + I_3^2}$ [3]) and I_{RMS} is a given RMS current value.

When no constraints are imposed on current and voltage amplitudes, the optimal values for three parameters θ_1, θ_3 and r are calculated by $dT_e/dr = 0$.

$$r_{opt} = \frac{\dfrac{\lambda_{m1}}{3\lambda_{m3}}}{\sqrt{1+\left(\dfrac{\lambda_{m1}}{3\lambda_{m3}}\right)^2}} = \frac{\dfrac{E_1}{E_3}}{\sqrt{1+\left(\dfrac{E_1}{E_3}\right)^2}} \qquad \theta_1=0 \qquad \theta_3=0 \qquad\qquad (6)$$

E_1 and E_3 are the first and the third harmonics of no load back-EMF respectively.

In Fig. 2, the no load back-EMFs and the back-EMF/Speed harmonic distribution of the studied machine are presented. Their shapes are clearly non-sinusoidal. The amplitude of the third harmonic is close to 22.7% of the amplitude of the fundamental harmonic. Considering a 5-phase machine, it allows us to generate a constant mean-value torque using the third harmonic. Therefore, according to the amplitude of the injected the third current harmonic should be 22% of the fundamental current harmonic.

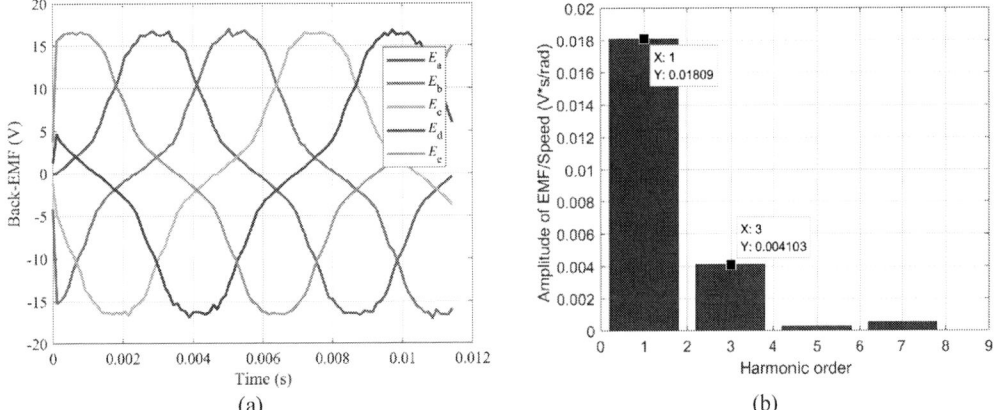

(a)   (b)

Fig. 2: Back-EMF at 750rpm (a) and back-EMF/Speed harmonic distribution (RMS value) (b).

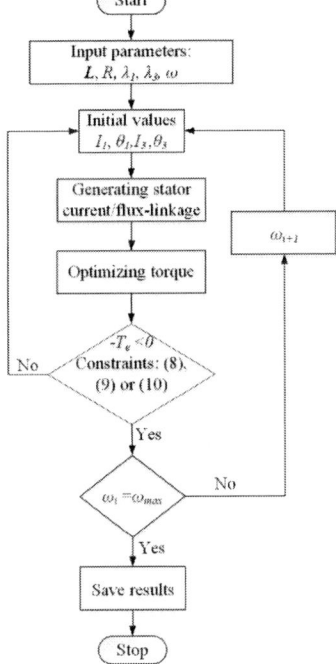

Fig. 3 Torque improvement algorithm

3. Torque Optimization algorithm

Finite element method is an effective tool to analyze and evaluate the design of electrical machine. During the analysis stage, optimal stator currents should be obtained in advance and supplied into the machine. Therefore, a program developed on MATLAB is used to obtain the optimal current shape that generates the maximum torque considering MTPA and MTPPC strategies. The torque optimizing problem is stated as following. The objective function is

$$\min_{(I_1, \varphi_1, I_3, \varphi_3)} \left(-T_e(I_1, \theta_1, I_3, \theta_3, L, R, \lambda_1, \lambda_3, \omega) \right) < 0 \tag{7}$$

With the voltage constraint

$$\max \left(U_{abcde} \right) \leq 60 \tag{8}$$

For the RMS current constraint

$$\max \left(\sqrt{\frac{I_1^2 + I_3^2}{2}} \right) \leq 71.4 \tag{9}$$

For the peak current constraint

$$\max \left(I_{abcde} \right) \leq 120 \tag{10}$$

The Flow chart of the optimization process is illustrated by Fig. 3. The parameters of this program are listed below:

The input inductance matrix L, resistance R and the first harmonic λ_1 and the third harmonic λ_3 of flux linkage without load are obtained by a no-load finite element simulation. The optimized parameters are the amplitudes (I_1 and I_3) and phases (θ_1 and θ_3) of the first and the third stator current harmonic. The optimizing algorithm is applied for the interested speed range to obtain the torque-speed characteristic. The objective function should be smaller than zero. If it is larger than zero, it means the machine is at generator mode.

The chosen constraints are associated to three components of the drive: the DC source, the Voltage Source Inverter and the machine. The value of the DC source will impact the maximum voltage which can be imposed to the machine. The OEW connection allows a maximum voltage per phase U_{abcde} equal to the voltage of the DC source.

The maximum RMS value, or the current density, which can be imposed in a machine is highly depending on thermal consideration. For example, we can consider that 71.4 Arms (6 A/mm²)) corresponds to steady-state operation (>30 minutes). As the thermal time constant constraints of transistor are less than 0.1s, the chosen peak value of the current I_{abcde} is associated to maximum allowable value 120A. If the MTPA strategy is applied, the corresponding stator currents can be optimized with the constraint (9). If the MTPPC strategy is considered, the corresponding optimal stator currents can be obtained with the constraint (10).

4.   Optimized results

a. Impact of the third harmonic injection for MTPA strategy

Firstly, the impact of the third harmonic injection is analyzed by the torque improvement algorithm with the current constraint (9). The first case (1h) is to only apply the fundamental current harmonic to optimized torque and the second case (1h+3h) injects the third current harmonic while the same maximum RMS value in both cases is considered. The results are presented in Fig. 4.

Comparing the curves of Fig. 4 (a), it is possible to conclude that the third harmonic current injection allows a torque increase for the same copper losses. The torque increase is quite small at low speed (2,4%). This low gain is expected because the amplitudes of the third harmonic of the current and of the back-EMF are equal to 22% of the fundamental ones. The obtained stator current (phase a) at low speed 750rpm is illustrated in Fig. 4 (b). The current of the second case (1h+3h) is no longer sinusoidal.

When the flux-weakening strategy is implemented, the torque increase overtakes 32% at 4500rpm as shown in Fig. 4 (a). The torque improvement at 5000rpm is over 100% because the available maximum speed is just 4800rpm for the first case (1h). It implies that injecting the third current harmonic can increase the flux-weakening region for the studied five-phase PMSM. The flux-weakening strategy is needed at high speed because the peak back-EMF exceeds the voltage limits of the drive. At high speed, over 2500rpm for this five-phase PMSM, the maximum peak voltage (60V) is a major constraint. The injection of the third harmonic increases from 2 to 4 the number of degrees of freedom (amplitudes and phases of first and third harmonics). By shifting the phase between the third and first harmonics, it is possible to significantly decrease the peak voltage. The resulting peak value of the voltage reference is less than or equal to the voltage limit. An example at 4000rpm is presented in Fig. 4 (c). The current shape is different from the sinusoidal stator current. The produced voltage and back-EMF are illustrated in Fig. 5. It is observed that back-EMF is 90V, but the voltage is reduced to 48V.

At last, the obtained optimal currents of these two cases are applied to the finite element model at speed 750rpm and 4000rpm. The torques are presented in Fig. 4 (a) by the red dots (case 1) and blue dots (case 2). At each speed, the torque is improved by the second case which is same as the conclusion summarized by analytical torque optimization algorithm. The difference is that the torques at 750rpm are smaller than the analytical results.

Because the inductances used in the algorithm are obtained at no load which means there is no saturation. But the finite element model has considered saturation.

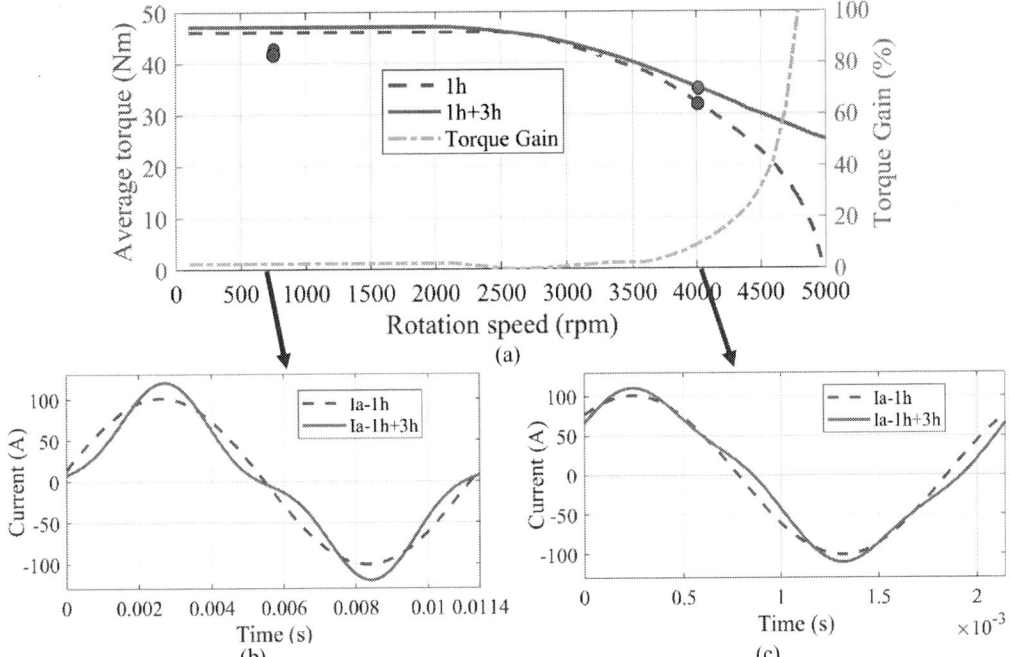

Fig. 4: Torque per speed and torque gain curves with and without third harmonic injection for a current of 71.4Arms (a), with examples of the current shape of the phase-a obtained at 750rpm (b) and 4000rpm (c).

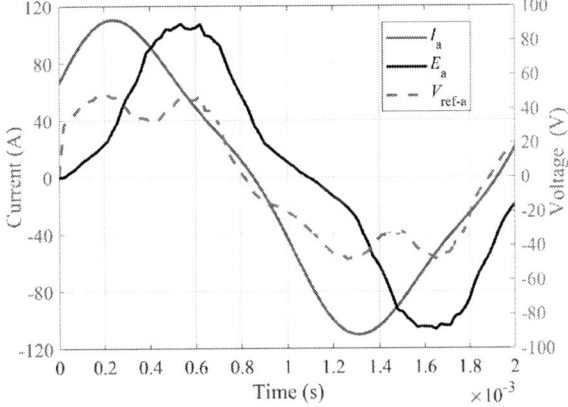

Fig. 5: Current, Back-EMF and Reference Voltage of phase a with third harmonic injection at 4000rpm.

b. MTPA and MTPPC strategies

In the previous section, the RMS current is used as a constraint for the machine characterization because this is directly related to the machine copper losses. However, when considering the whole drive, the peak-current is the limitation for the transistors. This is the reason why an analysis will be carried out by not limiting the RMS current but the peak one. This strategy is called MTPPC.

This analysis is only possible when the currents are composed by more than one harmonic, otherwise the peak and the RMS values are directly related. The chosen peak value for the MTPPC strategy is 120Â because it is the maximum peak-current value obtained with the MTPA strategy for the RMS current of 71.4Arms.

During the optimization, the peak current constraint (10) is used and the results are obtained. Fig. 6 and Fig. 7 present the maximum torque, RMS and peak current per speed curves for the both strategies. These figures confirm that the MTPPC strategy allows the drive to attain higher values of torque (Fig. 6) without increasing the current peak value (Fig. 7 (b)). At low speed, below 1700rpm, it is possible to have 31% of torque gain for the same peak current, despite a loss increase of 38,6% (Fig. 7 (a)). Different from the comparison of the previous section, the major gain is not in the flux-weakening operation zone.

EPE'20 ECCE Europe

Assigned jointly to the European Power Electronics and Drives Association & the Institute of Electrical and Electronics Engineers (IEEE)

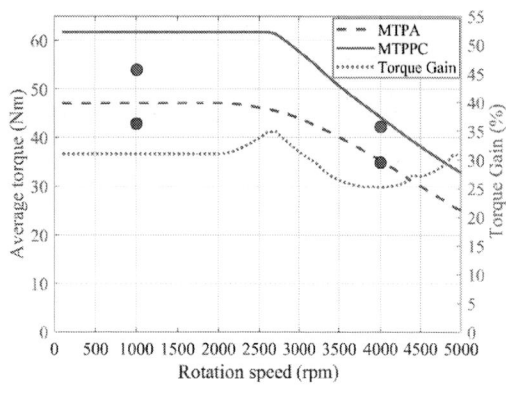
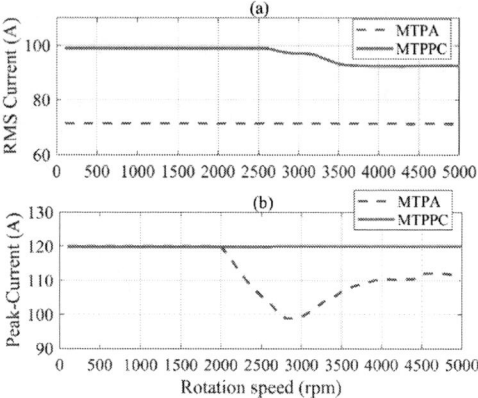

Fig. 6: Torque per speed curves for MTPA for a RMS current of 71.4Arms and MTPPC for a peak value of 120Â and torque gain percentage of MTPPC in comparison to MTPA.

Fig. 7: RMS current (a) and peak-current (b) per speed curves for MTPA for a RMS current of 71.4Arms and MTPPC for a peak value of 120Â.

Fig. 8 shows how different the current shapes are when comparing both strategies. The major difference is the phase shift of the third harmonic. Roughly speaking, the peak-current value when the MTPA strategy is applied is the sum of the first and third harmonics amplitude. This behavior is due to the shape of the back-EMF, intrinsic to the machine. Due to the back-EMF distribution, the third harmonic generates much less torque than the first harmonic. When applying the MTPPC strategy, the third harmonic of the current will be phase-shifted of π from the relative back-EMF harmonic, generating then a negative torque. However, it allows to have a higher first harmonic, consequently a much higher torque, without overtaking the peak-current limit.

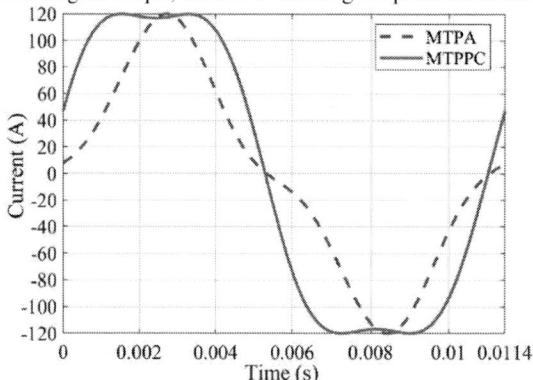

Fig. 8: Current shape of the phase a for both strategies at 750rpm.

This behavior is interesting because, in applications such as automotive traction, the higher torque values are usually used during the accelerations. As the duration of those accelerations are relatively short, their impact on machine losses is limited. Finally, torque/speed characteristics of Fig. 6 define which is the most adaptable strategy for each functioning point.

Lastly, these two strategies are simulated at 750rpm and 4000rpm by finite element method. The torques are presented in Fig. 6 by the red dots (MTPPC) and blue dots (MTPA). The torques are larger when MTPPC is applied which is same as the analytical result. The saturation reduces the torques obtained by finite element method. It can be observed the torque is much reduced in MTPPC (12.9% at 750rpm and 4.23% at 4000rpm) than in MTPA (9.8% at 750rpm and 0.23% at 4000rpm). Because the peak currents are same for these strategies, but the RMS current are different. So higher RMS current produces more saturation.

5. Conclusion

The impact of the third harmonic injection on torque generation of a five-phase open-winding PMSM with two different strategies was studied. This third harmonic injection has been optimized in order to generate the maximum torque considering the current limitations of the transistors and the machine. At first, the injected third current harmonic could enlarge the flux-weakening region for the RMS current constraint (MTPA). Besides, the analysis of two constraints with injection of current harmonic allowed us to adapt the control of the drive, in order to increase the torque density of the drive without increasing its sizing. For the same peak current constraint,

MTPPC could produce higher torque than MTPA. But the copper losses could also be larger because of higher RMS current. This control was highly relevant in an industrial context in which the sizing of the machine and the transistors of the inverter are defined together. A possible control for the automobile application was to apply MTPPC for short time such as acceleration and to use MTPA for steady state.

Acknowledgment

This work has been achieved within the framework of CE2I project. CE2I is co-financed by European Union with the financial support of European Regional Development Fund (ERDF), French State and the French Region of Hauts-de France.

References

[1] F. Barrero and M. J. Duran, "Recent Advances in the Design, Modeling, and Control of Multiphase Machines-Part I," *IEEE Transactions on Industrial Electronics*, vol. 63, no. 1, pp. 449–458, Jan. 2016.

[2] F. Scuiller, H. Zahr, and E. Semail, "Maximum Reachable Torque, Power and Speed for Five-Phase SPM Machine With Low Armature Reaction," *IEEE Transactions on Energy Conversion*, vol. 31, no. 3, pp. 959–969, Sep. 2016.

[3] J. Gong, H. Zahr, E. Semail, M. Trabelsi, B. Aslan, and F. Scuiller, "Design Considerations of Five-Phase Machine with Double p/3p Polarity," *IEEE Transactions on Energy Conversion*, vol. 34, pp. 12-24, 2018.

[4] P. Zhao and G. Yang, "Torque Density Improvement of Five-Phase PMSM Drive for Electric Vehicles Applications," *Journal of Power Electronics*, vol. 11, no. 4, pp. 401–407, Jul. 2011.

[5] K. Wang, Z. Gu, Z. Zhu, and Z. Wu, "Optimum injected harmonics into magnet shape in multiphase surface-mounted PM machine for maximum output torque," *IEEE Transactions on Industrial Electronics*, vol. 64, pp. 4434-4443, 2017.

[6] M. Slunjski, M. Jones and E. Levi, "Control of a Symmetrical Nine-phase PMSM with Highly Non-Sinusoidal Back-Electromotive Force Using Third Harmonic Current Injection," IECON 2019 - 45th Annual Conference of the IEEE Industrial Electronics Society, Lisbon, Portugal, 2019, pp. 969-974

[7] D. T. Vu, N. K. Nguyen, E. Semail, and T. J. dos Santos Moraes, "Control strategies for non-sinusoidal multiphase PMSM drives in faulty modes under constraints on copper losses and peak phase voltage," *IET Electric Power Applications*, 2019.

[8] Y. Sui, P. Zheng, Y. Fan and J. Zhao, "Research on the vector control strategy of five-phase permanent-magnet synchronous machine based on third-harmonic current injection," 2017 IEEE International Electric Machines and Drives Conference (IEMDC), Miami, FL, 2017, pp. 1-8

[9] G. Feng, C. Lai, M. Kelly, and N. C. Kar, "Dual three-phase PMSM torque modeling and maximum torque per peak current control through optimized harmonic current injection," *IEEE Transactions on Industrial Electronics*, vol. 66, pp. 3356-3368, 2018.

[10] D. T. Vu, N. K. Nguyen, E. Semail, and T. J. dos Santos Moraes, "Torque optimization of seven-phase BLDC machines in normal and degraded modes with constraints on current and voltage," *The Journal of Engineering*, vol. 2019, pp. 3818-3824, 2019

Comparative study of two control techniques of regenerative braking power recovering inverter based DC railway substation

Youssef Krim[1], <u>Khaled Almaksour</u>[1], Hervé Caron[2], Tony Letrouvé[3], Christophe Saudemont[1], Bruno Francois[1] and Benoit Robyns[1]

[1]Univ. Lille, Arts et Metiers Institute of Technology, Centrale Lille, Yncrea Hauts de France, ULR 2697 L2EP, F 59000 Lille, France

[2]SNCF Réseau, 6 Avenue François Mitterrand, 93574 La Plaine Saint Denis, France

[3]SNCF- Direction Innovation & Recherche, 1-3 Avenue François Mitterrand, 93574 La Plaine Saint Denis, France

Tel. : +33 328 384 858

E-mail : khaled.almaksour@yncrea.fr

Keywords

« Electrical train », « Regenerative power », « Efficiency », « Voltage Source Converter », «Regulation», « Control methods for electrical systems ».

Abstract

This work is focused on the improving of the energy efficiency of electrical train in braking mode. The improvement is explored by integration and control of a reversible inverter in a DC power substation to inject the braking power from the DC railway electrical network to the AC transmission grid. This solution makes the power substation reversible. The objective of this study is to investigate a control strategy by adjustment of the catenary DC voltage by the inverter in order to recover the maximum of electric braking energy by keeping the DC voltage of the substation stable with a reference value. A comparative study of simulation results of the proposed control scheme with a droop control technique for the railway substation "Massena" in Paris is presented.

I. Introduction

In recent years, the growth of railway traffics begets a significant increase in energy consumption [1]. In this context, many industrial and research projects are currently being conducted to identify strategies to reduce the electricity demand and improve the energy efficiency [2]. In general, an electric train can operate in four modes: acceleration, cruising, coasting and braking [3]. In the first, traction mode, the train accelerates until reaching the reference speed. Once the reference speed is reached, it goes to the cruising mode, which allows keeping this speed stable. The next step is the coasting, which sets a traction force equal to zero. Finally, comes the braking mode, the train decelerates until the total stop. There are many techniques to make a train braking, including mechanical braking, regenerative braking, resistance braking and combination of them [3]. The regenerative braking is implemented by reversing the train machine operation. In braking mode, the machine operates as a generator and, then, converts the mechanical energy to electrical energy. Neighboring trains that are in acceleration or cruising modes can consume the recovered energy injected to the catenary. When the distance between stations is short, the acceleration / braking cycle is repeated several times, which provides a significant amount of energy. However, under low traffic conditions, a braking could take place without a synchronized acceleration of other trains. If there is no train in accelerating mode, the injection of the recovered braking power tends to increase the catenary voltage [4]. There is a voltage limit that must not be exceeded to protect the railway system infrastructure against overvoltage. To satisfy this limitation, excess of energy must be dissipated in the braking resistors or in mechanical brakes. Consequently, the dissipation of the recovered energy is considered as a loss of energy.

One of the solutions to save energy is the power reinforcement of the railway power grid and the improvement of braking energy recovery rate [5]. The energy excess is stored in electrical storage such as, batteries [6], flywheels [7] and supra-capacitors [8]. The synchronization of trains circulation in the same rail is exploited to increase the reuse of braking energy [9]. When a train brakes and injects the recovered energy to the catenary, another train accelerates simultaneously to consume this energy. In addition, making the substation reversible is another promising solutions to improve the energy efficiency, because the recovered energy is sent back to the main AC grid by using power electronic converters [10]. In this framework, the French national railway company "SNCF" has decided to choose the last solution by connecting a reversible inverter in parallel with the rectifier that supplies Massena power substation. The chosen inverter is installed to make the power substation reversible, and consequently, improve the energy efficiency by reducing losses and the total energy consumed. This efficiency could be highly impacted by the control strategy applied to the inverter. A droop control is developed in [10] to control the inverter. With this control method, 22.5 % of the maximum braking power available is recovered. This limitation is caused by the train braking strategy and the characteristic of the droop control. For this reason, a control strategy with regulation of Massena DC voltage is suggested in this work.

Thus, the goal of this paper is to explore a control strategy of the inverter, which could maximize the recovered regenerative braking energy and keep the DC substation voltage stable with a reference value. The proposed control method operates in two modes. In braking mode, this control injects the total of the energy available in DC side to the AC grid. In traction mode, the inverter is blocked by its control strategy to prevent the power transit from the AC grid to the DC side. The power reference should therefore be zero if the measured voltage is lower than the reference voltage of the voltage regulation loop.

The remaining of this paper is organized as follows. In section II, a description of the studied railway system is developed. Section III focuses on the control of the braking energy recovery converter. A constant voltage control with a PI controller is proposed. In section IV, a comparative analysis between results obtained with the constant voltage controller and a droop controller is discussed. Finally, a conclusion is presented in section V.

II.   Description of the studied railway system

In the particular case of the Massena project, three electric power substations located on the line C of the Paris suburban rail are studied: 'Quai de La Gare', 'Massena' and 'Les Ardoines'. As depicted in Fig. 1, 'Massena' and 'Les Ardoines' power substations are supplied by the national 63 kV transmission grid. In traction mode, power conversion of each substation consists of a 12-phase rectifier powered by a double-winding transformer. The recovery of the braking energy is done by connecting an inverter in parallel with the diode rectifier of Massena substation. The train injects the braking power to the catenary to consume it by other trains in acceleration, but in case of low traffic conditions, the reversible substations reinjects this power to the AC grid. The advantage of the used inverter is to give the possibility to inject the total available power at DC grid by keeping the DC voltage stable with its reference. In this work, the inverter has a rated power of 1 MW and a maximum power of 2.25 MW.

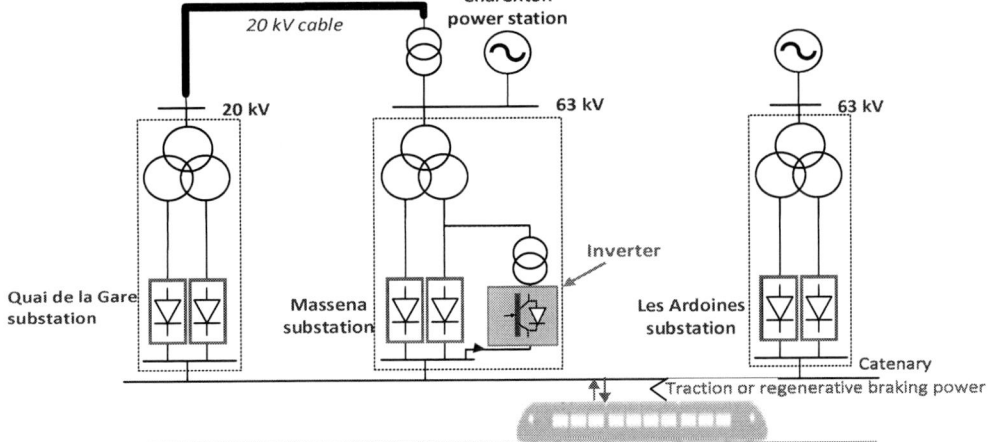

Fig. 1: A schematic diagram of the supply substations

This work presents a control strategy of the inverter. This control system offers the following services:
- Recover the maximum braking energy,
- Keep stable the Massena substation voltage with its reference value during braking modes,
- Does not allow power flow from the AC power grid to the DC catenary through the inverter in traction modes.

III. Control of the installed inverter

In this part, the inverter's control method is presented. This control design is based on a constant voltage control technique to adjust the DC voltage despite the re-injection of regenerative braking power. In order to establish a comparison framework, a design of a droop control is proposed. The constant voltage control is commonly used in renewable distributed generators [11] and for DC micro-grid systems [12], but rarely proposed in the context of recovering braking energy.

The control law of the inverter allows controlling AC voltages, V_{inv1}, V_{inv2} and V_{inv3}. The control scheme includes a DC voltage controller and a current loop (Fig. 2). The DC voltage controller defines the reference values for currents loop that will be injected to the AC power grid. These references are obtained thanks to either the droop controller or the constant voltage controller.

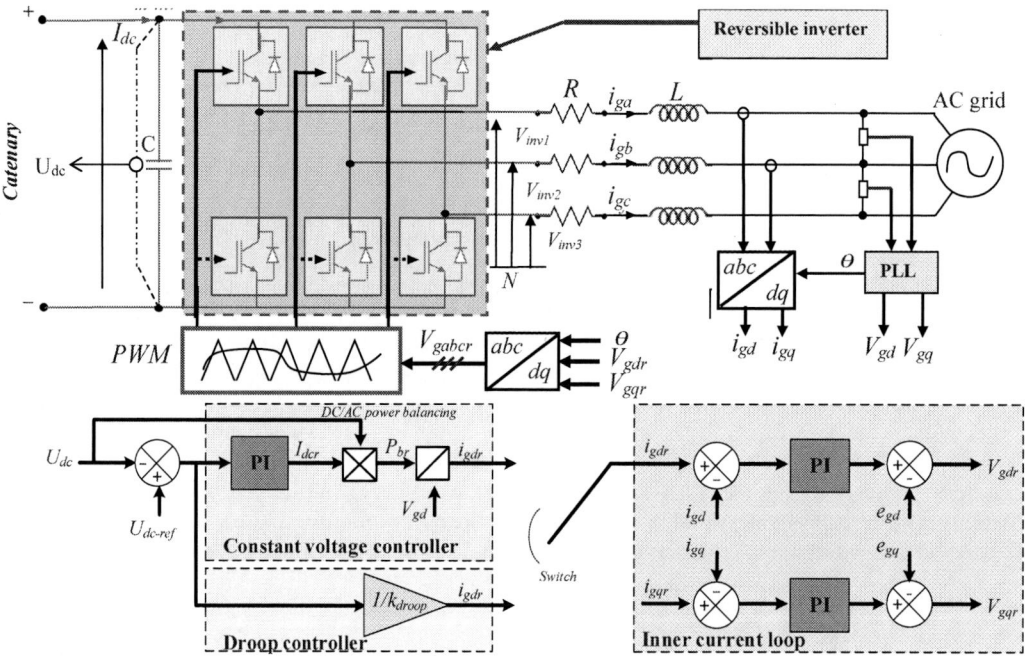

Fig. 2: Control structure of the inverter

III.1. Droop control method

The operating principle of the droop control is as follows: a variation of the DC voltage U_{dc} relative to the reference operating voltage "U_{dc-ref}=1750V" (no-load voltage of Massena substation) will generate a linear variation of the current reference, by moving on the linear current-voltage characteristic (Fig. 3). The slope of this characteristic is $1/k_{droop}$ where k_{droop} is the droop coefficient, as described in the following equation:

$$i_{gdr} = -\frac{1}{k_{droop}}\left(U_{dc-ref} - U_{dc}\right) \tag{1}$$

With:

$$k_{droop} = \frac{U_{dc\_max} - U_{dc-ref}}{-I_{inv\_max}} \tag{2}$$

where I_{inv_max} and U_{dc_max} are respectively the maximum DC current and the maximum DC voltage of the inverter.

The value of the droop coefficient considered in this study is 0.1733. This value is calculated by using the following inverter parameters: U_{dc_max}= 1950V, U_{dc-ref}= 1750V and I_{inv_max}= 1153.8A.

In braking mode, the U_{dc} voltage starts to increase. When this voltage exceeds U_{dc-ref}, the reference of the inverter current will be changed as a linear function according to the DC voltage as depicted in Fig. 3. When the voltage U_{dc} is lower than U_{dc-ref}, the reference current i_{gdr} remains equal to zero.

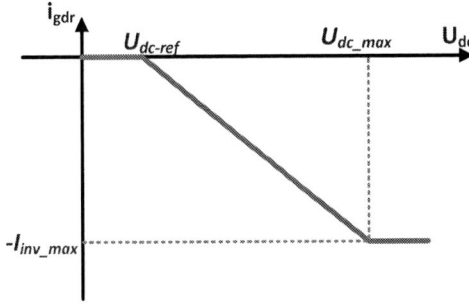

Fig. 3: Characteristic curve of the droop control

III.2. Constant DC voltage control method

Fig. 4 illustrates the regulation loop of the DC voltage. A Proportional Integral (PI) controller is implemented to keep the U_{dc} voltage stable at its reference value "U_{dc-ref}".

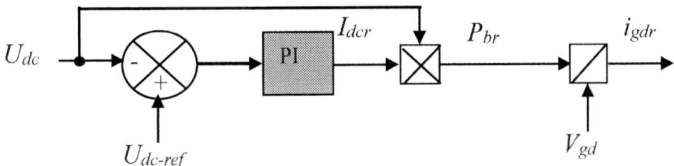

Fig. 4: Regulation loop of the constant DC voltage controller

The transfer function of the DC bus is deduced from Fig. 4 as follows:

$$F_{dc}(p) = \frac{U_{dc}}{i_{dc}} = \frac{1}{C.p} \tag{3}$$

Where p is the Laplace variable.

The following form defines the PI controller:

$$PI_{dc}(p) = k_p + \frac{k_i}{p} = G\frac{(1+\tau.p)}{\tau.p} \tag{4}$$

Thus, the Closed Loop Transfer Function (CLTF) is given by the following expression:

$$CLTF_{dc}(p) = \frac{1+\tau.p}{1+\tau.p+\dfrac{\tau.C}{G}p^2} \tag{5}$$

The PI controllers are adjusted by considering the pole placement method. The denominator of CLTF is compared with a desired characteristic equation whose dynamics is known as:

$$P = 1 + \frac{2\zeta}{\omega_n}p + \frac{1}{\omega_n^2}p^2 \tag{6}$$

where ζ is the damping coefficient and ω_n is the natural frequency. For $\zeta = 0.7$, this polynomial has a response time set as: $\omega_n.t_r = 3$.

The controller parameters are set to match the desired second order polynomial and they are defined by the following equations:

$$G = k_p = \frac{6\zeta}{t_r}C$$

$$\tau = \frac{k_p}{k_i} = \frac{2\zeta}{3}t_r$$

(7)

The output of the DC voltage controller is the reference current I_{dcr}. The multiplication of this current by the measured DC voltage gives the power P_{br} to be injected to the AC power grid.

III.3. Current control loop

The electrical equations of the filter for the connection of the inverter with the AC grid in the Park frame can be expressed as follows [13]:

$$V_{gd} = R.i_{gd} + L.\frac{di_{gq}}{dt}\underbrace{-L.\omega.i_{gq} + V_{inv\_d}}_{e_{gd}} = R.i_{gd} + L.\frac{di_{gq}}{dt} + e_{gd}$$

$$V_{gq} = R.i_{gq} + L.\frac{di_{gq}}{dt}\underbrace{+L.\omega.i_{gd} + V_{inv\_q}}_{e_{gq}} = R.i_{gq} + L.\frac{di_{gq}}{dt} + e_{gq}$$

(8)

Where R and L are respectively the resistance and inductance of the filter, i_{gd} and i_{gq} are direct and quadratic currents injected to the grid, V_{gd} and V_{gq} are the direct and quadratic voltages of the grid, V_{inv_d} and V_{inv_q} are the direct and quadratic voltages of the inverter output, and ω is the grid frequency. The coupling terms e_{gd} and e_{gq} are assimilated as measurable disturbances. Therefore, the studied filter can be presented by the following transfer function:

$$G_g(p) = \frac{i_{gd,q}}{V_{gd,q} - e_{gd,q}} = \frac{1}{R\left(1 + \frac{L}{R}p\right)}$$

(9)

In order to impose the reference currents i_{gdr} and i_{gqr}, we have used two Proportional Integral (PI) regulators. These regulators are identical because the transfer function is the same in the axis d and q. Fig. 5 presents the current control loop in the dq axis.

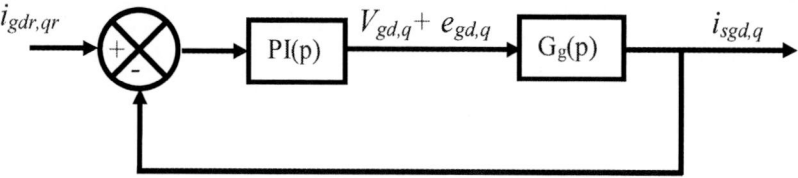

Fig. 5: Current control loop

The form of the used PI regulator is as follows:

$$PI(p) = k_p + \frac{k_i}{p} = G_R\frac{1 + \tau_R p}{\tau_R p}$$

(10)

where τ_R is the time constant and G_R is the regulator gain.
The following relation gives the open loop transfer function:

$$FTBO(p) = \frac{G_R\left(1 + \tau_R p\right)}{R.\tau_R p\left(1 + \dfrac{L}{R}p\right)} \tag{11}$$

Using the poles compensation method, the time constant τ_R of the studied regulator can be calculated as follows:

$$\tau_R = \frac{L}{R} \tag{12}$$

Consequently, the expression of *FTBO(p)* becomes:

$$FTBO(p) = \frac{G_R}{L.p} \tag{13}$$

Thus, the following relation gives the closed loop transfer function:

$$FTBF(p) = \frac{1}{1 + \dfrac{L}{G_R}p} \tag{14}$$

For a first order function, the response time at 5% is equal to triple of the time constant. In this case:

$$t_r = 3\frac{L}{G_R} \Rightarrow G_R = 3\frac{L}{t_r} \tag{15}$$

IV. Results and discussion

A circulation of one train is considered in this work for the validation of the proposed control. The speed profile is depicted in Fig. 6. According to this profile, the train starts its trajectory from "Quai de la Gare" substation (located at position 0.975 km) and passes by "Masséna" substation (located at position 2.495 km), then begins the braking phase and stops at "Les Ardoines" substation (located at position 6.8 km).

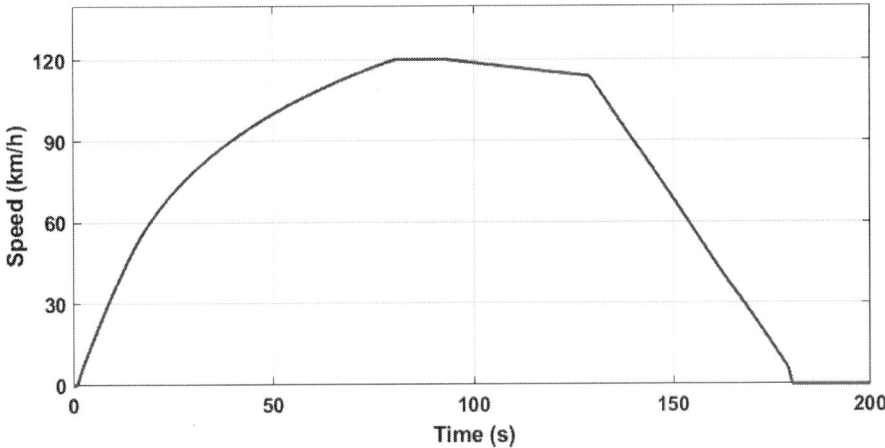

Fig. 6 : Train speed profile

The control of the inverter allows the transit of power from the catenary to the AC electrical network. When the measured DC voltage is lower than the inverter reference voltage (Fig. 7), the recovered power should therefore be zero (Fig. 9). In deceleration mode, the train injects the braking power to the catenary, which leads to an increase of the DC voltage (Fig. 7). The inverter droop control modifies the reference power to allow the transmission of power from the catenary to the AC network (Fig. 9).

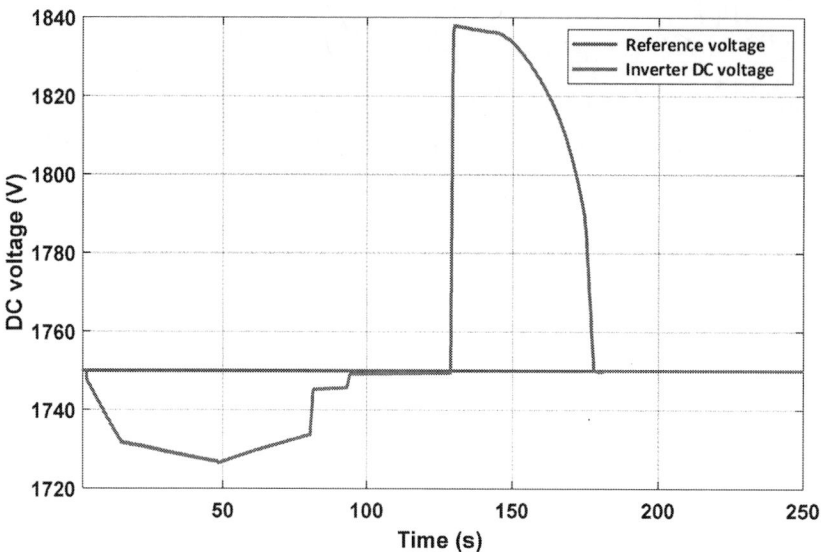

Fig. 7: Inverter DC voltage with a droop control method

Fig. 8 shows that the balance between the powers before and after the DC bus keeps a DC bus voltage stable with its reference value '1750V'. Without braking mode, the input error of the 'DC regulation loop' should be zero to prevent the passage of energy from the main AC network to the DC side. In braking mode, the DC voltage control loop makes it possible to transfer all of the braking energy available at the DC side to the AC electrical network to keep the DC voltage stable at its reference and therefore improves the recovery rate of braking energy.

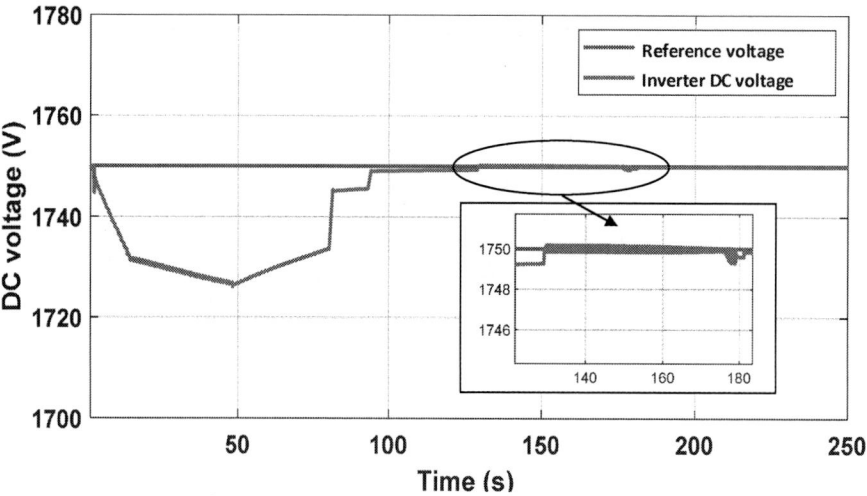

Fig. 8: Inverter DC voltage with a PI voltage control method

According to Fig. 9, it is shown that the maximum recovered power is well optimized thanks to the constant DC voltage controller comparing to the maximum recovered power using a droop controller. In addition, the DC voltage of "Massena" substation remains stable at its reference value "U_{dc-ref}" which demonstrates the feasibility of the proposed PI-based regulation control loop (Fig. 10). From Fig. 10, it is shown that the train voltage is very high with the droop control, which generates a fast limitation of the recovered power.

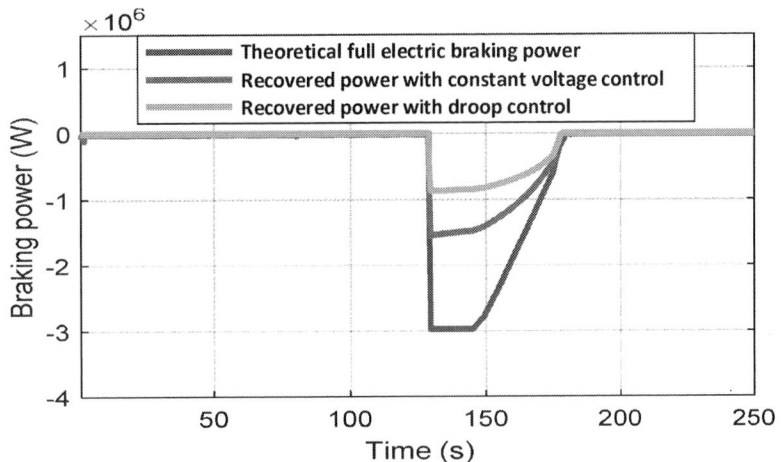

Fig. 9: Recovered power with a droop control and PI voltage control

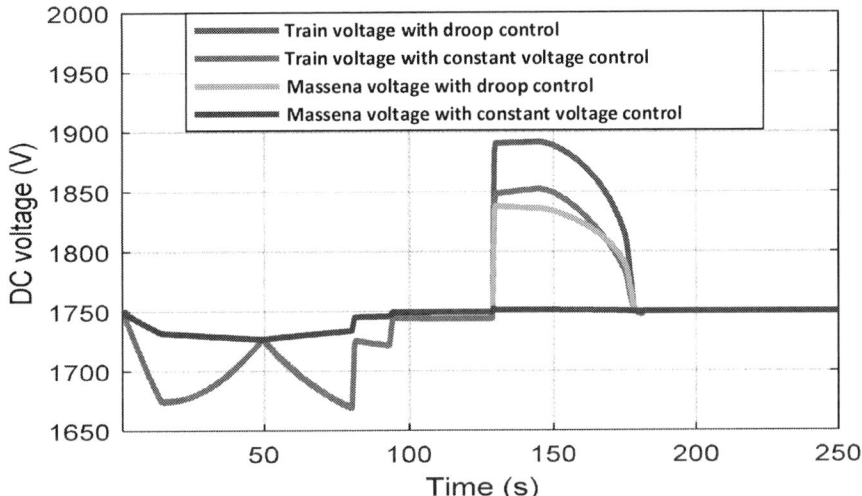

Fig. 10: DC voltage at the train and at Massena substation

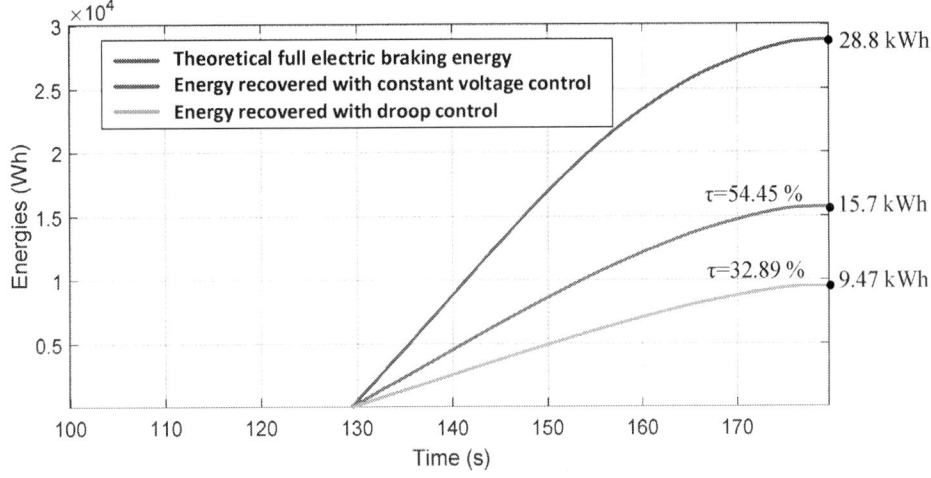

Fig. 11: Recovery rate with the two-inverter control strategies

Fig. 11 shows that the control with a constant voltage controller offers a significant improvement in energy efficiency with a braking energy recovery rate of 54.45 % of the theoretical full electric braking power against a rate of 32.89 % for a droop control. The difference between the recovered energy by the constant voltage control and the maximum braking energy is due to the limitation of the injection of braking energy imposed by the train electric brake controller. When the train voltage exceeds 1780 V, the train degrades the rate of power injection to the catenary to prevent the increase of the voltage in braking mode. However, with the constant voltage controller, the train voltage is well minimized, which gives more margin of energy recovery without degradation.

V.   Conclusion

In this paper, the presented solution to increase the energy efficiency of the railway system consists to control the installed power inverter in the reversible Massena substation. A control method based on DC voltage control loop is proposed and compared with a droop control method. It is highlighted that the proposed control scheme is preferred to maximize the recovered braking power, such as the recovery rate is optimized from 32.89 % with the droop control method to 54.45 % with the control with a PI DC voltage control. A droop control of the inverter with a modification of regenerative braking controller limitations of the train will be investigated in the future work.

References

[1]   M. Popescu, A. Bitoleanu, "A Review of the Energy Efficiency Improvement in DC Railway Systems", Energies, 2019, vol. 12, pp. 1-25.

[2]   K. Almaksour, Y. Krim, N. Kouassi et al., Comparison of dynamic models for a DC railway electrical network including an AC/DC bi-directional power station, Mathematics and Computers in Simulation (2020), https://doi.org/10.1016/j.matcom.2020.05.027.

[3]   M. Khodaparastan, A. A. Mohamed, W. Brandauer, "Recuperation of Regenerative Braking Energy in Electric Rail Transit Systems", IEEE Transactions on Intelligent Transportation Systems, 2019, vol. 20, n°8, pp. 2831 - 2847.

[4]   K. Holmes, "Smart grids and wayside energy storage". Passenger Transport, 2008, vol. 66, n°40.

[5]   P. Arboleya, I. El-Sayed, B. Mohamed, C. Mayet, "Modeling, Simulation and Analysis of On-Board Hybrid Energy Storage Systems for Railway Applications", Energies 2019, vol. 12, n°11, pp, 1-21.

[6]   X. Luo, J. Wang, M. Dooner, J. Clarke, "Overview of current development in electrical energy storage technologies and the application potential in power system operation", Applied Energy, 2015, vol. 137, pp. 511–536.

[7]   A. Rupp, H. Baier, P. Mertiny, and M. Secanell, "Analysis of a flywheel energy storage system for light rail transit", Energy, 2016, vol. 107, pp. 625–638.

[8]   M. Khodaparastan and A. Mohamed, "supercapacitors for electric rail transit system", in 6th International Conference on Renewable Energy Research and Application, 2017, vol. 5, pp. 1–6.

[9]   D. Fournier, D. Mulard, D. Fournier, D. Mulard, and A. G. Heuristic, "A greedy heuristic for optimizing metro regenerative energy usage" in Proceedings of the second international conference on railway technology: research, development and maintenance, 2015.

[10] K. Almaksour, H. Caron, N. Kouassi, Tony Letrouvé, Nicolas Navarro, Christophe Saudemont, Benoit Robyns, "Mutual impact of train regenerative braking and inverter based reversible DC railway substation", EPE'19 ECCE Europe conference, Genova, Italy, 2019.

[11] Y. Krim, D. Abbes, S. Krim, M. F. Mimouni, "Intelligent droop control and power management of active generator for ancillary services under grid instability using fuzzy logic technology", Control Engineering Practice, 2018, vol. 81, pp. 215-230.

[12] Z. Cabrane, M. Ouassaid, M. Maaroufi, "Battery and supercapacitor for photovoltaic energy storage: a fuzzy logic management", IET Renewable Power Generation, 2017, vol. 11, n°.8, pp. 1157-1165.

[13] Y. Krim, D. Abbes, S. Krim, M. F. Mimouni, "Classical vector, first-order sliding mode and high-order sliding-mode control for a grid-connected variable speed wind energy conversion system: a comparative study", Wind Engineering, 2017, vol. 42, n°1, pp. 16–37.

Junction Temperature Control Strategy for Lifetime Extension of Power Semiconductor Devices

Johannes Ruthardt, Hendrik Schulte, Philipp Ziegler, Manuel Fischer, Maximilian Nitzsche,
Jörg Roth-Stielow
INSTITUTE FOR POWER ELECTRONICS AND ELECTRICAL DRIVES
University of Stuttgart
Pfaffenwaldring 47
70569 Stuttgart, Germany
Tel.: +49 / (711) – 685.67387
E-Mail: johannes.ruthardt@ilea.uni-stuttgart.de
URL: http://www.ilea.uni-stuttgart.de

Keywords

«Reliability», «Power Semiconductor Devices», «Converter Control», «Thermal Stress», «Control Methods for Electrical Systems»

Abstract

The lifetime of power semiconductor devices mainly depends on their thermal stress. In particular, temperature swings cause damage due to different coefficients of thermal expansion of the different materials, which are used in power semiconductor devices. These temperature swings occur when the environmental temperature or the load conditions and with them the power losses change. Junction temperature controllers are able to extend the expected lifetime by reducing the occurring temperature swings. This paper proposes two control strategies to calculate a suitable set value for junction temperature controllers, which leads to a smoother temperature course with fewer swings. Both strategies do not affect the normal operation by for example limiting the output parameters. They affect the efficiency of the power electronic circuit to influence the power losses and thermal conditions.

Introduction

Reliability and lifetime are some of the important issues in developing power electronics. Especially the lifetime of power semiconductor devices becomes more important due to higher integration, which leads to more thermal stress. According to common lifetime models and studies, junction temperature swings have a huge impact on the lifetime of power semiconductor devices [1, 2]. These temperature swings occur for instance when the load condition changes due to different power losses, which are generated in the power semiconductor devices. Junction temperature control systems are developed which reduce the amplitude of the occurring temperature swings leading to an extended lifetime. These controllers require a possibility to affect and to measure the junction temperature. Besides that, a set value generator is required, which calculates a suitable junction temperature set value course with reduced swings.

In [3] a variable cooling system is used to affect the semiconductor device's junction temperature. Other approaches use the power losses of the power semiconductor devices, which have an impact on the junction temperature. The power losses can be affected by changing the modulation technique [4, 5], varying the switching frequency [6–10], applying a reactive current [11–13] or changing the switching and conduction characteristics of the power semiconductor devices [14–17]. The real-time junction temperature, which is fed back to the temperature controller, can be estimated with a thermal model in combination with a power loss model [13, 14, 18], measured by temperature sensors or measured via temperature sensitive parameters [19].

Many approaches for the set value generation were proposed. In [7] the occurring junction temperature course is calculated by a power loss and thermal model. The junction temperature controller limits the change rate of the temperature course. Therefore, the occurring change rate has to be determined. This

EPE'20 ECCE Europe

Assigned jointly to the European Power Electronics and Drives Association & the Institute of Electrical and Electronics Engineers (IEEE)

method is not robust if the junction temperature is measured instead of calculated by power loss and thermal models due to noise in the measured signal. The noise is amplified by the change rate calculation and can lead to instability of the control system. Besides, very accurate power loss and thermal models are required for this method. In [8, 20–22] the set value is also calculated by accurate power loss and thermal models. Two real-time junction temperatures are calculated – one temperature which would occur if the junction temperature controller was inactive and one which would occur if the junction temperature controller was active. Based on this data, a suitable set value is calculated. The drawbacks of this method are high complexity and the need for accurate power loss and thermal models.

In this work, two strategies are proposed, which are suitable for the use with noisy measured junction temperatures because no derivation of the measured value is required. In addition, the proposed strategies do not need temperature models and only a simple power loss model. Both strategies are evaluated with simulations and the results are compared. In a simulation, a junction temperature control system is built up for a two-level three-phase voltage-source inverter equipped with IGBTs. The switching frequency is used as the correction variable of the control system due to simple realization and few limitations. With the use of a variable switching frequency in an application, the whole frequency range has to be considered in the EMI-filter design, which is not considered in this paper. The proposed control strategies can also be used to calculate junction temperature set values for control systems using any other correction variables, which are mentioned before.

Junction Temperature Controller

The junction temperature control system consists of a PI-controller, a power loss model, a set value generator unit and the control path, see Fig. 1. The PI-controller calculates the power losses P_{loss}, which are necessary to adjust the measured junction temperature T_j to its set value $T_{j,\text{set}}$. The calculated power losses and the load conditions (DC-link voltage $V_{\text{DC,link}}$, modulation index m, phase shift φ between load current and the inverter's output voltage, amplitude of the load current $\hat{\imath}_{\text{load}}$) are fed into the power loss model to calculate the correction variable, which is the switching frequency f_{PWM} in this work. The switching frequency and the load conditions result in power losses in the inverter's IGBTs. This leads to a certain junction temperature, which is measured and fed back to the PI-controller. The set value generator calculates a suitable junction temperature set value, which reduces occurring temperature swings to extend the expected lifetime of the IGBTs.

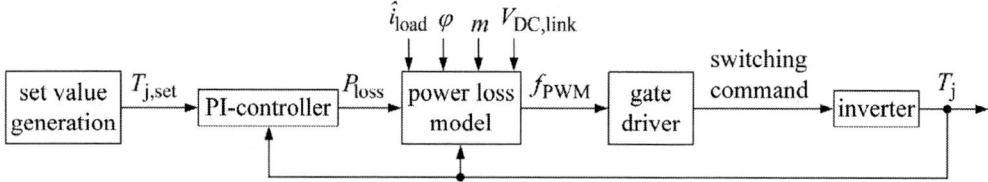

Fig. 1: Junction temperature control system

Control Path and PI-Controller

The control path consists of thermal resistances and capacitances between the semiconductor chips, in which the power losses occur, and the heatsink. An IGBT module (FF300R12ME4 from Infineon) is mounted on a heat sink to determine these elements. The IGBT, which junction temperature is controlled, is heated up by applying a high load current. Then, the load current is set to zero and the junction temperature is recorded with an infrared camera while the chip cools down. Therefore, opened and blackened modules are used which enables the use of an infrared camera. The cooling curve represents the step response of the control path. It is fitted with a second-order Foster model, which is accurate enough to tune the PI-controller [5, 12]:

$$g(t) = r_1 \cdot \left(1 - e^{-\frac{t}{\tau_1}}\right) + r_2 \cdot \left(1 - e^{-\frac{t}{\tau_2}}\right) \tag{1}$$

The parameters are fitted to $r_1 = 62.96$ mK/W, $\tau_1 = 39.12$ ms, $r_2 = 52.34$ mK/W and $\tau_2 = 1.037$ s. The thermal step response is differentiated with respect to time t and transferred into the Laplace domain to get the transfer function of the thermal system:

$$h(t) = \frac{d}{dt} g(t) \xrightarrow{\mathcal{L}} H(s) = \frac{T_j}{P_{loss}} = \frac{r_1}{1 + \tau_1 s} + \frac{r_2}{1 + \tau_2 s} \tag{2}$$

Fig. 2 shows the block diagram of the control system representing the transfer function of the control path and the PI-controller. The heat sink temperature T_h is a disturbance variable in the control system, which can be considered as constant due to a very slow change rate compared to the junction temperature's change rate. Another disturbance variable is the thermal cross-coupling between different semiconductor chips. The use of a PI-controller enables the elimination of the effect of disturbance variables in steady state due to the integral part.

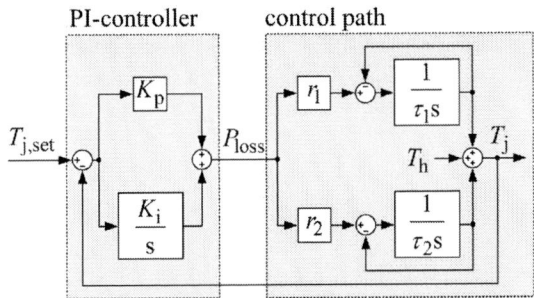

Fig. 2: Block diagram of the PI-controller and the control path

A MATLAB Simulink simulation is used to tune the PI-controller. The control parameters K_p and K_i are determined with the Ziegler-Nichols method [23, 24]. The result is $K_p = 454.5$ W/K and $K_i = 54945$ W/Ks. These parameters lead to a settling time of 17.4 ms and a bandwidth of 849.8 Hz. This relatively low bandwidth of the control system filters the noise of the measured junction temperature, as the noise's frequency is in the range of the switching frequency. That means a noisy temperature feedback does not affect the controller.

Power Loss Model

The power loss model is used to calculate the switching frequency, which is necessary to generate the power losses calculated by the PI-controller. Therefore, either only switching losses can be considered or both, switching and conduction losses. If only switching losses are considered, the conduction losses act as a disturbance variable in the control system. The main advantage of that is that the modulation index m and the phase shift φ are not required in the power loss model because they do not influence the switching losses. Inaccurate power loss models lead to a wrong calculated switching frequency, which also affects the control system as a disturbance variable. Due to manufacturing variations of the semiconductor devices, a power loss and a thermal model for every single device have to be determined and implemented. To keep the effort and calculations simple, only one power loss model for all devices is used. Besides, the junction temperature dependency of the losses is neglected, which makes the models also simpler but less accurate. In addition, only the switching losses are considered. All this leads to an error in the switching frequency calculation. The PI-controller can counteract the effect of all disturbance variables such as inaccurate and simple power loss models, which do not consider conduction losses. Simulation results show a better dynamic behavior when conduction losses are not considered for the switching frequency calculation using less accurate power loss models.

The basis for the power loss model is the power losses of the power semiconductor devices within one switching period. As the conduction losses are neglected to keep the calculation simple and easy-to-implement, these losses consist only of the turn-on and turn-off losses. Those are determined with the double pulse test at different load currents [15, 16]. The DC-link voltage is kept constant in this paper. The switching losses at different voltages have to be considered for applications with variable DC-link voltage. The experimental results of the switching losses are used to feed the power loss equations, which calculate the switching frequency required to adjust the demanded lower losses at certain operation conditions for the IGBTs and diodes.

$$f_{PWM} = \frac{P_{loss,IGBT}}{\left(\dfrac{m_{on} + m_{off}}{\pi}\right) \cdot \hat{\imath}_{load} + \left(\dfrac{E_{0,on} + E_{0,off}}{2}\right)} \tag{3}$$

$$f_{PWM} = \frac{P_{loss,diode}}{\left(\dfrac{m_{diode}}{\pi}\right) \cdot \hat{\imath}_{load} + \left(\dfrac{E_{0,diode}}{2}\right)} \tag{4}$$

The equations contain a linear relationship between the switching loss energy and the load current with the slope m and the offset E_0. Both equations are based on the power loss equations to calculate the mean switching losses within one period of the output current of a two-level three-phase inverter using pulse-width-modulation technique [8, 25]. Depending on whether the IGBT's or the diode's junction temperature is controlled, the corresponding equation has to be used. In this work, the proposed control strategies are explained and tested controlling the IGBT's junction temperature.

Control Strategy

The set value generator has to calculate a suitable junction temperature set value, which reduces the occurring temperature swings effectively. Therefore, a constant set value is not suitable. One reason is the limitation of the correction variable, which is the switching frequency in this work. Here, it is limited to 10 kHz $\leq f_{PWM} \leq$ 50 kHz as an example to show the control strategies. The limitation is in a frequency range, which is commonly used in many applications. It is a trade-off between e.g. efficiency, output current ripple, electro-magnetic interference, acoustic noise and volume of passive components. Depending on the application and the correction variable, it can be limited even more [15, 16, 20–22]. If the correction variable is in saturation, the control system will not be able to adjust the junction temperature to its set value anymore, which means that occurring junction temperature swings are not reduced or the reduction is minor. In this case, the set value has to be changed. In addition, it is not suitable to set a very high junction temperature for a long time during low load conditions due to a significantly decreasing efficiency of the whole system. In order to reduce junction temperature swings, the set value decreases with a predefined slope. The steeper the slope, the bigger the occurring junction temperature swings. However, this leads to more efficiency of the whole system due to lower mean junction temperature, which means less power losses occur. With a flatter slope, the junction temperature swings are more reduced, but the system is less efficient. Due to a continuously decreasing set value, it has to be reset to the measured junction temperature occasionally. Therefore, two trigger strategies are developed and tested by simulations. Both strategies decrease occurring junction temperature swings caused by changing load conditions. Occurring swings within a period of the load current of an inverter are not considered. These junction temperature swings are small and can be neglected when the load current's frequency is high [26].

Deviation of the Junction Temperature as a Trigger

The first strategy uses the deviation between the measured junction temperature and its set value as a trigger to reset the set value. When the deviation reaches a certain value, the reset is triggered, because it is assumed that the control system is in saturation. That means, it is a static deviation and not only dynamic for a short time. Fig. 3 shows the principle of this method. The predefined slope is set to $dT_{j,set}/dt = -0.2$ K/s. The occurring junction temperature T_j, its set value $T_{j,set}$ and a reference junction temperature $T_{j,ref}$, which would occur with inactive junction temperature control and a fixed switching frequency of $f_{PWM} = 10$ kHz, are depicted in the top diagram. All temperatures are related to the heat sink temperature, which is fixed to $T_h = 30°C$ in this simulation. The middle diagram shows the amplitude of the load current and the bottom diagram shows the adjusted switching frequency when the temperature control is active. The main advantage of this strategy is that no other measured parameter is required than the junction temperature, which has to be measured for the controller anyway. A drawback is that the trigger is not always set at the optimum point of time, see at $t = 2$ s for instance. A better point of time would be at $t = 3$ s, when the sign of the temperature change rate changes. By decreasing the allowed deviation this effect can be minimized. If the allowed deviation is too small, junction temperature measurement errors or a noisy measurement signal in a real application can lead to false trigger, which has to be prevented. Experiments with a real-time junction temperature measurement in combination with signal filters show that the amplitude of the noise is usually less than 2 K if a suitable measurement method is used [27, 28]. A dynamic deviation caused by a fast change of the load current or by one of the disturbance variables can also lead to a false trigger. Therefore, a trade-off between these effects has to be found. In this work, the allowed deviation is set to $\Delta T_j = 5$ K. This is a good compromise, which leads to a significant reduction of the occurring junction temperature swings and which does not cause any false triggers in all simulations, which were done in this work. The allowed deviation depends on the quality of the measured junction temperature signal and the dynamic of the controller. It has to be optimized for each application and setup.

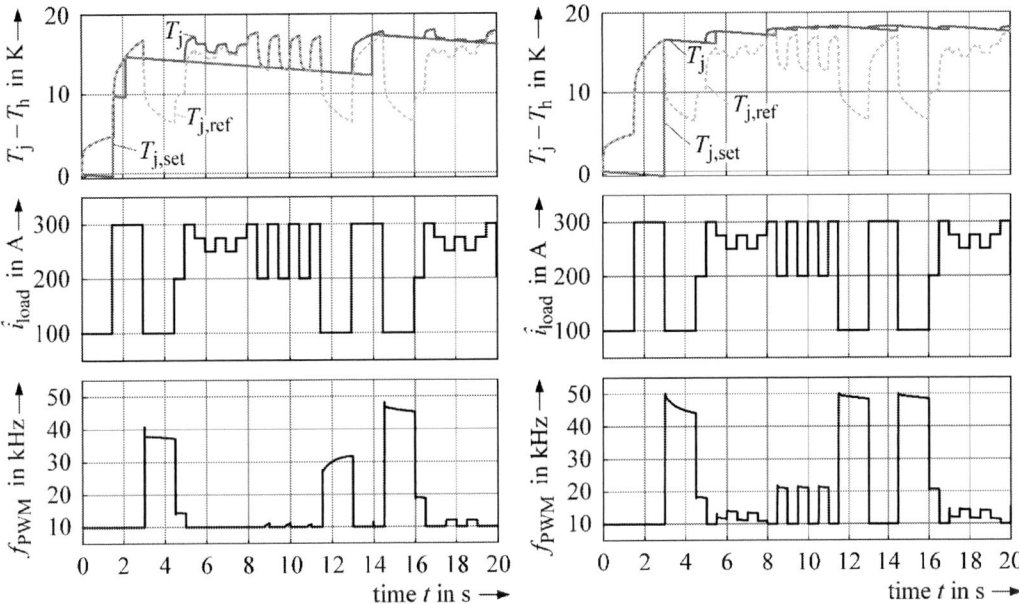

Fig. 3: Control strategy with junction temperature   Fig. 4: Control strategy with load current
trigger   amplitude trigger

Amplitude of the Load Current as a Trigger

This strategy uses the load current conditions as a trigger. It is assumed that the junction temperature increases when the amplitude of the load current increases and vice versa. This leads to the following trigger conditions: The trigger is set either when the amplitude of the load current increases and the switching frequency is in the upper limitation or when the amplitude of the load current decreases and the switching frequency is in lower limitation. Fig. 4 shows how this strategy works. The conditions and the load profile are the same as in Fig. 3. The top diagram shows that the trigger is mostly set at the optimal point of time in order to reduce occurring swings. This results in a significant reduction of junction temperature swings. A side effect is the higher mean junction temperature due to more overall power losses. A drawback of this method is the need of the amplitude of the load current. In many applications, the load current is controlled. In these cases, the load current set value can be used to extract the trigger signal, which is less noisy than the measured current value. As the response time of the electrical system is much smaller than of the thermal system, it is assumed that the current controller is always in steady state from the junction temperature control system's point of view. Another drawback of this strategy is that the reset is not triggered when the environmental conditions change such as the heat sink temperature. The advantage is that this strategy is very robust against false trigger when the load current set value is used to extract the trigger signal. In addition, the reset point is usually better than when using the other strategy which leads to a better reduction of the temperature swings.

Evaluation and Comparison

The load cycle depicted in Fig. 3 and Fig. 4 between $0 \leq t \leq 11.5$ s is repeated 87 times resulting in a 1000 s long load cycle to evaluate both strategies and to compare it with the operation with inactive temperature control. That means all simulations have the same load cycle in common, which makes a comparison possible. The occurring junction temperature swings are counted for both strategies and an inactive junction temperature control. Therefore, the rainflow algorithm is used, which considers the magnitude ΔT_j of the swings and the mean value within each swing $T_{j,m}$ [29]. The results are depicted in Fig. 5, where different swings are merged into groups of 3 K each for a better survey.

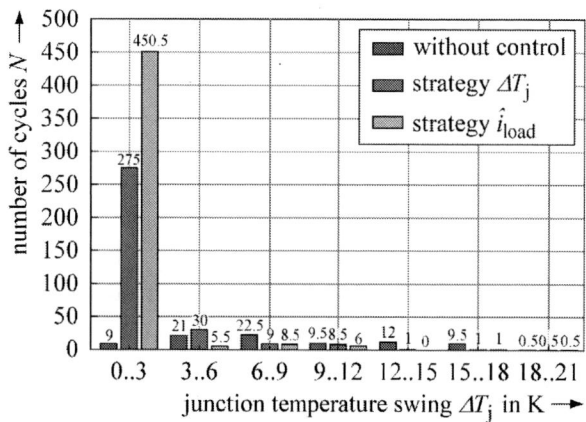

Fig. 5: Occurring temperature swings

With active junction temperature control fewer large temperature swings occur while the number of small swings increases. As large temperature swings have a much bigger impact on the lifetime than smaller ones the lifetime is increased according to common lifetime models. To prove this, the lifetime model from the LESIT study [1, 2] is used. This lifetime model calculates the number of cycles to failure N_f for a particular junction temperature swing ΔT_j and the corresponding mean temperature $T_{j,m}$:

$$N_f = A \cdot \Delta T_j^{\alpha} \cdot e^{\frac{E_a}{k_B \cdot T_{j,m}}}$$
(5)

The statistical lifetime parameters $A = 302500$ K$^{-\alpha}$ and $\alpha = -5.039$ are used from the LESIT study [1]. The activation energy is $E_a = 9.98 \cdot 10^{-20}$ J and the Boltzmann constant is $k_B = 1.38 \cdot 10^{-23}$ J/K. The damage caused by each temperature swing Q_v is calculated by dividing the number of the occurred swings N_v by the number of cycles to failure N_{fv}. As the damage caused by different temperature swings are independent of each other, the overall damage Q is calculated by summing up all damages [12]:

$$Q = \sum_v Q_v = \sum_v \frac{N_v}{N_{fv}}$$
(6)

The increase of the lifetime l with active junction temperature control is calculated by the ratio between the damage Q_{nc} with inactive control and the damage Q_c with active control:

$$l = \frac{Q_{nc}}{Q_c}$$
(7)

The results are shown in Table I. It is seen that both strategies increase the expected lifetime of the IGBTs significantly. The strategy, which uses the junction temperature deviation as a trigger, can't reduce all temperature swings due to not optimal reset of the temperature set value. Nevertheless, the control system with that strategy eliminates most of the large swings, which damage the IGBTs the most [2]. The strategy, which triggers on the change of the load conditions, is able to eliminate almost all large swings and increases the expected lifetime a lot.

Table I: Results

Strategy	Not controlled	ΔT_{j}	$\hat{\imath}_{\mathrm{load}}$
Lifetime l	100 %	433 %	567 %
Power Losses p_{loss}	100 %	121 %	131 %

While manipulating the switching frequency to affect the junction temperature, the switching losses of the simulated inverter is influenced. The relative increase of the power losses p_{loss} with active control is calculated with the ratio between the power loss energy of the whole test cycle with active control and inactive control:

$$p_{loss} = \frac{\int P_{\mathrm{loss,c}}(t)\,\mathrm{d}t}{\int P_{\mathrm{loss,nc}}(t)\,\mathrm{d}t} \qquad (8)$$

As the mean junction temperature is higher with active control compared to an inactive control, there are more power losses, see Table I. Using the strategy with the load current trigger results in more losses due to a better reduction of junction temperature swings, which leads to higher mean temperature. Table I shows exemplary results, which are generated in this work with a simulation model to show how the control strategies work and to evaluate them. However, the increase of the lifetime and power losses significantly depend on the used power electronic devices, the lifetime parameters, the limitation of the correction variable of the control system and the predefined slope, which is adjusted by the set value generator.

Conclusion

The expected lifetime of power semiconductor devices can be extended by a junction temperature control system. These control systems reduce the occurring temperature swings, which decreases damage. Therefore, the junction temperature controllers require a set value to adjust the junction temperature suitably. This paper presents two methods to generate a set value without the need for thermal models and accurate power loss models. Thus, these two methods are robust against parameter variations of the semiconductor devices and the cooling system. Simulations evaluate both methods. Therefore, a load cycle to a two-level three-phase voltage-source inverter equipped with IGBTs is simulated. The occurring junction temperature swings using both methods and with inactive temperature control are counted and the damage is calculated with the help of a lifetime model. The increased lifetime and increased power losses are discussed.

References

[1] J. Lutz, H. Schlangenotto, U. Scheuermann and R. de Doncker, *Semiconductor Power Devices: Physics, Characteristics, Reliability*. Berlin, Heidelberg: Springer-Verlag Berlin Heidelberg, 2011.

[2] M. Held, P. Jacob, G. Nicoletti, P. Scacco and M.-H. Poech, "Fast power cycling test of IGBT modules in traction application", in *International Conference on Power Electronics and Drive Systems*, Singapore, 1997, pp. 425–430.

[3] Y. Yerasimou, V. Pickert, B. Ji and X. Song, "Liquid Metal Magnetohydrodynamic Pump for Junction Temperature Control of Power Modules",, *IEEE Trans. Power Electron.*, 2018.

[4] J. Ruthardt, J. Wölfle, M. Zehelein and J. Roth-Stielow, "A New Modulation Technique to Control the Switching Losses for Single Phase Three-Level Active-Neutral-Point-Clamped-Inverters", in *International Exhibition and Conference for Power Electronics, Intelligent Motion, Renewable Energy and Energy Management (PCIM)*, Nürnberg, Germany, 2017.

[5] J. Wolfle, M. Nitzsche, J. Weimer, M. Stempfle and J. Roth-Stielow, "Temperature control system using a hybrid discontinuous modulation technique to improve the lifetime of IGBT power modules", in *18th European Conference on Power Electronics and Applications (EPE)*, Karlsruhe, Germany, 2016, pp. 1–10.

[6] M. Weckert and J. Roth-Stielow, "Lifetime as a control variable in power electronic systems", in *Emobility - electrical power train: International VDE Congress*, Leipzig, Germany, 2010.

[7] M. Weckert and J. Roth-Stielow, "Chances and limits of a thermal control for a three-phase voltage source inverter in traction applications using permanent magnet synchronous or induction machines", in *14th European Conference on Power Electronics and Applications (EPE)*, Birmingham, United Kingdom, 2011.

[8] J. Wolfle, J. Roth-Stielow, O. Koller and B. Bertsche, "Control Method to Increase the Reliability of IGBT Power Modules Validated on a Three Phase Inverter", in *IEEE Vehicle Power and Propulsion Conference (VPPC)*, Montreal, QC, Canada, 2015, pp. 1–6.

[9] J. Falck, M. Andresen and M. Liserre, "Active thermal control of IGBT power electronic converters", in *41st Annual Conference of the IEEE Industrial Electronics Society (IECON)*, Yokohama, Japan, 2015, pp. 1–6.

[10] D. A. Murdock, J.E.R. Torres, J. J. Connors and R. D. Lorenz, "Active thermal control of power electronic modules",, *IEEE Trans. on Ind. Applicat.*, vol. 42, no. 2, pp. 552–558, 2006.

[11] K. Ma, M. Liserre and F. Blaabjerg, "Reactive power influence on the thermal cycling of multi-MW wind power inverter", in *27th Annual IEEE Applied Power Electronics Conference and Exposition (APEC)*, Orlando, FL, USA, 2012, pp. 262–269.

[12] J. Wolfle, T. Roser, M. Nitzsche, N. Troster, M. Stempfle *et al.*, "Model based temperature control system to increase the expected lifetime of IGBT power modules executed on a neutral point diode clamped three level inverter", in *19th European Conference on Power Electronics and Applications (EPE)*, Warsaw, Poland, 2017, P.1-P.9.

[13] J. Wolfle, O. Lehmann and J. Roth-Stielow, "A novel control method to improve the reliability of traction inverters for permanent magnet synchronous machines", in *11th International Conference on Power Electronics and Drive Systems (PEDS)*, Sydney, Australia, 2015, pp. 379–384.

[14] C. H. van der Broeck, L. A. Ruppert, R. D. Lorenz and R. W. de Doncker, "Active thermal cycle reduction of power modules via gate resistance manipulation", in *33th Annual IEEE Applied Power Electronics Conference and Exposition (APEC)*, San Antonio, TX, USA, 2018, pp. 3074–3082.

[15] J. Ruthardt, J. Wölfle, N. Tröster, M. Fischer and J. Roth-Stielow, "Dead Time as a Correction Variable for Junction Temperature Control", in *20th European Conference on Power Electronics and Applications (EPE)*, Riga, Latvia, 2018.

[16] J. Ruthardt, C. Hermann, J. Wölfle, M. Fischer and J. Roth-Stielow, "Gate Driver Circuit with a Variable Supply Voltage to Influence the Switching Losses", in *9th International Conference on Power Electronics, Machines and Drives (PEMD)*, Liverpool, United Kingdom, 2018.

[17] J. Ruthardt, M. Fischer, J. Wölfle, N. Tröster and J. Roth-Stielow, "Three-Level-Gate-Driver to Run Power Transistors in the Saturation Region for Junction Temperature Control", in *International Exhibition and Conference for Power Electronics, Intelligent Motion, Renewable Energy and Energy Management (PCIM)*, Nürnberg, Germany, 2018.

[18] J. Ruthardt, P. Ziegler, M. Fischer and J. Roth-Stielow, "Model Based Junction Temperature Control Using the Gate Driver Voltage as a Correction Variable", in *21th European Conference on Power Electronics and Applications (EPE)*, Genova, Italy, Sep. 2019 - Sep. 2019, P.1-P.8.

[19] N. Baker, M. Liserre, L. Dupont and Y. Avenas, "Junction temperature measurements via thermo-sensitive electrical parameters and their application to condition monitoring and active thermal control of power converters", in *39th Annual Conference of the IEEE Industrial Electronics Society (IECON)*, Vienna, Austria, 2013, pp. 942–948.

[20] J. Wölfle, M. Nitzsche, N. Tröster, M. Stempfle and J. Roth-Stielow, "Comparison of Three Model Based Junction Temperature Control Systems to Increase the Lifetime of IGBT-Power- Modules", in *International Exhibition and Conference for Power Electronics, Intelligent Motion, Renewable Energy and Energy Management (PCIM)*, Nürnberg, Germany, 2017.

[21] J. Wölfle, M. Nitzsche, N. Tröster, J. Ruthardt, M. Stempfle *et al.*, "Combination of two variables in a junction temperature control system to elongate the expected lifetime of IGBT-power-modules", in *12th International Conference on Power Electronics and Drive Systems (PEDS)*, Honolulu, HI, USA, 2017, pp. 41–47.

[22] J. Wölfle, M. Pitters, J. Ruthardt, J. Schuster, M. Stempfle *et al.*, "Comparison of Two Model based Temperature Control Systems Implemented on a Three Level T-Type Inverter", in *International Exhibition and Conference for Power Electronics, Intelligent Motion, Renewable Energy and Energy Management (PCIM)*, Nürnberg, Germany, 2018.

[23] J. G. Ziegler and N. B. Nichols, "Optimum Settings for Automatic Controllers",, *Journal of Dynamic Systems, Measurement, and Control*, vol. 115, no. 2B, pp. 220–222, 1993.

[24] J. Lunze, *Control Theory of Digitally Networked Dynamic Systems*. Dordrecht: Springer, 2013.

[25] A. Wintrich, U. Nicolai, W. Tursky and T. Reimann, *Application Manual Power Semiconductors*, 2nd ed. Ilmenau: ISLE, 2015.

[26] M. Weckert, "Neuartige Regelung eines dreiphasigen Pulswechselrichters zur Verlängerung der Lebensdauer der Leistungshalbleitermodule", Dissertation, Universität Stuttgart, 2014.

[27] J. Ruthardt, K. Muñoz Barón, P. Marx, K. Sharma, M. Nitzsche *et al.,* "Online Junction Temperature Measurement via Internal Gate Resistance Using the High Frequency Gate Signal Injection Method", in *International Exhibition and Conference for Power Electronics, Intelligent Motion, Renewable Energy and Energy Management (PCIM)*, Nürnberg, Germany, 2019.

[28] M. Denk and M.-M. Bakran, "Junction Temperature Measurement during Inverter Operation using a TJ-IGBT-Driver", in *International Exhibition and Conference for Power Electronics, Intelligent Motion, Renewable Energy and Energy Management (PCIM)*, Nürnberg, Germany, 2015.

[29] K. Mainka, M. Thoben and O. Schilling, "Lifetime calculation for power modules, application and theory of models and counting methods", in *14th European Conference on Power Electronics and Applications (EPE)*, Birmingham, United Kingdom, 2011.

High Dynamic Power Balancing for Dual Two-Level Inverters during High-Speed Machine Operation

Johannes Büdel, Johannes Teigelkötter, *Senior Member, IEEE*, Alexander Stock,
Member, IEEE, Christian Herkommer and Kai Kuhlmann
University of Applied Sciences Aschaffenburg
Würzburger Straße 45
Aschaffenburg, Germany
Phone: +49 (0) 6021-4206837
Email: johannes.buedel@th-ab.de
URL: http://www.th-ab.de

Keywords

<<Control of drive>>, <<Doubly fed induction motor>>, <<Direct torque and flux control>>, <<Modulation strategy>>, <<Robust control>>

Abstract

This paper aims to present optimized square-wave modulation strategies for the dual two-level inverter, fed by two separate and galvanically isolated energy sources, to balance and control the energy distribution of both inverter systems. Especially for high-speed operation, the methods keep the machine in the required state while decreasing the switching operations to a minimum. Method one aims to compensate voltage fluctuations of variable DC-link voltages, while the second method ensures a symmetrical power distribution for the two inverter systems. In addition, both methods offer the possibility of adapting the amplitude of the usually fixed fundamental load voltage during square-wave modulation.

1 Introduction

Especially for applications in the field of traction drives, electrical machines and drive inverters need to have a high power density, maximum machine utilization and low current and torque fluctuations. Moreover, a small and simple system design, low weight and standard components are often required. For a high power demand, conventional topologies like the two-level inverter reach their limits and do not have the ability to achieve the claimed goals. The demands for adequate voltage and current quality require a high switching frequency. This leads to a deterioration of the overall system efficiency. The two-level inverter is therefore unattractive for mobile applications, where the stored energy is very limited. The use of multilevel inverters could be a promising option to meet those high requirements. Where two-level inverters can switch between only two output potentials, multilevel inverters can provide more than two voltage levels at the output. Thus, smaller voltage steps can be created, providing the possibility of reducing the switching frequency and improving the voltage and current quality. Due to the smaller voltage steps, the power semiconductors can also be designed for lower blocking voltages, which leads to a further reduction of the inverter losses.

For applications in mobile high-performance drives, the topology of the dual two-level inverter could be a promising option, as it combines the high requirements for traction application. The dual two-level inverter consists of two two-level inverters connected through an open-end winding machine. Each inverter is fed by an independent energy source. Compared to other multilevel topologies, the dual two-level inverter can generate a load voltage that is higher than the DC-link voltages of the single inverters. Consequently, the power electronics can be designed for a lower voltage. The system can be equipped with different energy sources. Also, the topology enables the controlling of the energy flow and power distribution of the sources in both directions. Thus, an energy exchange can be accomplished across the machine, from one source to the other, without affecting the normal machine operation. In addition, the dual two-level inverter comes with increased failure safety. If one inverter fails, the other one can maintain the operation of the machine with reduced power. The ability to control the power distribution and energy flow represents a further degree of freedom, which must be considered for optimal operation of the system. Even small differences in the switching times or fluctuations in the DC-Link voltages can lead to unbalanced power distribution. Also, for different energy sources, the power distribution has to be controlled to ensure optimal operation. If not taken into account, this can lead to voltage and power fluctuations and therefore to decreased efficiency, and to the problem of over- or discharging the energy sources during normal operation. To prevent this and to ensure an efficient and safe operation of the overall system, this paper presents different modulation strategies for the high-speed region taking the possibility of energy-flow control into account. The methods aim to compensate voltage fluctuations of variable DC-link voltages and ensure an implicit symmetrical power distribution for the two inverter systems respectively the energy sources for the entire operation range.

2 Dual Two-Level Inverter

The dual two-level inverter topology is shown in Figure 1. The setup consists of two inverter systems fed by two separate and galvanically isolated energy sources, energizing an electrical machine with open-ended windings (in this case, a permanent magnet synchronous machine or PMSM). The proposed topology is derived from the dual two-level inverter topology exemplarily described in [1–4]. The two isolated inverter-feeding energy sources, can differ in terms of their technology and characteristics. With special control algorithms, these different properties can be optimally used for enhancing the overall system behaviour. For example, if an energy source with a very slow dynamic behaviour is combined with a highly dynamic source, it is necessary to consider the dissimilar behaviour of the sources, especially in the case of operation point changes. If that is not considered, situations can occur, when the required machine torque and speed cannot be preserved, which could lead to a total system failure. Figure 1 shows the basic schematic with the corresponding voltage and current conditions. For the reference

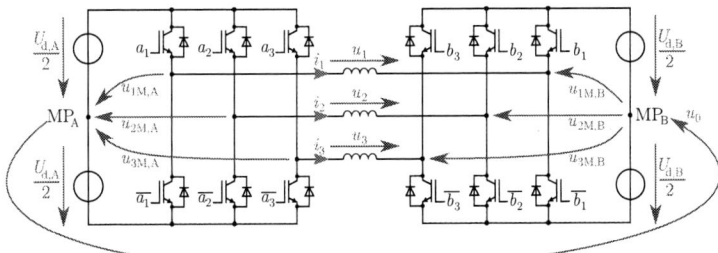

Figure 1: Basic dual two-level inverter schematic

point of the inverter voltages, the midpoint potentials of the DC-links MP_A and MP_B are chosen, with which, the inverter voltages can be calculated. The phase voltages u_1, u_2 and u_3 form a three-phase voltage system with a superimposed zero voltage u_0. Due to the complementary switching of the power semiconductors of each half-bridge, each inverter output voltage $u_{nM,A}$ and $u_{nM,B}$ has two output states ($n = 1, 2, 3$). Thus, the phase voltages u_n can have up to four different voltage levels. The number and size of the voltage levels depend on the ratio r of the DC-link voltage levels $U_{d,A}$ and $U_{d,B}$:

$$r = \frac{U_{d,A}}{U_{d,B}}, \text{with } U_{d,A} \geq U_{d,B} \tag{1}$$

For achieving legitimate multilevel behaviour, the adjacent voltage levels have to be equidistant. This can only be accomplished when $r = 1$ and $r = 2$. The corresponding voltage space vectors (see Figure 2a) reveal, that for $r = 1$, a three-level, and for $r = 2$, a four-level behaviour can be realized.

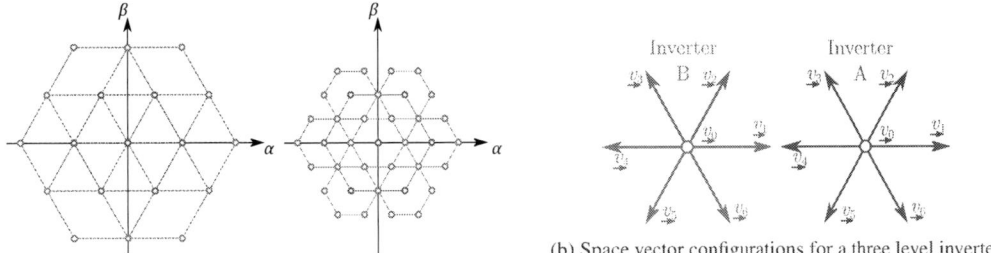

(a) Basic space vectors for $r = 1$ (left) and $r = 2$ (right)

(b) Space vector configurations for a three level inverter

Figure 2: Square-wave modulation for multilevel inverters

The superposition of the two inverter voltages also leads to redundant switching states [1]. This states can be optimally used for the energy flow control without affecting the machine torque and speed.

For an exact control of the energy flow and the power distribution of each inverter, the output power of the system has to be analysed. The space vector power is calculated as follows:

$$p = \underbrace{\frac{3}{2} \cdot \Re\left\{ \underset{\rightarrow}{u}_A \cdot \underset{\rightarrow}{i}^* \right\}}_{p_A} + \underbrace{\frac{3}{2} \Re\left\{ \underset{\rightarrow}{u}_B \cdot \underset{\rightarrow}{i}^* \right\}}_{p_B} \tag{2}$$

where $\underset{\rightarrow}{i}$ is the load current space vector and $\underset{\rightarrow}{u}_A$ and $\underset{\rightarrow}{u}_B$ are the inverter output voltage space vectors. Therefore, the active power of the inverters can be calculated:

$$p = \frac{3}{2} \cdot \left| \underset{\rightarrow}{i} \right| \cdot \left(\left| \underset{\rightarrow}{u}_A \right| \cdot \cos\left(\varphi_{u,A} - \varphi_i \right) + \left| \underset{\rightarrow}{u}_B \right| \cdot \cos\left(\varphi_{u,B} - \varphi_i \right) \right) \tag{3}$$

This shows that the dual two-level inverter output power is composed of both single-inverter powers. Because there is the same current space vector $\underset{\rightarrow}{i}$, the power distribution is dependent only on the output voltage space vectors $\underset{\rightarrow}{u}_A$ and u_B. Thus, the power distribution can be controlled by adapting both voltage space vectors by changing the corresponding amplitudes and phase angles. Furthermore, for symmetrical power distribution, the equality of the amplitudes and identical phase angels of both voltages is imperative.

3 Modulation Methods for High-Speed Operation with Power Balancing

This section presents the fundamental principle of the square-wave modulation for multilevel inverter systems. After that, the basic square-wave modulation characteristics for three-level inverters are explained. Next, two special square wave control methods for the dual two-level inverter with the objective of output power control and control of the power distribution are presented in detail. Both methods offer the possibility of adapting the amplitude of the fundamental load voltage during square-wave modulation.

3.1 High-Speed Machine Operation with Square-Wave Modulation

The so-called square-wave modulation only uses the active space vectors located on the outermost hexagon of the $\alpha - \beta$–trajectory (see Figure 3a). This space vectors are switched sequentially with fixed switching times synchronous to the fundamental frequency. This leads to a minimum of switching operations and therefore to lowest switching losses. As only the outermost space vectors are used, the maximum possible voltage amplitude is applied to the load. Therefore, this control method is optimally suited for application in the maximum-load region. Figure 3a displays the corresponding characteristic phase voltage shapes for square-wave modulation. Especially for a two-level inverter, the deviation from the ideal sinusoidal shape leads to a high current distortion and therefore to torque fluctuations. Multi-level inverters can reduce this influence. The additional space vectors lead to a great improvement in approaching the ideal sinusoidal voltage. As explained, for $r = 1$ and $r = 2$, the dual two-level inverter represents a three- or four-level inverter. Figure 3a shows the corresponding space vectors and voltage shapes for this combinations, as compared to a two-level inverter.

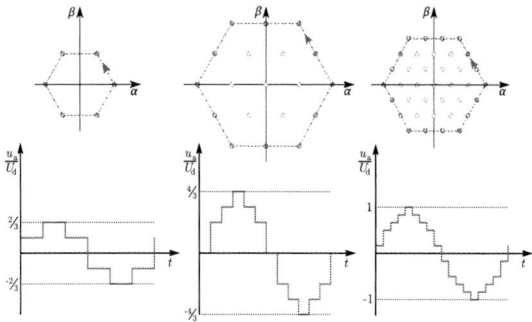
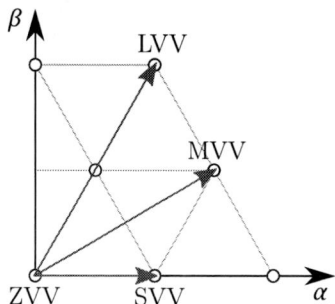

(a) Voltage space vectors and appropriate output phase voltages for a 2-, 3-, and 4-level inverter and different DC-link voltages

(b) Space vector configurations for a three level inverter

Figure 3: Square-wave modulation for multilevel inverters

The absolute lengths of the space vectors for $r = 1$ are shown in Figure 3b. While applying square-wave modulation, only the space vectors LVV and MVV are used to produce the trajectory. Thus, in comparison to a two-level inverter, the switched amplitudes of the space vectors are not constant and fluctuate between $\frac{2}{\sqrt{3}}U_d$ and $\frac{4}{3}U_d$. Therefore, this topology is able to produce a lower harmonic distortion. The modulation methods presented below refer to a configuration with $r = 1$. For $r = 2$, the functions have to be adapted accordingly.

EPE'20 ECCE Europe

Assigned jointly to the European Power Electronics and Drives Association & the Institute of Electrical and Electronics Engineers (IEEE)

3.2 Phase-Shifted Square-Wave Modulation

Because the dual two-level inverter is composed of two single inverters, both inverters can be operated with a two-level square wave modulation. According to the principle of superposition, the inverter output voltages are superimposed, resulting in the load voltage. Basically, both inverters can generate their switching sequence independent of each other, but the phase angles have to be linked. The resulting fundamental load space vector $\underrightarrow{u}_\text{f}$ is calculated by the sum of the two inverter output space vectors.

$$\underrightarrow{u}_\text{f} = \hat{u}_{\text{f,A}} \cdot e^{j\varphi_\text{A}} + \hat{u}_{\text{f,B}} \cdot e^{j\varphi_\text{B}} \tag{4}$$

By creating the Fourier series for the typical voltage waveform for square-wave modulation, the fundamental amplitude of both inverter voltages can be calculated. For a single two-level inverter, a constant amplitude of $\frac{2}{\pi}U_\text{d}$ is achieved. With $r = \frac{U_{\text{d,A}}}{U_{\text{d,B}}}$ and $\varphi_{\text{AB}} = \varphi_\text{B} - \varphi_\text{A}$, the equation can be solved to:

$$\hat{u}_\text{f} = \frac{2}{\pi} \cdot U_{\text{d,A}} \cdot \sqrt{\left(1 + \frac{1}{r} \cdot \cos\left(\varphi_{\text{AB}}\right)\right)^2 + \left(\frac{1}{r} \cdot \sin(\varphi_{\text{AB}})\right)^2} \tag{5}$$

With $r = 1$, the possibility of adapting the phase shift φ_{AB} enables a further degree of freedom to control the resulting fundamental voltage amplitude. Exemplarily, Figure 4a shows the relation of the single inverter voltages with the resulting load voltage for one phase. According to Figure 2b, the numbers 1-6 represent the corresponding applied inverter voltage space vectors. It can be seen that for identical switching sequences for both inverters, the resulting voltage amplitude is controlled only by changing the phase angle between both inverter-switching sequences. Using this control technique, the power distribution of both inverters is strongly dependant on the phase angle between both output voltage fundamentals. With (2), the contributed instantaneous power of each inverter can be expressed as:

$$p_\text{A} = \frac{2}{\pi} \cdot U_{\text{d,A}} \cdot \hat{i} \cdot \cos\left(\varphi_{\text{iA}}\right)$$
$$p_\text{B} = \frac{2}{\pi} \cdot U_{\text{d,B}} \cdot \hat{i} \cdot \cos\left(\varphi_{\text{iB}}\right) \tag{6}$$

Figure 4b visualizes the relationships in the complex plane. Only if the resulting fundamental load voltage space vector is in-phase with the current space vector ($\varphi_{\text{iA}} = \varphi_{\text{iB}}$), an equal power distribution can be achieved.

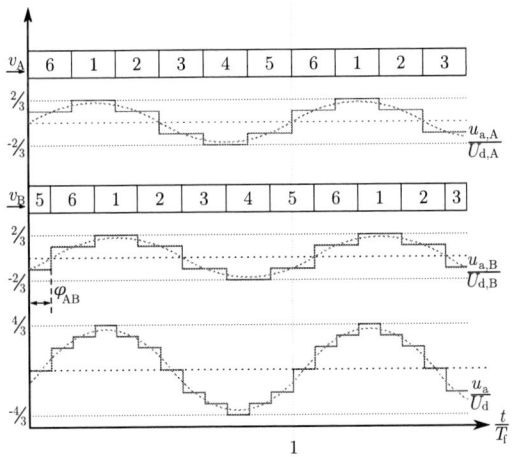

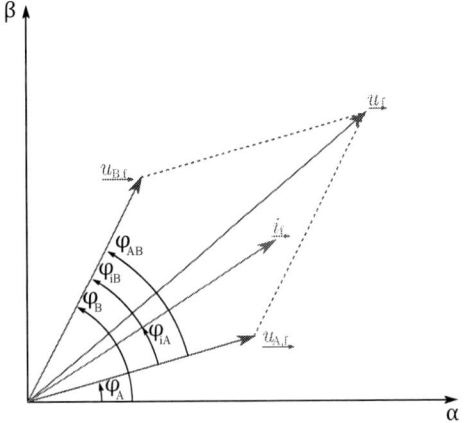

(a) Single inverter space vector sequences and voltages for phase shifted square-wave modulation

(b) Space vector constellation of the phase shifted square-wave modulation

Figure 4: Phase-Shifted Square-wave modulation for the dual two-level inverter

For $r = 1$, the amplitude of the load voltage space vector can continuously be adjusted in the range from 0 to $\frac{2}{\pi} \cdot (U_{\text{d,A}} + U_{\text{d,B}})$ by applying a phase offset between 0 and $180°$. Figure 5 furthermore displays the dependency of different DC-link voltage relations r for phase-shifted square-wave modulation. In principle, Equation 5 and Figure 5 show that the possibility of controlling the fundamental voltage amplitude by adapting the phase shift φ_{AB}, can be used to compensate voltage fluctuations for variable DC-link voltages. For a DC-link voltage change in one inverter, the output amplitude can be maintained by adapting the phase angle if the second inverter is not

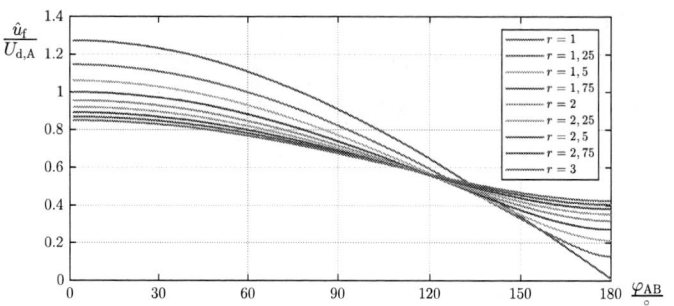

Figure 5: Control range of the resulting fundamental voltage amplitude for several values of r

operating at its limit. Further, the method can be used to control the power distribution within certain limits. As a result, different combined energy sources can be optimally used. As a symmetrical power distribution is given only for $\varphi_{AB} = 0$, this method can only be used where both inverters and energy sources tolerate an asymmetrical operation. For applications where a balanced power distribution is inevitable, the method has to be adapted. The following paragraph describes an approach for achieving this control objective.

3.3 Power Balance during Square-Wave Modulation

Under the requirement of equal DC-link voltage levels ($r = 1$), a symmetrical power distribution over a certain amount of the control range can be achieved by modifying the switching sequences for both inverters. As previously discussed, an equal phase angle between both inverter output voltages and the resulting load voltage is crucial for symmetrical power distribution. The maximum efficiency can be reached, when there is no phase shift between both inverter output voltage fundamentals. In order to maintain these criteria, the new switching sequences have to offer a degree of freedom to control the fundamental inverter output voltages without changing the phase angle between them.

In contrast to the previously presented control method, both inverters have unique switching sequences, which are constructed by periodically exchanging the roles of both inverters. Figure 6 shows an example, how such sequences are generated. Inverter B has to switch two consecutive space vectors, while Inverter A remains in its current state and does not change its output space vector (I to III). In the following states, the inverters change their roles. Now, Inverter A switches two consecutive space vectors while Inverter B remains in its current state (III to I). This way, there exist time segments, where one inverters instantaneous voltage space vector has a phase lead to the resulting space vector, while the other inverters output has a phase lag to the resulting space vector. Due to the role switching behaviour of the sequences, in the next time segment, the phase lead/lag relationships will be opposite. Therefore, an average symmetrical power distribution is achieved implicitly for equal DC-link voltage levels.

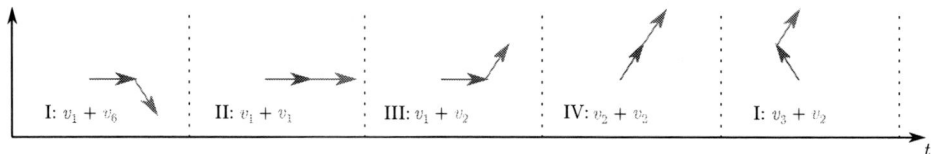

Figure 6: Construction of the alternating subsequent switching sequences

The complete switching sequence is shown in Figure 7a. As the resulting voltage waveform consists of large voltage vectors (LVV), as well as middle voltage vectors (MVV), the amplitude of the output fundamental voltage can be controlled using the relative on-times of these states. A large voltage vector is generated, when both inverters share the same state. In this paper, the control variable χ is introduced as the relative on-time for the large voltage vector.

By computing the Fourier series of the single inverter voltage waveforms, it can be seen, that both fundamental voltages are represented by the same Fourier coefficients. While χ does influence the amplitudes of the single inverter voltages in the same way, the phase angle between them is unaffected.

$$ \hat{u}_{f,A} = \frac{1}{\pi} \cdot U_{d,A} \cdot \left(\sin\left(\frac{\chi}{2}\right) + \sqrt{3} \cdot \cos\left(\frac{\chi}{2}\right) \right) \quad ; \quad \hat{u}_{f,B} = \frac{1}{\pi} \cdot U_{d,B} \cdot \left(\sin\left(\frac{\chi}{2}\right) + \sqrt{3} \cdot \cos\left(\frac{\chi}{2}\right) \right) \tag{7} $$

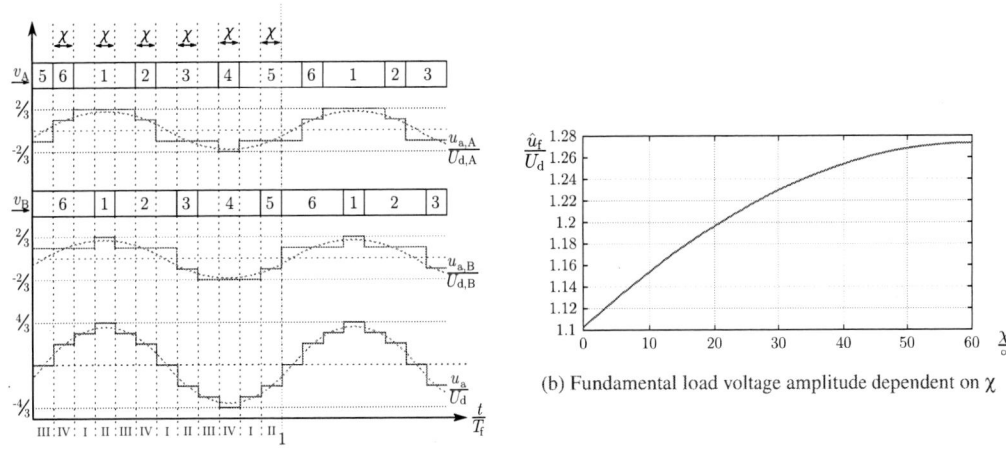

(a) Single inverter space vector sequences and voltages for power balanced square-wave modulation

(b) Fundamental load voltage amplitude dependent on χ

Figure 7: Phase-Shifted Square-wave modulation for the dual two-level inverter

With (7) and $U_{d,A} = U_{d,B} = U_d$, the resulting fundamental load voltage amplitude can be expressed as:

$$\hat{u}_f(\chi) = \frac{2 \cdot U_d}{\pi} \cdot \left(\sin\left(\frac{\chi}{2}\right) + \sqrt{3} \cdot \cos\left(\frac{\chi}{2}\right) \right) \tag{8}$$

As shown in Figure 7b, through variation of χ in the range of $[0°; 60°]$, the resulting fundamental voltage amplitude can be adjusted in the range of $[\frac{2 \cdot \sqrt{3} \cdot U_d}{\pi}; \frac{4 \cdot U_d}{\pi}]$. Assuming that the main part of the power is delivered by the fundamental components of the load voltage and current space vectors, it can be seen that this method provides a practical way to split the power evenly between both power sources.

A more precise description of the power distribution can be achieved, if the instantaneous power components of both inverters are considered. As previously described, the whole systems switching sequence consists of four repeating time intervals (I to IV, see Figure 6). For this explanation, the current is considered to be sinusoidal. In each interval, the corresponding voltage space vectors of both inverters are constant.

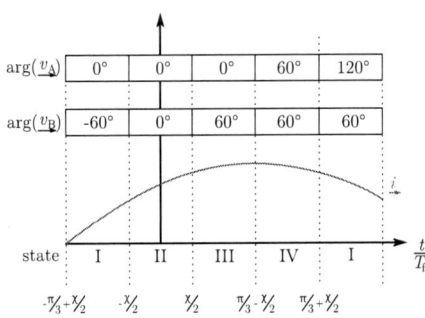

Figure 8: Mechanical and electrical quantities for a torque step with DSC

Therefore, the instantaneous inverter powers can individually be calculated for each state using (2):

$$p_A(t) = U_d \cdot \hat{i} \cdot \begin{cases} \cos(\omega t + \varphi) & -\pi/3 + \chi/2 \leq \omega t < \pi/3 - \chi/2 & \text{(states I to III)} \\ \cos(\omega t + \varphi - 2\pi/3) & \pi/3 - \chi/2 \leq \omega t < \pi/3 + \chi/2 & \text{(state IV)} \end{cases} \tag{9}$$

$$p_B(t) = U_d \cdot \hat{i} \cdot \begin{cases} \cos(\omega t + \varphi + \pi/3) & -\pi/3 + \chi/2 \leq \omega t < -\chi/2 & \text{(state I)} \\ \cos(\omega t + \varphi) & -\chi/2 \leq \omega t < \chi/2 & \text{(state II)} \\ \cos(\omega t + \varphi - \pi/3) & \chi/2 \leq \omega t < \pi/3 - \chi/2 & \text{(state III)} \\ \cos(\omega t + \varphi - 2\pi/3) & \pi/3 - \chi/2 \leq \omega t < \pi/3 + \chi/2 & \text{(state IV)} \end{cases} \tag{10}$$

With that, both inverters average power contributions for each time interval can be determined.

$$
\bar{p}_A = U_d \cdot \hat{i} \cdot
\begin{cases}
\dfrac{1}{\frac{\pi}{3}-\chi} \cdot \left(\ \sin\left(-\dfrac{\chi}{2}+\varphi\right) \quad -\sin\left(\ \dfrac{\chi}{2}+\varphi-\dfrac{\pi}{3}\right) \ \right) & \text{(state I)} \\[2ex]
\dfrac{1}{\chi} \cdot \left(\ \sin\left(\ \dfrac{\chi}{2}+\varphi\right) \quad -\sin\left(-\dfrac{\chi}{2}+\varphi\right) \ \right) & \text{(states II and IV)} \\[2ex]
\dfrac{1}{\frac{\pi}{3}-\chi} \cdot \left(\ \sin\left(-\dfrac{\chi}{2}+\varphi+\dfrac{\pi}{3}\right) \quad -\sin\left(\ \dfrac{\chi}{2}+\varphi\right) \ \right) & \text{(state III)}
\end{cases}
\tag{11}
$$

$$
\bar{p}_B = U_d \cdot \hat{i} \cdot
\begin{cases}
\dfrac{1}{\frac{\pi}{3}-\chi} \cdot \left(\ \sin\left(-\dfrac{\chi}{2}+\varphi+\dfrac{\pi}{3}\right) \quad -\sin\left(\ \dfrac{\chi}{2}+\varphi\right) \ \right) & \text{(state I)} \\[2ex]
\dfrac{1}{\chi} \cdot \left(\ \sin\left(\ \dfrac{\chi}{2}+\varphi\right) \quad -\sin\left(-\dfrac{\chi}{2}+\varphi\right) \ \right) & \text{(states II and IV)} \\[2ex]
\dfrac{1}{\frac{\pi}{3}-\chi} \cdot \left(\ \sin\left(-\dfrac{\chi}{2}+\varphi\right) \quad -\sin\left(\ \dfrac{\chi}{2}+\varphi-\dfrac{\pi}{3}\right) \ \right) & \text{(state III)}
\end{cases}
\tag{12}
$$

It can be seen, that in the states II and IV, the average inverter powers are equal, whereas in the states I and III, the inverter powers differ from each other. Because the average power $\bar{p}_A$ in the state I equals $\bar{p}_B$ in the state III, and vice versa, the temporary power fluctuations compensate each other after two subsequent odd-numbered states.

4 Simulation results

In this section, first simulation results implementing the power-balancing square-wave modulation are presented. In the first simulation, a static ohmic-inductive load is used to characterize the modulations behaviour without disturbances from external systems. In a second run, the modulator was implemented within a direct self-control strategy for high speed motor drives. Both simulations use a precise inverter hardware model in order to determine the power contributions from both inverters.

4.1 Static load

In a first approach, the power-balancing properties of the proposed method were simulated using a transistor-based model of the dual two-level inverter. Both inverters are interconnected by a three-phase symmetric ohmic-inductive load, which results in a similar configuration as shown in figure 1. The load is driven by the characteristic voltage waveform with a fundamental frequency of 500 Hz. Using the DC-Link currents and voltage levels, the instantaneous inverter powers can be estimated. In the simulation, the DC-Link voltages are assumed to be constant. In figure 9a, the instantaneous power distribution of the power-balancing square-wave modulation can be seen. Clearly visible is the periodicity of the previously described modulation states with a frequency of 1500 Hz. Furthermore, the alternating power sequences can be observed in the time intervals, when both inverters instantaneous powers differ from each another. Also, the equality of the average inverter powers can easily be proven by comparison of the normalized powers.

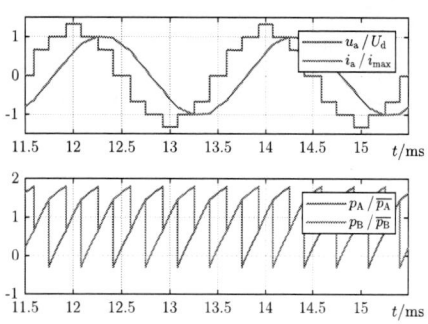

(a) Distribution of the instantaneous power in steady-state

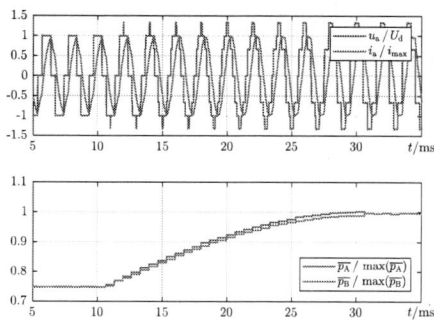

(b) Distribution of the average inverter powers during the variation of χ

Figure 9: Simulation of the power distribution at a static load

In a second experiment, the average power distribution was analysed during a continuous variation of the fundamentals voltage amplitude. Figure 9b shows the simulated results for that test setup. The control variable χ

continuously sweeps over the interval $[0; 60°]$ in the time interval $[10\,\text{ms}; 30\,\text{ms}]$. During that timespan, the modification of the voltage shape, which changes the fundamentals amplitude, can be observed. It is also noticeable, that minor fluctuations of both average inverter powers occur, when the fundamental amplitude changes rapidly. Both simulations show, that the proposed method is capable of supplying a three-phase load with a variable amplitude, high-frequency and loss-optimal voltage waveform, while maintaining the equal power balance over both inverters. Therefore, the modulation technique can be used to further increase the operational range of dual two-level inverter driven high power drives.

4.2 Application example : Direct self-control

As an application example, the power-balancing square-wave modulation can be used as an actor inside a direct self-control (DSC) algorithm for high power drives. The basic concept of a direct self-control, as well as the adaption for multilevel-inverters is well known and comprehensively described in [5] [6]. In short terms, the DSC algorithm provides a good approximation of the stator flux linkage trajectory and a highly dynamic torque control while maintaining a minimum of switching operations. The ideal path of the flux linkage space vector is described by a circle in the complex plane. Using the dual two-level inverter topology as a three-level inverter, a dodecagonal stator flux linkage trajectory can be achieved with a minimum of required switching operations. This leads to the characteristic output voltage waveform of a square wave modulation in steady-state. Furthermore, the previously

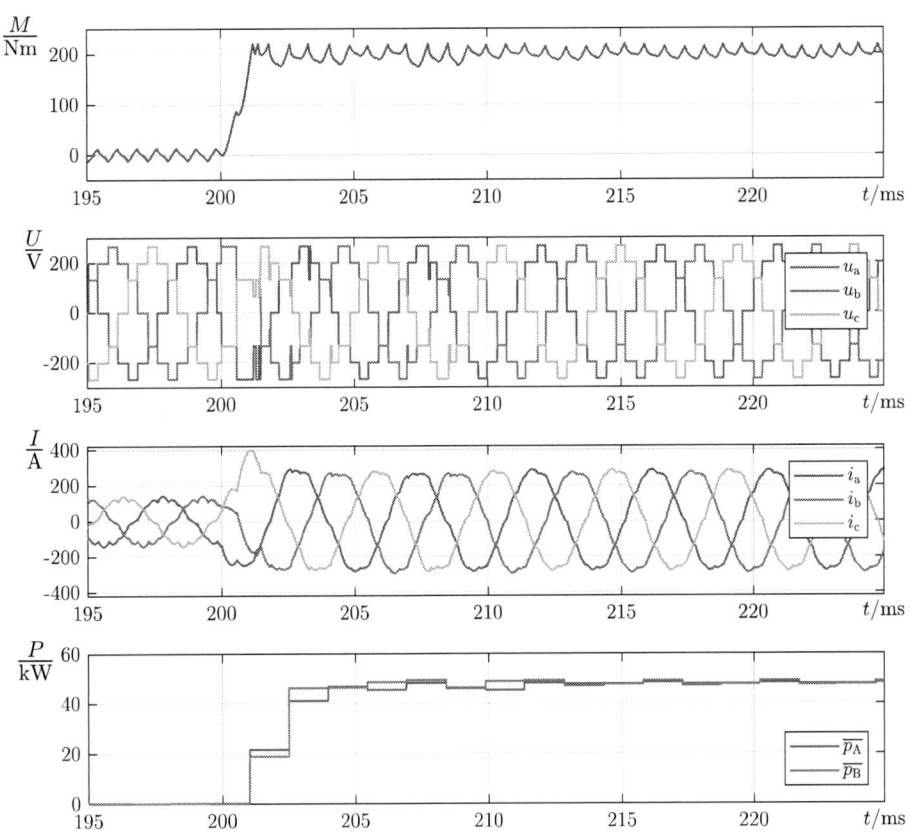

Figure 10: Mechanical and electrical quantities for a torque step with DSC

discussed possibility to modify the amplitude of the fundamental stator voltage enables several degrees of freedom, which are used to optimize the torque control loop of the DSC. The torque control basically works as a phase shift controller. In a small-signal model, the motor torque is determined by the angle between the stator and rotor flux linkage space vectors. If a positive torque is required, the stator flux linkage space vectors phase is advanced by short-cutting the flux linkage trajectory. A negative torque request leads to the output of a half- or zero-vector, which results in a decreased phase angle between the stator and rotor flux linkage vectors. This way, very fast torque step responses can be achieved, while keeping the required switching operations to a minimum. In order to meet the requirement of minimal switching losses, some voltage and flux distortions are intrinsically introduced, which means that an increased torque ripple has to be expected. In the high-speed operation range of traction drives, this torque ripple is acceptable in most cases. In general, a trade-off has to be made between high torque quality and increased switching losses.

In the simulation setup, the DSC algorithm for three-level inverters was implemented. In order to provide the switching sequences for the power-balancing square-wave modulation, a state machine, which selects the correct output voltage space vectors, was implemented. In figure 10, the most important mechanical and electrical quantities are shown. In this scenario, a torque step is applied to the machine while the rotational speed is held at a constant value.

It can be observed, that the required torque is reached at about 2 ms after the step is applied. The short-cutting process of the stator flux linkage trajectory, as well as minor corrections due to the increased torque can be noticed as the disturbances in the phase voltage waveforms, as the controller approaches the steady-state.

The plot at the bottom of figure 10 shows the average power distribution between both inverters. It can be verified, that the power-balancing square-wave modulation is capable of ensuring equal power contributions of both inverters for dynamic operation point changes. This confirms, that the proposed method is applicable as a modulation technique for dual two-level inverter high-power drives in the operating range of high speeds, where a symmetric power distribution is desired.

5   Conclusion

In this paper, the dual two-level inverter topology is described and new optimized control methods for the high speed operating range are presented. It is shown, that under the requirement of equal DC-link voltages, the dual two-level inverter can be understood as a single three-level topology. Two square-wave modulation techniques are discussed, which utilize the three level properties in order to apply a square wave modulation at the load. The described power balancing square-wave modulation technique is capable of ensuring an equal average power distribution between both inverters, while maintaining a minimum of required switching operations. It is furthermore shown, that in combination with a direct-self control strategy, the proposed method produces very promising results in the simulation approaches. As further objectives, the proposed control method will be implemented in a real dual two-level inverter system and final measurements will be made in order to confirm the methods functionality and performance.

References

[1] Kowalski, T.: Mess- und Betriebsverfahren von stromrichtergespeisten Drehfeldmaschinen mit supraleitender Statorwicklung. 1. Edition. Herzogenrath: Shaker (Forschungsberichte Leistungselektronik und Steuerungen, 12), 2019

[2] Stemmler, H; Guggenbach, P.: Configurations of high-power voltage source inverter drives. In: *1993 Fifth European Conference on Power Electronics and Applications*, Brighton, UK, 1993, pp. 7-14 vol. 5.

[3] Lega, A.: Multilevel Converters: Dual Two-Level Inverter Scheme, University of Bologna, Diss., 2007.

[4] Welchko, B. A.: A double-ended inverter system for the combined propulsion and energy management functions in hybrid vehicles with energy storage. In: *31st Annual Conference of IEEE Industrial Electronics Society, 2005. IECON 2005.*, Raleigh, NC, 2005, pp. 6 pp.-.

[5] Depenbrock M.: Direct Self-Control(DSC) of Inverter-Fed Induction Machine. In: *IEEE Transactions on Power Electronics, Vol 3, IEEE 1988.*.-.

[6] Springmeier F.: Direkte Ständergrößen-Regelung von Induktionsmaschinen am Dreipunktwechselrichter In: *VDI Verlag, 1993*, ISBN 3-18-143321-7

Charging High Voltage capacitors in pulsed power applications with a Capacitor Diode Voltage Multiplier of reduced size and lower ripple currents

Tristan Weinert and Wolfgang Oberschelp
Westphalian University of Applied Sciences
Neidenburger Straße 43
45897 Gelsenkirchen, Germany
Phone: +49 (0) 209-9596863
Fax: +49 (0) 209-9596544
Email: tristan.weinert@studmail.w-hs.de
URL: https://www.w-hs.de

Günter Schröder
University of Siegen
Hölderlinstraße 3
57068 Siegen, Germany
Phone: +49 (0) 271-7403356
Fax: +49 (0) 271-7402499
Email: guenter.schroeder@uni-siegen.de
URL: https://www.eti.uni-siegen.de/emas/

Keywords

≪DC power supply≫, ≪High voltage power converters≫, ≪Power supply≫, ≪Pulsed power≫, ≪Pulsed power converter≫

Abstract

The Cockroft-Walton multiplier is an established converter to generate high voltage, that can be used to charge capacitors for pulsed power applications. In this paper the drawbacks for this use case are shown and with the Two Phase Capacitor Diode Voltage Multiplier Type A a promising alternative solution is presented. While this topology seems to be the less obvious choice between both converters under steady state conditions, it shows its strength at repetitive capacitor charging. One of the benefits is a great reduction of the ripple currents overlayed to the desired constant DC current, by reducing reverse flowing charges and doubling the output frequency. Thus the pre-resistor for diode protection has lower losses or could be increased to improve the robustness. Filtering on the current measurement becomes simpler and disturbances are easier to avoid. The capacitor values can be greatly downscaled leading to reduced weight, size and cost. With smaller values, the power supply requirements are lowered, since the multipliers capacitance acts as an additional load to it. The reduction of energy stored inside the multiplier by a factor of 17 enhances the efficiency.

Introduction

The generation of high voltages is a key part in pulsed power applications. Since the first use-cases in radar equipment and the X-ray radiography, pulsed power generators had constantly evolved into many technical fields. Today the main applications of pulsed power are particle accelerators, lasers, nuclear fusion research, magnetic forming, plasma cleaning, medical equipment and treatment of food. The main concept of pulsed power technology is slow accumulation of energy, which can be discharged with high power in short time. This can be achieved with large capacitor banks and voltage multipliers charging them. The available charging modes are constant voltage, constant current and constant power, while constant current is popular especially for high repetition rates [1]. Modern capacitor chargers are a combination of switched-mode power supplies and voltage multipliers to keep blocking voltages over diodes low. The switched-mode power supplies are driven with high frequency which reduces the size of the voltage multipliers capacitors. A transformer is the link between power supply and multiplier, boosting the voltage to a level that the multiplier parts can withstand. The common AC-fed voltage multiplier in high voltage engineering is the Cockroft-Walton multiplier (C-W).

Since more and more applications come with restricted space requirements, e. g. mobile applications [2][3], there are increasing endeavors to become more efficient and volume optimized. In these applications the C-W is sometimes dismissed because of the heavy voltage drop under load and their huge amount of stored energy [4]. The key parameters for size reduction are frequency and capacitance. The voltage drop depends also on these parameters and additionally the number of multiplier stages. While the increase of the frequency improves volume and voltage drop, the increase of capacitance just advances the voltage drop. Higher capacitance leads to increased size and more stored energy inside of the multiplier. Another approach to reduce voltage drop is the research on new multiplier topologies.

In [5] a hybrid typology of the C-W and a Dickson charge pump is presented. While the hybrid topology reduces the voltage drop, unwanted features of the Dickson topology are carried over. These are unequally distributed voltages on the capacitors with the last one burdened with the whole output voltage. Higher voltages also lead to higher stored energy in the multiplier and therefore rising losses and space requirements.

With the rising popularity of wide-bandgap semiconductors the power supplies frequency can be increased without raising switching losses. The high frequency operation of the C-W is analyzed in [6] and [7]. The reverse recovery effect of the multipliers diodes can not be neglected at high frequencies. The authors of [6] suggest the use of SiC schottky rectifiers in the first stage of the multiplier. In [7] all semiconductors are SiC and the multiplier is integrated into the resonant circuit. However, due to the price of SiC devices the increase of frequency leads to additional costs, also on the power supply. In [8] a split-source multiplier and mixed capacitor distribution is proposed with comparable total capacitance to a common C-W. But the transformer isolation design will become more complex due to the multi-sectioning.

The Cockroft-Walton multiplier

A well known voltage multiplier used in pulsed power applications is the Cockroft-Walton generator (C-W). The circuit is depicted in Fig. 1 with a load capacitor C_L, a transformer T_1 on the input and a quantity n of five stages. This high-voltage generator is a cascade of multiple Villard voltage doublers (e.g. C_1 and D_1) combined with half-wave rectifiers (e.g. C_2 and D_2). It is normally fed with a constant AC voltage resulting in constant DC voltage of two times n the AC amplitude in steady state without load current, see (1).

$$V_{o,max} = 2 \cdot n \cdot \hat{v}_T \tag{1}$$

The odd numbered capacitors C_1, C_3, C_5, C_7 and C_9 form the coupling column. In the negative half-wave of the alternating voltage v_T, charge is transferred into the coupling capacitor C_1 shifting the voltage across D_1 with an offset. Likewise the alternating voltages across the rectifier diodes D_2, D_4, D_6 and D_8 charge the remaining coupling column capacitors and the voltage across the other diodes is shifted. Even numbered capacitors C_2, C_4, C_6, C_8 and C_{10} form the smoothing column responsible for less ripple on the rectified voltages and on the output.

Adding a significant load on the output of the multiplier will result in a current-dependent voltage drop inside the cascade minimizing the achievable output voltage. This drop is based on the charge pumping behavior of the circuit and will increase with the addition of further stages. The voltage drop can be separated into two components, a steady drop ΔV limiting the average voltage on the output and a dynamic drop δV seen as an overlayed ripple. The resulting voltage waveform is depicted in Fig. 2.

The constant current I_L equals a constant amount of charge $Q = I_L \cdot T$ that needs to be supplied by the smoothing column capacitors. Therefore each even numbered capacitor is discharged at least with one Q per period. Capacitor C_{10} is only discharged with one Q, but this charge has to be delivered by the capacitor C_9 during one half of the charge process. As a consequence the capacitor C_8 has to provide an additional charge Q intended for C_9, resulting in a total discharge of two times Q per period. In conclusion each added stage increases the charges supplied by the lower stage capacitors, leading to a discharge of $n \cdot Q$ from the first stage.

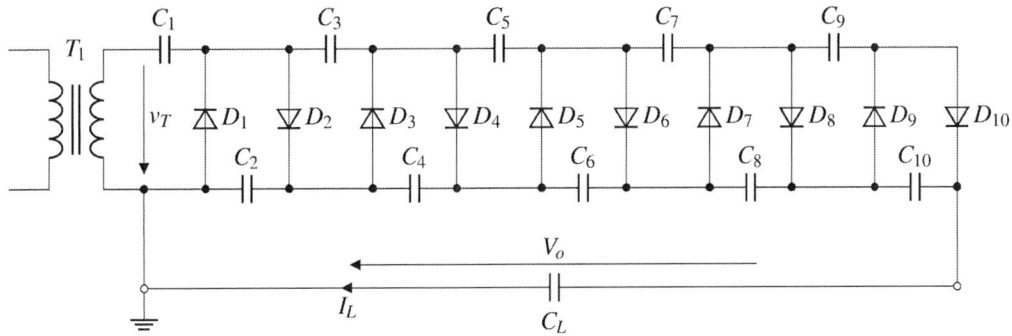

Fig. 1: Cockroft-Walton multiplier with capacitive load

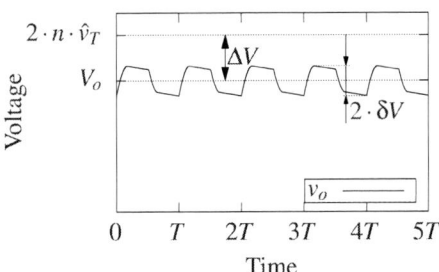

Fig. 2: Voltage drop of the Cockroft-Walton multiplier under load current

On closer examination the resulting voltages across the capacitors differ from the ideal voltages by a charge flow dependent voltage drop. This voltage drop is varying from stage to stage and will be higher on the lower stages if capacitors of equal value are considered. The impact on the output can be described by the sum of these voltages and will be minimized by choosing larger capacitance values for the lower stages. As it is desired to build a weight and cost optimized converter it is common to design a multiplier with equal capacitors of the value C_B and only increase the first coupling capacitor C_1 into $2C_B$. This capacitor has the highest effect on the resulting voltage drop and it is the capacitor burdened with the lowest voltage (Half of the voltage compared to the remaining capacitors). The voltage drop ΔV and the ripple δV for this configuration can be calculated via (2) and (3) for an ohmic load [9].

$$\Delta V = \frac{I_L}{f \cdot C_B} \cdot \left(\frac{2n^3}{3} - \frac{n}{6}\right) \qquad C_1 = 2C_B \qquad C_{k>1} = C_B \tag{2}$$

$$\delta V = \frac{I_L}{f \cdot C_B} \cdot \frac{n(n+1)}{4} \qquad C_{k(even)} = C_B \tag{3}$$

With the graphic view of (2) in Fig. 3, it can be seen that the voltage drop is a serious problem for the design of a C-W generator. As an example the voltage drop ΔV and the resulting output voltage V_o are plotted for a generator seeking a total output voltage of 50 kV with a load current of 40 mA. The plot of ΔV is independent of the input voltage v_T and it becomes obvious that the amount of stages is limited to only a small number. At stage numbers above 3 the linear part can be neglected and the voltage drops can be seen nearly as a cubic function of the amount of stages [9]. With 7 stages the voltage drop is already over the half of the desired output voltage and with 9 stages the drop becomes actually higher than the wanted output voltage. An opportunity to compensate the voltage drop is to increase the input voltage, which is chosen at an amplitude of 6 kV in the example suited for a five stage multiplier. This leads to a $V_{o,max}$ of 60 kV considering (1) and a voltage drop of nearly 10 kV at full load current resulting in the required 50 kV.

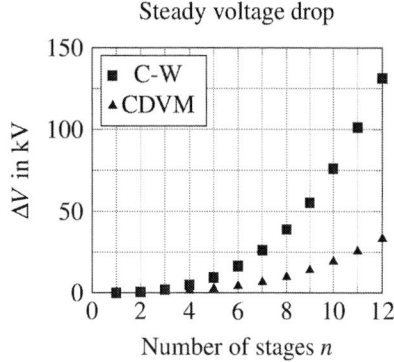
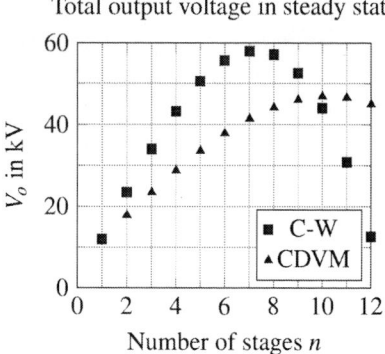

Fig. 3: Both multipliers depending on the amount of stages ($\hat{v}_T = 6\,\text{kV}$, $I_L = 40\,\text{mA}$, $f = 50\,\text{kHz}$, $C_B = 7\,\text{nF}$)

The second part of Fig. 3 shows the output voltage assuming this fixed $\hat{v}_T$ of 6 kV at full load. It can be seen again that the increment of stages will increase the voltage drop, limiting the voltage rise with every additional stage and also becomes disadvantageous at a certain point. The dynamic voltage ripple in this example is calculated via (3) and is approximately 860 V. Other options to minimize the voltage drop of the C-W generator are the increase of the frequency or the capacitance.

Boosting the switching frequency of the preliminary stage feeding the C-W multiplier is limited by several factors. Obvious factors are the higher switching losses of transistors, higher magnetization losses, the increase of the skin effect and limitations in the PWM resolution. Another problem is the diode reverse recovery becoming more significant at higher frequencies [6]. An increase of the multiplier capacitance is undesirable in terms of weight, size and cost [10]. Furthermore the multiplier capacitors form a large energy storage which can be dangerous or troublesome in several applications.

However, there are also drawbacks that need to be concerned in design of a C-W in capacitor charging applications. The capacitors of the voltage multiplier are in parallel to the load capacitance acting as a current divider. Therefore the preliminary power supply needs to provide a higher current which depends on the ratio of the capacitance values between load and multiplier. In case of an equivalent multiplier capacitance of the same value as the load capacitance, the supplied current needs to be doubled. Considering this, the optimal performance is achieved by using small capacitors in comparison to the load, with the side benefits of small size and weight. In addition the voltage multiplier acts as an energy storage which is also short-circuited during power pulse. For this reason it is necessary to add a precharging resistor between multiplier and load for the protection of the diodes against high currents. The energy stored in the voltage multiplier at full charge will be lost after every cycle, reducing the efficiency. Due to the charge reversal from the load back into the multiplier (see δV in Fig. 2) the ripple of the charging current will be high, causing increased losses at the precharging resistor.

For the control, a measurement of the output current is needed which is also placed in the discharge loop of the multipliers capacitance. Challenges in the measurement have to be solved regarding the high ripple current caused by the charge reversal. The ripple part of the current can be significantly higher than the current mean value, thus complex filtering is needed. Shunt resistors need to handle high current spikes with a sharp edge during discharge and high thermal losses due to the increased RMS current. To sum up, if smaller capacitors are used, voltage multipliers in pulsed power applications are much more efficient and a better power-to-weight ratio is achieved. Nevertheless, the voltage drop ΔV will rise with less capacitance in the multiplier, so choosing capacitors will become a trade-off.

The Two Phase Capacitor Diode Voltage Multiplier Type A

There are several other topologies of voltage multipliers that are less frequently used than the C-W. One of those topologies evaluated in [11], [12] is classified as the Two Phased Capacitor Diode Voltage

Multiplier (CDVM) Type A. It can be derived from the Full-Wave C-W by eliminating the smoothing column capacitors [13]. The proposed five stage CDVM is shown in Fig. 4, using the same amount of capacitors compared to the C-W in Fig. 1. A key difference is the insertion of a full-bridge rectifier on the output (D_{11}, D_{12}, D_{13} and D_{14}) and identical columns.

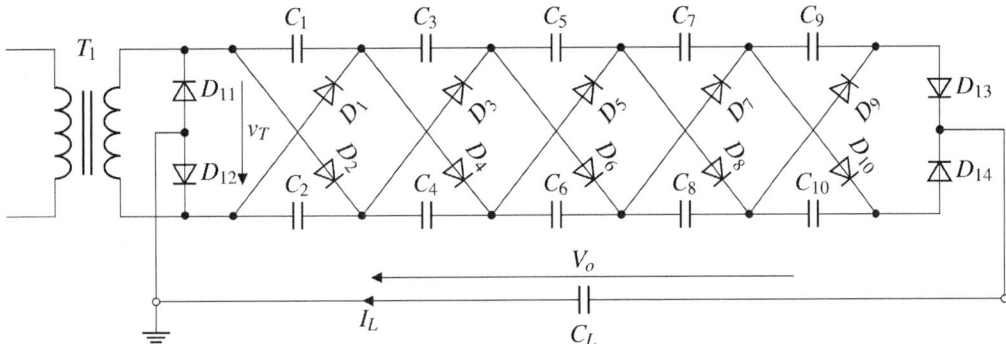

Fig. 4: Two Phase Capacitor Diode Voltage Multiplier Type A with capacitive load

The working principle of the CDVM differs from the C-W, as both half-waves of v_T are used symmetrically for the voltage shift. D_1 and C_1 act as a half-bridge rectifier in the positive half-wave of v_T, while D_2 and C_2 rectify the negative half-wave. In steady state without load both capacitors will be charged to the amplitude $\hat{v}_T$ of the input voltage. The sinusoidal input voltage will be reflected onto the diodes with an offset. When determining the voltage difference between the cathodes of D_1 and D_2 the offset becomes eliminated and an alternating voltage can be seen. With the neglection of losses the voltage is equivalent to v_T, which provides the ability to add multiple stages. When reaching steady state the capacitors in both legs of the CDVM are charged to $\hat{v}_T$. Each leg has a total voltage of $n \cdot \hat{v}_T$.

At the last stage the reflected input voltage will be again rectified with the full-bridge onto the load. To identify the overall maximum of the output voltage it is necessary to consider the load. The output voltage can be seen as a pulsating voltage with the DC offset of the legs capacitors. An output capacitor has a smoothing effect on the voltage, while a pure ohmic load does not affect the waveform. The calculation is done with (4), showing that a capacitor on the output offers the higher average output voltage.

$$V_{o,max} = \begin{cases} (n + \frac{2}{\pi}) \cdot \hat{v}_T & C_L \to 0 \\ (n+1) \cdot \hat{v}_T & C_L \to \infty \end{cases} \tag{4}$$

When comparing (4) with (1) it is noteworthy that the voltage doubling effect of each stage is lost with the proposed CDVM. Instead the output capacitor works as an additional stage without the need of internal capacitors. Observing only the steady state with negligible load current, the C-W would always be preferred over the Two Phase CDVM Type A. A look at the internal charges flowing while providing a load current, shows significant reductions in voltage drop. Since the structure of the considered CDVM is symmetric, both legs have the same charge flow. None of the multipliers legs supplies the load the entire period, so the charge $Q = I_L \cdot T$, due to the constant current I_L, is provided by both legs in parallel. The capacitors C_9 and C_{10} are discharged with only half of Q while the other capacitors are also burdened with additional charges. C_7 and C_8 are discharged with one Q because they need to deliver the output current and charges for the following capacitors C_9 and C_{10}. This results in a workload of $\frac{n}{2} \cdot Q$ on the first stage capacitors. In conclusion the capacitors of the proposed CDVM are only half discharged, compared with their counterpart in the C-W multiplier.

The static voltage drop (5) for a design with equal capacitor values has the same dependence on the load current, frequency and capacitance as in case of the C-W. In comparison the cubic part is greatly reduced, while the linear term changed the sign to positive, increasing the drop. Nevertheless, in the left part of Fig. 3 the voltage drop for both multipliers is depicted, showing a reduction of ΔV for all stage numbers.

As the plot was made with equal conditions, it can be considered, that smaller capacitance values are needed for the Two Phase CDVM Type A to achieve the same ΔV.

$$\Delta V = \frac{I_L}{f \cdot C_B} \cdot \left(\frac{2n^3 + n}{12} \right) \qquad C_k = C_B \tag{5}$$

Analyzing the dynamic ripple voltage over one of the legs δV_{leg} while neglecting the output smoothing (6) becomes half of the C-W voltage ripple (3). In addition δV becomes even smaller, because it will cancel out at the output since both legs operate phase shifted at $180°$. The resulting ripple voltage will be mainly influenced by the load capacitors interaction.

$$\delta V_{leg} = \frac{I_L}{f \cdot C_B} \cdot \frac{n(n+1)}{8} \qquad C_k = C_B \tag{6}$$

In the right part of Fig. 3 the constant voltage drop ΔV is subtracted from the maximal output $V_{o,max}$ with capacitive load and compared to the C-W. In steady state the C-W multiplier will still have the higher output voltage under the same conditions for a small amount of stages. Although the proposed CDVM will become better than the C-W exceeding ten stages, the benefit of additional stages has already reached its maximum. As the result of the steady state comparison of both voltage multipliers, the C-W seems to be the better choice. On the other hand, the use case of a HV capacitor charger for pulsed power applications is not similar to the steady state operation.

Comparing the charging process of the proposed CDVM and the C-W

When looking at the constant current controlled charging of HV-capacitors, (5) and (6) are not applicable. As the power supply, a Phase-Shifted Full Bridge converter (PSFB) with a series inductor on the primary side of the transformer is used, while the voltage multiplier acts as the rectifier. The supply is fed by a voltage of $400\,\text{V}$ and the transformer is wound with a turns ratio of 20. A switching frequency of $50\,\text{kHz}$ is implemented, which offers a great reduction of the multipliers capacitor size. The constant current is controlled by phase and duty cycle of the full-bridge and the measurement is done with a shunt resistor in series to C_L. While only the average of the current is constant, the waveform in the time domain is superimposed with an alternating part caused by the charge distribution inside the multiplier. Compared against the average value, the amplitude can be significantly higher, leading to a high RMS value of the current.

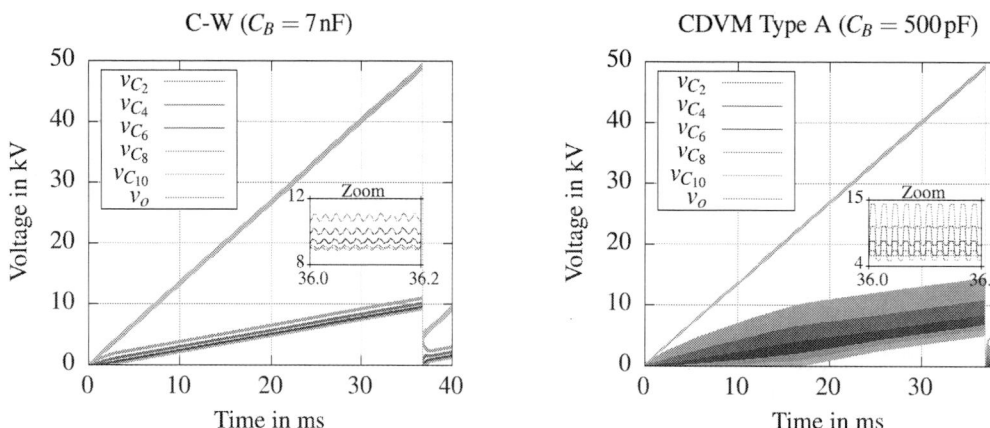

Fig. 5: Charging cycles with constant output current compared ($I_L = 40\,\text{mA}$, $f = 50\,\text{kHz}$, $C_L = 30\,\text{nF}$)

On the left side of Fig. 5 one charging cycle with the current-controlled C-W generator is depicted. The smoothing capacitor voltages v_{C_2} to $v_{C_{10}}$ are shown, which sum up to the output voltage v_o. After 37.5 ms the constant current has charged the 30 nF load capacitor C_L to the targeted voltage of 50 kV. The energy inside C_L is then discharged with a high power pulse emulated by a spark gap. After this a new cycle begins, which leads to the sawtooth-shaped profile of the voltages over all capacitors in the multiplier. The C-W generator was designed with a base capacitance of 7 nF. The energy of the load capacitor before discharge amounts to 37.5 J. Calculation of the energy inside the multiplier can be done with (7). With 3.3 J the multipliers energy is about 8.8 % of the load energy at discharge. Based on the energy content of the C-W generator an equivalent capacitance of 2.64 nF can be assumed.

$$E = \frac{1}{2} \cdot C_B \cdot \left(\frac{V_o}{n}\right)^2 \cdot \left(2n - \frac{1}{2}\right) \qquad C_1 = 2C_B \qquad C_{k>1} = C_B \tag{7}$$

Since the voltage on the load is coupled with the multiplier capacitor voltages, the power supply needs to provide an additional constant current. Load and multiplier form a current divider fed by a total current of the supply. This effect will become a problem at smaller load capacitors. The supply needs to serve double the current, if the load capacitor is in the same value range as the multipliers equivalent capacitance. In this case the power supply has to be designed with double power, following an increase in size, weight and cost.

At discharge the current of the multiplier is only limited with a high-ohmic resistor in series, needed to protect the multiplier diodes against destruction. Since the shunt resistor for current measurement is also located in the discharge circuit, it has to be tolerant to high current spikes. Indeed this resistor can be a gateway for disturbances causing latch-up of control circuits. Thus protection is needed, in addition to the extensive filtering against the high ripple currents of the back and forth traveling charges.

In the left of Fig. 6 the current waveform on the output with an average of 40 mA is outlined. It is obvious that the peak-to-peak value of the current is significantly higher with nearly 500 mA. Forming the RMS of the depicted waveform results in about 167 mA. Since the current acts squared on resistors power stress, the losses on pre-resistor and shunt are both around the factor 17 higher, compared with charging at pure DC current. The areas between the positive part of the waveform and the zero line represent charges going into the load, while the areas under zero are charges flowing back into the multiplier. Comparing both areas illustrates that the extensive charge travel is responsible for this behavior.

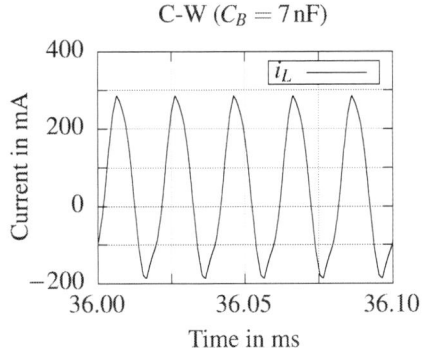 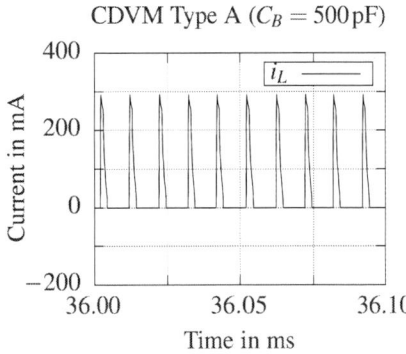

Fig. 6: Output current waveforms compared ($I_L = 40$ mA, $f = 50$ kHz, $C_L = 30$ nF)

This constant-current capacitor charging application is a use case in which the proposed CDVM performs much better than the C-W multiplier. While the C-W charges the load mainly via the smoothing column, booth legs of the Two-Phase CDVM Type A are involved in the charging process. Therefore the frequency seen on the charging current doubles and the dynamic ripple δV at the output cancels out. It can only be seen on the legs as δV_{leg} according to (6) in steady state. In the process of charging, steady state isn't reached and a higher amplitude of the legs ripple can be used for a higher voltage lift, while charging the following stage. Since δV cancels out on the output there is no need for high capacitance in the legs with regard to the load. Recapitulating the insights from the capacitor charging with the C-W, it seems to be attractive to reduce the leg capacitance and also benefit from higher voltage in amplitude.

In simulation it has been shown that a CDVM with significant smaller base capacitance of 500 pF and a slightly increased transformer turns ratio from 20 to 25 also fulfills the requirements. Fig. 5 shows a comparison with the proposed CDVM on the right, providing an even smoother waveform of the requested output voltage. The higher ripple on the capacitors can not be seen on the output, since both legs are operated phase-shifted and the ripple nearly cancels out after the output diodes. The distribution of the voltages on both of the capacitors legs is not consistent as before, boosting the voltage at the lower stages. On the higher stages the full ripple is limited at the zero line at the beginning, since the multiplier diodes are blocking voltages below. When the voltages over capacitor C_9 and C_{10} finally leave the discontinuous range, the boosting effect is reduced and the slope of all other multiplier capacitor voltages decreases. At this moment the control needs to increase the duty cycle of the power supply, to hold the constant current on the output. The sum of the multipliers energy at discharge is calculated via (8).

$$E = \frac{1}{2} \sum_{k=1}^{n} \left(C_k \cdot v_k^2 \right) \tag{8}$$

Analyzing the charging state of every capacitor in the simulation sums up to a energy of 0.19 J in the multiplier before discharge, much lower than with the C-W. Thereby the multiplier capacitors become neglectable also for smaller load capacitors reducing the power requirements for the power supply. The equivalent capacitance of the multiplier is around 152 pF.

When looking at the current waveform of the proposed multiplier in the right of Fig. 6, it is preferable to the C-W. Charges are only transported in one-way because reverse currents are blocked with the diodes D_{11} to D_{14}. One can think about blocking reverse currents with a simple diode on the output, also on the C-W. But the Two Phase CDVM Type A doubles the output frequency by providing fragments of charges during both half-waves of the transformer voltage. As a result the current spikes can be smaller in amplitude and filtering of the average value will be simpler. The RMS value of the depicted current is about 93 mA which is nearly the half of the C-W value. Accordingly, the losses on the resistive parts are almost quartered. Another option is to increase the pre-resistor and keep the losses, so a better protection against the application induced high current pulses is achieved.

Experimental verification

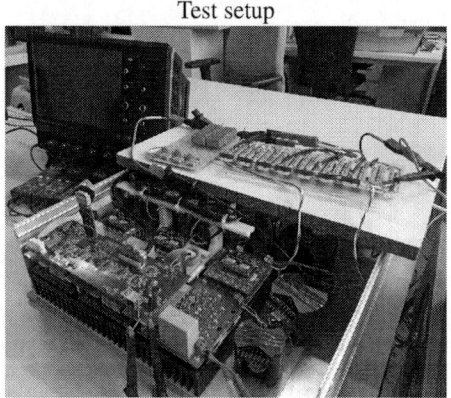

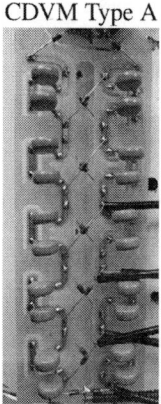

Fig. 7: Pictures of test setup and both voltage multipliers

To test the function of the CDVM Type A for pulse capacitor charging, a prototype (see Fig. 7) was built and compared with the C-W. A PSFB with primary side inductance was used to drive the voltage multiplier. The circuit is shown in Fig. 8 and the parameters are listed in Table I. To enable easier measurements on the multipliers capacitors, lower voltages were used for this purpose. Therefore, a 1:1 transformer was chosen and primary inductance L_D and load current I_L were adjusted to remain comparable to the high voltage version presented.

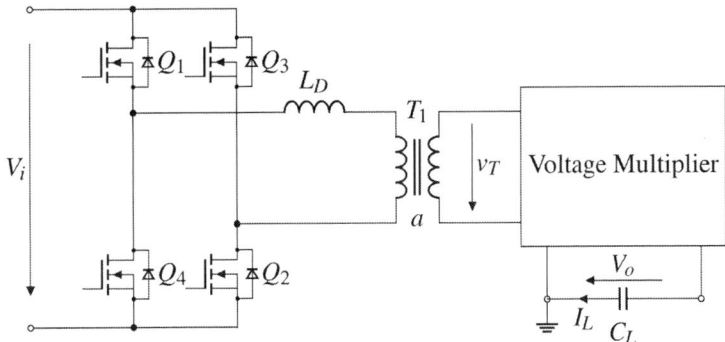

Fig. 8: Phase-Shifted Full Bridge Converter with primary inductor and Voltage Multiplier

By choosing a lower output voltage, the forward voltages would be much more significant when using high voltage diodes. For this reason, the multipliers were equipped with other diodes that have a blocking voltage of 1000 V and therefore also lower forward voltages. The capacitors in the voltage multiplier were dimensioned as in the theoretical comparison, which means that the CDVM Type A has a much smaller overall capacitance than the C-W.

Table I: Parameters of experimental verification

Parameter	Symbol	Value
Diodes	D_1 - D_{14}	UF 4007
MOSFETs	Q_1 - Q_4	STW55NM60ND
Capacitors C-W	C_1 , C_2 - C_{10}	14 nF, 7 nF
Capacitors CDVM Type A	C_1 - C_{10}	500 pF
Load Capacitor	C_L	34 nF
Transformer T_1 Turn ratio	a	1:1
Choke	L_D	12 mH
Switching frequency	f	50 kHz
Load Current	I_L	0.4 mA
Input voltage	V_i	200 V

Fig. 9 shows the output voltage and additionally the first three capacitor voltages of a column near the end of the charging process for both multipliers compared. It can be seen that the capacitors of the CDVM Type A are subject to a much higher ripple than in the C-W, as already shown in the simulation. This ripple results in a much higher output voltage than in no-load state and a performance comparable to C-W. As predicted, an equally smooth output voltage is achieved. The proposed functional principle of a CDVM Type A for capacitor charging could thus be verified.

Conclusion

In this paper the Two Phase Capacitor Diode Voltage Multiplier (CDVM) Type A was compared with the well-known Cockroft-Walton multiplier (C-W) for the charging of HV capacitors in pulsed power applications. It was shown that the voltage drop of the proposed multiplier depending of the number of stages is much lower than on the C-W. Caused by the voltage doubling of the C-W, the total output voltage will be higher than with the discussed CDVM in steady state. But in capacitor charging applications a voltage boosting effect of smaller capacitors ripple can be used to achieve the same performance with

C-W ($C_B = 7\,\mathrm{nF}$)

CDVM Type A ($C_B = 500\,\mathrm{pF}$)

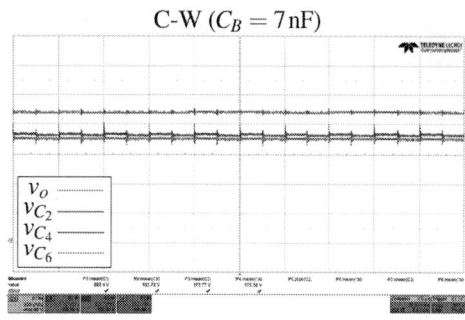 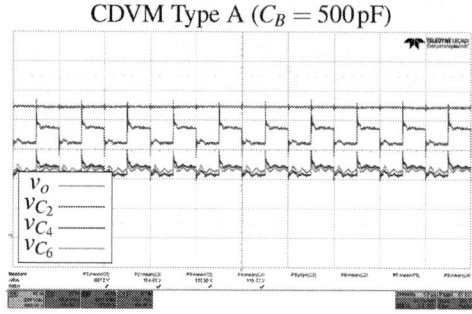

Fig. 9: Measurement on multipliers prototypes compared

less size, weight and cost. A reduction of the capacitors size also leads to further benefits. At first, reverse currents are blocked and the charging frequency on the load will be doubled by the multiplier. Therefore the ripple current is reduced significantly, decreasing losses and filtering requirements. Due to reduced discharged energy, the converter can be designed more robust with higher pre-resistance. Furthermore, the capacitive load is reduced by smaller multiplier capacitors, lowering the power demand for the feeding power supply. A prototype was built and first measurements confirm the function of the CDVM as capacitor charger. Further research on the topology is still pending.

References

[1] J. Lehr and P. Ron, Foundations of Pulsed Power Technology. Wiley-IEEE Press, 2017.

[2] A. V. Bilbao and S. B. Bayne, "Compact Rapid Capacitor Charger for Mobile Marx Generator Applications," 2019 IEEE Pulsed Power & Plasma Science (PPPS), Orlando, FL, USA, 2019, pp. 1-4, doi: 10.1109/PPPS34859.2019.9009680.

[3] T. Weinert, W. Oberschelp and G. Schroder, "A current-fed DC/DC converter for the efficient charging of HV capacitors in mobile applications," 2016 18th European Conference on Power Electronics and Applications (EPE'16 ECCE Europe), Karlsruhe, 2016, pp. 1-10, doi: 10.1109/EPE.2016.7695412.

[4] R. E. P. Frost, J. A. Pilgrim, P. L. Lewin and M. Spong, "An Investigation into the Next Generation of High Density, Ultra High Voltage, Power Supplies," 2018 IEEE International Power Modulator and High Voltage Conference (IPMHVC), Jackson, WY, USA, 2018, pp. 156-161, doi: 10.1109/IPMHVC.2018.8936830.

[5] S. Park, J. Yang and J. Rivas-Davila, "A Hybrid Cockcroft–Walton/Dickson Multiplier for High Voltage Generation," in IEEE Transactions on Power Electronics, vol. 35, no. 3, pp. 2714-2723, March 2020, doi: 10.1109/TPEL.2019.2929167.

[6] Saijun Mao, Pengcheng Zhang, J. Popovic and J. A. Ferreira, "Diode reverse recovery analysis of Cockcroft-Walton voltage multiplier for high voltage generation," 2017 IEEE 3rd International Future Energy Electronics Conference and ECCE Asia (IFEEC 2017 - ECCE Asia), Kaohsiung, 2017, pp. 1765-1770, doi: 10.1109/IFEEC.2017.7992315.

[7] U. Müter, K. F. Hoffmann, B. Wagner and A. Lunding, "Compact 80 kW silicon carbide high voltage generator," 2017 19th European Conference on Power Electronics and Applications (EPE'17 ECCE Europe), Warsaw, 2017, pp. P.1-P.8, doi: 10.23919/EPE17ECCEEurope.2017.8099198.

[8] L. Katzir and D. Shmilovitz, "Effect of the capacitance distribution on the output impedance of the Half-Wave Cockcroft-Walton voltage multiplier," 2016 IEEE Applied Power Electronics Conference and Exposition (APEC), Long Beach, CA, 2016, pp. 3655-3658, doi: 10.1109/APEC.2016.7468395.

[9] E. Kuffel, W. S. Zaengl, and J. Kuffel, High Voltage Engineering Fundamentals, 2nd ed. Oxford: Newnes, 2000.

[10] I. C. Kobougias and E. C. Tatakis, "Optimal Design of a Half-Wave Cockcroft–Walton Voltage Multiplier With Minimum Total Capacitance," in IEEE Transactions on Power Electronics, vol. 25, no. 9, pp. 2460-2468, Sept. 2010, doi: 10.1109/TPEL.2010.2049380.

[11] J. J. Kisch and R. M. Martinelli, "High frequency capacitor-diode voltage multiplier dc-dc converter development," Culver City, Tech. Rep., Sep. 1977.

[12] R. M. Martinelli and A. F. Ahrens, "Multiphase capacitor-diode voltage multiplier," 1978 IEEE Power Electronics Specialists Conference, Syracuse, NY, 1978, pp. 275-284, doi: 10.1109/PESC.1978.7072364.

[13] H. van der Broeck, "Analysis of a current fed voltage multiplier bridge for high voltage applications," 2002 IEEE 33rd Annual IEEE Power Electronics Specialists Conference. Proceedings (Cat. No.02CH37289), Cairns, Qld., Australia, 2002, pp. 1919-1924 vol.4, doi: 10.1109/PSEC.2002.1023094.

Review of optimization methods for the design of power electronics systems

Mylène Delhommais

Univ. Grenoble Alpes, CEA, LITEN, DEHT, LAEH, 38000 Grenoble, France

Tel.: +33/(0) 438 78 21 54

mylene.delhommais@cea.fr

liten.cea.fr

Keywords

«optimization», «design»

Abstract

Optimization methods for the design of power electronics are increasingly used. The variety of methods is rich but can be confusing for designers who wish to start using them. This article is an introduction to optimization methods that can be applied in power electronics through a review of the literature.

Introduction

Thanks to a growing goal to decarbonize energy production and consumption and the objective to improve system performance with reduced maintenance costs, the need for power electronics systems is significantly increasing.

In the same time, power converters are still a little, but necessary, function of a system. So generally, the system integrator will wish a transparent behavior of power electronics, the transparent term depending on the application. For example in more electrical aircrafts, the power converters should be lightweight and redundant. In hybrid or full electric vehicles, the power supplies must be small and cheap. Of course, a high efficiency is generally required for energy saving. In a stationary application as solar or wind farms, converters must be also efficient, robust and middle cost to make the renewable energies as competitive as possible with other power sources. These criteria are hardly compatible between them: an improvement of one criterion will degrade the others (if the design was optimal). All designs fulfilling this condition will draw the Pareto front of the system.

Besides, with the relatively recent arrival of Wide Band Gap (WBG) semiconductors on market, the designer had to apprehend the new limits of power converters (i.e. acquiring experience knowledge). Because there are a priori no new breakthrough technologies, designers must exploit the available technologies in their limits. The use of optimization algorithm helps them into finding those limits.

In the early 80's, Balachandran and Lee already showed the interests of using algorithms for power converter design optimization [1]. They also presented the different methods available on that time, which were quite poor compared with nowadays methods and tools. In 2014, a review on metaheuristic optimization methods applied to power converters has been made [2]. The authors presented the population-based methods as new optimization algorithms well adapted to power electronics design.

Figure 1 shows the number of papers dealing with optimization methods for power electronics obtained with (1) in Scopus database [3].

$$
\begin{aligned}
&SUBJAREA\,(\,ener\;OR\;engi\;OR\;math\;OR\;phys\,)\;AND\;(\,TITLE\,(\,(\,optimi?ation\;OR\;automation\;OR\;pareto\,)\;AND\;(\,converter^* \\
&OR\;"power\;supply"\;OR\;"power\;electronic^{*"}\;OR\;dc\_dc\;OR\;inverter^*\;OR\;pfc\;OR\;dc\_ac\;OR\;ac\_dc\,)\,)\;AND\;KEY\,(\,(\\
&optimi?ation\;OR\;automation\;OR\;pareto\,)\;AND\;(\,converter^*\;OR\;"power\;supply"\;OR\;"power\;electronic^{*"}\;OR\;dc\_dc\;OR \\
&sizing\;OR\;design\;OR\;presizing\;OR\;inverter^*\;OR\;pfc\;OR\;dc\_ac\;OR\;ac\_dc\,)\,)\,)\;AND\;(\,algorithm\;OR\;gradient\;OR \\
&programming\;OR\;tool\,)\;AND\;(\,sizing\;OR\;design\;OR\;presizing\,)\;AND\;NOT\;(\,mechanic^*\;OR\;fluid^*\;OR\;micro^*\,)\;AND\;(\\
&LIMIT\text{-}TO\,(\,LANGUAGE\,,\;"English"\,)\,)
\end{aligned} \tag{1}
$$

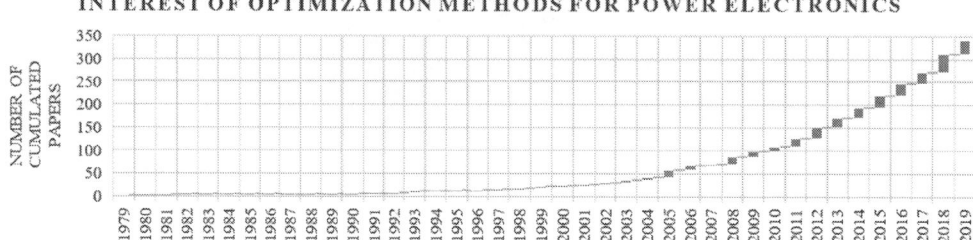

Figure 1: Number of cumulated papers dealing with power electronics optimization : these 40 last years, a total of 343 papers treated about power electronics optimization based on (1)

A first explanation of the growing interest in this subject in these last 15 years may be the increasing demand of highly constraint converters (together with an absence of new breakthrough technologies since WBG devices) combined with an easier access of efficient optimization algorithm, tools and powerful computing capabilities. This paper offers a review of the different optimization methods currently used in power electronics systems. It will be shown that the discretized and multi-physical aspects of power electronics conduct designers to a particular attention when using algorithms. The analysis will also be performed on the method choice made regarding the design phase in which the designer is.

First section reminds the design process and what is involved in each sketch phase. Second section will present the main categories of optimization methods and for which design phase they are generally used. The last section will give some detailed examples took in the literature to concretely show the differences between these methods. Finally, the conclusion will draw some perspectives for the optimization methods in power electronics design.

Design process

In this paper, the design process is cut in two stages. First stage is preliminary design stage which main objective is the definition of the specifications of the power electronics according to the system integrator needs and the feasibility of the requirements tested by pre-sizing the power electronics. The second objective of the preliminary design phase is the definition of the development plan for the detailed design phase that is stage 2. In this second stage, designers must deeply design the system (components selection, sub-functions design, simulations, schematics and circuits layout drawings) and conduct the necessary tests according to V-cycle to validate the design by experiments. The objective is to fulfill the requirements of the product in the allocated time and budget.

Power electronics optimization methods review

Power electronics systems are mainly an assembly of components taken off the shelf. It means that most of the design variables are discrete (i.e. choice of the components among catalogs). The other design variables may be continuous (switching frequencies, control parameters) or also discrete (technology or topology choice). These facts conduct power electronics design to be a combinatory problem (complex) to solve.

There are so several strategies to solve this problem: keeping it as a combinatory problem or not: either by setting the problem in a continuous area (by assuming all design variables continuous), either by using stochastic strategies. Then the optimization method must be chosen among: a heuristic method, a mathematic method or a metaheuristic method. It is not rare that designers hybrid these methods.

A heuristic method is based on rules: as trying every combination (enumeration more or less smart, branch & bound technic [4]) or define a smart sequence of design equation solving according to the optimization problem. A mathematic method is based on mathematical theories: as using the gradient to find the optimal point [5], solving linear equations [6] or geometric programming (solve convex problems). A metaheuristic method is "an over rules method" and is generally stochastic as evolutionary algorithms [2], [7] or neural network methods [8]. Figure 2 summarizes the three methods with a non-exhaustive list of specific methods. It should be noticed that there exists hundreds of different optimization methods with several variants and that Figure 2 is only a proposition of classification.

For the power electronics designer, the major difficulty is to know the most adapted method depending on his optimization problem. The method will be indeed different if the specifications of the system have to be negotiated; or if most questions are about the technologies or topology to select; or if it is about finding the optimal configuration of a converter.

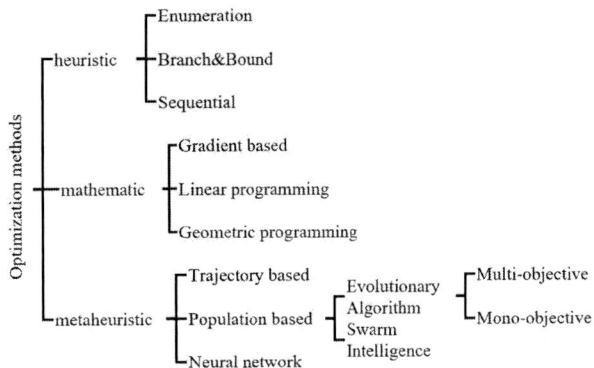

Figure 2: proposition of optimization methods classification

Table I is an example of the diversity of used methods to solve power electronics problems and is therefore non-exhaustive. As it can be seen the population based methods have more success (actually 108 references among the 343 papers from research (1)) thanks to their easier implementation for power electronics (accept the discrete variables with a lower computation cost than for enumeration technics). Secondly, genetic algorithm and sequential method are really appreciated for stage 2. During this stage, the customer needs are well known (specifications are fixed) and the designer has already identified the main issues to focus on (a part of the converter or a specific constraint): the optimization problem size is reduced and the objective is to well understand the design (sequential method) or find the global optimal design (with an evolutionary algorithm).

For stage 1, since the degree of freedom is larger than for stage 2, methods handling large optimization problem size such as mathematic ones (and in a second place NSGA-II algorithm which is quite performant) are preferred. They will help to negotiate the specifications through Pareto fronts and in finding the future issues to work on during stage 2.

Finally, Figure 3 shows the occurrence of the design objectives of papers of Figure 1. It is not surprising that efficiency is the most important criteria to maximize followed by the current or voltage ripple to minimize since the renewable energies drive the power electronics market. Then electric embedded systems require high power density converters.

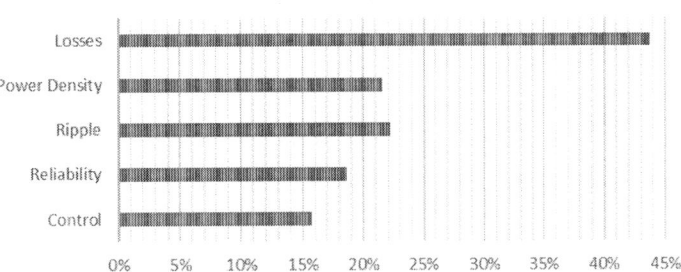

Figure 3: Percentage of paper dealing with a particular subject in (1)

The reliability and control are less threated. The methods used for these subjects are less widespread (Table I). It can be explained by the difficulty to set the models for them, moreover, to set models adapted to optimization methods.

Table I: Review of the different methods used to optimize power electronics

Paper	Stage	Component or system optimization	Discrete	Algorithm family	Optimization algorithm?	Objective(s)
[9]	1	System	No	gradient based	SQP [10] & modified MFD [11]	cost
[12]	1	System	No	gradient based	SQP [13]	weight
[14]	1	System	No	gradient based	SQP [13]	volume & losses
[15]	1	System	No	geometric programming	GP [16]	volume & losses
[17] [18]	1	System	Yes	population based	NSGA-II [19]	weight & losses
[20] [21]	1	System	Yes	population based	NSGA-II [19]	weight & losses & cost
[22] [23]	1	System	Yes	sequential	homemade	volume & losses
[24]	1	System	Yes	population based	GA	volume
[25]	1	System	Yes	population based	NSGA-II [19]	weight
[26]	1	System	Yes	population based and trajectory search	GA or PSO [27] (1) or SA (2)	weight
[28]	1/2	System	Yes	artificial intelligence	neural network	reliability
[29]	2	System	No	population based	GA	reliability
[30]	2	System	No	machine learning (Gaussian)	Bayesian Optimization [31]	losses
[32]	2	System	No	sequential & population based	GA combined with PSO	losses
[33]	2	System	Yes	population based	GA	cost
[34]	2	System	Yes	sequential	homemade	volume & weight
[35]	2	System	Yes	sequential	homemade	volume & losses
[36]	2	System	Yes	population based	GA or PSO [27]	ripple
[37]	2	System	Yes	sequential &/or SQP	homemade + SQP	volume & losses
[38], [39]	2	System	Yes	Enumeration	Chen's rearrangement for explicit enumeration [40]	weight & losses
[41]	2	System	Yes	gradient based	Conjugate Gradient Method [42] with Differential Evolution [43]	control
[44]	2	Printed spiral coil	No	population based	GA	losses
[45]	2	Regulator	No	population based	DE [43]	control
[46]	2	Regulator	No	population based & trajectory search	OLGSA (hybridization of GA and orthogonal search) → homemade	control
[47]	2	AC Filtering Inductor	No	gradient based	SQP	weight
[48]	2	filter	No	population based	MOEA [49]	ripple
[50]	2	LLC transformer	Yes	gradient based	sweeping and Levenberg-Marquard	losses
[51]	2	Thin film integrated magnetics	Yes	(1) trajectory search (2) population based	Pattern search (1) [52] or PSO [27] or DE [43] (2)	area or ripple or losses
[53]	2	DC link capacitors	Yes	sequential	homemade	volume & reliability & cost

Detailed examples of four optimization methods

In this section, four optimization used for the (pre)-design of interleaved Buck or Boost converters are more deeply described. The objectives were similar: maximizing the power density and efficiency. Table II summarizes the four examples.

Table II: Interleaved Buck or Boost converters design methods review

Paper	application	converter spec	method	Optimization algorithm	Problem size	computation time
[54], 2014	?	Boost, 6 kW, [290-480] → 650 V	heuristic	sequential (homemade)	3 input design variables	?
[55], 2015	automotive	Buck [250-1750] W, 60 → 14 V	metaheuristic	NSGA-II	9 input design variables ? constraints 2 objectives	30 min
[38], 2016	automotive	Boost of 40 kW, 220 → 600 V	heuristic	enumeration [40]	11 input design variables ? constraints 2 objectives	15 min
[56], 2017	aeronautic	Buck, 5 kW, [450-800] → [200-400] V	mathematic	SQP	36 input design variables 40 constraints 2 objectives	10 min

Brief description of Interleaved Buck or Boost Converter (IBC)

An IBC is made of several Buck or Boost converters set in parallel, each one duty-cycle being equally shifted. It presents several advantages among reducing current and voltage ripples (requiring smaller filters), current sharing between power components allowing a proportional increase of the converter power and natural redundancy for uninterrupted service.

To gain on the size of the inductors, it is possible to couple them. The choice of coupling the inductors and the coupling topology is function of the operating of the converter and of the criteria to optimize. Then the sizing decisions of this topology is close to the simple Buck or Boost converter: choice of the switching frequency and devices, conduction mode, inductors and capacitors for example. There are nevertheless additional design parameters that are the number of phases and the phase inductors coupling topology.

Detailed description of the four methods

IBC optimized with a heuristic sequential method

The main hypothesis of the designers was to limit the filters size by chosen 4-phases (based on Figure 4) and selecting Continuous Conduction Mode (CCM) and an apparent frequency at 0.5 MHz. Then the authors plotted four surfaces: converter losses and semiconductors junction temperature as functions of the switching frequency and the heatsink thermal resistance and as functions of the switching frequency and the phase inductor current ripple (as in Figure 5). Thanks to those plots, they re-adjusted the switching frequency, the required thermal resistance of the heat sink and the phase inductor value.

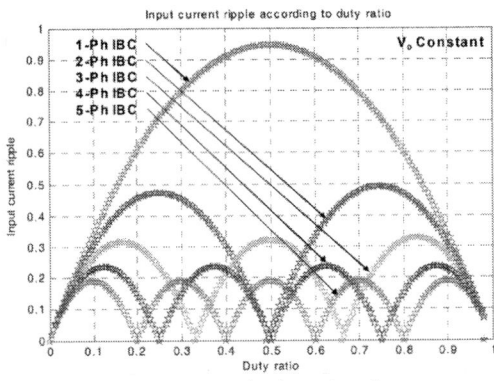

Figure 4: Input current ripple ratio of an Interleaved Boost Converter according to its number of phases and duty-cycle (from [57])

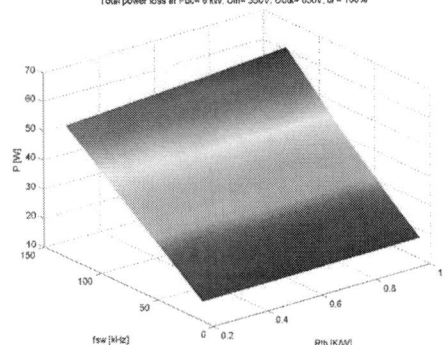

Figure 5: Constraints and objective criteria trade-off surface from [54]

IBC optimized with a metaheuristic population based method

Authors of [55] chose also continuous conduction mode to limit the constraints on active components and on the phase core losses. They used NSGA-II that had to select active, passive and mechanical components in a database while the switching frequency has been defined as a continuous input variable and the number of phases as discrete input variable. They could not optimize the converter on the all power operating range so they ran several optimizations at different power (Figure 6). Based on this, they had to select the design that offers a good compromise for the converter real operation. The main advantage of this method is the direct obtainment of the bill of material of power devices.

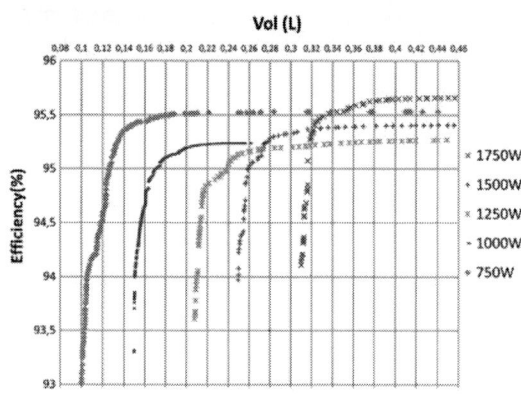

Figure 6: Pareto front of an IBC from [55]

IBC optimized with a heuristic enumeration method

To evaluate the gain of the use of coupled inductors, the authors of [38] compared the power density of three configurations: three-phase discrete, four-phase discrete and four-phase coupled. These optimizations have been performed with an explicit enumeration for integer programming problems [40]. So all variables had to be discretized including the switching frequency.

Figure 7 presents all feasible and Pareto-optimal designs for this problem. While the Pareto front of the three-phase discrete inductors is nearly continuous, the four-phase coupled inductor topology one is discontinuous. The authors explained this discontinuity by the fact that the semiconductors module and the heat sink are not set as design variables and besides drives the converter volume. Indeed, even if the switching frequency increases, the efficiency drop but the power density stays constant to 25 kW/kg.

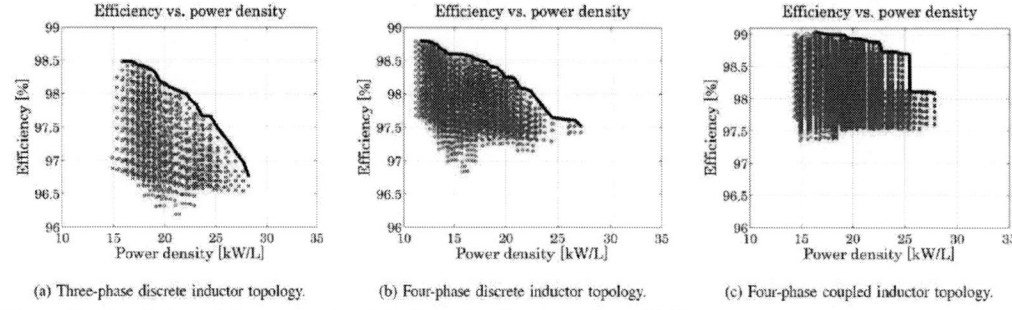

(a) Three-phase discrete inductor topology.

(b) Four-phase discrete inductor topology.

(c) Four-phase coupled inductor topology.

Figure 7: Feasible and Pareto-optimal designs for the three from [38]

IBC optimized with a mathematic gradient-based method

In contrary to the previous methods that were set in the second design stage, the optimization of the IBC in [56] has been performed in the first stage to negotiate the converter specifications. In that purpose, first, no a priori decision has been taken regarding the conduction mode, and secondly, several operating points of the converter have been optimized simultaneously. The number of constraints increased proportionally with the number of studied points. This increase has been made possible by using the gradient-based SQP algorithm that implied to make all the design variables continuous. Figure 8 shows the obtained Pareto front.

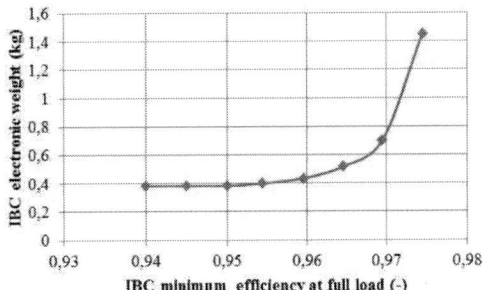

Figure 8: Pareto front of the IBC optimized with mathematic method

Comparison of the four methods for the optimization of IBC

Due to a lack of data about the optimization problem size and because the set of specification is not the same, it has no senses to compare the results of the four methods. But from that review, it is possible to give the advantages of each method: the understanding in the heuristic sequential [54], the possibility to use database in the metaheuristic [55], the ability to optimize the topology in the heuristic enumeration [38] and finally the ability to pre-size on a wide operating range in the mathematic one [56].

Conclusion

The interest of using optimization method for power electronics sizing has not to be proved anymore. There are still nevertheless some barriers for the designer to use it.

First, there are no ideal methods to optimize power converters, but several ones available to explore a large number of possible solutions. The choice of the method will depend on the project stage (1 or 2), if the problem can be made continuous or if it has to be very discretized, if the objective is to find the global optimal or just a good solution.

Secondly, power electronics designers should receive a basic course about optimization method. The aim of this paper was to introduce the different methods.

Third, the access to different numerical tools may be difficult for designers. It is associated with another constraint: the absence of a furnished database of models adapted to optimization methods. Then, the time spent into linking the algorithms to the models may be deterrent. This subject has been partially discussed in [58] in which the authors concluded that "multi-objective optimization is a multidisciplinary hierarchical method that guarantees full utilization of design space" that are among "the three trends emerging in the tools arena for design automation". In a close future, a roadmap should be constructed within power electronics community to specify the requirements of design by optimization tools.

References

[1] S. Balachandran et F. Lee, « Algorithms for Power Converter Design Optimization », *IEEE Trans. Aerosp. Electron. Syst.*, vol. AES-17, n° 3, p. 422-432, mai 1981, doi: 10.1109/TAES.1981.309070.

[2] S. E. De Leon-Aldaco, H. Calleja, et J. Aguayo Alquicira, « Metaheuristic Optimization Methods Applied to Power Converters: A Review », *IEEE Trans. Power Electron.*, vol. 30, n° 12, p. 6791-6803, déc. 2015, doi: 10.1109/TPEL.2015.2397311.

[3] Elsevier, *Scopus*. 2019.

[4] A. Chauvin, A. Hijazi, E. Bideaux, et A. Sari, « Combinatorial approach for sizing and optimal energy management of HEV including durability constraints », in *2015 IEEE 24th International Symposium on*

Industrial Electronics (ISIE), Buzios, Rio de Janeiro, Brazil, juin 2015, p. 1236-1241, doi: 10.1109/ISIE.2015.7281649.

[5] J. A. Snyman et D. N. Wilke, *Practical mathematical optimization*. New York, NY: Springer Berlin Heidelberg, 2018.

[6] A. Chauvin, A. Hijazi, E. Bideaux, et A. Sari, « Cost Optimization for Plug-In Integration in a Hybrid Electric Mini-Excavator with Mixed-Integer Linear Programming », in *2015 IEEE Vehicle Power and Propulsion Conference (VPPC)*, Montreal, QC, Canada, oct. 2015, p. 1-6, doi: 10.1109/VPPC.2015.7352923.

[7] S. Bechikh, R. Datta, et A. Gupta, Éd., *Recent Advances in Evolutionary Multi-objective Optimization*, vol. 20. Cham: Springer International Publishing, 2017.

[8] K. Hornik, M. Stinchcombe, et H. White, « Multilayer feedforward networks are universal approximators », *Neural Netw.*, vol. 2, n° 5, p. 359-366, janv. 1989, doi: 10.1016/0893-6080(89)90020-8.

[9] S. Busquets-Monge *et al.*, « Design of a Boost Power Factor Correction Converter Using Optimization Techniques », *IEEE Trans. Power Electron.*, vol. 19, n° 6, p. 1388-1396, nov. 2004, doi: 10.1109/TPEL.2004.836638.

[10] G. N. Vanderplaats, *Numerical optimization techniques for engineering design*. Colorado Springs, CO: Vanderplaats Research & Development, 2005.

[11] R. T. Haftka et Z. Gürdal, *Elements of structural optimization*, 3rd rev. and expanded ed. Dordrecht ; Boston: Kluwer Academic Publishers, 1992.

[12] M. Delhommais, G. Dadanema, Y. Avenas, J. L. Schanen, F. Costa, et C. Vollaire, « Using design by optimization for reducing the weight of a SiC switching cell », in *2016 IEEE Energy Conversion Congress and Exposition (ECCE)*, Milwaukee, WI, USA, sept. 2016, p. 1-8, doi: 10.1109/ECCE.2016.7855545.

[13] P. T. Boggs et J. W. Tolle, « Sequential quadratic programming for large-scale nonlinear optimization », *J. Comput. Appl. Math.*, vol. 124, n° 1-2, p. 123-137, déc. 2000, doi: 10.1016/S0377-0427(00)00429-5.

[14] C. Larouci, M. Boukhnifer, et A. Chaibet, « Design of Power Converters by Optimization Under Multiphysic Constraints: Application to a Two-Time-Scale AC/DC–DC Converter », *IEEE Trans. Ind. Electron.*, vol. 57, n° 11, p. 3746-3753, nov. 2010, doi: 10.1109/TIE.2010.2041134.

[15] A. Stupar, T. McRae, N. Vukadinovic, A. Prodic, et J. A. Taylor, « Multi-Objective Optimization of Multi-Level DC–DC Converters Using Geometric Programming », *IEEE Trans. Power Electron.*, vol. 34, n° 12, p. 11912-11939, déc. 2019, doi: 10.1109/TPEL.2019.2908826.

[16] S. Boyd, S.-J. Kim, L. Vandenberghe, et A. Hassibi, « A tutorial on geometric programming », *Optim. Eng.*, vol. 8, n° 1, p. 67-127, mai 2007, doi: 10.1007/s11081-007-9001-7.

[17] O. Deblecker, C. Versele, Z. De Greve, et J. Lobry, « Multiobjective optimization of a power supply for space application using a flexible CAD tool », in *2013 15th European Conference on Power Electronics and Applications (EPE)*, Lille, France, sept. 2013, p. 1-10, doi: 10.1109/EPE.2013.6631797.

[18] H. Suryanarayana et S. D. Sudhoff, « Design Paradigm for Power Electronics-Based DC Distribution Systems », *IEEE J. Emerg. Sel. Top. Power Electron.*, vol. 5, n° 1, p. 51-63, mars 2017, doi: 10.1109/JESTPE.2016.2626458.

[19] K. Deb, A. Pratap, S. Agarwal, et T. Meyarivan, « A fast and elitist multiobjective genetic algorithm: NSGA-II », *IEEE Trans. Evol. Comput.*, vol. 6, n° 2, p. 182-197, avr. 2002, doi: 10.1109/4235.996017.

[20] C. Versèle, O. Deblecker, et J. Lobry, « Multiobjective optimal choice and design of isolated dc-dc power converters », in *Proceedings of the 2011 14th European Conference on Power Electronics and Applications*, août 2011, p. 1-10.

[21] H. Visairo, M. A. Medina, et J. M. Ramirez, « Use of evolutionary algorithms for design optimization of power converters », in *CONIELECOMP 2012, 22nd International Conference on Electrical Communications and Computers*, Cholula, Puebla, Mexico, févr. 2012, p. 268-272, doi: 10.1109/CONIELECOMP.2012.6189922.

[22] J. Huber, G. Ortiz, F. Krismer, N. Widmer, et J. W. Kolar, « Pareto optimization of bidirectional half-cycle discontinuous-conduction-mode series-resonant DC/DC converter with fixed voltage transfer ratio », in *2013 Twenty-Eighth Annual IEEE Applied Power Electronics Conference and Exposition (APEC)*, Long Beach, CA, USA, mars 2013, p. 1413-1420, doi: 10.1109/APEC.2013.6520484.

[23] B. Strothmann, F. Schafmeister, et J. Bocker, « Pareto Design and Switching Frequencies for SiC MOSFETs Applied in an 11 kW Buck Converter for EV-Charging », in *2019 IEEE Applied Power Electronics Conference and Exposition (APEC)*, Anaheim, CA, USA, mars 2019, p. 2234-2240, doi: 10.1109/APEC.2019.8721850.

[24] K. Ejjabraoui, C. Larouci, P. Lefranc, et C. Marchand, « Pre-sizing of dc-dc converters by optimization under constraints; influence of the control constraint on the optimization results », in *2010 IEEE International Conference on Industrial Technology*, Vi a del Mar , Chile, 2010, p. 800-806, doi: 10.1109/ICIT.2010.5472627.

[25] H. Ounis, B. Sareni, X. Roboam, et A. De Andrade, « Multi-level integrated optimal design for power systems of more electric aircraft », *Math. Comput. Simul.*, vol. 130, p. 223-235, déc. 2016, doi: 10.1016/j.matcom.2015.08.016.

[26] Q. Li, R. Burgos, et D. Boroyevich, « Hierarchical Weight Optimization Design of Aircraft Power Electronics Systems Using Metaheuristic Optimization Methods », in *2018 AIAA/IEEE Electric Aircraft Technologies Symposium*, Cincinnati, Ohio, juill. 2018, doi: 10.2514/6.2018-4980.

[27] R. Eberhart et J. Kennedy, « A new optimizer using particle swarm theory », in *MHS'95. Proceedings of the Sixth International Symposium on Micro Machine and Human Science*, Nagoya, Japan, 1995, p. 39-43, doi: 10.1109/MHS.1995.494215.

[28] T. Dragicevic, P. Wheeler, et F. Blaabjerg, « Artificial Intelligence Aided Automated Design for Reliability of Power Electronic Systems », *IEEE Trans. Power Electron.*, vol. 34, n° 8, p. 7161-7171, août 2019, doi: 10.1109/TPEL.2018.2883947.

[29] H. Tripathi et N. Pradhan, « Reliability optimization of electronics module by derating using genetic algorithm », in *2016 International Conference on Control, Computing, Communication and Materials (ICCCCM)*, Allahbad, India, oct. 2016, p. 1-6, doi: 10.1109/ICCCCM.2016.7918264.

[30] G. K. Y. Ho, Y. Fang, et B. M. H. Pong, « A Multiphysics Design and Optimization Method for Air-Core Planar Transformers in High-Frequency *LLC* Resonant Converters », *IEEE Trans. Ind. Electron.*, vol. 67, n° 2, p. 1605-1614, févr. 2020, doi: 10.1109/TIE.2019.2910023.

[31] P. I. Frazier, « A Tutorial on Bayesian Optimization », *ArXiv180702811 Cs Math Stat*, juill. 2018, Consulté le: nov. 29, 2019. [En ligne]. Disponible sur: http://arxiv.org/abs/1807.02811.

[32] B. Zhao, X. Zhang, et J. Huang, « AI Algorithm-Based Two-Stage Optimal Design Methodology of High-Efficiency *CLLC* Resonant Converters for the Hybrid AC–DC Microgrid Applications », *IEEE Trans. Ind. Electron.*, vol. 66, n° 12, p. 9756-9767, déc. 2019, doi: 10.1109/TIE.2019.2896235.

[33] H. Helali *et al.*, « Power converter's optimisation and design. Discrete cost function with genetic based algorithms », in *2005 European Conference on Power Electronics and Applications*, Dresden, Germany, 2005, p. 7 pp.-P.7, doi: 10.1109/EPE.2005.219520.

[34] I. Laird, X. Yuan, J. Scoltock, et A. J. Forsyth, « A Design Optimization Tool for Maximizing the Power Density of 3-Phase DC–AC Converters Using Silicon Carbide (SiC) Devices », *IEEE Trans. Power Electron.*, vol. 33, n° 4, p. 2913-2932, avr. 2018, doi: 10.1109/TPEL.2017.2705805.

[35] H. Uemura, D. Yoshida, H. Fujimoto, Y. Okuma, et J. W. Kolar, « Benefits and challenges of design process that employs a multi-objective optimization approach for industrial applications », in *2015 IEEE 2nd International Future Energy Electronics Conference (IFEEC)*, Taipei, Taiwan, nov. 2015, p. 1-6, doi: 10.1109/IFEEC.2015.7361525.

[36] G. Marsala et A. Ragusa, « A Design Method of Zero Ripple Input Current DC-DC Converter Assisted by Constrained GA&PSO Algorithms to Minimize Power Losses », p. 6.

[37] L. Schrittwieser, M. Leibl, et J. W. Kolar, « 99% Efficient Isolated Three-Phase Matrix-Type DAB Buck–Boost PFC Rectifier », *IEEE Trans. Power Electron.*, vol. 35, n° 1, p. 138-157, janv. 2020, doi: 10.1109/TPEL.2019.2914488.

[38] J. Scoltock, G. Calderon-Lopez, Y. Wang, et A. J. Forsyth, « Design optimisation and trade-offs in multi-kW DC-DC converters », in *2016 IEEE Energy Conversion Congress and Exposition (ECCE)*, Milwaukee, WI, USA, sept. 2016, p. 1-8, doi: 10.1109/ECCE.2016.7855029.

[39] J. Scoltock, G. Calderon-Lopez, et A. J. Forsyth, « Topology and Magnetics Optimisation for a 100-kW Bi-Directional DC-DC Converter », in *2017 IEEE Vehicle Power and Propulsion Conference (VPPC)*, Belfort, déc. 2017, p. 1-6, doi: 10.1109/VPPC.2017.8330900.

[40] S. G. Chen, « Efficiency improvement in explicit enumeration for integer programming problems », in *2013 IEEE International Conference on Industrial Engineering and Engineering Management*, Bangkok, Thailand, déc. 2013, p. 98-100, doi: 10.1109/IEEM.2013.6962382.

[41] S. Pam, Y. Satija, P. Chawda, M. Mansour, R. Hanrahan, et J. Perry, « A web-based tool for compensation design of power converters using hybrid optimization », in *2016 IEEE Applied Power Electronics Conference and Exposition (APEC)*, Long Beach, CA, mars 2016, p. 3266-3272, doi: 10.1109/APEC.2016.7468334.

[42] S. S. Rao, *Engineering optimization: theory and practice.* New York: Wiley, 1996.

[43] R. Storn et K. Price, « Differential Evolution – A Simple and Efficient Heuristic for global Optimization over Continuous Spaces », *J. Glob. Optim.*, vol. 11, nº 4, p. 341-359, 1997, doi: 10.1023/A:1008202821328.

[44] G. Wang, X. Li, M. Wang, H. Yuan, et W. Xu, « Optimization of a Dual-Band Wireless Power and Data Telemetry System Using Genetic Algorithm », *IEEE Des. Test*, vol. 35, nº 3, p. 54-65, juin 2018, doi: 10.1109/MDAT.2016.2582828.

[45] Z.-H. Zhan et J. Zhang, « Differential evolution for power electronic circuit optimization », in *2015 Conference on Technologies and Applications of Artificial Intelligence (TAAI)*, Tainan, Taiwan, nov. 2015, p. 158-163, doi: 10.1109/TAAI.2015.7407129.

[46] Tao Huang, Jian Huang, et Jun Zhang, « An orthogonal local search genetic algorithm for the design and optimization of power electronic circuits », in *2008 IEEE Congress on Evolutionary Computation (IEEE World Congress on Computational Intelligence)*, Hong Kong, China, juin 2008, p. 2452-2459, doi: 10.1109/CEC.2008.4631126.

[47] A. Voldoire, Jl. Schanen, Jp. Ferrieux, C. Gautier, et C. Saber, « Optimal Design of an AC Filtering Inductor for a 3-Phase PWM Inverter Including Saturation Effect », in *2019 10th International Power Electronics, Drive Systems and Technologies Conference (PEDSTC)*, Shiraz, Iran, févr. 2019, p. 595-599, doi: 10.1109/PEDSTC.2019.8697246.

[48] L. Jianfeng, L. Meiyu, Y. Guangzheng, et C. Rusi, « Parameter optimization design method of LCL filter based on harmonic stability of VSC system », in *2019 4th International Conference on Intelligent Green Building and Smart Grid (IGBSG)*, Hubei, Yi-chang, China, sept. 2019, p. 226-231, doi: 10.1109/IGBSG.2019.8886239.

[49] X. Ma, Q. Zhang, G. Tian, J. Yang, et Z. Zhu, « On Tchebycheff Decomposition Approaches for Multiobjective Evolutionary Optimization », *IEEE Trans. Evol. Comput.*, vol. 22, nº 2, p. 226-244, avr. 2018, doi: 10.1109/TEVC.2017.2704118.

[50] L. Keuck, F. Schafmeister, et J. Bocker, « Computer-Aided Design and Optimization of an Integrated Transformer with Distributed Air Gap and Leakage Path for an LLC Resonant Converter », in *2019 IEEE Applied Power Electronics Conference and Exposition (APEC)*, Anaheim, CA, USA, mars 2019, p. 1415-1422, doi: 10.1109/APEC.2019.8722077.

[51] C. Feeney et N. Wang, « A new Electronic Design Automation tool for the optimization of PwrSoC/PwrSiP DC-DC converters », in *2018 IEEE Applied Power Electronics Conference and Exposition (APEC)*, San Antonio, TX, USA, mars 2018, p. 2905-2909, doi: 10.1109/APEC.2018.8341430.

[52] R. Hooke et T. A. Jeeves, « `` Direct Search'' Solution of Numerical and Statistical Problems », *J. ACM*, vol. 8, nº 2, p. 212-229, avr. 1961, doi: 10.1145/321062.321069.

[53] J. M. Lenz, D. Zhou, H. Wang, et J. R. Pinheiro, « Optimization Tool for Dc-Link Capacitor Bank Design in PV Inverters », p. 7.

[54] M. Zdanowski, J. Rabkowski, et R. Barlik, « Design issues of the high-frequency interleaved DC/DC boost converter with Silicon Carbide MOSFETs », in *2014 16th European Conference on Power Electronics and Applications*, Lappeenranta, Finland, août 2014, p. 1-10, doi: 10.1109/EPE.2014.6910918.

[55] C. Marchand, G. Coquery, C. Larouci, M. Bendali, et T. Azib, « Design methodology of an interleaved buck converter for onboard automotive application, multi-objective optimisation under multi-physic constraints », *IET Electr. Syst. Transp.*, vol. 5, nº 2, p. 53-60, juin 2015, doi: 10.1049/iet-est.2013.0045.

[56] M. Delhommais, J.-L. Schanen, F. Wurtz, C. Rigaud, et S. Chardon, « Design by optimization methodology: Application to a wide input and output voltage ranges interleaved buck converter », in *2017 IEEE Energy Conversion Congress and Exposition (ECCE)*, Cincinnati, OH, oct. 2017, p. 3529-3536, doi: 10.1109/ECCE.2017.8096629.

[57] G.-Y. Choe, J.-S. Kim, H.-S. Kang, et B.-K. Lee, « An Optimal Design Methodology of an Interleaved Boost Converter for Fuel Cell Applications », *J. Electr. Eng. Technol.*, vol. 5, nº 2, p. 319-328, juin 2010, doi: 10.5370/JEET.2010.5.2.319.

[58] A. Bindra et A. Mantooth, « Modern Tool Limitations in Design Automation: Advancing Automation in Design Tools is Gathering Momentum », *IEEE Power Electron. Mag.*, vol. 6, nº 1, p. 28-33, mars 2019, doi: 10.1109/MPEL.2018.2888653.

A Flexible Power Crossbar-based Architecture for Software-Defined Power Domains

Francesco Di Gregorio, Gilles Sassatelli, Abdoulaye Gamatié, Arnaud Castelltort
Laboratoire d'Informatique, de Robotique et de Microélectronique de Montpellier (LIRMM)
CNRS and University of Montpellier
Montpellier, France
firstname.lastname@lirmm.fr

Acknowledgments

This work was carried out as part of the ICARE project funded by the Occitanie region, France.

Keywords

≪Smart grids≫, ≪Microgrid≫, ≪Distributed Power≫, ≪Energy system management≫, ≪Software-Defined Power Domains≫.

Abstract

This paper proposes a novel approach referred to as "Software-Defined Power Domains", relying on a power crossbar component that makes it possible to setup arbitrary electrical topologies onto rather densely connected physical micro-grids. It presents the proposed crossbar architecture together with its topology switching operation and efficiency analysis.

Introduction

Energy transition and design of sustainable energy systems will require incremental yet profound long-term structural changes in energy production, storage and delivery. Besides the obvious expected role of renewable energies as sources, one prominent challenge lies in finding solutions capable of supporting the required gradual changes to power grids, from centralized generation towards distributed, heterogeneous and hybrid systems [1]. Smart grids and much of the work conducted in that area are meant to mature a portfolio of technologies enabling the advent of resilient self-regulated multi-scale power infrastructures capable of optimizing the overall efficiency and lifting dependence on fossil fuel energies.

Distributed generation exploiting intermittent renewable energy sources, distributed energy storage and smart scheduling decrease power losses by favoring the use of locally produced or stored energy, thereby significantly decreasing the average energy travel distance as well as the number of required current/voltage conversions. The main challenges are then to design technologies and control systems amenable to the integration into existing grids whatever their nature. We regard flexibility, in particular in terms of energy transport, as a primary concern that requires specific attention.

This paper therefore proposes a novel DC micro-grid (DCMG) crossbar architecture that makes it possible to install time-changing electrical topologies onto the existing physical architectures. We improve the state-of-the-art by proposing a system organization centered around a power crossbar that enables software-defined electrical topologies with the corresponding digital control architecture. This enables to handle dynamic and transparent switching (handover operation) from one set of energy actors to another. Our approach provides notable advantages through its flexibility, such as minimization of distribution / conversion losses via selection of appropriate routes between sources and loads. As the digital control architecture performs constant, high-frequency monitoring of multiple voltages and currents, the approach enables to setup a reactive failure detection and mitigation capable of recovering from various failures without interruption of energy delivery.

Related Works

DCMG Architectures

The choice of the architecture is very important in the design of DCMG systems [5] [6] [7] [10]. It influences many factors such as the cost, resiliency, controllability, reliability, availability, resource utilization, and flexibility. DCMG architectures have been surveyed according to the number of available paths between loads and the AC grid in [6]. In this survey, it is shown that system reliability heavily depends on path diversity: in radial and ring (see Fig. 1.b) configurations for example, there is a single AC feeder. The architecture topology then exhibits more or less paths to the AC grid interface. The limitation of these architectures is in case a fault occurs on the AC feeder, the whole micro-grid functionality will be compromised because of the impossibility to access this always present energy resource, thereby decreasing reliability.

To overcome these limitations, interconnected architectures such as ladder-type [10] and zonal-type DCMG [9] can be used. These two architectures solve the reliability issue by adding AC feeders and additional redundant paths to the power network. However, this addition increases the architecture implementation costs and decreases the resource utilization because of the rather static [8] [10] nature of most current architectures. To enhance the architecture flexibility, other redundant paths are usually added to the system to connect a point A with a remote point B, i.e. a load with the AC feeder. This adds complexity and costs to the whole system. In this paper, we use the term Energy Player (EP) to denote all energy devices involved in a micro-grid including distributed energy resources (DERs), loads, Energy Storage Systems (ESS) and AC grid. Even though the concept of energy routing already exists in literature [4] [3], there is no, to the best of our knowledge, physical implementation of a device which enables to proactively route the energy between EPs.

The proposed power crossbar-based architecture aims to tackle these issues by adding the possibility to dynamically and electrically reconfigure the power network by software. This greatly enhances the flexibility of the system compared to the existing static architectures. Moreover, power crossbars enable energy routing through the power network and increase the number of redundant paths between loads and the AC feeder, improving reliability without compromising resource utilization. Finally, this architecture further enables dynamic topology switching, opening an avenue for various run-time optimizations such as minimizing distribution losses or increased resilience through instant failure detection and recovery.

Crossbar Architectures

The architecture of an energy router is proposed in [15] which interfaces 2 AC Microgrids (ACMG) to the electrical grid. This energy router controls the power flow between the three ports using power converters. The limitation of this configuration is that it is static and not scalable. In fact, the greater the number of ports, the higher the complexity in its control. Moreover, this architecture has a single point of failure. In fact, a fault on the common DC Bus (DCB) determines the inability to interconnect the three ports.

Data crossbars as found in telecommunication networks are on their side specifically designed to create independent point to point connections between a set of inputs and a set of outputs. In a crossbar architecture, each input is connected to a dedicated bus which can be connected to any output bus by using one of the switches present at every X-Y cross-point. This allows to dynamically create data path where information flow from the input to the output. This technique is well known in computer science as circuit switching [16]. The advantage of this architecture is that a failure on a bus does not compromise the functioning of the whole crossbar, but only of the connected units. Such crossbars are however oriented, therefore limiting the applicability should one draw inspiration from such communication crossbars towards a power switching architecture where input and output concepts are irrelevants. Indeed a connection between two "outputs" is possible at the cost of allocating an input bus, thereby limiting the flexibility of the architecture.

Table I: Comparison of crossbar architectures.

Architecture	Cost	Flexibility	Reliability
Energy Router in [15]	Low	Low	Low
Crossbar (communication network)	High	Medium	Medium
Crossbar (proposed architecture)	Medium	High	High

The proposed crossbar architecture favours connections between EP groups and lifts this limitation. Each EP has the possibility to select among M independent busses (see Fig. 3.b) of the local crossbar which, in turns, can connect to the busses of the adjacent crossbars. This permits to dynamically create, destroy and bind independent groups of EPs which share energy without assumption on power flow direction. Moreover, the proposed architecture is more reliable because a failure on a DCB does not compromise the functioning of any of the connected EPs which still can exploit the $M-1$ remaining busses of the crossbar. Table I qualitatively compares the three architectures in terms of cost, flexibility and reliability.

Crossbar-based DCMG Architecture

The architecture of the proposed crossbar-based DCMG is shown in Fig. 1.c. It consists of a set of DERs (i.e. solar panel, wind pole), loads (i.e. servers, electric vehicles etc.) and ESSs (i.e. batteries or super-capacitors) interconnected together by using power crossbars. In order to let this heterogeneous set of energy sources and loads interact together, power converters are used. Fig. 2 shows the structure of the conventional bidirectional DC/DC converter used to interface all the EPs. Buck / Boost modes are supported in both directions thanks to the symmetrical structure. The power crossbar, as will be explained in the next section, guarantees galvanic isolation for each EP. Then, it allows the dynamic reconfiguration of the DCMG by software. Indeed, it permits to dynamically create, decompose or even connect independent clusters of EPs, here referred as Power Domains (PDs). It allows resource sharing between sources and loads connected to different crossbars. Finally, it acts as a bridge in order to connect any two adjacent power crossbars without involving the local EPs. These EPs can still share their local resources by using an alternative independent bus of the crossbar to which they are physically connected.

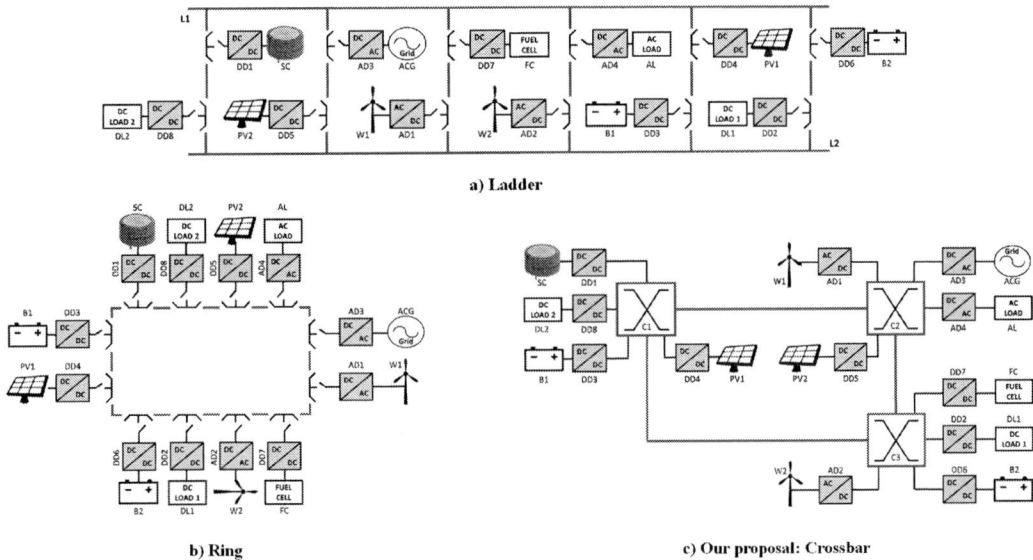

Fig. 1: DCMG implementation example: a) Ladder, b) Ring, and c) Crossbar-based architectures.

The control of the whole system and its associated crossbar is assigned to one or more Systems-on-Chip (SoCs) integrated on each crossbar. In particular, the Master-Slave control technique [2] [10] is

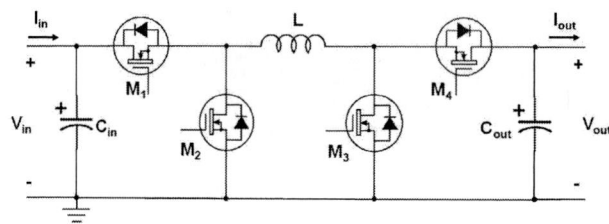

Fig. 2: Schematic of the Bidirectional DC/DC Converter configuration used in this work.

used for the local power management. In Master-Slave technique a Master controller is chosen with the objective to keep the voltage on the common DCB constant. All the other controllers of the same PD act as Slaves either injecting current into the DC Bus or plain loads. PI or PID controllers, which drive power converters in Constant Current (CC) or Constant Voltage (CV) mode, are implemented directly in software by exploiting the peripherals of the SoCs. Furthermore, diagnosis and fault recovery functionalities are handled by using local sensors and control. Finally, the addition of a communication interface allows different power crossbars to communicate, enabling the application of global and smart energy management strategies [8] to either increase the overall energy efficiency, reduce the energy cost or increase the reliability for critical loads, etc.

Power Crossbar

The power crossbar is the device that routes energy through the power network and defines PDs in software. It consists of a set of power switches (i.e. relays, MOSFET switches, etc.) which are controlled by software in order to physically define electrical paths on the Power Network for energy transfer between two or more EPs. Power crossbars have an arbitrary number of external ports and independent busses. Depending on this number, some connection combinations between external ports are achieved. The bigger the number of ports and independent busses, the greater the flexibility, at the expense of more complexity in the crossbar topology and control.

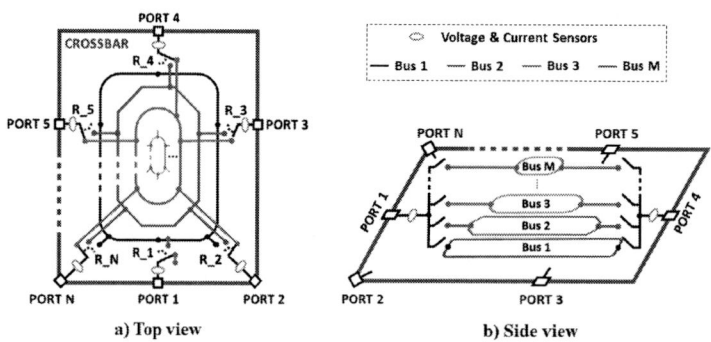

Fig. 3: N-ports, M-busses crossbar: a) Top and b) Side view.

Fig. 3 shows the architecture of an N-ports, M-busses power crossbar. Fig. 3.a represents a top view, outlining the number of ports while Fig. 3.b shows a side view, where the busses with their power switches are highlighted. The number of ports defines the maximum number of external connections of the crossbar. The number of busses defines both the maximum number of independent paths that can be created on a crossbar, and the maximum power capacity the crossbar can support. The maximum transmission power can be dynamically adjusted by connecting more busses in parallel.

Implementation Costs

Another important aspect to consider in the architecture is the implementation cost, notably in relation with the number of power switches. Here, we use the number of switches normalized to the number of

EPs, denoted by N_{sw_EP}, as a metric to estimate the implementation cost and compare different DCMG architectures. For instance, in Ring and Ladder ones (see Fig. 1.b and Fig. 1.a respectively), N_{sw_EP} is always equal to 3. In the proposed DCMG architecture, assuming identical crossbar instances, N_{sw_EP} is calculated with the following formula (1):

$$N_{sw\_EP} = M * (1 + N_{CR\_INT}/N_{EP}) \tag{1}$$

where M is the number of independent busses of the crossbar, N_{CR_INT} is the total number of crossbar ports used for interconnections, N_{EP} is the total number of EPs in the DCMG. For instance, in Fig. 1.c $N_{CR\_INT} = 6$ and $N_{EP} = 12$. From (1), using 2-busses crossbars ($M = 2$), one obtains an implementation cost of 3 switches per EP, which is equal to that of Ring and Ladder architectures. However, for our crossbar-based architecture, the cost can be reduced by reducing the ratio N_{CR_INT}/N_{EP}, which is obtained by adding more EPs per crossbar.

Control Strategy

Many control techniques for microgrids can be found in the literature [10][13][14]. A two-level hierarchical control has been chosen for the proposed crossbar-based DCMG. In particular, the first level of control manages the power balance between EPs connected to the same PD. Master-Slave technique is used to reach this objective for its simplicity and effectiveness in small and medium-sized power networks. Then, the second level of control defines PDs, assigns Master role to a power converter in each PD and manages the dynamic topology switching operation.

Dynamic Topology Switching

Having the ability to match sources to current load is here regarded as a prominent advantage. We here demonstrate the ability of performing dynamic (run-time) topology switching, which we refer to as handover operation.

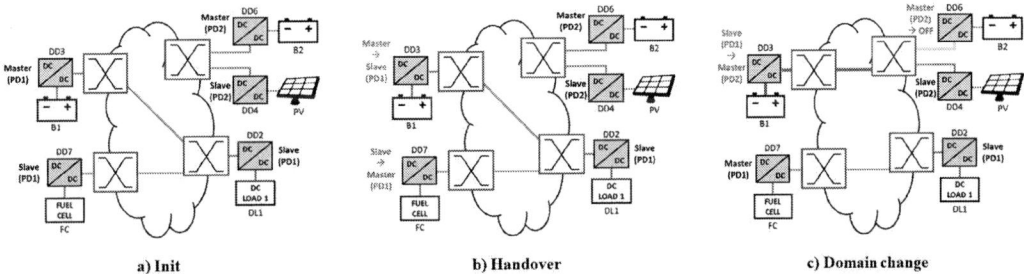

Fig. 4: Dynamic topology switching protocol: a) Init, b) handover and c) power domain change.

This makes it possible to change the Master control of a PD. In a crossbar-based DCMG each PD requires a Master that may turn into a Slave after a handover operation. We here describe a reliable, experimentally validated technique for handover operation. The technique does not alter the power quality that we here define as maximum acceptable voltage ripple, arbitrarily set to 6%. The method is simulated with the Powersim simulator. We consider the electromechanical relays that provide galvanic isolation for each EP. The simulated scenario, depicted in Fig. 4, involves two generic voltage sources of 20V, represented by a battery B1 and a fuel cell FC that reach a DC load DL1 by using the power crossbar network. At the beginning, the B1 converter DD3 is the Master of PD1, guarantying a constant voltage of 36V on the DCB and FC converter DD7 is in slave mode helping feeding DL1. On the other domain PD2, B2 converter DD6 is the Master, keeping a constant voltage of 30V, while receiving the maximum power the solar panel PV, which is in Slave mode, provides. Then, the handover protocol permits to change the Master control of PD1 between the two sources and lets B1 change domain, switching from discharging mode to charging mode.

Table II: Description of the dynamic topology switching steps.

Time	Description	Comments
0ms	Initial state.	The initial configuration is depicted in Fig. 4.a.
50ms	The DCMG controller gradually changes the current set point of FC in order to reach the current contribution of B1.	This reduces oscillations during the handover operation.
55ms	The handover operation takes place by reversing the control of the Master B1 and the Slave FC. Then, the re-configuration process starts by setting relays.	B1 current is set according to the last measurement of its current sensor in order to reduce perturbations on the DCB. Note that chosen relays [12] need 25ms to operate and 5ms to release.
60ms	B1 set point is gradually brought to 0A.	This permits to disconnect B1 without perturbing the DCB of PD1.
80ms	Power relays close and the reconfiguration process is completed. B2 disconnects from the PD2 domain and is turned OFF so as to let B1 charge and take all of the power coming from the solar panel. Master control of PD2 pass to B1.	The controller of B1 adapts its input current in order to keep the DCB to a constant voltage of 30V.

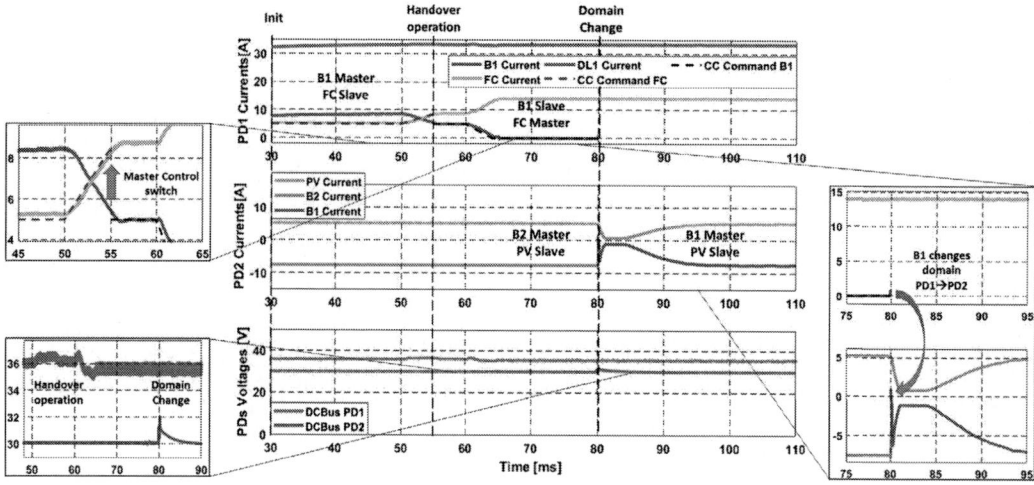

Fig. 5: PSIM simulation of handover and domain change.

The main steps of the handover and domain change operation are summarized in Table II and the simulation results depicted in Fig. 5. A ramp function with a slope of 2.6A/ms is used in the simulation to gradually change the set points to the PI controllers. As shown in Fig. 5 the handover operation does not affect the voltage of the PD1 DCB, which undergoes a slight voltage oscillation of 1V corresponding to 2,78% of its nominal value. A greater voltage peak of 2V corresponds to the activation of relays due to DD3 capacitances already charged from previous operation.

Fault Handling

The proposed DCMG architecture offers better fault handling capabilities with respect to the previous architectures. The presence of M independent busses per crossbar permits to overcome to fault conditions by simply switching the operating bus. Path diversity and dynamic reconfiguration permit to isolate the fault and to perform recovery by selecting another redundant path. Different types of fault can occur in a DCMG architecture. In this work, we consider two main types of fault:

- fault on a DCB (short circuit)
- fault on a terminal relay

Fig. 6 presents a schematic view of the presented architectures highlighting the interconnection components such as terminal relays, internal and external busses. A 2-busses 6-ports power crossbar has been chosen in this work which represents a good compromise between implementation costs and flexibility offered.

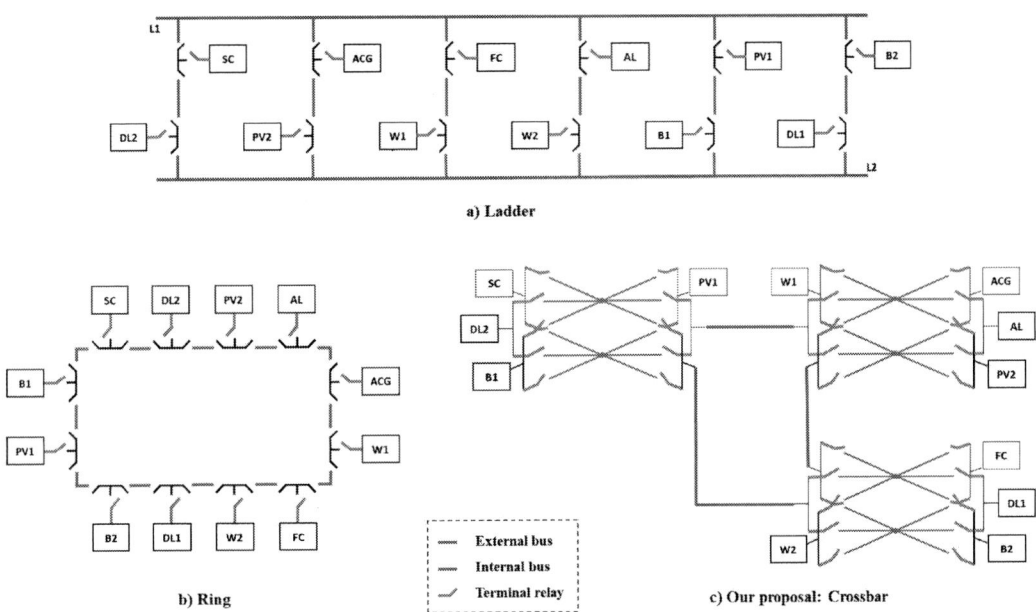

Fig. 6: Bus segmentation in the three DCMG architectures: a) Ladder, b) Ring and c) Crossbar-based.

A fault on a DCB can be more or less critical depending on its position. In ring architecture, for example, the bus is segmented along all the ring (see Fig. 6.b) and all the segments are equally critical. A fault on a segment does not have serious consequences, but two consecutive faults highly limit the connectivity of the EPs.

On the other hand, ladder architecture presents two types of busses (see Fig. 6.a): L1 / L2 external busses and the internal busses in each step of the ladder. Faults on the external busses are more critical especially when the DCMG scale up. In fact, two consecutive faults on the two external busses highly affect the connectivity of the whole DCMG.

Finally, the proposed crossbar-based DCMG presents two types of busses (see Fig. 6.c): internal busses which permit the connection of the EPs attached to the same crossbar and external busses allowing the interconnection between crossbars. A fault on an external bus is less critical than internal busses. In fact, consecutive faults on the external busses of the same crossbar limits the connection of the local EPs to the other crossbars. However, they still have the possibility to exchange energy locally. A fault on an internal bus does not compromise the functioning of any EP which can use the others $M - 1$ busses to interconnect. Consecutive faults on the busses of the same crossbar can limit the connectivity of the EPs connected to that crossbar but the others EPs can still function properly. Moreover, consecutive faults on busses of different crossbars does not compromise any of the EPs.

Terminal relays are the most critical parts in a connection between two EPs. In fact, a fault on these makes it impossible to access the power network by the associated EP. While ring and ladder architectures have a single terminal relay per EP, the proposed crossbar-based architecture has M terminal relays per EP. This redundancy permits an higher level of reliability for the crossbar-based architecture with respect to the others. To summarize, the bus segmentation redundancy and the dynamic topology reconfiguration offered by the crossbar-based DCMG provides better fault handling capability as well as fault isolation and recovery.

Efficiency Analysis

We analyze and compare the energy-efficiency of the three architectures shown in Fig. 1. To estimate the energy transport efficiency, the number of crossed relays is considered. We assume that cable losses can be neglected due to short distance and only account for the contribution to the transmission losses given by the static power loss needed to activate a relay and the power losses contact resistance. Unit loss figures are extracted from a commercial power relay for ensuring realistic assessment even though relative units such as number of relays crossed would yield to similar conclusions. The chosen relay [12] has a static power consumption of 1.7W, a maximum DC current of 50A and uses AgSnO2 as contact material that presents an average contact resistance of 2mΩ [11]. The comparison is made by analyzing the shortest paths allowing to connect a pair of source-load. Then, depending on the number of relays found on that path, an estimation of transmission efficiency is made.

Table III: Shortest paths ranking for the source-load pair B1 → DL1.

Shortest Path	Path Ring	Path Ladder	Path Crossbar-based
1st	B1→PV1→B2→DL1	B1→L2→DL1	B1→C1(1)→C2(1)→DL1
2nd	B1→SC→DL2→PV2→AL→ ACG→W1→FC→W2→DL1	B1→PV1→L1→B2→DL1	B1→ C1(2)→C2(1)→DL1
3rd	Does not exist.	B1→L2→W2→AL→L1→ B2→DL1	B1→C1(1)→C2(2)→DL1

In Table III, an example of comparison between the three architectures is made, by considering the pair battery B1 with the DC load DL1. In particular, the table presents the three shortest paths allowing to connect B1 with DL1. Note that the ring architecture always has two paths to connect a source with a load whose sum equals the number of serial relays on the ring. Ladder architecture has a better organization of its resources thanks to its two common busses L1 and L2 which overall provide higher efficiency and path redundancy. Crossbar-based architecture, thanks to the bus segmentation and configurability makes for a higher number of redundant paths with higher efficiency.

Fig. 7 shows an estimation of transmission efficiency obtained in the three architectures. In particular, four samples of source-load pairs were selected and represented, next to the average on all possible pairs, represented in the fifth columns. Results show that the crossbar-based architecture has a higher energy efficiency compared to Ring and Ladder. A transmission power of 1051W is used for the simulation, which corresponds to the optimal transmission power calculated from the relay characteristics: static power losses in the coil and contact resistance.

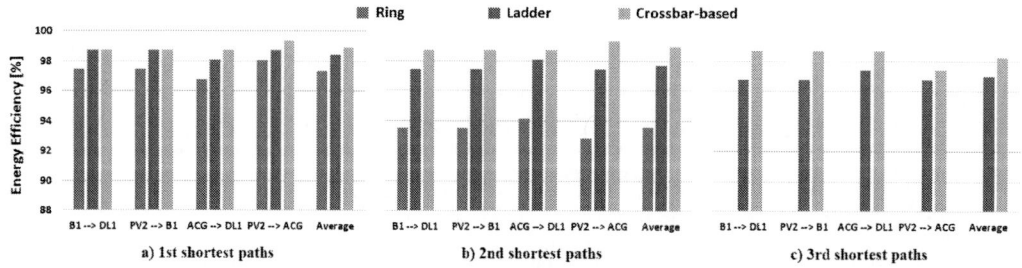

Fig. 7: Energy efficiency comparison between Ring, Ladder and Crossbar-based architectures for several source-load pairs: a) 1st, b) 2nd and c) 3rd shortest paths.

Conclusion

Energy transition is imminent and the role of renewable energy sources will be fundamental. The heterogeneity and intermittent nature makes static topology / architectures unsuited. Dynamic, software-defined architectures should be considered to facilitate this transition. This paper proposes a novel approach referred to as "software-defined power domains" that relies on an energy crossbar component. The solution makes it possible to setup arbitrary electrical topologies onto densely connected physical micro-grids. We improve state-of-the-art by proposing a system organization centered around this crossbar that enables software-defined electrical topologies with the corresponding digital control architecture that handles dynamic and transparent handover from one set of energy actors to another. This approach provides distinct advantages thanks to its flexibility, such as minimization of distribution / conversion losses through selection of appropriate routes between sources and loads. It provides a greater reliability, resource utilization and resilience. Finally, the dynamic switching operation together with an efficiency analysis of the proposed architecture are covered in the paper. Measures of the handover operation are underway on a recently realized prototype of crossbar-based DCMG.

References

[1] R. Salas-Puente, S. Marzal, R. Gonzlez-Medina, E. Figueres, and G. Garcer: Efficient management strategy of the power converters connected to the DC bus in a hybrid microgrid of Distributed Generation, EPE 2017-ECCE Europe, pp. P.1-P.10

[2] D. E. Olivares et al.: Trends in Microgrid Control, IEEE Transactions on Smart Grid, Vol. 5 no 4, pp. 1905-1919, July 2014

[3] M. H. Amini, K. G. Broojeni, T. Dragievi, A. Nejadpak, S. S. Iyengar, and F. Blaabjerg: Application of cloud computing in power routing for clusters of microgrids using oblivious network routing algorithm, EPE 2017-ECCE Europe, pp. P.1-P.11

[4] A. Brocco: Fully distributed power routing for an ad hoc nanogrid, 2013 IEEE International Workshop on Intelligent Energy Systems (IWIES), pp. 113-118, 2013

[5] D. Burmester, R. Rayudu, W. Seah, and D. Akinyele: A review of nanogrid topologies and technologies, Renewable and Sustainable Energy Reviews, Vol. 67, pp. 760-775, 2017

[6] D. Kumar, F. Zare, and A. Ghosh: DC Microgrid Technology: System Architectures, AC Grid Interfaces, Grounding Schemes, Power Quality, Communication Networks, Applications, and Standardizations Aspects, IEEE Access, Vol. 5, pp. 12230-12256, 2017

[7] D. Magdefrau, T. Taufik, M. Poshtan, and M. Muscarella: Analysis and review of DC microgrid implementations, 2016 International Seminar on Application for Technology of Information and Communication (ISemantic), pp. 241-246, 2016

[8] M. Gunasekaran, H. Mohamed Ismail, B. Chokkalingam, L. Mihet-Popa, and S. Padmanaban: Energy Management Strategy for Rural Communities DC Micro Grid Power System Structure with Maximum Penetration of Renewable Energy Sources. Applied Sciences 2018, Vol. 8, no. 4, pp. 585

[9] E. Tironi, M. Corti, and G. Ubezio: Zonal electrical distribution systems in large ships: Topology and control, 2015 AEIT International Annual Conference (AEIT), pp. 1-6, 2015

[10] J. Kumar, A. Agarwal, and V. Agarwal: A review on overall control of DC microgrids, Journal of Energy Storage, Vol 21, pp. 113-138, 2019

[11] A. Ksiazkiewicz, G. Dombek, and K. Nowak: Change in Electric Contact Resistance of Low-Voltage Relays Affected by Fault Current, Materials (Basel, Switzerland) Vol. 12, pp. 1-11, July 2019

[12] Finder: High Power relay 50 A 67 Series, 67.23.9.012.4300 datasheet, 2016

[13] P. Borazjani, N. I. A. Wahab, H. B. Hizam, and A. B. C. Soh: A review on microgrid control techniques, 2014 IEEE Innovative Smart Grid Technologies - Asia (ISGT ASIA), Kuala Lumpur, pp. 749-753, 2014

[14] J. M. Guerrero, M. Chandorkar, T. Lee and P. C. Loh: Advanced Control Architectures for Intelligent MicrogridsPart I: Decentralized and Hierarchical Control, IEEE Transactions on Industrial Electronics, Vol. 60 no 4, pp. 1254-1262, April 2013

[15] Y. Liu, Y. Fang, J. Li: Interconnecting Microgrids via the Energy Router with Smart Energy Management, Energies, Vol. 10 no 9, pp. 1297, 2017

[16] G. Broomell and J. R. Heath: Classification Categories and Historical Development of Circuit Switching Topologies. ACM Comput. Surv., Vol. 15 no 2, pp. 95133, 1983

Impact of grid-forming control on the internal energy of a modular multi-level converter

Ebrahim Rokrok[1], Taoufik Qoria[1], Antoine Bruyere[1], Bruno Francois[1], Haibo Zhang[1], Moez Belhaouane[1], Xavier Guillaud[1]

[1] L2EP, Univ. Lille, Arts et Metiers Institute of Technology, Centrale Lille, Yncrea Hauts-de-France, ULR 2697 - L2EP - Laboratoire d'Electrotechnique et d'Electronique de Puissance, F-59000 Lille, France

E-Mail: ebrahim.rokrok@centralelille.fr

Acknowledgements

This work is supported by the project "HVDC Inertia Provision" (HVDC Pro), financed by the ENERGIX program of the Research Council of Norway (RCN) with project number 268053/E2, and the industry partners; Statnett, Statoil, RTE and ELIA.

Keywords

«Grid-forming control», «Modular multi-level converter», «MMC internal energy».

Abstract

This paper presents a comparative analysis regarding the dynamic behavior of a Modular Multi-level Converter (MMC) with grid-forming control either with or without controlling the MMC internal energy. It has been demonstrated that the internal energy of the MMC in low-level control interacts with the high-level control, which is performed by a grid-forming scheme. In case of controlling the internal energy of the MMC, this interaction is mitigated. Moreover, it has been shown that the dynamic behavior of a grid-forming controlled MMC with internal energy control is identical to an equivalent 2-level Voltage Source Converter (VSC).

Introduction

The MMC has become the most preferred converter topology for high-power applications such as High Voltage Direct Current (HVDC) transmission systems. The MMC control has been widely studied in the literature mostly with the grid-following scheme in the high-level control [1]–[3]. In this case, the control relies on the existence of the instantaneous voltage formed by the AC grid. Compared to the two or three-level VSC, there are some internal capacitors, which store some energy. Two different types of control exist for the MMC: non-energy-based or energy-based control schemes [4].

With the recent interest of using grid-forming control schemes for providing the ancillary services to the power system, it is expected to adapt these kind of controls to the MMCs that will have a major share of power converters in the future power systems. The first examples of studying the utilization of the grid-forming control for the MMC have been published in [5], [6] with the aim of providing virtual inertia and frequency support for an MMC-based HVDC terminal. More recently, in [7] a Virtual Synchronous Machine (VSM) based grid-forming control of MMC in order to provide an inertia support to the grid has been examined with a laboratory demonstration. Authors in [8] have presented an evaluation of different inertia emulation strategies with grid-forming control based on a laboratory-scale prototype of a point-to-point HVDC transmission system with MMC. In these previous publications, the internal energy of the MMC is not controlled and therefore its effect and interactions with the grid-forming control has not been studied.

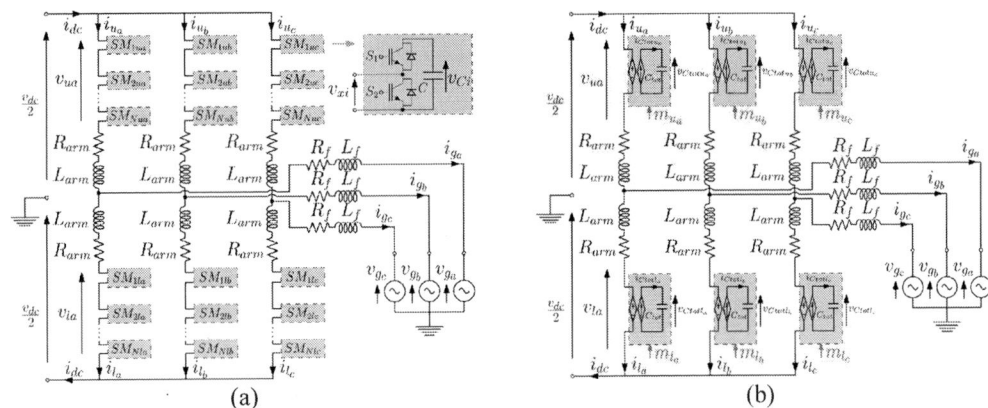

Fig. 1: MMC topology (a) and its arm average model (b)

The contribution of this work is firstly to investigate on the effect of internal energy control of the MMC on the general dynamic behavior of the system with a grid-forming control. Secondly, it has been demonstrated that by controlling MMC internal energy, the dynamic behavior of the grid-forming controlled MMC is identical to an equivalent 2- level VSC. To validate the contributions, the non-energy-based MMC, energy-based MMC and the equivalent 2-level VSC are implemented in Matlab-Simulink environment. Simulation results give the possibility of observing the MMC internal energy level in order to compare the obtained performance by the MMC with the equivalent VSC.

Recall on the MMC model

The MMC with half-bridge switching cells as the Sub-Modules (SMs), which is shown in Fig. 1-a, is the preferred topology for VSC-based HVDC systems in high power applications due to the low power losses, low harmonics and high scalability [9]. To simplify the analysis, by assuming that the sub-module voltages are equal and balanced, an average arm model can be used. This model, which is drawn in Fig. 1-b, allows the study of the overall energy transferred between the DC and AC side or the energy stored in the internal capacitors [4]. For each phase j ($j = a, b, c$), the model presents a leg containing an upper and a lower arm. Each arm consists of an aggregated capacitance C_{tot}, an inductance L_{arm} and an equivalent resistance R_{arm} [2]. The modulated voltages v_u and v_l as well as the current for each arm j are expressed as follow:

$$v_{u_j} = m_{u_j} v_{Ctot\,u_j}, \quad v_{l_j} = m_{l_j} v_{Ctot\,l_j}, \tag{1}$$

$$i_{Ctot\,u_j} = m_{u_j} i_{u_j}, \quad i_{Ctot\,l_j} = m_{l_j} i_{l_j}, \tag{2}$$

where m_{u/l_j} are the corresponding modulation indices, and v_{Ctotu/l_j} are the voltages across the upper/lower equivalent arm capacitance C_{tot}. By applying the Kirchhoff's Voltage Law (KVL), following equations can be obtained for each phase of the MMC:

$$\frac{v_{dc}}{2} = v_{u_j} + L_{arm}\frac{di_{u_j}}{dt} + R_{arm}i_{u_j} + L_f\frac{di_{g_j}}{dt} + R_f i_{g_j} + v_{g_j}, \tag{3}$$

$$\frac{v_{dc}}{2} = v_{l_j} + L_{arm}\frac{di_{l_j}}{dt} + R_{arm}i_{l_j} - L_f\frac{di_{g_j}}{dt} - R_f i_{g_j} - v_{g_j}, \tag{4}$$

where L_f, R_f are the inductance and resistance of the MMC transformer, respectively. v_{g_j} is the voltage at the point of common connection (PCC) and i_{g_j} is the grid current. In order to have a decoupled control on the inner dynamics, the DC and AC part of the above equations can be extracted by adding and subtracting (3) and (4), respectively:

$$\frac{v_{dc}}{2} - v_{diff_j} = L_{arm}\frac{di_{diff_j}}{dt} + R_{arm}i_{diff_j}, \tag{5}$$

$$v_{v_j} - v_{g_j} = L_{eq}\frac{di_{g_j}}{dt} + R_{eq}i_{g_j}, \tag{6}$$

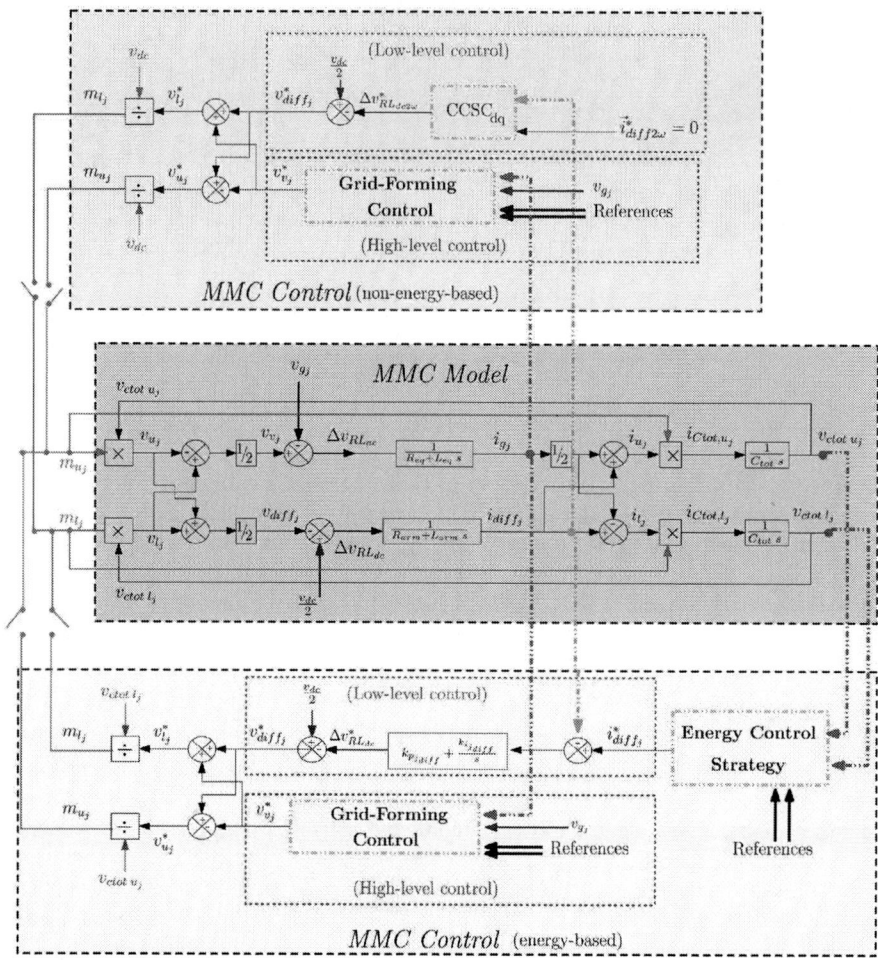

Fig. 2: General structure of the MMC model and control

where $L_{eq} = \frac{2L_f + L_{arm}}{2}$ and $R_{eq} = \frac{2R_f + R_{arm}}{2}$. The voltage $v_{diff_j} \triangleq \frac{v_{u_j} + v_{l_j}}{2}$ is the differential voltage that would internally drive the differential current $i_{diff_j} \triangleq \frac{i_{u_j} + i_{l_j}}{2}$. The voltage $v_v \triangleq \frac{v_{l_j} - v_{u_j}}{2}$ is the external voltage that drives the grid current $i_{g_j} = i_{u_j} - i_{l_j}$.

MMC control system overview

In the literature, two main control solutions for MMC are introduced: non-energy based control strategies and energy-based control strategies. In this section, the basics of these control schemes are recalled.

The non-energy based control strategies are based on the inner current controllers, which produce the voltage references $v^*_{diff_j}$ for the modulators to communicate with the low-level control. In this concept, a circulating current suppression controller (CCSC) strategy is widely used in the literature. In CCSC, by means of a Park's transform at -2ω the second harmonic component of the circulating current that has a large influence on the system losses and voltage ripple of the capacitors is eliminated. By considering (5), (6), the MMC control can be designed.

Fig. 2 illustrates the general structure of the MMC control for both non-energy-based and energy-based schemes. In the non-energy-based control, the DC part of the MMC (equation (6)) is controlled with the CCSC strategy. More details about the CCSC are given in [10].

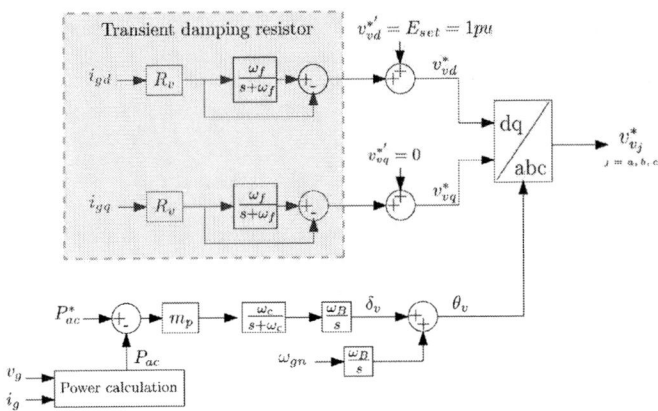

Fig. 3: Classical droop-based grid-forming control for the AC part control of the MMC

To have a full control of the MMC dynamics and the stored energy in the system, the energy-based control has been developed. In this scheme, all the system state variables are controlled. Unlike the non-energy based control, in which the differential current is not fully controlled and only a second harmonic component is suppressed, here the differential current is fully controlled. To control the DC part of the MMC, the energy-based control includes a feedback control loop on the energy of capacitors in each arm, which provides the differential current reference thanks to the energy control. The energy control strategy is explained in the next section.

The differential current is controlled with three PI controllers (one for each phase). The parameters of these PI controllers are designed with the desired response time of the transfer function $i_{diff_j}/i^*_{diff_j}$, which can be simply calculated from the Fig. 2.

In order to control the AC part of the MMC (equation (5)), a grid-forming control scheme provides the reference of the external voltage $v^*_{v_j}$. Several grid-forming control strategies have been proposed in the literature [11]–[13]. Among these grid-forming control solutions, the droop-based grid-forming control is considered as an extensively accepted baseline solution (Fig. 3).

The grid-forming control is based on the phase and magnitude control of the external voltage v_v. Fig. 3 shows the scheme of the droop-based grid-forming control [14], [15]. According to this figure, the converter active power P_{ac} is controlled through the voltage angle θ_v. The parameter m_p is the droop gain. A low-pass filter with $\frac{\omega_{gn}.\omega_B}{20} < \omega_c < \frac{\omega_{gn}.\omega_B}{5}$ is applied to the mismatch of the reference power and measured power in order to filter the measurement noise and to avoid frequency jump. This filter also brings the benefit of virtual inertia effect to this control. ω_{gn} is the grid nominal frequency and ω_B is the base frequency. It should be noted that the voltage set-point E_{set} is directly applied to the d-axis of the converter voltage with no internal cascaded voltage and current control loops. A damping resistor R_v that operates only during the transient, thanks to a high-pass filter tuned by ω_f, is acting on the voltage references in order to damp the oscillation on the grid current in case that there is not enough resistive effect in the grid. More details about this grid-forming control are given in [14], [15].

Energy control strategy

By considering the equivalent arm capacitors as the main energy storage elements in the MMC, the stored energy in upper/lower arm ($W_{Ctot\,u/l_j}$) can be calculated as:

$$\frac{W_{Ctot\,u/l_j}}{dt} = v_{Ctot\,u/l_j}i_{Ctot\,u/l_j}. \tag{7}$$

Based on (1) and (2):

$$v_{Ctot\,u/l_j}i_{Ctot\,u/l_j} = v_{u/l_j}i_{u/l_j}. \tag{8}$$

By expressing v_{u/l_j} and i_{u/l_j} based on the external voltage and differential voltage/current, following equations are obtained for the MMC energy in upper/lower arm:

$$\frac{W_{Ctot\,u_j}}{dt} = v_{Ctot\,u_j}i_{ctot\,u_j} = v_{u_j}i_{u_j} = \left(v_{diff_j} - v_{v_j}\right)\left(i_{diff_j} + \frac{i_{g_j}}{2}\right). \tag{9}$$

$$\frac{W_{Ctot\,l_j}}{dt} = v_{Ctot\,l_j}i_{ctot\,l_j} = v_{l_j}i_{l_j} = \left(v_{diff_j} + v_{v_j}\right)\left(i_{diff_j} - \frac{i_{g_j}}{2}\right). \tag{10}$$

In [4] it has been demonstrated that a simpler and decoupled control of the stored energy can be achieved by defining the energy variables as follow:

Per-phase energy sum (W_j^Σ): $\qquad W_j^\Sigma = W_{Ctot\,u_j} + W_{Ctot\,l_j} = \frac{1}{2}C_{tot}\left(v_{ctot\,u_j}^2 + v_{ctot\,l_j}^2\right).$ $\qquad$ (11)

Per-phase energy difference (W_j^Δ): $\quad W_j^\Delta = W_{Ctot\,u_j} - W_{Ctot\,l_j} = \frac{1}{2}C_{tot}\left(v_{ctot\,u_j}^2 - v_{ctot\,l_j}^2\right).$ $\qquad$ (12)

The per-phase energy sum mainly comprises a DC and a double frequency component $2\omega_g$. The per-phase energy difference mainly comprises an oscillation with fundamental component ω_g [4], [16]. In order to control the energy stored per MMC arm while ensuring the energy balance between the upper and lower arm, both W_j^Σ and W_j^Δ have to be controlled. The control task is to master only the average value of both previously defined energy variables.

By considering an AC (with the fundamental grid frequency) and a DC part in the differential current (i.e. $i_{diff_j} = i_{diff_j}^{ac} + i_{diff_j}^{dc}$), the variation in the average value of the energy variables ($\bar{W}_j^\Sigma$, $\bar{W}_j^\Delta$) are expressed as follow [4], [17]:

$$\frac{d\bar{W}_j^\Sigma}{dt} \approx v_{dc}i_{diff_j}^{dc} - \frac{1}{3}P_{ac} \tag{13}$$

$$\frac{d\bar{W}_j^\Delta}{dt} \approx -2 * < i_{diff_j}^{ac}v_{g_j} >_T, \tag{14}$$

where $P_{ac} = v_{g_d}i_{g_d}$ is the total injected AC power of the MMC. The average value of the per-phase energy sum and per-phase energy difference are controlled through the DC and AC component of the differential current, respectively.

For controlling $\bar{W}_j^\Sigma$, as depicted in Fig. 4, three independent PI controllers are implemented. The $\bar{W}_j^\Sigma$ is obtained with a notch filter tuned at the frequency $2\omega_g$. The output of this control loop gives the reference for the DC component of the differential current.

The energy difference controller is shown in Fig. 5, where the matrices **R** and **K** are defined in (15) and (16), respectively. The matrix **R** produces a three-phase AC current with the internal phase angle θ_v, which is produced by the grid-forming control. The frequency of this three-phase signal in steady state is equal to the grid nominal frequency ω_{gn} [14].

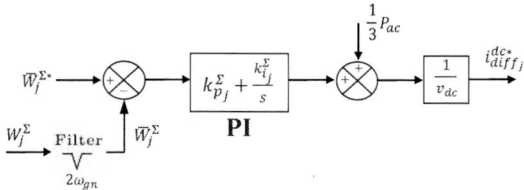

Fig. 4: Scheme of the per-phase energy sum control

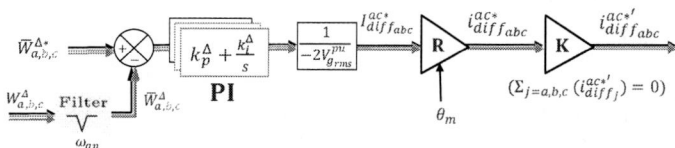

Fig. 5: Scheme of the per-phase energy difference control

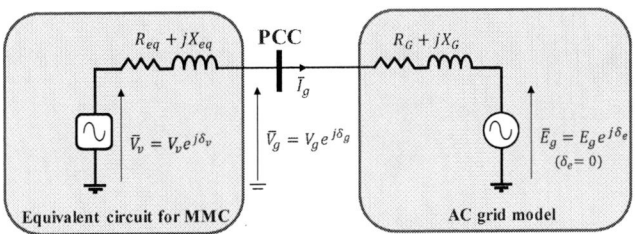

Fig. 6: Single-phase equivalent model of the system

The matrix **K** is used to cancel the sum of the AC differential component in order to mitigate this AC component in the DC current during transient. More details about this control can be found in [18].

$$\mathbf{R} = \begin{bmatrix} \cos(\theta_v) & 0 & 0 \\ 0 & \cos(\theta_v - \frac{2\pi}{3}) & 0 \\ 0 & 0 & \cos(\theta_v + \frac{2\pi}{3}) \end{bmatrix}. \tag{15}$$

$$\mathbf{K} = \begin{bmatrix} 1 & -\frac{1}{2} & -\frac{1}{2} \\ -\frac{1}{2} & 1 & -\frac{1}{2} \\ -\frac{1}{2} & -\frac{1}{2} & 1 \end{bmatrix}. \tag{16}$$

Case studies

Assuming that the MMC is connected to an AC grid, which is modeled with the Thevenin equivalent impedance ($Z_G = R_G + jX_G$) in series with an ideal AC source, Fig. 6 shows the single-phase equivalent circuit of the system. The active power flow is controlled through the phasor angle of the external MMC voltage δ_v. This angle can be observed in the grid-forming control scheme that is presented in Fig. 3. Time-domain angle of the MMC external voltage is given by $\theta_v = \omega_B \omega_{gn} t + \delta_v$.

In order to assess the dynamic performance of the grid-forming MMC, as it is shown in Fig.7, two cases are considered as follow:

- Non-energy-based control: the switches S1 and S2 are set to the position "a".
- Energy-based control: the switches S1 and S2 are set to the position "b".

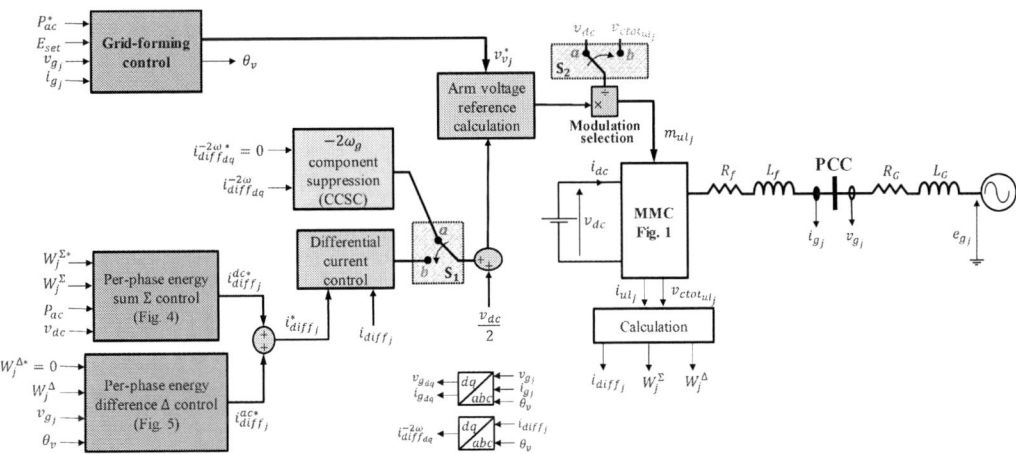

Fig. 7: General structure of the MMC control

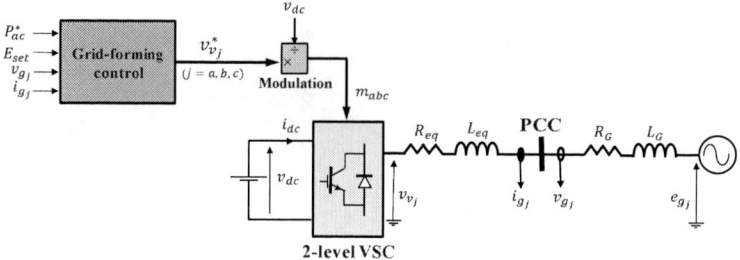

Fig. 8: General structure of the equivalent 2-level VSC

To tune the PI controllers for various control loops of the Fig. 7, we consider following response times: 5ms for the CCSC, 5 ms for the differential current control loop, 100 ms for the per-phase energy sum and 200 ms for the per-phase energy difference.

These two case studies could be compared to a base case with an equivalent 2-level voltage source converter (VSC). In this case, there is no internal stored energy and the grid-forming converter is simply providing the external voltage reference v_v^*. The structure of the equivalent 2-level VSC is illustrated in Fig. 8. It should be noted that the connection impedance is $Z_{eq} = R_{eq} + jX_{eq}$.

Simulation results

The system parameters are given in Table I. Two steps are applied to the active power reference of the power converter including 1 pu at $t = 0.5\ s$ and -0.5 at $t = 1.5\ s$. Fig. 9 shows the active power response of the power converter for different cases. It can be seen that the grid-forming control is properly implemented to the MMC and the dynamics of the power response are quite well-damped. The power response of the energy-based MMC is perfectly matched to the equivalent 2-level VSC, however for the non-energy-based MMC a slight difference is observed. Therefore, it seems that there is an interaction between the power control and energy control in the MMC. To ensure this fact, the average of per-phase energy sum and per-phase energy difference are presented in Fig. 10 and Fig. 11, respectively. From these figures, it can be clearly seen that although the stored energy in non-energy based MMC is stabilized naturally, however the energy level is not controlled at 1 pu. The energy difference is settled to zero in steady-state for both cases.

Table I: System parameters

Parameter	Value	Parameter	Value
Base power S_B	1 GW	E_{set}	1 pu
Converter nominal power P_n	1 GW	L_{arm}	0.18 pu
Base voltage V_B	320 kV	R_{arm}	0.005 pu
Grid voltage E_g	1 pu	Transformer inductance X_f	0.15 pu
Base frequency ω_B in rad/s	314.16	Transformer resistance R_f	0.005 pu
Grid nominal frequency ω_{gn}	1 pu	v_{DC}	640 kV
AC line-line voltage	320 kV	$k_{p_{diff}}$	0.17
Grid thevenin inductance X_G	0.1 pu	$k_{i_{diff}}$	100
Grid thevenin resistance $R_G = X_G/10$	0.01 pu	k_p^{Σ}	0.59
m_p	2%	k_i^{Σ}	12.72
ω_c	31.4 rad/s	k_p^{Δ}	0.3
R_v	0.09 pu	k_i^{Δ}	3.18
ω_f	60 rad/s		

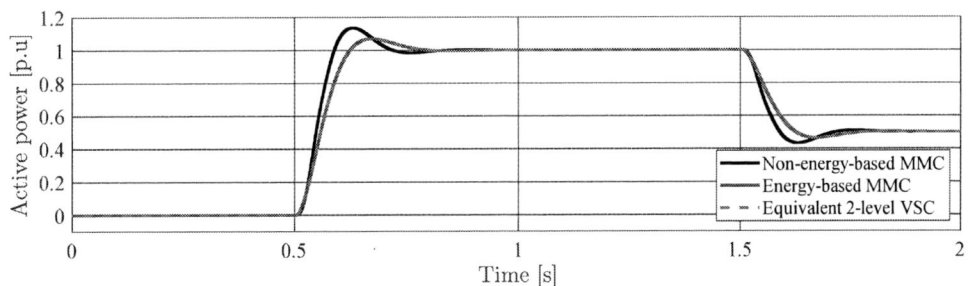

Fig. 9: Active power response of various case studied to the steps in the converter power set-point

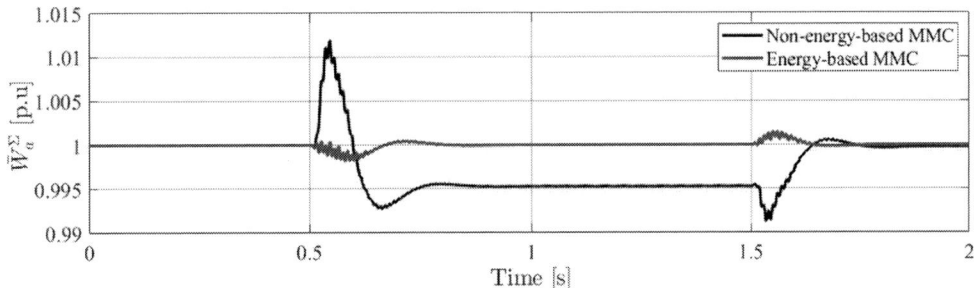

Fig. 10: Average of per-phase energy sum (phase a)

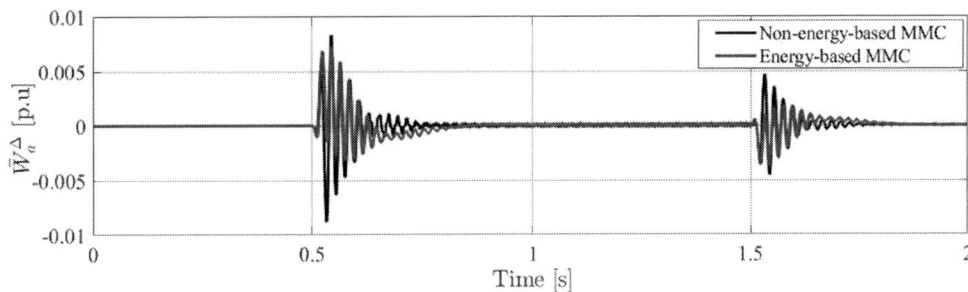

Fig. 11: Average of per-phase energy difference (phase a)

In order to find the origin of this difference, it is usefull to have a look to the phasor angle of the voltage δ_v at the terminal of the converter, which is ilustrated in Fig. 12. It can be observed that the behaviour of the angle is different in non-energy-based MMC. As it has been already seen, the active power reached 1 pu in steady-state. However, the converter based on the non-energy-based MMC transfers the same amount of active power with a lower voltage angle. By ignoring the resistances, a simple static consideration of the equivalent circuit depicted in Fig. 6 gives the relation between the active power and voltage angle of the power converter as follows:

$$P_{ac} = \frac{V_v.E_g}{X_{eq}+X_G} sin(\delta_v) \qquad (17)$$

According to (17), for non-energy-based MMC the magnitude of the external voltage V_v must be higher to inject 1 pu active power with a lower angle. Fig. 13 shows the d-component of the external voltage for all case studies. Although the reference value of this voltage is set to $E_{set} = 1\ pu$ in the control, the MMC with a non-energy-based control is not tracking the set-point in steady-state.

The reason why non-energy-based MMC presents a different dynamic behavior compared to the energy-based MMC and equivalent VSC is linked to the control of the modulated external voltage. In order to justify this result, a focus should be put on the general structure of the MMC control that is presented in

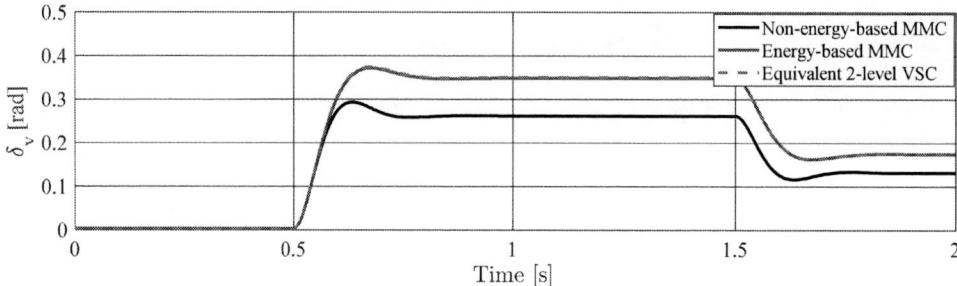

Fig. 12: Phasor angle of the voltage v_v

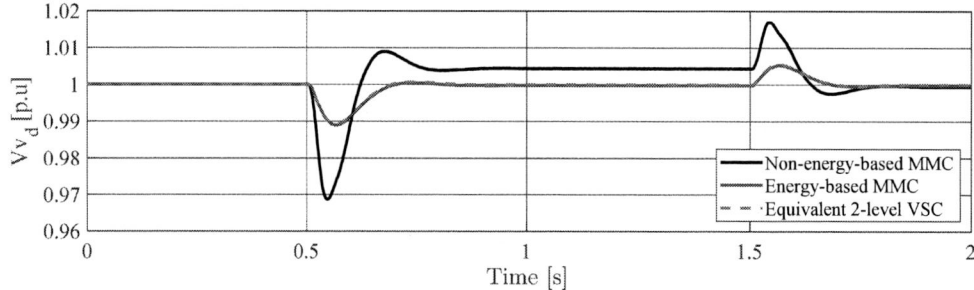

Fig. 13: d-component of the external voltage v_v

Fig.2. It can be seen that the obtained voltages v^*_{u/l_j} are divided to different voltages (i.e. $v_{Ctot\,u/l_j}$ and v_{dc}) to obtain the modulation indices. Contrary to the energy-based control, since the capacitor voltages are not sensed in the non-energy-based control, the only possibility to obtain the modulation indices is to divide the voltage voltages v^*_{u/l_j} to v_{dc}, which is a constant value. This is a strict adherence to the principle of the model inversion. Therefore, the voltage magnitude in the presented non-energy-based scheme is not under control. In the equivalent VSC, as it is shown in Fig. 8, since there is no internal capacitor, the external voltage reference is divided to v_{dc} in order to obtain the modulation indices. In this case, due to the fact that the DC bus voltage is generally kept very closed to the nominal value of v_{dc}, the model inversion is met. That is why the energy-based MMC has similar dynamic behavior as the equivalent VSC.

Conclusion

This paper has investigated the effect the grid-forming control on the internal energy of an MMC. It has been demonstrated that the stored energy in the MMC interacts with the active power response when the internal energy is not controlled. Moreover, it has been highlighted that the dynamic behavior of the active power in an energy-controlled MMC is similar to an equivalent 2-level VSC. This means that in the studies where the focus is on the dynamic behavior of the power converters at the AC side with no specific consideration about the DC bus and internal energy of the converter, a 2-level VSC can be used instead of MMC that results in reducing the complexity of the analysis.

References

[1] S. Samimi, F. Gruson, P. Delarue, F. Colas, M. M. Belhaouane, and X. Guillaud, "MMC Stored Energy Participation to the DC Bus Voltage Control in an HVDC Link," *IEEE Trans. Power Deliv.*, vol. 31, no. 4, pp. 1710–1718, Aug. 2016.

[2] J. Freytes *et al.*, "Improving Small-Signal Stability of an MMC With CCSC by Control of the Internally Stored Energy," *IEEE Trans. Power Deliv.*, vol. 33, no. 1, pp. 429–439, Feb. 2018.

[3] T. Qoria, F. Gruson, P. Delarue, P. Le Moigne, F. Colas, and X. Guillaud, "Modeling and Control of the Modular Multilevel Converter connected to an inductive DC source using Energetic Macroscopic Representation," in *20th European Conference on Power Electronics and Applications, EPE 2018 ECCE*

Europe, 2018.

[4] S. Samimi, "Modélisation et Commande des Convertisseurs MMC en vue de leur Intégration dans le Réseau Electrique," PhD thesis, ECOLE CENTRALE DE LILLE, 2016.

[5] C. Verdugo, J. I. Candela, and P. Rodriguez, "Grid support functionalities based on modular multilevel converters with synchronous power control," in *2016 IEEE International Conference on Renewable Energy Research and Applications, ICRERA 2016*, pp. 572–577.

[6] O. D. Adeuyi *et al.*, "Frequency support from modular multilevel converter based multi-terminal HVDC schemes," in *IEEE Power and Energy Society General Meeting*, September 2015.

[7] S. D'Arco, G. Guidi, and J. A. Suul, "Operation of a Modular Multilevel Converter Controlled as a Virtual Synchronous Machine," in *2018 International Power Electronics Conference (IPEC-Niigata 2018 -ECCE Asia)*, pp. 782–789, 2018.

[8] S. D'Arco, T. T. Nguyen, and J. A. Suul, "Evaluation of Virtual Inertia Control Strategies for MMC-based HVDC Terminals by P-HiL Experiments," *IECON Proc. (Industrial Electron. Conf.*, pp. 4811–4818, October, 2019.

[9] Q. Song, W. Liu, X. Li, H. Rao, S. Xu, and L. Li, "A Steady-State Analysis Method for a Modular Multilevel Converter," *IEEE Trans. Power Electron.*, vol. 28, no. 8, pp. 3702–3713, Aug. 2013.

[10] G. Bergna *et al.*, "An energy-based controller for HVDC modular multilevel converter in decoupled double synchronous reference frame for voltage oscillation reduction," *IEEE Trans. Ind. Electron.*, vol. 60, no. 6, pp. 2360–2371, 2013.

[11] J. G. Webster, T. Dragičević, L. Meng, F. Blaabjerg, and Y. Li, "Control of Power Converters in ac and dc Microgrids," *Wiley Encycl. Electr. Electron. Eng.*, vol. 27, no. 11, pp. 1–23, 2012.

[12] T. Qoria, F. Gruson, F. Colas, X. Guillaud, M. S. Debry, and T. Prevost, "Tuning of cascaded controllers for robust grid-forming voltage source converter," *20th Power Syst. Comput. Conf. PSCC 2018*, pp. 1–7, 2018.

[13] A. Tayyebi, F. Dörfler, Z. Miletic, F. Kupzog, and W. Hribernik, "Grid-Forming Converters – Inevitability, Control Strategies and Challenges in Future Grids Application," *CIRED Work.*, 2018.

[14] E. Rokrok, T. Qoria, A. Bruyere, B. Francois, and X. Guillaud, "Effect of Using PLL-Based Grid-Forming Control on Active Power Dynamics Under Various SCR," in *IECON 2019 - 45th Annual Conference of the IEEE Industrial Electronics Society*, 2019, pp. 4799–4804.

[15] E. Rokrok, T. Qoria, A. Bruyere, B. Francois, and X. Guillaud, "Classification and Dynamic Assessment of Droop- Based Grid-Forming Control Schemes: Application in HVDC Systems," in *21th Power Systems Computation Conference (PSCC)*, Porto, June 2020.

[16] S. Samimi, X. Guillaud, F. Gruson, and P. Delarue, "Synthesis of different types of energy based controllers for a Modular Multilevel Converter integrated in an HVDC link," in *11th IET International Conference on AC and DC Power Transmission*, 2015.

[17] P. Delarue, F. Gruson, and X. Guillaud, "Energetic macroscopic representation and inversion based control of a modular multilevel converter," in *15th European Conference on Power Electronics and Applications, EPE 2013*, 2013.

[18] K. Shinoda, J. Freytes, A. Benchaib, J. Dai, H. Saad, and X. Guillaud, "Energy Difference Controllers for MMC without DC Current Perturbations," in *2nd International Conference on HVDC (HVDC2016 - CIGRE)*, 2016.

Combining multiple temperature-sensitive electrical parameters using artificial neural networks

Daniel Herwig, Torben Brockhage, and Axel Mertens
LEIBNIZ UNIVERSITY HANNOVER
Institute for Drive Systems and Power Electronics
Hanover, Germany
Email: daniel.herwig@ial.uni-hannover.de
URL: https://www.ial.uni-hannover.de

Acknowledgments

The project on which this report is based was funded by the German Federal Ministry of Education and Research under the funding code 16EMO0325. The authors are responsible for the content of this publication.

Keywords

≪Circuits≫, ≪Component for measurements≫, ≪Data analysis≫, ≪Device modeling≫, ≪Estimation technique≫, ≪Hardware≫, ≪IGBT≫, ≪Industrial application≫, ≪Maintenance≫, ≪Measurement≫, ≪Neural network≫, ≪Reliability≫, ≪Thermal stress≫,

Abstract

Temperature-Sensitive Electrical Parameters (TSEPs) are often discussed for on-line determination of the junction temperature of semiconductors, and as key parameters for condition monitoring. This paper focuses on the combination of several simultaneously captured TSEPs using Artificial Neural Networks (ANNs) to reduce cross-dependencies and improve accuracy.

Introduction

Temperature-Sensitive Electrical Parameters (TSEPs), like collector-emitter saturation voltages or switching times [1], can be used to estimate a power module's junction temperature during operation [2]. Knowledge of the junction temperature can be used to improve existing thermal models or to detect common degradation effects, like bond wire lift-offs [3] or solder degradation. Evaluating a TSEP usually also requires knowledge of the collector current or the dc-link voltage. This information can be derived from the inverter's sensors. Additionally, parasitic effects can affect the TSEPs, e.g., temperature dependencies of the external gate resistor or discrete resistors within the power module may lead to significant temperature estimation deviations. In the first part of this work, we will discuss the selection of a suitable subset of TSEPs that can be used to eliminate cross-dependencies to external sensors or unwanted parasitic influences. A dedicated measurement PCB has been designed, capable of measuring four TSEPs at the same IGBT turn-on event: the IGBT's saturation voltage, the diode's forward voltage, the turn-on delay time and the current rise time. The PCB is validated in double-pulse experiments, mapping the TSEPs for a wide range of temperatures, currents and dc-link voltages. In the second part, the four TSEPs are then combined using ANNs to obtain a single temperature estimate. The results of the ANNs are compared to a conventional approach based on analytical models of the TSEPs. An extended calibration concept of the TSEPs for regular inverter operation is outlined shortly. The described methods were applied to an Infineon FF1000R17IE4D B2 Primepack IGBT Module, $I_N = 1000\,\text{A}$, $U_{CES} = 1700\,\text{V}$ and a Power Integrations 2SP0320 Dual-Channel SCALE driver.

EPE'20 ECCE Europe

Assigned jointly to the European Power Electronics and Drives Association & the Institute of Electrical and Electronics Engineers (IEEE)

Selection of a subset

The possibility of combining different TSEPs in order to eliminate unwanted dependencies has been shown in [4], [5] for single dependencies which can be analyzed during calibration. In this section, the selection of several TSEPs that will be combined is discussed, focusing on reducing the impact of unknown parameters on the TSEP analysis. The key challenges are:

- improving the transferability of TSEP calibration data created in double-pulse experiments to regular inverter operation,

- reducing the influence of usually unknown parameters, e.g., variations in the external gate resistance,

- low temperature dependency of the measurement circuit itself,

- independence from inverter sensors, if possible.

The gate drive's voltage rails are assumed to be stabilized, but the overall gate loop resistance can vary significantly. Combining TSEPs that are measured at different times, e.g., turn-on delay time and turn-off delay time [6], are avoided, as system parameters like load current change between the measurements. Furthermore, additional TSEPs should not introduce unnecessary new dependencies. Timing measurements were generally preferred, as the required measurement circuits can be implemented mostly digitally and are therefore less dependent on the temperature. As a result, the aforementioned subset of four TSEPs was chosen. Table I shows the dependencies of these TSEPs on known and unwanted parameters.

Table I: Most significant dependencies of the four proposed TSEPs

Dependency \ TSEP	t_{d}	t_{ri}	$U_{\mathrm{ce,sat}}$	U_{f}
Junction temperature IGBT	✓	✓	✓	
Junction temperature diode		✓		✓
Load current		✓	✓	✓
DC-link voltage	✓	✓		
Bond wire lift-off			✓	✓
Solder degradation	✓	✓	✓	✓

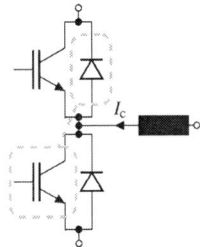

Fig. 1: One evaluated switching pair consisting of an IGBT and its complementary diode

The IGBT's saturation voltage $U_{\mathrm{ce,sat}}$ is included, as it contains information on the degradation of bond wires [7]. Distinguishing changes in voltage due to junction temperature changes and bond wire degradation is challenging without additional information. The turn-on delay time t_{d} can be used as a second reference for junction temperature. It does not depend on the load current, but its temperature sensitivity is not pronounced. The current rise time t_{ri} was introduced to subset, which can be measured with the same circuit. In practice, only the time until the current peak is reached, $t_{\mathrm{d}} + t_{\mathrm{ri}}$, can be measured with reasonable effort. This introduces a new dependency on the complementary diode's temperature, which is why its forward voltage U_{f} is incorporated as well. The diode's forward voltage can be measured by the same circuit that captures the saturation voltage of the complementary IGBT.

These TSEPs can be captured at the same turn-on switching process, therefore the load current and dc-link voltage can be assumed to be constant for all TSEP measurements. Fig. 2 shows the implemented measurement circuit's concept, as well as the timing of these measurements.

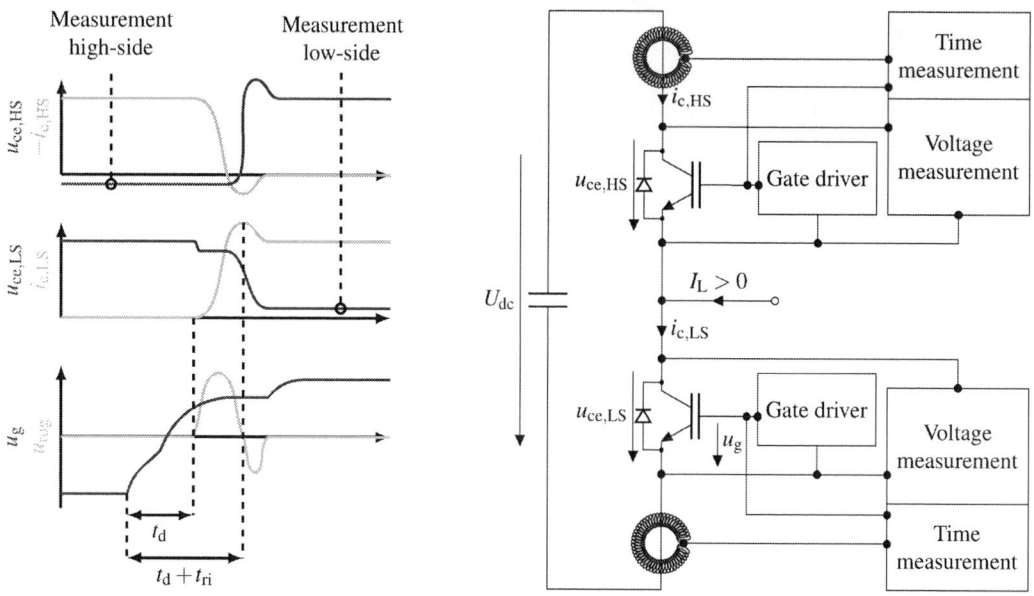

Fig. 2: Capturing times for the low-side IGBT and the high-side diode; concept of the TSEP measurement circuit

On-line TSEP measurement hardware

The PCB can be retrofitted on top of an industrial gate driver. A Rogowski coil located at the power terminals of the IGBT module generates the stop signal for the time measurements. The stop signal could also be generated by directly contacting the power emitter and using the induced voltage during switching [6]. Using a Rogowski coil is more complex and not recommended in general, but it was chosen here to test contactless design options with a more flexible placement.

The time measurement circuitry is based on a TDC7200 by Texas Instruments with a resolution of 55 ps. The IC has a separate start and stop pin, but multiple stop signals must be generated as pulses on the single stop pin. The analog front end must interface the high input voltage to CMOS level with low latency, but also act as a barrier for EMI. The core concept of the analog front end for one stop signal is depicted in Fig. 3. An open collector concept with a pull-up resistor was chosen to allow the parallel

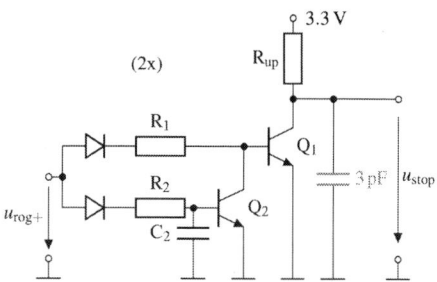
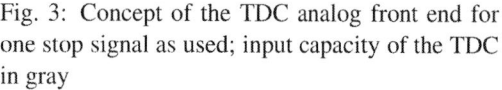
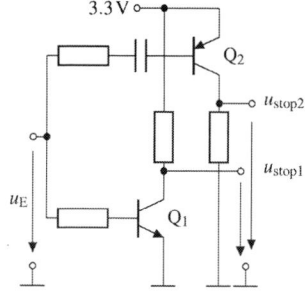

Fig. 3: Concept of the TDC analog front end for one stop signal as used; input capacity of the TDC in gray

Fig. 4: Recommended concept for direct connection of the power emitter and two separate stop signals; input capacities omitted

connection of multiple front ends, creating a pulse train for the TDC. When the input voltage u_{rog+} rises, the transistor Q_1 shorts the TDC's pin to ground. The latency and fall time of the resulting signal u_{stop} is mostly defined by the gain-bandwidth product of Q_1 and the input capacitance of the TDC's input pin ($\leq 3\,\mathrm{pF}$). R_1 can be chosen very small, as the pulse is short. A bipolar transistor was chosen over a MOSFET, as the voltage drop on R_1 protects the front end from large voltage spikes, and the almost fixed voltage at the transistor's base reduces capacitive coupling through the transistor. In order to release the stop signal afterwards, a secondary signal path via R_2 and Q_2 is introduced. In this path, the input signal is delayed, shorting the base of Q_1 after a time defined by $R_2 \cdot C_2$. In order to measure t_d and $t_\mathrm{d} + t_\mathrm{ri}$ during turn-on, two of these front ends are paralleled at the open collector output and fed with the positive and negative signal from the Rogowski coil. When directly connecting the power emitter and using two separate stop signals, a simplified circuitry can be used, see Fig. 4.

The voltage measurement uses a high voltage MOSFET Q_1 to block the dc-link voltage while the IGBT is turned off, see Fig. 5. When the IGBT is turned on, Q_1 and Q_2 are conducting and allow the direct measurement of the saturation voltage. An alternative concept using diodes is described in [9]. The voltage limiting of the first stage, Q_1 and C_1, is not sufficient for direct use with an ADC, therefore a second stage, Q_2 and C_2, at low voltage was added. The MOSFETs have to be chosen with adequately low on-resistance, so that the low-pass effect of the circuitry allows measurements while the IGBT is turned on. A comparator can be used to reduce leakage current through Q_1 while the IGBT is turned off. The circuit can also be used to measure the forward voltage of the IGBT's antiparallel diode.

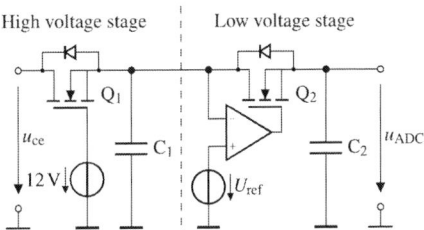

Fig. 5: Concept of the saturation voltage measurement

Experimental results in double-pulse experiments

The four TSEPs were measured and mapped in double-pulse experiments for temperatures ϑ_j between $25\,^\circ\mathrm{C}$ and $130\,^\circ\mathrm{C}$, collector currents I_c from $50\,\mathrm{A}$ to $1000\,\mathrm{A}$ and dc-link voltages U_dc from $500\,\mathrm{V}$ to $900\,\mathrm{V}$. The dc-link voltage drop during the first pulse of the experiment was compensated by increasing the dc-link voltage in advance, so that the voltage at the switching instance was at the desired setpoint. Due to the setup's ratio between the dc-link capacity and the load inductance and the voltage limit of the dc-link, not all large currents could be sampled at high dc-link voltages. Fig. 6 shows the experimental results acquired using the previously shown measurement PCB.

In order to obtain the temperature from TSEP measurements, a direct fit from the measured TSEPs can be used, e.g., $\vartheta_\mathrm{j} := f_1(U_{\mathrm{ce,sat}}, I_\mathrm{c}, U_\mathrm{dc})$. Simply fitting the data this way with a linear least squares method resulted in unusable fits. The fit results were mostly defined by their behavior near the temperature independent point of the saturation voltage $U_\mathrm{ce,sat}$ and forward voltage U_f. This strategy requires a weighting concept to handle the temperature independencies, e.g., using the inverse of the temperature sensitivity. An alternative is to fit the TSEP as a function of the temperature first, e.g., $U_\mathrm{ce,sat} := f_2(\vartheta_\mathrm{j}, I_\mathrm{c}, U_\mathrm{dc})$, and then use its inverse function to estimate the temperature. The latter was chosen in context the of this paper.

Based on physical models of the semiconductors, several analytical fit functions with varying level of detail were formed. To evaluate their performance without overfitting, the sample points were randomly binned into a training, and a validation dataset ($80\,\%/20\,\%$). A test dataset was not used, since the dominant nonlinearities of the TSEPs are expected to behave like low-order polynomials or logarithmic functions. Sample points at the edges of the investigated range, e.g., at maximum or minimum temperature, were forced into the training dataset. Each function was fitted using the training dataset only. Afterwards, the resulting root-mean-square deviation (RMSD) of the fit was evaluated on the validation

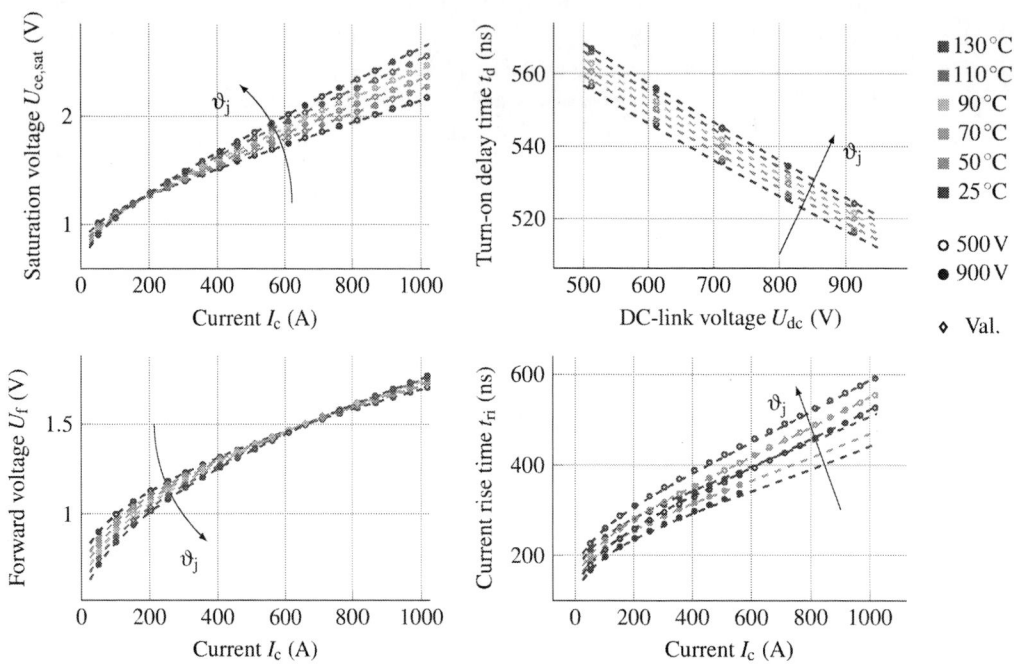

Fig. 6: TSEP measurement results of the turn-on times and the forward voltages of IGBT and diode; the regression (dashed) is based on analytical models of the TSEP. Excluded validation data marked with diamonds

dataset. The analytical model with the lowest RMSD on the validation dataset was chosen as the fitting function to avoid overfitting, giving (1) to (4):

$$U_{\text{ce,sat}}(I_c, \vartheta_j) \approx a_1 + a_2 \cdot \vartheta_j + a_3 \cdot \vartheta_j^2 + (a_4 + a_5 \cdot \vartheta_j) \cdot I_c + (a_6 + a_7 \cdot \vartheta_j) \cdot \ln(I_c/(1\,\text{A})) \tag{1}$$

$$U_f(I_c, \vartheta_j) \approx b_1 + b_2 \cdot \vartheta_j + b_3 \cdot \vartheta_j^2 + (b_4 + b_5 \cdot \vartheta_j) \cdot I_c + (b_6 + b_7 \cdot \vartheta_j) \cdot \sqrt{I_c/(1\,\text{A})} \tag{2}$$

$$t_d(U_{dc}, \vartheta_j) \approx c_1 + c_2 \cdot U_{dc} + c_3 \cdot \vartheta_j + c_4 \cdot U_{dc}^2 + c_5 \cdot U_{dc} \cdot \vartheta_j \tag{3}$$

$$t_{ri}(I_c, U_{dc}, \vartheta_j) \approx d_1 + d_2 \cdot I_c + d_3 \cdot I_c^2 + (d_4 + d_5 \cdot I_c) \cdot \vartheta_j$$
$$+ (d_6 + d_7 \cdot I_c) \cdot U_{dc} + (d_8 + d_9 \cdot \vartheta_j) \cdot \sqrt{I_c/(1\,\text{A})}. \tag{4}$$

Calculating the inverse of (1) or (2) results in two separate solutions, due to their parabolic properties with regard to temperature ϑ_j. Inverting (1) yields

$$\vartheta_{j,U\text{ce,sat}} = -\frac{a_2 + a_5 \cdot I_c + a_7 \cdot \ln(I_c/(1\,\text{A}))}{2 \cdot a_3}$$
$$\pm \sqrt{\left(\frac{a_2 + a_5 \cdot I_c + a_7 \cdot \ln(I_c/(1\,\text{A}))}{2 \cdot a_3}\right)^2 - \frac{a_1 - U_{\text{ce,sat}} + a_4 \cdot I_c + a_6 \cdot \ln(I_c/(1\,\text{A}))}{a_3}}. \tag{5}$$

At large or small currents, the applicable solution can be easily identified. Close to the temperature-independent point of $U_{\text{ce,sat}}$, both solutions are plausible. Fig. 7 shows the location of the temperature-independent point according to (5), which changes with temperature. The determined position of this point strongly depends on the chosen fitting function. Equation (5) provides two solutions, as long as both are within the range of viable temperatures, the appropriate temperature estimate cannot be selected solely on the measured current and saturation voltage. There are several possibilities to handle these ambiguous results in application. Based on a chosen range of expected temperature, the TSEP could be

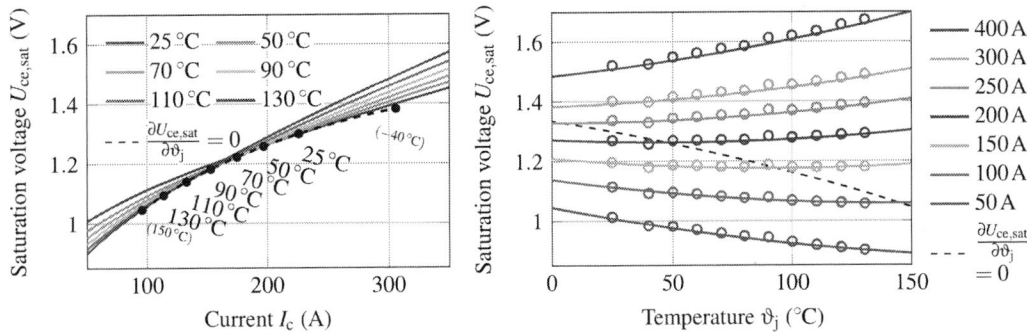

Fig. 7: Saturation voltage $U_{ce,sat}$ depending on the collector current I_c and the temperature ϑ_j; fit results as solid line; measured values as unfilled circles; dashed line marks the location of the temperature-independent operating points according to (1); extrapolation of the location is not recommended

discarded if both solutions are viable. When combining TSEPs with a thermal model, the temperature estimated by the model could be used to select the appropriate TSEP temperature estimate. In the context of this paper, the solution closest to the real temperature will be selected to improve comparability to the ANNs.

Combination of TSEPs

The description of the behavior of a TSEP by an analytical model can be used to create a fitting function for temperature estimation. Though this concept can be successfully applied to evaluate individual TSEPs, finding an analytical model that also allows the combination of several TSEPs into a single temperature is not possible analytically. In order to combine the analytical models without the use of ANNs, a reference model has been developed. In the reference model, the four TSEPs are evaluated individually, resulting in four different temperature estimates $\vartheta_{j,Uce,sat}$, $\vartheta_{j,Uf}$, $\vartheta_{j,td}$ and $\vartheta_{j,tri}$. The four estimated temperatures are then weighted inversely proportional to their Root-Mean-Square Deviation (RMSD) at the given operating conditions, see (6). Fig. 8 shows the RMSD of the fitting function of each individual TSEP depending on the collector current and the resulting weighting factor w.

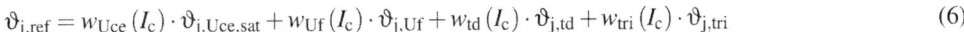

$$\vartheta_{j,ref} = w_{Uce}\left(I_c\right) \cdot \vartheta_{j,Uce,sat} + w_{Uf}\left(I_c\right) \cdot \vartheta_{j,Uf} + w_{td}\left(I_c\right) \cdot \vartheta_{j,td} + w_{tri}\left(I_c\right) \cdot \vartheta_{j,tri} \qquad (6)$$

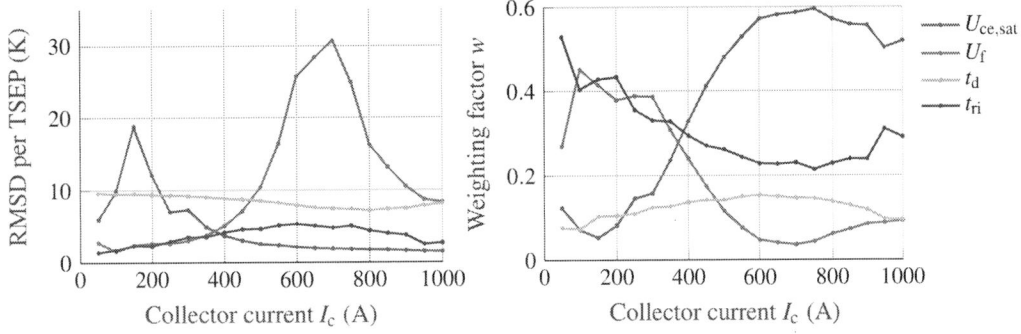

Fig. 8: RMSD of the temperature estimation of the individual TSEPs and the resulting weighting factor in the reference model

As an alternative to analytical functions, an ANN is used to create a mapping from the four selected TSEPs, the load current and dc-link voltage to the junction temperature. As all TSEPs are continuous functions with little nonlinearity, an ANN with a single hidden layer of six neurons and a hyperbolic tangent sigmoid as the activation transfer function was used, in accordance with [8]. In addition to the full ANN $\vartheta_{j,\text{ANN,full,6}}$ including the current and voltage information, a reduced ANN $\vartheta_{j,\text{ANN,red,6}}$ is evaluated based solely on the four TSEPs. The ANNs use the identical set of training and validation data, like the analytical functions used. The ANN with six neurons and the analytical reference model have a different number of fitting parameters. To compare the performance of ANNs and the reference model at a similar amount of fitting parameters, another full ANN with only three neurons and the dc-link voltage and collector current information was generated, $\vartheta_{j,\text{ANN,full,3}}$, see Table II.

Table II: Overview of the compared ANNs

	Identifier	Inputs	Scope
ANN with six neurons	$\vartheta_{j,\text{ANN,full,6}}$	$U_{\text{ce,sat}}, U_{\text{f}}, t_{\text{d}}, t_{\text{ri}}, I_{\text{c}}, U_{\text{dc}}$	compared throughout
Analytical reference model	$\vartheta_{j,\text{ref}}$	$U_{\text{ce,sat}}, U_{\text{f}}, t_{\text{d}}, t_{\text{ri}}, I_{\text{c}}, U_{\text{dc}}$	the paper
ANN with three neurons	$\vartheta_{j,\text{ANN,full,3}}$	$U_{\text{ce,sat}}, U_{\text{f}}, t_{\text{d}}, t_{\text{ri}}, I_{\text{c}}, U_{\text{dc}}$	short, isolated
ANN with six neurons without inverter sensors	$\vartheta_{j,\text{ANN,red,6}}$	$U_{\text{ce,sat}}, U_{\text{f}}, t_{\text{d}}, t_{\text{ri}}$	comparisons

Fig. 9 shows the error histogram of the combination concepts. Fig. 10 depicts the RMSD of the different combination concepts over the collector current I_{c}. Fig. 11 presents the resulting overall RMSD across the entire fitting range for the individual TSEPs, the reference model and the ANNs with and without the current and voltage measurement. The analytical reference model $\vartheta_{j,\text{ref}}$ has the largest RMSD. The full ANN with six neurons $\vartheta_{j,\text{ANN,full,6}}$ has the lowest temperature deviation for the given data. The ANN with only three neurons also shows less deviation than the reference model, except for small currents. The reduced ANN without the collector current and dc-link voltage measurements also shows an RMSD below 2 K.

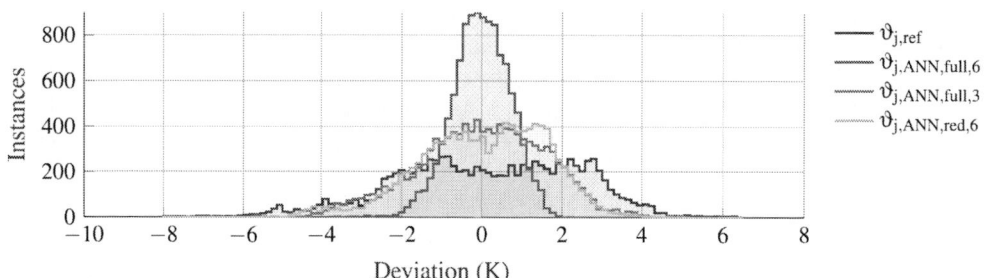

Fig. 9: Error histogram of the different combination concepts in 100 bins

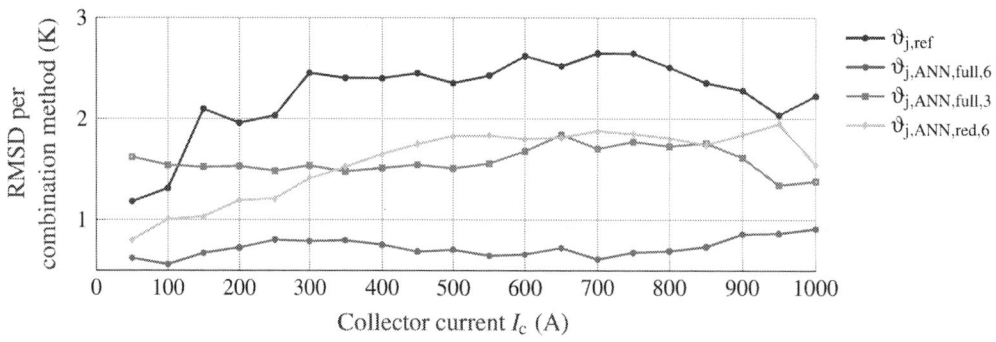

Fig. 10: RMSD of the temperature estimation of the different combination concepts; the reference model is based on analytical models; the reduced ANN does not have dc-link voltage or current information

The large deviations of the collector-emitter saturation voltage and the diode's forward voltage are a result

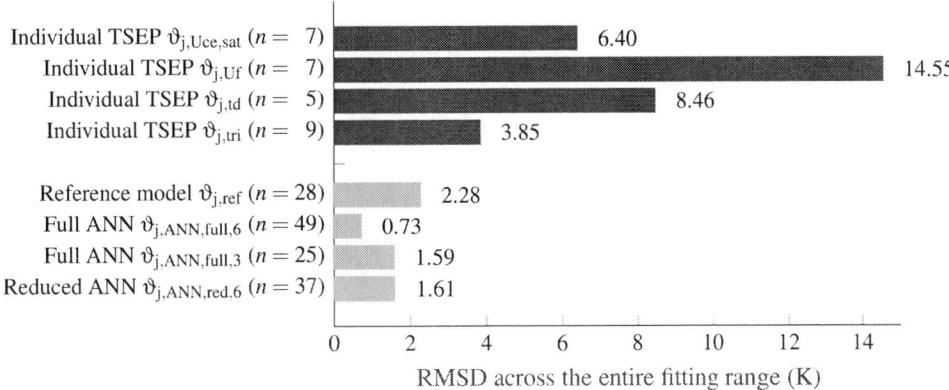

Fig. 11: Comparison of the overall RMSD of the different combination concepts and individual TSEPs; n is the number of fitting parameters used

of the poles in Fig. 8. Therefore, for those TSEPs, the overall RMSD misrepresents their performance at other operating conditions. Combining the TSEPs significantly reduces the RMSD. The full ANN shows the smallest RMSD of below 1 K. It should also be possible to obtain the dc-link voltage and the collector current from the four TSEPs. Therefore, additional ANNs have been trained which estimate the dc-link voltage and the collector current only from the four TSEPs. These ANNs showed an overall RMSD of 3.1 A and 6 V, respectively.

Required number of calibration points

The results in Fig. 11 show that ANNs result in low overall RMSD, but there are several other aspects that may affect the choice of the combination concept. The results presented so far have used measurements at 20 current levels, eleven temperatures and nine voltage levels. The parameters were chosen in a grid for graphical representation. Even though not all calibration points at high currents could be measured at high voltages, the fits are based on a total of 1620 calibration points. Each point was measured several times. In order to estimate the influence of the number of calibration points on the analytical reference model and the ANN with six neurons, the fits have been redone on a reduced number of calibration points.

The highest and lowest level of current, temperature and voltage were kept in the training data, while the density of calibration points in between was reduced. The used ANN is prone to be stuck in local minima during training. For each set of calibration points, 100 ANNs were trained with identical settings, but different weight initializations. Fig. 12 shows the resulting overall RMSD in a log-log scale. The results of the ANNs are shown as box plots. The ANNs were trained according to [10], with Bayesian regularization enabled [11]. The fitting of the analytical reference model (1) to (4) is a linear least square problem, therefore the solver can be expected to converge to the global minimum and was fitted only once. If several combinations of the used number of current, voltage and temperature calibration points result in the same total number of calibration points, the ANN results are merged into one box plot.

The interquartile range of the ANNs shows results comparable to the analytical reference model, but the number of outliers with extreme deviation rises below 600 calibration points. The ANN fit with the lowest RMSD is consistently below the analytical reference model. These results show that the ANN can have similar performance, even with few calibration points, but more sophisticated methods for finding a good solution are needed.

Another aspect to consider is extrapolation. Fig. 13 shows the RMSD of the analytical reference model and a representative ANN fit with three neurons when extrapolating. For this comparison, new fits were created based on a reduced range of currents and temperatures $\vartheta'_{j,ANN,full,3}$ and $\vartheta'_{j,ref}$. The ANN shows significantly larger RMSD outside the fitted area, especially at low temperatures. A disadvantage when

Combining multiple temperature-sensitive electrical parameters using artificial neural networks

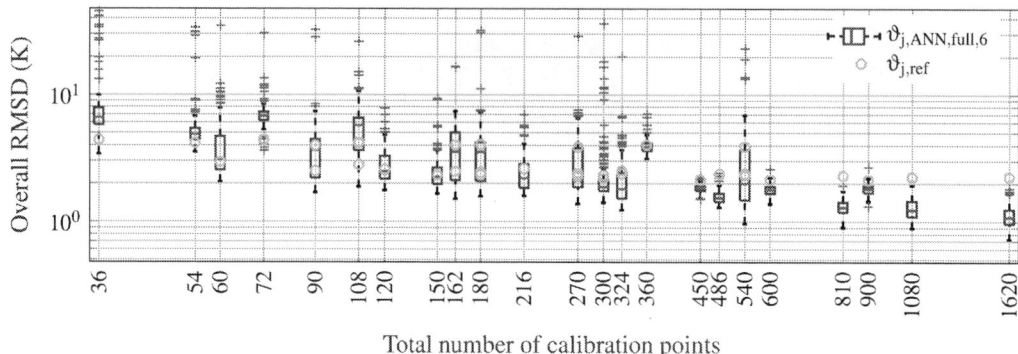

Fig. 12: Overall RMSD of the fits depending on the available amount of calibration points; box plot of 100 random ANN fits; green circles show the result of the reference model; 25th to 75th percentile in blue box; whiskers mark extreme data, not considered outliers; outliers as red crosses; if the number of calibration points are matched by multiple combinations, the ANN results are merged

extrapolating with ANNs is expected, as they do not incorporate an underlying analytical model of the physical system. Furthermore, the ANN showed strongly varying extrapolation performance, depending on the training result. The better extrapolation performance of the analytical reference model may allow the use of a smaller range during calibration than in regular inverter operation.

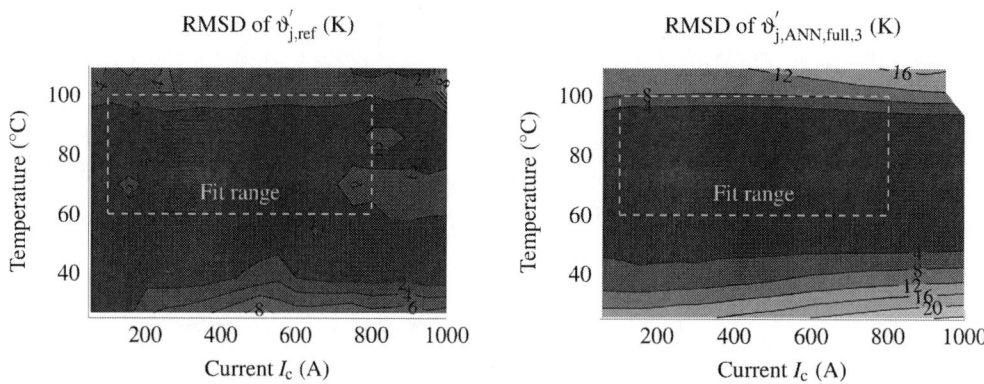

Fig. 13: RMSD of the reference model and comparable ANN when extrapolating data as contours; only the measured calibration points within the green rectangle were used for the fits. The current range was not chosen smaller to span beyond the temperature-independent points of $U_{ce,sat}$ and U_f. The measured temperature of the module's NTC is used as the temperature reference.

Extended calibration concept

Usually, TSEP calibration is done in double-pulse experiments. The device is heated to a known (homogenous) temperature, and the TSEP's value is captured. The previous sections demonstrated the viability of the combination of several TSEPs in order to eliminate unwanted dependencies with double-pulse experiments. Transferring TSEP calibration data from double-pulse experiments to regular inverter operation under load is challenging. Of the proposed TSEPs, the time until the peak current is reached during turn-on, $t_d + t_{ri}$, depends on the temperature of the IGBT, as well as on its complementary diode's temperature. During operation, these temperatures are expected to be different, e.g., when the power factor of the load changes. In order to calibrate the TSEP, regular double-pulse experiments are used to calibrate the IGBT's saturation voltage, the diode's forward voltage and the turn-on delay time. Afterwards, a full bridge setup is proposed to drive a constant dc current at the phase terminal in PWM operation. Introducing a duty cycle offset in both half bridges does not affect the constant load current, but it affects the ratio at which the current is conducted through IGBT or diode chips of the module. This concept allows sample points with different temperatures for IGBT and diode. The previously calibrated TSEPs $U_{ce,sat}$, U_f and t_d can be used as temperature references for the IGBT and the diode, enabling the calibration of the current rise time t_{ri}.

Conclusion

In this work, the combined evaluation of several TSEPs was proposed. A dedicated measurement circuit was designed, capable of capturing four TSEPs for the low- and high-side switch of a power module while operating. Different methods for combining the measured TSEPs into a single temperature estimate were compared. An analytical reference model was formed, incorporating an underlying physical model of the system. This model was compared to fitting the data using a shallow neural network with six neurons, without any additional information of the physical system.

The chosen set of TSEPs includes a parameter which depends on the temperature of the IGBT and the diode, thus making an extended calibration for regular inverter operation necessary. Even though the TSEPs enable temperature estimation without the use of the inverter's sensors, it is suggested to include both to increase robustness against parasitic influences, e.g., changes in the gate driver's resistance and to detect degradation.

The use of an ANN has proven to enhance the quality of temperature estimation significantly and outperforms approaches based on analytical models of the TSEPs at high numbers of calibration points. When the density of the calibration points is reduced, the performance of the trained ANNs varies widely depending on the weight initialization. More sophisticated initialization strategies are required to find an appropriate fit. However, at any number of calibration points, an ANN could be found with a lower RMSD than with the analytical reference model. The underlying physical model incorporated in the analytical model improves precision when extrapolating compared to ANNs, which enables the use of a reduced range of temperatures, voltages and currents during calibration.

References

[1] Kuhn H. and Mertens A.: On-line junction temperature measurement of IGBTs based on temperature sensitive electrical parameters, 2009 13th European Conference on Power Electronics and Applications, pp 1-10
[2] Avenas Y., Dupont L. and Khatir Z.: Temperature Measurement of Power Semiconductor Devices by Thermo-Sensitive Electrical ParametersA Review, IEEE Transactions on Power Electronics, Vol. 27 no 6 pp 3081-3092, June 2012
[3] Krone T., Dang Hung L., Jung M. and Mertens A.: Advanced condition monitoring system based on on-line semiconductor loss measurements, 2016 IEEE Energy Conversion Congress and Exposition (ECCE), Milwaukee, WI, 2016, pp 1-8
[4] Wang X., Zhu C., Luo H., Li W. and He X.: Elimination of collector current impact in TSEP-based junction temperature extraction method for high-power IGBT modules, Chinese Journal of Electrical Engineering, Vol. 2 no 1 pp 85-90, June 2016
[5] Simon S.: Beitrag zur Zustandsberwachung von IGBT-Modulen mit temperatursensitiven Parametern, Hannover: Gottfried Wilhelm Leibniz Universitt, Dissertation, 2018
[6] Weber S., Schlter M., Borowski D. and Mertens A.: Simple analog detection of turn-off delay time for IGBT junction temperature estimation, 2016 IEEE Energy Conversion Congress and Exposition (ECCE), Milwaukee, WI, 2016, pp 1-7
[7] Gonzalez-Hernando F.,San-Sebastian J.: Garcia-Bediaga A., Arias M., Iannuzzo F. and Blaabjerg F., On-line Condition Monitoring of Bond Wire Degradation in Inverter Operation, 2018 IEEE Energy Conversion Congress and Exposition (ECCE), pp 4115-4121
[8] Dias Pereira J. M., B. Silva Girao P. M. and Postolache O.: Fitting transducer characteristics to measured data, IEEE Instrumentation & Measurement Magazine Vol. 4 no 4 pp 26-39, Dec. 2001
[9] Ghimire, Pramod et al.: A review on real time physical measurement techniques and their attempt to predict wear-out status of IGBT, 2013 15th European Conference on Power Electronics and Applications (EPE), pp 1-10
[10] Fit Data with a Shallow Neural Network, Accessed on: April 4, 2020, [Online] Available: https://www.mathworks.com/help/deeplearning/gs/fit-data-with-a-neural-network.html
[11] Goodfellow I., Bengio Y., and Courville A.: Deep Learning, 2016, The MIT Press

Single-Phase Measurement of the Output Impedance of the Four-Quadrant Cascaded H-Bridge Converter Cell Using Wideband Signals

Marko Petković and Dražen Dujić
Power Electronics Laboratory - PEL, EPFL
Route Cantonale 15
CH-1015 Lausanne, Switzerland
Phone: +41 21 693 56 25
Email: marko.petkovic@epfl.ch; drazen.dujic@epfl.ch
URL: https://www.epfl.ch/labs/pel

Keywords

≪Impedance measurement≫, ≪Modelling≫, ≪Multilevel converters≫.

Abstract

The development and integration of power electronics equipment and converters for medium voltage ac and dc application has created different subsystem interactions that require proper investigation, understanding, description and estimation of global system stability through impedance-admittance measurements and identification. Four-quadrant Cascaded H-Bridge topology features high output voltage resolution and high effective switching frequency which enables high-dynamic, high-fidelity voltage perturbation injection for medium voltage impedance and admittance measurement. While being a perturbation injection and a measurement instrument it also needs to assure that it does not cause instabilities in the device under test. Hence, a mechanism to determine the output impedance of a single cell of a Cascaded H-Bridge converter is needed. To validate the theoretical developments, a new method to measure the single-phase output impedance in the *dq*-frame is proposed.

Introduction

The increase in energy demand combined with the advances in the field of power electronics has incentivized the expansion of renewable energy system as well as extensive placement into service of power conversion devices. The interaction of different elements in the grid, on one side leads to the change in power network behaviour that is becoming increasingly complex and on the other side results in system instabilities [1], [2]. In order to overcome this issue and prevent future instabilities, an early identification and characterization of the present system and its future potential components would reduce the risk of encountering unpredictable behavior which would, in turn, provide a stable network with uninterrupted operation. Some of the criteria used to assess the stability are presented in [3] and they rely either on analysing the minor-loop gain or the passivity of the system. Whatever the criteria are, the knowledge of the impedance of the network being analysed is required.

Unfortunately, the problem of high-power and medium voltage (MV) impedance/admittance measurement and system identification had not yet been fully addressed. Measurements in MV application require equipment designed for operation at MV level and a possibility to inject perturbation into MV system, while at the same time having sufficiently wide bandwidth in order to characterize the unknown object over a wide frequency range. These requirements combined are not easily fulfilled at the MV level and thus the research performed in this field is scarce and the devices developed for that purpose are few. The MV impedance estimators developed so far have either limited bandwidth, up to 1 kHz or comprise an output side transformer in order to step-up the voltage to the MV levels, which in turn also limits the bandwidth. Still, the demand for such equipment is increasing due to the need to assist the development of recent medium voltage dc (MVdc) and medium voltage ac (MVac) applications, grid integration of renewable energy sources and storage devices, energy transmission and distribution in the MV range [4]–[6] and assure safe integration with the existing apparatus.

The MV converters based on the Cascaded H-Bridge (CHB) fall into the category of devices capable of having sufficiently high voltage output and bandwidth for MV impedance/admittance measurements. A variation of such topology that makes use of an active input converter stage interfaced to a multi-winding transformer (MWT) is presented in Fig. 1. The topology consists of step-down MV–MWT with 15 secondary phase-shifted outputs permitting to stack up to 5 cells per phase and thus effectively increase the output voltage levels and switching frequency allowing higher frequency voltage perturbation injection and impedance measurement. As a matter of fact, the presence of the MWT on the input side is one of the advantages of this topology. Having the MWT at the input means that there is no need to have a step-up transformer on the output side to elevate the voltage to MV level. As a result, the output stage high-frequency bandwidth is not limited by the output transformer. The CHB cells outputs are interfaced to a three-phase LC-type filter on which the voltage output is controlled. A single cell of the CHB converter is presented in Fig. 2. It consists of an Active Front End (AFE) as the input conversion stage and an H-bridge (HB) inverter as the output conversion stage. It is taken as an assumption that the dynamics of a single-phase leg consisting of five cells can be represented by the dynamics of one cell, which is why the cell in Fig. 2 is represented with an LC-type filter at the output.

The initial study on the feasibility of this solution was performed in [7], where there was a need to superimpose a high-frequency small-signal in addition to the fundamental component. When this is performed in a closed-loop, there exist possible interactions of synchronization and closed control loops of AFE and HB. Our work seeks to reveal if the presence of an active input element, such as AFE, impacts the shape of the impedance of the output stage HB and subsequently the cascaded connection of multiple HBs.

The rest of the paper is organised as follows. Section II describes the main features of a single cell of the CHB converter. Section III provides the state-space model of the cell in the dq-frame. Section IV characterizes the source-affected dynamics of the cell output. Section V proposed a new method to measure the output impedance of a single-phase inverter in the dq-frame using wideband pseudo-random binary sequence (PRBS) signal perturbation injection. Finally, Section VII concludes the work.

Cascaded H-Bridge Cell

The MWT (cf. Fig. 1) supplies three-phase isolated voltages from its secondary, low-voltage, side to the cell. Isolated three-phase supply makes the cells and the dc-links largely independent from each other [8]. The input filter of the cell is the L-type filter with its parasitic resistance. The rectification of the input voltage is performed by the three-phase AFE, while the dc-link serves as an energy buffer from which the output stage, a single-phase HB, is supposed to inject perturbation voltages into an unknown network or device under test (DUT). The AFE control system comprises a PLL that performs synchronisation, as it is essentially a grid-connected converter through the MWT. Apart from this, it also performs the cascaded control of the dc-link voltage through the input current control. On the output side the HB performs cascaded control of its output voltage through the control of the filter inductor current.

The knowledge of the cell output impedance is beneficial due to the fact that the CHB should stay passive as seen by the DUT and not violate the stability criteria [3]. The information of the output impedance in this work is obtained by modelling the converter cell and by measuring it using single-phase measurements in the dq-frame combined with the application of wideband signals. Moreover, the influence of the active input element on the shape of the output impedance is investigated in order to reveal whether the input and the output stage can be considered decoupled for any future analysis.

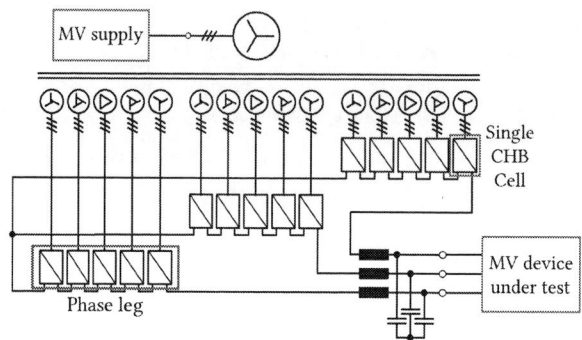

Fig. 1: Cascaded H-bridge topology for high dynamic medium-voltage perturbation injection and impedance measurement interfaced to a multi-winding transformer.

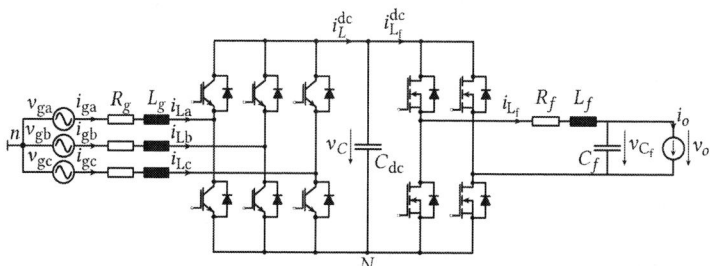

Fig. 2: CHB converter single cell comprising three-phase supplied Active Front End (AFE) interfaced to an HB phase inverter with an *LC* filter.

Single Cell Modelling

Active Front End Modelling

Closed-loop modelling of the AFE follows the state-space modelling approach outlined in [9], [10] and the intermediate developments are not given in this work, but only the final open-loop and closed-loop models in the Synchronous Reference Frame (SRF) are presented. State-space averaged model is presented in Eq. (1), where V_{dc}, D_d^{AFE}, D_q^{AFE}, I_{gd} and I_{gq} are steady-state values of dc-link voltage with duty cycles and currents represented in dq-frame. To obtain the open-loop dynamics the model in Eq. (1) can be solved using Eq. (2) and the dynamics is represented by Eq. (3). State and output variables are given in the vector on the left hand side side, while the input variables are given in the right hand side vector of Eq. (3).

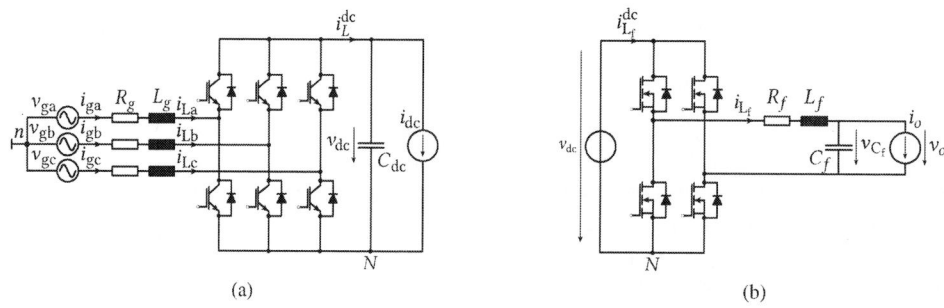

(a)   (b)

Fig. 3: Partitioned single cell of the CHB converter for the purposes of input and output stage modelling.

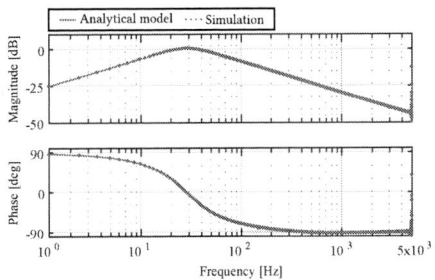

Fig. 4: Output impedance of the AFE under the input current and output voltage control loops closed.

$$
s \begin{bmatrix} \tilde{i}_{gd} \\ \tilde{i}_{gq} \\ \tilde{v}_{dc} \end{bmatrix} = \underbrace{\begin{bmatrix} -\dfrac{R_g}{L_g} & \omega_g & -\dfrac{D_d^{\text{AFE}}}{L_g} \\ -\omega_g & -\dfrac{R_g}{L_g} & -\dfrac{D_q^{\text{AFE}}}{L_g} \\ \dfrac{3}{2}\dfrac{D_d^{\text{AFE}}}{C_{dc}} & \dfrac{3}{2}\dfrac{D_q^{\text{AFE}}}{C} & 0 \end{bmatrix}}_{\mathbf{A}} \begin{bmatrix} \tilde{i}_{gd} \\ \tilde{i}_{gq} \\ \tilde{v}_{dc} \end{bmatrix} + \underbrace{\begin{bmatrix} \dfrac{1}{L_g} & 0 & 0 & -\dfrac{V_{dc}}{L_g} & 0 \\ 0 & \dfrac{1}{L_g} & 0 & 0 & -\dfrac{V_{dc}}{L_g} \\ 0 & 0 & -\dfrac{1}{C_{dc}} & \dfrac{3}{2}\dfrac{I_{gd}}{C} & \dfrac{3}{2}\dfrac{I_{gq}}{C} \end{bmatrix}}_{\mathbf{B}} \begin{bmatrix} \tilde{v}_{gd} \\ \tilde{v}_{gq} \\ \tilde{i}_{dc} \\ \tilde{d}_d^{\text{AFE}} \\ \tilde{d}_q^{\text{AFE}} \end{bmatrix}
\tag{1}
$$

$$
\begin{bmatrix} \tilde{i}_{gd} \\ \tilde{i}_{gq} \\ \tilde{v}_{dc} \end{bmatrix} = \underbrace{\begin{bmatrix} 1 & 0 & 0 \\ 0 & 1 & 0 \\ 0 & 0 & 1 \end{bmatrix}}_{\mathbf{C}} \begin{bmatrix} \tilde{i}_{gd} \\ \tilde{i}_{gq} \\ \tilde{v}_{dc} \end{bmatrix} + \underbrace{\begin{bmatrix} 0 & 0 & 0 & 0 & 0 \\ 0 & 0 & 0 & 0 & 0 \\ 0 & 0 & 0 & 0 & 0 \end{bmatrix}}_{\mathbf{D}} \begin{bmatrix} \tilde{v}_{gd} \\ \tilde{v}_{gq} \\ \tilde{i}_{dc} \\ \tilde{d}_d^{\text{AFE}} \\ \tilde{d}_q^{\text{AFE}} \end{bmatrix}
$$

$$
\mathbf{G} = \mathbf{C}(s\mathbf{I} - \mathbf{A})^{-1}\mathbf{B} + \mathbf{D}
\tag{2}
$$

$$
\begin{bmatrix} \tilde{\mathbf{i}}_{g,dq} \\ \tilde{v}_{dc} \end{bmatrix} = \begin{bmatrix} \mathbf{Y}_{\text{in,o}} & \mathbf{G}_{\text{io,o}} & \mathbf{G}_{\text{ci,o}} \\ \mathbf{T}_{\text{oi,o}} & -Z_{\text{out,o}} & \mathbf{G}_{\text{co,o}} \end{bmatrix} \begin{bmatrix} \tilde{\mathbf{v}}_{g,dq} \\ \tilde{i}_{dc} \\ \tilde{\mathbf{d}}_{dq}^{\text{AFE}} \end{bmatrix}
\tag{3}
$$

Direct link between the AFE and HB is the output impedance of the AFE and thus is the element that could influence the dynamics of the HB. For this reason the next important step is to find the closed-loop output impedance of the AFE. In the closed-loop the AFE is controlled in cascaded manner, with the inner loop being the input current control using proportional integral (PI) controllers in the SRF and the outer controller being the dc-link voltage PI controller. Under the current control loop closed, the output impedance is given as

$$
Z_{\text{out,cl}}^{\text{gcc}} = -Z_{\text{out,o}} + \mathbf{G}_{\text{co,o}}\left(\mathbf{G}_{\text{dec}} - \mathbf{G}_{\text{PI}}^{\text{i}}\right)\mathbf{G}_{\text{io,cl}}^{\text{gcc}}
\tag{4}
$$

Closing the dc-link voltage control loop reshapes the output impedance as

$$
Z_{\text{out,cl}}^{\text{dvc}} = -\left[\mathbf{I} + \mathbf{G}_{\text{co,cl}}^{\text{gcc}}\mathbf{G}_{\text{PI}}^{\text{v}}\right]^{-1} Z_{\text{out,cl}}^{\text{gcc}}
\tag{5}
$$

where $\mathbf{G}_{\text{dec}}$, $\mathbf{G}_{\text{PI}}^{\text{i}}$ and $\mathbf{G}_{\text{PI}}^{\text{v}}$ are the current controller decoupling matrix, current controller PI regulator and voltage controller PI regulator. AFE with parameters in Table I was simulated in PLECS and its analytical output impedance from Eq. (5) was compared to the impedance obtained from the frequency response. The result is presented in Fig. 4 and it shows good match. Several things can be noted in the impedance shape. The impedance magnitude has a low value in low frequency range due to a high gain of the PI controller in the same region. The breaking point at 35 Hz where the impedance has almost unity gain denotes the end of the closed-loop dc-link control bandwidth.

$$
s\begin{bmatrix}\tilde{i}_{\text{Ld}}\\\tilde{i}_{\text{Lq}}\\\tilde{v}_{\text{Cd}}\\\tilde{v}_{\text{Cq}}\end{bmatrix}=\underbrace{\begin{bmatrix}-\dfrac{R_f}{L_f} & \omega_s & -\dfrac{1}{L_f} & 0\\[2mm]-\omega_s & -\dfrac{R_f}{L_f} & 0 & -\dfrac{1}{L_f}\\[2mm]\dfrac{1}{C_f} & 0 & 0 & \omega_s\\[2mm]0 & \dfrac{1}{C_f} & -\omega_s & 0\end{bmatrix}}_{\mathbf{A}}\begin{bmatrix}\tilde{i}_{\text{Ld}}\\\tilde{i}_{\text{Lq}}\\\tilde{v}_{\text{Cd}}\\\tilde{v}_{\text{Cq}}\end{bmatrix}+\underbrace{\begin{bmatrix}-\dfrac{D_d^{\text{HB}}}{L_f} & 0 & 0 & -\dfrac{V_{\text{dc}}}{L_f} & 0\\[2mm]-\dfrac{D_q^{\text{HB}}}{L_f} & 0 & 0 & 0 & -\dfrac{V_{\text{dc}}}{L_f}\\[2mm]0 & -\dfrac{1}{C_f} & 0 & 0 & 0\\[2mm]0 & 0 & -\dfrac{1}{C_f} & 0 & 0\end{bmatrix}}_{\mathbf{B}}\begin{bmatrix}\tilde{v}_{\text{dc}}\\\tilde{i}_{\text{od}}\\\tilde{i}_{\text{oq}}\\\tilde{d}_d^{\text{HB}}\\\tilde{d}_q^{\text{HB}}\end{bmatrix}
\tag{6}
$$

$$
\begin{bmatrix}\tilde{i}_{\text{Ld}}\\\tilde{i}_{\text{Lq}}\\\tilde{v}_{\text{Cd}}\\\tilde{v}_{\text{Cq}}\\\tilde{i}_L^{\text{dc}}\end{bmatrix}=\underbrace{\begin{bmatrix}1 & 0 & 0 & 0\\0 & 1 & 0 & 0\\0 & 0 & 1 & 0\\0 & 0 & 0 & 1\\D_d^{\text{HB}} & D_q^{\text{HB}} & 0 & 0\end{bmatrix}}_{\mathbf{C}}\begin{bmatrix}\tilde{i}_{\text{Ld}}\\\tilde{i}_{\text{Lq}}\\\tilde{v}_{\text{Cd}}\\\tilde{v}_{\text{Cq}}\end{bmatrix}+\underbrace{\begin{bmatrix}0 & 0 & 0 & 0 & 0\\0 & 0 & 0 & 0 & 0\\0 & 0 & 0 & 0 & 0\\0 & 0 & 0 & 0 & 0\\0 & 0 & 0 & I_{\text{Ld}} & I_{\text{Lq}}\end{bmatrix}}_{\mathbf{D}}\cdot\begin{bmatrix}\tilde{v}_{\text{dc}}\\\tilde{i}_{\text{od}}\\\tilde{i}_{\text{oq}}\\\tilde{d}_d^{\text{HB}}\\\tilde{d}_q^{\text{HB}}\end{bmatrix}
$$

$v_{\text{g,d}} = 580\,\text{V}$	$f_g = 50\,\text{Hz}$
$v_{\text{g,q}} = 0\,\text{V}$	$L_g = 5.4\,\text{mH}$
$i_{\text{g,d}} = 41.5\,\text{A}$	$R_g = 10\,\text{m}\Omega$
$i_{\text{g,q}} = 0\,\text{A}$	$v_{\text{dc}} = 1.2\,\text{kV}$
$K_{\text{p,gcc}} = 18$	$K_{\text{i,gcc}} = 33$
$K_{\text{p,dvc}} = 1.4$	$K_{\text{i,dvc}} = 192$

Table I: AFE parameters

$v_{\text{C,d}} = 600\,\text{V}$	$f_s = 50\,\text{Hz}$
$v_{\text{C,q}} = 0\,\text{V}$	$L_f = 1.4\,\text{mH}$
$i_{\text{o,d}} = 60\,\text{A}$	$R_f = 100\,\text{m}\Omega$
$i_{\text{o,q}} = 0\,\text{A}$	$C_f = 24\,\mu\text{F}$
$K_{\text{p,lcc}} = 5$	$K_{\text{i,lcc}} = 5$
$K_{\text{p,cvc}} = 1$	$K_{\text{i,cvc}} = 52$

Table II: HB parameters

Single-Phase Inverter Modelling

In order to find the closed-loop model, first the open-loop model and dynamics have to be defined and solved. As the dc-link voltage control is performed on the AFE side, the inverter input is initially considered to be an ideal voltage source (c.f. Fig. 3b). The modelling is performed in the *dq*-frame and it follows similar approach as the modelling of the AFE. The state-space averaged open-loop model is given in Eq. (6) whose open-loop dynamics is again solved using Eq. (2). The open-loop dynamics is represented with Eq. (7). The inverter control is performed as well in a cascaded manner in SRF. The inner loop controls the filter inductor current while the outer one controls the filter capacitor voltage, or the output voltage. Closed-loop output impedance under inductor current loop closed is given as Eq. (8). Closing the capacitor voltage control loop reshapes the impedance as presented in Eq. (9). The denominators Δ_{i} and Δ_{v} are calculated using Eqs. (10) and (11).

$$
\begin{bmatrix}\tilde{\mathbf{i}}_{\text{L,dq}}\\\tilde{\mathbf{v}}_{\text{C,dq}}\\\tilde{i}_L^{\text{dc}}\end{bmatrix}=\begin{bmatrix}\mathbf{G}_{\text{iL,o}} & \mathbf{T}_{\text{oL,o}} & \mathbf{G}_{\text{cL,o}}\\\mathbf{G}_{\text{iC,o}} & \mathbf{Z}_{\text{out,o}} & \mathbf{G}_{\text{co,o}}\\\mathbf{Y}_{\text{in,o}} & \mathbf{T}_{\text{oi,o}} & \mathbf{G}_{\text{ci,o}}\end{bmatrix}\begin{bmatrix}\tilde{v}_{\text{dc}}\\\tilde{\mathbf{i}}_{\text{o,dq}}\\\tilde{\mathbf{d}}_{\text{dq}}^{\text{HB}}\end{bmatrix}
\tag{7}
$$

$$
\mathbf{Z}_{\text{out,cl}}^{\text{lcc}}=-\mathbf{Z}_{\text{out,o}}+\mathbf{G}_{\text{co,o}}\Delta_{\text{i}}^{-1}\left(\mathbf{G}_{\text{dec}}^{\text{i}}-\mathbf{G}_{\text{PI}}^{\text{i}}\right)\mathbf{T}_{\text{oL,o}}
\tag{8}
\qquad
\mathbf{Z}_{\text{out,cl}}^{\text{cvc}}=-\Delta_{\text{v}}^{-1}\mathbf{Z}_{\text{out,cl}}^{\text{lcc}}
\tag{9}
$$

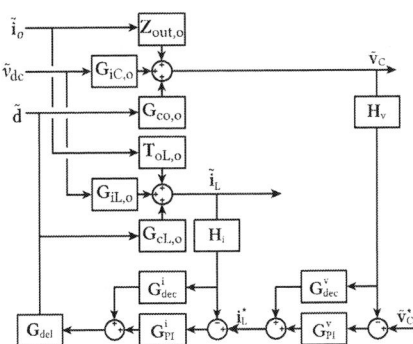

Fig. 5: Closed-loop dynamics of the HB inverter with cascaded inductor current and capacitor voltage control.

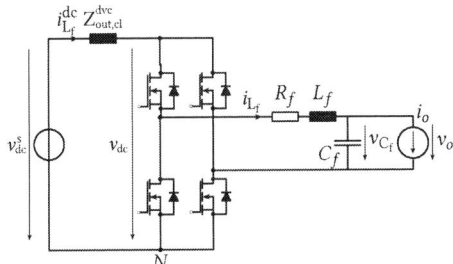

Fig. 6: HB with an output LC filter and internal source impedance.

$$\Delta_i = \left[\mathbf{I} - \mathbf{G}_{cL,o} \left(\mathbf{G}_{dec}^i - \mathbf{G}_{PI}^i \right) \right]^{-1} \qquad (10) \qquad \Delta_v = \left[\mathbf{I} - \mathbf{G}_{co,cl}^{cvc} \left(\mathbf{G}_{dec}^i - \mathbf{G}_{PI}^i \right) \right]^{-1} \qquad (11)$$

Source-Affected Dynamics Characterization

In previous section it was stated that the dc-link voltage source is ideal with respect to the fact that the voltage was controlled on the AFE side. However, in practice the control loops do not have an infinite bandwidth and as such the voltage source cannot be considered ideal. This means that the source contains a non-zero finite internal impedance which is the closed loop output impedance $Z_{out,cl}^{dvc}$ of the active input element, i.e. the AFE. From Fig. 6, the dc-side current of the inverter can be defined as in Eq. (12). Replacing the open-loop dynamics from Eq. (7) with the closed-loop one, the dc-side current of the single-phase inerter, i_{dc}^L, can be obtained using Eq. (13).

$$i_{dc}^L = \frac{v_{dc}^S - v_{dc}}{Z_{out,cl}^{dvc}} \qquad (12) \qquad i_{dc}^L = Y_{in,cl}^{cvc} v_{dc} + \mathbf{G}_{ci,cl}^{cvc} \tilde{\mathbf{v}}_{C,dq}^{\star} + \mathbf{T}_{oi,cl}^{cvc} \tilde{\mathbf{i}}_{o,dq} \qquad (13)$$

Furthermore, replacing Eq. (12) into Eq. (13) gives the v_{dc}^S expressed in terms of v_{dc} as

$$\tilde{v}_{dc} = \frac{\tilde{v}_{dc}^S - Z_{out,cl}^{dvc} \mathbf{G}_{ci,cl}^{cvc} \tilde{\mathbf{v}}_{C,dq}^{\star} - Z_{out,cl}^{dvc} \mathbf{T}_{oi,cl}^{cvc} \tilde{\mathbf{i}}_{o,dq}}{1 + Z_{out,cl}^{dvc} Y_{in,cl}^{cvc}} \qquad (14)$$

Replacing $\tilde{v}_{dc}$ in Eq. (7) by Eq. (14) gives the closed-loop source affected dynamics expressed Eq. (15).

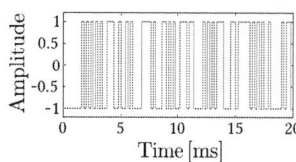 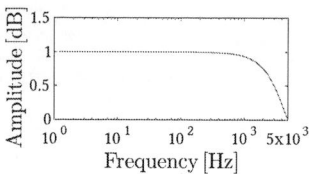

Fig. 7: PRBS-12 signal in time and frequency domain

$$
\begin{bmatrix} \tilde{\mathbf{i}}_{\mathrm{L,dq}} \\ \tilde{\mathbf{v}}_{\mathrm{C,dq}} \\ \tilde{i}_L^{\mathrm{dc}} \end{bmatrix} = \begin{bmatrix} \dfrac{\mathbf{G}_{\mathrm{iL,cl}}^{\mathrm{cvc}}}{1+Z_{\mathrm{out,cl}}^{\mathrm{dvc}} Y_{\mathrm{in,cl}}^{\mathrm{cvc}}} & \mathbf{T}_{\mathrm{oL,cl}}^{\mathrm{cvc}} - \dfrac{\mathbf{G}_{\mathrm{iL,cl}}^{\mathrm{cvc}} Z_{\mathrm{out,cl}}^{\mathrm{dvc}} \mathbf{T}_{\mathrm{oi,cl}}^{\mathrm{cvc}}}{1+Z_{\mathrm{out,cl}}^{\mathrm{dvc}} Y_{\mathrm{in,cl}}^{\mathrm{cvc}}} & \mathbf{G}_{\mathrm{cL,cl}}^{\mathrm{cvc}} - \dfrac{\mathbf{G}_{\mathrm{iL,cl}}^{\mathrm{cvc}} Z_{\mathrm{out,cl}}^{\mathrm{dvc}} \mathbf{G}_{\mathrm{ci,cl}}^{\mathrm{cvc}}}{1+Z_{\mathrm{out,cl}}^{\mathrm{dvc}} Y_{\mathrm{in,cl}}^{\mathrm{cvc}}} \\[3ex] \dfrac{\mathbf{G}_{\mathrm{iC,cl}}^{\mathrm{cvc}}}{1+Z_{\mathrm{out,cl}}^{\mathrm{dvc}} Y_{\mathrm{in,cl}}^{\mathrm{cvc}}} & \mathbf{Z}_{\mathrm{out,cl}}^{\mathrm{cvc}} - \dfrac{\mathbf{G}_{\mathrm{iL,cl}}^{\mathrm{cvc}} Z_{\mathrm{out,cl}}^{\mathrm{dvc}} \mathbf{T}_{\mathrm{oi,cl}}^{\mathrm{cvc}}}{1+Z_{\mathrm{out,cl}}^{\mathrm{dvc}} Y_{\mathrm{in,cl}}^{\mathrm{cvc}}} & \mathbf{G}_{\mathrm{co,cl}}^{\mathrm{cvc}} - \dfrac{\mathbf{G}_{\mathrm{iC,cl}}^{\mathrm{cvc}} Z_{\mathrm{out,cl}}^{\mathrm{dvc}} \mathbf{G}_{\mathrm{ci,cl}}^{\mathrm{cvc}}}{1+Z_{\mathrm{out,cl}}^{\mathrm{dvc}} Y_{\mathrm{in,cl}}^{\mathrm{cvc}}} \\[3ex] \dfrac{Y_{\mathrm{in,cl}}^{\mathrm{cvc}}}{1+Z_{\mathrm{out,cl}}^{\mathrm{dvc}} Y_{\mathrm{in,cl}}^{\mathrm{cvc}}} & \dfrac{\mathbf{T}_{\mathrm{oi,cl}}^{\mathrm{cvc}}}{1+Z_{\mathrm{out,cl}}^{\mathrm{dvc}} Y_{\mathrm{in,cl}}^{\mathrm{cvc}}} & \dfrac{\mathbf{G}_{\mathrm{ci,cl}}^{\mathrm{cvc}}}{1+Z_{\mathrm{out,cl}}^{\mathrm{dvc}} Y_{\mathrm{in,cl}}^{\mathrm{cvc}}} \end{bmatrix} \begin{bmatrix} \tilde{v}_{\mathrm{dc}}^{\mathrm{S}} \\ \tilde{\mathbf{i}}_{\mathrm{o,dq}} \\ \tilde{\mathbf{v}}_{\mathrm{C,dq}}^{\star} \end{bmatrix} \tag{15}
$$

Single Phase Measurement in *dq*-frame Using Wideband Signals

Broadly used method in system identification measurements is the sine sweep method requiring one by one frequency small-signal perturbation injection and collection of responses, thus requiring large number of measurements and time to be executed. In the case where the measurement time is excessive, the measured system may change its operating point and subsequently the characteristics measured would change. This is improved to a degree with multi-tone signals but the energy of this signal is reduced with the number of tones it consists of. On the other hand, the PRBS signals (cf. Fig. 7) have lately been used substantially for system identification and impedance measurement [11]–[13]. The advantage of these signals lies in the fact that they are well suited for the characterization of dynamics systems where rapid measurement is necessary due to the possibility of variation of the system state in time. What is common for measurements in [11]–[13] is that they are performed in dc or three-phase ac systems after which the results are represented in *dq*-frame even though they represent ac measurements. Hilbert transform was successfully applied to single-phase measurement and representation in *dq*-frame in [14]. However, the transform was applied to a single tone perturbation and frequency sweep was performed. Here, a method that applies Hilbert transform to wideband PRBS signal is given. The reason for using the Hilbert transform is to create an orthogonal signal, containing all frequencies, to the one measured in single-phase. As the output impedance matrix in the *dq*-frame is defined as

$$
\mathbf{Z}_{\mathrm{out,cl}}^{\mathrm{cvc}} = \begin{bmatrix} Z_{\mathrm{out,cl,dd}}^{\mathrm{cvc}} & Z_{\mathrm{out,cl,qd}}^{\mathrm{cvc}} \\ Z_{\mathrm{out,cl,dq}}^{\mathrm{cvc}} & Z_{\mathrm{out,cl,qq}}^{\mathrm{cvc}} \end{bmatrix} = \begin{bmatrix} \tilde{i}_{dd} & \tilde{i}_{dq} \\ \tilde{i}_{qd} & \tilde{i}_{qq} \end{bmatrix}^{-1} \begin{bmatrix} \tilde{v}_{dd} & \tilde{v}_{dq} \\ \tilde{v}_{qd} & \tilde{v}_{qq} \end{bmatrix} \tag{16}
$$

Where $\tilde{i}$ and $\tilde{v}$ are the small-signal current perturbation and the resulting current response. Measurement process of $\mathbf{Z}_{\mathrm{out,cl}}^{\mathrm{cvc}}$ requires four perturbation injections and four measurement sets. First perturbation is created in the *dq*-frame as $[p(t),0]$ and then transformed into the αβ-frame using inverse Clarke transform. Afterwards only the α-axis is injected into the current source in Fig. 3b, where the $p(t)$ is the PRBS signal in time domain. The resulting voltage measured would be equal to Eq. (17). If a second perturbation is created as $[-\hat{p}(t),0]$ and injected in the same manner as before the resulting voltage leads to to Eq. (18).

$$
v_C^s = V_m \cos \omega_s t + v_p(t) \cos \omega_s t \tag{17} \qquad v_C^s = V_m \cos \omega_s t - \hat{v}_p(t) \cos \omega_s t \tag{18}
$$

Applying the Hilbert transform to Eq. (18) and combining the resulting transformed equation with Eq. (17) provides two voltage measurements, one in α-axis and one in β-axis as

$$v^s_{C,\alpha} = V_m \cos \omega_s t + v_p(t) \cos \omega_s t \qquad (19) \qquad v^s_{C,\beta} = V_m \sin \omega_s t + v_p(t) \sin \omega_s t \qquad (20)$$

Similarly, the third and fourth perturbations are created as $[0, -p(t)]$ and $[0, -\hat{p}(t)]$, respectively. After the injection the resulting voltages are obtained as

$$v^s_C = V_m \cos \omega_s t + v_p(t) \sin \omega_s t \qquad (21) \qquad v^s_C = V_m \cos \omega_s t + \hat{v}_p(t) \sin \omega_s t \qquad (22)$$

If the Hilbert transform is now applied to Eq. (22) and the resulting equation is combined with Eq. (21) two voltage measurements, one in α-axis and one in β-axis are obtained as

$$v^s_{C,\alpha} = V_m \cos \omega_s t + v_p(t) \cos \omega_s t \qquad (23) \qquad v^s_{C,\beta} = V_m \sin \omega_s t - v_p(t) \sin \omega_s t \qquad (24)$$

Subsequently, Clarke transformation can be applied to Eqs. (19) and (20) and Eqs. (23) and (24) and the voltages can be represented in the dq-frame. Upon obtaining the measurements the frequency spectrum of the measurements is extracted using discrete Fourier Transform (DFT) algorithm from where Eq. (16) can be applied.

Simulation Results

The performance of the proposed modelling and measurement method is put under test using PLECS simulation environment. A 5 kHz PRBS-12 signal is injected into on top of the fundamental load current and the resulting output voltage is measured. In order to investigate the effect of presence of an active converter stage in the cell, two types of CHB cell were simulated. The first cell type is a standalone HB-inverter with an ideal voltage source as an input, while in the second case the AFE control system controls the capacitor dc-link voltage which serves as an input voltage source to the HB-inverter. In both cases, the HB-inverter is operating with parameters given in Table II. The d-axis closed-loop output impedances $Z^{cvc}_{out,cl}$ of the HB-inverter for with parameters in Table II are presented in Fig. 9. By comparing the two frequency responses it is easy to notice that the proposed modelling holds ground as the analytical characteristics matches the simulation results that are obtained by measuring the output impedance using the method proposed in this work. Besides, it is also noticeable that the output impedance of the HB-inverter does not change when there is an active element on the input side.

Conclusion

This paper presented a promising solution for the problem of MV impedance/admittance measurement based on the MV-CHB converter. To confirm the suitability of the proposed idea, the model of the converter is developed systematically starting with the model of one single-phase cell of the converter and by confirming that the AFE input stage of a single cell does not affect the shape of the HB output impedance. The conclusion is reached through the wideband measurement of the output impedance using PRBS signals, while the results are shown methodically through analytical modelling and simulations. This type of measurement can successfully be performed in dq-frame by using the Hilbert transform of wideband signals. Compared to previous measurement methods, such as the much slower ac-sweep method, the wideband measurements allow rapid verification of concepts put forward. Moreover, the use of wideband measurements, namely the use of PRBS signals, reduce the risk of the measured system changing state while being measured.

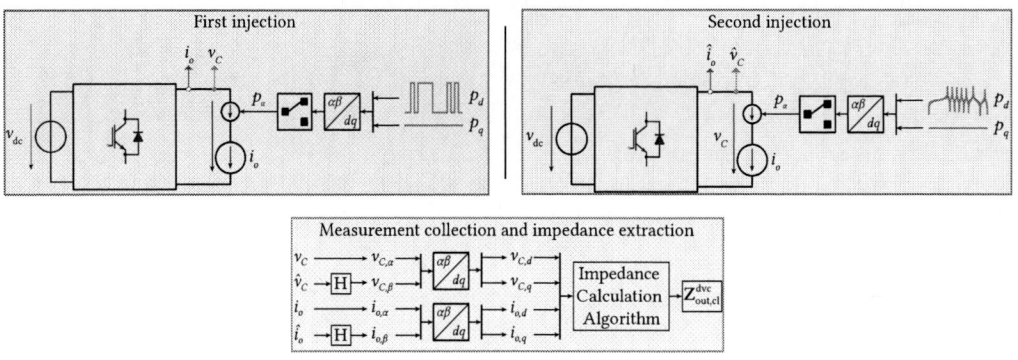

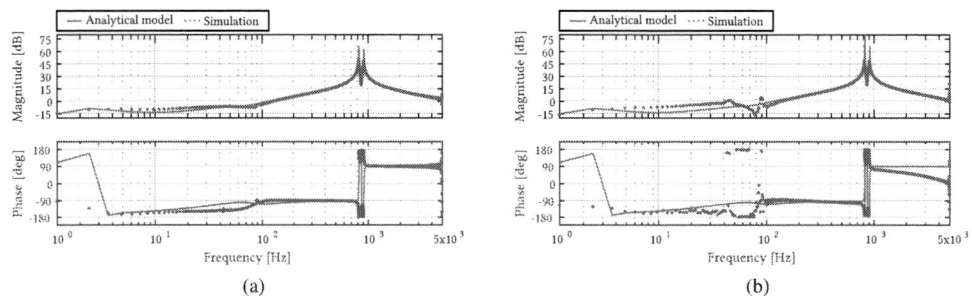

Fig. 8: Output impedance measurement of a single-phase inverter in the *dq*-frame

Fig. 9: *d*-axis output impedance $\mathbf{Z}_{\text{out,cl,dd}}^{\text{cvc}}$ with a) standalone-operation of the HB-inverter compared to the same characteristics, b) when the AFE is operating at the input stage.

References

[1] X. Wang and F. Blaabjerg, "Harmonic stability in power electronic based power systems: Concept, modeling, and analysis," *IEEE Transactions on Smart Grid*, pp. 1–1, 2018.

[2] U. Javaid, F. D. Freijedo, D. Dujic, and W. Van Der Merwe, "Dynamic assessment of source–load interactions in marine MVDC distribution," *IEEE Transactions on Industrial Electronics*, vol. 64, no. 6, pp. 4372–4381, 2017.

[3] A. Riccobono and E. Santi, "Comprehensive review of stability criteria for dc power distribution systems," *IEEE Transactions on Industry Applications*, vol. 50, no. 5, pp. 3525–3535, Sep. 2014.

[4] U. Javaid, F. D. Freijedo, W. van der Merwe, and D. Dujic, "Stability analysis of multi-port mvdc distribution networks for all-electric ships," *IEEE Journal of Emerging and Selected Topics in Power Electronics*, pp. 1–1, 2019.

[5] M. Mogorovic and D. Dujic, "100 kw, 10 khz medium-frequency transformer design optimization and experimental verification," *IEEE Transactions on Power Electronics*, vol. 34, no. 2, pp. 1696–1708, Feb. 2019.

[6] M. Utvic, S. Milovanovic, and D. Dujic, "Flexible medium voltage dc source utilizing series connected modular multilevel converters," in *[Detailed technical programme of the 21th European Conference on Power Electronics and Applications (EPE ECCE Europe)]*, 2019.

[7] M. Petković, N. Hildebrandt, F. D. Freijedo, and D. Dujić, "Cascaded H-Bridge Multilevel Converter for a High-Power Medium-Voltage Impedance-Admittance Measurement Unit," in *2018 International Symposium on Industrial Electronics (INDEL)*, Nov. 2018, pp. 1–8.

[8] Y. Li, Y. Wang, R. Wang, J. Wu, H. Zhang, Y. Feng, S. Li, W. Yao, and B. Q. Li, "Distributed Generation Grid-Connected Converter Testing Device Based on Cascaded H-Bridge Topology," *IEEE Transactions on Industrial Electronics*, vol. 63, no. 4, pp. 2143–2154, Apr. 2016.

[9] R. W. Erickson and D. Maksimovic, *Fundamentals of power electronics*. Springer Science & Business Media, 2007.

[10] T. Suntio, T. Messo, and J. Puukko, *Power Electronic Converters: Dynamics and Control in Conventional and Renewable Energy Applications*. Wiley, 2017.

[11] R. Luhtala, T. Roinila, and T. Messo, "Implementation of real-time impedance-based stability assessment of grid-connected systems using mimo-identification techniques," *IEEE Transactions on Industry Applications*, vol. 54, no. 5, pp. 5054–5063, Sep. 2018.

[12] T. Roinila, H. Abdollahi, S. Arrua, and E. Santi, "Real-time stability analysis and control of multiconverter systems by using mimo-identification techniques," *IEEE Transactions on Power Electronics*, vol. 34, no. 4, pp. 3948–3957, Apr. 2019.

[13] T. Roinila, T. Messo, R. Luhtala, R. Scharrenberg, E. C. W. de Jong, A. Fabian, and Y. Sun, "Hardware-in-the-loop methods for real-time frequency-response measurements of on-board power distribution systems," *IEEE Transactions on Industrial Electronics*, vol. 66, no. 7, pp. 5769–5777, Jul. 2019.

[14] Y. Liao, Z. Liu, H. Zhang, and B. Wen, "Low-frequency stability analysis of single-phase system with dq-frame impedance approach—part i: Impedance modeling and verification," *IEEE Transactions on Industry Applications*, vol. 54, no. 5, pp. 4999–5011, Sep. 2018.

A Novel Three-Phase PFC Diode Rectifier by LC Network Circuits for High Frequency Generator

Shin-ichi MOTEGI[†] and Yasuyuki NISHIDA[††]
[†]Kobe City College of Technology, 8-3, Gakuen-Higashi-Machi, Nishi-ku, Kobe, Japan
[††]Chiba Institute of Technology, 2-17-1, Tsudanuma, Narashino, Chiba, Japan
Tel.: +81 / (78) – 795 – 3311
Fax: +81 / (78) – 795 – 3314
E-Mail: motegi@mem.iee.or.jp

Keywords

Three-phase system, High frequency power converter, Power factor correction, Harmonics, Passive component, Transformer

Abstract

This study deals with a three-phase diode rectifier for high frequency generator and it offers very fine PFC nature. The rectifier has an L-C network circuit in each phase of the input. Performance of the rectifier with most of all practically available combinations of inductance and capacitance of the L-C network circuit has been investigated through simulations. As a result, useful combinations of the constants to obtain desirable performance of the new rectifier are found. Referring to the simulation study, experimental setup was build and tested. The Total-Power-Factor of the input and Total-Harmonic-Distortion of input current in the experiments are 99% and 5%, respectively, at rated condition (4kW). It is shown in this paper that the new rectifier is practically advantageous than ordinary rectifier, e.g., three-phase bridge diode rectifier, and even well comparable to double three-phase bridge diode rectifier.

Introduction

Three-phase rectifiers have been widely applied in practice and those PFC performance is essential to obtain dc power from a high frequency generator with a high Total-Power-Factor to reduce size, weight and cost of the facility and a high quality input current to mitigate harmonic problem in the generator. Although the PWM rectifiers are currently the most common topology in the three-phase PFC rectifier, they have disadvantages in reliability or environmental durability (especially taking the necessity of controllers into account) and producing EMI noises comparing with passive scheme. On the other hand, conventional multi-pulse (e.g., 12-pulse, 18-pulse and etc.) diode rectifiers [1]-[6] are alternative to obtain three-phase PFC rectifier but an isolation or non-isolation transformer (or an auto-transformer) is required in the rectifiers and they results in a bulky, heavy, and expensive solution.

The authors have been studied a single-phase PFC diode rectifier without active switching devices and achieved a very fine topology [7]. The proposed single-phase passive PFC rectifier is characterized by a very high total power factor (>98% under rated load conditions) and very low total harmonic distortion of the input current (<10% under rated load conditions). In this paper, the authors apply the idea of the proposed single-phase passive PFC rectifier [7] into a three-phase PFC rectifier [8], [9] (this paper calls it "KOBE rectifier"). The performance of the proposed KOBE rectifier is compared with that of ordinary capacitor-input-type (CIT) diode rectifier through simulations.

Topology

Proposed KOBE Rectifier

The proposed three-phase PFC diode rectifier (KOBE rectifier) without active switching devices or transformers and for high frequency generator [10] is shown in Fig. 1. As shown in the figure, the proposed KOBE rectifier consists of three of single-phase PFC diode rectifier [7] connected in Y-form and each of them has an L-C network circuits (L_S, L_A + C_A, L_B) on the AC-line. The performance strongly depends on the inductance and the capacitance of the L-C network. So, most of all practically available combinations of inductance and capacitance of the L-C network circuit have been investigated through simulations to determine desirable combination of the L and C to achieve a fine performance. As a result, a useful combination, i.e., L_S=1 mH, L_A=8 mH, L_B=3 mH, and C_A=1.72 μF shown in Table I, to obtain desirable performance of the new rectifier is found.

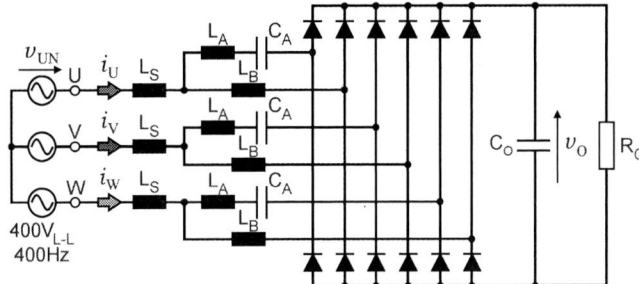

Fig. 1: Proposed three-phase PFC diode rectifier (KOBE Rectifier).

Conventional Double Three-Phase-Bridge Capacitor-Input-Type Diode Rectifier

The single three-phase-bridge Capacitor-Input-Type (CIT) diode rectifier offers the best simple topology in the three-phase rectifiers and so it's widely applied in practice. Though, significant distortion of the input current is only one disadvantageous feature of it. To overcome the problem but remaining the advantageous features of the diode rectifier, the double three-phase-bridge CIT diode rectifier is often applied in practice. Although there are variety of topologies of the double three-phase-bridge CIT diode rectifier (12P-CIT rectifier), the best orthodox one shown in Fig. 2 is chosen to compare the performance with that of the proposed rectifier in this study. Discussion of the proposed rectifier operation is omitted in the paper since it's very complicated but it's not the focus in the paper.

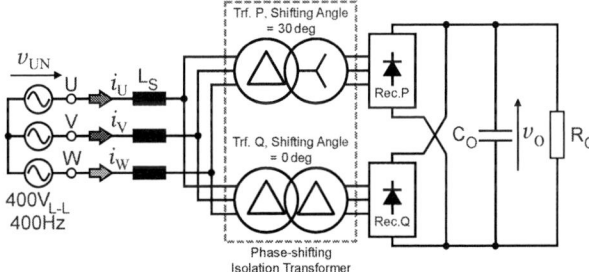

Fig. 2: Double 3-phase-bridge capacitor input type (12P-CIT) rectifier.

Table I: Operating condition of rectifiers in simulation and experiment.

			KOBE Rectifier (Fig.1)	12-Pulse Capacitor Input-Type Rectifier (Fig.2)
AC source voltage		V_S [V_{RMS}]	400V, 400Hz	400V, 400Hz
AC-side	Inductor	L_S [mH]	1	3
	Indcutor	L_A [mH]	8	Not Connected
	Inductor	L_B [mH]	3	Not Connected
	Capacitor	C_A [μF]	1.72	Not Connected
	Isolation Transformer	Trf. P	Not Connected	$1:1/\sqrt{3}$
		Trf. Q	Not Connected	1:1
DC-side	Capacitor	C_O [μF]	600	600
DC Output Power		P_O [kW]	4	4

Simulation Study

The performances of the proposed KOBE rectifier shown in Fig. 1 and the 12P-CIT rectifier shown in Fig. 2 have been investigated through simulations. The operating condition of the rectifiers in the simulations is shown in TABLE I. All the components in the rectifiers are ideal in the simulation study. The capacitance C_O of the dc capacitor is set to $600\,\mu F$ and is enough to reduce the dc voltage ripples so that the ripples do not affect the input performance of the rectifiers. Most of all practically available combinations of inductance and capacitance of the L-C network circuit have been investigated through simulations. As a result, useful combinations of the constants to obtain desirable performance of the new rectifier are found. Here, the constants for the L-C network circuits are $L_S=1\,mH$, $L_A=8\,mH$, $L_B=3\,mH$, and $C_A=1.72\,\mu F$, as shown in Table I.

Operating Waveforms and Input Current Quality

Figure 3 (a) shows operating waveforms while (b) shows spectrum of the input line current i_U of the proposed KOBE rectifier. As seen in Fig. 3(a), the input line currents draw high quality waveforms and the Total-Harmonic-Distortion THD of the input line current i_U is only 5.27%. Since the phase voltage v_{UN} of the high frequency generator and the input line current i_U are almost in-phase each other, the TPF is almost unity (i.e., 99.70%). As seen in Fig. 3(b), the input current i_U involves all the orders of harmonics but the amplitudes are all very low. From the waveforms and spectrum, it has been confirmed that the proposed rectifier offers a very high quality performance in input Total-Power-Factor and input current waveform.

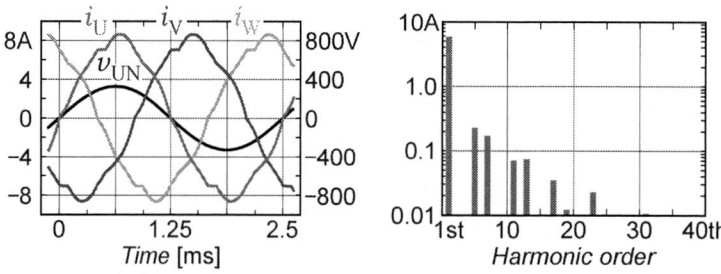

(a) Operating waveforms (i_U, i_V, i_W, v_{UN})   (b) Spectrum of input line current i_U

Fig. 3: Operating waveforms (a) and spectrum of input current i_U (b) of KOBE rectifier (simulation).

Figure 4 (a) shows operating waveforms and (b) shows spectrum of the input line current i_U of the 12P-CIT rectifier under condition with $L_S=3$ mH. It is known from the waveforms in Fig. 4(a) that the input line current draws a fine waveform as similar as those of the proposed rectifier and the $THD =$ 5.07%, i.e., slightly better than that of the proposed KOBE rectifier. The input current i_U is slightly delayed for the high frequency generator phase voltage v_{UN} due to the line inductor L_S and thus, the TPF is slightly lower (95.32%) than that of the proposed one. As seen in Fig. 4(b), the input line current i_U does not involves particular orders (i.e., 5th, 7th, 17th, 19th and etc.) of harmonics due to double three-phase-bridge topology.

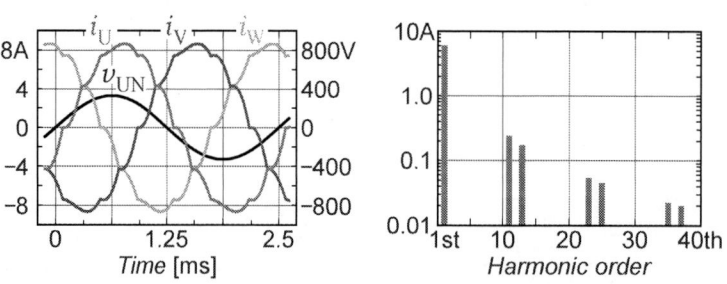

(a) Operating waveforms (i_U, i_V, i_W, v_{UN})   (b) Spectrum of input line current i_U

Fig. 4: Operating waveforms (a) and spectrum of input current i_U (b) of the 12P-CIT rectifier (simulation).

Comparing the performance of the proposed KOBE rectifier and 12P-CIT rectifiers obtained from simulation study and shown in Fig. 3 and Fig. 4, respectively, it is known that the two rectifiers offer almost the same performance or that of the proposed rectifier is slightly better.

Capacity (VA) of Magnetic Components

As same as the electrical characteristics, "volume, weight and cost" are essential to evaluate the performance of the rectifiers, and thus those of the inductors applied on the two rectifiers are tried to evaluate and compare in the study.

Although the "kVA" rating is a convenient and reasonable index to evaluate the "volume, weight and cost" of inductors, it is good only in condition with sinusoidal voltage and current, in a strict sense. Though, the current flowing through the inductor is not much distorted in the rectifiers and thus, the ordinary kVA is applied in this study to evaluate and compared the "volume, weight and cost" of the inductors.

$$S_L = 2\pi f L \left(I_{L(RMS)}\right)^2 \text{ [VA]} \tag{1}$$

f: Frequency of generator [Hz] ($f = 400\,\text{Hz}$)

L: Inductance of the inductor [H]

$I_{L(RMS)}$:   RMS value of current flowing through the inductor [A_{RMS}]

TABLE II shows S_L of the inductors applied on the two rectifiers for $P_O=4\text{kW}$ condition. The RMS value of the currents flowing through the inductors are shown in TABLE II(b). Using the inductance and the RMS value of the currents shown in TABLE II into Eq. (1) and multiplied by "3" to get three-phase total values, the three-phase total "VAs" (i.e., $S_{L\text{-}3P}$) are obtained as shown in TABLE II(c). Total sum of the "VA" $S_{L\text{-}3P}$ of all the nine inductors is shown in TABLE II(d) and it is 1119 VA or 27.97% of maximum P_O (i.e., 4 kW). The 12P-CIT rectifier employs only three inductors L_S and total sum of the "VA" is 831 VA (21% of max. P_O) as shown in TABLE II(d). Thus, the total sum of the "VAs" of the inductors of the 12P-CIT rectifier is 7% less than that of the proposed KOBE rectifier. However, the 12P-CIT rectifier requires isolation transformers or auto-transformers in addition to the inductors L_S. The kVA of the transformer depends on type of it but the minimum value is offered by the auto-transformers scheme and is 15 to 20% of the output power P_O or so [3], [6]. The kVA of inductors and transformer are based on different rule and the detail discussion is omitted here but half of the "inductor kVA" nearly equals to the "transformer kVA." [3], [6] Considering this, the modified total kVA of the KOBE rectifier is 14% (i.e., 28%/2) while that of 12P-CIT rectifier is 25.5 to 30.5% (i.e., 21%/2 + 15 to 20%). Thus, the proposed KOBE rectifier is advantageous in the "volume, weight and cost" than the 12P-CIT rectifier.

Table II: Capacity (VA) of Inductors (at P_O = 4kW)

		KOBE Rectifier (Fig.1)			12-Pulse CIT (Fig.2)
		L_S	L_A	L_B	L_S
(a)	Inductance L [mH]	1	8	3	3
(b)	I [A_{RMS}]	5.79	2.10	5.15	6.06
(c)	$S_{L\text{-}3P}$ [VA]	252.77	266.00	599.92	830.67
(d)	Total $S_{L\text{-}3P}$ [VA]	**1118.69**			**830.67**
(e)	TPF [%]	99.70			95.32
(f)	THD [%]	5.27			5.07

Experimental Results

An experiment was conducted using a prototype to measure the TPF, THD, V_O, and efficiency of the proposed KOBE rectifier. The operating condition in the experiments is shown in TABLE I. In the following experiments, an amplifier sinusoidal power source (three-phase ES2000 system by NF

Corporation) is applied to emulate the high frequency generator and a high accuracy power meter (WT3000E, Yokogawa Test & Measurement Corporation) is employed to measure the *TPF*, *THD*, V_O, and efficiency.

operating Waveforms and Input Current Quality

Figure 5 (a) shows operating waveforms while (b) shows spectrum of the input line current i_U, respectively, of the proposed KOBE rectifier. As seen in Fig. 5(a), the input line currents draw high quality waveforms and the *THD* of the input line current i_U is only 5.04%. Since the phase voltage v_{UN} of the mains and the input line current i_U is almost in-phase each other, the *TPF* is almost unity (i.e., 99.33%). As seen in Fig. 5(b), the input current i_U involves all the orders of harmonics but the amplitudes are all very low (the largest was the 5th harmonic component of approximately 210mA). From the waveforms and spectrum, it has been confirmed that the proposed rectifier offers a very high quality performance in input *TPF* and input current waveform.

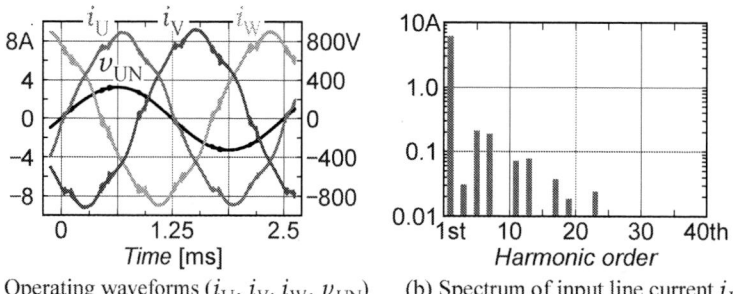

(a) Operating waveforms (i_U, i_V, i_W, v_{UN})   (b) Spectrum of input line current i_U

Fig. 5: Operating waveforms (a), spectrum of input current i_U (b) of the proposed KOBE rectifier (experimental results).

Static Characteristics

Figure 6 and Fig. 7 show several static characteristic curves of the proposed rectifier. The *TPF* of input, the *THD* of input line current i_U, the dc output voltage V_O, and conversion efficiency η, for all the output power P_O are shown in Fig. 6(a), (b), Fig. 7(a), and (b), respectively.

The *TPF* offers a high value for middle to high wider P_O range as shown in Fig. 6(a). The *TPF* is over 95% at P_O = 0.6kW and above and the maximum value is 99.51% at P_O=2.8kW. The *THD* offers a low value for middle to high wider P_O range as shown in Fig. 6(b). The *THD* is under 10% at P_O = 2.1 kW or above and the lowest value is 4.97% at P_O=3.8kW.

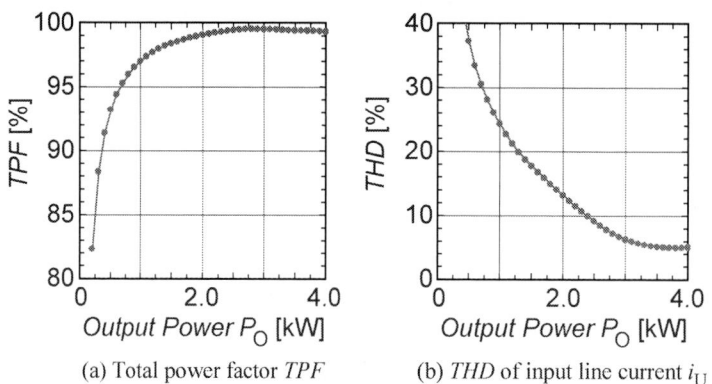

(a) Total power factor *TPF*   (b) *THD* of input line current i_U

Fig. 6: Static characteristic curves of the proposed rectifier (experimental results).

The output voltage V_O decreases rapidly as P_O increases for very lower P_O range (not shown in Fig. 7(a)) but the decreasing rate is lower excepting the very low P_O range as shown in Fig. 7(a). The output voltage V_O reaches to $498.9\,V_{AVG}$ at $P_O = 4.0\,\text{kW}$. Rated voltage dropping rate is 13.4% in the case. Although the conversion efficiency η slightly decreases as P_O increases, excepting very light load range, the efficiency η remains in a high value (i.e., 98.5%) even in the maximum output power point of $P_O = 4.0\,\text{kW}$.

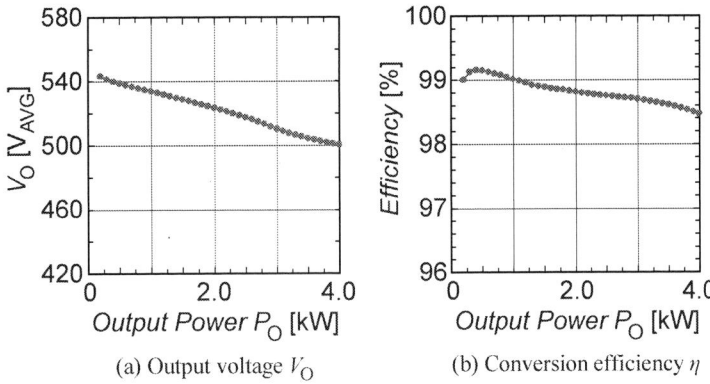

(a) Output voltage V_O (b) Conversion efficiency η

Fig. 7: Static characteristic curves of the proposed rectifier (experimental results).

Conclusion

A novel three-phase PFC diode rectifier without active switching devices or transformers (this paper calls it the KOBE rectifier) is proposed and the performance is evaluated and compared with ordinary double three-phase bridge diode rectifier of Capacitor-Input-Type (12P-CIT rectifier) through simulations and experiments. The proposed KOBE rectifier consists of three of single-phase PFC diode rectifier connected in Y-form and each of them has an L-C network circuit on each the ac input side. Since the characteristics depend on L-C constants and operating condition of the rectifier, appropriate combination of the constants and operating condition is explored by means of simulations. As a result, practically reasonable characteristics are obtained and shown in this paper. Comparing the characteristics of the proposed KOBE Rectifier and of the 12P-CIT rectifiers, it is confirmed that the new rectifier offers almost the same or slightly better characteristic than the ordinary one.

The kVA of the inductors applied on the two rectifiers are calculated and the size, weight and cost of the rectifier are evaluated by taking the transformer employed in the 12P-CIT rectifier into account. It is known as a result that the proposed rectifier is advantageous than the 12P-CIT rectifier.

It is concluded that the proposed KOBE rectifier is better than the 12P-CIT rectifier.

Further studies regarding the following subjects remain and those will be studied future.

- Evaluation of "volume, weight and cost" of phase shifting transformer (isolation or non-isolation ones) in the ordinary rectifier and ac capacitor C_A in the new rectifier
- Dynamic characteristic

References

[1] B. M. Bird, J. F. Marsh, and P. R. McLellan: "Harmonic reduction in multiplex convertors by triple-frequency current injection," *Proc. IEE*, Vol. 116, No. 10, pp. 1730-1734, 1969

[2] A. Ametani: "Generalized method of harmonic reduction in a.c.-d.c. converters by harmonic current injection," *Proc. IEE*, Vol. 119, No. 7 pp. 857-864, 1972

[3] C. Niermann: "New Rectifier Circuits with Low Main Pollution and Additional Low Cost Inverter for Energy Recovery," *EPE'89*, Vol. III, pp. 1131-1136, 1989

[4] A. Uan-Zo-li, F. Lacaux, R. P. Burgos, F. Wang, and D. Boroyevich: "Multi-level Modeling of Autotransformer Rectifier Units for Air-craft Applications," *Conf. Proc. of IPEC-Niigata 2005*, pp. 1831-1837, 2005

[5] H. Mochikawa, J. Tsuda, M. Uesugi, T. Yamashita, and T. Kobayashi: "A New 18-Pulse Rectifier Circuit with Small Phase-Shifting Transformer," *Conf. Proc. of IPEC-Niigata 2005*, pp. 1838-1842, 2005

[6] Y. Nishida: "Passive and Hybrid PFC Rectifiers, -A Survey and Exploration of New Possibilities-," *T. IEE Japan*, Vol. 126-D, No. 7, pp. 927-940, 2006

[7] S. Motegi: "A 3.5kW Single-Phase Voltage Doubler Rectifier without Switching Devices for European Use that Meets IEC 61000-3-2 (Class A)," *T. IEE Japan*, Vol. 138-D, No. 11, pp. 841-847, 2018 (in Japanese)

[8] S. Motegi and Y. Nishida: "Three-Phase PFC Rectifier by Means of Three Single-Phase Passive PFC Units," 2019 10th International Conference on Power Electronics and ECCE Asia (ICPE 2019 - ECCE Asia), WeF1-4, 2019

[9] S. Motegi: "A Three-Phase High Power Factor Diode Rectifier without Active Switching Devices or Transformers," *T. IEE Japan*, Vol. 140-D, No. 6, pp. 495-496, 2020 (in Japanese)

[10] http://www.yanmar-es.com/products/chp/

Frequency-domain simulation of power electronic systems based on multi-topology equivalent sources modelling method

Stephane VIENOT [1,3], Arnaud VIDET [1], Nadir IDIR [1], Lamine KONE [2],
Sébastien WEISS [3], Frederic LAFON [3]

[1] Univ. Lille, Arts et Metiers ParisTech, Centrale Lille, HEI,
L2EP - Laboratoire d'Electrotechnique et d'Electronique de Puissance,
59000 Lille, France
[2] IEMN, CNRS UMR 8520, 59000 Lille, France
[3] Valeo - Electrical Systems, 62630 Etaples, France,
stephane.vienot@valeo.com
nadir.idir@univ-lille.fr

Keywords

EMC/EMI , Modelling, Frequency-Domain Analysis.

Abstract

ElectroMagnetic Interference (EMI) simulation of power converters helps engineers in the design process. In this paper, we describe a frequency-domain simulation method based on the Multi-Topology Equivalent Sources (MTES) model. The aim is to reproduce the non-linear behavior of power switches for a fast evaluation of the conducted EMI. The method performance is validated on a DC-DC converter.

Introduction

Car manufacturers have integrated more and more electronic systems in vehicles. To comply with ElectroMagnetic Compatibility (EMC) constraints, simulations are used to reduce cost and development time [1]. Electrical systems can be modelled in time and/or frequency domains. Frequency-domain models are usually established from an impedance network associated with generators that represent ElectroMagnetic Interference (EMI) sources [2,3]. Simulations in frequency-domain are faster than in time-domain, so they can be evaluated many times in an optimization process in order, for example, to design an EMI filter [4,5].

However, models in frequency-domain do not take into account the evolution over time of the power switch states and therefore their impedances. Thus, we propose to use the Multi-Topology Equivalent Source (MTES) modelling method [6] to associate each transient with its equivalent source and impedance. The MTES method uses the simulation of several topologies that are then associated to reconstruct current or voltage. Previous work showed that the overall results were sensitive to the model association process [7], we propose an improvement thanks to a specific modeling of the EMI sources.

This paper presents how to perform a frequency-domain simulation with several topologies in order to take into account the different states of semiconductor components during a switching period. In addition, a new approach to model EMI sources is given to improve the result robustness. The MTES method is finally applied to a Buck converter, the comparison with a reference time-domain simulation allows to verify the validity of the proposed method as well as the improvement of the simulation duration.

Impedances and sources of EMI in power electronics

EMI sources are related to high values of di/dt and dv/dt [8] which occurs, in power electronics, during switching of semiconductor components. From a functional point of view, the current and voltage are varying with the semiconductor states and so to their impedances. With a simple example, we can show how these states can impact the propagation path of EMI.

Impact of the state of semiconductor on the propagation path

Figure 1(a) shows a simplified leg structure that can operate as a DC/DC converter. Two Line Impedance Stabilization Networks (LISN) are used at the input and the output is connected via a power cable represented by L_2 and C_{cm}. In high frequency, the LISN capacitors and C_{bus} can be approximated with short circuits because the 50Ω load of the LISN is much higher. Finally, we can see in figure 1(b) and figure 1(c) that the states of power switches K_1 and K_2 have a direct impact on the common mode impedance and thus on the current path and its resonance frequency.

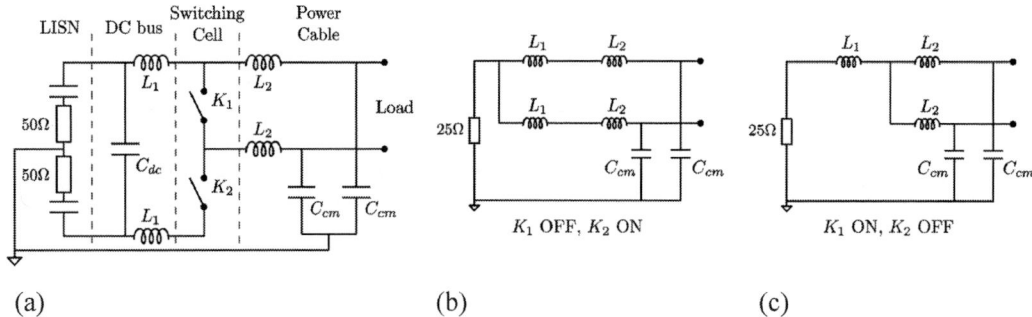

(a)   (b)   (c)

Fig. 1: Simplified case study of the power converter impedance

However, EMI are generated during commutation, so we need a method to associate sources and impedances in order to have a good representation of the EMI generation and coupling paths. In the next section, we describe how to achieve this goal with the MTES.

Multi-Topology Equivalent Sources (MTES)

To illustrate the method, we consider that the power switches K_1 and K_2 are two MOSFETs. Depending on their states, the MOSFET impedance is different. During conduction, the MOSFET impedance is low and can be approximated with its on-state resistance R_{dsON}. On the other hand, when the MOSFET is in off-state, its impedance is high and is related to the MOSFET output capacitance, C_{oss}, under the DC voltage bias.

The EMI sources correspond to the transient of current and voltage during switching. The oscillation resulting from this transient depends on the circuit loop impedance. When the transistor turns on, the source impedance becomes low so we can use a voltage source to model the transient and the resulting propagation path for EMI. Similarly, when it turns off, its impedance is high therefore a current source can be used.

In addition, conduction of high side and low side MOSFETs are complementary. The equivalent source topologies to model a converter leg are given in figure 2. So, the Low Current and High Voltage (LCHV) topology and Low Voltage and High Current (LVHC) topology represent the sources and impedances of each transient occuring during a switching period.

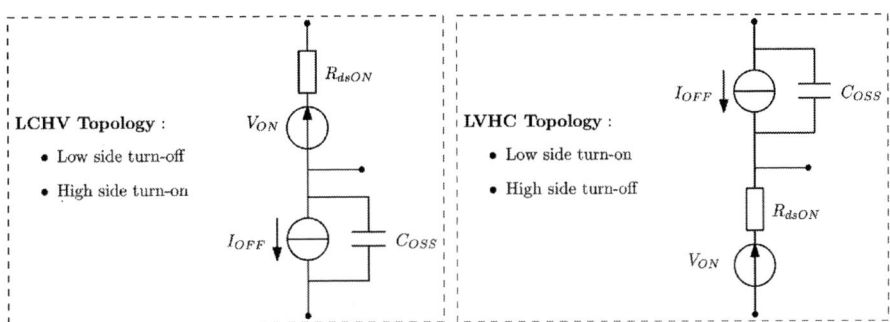

Fig. 2: LCVH and LVHC topologies [7]

The final result is a combination of both topologies. Previous work [7] made use of validity functions. However, the obtained results were dependent on the functions definition. Two improvements are proposed in this paper.

First, for each commutation, we propose to split the current and voltage transients into two topologies. This allows using different impedances during the commutation process in order to accurately represent voltage and current. Secondly, to simplify and improve the robustness, a mathematical expression is derived to model EMI sources. It represents a single transient to avoid needing validity function. As a consequence, the global result corresponds to the sum of the simulated topologies.

Modelling topologies to represent switching process

In this section, we describe the modelling method used to represent the commutation in the frequency-domain. Based on simplified switching curves, we define the appropriate topologies that are used to simulate a leg structure. The analysis is focused on a semi-conductor hard switching behavior, the associated component of the leg is represented by a single impedance.

Model for turn-on

The simplified current and voltage waveforms during turn-on are given in figure 3. First the current rise to the load current, during this phase the voltage stays constant. We can deduce that the equivalent source is neither high or low because there is a significant voltage and current. Then the drain to source voltage falls to zero, the MOSFET state is on so its impedance is low and is directly related to R_{dsON}.

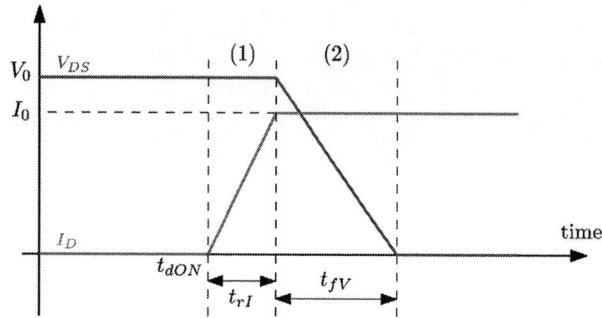

Fig. 3: Simplified MOSFET current and voltage waveforms during turn-on

To represent this commutation, we use two topologies given figure 4. On the left, the current switching topology allows to simulate the first transient. We use the resistance $R_{Transient}$ to simulate the source impedance, its value is derived from DC bus voltage V_0 and load current I_0 : $R_{Transient} = V_0/I_0$. For the voltage switching, we use the LVHC with the high side current source equal to zero. This topology induces ringing in the circuit due to the low impedance voltage source.

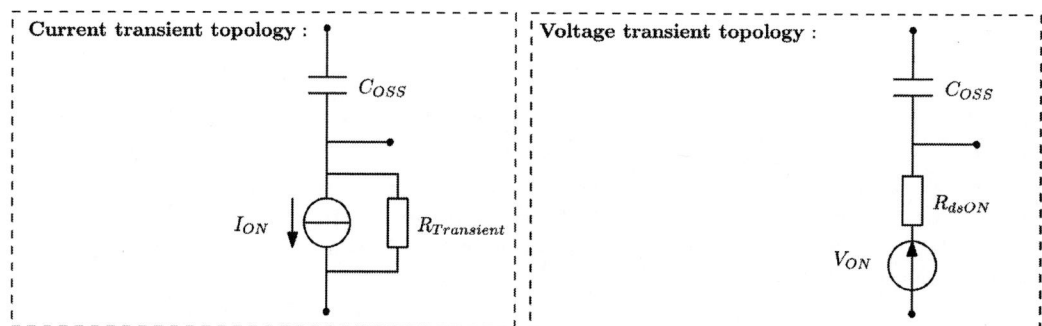

Fig. 4: Modelling topologies during turn-on

Model for turn-off

For the turn-off, the voltage transient occurs before the current as shown in figure 5. As a result, the topologies used to simulate this commutation includes a voltage source with the same resistance $R_{Transient}$. Then the current transient is modelled with a high impedance source as shown in figure 6.

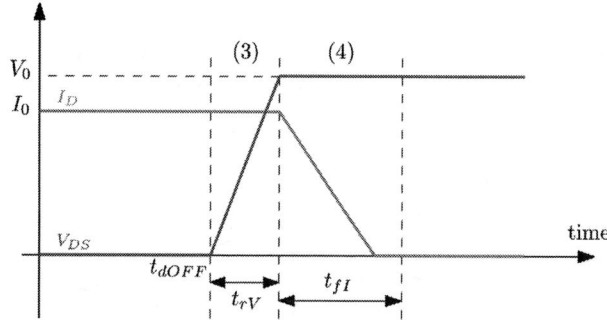

Fig. 5: Simplified MOSFET current and voltage waveforms during turn-off

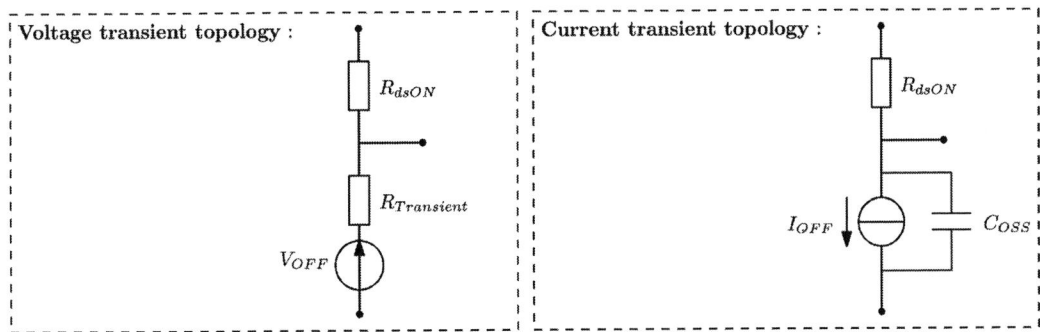

Fig. 6: Modelling topologies during turn-off

Frequency model of time-domain signal

The EMI sources, MOSFET drain current and drain to source voltage, can be modeled with trapezoidal waveforms [9]. The aim of this section is to show how to isolate each switching transient in order to have the appropriate source associated with its topology.

Extraction of the rise and fall transients of a trapezoidal signal

A trapezoidal signal $x(t)$ can be decomposed into two periodic signals $x(t) = x_r(t) + x_f(t)$ as shown in figure 7. The signals $x_r(t)$ and $x_f(t)$ allow to respectively model the rise and fall transients, these signals contain a slow ramp to ensure the signals periodicity. The frequency-domain expression of these signals allows to simulate a single transient of a waveform, and thus to adapt the topology for each transient.

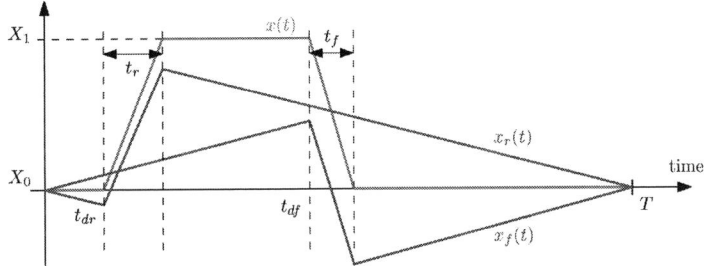

Fig. 7: Decomposition of a trapezoidal signal to extract rise and fall transients

EMI transient characteristics in frequency-domain

Fourier series are used to determine the spectrum of $x_r(t)$. The expression of Fourier coefficients $c_n[x_r]$ are given by the equation (1).

$$c_n[x_r(t)] = \frac{1}{T} \frac{X_1 - X_0}{t_r} \frac{1 - e^{-i2\pi f_n t_r}}{(i2\pi f_n)^2} e^{-i2\pi f_n t_{dr}}$$

(1)

Where X_0 and X_1 are respectively the initial and final values. T is the signal period, t_{dr} the time delay and t_r the rise time. f_n are the harmonic frequencies : $f_n = n/T$. With the adequate parameters, this expression can also model a fall transient.

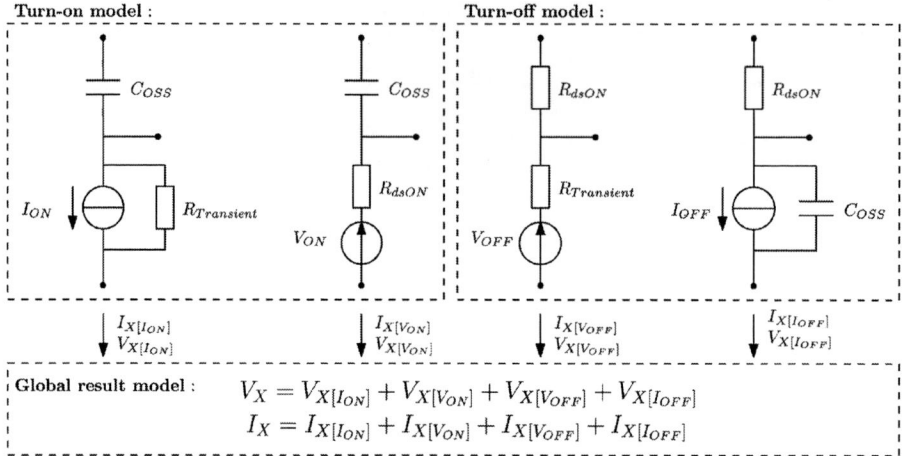

Fig. 8: Workflow for the proposed MTES method

We have defined how to model a single transient for a MTES simulation with $c_n[x_r]$. As a result, we can simulate each transient with its associated topology. The workflow is illustrated in figure 8. The switching cell is substituted by the proposed equivalent sources and impedances. The global result for a current I_x or voltage V_x in the simulation corresponds to the sum of the responses of current and voltage sources for turn-on and turn-off. This allows to represent the appropriate propagation path associated with each switching transient.

Application to the Buck converter

In this section, the MTES method is applied to a Buck converter using a leg structure. The source characteristics are extracted in a simple time-domain simulation. Then we validate the modelling method with the commutation waveforms. Finally, conducted EMI results in common and differential mode are compared to the reference simulation.

The figure 9 gives the reference time-domain simulation schematic. The electrical parameters of the Buck converter are : 100V input voltage and 65A output current. The high side MOSFET is used as a freewheeling diode, the reverse recovery is not considered. The low side MOSFET is driven with a 10kHz switching frequency and 0.4 duty cycle.

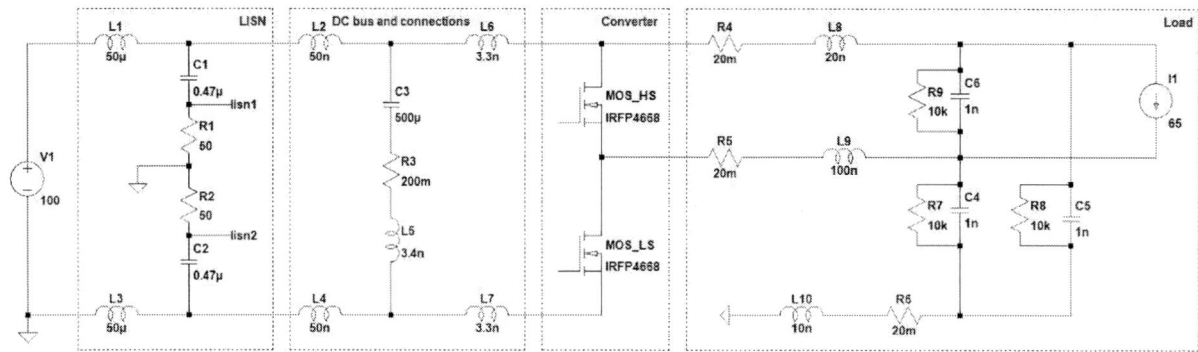

Fig. 9: Reference time-domain simulation circuit

The model includes a DC bus capacitor and its high frequency model. On the load side, a power cable is modeled with a RLC cell and a common mode capacitive coupling to the ground. Line Impedance Stabilisation Networks (LISN) are used to extract the conducted emissions. From this schematic, we can compute the differential and common mode voltage at the LISN.

Extraction of EMI source characteristics

To avoid the influence of external parasitic elements, the characteristics of the commutation waveforms are extracted from the simplified simulation circuit as shown in figure 10. The simulation is faster than the total converter because slow response times related to the LISN and DC bus capacitor are removed.

The input parasitic inductance, 10nH, corresponds to capacitor stray inductance combined with the connection inductances. This input parasitic inductance and MOSFET capacitances are kept in the model to determine di/dt and dv/dt. The circuit does not include damping elements such as resistance, these have an impact on the ringings but are not necessary to extract the rise and fall time.

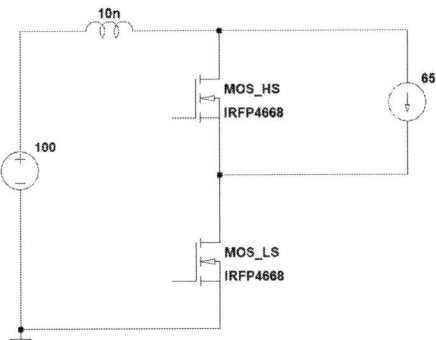

Fig. 10: Simplified circuit for extraction of EMI source characteristics

Figure 11 gives, in dashed lines, the results of the simulation of the simplified model. The trapezoidal waveforms are drawn in solid lines for the current and voltage transition models. The trapezoidal signals are set with a time delay t_{dr}, and a rise time t_r, in order to obtain a good correlation with the simulated signals. During the turn-on, the voltage falls due to the Ldi/dt. We use the current waveform to determine the voltage time delay : the voltage commutation starts when the current transient is over. Then, the rise time is identified when the voltage reaches zero at approximately 90ns.

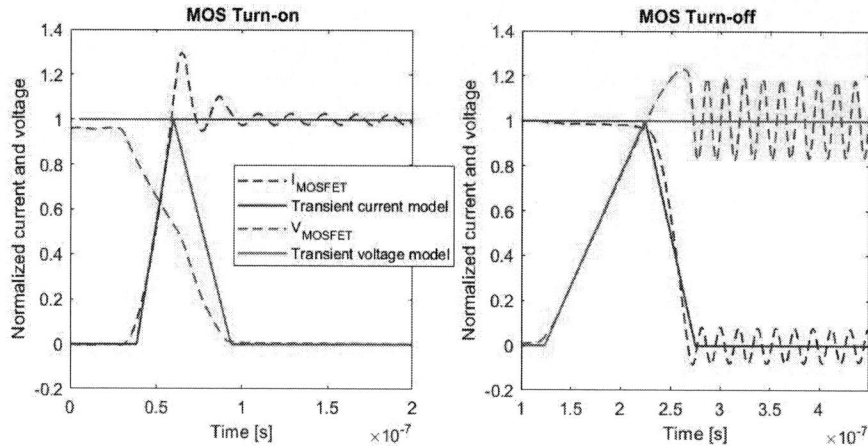

Fig. 11: MOSFET commutation voltage and current waveforms for MTES simulation

Time-domain validation

To evaluate the model performance, we compare the commutation curves from the time-domain simulation to the proposed MTES method with equivalent sources. The EMI characteristics previously extracted are integrated to the four topologies. The results are given in figure 12.

During turn-on, the *di/dt* is well represented and the source impedance allows to represent the voltage drop across the MOSFET. During the voltage transient, ringings are less prominent in the MTES simulation, but once the commutation is over, we retrieve the appropriate ringing. The turn-off results show a good correlation with the reference simulation.

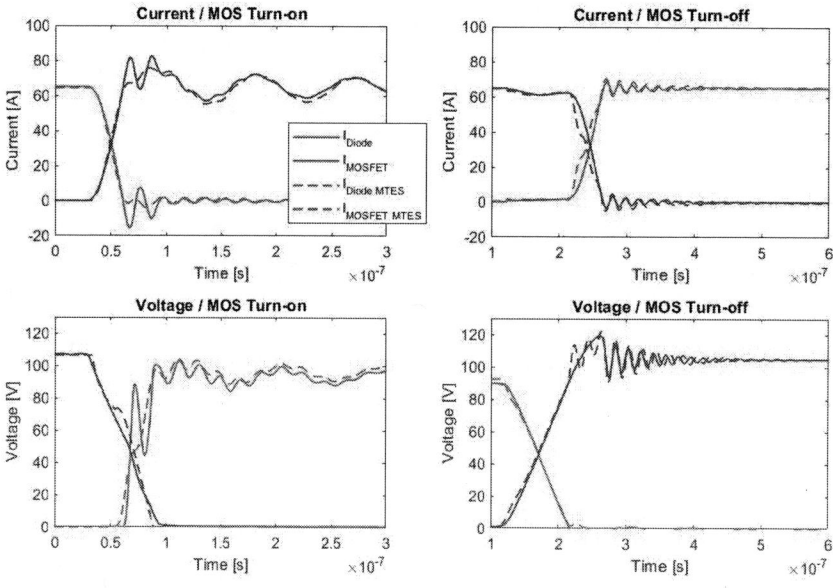

Fig. 12: Switching curves from time-domain compared to MTES method

Frequency-domain validation

The EMI characteristics are now integrated to a frequency-domain simulation. The four topologies are simulated simultaneously and the global result corresponds to the sum of the LISN voltage of each topology. Results for differential mode and common mode emission are provided in figure 13. The trapezoidale source model allows to obtain a good correlation on a wide frequency band. The low frequency harmonics are correctly simulated by the MTES. Then, in high frequency, we identify the resonance frequencies related to the impedance network at 11MHz and 52MHz.

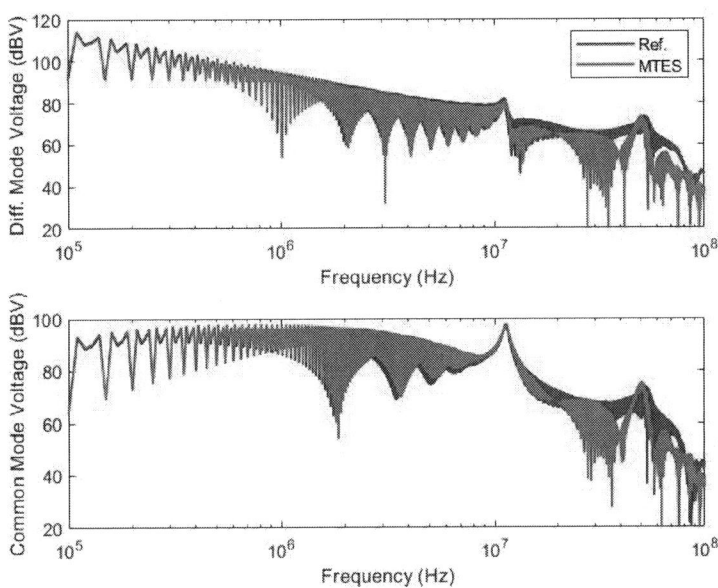

Fig. 13: Differential mode and common mode conducted emissions

The simulation time for the proposed frequency-domain modeling method is reduced by a factor of 100 (from 220s to 1.7s), these results could be improved with an optimized multi-core processing that is suitable for frequency-domain simulation. The duration of time-domain simulation is related to the small time step required for high frequency simulation coupled to the long simulation needed to reach steady state.

Since the MTES takes into account the source impedance, the model stays valid with a modification of the propagation path around the switching cell. As a result, the MTES model can be used to perform EMC filter design in the frequency-domain.

Conclusion

Based on the MTES approach, a new method allowing to determine the EMI sources in the frequency-domain and taking into account the non-linearity is proposed. The obtained results show that this method is accurate in a wide frequency band. In addition, the proposed source model can be easily adjusted to represent the EMI source characteristics. This will allow to simulate more complex converters with time dependent duty cycle and current load, such as inverters.

The MTES requires to simulate several topologies in the frequency-domain. As shown, it remains much faster than time-domain simulation. It allows the evaluation of the conducted emission levels of power converters during the design process. Thanks to its fast simulation speed, MTES model can be used to optimize the performances of EMI filters.

References

[1] Egot-Lemaire, S., Klingler, M., Lafon, F., Marot, C., Koné, L., Baranowski, S., & Démoulin, B. (2012). Modeling methodology of automotive electronic equipment assessed on a realistic subsystem. IEEE Transactions on Electromagnetic Compatibility, 54(6), 1222-1233.

[2] Gubia, E., Sanchis, P., Ursúa, A., López, J., & Marroyo, L. (2005). Frequency domain model of conducted EMI in electrical drives. IEEE Power Electronics Letters, 3(2), 45-49.

[3] Labrousse, D., Revol, B., & Costa, F. (2010). Common-mode modeling of the association of N-switching cells: Application to an electric-vehicle-drive system. IEEE Transactions on Power Electronics, 25(11), 2852-2859.

[4] Zaidi, B., Videt, A., & Idir, N. (2018). Optimization method of CM inductor volume taking into account the magnetic core saturation issues. IEEE Transactions on Power Electronics, 34(5), 4279-4291.

[5] Robert, F., Bensetti, M., dos Santos, F. V., Dufour, L., & Dessante, P. (2017). Multiphysics modeling and optimization of a compact actuation system. IEEE Transactions on Industrial Electronics, 64(11), 8626-8634.

[6] Marlier, C., Videt, A., Idir, N., Moussa, H., & Meuret, R. (2012, September). Modeling of switching transients for frequency-domain EMC analysis of power converters. In 2012 15th International Power Electronics and Motion Control Conference (EPE/PEMC) (pp. DS1e-1). IEEE.

[7] Marlier, C., Videt, A., Idir, N., Moussa, H., & Meuret, R. (2013, September). Hybrid time-frequency EMI noise sources modeling method. In 2013 15th European Conference on Power Electronics and Applications (EPE) (pp. 1-9). IEEE.

[8] Paul, C. R. (2006). Introduction to electromagnetic compatibility (Vol. 184). John Wiley & Sons.

[9] Lai, J. S., Huang, X., Pepa, E., Chen, S., & Nehl, T. W. (2006). Inverter EMI modeling and simulation methodologies. IEEE Transactions on Industrial Electronics, 53(3), 736-744.

Modular Multilevel Converter with Distributed Galvanic Isolation: A Decentralized Voltage Balancing Algorithm with Smart Gate Drivers

DARBAS Corentin[1], GINOT Nicolas[1], OLIVIER Jean-Christophe[2], POITIERS Frédéric[1]
Univ Nantes, CNRS, IETR UMR 6164 [1], IREENA[2]
Rue Christian Pauc, 44300 Nantes, France[1]
37 boulevard de l'université – BP406 – 44602 Saint-Nazaire Cedex, France[2]
Tel.: (+33) 02 49 14 20 80
E-Mail: corentin.darbas@univ-nantes.fr, nicolas.ginot@univ-nantes.fr,
jean-christophe.olivier@univ-nantes.fr, frederic.poitiers@univ-nantes.fr

Keywords

"Modular Multilevel Converter (MMC)", "Distributed Galvanic Insulation (DGI)", "Voltage Balancing Algorithm (VBA)", "Smart Gate Driver", "Cascaded Drivers"

Abstract

This work introduces a novel concept of Distributed Galvanic Insulation (DGI) for multilevel converters. By daisy chaining the submodule gate drivers, the requirement in terms of isolation voltage is highly decreased. The use of Smart gate drivers, which implement advanced communication features, allows communication signals to flow along the chain. Applied to MMC, the DGI forbids the use of classical Voltage Balancing Algorithms. A new algorithm, directly integrated on the gate drivers, is proposed for DGI-MMC and is validated by simulation.

1. Introduction

Multilevel converters had become appealing solutions for medium and high power conversion due to their high voltage capability, lower output harmonics and reduced stress on power devices [1] [2]. Recently, a highly attractive multilevel topology had been introduced by Marquardt & Lesnicar [3]. The Modular Multilevel Converter (MMC) is based on the assembly of smaller conversion units, often called submodules (SM) or cells. One MMC phase is composed of two "arms", each of them is constituted of a chain of N submodules terminated by an inductor L_{arm}. The three-phase MMC is given in Figure 1 (a). Various submodule topologies have been introduced in the literature. The most common submodule topology is also the simplest one, namely the "Half-Bridge submodule", given in Figure 1(b).

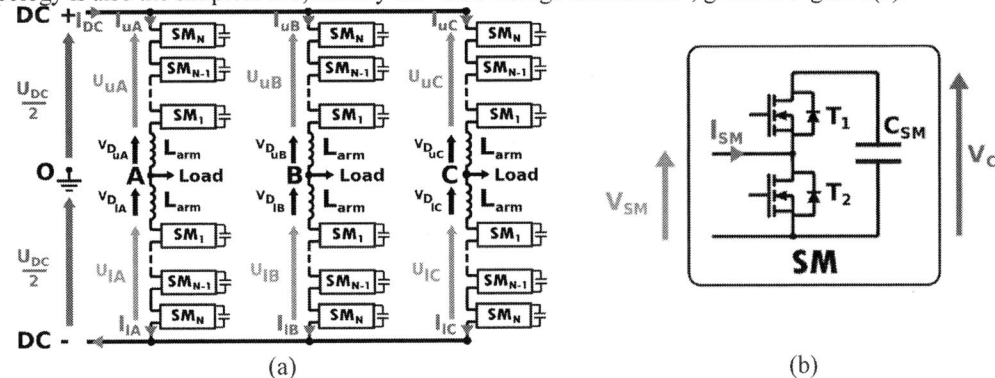

| (a) | (b) |

Figure 1 : (a) Three-phase MMC Topology and (b) a Half-Bridge Submodule

The submodule is said "ON" or "inserted" when T_1 is conducting, the output voltage V_{sm} is then equal to the capacitor voltage V_C. The submodule is said "OFF" or "bypassed" when T_2 is conducting, implying $V_{SM} = 0$ V. The half bridge submodule exhibits the lowest losses among all submodule topologies while allowing very simple control, as V_{SM} only takes two values.

EPE'20 ECCE Europe

Assigned jointly to the European Power Electronics and Drives Association & the Institute of Electrical and Electronics Engineers (IEEE)

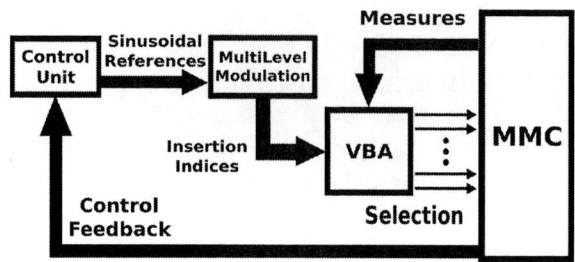

Figure 2 : MMC main control architecture with external voltage balancing algorithm

Due to the high number of capacitors connected in series, the voltage balancing of multilevel topologies remains a challenge [4]. Various control and modulation methods have been introduced in the literature concerning the MMC [5], [6], [7], where the switching orders are directly generated for each submodule. Another idea to balance the capacitor voltages consists on the use of hardware acceleration. An external system, usually based on FPGA, is dedicated to balance the capacitor voltages. The control stage is in charge of the generation of the references, which are fed to the modulation stage. The insertion indices are the number of submodules N_{ON} to insert in the arms. They are generated by the modulation stage. The Voltage Balancing Algorithm (VBA) is executed on the external hardware. It selects the correct number of submodules to insert in each arm N_{ON}. Such control structure is given in Figure 2.

Various types of external algorithm had been proposed in the literature [8] [9]. They are generally based on the measurement or the estimation of the arm current and the submodule capacitor voltages. The inserted capacitors are charged or discharged by the arm current I_{arm}, depending on its sign. The correct way to balance the capacitor voltages is to insert in the arm the N_{ON} capacitors with the highest voltages when $I_{arm} < 0$. Similarly, the capacitors with the lowest voltages are inserted if $I_{arm} > 0$.

Although highly reducing the computational burden on the main controller, external voltage balancing algorithm has the general drawback of increasing the switching frequency. It is however not the case of the one presented in [10] and referred as "Reduced Switching Frequency" (RSF). RSF waits for a change of the insertion index ΔN_{ON} before activating. ΔN_{ON} is defined as the difference between N_{ON} the insertion index at time (t), and $N_{ON_{OLD}}$ the insertion index at time $(t-1)$. While $\Delta N_{ON} = 0$, the gate signals of the transistor submodules are kept constant.

If $\Delta N_{ON} > 0$, then ΔN_{ON} submodules are inserted in the arm. Only the submodules already bypassed are allowed to switch. As explained before, the arm current's sign determines which capacitors need to be inserted, depending on their voltages. If $\Delta N_{ON} < 0$, then ΔN_{ON} submodules are bypassed among the already inserted ones, without any other switching allowed. If $I_{arm} < 0$, the submodules with the highest voltages are bypassed, otherwise the submodules with the lowest voltages are bypassed.

That way, only ΔN_{ON} switching are executed and all other switching are forbidden. The overall switching frequency remains unchanged by the VBA stage and is only imposed by the modulation stage.

For this type of algorithm to be executed properly, the capacitor voltages need to be sampled regularly, and sorted by the FPGA. The challenge is then the practical implementation of such algorithm on HVDC systems, with a large number of submodules and a high DC-bus voltage, which requires very high galvanic insulation barriers. In this paper, a distributed galvanic isolation coupled to a daisy chained Voltage Balancing Algorithm is proposed.

2. The Distributed Galvanic Isolation with smart gate-drivers

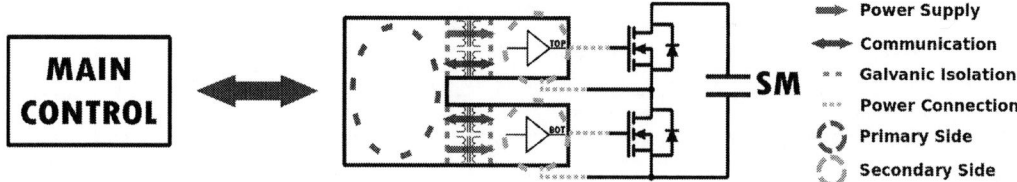

Figure 3 : Typical built-in two channel driver

The MMC half-bridge submodule transistors are typically driven by a built-in two-channel driver, as presented on Figure 3. It is composed of a primary side separated from two secondary channels by a galvanic isolation. The secondary channels are usually called "Top" and "Bottom" (Bot). The primary side is communicating with the main control unit through a link usually referenced to ground, for safety reasons. However, the secondary channels are referenced to the emitter/source of the power transistors, which are usually at high potential compared to the ground. In order to maintain the control side isolated from the power side, the galvanic isolation needs to withstand the voltage between the ground and the converter potentials.

The gate signals are usually fed up to the secondary sides via optical component, which can withstand extremely high voltages but can't provide power. However, the power supply of each secondary side must also be isolated from the primary. That isolation is usually realized with magnetic components, which can be difficult to integrate on boards. Usually, insulation voltages of built-in gate drivers proposed by main manufacturers don't reach more than 10 kV. Such drivers are designed for power modules operating at voltages up to 6.5 kV [11]. High power converters can reach several hundreds of volts, preventing the use of such built-in drivers. With the increase of the isolation voltage, the magnetic component become very bulky, hence costly.

Common drivers only exchange gate signals and error messages between their primary and secondary sides. Their only purpose is to execute the commutation from the control unit. Recently, "Smart gate drivers" have been introduced, which are able to realize more complex tasks and exchange more complex information between their sides. A typical smart gate driver is presented in [12] & [13], which features a bidirectional channel dedicated to data exchanges between the two sides (measurements, monitoring, etc). The gate driver presented in [14] allows the same features without the need of an additional communication channel. The messages are sent in the form of binary words at high bit rates.

This work presents a new approach to distribute the galvanic insulation between an MMC arm's drivers, allowing the use of classical built-in drivers regardless of the application voltages. One arm regular connection structure between the submodules' drivers and the control unit is showed Figure 4 (a). All drivers' primaries are referred to the safe potential of the control unit (in black), while each secondary are referred to their power modules' emitters/sources. This allows direct communication between the drivers and the control unit, but implies a high voltage isolation between the safe ground and the power grounds.

The structure introduced in that work is illustrated on Figure 4 (b) and called Distributed Galvanic Insulation. Only one driver is linked to the control unit, its primary being referred to the signal ground. The corresponding submodule is named SM_1, then all drivers are named from 2 to N, the Nth driver being the closet to the DC bus. All the drivers are cascaded: the primary of SM_2 is referred to the top channel of SM1's driver, SM_3's driver primary being referred to SM_2's driver top channel, etc. The primary side of SM_N's driver is referred to SM_{N-1}'s driver top channel.

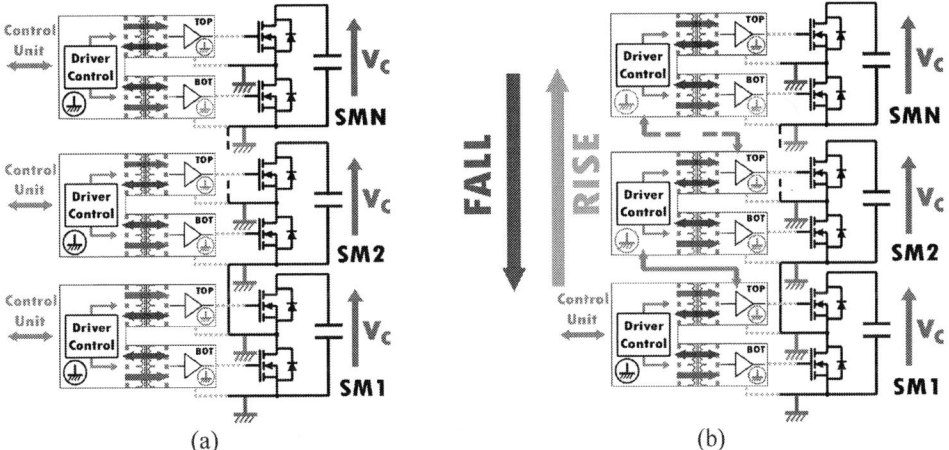

(a)   (b)

Figure 4 : MMC with regular architecture (a) & MMC with Distributed Galvanic Isolation (DGI) (b)

The MMC submodules are connected such as SM$_p$'s bottom channel and SM$_{p+1}$'s top channel shares the same voltage reference. That implies that the bottom channels of each chained driver shares the same references as their primaries. The physical galvanic isolation barrier must however not be suppressed in order to avoid EMI issues caused by the huge $\frac{dv}{dt}$ and $\frac{di}{dt}$ of the power side. But the lower the voltage across the barrier, the easiest it is to design and integrate to a board.

Furthermore, the voltage between the references of SM$_p$'s channels is equal to the capacitor voltage V_C. The galvanic isolation between the top channel and the primary side of the chained drivers needs only to withstand the capacitor voltage V_C. In most cases, the capacitor voltage V_C is equal to the DC bus voltage divided by the number of submodule in each arm as in eq. 1 [15].

$$V_C = \frac{U_{DC}}{N} \qquad \text{eq. 1}$$

That way, increasing the number of submodules in the system allows decreasing the value of V_C, thus the voltage across the galvanic isolation. Compared to the conventional structure, the need of galvanic isolation is decreased by a factor N for N-1 drivers. That is a critical improvement compared to the conventional structure, which allows regular drivers to be used in high voltage power converters. One more asset of that architecture is the need of communication wires between the control unit and the drivers. In conventional MMC, one wire per submodule is required and the number of wire N$_{wire}$ is given in eq. 2 :

$$N_{wire} = 3 \times 2 \times N = 6\,N \qquad \text{eq. 2}$$

With the proposed structure, there is only one data wire per arm, leading to only 6 wires required for a three phase MMC, which is once again a reduction by a factor N. This DGI approach can be applied to any type of multilevel structure.

The lack of direct communication between the drivers and the control unit in the proposed DGI imposes new communication channels between each driver. Only Smart gate drivers could handle such advanced communication features. A message sent from one driver to another needs to cross an insulation barrier. That process usually takes a short time, especially when the isolation is realized with magnetic component [14]. Thus, the amount of information send along the chain needs to be kept as low as possible, because the time for an information to propagate along the chain becomes proportional to the number of driver. Messages sent from low index to higher index drivers are "*Rising*", and messages from high index to lower index drivers are "*Falling*" (see Figure 4).

Another issue is the reliability of the structure. In high voltage MMC, the gate driver power is usually drawn from the submodule capacitor [16]. However, if a submodule fails, and its capacitor voltage goes low, the driver won't be powered anymore. It is thus not a viable solution to use self-power for the gate driver units in the DGI. Several proposition have been introduced in the literature to power gate drivers (Double Galvanic Insulation Transformer [17] which allows very high insulation capability while powering many drivers simultaneously; cascaded power supplies [18] which highly increase the driver noise immunity while being well adapted to the proposed structure, etc). The gate drivers power supplies are not further developed in this work and are then not considered.

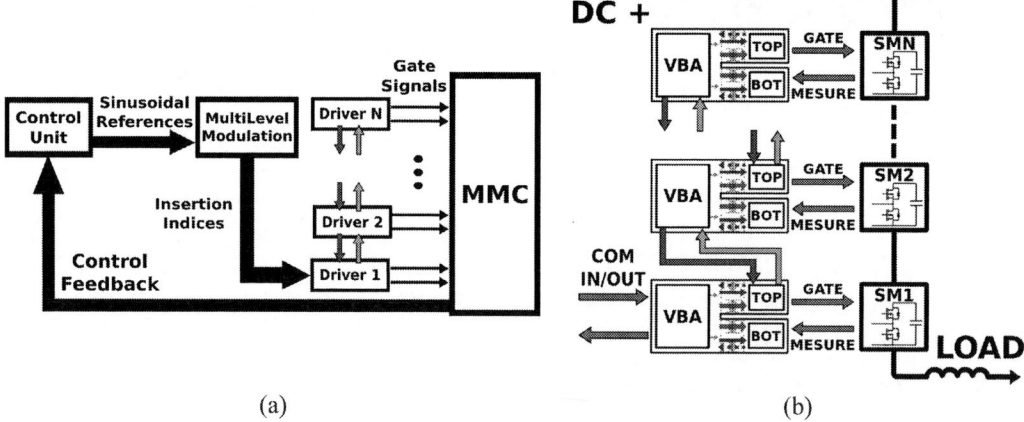

(a) (b)

Figure 5 : MMC with DGI structure and proposed VBA inside the driver chain : overall control architecture (a) and a detailed architecture of one MMC arm (b)

One of the MMC's features is the voltage balancing algorithm (VBA) discussed earlier in the paper. This algorithm needs to sort the capacitor voltages and select the right submodule to insert. The proposed idea is to use a distributed control where each driver is responsible of its own submodule switching. Conventional VBA allows simultaneous switching.

In the DGI structure, that must be avoided in order to limit the flow of information between the drivers. A voltage balancing algorithm is proposed in the next section which is directly implemented in the chained driver structure and overcomes the previous difficulties. The overall topology of the MMC with the proposed structure together with the voltage balancing algorithm are presented Figure 5 (a).

3. Voltage Balancing Algorithm for DGI structures

Each driver of the chain is numbered as its submodule index. The proposed VBA is structurally similar to the RSF algorithm presented in [10]. The insertion index is sent to the first driver of the chain, Driver 1 (D1). The algorithm is executed by the electronics on the driver primary sides, typically by a fast component like a FPGA, each time the insertion index changes its value ($\Delta N_{ON} \neq 0$).

For the execution of the proposed VBA, each driver must have a direct access to the measurement of their own capacitor voltage V_C. It is also an interesting feature to measure the sign of the arm current directly from the submodule. The channel between the drivers can be used in both directions. One-phase of the MMC with the proposed structure is depicted Figure 5 (b).

The key to minimize the time delay introduced by the voltage balancing algorithm on the cascaded structure is to limit the amount of data exchanged between drivers. Indeed, the time for a bit to cross the galvanic isolation of one module is relatively high compared to the characteristic time of a FPGA.

The proposed algorithm will thus avoid to send long binary words on the channels and limits the communication to short words. The algorithm presented hereafter is only able to produce a single module removal or insertion at each execution. If $|\Delta N_{ON}| > 0$, which represents the number of switching to execute, the algorithm is run $|\Delta N_{ON}|$ times. That way, there is no need to sort all the capacitor voltages, but only to find the highest one, or lowest one, depending on the operation to fulfill.

3.1. Proposed Voltage Balancing Algorithm

The main idea of the algorithm is to realize all the capacitor voltages comparisons at the same time. For that purpose, a unique token is passed between the chained drivers during the procedure, depending on their priority to switch during the procedure. The last "token bearer" will switch its state.

An internal counter is launched in each driver. Its duration is directly determined by its priority to switch. This allows an efficient selection of the right submodule to switch, detailed in the following. The counter duration depends on the signs of the insertion index transition $sign(\Delta N_{ON})$ and of the arm current $sign(I_{arm})$, and on the capacitor voltage V_C. It will be detailed later in the paper.

Different working examples are given Figure 6, with "rising" messages depicted in orange, "falling" messages in scarlet, and the internal counter in blue.

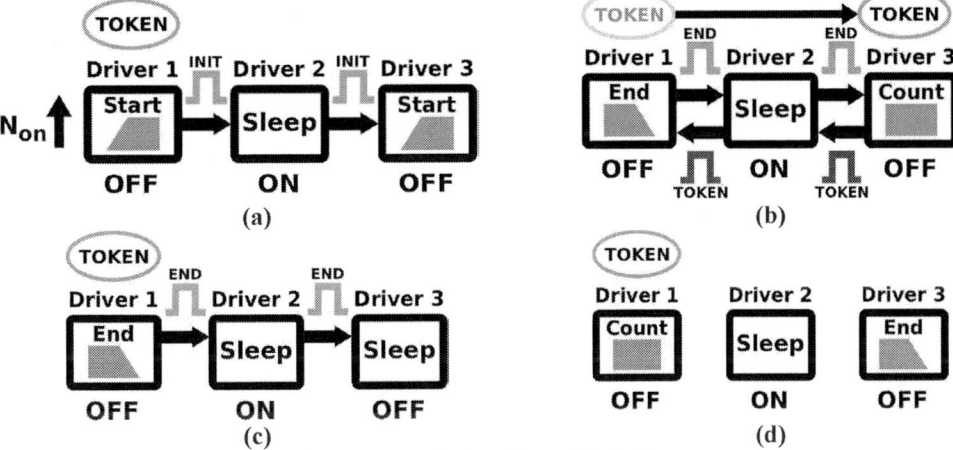

Figure 6 : Different scenarios of the proposed algorithm with 3 drivers

During the procedure, some drivers will be removed from the selection process as their priority to switch are low. They enter a state referred as "sleep mode". In sleep mode, the driver will propagate the incoming messages in both directions as a member of the driver chain without interfering. It does not interpret them nor it runs the algorithm. The final token bearer will be the only out of "sleep mode" at the end of the procedure.

In accordance with RSF, some switchings are not allowed. For example, if $\Delta N_{ON} > 0$, all the drivers already inserted in the arm must remain in their current state. Let define S as the state of the submodule, with S = 1 meaning that the submodule is ON, and S = 0 that the submodule is OFF. A driver is concerned by the voltage balancing algorithm if and only if Crit. 1 is met :

$$S \neq \text{Sign}(\Delta N_{ON}) \qquad \text{Crit. 1}$$

If Crit. 1 is not met, the submodule must remain in its current state and enters "sleep mode".

The initialization step is schematized Figure 6 (a), when an increase in N_{ON} is detected, meaning that a submodule needs to be switched ON. Driver 2 is already ON. As soon as Driver 1 detects a change in N_{ON}, it instantly sends an initialization code which rises along the chain of drivers from D1 to DN.

That code contains the information of the sign of ΔN_{ON}, which is referred as the *transition bit*. The initialization message length may differ depending on the application, but it has little impact on the procedure length itself.

Driver 1 takes the token by default. If D1 cannot switch due to Crit. 1, the token is given to the first driver of the chain which satisfies Crit. 1. In the next examples, it is assumed that D1 meets Crit. 1.

The *transition bit* allows each driver to check for Crit. 1. As can be seen on the example Figure 6 (a), D2 was already ON and thus enters sleep mode. It propagates the initialization code anyway.

After the reception of the initialization code, each driver which is not in sleep mode will measure the arm current sign and its capacitor voltage in order to generate its counter and starts their counters accordingly.

Nothing happens during t_{count_1}, the duration of driver 1's counter. It's an image of its capacitor voltage, thus its priority to switch. As shown on Figure 6 (b), at the end of its counter, D1 sends a "rising" *ending bit* to the upper drivers of the chain. The driver which are already in sleep mode just propagates this *ending bit*, like D2 in the example. All the drivers whose counters are already OFF enter sleep mode and propagate the ending bit along the chain.

The *ending bit* stops when it crosses a driver whose counter is still ON, which is D3 on Figure 6 (b). D3 has higher priority to switch and thus takes the token. It sends a "falling" *token bit* to Driver 1. That bit is simply "falling" through the chain, as all the drivers between D1 and D3 are now in sleep mode (only D2 in the example, but that is true regardless of the position of the previous and current token bearers). After the reception of the *token bit*, D1 enters sleep mode itself and release its token.

If no *token bit* is received by the token bearer after the "rise" of its *ending bit*, it means that all the other drivers had already stopped counting. The current token bearer has the highest priority for switching. It keeps its token and will consequently switch its state at the end of the procedure. That scenario is illustrated on Figure 6 (c), where D2 and D3 are already in sleep mode when D1 ends its counter. It gets not *token bit* back. As all the counters are run at the same time, there is no need to compare all the voltages together.

In order to limit the number of comparison, thus the number of bits exchanged between the drivers, only the driver which is holding the token is allowed to use the channel. The other drivers can't send any "rising" or "falling" message but the *token bit* in the case previously described (Figure 6 (b)). As illustrated on Figure 6 (d), when D3 ends its counter, it does not send any *ending bit* to any driver. It simply enters "sleep mode", as it wasn't holding the token. That strict restriction to access the communication channels allows avoiding issues of messages crossing each other, which could lead to loss of information, and threatens the stability of the process.

With the proposed algorithm, the token can never be lost, nor can it be multiplied. Only one token is passed from drivers to other drivers until the end of the procedure. Furthermore, the delays introduced by the barriers don't cause synchronization issues between drivers, as they are automatically and naturally counterbalanced. Even if all the drivers have different barriers' delays, the synchronization won't be impacted.

3.2. Counter generation

The counter maximum value is determined by the priority of the submodule to switch, according to the rules of RSF. It is a fraction of the overall maximum counter period T. T is selected by the converter designer as a trade-off between voltage precision and counter duration. Let q be the voltage resolution of the algorithm. If q = 1, then each counter increment corresponds to 1 V of the capacitor voltage.

In a majority of applications, the capacitor voltages are kept between two voltage values. Those values, V_{max} and V_{min}, are usually designed in order to keep the capacitor voltage around 10% of its reference value $V_{C_{ref}}$. This is a common design parameter, as it is important that all capacitor voltages are kept close to the rated voltage in order to limit the output voltage THD and the magnitude of the circulating current [15]. Then, as V_C only takes values between V_{min} and V_{max}, if the counter is incremented by 1 each clock cycles, which frequency is f_{clk}, then T is given in eq. 3:

$$T = \frac{V_{max} - V_{min}}{q} \frac{1}{f_{clk}} \qquad \text{eq. 3}$$

Let's assume that under safe operating conditions, the capacitor voltages are kept between 1440 V and 1760 V, while the voltage resolution is q = 3V and the clock frequency is 10 MHz. Eeq. 4 gives the value of T:

$$T = \frac{320}{3} \frac{1}{10\,10^6} = 10.7\ \mu s \qquad \text{eq. 4}$$

There is a trade-off between the voltage resolution and the duration of the algorithm. Increasing the clock rate is also a solution for decreasing the value of T.

The counter maximum value will depend not only on the capacitor voltage V_c, but also on the signs of ΔN_{ON} and I_{arm}. Indeed, the capacitor voltage V_c with the highest priority is sometimes the highest, sometimes the lower. If $\Delta N_{ON} > 0$ and $I_{arm} < 0$ or $\Delta N_{ON} < 0$ and $I_{arm} > 0$, the highest V_c is selected. Otherwise, it is the lowest V_c which is selected. That criterion Crit. 2 can be written as:

$$\text{Sign}(\Delta N_{ON}) \neq \text{Sign}(I_{arm}) \qquad \text{Crit. 2}$$

If Crit. 2 is true, then the highest voltage needs to be switched. If Crit. 2 is false, it is the lowest voltage that is switched. There are then two definitions for the counter duration t_{compt} given in eq. 5:

$$\begin{cases} t_{compt} = \dfrac{V_c - V_{min}}{q} \dfrac{1}{f_{clk}} & \text{If Crit. 2 is true} \\[2mm] t_{compt} = \dfrac{V_{max} - V_c}{q} \dfrac{1}{f_{clk}} & \text{If Crit. 2 is false} \end{cases} \qquad \text{eq. 5}$$

Duration of the proposed algorithm

The overall duration of the algorithm t_{algo_1} needs to be larger than the maximum counter period T. It depends on T but also on propagation time of one bit from a driver to its neighbor : t_{driver}. To evaluate t_{algo_1}, the worst case scenario must be taken into account. It corresponds to the situation when driver 1 holds the token during almost all the counting period T, but driver N takes the token at T. It happens when the duration of driver 1's counter t_{compt_1} and the duration of driver 1's counter t_{compN} are as shown in eq. 6:

$$\begin{cases} V_{c1} = V_{max} - q \Leftrightarrow t_{compt_1} = T - \dfrac{1}{f_{clk}} \\[2mm] V_{cN} = V_{max} \Leftrightarrow t_{compt_N} = T \end{cases} \qquad \text{eq. 6}$$

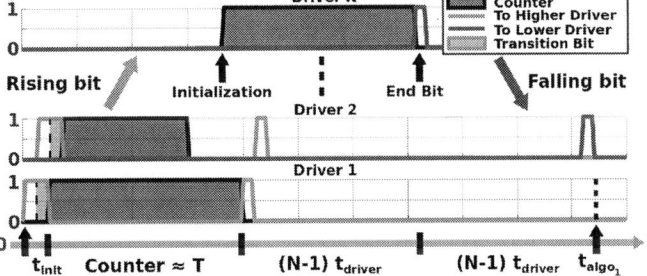

Figure 7 : Chronogram of the proposed VBA under worst case scenario

In that case, driver 1's counter ends just before the maximum period T. An *ending bit* is sent by driver 1 along the driver chain and it reaches driver N just before the end of its own counter. Driver N takes the token and sends back its *token bit* to driver 1 for him to release its token. The chronogram of the chain's different internal and external signals under the worst case scenario is shown Figure 7.

Each bit of Figure 7 has a length of t_{driver}, which represents the time taken for one bit to reach the next driver. t_{driver} is mainly due to the crossing of the galvanic isolation. Let's assume that the algorithm is executed by devices able to simultaneously receipt and send bits. It is typically the case of FPGAs. This way, the length of the initialization code is not important, as the FPGA will propagate any entering message instantly. The initialization code of Figure 7 is only 2-bits long. Its time t_{init} is in that case given in eq. 7:

$$t_{init} = 2\, t_{driver} \qquad\qquad \text{eq. 7}$$

After the initialization, driver 1 counter starts. Its duration is almost T. The *ending bit* "rises" through N - 1 driver to reach driver N. The time to reach driver N is then $(N-1)\, t_{driver}$. The *token bit* "falls" through N - 1 drivers as well and reaches driver 1. The time to reach driver 1 is once again $(N-1)\, t_{driver}$. The overall algorithm duration t_{algo_1} is finally given by eq. 8, assuming $t_{compt_1} \cong T$:

$$t_{algo_1} = 2(N-1)\, t_{driver} + T + t_{init} = 2N\, t_{driver} + T \qquad\qquad \text{eq. 8}$$

t_{algo_1} is the time between the reception of ΔN_{ON} by driver 1 and the end of the algorithm. t_{algo_2} can then be defined as well. It represents the time between the reception of the first bit of the initialization code by driver 2 and the end of the procedure as in eq. 9:

$$t_{algo_2} = t_{algo_1} - t_{driver} \qquad\qquad \text{eq. 9}$$

It is the overall procedure duration as seen by driver 2. More generally, the procedure duration t_{algo_p} for any driver p of the chain is:

$$t_{algo_p} = t_{algo_1} - (p-1)\, t_{driver} \qquad\qquad \text{eq. 10}$$

Each value of t_{algo_p} is pre-implemented in the corresponding driver p. The token bearer will switch its state "t_{algo_p} seconds" after the reception of the initialization code first bit. This allows the algorithm to be a deterministic procedure with a constant time t_{algo_1}, without the need of synchronizing the drivers at any moment. Additional delays may be introduced in the procedure if the measurement of V_c or $sign(I_{arm})$ requires some time.

4. Simulation of the proposed algorithm

A three-phase MMC inverter with N = 30 submodules per arms is simulated via Matlab/Simscape. The converter electrical model is built with Simscape blocks, allowing accurate results while simulating both of the power converter and the high speed devices. The model is easily scalable to any number of submodules. The DC bus voltage is 48 kV pole to pole. Thus the mean value of the capacitor voltages are 1.6 kV according to eq. 1. Their capacitance value is selected as C = 2.5 mF and the arm inductor value is $L_{arm} = 240\ \mu H$.

The circulating current is regulated with the Circulating Current Suppression Control introduced in [10]. The converter output is connecter to a three-phase grid with each phase at 20 kV_{MAX} from the converter ground. The power delivered to the grid is regulated with a typical vector control in Park reference. The modulation stage uses PDPWM (Phase-Disposition Pulse Width Modulation) with a carrier frequency of 6.2 kHz. The whole simulation architecture follows the previous schematic on Figure 5(a).

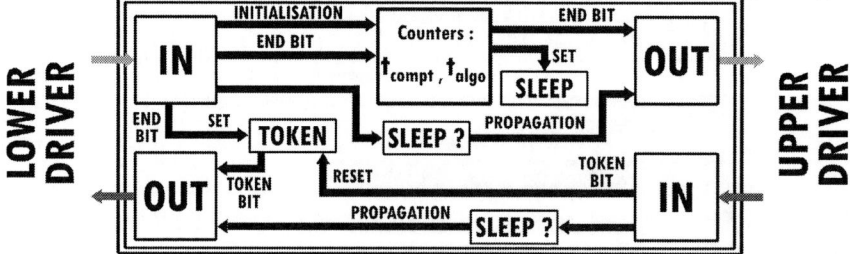

Figure 8 : Functional diagram of one driver for the simulation of the proposed algorithm

A simplified schematic of the main functions, bit setting/resetting and logical tests performed in the driver is shown Figure 8. Each driver is modeled with Simulink logic blocs (numerical and logical operators, mux, etc), like it would typically be synthetized in a FPGA. The drivers' clock frequency is $f_{clk} = 10$ MHz. The selected counter parameters are q = 2, $V_{min} = 1440$ V and $V_{max} = 1760$ V, i.e $\pm 10\%$ of the capacitor rated voltage. The time for one bit to propagate from one driver to the other t_{driver} is 200 ns (two clock cycles). In those conditions, the algorithm duration is $t_{algo_1} = 28.5$ µs.

The output apparent power is $S_{out} = 10$ MVA with a power factor of $\frac{\pi}{4}$ at a fundamental frequency of 60 Hz. The active and reactive powers P_{out} and Q_{out} injected in the grid are then:

$$\begin{cases} \text{Pout} &= 7.07 \text{ MW} \\ \text{Qout} &= 7.07 \text{ MVAR} \end{cases}$$

The simulation results are given on Figure 9 and Figure 10. Figure 9 shows the converter three phase output voltages and currents. The RMS values of the output voltages and currents are $V_{out_{RMS}} = 15.6$ kV and $I_{out_{RMS}} = 235$ A.

The u-phase 60 capacitor voltages are shown on Figure 10. The two dash-dotted lines show the upper and lower limit of $1600\ V \pm 10\%$. It can be seen that the algorithm balances the capacitor voltages efficiently. The 30 capacitor voltages of each arm are maintained close together. Furthermore, they are kept between the two voltage limits. The voltages behave typically the same as with standard voltage balancing algorithms [10] [9].

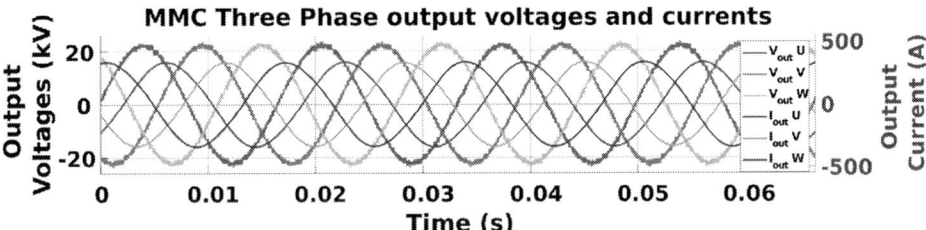

Figure 9 : Output voltages and currents

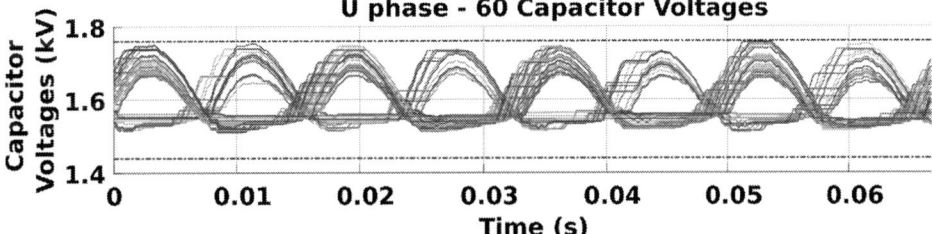

Figure 10 : U phase capacitor voltages (upper and lower arms)

5. Conclusion

That work introduces a new cascaded architecture for multilevel converters which allow the use of common medium voltage built-in drivers, regardless of the voltage of the application. Using smart gate drivers, messages from the control unit are cascaded through each gate driver one by one. Applied to MMC, that structure imposes to find new ways to perform voltage balancing algorithm. A solution is proposed which minimizes the number of exchanged bits between the drivers, reducing delays. Increasing the number of submodule will increase the algorithm execution time but not its complexity, as there is no direct sorting of the voltages. That scalability is an important asset, especially for a converter like the MMC. If submodules are added or removed, the driver need not to be reprogrammed. The structure of the algorithm is the same as the "RSF" algorithm. It features a fixed and controlled execution time, and a tunable resolution. A trade-off must be found between the execution time and the accuracy of the sorting depending on the application.

The simplicity of the proposed solution is one main advantage. Any driver able to communicate with its close environment can execute the algorithm as only few bits' generation, tests and counters are required. This can be easily realized using elementary analog circuits with little number of logical gates, or standard FPGA. A simulation tool on Matlab/Simscape allowed to verify the ability of the algorithm to balance the capacitor voltages of a MMC with 30 submodules per arms. The proposed VBA is implemented using logic blocks which mimics fast analog circuit behavior. Later works will focus on the algorithm dynamics and implementation on real physical systems.

References

[1] M.Tolbert and X.Shi, "Multilevel Power Converters," in *Power Electronics Handbook, 4th Edition*, M.H.Rashid, Ed. 2017.

[2] I. Colak, E. Kabalci, and R. Bayindir, "Review of multilevel voltage source inverter topologies and control schemes," *Energy Convers. Manag.*, vol. 52, no. 2, pp. 1114–1128, 2011.

[3] A. Lesnicar and R. Marquardt, "An innovative modular multilevel converter topology suitable for a wide power range," *2003 IEEE Bol. PowerTech - Conf. Proc.*, vol. 3, pp. 272–277, 2003.

[4] A. Mahé, A. Houari, J.-C. Olivier, M. Machmoum, and J. Deniaud, "Modulation Technic Highlight for State of Charge Balancing on a Series Cascaded Converter," *Electrimacs*, no. July, pp. 107–118, 2017.

[5] H. Akagi, S. Inoue, and T. Yoshii, "Control and performance of a transformerless cascade PWM STATCOM with star configuration," *IEEE Trans. Ind. Appl.*, vol. 43, no. 4, pp. 1041–1049, 2007.

[6] K. Ilves, A. Antonopoulos, S. Norrga, and H.-P. Nee, "A new modulation method for the modular multilevel converter allowing fundamental switching frequency," in *8th International Conference on Power Electronics - ECCE Asia*, 2011, pp. 991–998.

[7] J. Qin and M. Saeedifard, "Predictive control of a modular multilevel converter for a back-to-back HVDC system," *IEEE Trans. Power Deliv.*, vol. 27, no. 3, pp. 1538–1547, 2012.

[8] A. Hassanpoor, L. Ängquist, S. Norrga, K. Ilves, and H. P. Nee, "Tolerance band modulation methods for modular multilevel converters," *IEEE Trans. Power Electron.*, vol. 30, no. 1, pp. 311–326, 2015.

[9] A. Hassanpoor, A. Roostaei, S. Norrga, and M. Lindgren, "Optimization-based cell selection method for grid-connected modular multilevel converters," *IEEE Trans. Power Electron.*, vol. 31, no. 4, pp. 2780–2790, 2016.

[10] Q. Tu, Z. Xu, and L. Xu, "Reduced Switching-frequency modulation and circulating current suppression for modular multilevel converters," *IEEE Trans. Power Deliv.*, vol. 26, no. 3, pp. 2009–2017, 2011.

[11] 1SC0450E2A, "Single-Channel Cost-Effective SCALE™-2 IGBT Driver Core with Customer- Specific Gate Drive Input for 4500V and 6500V IGBTs," *Power Integration*, 2015. [Online]. Available: https://gate-driver.power.com/sites/default/files/product_document/application_manual/1SC0450E2A0_Manual.pdf.

[12] C. Bouguet, N. Ginot, and C. Batard, "Communicating gate driver for SiC MOSFET," *PCIM Eur. 2016; Int. Exhib. Conf. Power Electron. Intell. Motion, Renew. Energy Energy Manag.*, no. May, pp. 712–719, 2016.

[13] C. Bouguet, N. Ginot, and C. Batard, "Communication Functions for a Gate Driver under High Voltage and High dv/dt," *IEEE Trans. Power Electron.*, vol. 33, no. 7, pp. 6137–6146, 2018.

[14] J. Weckbrodt, N. Ginot, C. Batard, and S. Azzopardi, "Short pulse transmission for sic communicating gate driver under high dv/dt," *PCIM Eur. Conf. Proc.*, no. 225809, pp. 1552–1557, 2018.

[15] K. Sharifabadi, L. Harnefors, H.-P. Nee, S. Norrga, and R. Teodorescu, *Design, Control and Application of Modular Multilevel Converters for HVDC Transmission Systems*. Wiley - IEEE Press, 2016.

[16] J. Wang *et al.*, "Design and Testing of 6 kV H-bridge Power Electronics Building Block Based on 10 kV SiC MOSFET Module," *2018 Int. Power Electron. Conf. IPEC-Niigata - ECCE Asia 2018*, pp. 3985–3992, 2018.

[17] S. Brehaut and F. Costa, "Gate driving of high power IGBT through a Double Galvanic Insulation Transformer," *IECON Proc. (Industrial Electron. Conf.*, pp. 2505–2510, 2006.

[18] V. S. Nguyen, L. Kerachev, P. Lefranc, and J. C. Crebier, "Characterization and Analysis of an Innovative Gate Driver and Power Supplies Architecture for HF Power Devices with High dv/dt," *IEEE Trans. Power Electron.*, vol. 32, no. 8, pp. 6079–6090, 2017.

Comparison and optimization of magnetically coupled and non-coupled magnetic devices in interleaved operation

Peter Zacharias and Alejandro Aganza-Torres
Dept. of Electric Power Supply Systems. Kassel University
Willhelmshöhe Allee 71 - 73, 34121
Kassel, Germany
Tel.: +49 / (0561) – 804-6344.
E-Mail: peter.zacharias@uni-kassel.de / alejandro.aganza@uni-kassel.de

Keywords

« Magnetic Device », « Interleaved Converters », « High Power Density Systems », « Passive Component », « Flux Separation ».

Abstract

Interleaved operating DC/DC converters with coupled magnetic elements are well known as an option for high power density conversion systems. However, the design of the magnetic coupling element is usually made as single device, considering the sum of DC and AC flux swing for sizing. The presented analysis shows that it is possible to separate the magnetic AC and DC flux to different magnetic core elements with specific materials to minimize the respective losses. The approach, modelling and achievable volume reduction for non-coupled and coupled schemes are presented and compared in this paper.

Introduction

Phase-shifted operation of DC/DC converters allow currents to be distributed over several switching stages. The individually switched current results from the total current divided by the number of phases 'p'. The phase shift is usually 360°/p. Lower currents allow higher commutation inductances at the same overvoltage at the switches. By the interleaved switching one has 'virtually' a recharging of the buffer capacity with the p-fold frequency, so that this capacity can be reduced. In principle, magnetically uncoupled and magnetically coupled inductors can be used for phase-shifted operation.

With magnetically coupled inductors, the leakage inductance is used as storage inductance, while the magnetization inductance is used for coupling. Fig. 1 shows two examples of magnetically coupled toroid cores with the same electrical performance. In Fig. 1 a), a central core with high permeability is used for coupling while the outer yellow toroid cores act as inductive storages. The setup represents itself a transformer equivalent circuit of the T type. In Fig. 1 b), a setup with separate toroid cores for storage (red/left and orange/right) and coupling (central, blue and yellow winding). As Fig. 2 shows, both setups can be considered electrically equivalent. It is interesting to note that the storage and coupling functions are each assigned their own core materials. These can therefore be optimized for the respective task.

The arrangement shown in Fig. 1 a) is indeed a manufacturing challenge. Due to the construction, however, a pure toroid field is induced in all 3 cores. Although the leakage inductances are comparatively high in Fig. 1 a), there is practically no leakage field in the space around the component. In Fig. 1 b), which represents the same electrical functionality, the storage chokes are separated. That means the coupling function and the storage functions are carried out with different windings. This creates a maximum flexibility.

a)

b)

Fig. 1: Magnetically coupled chokes: a) transformer with augmented stray inductance, b) separate storage chokes with 1:1 transformer as coupling core.

For a better understanding Fig. 2 shows magnetic and electrical equivalent circuit diagrams. It is also clear that the different design represents the same electrical properties.

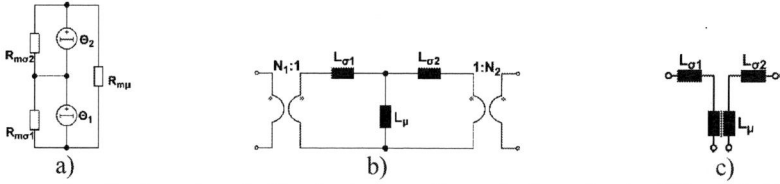
a)   b)   c)

Fig. 2: Model approaches for the setups in Fig. 1: a) magnetic equivalent circuit of Fig. 1a), b) electric equivalent circuit of a), c) electric equivalent circuit of Fig. 1b).

Non-coupled and coupled multiphase converters in interleaved mode

The fundamental problems of coupled and uncoupled inductive elements in the phase-shifted operation of multi-phase converters are examined below using 2 two-phase circuit examples. Fig. 3 a) shows the known design with non-coupled chokes. The two pairs of switches are operated 180° out of phase. The current ripple of a choke (red) does not differ from the usual step-down converter as can be seen in Fig. 4 a) and becomes maximum at D = 0.5. The ripple of the output current (orange) disappears due to the shift of the two phases of 180° but at D = 0.5. The frequency of the output current doubles compared to the clock frequency of a pair of switches. The frequency of the current (red) in a choke is equal to the switching frequency, as shown in Fig. 5 a). Here it becomes clear that the different design represents the same electrical properties.

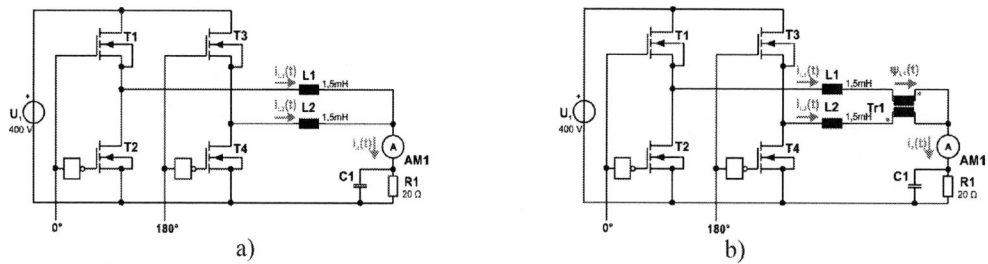
a)   b)

Fig. 3: Buck converter, 2-phase interleaved: **a)** non-coupled, **b)** magnetically coupled.

In the case of the buck converter with magnetic coupling in Fig. 3 b), the current ripple is also reduced by the chokes (red), as can be seen in Fig. 4 b). This and the reduction of the total ripple current amplitude is due to the disappearance of the fundamental oscillation in the current amplitude (Fig. 5 a)). The reason for this is that the coupling transformer absorbs the fundamental frequency in its voltage curve. The main frequency of the flux linkage is accordingly the switching frequency, 10 kHz in the given example, as shown in Fig. 5 b).

Comparison and optimization of magnetically coupled and non-coupled magnetic devices in interleaved operation

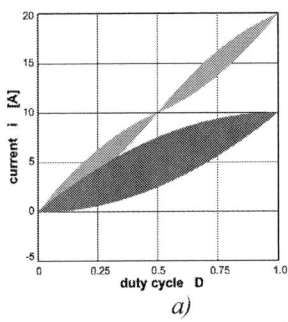

a)

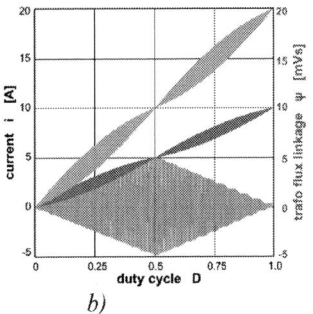

b)

Fig. 4: Total current (orange) and choke current for a) non-coupled chokes and b) for magnetically coupled chokes.

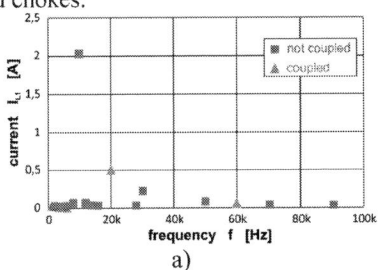

a)

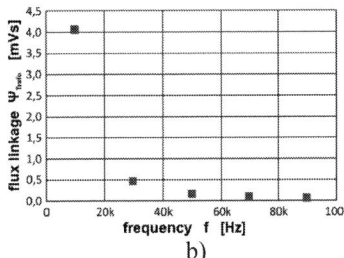

b)

Fig. 5: a) FFT of choke current and b) FFT of flux linkage at switching frequency of 10 kHz.

The dependencies can be schematized (Fig. 6). The blue curve (a) shows the current ripple (ripple of the flux linkage) of the chokes in the uncoupled case as an envelope curve. Here is:

$$\frac{\psi_x}{U_1} \cdot f = \frac{\Delta\psi_x}{2\cdot U_1} \cdot f = \frac{\left(i_{\max}-i_{\min}\right)\cdot L_x}{2\cdot U_1} \cdot f \tag{1}$$

the relative amplitude of the alternating component of the flux linkage on component L_x. Dividing this value by the effective magnetization cross section A_{Fe} and the number of windings N gives the amplitude of the alternating component of the magnetic flux density in this component as:

$$B_x = \frac{\Delta\psi_x}{2\cdot N\cdot A_{Fe}} = \frac{\psi_x}{N\cdot A_{Fe}} \tag{2}$$

The curve (Fig. 6 a), magenta) shows the known dependence for the amplitude of the ripple current on the duty cycle D for uncoupled chokes L_1 and L_2:

$$\frac{\psi_{L1}}{U_1} \cdot f_{sw} = 0{,}5\cdot D\cdot\left(1-D\right) \tag{3}$$

In the case of magnetic coupling, an autotransformer with the magnetization inductance L_μ acts as a coupling element and absorbs a large part of the voltage differences with its windings (Fig. 4b, Fig. 6 b), red). The basic frequency of the flux linkage is equal to the switching frequency (Fig. 5 b)).

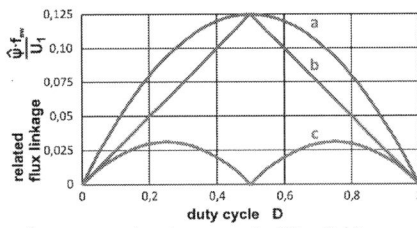

Fig. 6: Related Flux linkage for the magnetic elements in Fig. 3 b).

The alternating component of the flux linkage stresses the coupling core. The envelope of the amplitudes can be described by:

$$\frac{\psi_{L\mu}}{U_1} f_{sw} = 0,25 \cdot D \quad \text{for } 0 \leq D \leq 0,5 \qquad \text{and} \qquad \frac{\psi_{L\mu}}{U_1} f_{sw} = 0,25 \cdot (D-1) \quad \text{for } 0,5 \leq D \leq 1 \qquad (4)$$

Accordingly, the amplitudes of the alternating component of the flux linkage in the two inductors L_1 and L_2 are obtained for the two aforementioned intervals for D

$$\frac{\psi_{L1,2}}{U_1} f_{sw} = 0,125 \cdot D \cdot (0,5-D) \qquad \text{or} \qquad \frac{\psi_{L1,2}}{U_1} = 0,125 \cdot (0,5-D) \cdot (1-D) \qquad (5)$$

This procedure can be extended by cascading, as shown in Fig. 7 a). Here again a smoothing choke is used for each phase. In pairs, the 180° out-of-phase currents are combined by the two autotransformers L_{12} and L_{34}, so that a 2-phase system is created from a 4-phase system. These two are in turn combined by a coupling transformer L_{56}. This produces a load current with a current ripple with 4 times the switching frequency.

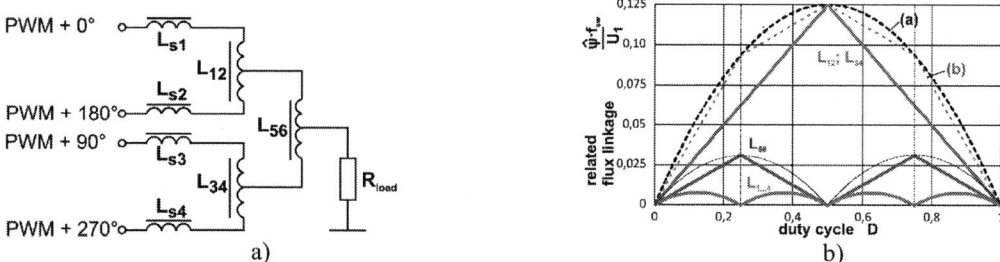

Fig. 7: a) Procedure for cascading phases operated in pairs in a 4-phase system and b) Related flux linkage for the magnetic elements in Fig. 7 a).

Fig. 7 b) shows the envelopes for the amplitudes of the flux chains on the inductive components to illustrate the method of operation. The maximum amplitude of the alternating component of the flux linkage and thus of the ripple current in the smoothing chokes $L_{s1...4}$ is only a quarter of the variant in Fig. 4 b). Their maximum amplitude of the flux linkage is only ¼ of the value of the variant without coupling. The frequency of flux linkage in autotransformers L_{12} and L_{34} is equal to the switching frequency of one phase. The autotransformer L_{56} carries twice the phase current of one of the other two transformers at twice the switching frequency. It becomes clear that apparently, at the same ripple current, the energy content and thus the volume of the required storage chokes decreases. However, coupling transformers are used as additional components.

If the total volume and the total weight of transformers and chokes is reduced by the coupled approach, advantages for the application can be achieved. Especially in mobile applications volume and mass play an important role in the evaluation of technical solutions. Depending on the objective of the application, frequency, voltage and the materials used for cores and windings, the decreasing trend for the choke volume is contrasted by an increasing trend for the volumes of the transformers. At high frequencies, the leakage inductances of the transformers can be used as smoothing inductances under certain circumstances [6, 7], resulting in a reduction of the number of components and an increase in compactness. For finding solutions with minimal volume a detailed loss analysis is required, the approaches are developed in the following sections.

Example of Transformer Optimization Problem

In order to achieve comparability of the results, in the selected examples, the same maximum ripple current is defined for a given maximum continuous power of a buck converter and the following given values: Input voltage: 400 V, Output voltage: 20 A, Maximum amplitude of ripple current: 0,5 A, Maximum output power: 8 kW and Switching frequency: 20 kHz. The maximum continuous apparent power of a coupling transformer is obtained for a sinusoidal voltage as:

$$S = \frac{\hat{i}_x \cdot \psi_x}{2} \omega \qquad (6)$$

In the range of the maximum values, however, in the present case the current is triangular and the voltage rectangular, so that the following applies to the effective values of current and voltage

$$I_x \approx I_{dx} \quad \text{and} \quad U_x = u_x = \psi_x \cdot 4 \cdot f. \tag{7}$$

The rated power of the coupling transformers is calculated accordingly with:

$$S_x = I_{dx} \cdot U_x = \psi_x \cdot 4 \cdot f \tag{8}$$

The values of current, voltage, frequency and flux linkage influence the size of the inductive components in a non-linear way. In the following sections, these influences are therefore examined in detail. We start with the transformer as the simpler optimization problem.

General Approach for Transformers

The influence of switching frequency and geometric parameters on the transmittable apparent power and transformer losses is to be investigated here on a transformer with its geometric and electrical characteristics. As specifications and principal dependencies are given or known: 1) The transformer has a primary and a secondary winding with $N_1 = N_2$. (This is only a simplification, not a limitation), 2) The excitation of the material is symmetrical and sinusoidal and 3) The core losses can be calculated approximately with knowledge of the core form and material with the Steinmetz formula as:

$$P_{vcore} = p_v\left(B, f\right) \cdot V_{Fe} = V_{Fe} \cdot k_1 \cdot \left(\frac{\hat{B}}{1T}\right)^{\alpha} \cdot \left(\frac{f}{1kHz}\right)^{\beta} \qquad \text{with:} \qquad \hat{B} = \frac{\sqrt{2} \cdot U_1}{2\pi f \cdot N_1 \cdot A_e} \tag{9}$$

where: B is the amplitude of bipolar induction, A_e is the effective magnetization cross section, l_{Fe} is the effective magnetic length, f is the frequency, k_1 is the coefficient of loss of core material, α, β depend on the material, U_1 is the primary voltage, V_{Fe} is the core volume and N_1 is the number of primary turns. Therefore, the core losses can be calculated as:

$$P_{vcore} = V_{Fe} k_1 \left(\frac{\sqrt{2} U_1}{2\pi f N_1 \, A_{Fe} 1T}\right)^{\alpha} \left(\frac{f}{1kHz}\right)^{\beta} = K_{Fe} \cdot N_1^{-\alpha} \tag{10}$$

And the winding losses are approximately for 2 windings given as:

$$P_{V.Cu} = \frac{2\rho N_1 l_m I_1^2}{k_f \dfrac{A_W}{N_1}} = \frac{2\rho N_1^2 l_m \left(\dfrac{S}{U_1}\right)^2}{k_f A_W} = K_{Cu} \cdot N_1^2 \tag{11}$$

where ρ is the specific resistance of the winding material, l_m is the mean winding length, S is the transformer apparent power, A_w is the cross-section area of winding window and k_f is the fill factor of the winding window. The specific resistance of the winding material is to be applied for larger frequency ranges than frequency-dependent. If the measured curves (without proximity effect) shown in Fig. 8 a) are applied, an approximation formula can be derived as:

$$\rho_{Cu}\left(f\right) = \rho_{Cu.0} \cdot \left(1 + \left(\frac{f}{f_0}\right)^{\varepsilon}\right) \tag{12}$$

For the winding losses one then gets the expression:

$$P_{V.Cu} = \frac{2\rho_{Cu.0}\left(1 + \left(\dfrac{f}{f_0}\right)^{\varepsilon}\right) N_1^2 l_m \left(\dfrac{S}{U_1}\right)^2}{k_f A_W} = K_{Cu} \cdot N_1^2 \tag{13}$$

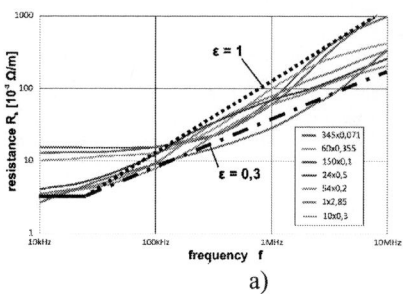 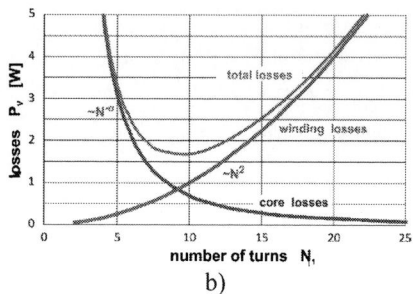

a) b)

Fig. 8: a) Measured resistances of straight conductors made of different strands [8] and b) Division of transformer losses into winding losses and core losses for a calculated example.

This gives at fixed conditions the expression for the total losses of the transformer depending on the number of windings:

$$P_{vtot} = P_{V.Fe} + P_{V.Cu} = K_{Fe} \cdot N_1^{-\alpha} + K_{Cu} \cdot N_1^2 \tag{14}$$

Both loss components are dependent on the number of turns N_1, but with the opposite tendency, as Fig. 8 b) shows. This results in a pronounced minimum for the core losses. So, an optimization task can be formulated in relation to the windings :

$$P_{v.tot} = K_{Fe} \cdot N_1^{-\alpha} + K_{Cu} \cdot N_1^2 \Rightarrow \min \tag{15}$$

By setting the first derivative of $P_v(N)$ to zero, one finds the necessary condition for a local extremum and

$$N_{1opt} = \left(\frac{\alpha \cdot K_{Fe}}{2 \cdot K_{Cu}} \right)^{\frac{1}{\alpha+2}} \tag{16}$$

if one inserts this value into the initial formula, one gets for the relation between copper losses and core losses as:

$$\frac{P_{V.Cu}}{P_{V.Fe}} = \frac{\alpha}{2} = \gamma(\alpha) \tag{17}$$

In the optimal case, the ratio of the loss components P_{VCu} and P_{VFe} can thus be determined by $\gamma(\alpha)$. For $\alpha = 2$, winding losses and core losses are the same when designed for minimum losses. This corresponds to a frequently used solution approach. At $\alpha = 1.6$, for example, copper losses contribute approximatively 80% of the core losses in the optimum case. This corresponds to 44% of the total losses of the transformer. Modern power ferrites for high frequencies have α values in the range of 2...3 (see Table I). Accordingly, in the best case the winding losses there are somewhat higher than the winding losses and are in the range 50...60% of the total losses. It should be noted that the Steinmetz formula is only an approximation of empirical data (see example Fig. 9 b)). In the case of power ferrites, a satisfactory approximation of the formula to the behaviour of the data is only given in a narrow working range. The parameters k_1, α and β must be determined for the corresponding range. Table I lists some materials with their values for α and the frequency exponent β.

Table I: Parameters of the Steinmetz formula, determined from the data sheet values.

Material	α @ 100°C, 100 kHz	β @ 100°C, 0,1 T	k_1 [mW/cm³]
N27	2,25 (@ 25 kHz)	1,28	97,7
N87	2,77	1,424...2,025	2,97...0,43
N97	2,65	1,35...2,22	2,99...0,0345
3C90	2,70	1,494	42,14
3C96	3,03	2,08	2,93
Fi339	3,00	1,285	370,7

A considerable tolerance of the characteristic values summarized in Table I is to be assumed. Especially for materials for higher frequencies, the specific losses $p_V(B, f)$ are not straight lines even on a double logarithmic scale. Calculating absolute values for losses will therefore be subject to greater uncertainty. However, the values should be useful for tendential investigations. With these considerations an easy to carry out optimization of an inductive component can be built up. For the expected total losses one can now write:

$$P_{V.tot} = P_{V.Fe} + P_{V.Cu} = P_{V.Fe} + P_{V.Fe} \cdot \gamma(\alpha)$$ (18)

The total losses of the transformer are then proportional as:

$$P_{V.tot} \sim P_{V.tot} = \left(1 + \left(\frac{f}{f_0}\right)^{\varepsilon}\right)^{\frac{\alpha}{\alpha+2}} \left(\frac{f}{1kHz}\right)^{\frac{2(\beta-\alpha)}{2+\alpha}}$$ (19)

A frequently unknown parameter for the core designs used is the thermal resistance for the release of the generated heat losses to the environment. If known, it enables the calculation of the possible transferable apparent power. For some core forms, information on the thermal resistance R_{th} can be found in literature. Here too, the maximum permissible temperature inside the core must not be exceeded. This can be used to formulate the condition of use such that: $P_{Vtot\ opt}\ R_{th} \leq (T_{max} - T_{amb})$; or the estimation of the R_{th} in [3] through the approximation formula:

$$R_{th} \approx 53 \frac{K}{W} \cdot \left(\frac{V_{core}}{1cm^3}\right)^{-0,54}$$ (20)

The maximum internal temperature will usually be about 100°C. Fig. 9 a) shows a plot of available R_{th} for different standardized core forms. They mainly follow the above fixed estimating formula.

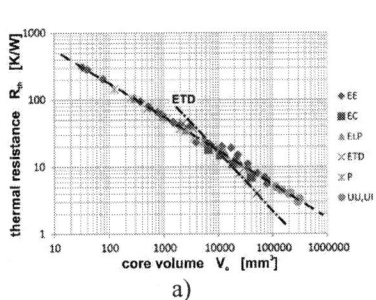
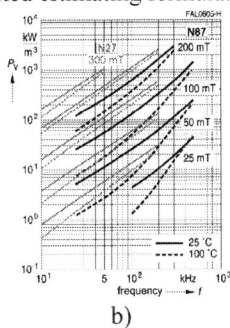

a)   b)

Fig. 9: a) Thermal resistance of ferrite core devices according to data provided by various manufacturers [7] and b) Comparison of ferrite materials N27 an N87 [2].

The example of ETD cores show, that the above formula is only for rough estimation purposes suitable. Looking on suitable transformer material, the ferrites N27 and N87 are compared in Fig. 9 b). Obviously N87 is of advantage in the frequency range of 20 and 40 kHz. The interest is focussed on toroid cores because of their low stray permeance. By interpolation of the characteristics for N87 @ 100°C the Steinmetz parameters for f=20 kHz can be found as: α =2,852, β = 2,301 and k_1 = 1,579 mW/cm³. These parameters are used for optimization of the transformer.

Optimization of a Choke

When optimizing magnetic storage chokes, there are basically 2 possibilities. In the case of toroidal cores, for example, the geometry and material of the cores used are completely fixed. The free parameters here are the number of turns and the fill factor. However, if you use an existing core shape with an adjustable air gap, you get an additional freely adjustable parameter. First the case of given toroidal cores is considered; in particular powder cores in which metal powder or carbonyl iron powder is used and which are characterized by a particularly high saturation induction. This makes particularly compact storage chokes with the corresponding parameter combinations possible under appropriate conditions. The procedure is basically the same as for transformers. The stress for the magnetic core comes from the time change of flux density. Converting this into stress parameters of the Steinmetz formula one can write:

$$P_{Vcore} = V_{Fe}k_1\left(\frac{0{,}5\cdot\Delta\Psi}{N_1 A_{Fe}\cdot 1T}\right)^{\alpha}\left(\frac{f}{1kHz}\right)^{\beta} = V_{Fe}k_1\left(\frac{0{,}5\cdot\Delta I\cdot L}{N_1 A_{Fe}\cdot 1T}\right)^{\alpha}\left(\frac{f}{1kHz}\right)^{\beta} = K_L\cdot N^{-\alpha} \tag{21}$$

At chokes we have only one coil in the winding window. At the same loss density K_{Cu}, within the above formula, has to be divided by 2 in the optimizing approach. Thus, the optimization approach is given as:

$$\frac{\partial P_V}{\partial N}\left(K_L\cdot N^{-\alpha} + \frac{K_{Cu}}{2}\cdot N^2\right) = 0 \tag{22}$$

This delivers the optimum number of winds as:

$$N_{opt} = \left(\frac{\alpha\cdot K_L}{K_{Cu}}\right)^{\frac{1}{\alpha+2}} \tag{23}$$

In the case of choke optimization too at optimum number of windings a special relation of copper losses and core losses can be considered as:

$$\frac{P_{V,Cu}}{P_{V,Fe}} = \frac{\alpha}{2} = \gamma(\alpha) \tag{24}$$

One important parameter of each core is the A_L value, which is measured as small-signal parameter at almost zero current. If the A_L value would be not depending on the current, the boundary condition of the optimization problem would be $N_{opt}\geq\sqrt{L/A_L}$. Since this is not the case, as [1] shows an approximating formula can be used as:

$$\mu_{diff}(H) = \frac{\mu_i}{100(a+b\cdot H^c)} \approx \frac{\mu_i}{100\left(a+b\cdot\left(\frac{N\cdot I}{l_m}\right)^c\right)} \tag{25}$$

The small signal permeability is depicted in Fig. 10 a) for some of the materials in [1] as examples.

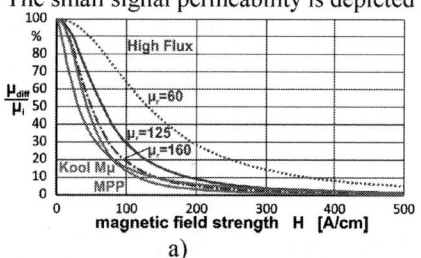
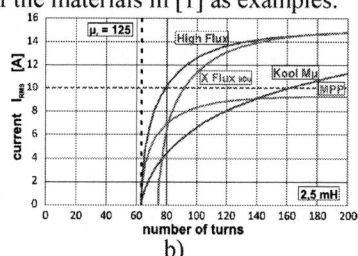

a)

b)

Fig. 10: a) Change of small signal permeability μ_{diff} with magnetic field strength and b) Maximum RMS current for 2,5 mH for different core materials at different number of turns, (Magnetics, 55070A2, $D_a = 68$ mm, $\mu_r = 125$).

With this dependency on the current the small-signal inductance is:

$$L_{diff}(I) = \frac{A_L\cdot N^2}{100\left(a+b\cdot\left(\frac{N\cdot I}{l_m}\right)^c\right)} \geq L_{min} \tag{26}$$

This equation delivers the implementation of the current depending differential inductance into a boundary condition:

$$I \leq \frac{l_m}{N}\left(\frac{A_L\cdot N^2 - L_{min}\cdot 100a}{L_{min}\cdot 100b}\right)^{\frac{1}{c}} \tag{27}$$

Fig. 10 b) shows the boundary condition for the same materials as used in Fig. 10 a) for the same size of toroid core. Design solutions only may be possible below the curves. Considering the target current

of maximum 10 A per choke, only High Flux for this core size is possible. That means High Flux 125 is a possible material with this core size. The lowest losses can be found at the boundary condition at 10 A and minimum number of turns. The main optimization problem considering these constraints is:
$P_{V\,tot}\,(I,N) \Rightarrow$ min; then: $P_{V\,tot}\,(I,N)\,R_{th} + T_{amb} \leq 100\,°C$; $L_{diff}\,(I,N) \geq L_{min}$ and $I \leq I_{max} = 10$ A.
If one enters the optimization goal and the constraints in a diagram, one gets for example Fig. 11 a), where the calculated maximum temperature as a function of the flowing current I_{RMS} and a certain number of turns N is entered with isotherms. For the given maximum current of 10 A, a minimum temperature of 120°C is found. But even this is outside the possible solution area marked by the green line (L = 2.5mH). This makes it impossible for the selected core to solve the optimization task.

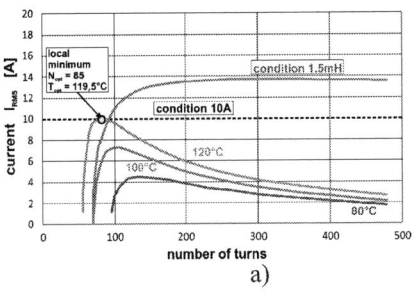
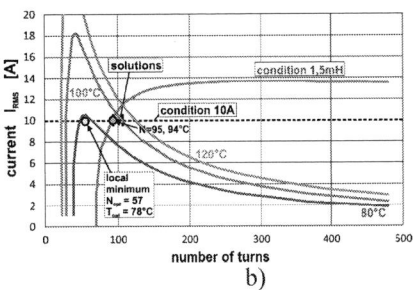

a) b)

Fig. 11: Maximum temperature and constraints of the optimization of a toroidal core for chokes in interleaved mode: a) non-coupled case and b) magnetically coupled case.

In Fig. 11 b) the same problem is visualized for the case of using a transformer for magnetic coupling of the phases. Here the local minimum of the maximum temperature for 10A is outside the green limited area. But the requirement T<100°C can be met at 10A on the edge. The corresponding solution is marked.

Discussion of Results

Table II summarizes the optimization results. Only existing core designs with existing material properties were used. In the case of chokes, the high induction is remarkable, which leads to compact cores. In the lower frequency range, the losses are dominated by the winding losses due to the low permeability values and the associated high number of turns.

Table II: Results of a discrete optimization considering only toroid existing core types of High Flux (chokes) and ferrite N87 (transformers).

	Non-Coupled		Coupled	
# Phases	1	2	2	4
Δi_{load}	0,5 A			
L_x / I_{RMS} Mass /Losses B_{max} / T_{max} fill factor	5 mH* / 20 A, 4862 g / 39,04W 1,37 T / 98°C 0,4	2,5 mH** /10 A 1100 g/12,4 W 1,18 T / 100°C 0,4	2,5mH / 10 A 557 g / 5 W 1,0 T / 77°C 0,3	1,25 mH / 5 A 85g / 2W 1,26 T / 103°C 0,4
W_{Lx}	1 Ws	0,25 Ws	0,25 Ws	0,0625 Ws
Δi_{Lx}	0,5 A @ 20kHz	2 A @ 20kHz	0,25 A @ 40kHz	0,125 A @ 80kHz
$\psi_{\mu 1}$	-	-	2,5 mVs @ 20kHz	2,5 mVs @ 20kHz
S_1;mass, loss	-	-	2kW; 255g; 10 W	2x 1kW ; 2x 129 g;10 W
$\psi_{\mu 2}$	-	-	-	625 µVs @ 40kHz
S_2; mass; loss	-	-	-	2kW; 74g; 0,266T
Total mass	4862 g	2200 g	1333 g	630 g
Total loss	38,9 W	24,8 W	20W+	18W+

With their relatively high permeabilities for induction, ferrite transformers show that at frequencies of 20 to 40 kHz are at the lower end of the typical N87 utilization range. The calculated mass is made up of the mass of the cores and the windings of all components. The first two implementation variants do not include coupling transformers. When the weight of all the components is compared, Fig. 12 a) is obtained. The weight shares for components with High Flux material and N87 are shown separately. In many applications, minimal weight and minimal losses are required in addition to only minimal weight. Fig. 12 b) shows the total losses of an arrangement as a function of the duty cycle at a load current of 20A. The winding losses practically do not change. Corresponding to their share in the total losses, the core losses as a function of the duty cycle lead to changes. The core losses in the storage chokes and the coupling transformers become maximum at different duty cycles D. In their superposition, the losses result in the diagram shown in Fig. 18 b).

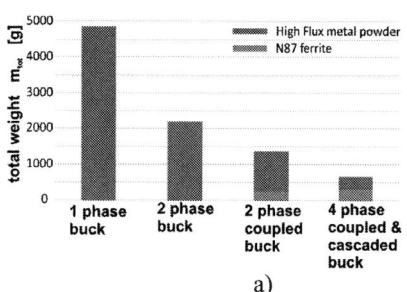
a)

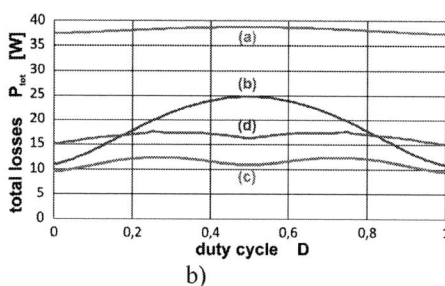
b)

Fig. 12: a) Comparison of the weight of the magnetic components for the investigated configurations and b) Comparison of the total weight of the magnetic components for the investigated configurations.

The variants in Fig 12 b) '(a)...(c)' show a decreasing tendency while variant '(d)' shows higher losses. Whether this is the case because of the discretization of the possible results or whether it is a real trend cannot be determined in the given case. One assumption is that the losses in a continuous solution tend towards a low limit value. The selection according to the lowest weight probably leads to a tendency to increase the losses.

Conclusion

For interleaved operation of several phases a new possibility of coupling and cascading inductive components for current smoothing at the output of DC/DC and DC/AC converters was presented. This approach shifts the magnetic AC flux load to the coupling transformers used. It enables the use of adapted materials. The result is a significant reduction in volume and weight of the magnetic components. Coupling of phases in pairs with phase shift of 180° leads to maximum suppression of harmonics. The reduction of harmonics shown in [10] with pairwise coupling of phases in three-phase systems is significantly improved.

References

[1] Catalogue Magnetics Co. Pittsburgh, 2017.

[2] Catalogue Ferrites and Accessories (EPCOS).

[3] http://encyclopedia-magnetica.com/doku.php/thermal_resistance#, consulted June 2020.

[4] Gallagher, J.: Coupled Inductors improve multiphase buck efficiency. power electronics.com, 2006,

[5] Wibowo, S. A.; Ting, Z.; Kono, M.; Taura, T.; Kobori, Y.; Onda, K.; Kobayashi, H.: Analysis of coupled inductors for Low-Ripple Fast-Response Buck Converter. IEICE Trans. Fundamentals, Vol. E92-A, No. 2, Februar 2006.

[6] Dick, Chr.: Ultra Compact 2 kW 12 V - 48 V Converter Using a 4-Phase Coupled Inductor. PCIM 2018, Nürnberg.

[7] Zacharias, P.: Magnetic Devices. (Magnetische Bauelemente, German). Springer Nature 2020.

[8] Fenske, F.: Characterization of winding goods under the influence of the skin and proximity effect (Charakterisierung von Wickelgütern unter Einfluss des Skin- und Proximity-Effekts, German), master thesis,University of Kassel 2012.

[9] Boillat, D. O.; Kolar, J. W.: Modeling and Experimental Analysis of a Coupling Inductor Employed in a High Performance AC Power Source. ICRERA 2012, Nagasaki, Japan.

[10]Schroeder, J. C.; Fuchs, F. W.: Detailed Characterization of Coupled Inductors in Interleaved Converters Regarding the Demand for Additional Filtering. 2012 IEEE Energy Conversion Congress and Exposition (ECCE), Raleigh, NC, 2012.

[11]Liu, J.: Investigation of Multiphase Power Converter using Integrated Coupled Inductor Regarding Electric Vehicle Applicaticon. PhD thesis, University of Kassel.

EPE'20 ECCE Europe

Assigned jointly to the European Power Electronics and Drives Association & the Institute of Electrical and Electronics Engineers (IEEE)

Experimental tuning and design guidelines of a dynamically reconfigured weighting factor for the predictive torque control of an induction motor

İlker ŞAHİN

Aselsan

Ankara, Turkey
ilkersahin@aselsan.com.tr
www.aselsan.com.tr

Ozan KEYSAN

Middle East Technical University

Ankara, Turkey
keysan@metu.edu.tr
www.metu.edu.tr

Eric MONMASSON

CY Cergy Paris University

Paris, France
eric.monmasson@u-cergy.fr
www.u-cergy.fr

Keywords

«MPC (Model-based Predictive Control)», «Adjustable speed drive», «Highly dynamic drive», «Induction motor», «Drive».

Abstract

In this study, an experimental approach for the weighting factor design of the predictive torque control of an induction motor (IM) is presented. The weighting factor is set to the value which yields a low total harmonic distortion (THD) figure in the phase currents. Then, this optimized value is dynamically adjusted with respect to the speed error term, which is defined as the magnitude of difference between the commanded speed and the actual speed. By doing so, superior transient performance is achieved without sacrificing steady-state THD values. The presented approach for dynamically adjusted weighting factor is tested on a two-pole, 500 W IM, driven by a two-level voltage source inverter (2L-VSI), via finite control set model predictive control (FCS-MPC). Experimental results show the effectiveness of the proposed method: by the adoption of dynamic lambda approach, speed reversal time from -3000 rpm to +3000 rpm is decreased by 38% under the same controller structure and settings.

Introduction

Model predictive control (MPC) has emerged as a promising alternative for the control of power electronics and received significant attention from the research community in the last two decades. Considering the finite control-set category of MPC (FCS-MPC), the methodology for cost function design and optimal weighting factor selection still remains as a challenge that requires further research [1]–[3].

Contrary to the conventional control methods that are typically based on PID compensators, FCS-MPC technique does not involve PID stages. Rather, system states are estimated, potential future states are predicted, and the best control action is decided by the minimization of a predefined cost function. Considering the nature of power electronic converters, this almost always corresponds to a multi-variable optimization. The relative importance of the controlled variables is set by weighting factors in the cost function. General guidelines for optimal weighting factor design are presented in [1], [3], [4] where the tuning of the weighting factors is accomplished through simulation and experiments.

A control algorithm tuned for a specific operating condition may not yield the optimum control over a wide range of operating points. Therefore, dynamic adjustment of the weighting factors according to the operating conditions are proposed for an IM drive [5] and 3L-NPC inverter [6] and rectifier [2]. In [5], torque ripples are estimated, and the weighting factor is optimized to reduce the ripples through a complicated formulation. In [2], weighting factors are calculated as continuous functions depending on the preexisting values of the respective variables. In [6], neural networks are utilized for the determination of weighting factors. There also exist variations of MPC that do not utilize weighting factors in their optimization process [7], [8]. Although these approaches may seem to be avoiding the design of the weighting factor, the effect of the weighting factor is implicitly included in their modified control schemes. In other words, the effort for the weighting factor design is just replaced by the effort for the design of structural modifications that allow the operation without weighting factors.

In this paper, FCS-MPC of an IM is presented. The cost function is tuned experimentally: the weighting factor (λ) is varied over a wide range to observe its influence on the current THD. Hence, an optimal λ is found. Secondly, a simple algorithm that updates the λ value according to the speed error is suggested such that the torque error term in the cost function is prioritized in the event of a speed transition. The main motivation of this study is to provide the research community a simple and effective algorithm for the dynamic adaptation of λ and to report an experimental cost function tuning procedure for the predictive torque control of an IM.

FCS-MPC of an IM

For the sake of completeness, the control structure of FCS-MPC of an IM will be presented first. State-space model of an IM is given as (1)-(4).

$$V_s = R_s I_s + \frac{d\Psi_s}{dt} \tag{1}$$

$$0 = R_r I_r + \frac{d\Psi_r}{dt} - jw_e \Psi_r \tag{2}$$

$$\Psi_s = L_s I_s + L_m I_r \tag{3}$$

$$\Psi_r = L_m I_s + L_r I_r \tag{4}$$

The flowchart of the FCS-MPC routine is depicted in Fig. 1. It is mainly composed of three steps: estimation, prediction, and optimization. First, rotor and stator fluxes are estimated by using the current measurement and machine parameters. A suitable way for this is to use the rotor current model of the IM as in (5)-(6). Discrete versions of rotor and stator flux estimations can be obtained by the Euler approximation as presented in (7)-(8).

$$\frac{d\Psi_r}{dt} = R_r \frac{L_m}{L_r} I_s - \left(\frac{R_r}{L_r} - jw_e\right)\Psi_r \tag{5}$$

$$\Psi_s = \frac{L_m}{L_r}\Psi_r + \sigma L_s I_s \tag{6}$$

$$\Psi_r(k) = \Psi_r(k-1) + \Delta t.\left(R_r \frac{L_m}{L_r} I_s(k) - \left(\frac{R_r}{L_r} - jw_e(k)\right)\Psi_r(k-1)\right) \tag{7}$$

$$\Psi_s(k) = \frac{L_m}{L_r}\Psi_r(k) + \sigma L_s I_s(k) \tag{8}$$

where $\sigma = 1 - L_m^2/L_s L_r$ is the leakage factor and Δt is the control period.

Having estimated the values of $\Psi_s(k)$ and $\Psi_r(k)$, next step in the FCS-MPC routine is to predict the values of stator flux (9), current (10) and torque (11) at the next control cycle, under the influence of a particular voltage vector $V_i(k)$. The superscript p is used to indicate predicted variables.

$$\Psi_s^p(k+1) = \Psi_s(k) + \Delta t.V_i(k) - \Delta t R_s I_s(k) \tag{9}$$

$$I_s^p(k+1) = \left(1 - \frac{\Delta t}{\tau_\sigma}\right)I_s(k) + \frac{\Delta t}{\Delta t + \tau_\sigma}\frac{1}{R_\sigma}\left(\left(\frac{k_r}{\tau_r} - k_r j\omega_e(k)\right)\Psi_r(k) + V_i(k)\right) \tag{10}$$

$$T_e^p(k+1) = 1.5p\Im\{\Psi_s^{p*}(k+1)I_s^p(k+1)\} \tag{11}$$

where, $k_r = L_m/L_r$ is the rotor coupling factor, $R_\sigma = R_s + k_r^2 R_r$ is the stator referred equivalent resistance, $\tau_\sigma = \sigma L_s/R_\sigma$ is the stator transient time constant and $\tau_r = \sigma L_r/R_r$ is the rotor time constant.

After the predictions for a voltage vector is obtained, the outcomes in terms of stator flux magnitude and torque can be evaluated with a cost function. This action is repeated for each admissible voltage vector as shown in Fig. 1. The voltage vector for which the cost function is minimized, is the optimum vector. Although different forms for the cost function can be proposed, the conventionally used type is given in (12). A weighting factor λ is included to adjust the relative importance of torque and flux terms hence to tune the cost function. It is also good practice to include a penalty factor in the cost function such that an infinite cost is assigned to a voltage vector if the current prediction of (10) foresees an overcurrent under the influence of that particular vector.

$$g = \left|T_e^{ref} - T_e^p(k+1)\right| + \lambda \left|\left|\Psi_s\right|^{ref} - \left|\Psi_s^p(k+1)\right|\right| \tag{12}$$

The voltage vector, for which the cost function of (12) yields the minimum value, is selected as the optimum and applied at the next switching instant.

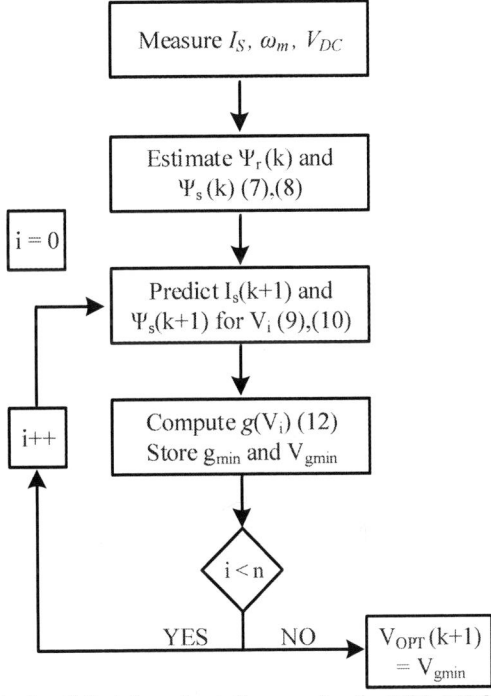

Fig. 1: A simplified flowchart diagram for the FCS-MPC routine.

Considering the different natures of the variables utilized in the cost function of (12), a per-unit approach would be appropriate because the rated values of different motors' torque and flux can largely vary. Similarly, the percent change in torque and flux for the same voltage vector can also be different. Those differences can be tuned for a satisfactory motoring action, with a careful selection of λ.

The experimental tuning of the weighting factor

The test setup is shown in Fig. 2. A modified IM is utilized in the experiments, which was originally a spindle motor intended for 200 Hz operation. It is driven via the motor drive development platform TMDXIDDK379D, which houses a 2L-VSI, a DSP, and the required sensor circuits. The DC bus voltage is set at 350 V, therefore the operating frequency is limited with 75 Hz, in the experiments. Re-defined rated values and circuit parameters of the motor is given in Table I. The control algorithm is run inside an interrupt service routine of 40 kHz, (i.e. sampling frequency is set as 40 kHz). For the interested reader, further details on the experimentation setup are provided in [9].

Fig. 2: The experimental setup

Table I: The IM Parameters

Name	Symbol	Value
Power	P	500 W
Stator voltage	V_{ab}	150 V
Stator current	I	3.5 A
Base frequency	f	50 Hz
Torque	T_L	1.2 N.m
Number of poles	p	2
Number of turns in one phase	N_s	104
Stator resistance	R_s	2.3 Ω
Rotor resistance	R_r	3.1 Ω
Magnetizing inductance	L_m	90 mH
Stator leakage inductance	L_{ls}	5 mH
Rotor leakage inductance	L_{lr}	1 mH
Stator flux magnitude ref.	$|\Psi_s|^{ref}$	0.4 Wb

For the experimental tuning of λ, THD values of stator phase currents are examined for various λ. A THD vs λ graph is obtained and depicted in Fig. 3 for both rated load and no-load cases. It is observed that the increasing values of λ provide better shaped sine waves hence lower THD values. However, increasing λ further beyond $\lambda = 50$ brings little to no improvement in THD. Therefore, $\lambda = 50$ value is chosen as the optimum. For the operating condition of $\omega = 3000\ rpm$ and $T_e = 1.2\ N.m$, current and torque waveforms are depicted in Fig. 4 for the selected values of λ.

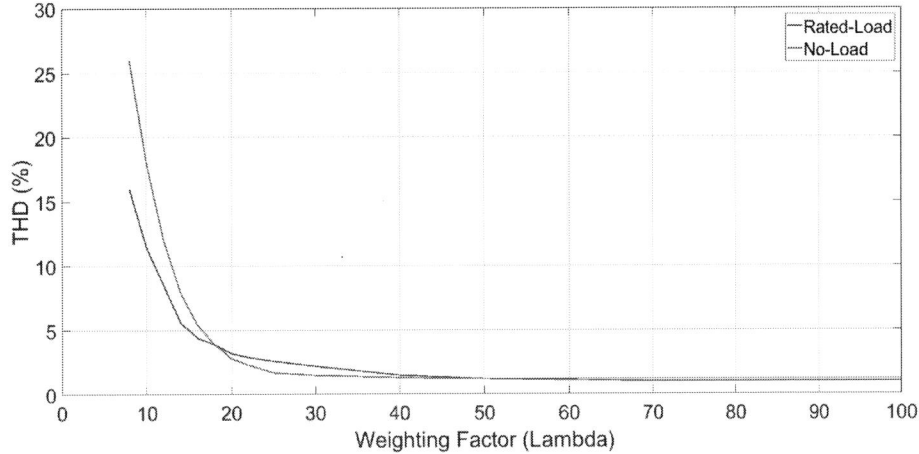

Fig. 3: Phase current THD vs weighting factor (λ) graph for rated load (blue) and no-load (red) cases.

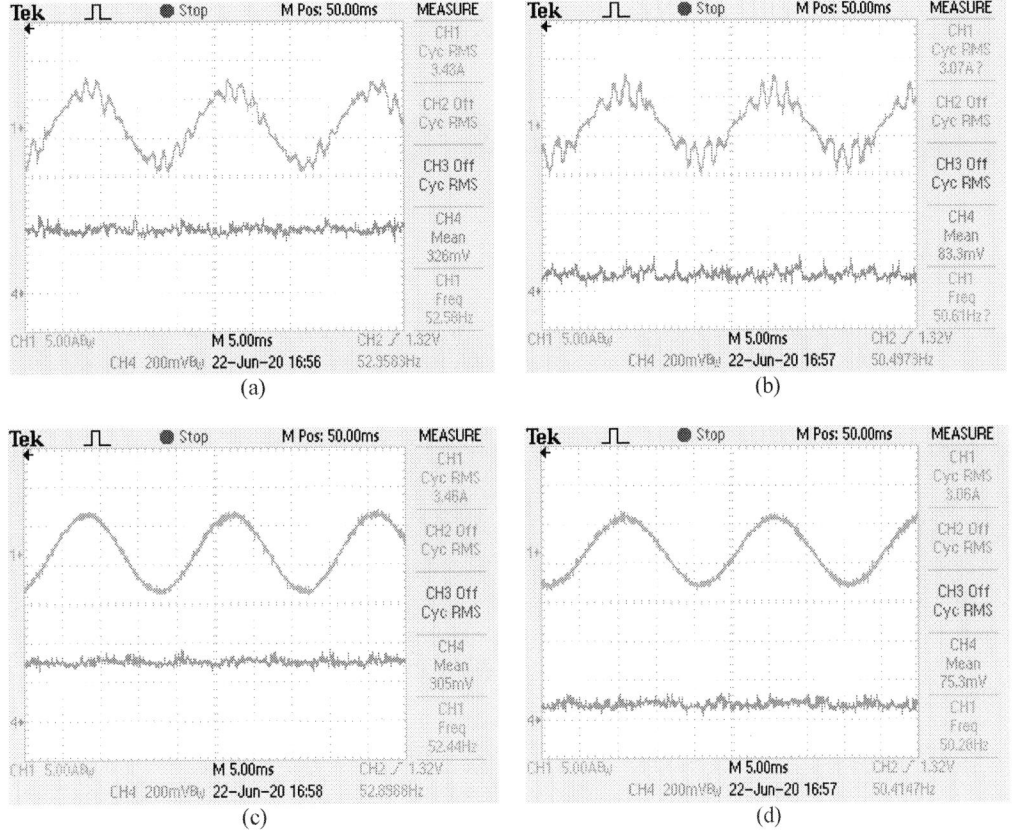

Fig. 4: Phase current (Ch. 1, orange) and torque (Ch. 4, green) waveforms for $\lambda = 8$ (a, b) and $\lambda = 50$ (c, d) cases, under rated-load $(\tau \cong 1.2\ N.m)$ (a, c) and no-load $(\tau \cong 0\ N.m)$ (b, d) conditions.

It is evident from (12) that increasing value of λ prioritizes flux over torque and the experimental THD figures depicted in Fig. 3 show that this corresponds to smoother current waveforms. To test the dynamic performance for different magnitudes of λ, a sudden transition from no-load to full (rated) load at rated speed (3000 rpm) is tested for different values of λ and results are depicted in Fig. 5. The speed information is included (purple waveform, Ch. 3) in the respective figures and the momentary decrease in speed due to sudden loading is also noted. It is interesting to note that the response of the drive against the sudden loading seems pretty much not affected by the value of λ. Because the controller is trying to preserve the sinusoidal shape of the current and this effort in turn enables a fast increase in torque even if the torque error term becomes less effective in the cost function. Nonetheless, smaller values for λ (hence, more emphasis on the torque error) enables faster transient during start-up and speed reversal, as will be shown in the next section.

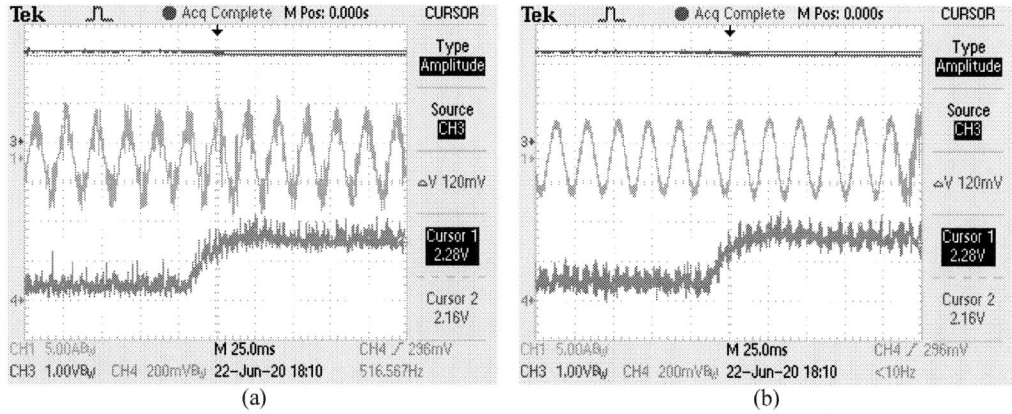

Fig. 5: Phase current (Ch. 1, orange), torque (Ch. 4, green), and speed (Ch. 3, purple) waveforms for $\lambda = 8$ (a) and $\lambda = 50$ (b) cases, under sudden load change from no-load to rated-load at $\omega = 3000 \, rpm$.

Dynamic weighting factor adjustment according to the speed error

Having chosen the optimum weighting factor as $\lambda = 50$ from the THD results presented above, the next step will be dynamically adjusting the λ value for a better transient performance. A very simple yet effective algorithm is suggested for the dynamic λ adjustment: by observing the speed error $\Delta\omega$ as defined in (14), the λ value can be lowered from its optimum value with increasing $\Delta\omega$ to prioritize the torque error and hence to enable faster transients.

$$\Delta\omega = \left| \omega^{ref} - \omega^{actual} \right| \qquad (14)$$

In the experiments regarding dynamic λ approach, weighting factor value is adjusted with respect to the speed error $\Delta\omega$ both linearly and quadratically as depicted in Fig. 6. A per-unit approach is adopted in the formulation of the $\Delta\omega$ variable and a saturation level is utilized as $\Delta\omega = 1$. For the increasing values of $\Delta\omega$ term, the λ value is not totally reduced to zero but to a rather small value ($\lambda = 5$) so as not to totally lose the flux magnitude controllability.

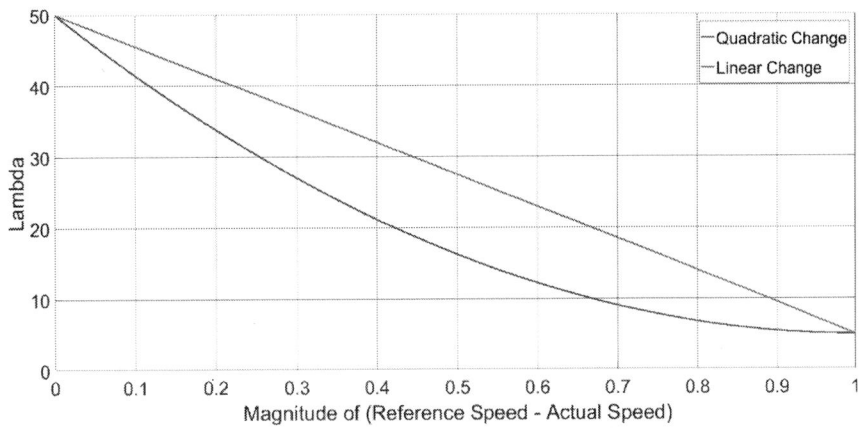

Fig. 6: The dynamic adjustment of λ with respect to $\Delta\omega$, linear (red) and quadratic (blue).

The effectiveness of the proposed λ variation technique is compared to the optimum value of λ found from minimum THD approach. A convenient way to test this is to observe the drive performance under sudden speed reversal command. The comparison is provided in Fig. 7 for the fixed case of $\lambda = 50$ and variable (linear and quadratic change with respect to $\Delta\omega$) λ cases for the speed transition from -3000 rpm to +3000 rpm under the rated load. Experimental results show up to 38% reduction in the transition time compared to the fixed λ case. The steady-state performance figures are not affected because λ converges to its minimum THD setting ($\lambda = 50$) as the speed error term $\Delta\omega$ vanishes.

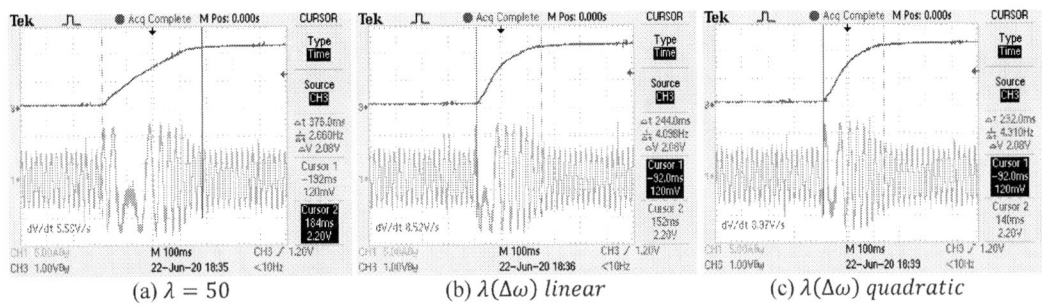

(a) $\lambda = 50$ (b) $\lambda(\Delta\omega)\ linear$ (c) $\lambda(\Delta\omega)\ quadratic$

Fig. 7: Current (Ch. 1, orange) and speed (Ch. 3, purple) waveforms for speed reversal command ($-3000\ rpm \rightarrow +3000\ rpm$) for constant and dynamic λ.

Similarly, the proposed method of dynamic λ adjustment is tested for start from stand-still command, and the experimental results are depicted in Fig. 8. The transition time is decreased up to 25%.

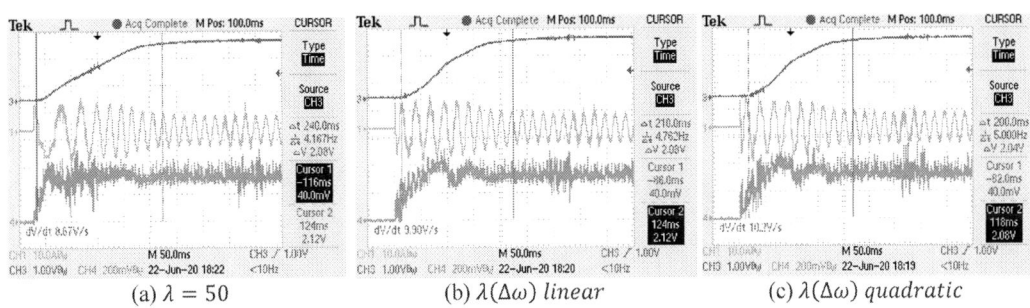

(a) $\lambda = 50$ (b) $\lambda(\Delta\omega)\ linear$ (c) $\lambda(\Delta\omega)\ quadratic$

Fig. 8: Current (Ch. 1, orange) and speed (Ch. 3, purple) waveforms for start from stand-still command, for constant and dynamic λ.

Conclusion

An experimental cost function tuning procedure is reported for a predictive torque controlled IM. The optimization is made based on the phase current THD values. Having found the optimized weighting factor value, a dynamic update method is proposed to prioritize torque error element in case of a speed transition, hence to improve transient performance. The proposed algorithm, which is based on the speed error term, is trivial to implement and does not introduce any significant computational burden. Experiment results show the effectiveness of the proposed approach. Transition time during speed reversal command has been reduced by 38% with the adoption of dynamic λ, without sacrificing steady-state THD figures.

References

[1] S. Vazquez, J. Rodriguez, M. Rivera, L. G. Franquelo, and M. Norambuena, "Model Predictive Control for Power Converters and Drives: Advances and Trends," *IEEE Trans. Ind. Electron.*, vol. 64, no. 2, pp. 935–947, Feb. 2017.

[2] L. M. A. Caseiro, A. M. S. Mendes, and S. M. A. Cruz, "Dynamically Weighted Optimal Switching Vector Model Predictive Control of Power Converters," IEEE Trans. Ind. Electron., vol. 66, no. 2, pp. 1235–1245, 2019.

[3] P. Cortes et al., "Guidelines for weighting factors design in Model Predictive Control of power converters and drives," IEEE Int. Conf. Ind. Technol., pp. 1–7, 2009.

[4] S. Thielemans, T. J. Vyncke, and J. Melkebeek, "Weight factor selection for model-based predictive control of a four-level flying-capacitor inverter," IET Power Electron., vol. 5, no. 3, pp. 323–333, 2012.

[5] S. A. Davari, D. A. Khaburi, and R. Kennel, "An improved FCS-MPC algorithm for an induction motor with an imposed optimized weighting factor," IEEE Trans. Power Electron., vol. 27, no. 3, pp. 1540–1551, 2012.

[6] O. Machado, F. J. Rodriguez, E. J. Bueno, and P. Martin, "A Neural Network-Based Dynamic Cost Function for the Implementation of a Predictive Current Controller," IEEE Trans. Ind. Informatics, vol. 13, no. 6, pp. 2946–2955, 2017.

[7] C. A. Rojas, J. Rodriguez, F. Villarroel, J. R. Espinoza, C. A. Silva, and M. Trincado, "Predictive torque and flux control without weighting factors," IEEE Trans. Ind. Electron., vol. 60, no. 2, pp. 681–690, 2013.

[8] Y. Zhang and H. Yang, "Two-Vector-Based Model Predictive Torque Control Without Weighting Factors for Induction Motor Drives," IEEE Trans. Power Electron., vol. 31, no. 2, pp. 1381–1390, 2016.

[9] İ. Şahin and O. Keysan, "A simplified discrete-time implementation of FCS-MPC applied to an IM drive," in 21st European Conference on Power Electronics and Applications, EPE 2019 ECCE Europe, 2019, pp. 1–8.

Compensation of Temperature Dependence in a Module Parasitic Based Current Measurement System

Frank Lautner, Mark-M. Bakran
UNIVERSITY OF BAYREUTH, DEPARTMENT OF MECHATRONICS
Centre for Energy Technology - ZET
Universitätsstraße 30
95447 Bayreuth, Germany
Tel.: +49 / (0)921 – 55-7815.
Fax: +49 / (0)921 – 55-7802.
E-Mail: frank.lautner@uni-bayreuth.de
URL: http://www.mechatronik.uni-bayreuth.de

Acknowledgements

This project was supported by the ZF Friedrichshafen AG. Special thanks go to the department of Engineering Electric Mobility & Mechatronics, Auerbach i. d. OPf.

Keywords

«Current sensor», «Component for measurements», «Measurement», «Sensor», «Module temperature measurement»

Abstract

This paper presents a current sensing method, which evaluates the voltage across the parasitic resistance and inductance in power modules. Such a sensing method requires an appropriate low pass filter to achieve a signal that is proportional to the current to be measured. In inverter operation, however, the parasitic resistance in the measurement section is variable due to its temperature dependence. This requires firstly a temperature estimation for the measurement section in the semiconductor module and a correction of the deviations arising from that circumstance. At a first glance, three different possible temperature estimation methods seem possible. The first approach by using an additional temperature sensor is only considered as a trivial solution and not examined further. Its disadvantage is the additional component, which is then necessary for current measurement. The second and third method are investigated in detail, which is on the one hand the use of mostly present negative-temperature-coefficient (NTC) thermistors in the module and on the other hand, a completely new approach that exploits the temperature dependence of the sensed signal waveform. Both approaches were tested in inverter operation. They improve current sensing errors by means of module parasitics from ± 25 % to ±2…5 %.

Introduction

Current sensing is an important task for various applications such as electrical or hybrid drives, energy production and distribution. Current sensors used for control purposes need an accurate and robust functionality. Up to now, different kinds of measurement approaches are discussed or already applied in power electronic systems. One simple method is the shunt resistor that, however, has either a limited sensitivity with a small resistance or high losses with higher resistances during operation. With this approach, also cooling issues have to be solved [1]. An advantage of this method is its relatively low price compared to alternatives like Hall-Effect sensors or sensors that are based on magneto resistive effects [1]. These mostly show higher costs, but better flexibility. In sum, for every current sensing method, a trade-off between different pros and cons has to be made. An ideal solution would be an approach, which is easy to implement, cheap and shows high accuracy. Such an approach only can be found, if existing components in the power electronic system can be used as current sensitive elements. For a SiC-MOSFET equipped inverter there is the possibility to assess the chip's drain-source-voltage [2] [3]. A further method is to measure the voltages across parasitic inductances in such a module, which is also one focus of this paper. In nearly every module a section consisting of parasitic inductance and resistance is accessible, which provides a signal proportional to the current and proportional to the current slope (proportional-derivative signal) in every switching pulse. To get the current itself it is necessary to find a method to re-calculate the current out of the measured signal. This can be done with a network that shows the exact reverse behaviour to the measurement section. The solution is a simple, first order low pass filter (proportional-integrating transfer function), see Fig. 1. This procedure seems to match all requirements for a feasible current sensing technique (simple to implement, no excessive side effects, cheap …) and should therefore be examined further, especially for accuracy.

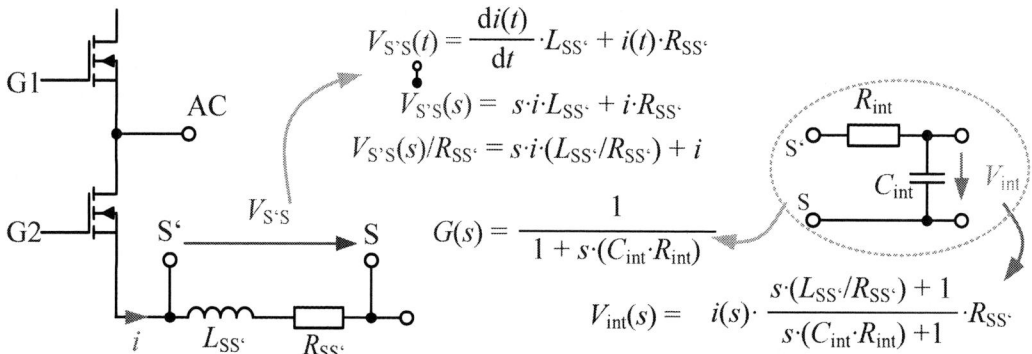

Fig. 1: Exemplary accessible measurement section of a MOSFET half bridge module. Across the section "auxiliary source (S') – load source(S)" the proportional-derivative of the current can be measured. With an exact fitting of a downstream first order low pass filter, the original current can be reconstructed (in reality, an active low pass filter has to be used, this is neglected due to only a proportional factor in this figure).

Challenges and Main Side Influence of the Proposed Method: Module Temperature

The only situation, why this method might not work, is a module that does not provide access to a measurement section (current traversed LR-network). Most of the actual power semiconductor modules, however, provide such networks or, like discussed in [4] provide sections between two semiconductor switches, which can be used for this method by correcting the measured signal afterwards with a software based solution. The most difficult challenge of this method, however, is the temperature dependence.

An exactly tuned low pass filter has to be found, to get an accurate representation of the current through the measurement section. This means, the time constant of the measurement section has to meet the time constant of the low pass filter. Considering this in detail, we can exclude any temperature influence on the parasitic inductance $L_{SS'}$. In contrast, temperature does effect the parasitic resistance of the section, which represents the resistance of bond wires and conductive planes in the power module. These parts consist primarily of aluminium or copper that both show an average temperature coefficient of 0.004 1/K. To give an example with characteristic values, a cold module ($L_{SS'}$ = 7 nH, $R_{SS'}$ = 4 mΩ, 20 °C) shall be compared to a hot one (80 °C). Here, the first case results in a to-be-compensated time constant of τ_{cold} = 1.75 µs, whereas the heated module shows τ_{hot} = 1.41 µs, with a now present parasitic resistance of 5.28 mΩ. This situation has two effects, according to the equations in Fig. 1: Firstly, the measurement section's time constant is not compensated correctly any longer. This means, the sensor output signal is no mere proportional representation of the current (P-behaviour), but now shows PDT1-behaviour (proportional-derivative element with first order lag). The second effect is the changed proportional gain of the sensor output signal, because $R_{SS'}$ changes. To illustrate this effect, Fig. 2 shows a simulation of this measurement principle for different bond wire temperatures.

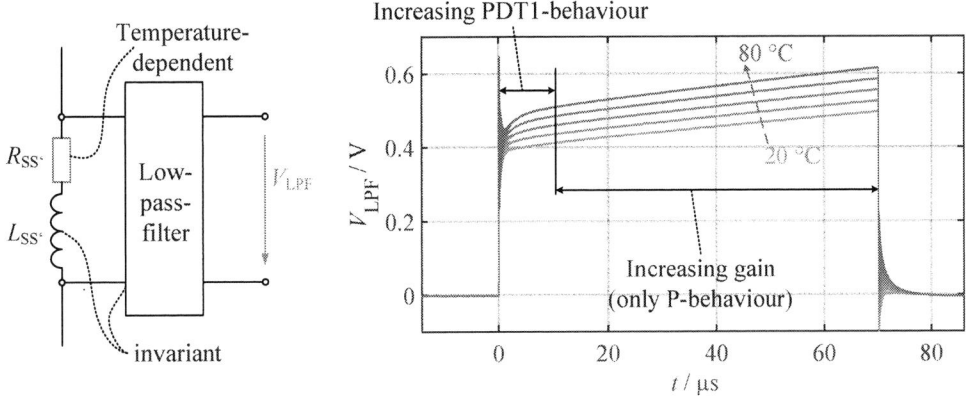

Fig. 2: Simulated waveforms of the sensing low pass filter for different measurement section temperatures (measured current pulse approx. 50 A). Due to invariant filter and parasitic inductance but changing value of $R_{SS'}$ different results are achieved for different temperatures.

With this characteristic, a sensor system calibrated to the 20 °C-case and sampling at approx. 10 µs after switching on, would not show 50 A at 80 °C, but up to 62 A as measurement value. If it is necessary to sample the sensor signal during that time period (because of only short current pulses) not only a proportional factor has to be corrected, but also the nonlinear change due to an increasing PDT1-behaviour. This can be simplified to a linear correction only for one single and exact sampling point for every switching pulse.

Thus, two main challenges have to be overcome: Firstly, identifying the actual bond wires' and conducting planes' temperature in the power module and, secondly, finding a feasible solution for compensating the resulting errors. Aim of this paper is presenting different approaches and their tests for these two main challenges.

Estimating the Appropriate Temperature

Exact temperature measurement is a wide topic in power electronics [5] and would lead to very high additional efforts for a low-cost current sensing technique. Moreover, most of the temperature sensors try to sense the semiconductor's junction temperature, which is not exactly the same as the here required temperature of bond wires and conducting planes. In literature, a deviation of 10 K and more is discussed [6]. It has yet to be clarified, if such errors affect the current measurement significantly, or only at special operating points. It seems easier (but more expensive) to find an appropriate additional temperature sensor than compensating temperature without such tools. Because of that, this paper focuses on the cheaper approaches.

Estimating Temperature by Means of NTC-Thermistor

Such a method can be found in the assessment of the usually inserted module-NTC-thermistor. This temperature sensor is located on the module baseplate and allows the measurement of the temperature at this position. However, this determined temperature is even more far away from the bond wires (Fig. 3). The copper conduction plates may be covered best with this method, but these state only a subordinate role in the overall resistance. To assess this temperature compensation method, a few points have to be considered:
(1) What time lag will occur between bond-wire temperature and possible measurement of this temperature at the NTC? Is this measurement principle for the bond wire temperature fast enough? (2) Which difference in absolute values between bond wires and NTC (even steady state) have to be

considered (3) Are there special situations or special operating points, when an accurate temperature estimation is not possible? What are the limitations of this method?

To answer these questions one has to consider the thermal path between bond wires and module-NTC. Figure 3 illustrates a worst-case scenario situation for a power semiconductor module.

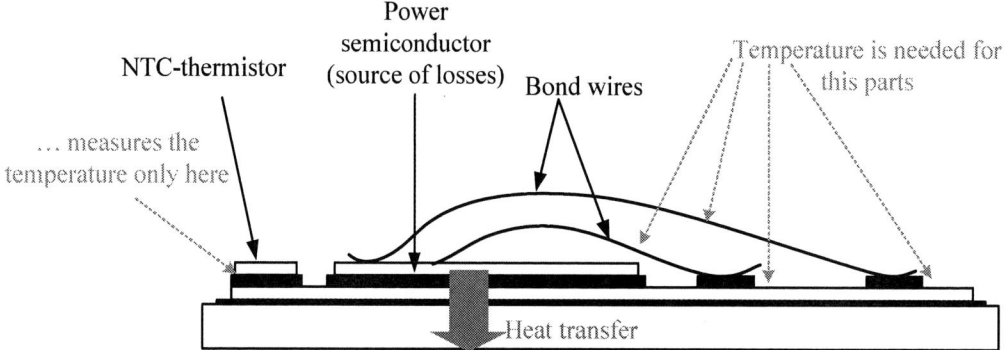

Fig. 3: Thermal situation in a power semiconductor module, where the NTC-thermistor is located far away from the evaluated parasitic resistance of the module.

Here, the temperature of the bond wires as well as the temperature of conducting planes can only be measured indirectly via the base plate. This thermal path, however, is crossed by the main thermal path of the semiconductor chip, and only a very rough temperature estimation may be possible. It has to be considered, however, that this depicted setup is (1) a worst-case situation and (2) the respective thermal resistances and capacitances of the considered path play a key role in temperature distribution of a module. Disadvantageous situations with low thermal capacitances and high thermal resistances lead to an uneven temperature distribution and will result in delayed and error prone temperature signals at the NTC-thermistor and therefore false measured current signals, at least for a certain time span. High thermal capacitances and low thermal resistances in the module may enable an acceptable solution for the given problem, because in this case a uniform temperature distribution is achieved. Such a situation can be seen in real measurement data, which is shown in Fig. 4. Here, one single operating point is considered ($\hat{I} = 50$ A, $f_{el} = 20$ Hz). In this situation, the module was heated to a temperature of approx. 40 °C. Then, the temperature coefficient leads to a deviation in the current measurement of max. 6 % (depending on the sampling, see. Fig. 2) when calibrated to 25 °C. This can be seen in the uncompensated mode (left). If the NTC-thermistor is taken into account, the deviation can be compensated accurately in steady state, which means after running a sufficient time span in the same operating point assuming that the temperature of the NTC thermistor and the temperature of the considered measurement section is the same (right plot, "Using NTC-thermistor temperature"). This method works also for a temperature swing back to approx. 23 °C, which is done in the experiment at $t = 28$ s. Due to the aforementioned thermal path (resistances and capacitances) between NTC-temperature and a resulting temperature difference between measurement and reality, this leads to a temporary deviation, or false-compensation of the temperature effect. In the present test, a maximum deviation of < 2 % was found. The fact that there is only a small deviation has different reasons: The electrical frequency of $f_{el} = 20$ Hz is high enough to provide a very stable loss injection into the module and therefore a uniform temperature distribution between module NTC thermistor and measurement section. As said above, the temperature in the bond wires and the conducting planes may not be the same for any current pulse, what means that a method considering the average module temperature (and the NTC-temperature estimation has undoubtedly such averaging effects) may be advantageous in real operation. The last reason for the good functionality is the chosen current path through the module: There may be other operating points, where current paths and the NTC thermistor are located far away from each other (like Fig. 3), which makes this approach less accurate.

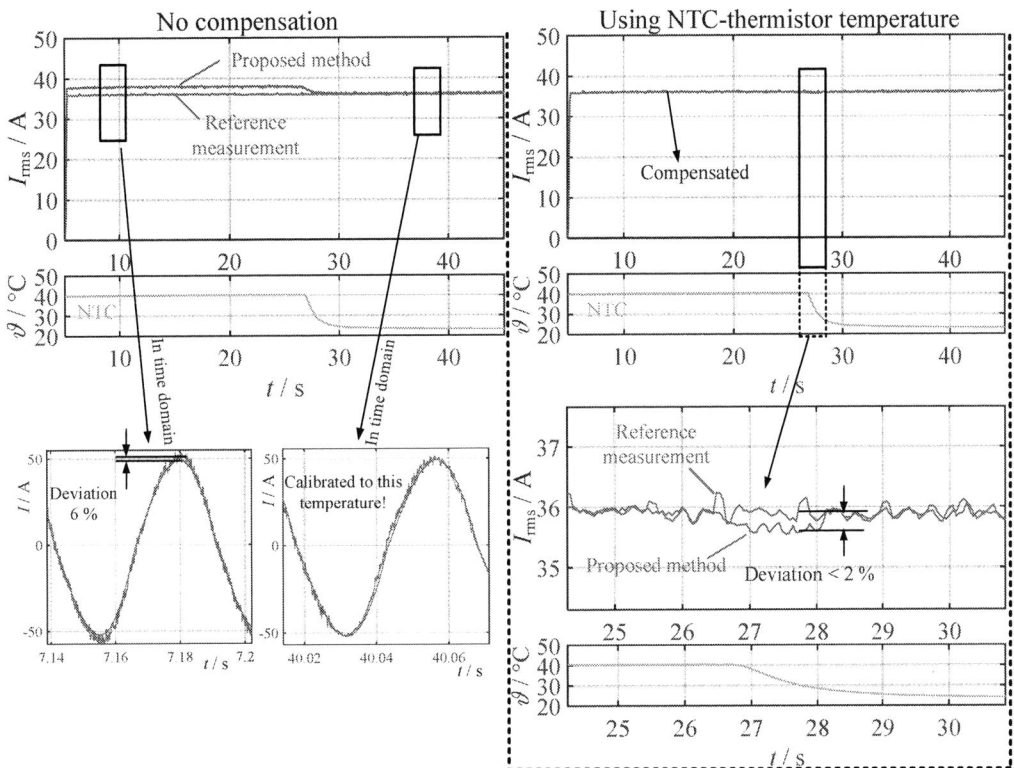

Fig. 4: Current measurement via module parasitics at operating point $\hat{I} = 50$ A. $f_{el} = 20$ Hz. Measurement voltage across S'(HSS) and D'(LSS) is low pass filtered, sampled (red) and compared to a simultaneously evaluated Hall-Effect sensor (magenta). The temperature is changed at $t = 28$ s by abruptly starting module cooling. Left: Without any temperature compensation; Right: Evaluating NTC-temperature and correcting the measured current signal by means of the temperature coefficient of copper/aluminium.

Compensating Before Measurement or Correcting the Temperature Afterwards?

In Fig. 4 the measured temperature value was corrected afterwards to process the accurate current value by software solution. This will be the method of choice for an easy and flexible elimination of the temperature impact. Another possibility would be the fast changing of low pass filter parameters for the next current pulse, so that another filter time constant is active and the measurement signal is filtered with the respective time constant for the valid temperature. Such a switching of time constants may anyway be necessary for special modules [4].

Correcting Temperature by Means of the Sensor Signal Waveform

Identifying a way to find the temperature in the measurement section itself and not in vicinity to it, the measured signal waveform was found to be a possible indicator. As seen in Fig. 2, the signal will show PDT1 behaviour, if it is not calibrated to the correct temperature. This offers a good possibility for a temperature estimation. If the signal is sampled, for example, exactly 4.0 µs after the switching pulse, and again maybe 10 µs later (when the waveforms for different temperatures run in parallel), a characteristic difference can be seen. This difference will be minimal if the low pass filter is calibrated to the present temperature. In this case, only the current rise due to the load inductivity can be measured. For higher temperatures, the PDT1-behaviour will increase and the difference between first and second sample grows. This difference is the temperature sensitive parameter, which has to be transferred to a correction factor for the second sample (which is then only proportional to current). To improve the sensitivity of this effect, it is possible to set the low pass filter time constant consciously to another value, so that the PDT1-behaviour is always present and a temperature deviation shows a higher effect. This procedure is outlined in Fig. 5. Here, two examples for possible difference values are depicted.

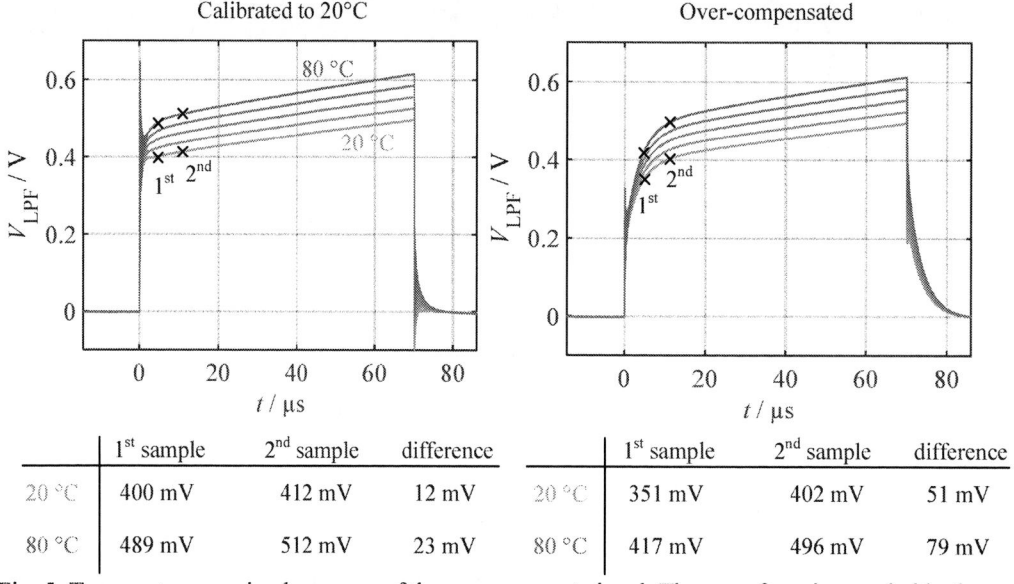

	1st sample	2nd sample	difference		1st sample	2nd sample	difference
20 °C	400 mV	412 mV	12 mV	20 °C	351 mV	402 mV	51 mV
80 °C	489 mV	512 mV	23 mV	80 °C	417 mV	496 mV	79 mV

Fig. 5: Temperature sensing by means of the measurement signal. The waveform is sampled in the PDT1-phase (1st sample) and in the P-phase (2nd sample). For the situation, when measurement time constant is fitted exactly to the low pass filter time constant at 20 °C (left plot), the temperature sensitive difference is only 23 mV at 80 °C. If the low pass filter is mistuned deliberately, this value can be increased, which makes this approach more sensitive to temperature.

These difference values can also be derived mathematically: Retransforming the equations in Fig. 1 to time domain again, a current step (which is nearly the real waveform of a current pulse) would give the following equation:

$$V_{\text{int}}(t) = I \cdot R_{\text{SS}'} \cdot \left(1 + \frac{\frac{L_{\text{SS}'}}{R_{\text{SS}'}} - C_{\text{int}} \cdot R_{\text{int}}}{C_{\text{int}} \cdot R_{\text{int}}} \cdot e^{-\frac{t}{C_{\text{int}} \cdot R_{\text{int}}}} \right) \tag{1}$$

If this signal is sampled at two different time instances, the difference of these values can be received. This difference can be determined exactly by:

$$\Delta V_{\text{int}}(t_1, t_2) = V_{\text{int}}(t_2) - V_{\text{int}}(t_1) = I \cdot R_{\text{SS}'} \cdot \frac{\frac{L_{\text{SS}'}}{R_{\text{SS}'}} - C_{\text{int}} \cdot R_{\text{int}}}{C_{\text{int}} \cdot R_{\text{int}}} \cdot \left(e^{-\frac{t_2}{C_{\text{int}} \cdot R_{\text{int}}}} - e^{-\frac{t_1}{C_{\text{int}} \cdot R_{\text{int}}}} \right) \tag{2}$$

Out of this equation different findings can be drawn. Firstly, the difference signal $\Delta V_{\text{int}}(t_1,t_2)$ is dependent on the actual current pulse. Secondly, it is dependent on the parasitic resistance proportionally and on the difference of the two time constants (measurement section and sensor) and third, it is dependent on the chosen sensor time constant itself. The main interesting figure out of these, however, is the temperature dependence ($R_{\text{SS}'}$). It has to be tested in the following, whether the influence of current and temperature can be separated. Nevertheless, such a temperature compensation method has different advantages:

- It measures the exact present temperature directly (and not indirectly like module NTC).
- It considers the effect of hot bond wires and not only its temperature at one special point, so that a significant uneven temperature distribution would not lead to false compensation.
- It can be applied without any additional sensor.

The challenge in this approach is, however, to get two sampled values shortly after the current pulse transient. Here, the sampling points have to be determined accurately to avoid too early sampling (ringing) and too late sampling (no temperature effect). This may lead to a special calibration of this method to every module where this approach is applied.

Accuracy and Applicability of Temperature Estimation by Means of the Sensor Signal

In a real inverter application, this temperature estimation approach has to be firstly accurate enough to get a proper temperature correction and secondly a minimal invasive intervention to keep this approach as simple and flexible as possible. At a second glance at the proposed temperature compensation, some additional challenges have to be solved to reach these goals. One point is the fact that the differential signal ("difference" in Fig. 5) is on the one hand dependent on the temperature, which is the intended influence, but on the other hand also grows with higher current values. This circumstance is depicted in Fig. 6.

In this plot, the problem of differentiating between higher current and higher temperature is shown. In reality, a higher measurement signal can have two causes: The first one is a higher current, which traverses the measuring section and of course leads to a higher measured signal when sampled after the sensor. The other cause, however, can be a higher parasitic resistance in the measuring section without the current having changed. Such an increase in parasitic resistance occurs naturally at a higher module temperature.

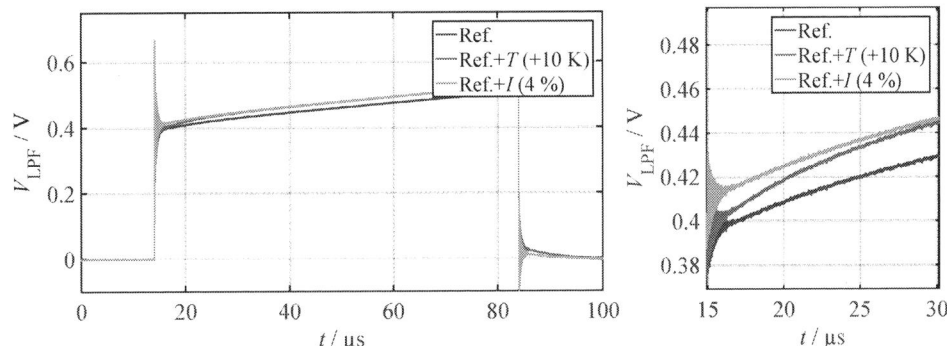

Fig. 6: Simulated sensor output for an exemplary current pulse. The three waveforms represent an (exemplary) reference pulse (dark blue), the same pulse at higher measurement section's temperature (+10 K, red) and a pulse at the same temperature as the reference but with 4 % more current (light blue). Right side: detailed section

Therefore, if only the proportional signal, long after the switching pulse is examined (e.g. in Fig. 6 at $t > 40$ µs), the distinction between higher current and higher resistance would not be possible. However, it can be seen in Fig. 6 that there is a different waveform for the curve representing higher current and the one representing higher temperature. Especially in the short period of time after switching, a difference between "higher temperature" and "higher current" can be seen.

With this effect, a distinction between higher current and higher temperature is possible. However, it requires a very accurate knowledge of the measurement section's time constant and in addition an adapted sampling procedure. This should contain a first sampling in the PDT1 dominated period shortly after the switching pulse (in Fig. 6: 15 µs $< t >$ 30 µs) and another one in the current-proportional period. In the present case, a realistic value of 6 µs was assumed for the measurement section. Therefore, after $5 \cdot \tau_{SS'} = 5 \cdot (L_{SS'}/R_{SS'}) = 30$ µs (rule of thumb), the time-dependent component of the PDT1 behaviour decays again and a distinction between higher current and higher temperature is no longer possible. In Fig. 6 the red line has completely changed to the light blue line after this time.

Another challenge is the practical implementation of this temperature estimation method. As shown in Fig. 6, the waveforms for higher temperature and higher current are obviously not very far apart. Further research on this aspect suggests that in worst case (i.e. at very low currents) the distinction must be made based on single millivolts at the sensor output.

Both, distinguishability and absolute difference between the temperature rise and the current change can, however, be influenced by the choice of the transfer behaviour of the sensor. For example, both values are higher for a second order lag element as sensor, whereby it must be objected that the entire measurement signal then no longer is a representation of the current, but a more or less detached output signal is obtained, which must again be specifically converted into the causal current.

Realisation of Temperature Estimation by Means of the Sensor Signal

In the following, measurements are carried out with a slight over- or under-compensation using a first-order-lag-sensor (PT1). The aim of the measurements is to show whether a temperature determination with this approach is possible at all, or if measurement inaccuracies and disturbances reduce the statements of the simulation. Two points per current pulse were sampled at the output of the sensor in different measurement series during a single-phase inverter operation. The first about 7 µs after the switching process and the second about 5 µs later.

The measuring section's time constant is approx. 6 µs ($R_{SS'}$ = 0.3 mΩ), which was already assumed above, and a significant overcompensation was chosen. This is advantageous, because the difference between the first and second sampling is relatively large (see Fig. 5, right), but has the disadvantage of a somewhat smaller distinguishability as with exact compensation. Just like in the previous simulation, a measurement at room temperature and a certain current value (the current amplitude was controllable in the test rig) was first taken as a measurement reference. Measurement series no. 2 then examined the sensing at a 4 % higher current, measurement series no. 3 at a 4 % higher $R_{SS'}$ due to a 10 K higher temperature. Out of the up-to-now simulation results, the current sensor should output a 4 % increased value for both, higher temperature and higher current, whereas the difference of the first two samples should yield in distinguishably values in those cases.

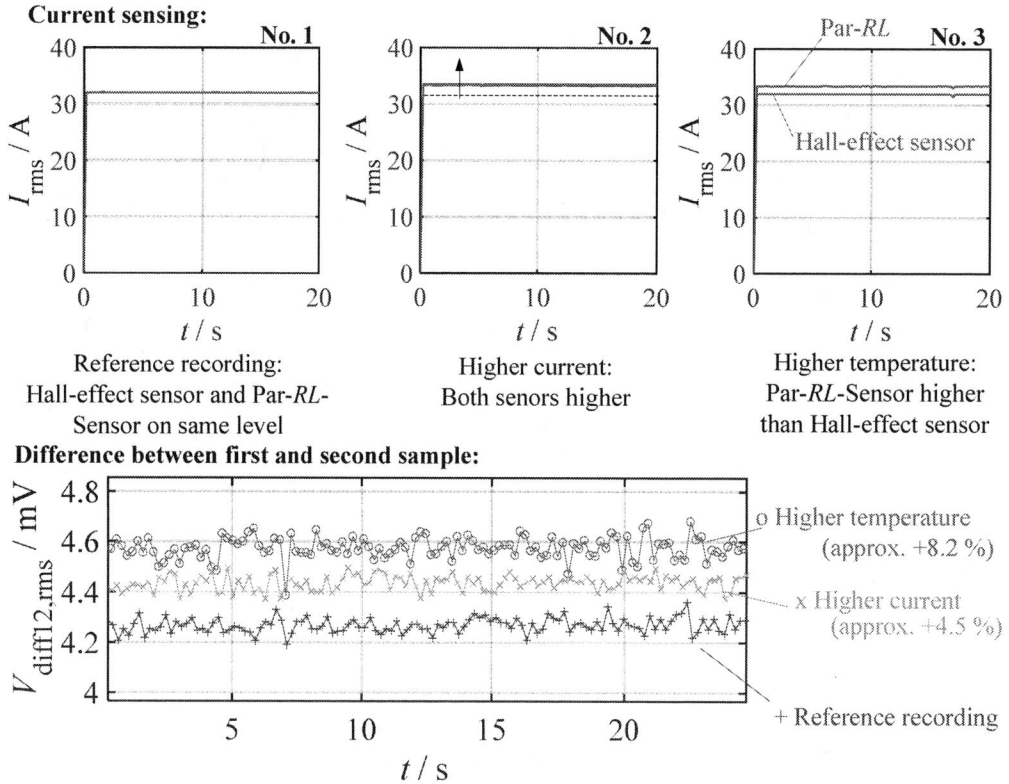

Fig. 7: Upper half: Three measurement series (reference, higher current and higher temperature) with Hall-effect sensor and Par-*RL*-Sensor in inverter operation (root-mean-square-values of the sensed sinusoidal signal). Lower half: Difference between first and second sample at a current pulse for the three measurement series (reference: marker "+", +10 K: marker "o", +4 % current: marker "x")

The investigation at higher temperature was carried out in thermal equilibrium, in order to guarantee that the whole module had a higher temperature. To get a good overview on the derived measurement values, the single situations were tested approx. half a minute in inverter operation with a set sinusoidal current waveform (f_{el} = 10 Hz and $\hat{I} \approx 45$ A). The results of the presented approach, the reference sensor and the difference signal were saved and processed offline. In the following, for all the signals their respective root-mean-square-value was calculated to make even small, proportional deviations visible. The signals processed in the aforementioned way are shown in Fig. 6 depicted for all measurement series.

It can be seen that there is an expected higher difference value $V_{diff,12,rms}$ between the reference recording and the situation with higher current. This increase is approx. 4.5 %, which is in accordance with the set current in this situation. The measurement series at higher temperatures does only once reach this $V_{diff,12,rms}$-level and stays above this value in all other recorded time periods (higher difference, cf. Fig. 6). It has to be admitted, that the absolute difference values yield only to a few millivolts, but as said in the beginning of this examination, it was outlined for a relatively small current (< 50 A for a 600 A module) and only a low PDT1-distortion of the sensor signal.

In the examined case, temperature correction can therefore be provided as follows: First, a calibration must be carried out for the evaluated module. The time constant of the active low-pass filter must already be determined. This calibration has to be performed at reference temperature (e.g. room temperature). For this purpose, all possible currents are tested in a test-bench and the sensor value (of the RL sensor) is noted. At the same time, all respective differential values must be assigned. Each current value at reference temperature is now assigned to a difference value $V_{diff,12,rms}$ that is valid for room temperature. In real operation, the difference value must be queried after each current measurement. If both match the calibration, the measurement was taken at reference temperature. If the values do not match, a different temperature is present and the measured current must be corrected.

Further measurements that try to get a higher grade of distinguishability were found by doing an under-compensation of the low pass filter (the opposite procedure that was done in Fig. 5). Here, the difference signal at higher temperatures stays closer to the reference values. In that case, the actual measurement accuracy makes it easier to decide, whether there is higher current or higher temperature. Therefore, a temperature estimation with the accuracy of ± 5 K is possible all the time. Nevertheless, at this point it has to be concluded out of Fig. 7 that the measurement of the two different values has to be more exact than in the actual setup. Only then, temperature estimation in the range of ± 5 K can be guaranteed, which will reduce temperature dependent errors below ± 2 %.

Conclusion

Current sensing by means of module parasitics is a very low cost alternative to today's current sensors. However, by saving additional sensors, available signals of the power modules have to be processed, which are not exclusively sensitive to current. In this case, the temperature of bond wires and conducting planes has significant impact on the measurement signal, which introduces great deviations between measured and present current. In this paper, different procedures for achieving the correct temperature value and compensating this effect were found.

Three different temperature estimation techniques were considered in this paper: The first one is to use an additional temperature sensor. This method was not examined further in this paper because it introduces another measurement circuit additional to the current sensing. Such a solution should only be used if others do not provide enough accuracy. Nevertheless, there might be interesting methods that are only sensible in combination with current sensing.
The second temperature estimation is the signal evaluation of an often-present module NTC thermistor. This method seemed obvious because the disadvantage of this temperature-dependent resistor, its limited accuracy for junction temperature measurement, is only of minor importance for the present application. Here, a measurement accuracy of ± 5 K is acceptable to gain a current accuracy of ± 2 %. However, it was found that such NTC thermistors might be situated far away from the interesting current

measurement section. Such a sensor provides the temperature information either for the wrong place or at least delayed.

The third method uses the low-pass-filtered signal itself as a temperature dependent signal. Two well considered sampling points of the sensor signal are sufficient to get the temperature information that is needed for compensation. In this case, the low-pass filter (sensor) has to be tuned properly to get a sufficiently high signal. It has also be paid attention to the fact that the temperature dependence is not overlaid by other effects. Moreover, a special calibration to the module has to be performed, which contains all possible currents and some model temperatures that might occur in real operation. With such a calibration, the temperature may be estimated up to now in an accuracy range of ± 5 K. It is intended to extend the approach and especially its accuracy so that temperature influence is no longer the limiting factor of the overall current sensing method.

References

[1] Vogel K., Gadermann M., Schmal A. and Urban C.: System Cost Reduction with Integration of Shunts in Power Modules in the Power Range above 75 kW, International Exhibition and Conference for Power Electronics, Intelligent Motion, Renewable Energy and Energy Management, PCIM Europe 2018, pp. 182-188, June 5-7, 2018.

[2] Pajer R., Rodič M. and Milanovič M.: Evaluation of MOS-FET RDSon Resistance for Current Measurement Purposes in Power Converter's Applications, 2015 International Conference on Electrical Drives and Power Electronics (EDPE), pp. 503-508, September 21-23, 2015.

[3] Denk M., Lautner F. and Bakran M.-M.: Accuracy Analysis of UCE(on)-based Measurement of the Inverter Output Current at Higher Motor Speeds, 2017 19th European Conference on Power Electronics and Applications (EPE'17 ECCE Europe), September 11-14, 2017.

[4] Lautner F. and Bakran M.-M.: Optimisation and Real Life Challenges of an Integrated Parasitics Based Current Measurement System, 21st European Conference on Power Electronics and Applications (EPE), Genoa, September 2-6, 2019.

[5] Avenas Y., Dupont L. and Khatir Z.: Temperature Measurement of Power Semiconductor Devices by Thermo-Sensitive Electrical Parameters—A Review, IEEE Transactions on Power Electronics, Vol. 27, no. 6, pp. 3081-3092, June 2012.

[6] Niu H.: The Effect of Load Properties on the Reliability of Machine Drives - The Temperature and Stress Analysis of Power Module Bond Wires, 2017 IEEE Energy Conversion Congress and Exposition (ECCE), October 1-5, 2017.

Development and Implementation of a Low-Cost Research Platform for Control Applications for Inverter-Based Generators

Jesus D. Vasquez Plaza, Juan F. Patarroyo-Montenegro, and Fabio Andrade
Dept. of Electrical and Computer Engineering
University of Puerto Rico at Mayagüez Campus
Mayagüez, Puerto Rico
E-Mail: jesus.vasquez@upr.edu, juan.patarroyo@upr.edu, and
fabio.andrade@upr.edu
URL: https://inec.uprm.edu/sec/

Keywords

«Research platform», «Microcontroller», «LC-Filter», «Resonant Controller»

Abstract

This document presents the develop and implementation of a low-cost research platform based on a microcontroller. The platform was validated through implementation of a resonant controller in $\alpha\beta$-frame for Inverter-Based Generators. Also, this work proposes a comparative analysis between a dSPACE 1006 and the embedded systems for control applications in inverter-based generators (IBGs). For the development of the research platform, a microcontroller from the Texas Instruments TMS320F28379D C2000 line was used. Our analysis revealed that control behavior for IBG applications through a low-cost platform based on a microcontroller is similar to that of a dSPACE.

Introduction

In recent times, the use of renewable energy sources is increasing dramatically worldwide. This has led to the questioning about sustainability of traditional electric power systems since they are mainly based on the generation of fossil fuels (oil, coal, natural gas). The environmental needs are increasingly demanding a change of the energetic paradigm due to the negative impact that fossil fuels cause. Thanks to the inclusion of renewable energy, the concept of microgrid is becoming popular. Thus, the use of Inverter-based Generators (IBG) is more common nowadays.

A microgrid can work in two operating modes: grid connected or in isolated mode. To achieve optimal operation, the microgrids need advanced control strategies to guarantee the voltage and frequency levels in any of the operating modes. A flexible platform that can operate in both operating modes without any significant changes in hardware is ideal for research and education issues

Currently, the control of microgrids is a source of continuous research, specifically in the control of IBG. That is why companies like OPAL RT, National Instruments, dSPACE Typhoon, RTSD, RT Box Plexim, etc., are increasing their offer with respect to Hardware-in-the-Loop (HIL) systems. These companies provide multiple modules to simulate real-time energy systems. These technologies allow research centers to improve their infrastructure and, in this way, they emulate and solve problems regarding to renewable energies. In [1], [2], [3] and [4] some works relating to the development, implementation and analysis of controllers of power electronic systems in which they use HIL technologies are presented.

Although these tools are very powerful, they are expensive as well. Therefore, they are difficult to access by low-income universities that belong to developing countries. This paper aims to present a low-cost microcontroller-based platform that allows for the implementation of advanced control strategies for IBG. Also, this project presents the development of a comparative analysis in terms of performance with respect to one of the HIL tools mentioned above, specifically a dSPACE 1006.

Microcontroller-Based Research Platform for IGB Control Applications

For the development of the research platform for IBGs, we developed a control box (CBOX) based on a microcontroller. This control box is composed of a Texas Instruments (TI) C2000 TMS320F28379D microcontroller (uC), custom made analog signal conditioning board and electrical signal to light pulses conversion board (optical fiber). Aditionally, a IBG Danfoss VLT-302 inverter and a SensorBox were used, the last two supplied by the Sustainable Enegy Center (SEC) of the University of Puerto Rico de Mayaguez (UPRM). The CBOX contains all the necessary ports to connect to the SensorBox and the IBG to perform the respective control. A set of experiments was developed to program and interact in real time with the uC through the blocks and tools provided by the Simulink-Matlab visual programming environment. These experiments validated the different peripherals of the uC (ADCs, Digital Input/Output, etc.) and implemented the desired control strategies for the control of IBGs. In addition, Matlab offers a work environment called GUIDE [5] allowed to design a custom user interface, a drag-and-drop environment to observe the values of the controller variables in real-time and to change the parameters of the controller on the fly, if necessary. A general description of the developed platform is shown in Fig 1.

The CBOX uses a C2000 TMS320F28379D microcontroller. It operates at 200MHz and is equipped with 24 analog-to-digital (ADC) conversion channels with a resolution of 12 bits, 16 independent 32-bit PWM channels, floating point unit (FPU), JTAG emulation tool and other features that make this microcontroller ideal for high capacity digital control purposes in IBG applications.

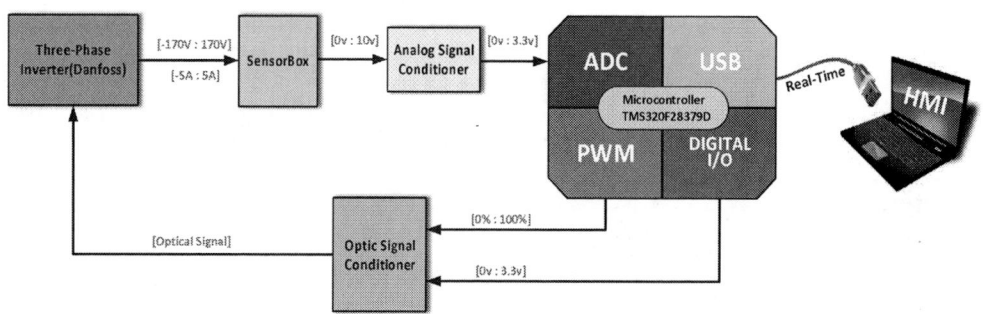

Fig 1. General system overview

System Assembly

The three boards contain in the CBOX are describe as follow: first, analog signal conditioning board is required to convert the voltage signal levels from the SensorBox into a range that the microcontroller can read (0-3.3V). The CBOX is packaged in an enclosure which was designed in SolidWorks and made by a 3D printer. It contains a port to connect it to the conventional grid (120Vac), a USB port to communicate and program the uC in Real-Time, 14 ports for analogue signal input from the SensorBox, 5 output ports and 1 input for optical signals to control the IBG. The CBOX was designed with 14 channels that allow analog signals to be read simultaneously, from the SensorBox. The SensorBox was originally designed to convert the measured signals into a range of voltage levels from -10V to + 10V. For this reason, a signal conditioner circuit is used to convert these voltage levels to a range of 0V to 3.3V, which is the operating range of the ADC unit of the uC. To achieve this objective, it was necessary to attenuate and add an offset value to the signals coming from the SensorBox in order to convert them into positive unipolar signals to read it properly with the uC. The OP482 amplifier from Analog Devices was used for this task because it has a high response speed (9V/µs), wide bandwidth (4 MHz), low noise and other features that make it ideal for this application. this amplifier has 4 independent channels with high response speeds.

To generate an attenuation of 0.15, resistors of RD=150Ω and RC=1kΩ were used Fig 2 shows the proposed design of the analog signal conditioning circuit. The output voltage is given by (1).

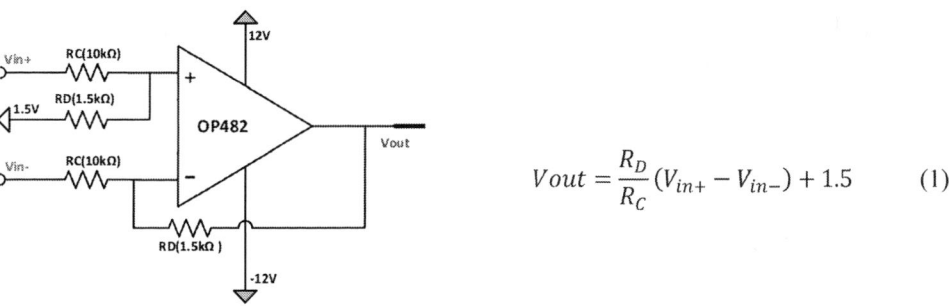

$$Vout = \frac{R_D}{R_C}(V_{in+} - V_{in-}) + 1.5 \qquad (1)$$

Fig 2. Analog signal conditioning circuit

Second, Optical Signal Conditioning Board is required to convert the PWM (electrical signals) into light signals, which will be responsible for carrying the control signals to the IBG, this board receives electrical signals and transforms them into light signals, has 6 optical signal ports, 5 outputs and 1 input. The 5 optical output signals are distributed as follows: 3 signals are from PWM responsible for applying the control signals to the IBG; a Reset signal to reset the IBG in case it is blocked a non-lethal fault; and an Enable signal for activating or deactivating the IBG. On the other hand, the input signal called TRIP informs the IBG had a fault and is blocked. This board is interconnected with the Analog Signal Conditioning Board through a ribbon cable.

Finally, BNC Signals Board is required to connect to 14 BNC channels of the SensorBox output ports. This board is connected to the Analog Signal Conditioning Board using a ribbon cable. In short, this card serves as an interface between the SensorBox and the Analog Signal Conditioning.

Fig 3 shows the final assembly of the CBOX. The CBOX is organized so that its main modules can be visually identified and used with little risk of damage.

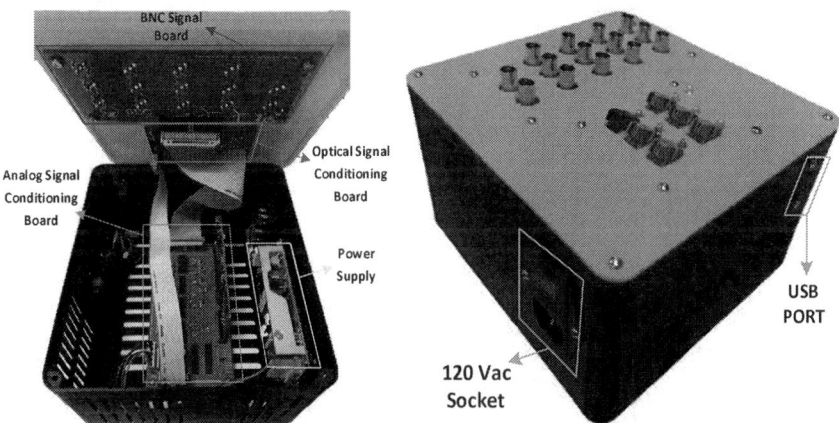

Fig 3. Final box assembly of the CBOX. **Left:** internal view. **Right:** external view

Graphic Interface

Grafic Interfaces allow easy control of software applications, which eliminates the need to learn a programming language and write commands in order to run an application. With the help of Matlab GUIDE, an interface was created that allows you to program the uC with files previously created from Simulink and at the same time observe and change the parameters of the controllers that were implemented. For each type of controller implemented in this platform, different interface were developed.

Fig 4 shows the interface developed for the αβ controller, which will be discussed later. As you can be seen, the interface consists of 4 sections that will be described below.

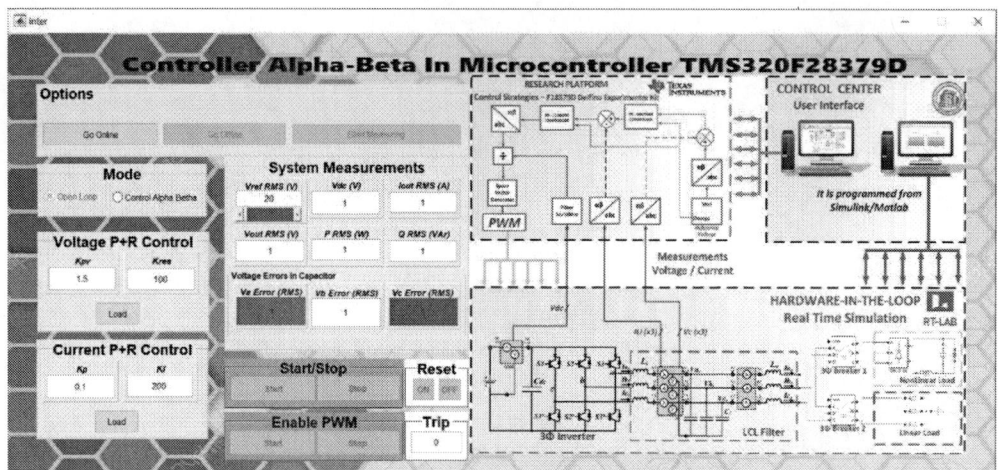

Fig 4.User interface for the microcontroller-based platform for IBG control applications

The ***Options*** section contains three buttons.

- The "*Go Online*" button allows you to program the uC with the control strategy and keep the system running in real-time.

- The "*Go Offline*" button allows you to reset the uC and finish the data transmission in Real-Time.
- The "*Start Measuring*" button allows you to display some variables of interest on the screen in real-time, such as the dc bus voltage (Vdc), RMS value of the IBG output current (Iout), RMS value of the voltage of IBG output (Vout), Ppower that is delivering the IBG (P RMS) and voltage errors with respect to the reference (Va, Vb, Vc RMS Error).

The ***Mode*** section allows you to switch between the type of control strategy or the open loop system without a controller.

The ***System Measurements*** section allows visualizing on display the variables of interest in real-time at the moment the "*Start Measurement*" button is pressed. It also contains a slider bar that allows you to vary the reference voltage for the IBG between 20 and 120 volts RMS as desired, it is also accompanied by a display (Vref) where you can enter numerically and observe the value of the voltage reference.

Finally, the section of ***Voltage and Current P+R*** Control allows to visualize and change the values of the volatage parameters and current controllers (*Kpv, Kres, Kp, Ki*) as desired, the selection of these values will be farther discussed.

Platform Validation

To validate the correct functioning of the platform, a three-phase inverter with an LCL filter was used. The platform was validated by controlling the voltage in the capacitors (Ci) and the current control in the input inductance (Li).

In order to control the inverter, it is necessary to measure the signals of interest, which is achieved with the SensorBox described above. Once the signals of interest are measured, they are sent to the uC, which is responsible for carrying out the respective control and generating the PWM signals that will switch the inverter's power transistors and obtain the desired output voltage.

In Fig 5, a general scheme of the inverter with the LCL filter and a load is observed. In addition, all the voltage and current measurements necessary to control the inverter are observed. On the other hand, the PWM block represents the optical signals from the optical signal conditioning board.

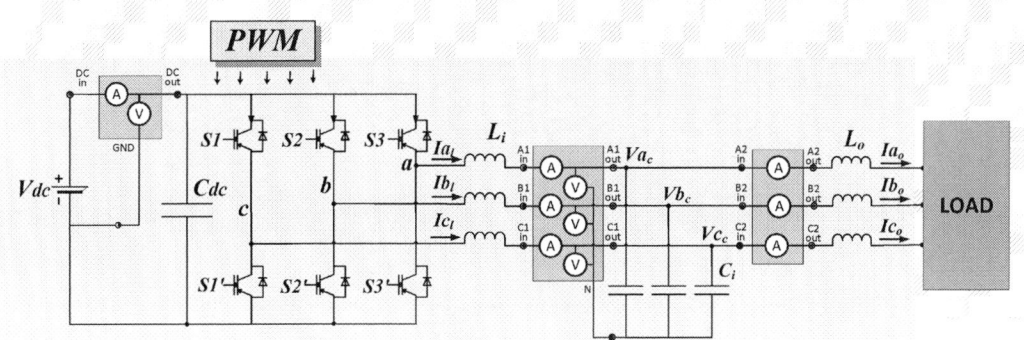

Fig 5. General scheme of the three-phase inverter with LCL filter

The model of the inverter is represented in equations of state. For simplicity of the model and control of the three phase inverters, the αβ transform is used in the reference frame aligned to the voltage in the capacitor of the LC filter. The LC filter model is given by the following equations:

$$\begin{bmatrix} \dot{V_c} \\ \dot{i_l} \end{bmatrix} = \begin{bmatrix} 0 & 1/C \\ -1/Li & -Ri/Li \end{bmatrix} * \begin{bmatrix} V_c \\ i_i \end{bmatrix} + \begin{bmatrix} 0 \\ 1/Li \end{bmatrix} * V_{ref} \tag{2}$$

To validate the platform the resonant control in αβ-frame was selected, work related to the implementation of these controllers is described in [6]–[9]. shows the general scheme of the system with PR control (αβ), in which there is an internal loop that controls the circulating current by Li and an external voltage loop that controls the voltage in the capacitor Vc.

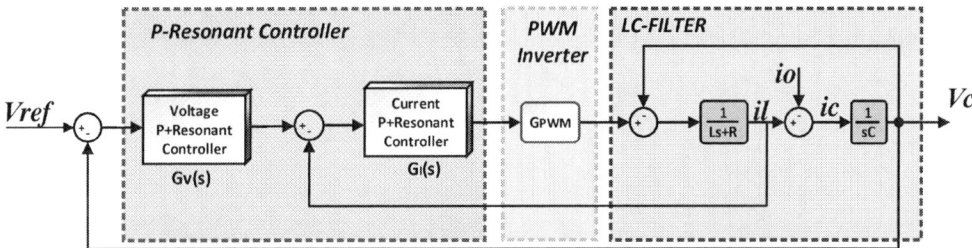

Fig 6. General schematic of the IBG with PR (αβ) current and voltage controller

where the GPWM transfer function corresponds to the delay due to the calculation device (Ts) and the PWM (0.5Ts), with Ts being the sample period. The GPWM transfer function is given by:

$$G_{PWM}(s) = \frac{1}{1 + 1.5T_s s} \tag{3}$$

The transfer functions GV (s) and GI (s) correspond to the PR controllers of voltage and current, respectively. The main characteristic of these controllers is that it has an infinite magnitude gain in its resonant frequency ω, and it presents an abrupt 180 phase change in the frequency ω. The equations G_V (s) and G_I (s) are given by (4) and (5), where K_{pv} - K_{pi} and K_{iv} - K_{ii} are the proportional and resonant constants for current and voltage controllers, respectively. ω0 is the desired resonant frequency, in this almost 377rad/s [10][11].

$$G_v(s) = K_{pv} + \frac{K_{iv}s}{s^2 + \omega_0{}^2} \qquad\qquad G_i(s) = K_{pi} + \frac{K_{ii}s}{s^2 + \omega_0{}^2} \tag{4}$$

For the controller mentioned above, the design and calculation of their parameters was carried out by means of the frequency response using the bode diagrams as in [10]. **Table 1** shows the values of the system parameters to be controlled.

Parameters	Symbol	Nominal Value
Switching frequency	fs	10kHz
Sampling Time	Ts	100μs
Filter inductance	Li - Lo	1.8mH
Filter capacitance	C	8.8μF
LC Filter resistance	R_i	2Ω
System frequency	ω_0	2*pi*60 rad/s

Table 1. System Parameters

The constants of the PR controllers obtained for the current and voltage controller were $K_{pi}=0.1$ - $K_{ii}=200$ and $K_{pv}=1.5$ - $K_{iv}=100$, respectively.

Experimental Results

Prior to experimental implementation, to carry out the comparative analysis of the platform developed in this investigation with respect to a HIL technology, it was necessary to validate the system with the simulation strategies described. For this reason, the MATLAB-Simulink tool was used, which allows executing complex system simulations with precision.

Subsequently, to evaluate the performance of the proposed research platform, one of the inverters of the SEC-UPRM microgrid test bench was used [12]. The experimental setup is represented in Fig 7 consisted of a 2.2kW Danfoss inverter, an LCL filter, a CBOX or dSPACE, and a SensorBox. The switching frequency of the inverter was adjusted to 10 kHz by symmetric spatial vector modulation. Finally, the voltage in the capacitors (Vc) was acquired using a Lecroy Wavesurfer 64Xs oscilloscope, which allows the measured signals to be converted into data that can be processed in Matlab. For validation, our platform was compared to the behavior of the dSPACE 1006. In the two systems, the same controller was implemented with the same parameters, the same inverter was controlled with the same LCL filter.

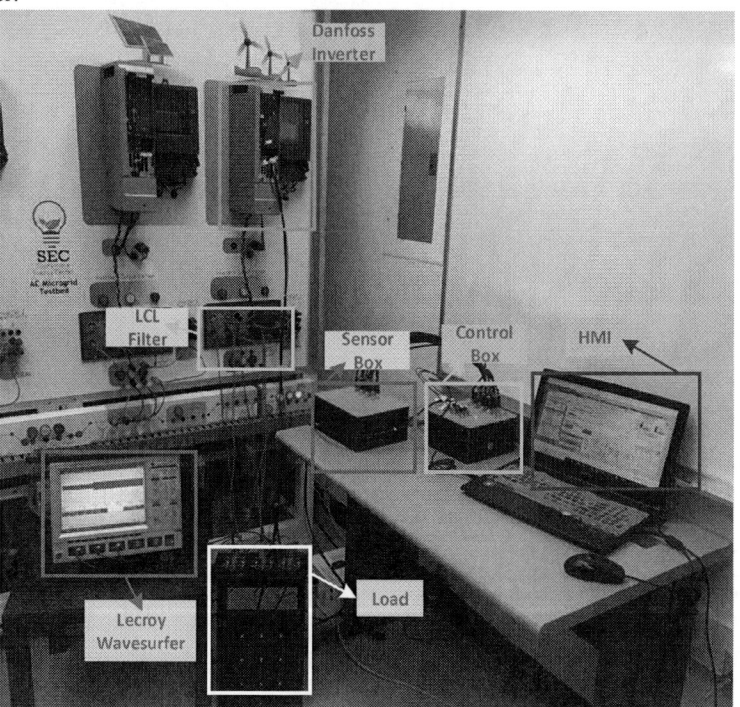

Fig 7. Experimental Low-Cost Research Platform

The experimental results of the PR (αβ-frame) control implemented was validated for both the CBOX and a the dSPACE. Fig 8 shows the voltage behavior in the capacitor for the controllerless system with our platform and the dSPACE.

The voltage reference was initially set at 60V RMS. After 1 second, a 60V step was used to reach the voltage of the conventional grid (120V RMS). It was observed that the transient in the simulated model had more harmonic components as in the open loop model because the simulation did not consider all damping factors. It is observed that the behavior of Vc controlled by our platform and by dSPACE had a similar dynamic. The simulated system had a settling time of 15ms, and a maximum exceeding of 7V with respect to the reference (less than 5%). The system controlled by our platform and the dSPACE behaved in a similar way and had a maximum overshoot in one of its 15V phases (less than 10%) and also presented a maximum error of 5V in steady state, that is, an error of less than 3%.The exact similarity values between our platform and the dSPACE when implementing the αβ controller will be presented in the next section.

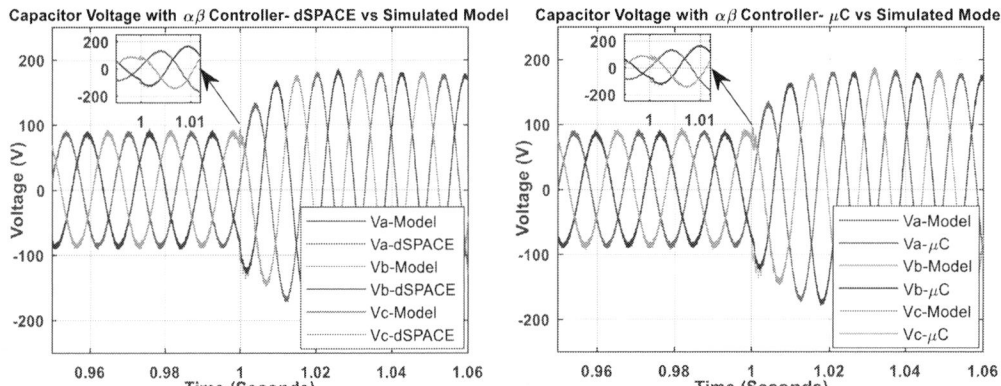

Fig 8. PR (αβ) Controller System - voltage capacitor. RIGTH: uC and Simulated Model. LEFT: dSPACE and Simulated Model

Comparative Analysis

For the PR (αβ-frame) controller implemented in our platform and in a dSPACE 1006 the voltage signals in the capacitor (Vc) is compared, which is superimposed and compared with respect to the Vc of the simulated model in a time interval determined.

For each platform with respect to the simulated model, the root mean normalized square error (NRMSE) is defined by [13].

$$NRMSE = 100 \times \left(1 - \frac{\|Vc_{ref} - Vc_{Pl}\|}{\|Vc_{ref} - mean(Vc_{ref})\|^2} \right).$$ (5)

where ‖.‖ Indicates Euclidian norm of a vector. The vectors Vc_{Pl} and Vc_{ref} represent the Vc for each platform and the Vc of the simulated model, respectively. The NRMSE is suitable for this validation because it considers the complete measurement during a specific time interval. This means that it is suitable for comparing responses with a high frequency component, such as signals from an IBG that operates at high frequencies. Fig 9 shows the NRMSE for the system with PR controller (αβ-frame) All adjustment values are above 96% in each of the phases of Vc with respect to the simulated model. In addition, the system controlled by the uC has an error rate of less than 1% with respect to the dSPACE when compared to the simulated model. These results demonstrate that our platform has the ability to keep the IBG stable and controlled through the implementation of the αβ controller, obtaining results similar to those of dSPACE.

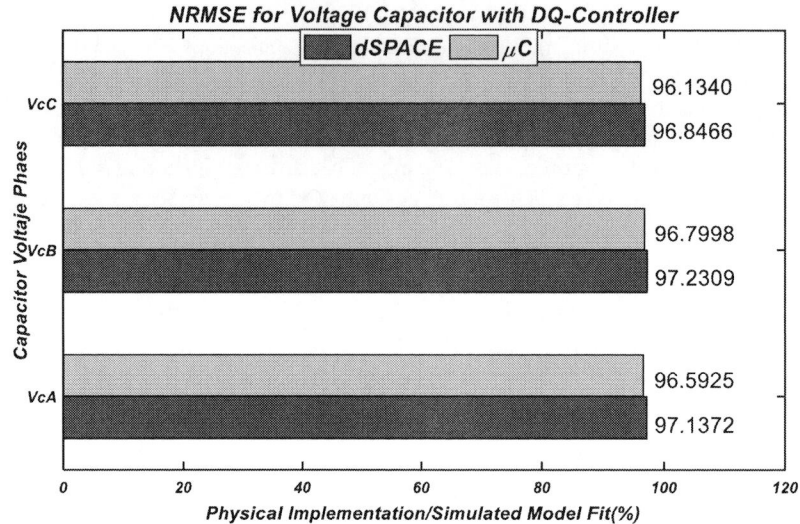

Fig 9. NRSME for capacitor voltage between simulated model and uC - dSPACE with PR (αβ-frame) controller

Through the results obtained when applying the NPSME to the Vc Voltage obtained with our uC-based platform, as well as when implementing PR (αβ-frame) controller in the dSPACE with a real IBG, we observed that the errors regarding the dSPACE in all the cases were less than 1% with respect to the simulated model. These results demonstrate that our platform has the capabilities to implement control strategies for IBG applications, obtaining results similar to that of the dSPACE.

On the other hand, when directly comparing our platform and the dSPACE, NRSME values of 97.25%, 96.79%, and 96.93 are obtained for the three phases of the voltage in the capacitor VcA, VcB, and VcC, respectively.

According to our criteria, it is considered that a value greater than 85% similarity between our platform and the dSPACE is sufficient to determine that the platform meets the requirements to implement control strategies for IBG applications. The results obtained demonstrate that our platform has the capacity to implement controllers for IBG applications, since through the implementation of the ab and dq controller there is a stable and contracted system with a percentage of similarity greater than the threshold (85%) established with respect to a dSPACE 1006.

Conclusion

By implementing a PR (αβ-frame) controller for IBG applications through a low-cost uC-based platform, a similar behavior is obtained as when implementing the controllers with high-performance technologies. The comparative analysis carried out demonstrates that the results obtained with our platform based on a uC and the dSPACE 1006 are similar. The platform proposed in this work allows for the research and implementation of advanced control strategies for IBG to be accessible to universities with low budgets that are in the process of development.

References

[1] M. Mobarrez, N. Ghanbari, and S. Bhattacharya, "Control Hardware-in-the-Loop Demonstration of a Building-Scale DC Microgrid Utilizing Distributed Control Algorithm," IEEE Power Energy Soc. Gen. Meet., vol. 2018-Augus, pp. 1–5, 2018.

[2] S. Acharya, M. S. El Moursi, and A. Al-Hinai, "Coordinated Frequency Control Strategy for an Islanded Microgrid With Demand Side Management Capability," IEEE Trans. Energy Convers., vol. 33, no. 2, pp. 639–651, 2018.

[3] A. Mortezaei, M. G. Simoes, A. S. Bubshait, T. D. C. Busarello, F. P. Marafao, and A. Al-Durra, "Multifunctional Control Strategy for Asymmetrical Cascaded H-Bridge Inverter in Microgrid Applications," IEEE Trans. Ind. Appl., vol. 53, no. 2, pp. 1538–1551, 2017.

[4] A. S. Bubshait, A. Mortezaei, M. G. Simoes, and T. D. C. Busarello, "Power Quality Enhancement for a

Grid Connected Wind Turbine Energy System," IEEE Trans. Ind. Appl., vol. 53, no. 3, pp. 2495–2505, 2017.

[5] Matlab, "Create a Simple App Using GUIDE - MATLAB & Simulink." [Online]. Available: https://www.mathworks.com/help/matlab/creating_guis/about-the-simple-guide-gui-example.html.

[6] A. G. Yepes, F. D. Freijedo, J. Doval-Gandoy, Ó. López, J. Malvar, and P. Fernandez-Comesaña, "Effects of discretization methods on the performance of resonant controllers," IEEE Trans. Power Electron., vol. 25, no. 7, pp. 1692–1712, 2010.

[7] F. D. Freijedo et al., "Grid-synchronization methods for power converters," IECON Proc. (Industrial Electron. Conf., no. December, pp. 522–529, 2009.

[8] L. Antonio De Souza Ribeiro, F. D. Freijedo, F. De Bosio, M. Soares Lima, J. M. Guerrero, and M. Pastorelli, "Full Discrete Modeling, Controller Design, and Sensitivity Analysis for High-Performance Grid-Forming Converters in Islanded Microgrids," IEEE Trans. Ind. Appl., vol. 54, no. 6, pp. 6267–6278, 2018.

[9] M. Bobrowska-Rafał, "Grid Synchronization and Control of Three-Level Three-Phase Grid-Connected Converters based on Symmetrical Components Extraction during Voltage Dips," PhD Thesis, p. 141, 2013.

[10] J. C. Vasquez, J. M. Guerrero, M. Savaghebi, J. Eloy-Garcia, and R. Teodorescu, "Modeling, analysis, and design of stationary-reference-frame droop-controlled parallel three-phase voltage source inverters," IEEE Trans. Ind. Electron., vol. 60, no. 4, pp. 1271–1280, 2013.

[11] R. Teodorescu, F. Blaabjerg, M. Liserre, and P. C. Loh, "Proportional-resonant controllers and filters for grid-connected voltage-source converters," vol. 150, no. 20030259, 2003.

[12] J. F. Patarroyo-Montenegro, J. E. Salazar-Duque, S. I. Alzate-Drada, J. D. Vasquez-Plaza, and F. Andrade, "An AC Microgrid Testbed for Power Electronics Courses in the University of Puerto Rico at Mayagüez," 2018 IEEE ANDESCON, ANDESCON 2018 - Conf. Proc., pp. 1–6, 2018.

[13] A. G. Barnston, "Correspondence among the correlation, RMSE, and Meidke Foresast verification measures; Refinement of the Neidke Score," Weather and Forecasting, vol. 7, no. 4. pp. 699–709, 1992.

Control of Parallel Connected Voltage Source Inverters in a Microgrid for Experimental Testing

Jesus D. Vasquez-Plaza[1], Jorge Tenorio[2], J. M. Ramírez-Scarpetta[2], *Member, IEEE*, Jose Alex Restrepo[2] and Fabio Andrade[1], *Member, IEEE*
[1]Dept. of Electrical Engineering, University of Puerto Rico,
Mayaguez, Puerto Rico
[2]Dept. of Electrical and Electronic Engineering, Universidad del Valle
Santiago de Cali, Colombia
jesus.vasquez@upr.edu, jorge.tenorio@correounivalle.edu.co,
jose.ramirez@correounivalle.edu.co, restrepo@usb.ve and
fabio.andrade@upr.edu

Keywords

«AC Microgrid», «Primary Control», «Secondary Control», «Load Sharing».

Abstract

This article describes the primary and secondary controls of voltage source inverters that are part of a custom-made AC microgrid for testing control strategies. The conceptual and physical structure of the microgrid is described. A cascade voltage-current control and a state feedback control are implemented for inverters with LCL output filters. Also, the conventional droop controls are implemented for power sharing. Simulation and experimentation results are presented of voltage, current and load sharing controls between two inverters of the microgrid, which validate these control strategies for teaching of microgrid control strategies.

Introduction

In recent decades, the increase in energy demand and the need for power generation with a low carbon footprint has led to the use of renewable energy sources such as solar, wind, biomass, and tidal power. To take advantage of these energy sources efficiently, safely, and reliably, various control strategies have been proposed making use of the flexibility of the electronic power converters used to manage energy. Initially these sources were integrated in hard-to-reach rural locations for energy suppliers and also in the industrial sector, where the power requirements are in the order of hundreds of kW to a few MW [1]. Nowadays the concept has been extended, not only to mitigate the environmental problem, but also for its commercial appeal. The flexibility of electronic power converters to deliver or absorb energy with fast dynamics, allows a quick control of power flows. In this context, electrical microgrids (μGs) are making the structure of power grids worldwide change their conception of generation centralized to distributed generation with an active energy trade where users at any given time can become generators electric power. Storage systems (such as flywheels, superconductors, lithium ion-based batteries and graphene in silicone gel, etc.) are fundamental elements in this conception [2][3]. Electric vehicles are proposed as important agents of the μRs storage systems [4]. Energy is not only stored in kinetics in large generators, but in a distributed way, which gives more flexibility and reliability to μGs, since many of the renewable energy sources such as wind and solar are not dispatchable. The μGs of urban, rural or industrial areas are not appropriate to research control strategies since they normally feed loads that do not support disconnection by testing.

To attend all these mentioned challenges, it is necessary to have small-scale laboratories of μGs that allow experimenting different control strategies for voltage and frequency regulation, and the power sharing between various intermittent generators. This document presents the voltage, current and power sharing controls between two generators that are part of the μG implemented at the laboratory level in the Department of Electrical and Electronic Engineering of the Universidad del Valle (*μGUV*), Fig 1 [5].

The structure of this document is as follows: Section II provides a brief description of the hierarchical control structure of the μGUV [5], Section III presents the results obtained with the controls at the simulation level and finally in Section IV presents the experimental results of the controls in the μGUV.

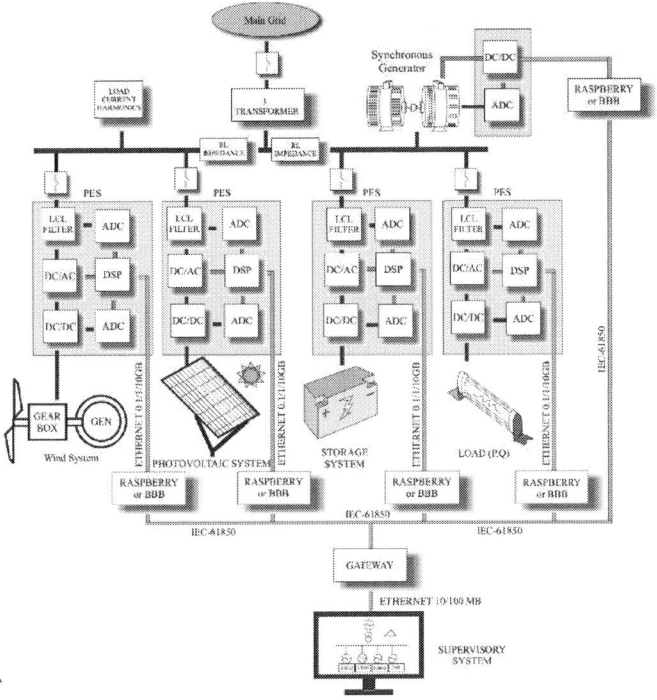

Fig 1. General scheme of the μG in the Universidad del Valle Colombia

Control Structure

Researchers have adopted the hierarchical control structure of traditional electric power systems for μGs [6], [7], [8]. These levels are presented in Fig 2. It must be ensured that each level of control has a low impact on the other, that means the control levels are decoupled. Each level must operate with different bandwidths, which decrease as the level rises.

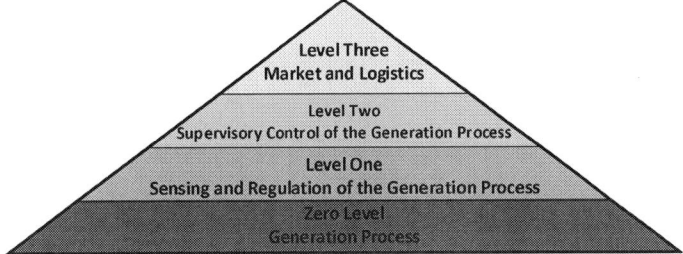

Fig 2. Hierarchical control levels according to the standard IEC/SAC 62264

The tertiary level is in charge of the management of electrical energy and control of the power flow between the μGs and the public grid, either injection or extraction of energy [9].
The secondary control restores the nominal values of frequency and voltage [10], this can also include synchronization with the electrical grid in order to connect or disconnect from it [11].
The primary control regulates the frequency and voltage, as well as the load sharing between generators maintaining the stability of the μG [12], [13].
The zero-level control is responsible for the inner voltage and current controls of the inverters [14], [15].

Zero Level Control

Fig 3 presents the general structure of the zero-level control. It has a cascade control structure with a voltage controller of the capacitor voltage in the output LC filter output. Also, a current controller of the inductor current of this output filter. Both controllers are implemented in the rotating coordinate system dq. A feedback of the filter states with Ki and Kv gains is added to the control effort of the internal current loop. Finally, the trigger signals to the IGBTs are generated, operating the power converter as a Voltage Source Inverter (VSI).

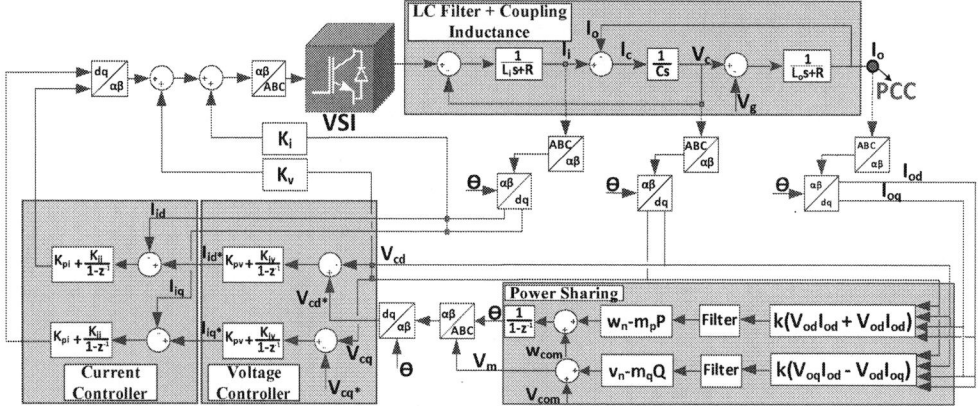

Fig 3. The whole control structure scheme adopted by this investigation

Discrete state feedback control brings the poles of the LC filter to an appropriate stability region. The state feedback gains are calculated using the LQR (Linear – Quadratic Regulator) optimal control technique. The LQR gains were calculated using the Matlab command *lqrd*, which has R and Q as input parameters and returns the gains for the states feedback. To determine optimal gains, a fixed value of Q = 1 was established and a sweep was made of R giving it different values. Fig 4 shows the poles of the system for each gain obtained by varying R. The state feedback gains are adjusted with the LQR control to obtain conjugate complex poles with real part of 0.1 and damping coefficient of 0.65.

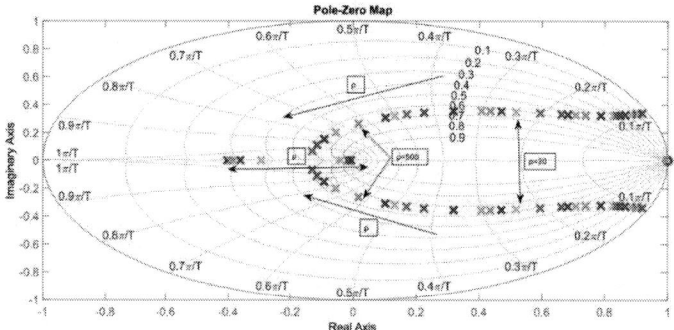

Fig 4. Optimal location of the poles of the LC filter by state feedback

Output capacitor voltage control $V_{Cd,q_{VSI}}$ is performed through a cascade control strategy of voltage and current using standard Proportional Integral (PI) type controllers in the dq rotational frame, which guarantees a zero steady state error. The voltage controller generates the current control reference for the inductor current controller. The capacitor voltage, $V_{Cd,q}$, and the current of the inductor, $I_{Cd,q}$ are measuring by a LEM sensor board. The controller functions for $V_{Cd,q}$ and $I_{Cd,q}$ are given by (1) and (2).

$$I_{d,q_{ref}}(z) = \left(K_{pV} + \frac{K_{iV}}{1 - z^{-1}} \right) \left(V_{d,q_{ref}} - V_{Cd,q} \right) \tag{1}$$

$$V_{d,q_{VSI}}(z) = \left(K_{pI} + \frac{K_{iI}}{1 - z^{-1}} \right) \left(I_{d,q_{ref}} - I_{Cd,q} \right) \tag{2}$$

The direct component of the voltage reference $V_{d,q_{ref}}$ comes from a conventional load sharing control strategy; this strategy is a classic droop control, which a frequency droop is performed with the filtered active power P and the voltage droop with the filtered reactive frequency Q.

Secondary Control

Droop controls allow power sharing between inverters, but load variations or connection and disconnection of other inverters change the system's frequency and voltage levels. The secondary control is implemented to eliminate the frequency and voltage deviations in steady state. VSI measure the signals of interest and send them to a single central controller, which produces control signals that adjust the outputs of the *Wcom* and *Vcom in the* droop controls. This centralized control is implemented and requires a communication network. The controller is an integrator control with a constant time of 0.4 seconds, in (3) the corresponding equation is observed:

$$C_S(z) = \left(\frac{K_{pc} + K_{ic} + K_{pc} * z^{-1}}{1 - z^{-1}} \right) \tag{3}$$

Simulated Results

Two inverters connected with coupling inductors and connected to a load are simulated, see Fig 5 and Table I with the system parameters. The model of the system corresponds to the upper right box in Fig 3 (LC Filter + Coupling Inductance). Both droop controls were adjusted to mp = 0.001255 and mq = 0.02; to get an equal load sharing; the system is affected with a load increase to observe the behavior associated with this disturbance. The common connection point (PCC) is the point where the two VSIs are connected. SW1 and SW2 are controlled by digital signals from one of the two VSIs. SW1 allows to connect or disconnect the inverters in parallel and SW2 allows to connect or disconnect a load.

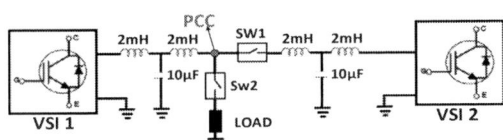

Fig 5. Simulation configuration scheme

Table I. Two inverter system parameters

System Parameters			
Three-Phase Load		**Inverters (VSI)**	
Voltage	311 V	P_{MAX}	1000 W
Frequency	60 Hz	Q_{MAX}	500 VAR
P	1000 W	V_{MAX}	311V ± 5 %
Q	0 VAR	F_{MAX}	60Hz±0.17 %

Voltage and Current Control

The performance of the external voltage controller is observed in Fig 6, the dotted waveforms are the references for the controller and the continuous line waveforms are the controlled voltages in the three-phase capacitor array. A settling time of 5ms was achieved and stationary voltage error is less than 10V (less than 4%). Additionally, the performance of the inner current controller is shown in Fig 7. The three-phase controlled currents that circulate through the inductor of the LC filter are observed, the current error with respect to the reference is less than 0.2A and a settling time was approximately 2.5ms. The inner current has a high frequency content that in this case are negligible.

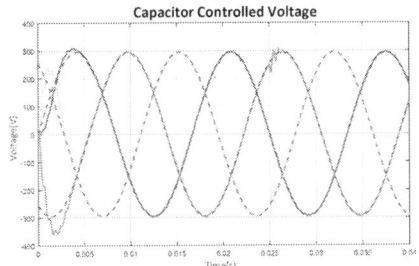

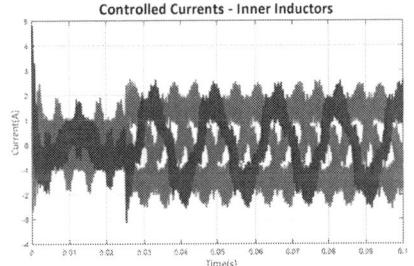

Fig 6. External voltage control result in simulation

Fig 7. Inner current control result in simulation

Power Sharing Control

To validate in simulation the performance of the droop control, an increase of the load is made every 0.1s to vary the active and reactive power delivered by the inverter. An increase in resistive load is made until reaching the Pmax value that each inverter can provide Fig 8 (a). Also, Fig 8 (b) shows how the frequency falls as the inverter delivers more active power P.

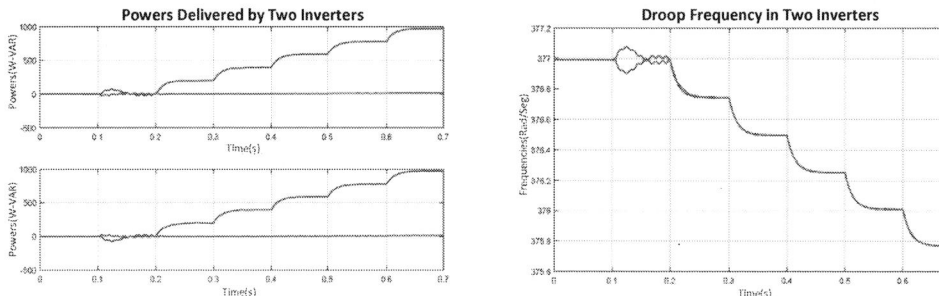

Fig 8. (a) The upper and lower image show the variation of the Active Power (blue) and Reactive Power (Red) delivered by inverters 1 and 2 respectively by varying the connected load - (b) Variation of Frequencies in inverters 1 (Blue) and 2 (Red) obtained with the Droop Control by increasing the power delivered by each Inverter

Secondary Control

Fig 9 (a) shows the effect of the droop control on the frequencies of the two inverters connected in parallel, where the blue signal is the frequency of the inverter that imposes the voltage at the PCC and the red signal is the frequency of the inverter that is synchronized to the PCC voltage. By making increments in the load, the frequency returns to the reference value due to the correct operation of the secondary control. Moreover, Fig 9 (b) shows the effect of the droop control in the magnitudes of the reference voltages of the two inverters in parallel, which the blue signal is the voltage magnitude of the inverter that imposes the voltage on the PCC and the red signal is the voltage magnitude of the inverter that is synchronized to the PCC voltage. After making several increases in the resistive load, reactive power circulates through the coupling inductances and inductances of the LC filters and with this the magnitude of the voltage has relatively small changes, with the help of the secondary voltage control it returns to its reference value.

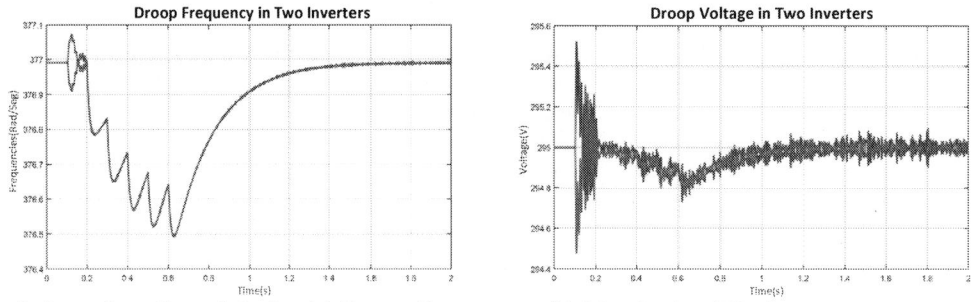

Fig 9. Secondary Control Acting (a) System Frequency - (b) Magnitude of Vref

Experimental Results

Experimental tests were performed on the μG prototype presented in [5]. Fig 10 presents a view of the circuit elements present in the converters of the *μGUV*. This platform has the following elements: Each inverter has an ADSP 21369 digital processing unit from Analog Devices, an interface board designed to fit directly on the bottom of the DSP. This board has an FPGA that performs the data transfer between the processing unit (DSP) and the power stage. This card receives the reference signals from the DSP and generates the PWM pulses for the activation of the IGBTs, third, it has two sensor boards composed of an ADC with a 16 bits resolution, and can sample 200kS/seg in eight channels. It has a trigger card in charge of sending the switching signal to each IGBT. These signals pass through a buffer and the output of this goes directly to the ACPL-332J driver, which is configured as recommended in the data sheet. This driver has advantages that allow switching of the IGBTs, makes the protection against failures of the collector-emitter voltage of the IGBT compact and performs a "soft" shutdown of the IGBT. Each sensor card can measure the signal of eight variables that can be voltage or current.

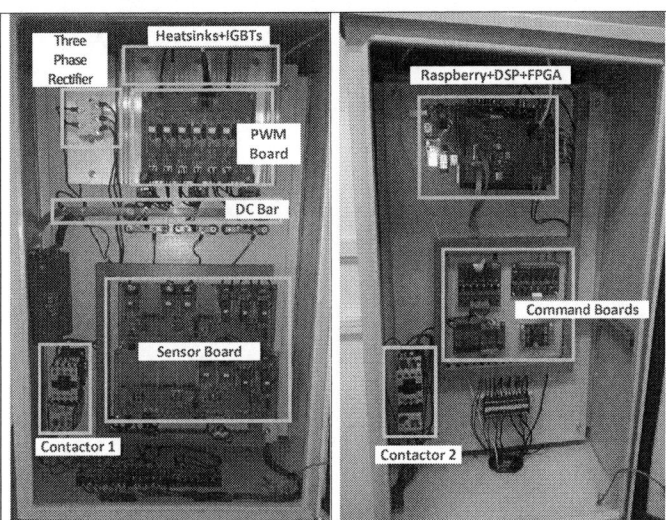

Fig 10. Experimental Assembly of the Power and Control Stages

Voltage and Current Controllers

First, the zero level controls, the inner current and the external voltage are checked. A 60V peak voltage reference is imposed for the test and a 6Ω load is connected. Fig 11 shows the current (two-phases) controlled in the inductor of the LC filter. It presents a settling time of 1.5ms without current peaks that could damage the system, which indicates a good performance of the inner current controller. Fig 12 shows the two-phase voltage signals of the capacitor of the LC filter (blue signals) and the current delivered to the load of one phase (green signal), these signals were captured in the oscilloscope Tektronix TDS3034C exactly during the time the inverter was in operation.

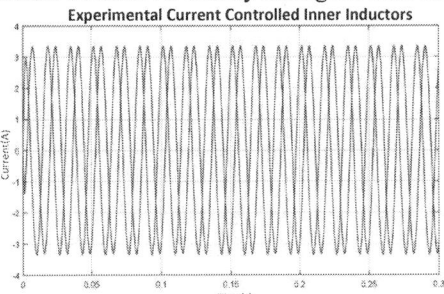

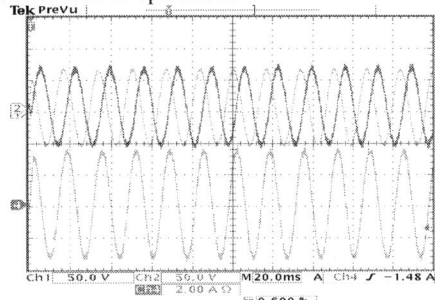

Fig 11. Experimental Current in Two Phases of the Filter Inductor

Fig 12. Experimental Voltages and Current Taken by Oscilloscope

Fig 13 shows the power sharing between the two inverters connected in parallel for a certain period of time (4s). The voltage on the DC bus of the second inverter is shown on channel 1, on the left. The DC bus falls at the moment that this inverter comes into operation since it is responsible for delivering the power to the load. Once the inverter stops injecting current into the load, it recovers its value. The channel 2 signal is the voltage controlled in a phase of the capacitors of the LC filter of the second inverter, which is not affected by the parallel operation of the inverters.

The signal of channel 3 indicates the time during which the second inverter starts to operate in parallel next to the first inverter. The load sharing between inverters is shown in the figure on the left, the current delivered by the first inverter (channel 4) starts with a value close to 2A, delivering only the current to the load. When the second inverter is connected in parallel, the current delivered by the first inverter decreases by half during the time that the second inverter is connected. The delivered current value that decreases in the first inverter while connected in parallel, is the current that the second inverter brings to the load and is shown in the graph on the right (blue and red). Once the second inverter stops injecting current into the load, the current of the first inverter (channel 4) increases again to its initial value, indicating that it provides all the power required by the load.

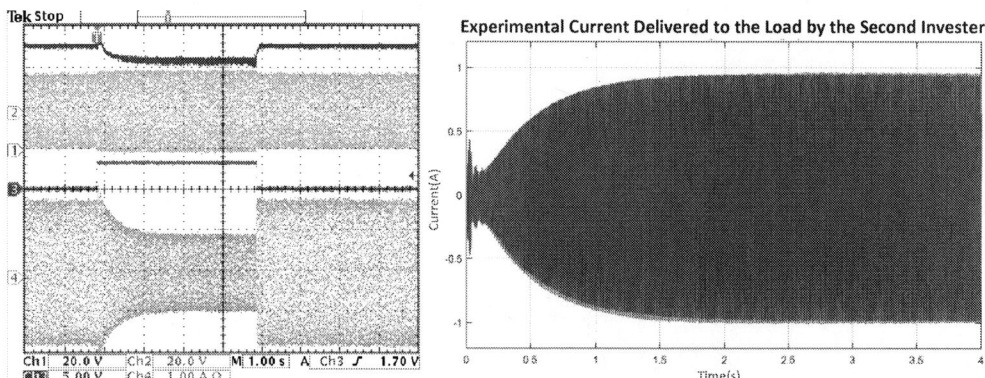

Fig 13. Experimental load sharing between two inverters connected in parallel

Conclusions

This work describes the design, simulation, construction and validation of the operation of an interconnected system of two inverters in parallel, the traditional droop control for load sharing was implemented. A symmetrical power distribution was obtained between the two inverters with a stabilization time of 40ms. In addition, the system has a cascade control structure for the capacitor voltage and the inductor current in dq coordinates. The performance of the current control loop has stabilization time of 0.5ms. Moreover, the performance of the voltage control loop has a stabilization time of 1.5ms. This concludes that the theory used in this application has been successfully validated through simulation tools and with a system built and validated experimentally. The platform used to control investors provides ample flexibility in the implementation of control laws because the DSP has a high processing capacity and a programming language that enables its use.

References

[1] J. Gutierrez-Vera, "Renewables for sustainable village power supply," in 2000 IEEE Power Engineering Society, Conference Proceedings, 2000, vol. 1, pp. 628–633.

[2] T. Kousksou, P. Bruel, A. Jamil, T. El Rhafiki, and Y. Zeraouli, "Energy storage: Applications and challenges," Sol. Energy Mater. Sol. Cells, vol. 120, pp. 59–80, Jan. 2014.

[3] X. Tan, Q. Li, and H. Wang, "Advances and trends of energy storage technology in Microgrid," Int. J. Electr. Power Energy Syst., vol. 44, no. 1, pp. 179–191, 2013.

[4] E. L. Karfopoulos, K. A. Panourgias, and N. D. Hatziargyriou, "Distributed Coordination of Electric Vehicles providing V2G Regulation Services," IEEE Trans. Power Syst., vol. 31, no. 4, pp. 2834–2846, Jul. 2016.

[5] J. A. Restrepo-Zambrano, J. M. Ramírez-Scarpetta, M. L. Orozco-Gutiérrez, and J. A. Tenorio-Melo, "Experimental framework for laboratory scale microgrids," Rev. Fac. Ing., vol. 2016, no. 81, pp. 9–23, 2016.

[6] A. Bidram and A. Davoudi, "Hierarchical structure of microgrids control system," IEEE Trans. Smart Grid, vol. 3, no. 4, pp. 1963–1976, 2012.

[7] J. M. Guerrero, J. C. Vasquez, J. Matas, L. G. De Vicuña, and M. Castilla, "Hierarchical control of droop-controlled AC and DC microgrids - A general approach toward standardization," IEEE Trans. Ind. Electron., vol. 58, no. 1, pp. 158–172, Jan. 2011.

[8] E. Planas, A. Gil-De-Muro, J. Andreu, I. Kortabarria, and I. Martínez De Alegría, "General aspects, hierarchical controls and droop methods in microgrids: A review," Renewable and Sustainable Energy Reviews, vol. 17. pp. 147–159, Jan-2013.

[9] L. Meng, J. M. Guerrero, J. C. Vasquez, F. Tang, and M. Savaghebi, "Tertiary control for optimal unbalance compensation in islanded microgrids," in 2014 IEEE 11th International Multi-Conference on Systems, Signals and Devices, SSD 2014, 2014.

[10] C. Ahumada, R. Cárdenas, D. Sáez, and J. M. Guerrero, "Secondary Control Strategies for Frequency Restoration in Islanded Microgrids With Consideration of Communication Delays," IEEE Trans. Smart Grid, vol. 7, no. 3, pp. 1430–1441, May 2016.

[11] I. Serban and C. Marinescu, "Battery energy storage system for frequency support in microgrids and with enhanced control features for uninterruptible supply of local loads," Int. J. Electr. Power Energy Syst., vol. 54, pp. 432–441, 2014.

[12] Y. Tan, K. M. Muttaqi, and L. G. Meegahapola, "A droop control based load sharing approach for management of renewable and non-renewable energy resources in a remote power system," in 2013 Australasian Universities Power Engineering Conference, AUPEC 2013, 2013.

[13] J. M. Guerrero, L. Hang, and J. Uceda, "Control of distributed uninterruptible power supply systems," IEEE Trans. Ind. Electron., vol. 55, no. 8, pp. 2845–2859, 2008.

[14] A. G. Yepes, F. D. Freijedo, Ó. Lopez, and J. Doval-Gandoy, "High-Performance Digital Resonant Controllers Implemented With Two Integrators," IEEE Trans. Power Electron., vol. 26, no. 2, pp. 563–576, Feb. 2011.

[15] D. Yang, X. Ruan, and H. Wu, "Impedance shaping of the grid-connected inverter with LCL filter to improve its adaptability to the weak grid condition," IEEE Trans. Power Electron., vol. 29, no. 11, pp. 5795–5805, 2014.

Optimization strategy for the sizing of passive magnetic components

Guillaume Devos[1,2], Maya Hage-Hassan[1], Philippe Dessante[1], Cyrille Gautier[2], Adrien Mercier[1], Éric Labouré[1]

[1]GEEPS – GROUP OF ELECTRICAL ENGINEERING, PARIS
11, rue Joliot Curie
Gif-sur-Yvette, France
Tél: 01.69.85.16.33
Fax: 01.69.41.83.18

[2]SAFRAN TECH
1, Rue Jeunes Bois
Châteaufort, France
Tél: 01 61 31 80 00
E-Mail: guilaume.devos@centralesupelec.fr

Keywords

Magnetic device, Airplane, Converter circuit, neural network, Device modeling, Energy converters for HEV, Passive component integration.

Abstract

This paper presents an optimization methodology for the design of magnetic components embedded in aeronautical structures. We want to realize a multi-objective optimization of an inductance, taking into account some physical phenomenon that are not easily described with mathematical formulations, and to set up a methodology that allows us to harvest optimization time. We will use, for that purpose, analytical formulations, Finite Element Analysis (FEA), and a neural network to reduce the time consumption of components evaluation.

Notations

A_e	m^2	core cross section
α	-	Steinmetz parameter for frequency
B_{max}	T	maximum induction level
β_F	-	Steinmetz parameter for induction
$\Delta\theta$	K	Temperature elevation
η	-	efficiency
f	Hz	frequency
h	W.m^{-2}.K^{-1}	Heating coefficient
J	A.m^{-2}	RMS current density
k_w	-	Winding coefficient
k_i	-	Mean current divided by RMS current
K_{ripple}	-	Current ripple divided by maximum of the current
K_P	-	Corrective Steinmetz parameter
K_{P_i/P_c}	-	Iron losses divided by copper losses
K_S	-	Exchange surface divided by $(A_e S_w)^{\frac{1}{2}}$
k_v	-	$1 -$ duty cycle
K_{V_c}	-	Volume of copper divided by $(A_e S_w)^{\frac{3}{4}}$

K_{V_i}	-	Volume of iron divided by $(A_e S_w)^{\frac{3}{4}}$
P_c	W	Copper losses
P_i	W	Iron losses
P_{max}	W	Power flowing through the component
R_{ac}/R_{dc}	-	AC resistance divided by DC resistance
ρ	$\Omega.m$	Resistivity at given temperature
S_w	m^2	Winding path area
N	-	Winding number of turns

Introduction

The integration of more electrical power in the aerospace systems will help redefine the development of future aircrafts. Mass optimization of power converters based on Wide Band Gap semiconductors is essential for facing the challenges in the aerospace field brought about by the concepts of More Electrical Aircraft (MEA) and More Electrical Propulsion (MEP). In Fig. 1, a view to the exponentially increasing MEA and MEP development is presented [1].

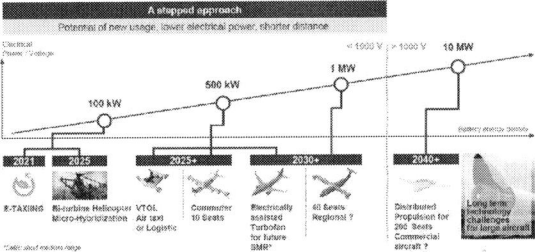

Fig. 1: a roadmap to hybrid electric energy & propulsion

In the field of More Electrical Propulsion, aircraft network architecture could include high-power propulsion electrical distribution network and storage equipment. In a configuration that includes a high voltage DC bus and a Li-ion battery, a non-isolated DC/DC converter can be inserted between the battery and the DC bus. The aim of this converter is to limit the voltage variations of the DC bus and to control the power flow of the battery. For a Li-ion battery of 540V, the voltage can vary from 400V to 630V depending on the state of charge. In the case of a DC bus voltage between 750V and 1000V, we choose to work on a multiphase synchronous boost converter, based on SiC transistors (fig. 2). In such a converter, weight optimization depends on the number of phases, the switching frequency and the magnetic components. To have an optimized sizing of the magnetic component according to the materials and other parameters is therefore most important. In addition, this optimization process must be fast enough to allow the evaluation of many converter configurations (frequency, number of phases, current ripple and magnetic material) and allow the computation of an overall optimum.

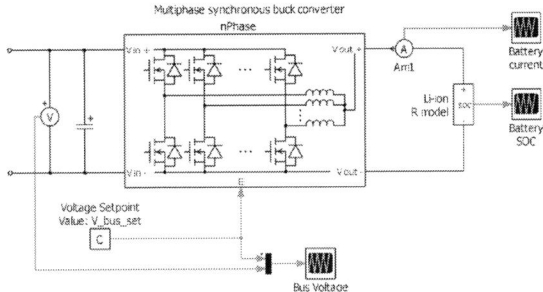

Fig. 2: Multiphase synchronous buck converter

This paper will present an optimization strategy. We used the design of a smoothing inductance located at the output of a DC/DC converter, carrying 1 kW at 25 kHz, as illustration of the global methodology. This strategy will take place in four stages:

1. A pre-sizing process with simplified analytical formulas (*coarse model*), gives a set of initial geometric and size solutions to the required specifications. We do not take into account AC losses in windings during this pre-sizing process.
2. Copper losses are refined on a few selected geometries using finite element software (FEMM 4.2 [2]) to calculate a R_{ac}/R_{dc} ratio taking into account eddy currents and proximity effects in the windings. This model based on finite element analysis is here after defined as fine model,
3. A surrogate model for copper losses is proposed, it is based on the correction of R_{ac}/R_{dc} ratio by means of a neural network model,
4. This substitute corrected model is then used in a stochastic multi-objective algorithm based on a "differential evolution algorithm" (DEA) [3].

1. Objective of the work

The general objective of this work is to optimize the previously described magnetic component. We want to minimize two variables:

- the weight
- the losses ratio (amount of losses divided by P_{max}, equal to 1 -η)

Geometries are defined by 6 parameters:

- 4 geometric parameters (length and width of the winding path, width and depth of the magnetic central leg),
- the number of turns,
- the winding coefficient.

The component is constrained by the saturation level of induction in the magnetic core.

The corresponding problem can be formulated as follow:

$$\min_{x} F(x) = \{Weight(x), 1 - \eta(x)\} \qquad x = \{\ell_1, \ell_2, \ell_3, \ell_4, N, k_w\}$$

$$\text{Domain } \Omega : \begin{cases} 0 < \ell_1 \leq L_{max} \\ 1 \leq N \leq N_{max} \\ 0 < k_{w\,min} < k_w < k_{w\,max} < 1 \end{cases}$$

Subject to $g(x) \leq 0 : g(x) = B(x) - B_{max}$

2. Analytical pre-sizing process

The developments of our strategy are confined to the following design specifications:
- The designed magnetic component is a filter inductor designed for a Buck converter as represented in Fig. 3.a., carrying 1 kW at 25 kHz.
- The inductor voltage is +/- 135 V with an average current I_o of 7.4 A, a current ripple rate of 10% and a duty cycle of ½ (Fig. 3.b.).
- The magnetic core is a gapped ferrite core made with a 3C90 material. Its geometry is defined in Fig. 3.c.

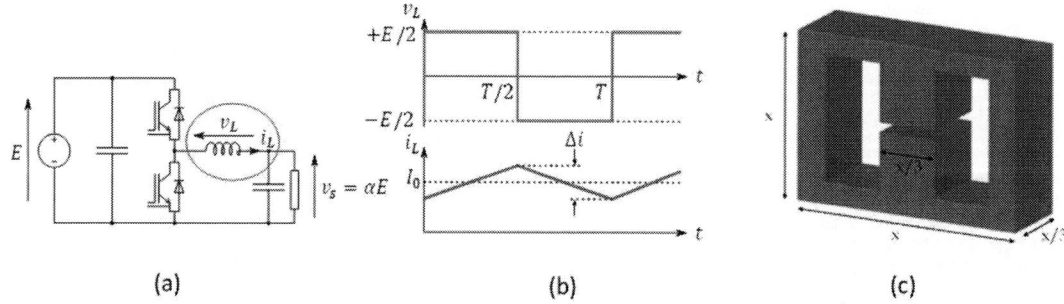

(a)   (b)   (c)

Fig. 3: a. schematic of the DC/DC converter. b. Current/voltage waveforms. c. Magnetic core shape.

Once the strategy is validated, it will be extended to others configurations. In the pre-sizing process, the geometry of all the components is assumed to be homothetic of a generic shape defined by the x parameter of Fig. 3.c.

The pre-sizing process is a mix between a thermally limited method [4] and a method using the area product "A_eS_w" [5]. This method is based on the following formulation:

$$P_{max} \propto A_eS_w \times f \times J \times B_{max} \tag{1}$$

Many calculation models for core losses exist [6]. In this pre-sizing process, we have chosen a formula based on the Steinmetz equation, adapted for ferrites under this kind of electrical waveforms:

$$P_i = K_P f^\alpha \left(K_{ripple} \times B_{max} \right)^{\beta_F} K_{V_i} (A_eS_w)^{\frac{3}{4}} \tag{2}$$

Electrical losses in windings are expressed by a formula that neglects AC component due to skin and proximity effects [7]:

$$P_c = \rho J^2 K_{V_c} (A_eS_w)^{\frac{3}{4}} \tag{3}$$

Considering a simple thermal model and introducing a ratio between iron and copper losses lead to the following formula for A_eS_w area product:

$$(A_eS_w)^{\frac{7\beta_F-2}{8\beta_F}} = \frac{k_v P_{max}}{f\, k_i k_w} \frac{\left(1 + K_{P_i/P_c}\right)^{\frac{1}{\beta_F}+\frac{1}{2}}}{K_{P_i/P_c}^{\frac{1}{\beta_F}}} \times \frac{\left(K_P f^\alpha K_{V_i}\right)^{\frac{1}{\beta_F}} \left(\rho\, K_{V_c}\right)^{\frac{1}{2}}}{(h\, K_S\, \Delta\theta)^{\frac{1}{\beta_F}+\frac{1}{2}}} \tag{4}$$

The calculation of the derivative with respect to K_{P_i/P_c} of this expression allows us to find that the volume of the component, when K_{P_i/P_c} increases, is decreasing then increasing. The minimum volume is reached for K_{P_i/P_c} equal to $\frac{2}{\beta_F}$.

However, this value of K_{P_i/P_c} can lead to a maximum induction level higher than the saturation level of the magnetic material. In this case, during the execution of the proposed algorithm, the value of K_{P_i/P_c} is decreased by steps of 10% while the saturation level is exceeded. A set of pre-sized components is computed using the implementation of this procedure (fig 4.a.) with different values for the temperature elevation ($\Delta\theta$). The circles on Fig. 4.b represents the Pareto front of pre-sized components according to the previous inductor specifications and for temperature elevation in the range of 1 to 80 °C.

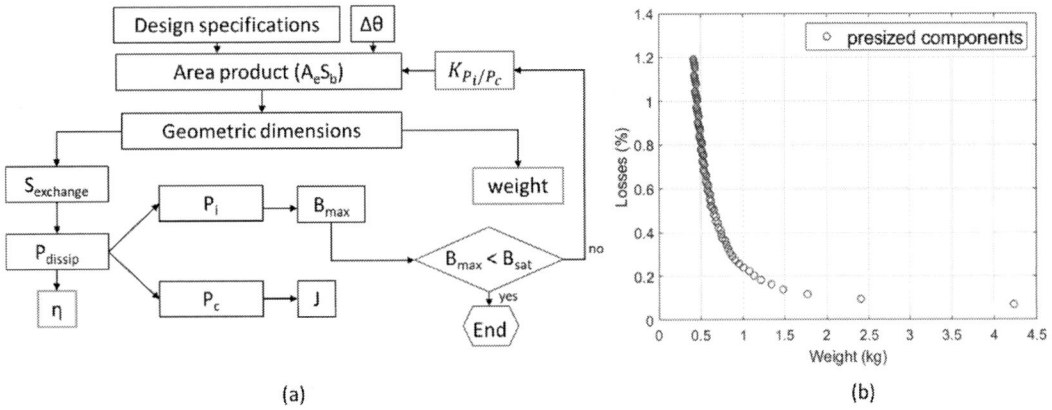

(a)　　　　　　　　　　　　　　　(b)

Fig. 4. (a) Flow chart of the pre-sizing process.　(b) Pareto front of computed results ($\Delta\theta$ between 1 and 80 °C).

The results of this analytical pre-sizing process is fed to a multiobjective DEA to find the Pareto based on this model. This analytical pre-sizing process gives a set of solutions based on the homothetic shape hypothesis presented in Fig.3. Computing a DEA with analytical formulations allows to quickly obtain a set of solutions for different geometric shape ratios.

3. Finite Element Analysis

The Analytical pre-sizing process doesn't consider eddy currents and proximity effects, a 2D Finite Element Method Analysis (FEA) is used to remedy this issue with the free software FEMM 4.2. Modeling a magnetic component with detailed windings is time consuming, to overcome this issue, a homogenized model of the windings [8] has been used. The implementation of the homogenized model for windings drastically reduces the time required to calculate losses for a given component, from several minutes to a few seconds on a desktop computing station. In the context of our work, we use a laptop with a four-core CPU Intel core i5 7440 HQ (2.8 GHz). Considering a component with 64 turns (presented on Fig.5.), the calculation of R_{ac}/R_{dc} with detailed windings takes 6300 seconds whereas it takes only 16.6 seconds with the homogenized model.

The detailed and homogenized models have been compared on a frequency range between 100 Hz and 1 MHz in the fig 5.a. Calculations of R_{ac}/R_{dc} with homogenized model is close to the values obtained with detailed windings. The relative error between the two models does not exceed 3 %. This is quite acceptable while the calculation time is, according to previously given values, divided by approximately 400.

At this time, the effect of the gap on the center leg on R_{ac} losses is not taken into account. This will be done in future work, however, this does not fundamentally change the method.

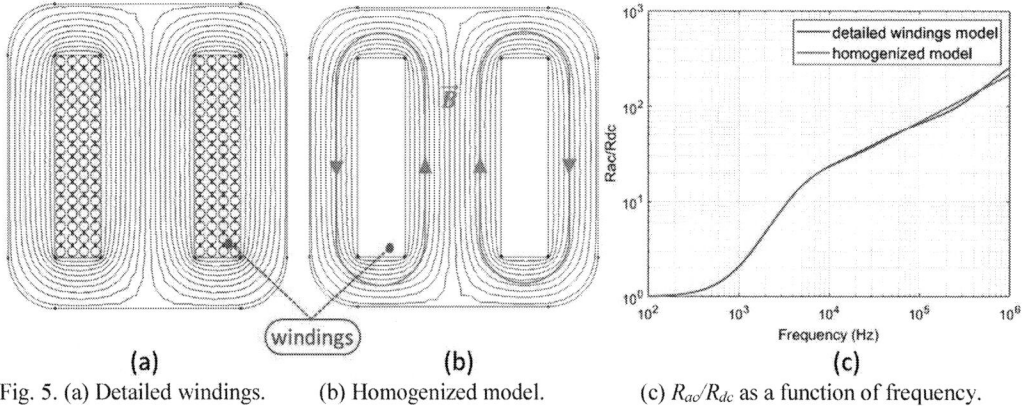

Fig. 5. (a) Detailed windings.   (b) Homogenized model.   (c) R_{ac}/R_{dc} as a function of frequency.

4. Corrected copper losses model using a neural network

Although the implementation of the homogenized model for windings drastically reduces the time required to calculate losses for a given component (between several minutes and a few seconds), the gain in computation time does not allow for the systematic use of finite element analysis in an optimizing routine. For example, considering 15 seconds for one calculation in a procedure of 100 iterations with a set of 500 different geometries leads to approximatively 208 hours exclusively dedicated to these analyses. As these results should be used in a more complex optimization process for the whole converter, it is therefore necessary to reduce the number of FEA evaluations, and then find a way to assign a R_{ac}/R_{dc} value for each geometry of the design space.

A methodology based on the Space Mapping Technique [9] is proposed and is described in Fig 6.a. The geometries were chosen with regular spacing on the Pareto front resulting from the optimization of the analytical pre-sizing model (fig 4b).

With this set of 124 geometries and their calculation via FEA, we construct an approximation of these data based on a neural network (using *newrb* function in Matlab software). Of this 124 geometries, 20 are used to train the neural network, and 104 are used for its validation.

Values of R_{ac}/R_{dc} has been calculated with FEMM for all these 124 points. In Fig 6.b we represent the value of R_{ac}/R_{dc} obtained by FEM calculation, and the approximation of R_{ac}/R_{dc} ratio that the neural network assigns to all the 124 points.

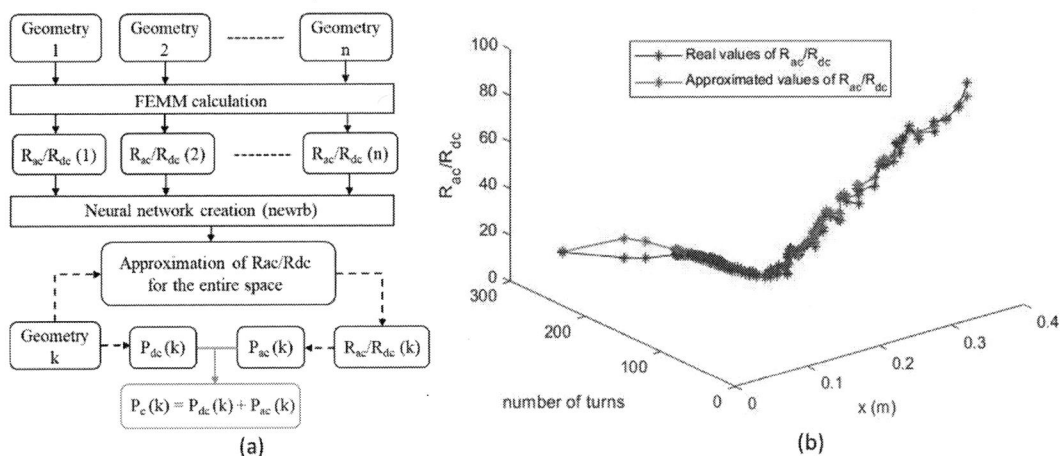

(a)   (b)

Fig. 6. (a) Space mapping: Flow chart of the creation and utilization of the neural network. (b) Construction of a neural network. In blue: FEM computed values of R_{ac}/R_{dc}. In red: approximated values of R_{ac}/R_{dc}.

This neural network gives a quite good approximation of R_{ac}/R_{dc}. We will use it in our optimization algorithm, and then verify its efficiency in this process.

5. Global optimization strategy

The adopted global strategy employed here is described by the flow chart in Fig.7.

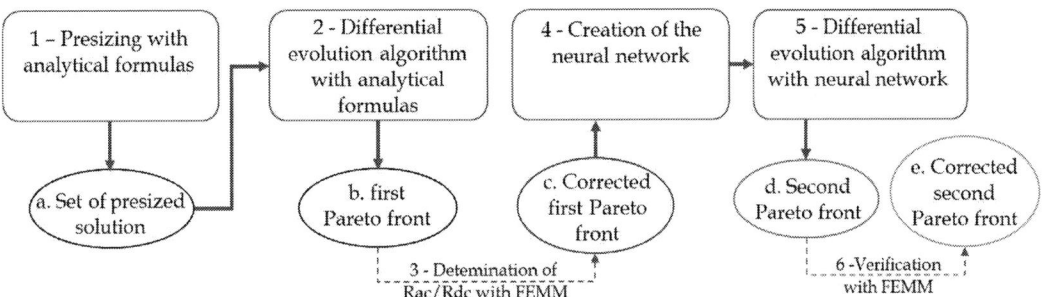

Fig. 7. Flow chart of the complete optimization strategy.

A first set of optimal geometries (Fig7.a) is computed using the analytical pre-sizing process (step 1 of fig.7, see also fig 4.). This set is fed to a first DEA (step 2) to compute a first optimal Pareto front (b). To take into account the AC effect in windings, we compute the FEA calculation of R_{ac}/R_{dc} of these geometries (step 3), and use the results (c) to train a neural network (step 4). This neural network approximation is used in a second DEA (step 5) to compute the second Pareto front (d). The final result (e) is then determined by a complete FEA of the results (step 6).

We represent the corrected first Pareto front (c, blue line), the one obtained after the second computation of DEA (d, red line) and its verification through FEMM (e, green line) on Fig.8.

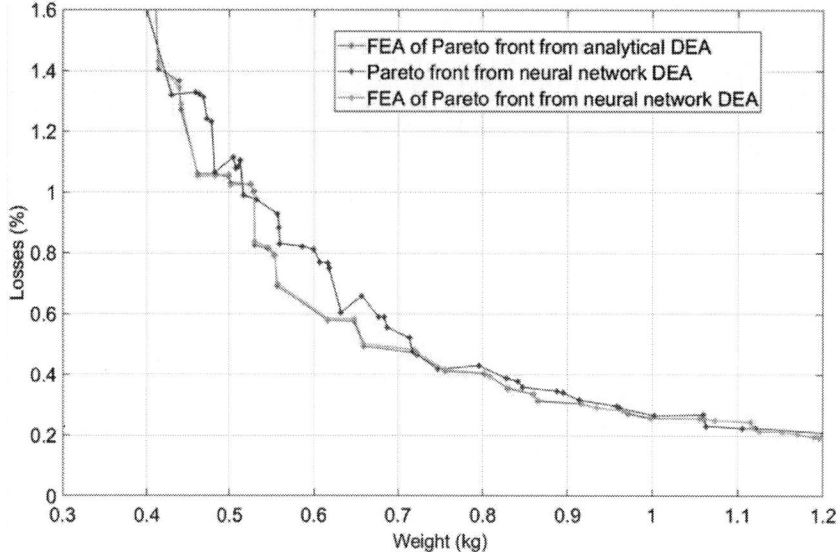

Fig. 8. Pareto optimal solutions obtained with analytical formulas and with neural network assignation

In order to verify the accuracy of the neural network and the need to consider AC effects during the optimization process, we represent three sets of solutions.

Indeed, the comparison between the Pareto front obtained after the second DEA (red line), and its correction with FEA analysis (green line) shows little differences between these two sets. It allows us to admit the good accuracy of the neural network in the attribution of R_{ac}/R_{dc} for these geometries.

The comparison between the Pareto front obtained after the first DEA, corrected by FEA (blue line) and the FEA correction of the second Pareto front (green line), shows that the geometries obtained after the second DEA present lower level of losses for a same weight, than those obtained without consideration of AC effects during DEA. We can conclude that it is important to take into account AC effects during the DEA in order to reach better optimized solutions.

Conclusion

The proposed optimization methodology provides an efficient way to design components in which physical phenomena cannot be estimated correctly with analytical models. The proposed solution uses few Finite Element calculations and a neural network to extend these calculations to the entire design space. In further works, we will apply this methodology to other specifications and at different frequencies. We will also try to adapt this methodology in order to consider the presence of an air gap on the electrical losses of the component.

References

[1] V. Garnier, Perspectives and activities on hybrid / electric propulsion, MEA 2019, Toulouse

[2] Crozier, R, Mueller, M., "A New MATLAB and Octave Interface to a Popular Magnetics Finite Element Code", Proceedings of the 22nd International Conference on Electric Machines (ICEM 2016), September 2016

[3] Storn, Rainer & Price, Kenneth. (1997). Differential Evolution - A Simple and Efficient Heuristic for Global Optimization over Continuous Spaces. Journal of Global Optimization. 11. 341-359. 10.1023/A:1008202821328.

[4] A. Van den Bossche et V. Cekov Valchev, « Inductors and Transformers for Power Electronics ». CRC Press, 2005.

[5] F. Forest, E. Laboure, T. Meynard and M. Arab, "Analytic Design Method Based on Homothetic Shape of Magnetic Cores for High-Frequency Transformers," in *IEEE Transactions on Power Electronics*, vol. 22, no. 5, pp. 2070-2080, Sept. 2007.

[6] Krings, Andreas & Soulard, Juliette. (2010). Overview and Comparison of Iron Loss Models for Electrical Machines. Journal of Electrical Engineering. 10. 162-169.

[7] C. R. Sullivan, "Computationally efficient winding loss calculation with multiple windings, arbitrary waveforms, and two-dimensional or three-dimensional field geometry," in *IEEE Transactions on Power Electronics*, vol. 16, no. 1, pp. 142-150, Jan. 2001.

[8] D. C. Meeker, "An improved continuum skin and proximity effect model for hexagonally packed wires", Journal of Computational and Applied Mathematics, Volume 236, Issue 18, 2012, Pages 4635-4644.

[9] Hage Hassan, M., Remy, G., Krebs, G. and Marchand, C. (2014), "Radial output space mapping for electromechanical systems design", COMPEL - The international journal for computation and mathematics in electrical and electronic engineering, Vol. 33 No. 3, pp. 965-975.

[10] MATLAB and deep learning Toolbox 2018b, The MathWorks, Inc., Natick, Massachusetts, United States.

Exploiting a Multi-Port Transformer for Minimal DC-Link Capacitance for an Automotive Onboard Charger

Franz Vollmaier, Alexander Connaughton, Thomas Langbauer, Klaus Krischan*
Silicon Austria Labs GmbH
Inffeldgasse 33
Graz, Austria
Phone: +43 / 664 852 95 00
Email: franz.vollmaier@silicon-austria.com
URL: https://www.silicon-austria.com

Acknowledgments

Acknowledgement goes to our partners: AT&S, AVL, Fronius, Infineon and TDK, and the Electric Drives and Machines Institute at Graz University of Technology*

Keywords

≪DC-Link capacitance≫, ≪Multiport LLC≫, ≪3 phase onboard charger≫

Abstract

In contrast to common 3-phase topologies, DC-link capacitance can be reduced by combining input power after the DC-link using a multi-port transformer and suitable control. Initial results of this approach demonstrate stable 12 kW output power using only 5 µF DC-link capacitance per phase, promising higher power density and reliability. Furthermore, a novel control scheme is proposed that allows high frequency reactive power flow through the multi-port transformer to be manipulated via gate-signal phase angles, in order to achieve maximum efficiency and minimum DC-link voltage ripple for any load condition.

Introduction

Automotive onboard chargers, as well as 3-phase power supplies, use an active power factor correction (PFC) stage to improve input current ripple, power factor, power density and filtering requirements [1] [2]. Nevertheless, significant high voltage DC-link capacitance on the output of such PFC stages may be needed for stable operation of a subsequent DC-DC stage [3] [4]. Automotive qualification may even require bulky film capacitors [5]. For example, Table I lists some commercial and academic prototypes and their corresponding DC-link capacitors. A large DC-link capacitance risks high inrush current at start-up and large reverse current during input voltage drops [6]. To avoid such effects, additional protection hardware prior to the PFC is often needed, limiting efficiency and further increasing volume, weight and cost. To sidestep these issues and to enable significant reduction of the DC-link capacitance, this paper proposes the multi-port topology shown in Fig. 1.

Proposed Multi-Port Topology

In contrast to the converter shown in [10] and the Vienna Rectifier, where three phase power combines at the DC-link [12], this topology enables very low DC-link capacitance (C_{DC}) by combining the power after the DC-link(s) using a single multi-port transformer. This structure allows the control to reduce the active-power taken from a DC-link capacitor when its respective input phase power decreases.

EPE'20 ECCE Europe

Assigned jointly to the European Power Electronics and Drives Association & the Institute of Electrical and Electronics Engineers (IEEE)

Table I: Examples of DC-link capacitors for PFCs of commercial & academic demonstrators including the proposed Multi-Port onboard charger topology

Source	Rated Input Power	PFC Topology	DC-link Capacitor	Quantity & Name	Total Volume & Capacitance
[7]	6.6 kW 1 Phase	Totem Pole PFC	400 V, 250 μF Electrolytic	18 (9 parallel, 2 series) EKXJ401ELL251MM50S	1130μF 229 cm^3
[8]	5.0 kW 3 Phase	Buck PFC	420 V, 47 μF Electrolytic	8 (8 parallel) EKXJ421ELL470MK35S	376 μF 34 cm^3
[9]	2.4 kW 3 Phase	Vienna Rectifier	450 V, 220 μF Electrolytic	4 (2 parallel, 2 series) TDK B43504C5227M	220 μF 115 cm^3
[10] [11]	9 kW 3 Phase	Isolated Full-Bridge LLC	450 V, 10 μF Film	2 (2 parallel) *volume estimated	20 μF *69 cm^3
	12 kW 3 Phase	Proposed Multi-Port	500 V, 1 μF Ceramic	3x 5 (5 parallel) B58031I5105M062	15 μF 3.4 cm^3

Furthermore, when the phase input power is close to zero, this DC-link capacitor will merely provide reactive-power to the subsequent LLC based DC-DC stages resonant tank and transformer. This means that the DC-link capacitors need to store much less energy than a conventional unified DC-link capacitor, as sufficient power is always available from the remaining two phases (2). This leads to double line frequency DC-Link power pulsing of each single phase connected to the transformer and exploits the constant power flow of three phase supply (2). The proposed topology has the advantages of:

- fewer parts than the topology in [10] resulting in 28 vs. 36 active switching devices or even less considering a non-interleaved PFC,
- inherent isolation & bidirectionality,
- flexibility for a real or virtual neutral point, or a delta connection for 120 V$_{rms}$ input voltage,
- a 400 V DC-link with one level of 650 V switches for 230 V$_{rms}$ grid voltage, unlike the Vienna Rectifier [13] or T-type PFC [19] it permits smaller creepage and clearance distance and higher achievable power density,
- optional output power reduction capabilities in case of either an input voltage drop or asymmetric phase voltages, ensuring resistive load to the grid and prevention of power pulsing to the battery,
- inherent voltage balancing between the DC-link capacitors of the PFC

Control and Limit to DC-link Capacitance Reduction

The DC-DC stage is a phase shift series resonant converter as presented in [14], switching slightly above the resonant frequency (f_{res}) of C_R and L_R with the output voltage controlled with the switching frequency, phase angle, PWM duty cycle, or burst mode. However, to ensure the constant overall power flow with minimal DC-capacitance and a minimum amount of transfered reactive power, a simple yet novel 'power pulsing' control scheme is required. Here, each input-side half-bridge changes the phase φ_m difference of the voltage vectors V'_m between its PWM and the output side PWM controlling V_B.

$$V_{ACm}(t) = \hat{V}_{ACm} \cdot \sin\left(\omega_{grid}t + \frac{2\pi}{3}(m-2)\right) \qquad m \in 1,2,3 \qquad (1)$$

$$P_{OUT} = \sum_{m=1}^{3} \hat{V}_{ACm}\hat{I}_{ACm} \cdot \sin^2\left(\omega_{grid}t + \frac{2\pi}{3}(m-2)\right) \qquad (2)$$

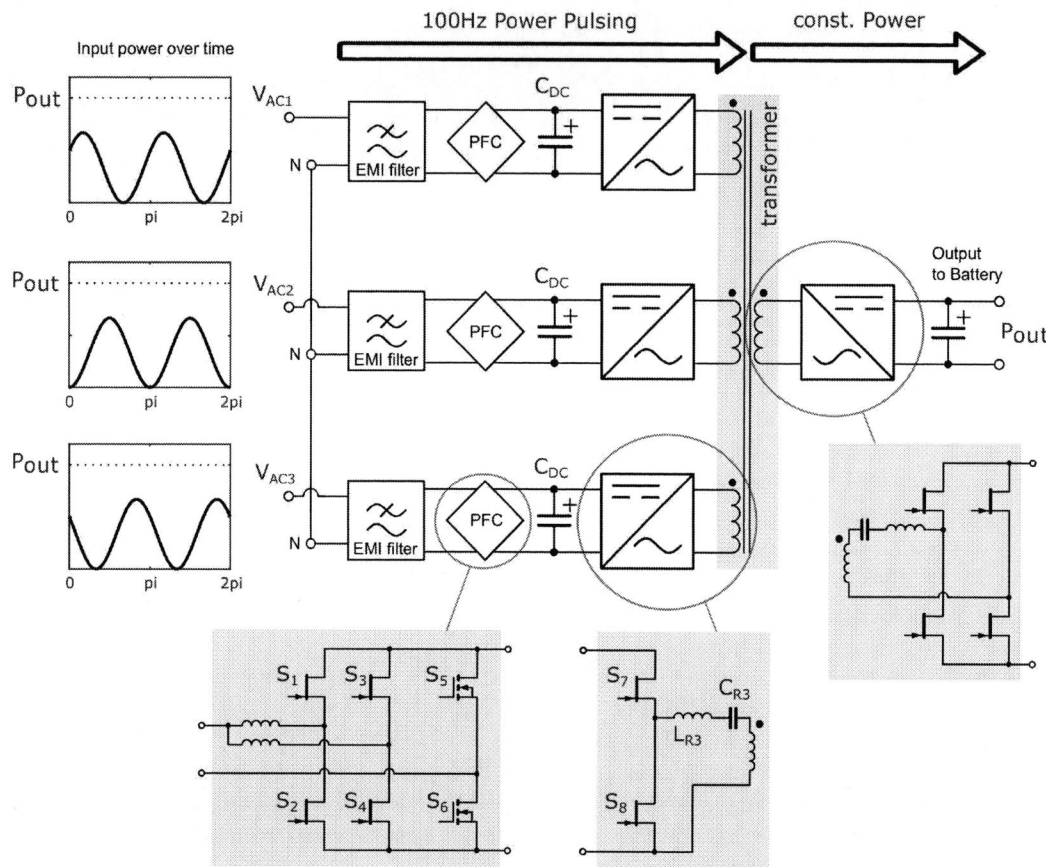

Fig. 1: Proposed Multi-Port converter structure and resulting power flow.

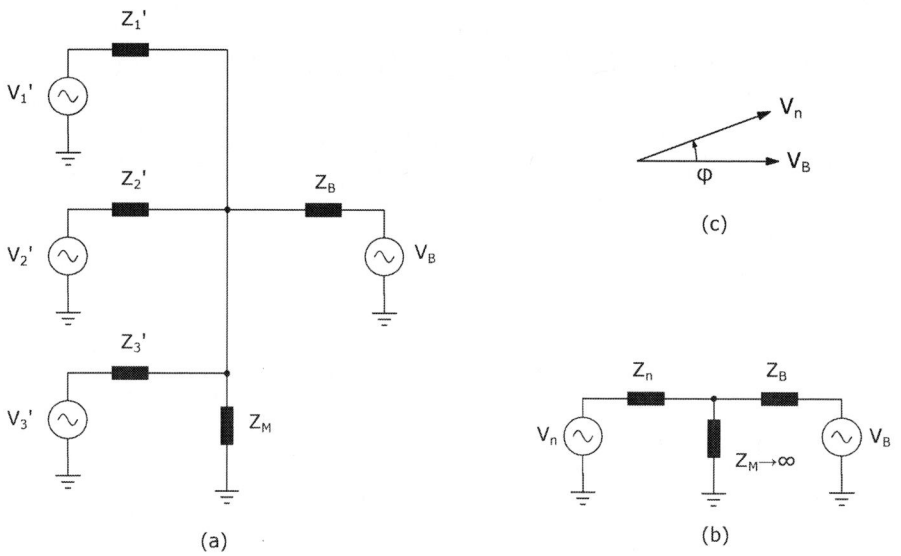

Fig. 2: Equivalent circuit diagram: (a) detailed (b) reduced (c) related voltage vectors

Fundamental tone approach for active and reactive power flow estimation

The resonant tanks input voltages of the DC-DC stage are set up completely symmetrically according to (3). The transformer's turns ratio a_m related to the output is 1:1, which will result in an equivalent circuit model shown in Fig. 2. This can be further reduced by paralleling the inputs (5). In contrast to [18] the transformer's main inductance will be neglected for this approach.

$$V_n \equiv V_1' \equiv V_2' \equiv V_3' \tag{3}$$

$$Z_m' = Z_m \cdot \frac{1}{a_m^2} \qquad Z_m = \sqrt{\frac{L_{Rm}}{C_{Rm}}} \qquad \begin{aligned} m &\in 1,2,3 \\ a_m &= 1 \end{aligned} \tag{4}$$

$$Z_k = Z_n + Z_B \qquad Z_n = \left(Z_1'^{-1} + Z_2'^{-1} + Z_3'^{-1}\right)^{-1} \qquad Z_M \to \infty \tag{5}$$

Neglecting the exact value for the main inductances will result in a transfer impedance Z_k for observing the overall power flow. Considering P_{out} from (2), the amount of power flow can be determined by using the operating principals for LLC circuits described in [14]. The active power P_B and reactive power Q_B related to the output port can be observed by (6), which represents the relation between active and reactive power transferred to the port B. Especially the difference between the voltages V_n and V_B at $0°$ phase shift will result mainly in a change in reactive power, also shown in Fig. 2c.

$$
\begin{aligned}
P_B &= \frac{V_n V_B}{|Z_k|^2}(R\cos(\varphi) + X\sin(\varphi)) - \frac{V_B^2}{|Z_k|^2}R \qquad X = \Im(Z_k) \\
Q_B &= \frac{V_n V_B}{|Z_k|^2}(-R\sin(\varphi) + X\cos(\varphi)) - \frac{V_B^2}{|Z_k|^2}X \quad R = \Re(Z_k)
\end{aligned}
\tag{6}
$$

However, the amount of transferred active power is related to the difference in the voltages V_n and V_B, the transfer impedance Z_k and the phase shift φ. In particular, the amount of reactive power will be lower the higher the phase shift φ for the same relation between the voltages V_n and V_B is [17]. However, a small amount of reactive power should be considered to ensure zero voltage switching is achieved.

Blocking reactive power during low grid voltage

According to (2) the power of each phase is pulsing proportionally to $\sin^2(\omega_{grid}t)$, resulting in low active power when V_{AC} is low. Considering $f_s > f_{res}$, the resonance tank behaves like an inductor between the ports. Therefore, mostly reactive power will be transferred between the ports during this state, resulting in unnecessary conduction losses in the switches. To reduce the amount of reactive power to an absolute minimum (6) will be transformed, so that Q_B becomes zero, causing a change of the voltage V_n. This will result in a relation between the voltages V_n and V_B according to (7) for $\Re(Z_k) = 0$. In addition, the phase angle φ has to be changed too, so that a minimum of reactive power will be transferred between the DC-links.

$$V_n = \frac{V_B}{\cos\varphi} \qquad \varphi = \arctan\left(\frac{P_B \cdot Z_k}{V_B^2}\right) \qquad \Re(Z_k) := 0 \tag{7}$$

Accurate power flow estimation using vector description

The equations mentioned in (7) describe the relation between transfered power, voltage vectors and the circuit impedance based on a simplified model. Considering asymetric supply voltages on the input, temporary events like line voltage drops, overvoltage transitions or a non-symetric circuit design of the input ports, the simplified model will no longer provide a correct relation. In addition, the resonant tanks and transformer will inevitably cause some losses which make the estimation of the correct voltage vector angles for a minimum reactive power transfer difficult. However, a calculation of the transfered complex power (9) can be performed by using the admittance matrix (8) of the equvalent circuit to calculate the correct phase angles for the voltage vectors of the fundamental tone approximation.

$$
\begin{pmatrix} I_1 \\ \vdots \\ I_m \end{pmatrix} = \begin{pmatrix} Y_{11} & \cdots & Y_{1m} \\ \vdots & \ddots & \vdots \\ Y_{m1} & \cdots & Y_{mm} \end{pmatrix} \begin{pmatrix} V_1 \\ \vdots \\ V_m \end{pmatrix}
\tag{8}
$$

$$
S_m = V_m \cdot I_m^* \qquad\qquad m \in 1,2,3 \tag{9}
$$

DC-Link capacitor limits

The DC-link capacitance C_{DC} should be dimensioned to stay charged above its instantaneous grid voltage to avoid unwanted body diode conduction in the PFC switches, resulting in non-sinusoidal input currents and potentially harming components with excessive current. In normal operation, and within a switching period T_S, the 'worst possible' DC-link voltage drop would occur if energy would be delivered to the DC-DC stage during peak phase output power. I.e. when $V_{AC}(t) = \hat{V}_{AC}$. Analysis of this scenario gives a safe estimate for minimum C_{DC} in (10). For the parameters in Table II and for a 400 V DC-link voltage (V_{DC}), (10) suggests a minimum $C_{DC} = 4.03$ µF for each phase. As such, stored DC-link energy can be orders of magnitudes smaller than the capacitors in Table I.

$$
C_{DC} \geq \frac{4 P_{OUT} T_S}{3 \left(\hat{V}_{DC}^2 - \hat{V}_{AC}^2 \right)} \tag{10}
$$

In case there is no power flow in the phase, shown in Fig. 4b and 5b respectively , the DC-link current is only the oscillating resonant current of the DC-DC stage, indicating that the remaining phases are providing the constant load. As predicted, Fig. 5b shows that at full load the DC-link voltage never drops below the peak grid voltage with a 5 µF DC-link. The resulting capacitor current peaks of up to 80 A reinforces the use of ceramic CeraLink capacitors due to their low ESR at frequencies above 100kHz [15][16].

Simulation Results / DC-link Performance

To evaluate the small DC-link of the multi-port topology, simulations are done with a transient circuit simulator using the settings in Table II. Fig. 3 shows power and DC-link voltage for one input phase, with zoomed waveforms during full and low input voltage shown in Fig. 4 and 5 respectively. Simulation results are shown using two modes of control.

1. Firstly, the phase shift φ of the voltage vectors is set to 0°, showing the inherent stability of the circuit. In this case the transferred active power will be affected by the difference in DC-link voltages related to the output voltage. Although such operation is inherently stable, it introduces a massive amount of reactive power.

2. Secondly, to reduce this reactive power transfer, a phase shift modulation is used for each input phase m based on (11) which combines (7) and (2) related to the input voltage (1).

Table II: Circuit simulation parameters and key modelled components from proposed topology.

Circuit Parameter	Value	*Circuit Symbol	*Component
Rated Power	12 kW	C_{DC}	5x 1μF CeraLink B58031I5105M062
Inp. Voltage	*230 VAC		
Out. Voltage	320 VDC	S_1-S_4	CoolGAN 70 mΩ IGO60R070D1
f_s (PFC)	215 kHz		
f_s(DC-DC)	73 kHz	S_5-S_6	CoolMOS 40 mΩ IPP60R040C7
Resonant L_r	*25 μH		
Resonant C_r	*200 nF	S_7-S_8	CoolGAN 70 mΩ IGO60R070D1
Magnetizing inductance	*100 uH		

*applies for each of the three input phases.

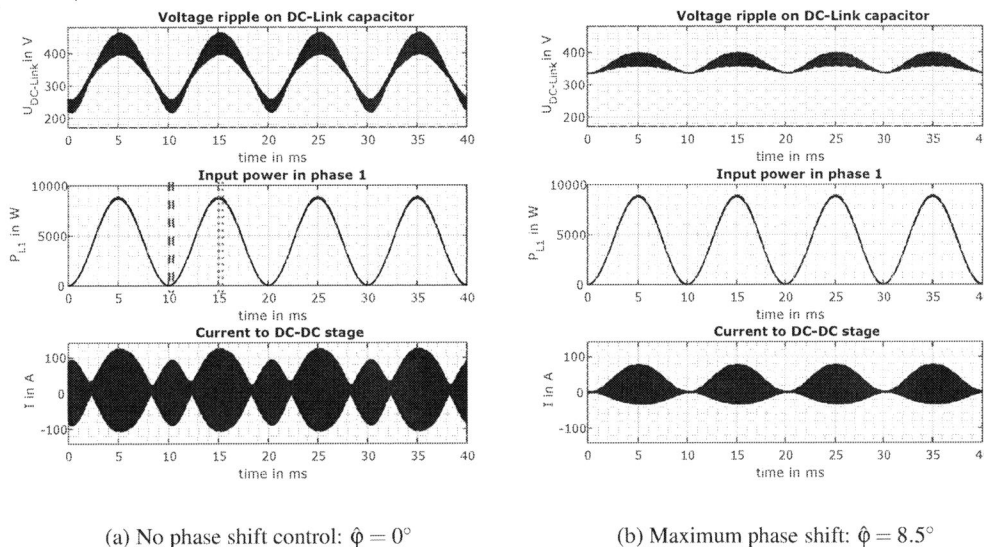

(a) No phase shift control: $\hat{\varphi} = 0°$ (b) Maximum phase shift: $\hat{\varphi} = 8.5°$

Fig. 3: DC-link voltage and power of one input phase, without phase-angle control (a) and with (b).

$$\varphi_{m,equ}(t) = \arctan\left(\frac{2}{3}P_{out} \cdot sin\left(\omega_{grid}t + \frac{2\pi}{3}(m-2)\right)^2 \cdot \frac{Z_k}{V_B^2}\right) \qquad m \in 1,2,3 \qquad (11)$$

Fig. 3a clearly shows the high amount of current to the DC-DC stage representing reactive power compared to Fig. 3b which uses the active phase shift control. During the zero crossing of the phase's input voltage, the respective phase's transformer current is minimized when phase angle modulation is enabled. Especially the amount of voltage swing on the DC-link capacitor will be massively reduced and thus reducing losses. Fig. 4a & 5a as well as Fig. 4b & 5b show the simulation results of the zoomed section from Fig. 3a and 3b respectively. The difference in DC-link capacitor current as well as the current to the DC-DC stage can clearly be seen.

Since (7) is based on the simplified equivalent circuit model, the phase angle $\varphi_{m,equ}(t)$ only gives an indication of the correct angle. The exact solution has to be calculated using a transformed version of (8) & (9) to solve for $\varphi_{m,equ}(t)$.

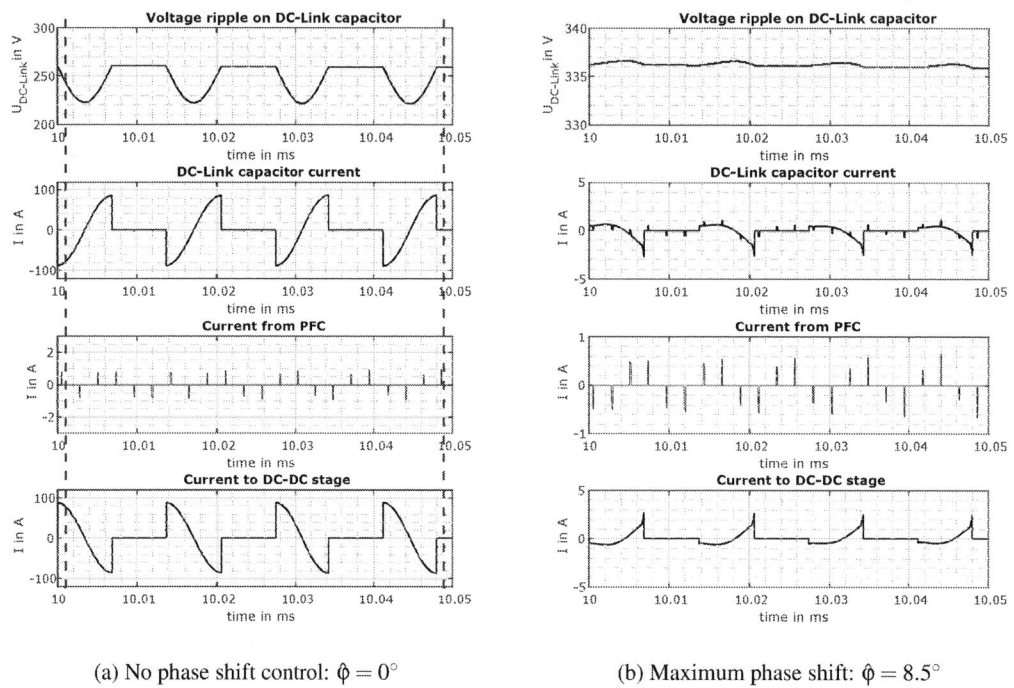

(a) No phase shift control: $\hat{\varphi} = 0°$ (b) Maximum phase shift: $\hat{\varphi} = 8.5°$

Fig. 4: Comparison of voltages & currents with and without phase shift control at **low** $V_{AC}(t)$.

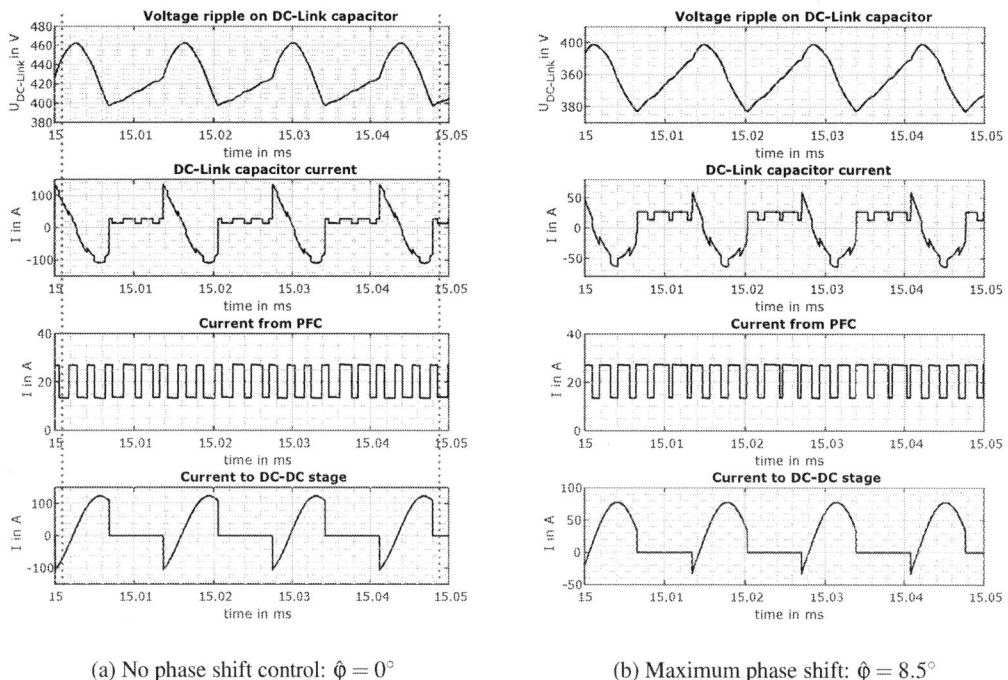

(a) No phase shift control: $\hat{\varphi} = 0°$ (b) Maximum phase shift: $\hat{\varphi} = 8.5°$

Fig. 5: Comparison of voltages & currents with and without phase shift control at **high** $V_{AC}(t)$.

Conclusions

This paper introduces a multi-port topology that, in contrast to the most common topologies used today, combines three phase input power after the DC-link(s) with a multi-port transformer. The contributions and key associated conclusions of this paper are summarized as followed:

- This topology can exploit the constant power available in 3-phase supply to achieve massive shrinkage of the DC-link capacitor.
- It has been shown that the topology is inherently stable, also with a disconnected neutral line.
- The proposed control scheme allows the multi-port topology to operate with minimal reactive power flow which gives the opportunity for high efficiency and improved power density when compared with conventional concepts.
- Additionally, it inherently provides the galvanic isolation needed for the onboard charger without requiring an additional stage.

References

[1] C. Saber, D. Labrousse, B. Revol and A. Gascher, "Challenges Facing PFC of a Single-Phase On-Board Charger for Electric Vehicles Based on a Current Source Active Rectifier Input Stage," in IEEE Transactions on Power Electronics, vol. 31, no. 9, pp. 6192-6202, Sept. 2016.

[2] C. Shi, Y. Tang and A. Khaligh, "A Single-Phase Integrated Onboard Battery Charger Using Propulsion System for Plug-in Electric Vehicles," in IEEE Transactions on Vehicular Technology, vol. 66, no. 12, pp. 10899-10910, Dec. 2017.

[3] A. Tokumasu, H. Taki, K. Shirakawa, and K. Wada, "AC/DC converter based on instantaneous power balance control for reducing DC-link capacitance," 2014 Int. Power Electron. Conf. IPEC-Hiroshima - ECCE Asia 2014, pp. 13791385, 2014.

[4] M. Kasper and G. Deboy, "GaN HEMTs Enabling Ultra-Compact and Highly Efficient 3kW 12V Server Power Supplies," 2018 IEEE Int. Power Electron. Appl. Conf. Expo., vol. 3, pp. 10591064, 2018.

[5] M. Rosekeit and R. W. De Doncker, "Smoothing power ripple in single phase chargers at minimized dc-link capacitance," 8th International Conference on Power Electronics - ECCE Asia, Jeju, 2011, pp. 2699-2703.

[6] A. Mallik, J. Lu, S. Zou, P. He, and A. Khaligh, "Minimum inrush start-up control of a single-phase interleaved totem-pole PFC rectifier," Conf. Proc. - IEEE Appl. Power Electron. Conf. Expo. - APEC, vol. 2018-March, pp. 754759, 2018.

[7] https://www.wolfspeed.com/media/downloads/CPWR-AN25.pdf - Wolfspeed. - Accessed on 13/07/2019

[8] A. Stupar, T. Friedli, J. Minibock, J. W. Kolar, "Towards a 99% efficient three-phase buck-type PFC rectifier for 400-V dc distribution systems", IEEE Trans. Power Electron., vol. 27, no. 4, pp. 1732-1744, Apr. 2012.

[9] http://www.ti.com/lit/ug/tiducj0b/tiducj0b.pdf -Texas Instruments. Vienna Rectifier-Based Three-Phase Power Factor Correction (PFC) Reference Design - Accessed on 22/06/2019

[10] K. Yoo, K. Kim and J. Lee, "Single- and Three-Phase PHEV Onboard Battery Charger Using Small Link Capacitor," in IEEE Transactions on Industrial Electronics, vol. 60, no. 8, pp. 3136-3144, Aug. 2013.

[11] S. -. Youn and J. -. Lee, "Electrolytic-capacitor-less AC/DC electric vehicle on-board battery charger for universal line applications," in Electronics Letters, vol. 47, no. 13, pp. 768-770, 23 June 2011.

[12] J. W. Kolar, U. Drofenik, and F. C. Zach, "VIENNA rectifier II - a novel single-stage high-frequency isolated three-phase PWM rectifier system," IEEE Trans. Ind. Electron., vol. 46, no. 4, pp. 674691, 1999.

[13] R. Greul, U. Drofenik, and J. W. Kolar, "Analysis and comparative evaluation of a three-phase three-level unity power factor y-rectifier," 25th Int. Telecommun. Energy Conf. 2003. INTELEC., pp. 421428, 2003.

[14] K. T. Manez, Z. Zhang, and Y. Xiao, "Three-Port Series-Resonant Converter DC Transformer with Integrated Magnetics for High Efficiency Operation," IEEE Energy Convers. Congr. Expo. ECCE, pp.56225629, 2018.

[15] D. Neumayr, D. Bortis, J. W. Kolar, M. Koini, and J. Konrad, "Comprehensive large-signal performance analysis of ceramic capacitors for power pulsation buffers," 2016 IEEE 17th Work. Control Model. Power Electron. COMPEL 2016, pp. 18, 2016.

[16] https://www.tdk-electronics.tdk.com/inf/20/10/ds/B58031_LP.pdf - TDK - Accessed on 15/07/2019

[17] M. Michel, Leistungselektronik. Berlin Heidelberg: Springer-Verlag, 2011, p. 211-215.

[18] K. T. Manez, Z. Zhang, and Y. Xiao, "Multi-Port Isolated LLC Resonant Converter for Distributed Energy Generation with Energy Storage," IEEE Energy Convers. Congr. Expo. ECCE, pp.2219-2226, 2017.

[19] R. Yamada, Y. Nemoto, S. Fujita and Q. Wang, "A battery charger with 3-phase 3-level T-type PFC," 2015 IEEE International Telecommunications Energy Conference (INTELEC), Osaka, 2015, pp. 1-5, doi: 10.1109/INTLEC.2015.7572419.

Design and optimization of high-efficiency 1W 500V-12V isolated low-cost DC/DC converter

Etienne Foray, Christian Martin, Bruno Allard
Laboratoire Ampere
University of Lyon, INSA Lyon, Université Lyon 1, Ecole Centrale Lyon, CNRS
Lyon, France
etienne.foray@insa-lyon.fr ; christian.martin@univ-lyon1.fr ; bruno.allard@insa-lyon.fr
URL: http://www.ampere-lab.fr

Keywords

≪ZVS converters≫, ≪Resonant converter≫, ≪High frequency power converter≫, ≪Converter circuit≫, ≪Transformer≫

Abstract

Conversion from high-input-voltage to low-output-voltage in low-power mode is challenging, especially with high-efficiency and small-size requirements. Standard Flyback solutions are not good enough for these applications. This works analyzes the design trends for a low-power Asymmetrical Half-Bridge Flyback, with special attention given to the power-stage and transformer design. A simplified model of the converter is built to compare a large number of solutions and to validate the design approach. Results from the model are analyzed, through the study of some key parameters and some interesting designs are identified. They show the benefits of this approach with respect to existing solutions, based on standard Flyback.

Introduction

In some applications, high-voltage DC buses are becoming more and more common, and some tiny systems need to be powered directly and continuously from the HV bus. In automotive for example, batteries of an electric vehicle are nowadays often around 400 V, and a system like a Pyroswitch [1] needs to be powered directly and continuously from the high-voltage input. This creates a need for High-Voltage Low-Power (HVLP) isolated DC/DC converters, with requirements on small size and high-efficiency brought by the embedded context.

A usual choice when designing an isolated low-power converter is the Flyback architecture, as it is a simple and low-cost solution. Classical Flyback [2], active clamp Flyback [3] or Flyback with snubber [4] have been experimented as low-power off-line converters. However, in the case of HVLP converters, efficiency is often not very good or the maximum input voltage supported is not high enough. Moreover, the Flyback power-stage has several drawbacks such as high voltage stress on the main switch and difficulty to operate in Zero Voltage Switching (ZVS) mode.

A solution proposed to improve the Flyback performances is the topology called "Asymmetrical Half-Bridge Flyback" (AHBF). It decreases the stress on the switch and it allows ZVS operation to improve efficiency, while offering the possibility to operate at high switching frequencies. It was already studied in past works [5, 6, 7, 8, 9] but with larger output power (typically 50 W to 100 W). To our best knowledge, ABHF design was never studied to operate at low output power (< 5 W) and high input voltage.

This work wishes to explore the possibilities offered by AHB Flyback topology (high switching frequency, ZVS, high efficiency...) when designing a converter with high input voltage, low output power under low output voltage. This work focuses on specificities due to low-power as it changes some important design considerations, especially when targeting high-efficiency.

The objective is to design a low-cost converter with general specifications summarized in Table I. Therefore, the targeted technology for the power switches belongs to standard Si HV-BCD processes. They offer easier and cheaper integration possibilities (of gate-driver, control, local supply...) than GaN or SiC solutions. [10] shows an example of HVLP solution, based on classical Si HV-BCD process, without isolation, but it still demonstrate the integration possibilities. As an isolation transformer is required in this case, PCB planar transformers solutions with reduced number of layers (max. 4) are considered in order to comply with the low-cost constraint.

Table I: Converter's general specifications

Input voltage	From 300 V to 500 V
Output voltage	12 V
Output power	~ 1 W
Isolation	Required
Efficiency	$> 85\%$

The paper presents the analysis of design trends specific to the high-voltage low-power conversion, from both power-stage and transformer points of view, using a simplified model of the converter to estimate losses.

First, the AHBF topology is described and its behavior is discussed. Then, a design approach for the power-stage is presented, taking into account the low-power specificity. Some important parameters of the transformer design are discussed and the simplified model used to explore the design space is introduced. Finally, results from models are analyzed to identify important design parameters and the proposed solution is compared to existing HVLP Flyback solutions.

Asymmetrical Half-Bridge Flyback

The selected AHBF topology is pictured in Fig. 1. It can also be seen as an isolated version of a Buck converter (where inductance is replaced by a transformer), which is why it is sometimes referred as the "Fly-Buck" [11].

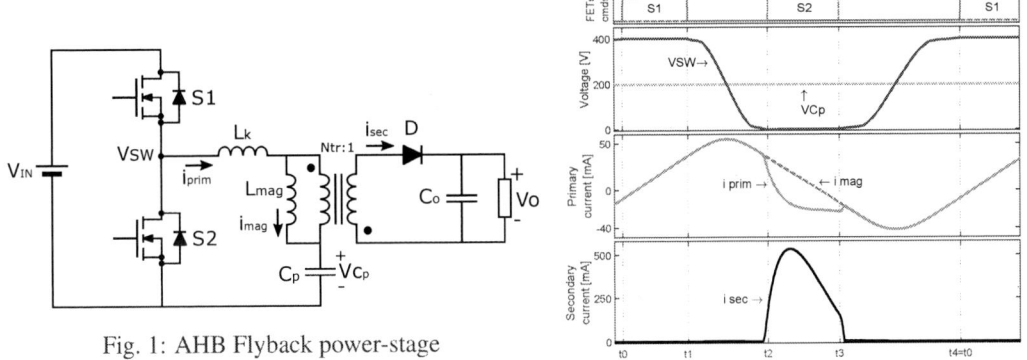

Fig. 1: AHB Flyback power-stage

Fig. 2: Simulation waveforms

General introduction to operating principle

The power-stage is composed of two complementary switches $S1$ and $S2$, on which maximum stress voltage is V_{IN}. The transformer is connected to the switching node V_{SW} and a capacitor C_p is connected in series with the transformer primary side. The secondary side is simply composed of a rectification diode

D and an output filter capacitor C_O. A detailed analysis of the converter operating modes is covered in [5, 6] and the main aspects will only be quickly described here.

The switching period is divided in four steps, as presented in Fig. 2. A first step ($t_0 - t_1$), when $S1$ is ON, during which the transformer gets charged. A second step ($t_1 - t_2$) starts when $S1$ turns-off and part of the energy stored inside the transformer is used to move the switching node V_{SW} from V_{IN} to ground, in order to operate a ZVS turn-on of $S2$. During the third step ($t_2 - t_3$), $S2$ is ON and the negative voltage on the primary side of the transformer translates into a positive voltage on the secondary side, allowing forward biasing of the secondary diode. As a result, the current i_{SEC} starts flowing in the secondary side and energy is transmitted to the output. The final step ($t_3 - t_4$) starts when $S2$ turns off. The negative magnetizing current moves the switching node V_{SW} from ground to V_{IN} and $S1$ can turn-on in ZVS conditions to start a new period.

A deep analysis of the converter operating mode and design overview were already given in past works but it was for a converter with much higher output power (>50W). Low-power will change some important considerations in the topology analysis and behavior.

Influence of the leakage inductance

The role of the leakage inductance L_k was analyzed in [5], showing how it stores current to allow ZVS. However, it is only possible for large output current, as the energy stored in the leakage inductance depends on the current flowing in the secondary side when the low-side switch turns off. When the output power is low, waveforms (Fig. 2) show that the current at this moment is small. More generally, the current flowing in the primary side is mostly the magnetizing current as the overal current is small, making it difficult to use this mechanism to operate in ZVS. Hence the role of the leakage inductance in the case of low-power opearation is less important than for high-power operation. Note that the leakage inductance still has an effect on the dynamics of the output current, as it limits the rising slope ($\frac{di_{sec}}{dt}$), but it should not be a problem in the case of low-power operation.

ZVS mechanism

When the current flowing to the output is small, the energy stored inside the magnetizing inductance is used to perform ZVS, as presented in [6][9] in the case of light-load. The magnetizing current indeed takes both positive and negative values, because its offset is very small in the case of low-power operation (the offset value is the ratio of the DC output current by the transformer turn-ratio). This offset leads to slightly different dead-time duration values (for low-side and high-side switches), but both dead-time values should be taken into account.

In order to operate in ZVS conditions, the magnetizing current during dead-time should be high enough to completely charge-discharge the capacitance connected to the middle node V_{SW}. This capacitance is actually the sum of several parasitic capacitors, due to the switches' output capacitance, the diode capacitance (as studied in [9]) and the transformer primary-secondary winding capacitance. If the previous analysis [9] made the assumption of a constant capacitance value, it is not the case under high-voltage, as the Si MOSFET output capacitance C_{OSS} is highly non-linear with voltage. Moreover the magnetizing current available to operate a resonance is mainly fixed by the magnetizing inductance value and the switching frequency.

The later statements are an introduction to a necessary trade-off between the switching frequency and the magnetizing current as detailed in following Sections.

Dead-time influence on main transfer function

In the previous analysis of the topology, one assumption is that the dead-time duration values can be neglected as they are shorter than the switches on-time values. However, in this case, due to the small current available for ZVS, dead-time duration values will become longer such that their influence on the converter main transfer function cannot be neglected.

The influence of the dead-time duration values on the output voltage can be explained by the current going through the series capacitor Cp during the transients and by the relation between V_{Cp} and V_{OUT}.

With the approximation that the voltage across the primary winding varies linearly during dead-time duration, volt-second balancing across the transformer may be evaluated according to (1).

$$V_{OUT} = D_{eff} * \frac{V_{IN}}{N_{tr}} - V_D = \frac{V_{IN}}{N_{TR}} * (t_{on\_S1} + \frac{1}{2} \sum dt) * f_{SW} - V_D \qquad (1)$$

Where V_D is the voltage drop across the secondary side diode, f_{SW} the switching frequency, N_{TR} the transformer turn-ratio (as shown in Fig. 1), D_{eff} the effective duty-cycle of $S1$, t_{on_S1} the on-time of switch $S1$ and $\sum dt$ the total duration of dead-times in a switching period.

Even if the later assumption is not very accurate, especially when dead-times become longer, (1) still illustrates the impact of dead-times on the relation between input and output voltages. Dead-times' influence can no longer be neglected, especially when their duration is similar or larger than the switches' on-times, as explained in the next Section.

Design approach of the power-stage

The goal is to target designs with high-efficiency. The main source of losses are therefore considered: conduction losses and core losses. The ZVS operation mode allows to significantly decrease the switching losses, therefore they are neglected in this analysis.

The difficulty when designing HVLP solution comes from losses that do not scale with the output power but with the input voltage. One example is the core losses that vary with the voltage applied to the transformer. Another example is the current on the primary side of the transformer, that is mainly due to the magnetizing current, which does not depend on the output power. The low output current condition will therefore not translate into a low current value on the primary side of the transformer.

The analysis of ZVS mechanism already showed that a minimum value of magnetizing current is required in order to perform a ZVS operation for both high-side and low-side switches. But as the current is also going to generate conduction losses, especially in the primary winding resistance and switches on-resistance, it should be minimized. Therefore, an important design step will be to determine the minimum quantity of current required to perform a ZVS operation, depending on the value of total parasitic capacitors connected to the middle node (C_{SW}). As a consequence of a small current during ZVS transition, dead-times between switches' conduction will become very long, as shown by simulation waveforms in Fig. 2 ($t_1 - t_2$ and $t_3 - t_4$).

However, dead-times' duration values will be limited by the switching frequency. If a minimum on-time for the switches is considered, there will be a limit to the maximum duration of the dead-times. An increased frequency will therefore lead to shorter dead-times. For dead-times to become shorter, if a same value of C_{SW} capacitance is considered, a higher current is required. This shows that there is a relation between frequency and the amount of magnetizing current, and so a relation between the switching frequency and the conduction losses on the primary side of the transformer.

But the operating frequency is not the only parameter impacting the magnetizing current: the input voltage and the magnetizing inductance also play a significant role. For a given input voltage, there is a maximum value for the magnetizing inductance depending on the operating frequency that minimizes the magnetizing current, but still keeps it high enough to perform ZVS transitions. Or, equivalently for a given magnetizing inductance value, there is a maximum switching frequency that allows to operate in ZVS mode. When considering an input voltage range, it will be necessary to select the operating frequency depending on the input voltage, as magnetizing inductance value will be fixed. A variable frequency over the input voltage range will then allow to keep at minimum the magnetizing current, and therefore minimize conduction losses on the transformer primary side.

Note that the transformer turn-ratio Ntr also impacts the maximum dead-times' duration values, as it is discussed in the next Section.

Planar transformer design

This section details an analysis of the impact of several parameters of the transformer on converter's performances.

Choice of the technology

The transformer is a critical device of a HVLP converter, as it greatly impacts performances and the size of the solution, as it is the bulkiest element of the converter. In this case, a planar transformer technology with tracks on the Printed Circuit Board (PCB) was selected as it is offering a low profile and better power density than any conventional transformers [12, 13, 14]. Besides, planar transformers are more appropriate for high switching frequencies since the parasitic components are more repeatable over several prototypes. Moreover, planar E-cores offer a good trade-off between volume and core cross-section, leading to lower core losses.

In order to respect low-cost and low-capacitance objectives, 4-layers-PCB designs are considered. Two layers are dedicated to the primary winding, one to the secondary winding and the last one is for inter-connections.

Transformer turn-ratio

As revealed with (1), a relation exists between D_{eff}, N_{TR}, V_{IN} and V_{OUT}. The analysis of the power-stage design showed that to minimize the primary current, the dead-times should be as long as possible. The maximum dead-time duration is determined by two parameters: D_{eff} needed to ensure the right output voltage and the minimum on-time values of high-side and low-side switches. For a given input and output voltage, the transformer turn-ratio can be optimized in order to get the lowest constraint possible on dead-times' duration, and therefore the lowest primary current. This optimal turn-ratio can be approximated with (2). In the case of 400 V input, 12 V output, with $V_D = 0.5\ V$ and minimum on-times of 200 ns for $S1$ and 300 ns for $S2$, the optimal turn-ratio varies from 15.8 at $f_{SW} = 300\ kHz$ to 14.4 at $f_{SW} = 1000\ kHz$.

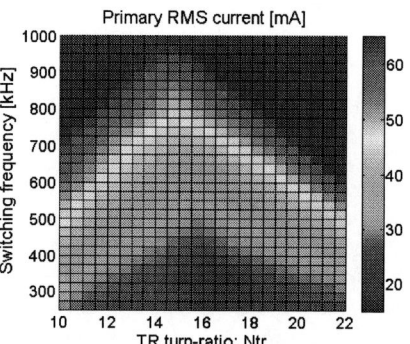

Fig. 3: Variation of primary current vs Ntr and f_{SW} at Vin = 400V

$$Ntr_{opti} = \frac{V_{IN}}{2 * (V_{OUT} + V_D)} * \left(1 - f_{SW} * (t_{on\_min\_S2} - t_{on\_min\_S1})\right) \tag{2}$$

These values correspond quite well to those shown in Fig. 3, obtained with the simplified model (discussed in the next Section). This figure also illustrates the trade-off between the switching frequency and the primary current discussed in previous Section. Note that this is valid only for a single input voltage value, the case of an input voltage range in discussed in Results' Section.

Number of primary turns

Number of turns on the primary side of the transformer is an important design parameter as it impacts both core losses and conduction losses.

To estimate core losses, one important quantity is the maximum flux density reached inside the core. It is inversely proportional to the number of turns and an increased number of turns will lower the flux density and thus the core losses. A maximum number of turns is determined by the geometry of the E-core considered in the planar design, but also by the accuracy of the PCB manufacturing, as tracks require a minimum width and a minimum space between them. A constraint of 0.1 mm is used for minimum width and minimum spacing what do not impact the cost of PCB. However, changing the number of turns is also changing the winding resistivity. A larger number of turn means a longer winding with smaller width, so its resistance rises greatly. This shows that a trade-off is to be found between core losses and conduction losses.

This trade-off is not straightforward because the operating frequency will also have an impact on both core losses and conduction losses. As shown previously, a higher switching frequency is linked to a larger primary current. But the switching frequency will also decrease the core losses, as it allows to decrease the maximum volt-second across the transformer. Since frequency and the number of turns will have an impact on these different parameters, a trade-off between the switching frequency and the number of primary winding turns will also exist. It is difficult to proceed analytically because frequency also impacts resistivity of the tracks, with skin and proximity effects. The influence of these phenomena can be modeled under a 1D-magnetic-field assumption like in [15][16], but it is questionable to get good results in the case of many tracks on each layer and a low porosity factor [14].

About parasitic capacitors

Parasitic capacitors need to be taken into account when designing the planar transformer: the primary-to-secondary capacitor (Cps) and the self-resonating capacitor.

The first one is connected to the middle node (V_{SW}) that will be charged/discharged during ZVS transients. Having a too large value for this capacitance will translate into longer ZVS transient or larger primary current, leading to extra conduction losses and lower efficiency, as shown in Fig. 4. To reduce this capacitance, the transformer design should have distant primary and secondary layers even though it will impact the coupling factor and the magnetizing inductance.

The second capacitor, called self-resonating capacitor, is the equivalent capacitor of the primary side. It sets the self-resonant frequency (SRF) of the transformer along with the magnetizing inductance. The Impact of SRF on transformer performances is not easy to anticipate but design should be such that the power-stage switching frequency is not too close to SRF to avoid increasing conduction losses.

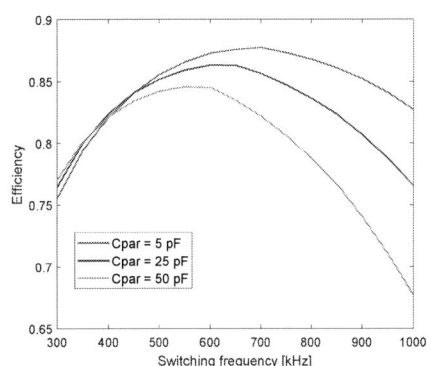

Fig. 4: Impact of transformer capacitance Cps on efficiency

Simplified models of power-stage and transformer

A simplified model of the power-stage was created to be able to compare many different designs with a reasonably good accuracy inside a Matlab routine. One benefit from using a model is that it allows to take into account different types of losses (core and conduction losses). Another benefit is that the computation can include some optimization routines in its flow represented in Fig. 5. This is useful to evaluate an optimized value for the L_{mag}/f_{SW} couple.

Short description of models

Model's input parameters are the specifications of the converter (input voltage, output voltage, power), the general characteristics of the transformer (E-core geometry and material, number of turn, parasitic capacitance) and some constraints from power switches (ON-resistance, output capacitance, minimum on-time).

It computes a relation between the magnetizing inductance and frequency. It optimizes the value of inductance over a wide range of switching frequency values making it as big as possible (to reduce the primary RMS current), while keeping the converter in desired operating conditions (ZVS, minimum on-time for FETs and duty-cycle value to keep output voltage at the right value). This is done through an iterative process, as it computes, for each inductance value, the primary current and primary voltage waveforms during several switching periods, until it gets to steady-state. From this step it gets values for L_{mag} and for primary/secondary currents and voltages at each frequency.

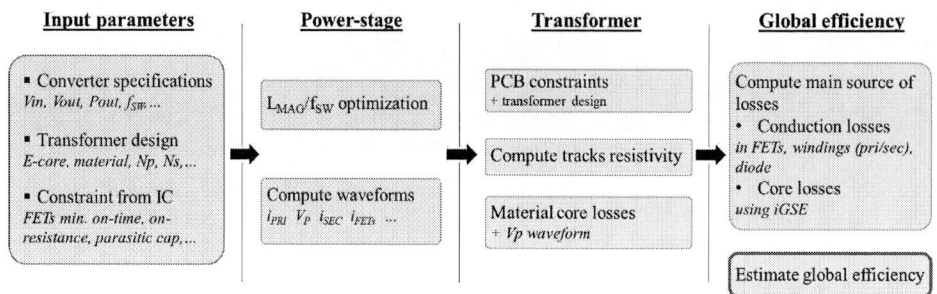

Fig. 5: Simplified representation of computation routine

Losses related to the transformer are then computed. PCB constraints and the number of primary winding turns are used to estimate the windings DC resistance. Conduction losses are computed, using currents RMS value and the estimated R_{AC} resistivity of windings. The model use to compute the AC resistance is a simple constant factor ($F_{AC} = 3$) with respect to the DC-resistance (the value of the factor was determined by measuring various planar transformer winding resistivity at their nominal frequency of operation). Core losses are calculated using the material parameters and the improved Generalized Steinmetz Equation (iGSE) [17] with a trapezoidal waveform of primary current.

It is then possible to evaluate an efficiency over the frequency range considered, for a large number of transformer designs, changing the number of turns on primary and secondary sides for example. Moreover most important sources of loss are considered: condution losses in MOSFETs, transformer windings, rectification diode and core losses in the transformer.

Results from pre-design phase

A large number of designs were tested using the Matlab routine to identify design tendencies and identify critical parameters. Few E-core were selected with different sizes (E22, E18, E32). The E+PLT association was used for all the geometries to decrease the solution's volume and height. The selected core material is finally 3F4 because of its large frequency range (200 kHz - 2 MHz) and low losses. The impact of some important design parameters are now discussed and some interesting designs are selected and compared to existing solutions.

Number of primary turns

The impact of the number of primary turns on efficiency is shown in Fig. 6. Three solutions are compared for the same turn-ratio $Ntr = 15$. Performances are computed for 400 V input and 12 V, 1 W output.

What can be observed is that a minimum number of turns is needed to get good performances as efficiency for $Np = 15$ is much lower than for the other values. This is because core losses are too high if the number of turns is too small. However, when comparing solutions for $Np = 30$ and $Np = 45$, no major improvement is observed. One interesting point is that the peak efficiency is reached at different frequency values for both solutions. The solution with a smaller number of turns ($Np = 30$) has to go to reach high operating frequency in order to get low core losses but conduction losses will increase which is the case when going higher than 900 kHz, where efficiency starts to drop.

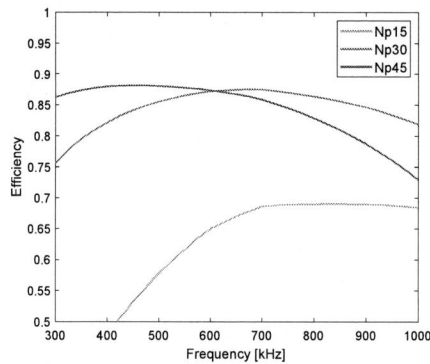

Fig. 6: Estimated efficiency for different number of primary turn

On the other hand, the solution with a larger number of turns ($Np = 45$) shows a satisfying efficiency at low frequency because core losses are small. But as soon as the frequency increases above 500 kHz, the efficiency drops due to larger conduction losses because a high number of turns translates into a high winding resistance.

Size of E-core

Size of the E-core will also impact the performances. One of the reason for this is because a bigger core will have a larger area for windings, offering the possibility to have more turns and to decrease the core losses (up to a certain limit, as shown previously). Another reason is because the bigger the E-core, the larger its cross section will be, and so flux density and core losses will decrease.

In Fig. 7 we can see the influence of the E-core size on global efficiency. It illustrates the well-known trade-off existing between the size of a solution and its performances. Note that these results were obtained using the same number of turns (for primary and secondary), such that the only difference comes from E-core cross section. The efficiency improvement is only due to the reduction of the core losses.

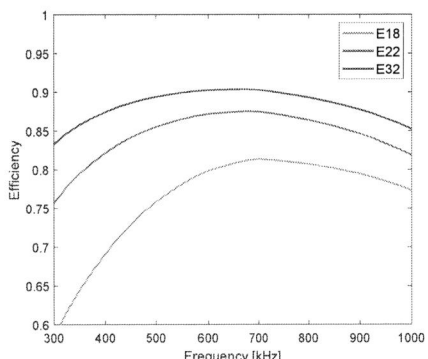

Fig. 7: Efficiency vs frequency for different E-core sizes

Input voltage range

What is the impact on the design and on performances if an input voltage range is considered instead of a fixed Vin value? When considering a complete range for the input voltage, the efficiency target should stay higher than a given value (85% in this case) over the range. Therefore, the design should be optimized for the worst case efficiency, what happens for a high Vin value (due to higher currents and higher core losses). It means for example that the transformer turn-ratio should be selected to get an optimal behavior at Vin = 500V using (2).

Fig. 8 illustrates that variations of the efficiency and the frequency over the input voltage range for two different designs. The first design (in red) is optimized for Vin = 400 V since Ntr = 15, while the second (in blue) is designed for Vin = 500 V with Ntr = 20. The efficiency of the first design is higher when the input voltage is below 400 V, but it significantly drops for higher voltages. On the other hand, the second design presents an efficiency that is more constant over the input voltage range and it becomes more interesting for Vin > 450 V.

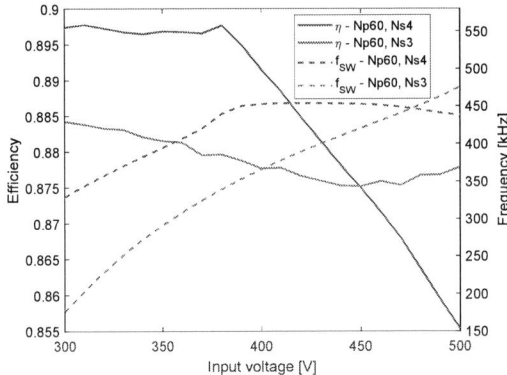

Fig. 8: Efficiency and frequency variations over Vin range

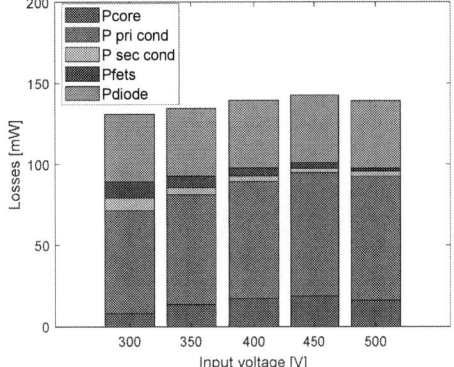

Fig. 9: Losses breakdown at different input voltages - Results for E32 - 3F4, Np60, Ns3

A non-optimal value of turn-ratio modifies the variations of the switching frequency. The first design reaches its maximum efficiency for 400 V input voltage. For higher input voltages, the magnetizing current increases as the frequency is constant. This extra current in the primary side leads to higher conduction losses hence a lower efficiency for high input voltages. For the second design, the switching frequency is evaluated with a change almost linear over the input voltage range, leading to almost constant conduction and core losses over the input voltage range as shown in Fig. 9. The input voltage increase is compensated by the frequency increase, so the primary current does not really change and conduction losses stay constant over the input voltage range.

Fig 9 presents a breakdown of losses for several input voltage values. It must be noted that a large part of losses is due to conduction losses in the primary side of the transformer. The rectifier diode has also an important impact on the efficiency as it could be reduced by choosing a diode with lower voltage drop or by using active rectification instead.

Interesting designs

From the design space exploration, few interesting designs are selected. Their expected performances are summarized in Table II and compared to current Flyback solutions.

Table II: Results obtained with model for different transformer designs

E-core ref ▶ Parameters ▼	E18/4/10R+PLT 3F4	E22/6/16R+PLT 3F4	E32/6/20R+PLT 3F4
Primary turns	40	50	56
Secondary turns	2	3	3
Frequency	230 kHz to 610 kHz	270 kHz to 450 kHz	160 kHz to 350 kHz
Efficiency @Vin	85% @300V 80% @400V 80% @500V	89 % @300V 88 % @400V 85 % @500V	91% @300V 90% @400V 89% @500V
Maximum L_{MAG}	1.3 mH	2.2 mH	3.7 mH
i_{prim} RMS	37 mA to 41 mA	28 mA to 37 mA	23 mA to 27 mA
Core footprint	1.8 cm^2	3.5 cm2	6.4 cm2

The second design using E22 core is a good trade-off between size (smaller than E32 solution) and performances (efficiency > 85% over Vin range). The solution with E18 is interesting at low Vin value but its efficiency drops for input voltage higher than 400 V.

When comparing these solutions with previous works on high-voltage and low-power solutions, the design identified so far here shows some superiority as it seems to be more efficient for a similar size. In [3]: peak efficiency is 75% for Vin>300 V, in [4] efficiency <71 % for Vin>300 V, and the transformer size is similar to E18. Finally [2] indicates an efficiency >80% but the overall solution size is larger as it uses large passive devices to compensate for a low switching frequency (below 20 kHz).

Some primary experimental results are pending because of COVID-19 lock-down.

Conclusion

In this paper, the impact of High-Voltage Low-Power conversion when designing an Asymmetrical Half-Bridge Flyback is analyzed. An approach is proposed for analyzing the design of the power-stage and of the transformer, studying the influence of several key parameters. A simplify model of the power-stage is built to explore against a large number of design parameter values, with the constraints of a low cost on switches and transformer. Finally, computation results based on the simple models are presented and show interests of this topology at low-power when compared to standard Flyback-based solutions. A transformer design based on small E18-core shows >80% efficiency at 500 V input and a solution based on E22-core reaches 85% efficiency over a large input voltage range.

References

[1] M. Koprivsek, "Advanced solutions in over-current protection of hvdc circuit of battery-powered electric vehicle," in *PCIM Europe 2018; International Exhibition and Conference for Power Electronics, Intelligent Motion, Renewable Energy and Energy Management*, pp. 1–4, June 2018.

[2] N. Nielsen, "An ultra low-power off-line APDM-based switchmode power supply with very high conversion efficiency," in *APEC 2001. Sixteenth Annual IEEE Applied Power Electronics Conference and Exposition (Cat. No.01CH37181)*, IEEE, 2001.

[3] P. Alou, O. Garcia, J. Cobos, J. Uceda, and M. Rascon, "Flyback with active clamp: a suitable topology for low power and very wide input voltage range applications," in *APEC. Seventeenth Annual IEEE Applied Power Electronics Conference and Exposition (Cat. No.02CH37335)*, IEEE, 2002.

[4] S. Zhao, J. Zhang, and Y. Shi, "A low cost low power flyback converter with a simple transformer," in *Proceedings of The 7th International Power Electronics and Motion Control Conference*, IEEE, jun 2012.

[5] T.-M. Chen and C.-L. Chen, "Analysis and design of asymmetrical half bridge flyback converter," *IEE Proceedings - Electric Power Applications*, vol. 149, no. 6, p. 433, 2002.

[6] S.-S. Lee, S.-K. Han, and G.-W. Moon, "Analysis and design of asymmetrical ZVS PWM half bridge forward converter with flyback type transformer," in *2004 IEEE 35th Annual Power Electronics Specialists Conference (IEEE Cat. No.04CH37551)*, IEEE, 2004.

[7] B.-R. Lin, H.-K. Chiang, C.-H. Tseng, and K.-C. Chen, "Analysis and implementation of an asymmetrical half-bridge converter," in *2005 International Conference on Power Electronics and Drives Systems*, IEEE, 2005.

[8] L. Huber and M. M. Jovanovic, "Analysis, design, and performance evaluation of asymmetrical half-bridge flyback converter for universal-line-voltage-range applications," in *2017 IEEE Applied Power Electronics Conference and Exposition (APEC)*, IEEE, mar 2017.

[9] M. Li, Z. Ouyang, and M. A. Andersen, "Analysis and optimal design of high frequency and high efficiency asymmetrical half-bridge flyback converters," *IEEE Transactions on Industrial Electronics*, pp. 1–1, 2019.

[10] M. Otsuka and O. Trescases, "A two-chip quasi-resonant buck converter with a 700v power-stage and mixed-signal current-mode control," in *2016 IEEE 17th Workshop on Control and Modeling for Power Electronics (COMPEL)*, IEEE, jun 2016.

[11] X. Fang and Y. Meng, "Isolated bias power supply for igbt gate drives using the fly-buck converter," in *2015 IEEE Applied Power Electronics Conference and Exposition (APEC)*, pp. 2373–2379, IEEE, March 2015.

[12] S. Ben-Yaakov, "The benefits of planar magnetics in of power conversion planar magnetics (pm) : The technology that meets the challenges of hf switch and resonant mode power conversion," 2006.

[13] C. Ropoteanu, P. Svasta, and C. Ionescu, "A comparative simulation analysis of toroid and planar magnetic cores near MHz region," in *2017 IEEE 23rd International Symposium for Design and Technology in Electronic Packaging (SIITME)*, IEEE, oct 2017.

[14] Z. Ouyang and M. A. E. Andersen, "Overview of planar magnetic technology—fundamental properties," *IEEE Transactions on Power Electronics*, vol. 29, pp. 4888–4900, sep 2014.

[15] P. Dowell, "Effects of eddy currents in transformer windings," *Proceedings of the Institution of Electrical Engineers*, vol. 113, no. 8, p. 1387, 1966.

[16] J. Ferreira, "Improved analytical modeling of conductive losses in magnetic components," *IEEE Transactions on Power Electronics*, vol. 9, no. 1, pp. 127–131, 1994.

[17] K. Venkatachalam, C. Sullivan, T. Abdallah, and H. Tacca, "Accurate prediction of ferrite core loss with nonsinusoidal waveforms using only steinmetz parameters," in *2002 IEEE Workshop on Computers in Power Electronics, 2002. Proceedings.*, IEEE.

Challenges in Calibrating an Unconventional Partial Discharge Measurement System for Pulsed Voltages

Markus Fürst, Mark-M. Bakran
UNIVERSITY OF BAYREUTH, DEPARTMENT OF MECHATRONICS
Centre for Energy Technology - ZET
Universitätsstraße 30
95447 Bayreuth, Germany
Phone: +49 (0) 921 55-7807
Fax: +49 (0) 921 55-847802
Email: Markus.Fuerst@uni-bayreuth.de
URL: http://www.mechatronik.uni-bayreuth.de

Acknowledgments

This project was supported by the ZF Friedrichshafen AG. Special thanks go to the department of Basic Development Electric Drives, Schweinfurt.

Keywords

≪Insulation≫, ≪Reliability≫, ≪Sensor≫, ≪Frequency-Domain Analysis≫, ≪Partial Discharge≫.

Abstract

In this paper, the signal of an unconventional partial discharge (PD) sensor is analysed on characteristic parameters both in time domain and frequency domain. This is done to find correlations to the standardised apparent charge q_a according to IEC 60270 and therefore make the used unconventional PD measurement system comparable to the conventional system. The benefit of the unconventional PD sensor is that it can be used even with pulsed voltage waveforms containing high switching noise due to high dV/dt . Therefore, a calibration of this system would enable a fair comparison to conventional sine wave test stands. The measurements for this investigation are mainly performed in a conventional sine wave test stand, where both PD measurement systems are mounted to compare their signals. As unconventional PD sensor a self-built PD Rogowski sensor with a high bandwidth is used. The characteristic parameters considered and analysed in the time domain are mainly the PD waveform, its time integral, the maximum and average peak amplitude of single PD events, the here introduced "average PD event" as well as the absolute PD quantity. In the frequency domain it is analysed if there are characteristic frequency bands, where the signal intensity correlates with the apparent charge. The derived findings are then examined further using only the unconventional sensor in a test stand for pulsed voltage waveforms.

Introduction

The design and dimensioning of the motor winding insulation in automotive electric motors is a difficult task because of contrary design goals with regard to thermal and electrical aspects. This is further complicated by the fact that the electric insulation properties are dependent on various parameters such as environmental conditions [1] including temperature [2] and humidity. The properties of the insulation determine the durability of the insulation, as it must withstand the applied voltage stress in the intended application. A poor design could otherwise lead to partial discharges (PD) occurring in insulation voids, which could lead to a reduced lifetime due to degradation of the insulation [3, 4]. In typical state-of-the-art automotive traction applications with 400 V the designers have already gained lots of experience about the insulation in the field to base their design decisions on. But increasing DC link voltages in battery electric vehicles and the desire for increasing switching speed to reduce switching losses are new challenges for an appropriate electric insulation design.

Because of the risk of reduced lifetime due to PD it is recommended in IEC 60034-18-41 to design the insulation completely free of PD during the whole lifespan [5]. To fulfil this requirement a sensitive and precise PD inception voltage (PDIV) measurement is necessary. This is possible with state-of-the-art 50 Hz sine wave PD tests, because of their low noise level below 1 pC. But because the PDIV also depends on the voltage waveform, the PDIV should be determined while the device under test (DUT) is stressed with the actual voltage waveform of the application [6]. The considered field of application of this paper are automotive applications. Therefore, a realistic PWM voltage waveform is needed to determine the PDIV in the application. Unfortunately, the high voltage slopes of this pulsed voltage waveform generates a lot of high frequency noise on the measured PD signal, which is why the sine wave PD measurement concept from IEC 60270 cannot be applied [7]. By using unconventional PD sensors like high frequency current transducers (HFCT) or ultra high frequency (UHF) antennas the detection of PDs is possible during pulsed voltage waveforms, like it is done in many publications. But it is challenging to achieve a high sensitivity especially in the presence of high switching noise [8]. Another common disadvantage of these unconventional measurement systems is that their PD output signal is not standardised. Most of these systems provide just the voltage signal of the respective PD sensor or the number of detected PD events. Due to the different sensor types or their spatial arrangement, the output signals of unconventional measurement systems are difficult to compare to each other and especially to the apparent charge q_a of state-of-the-art 50 Hz measurement systems. In literature some elaborate calibration methods are proposed to get a good reproducibility and comparability between different measurement systems or sensor types, e.g. for UHF antennas in [9].

In this paper it is investigated, whether there are parameters of the PD signal of a self-built PD sensor in the time domain or in the frequency domain, which show a correlation to the apparent charge q_a and could therefore be used to calibrate this unconventional not-standardised PD measurement system to the apparent charge q_a in an easy way. The applied PD sensor is a self-built Rogowski coil with a bandwidth of about 1 GHz. The measurements are done by using both a standardised PD measurement systems according to IEC60270 and the unconventional one simultaneously in one sine wave test stand and comparing their output signals. The transferability to pulsed measurements is also investigated with a pulsed voltage test stand.

Test Stands and the Unconventional PD Measurement System

Unconventional PD measurement system

The PD detection and measurement with the unconventional sensor is done with a self-build Rogowski coil, which is placed around one of the supply lines of the DUT, as it can be seen in both schematics of Fig. 1. In [10] this sensor was first introduced. The measurement method is based on the principle of a HFCT. Due to its high bandwidth of up to 1 GHz, it is able to measure the high frequency components of PD events. The voltage output of the PD sensor is proportional to the di/dt of the DUT's supply. Depending on the switching noise intensity, the sensor output has to be filtered with an analogue high pass filter to damp the switching noise. With this method the PD measurement system is able to detect PD even in the presence of high switching noise, which was already shown in previous work [10]. Due to the lack of high frequency noise in the sine wave test stand the measurements were conducted without the mentioned high pass filter. For better comparability between both test stands the switching noise on the PD sensor signal with pulsed voltages was reduced by limiting the dV/dt to about 2 V/ns and with the generally used high pass filter with a cut-off frequency of 150 MHz.

The PD measurement system uses a 1 GHz Lecroy HDO4104 oscilloscope to measure and analyse the PD sensor signal. The oscilloscope triggers on events in the unconventional PD sensor output signal V_{TE} that exceed a defined threshold value. The threshold value is set slightly above the remaining noise level.

Because of the aforementioned different noise levels of the two test stands used, the threshold has to be set on slightly different values.

One issue the introduced measurement system has is that the used oscilloscope is not suitable for a continuous measurement at high sampling rates. The maximum sampling rate of $2.5\,\mathrm{GS/s}$ for a single channel is needed to get a good representation of the high frequency components of the PD events. Unfortunately, this leads to a relatively short measurement time frame which the oscilloscope is able to capture as maximum. But the actual issue here is the oscilloscope's dead time before a new frame can be captured. This update rate can lead to a couple of not detected PDs during the dead time. For single pulse tests this is not an issue, but all of the conducted PD measurements were performed in a time range of several seconds. Depending, for example, on the sampling rate, the captured time range and the trigger settings, the ratio of actually captured time of the oscilloscope t_{cap} to the time which has passed t is approximately just $t_{\mathrm{cap}}/t = 0.01\,\%$.

But despite this low ratio, it is important to know that this issue only affects situations with a lot of PD events occurring in a very short time frame and does not affect the ability to detect single PDs and therefore of course the PD inception voltage (PDIV).

The aim of the performed measurements in this work includes the comparison of the two PD measurement systems at different PD activity levels. Therefore not only the PDIV was of interest, but also the PD events themselves. This requires testing at voltages where many PDs are generated, which is why the issue described above has to be considered in this case. Therefore, the captured PDs during a fixed measurement time in this paper are just a smaller sample out of all occurred PDs. But because the recorded PD events are almost evenly distributed during the measurement time, it is assumed that they are a good representation of all occurred PDs during the measurement time. To ensure the statistical representativity a minimum of 400 PD event samples was acquired for every measurement in this work.

Test stands for sine wave and pulsed voltage waveforms

For the investigations in this paper two different test stands were used. One for standard 50 Hz sine wave voltage waveforms and another for highly adjustable pulsed voltage waveforms.

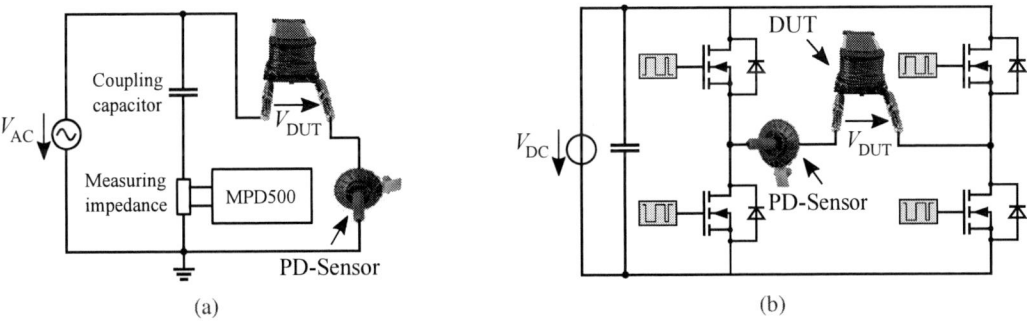

Fig. 1: Simplified schematic of the sine wave grid voltage test stand with included PD sensor in the ground path (a) and the pulsed voltage test stand (b).

In Fig. 1 (a) the simplified schematic of the applied conventional 50 Hz sine wave PD test stand is shown. It includes a conventional PD measurement system consisting of a coupling capacitor and a measuring impedance in parallel to the DUT. The conventional PD analysis system used in this work is the Omicron MPD500. It has a very high continuous sampling rate for PD measurement but the apparent charge q_{a} according to IEC 60270 is only calculated every 300 ms. In this time frame there are 15 full sine wave periods and therefore a lot of possible PD spots. Hence, q_{a} most likely includes many single PD events with possibly strongly deviating intensity. Therefore, it is of course challenging to calibrate single PD events or an average of them to this apparent charge.

To compare the measurement results of the MPD500 measurement system (designed in compliance to IEC 60270) with the used unconventional Rogowski PD sensor, the sensor is placed also in the existing sine wave test stand (see Fig. 1 (a)), which makes it possible to measure and collect the PD data simultaneously and analyse them afterwards.

The schematic of the second test stand used in this paper is shown in Fig. 1 (b). It is an H-bridge setup consisting of four CREE C2M0045170D MOSFETs with a break down voltage of 1700 V. In this configuration the insulation of the investigated DUT can be stressed with highly adjustable pulsed voltage waveforms. By changing the gate resistors, the dV/dt and therefore the rise and fall times of the voltage pulses can be adjusted. These high voltage slopes produce much more switching noise which ranges to much higher frequencies than the sinusoidal test stand. Therefore, the conventional MPD500 measurement system is not applicable in this test stand. So it is not possible to compare the both measurement systems in this setup, but the PD signals from the proposed unconventional PD sensor can be compared in both test stands with their different voltage waveforms. This may reveal potentially different PD behaviour at the given voltage waveforms.

All investigations were performed with a single tooth winding as DUT. It is cut open after a few windings, so there is no electrical connection, which is why the DUT acts as a capacitive load. Thus, by applying V_{DUT} to the terminals the winding-winding insulation is stressed.

Comparison of PD Measurement Systems in 50 Hz Sine Wave Test Stand

In the sine wave test stand multiple measurements were carried out at different apparent charge levels. Because it is not possible to set a specific q_a value, a nearly constant level of charges is achieved by slowly adjusting the applied voltage stress V_{AC}. Even at a very steady voltage level the apparent charge can fluctuate a lot. Therefore, several measurement attempts were made for different charge levels, but only those in which q_a was nearly constant for a measurement time of about 30 s were taken into account for the following investigations.

Because of the mentioned issue at achieving a steady apparent charge level, especially at a precise predefined value, the actual measured q_a values are not evenly distributed over the measuring range. Therefore, the measurements for all investigations were measured at 30 pC, 180 pC, 1050 pC and 2750 pC. The fact that it is possible to take measurements with the presented unconventional sensor at an average of about 30 pC shows the good sensitivity of the sensor. Due to this the self-built PD sensor is able to detect the PDIV in the sine wave setup at a very low apparent charge level as good as the conventional PD measurement system.

During the four measurements of nearly constant PD activity the oscilloscope triggered on PD events with a trigger value of 10 mV just above the noise level of about 4 mV. The difference of 6 mV was chosen as a safety margin for unexpected disturbances. This ensures that all triggered events are PDs. During the measurement time of about 30 s about 400 to 600 single PD events were recorded at every q_a level. As mentioned before, these events are just a smaller sample out of all occurred PDs, but it is assumed that these samples are statistically relevant.

The examination and comparison of the measured data is divided into two main evaluation methods in this paper, the time domain and frequency domain by means of the Fourier transformation.

PD characteristics in the time domain

Average PD signal

First the correlation between q_a and the direct voltage output of the unconventional PD sensor V_{PD} is examined. To enable a good comparison between the kind of averaged apparent charge value and the unconventional sensor it is reasonable to compare it to an averaged signal instead of the single recorded PD events of the sensor. For this purpose, in this paper an "average PD event" is introduced. It is calculated by averaging the absolute voltage signal of all recorded PD events during the measurement time of 30 s at one q_a level (approx. 500 PD events). By using the absolute value of V_{PD} ($|V_{PD}|$), a possible cancellation due to PDs with different polarity is avoided.

In Fig. 2 (a) the average PD events $|V_{PD}|_{avg}$ for all measurement series are shown. As it can be seen, the average PD events differ between different charge levels: The signal amplitude changes noticeably, but

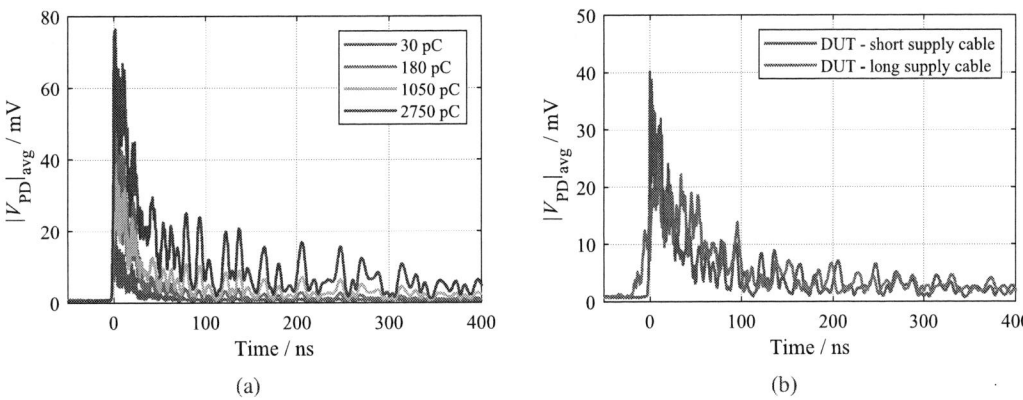

Fig. 2: (a): "Average PD events" measured with the unconventional PD sensor for different q_a values. (b): "Average PD events" at approximately the same q_a level with different supply cable lengths.

the signal form itself does not change significantly. Overall, the peak voltage amplitude (approximately 0 ns to 30 ns) and the level of the following tail increases with q_a. Interestingly, the increase of the peak amplitude is not monotonous, as the 180 pC curve is slightly higher than the one with 1050 pC. Whereas this is not the case in the tail of the average PD events, because there the signal value rises monotonically. Maybe the deviation from the monotonous increase is just due to statistical fluctuations and a small PD sample size. Nevertheless, this result is promising as the amplitudes of the average PD event have a recognisable correlation to the apparent charge. Especially the signal amplitude of the tail could be used as a calibration parameter between both PD measurement systems.

But there are also a few drawbacks of this calibration method. Firstly, the absolute voltages are quite low, with just a few mV, especially when calibrating to the tail of the average PD event. Therefore, the signal is prone to noise, particularly in the range of low apparent charge values of up to 1050 pC. But achieving a high sensitivity at low apparent charge levels is crucial to detect a low and therefore more realistic PDIV. Secondly, this method requires a recording of PDs in advance and calculate the average PD event in the post processing afterwards, which is a lot of effort and has the disadvantage, that the PD activity is not known during the measurement. In the following sections other parameters are investigated, which could be evaluated in real time.

Another general difficulty in the calibration of the PD sensor output voltage to the apparent charge is the fact that the PD signal depends among others on the used DUT as well as the impedance of the supply cables. In Fig. 2 (b) a comparison of two average PD events with similar apparent charge levels is given as an example for this issue. One of the measurements was recorded with a short supply cable. For the second measurement an additional cable of about 1 m extended the supply cable between the sensor and the DUT. As it can be can be seen, the signal waveform is deformed and damped in the latter case. The same issue can be found by comparing different DUTs. This is caused most likely by the change of the attenuation in the PD signal path to the PD sensor. For a similar reason a calibrator is required in the conventional 50 Hz PD measurement setup, which can eliminate such effects.

It is therefore unreasonable to assume that the unconventional PD sensor can be calibrated in the sine wave test stand with the MPD500 and then easily be used in another completely different test setup, such as the pulse test stand. A PD calibrator could very likely solve this issue also in the unconventional measurement setup. There is just one problem, typical PD calibrators discharge their calibration charge at a considerably lower frequency compared to some frequency components of the PDs. Depending on the transfer behaviour and frequency range of the used PD sensor the calibrator pulse could lead to a different sensor response than a real PD with the same charge. A custom calibrator with a suitable frequency range and a good known transfer behaviour is therefore necessary to fully calibrate the unconventional PD measurement system for different test stands.

Peak amplitude of PD events

As described in the last section the amplitude of the average PD event is a potential parameter for calibration. Therefore, the peak amplitude of the single PD events is evaluated in more detail (see Fig. 3 (a)). Each bar shows the minimum, maximum and the average value of the absolute peak voltage $|V_{PD}|_{max}$ of

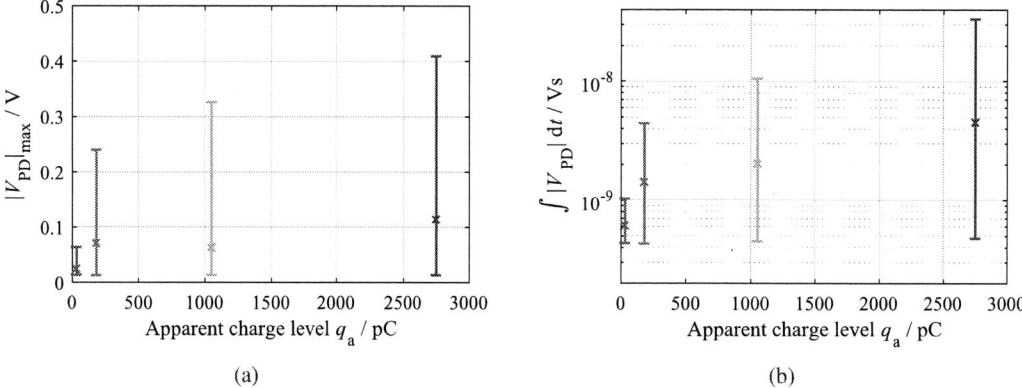

(a)   (b)

Fig. 3: Minimum, maximum and the average of the absolute peak voltage (a), and the time integral of the absolute PD events (b) at specific q_a levels.

every single recorded PD at the respective apparent charge level. The trend of the average values with increasing q_a is of course the same as the peak values showed in Fig. 2 (a). It can also be observed that in every measurement at a constant apparent charge level there are always very small PD events present which are just slightly higher than the threshold value of $10\,mV$. Therefore, the minimal peak values in a measurement can not be used for a calibration. Whereas the highest peak values of the measurement series does show a monotonous increasing trend correlating with the apparent charge. Due to a higher sensitivity compared to the average PD evaluation, and the possibility to analyse the highest peak values more easily in real time, the maximum peak voltage of the single PD events seems to be a better calibration parameter according to these measurements.

Time integral of PD events

Another intuitive way of analysing the PD signals in the time domain is to compare the voltage-time area of the single PD events. Because it is expected that at higher apparent charge levels there are PD events with more energy present, which should therefore be reflected by larger voltage-time areas of the sensor signal. Therefore, the absolute sensor signal $|V_{PD}|$ of every single measurement is integrated over the time duration of the respective PD event. In Fig. 3 (b) the results of these integrations are shown. The error bars again indicate the minimal and maximal occurred integral of a single PD event for each measurement row. Additionally, the x displays the average value of all recorded PD events at their respective apparent charge level.

Just as in the previous evaluation, the minimal values of the integral of single PD events are almost identical for the different q_a levels. But the average and maximum values show a significant and monotonous dependency on the apparent charge.

Therefore, the time integral of the absolute PD sensor signal is a potentially good calibration parameter for comparisons between the sine wave and pulsed voltage test stands. But it should also be mentioned that performing the integration in real time is much more difficult than evaluating the peak voltage, for example.

Number of PD events

Another possibility to analyse the PD activity is to count the occurring number of PD events in a given period of time. The conventional measurement system MPD500 has a build in feature for this purpose and also for the proposed unconventional system it is possible to design an analogue counting circuit for example to fulfil this purpose. But in the following the characteristics should only be examined with the MPD500. Therefore, the number of PD events at different apparent charge levels was measured for 30 s each. The results are shown in Fig. 4 on a logarithmic scale. As it can be seen in the figure, the slope was compared at two different threshold values 50 pC and 100 pC, above which PD events were counted.

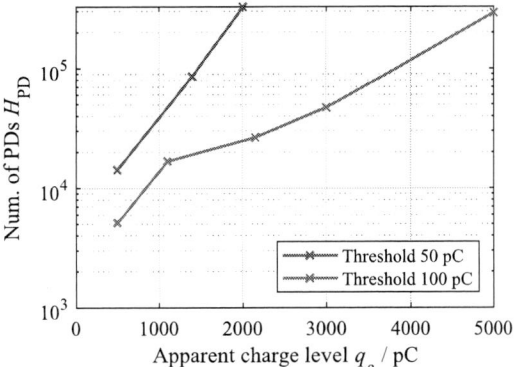

Fig. 4: Number of PD events in 30 s at different apparent charge levels for two tested threshold values.

The data shows an almost linear trend on the logarithmic scale for both lines but with different slopes. Therefore, the PD numbers increase exponentially with higher apparent charge, at least in the tested sine wave test stand. Due to the slower rise of the measurement with the higher threshold value of 100 pC it can be concluded that there is also an increase in PD event numbers with low charge at higher q_a levels, which are only taken into account by the lower threshold value of 50 pC. Therefore, it is not the case that the same PD events which occur at low apparent charge levels just have more charge at higher levels, but there are also an increasing amount of PD events with small charge. This is in accordance to the findings in the previous sections where there are also always small PDs present even at higher apparent charge levels.

Due to the exponential dependency between the PD event numbers and the apparent charge, as it can be seen in Fig. 4, it seems to be a good calibration parameter. However, a slight deviation of the threshold value could lead to a significant error in the calculated apparent charge value in the unconventional system. Additionally, it could be more difficult in a pulsed voltage test to separate single different PD events from each other, because they mostly occur in the much shorter voltage rise times [10] compared to a 50 Hz sine wave test, which could make the calibration invalid.

PD characteristics in the frequency domain

Searching for suitable calibration parameters, the frequency range will also be investigated. For example, it is conceivable that the detected partial discharges with the unconventional PD measurement system take place mainly in certain frequency bands, where a correlation to the apparent charge may be found. This could justify a bandpass filter, which effectively filters other frequency noise. Moreover, it is possible that characteristic PD frequencies shift at higher apparent charge levels and therefore the frequency itself could enable a calibration of the unconventional sensor to the apparent charge.

To analyse the frequency spectrum of the PDs the fast Fourier transformation was applied to all recorded PD events of the same measurement series analysed before which were measured with the unconventional measurement system. In Fig. 5 all frequency components are averaged individually over all measured PD events at a specific apparent charge level, similar to the average PD event.

The averaged frequency spectrum from Fig. 5 (a) for the whole bandwidth shows some characteristic frequencies where the magnitude is clearly dependent on the apparent charge level. One wide frequency band is from about 400 MHz to 500 MHz. This information can be used to design a bandpass filter which for example damps everything except a small band at about 450 MHz. That design could be beneficial especially in the pulsed voltage test stand, where a lot of switching noise is present. But just like in the

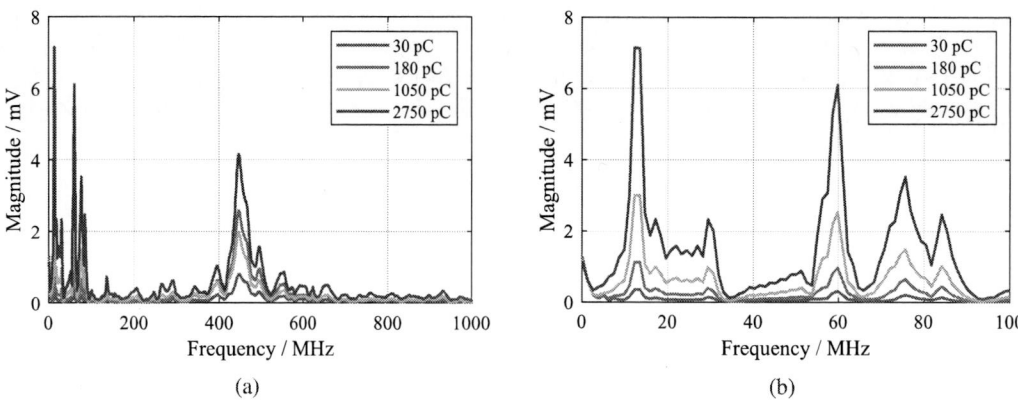

(a)   (b)

Fig. 5: (a): Averaged Fourier spectrum of PD events at different q_a levels. (b): Zoom on the frequency band between 0 MHz to 100 MHz.

evaluation of the average PD the magnitude does not increase monotonically with the apparent charge level.

Moreover, especially the frequency band from 400 MHz to 500 MHz with a high mid-frequency is dependent on the used DUT and supply cable length as it was already seen in Fig. 2 (b). A comparison between two different DUTs, a wire twist and the used single tooth, is given at approximately the same apparent charge level in Fig. 6 (a). It is obvious that the frequency spectrum is quite different. This is also the fact in Fig. 6 (b), where the same DUT is measured with two different supply cable lengths. Especially, because the signal in this frequency range almost completely disappeared in Fig. 6 (b), a bandpass in this frequency range is not suitable.

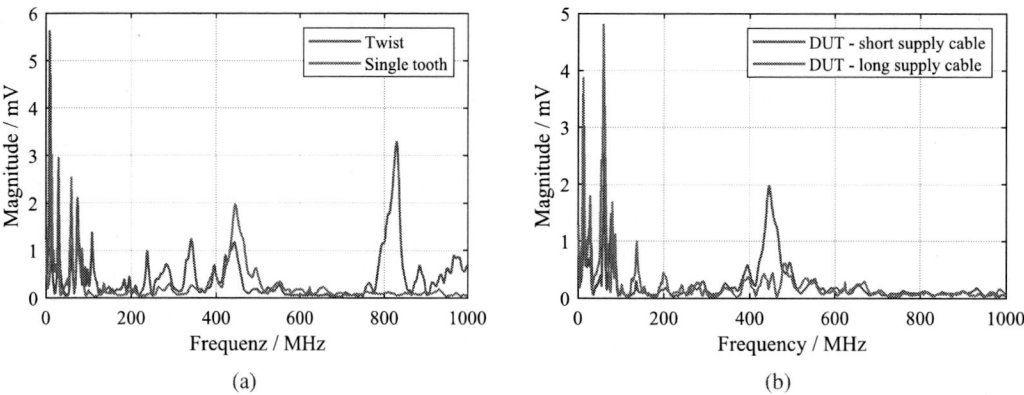

(a)   (b)

Fig. 6: Averaged Fourier spectrum for different DUTs (a) and different supply cable lengths (b).

Another frequency band which is clearly dependent on the apparent charge level is from 0 MHz to 100 MHz. In Fig. 5 (b) this frequency band is zoomed for better visualisation. As it can be seen, in this range the magnitude is monotonous increasing with the apparent charge. Therefore, this frequency range would be best for a calibration. But unfortunately in this low frequency range the switching noise in the pulsed voltage tests is quite high and even higher than the PD signals at low apparent charge levels. In [8] it was already shown that the switching noise at about 50 V/ns in the presented pulsed voltage test bench is dominant up to a frequency of about 150 MHz. Therefore, this frequency range has to be damped in the intended test stand and filtered which is why it is not suitable for the required calibration.

Overall, the frequency domain analysis shows that the occurring frequency components of the PDs do not qualify itself as a calibration parameter to the apparent charge in the given tests. The characteristic frequency in the PD signal spectrum turned out to correlate to the DUT and not to the apparent charge, due to a change in the resonance frequency. But, the frequency spectrum can be helpful to identify frequency ranges where noise or PD signals are present and therefore facilitate the decision making which frequency filters should be used in the measurement setup.

Unconventional PD Measurement System in the Pulsed Voltage Test Stand

In the following, a series of measurements conducted in the pulsed test stand (see Fig. 1 (b)) is presented to confirm in principle the applicability of some parameters found in the sine wave test stand investigations. Due to the lack of the MPD500 reference and thus the respective apparent charge value in this test stand the measurements are not conducted at specific apparent charge levels but at discrete DC-link voltages. The voltage was increased from below the PDIV of about 800 V to 1250 V in steps of 50 V. At each step about 600 single oscilloscope measurements were recorded. As mentioned before in the pulsed voltage test stand an analogue high pass filter of 150 MHz is used to damp the switching noise. Of course these measurements at fixed V_{DC} levels are not ideal in order to confirm the trends of the calibration parameters found in the aforementioned investigations. But assuming an overall increase of PD activity with higher voltage stress above the PDIV, it will at least give an indication if the trends are also valid in the pulsed voltage test stand.

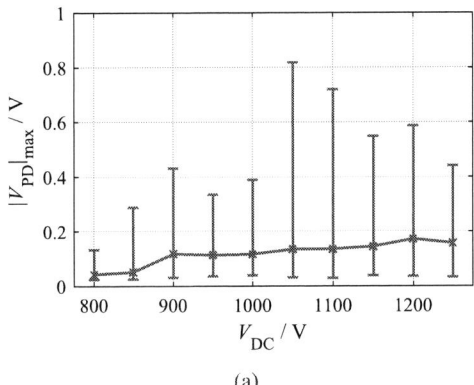
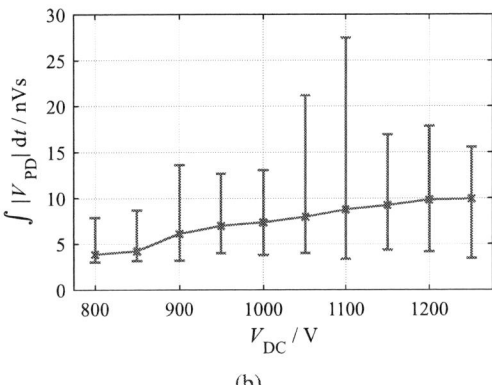

(a) (b)

Fig. 7: Minimum, maximum and the average of the absolute peak voltage (a), and the time integral of the absolute PD events (b) at increasing V_{DC} levels.

In Fig. 7 (a) the average value as well as the variations of the absolute peak voltage $|V_{PD}|_{max}$ which occurred in the approximate 600 recorded PD events at every V_{DC} voltage is displayed. Similar to the analysis in Fig. 3 (a) the average values are affected by the increasing applied voltage stress, but do not increase monotonically. However, the maximum voltages of the single PD events behave differently as they also do not increase monotonously. This is most likely due to statistical fluctuations in the occurred PD intensities and possibly by a insufficient sample quantity. Because of these uncertainties the single maximum peak voltage of the recorded samples is no reliable calibration parameter.

Because the integral of the absolute PD signal was the most promising calibration parameter in the comparison at the sine wave test stand, this parameter is also analysed in the pulsed test stand. Fig. 7 (b) shows the results. Like before in Fig. 7 (a) the maximum values fluctuate a lot, but the average integral values in fact rise monotonically as in the previous investigation in Fig. 3 (b).

The results in the pulsed voltage test stand show that even with the analogue filter the average absolute integral value of a reasonable number of PD events seems to be the most accurate one of the investigated parameters to calibrate the unconventional PD measurement system to the apparent charge. But the acquisition and calculation of this value in real time during the PD testing needs some effort. A possible alternative parameter at least for detecting the PDIV at low q_a is the average peak voltage $|V_{PD}|_{max}$, because in this range the data shows an monotonous increase with a high sensitivity over the PD activity.

Conclusion

Precise and sensitive PD measurements are needed to ensure long lifetime of electrical insulation of electric motors especially with higher system voltages and increasing dV/dt stress. Because the PD behaviour is dependent on the voltage waveform, tests have to be conducted with realistic pulsed voltage waveforms. Conventional PD measurement systems cannot be used under these high dV/dt pulse patterns, which makes it crucial to introduce unconventional PD sensors and measurement systems. To compare results from both measurement systems and test stands a calibration is needed. But calibrating an unconventional PD measurement system to the apparent charge q_a according to IEC 60270 is a big challenge. In a first step this paper tried to find correlations between both measurement system types to enable a possible calibration which is valid even for different test environments. A basic requirement for the desired calibration parameter is that it changes monotonically at increasing apparent charge values. Various parameters of the used unconventional PD sensor in the time domain and frequency domain were investigated.

The measurements in frequency domain show no useful correlations. But analysing the PD sensor signal in the frequency domain is still beneficial for choosing the right filters for the measurement system, so that the switching noise can be damped sufficiently while enable enough of the PD signal to pass in the pulsed voltage test stand. On the contrary, the analysis in the time domain did show some promising results. Especially the maximum peak voltage of the recorded PD events $|V_{PD}|_{max}$ as well as the time integral of the absolute PD sensor signal did show a monotonically increase with rising apparent charge levels at least in the sine wave test stand. In the subsequent measurements conducted in the pulsed voltage test stand, only the average time integral of the sampled PD events could be confirmed as a promising calibration parameter.

This paper showed in principle, that it is possible to use an unconventional PD measurement system to get sufficient information on the qualitative PD activity in a fixed test setup. But the quantitative calibration on the apparent charge q_a according to IEC 60270 remains a huge challenge. As the results show, a parameter which integrates the PD signal of a reasonable number of PD events is most suitable as a calibration parameter, because it is less prone to statistical fluctuations and modifications in the overall test setup, but it is challenging to realise in an pulsed voltage test stand.

References

[1] K. Younsi, D. Snopek, J. Hayward, P. Menard, and J. C. Pellerin, "Seasonal changes in partial discharge activity on hydraulic generators," in *Proceedings: Electrical Insulation Conference and Electrical Manufacturing and Coil Winding Conference (Cat. No.01CH37264)*, 2001, pp. 423–428.

[2] R. Schifani, R. Candela, and P. Romano, "On PD Mechanisms at High Temperature in Voids Included in an Epoxy Resin," *IEEE Transactions on Dielectrics and Electrical Insulation*, vol. 8, no. 4, pp. 589–597, 2001.

[3] A. Cavallini, G. C. Montanari, D. Fabiani, and M. Tozzi, "The influence of PWM voltage waveforms on induction motor insulation systems: Perspectives for the end user," in *8th IEEE Symposium on Diagnostics for Electrical Machines, Power Electronics & Drives*, 2011, pp. 288–293.

[4] A. Küchler, *Hochspannungstechnik*. Berlin, Heidelberg: Springer Vieweg, 2017.

[5] *IEC 60034-18-41:2014: Rotating electrical machines – Part 18-41: Partial discharge free electrical insulation systems (Type I) used in rotating electrical machines fed from voltage converters – Qualification and quality control tests (IEC 60034-18-41:2014)*.

[6] *IEC 61934:2011: Electrical insulating materials and systems – Electrical measurement of partial discharges (PD) under short rise time and repetitive voltage impulses (IEC/TS 61934:2011)*.

[7] *IEC 60270:2000: High-voltage test techniques - Partial discharge*.

[8] M. Fürst and M.-M. Bakran, "An Advanced Filtering Method for Partial Discharge Measurement in the Presence of High dV/dt," in *PCIM Europe 2019; International Exhibition and Conference for Power Electronics, Intelligent Motion, Renewable Energy and Energy Management*, 2019, pp. 1855–1861.

[9] M. Siegel, S. Coenen, M. Beltle, S. Tenbohlen, *et al.*, "Calibration Proposal for UHF Partial Discharge Measurements at Power Transformers," *Energies*, vol. 12, no. 16, p. 3058, 2019.

[10] M. Denk and M.-M. Bakran, "Partial Discharge Measurement in a Motor Winding Fed by a SiC Inverter - How Critical is High dV/dt Really?" In *PCIM Europe 2018; International Exhibition and Conference for Power Electronics, Intelligent Motion, Renewable Energy and Energy Management*, 2018, pp. 685–690.

Electrothermal Modeling of GaN Power Transistor for High Frequency Power Converter Design

Loris Pace[1], Florian Chevalier[1], Arnaud Videt[1], Nicolas Defrance[2], Nadir Idir[1], Jean-Claude De Jaeger[2]

[1]LABORATORY OF ELECTRICAL ENGINEERING AND POWER ELECTRONICS
[2]INSTITUTE OF ELECTRONICS, MICROELECTRONICS AND NANOTECHNOLOGIES
University of Lille, Avenue Henri Poincaré, Cité Scientifique
Villeneuve d'Ascq, France
E-Mail: loris.pace@centralelille.fr, nadir.idir@univ-lille.fr

Keywords

«Gallium Nitride», «High frequency power converter», «High power density systems», «Modelling», «Radio frequency».

Abstract

This work proposes the electrothermal modeling of a packaged GaN power transistor in order to evaluate by simulation its performances in a 200 W – 1 MHz DC/DC converter. The complete electrical modeling of the high frequency converter using EM-circuit co-simulations is presented. After validation of the GaN transistor switchings waveforms and estimation of power losses, the operating temperature of the device is simulated and experimentally validated.

Introduction

High frequency operation leads to the reduction of volume and weight of power converters as it is required by many embedded systems. Increased carriers mobility and reduced inter-electrode capacitances in Wide BandGap (WBG) power semiconductors such as GaN HEMT or SiC Schottky diodes makes these devices good candidates for the development of efficient high frequency power converters [1]. Designing power converter by simulation requires accurate device models. The first section of this paper details the high frequency characterization and modeling work carried out in order to get complete accurate electrical models at ambient temperature for the WBG devices that will be used as power switches in a 200 W – 1 MHz Buck converter. The complete electrical modeling of the high frequency power converter is proposed in Keysight ADS® software using EM model for the PCB. The switching waveforms of the GaN transistor are simulated and compared to experimental results in short time operation of the converter to avoid self-heating effects on switchings. In the third section, a methodology based on dissipated power measurements is presented to get the thermal model of the GaN power transistor. The temperature dependencies of the drain current source and the access resistances are highlighted. A method is also detailed to get thermal models for the cooling system. Then, thermal performances of the GaN device are estimated by simulation in ADS® and experimental validations of the thermal aspects are carried out in short and long operating times. Measurements are compared with simulation results.

Electrical Characterization and Modeling of Packaged WBG Power Devices

Electrical Characterization and Modeling of the GaN Power Transistor

In this work, the use of the GaN transistor (GS66502B form GaN Systems® - 650 V -7.5 A) is considered as the main power switch in the high frequency power converter. The transistor, presented in Fig. 1(a), is constituted of three terminals located at the bottom side of the device. The source connection is used for thermal dissipation. The electrical model considered for the GaN transistor is shown in Fig. 1(b). In this equivalent circuit, the presence of access resistances and inductances at each electrode is taken into account. The intrinsic part of the transistor is constituted of voltage-dependent inter-electrode capacitances, gate and drain current sources.

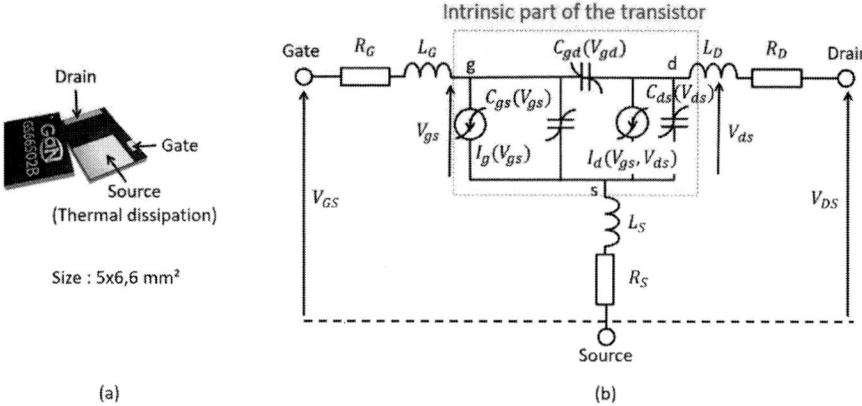

(a)

(b)

Fig. 1: GaN power transistor GS66502B : (a) packaging (b) electrical model

This work proposes the model identification (Fig. 1(b)) using high frequency techniques such as S-parameter and pulsed I-V characterizations. For S-parameters, as the device is packaged, a specific test fixture on PCB is fabricated in order to perform accurate measurements with the test equipment. As shown in Fig. 2(a), the testing board includes grounded coplanar 50 Ω transmission lines and on-board coaxial connectors permitting high frequency measurements such as S-parameters without reflection. Fig. 2(b) gives the equivalent circuit of the testing board in Fig. 2(a). $[Z_{DUT}]$ is the impedance matrix related to the GaN transistor, Y_0 represents the capacitive coupling between a transmission line and the ground plane, $Z_1 - Z_3$ are the impedances of the transmission lines and $Y_4 - Y_6$ represent the capacitive coupling between transmission lines.

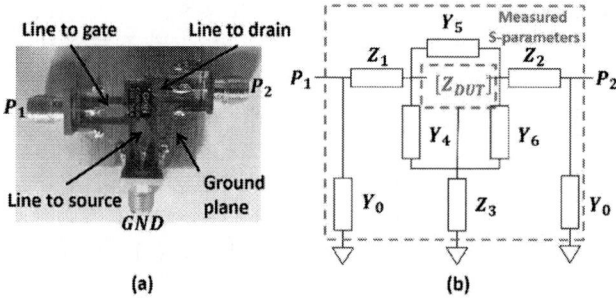

(a)

(b)

Fig. 2: S-parameter testing board for the GaN transistor (a) PCB (b) equivalent circuit

According to Fig. 2(b), to get the impedance parameters of the GaN transistor $[Z_{DUT}]$ from the measured S-parameters, a calibration procedure is required [2]. Y_0 parameter can be obtained by performing 1-port S-parameter measurement on the calibration fixture shown in Fig. 3(a) which consists in an open transmission line. Then, $Z_1 - Z_3$ parameters are obtained by performing 2-port S-parameter measurement on the calibration fixture shown in Fig. 3(b) which consists in three shorted transmission lines. Finally, $Y_4 - Y_6$ parameters are obtained by performing 2-port S-parameter measurement on the calibration fixture shown in Fig. 3(c) which consists in the complete characterization board on PCB without connecting the transistor. Once these parameters are known, the impedance matrix of the device can be easily obtained from the measured S-parameters.

The access parasitic resistances and inductances of the transistor are extracted using the Cold FET technique which consists in biasing the device at V_{GS} higher than the threshold voltage V_{TH} (around 1.3 V) to make the channel conductive and $V_{DS} = 0V$ to avoid thermal issues. In this configuration, the equivalent circuit of the GaN transistor can be represented as given in Fig. 4 [2]. From this figure, the real parts of the impedance parameters can be expressed as given by equations (1) to (3) according to [2]. K_1 and K_2 are fitting parameters [2].

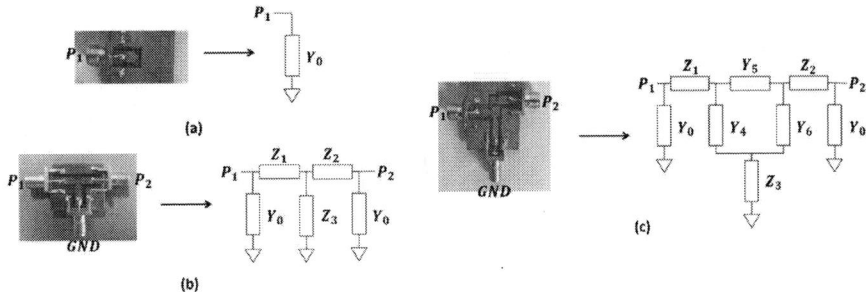

Fig. 3: Calibration fixtures (a) 1-port open transmission lines (b) 2-port short standard (c) 2-port open standard

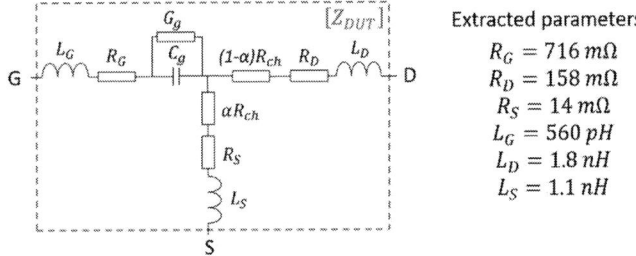

Extracted parameters :
$R_G = 716\ m\Omega$
$R_D = 158\ m\Omega$
$R_S = 14\ m\Omega$
$L_G = 560\ pH$
$L_D = 1.8\ nH$
$L_S = 1.1\ nH$

Fig. 4: Equivalent circuit of the GaN transistor in Cold FET conditions

$$Re\left(Z_{11_{DUT}} - Z_{12_{DUT}}\right) = R_G + \frac{G_g}{G_g^2 + C_g^2 \omega^2} \qquad (1)$$

$$Re\left(Z_{22_{DUT}}\right) = R_D + R_S + \frac{1}{K_1(V_{GS} - V_{TH})} \qquad (2)$$

$$Re\left(Z_{12_{DUT}}\right) = R_S + \frac{1}{K_2(V_{GS} - V_{TH})} \qquad (3)$$

By performing S-parameters in the range 1 MHz- 100 MHz at several value of V_{GS}, the access resistances R_G, R_D and R_S of the GaN device can be obtained by fitting methods [2]. Fig. 5(a) shows the proposed method to determine R_D and R_S from $Re\left(Z_{22_{DUT}}\right)$ and $Re\left(Z_{12_{DUT}}\right)$ vs. V_{GS}.

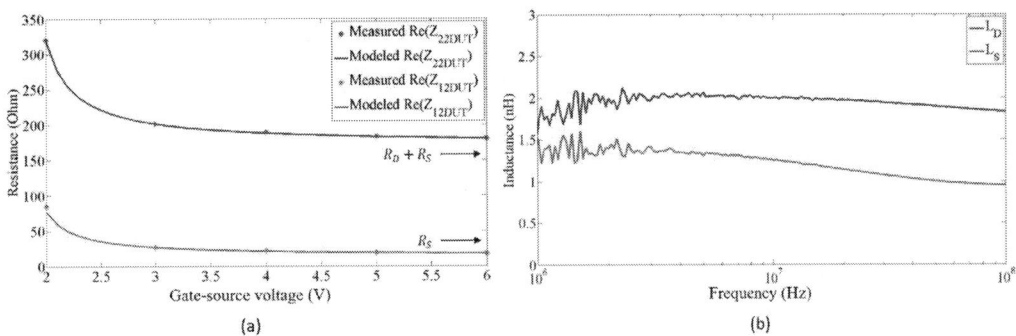

Fig. 5: Determination of drain and source access elements (a) $Re\left(Z_{22_{DUT}}\right)$ and $Re\left(Z_{12_{DUT}}\right)$ vs. V_{GS} (b) L_D and L_S extraction versus frequency

From Fig. 4, the access inductances of the GaN transistor can be obtained from the imaginary parts of the impedance parameters as given by equations (4) to (6) according to [2], [3]. Fig. 5(b) shows measurement results of L_D and L_S obtained by performing S-parameters between 1 MHz and 100 MHz at $V_{GS} = 6V$. L_G is determined as the slope of the curve $Im(Z_{11_{DUT}} - Z_{12_{DUT}}).\omega$ vs. ω^2 [3].

$$Im(Z_{11_{DUT}} - Z_{12_{DUT}}).\omega = L_G.\omega^2 - \frac{1}{C_g} \tag{4}$$

$$L_D = \frac{Im(Z_{22_{DUT}} - Z_{12_{DUT}})}{\omega} \tag{5}$$

$$L_S = \frac{Im(Z_{12_{DUT}})}{\omega} \tag{6}$$

The evolution of inter-electrode capacitances versus their respective voltage are obtained by 2-port S-parameter measurements according to the proposed method in [2]. In order to apply high voltage to the device, specific biasing fixtures are fabricated on PCB. These bias Tee allow accurate S-parameter measurements in the range 1 MHz – 100 MHz. After calibration process and subtracting the access parasitic, the inter-electrode capacitances are directly obtained from admittance parameters of the intrinsic part in off-state conditions [2]. Fig. 6(a) compares the measured capacitances C_{iss}, C_{oss} and C_{rss} versus V_{DS} voltage to the capacitances given by manufacturer's datasheet. The generic non linear model given by equation (7) is then used to model the capacitances evolution versus their respective voltages and modeling results are shown in Fig. 6(b).

$$C_{ij}(V_{ij}) = C_{0_{ij}} + \sum_{k=1}^{3} C_{k_{ij}}\left(1 + \tanh\left(\frac{V_{ij} + V_{k1_{ij}}}{V_{k2_{ij}}}\right)\right), \ ij = \text{gs, gd, ds.} \tag{7}$$

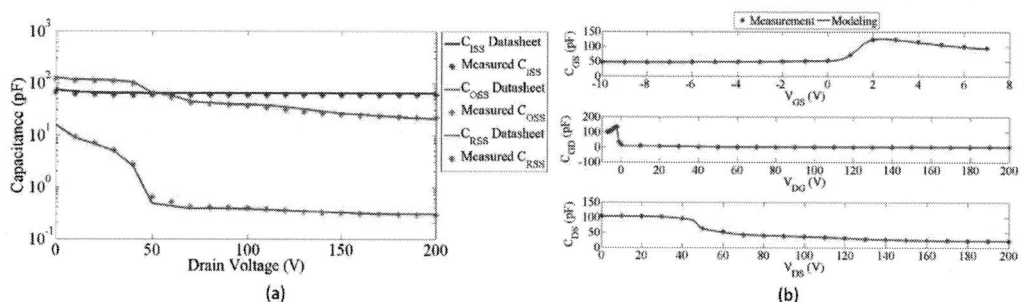

Fig. 6: Inter-electrode capacitances (a) extraction at 10 MHz versus V_{DS} (b) modeling of the inter-electrode capacitances

The static characteristics of the GaN transistor are measured by the curve tracer B1505A. The evolution of the drain current I_d versus V_{gs} and V_{ds} is obtained in pulsed mode in order to limit self-heating of the device [4]. An optimization procedure is then implemented in ADS® Software, as shown by the simplified scheme in Fig. 7(a), to model the transistor characteristics using non-linear equations given by (8) and (9) for the direct and reverse part of the characteristic respectively [5]. Modeling results versus V_{gs} and V_{ds} are presented in Fig. 7(b) and Fig. 7(c) respectively.

$$I_d(V_{gs}, V_{ds}) = K_d.\ln\left(1 + \exp\left(\frac{V_{gs} - b_d}{c_d}\right)\right)\frac{(m_d + n_d.V_{gs}).V_{ds}}{1 + (d_d + e_d.V_{gs}).V_{ds}} \tag{8}$$

$$I_d(V_{gs}, V_{ds}) = K_i.\ln\left(1 + \exp\left(\frac{V_{gd} - b_i}{c_i}\right)\right)\frac{V_{ds}}{1 + (d_i + e_i.V_{gd}).V_{ds}} \tag{9}$$

Finally, the gate current I_g evolution versus V_{gs} is measured in open-drain configuration and modeled according to [4]. Fig. 7(d) shows the modeling results of the gate current versus gate-source voltage.

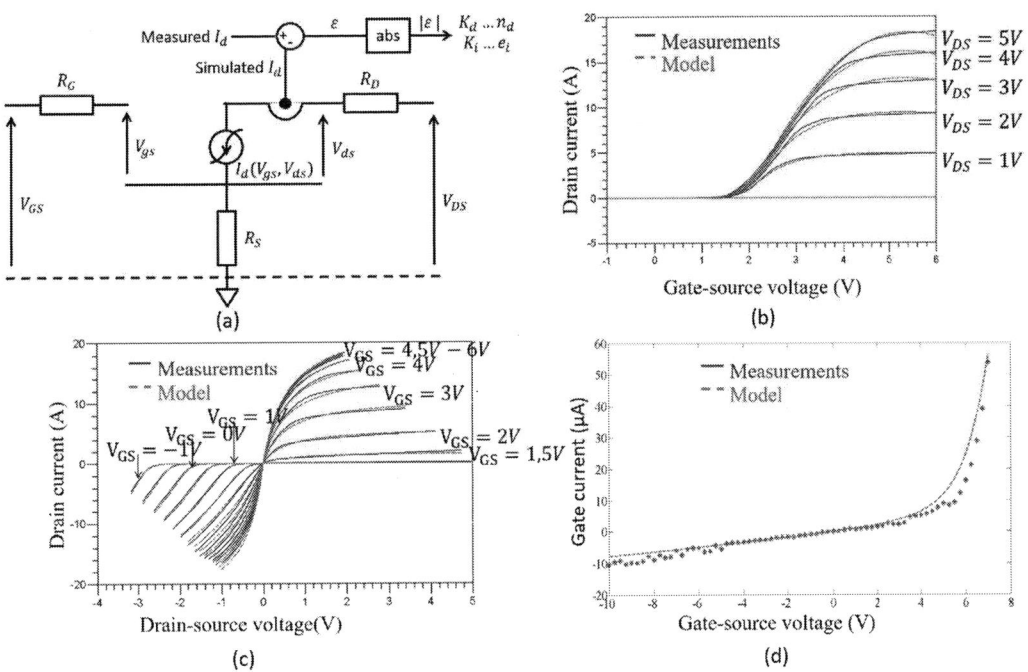

Fig. 7: Modeling of the static characteristics (a) optimization procedure for $I_d(V_{gs}, V_{ds})$ (b) $I_d(V_{gs})$ (c) $I_d(V_{ds})$ (d) $I_g(V_{gs})$

Electrical Characterization and Modeling of the SiC Schottky Diode

The SiC Schottky diode (IDDD04G65C6XTMA1 from Infineon® - 650 V - 8 A), presented in Fig. 8(a), is used as the freewheeling diode of the high frequency power converter. It is selected mainly for its low capacitive charge (7 nC) allowing efficient high frequency operation and its top-side cooling system permitting to optimize cooling requirements of the whole converter. The electrical modeling of the diode is necessary to accurately estimate the performances of the GaN transistor in the power converter. The considered model for the SiC diode is given in Fig. 8(b).

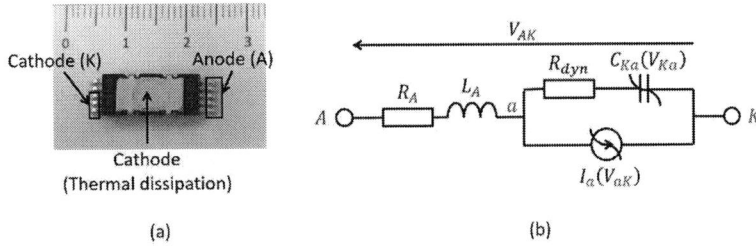

Fig. 8: SiC Schottky diode (a) presentation of the diode (b) considered electrical model

The access parasitic and the inter-electrode capacitance of the SiC diode are extracted using 1-port S-parameter measurements. Similarly to the characterization of the GaN transistor, a specific test fixture is made on PCB for accurate measurements as shown in Fig. 9(a). The equivalent circuit of the characterization is presented in Fig. 9(b). Using the previously presented calibration procedure, parameters Y_0, Z_1, Z_2 and Y_3 can be determined and then the impedance of the diode Z_D is obtained from the S-parameter S_{11}. According to the previous method, the parameters $R_A = 140~m\Omega$, $R_{dyn} = 148~m\Omega$ and $L_A = 7.2~nH$ are obtained for the device [3]. The evolution of the diode capacitance C_{Ka} versus the applied reverse voltage V_{Ka} is determined by adapting the previous method used for the GaN transistor. Fig. 10(a) gives the modeling results of the diode capacitance versus reverse voltage using the non-linear model in (7). Finally, the diode current source I_a versus V_{aK} is experimentally determined using the curve tracer B1505A in pulsed mode. The resulting static characteristic is then modeled by a diode non-linear model described by authors in [6] and results are shown in Fig. 10(b).

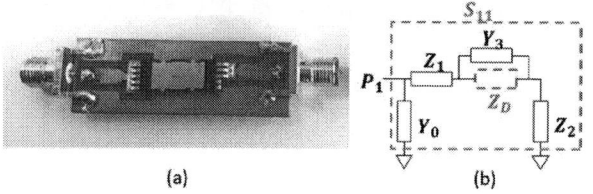

Fig. 9: Characterization board for the SiC diode (a) presentation (b) equivalent circuit

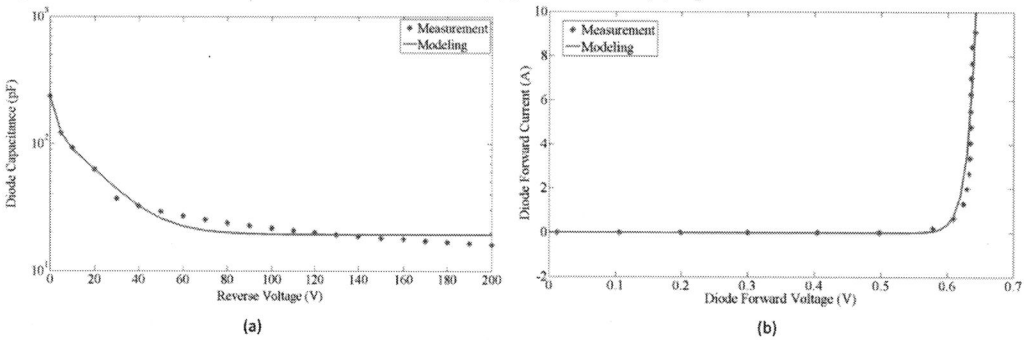

Fig. 10: SiC diode intrinsic characteristics (a) capacitance modeling (b) current source modeling

Evaluation of the GaN Transistor Electrical Performances in the High Frequency Power Converter

Power Converter Design

In this work, a high frequency Buck converter is designed using the modeled GaN power transistor and SiC Schottky diode. A schematic representation of the converter is given in Fig. 11(a). The converter operates at 1 MHz with an input voltage of 200 V, an output current of 2 A and a duty cycle of 50 %. Therefore, the output power is set to 200 W. It is fabricated on a two-sided 1.6 mm FR4 PCB with 105 µm Copper each side. The transistor's drain current is measured by a current shunt SDN-414-025 (25 mΩ, 2 W) and the V_{DS} voltage is measured by high voltage passive probe PPE 4kV. The layouts on top and bottom sides are presented in Fig. 11(b) and Fig. 11(c) respectively.

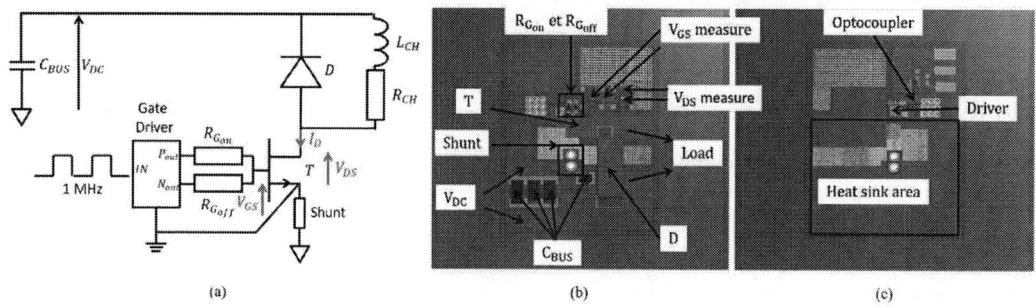

Fig. 11: Presentation of the converter (a) schematic (b) top side layout (c) bottom side layout

Electrical Modeling of the Power converter

In order to accurately evaluate the electrical performances of the GaN power transistor, a complete modeling of the converter is proposed. Electrical models of the passive components (DC bus capacitors, load inductors...) are obtained from measurements using an impedance analyzer HP 4294A. The gate driver model is built according to measured rise and fall times. The current shunt is modeled according to [7] from measurements using the impedance analyzer and the high voltage probe model is estimated according to the work proposed by authors in [8]. All these electrical models are then implemented in ADS® software. The fabrication files of the converter PCB are imported in ADS® Momentum to get an ElectroMagnetic (EM) model that can then be used in circuit simulations. EM simulations are set up to 2 GHz. As shown in Fig. 12(a), the power loop is designed on a single side of the PCB, while the gate loop is designed using both sides of the PCB. Fig. 12(b) shows the power loop and control loop inductances evolution versus frequency obtained from the EM model. Relatively low loop inductance of around 3 to 4 nH are obtained for the proposed converter design.

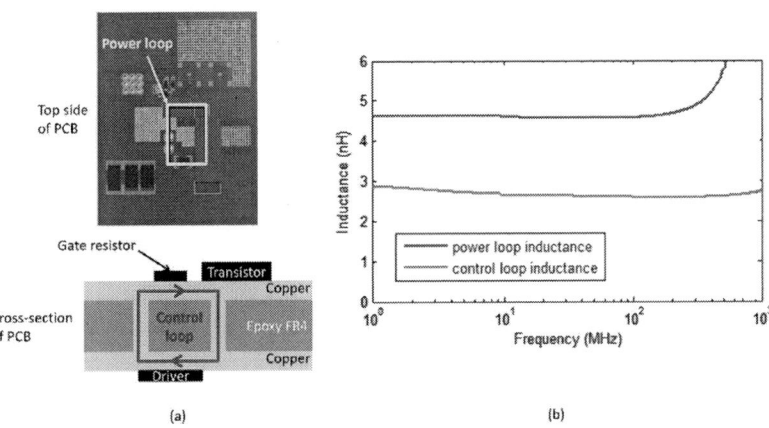

(a)   (b)

Fig. 12: Power and control loop (a) proposed design (b) inductances evolution versus frequency

Experimental Validation of the Proposed Electrical Modeling

Operating tests have been carried out on the high frequency power converter at 1 MHz. The transistor's drain current I_D and V_{DS} voltage are measured after 10 μs of operation (10 cycles in burst mode), once the electrical steady state is reached and to avoid self-heating effects on waveforms. To validate the electrical modeling, simulated and measured current and voltage waveforms are compared during turn-on and turn-off transitions of the GaN transistor in Fig. 13.

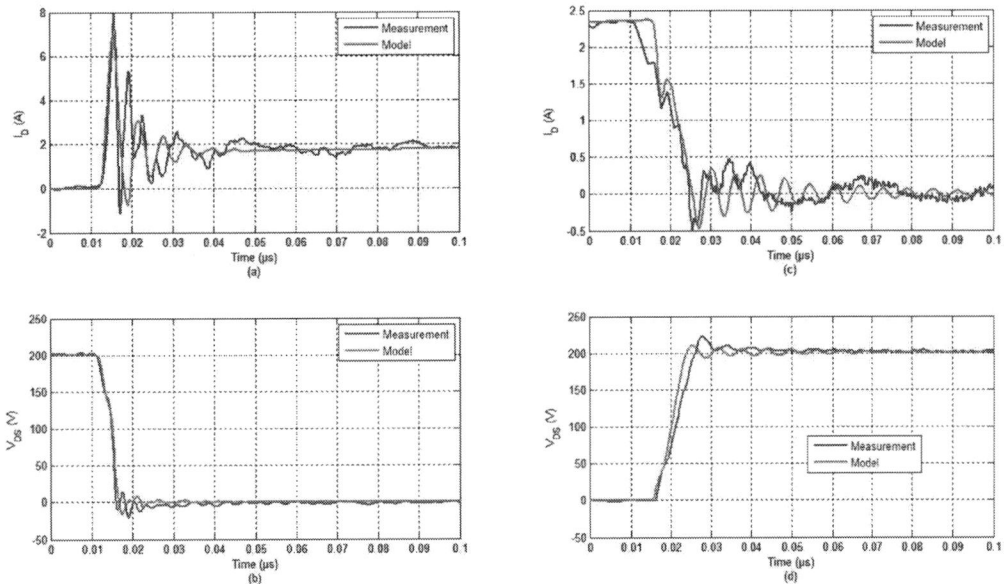

Fig. 13: Current and volatge switching waveforms comparison between simulation and measurement at turn-on ((a) and (b)) and turn-off ((c) and (d)) of the GaN power transistor

Good agreement is observed between simulation and experimental results in terms of rise and fall times and high frequency dynamics. Slight differences in high frequencies may be due to non-controlled capacitive and inductive couplings in the experimental test bench.

Evaluation of the GaN Transistor Thermal Performances in the High Frequency Power Converter

Estimation of Power Losses in the GaN Transistor

In order to evaluate the temperature of the GaN device in operation in the high frequency converter, the power losses model $P(T_j)$, depending on the junction temperature T_j, is considered as given by equation (10).

$$P(T_j) = P_0 + k_{\Delta R}(T_j - T_a)I_{D0}^2, \text{ with } P_0 = 3.4\ W, k_{\Delta R} = 2\ m\Omega/°C, T_a = 23\ °C, I_{D0} = 1\ A \qquad (10)$$

The average power losses on one cycle at ambient temperature, P_0, is estimated from simulated waveforms shown in Fig. 13. These power losses include conduction and switching losses in one period of operation. The coefficient $k_{\Delta R}$ is used to model the rise of the on-state resistance $R_{DS_{on}}$ according to the junction temperature T_j. This coefficient is determined by a linearization of the $R_{DS_{on}}(T_j)$ experimental characteristic given in [2]. I_{D0} is the average value of the drain current. T_a is the room temperature during tests.

Thermal Modeling of the GaN Power Transistor and its Cooling System

The proposed method to determine the thermal model of the packaged GaN power transistor is based on dissipated power measurements during 1 ms as shown in Fig. 14(a) for $V_{GS} = 4V$ and different V_{DS} values. The power drop during measurements observed in Fig. 14(a) is related to the effect of temperature rise on the static characteristic of the transistor. In order to take into account this self-heating phenomena, an electrothermal coefficient γ is used in the modeling. Equation (11) gives the proposed model for the dissipated power curves in Fig. 14(a) [9]. $p_D(\Delta T(t))$ is the dissipated power, P_{D0} is the initial power at the beginning of pulses, $\Delta T(t)$ is the GaN device temperature rise and T_a is the room temperature in K. Then, the optimization procedure illustrated in Fig. 14(b) is implemented in ADS® using a second order Foster model with parameters R_{TH1}, R_{TH2}, C_{TH1} and C_{TH2}.

$$p_D(\Delta T(t)) = P_{D0}\left(\frac{T_a}{T_a + \Delta T(t)}\right)^{\gamma} \qquad (11)$$

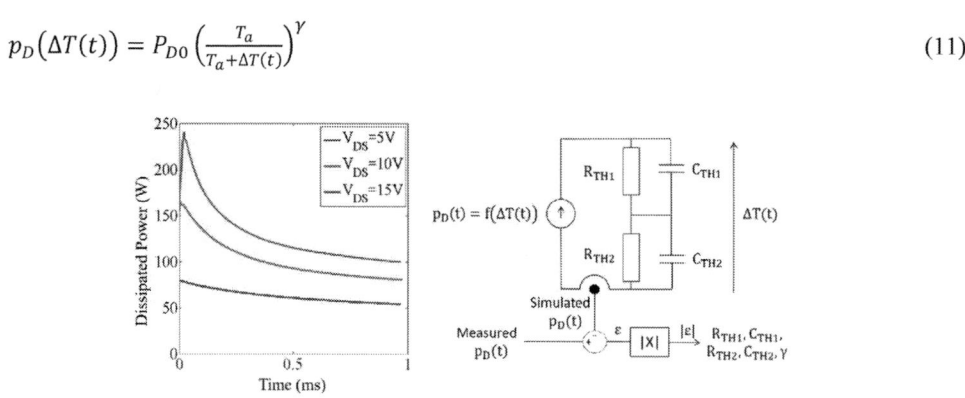

(a)   (b)

Fig. 14: GaN transistor thermal modeling (a) dissipated power measurements (b) optimization procedure for thermal model extraction

Simulation results obtained after optimization are presented in Fig. 15(a). The Foster model is then transformed into a Cauer model for next thermal simulations [10]. The resulting parameters R_{TH3}, R_{TH4}, C_{TH3} and C_{TH4} of the Cauer model as well as the γ parameter are given in Table I. Finally, in order to complete the electrothermal modeling of the GaN transistor, the influence of the device's junction temperature on the drain and source resistances is modeled regarding previous work detailed in [2].

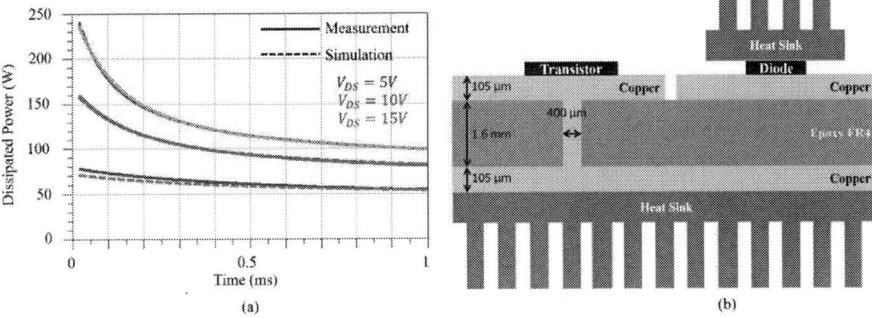

Fig. 15: (a) Simulation results after optimization for thermal model extraction (b) cooling system presentation for the high frequency power converter (cross-section)

Fig. 15(b) illustrates the cooling systems designed for the WBG devices in the power converter. As the GaN transistor dissipates heat from its bottom side, thermal via-holes are designed to conduct heat to a heat sink set to the bottom side of the converter PCB. The rigorous thermal modeling of the vias requires 3D simulations in Multiphysics software which is an outlook of this work. Here, the thermal model of via-holes is estimated as a first order Cauer model with parameters $R_{TH_{vias}}$ and $C_{TH_{vias}}$. Knowing the via-holes geometry and copper physical properties, these parameters are estimated for an equivalent area including 12 thermal via-holes directly located under the thermal source of the transistor. Using the same approach, knowing the heat sink geometry and aluminium physical properties, an equivalent thermal resistance $R_{TH_{HS}}$ and capacitance $C_{TH_{HS}}$ can be deduced. The thermal model parameters for via-holes and heat sink are given in Table I.

Table I: Thermal model parameters for the GaN transistor and its cooling system

R_{TH3} ($°C/W$)	R_{TH4} ($°C/W$)	C_{TH3} ($mJ/°C$)	C_{TH4} ($mJ/°C$)	γ	$R_{TH_{vias}}$ ($°C/W$)	$C_{TH_{vias}}$ ($mJ/°C$)	$R_{TH_{HS}}$ ($°C/W$)	$C_{TH_{HS}}$ ($J/°C$)
0.66	0.56	0.57	2.1	4.3	14	1.6	2.9	135

Experimental Validation of the Proposed Thermal Modeling

Using the previous power losses model as well as the transistor, via-holes and heat sink thermal models, the junction temperature T_j and the case temperature T_c of the GaN device are estimated by simulation using the circuit given in Fig. 16(a). To validate the temperature estimations, measurements are performed within one hour of the power converter operation in nominal conditions 200 W, 1 MHz. The temperature of the GaN transistor is measured on the top side by mean of a thermal camera as shown in Fig. 16(b). Fig. 16(c) gives the comparison between simulated and measured temperatures of the transistor versus operating time. These results show that the proposed thermal modeling can correctly estimate the temperature of the GaN device in a wide range of operating time of the power converter. However, fast temperature transient estimation cannot be validated with the test equipment.

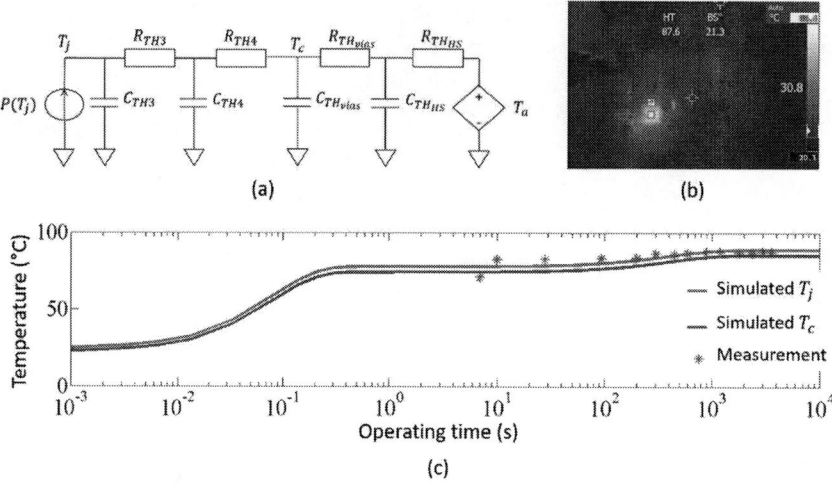

Fig. 16: Evaluation of the GaN transistor temperature (a) simulation circuit (b) measured temperature after one hour (c) comparison between simulated and measured temperatures

Conclusion

In this paper, a methodology is presented to electrically and thermally model a packaged GaN power transistor in order to accurately evaluate by simulation its performances in a 1 MHz – DC/DC converter. The electrical modeling of the GaN device is based on S-parameters and pulsed IV measurements, while the thermal modeling is based on dissipated power analysis. Several optimization procedure are then proposed to extract parameters for non-linear elements of the models. A complete electrothermal model of the GaN transistor is implemented in ADS® software with the aim to perform EM/circuit co-simulations of the power converter in operation. The temperature of the device is then analysed in short and long operating time of the power converter. The experimental validations show the potential of the proposed method for the design of efficient high frequency power converter design.

References

[1] A. M. S. Al-bayati, S. S. Alharbi, S. S. Alharbi and M. Matin, "A comparative design and performance study of a non-isolated DC-DC buck converter based on Si-MOSFET/Si-Diode, SiC-JFET/SiC-schottky diode, and GaN-transistor/SiC-Schottky diode power devices," *2017 North American Power Symposium (NAPS)*, Morgantown, WV, 2017, pp. 1-6.

[2] L. Pace, N. Defrance, A. Videt, N. Idir, J. -C. De Jaeger and V. Avramovic, "Extraction of Packaged GaN Power Transistors Parasitics Using S-Parameters," in *IEEE Transactions on Electron Devices*, vol. 66, no. 6, pp. 2583-2588, June 2019.

[3] L. Pace, N. Defrance, J. D. Jaeger, A. Videt and N. Idir, "A Method to Determine Wide Bandgap Power Devices Packaging Interconnections," *2019 IEEE 23rd Workshop on Signal and Power Integrity (SPI)*, Chambéry, France, 2019, pp. 1-4.

[4] L. Pace, N. Defrance, A. Videt, N. Idir and J. Dejaeger, "S-Parameter Characterization of GaN HEMT Power Transistors for High Frequency Modeling," *PCIM Europe 2018; International Exhibition and Conference for Power Electronics, Intelligent Motion, Renewable Energy and Energy Management*, Nuremberg, Germany, 2018, pp. 1-8.

[5] A. M. Bouchour, P. Dherbécourt, A. Echeverri, A. E. Oualkadi and O. Latry, "Modeling of Power GaN HEMT for Switching Circuits Applications Using Levenberg-Marquardt Algorithm," 2018 International Symposium on Advanced Electrical and Communication Technologies (ISAECT), Rabat, Morocco, 2018, pp. 1-6.

[6] S. Khannal, A. Noor, S. Neeleshwar and M. S. Tyagi, "Modeling aspects of current calculation of 4H-SiC Schottky diode," *2008 International Conference on Recent Advances in Microwave Theory and Applications*, Jaipur, 2008, pp. 857-860.

[7] Z. Liu, X. Huang, F. C. Lee and Q. Li, "Package Parasitic Inductance Extraction and Simulation Model Development for the High-Voltage Cascode GaN HEMT," in IEEE Transactions on Power Electronics, vol. 29, no. 4, pp. 1977-1985, April 2014.

[8] K. Ammous, H. Morel and A. Ammous, "Analysis of Power Switching Losses Accounting Probe Modeling," in IEEE Transactions on Instrumentation and Measurement, vol. 59, no. 12, pp. 3218-3226, Dec. 2010.

[9] C. Anghel, A. M. Ionescu, N. Hefyene and R. Gillon, "Self-heating characterization and extraction method for thermal resistance and capacitance in high voltage MOSFETs," ESSDERC '03. 33rd Conference on European Solid-State Device Research, 2003., Estoril, Portugal, 2003, pp. 449-452, doi: 10.1109/ESSDERC.2003.1256910.

[10] K. Murthy and R. Bedford, "Transformation between Foster and Cauer equivalent networks," in IEEE Transactions on Circuits and Systems, vol. 25, no. 4, pp. 238-239, April 1978.

Modeling and fault detection in photovoltaic systems using the I-V signature

Abdelhadi BENZAGMOUT[abc], Thierry TALBERT[a], Olivier FRUCHIER[a], Thierry MARTIRE[b], Philippe ALEXANDRE[c], Carolina PENIN[c]

[a]PROMES-CNRS Laboratory, University of Perpignan Via Domitia / Perpignan, France
[b]IES CNRS, University of Montpellier / Montpellier, France
[c]ENGIE Green / 215 Rue Samuel Morse, Montpellier, France
E-Mail: abdelhad.benzagmout@promes.cnrs.fr

Acknowledgements

PROMES-CNRS and Institut d'Electronique of Montpellier want to thanks the region of occitanie and ENGIE-Green for their funding of this project, and to thank ENGIE-Green for sending and permit us to use the PV panels from the "Rivesaltes-Grid" project. Also, for the opportunity to carry out tests on a PV power plant in full operation conditions.

Keywords

«Photovoltaic», «Modeling», «Electrical signature», «Fault detection».

Abstract

Photovoltaic power plants are more and more numerous all over the planet. In France, profit of PV power plants is done by a fixed buying back price of electricity. The repurchase price of electricity is constant until 2020, then variability of electricity will begin. Thus, to keep PV power plant profitable, a new management model is necessary. This model includes resource prediction, fault detection and management of maintenance and servicing to maximize energy production. This paper focuses only on the fault detection. In the first part, we will present the I-V signature of the panel in healthy and faulty mode. Then, we will focus on the electrical model. This model permits us to simulate the cell, panel and string in healthy and faulty mode. Finally, we present the fault detection algorithm using the I-V signature.

Introduction

One of the most relevant challenges for a cleaner and more efficient energy is high penetration rates from renewable sources on electricity grid considering their inconveniences: variability and intermittency of the produced power.

Photovoltaic power plant (several MW) are more and more numerous. But the management remain classic due to fixed price of the electricity. Moreover, the decrease of the electricity price and price variations implies new management model of power plants. Those improvements can be divided into (a) resource prediction and (b) fault detection.

Actually, fault detection method used in PV power plants is very simple. Full disconnection from the grid is the only default actually managed in real time. This includes, faults from transformer(s) and inverter(s). All the other points, permanent or temporary shading, hot spots, etc. are not verified in real time. All the defects : cells, panels, strings defects, connections etc. are managed from the energy point of view. ie the impact of this defaults on power are verified off-line. The fault is sometimes detected several days after it appears. Currently, the cause for triggering an alarm to the owner of the power plant is done as follows. The contract between the owner and the electricity manager define the number of kWh or MWh per year that needs to be generated. From this value, the owner define the value of kWh or MWh per months and per weeks. Thus, an alarm is sent if the produced energy is too far from the previous value. The variation of energy is given at the contract signature. This methodology is good if repurchase price of electricity is constant. But, if the power plant owner wants to respond for example to spot market (where the price fluctuates depending on the time), it's clearly not the best way to proceed.

EPE'20 ECCE Europe

Assigned jointly to the European Power Electronics and Drives Association & the Institute of Electrical and Electronics Engineers (IEEE)

There are few PV power plants in France that are interested to spot market. But in 2020, integration of the variability of the price is an obligation. A solution to the problem can be the integration of fault detection to anticipate maintenance and servicing of the PV installation. Then it's possible to increase performance ratio of the power plant[1].

A fault detection system needs (a) an electrical device that can measure current and voltage for different loads, (b) a full list of fault defects and the corresponding signatures, and of course a fault detection algorithm. On this article, we first present the electrical signature IV, this signature which represents the basic element to make the fault detection. Then, we will show the signature of some common defects. In the second part, we describe the work done to develop the electrical models. Then, in the next part, we explain how those models will be used to make simulations of panel / string / power plant installation in healthy and faulty mode. In the fourth part, we will propose a fault detection algorithm. Finally, we conclude and discuss on the performance of the proposed solution and the future perspectives of this device integration in a complete detection of defects in photovoltaic power plants.

Electrical signature of PV panels

The I-V signature allows to show the existence of defects, each defect has its own signature. As we note in (Figure 1, 2), the panel signature in healthy mode represents the reference (green curve) under a given value of illumination and temperature. All the signatures of the defects (T_c, R_s, R_{sh}, shading, soiling) have their own behavior:[2], [3]

- T_c: decrease in V_{oc} (open circuit voltage);
- R_s: decrease of the slope (*);
- R_{sh}: increase of the slope (**);
- Shading: inflection point (***);
- Soiling: dimming of I_{cc} (short circuit current).

Therefore, the I-V signature is the basic element that will be analyzed to do the fault detection. However, fault detection requires first a reliable electrical model to simulate the signatures of these defects under different constraints and at different scales.

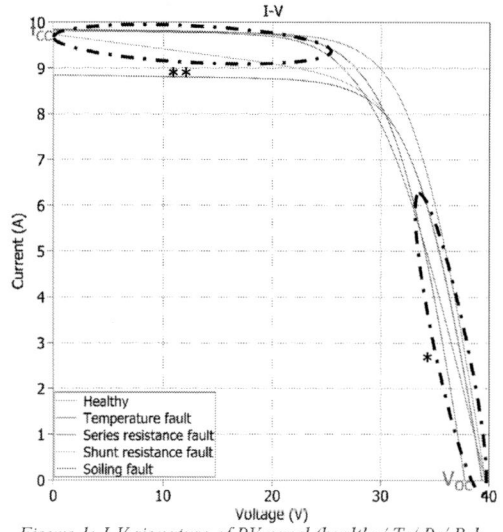

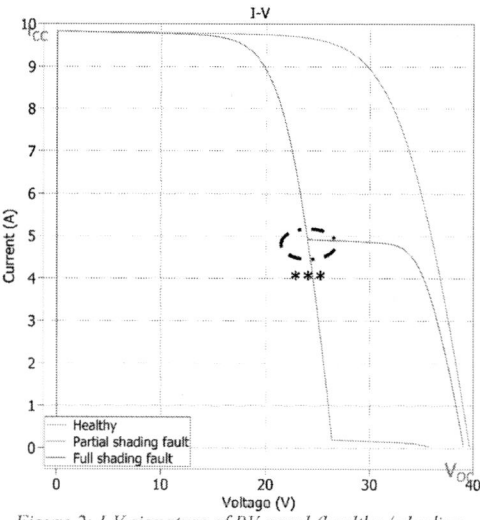

Figure 1: I-V signature of PV panel (healthy/ Tc/ Rs/ Rsh and soiling defect)

Figure 2: I-V signature of PV panel (healthy / shading defect)

Modeling of PV cell, panel and string

The model used for the photovoltaic cell, is the model known in the literature: single-diode model. This model is described by the following equation: [4], [5]

$$I = I_L - I_0 \left(\exp\left(\frac{q(V + IR_s)}{AKT} \right) - 1 \right) - \frac{V + IR_s}{R_{sh}} \qquad (1)$$

- I_L: Photo-current (A)
- I_0: Saturation current
- R_s: Series resistance (Ω)
- R_{sh}: Shunt resistance (Ω)
- A: Ideality factor (without unity)

Based on the electrical model of a photovoltaic cell (Figure 6), we can build the model of the photovoltaic panel (Figure 7). This consists of a serial-parallel set of several cells. When a cell produces less current due to a fault, the associated bypass diode short-circuits the corresponding block of cells. Thanks to this representation, we can now model all the defects that can affect a photovoltaic system by acting on the parameters of the model (Table I). For example, if we want to model a faulty mode panel with a series resistance fault, we must increase the value of the series resistance in the model.

Table I: Fault modeling

Fault	Modeling	Effect	Demonstration and effect explanation
Temperature (T_c)	Increase the T_c	Decrease in V_{oc}	$$I_{cc} = \frac{I_L}{1 + \frac{R_s}{R_{sh}}}$$ $$I_{cc} = \frac{\frac{GTI}{1000} \cdot \left(I_{L_{r\acute{e}f}} + \mu_{I_{cc}} \cdot (T_C - T_{r\acute{e}f}) \right)}{1 + \frac{R_s}{R_{sh}}}$$ $$\frac{\partial \left(R_s / R_{sh} \right)}{\partial T_c} \approx 0 \;;\; \frac{\partial I_L}{\partial T_c} > 0$$ $$\frac{\partial I_{cc}}{\partial T_c} \ll \Rightarrow \frac{\partial I_{cc}}{\partial T_c} \approx 0$$ $$V_{oc} = \frac{nKT_c}{q} \cdot \ln\left(1 + \frac{I_L}{I_0} \right)$$ $$I_0 = I_{0_{r\acute{e}f}} \cdot \left(\frac{T_c}{T_{r\acute{e}f}} \right)^3 \cdot \exp\left(\left(\frac{qE_g}{nT_{r\acute{e}f}} \right)\left(\frac{1}{T_{r\acute{e}f}} - \frac{1}{T_c} \right) \right)$$ $$V_{oc} = a_0 \cdot T_c \cdot \ln\left(1 + \frac{a_1}{a_2 \cdot T_c^3 \cdot \exp\left(a_3 \left(\frac{1}{a_4} - \frac{1}{T_c} \right) \right)} \right)$$ The study of the variation of the function $V_{oc}(T_c)$ allows us to show $$\frac{\partial V_{oc}}{\partial T_c} < 0$$ **When T_c increases, I_{cc} remains roughly constant and V_{oc} decreases[6]**
Series resistance (R_s)	Increase the R_s	Decrease of the slope (*)	$$I_{cc} = \frac{I_L}{1 + \frac{R_s}{R_{sh}}} \Rightarrow \frac{\partial I_{cc}}{\partial R_s} < 0$$ Note : $\frac{R_s}{R_{sh}} \ll$ example (for *sunmodule plussw285* $R_s = 0{,}369\Omega$, $R_{sh} = 318{,}18\Omega$, $\frac{R_s}{R_{sh}} = 1{,}1 \cdot 10^{-3}$) $\Rightarrow \frac{\partial I_{cc}}{\partial R_s} \ll$ $$V_{oc} = V_t \times \ln\left(1 + \frac{I_L}{I_0} \right) = V_{R_{sh}} \big|_{I=0}$$ V_{oc} is independent of $R_s \Rightarrow \frac{\partial V_{oc}}{\partial R_s} = 0$ **When R_s increases, V_{oc} remains constant and I_{cc} decreases[7]**
Shunt resistance (R_{sh})	Decrease the R_{sh}	Increase of the slope (**)	$$I_{cc} = \frac{I_L}{1 + \frac{R_s}{R_{sh}}} \Rightarrow \frac{\partial I_{cc}}{\partial R_{sh}} > 0$$

			$$V_{oc} = V_{R_{sh}}\big|_{I=0} = R_{sh} \cdot I_{R_{sh}} \Rightarrow \frac{\partial V_{oc}}{\partial R_{sh}} > 0$$ **When R_{sh} decreases, I_{cc} and V_{oc} will also decreses[8]**
Shading	Decrease the GTI at the cell level	Inflection point (***)	**When a cell is affected by shading**, the current produced decreases significantly. Other non-shaded cells continue to produce a current value that is higher than the shaded cell. These non-shaded cells, if they are in series with the shaded cell, will debit a higher current in the shaded cell. This will force the shaded cell to switch to receiver mode. The negative voltage of the shaded cell can therefore reverse the voltage of the all series of cells and thus **activate the bypass diode and make the inflection point appear**[9]
Soiling	Decrease the GTI at the panel level	Decrease of the I_{cc}	$$I_{cc} = \frac{I_L}{1 + \frac{R_s}{R_{sh}}}$$ $$I_{cc} = \frac{\frac{GTI}{1000} \cdot \left(I_{L_{réf}} + \mu_{I_{cc}} \cdot (T_C - T_{réf})\right)}{1 + \frac{R_s}{R_{sh}}}$$ $$\frac{\partial I_{cc}}{\partial GTI} > 0$$ $$V_{oc} = V_t \times \ln\left(1 + \frac{I_L}{I_0}\right) \Rightarrow \frac{\partial V_{oc}}{\partial GTI} > 0$$ **When GTI decreases, I_{cc} and V_{oc} will also decreses[10]**

The electrical model of the panel have been used under the software PLECS®. The first simulations are made under a constant illumination and constant temperature. Then, a model of the complete system (photovoltaic string + inverter) (Figure 8) was developed in order to make simulations under variable illumination and variable temperature and a power evolution managed by the MPPT of the inverter.

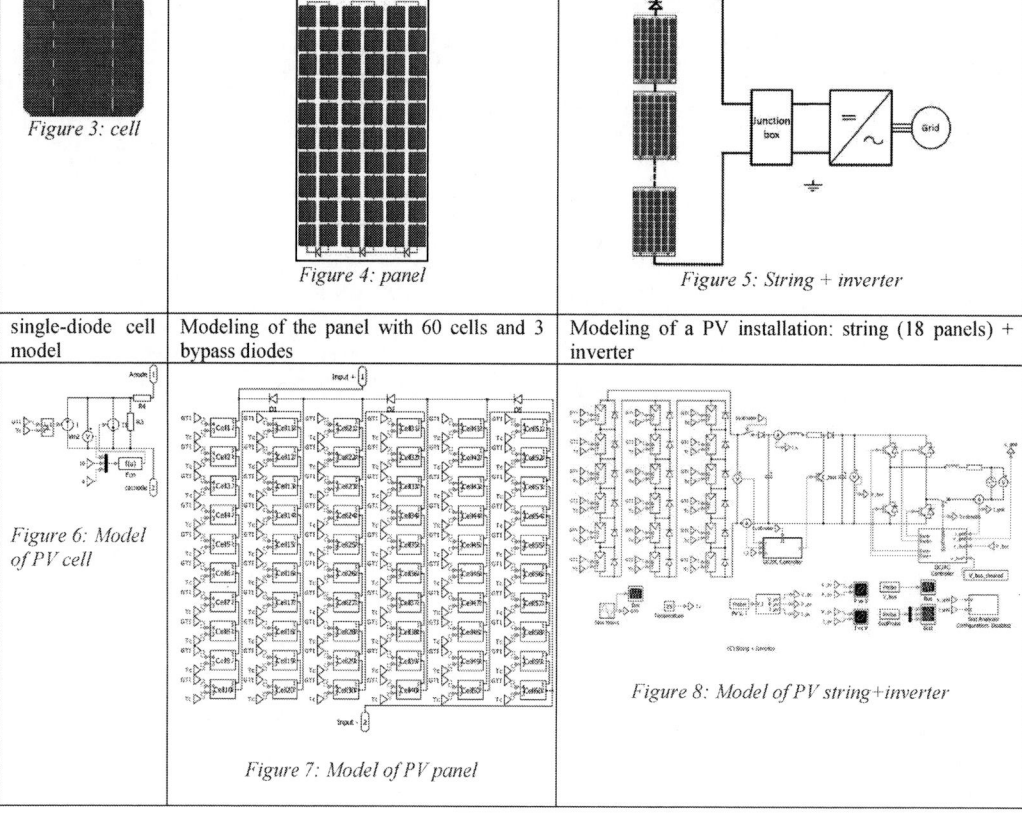

Figure 3: cell

Figure 4: panel

Figure 5: String + inverter

single-diode cell model	Modeling of the panel with 60 cells and 3 bypass diodes	Modeling of a PV installation: string (18 panels) + inverter

Figure 6: Model of PV cell

Figure 7: Model of PV panel

Figure 8: Model of PV string+inverter

Simulation

Simulation under constant illumination

The electrical model presented in this section corresponds to the Rivesaltes-grid installation (Sun Module Plus SW285). This panel has the characteristics given in (Table II).

Table II: characteristics of PV panel and PV car parks

Characteristic of PV panel		characteristic of PV car park	
Maximal Power	285 W	Maximal Power	20.5 kW
Short-circuit current	9.84 A	Short-circuit current	39.36 A
Open circuit voltage	39.70 V	Open circuit voltage	714.60 V
MPPT Current	9.20 A	Number of string	4
MPPT Voltage	31.30 V	Number of panels/string	18
Number of cells in series	60	Number of inverters	1
Number of bypass diodes	3	Number of inverter input	2

In this part, we performed simulations under constant illumination and constant temperature (GTI=723W/m2, Tc=25° C) using the model shown in (Figure 7). The particularity of this representation is the possibility of assigning different values of GTI and T_c for the sixty cells.

The parameters of the PV cell used in the model come from the SAM software database of the NREL laboratory. This simulation made it possible to trace the I-V signature of the panel in healthy and faulty mode (Figure 1, 2).

Simulation under variable illumination

In this part, we have modeled a small PV power plant (photovoltaic car parks) of the Rivesaltes-grid installation. This photovoltaic car parks includes 4 string, each string includes 18 panels. Table I shows the characteristics of the photovoltaic car parks. Figure 8 shows the simulated structure (string+inverter). The daily illumination profile is modeled by a Gaussian function. Figure 9 shows the evolution of the power (search for maximum power point) at the input of the inverter for half a day.

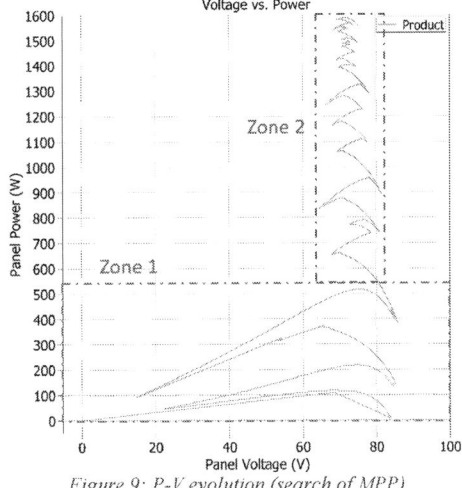

Figure 9: P-V evolution (search of MPP)

As shown in the curve P-V in Figure 9, we can divide the curve into two zones:

1. Zone 1 (Start of the inverter operation): in this zone we have almost all of the P-V curve (from V=0 to V=V_{oc}). Therefore, the I-V curve can be plotted directly for fault detection. However, some defects may be hidden, such as the partial shading fault, series resistance fault and soiling defect.

2. Zone 2 (Inverter operation): In this zone, we only have P-V curve elements around the maximum power point (P_{mpp}, V_{mpp}). For this, we have developed methods to reconstruct the complete curve from the partial curve.

Fault detection

Detection algorithm

The detection algorithm is divided into two parts:

1. Detection: This part makes it possible to compare the signature measured by the I-V measurement system, with a reference signature obtained by the simulation software under the same meteorological conditions (illumination and temperature). The comparison between the two signatures is done by means of the integrals of the two functions (I_m, I_e). When the residual (deviation Δ) exceeds a certain threshold, the fault alarm is activated (Figure 10). Right away, we start the identification part.

2. Identification: this part uses the process of calculating the 5 parameters of the electrical model. The calculation process is presented below. When we finish the evaluation of these parameters, we proceed to the isolation of the fault to determine its nature (Figure 10).

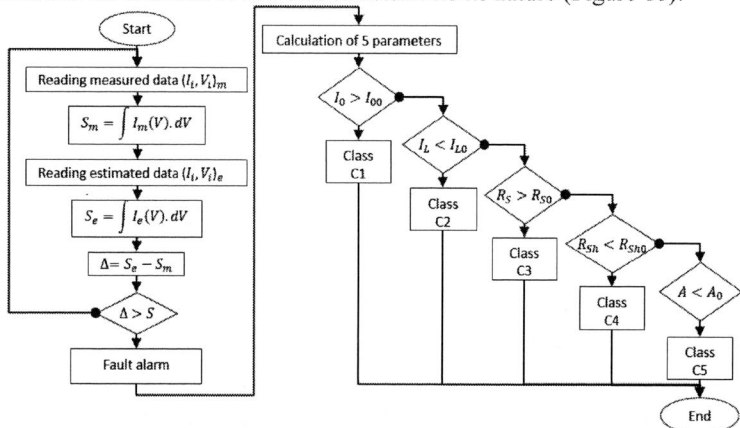

Figure 10: Detection algorithm

Identification of the 5 PV cell/panel parameters

The mathematical developments of equation (1) make it possible to extract the 5 parameters[11]–[13]. The input elements necessary to make this calculation are : I_{cc}, V_{oc}, R_{s0}, R_{sh0}, I_{mpp} and V_{mpp} sequence (Figure 11).

The replacement of equation (1) in the boundary conditions ($I = I_{cc}$, $V = 0$) and ($I = 0$, $V = V_{oc}$) allows to find:

$$\begin{cases} I_{cc} = I_L - I_0(\exp(BI_{cc}R_s) - 1) - \dfrac{I_{cc}R_s}{R_{sh}} \\[2mm] 0 = I_L - I_0(\exp(BV_{oc}) - 1) - \dfrac{V_{oc}}{R_{sh}} \end{cases} \quad (2)$$

We pose: $B = \dfrac{q}{AkT}$, $A_1 = \exp(BV_{oc})$ and $A_2 = \exp(BI_{cc}R_s)$

Thus:

$$\begin{cases} I_0 = \dfrac{I_{cc}\left(1 + \dfrac{R_s}{R_{sh}}\right) - \dfrac{V_{oc}}{R_{sh}}}{(A_1 - A_2)} \\[4mm] I_L = I_s(A_1 - 1) + \dfrac{V_{oc}}{R_{sh}} \end{cases} \quad (3)$$

In order to calculate R_s and R_{sh} we will derive equation (1):

$$\frac{d}{dI}\left(I = I_{Ph} - I_0\big(\exp\big(B(V + IR_s)\big) - 1\big) - \frac{V + IR_s}{R_{sh}}\right) \quad (4)$$

Thus:

$$1 = -I_0\left[B\frac{dV}{dI} + BR_s\right]\big(exp\big(B(V + IR_s)\big) - 1\big) - \frac{1}{R_{sh}}\left[\frac{dV}{dI} + R_s\right] \quad (5)$$

We replace equation (2) in the boundary conditions, we get the following two equations:

$$
\begin{cases}
1 = -I_0\left[B\left(\dfrac{dV}{dI}\right)_{V=V_{oc}} + BR_s\right](\exp(BV_{oc}) - 1) - \dfrac{1}{R_{sh}}\left[\left(\dfrac{dV}{dI}\right)_{V=V_{oc}} + R_s\right] \\[3mm]
1 = -I_0\left[B\left(\dfrac{dV}{dI}\right)_{I=I_{cc}} + BR_s\right](\exp(BI_{cc}R_s) - 1) - \dfrac{1}{R_{sh}}\left[\left(\dfrac{dV}{dI}\right)_{I=I_{cc}} + R_s\right]
\end{cases}
\tag{6}
$$

We pose:

$$
\begin{cases}
R'_s = -\left(\dfrac{dV}{dI}\right)_{V=V_{oc}} \\[3mm]
R'_{sh} = -\left(\dfrac{dV}{dI}\right)_{I=I_{cc}}
\end{cases}
\tag{7}
$$

Replacing (7) in (6), we get:

$$
\begin{cases}
R_s = R'_s - \dfrac{1}{\dfrac{1}{R_{sh}} + BI_0A_1} \\[5mm]
R_{sh} = \dfrac{1}{\dfrac{1}{R'_{sh} - R_s} - BI_0A_2}
\end{cases}
\tag{8}
$$

In order to calculate A, we replace equation (1) at the point of maximum power (I_{mpp}, V_{mpp}):

$$
I_{MPP} = I_L - I_0\left(exp\left(B(V_{MPP} + I_{MPP}R_s)\right) - 1\right) - \frac{V_{MPP} + I_{MPP}R_s}{R_{sh}}
\tag{9}
$$

Thus:

$$
A = \frac{q(V_{MPP} + I_{MPP}R_s)}{kT \ln\left(\dfrac{I_L + I_0 - I_{MPP} - \dfrac{V_{MPP} + I_{MPP}R_s}{R_{sh}}}{I_0}\right)}
\tag{10}
$$

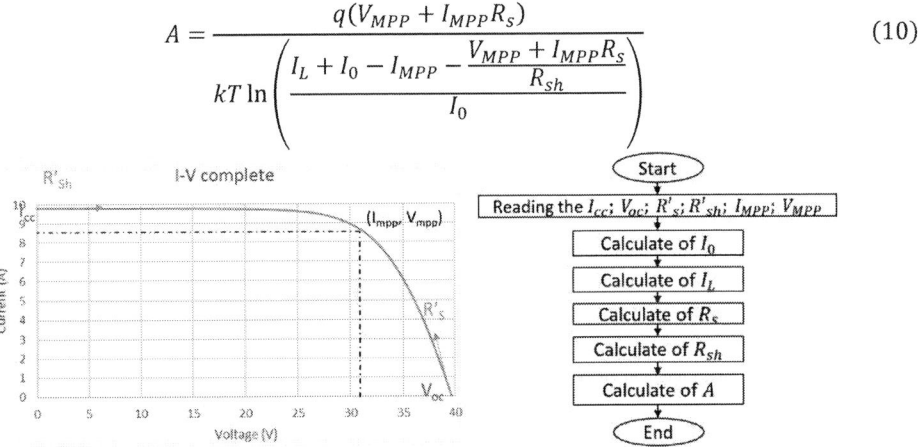

Figure 11: calculating processes

Reconstruction methods

The industrials always have a preference for non-intrusive measurement systems. Indeed, to trace the complete I-V curve we need disconnect the panel or the string of the network, it would then be necessary to switch to a non-intrusive system, which we recover that data from the inverter (P-V, power evolution during the day, Figure 4). However, we have incomplete curves (elements around the maximum power point) and then we can not directly apply the detection algorithm. Therefore, it is firstly be necessary to reconstruct the complete measurement curve $I_m = f(V_m)$. This reconstruction can be done by two different methods:

White Box Method: This method uses the previously presented electrical model in the modeling section. This method of reconstruction involves the calculation of the five parameters of the model (I_0, I_L, R_s, R_{sh}, A). Doing this, we use the same process presented in the previous subsection (Figure 6). So, the application of the white box method requires a prior knowledge several parameters (I_{cc}, V_{oc}, R_{s0}, R_{Sh0}, I_{mpp}, V_{mpp}). Missing one of these parameters increases the complexity of algorithm and running time. In the case where we only have a partial signature (elements around the maximum power point), we only have 2 information about maximum power point (I_{mpp}, V_{mpp}) and we miss 4 (I_{cc}, V_{oc}, R's, R'$_{Sh}$), making the application of this method much more complex. As a consequence, we chose another method (black box).

Black Box Method: This method relies on the extrapolation of the incomplete signature by mathematical functions without needs to know the physical structure of the model. This extrapolation was done by several models:

- Linear model :
  - Linear: we extrapolate both sides of the signature by a simple affine function (11), the slopes of this equation are determined by the two extreme points of each side of the curve (incomplete curve). Once the slopes are determined the coefficient (b) is calculated by replacing a point (x_i,y_i) in the equation.

$$y = a * x + b \tag{11}$$

$$a_{oc} = \left(\frac{\Delta I}{\Delta V}\right)_{oc} ; a_{cc} = \left(\frac{\Delta I}{\Delta V}\right)_{cc} \tag{12}$$

$$b = y_i - a * x_i \tag{13}$$

  - Linear-Max-Min: in this model we use the same approach and the same equation of the linear model, except that in this case, we calculate the variation (dI/dV) for all the point cloud of the curve to be extrapolated, we take the max value to extrapolate the curve on the V_{oc} side, and we take the min value to extrapolate on the I_{cc} side.

$$a_{oc} = max\left(\frac{\Delta I}{\Delta V}\right)_i ; a_{cc} = min\left(\frac{\Delta I}{\Delta V}\right)_i \tag{14}$$

- Exponential model: in this model we extrapolate by an exponential function (15), we take 3 points (x_i,y_i) of the cloud and replace them in the equation (15), we obtain a system of 3 equations and 3 unknowns. Therefore, the resolution of the system allows us to determine the coefficients a, b and c.

$$y = a - b * \exp(x) - c * x \tag{15}$$

Results and discussion

The algorithm was tested on several defaults (series resistance, shunt resistance, temperature, shading and soiling), with different values. For example, a 10% step change was applied to the series resistance fault (Table 4). « d_i » represents the algebraic variation value of each parameter.

- *Series resistance fault*: $R_s(fault) = R_{s_0}(healthy) * (1 + d_1); d_1 > 0$
- *Shunt resistance fault*: $R_{sh}(fault) = R_{sh_0}(healthy) * (1 + d_2); d_2 < 0$
- *Temperature fault* : $T_c(fault) = T_{c_0}(healthy) * (1 + d_3); d_3 > 0$
- *Shading fault* : $GTI_{at\ cell\ level}(fault) = GTI_{0\ at\ cell\ level}(healthy) * (1 + d_4); d_4 < 0$
- *Soiling fault* : $GTI(fault) = GTI_0(healthy) * (1 + d_5); d_5 < 0$

The detection part is based on the evaluation of the deviation Δ, when the deviation exceeds the threshold S which we have set at 1 in our case, the alarm is activated. « A! » is the binary state of the alarm. When the alarm is activated (A!=1), the identification process starts by applying the algorithm presented in the paragraph "Identification of the 5 PV cell/panel parameters".

Table 4, presents the results of the tests carried out. For each defect, we calculate the percentage variation of the 5 parameters as well as Icc and Voc. Deviations below 0.01% are neglected.

- Series resistance fault: the variation R_s% is always negative because it is an increase in its value compared to the healthy case. All the variations d% tested, have led to deviations Δ that exceed the set threshold (S=1). Therefore, the alarm is activated and we notice that the calculated variations R_s% are close to the fault values introduced. All the other parameters have almost zero variation.
- Shunt resistance fault: the variation R_{sh}% this time is positive because it is a decrease in its value. However, this time some variations (R_{sh}%=-10%; R_{sh}%=-20%) are not detected by the algorithm. These variations have Δ deviations below the threshold (S=1).
- Temperature fault: contrary to the other parameters, for this fault the variations is in 20% steps (corresponding to 5°C steps). The variations obtained for R_s% and R_{sh}% are due to variations in the extreme points I_{cc} and V_{oc} which are used to calculate $R'_s = -\left(\frac{dV}{dI}\right)_{V=V_{oc}}$ and $R'_{sh} =$

$-\left(\dfrac{dV}{dI}\right)_{I=I_{cc}}$. On the other hand, the variation of I_L% is only due to the variation of V_{oc}. It turns out that in some models $I_{cc} \approx I_L$ and $I_{cc} = I_L$, we can deduce that it is not a illumination fault.

- Shading fault: for this fault, we tested partial shading cases, by reducing the illumination of the shaded cell by 25% steps. The value [1] indicates the number of hidden cells. Then we reproduced the same variations on two cells simultaneously to activate two bypass diodes. The indicative element for this defect is the existence of the inflection point (IP=1). The cases where we have * are the cases where we find A% variations in complex numbers, this is because the term included in the function $ln\left(\dfrac{I_L+I_0-I_{MPP}-\frac{V_{MPP}+I_{MPP}R_s}{R_{sh}}}{I_0}\right)$ becomes negative.

- Shading fault: the variations obtained this time show the good coherence of the two parameters IL and Icc (IL%= Icc%). These variations correspond to the fault values entered. The V_{oc}% variation is due to the slight shift of V_{oc} on the signature. R_s% and R_{sh}% are always due to variations in the extreme points.

Table III: algorithm test results

Fault	I-V signature	%d	A!	Δ	% R_s	% R_{sh}	% I_L	% I_0	% A	% I_{cc}	% V_{oc}	IP
R_s		10	1	1.9	**-9.8**	0	0	0	0	0	0	0
		20	1	3.8	**-19.7**	0	0	0	0	0	0	0
		30	1	5.7	**-29.6**	0	0	0	0	0	0	0
		40	1	7.6	**-39.5**	0	0	0	0	0	0	0
		50	1	9.5	**-49.4**	0	0	0	0	0	0	0
		60	1	11.4	**-59.3**	0	0	0	0	0	0	0
		70	1	13.4	**-69.2**	0	0	0	0.1	0	0	0
		80	1	15.3	**-79.2**	0	0	0	0.1	0	0	0
		90	1	17.2	**-89.1**	0	0	0	0.1	0.1	0	0
		100	1	19.1	**-99**	0	0	0	0.1	0.1	0	0
R_{sh}		-10	0	0.2	-	-	-	-	-	-	-	-
		-20	0	0.6	-	-	-	-	-	-	-	-
		-30	1	1	-0.1	**30**	-0.5	0	0.4	0	0	0
		-40	1	1.6	-0.2	**40**	-0.8	0	0.7	0	0	0
		-50	1	2.4	-0.3	**50**	-1.2	0	1	0.1	0	0
		-60	1	3.6	-0.7	**60**	0.4	0	-0.5	0.1	0.1	0
		-70	1	5.7	-1	**70**	-0.5	0	0.4	0.2	0.1	0
		-80	1	9.7	-1.2	**80**	-0.3	0	0.1	0.4	0.2	0
		-90	1	21.9	-4.8	**90**	-0.1	0	-0.5	1.1	0.5	0
T_c		20	1	6.2	-0.4	0.1	28.5	0	*	0	**1.4**	0
		40	1	12.4	-2.3	0.2	52.8	0	*	-0.1	**3.2**	0
		60	1	18.7	-2.9	0.3	65.9	0	*	-0.1	**4.6**	0
		80	1	24.9	-3.9	0.5	75.2	0	*	-0.2	**6.1**	0
		100	1	31.2	-4.1	0.7	81.7	0	*	-0.2	**7.5**	0
		120	1	37.4	-6.1	0.9	87.3	0	*	-0.3	**9.3**	0
		140	1	43.7	-6.4	1.2	90.3	0	*	-0.3	**10.8**	0
		160	1	50	-8	1.6	92.5	0	*	-0.4	**12.2**	0
		180	1	56.3	-10	2.1	94.3	0	*	-0.4	**14**	0
Shading		[1]-25	1	29.8	-5	33.3	9.4	0	2.1	0	0.4	**1**
		[1]-50	1	60.1	-14.6	33.3	19.9	0	*	0	0.9	**1**
		[1]-75	1	90.5	-44.8	33.4	35.9	0	*	0	1.9	**1**
		[2]-25	1	59.7	-9.6	66.7	14.8	0	2.2	0	0.8	**1**
		[2]-50	1	120	-29.1	66.7	34.1	0	3.8	0	1.9	**1**
		[2]-75	1	181.1	-91.2	66.8	58.2	0	*	0	3.9	**1**
Soiling		-10	1	35.3	0	0	**9.1**	0	0.8	**10**	0.4	0
		-20	1	70.9	-0.2	0	**19.4**	0	0.6	**20**	0.9	0
		-30	1	106.7	-0.3	0	**30**	0	-0.3	**30**	1.51	0
		-40	1	142.6	-1	0	**39.4**	0	0.7	**40**	2.11	0

		-50	1	178.6	-2.1	0	**49.9**	0	0	**50**	2.9	0
		-60	1	214.7	-3	0	**59.5**	0	0.9	**60**	3.8	0
		-70	1	250.7	-7.4	0	**70.2**	0	-1.2	**70**	5.1	0
		-80	1	286.5	-14.5	0	**79.9**	0	0.1	**80**	6.8	0
		-90	1	321.8	-57.5	0	**89.9**	0	-0.2	**90**	9.9	0

Conclusion

In this paper, we present the basic building blocks (signature, modeling, simulation and algorithm) for detecting and identifying faults in a PV installation. In First, we present the I-V signatures of the panel in healthy mode and faulty modes. The presented signatures show that each fault has its own indices (slope, inflection point...). In order to build a database of the module behaviors for each defect, a reliable model is essential. The model we present allows us to describe the behavior of the cell, panel or string. This model can be used to perform simulations under constant and variable illumination. Based on the results obtained, a fault detection algorithm has been developed. This one was tested on several defects (series resistance, shunt resistance, temperature, shading and soiling). The analysis of the results allowed to validate the algorithm, for some faults we had parasites on the variation of some parameters, such as the temperature fault, its parasites will be treated so as not to disturb the results and lead to false diagnostics. The next work will also focus on the sequence of faults (snail trail, insulation fault...) not yet tested, and we will concentrate on the combination of several faults simultaneously.

References

[1] A. Benzagmout, T. Martire, G. Beaufils, O. Fruchier, T. Talbert, et D. Gachon, « Measurement of the i (v) characteristics of photovoltaic arrays by the capacitive load method for fault detection », in *2018 IEEE International Conference on Industrial Technology (ICIT)*, 2018, p. 1031–1036.

[2] M. Bressan, « Développement d'un outil de supervision et de contrôle pour une installation solaire photovoltaïque », PhD Thesis, Université de Perpignan, 2014.

[3] L. Bun, « Détection et Localisation de Défauts pour un Système PV », PhD Thesis, 2011.

[4] R. Chenni, M. Makhlouf, T. Kerbache, et A. Bouzid, « A detailed modeling method for photovoltaic cells », *Energy*, vol. 32, n° 9, p. 1724-1730, sept. 2007, doi: 10.1016/j.energy.2006.12.006.

[5] W. C. Benmoussa, S. Amara, et A. Zerga, « Etude comparative des modèles de la caractéristique courant-tension d'une cellule solaire au silicium monocristallin », *Rev. Energ. Renouvelables ICRESD-07*, p. 301–306, 2007.

[6] Y. Hishikawa *et al.*, « Voltage-Dependent Temperature Coefficient of the I–V Curves of Crystalline Silicon Photovoltaic Modules », *IEEE J. Photovolt.*, vol. 8, n° 1, p. 48-53, janv. 2018, doi: 10.1109/JPHOTOV.2017.2766529.

[7] E. E. van Dyk et E. L. Meyer, « Analysis of the effect of parasitic resistances on the performance of photovoltaic modules », *Renew. Energy*, vol. 29, n° 3, p. 333-344, mars 2004, doi: 10.1016/S0960-1481(03)00250-7.

[8] E. L. Meyer et E. Ernest van Dyk, « The effect of reduced shunt resistance and shading on photovoltaic module performance », in *Conference Record of the Thirty-first IEEE Photovoltaic Specialists Conference, 2005.*, Lake buena Vista, FL, USA, 2005, p. 1331-1334, doi: 10.1109/PVSC.2005.1488387.

[9] S. Guo, T. M. Walsh, A. G. Aberle, et M. Peters, « Analysing partial shading of PV modules by circuit modelling », in *2012 38th IEEE Photovoltaic Specialists Conference*, Austin, TX, USA, juin 2012, p. 002957-002960, doi: 10.1109/PVSC.2012.6318205.

[10] C. Schill, S. Brachmann, et M. Koehl, « Impact of soiling on IV-curves and efficiency of PV-modules », *Sol. Energy*, vol. 112, p. 259-262, févr. 2015, doi: 10.1016/j.solener.2014.12.003.

[11] J. P. Charles, M. Abdelkrim, Y. H. Muoy, et P. Mialhe, « A practical method of analysis of the current-voltage characteristics of solar cells », *Sol. Cells*, vol. 4, n° 2, p. 169-178, sept. 1981, doi: 10.1016/0379-6787(81)90067-3.

[12] D. S. H. Chan, J. R. Phillips, et J. C. H. Phang, « A comparative study of extraction methods for solar cell model parameters », *Solid-State Electron.*, vol. 29, n° 3, p. 329-337, mars 1986, doi: 10.1016/0038-1101(86)90212-1.

[13] D. Laplaze et I. Youm, « Modélisation d'une cellule photovoltaïque I: Détermination des paramètres à partir de la caractéristique courant-tension sous éclairement », *Sol. Cells*, vol. 14, n° 2, p. 167–177, 1985.

Efficiency Requirements for Passively Cooled Converters with Thermal Measurement Based 3D-FEM Simulation

Julian Weimer[1], Dominik Koch[1], Maximilian Nitzsche[2], Matthias Zehelein[2], Ingmar Kallfass[1]

[1] INSTITUTE OF ROBUST POWER SEMICONDUCTOR SYSTEMS
[2] INSTITUTE FOR POWER ELECTRONICS AND ELECTRICAL DRIVES
University of Stuttgart
Pfaffenwaldring 47
70569 Stuttgart, Germany
Email: julian.weimer@ilh.uni-stuttgart.de
URL: http://www.ilh.uni-stuttgart.de/

Keywords

≪High power density systems≫, ≪Thermal design≫, ≪Modeling≫, ≪Cooling≫, ≪Natural convection≫

Abstract

Passively cooled housings for modern power electronic converters with high power density require maximum heat dissipation over little surface area. To determine the power dissipation budget, this paper presents thermal measurement based convection modeling with temperature dependent heat transfer coefficients and the derivation of efficiency requirements for given housing dimensions.

Introduction

Modern GaN and SiC power semiconductors allow power electronic applications with ever-higher power densities. In particular, soft switching can significantly reduce power dissipation [1], [2] and therefore reduce the size of the corresponding housing. This trend also results in a more even distribution of losses across all components and helps to avoid the problem of internal hot spots at the semiconductor junction layers. Especially for passively cooled plastic housings, hot spots on the housing surface thus become the limiting factor for the thermal heat dissipation concept. The level of natural surface convection depends on the surface temperature of the housing that is limited by the maximum contact temperature. External hot spots hence determine the power dissipation budget of the entire system.

Purpose of this work is the prediction of the minimum efficiency requirement for the target application of a high frequency DC/DC converter [3]. In addition, a accurate thermal model of surface convection is required for the virtual prototyping process to optimize component position and heat spread within the enclosure. The dimensions of the smooth plastic housing were determined based on component sizes and application requirements of the plug integrated power supply.

The common approach for determining the maximum power dissipation via the housing surface is by means of complex analytical approximations of the airflow dependent convection [4]. For this purpose, we assume simplified geometries and temperature distributions and use large parameter tables for the thermophysical properties of different materials. The so determined power dissipation budget can therefore only serve as a first approximation. Alternatively, also complex 3D-FEM simulations with fluid dynamics can predict surface convection [5]–[12]. Both methods are very error-prone and require a high degree of experience.

This paper presents thermal measurements of the surface convection of a smooth plastic housing and the resulting maximum system losses at a maximum contact temperature. Based on the temperature

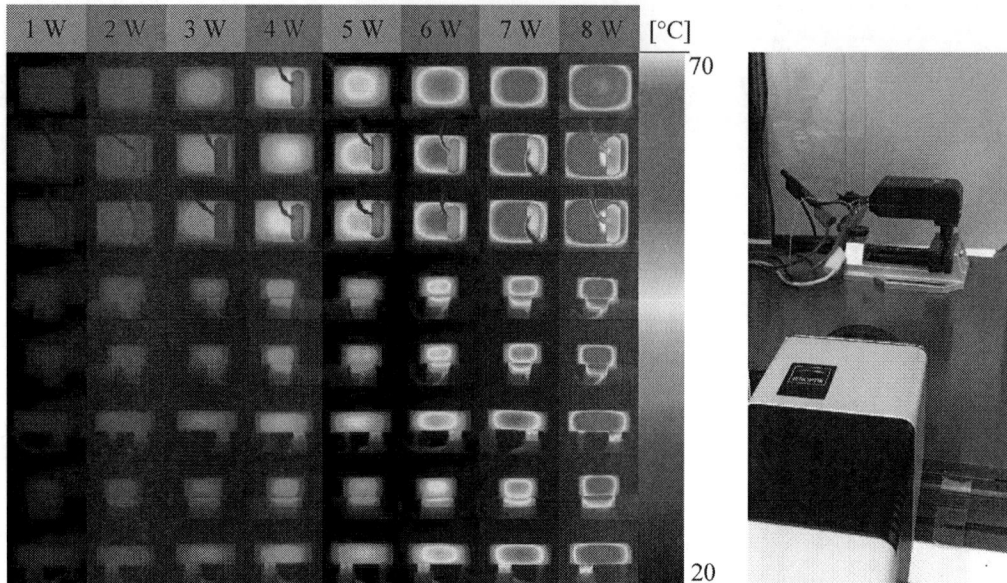

Fig. 1: Infrared images of each side of the target housing in steady-state with a heating power of 1 W to 8 W (left). Measurement setup with infrared camera and housing (right)

measurements for the different housing surfaces, a 3D-FEM model with temperature-dependent heat transfer coefficients was created which allows an accurate prediction of the maximum temperature at different temperature distributions and power losses. This approach enables more accurate models than with analytical modeling concepts, but is not as error-prone as complex fluid simulation. With an improved distribution of losses within the simulation, the possible power dissipation budget for a more homogeneous surface temperature is determined.

Measurement of the surface convection of the target housing

Since the natural convection is dependent on many parameters, such as surface material, surface texture, loss distribution or the housing orientation [11], [12], the thermal behavior of the housing is determined experimentally. The measurement setup consists of two ceramic resistors positioned in the center of the silicone filled polylactide (PLA) housing. These resistors allow to apply a wide range of power dissipation. By a central positioning and by molding with a high thermal conductive silicone, the temperature distribution at the surface is largely homogeneous. Therefore, higher power losses are achievable with the same maximum surface temperature. Usual housing concepts, especially with lower power density, are not filled and therefor show convection inside the enclosure [6]–[9]. Due to the inner convection obvious hot spots above the heat sources can be seen [8]. In addition to the housing with the target dimensions for the application, two other larger housings are thermally measured and simulated. However, the following analysis focuses on the target housing.

Using an infrared camera, the average and maximum surface temperatures of the thermal steady state are determined for all sides of the housing for different power losses. The infrared camera is calibrated to the emission factor of the plastic with a thermocouple measurement on the surface. Fig. 1 shows the infrared camera measurements for the target housing with a power dissipation from 1 W to 8 W. At a thermal settling time of up to 2 h 45 min for the largest housing, almost three days of thermal settling were measured for all housings and operating points. The concept's lower risk of error is thus at the expense of a high measurement effort. With exact measurements of the partial geometry areas and corresponding weighting of the infrared camera surface temperatures, the average temperature increase of the total surface $\Delta T = T_{\text{surf}} - T_{\text{amb}}$ can be calculated. Partial surfaces of the housings evaluated from several sides

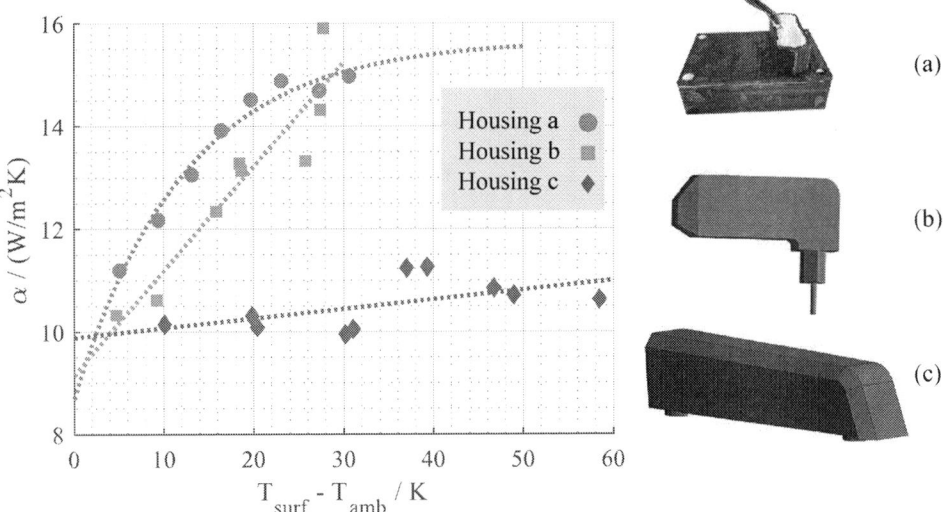

Fig. 2: Measured and fitted heat transfer coefficient α for different housings of different sizes. All Housings are filled with a thermal silicone and have a smooth plastic surface. The volume of the target housing (a) is $0.1\,\text{dm}^3$ with a surface of $150\,\text{cm}^2$.

like the connector surface are weighted so that their temperature is not taken into account too much. This is especially necessary for the housings (b) and (c) with their curved shape. The heat transfer coefficient α of the natural convection at a surface A is calculated according to [4]:

$$\alpha(\Delta T) = \frac{1}{R_{\text{th}}(\Delta T)A} = \frac{P_{\text{Loss}}}{(T_{\text{surf}} - T_{\text{amb}})A} \tag{1}$$

For the calculation, special attention must be paid to the ambient temperature at the time of measurement. For this purpose, a thermocouple is positioned close to the setup. Fig. 2 shows the measurement results of the temperature-dependent heat transfer coefficient for different housings for different power losses between $1\,\text{W}$ and $14\,\text{W}$ and thus for different surface temperatures. The housings with $0.2\,\text{dm}^3$ (b) and $0.4\,\text{dm}^3$ (c) volume show a linear dependence on the temperature in the examined range. The largest housing (c) was measured with ground contact and indicates only a small temperature dependence of the heat transfer coefficient α. The target housing with only $0.1\,\text{dm}^3$ (a) demonstrates a saturation of the temperature dependency at lower surface temperatures like in comparable measurements [5]. Under the simplified assumption of a homogeneous distribution of losses over the entire surface, the heat transfer coefficients of individual regions can be determined and compared. For this, the power losses P_{Losses} from equation (1) are weighted with the total surface area A in relation to the sub-areas of top, bottom, side and connector $A_{\text{Top/Bot/Side/Con}}$. The results show, that the assumption of an uniform convection over the entire surface A is too simplified and would lead to inaccurate results in a corresponding simulation (see Fig. 3). In the following, a distinction is thus made between the heat transfer coefficients of the surfaces Top, Bottom (Bot), Side and Connector (Con). The orientation of the housing also plays a role here. Although the measurements showed only a small orientation dependence for the housing (a), this cannot be generally assumed.

Thermal simulation model and fitting of heat transfer parameters

For the simulation of the free surface convection, a thermostatic 3D-FEM model of the measurement setup is used. The geometry matches the dimensions of the thermally measured target housing with the

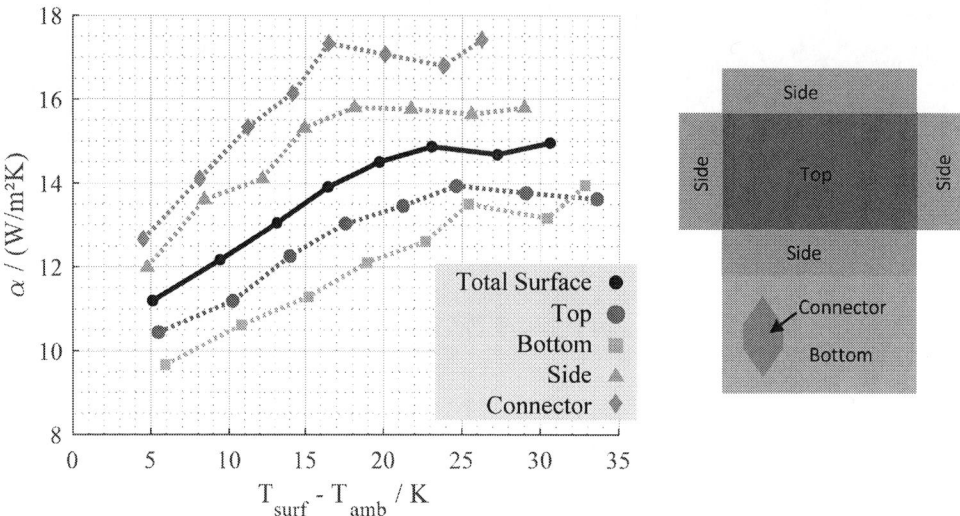

Fig. 3: Measurement of the heat transfer coefficient of the different surfaces of the target housing and their geometric definition.

two ceramic resistors. The thermal conductivity of the PLA plastic housing and the thermal silicone are selected according to the material data sheets at $0.23\,\mathrm{W\,m^{-1}\,K^{-1}}$ and $1.25\,\mathrm{W\,m^{-1}\,K^{-1}}$. Due to air bubbles in the silicone, a value reduced by 40 % is assumed. In order to achieve better results, a more accurately measured epoxy resin can alternatively serve as a filling material. Since the FEM simulation of temperature-dependent heat transfer coefficients always refers to the temperature on a mesh node and not to average surface temperatures, the heat transfer coefficient values cannot directly be transferred from the measurement. The evaluation of the measurement of the surface temperatures allows only the calculation of the convection of the entire surface. Even the differentiation into heat transfer coefficients of different sub-areas is not accurate enough due to the simplification to a homogeneous distribution of the losses. Therefore, the heat transfer coefficients of the different surfaces are only defined with direct measured values as start parameters and are adapted in the course of an iterative optimization based on the measurement temperature results. For this purpose a co-simulation between the numerical tool Matlab and the thermal FEM simulation tool ANSYS was developed. Fig. 4 shows the flow diagram of the fitting process and the definition of the heat transfer coefficient based on exponential functions. The heat transfer coefficient of a single mesh node has the same characteristic curve shape as the heat transfer coefficient of the whole surface defined by the parameters α_m, α_c and α_τ:

$$\alpha(\Delta T) = \alpha_{m,\mathrm{Top/Bot/Side/Con}}\left(1 - exp(-\frac{1}{\alpha_\tau}\Delta T)\right) + \alpha_c \qquad (2)$$

In order to limit the degrees of freedom of the optimization, the heat transfer coefficients definitions for the four different surfaces are only varied with two parameters. The shape parameter α_τ is equally for all surfaces to $\alpha_\tau = 12.35\,\mathrm{K}$ according to the measurement results of the total convection. The fitting process is divided into two phases. In phase 1 all surface convection parameters are varied together. The total convection of the simulated housing is determined from the simulation according to equation (1) ($\alpha_{m,\mathrm{sim}}$ and $\alpha_{c,\mathrm{sim}}$) and compared with the measurement ($\alpha_{m,\mathrm{meas}}$ and $\alpha_{c,\mathrm{meas}}$). If the parameters of the total convection are above or below the measured parameters, they are reduced or increased for all surfaces. In phase 2, the individual maximum surface temperatures $\Delta T_{\mathrm{max,x,sim}}$ are compared with the measured ones ($\Delta T_{\mathrm{max,x,max}}$). Based on a higher or lower maximum surface temperature the parameter $\alpha_{c,x}$ is adapted for each surface. Parameter α_m is kept constant in this phase based on the measurements of Fig. 3.

Efficiency Requirements for Passively Cooled Converters with Thermal Measurement Based 3D-FEM Simulation

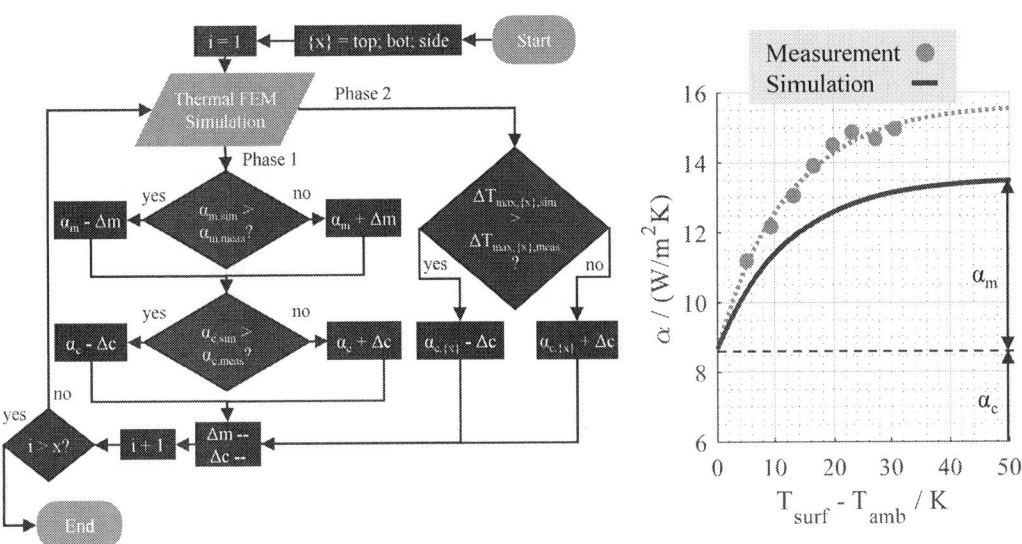

Fig. 4: Flow diagram for the two phases of heat transfer coefficient optimization (left) and exemplary still unfitted parameters for heat transfer coefficient definition of a surface in the simulation (right).

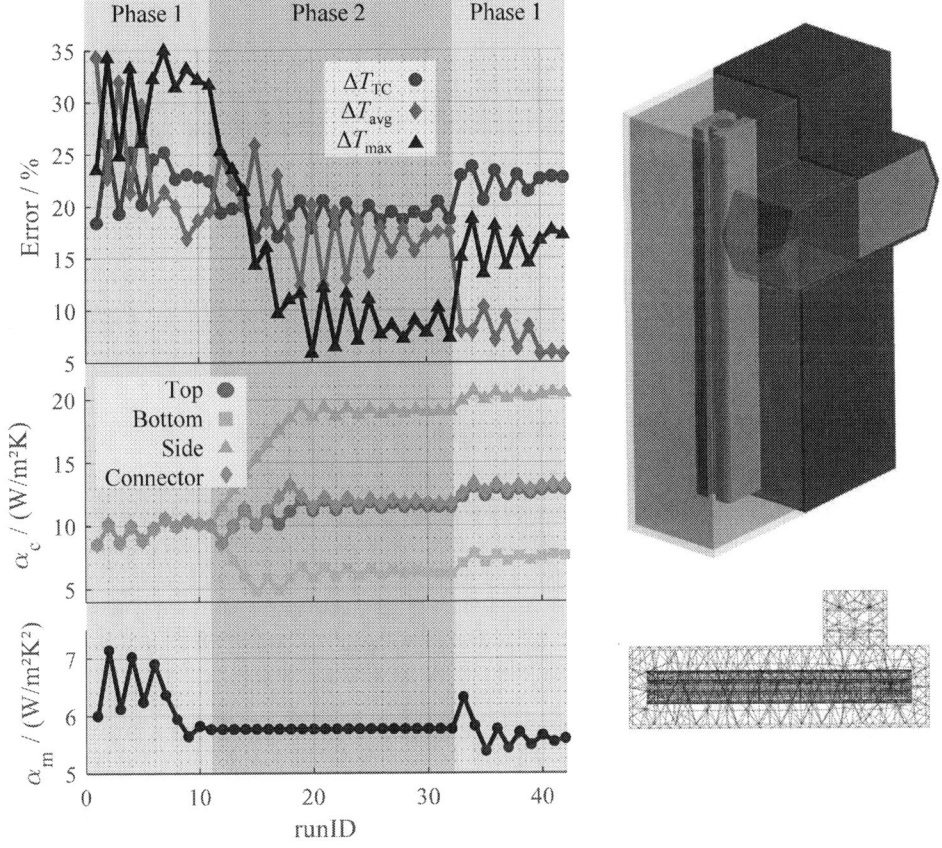

Fig. 5: Convection parameters and errors of the 3D-FEM simulation compared to the measurement for the iterations steps of the fitting process (left) and geometry and mesh of the thermal model utilizing symmetry (right).

Fig. 5 shows the results of the optimization process over 42 iterations. At the end of the optimization process of phase 2, the maximum error for all individual surfaces with respect to the maximum surface temperature can be reduced to less than 10 %. A prediction for similar temperature distributions with the help of this model is thus possible. If phase 2 is followed by another optimization of the total convection according to phase 1, the error of the average surface temperature can also be reduced to below 10 %. Overall, the error for thermal measurements is less than 20 % for all measured parameters. Only the internal temperature determined by a thermocouple shows a slightly larger error. This shows that a further reduction of the thermal conductivity of the filling material might be necessary. In order to achieve more accurate results, the thermal conductivity of the different materials must also be determined experimentally. Fig. 6 shows the direct comparison of the surface temperatures after the iterative optimization (runID 31) for a power dissipation of 8 W.

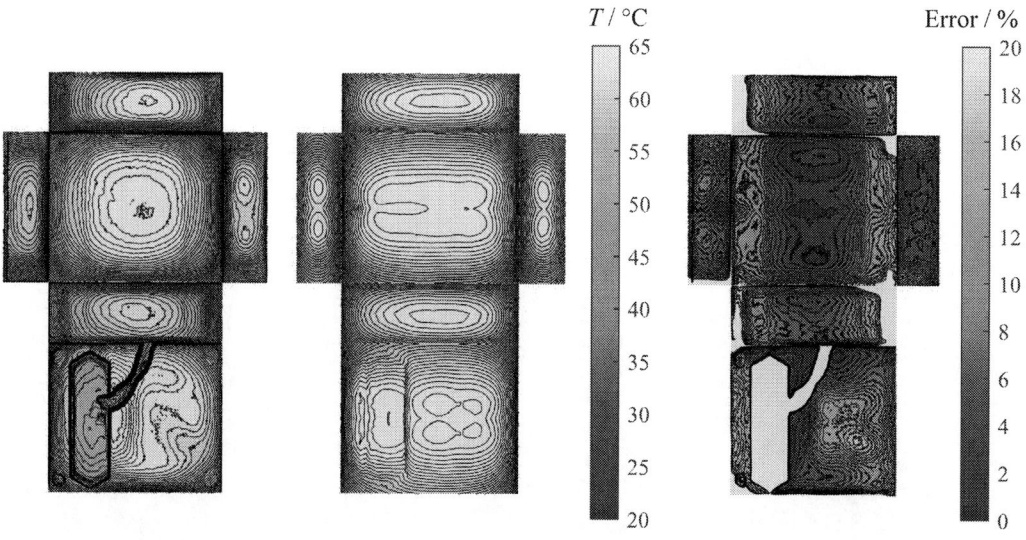

Fig. 6: Comparison of the surface temperature of the target housing measured (left) and simulated (center) for a power loss of 8 W at an ambient temperature of 25 °C and the corresponding error of the FEM simulation regarding the measured surface temperature also at 8 W (right).

Estimation of the power dissipation budget

The power dissipation budget of the housing can be determined directly from the measurements of the surface convection. Fig. 7 (right) shows the maximum surface temperature as a function of the applied power dissipation. Using linear approximation, the power dissipation budget of 6.5 W is obtained for a maximum contact temperature of 65 °C at an ambient temperature of 25 °C. However, this does not represent the maximum limit for the housing. Assuming that the temperature dependence of the convection changes only slightly, a power dissipation budget of 9.4 W can be determined on the basis of an adapted thermal model for a more homogeneous distribution of the surface temperature (see Fig. 7).

If the power dissipation budget is not sufficient for the target application, the simulation model can also determine the power dissipation for a slightly larger housing. However, significant changes to the dimensions would change the airflow and thus the convection too much. With the target application output power of 180 W, a minimum efficiency of 95 % must be achieved assuming optimal distribution of power loss for the target housing.

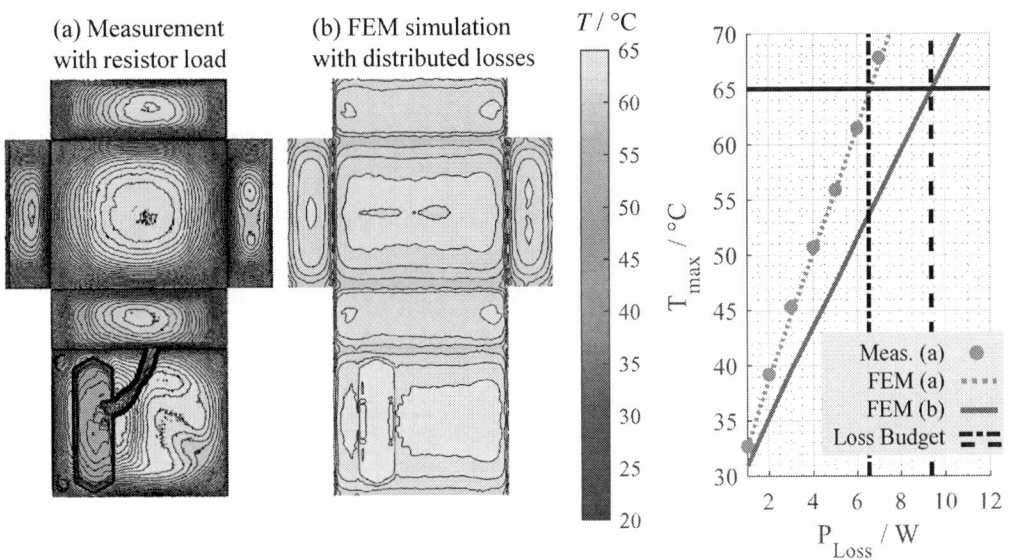

Fig. 7: Surface temperature comparison (left) between measurement with a load resistor in the middle (a) and surface temperature simulation with the fitted model with distributed losses (b). Linear power dissipation budget approximation based on measurement and simulation (right)

Conclusion

This paper presents thermal measurements of surface temperatures of a passively cooled plastic housing for a DC/DC converter target application. A thermostatic 3D-FEM simulation with temperature dependent heat transfer coefficients for sub-areas was fitted to infrared camera measurements. With an error of less than 10 % for all operating points, maximum surface temperatures can be determined for similar temperature distributions. Thus, a maximum possible thermal power dissipation budget of 9.4 W is estimated. With the target housing an efficiency of at least 95 % must be achieved with an output power of 180 W. The presented simulation model is suitable for further optimization of the thermal converter concept.

References

[1] J. Weimer and I. Kallfass, "Soft-Switching Losses in GaN and SiC Power Transistors Based on New Calorimetric Measurements", in *2019 31st International Symposium on Power Semiconductor Devices and ICs (ISPSD)*, 2019, pp. 455–458.

[2] D. Koch, S. Araujo, and I. Kallfass, "Accuracy Analysis of Calorimetric Loss Measurement for Benchmarking Wide Bandgap Power Transistors under Soft-Switching Operation", in *2019 IEEE Workshop on Wide Bandgap Power Devices and Applications in Asia (WiPDA Asia)*, 2019, pp. 1–6.

[3] M. Nitzsche, M. Zehelein, D. Koch, J. Weimer, and J. Roth-Stielow, "Design Flow of a Compact High-Frequency DC/DC Converter with Optimum Average Efficiency in a Wide Operation Range", in *2020 22nd European Conference on Power Electronics and Applications (EPE20 ECCE Europe)*, Submitted, 2020.

[4] V. Gesellschaft, *VDI Heat Atlas*, ser. VDI-Buch. Springer Berlin Heidelberg, 2010.

[5] A. Samson, M. Janicki, T. Raszkowski, and M. Zubert, "Determination of average heat transfer coefficient value in compact thermal models", in *2016 17th International Conference on Thermal, Mechanical and Multi-Physics Simulation and Experiments in Microelectronics and Microsystems (EuroSimE)*, 2016, pp. 1–4.

[6] H. Mi, Ye Gao, Pingan Liu, and Jianfeng Zou, "Numerical simulations of unsteady natural convection in an enclosure with two hot circular parts at bottom", in *2010 Sixth International Conference on Natural Computation*, vol. 8, 2010, pp. 4105–4108.

[7] M. A. Moussaoui, E. B. Lahmer, Y. Admi, and A. Mezrhab, "Natural Convection Heat Transfer in a Square Enclosure with an Inside hot Block", in *2019 International Conference on Wireless Technologies, Embedded and Intelligent Systems (WITS)*, 2019, pp. 1–6.

[8] M. N. Nikitin, "Modeling of natural convection", in *2016 2nd International Conference on Industrial Engineering, Applications and Manufacturing (ICIEAM)*, 2016, pp. 1–4.

[9] N. Y. Zhan and M. Yang, "The study of natural convection and heat transfer in the enclosure with heat transfer coupled in natural convection", in *2011 International Conference on Materials for Renewable Energy Environment*, vol. 2, 2011, pp. 1851–1855.

[10] A. Bouknadel, I. Rah, H. El Omari, and H. El Omari, "Comparative study of fin geometries for heat sinks in natural convection", in *2014 International Renewable and Sustainable Energy Conference (IRSEC)*, 2014, pp. 723–728.

[11] K. K. Sikka, T. S. Fisher, K. E. Torrance, and C. R. Lamb, "Effects of package orientation and mixed convection on heat transfer from a PQFP", *IEEE Transactions on Components, Packaging, and Manufacturing Technology: Part A*, vol. 20, no. 2, pp. 152–159, 1997.

[12] K. J. Kim, "Orientation effects on the performance of natural convection cooled hybrid fins", in *20th International Workshop on Thermal Investigations of ICs and Systems*, 2014, pp. 1–3.

Generic control law for DC and AC machines

Pierre-Philippe Robet, Maxime Gautier, Yannick Aoustin
UNIVERSITY OF NANTES, LS2N
1, rue de la Noë - BP 92101 - 44321 Nantes Cedex 03,
Nantes, France
Tel.: +33 (0)2 28 09 21 53
E-Mail: pierre-philippe.robet@univ-nantes.fr
maxime.gautier@univ-nantes.fr
yannick.aoustin@univ-nantes.fr
URL: https://www.ls2n.fr/

Keywords

«DC machine», «AC machine», «Asynchronous», «Synchronous», «Drives»

Abstract

In this paper we propose a generic method to get the gains of the controller of nested loops of current, velocity and position with an *IP-IP-P* structure. The literal expressions of the coefficients of the controller are identical for each three closed loop by changes of variables. By another changes of variables, the proposed method gives the controller coefficients calculation for a synchronous machine or an asynchronous machine.

Introduction

The position control of a rotor of an electrical motor is often needed for industrial applications. The three nested closed loop system is widely used in industrial application because of its simplicity [1]. We find the position control loop, the velocity control loop and the current control loop. The current control loop is also called torque control loop because of linear relation between the motor current and the generated torque. Frequently poles placement methods are used [2] [3] [4]. Significant interdependence occurred when the ratio between the PWM sampling frequency and the -3dB cutoff frequency of the closed loop of current is less than seven [5]. A ratio less than seven also brings significant interaction between the closed loop of current and the closed loop of speed. In that case, to adjust the interdependence of the three closed loops, a global state synthesis must be performed to obtain better results of gains controller calculation [6] or automatic tuning algorithm [7]. In this paper we work under the hypothesis that ratios are greater than seven; this is mostly the case on actual systems. This assumption gives a good ratio stability performance of the control of an electrical machine. It also provides simplifications on the transfer function so that generics expressions of the controller coefficients can be used. In other words, the calculation of the closed loop of current is sufficient. Coefficients calculation for the closed loop of speed and position can be deduced by changes of variables. By the way, another change of variables makes easy to calculate the controllers of AC machines from the DC machine.

The paper is outlined as follows. First we describe the nested loops of current, speed, position with the IP-IP-P structure. Second, we explain the calculation method to get the controller coefficients for the speed loop, the position loop and the current loop. Third, the limitation base band is discussed. Fourth, we give an example of calculation. Fifth, an experimental validation is proposed with a DC motor. Sixth, we explain how the method can be used with AC machines. Seventh, a validation on actual AC motors is done. Eighth, we discussed practical issues.

Position loop with *IP-IP-P* structure applied to a DC machine

A three cascaded-loops controller is used to control the position with an *IP-IP-P* structure giving no tracking error on a step response of position. The control law is realized with an *IP* controller for the current and the velocity loop and with a *P* controller for the position loop.

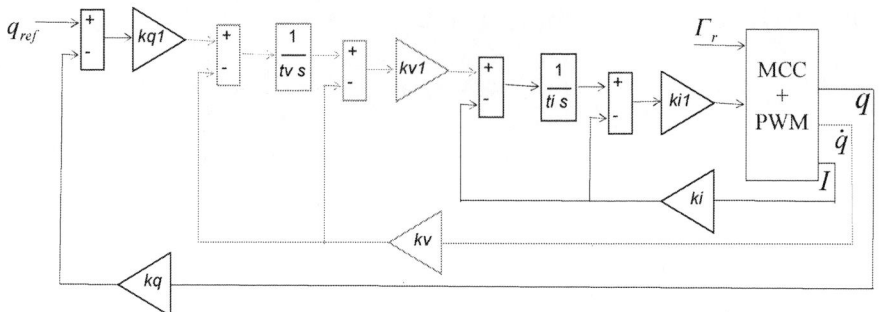

Fig. 1 Nested loops of current, of velocity, of position *IP-IP-P* structure

Let us consider a joint driven by a voltage source amplifier and a DC permanent magnet motor. Electrical and mechanical equations are the following:

$U = L\ \dot{I} + R\ I + kt\ \dot{q}$

$\Gamma_{em} = kt\ I = J\ \ddot{q} + fv\ \dot{q} + fs\ sign(\dot{q}) + \Gamma r$

where I, U, R, L, kt are respectively the armature current and voltage, the resistance, the inductance and the torque constant of the motor. For the mechanical dynamics, $\Gamma_{em}, J, fv, fs, \Gamma r, \dot{q}, \ddot{q}$, , are respectively the electromagnetic motor torque, the inertia moment, the viscous friction coefficient, the coulomb friction torque, the torque disturbances, the motor velocity and acceleration. The viscous friction coefficient $fs\ sign(\dot{q})$ and Γr are considered as perturbations.

The calculation method

Closed loop of speed with a proportional controller (*P*)

A good compromise between stability and high bandwidth comes with $-3dB$ bandwidth $w_{\dot{q}_d}$ of the speed loop and $-3dB$ bandwidth w_{I_d} of the current loop such that: $w_{\dot{q}_d} \leq \dfrac{w_{I_d}}{\alpha}$ and $\alpha \geq 7$. In that case, it possible to model the closed loop transfer of the current by a gain : $T_{bfI_{ref}I} \underset{w_{I_d} \gg w_{\dot{q}_d}}{\simeq} \dfrac{1}{ki}$, where ki is the current sensor gain. Introducing $G_\Gamma = kt / ki$ as the gain of the torque actuator, the closed loop transfer function of velocity is given by:

$$T_{bf\dot{q}_{ref}^P \dot{q}} = \frac{1}{kv}\frac{b}{b + \left(1 + \dfrac{J}{fv}s\right)} = \frac{1}{kv}\frac{\dfrac{b}{b+1}}{1 + \dfrac{tm}{(b+1)}s} \tag{1}$$

with $b = \dfrac{kv\ kv1\ G_\Gamma}{fv}$ and $tm = \dfrac{J}{fv}$ (2)

Where tm is the mechanical time constant of the DC machine

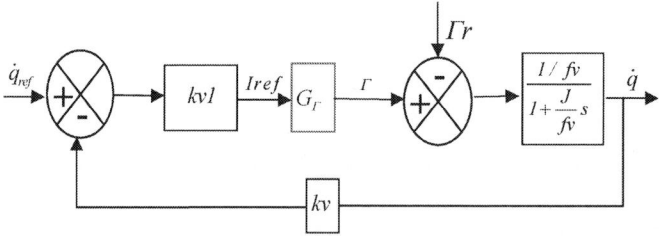

Fig. 2 Closed loop of velocity with *P* structure

Closed loop of speed with an integral proportional controller (*IP*)

The (*IP*) structure depends on the (*P*) structure given Fig. 2 as show on Fig. 3.

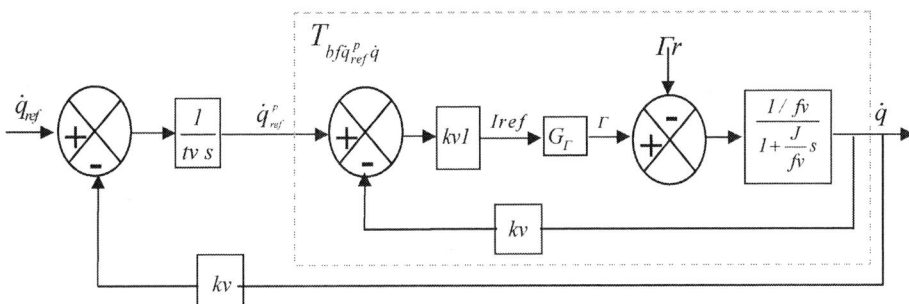

Fig. 3 Closed loop of velocity with *IP* structure

We have shown that the expression $T_{bf\dot{q}_{ref}^p\dot{q}}$ is given by Eq(1). In closed loop, it comes:

$$T_{bf\dot{q}_{ref}\dot{q}} = \frac{1}{kv}\frac{1}{1+\dfrac{b+1}{b}tv\,s+\dfrac{tm\,tv}{b}s^2} = \frac{1/kv}{\dfrac{s^2}{w_{\dot{q}}^2}+\dfrac{2\zeta_{\dot{q}}}{w_{\dot{q}}}s+1} \tag{3}$$

With $\dfrac{tv\,tm}{b}=\dfrac{1}{w_{\dot{q}}^2}$; $\zeta_{\dot{q}}=\dfrac{1}{2}\sqrt{\dfrac{(b+1)\,tv}{tm}}=\dfrac{1}{2}w_{\dot{q}}\dfrac{b+1}{b}tv$ (4)

where $w_{\dot{q}}$ is the natural frequency in radian of the second-order system of velocity and $\zeta_{\dot{q}}$ the damping ratio.

Moreover, we notice that: $\dfrac{b}{tm}(\dfrac{b+1}{b})=\dfrac{b+1}{tm}=2\zeta_{\dot{q}}w_{\dot{q}}$ (5)

Let us recall that the cutoff frequency $w_{\dot{q}_d}$ at $-3dB$, compared to the reference level $0dB$, of a second order is given by : $w_{\dot{q}_d}=\beta_{\zeta\dot{q}}w_{\dot{q}}$ (6)

where $\beta_{\zeta\dot{q}}=\left((1-2\zeta_{\dot{q}}^2)+\sqrt{(4\zeta_{\dot{q}}^4-4\zeta_{\dot{q}}^2+2)}\right)^{\frac{1}{2}}$

Usually $\dfrac{\sqrt{2}}{2}\le\zeta_{\dot{q}}\le 1$ for the velocity loop. Consequently for a given $w_{\dot{q}_d}$ at $-3dB$ base band, the natural pulsation is calculated with $w_{\dot{q}}=w_{\dot{q}_d}/\beta_{\zeta\dot{q}}$. With (5) and (6), the $-3dB$ cutoff frequency is given by:

$$w_{\dot{q}_d}=\beta_{\zeta\dot{q}}\frac{1}{2\zeta_{\dot{q}}}(\frac{b+1}{b})\frac{b}{tm}=\beta_{\zeta\dot{q}}\frac{1}{2\zeta_{\dot{q}}}\frac{b+1}{tm}. \tag{7}$$

Let us put $tmq=\dfrac{2\zeta_{\dot{q}}}{\beta_{\zeta\dot{q}}}tm$, so that $w_{\dot{q}_d}=\dfrac{b+1}{tmq}$ (8)

An approximation of (3) is to consider the second order as a first order transfer function approximated by a transfer of the same cutoff frequency $w_{\dot{q}_d}$ at $-3dB$:

$$T_{b\dot{q}_{ref}^{IP}\dot{q}}\simeq\frac{1/kv}{1+\dfrac{tmq}{b+1}s} \tag{9}$$

Closed loop of position with a proportional controller (*P*)

The (*P*) structure depends on the (*IP*) structure given Fig. 3 as show on Fig. 4.

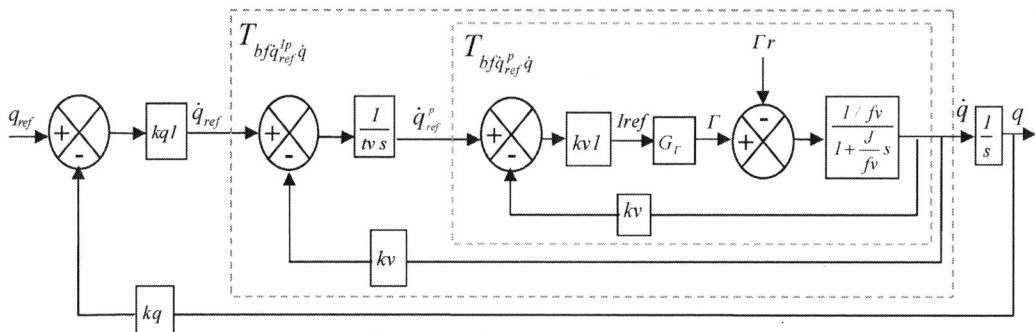

Fig. 4 Closed loop of position with *IP* structure

As we say in the previous section, a ratio of 7 between two $-3dB$ bandwidth allow to considered the highest one as a gain. For the calculation of the position loop, we improved the approximation of the velocity loop by taking the first order transfer function given by (9). To get generic calculation let us introduced tq such that :

$$kq1 = \frac{kv}{kq}\frac{1}{tq}\frac{b}{b+1} \qquad (10)$$

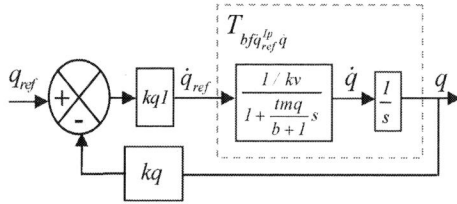

Fig. 5 Simplified closed loop of position

The closed position loop transfer function is given with (9) and (10) by:

$$T_{bf\dot{q}_{ref}q} = \frac{1}{kq}\frac{1}{1+\frac{b+1}{b}tq\ s+\frac{tmq\ tq}{b}s^2} = \frac{1/kq}{\frac{s^2}{w_q{}^2}+\frac{2\zeta_q}{w_q}s+1} \qquad (11)$$

With $\quad \dfrac{tq\ tmq}{b} = \dfrac{1}{w_q{}^2}; \quad \zeta_q = \dfrac{1}{2}\sqrt{\dfrac{(b+1)\ tq}{tmq}} = \dfrac{1}{2}w_q\dfrac{b+1}{b}tq \qquad (12)$

Where w_q is the natural frequency in radian of the second-order system of position and ζ_q the damping ratio.

Moreover, we notice that: $\dfrac{2\zeta_q}{w_q} = (\dfrac{b+1}{b})tq \Leftrightarrow \dfrac{b}{tmq}(\dfrac{b+1}{b}) = \dfrac{b+1}{tmq} = 2\zeta_q w_q \qquad (13)$

Base on (6), the cutoff frequency in radian w_{q_d} at $-3dB$ is given by :

$$w_{q_d} = \beta_{\zeta q}w_q \qquad (14)$$

Thus with (13) and (14), it comes :

$$w_{q_d} = \beta_{\zeta q}\frac{1}{2\zeta_q}w_q = \beta_{\zeta q}\frac{1}{2\zeta_q}\frac{b+1}{tmq} \qquad (15)$$

From previous relationships (6), (7), (8), (15), it come the relations between the natural pulsations and bandwidth pulsations at $-3dB$:

$$w_q = \frac{\beta_{\zeta\dot{q}}}{2\zeta_q}w_{\dot{q}}, \qquad w_{q_d} = \frac{\beta_{\zeta q}}{2\zeta_q}w_{\dot{q}_d}, \quad \text{where} \quad w_{q_d} = \beta_{\zeta q}w_q \quad \text{and} \quad w_{\dot{q}_d} = \beta_{\zeta\dot{q}}w_{\dot{q}} \qquad (16)$$

Closed loop of current with an integral proportional controller (*IP*)

The pulse width modulation amplifier (PWM) can be considered as gain G, because of the high sampling frequency Fs relative to the $-3dB$ bandwidth w_{I_d} of the current loop [8]. Experimentation have shown that this assumption is good as soon as $w_{I_d} \leq \dfrac{2\pi Fs}{\beta}$ and $\beta \geq 7$.

The *IP* structure of the current loop is identical to that one of the speed loop. By a simple change of variables, it come the correspondence between Fig. 3 and Fig. 6:

$$fv \rightarrow R,\ J \rightarrow L,\ G_\Gamma \rightarrow G,\ kv1 \rightarrow ki1,\ tv \rightarrow ti,\ w_{\dot q} \rightarrow w_I,\ tm \rightarrow te,\ b \rightarrow a,$$

$$\zeta_{\dot q} \rightarrow \zeta_i,\ \beta_{\zeta_q} \rightarrow \beta_{\zeta_I},\ w_{\dot q} \rightarrow w_I,\ w_{\dot q_d} \rightarrow w_{I_d} \tag{17}$$

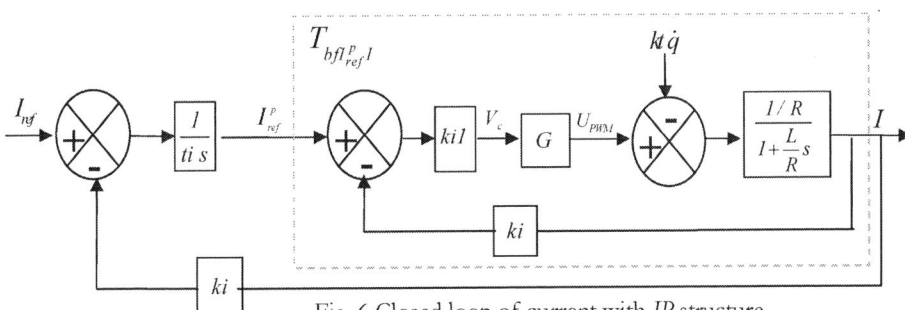

Fig. 6 Closed loop of current with *IP* structure

Thus, the $-3dB$ bandwidth of the current closed loop is given by (6), (7), (17) and (2):

$$w_{I_d} = \beta_{\zeta_I} w_I = \beta_{\zeta_I} \frac{1}{2\zeta_I} \frac{a+1}{te} \tag{18}$$

With $a = \dfrac{ki\ ki1\ G}{R}$ and $te = \dfrac{L}{R}$ $\qquad$ (19)

Calculation example

Experimentation have shown that a good compromise between stability and high bandwidth is to have a $-3dB$ bandwidth of the closed loop of current given by: $w_{I_d} \leq \dfrac{2\pi Fs}{\beta}$ and $\beta \geq 7$ where Fs is the sampling frequency of the PWM.

With (19) and (18) and choosing a damping ratio: $\zeta_I = 0.5 \rightarrow \beta_{\zeta_I} = 1.27$, the $ki1$ gain of the closed loop of current is calculated by:

$$\frac{2\zeta_I w_{I_d}}{\beta_{\zeta_I}} = \frac{a+1}{te} \Leftrightarrow \frac{ki\ ki1\ G}{R} = (\frac{2\zeta_I w_{I_d} te}{\beta_{\zeta_I}} - 1) \Leftrightarrow ki1 = \frac{2\zeta_I w_{I_d} L}{\beta_{\zeta_I} ki\ G} - \frac{R}{ki\ G} \tag{20}$$

A this level the value a given by (19) can be calculated.

With (4) and (17), it comes : $\zeta_I = \dfrac{1}{2} w_I \dfrac{a+1}{a} ti \Leftrightarrow ti = \dfrac{2\zeta_I}{w_I} \dfrac{a}{a+1} = \dfrac{2\zeta_I \beta_{\zeta_I}}{w_{I_d}} \dfrac{a}{a+1}$ $\qquad$ (21)

Where w_I is the natural frequency in radian of the second-order system of current and ζ_I the damping ratio.

A good compromise between stability and high bandwidth is to have $-3dB$ bandwidth $w_{\dot q_d}$ of the closed loop of velocity given by: $w_{\dot q_d} \leq \dfrac{w_{I_d}}{\alpha}$ and $\alpha \geq 7$.

Let us choose a damping ratio $\zeta_{\dot{q}} = 1/\sqrt{2} \rightarrow \beta_{\zeta\dot{q}} = 1$ and no over shoot on the closed loop of position $\zeta_q = \sqrt{2} \rightarrow \beta_{\zeta_q} = 0.4028$.

From (6), (5) and (2), it comes :

$$\frac{2\zeta_{\dot{q}}w_{\dot{q}_d}}{\beta_{\zeta\dot{q}}} = \frac{b+1}{tm} \Leftrightarrow \frac{kv\,kv1\,G_\Gamma}{Fv} = (\frac{2\zeta_{\dot{q}}w_{\dot{q}_d}tm}{\beta_{\zeta\dot{q}}} - 1) \Leftrightarrow kv1 = \frac{2\zeta_{\dot{q}}w_{\dot{q}_d}J}{\beta_{\zeta\dot{q}}\,kv\,G_\Gamma} - \frac{Fv}{kv\,G_\Gamma} \tag{22}$$

A this level the value b given by (2) can be calculated.

With (4), it comes : $\zeta_{\dot{q}} = \frac{1}{2}w_{\dot{q}}\frac{b+1}{b}tv \Leftrightarrow tv = \frac{2\zeta_{\dot{q}}}{w_{\dot{q}}}\frac{b}{b+1} = \frac{2\zeta_{\dot{q}}\beta_{\zeta\dot{q}}}{w_{\dot{q}_d}}\frac{b}{b+1}$ \hfill (23)

The $-3dB$ bandwidth of the closed loop of position is given by (16): $w_{q_d} = \frac{\beta_{\zeta_q}}{2\zeta_q}w_{q_d} \simeq \frac{w_{\dot{q}_d}}{7}$.

With (12) it comes: $tq = \frac{2\zeta_q}{w_q}\frac{b}{b+1} = \frac{2\zeta_q\beta_{\zeta q}}{w_{q_d}}\frac{b}{b+1}$ \hfill (24)

The coefficient $kq1$ is deduced using (10) : $kq1 = \frac{kv}{kq}\frac{w_{q_d}}{2\zeta_q\beta_{\zeta q}}$ \hfill (25)

Equations (21),(23),(24) show that ti, tv, tq expressions are similar but with coefficient indices related to the loop of current, speed and position. It is the same with (20) and (22) with coefficient indices related to the loop of current and speed.

Base band limitation due to saturation

The saturation can be study with a transient or harmonics point of view [9]. Both methods give similar results. In this paper, we used the harmonic approach. With Fig. 6, the transfer function between the PWM input voltage and the reference current is:

$$\frac{V_c}{I_{ref}} = \frac{V_c}{I}\frac{I}{I_{ref}} = \frac{1+te\,s}{G/R}\frac{1/ki}{\frac{s^2}{w_I^2} + \frac{2\zeta_I}{w_I}s + 1}$$

To avoid saturations, the maximum input PWM voltage $V_{c_{max}}$ is given by:

$$V_{c_{max}} \geq I_{ref}\frac{\|1+te\,s\|}{\frac{G\,ki}{R}\left\|\frac{s^2}{w_I^2} + \frac{2\zeta_I}{w_I}s + 1\right\|} \tag{26}$$

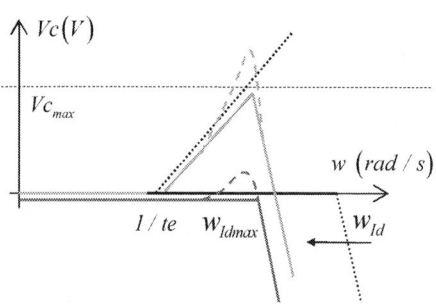

Fig. 7 Bode diagram, $Vc/Iref(-)$, $I/Iref(-)$

Generally for the current closed loop, with $0.5 \leq \zeta_I \leq 0.707$, the maximum over shoot is $Q_I = \frac{1}{2\zeta_I\sqrt{1-\zeta_I^2}} \leq 1.155$ and we have the following ratio: $1 \geq \frac{w_r}{w_I} = \sqrt{1-\zeta_I^2} \geq 0.866$. Thus, the effect of the over shoot, which is near w_I can be neglected later.

Approximation of (26) at $-3dB$ cut off frequency is given by: $V_{c_{max}} \geq I_{ref}\ \frac{R}{G.ki}te\,w_{Id}\frac{1}{\sqrt{2}}$

Finally the maximum $-3dB$ base band is given by: $w_{Id} \leq w_{Id\,max} = \sqrt{2}\,\dfrac{V_{c_{max}}\,G\,ki}{I_{ref_{max}}\,L}$ (27)

Where $I_{ref_{max}} \leq I_{max}^{motor}$ is the maximum current given by the chosen security of the closed loop of current.

For the closed loop of velocity and position ζ_q, $\zeta_{\dot q}$ are chosen higher than 0.707, thus there is no over shoot. Looking Fig. 3 and Fig. 6, $Iref$ in the velocity loop play the same role than V_c into the current loop. By a change of variables given with (17), the previous result can be used to get the maximum $-3dB$ base band of the velocity.

$$w_{qd} \leq w_{qd\,max} = \sqrt{2}\,\frac{I_{ref_{max}}\,Gt\,kv}{\dot{q}_{ref_{max}}\,J}$$ (28)

Where $\dot{q}_{ref_{max}} \leq \dot{q}_{ref_{max}}^{motor}$ is the maximum velocity given by the chosen security of the closed loop of velocity.

The maximum position $q_{ref_{max}}$ is obtained with (11) and using Fig. 5:

$$\dot{q}_{ref_{max}} \geq q_{ref}\,\frac{\|s\|\,\left\|1+\dfrac{tmq}{b+1}\,s\right\|}{\dfrac{kq}{kv}\,\left\|\dfrac{s^2}{w_q^2}+\dfrac{2\zeta_q}{w_q}s+1\right\|}$$ With $w_{\dot{q}_d} = \dfrac{b+1}{tmq} > w_{q_d}$, this result can be approximated at

$-3dB$ cut off frequency by: $\dot{q}_{ref_{max}} \geq q_{ref}\,\dfrac{kv}{kq}w_{qd}\,\dfrac{1}{\sqrt{2}}$,

Thus the maximum base band is: $w_{qd} \leq w_{qd\,max} = \sqrt{2}\,\dfrac{\dot{q}_{ref_{max}}\,kq}{q_{ref_{max}}\,kv}$ (29)

It is important to note that (27), (28), (29) cannot be mixed, because each approximation to calculate the bandwidth limit has been made at different frequencies.

Experimental case study with the DC machine

The linear positioning system EMPS300 connected to the control system from dSPACE™ is used. The control system is based on a TMS 320C31 Texas Instruments™ processor and Matlab-Simulink software, in order to get a high rate numerical control with a big computational capacity. The EMPS300 main components are a DC permanent magnet motor with DC tachometer and current controlled four quadrant PWM chopper, a ball screw drive the positioning unit and an incremental encoder [10]. All analog and digital signals are directly accessible between EMPS300 and Simulink using the "C" code generator RTW Matlab toolbox [11] and the RTI program from dSPACE, to start experiments easily and immediately.

The proposed experimental validation is realized with the following parameters:
$L = 3.2E-3\,(H)$, $R = 2.2\,(\Omega)$, $kt = 5.13E-2\,(V/rad/s)$,
$J = 1.61E-5\,(kgm^2)$, $fv = 9.16E-5\,(Nm/rd/s)$, $fs = 7.28E-3\,(Nm)$.
The sampling frequency is 4.444kHz. The simulated current, velocity and position matches the actual one as shown on Fig. 10. The theoretical calculation gives a $-3dB$ bandwidth of the closed loop of position for $82\,rad/s$. On Fig. 9, the Bode diagram obtained in simulation gives $-3dB$ bandwidth for $89\,rad/s$

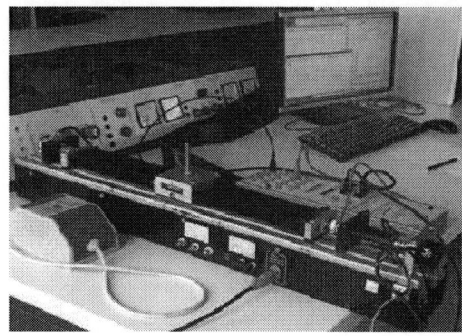

Fig. 8 Linear positioning system EMPS300

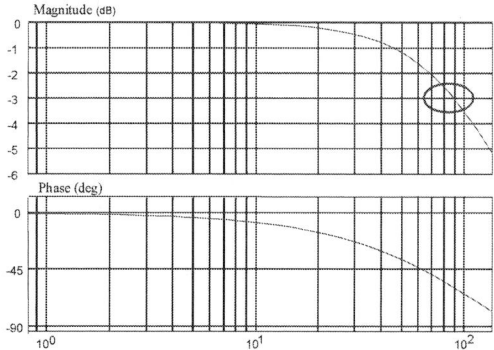

Fig. 9 Bode diagram of q / q_{ref}

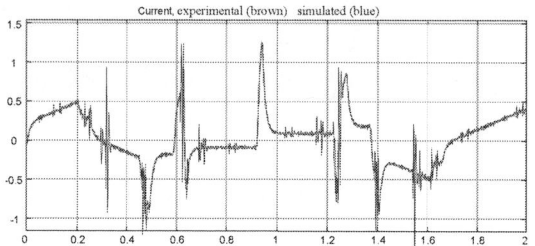

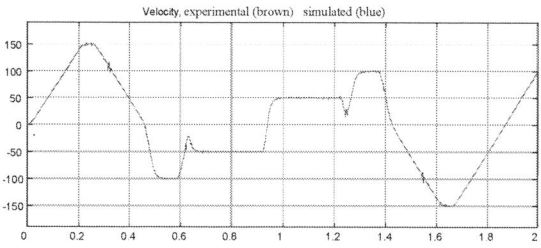

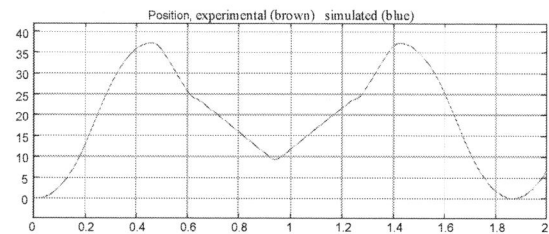

Fig. 10 Trajectories of current, velocity and position

Position loop with *IP-IP-P* structure applied to AC machines

Let us recall equations of the synchronous machine into the *dq* frame [12], with a control law such that $I_d = 0$ and $\phi_{fd} = 0$, $\phi_{fq} = C^{te}$:

$$
\begin{bmatrix} V_d \\ V_q \end{bmatrix} = \begin{bmatrix} R_s & -p\dot{q}L_q \\ p\dot{q}L_d & R_s \end{bmatrix} \begin{bmatrix} I_d \\ I_q \end{bmatrix} + \begin{bmatrix} L_d & 0 \\ 0 & L_q \end{bmatrix} \begin{bmatrix} \dot{I}_d \\ \dot{I}_q \end{bmatrix} + \begin{bmatrix} p\,\dot{q}\,\phi_{fd} \\ p\,\dot{q}\,\phi_{fq} \end{bmatrix}
\tag{30}
$$

$$
\Gamma_{em} = p\phi_{fq}I_q = J\,\ddot{q} + fv\,\dot{q} + fs\,sign(\dot{q}) + \Gamma\,r
\tag{31}
$$

where $V_{dq} = (V_d, V_q)$, $I_{dq} = (I_d, I_q)$, R_s, $L_{dq} = (L_d, L_q)$, p, $\phi_{fd} = 0$, $\phi_{fq} = C^{te}$, are respectively into the *dq* frame : the voltages, the currents, the resistance, the inductances, the number of pole pairs, the rotor flux constants.

Let us recall equations of the asynchronous machine [13].

$$
\begin{bmatrix} V_d \\ V_q \end{bmatrix} = \begin{bmatrix} R_s\sigma_R & -p\dot{q}_s L_{cs}\sigma_L \\ p\dot{q}_s L_{cs}\sigma_L & R_s\sigma_R \end{bmatrix} \begin{bmatrix} I_{ds} \\ I_{qs} \end{bmatrix} + \begin{bmatrix} L_{cs}\sigma_L & 0 \\ 0 & L_{cs}\sigma_L \end{bmatrix} \begin{bmatrix} \dot{I}_{ds} \\ \dot{I}_{qs} \end{bmatrix} + \begin{bmatrix} -\sigma_{sr}/\tau_r & -p(\dot{q}_s - \dot{q}_r)\sigma_{sr} \\ p(\dot{q}_s - \dot{q}_r)\sigma_{sr} & -\sigma_{sr}/\tau_r \end{bmatrix} \begin{bmatrix} \psi_{dr} \\ \psi_{qr} \end{bmatrix}
\tag{32}
$$

$$
\Gamma_{em} = (\sigma_{sr}\,p\,\Psi_{dr})I_{qs} = J\,\ddot{q} + fv\,\dot{q} + fs\,sign(\dot{q}) + \Gamma\,r
\tag{33}
$$

Where $V_{dq} = (V_d, V_q)$, $I_{dqs} = (I_{ds}, I_{qs})$, R_s, are respectively into the *dq* frame, the stator voltages, the stator currents, the stator resistance. With L_{sp}, L_{rp} the self inductance of stator and rotor and M_s, M_r the mutual inductance of stator and rotor phase, it becomes: $L_{cs} = L_{sp} - M_s$ and $L_{cr} = L_{rp} - M_r$. Let

us define $M = 2M_{sr} / 3$ the stator to rotor mutual inductance, p the number of pole pairs, $p\,q_s$, $p\,q_r$ the stator and rotor electric angles into the *dq* frame, $(q_s - q_r)$ the rotor position. Let us define:

$$
\tau_r = \frac{L_{cr}}{R_r}, \quad \tau_s = \frac{L_{cs}}{R_s}, \quad \sigma_{sr} = M_{sr}/L_{cr}, \quad \sigma_L = 1 - M_{sr}^2/L_{cr}L_{cs}, \quad \sigma_R = 1 + \frac{\tau_s}{\tau_r}(1 - \sigma_L) = 1 + \sigma_{sr}^2\frac{R_r}{R_s}.
$$

Different methods are used to control the asynchronous machine [13][14]. In this paper, the control law is such that $\Psi_{qr} = 0$, $I_{ds} = cte$, so that $\Psi_{dr} = I_{ds} M_{sr} = C^{te}$.

Usually the dynamic of the *d*-axis is chosen equal to the one of the *q*-axis for AC machines. The electrical equations on the *q*-axis of the AC machines compared to the DC machine are:

$$V^{DC} \quad = \quad R \quad I \quad + \quad L \quad \dot{I} \quad + \quad k_t \quad \quad \dot{q}$$

$$V_q^{synchronous} \quad \simeq \quad R_s \quad I_q \quad + \quad L_q \quad \dot{I}_q \quad + \quad p\phi_{fq} \quad \quad \dot{q}$$

$$V_{qs}^{asynchronous} \simeq R_s \sigma_R \quad I_{qs} \quad + L_{cs}\sigma_L \quad \dot{I}_{qs} \quad + p \ \sigma_{sr} I_{ds} M_{sr} \ (\dot{q}_s - \dot{q}_r) \quad \text{with } p\dot{q}_s L_{cs}\sigma_L I_{ds} << R_s\sigma_R \ I_{qs}$$

Torques equations of all machines are:

$$\Gamma_{em}^{DC} = k_t \ I \quad ; \qquad \Gamma_{em}^{synchronous} = p \ \phi_{fq} \ I_q \ ; \qquad \Gamma_{em}^{asynchronous} = p \ \sigma_{sr} \ I_{ds} M_{sr} \ I_{qs}$$

The five (*ki1, kv1, ti, tv, tq*) coefficients of the controller calculated in previous section are easily obtained by doing a change of variables with the parameters of the DC machine.

For the synchronous machine: $R \rightarrow R_s$, $L \rightarrow L_q$, $kt \rightarrow p\phi_{fq}$

For the asynchronous machine: $R \rightarrow R_s\sigma_R$, $L \rightarrow L_{cs}\sigma_L$, $kt \rightarrow p \ \sigma_{sr} \ I_{ds} M_{sr}$.

Experimental case study with the AC machine

The control system is based on a TMS 320C31 Texas Instruments™ processor and Matlab-Simulink software. All analog and digital signals are directly accessible between the AC machines and Simulink using the "C" code generator RTW Matlab toolbox [15] and the RTI program from dSPACE.

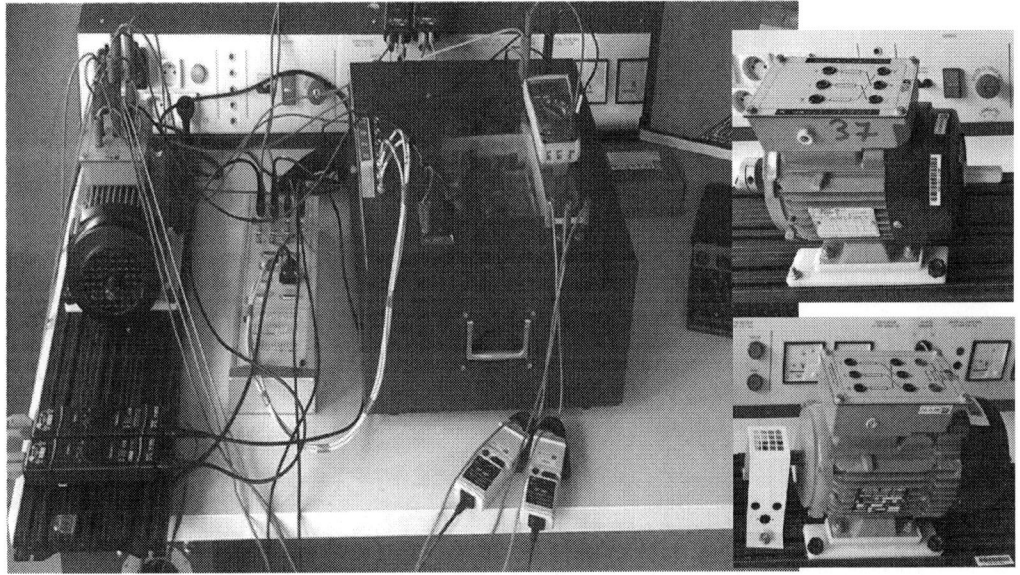

Fig. 11 Nested loops of current velocity and position (Asynchronous / Synchronous)

The parameters for the synchronous machine are given by:

$L_d = 0.496\,H$, $L_q = 0.186\,H$, $R_s = 31.3\,\Omega$, $\phi_{fq} = 0.749\ V/rad/s$, $p=2$,

$J=1.95E-3 kgm^2$, $fv=2.92E-3 m/rd/s$, $fs=7.74E-2 Nm$

The parameters for the asynchronous machine are given by:

$L_{cs} = 1.2\,H$, $R_s = 25\,\Omega$, $\tau_r = 0.067s$, $\sigma_L = 0.25$, $p=2$

$J=3.1E-3 kgm^2$, $fv=2.25E-3 m/rd/s$, $fs = 2.77E-2 Nm$

For both synchronous and asynchronous machine, the simulated signals match the experimental one.

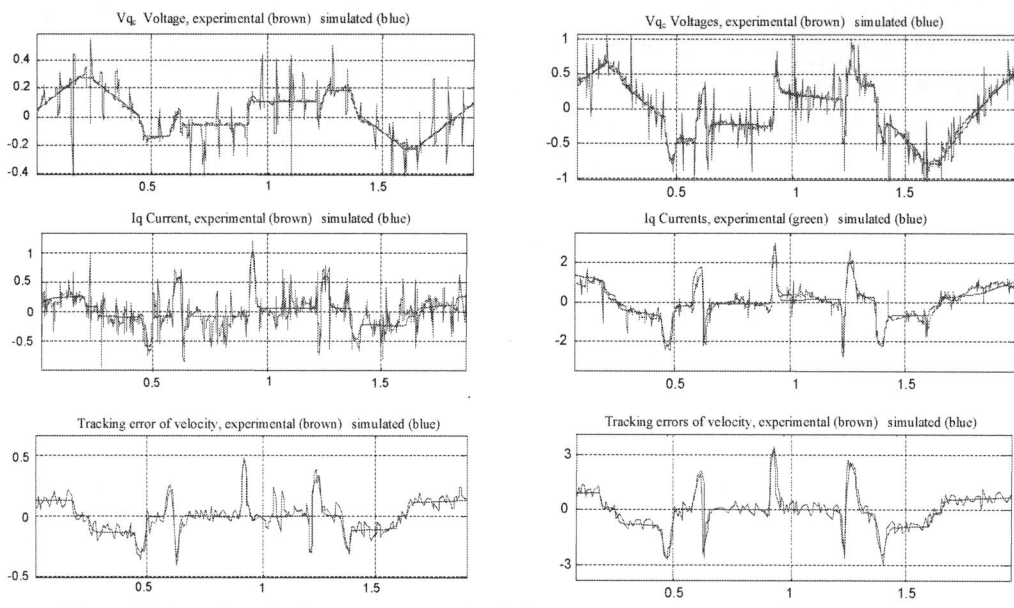

Fig. 12 Trajectories of voltage, current and velocity errors (Synchronous / Asynchronous)

Practical issues with the DC machine and the AC machine

* Accuracy around 10% is obtained on the calculation of the various bandwidths at $-3dB$ if there is a ratio > 7 between the various bandwidths. To obtain a ratio of 7 between the $-3dB$ base band of the closed position loop and that of speed, it is necessary to take a damping coefficient $\zeta_q \geq \sqrt{2}$. Otherwise, the theoretical calculation of the $-3dB$ based band introduced an error greater than 10% compared to the practice. Nevertheless, the proposed method provides a good compromise between performance and stability. To maximize the performance of all three loops a global state synthesis has to be perform [13], [14].

* In a position-controlled system, there is a natural saturation in the control of the PWM [9], which can be of analog or digital type. On almost all systems, there is also a current limitation and sometimes a speed limitation. We have discussed that in a previous section. Some algorithm even takes into account the sensor and the environment constrains [16]. Taking all these limits into account, the bandwidths of the different loop are limited.

* In digital application, the input control voltage of the Pulse with Modulated Wave is given by the computer zero order hold output. Excluding computing delay constraints, the transfer function between the calculated control voltage and the PWM output voltage depend on the modelisation and is currently given by : $G(F) \underset{F_s/F > 4}{\simeq} G \ exp(-j\pi F/Fs)$. With $Fs/F \geq 7$, a maximum phase shift of $26°$ is introduced. But for an ideal second order, the phase margin of the current closed loop of $-3dB$ bandwidth is greater than $60°$ with $0.5 \leq \zeta_1 \leq 0.707$. Consequently, in the worst case the phase margin is still $34°$.

With sensorless AC machines, the delay should be taken into account into real time calculation of the rotation angle with the Park transform [17][18]. But with the presented method, the ratio between the $-3dB$ velocity bandwidth and the sampling frequency is $Fs/F \geq 49$. That means, a maximum delay of $3.6°$ which can be neglected because it introduced 2% error on a 360° rotation angle.

Conclusion

Given a maximum $-3dB$ bandwidth due to saturations on the different loops, the proposed method calculates the control law from a conventional three nested loop. The main idea is to say that a good compromise between stability and high dynamic is to take at least a sampling PWM frequency seven times higher than the $-3dB$ cutoff frequency of the current closed loop. And a $-3dB$ cutoff frequency of the speed closed loop at least seven times lower than the one of the current closed loop.

In that case with the *IP-IP-P* structure, the equations show that *ti, tv, tq* expression are similar with coefficient indices related to the loop of current, velocity and position. This is the same with *ki1, kv1*. In addition with *IP-IP-P* structure applied to AC machines, we conclude that the calculation of the five coefficients (*ki1, kv1, ti, tv, tq*) of the controller can easily be obtained by a changing of variables.

The analytical expression of *ki1, ti* is sufficient to calculate all the other coefficients with a DC machine or with AC machines. This is why the method presented is generic for the *IP-IP-P* structure.

References

[1] Takesue N., Zhang G., Furusho J. ; Sakaguchi, M., "High stiffness control of direct-drive motor system by a homogeneous ER fluid", Robotics and Automation, 1999. Proceedings. 1999 IEEE International Conference on Robots Publication Year: 1999 , Page(s): 188-192 vol.1.

[2] Umland J.W., Safiuddin M., "Magnitude and symmetric optimum criterion for the design of linear control systems: what is it and how does it compare with the others", IEEE Transactions on Industry Applications, Volume: 26 , Issue: 3 Publication Year: 1990 , Page(s): 489 – 497.

[3] Keviczky L., Banyasz C., "A new pole-placement regulator design method", Proceedings of the 34th IEEE Conference on Decision and Control, Volume: 4 Publication Year: 1995 , Page(s): 3408 - 3413 vol.4.

[4] Datta S., Chakraborty Debraj., Belur M.N., "Low order controller with regional pole placement" American Control Conference (ACC), 2012 , Publication Year: 2012, Page(s): 6702 – 6707.

[5] Robet P.Ph., Gautier M., Bergmann C., " A Frequency approach for current loop modeling with a PWM converter ", IEEE Transaction on Industry Applications, pp.1000-1014, Vol.34, sept / oct 1998.

[6] P.P. Robet, M. Gautier, "Conventional control with nested loops including an acceleration loop". Conference on Power Electronics and Applications (EPE14-ECCE Europe), Aug 2014, Lappeenranta, Finland

[7] A. Sarca, "A New Approach for Electrical Drives Controllers Tuning Based on Optimal Closed-Loop Bandwidth Detection," 2019 11th International Symposium on Advanced Topics in Electrical Engineering (ATEE), Bucharest, Romania, 2019, pp. 1-5.

[8] Robet P.Ph., Gautier M., Bergmann C., " A Frequency approach for current loop modeling with a PWM converter ", IEEE Transaction on Industry Applications, pp.1000-1014, Vol.34, sept / oct 1998.

[9] P.Ph Robet, M.Gautier, C.Bergmann, " Control Design with Technological Constraints ", EPE 95, pp 3.869-3.873, Sevilla, Spain, 19 september 1995.

[10] dSPACETM , "EMPS300, RTI, DS1102 guides", dSPACE GmbH, D 33100, Paderborn, Germany, 1995.

[11] MathWorks, Matlab, Simulink "reference guides and Signal, Optimization, RTW toolboxes User's Guides", The MathWorks, Inc. Natick, Mass, USA, 2008.

[12] P.Ph Robet, M.Gautier, C.Bergmann, " Synchronous Machine Global Dynamic Control Law ", IEEE-SMC, CESA, pp 360-364, Lille, France, 9 July 1996.

[13] P.Ph Robet, M.Gautier, C.Bergmann, "Asynchronous machine global dynamic control law ", Lausanne, Switzerland, september EPE 99.

[14] P. J. Koratkar and A. Sabnis, "Comparative analysis of different control approaches of direct torque control induction motor drive," 2017 International Conference on Intelligent Computing, Instrumentation and Control Technologies (ICICICT), Kannur, 2017, pp. 831-835.

[15] MathWorks, Matlab, Simulink "reference guides and Signal, Optimization, RTW toolboxes User's Guides", The MathWorks, Inc. Natick, Mass, USA, 2008.

[16] L. N. Rassudov and A. P. Balkovoi, "Optimisation in servo motion control: Considering hardware constraints," 2018 25th International Workshop on Electric Drives: Optimization in Control of Electric Drives (IWED), Moscow, 2018, pp. 1-5.

[17] W. Lv, K. Huang, H. Wu, X. MO and M. Shen, "A Dynamic Compensation Method for Time Delay Effects of High-Speed PMSM Sensorless Digital Drive System," 2019 22nd International Conference on Electrical Machines and Systems (ICEMS), Harbin, China, 2019, pp. 1-5.

[18] H. Zhang et al., "A Time-Delay Compensation Method for IPMSM Hybrid Sensorless Drives in Rail Transit Applications," in IEEE Transactions on Industrial Electronics, vol. 66, no. 9, pp. 6715-6726, Sept. 2019.

A High Performance 48-to-8 V Multi-Resonant Switched-Capacitor Converter for Data Center Applications

Rose A. Abramson, Zichao Ye, Robert C. N. Pilawa-Podgurski
UNIVERSITY OF CALIFORNIA, BERKELEY
2626 Hearst Ave
Berkeley, CA 94720, USA
Email: rose_abramson@berkeley.edu, yezichao@berkeley.edu, pilawa@berkeley.edu

Acknowledgments

The information, data, or work presented herein was funded in part by the Advanced Research Projects Agency-Energy (ARPA-E), U.S. Department of Energy, under Award Number DE-AR0000906 in the CIRCUITS program monitored by Dr. Isik Kizilyalli. The views and opinions of authors expressed herein do not necessarily state or reflect those of the United States Government or any agency thereof. Rose Abramson was supported by the Department of Defense (DoD) through the National Defense Science & Engineering Graduate (NDSEG) Fellowship Program.

Keywords

≪Converter circuit≫, ≪DC power supply≫, ≪Efficiency≫, ≪High power density systems≫

Abstract

Resonant switched-capacitor (SC) converters are becoming increasingly attractive for high performance applications due to their efficient utilization of switches and use of high energy density capacitors. Multi-phase SC converters can offer further improvements in performance, as they can achieve the same conversion ratio as traditional two-phase SC converters with fewer capacitors and switches, making them ideal for large conversion ratio applications. This paper presents theoretical analysis and experimental results for a resonant multi-phase SC converter comprising a cascaded series-parallel topology derived from the conventional 4-to-1 series-parallel converter. The proposed converter can achieve a 6-to-1 conversion ratio with the same number of switches and capacitors as the 4-to-1 series-parallel converter. A 48-to-8 V prototype designed for data center applications was built and tested with 40 A maximum output current. The prototype achieved 99.0% peak efficiency (98.5% including gate drive loss) and 2230 W/in^3 power density, demonstrating one of the best in-class performances.

Introduction

The power demands of data centers have grown rapidly due to the development of cloud computing, high-performance computing, and big data analysis. In 2014, data centers used more than 1.8% of all electricity in the US, and it is projected that they will use 10% by 2020 [1]. Therefore, much work has been done to increase the efficiency and reduce the physical footprint of data center power delivery networks. One of these efforts is to distribute power to servers at a higher voltage (e.g. 48 V) compared to the traditional 12 V bus to reduce resistive losses. One approach for converting the 48 V bus to a voltage that the server equipment can use is through a two-stage architecture. Here, the first stage converts the 48 V to an intermediate voltage (e.g. 5-12 V) and the second stage converts the intermediate voltage to the point-of-load (PoL) voltage. The intermediate bus voltage converter can be unregulated, as the second-stage buck converter can provide regulation. Recent research has indicated that using an intermediate bus voltage lower than 12 V can result in better overall system efficiency once the efficiency of the second-stage buck is considered as well [2].

Due to their high efficiency and high power density, resonant switched-capacitor converters (ReSC) have received increased attention for use in data center applications. Resonant switched-capacitor circuits not only demonstrate an efficient utilization of switches [3] and the use of high energy density capacitors [4], but also allow for soft-switching and soft-charging operations [5–8]. Recent hardware demonstrations of ReSC converters have demonstrated excellent efficiency and power density, in applications ranging from high power discrete implementations [9–15] to CMOS integrated solutions [16, 17].

This work proposes and explores the performance of a new multi-resonant cascaded series-parallel 6-to-1 topology that can achieve very high efficiency and power density. Multi-phase switched-capacitor converters are especially attractive for data center applications, as they can achieve the same conversion ratio as traditional two-phase switched-capacitor circuits with fewer capacitors and switches, further improving efficiency and power density.

A 48-to-8 V, 40 A, fixed ratio (unregulated) converter prototype was designed and implemented. The prototype achieved 99.0% peak efficiency (98.5% with gate drive loss) and 2230 W/in^3 power density, both of which are among the highest of existing work.

Multi-Resonant Cascaded Series-Parallel Converter

Proposed topology and operating principle

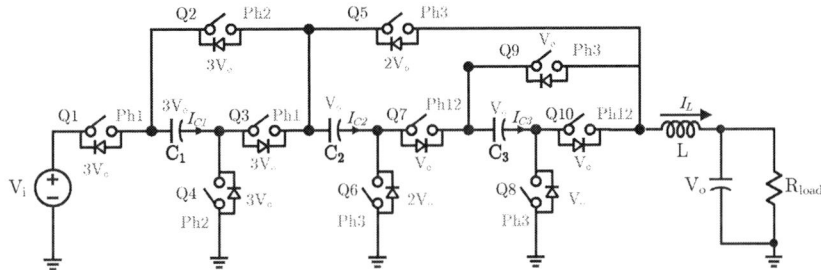

Fig. 1: Schematic of the proposed 6-to-1 cascaded series-parallel converter with device ratings labeled.

Fig. 1 shows the schematic for the proposed 6-to-1 cascaded series-parallel converter, with the output given by $V_{out} = \frac{V_{in}}{6}$. This circuit topology can be derived from the classic 4-to-1 series-parallel topology, by moving the source terminal of Q_2 in Fig. 1 from the left-side of the inductor to the positive-side of C_2 and adding an additional operating phase. The converter achieves a 2-to-1 step-down during the first $\frac{1}{3}$ of the switching period through a series-mode operation, followed by a 3-to-1 parallel-mode operation during the last $\frac{2}{3}$ of the switching period. This converter therefore can achieve a 6-to-1 conversion ratio for the same number of switches and capacitors as the 4-to-1 converter, although the device ratings on one switch and one capacitor are increased from V_o for the 4-to-1 converter to $3V_o$ for the 6-to-1 converter.

The current waveforms in the inductor and flying capacitors C_1-C_3, gating signals (matching switch labels in Fig. 1), and equivalent circuits for each phase are shown in Fig. 2. During the time 0 to $\frac{T}{3}$ (*Phase* 1 and *Phase* 2), the resonant frequency is determined by the series combination of C_1, C_2, C_3, and the inductor, L. Then, $f_{res,1} = f_{res,2} = \frac{1}{2\pi\sqrt{LC_{eq}}}$, where $\frac{1}{C_{eq}} = \frac{1}{C_1} + \frac{1}{C_2} + \frac{1}{C_3}$.

During the time $\frac{T}{3}$ to T (*Phase* 3), the resonant frequency is determined by the parallel combination of C_2 and C_3, so that $f_{res,3} = \frac{1}{2\pi\sqrt{L(C_2||C_3)}}$.

The time duration of each phase can then be derived from its respective resonant frequency. The duration of *Phase* 1 and *Phase* 2 can be written as:

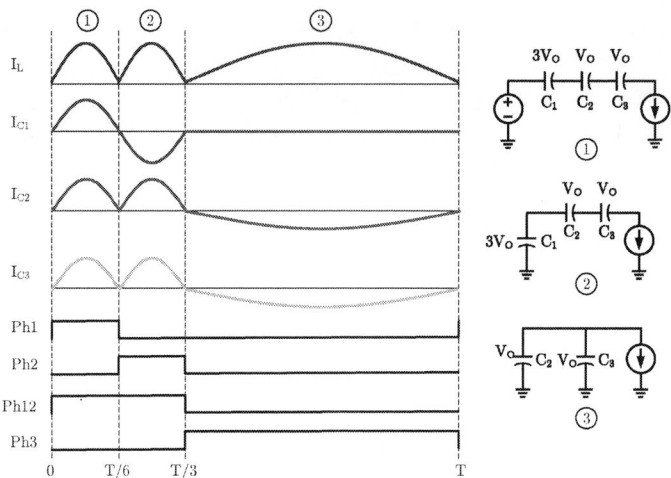

Fig. 2: Key current waveforms and control signals for the proposed converter. The converter state for each of the three phases is also shown.

$$T_1 = T_2 = \frac{T}{6} = \frac{1}{2} \cdot 2\pi \sqrt{L\left(\frac{1}{\frac{1}{C_1} + \frac{1}{C_2} + \frac{1}{C_3}}\right)} \tag{1}$$

where T is the switching period. T_1 has a factor of π rather than 2π at it represents a half cycle of the full resonant period.

The duration of *Phase* 3 can similarly be written:

$$T_3 = \frac{2T}{3} = \frac{1}{2} \cdot 2\pi \sqrt{L(C_2 + C_3)} \tag{2}$$

Due to the parallel operation mode, it is necessary that $C_2 = C_3 = C$. Substituting this into (1) and (2) and solving for $4 \cdot T_1 = T_3$ (as the duration of *Phase* 3 is $4\times$ that of *Phase* 1 or *Phase* 2), the minimum C_1 to achieve soft charging can be found to be $C_{1,\min} = \frac{1}{6}C$. As can be seen from Fig. 1, C_1 sees $3V_o$ while C_2 and C_3 only see V_o, so the highest voltage rated capacitor is also the lowest valued capacitor. This can allow C_1 to take up a similar volume to C_2 and C_3 despite its higher voltage rating.

If operated at resonance, the converter can achieve zero-current switching (ZCS) as the current reaches zero at the switch transitions. The resonant frequency represents the minimum frequency to achieve soft-charging [7]. Here, the hardware prototype is operated above resonance to account for component non-idealities and reduce the output impedance and conduction loss. This also allows for the use of Class 2 ceramic capacitors, as the flying capacitors do not have to be precisely tuned to an exact ratio. In this operation mode, the converter can also achieve partial zero-voltage switching (ZVS) at switch turn-on.

Table I compares the component count and voltage ratings for several different switched-capacitor topologies that can be augmented with resonant inductor(s) and operated in a resonant mode. The proposed topology has the lowest number of components of all the 6-to-1 converters in Table I, and the same number of components as the lower conversion-ratio 5-to-1 Fibonacci converter. This is of note as the 5-to-1 Fibonacci converter demonstrates the maximum gain possible in a two-phase switched-capacitor converter, with 10 switches and 3 flying capacitors [18] [19]. The proposed topology also requires lower voltage capacitors compared to the FCML and Switched Tank (Dickson) converters.

Table I: Comparison of number and voltage rating of components for several resonant switched-capacitor converters

Topology	Conversion Ratio	Number of Switches	Switch Rating	Number of C_{fly}	C_{fly} Rating	Number of Inductors
Proposed Topology	6-to-1	10	$4 \times 3V_o$ $2 \times 2V_o$ $4 \times V_o$	3	$1 \times 3V_o$ $2 \times V_o$	1
Series-Parallel	6-to-1	16	$3 \times 5V_o$ $2 \times 4V_o$ $2 \times 3V_o$ $2 \times 2V_o$ $7 \times V_o$	5	$5 \times V_o$	1
Switched Tank (Dickson)	6-to-1	16	$2 \times 2V_o$ $14 \times V_o$	5	$1 \times 5V_o$ $1 \times 4V_o$ $1 \times 3V_o$ $1 \times 2V_o$ $1 \times V_o$	3
FCML	6-to-1	12	$12 \times V_o$	5	$1 \times 5V_o$ $1 \times 4V_o$ $1 \times 3V_o$ $1 \times 2V_o$ $1 \times 1V_o$	1
Fibonacci	5-to-1	10	$2 \times 3V_o$ $4 \times 2V_o$ $4 \times V_o$	3	$1 \times 3V_o$ $1 \times 2V_o$ $1 \times V_o$	1

Two useful metrics for comparing topologies are the switch stress (VA_{RMS}) rating and the total passive component volume. The switch stress relates to how much voltage the switches must block and how much current they must conduct, and is a good indication of the efficiency of the topology. The passive component volume is proportional to the amount of reactive energy that needs to be processed and stored in the converter, and reflects the power density of the topology. Fig. 3 plots the passive volume vs. the switch stress rating, using the method outlined in [20] assuming $\frac{\rho_C}{\rho_L} = 100$, where ρ_C is the energy density of the capacitor(s) and ρ_L is the energy density of the inductor(s). The switch stress and passive volume are normalized to the theoretical lowest possible value to allow for a comparison across topologies. In this plot, the ideal converter would be situated near the origin, and exhibit both low passive volume and low switch stress. The buck converter operating at a conversion ratio of 6-to-1 is also plotted to provide a benchmark for comparison with the presented hybrid converters.

Looking at Table I and Fig. 3, it can be seen that the proposed converter has much lower passive volume and therefore potentially higher power density compared to the Switched Tank (Dickson) converter, though it has higher switch stress. The proposed work has lower switch stress compared to the series-parallel topology, while its passive volume is only slightly higher. This is of note as the passive volume of the series-parallel topology is known to be at the theoretical lower limit [20]. This means that the proposed topology can achieve similar theoretical performance to the series-parallel converter even with a greatly reduced number of capacitors and switches (and their associated gate drive circuitry). Furthermore, as devices might not be able to be sized exactly to the theoretical VA rating due to limited availability of different voltage ratings, in a physical implementation topologies with different theoretical VA ratings may end up using the same device. This practical concern is most apparent in low-voltage Silicon power transistors, where discrete transistor products below 25 V are not readily available. In applications such as 48 V conversion, the series-parallel converter is very attractive despite its high theoretical switch stress, due to both the limited available switch voltage ratings in this application space as

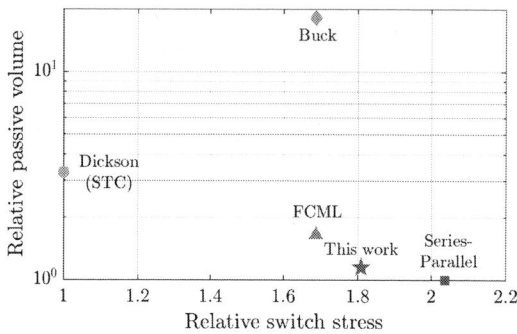

Fig. 3: Relative passive volume vs. normalized switch stress for several 6-to-1 converters.

well as the low output impedance exhibited by the converter due to its many parallel current paths [10]. The proposed topology operates in a similar manner to the series-parallel topology, and has the potential to exhibit very high performance in this application space compared to Dickson-based topologies due to its low passive volume and low number of components, as the disadvantage of switch utilization is relatively mild considering the actual switches available.

Hardware Implementation and Experimental Results

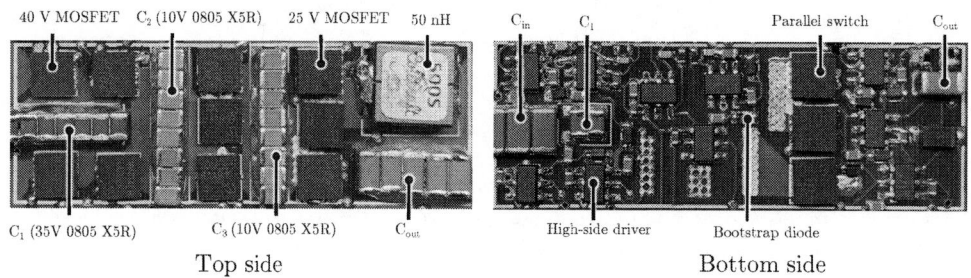

Fig. 4: Photograph of the converter. Dimensions: $1.38 \times 0.46 \times 0.22$ inch ($3.5 \times 1.17 \times 0.56$ cm).

Table II: Components for Prototype Converters

Component	Description	Device	Value
Q1-Q4	40 V Si switches	Infineon BSZ018N04LS6	40 V, 40 A, 1.8 mΩ
Q5-Q10	25 V Si switches	Infineon BSZ010NE2LS5	25 V, 40 A, 1.0 mΩ
L	Resonant inductor	Coilcraft SLC7530S-500MLB	50 nH, 50 A I_{sat}, 0.123 mΩ
C1	Flying Capacitors	TDK C2012X5R1V226M125AC	$16 \times 22 \mu F^*$ 35 V X5R 0805
C2	Flying Capacitors	Murata GRM21BR61A476ME15L	$16 \times 47 \mu F^*$ 10 V X5R 0805
C3	Flying Capacitors	Murata GRM21BR61A476ME15L	$16 \times 47 \mu F^*$ 10 V X5R 0805
	Gate driver	Analog Devices LTC4440-5	80 V, 1.1 A peak output current
	Bootstrap diode	ON Semiconductor NSR0340V2T1G	40V, 250 mA, Schottky
	Controller	TI TMS320F28069	

* The capacitance listed here is the nominal value before dc derating.

Fig. 4 shows an annotated photograph of the hardware prototype, with key components labeled. The PCB stack-up consists of 4 layers, with 4 oz copper on the outer layers (where the critical conduction path is) and 3 oz copper on the inner layers. As the maximum theoretical voltage a switch can see in this topology is $3V_o$, relatively low-voltage switches can be used (40 V and 25 V). Silicon devices are used for

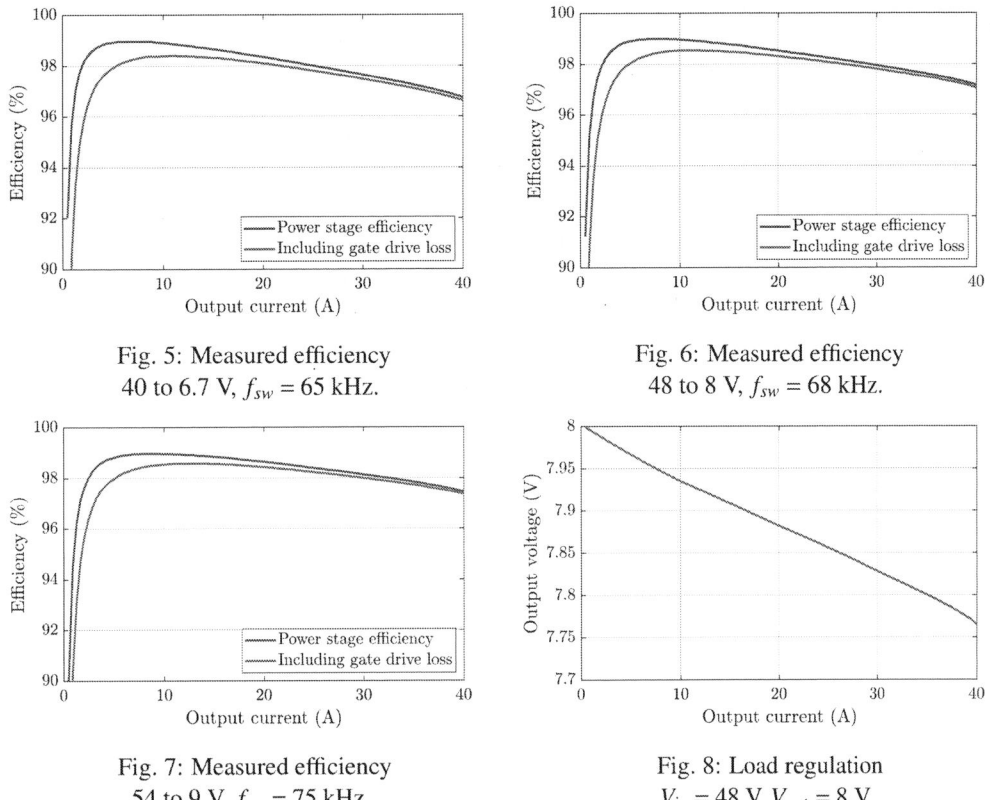

Fig. 5: Measured efficiency
40 to 6.7 V, f_{sw} = 65 kHz.

Fig. 6: Measured efficiency
48 to 8 V, f_{sw} = 68 kHz.

Fig. 7: Measured efficiency
54 to 9 V, f_{sw} = 75 kHz.

Fig. 8: Load regulation
V_{in} = 48 V, V_{out} = 8 V.

this prototype, as at these low voltages the performance of Si can match that of GaN. The high-current path devices ($Q_8 - Q_{10}$) are paralleled to reduce conduction losses. Each floating switch is driven by a floating high-side gate driver powered using the cascaded bootstrap method [21], from a 9 V source. The total gate drive current was measured with a Yokogawa WT310 digital power meter, while the power stage voltage, current, and efficiency was measured with a Yokogawa WT3000E precision power meter for the most accurate results at the high efficiencies obtained. Table II lists the components used in the prototype.

Table III lists the operating conditions. The gate drive signals were programmed with a constant deadtime of 44 ns. The converter was tested up to 40 A output current, and achieved a peak efficiency of 99.0% and a full-load efficiency of 97.1% (98.5% and 97.0% with gate drive loss included, respectively) for a 48-to-8 V step-down conversion operating at 68 kHz. The power density at full-load was 2230 Win^3 with a box volume of 0.139 in^3 (2.29 cm^3). Efficiency curves for 48-to-8 V operation from 0 A to 40 A are given in Fig. 6. Efficiency curves were also taken for additional voltage levels within the expected

Table III: Converter Operating Conditions

Parameter	Value
Input Voltage	48 V (40 - 54 V)
Output Voltage	8 V (6.7 - 9 V)
Output Current	40 A
Power (Measured)	310 W (260 - 350 W)
Switching Frequency	68 kHz (65 - 75 kHz)
Dimensions	1.38 inch × 0.46 inch × 0.22 inch
	(3.5 cm × 1.17 cm × 0.56 cm)
Box Volume	0.139 in^3 (2.29 cm^3)

Table IV: Measured Converter Performance

Metric	$V_{in} = 40$ V $f_{sw} = 65$ kHz	$V_{in} = 48$ V $f_{sw} = 68$ kHz	$V_{in} = 54$ V $f_{sw} = 75$ kHz
Peak Efficiency	99.0% (98.4% with gate loss)	99.0% (98.5% with gate loss)	99.0% (98.5% with gate loss)
Full-Load Efficiency	96.7% (96.6% with gate loss)	97.1% (97.0% with gate loss)	97.4% (97.3% with gate loss)
Power Density	1840 W/in^3	2230 W/in^3	2510 W/in^3

range of a 48 V nominal intermediate bus for datacenter applications. Efficiency sweeps for 40-to-6.7 V and 54-to-9 V operation are shown in Fig. 5 and Fig. 7. Table IV lists the efficiency and power density of the converter for all tested input voltage and frequency conditions.

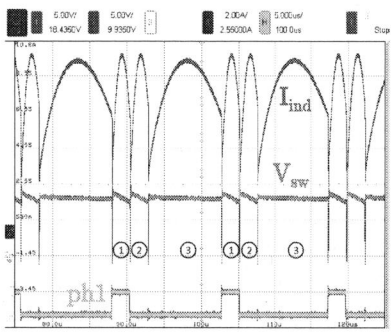

Fig. 9: Inductor current, switch node voltage, and ph1 gate signal waveforms for 48-to-8 V.

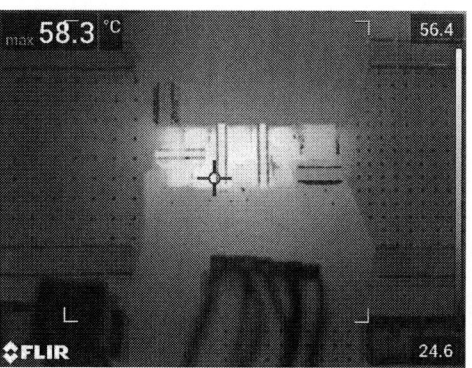

Fig. 10: Thermal image at full-load (40 A) for 48-to-8 V operation with fan cooling.

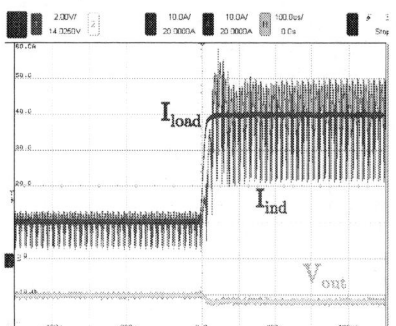

Fig. 11: Load-step from 10 A to 40 A for 48-to-8 V.

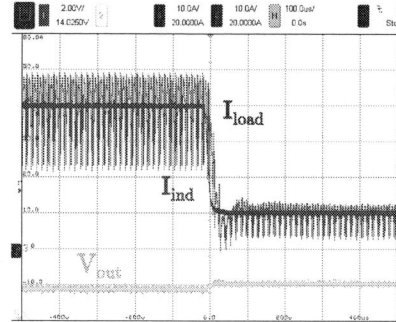

Fig. 12: Load-step from 40 A to 10 A for 48-to-8 V.

The high efficiency achieved by the converter decreases the impact of load regulation, as even though the converter operates in an open-loop fixed-ratio mode, it exhibits an output droop of only 234 mV (2.9% of V_{out}) at full-load as shown in Fig. 8.

Fig. 9 shows inductor current, I_L, switch node voltage, V_{sw}, and the ph1 gate drive signal for a 48 V input voltage. The three operating phases in Fig. 2 are labeled over two switching periods. As mentioned previously, the converter is operated above resonance to account for component tolerances and reduce conduction loss, as can be seen by the non-zero current at phase transitions. Partial ZVS at switch turn-on can be observed by noting that the switch node voltage goes below zero due to body diode conduction at the switch transitions. Future areas of research include optimizing the gate drive signals to improve the

ZVS performance of the converter.

Fig. 10 shows a thermal image of the converter operating at full-load at 48 V input. Even at 40 A output, the converter temperature did not exceed 59°C due to the low loss of the converter over its entire operating range. A bench-top fan was used to supply air cooling over the PCB.

Fig. 11 and Fig. 12 show the transient response of the converter for a step-up and step-down current load-step. Fig. 11 shows a transient from 10 A to 40 A at 48 V input. The inductor current stabilizes after approximately 200 μs after the initial ramp up in load current, while the voltage undershoot lasts less than 100 μs and then reaches its new steady state after approximately 160 μs.

Fig. 12 shows a transient from 40 A to 10 A at 48 V input. The inductor current stabilizes after approximately 120 μs after the initial ramp down in load current, while the voltage overshoot lasts less than 100 μs and then reaches its new steady state after approximately 200 μs.

Table V compares this work with some of the best existing works. The efficiencies given (including this work) include gate drive loss unless otherwise noted. The converter was tested at both 48-to-8 V and 54-to-9 V operating conditions to allow for a more equal comparison with prior work. The proposed topology has the highest power density of all the listed converters, and at 54-to-9 V operation the power density is almost 2× higher than the next most power dense converter (the EPC2905 buck [22]). The proposed topology's peak and full-load efficiencies are also higher than the EPC9205 buck converter, highlighting the potential efficiency advantages of the ReSC approach compared to more conventional approaches. The proposed topology also achieves a higher full-load efficiency compared to the Vicor VTM Current Multiplier [23], even at at higher full-load current. The power density and peak efficiency of the proposed converter are also higher as well, although the Vicor is a highly integrated product that may have more auxiliary circuitry compared to a prototype converter.

The proposed topology also has considerably higher power density and similar efficiencies compared to the 6-to-1 switched tank converter in [13]; however, [13] does not explicitly state whether their efficiencies include gate drive losses, which are often excluded in reported efficiencies for ReSC converters. Compared to the Google Switched Tank 4-to-1 converter [12], the proposed topology achieves roughly the same full and peak efficiencies for a larger conversion ratio (6-to-1 compared to 4-to-1).

Conclusion

In this paper, a multi-resonant cascaded series-parallel converter is proposed for 48 V to intermediate bus applications in datacenter power delivery architectures. The converter can achieve the same conversion ratio as traditional two-phase switched-capacitor converters with fewer capacitors and switches, allowing for very high power density and efficiency. A 48-to-8 V, 40 A converter prototype was built and tested, with 99.0% peak efficiency (98.5% with gate drive loss) and 2230 W/in^3 power density.

Table V: Comparison of this work and existing high step-down ratio bus converters

Reference	Topology	Voltage ratio	Output current (A)	Power density (W/in³)	Efficiency	Notes
This Work	Multi-Resonant Cascaded Series-Parallel SCC	48-to-8 V	40	2230	full-load: 97.0%, peak: 98.5%	6-to-1 fixed-ratio, Si MOSFET
		54-to-9 V	40	2510	full-load: 97.3%, peak: 98.5%	
6-to-1 Switched Tank [13]	Resonant Dickson SCC	54-to-9 V	50	750	full-load: 97.18%*, peak: 98.55%*	6-to-1 fixed-ratio, GaN FET
EPC9205 [22]	Buck	48-to-8 V	14	1300	full-load: 93.2%, peak: 94.7%	GaN FET
Vicor VTM Current Multiplier [23]	Sine Amplitude Converter	48-to-8 V	30	~900	full-load: 95.7%, peak: 95.8%	6-to-1 fixed-ratio
		55-to-9.2 V	30		full-load: 95.8%, peak: 95.9%	
Google 4-to-1 Switched Tank [12]	Resonant Dickson SCC	54-to-13.5 V	50	500	full-load: 97.41%, peak: 98.61%	4-to-1 fixed-ratio, Si MOSFET

* Not explicitly stated if efficiency number include gate drive loss

References

[1] A. Shehabi *et al.*, "United states data center energy usage report," *Lawrence Berkeley National Laboratory, Berkeley, CA, Tech.Rep. LBNL-1005775*, Aug, 2016. [Online]. Available: https://datacenters.lbl.gov/sites/default/files/DCDWebscale_Shehabi_072016.pdf

[2] T. Lee, "Powering graphics processors from a 48-v bus," 2019. [Online]. Available: https://www.powerelectronictips.com/powering-graphics-processors-from-a-48-v-bus/

[3] M. D. Seeman and S. R. Sanders, "Analysis and optimization of switched-capacitor dcdc converters," *IEEE Transactions on Power Electronics*, vol. 23, no. 2, pp. 841–851, 2008.

[4] C. B. Barth, T. Foulkes, I. Moon, Y. Lei, S. Qin, and R. C. N. Pilawa-Podgurski, "Experimental evaluation of capacitors for power buffering in single-phase power converters," *IEEE Transactions on Power Electronics*, vol. 34, no. 8, pp. 7887–7899, 2019.

[5] R. C. N. Pilawa-Podgurski, D. M. Giuliano, and D. J. Perreault, "Merged two-stage power converterarchitecture with softcharging switched-capacitor energy transfer," in *2008 IEEE Power Electronics Specialists Conference*, 2008, pp. 4008–4015.

[6] R. C. N. Pilawa-Podgurski and D. J. Perreault, "Merged two-stage power converter with soft charging switched-capacitor stage in 180 nm cmos," *IEEE Journal of Solid-State Circuits*, vol. 47, no. 7, pp. 1557–1567, 2012.

[7] Y. Lei and R. C. N. Pilawa-Podgurski, "A general method for analyzing resonant and soft-charging operation of switched-capacitor converters," *IEEE Transactions on Power Electronics*, vol. 30, no. 10, pp. 5650–5664, 2015.

[8] S. Pasternak, C. Schaef, and J. Stauth, "Equivalent resistance approach to optimization, analysis and comparison of hybrid/resonant switched-capacitor converters," in *2016 IEEE 17th Workshop on Control and Modeling for Power Electronics (COMPEL)*, 2016, pp. 1–8.

[9] Z. Ye, Y. Lei, and R. C. N. Pilawa-Podgurski, "The cascaded resonant converter: A hybrid switched-capacitor topology with high power density and efficiency," *IEEE Transactions on Power Electronics*, vol. 35, no. 5, pp. 4946–4958, 2020.

[10] W. C. Liu, Z. Ye, and R. C. N. Pilawa-Podgurski, "A 97% peak efficiency and 308 A/in3 current density 48-to-4 V two-stage resonant switched-capacitor converter for data center applications," in *2020 IEEE Applied Power Electronics Conference and Exposition (APEC)*, 2020, pp. 468–474.

[11] Z. Ye, R. A. Abramson, and R. C. N. Pilawa-Podgurski, "A 48-to-6 V multi-resonant-doubler switched-capacitor converter for data center applications," in *2020 IEEE Applied Power Electronics Conference and Exposition (APEC)*, 2020, pp. 475–481.

[12] S. Jiang, S. Saggini, C. Nan, X. Li, C. Chung, and M. Yazdani, "Switched tank converters," *IEEE Transactions on Power Electronics*, vol. 34, no. 6, pp. 5048–5062, 2019.

[13] Y. Li, X. Lyu, D. Cao, S. Jiang, and C. Nan, "A 98.55% efficiency switched-tank converter for data center application," *IEEE Transactions on Industry Applications*, vol. 54, no. 6, pp. 6205–6222, 2018.

[14] Y. Chen, J. Baek, and M. Chen, "LEGO-boost: A merged-two-stage resonant-switched-capacitor converter with high voltage conversion ratio," in *2020 IEEE Applied Power Electronics Conference and Exposition (APEC)*, 2020, pp. 48–55.

[15] T. Urkin, G. Sovik, E. E. Masandilov, and M. M. Peretz, "Digital lock-in controller IC for optimized operation of resonant SCC," in *2020 IEEE Applied Power Electronics Conference and Exposition (APEC)*, 2020, pp. 1236–1243.

[16] W. Liu, P. Assem, Y. Lei, P. K. Hanumolu, and R. Pilawa-Podgurski, "A 94.2%-peak-efficiency 1.53A direct-battery-hook-up hybrid dickson switched-capacitor DC-DC converter with wide continuous conversion ratio in 65nm CMOS," in *2017 IEEE International Solid-State Circuits Conference (ISSCC)*, 2017, pp. 182–183.

[17] C. Schaef *et al.*, "A 93.8% peak efficiency, 5V-input, 10A max ILOAD flying capacitor multilevel converter in 22nm CMOS featuring wide output voltage range and flying capacitor precharging," in *2019 IEEE International Solid- State Circuits Conference - (ISSCC)*, 2019, pp. 146–148.

[18] M. S. Makowski and D. Maksimovic, "Performance limits of switched-capacitor dc-dc converters," in *Proceedings of PESC '95 - Power Electronics Specialist Conference*, vol. 2, 1995, pp. 1215–1221 vol.2.

[19] M. S. Makowski, "Realizability conditions and bounds on synthesis of switched-capacitor dc-dc voltage multiplier circuits," *IEEE Transactions on Circuits and Systems I: Fundamental Theory and Applications*, vol. 44, no. 8, pp. 684–691, 1997.

[20] Z. Ye, S. R. Sanders, and R. C. N. Pilawa-Podgurski, "Modeling and comparison of passive component volume of hybrid resonant switched-capacitor converters," in *2019 20th Workshop on Control and Modeling for Power Electronics (COMPEL)*, 2019, pp. 1–8.

[21] Z. Ye, Y. Lei, W. Liu, P. S. Shenoy, and R. . Pilawa-Podgurski, "Improved bootstrap methods for powering floating gate drivers of flying capacitor multilevel converters and hybrid switched-capacitor converters," *IEEE Transactions on Power Electronics*, vol. 35, no. 6, pp. 5965–5977, 2020.

[22] EPC Inc, "Building the Smallest and Most Efficient 48 V to 5 - 12 V DC to DC Converter using EPC2045 and ICs," 2018. [Online]. Available: https://epc-co.com/epc/Portals/0/epc/documents/application-notes/How2AppNote001n\%2048n\%20Vn\%20ton\%205-12n\%20V.pdf/

[23] Vicor Power, "VTM Current Multiplier VTM48Ex080y030A00," 2018. [Online]. Available: http://www.vicorpower.com/documents/datasheets/VTM48E_080_030A00.pdf

SISO Control Strategy of Resonant Dual Active Bridge with a Tuned CLC Network

Meiqi Wang[1], Bo Yang[2], Lie Xu[3], Jing Li[1], David Gerada[4], Chunyang Gu[1], He Zhang[1], Chris Gerada[1,4], Yongdong Li[3]

[1]THE UNIVERSITY OF NOTTINGHAM NINGBO CHINA
[2]XI'AN UNIVERSITY OF TECHNOLOGY
[3]TSINGHUA UNIVERSITY
[4]THE UNIVESITY OF NOTTINGHAM
Ningbo, China
Xi'an, China
Beijing, China
Nottingham, United Kingdom
E-Mail: Meiqi.Wang@nottingham.edu.cn

Acknowledgements

This work was supported by the Zhejiang Natural Science Foundation under Grant LY19E070002 and LQ19E070002 and by the Zhejiang Provincial Department of Human Resources and Social Security under Grant QJD1803013.

Keywords

«Bidirectional DC/DC converter», «RDAB», «Control Design», «Modeling», «System Efficiency»

Abstract

This paper proposed a linear state-space model for a resonant dual active bridge with a tuned capacitor-inductor-capacitor network which is applied to an energy storage system. The proposed model is used for predicting the behavior of the proposed system and estimated the state variables under large signal variation. Using the proposed model, a decoupled control strategy was designed which realizes the high frequency link currents control and also improves the efficiency of the converter. By applying the steady-state relationship between state variables and the system inputs, the controller was simplified to a single proportional integral (PI) controller instead of from three PI controllers, which was implemented on a low cost digital signal processor and verified experimentally.

Introduction

Recently, dual active bridges (DABs) have received an increasing attention due to their bidirectional power transmission and galvanic isolation capabilities, which can provide a high power density and accommodate a wide range of voltages by operating in both buck and boost modes. Early DAB converters were controlled using single-phase-shift (SPS) control to allow for bidirectional power transfer at variable power levels, which can easily achieve soft switching when the magnitudes of the two ac voltages are matched by the transformer ratio. However, when the converter is operated outside of the proper voltage range, the zero-voltage switching (ZVS) can hardly be achieved and the peak current will increase which then lead to an increase of switching and conduction losses [1]. In order to reduce the switch current stresses and the attendant switching and conduction losses, many other complicated control strategies have been proposed [2] [3] which have shown an improved soft switching range and lower conduction losses, however, these methods usually require more complicated control systems and still have a large reactive current at rated power.

In order to improve the soft switching range, a number of quasi-resonant type DAB topologies, consisting of series resonant networks, have been investigated [4] [5]. The most common resonant dual active bridge (RDAB) topology is the series RDAB (SRDAB) [6] - [8], which has received an increasing attention due to it low requirement of resonant network components and its inherent dc blocking capability. However, it needs a wide switching frequency range to modulate the power transfer, which make the design of controller and filter being more complicated.

Regardless of control and resonant schemes employed, the existing DABs typically requires a more complex control system, particularly when they operate with wide load and supply voltage variations. In this paper, based on linear state-space model and mathematical analysis, a simplified decoupled control strategy for a capacitor-inductance-capacitor (CLC) tuned resonant DAB is proposed and implemented with a single PI controller which achieves the control of both the primary and secondary ac current amplitudes while ensuring soft switching of the switches at a wide voltage range. An experimental prototype is designed and operated in various conditions to investigate the performance of the control scheme.

Basic operation of the RDAB

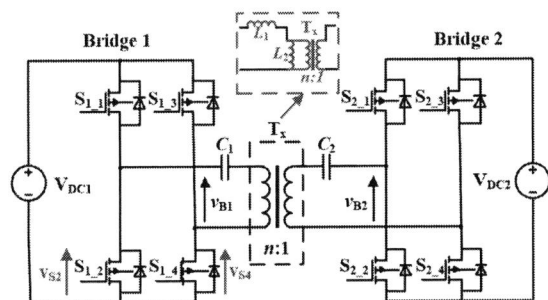 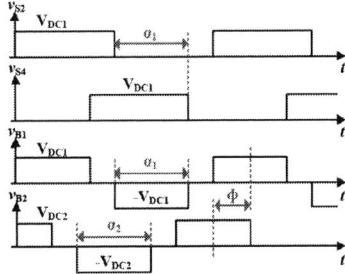

Fig. 1 Resonant dual active bridge   Fig. 2 TPS modulation strategy

The structure of capacitor-inductor-capacitor (CLC) RDAB is shown in Fig. 1. Two additional capacitors C_1 and C_2 is added to the primary side and secondary side of transformer T_x, respectively, which then act as an integral part of resonant network with T_x's leakage and mutual inductance L_1 and L_2. The two full-bridge converters are controlled using a Three Phase Shift (TPS) modulation scheme as demonstrated in Fig.2. The switches of each leg are controlled to have a duty cycle of 50% and the output of the first leg is phase shifted from the output of the second leg by α_1 for the primary bridge and α_2 for the second bridge. The output of the secondary full-bridge v_2 is phase shifted from the output of the primary full-bridge v_1, to lead v_1 by ϕ. The outputs of the full-bridge converter have been simplified into voltage source v_{B1} and v_{B2}, which are represented by the Fourier series in (1) and (2).

$$v_{B1} = V_{DC1} \frac{4}{\pi} \sum_{b_1=1,3\ldots}^{n} \frac{1}{b_1} \cos(b_1 \omega_s t) \sin\left(\frac{b_1 \alpha_1}{2}\right) \tag{1}$$

$$v_{B2} = V_{DC2} \frac{4}{\pi} \sum_{b_2=1,3\ldots}^{n} \frac{1}{b_2} \cos(b_2 \omega_s t + b_2 \varphi) \sin\left(\frac{b_2 \alpha_2}{2}\right) \tag{2}$$

The magnitude of the power transferred has been quantified in a previous paper by the author [9], which was shown that majority of the power was transferred at the fundamental frequency of the tuned resonant network. The tuned network presents a high impedance to harmonics generated by the converters. Consequently, the effects of these harmonics on the power transfer will be insignificant and therefore ignoring harmonics, the magnitude of the power can be simplified to (3):

$$P = \frac{8n V_{DC1} V_{DC2}}{\pi^2 L_2} \sin(\varphi) \sin\left(\frac{\alpha_2}{2}\right) \sin\left(\frac{\alpha_2}{2}\right) \tag{3}$$

Mathematical analysis of RDAB

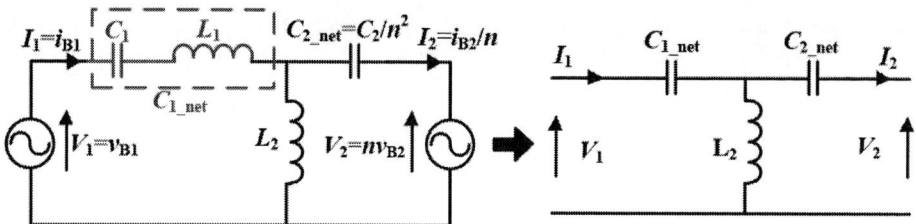

Fig. 3 Equivalent circuit model of RDAB (a) at fundamental frequency (b) at HF ac link

In order to simplify the mathematical analysis, the converter works at fundamental frequency is simplified as shown in Fig.3 (a), where the voltage across C_1 and L_1 behaves as a capacitor (C_{1_net}) voltage. Then the fundamental circuit at switching frequency is illustrated in Fig.3 (b) while circuit resistance have a relatively insignificant effect on the current and will, therefore, be ignored and their losses can be allowed for late, then the relationships between original circuit and the equivalent circuit are shown as follows:

$$Z_1 = -jX_{C1_{net}} = -j\left(X_{C_1} - X_{L_1}\right) \tag{4}$$

$$Z_2 = -jX_{C2_{net}} = -jn^2 X_{C_2} \tag{5}$$

$$Z_3 = jX_{L_2} \tag{6}$$

The resonant network consisted with C_{1_net}, C_{2_net} and L_2 is tuned to the fundamental of the switching frequency f_s as follow:

$$X_{1_{net}} = X_{2_{net}} = \frac{1}{\omega_s C_{1_{net}}} = \frac{1}{\omega_s C_{2_{net}}} = \omega_s L_2 = X_{L_2} \tag{7}$$

Substituting (7) to (4) to (6), the following relationship among T-equivalent circuit (shown in Fig. 4(a)) was found:

$$Z_1 = Z_2 = -Z_3 = -j\omega_s L_2 = \frac{n^2}{j\omega_s C_2} \tag{8}$$

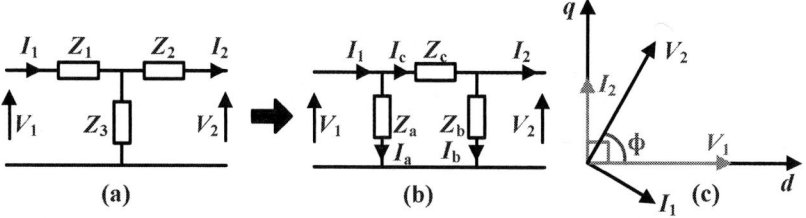

Fig. 4 Equivalent circuit and steady-state voltage and current phasors at fundamental frequency of resonant DAB (a) T-equivalent circuit (b) delta-equivalent circuit (c) voltage and current phasors

To recognize the relationship between bridge currents, i_1 and i_2, and voltages, v_1 and v_2, a delta-equivalent circuit of resonant network is illustrated in Fig. 4(b), where the independences are calculated as follows:

$$Z_a = Z_b = -Z_c = Z_1 + Z_3 + \frac{Z_1 Z_3}{Z_2} = j\omega_s L_2 \tag{9}$$

According to KVL and KCL of delta network and the inductances relationship (9), at the switching frequency and steady-state, the bridge current can be calculated as shown in (10) and (11), which shows that i_1 is decoupled with v_1 and i_2 is decoupled with v_2 (as shown in Fig. 4c).

$$I_1 = I_a + I_c = (\frac{1}{Z_a} + \frac{1}{Z_c})V_1 - \frac{1}{Z_c}V_2 = \frac{1}{j\omega_s L_2}V_2 \tag{10}$$

$$I_2 = I_c + I_b = -(\frac{1}{Z_b} + \frac{1}{Z_c})V_2 + \frac{1}{Z_c}V_1 = -\frac{1}{j\omega_s L_2}V_1 \tag{11}$$

To investigate the effects that the modulation variables have on the current, (1) and (2) have been substituted into (10) and (11), the magnitudes of the high frequency link current i_1 and i_2 are abstained as (12) and (13), where it can be found that the magnitudes of currents i_1 and i_2 are controlled by the phase shift α_2 and α_1 respectively.

$$|I_1| = \frac{4nV_{DC2}}{\pi\omega_s L_2}\sin\left(\frac{\alpha_2}{2}\right) \tag{12}$$

$$|I_2| = -\frac{4nV_{DC1}}{\pi\omega_s L_2}\sin\left(\frac{\alpha_1}{2}\right) \tag{13}$$

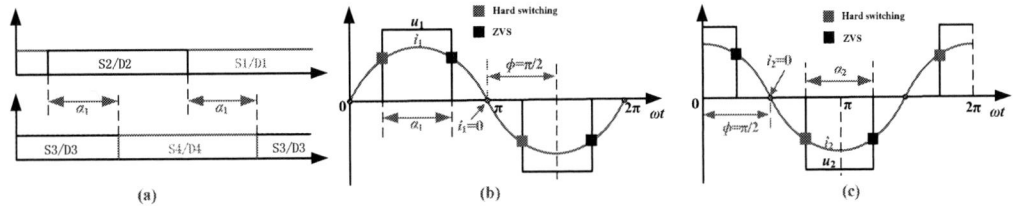

Fig.5 Relationship among pulse width, voltages and currents (a) primary switch states (b) primary voltage and currents waveforms (c) secondary voltage and currents waveforms

The relationships between the voltages and currents in (10) and (11) are illustrated in Fig. 4(c), where i_2 is fixed to lead v_1 by 90° and v_2 is fixed to lead i_1 by 90°. As it can be seen that when ϕ is fixed at 90° or -90° (depending on the power flow direction), the currents of the bridges (i_1 and i_2) will align to the voltages (v_1 and v_2), which then results in unity power factor (UPF) at the ac terminals of both full-bridge converters. The relationship among pulse width, voltage and current is illustrated in Fig. 5, where α_1 controls the pulse width of v_1 and therefore the magnitude of the fundamental component of v_1, α_2 controls the pulse width of v_2, while ϕ controls the phase difference between v_1 and v_2.

According to the relationships between currents and voltages shown in Fig. 4(c), the voltage midpoint corresponds to the current starting point, it can be expressed as ϕ also controls the phase difference between i_1 and v_1, as well as the phase different between i_2 and v_2, which is illustrated in Fig 5. In UPF applications, ϕ is fixed at 90° or -90°, the power transfer is modulated using α_1 and α_2, the freedom of the system is simplified to two degrees.

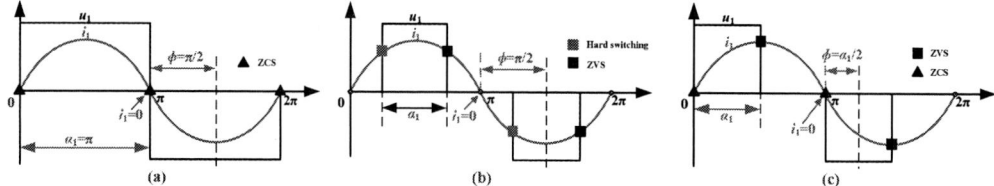

Fig. 6 Voltage and current under different modulations (a) at maximum power transfer capability (b) with modulation of α_1 and α_2, while φ fixed at $\pm\pi/2$ (c) with proposed modulation.

As shown in Fig. 6(a), in the maximum power transmission, the converter is zero-current switching or near ZCS, however, at the magnitude of power transfer is modulated down using α_1 and α_2, some of the switches start hard switching as can be seen in Fig. 6(b). This leads to an increase of the switching loss in high frequency link and a decrease of the power transmission efficiency of the system. From the analysis of Fig. 5, when the relationship among the three freedoms of the system can be implemented as (14), a wide power range of soft-switching can be achieved as shown in Fig. 6(c).

$$\varphi = \frac{\alpha_1}{2} = \frac{\alpha_2}{2} \tag{14}$$

Control scheme design of RDAB

The purpose of the RDAB control is to control the amplitudes of the currents in the resonant tank, i_1 and i_2, by α_1 and α_2, while ensuring soft switching of the switches by modulating ϕ. In order to simplify the control algorithm, the d-q decoupling control is introduced, and the d-axis is oriented in the direction of v_1, the relationship between the steady-state voltage and the phase angle in the d-q frame is derived as follows:

$$v_{1q} = 0 \tag{15}$$

$$v_{1d} = \frac{4}{\pi}V_{DC1}\sin\left(\frac{\alpha_1}{2}\right) = v_{1max}\sin\left(\frac{\alpha_1}{2}\right) \tag{16}$$

$$v_{2q} = \frac{4}{\pi}V_{DC2}\sin\left(\frac{\alpha_1}{2}\right)\sin\varphi = v_{2max}\sin(\frac{\alpha_1}{2})\sin\varphi \tag{17}$$

$$v_{2d} = \begin{cases} \frac{4}{\pi}V_{DC2}\sin\left(\frac{\alpha_1}{2}\right)\cos\varphi & (\varphi \geq 0) \\ \frac{-4}{\pi}V_{DC2}\sin\left(\frac{\alpha_1}{2}\right)\cos\varphi & (\varphi < 0) \end{cases} = v_{2max}\sin\left(\frac{\alpha_1}{2}\right)\cos\varphi \tag{18}$$

Since v_{1q} is zero due to the selection of v_{1d} as the reference frame (as shown in Fig.4 (c)), v_{1d} is the only input that affects i_2 and the current i_2 lead v_1 by 90, the following assumptions can be made:

$$i_{2d} = 0 \tag{19}$$

$$i_{2q} = i_{2max} \sin(\frac{\alpha_1}{2}) \tag{20}$$

$$i_{1d} = i_{1max} \sin(\frac{\alpha_1}{2}) \sin\varphi \tag{21}$$

$$i_{1q} = i_{1max} \sin(\frac{\alpha_1}{2}) \cos\varphi \tag{22}$$

By submitting (14) to (15) - (22), the relationships between primary side and secondary side can be found as follows:

$$v_{2q} = \frac{v_{2max}}{v_{1max}^2} v_{1d}^2 \tag{23}$$

$$v_{2d} = \frac{v_{2max}}{v_{1max}} v_{1d} \sqrt{1 - \left(\frac{v_{1d}}{v_{1max}}\right)^2} \tag{24}$$

$$i_{1d} = \frac{i_{1max}}{i_{2max}^2} i_{2q}^2 \tag{25}$$

$$i_{1q} = \frac{i_{1max}}{i_{2max}} i_{2q} \sqrt{1 - \left(\frac{i_{2q}}{i_{2max}}\right)^2} \tag{26}$$

As it can be seen, v_{2q} and v_{2d} can be calculated by v_1, i_{1d} and i_{1q} can be calculated by i_{2q}, and the amplitude of i_{2q} or i_2 depends on v_1, the system transmission power can be adjusted by controlling the amplitude of α_1, whereby the degree of freedom of the system is reduced to 1. At the same time, the maximum soft switch control will be achieved. According to the above analysis, the input and output relationship of the AC system is directly related, the single-in-single-out (SISO) modulation algorithm is proposed, as shown in Fig. 7. α_1 and α_2 are varied to control the magnitudes of the currents i_1 and i_2, while ϕ will be varied along with the change to improve the soft switching range of the converter.

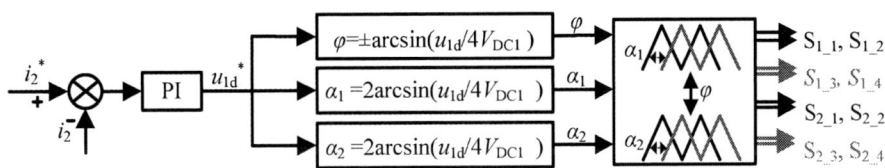

Fig. 7 SISO control scheme of CLC tuned RDAB

As mentioned before, the input of the AC system is directly related to the output, therefore, only the amplitude of i_{2q} which is the same as the amplitude of i_2 is needed for the control. This measurement can be easily realized through the use of an inexpensive current transformer (CT). Additionally, only one PI controller is implemented, the inputs for the other loops are derived from (23) - (26), which make the RDAB being more flexible for complicated applications like hybrid DC-AC microgrid which is enabled by energy storage systems.

Experimental results

An experimental prototype of CLC tuned RDAB was designed to investigate the performance of the control strategy and the mathematical analysis (as shown in Fig. 8). In Fig. 10, the voltage and current waveforms of the primary and secondary H bridges with varying phase shifts are illustrated. From Fig.10, all the switches in the primary and secondary full bridges realize the soft switching. The power throughput and efficiency of the prototype RDAB under various modulation levels are shown in Fig. 9. The maximum efficiency of the system is observed at rated power and is approximately 96%. Furthermore, the system efficiency is maintained well above 82% even at lower modulation factors, which suggest a significant improvement in performance in comparison to conventional DAB converters.

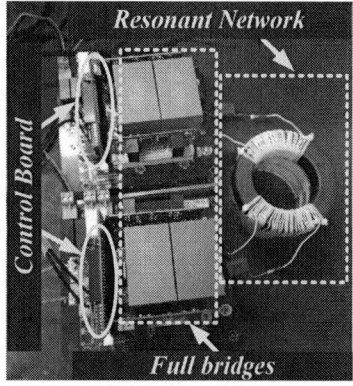

Fig. 8 Experimental prototype of CLC tuned RDAB

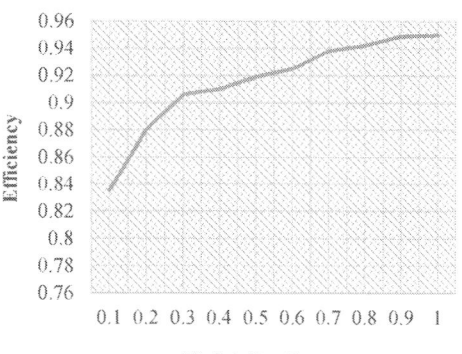

Fig. 9 Converter efficiency over the modulation range

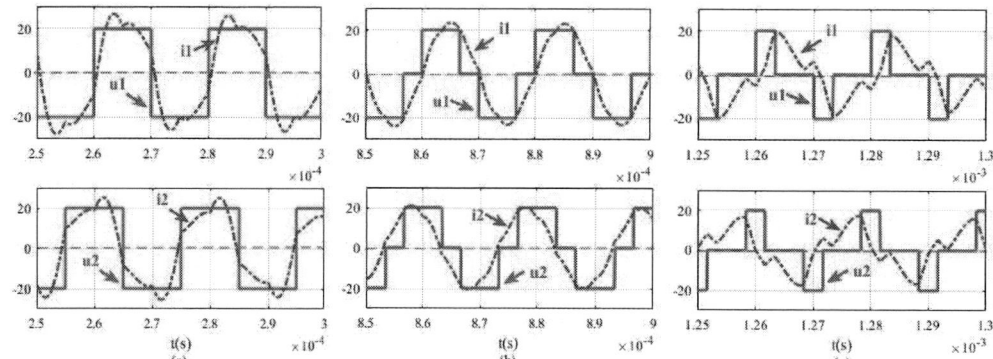

Fig. 10 Voltage and current waveforms under different modulation (a) $\alpha_1 = \alpha_2 = 2$ $\phi=180°$ (M=1) (b) $\alpha_1 = \alpha_2 = 2$ $\phi=120°$ (M=0.65) (c) $\alpha_1 = \alpha_2 = 2$ $\phi=60°$ (M=0.125)

Conclusion

In this paper, a mathematical analysis has been presented to accurately predict the performance of the proposed SISO control strategy. Experimental results of a prototype RDAB, operated under various conditions, have been presented to demonstrate the improved performance of the converter. Results indicate that the proposed SISO control strategy for resonant DAB topology has a good closed-loop dynamic response and steady-state performance, while indicates a potential to further increase the operating efficiency by employing a purpose-designed transformer, which will be smaller than that of a conventional DBA converter on account of the lower operating currents.

References

[1] S. Inoue and H. Akagi.: A Bidirectional DC–DC Converter for an Energy Storage System with Galvanic Isolation, in IEEE Transactions on Power Electronics, vol. 22, no. 6, pp. 2299-2306, Nov. 2007.

[2] F. Krismer.: Accurate Small-Signal Model for the Digital Control of an Automotive Bidirectional Dual Active Bridge, in IEEE Transactions on Power Electronics, vol. 24, no. 12, pp. 2756-2768, Dec. 2009.

[3] W. Han.: Wide-Range ZVS Control Technique for Bidirectional Dual-Bridge Series-Resonant DC–DC Converters, in IEEE Transactions on Power Electronics, vol. 34, no. 10, pp. 10256-10269, Oct. 2019.

[4] R. P. Twiname.: A New Resonant Bidirectional DC–DC Converter Topology, in IEEE Transactions on Power Electronics, vol. 29, no. 9, pp. 4733-4740, Sept. 2014.

[5] M. Yaqoob.: Extension of Soft-Switching Region of Dual-Active-Bridge Converter by a Tunable Resonant Tank, in IEEE Transactions on Power Electronics, vol. 32, no. 12, pp. 9093-9104, Dec. 2017.

[6] J. Jung, H. Kim, M. Ryu and J. Baek, Design Methodology of Bidirectional CLLC Resonant Converter for High-Frequency Isolation of DC Distribution Systems, in IEEE Transactions on Power Electronics, vol. 28, no. 4, pp. 1741-1755, Apr. 2013.

[7] H. Kim, M. Ryu, J. Baek and J. Jung, High-Efficiency Isolated Bidirectional AC–DC Converter for a DC Distribution System, in IEEE Transactions on Power Electronics, vol. 28, no. 4, pp. 1642-1654, Apr. 2013.

[8] F. Ibanez, J. M. Echeverria and L. Fontan, Novel technique for bidirectional series-resonant DC/DC converter in discontinuous mode, in IET Power Electronics, vol. 6, no. 5, pp. 1019-1028, May 2013.

[9] R. Twainame, D. THrimawithana, U. Madawala, and C. Baguley, A dual active bridge topology with a tuned CLC network, IEEE Transaction on Power Electronics, vol.30, no. 12, pp. 6543-6540, Dec. 2015.

Impact of steady-state grid-frequency deviations on the performance of grid-forming converter control strategies

Anant Narula[1], Massimo Bongiorno[1], Mebtu Beza[1], Jan R Svensson[2], Xavier Guillaud[3], Lennart Harnefors[4]

[1] Chalmers University of Technology, Dept. of Electrical Engineering, Gothenburg, Sweden
[2] Hitachi ABB Power Grids, Power Grids Research, Västerås, Sweden
[3] Univ. Lille, Arts et Metiers Institute of Technology, Centrale Lille, Yncrea Hauts-de-France, ULR 2697-L2EP, F-59000 Lille, France
[4] ABB Corporate Research, Västerås, Sweden
anant.narula@chalmers.se

Keywords

≪Renewable energy systems≫, ≪Converter control≫, ≪Grid-forming converters≫, ≪Passivity≫, ≪Input admittance≫.

Abstract

The aim of this paper is to investigate the impact of steady-state deviations in the grid-frequency on the performance of grid-forming converter control strategies. In particular, the virtual synchronous machine (VSM) control structure has been the focus of study in this paper. Two alternative solutions to address the issue with the conventional VSM structure are suggested, and their impact on the active power response and the power dissipation properties of the converter are compared. For the latter, the ac-side input admittance of the grid-connected converter is derived for the different control structures under consideration. Finally, the dynamic performance of the different control structures is verified through the time-domain simulations.

Introduction

Electric power generation using renewable energy sources (RES), particularly wind and solar, is continuously increasing as utility and power providers are turning to cleaner, more sustainable and abundant sources of energy. This in turn has led to a substantial increase in the converter-interfaced generation. In order to allow large penetration of RES and at the same time guarantee a stable operation of the power system, different converter control strategies have been developed in the last years, which are aimed at providing various functionalities such as inertia and frequency support, black-start and synchronization capabilities, and are often termed as grid-forming converter control strategies. Virtual synchronous machine (VSM) control is one of the variants of grid-forming converter control strategy. The VSM control offers the ability to provide the power system a virtual (or "synthetic") inertia and thereby reduce the rate-of-change-of-frequency (RoCoF) during normal operating conditions as well as large power system contingencies. This control strategy was first suggested in [1], and since then many variants have appeared in the literature [2–6]. All these control strategies share one main feature similar to the conventional synchronous machine, i.e. the active power output of the converter is controlled by directly regulating the converter-voltage angle; thus, the active power loop provides the angle for grid-synchronization, meaning that the phase-locked loop (PLL) is not needed during normal operating conditions [2,5,6]. Although the effectiveness of these control strategies has been proved under different conditions, little attention has been given to the operation of VSM control when the grid-frequency deviates from its nominal value, while still being within the limits accepted by the Grid Codes.

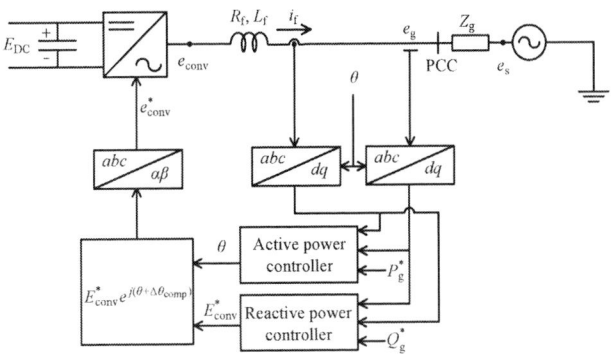

Fig. 1: Grid-connected VSC single-line diagram and its control structure.

The aim of this paper is to investigate the impact of steady-state grid-frequency deviations on the performance of the VSM control and investigate two alternative solutions to guarantee proper operation of the converter system. The tuning criteria for the investigated control alternatives is also presented. In order to study the power dissipation properties of the converter and thereby the robustness of the control system, the converter's ac-side input admittance is derived for the different control structures under consideration. Finally, the theoretical findings are verified through the time-domain simulations.

Controller design and system modeling

The single-line diagram of a generic grid-forming converter system together with its control structure is shown in Fig. 1, where a voltage-source converter (VSC) is connected to a generic ac grid through an RL filter of resistance, R_f, and inductance, L_f (inductive reactance, X_f). The grid model comprises of an impedance, Z_g, behind an infinite bus of voltage, e_s. The voltage at the point-of-common-coupling (PCC) or the grid-voltage is denoted by e_g. The voltage at the converter's terminal is denoted by e_{conv}, and i_f is the current exchanged between the converter and the grid.

One main feature of the grid-forming converter control strategy is that the active power is controlled by directly regulating the converter-voltage angle, θ; thus the active power loop provides the angle for grid-synchronization. The PCC voltage (or alternatively the reactive power) is controlled by regulating the converter-voltage magnitude, E_{conv}. An additional current controller can be included in the control structure in order to allow proper converter operation in case of fault-ride-through; however, as this is beyond the scope of this paper, this control loop is not considered in the derived model.

In the following, the conventional structure of the VSM control is first described and the issue related to its operation under grid-frequency deviations is then identified. Two alternative solutions to address this issue are described next, and the corresponding converter's ac-side input admittance models are derived in this section.

Conventional virtual synchronous machine control

In a conventional VSM control structure, the active power is controlled by adjusting the converter-voltage angle, θ, in order to emulate the mechanical dynamics of a synchronous machine, as

$$\theta = \frac{1}{s}\omega_{conv} = \frac{1}{s}\left[\omega_1 + \frac{1}{sM + K_D}(P_g^\star - P_g)\right], \tag{1}$$

where s is the Laplace transform variable and shall be interpreted as d/dt wherever appropriate, ω_1 is the rated angular synchronous frequency, ω_{conv} is the angular frequency of the converter-voltage, which can be interpreted as the angular speed of the virtual machine, M is the ω_1-scaled virtual inertia, K_D is the virtual mechanical damping constant, and $P_g^\star$ is the reference active power. M can be expressed in terms of inertia constant, H (expressed in seconds), and rated apparent power of the converter, S_{base}, as $M = \frac{2HS_{\text{base}}}{\omega_1}$. Neglecting the losses in the filter reactor, the steady-state active power injected by the converter, P_{g}, can be expressed as

$$P_{\text{g}} = \frac{E_{\text{conv}} E_{\text{g}} \sin(\delta_{\text{L}})}{X_{\text{f}}}, \tag{2}$$

where E_{g} represents the magnitude of the grid-voltage phasor, δ_{L} represents the difference between the converter-voltage angle, $\theta = \frac{\omega_{\text{conv}}}{s}$, and the estimated grid-voltage angle, $\hat{\theta}_{\text{g}} = \frac{\omega_1}{s}$. Note that the angle, δ_{L}, corresponds to the load-angle in the synchronous machine, which using (1) can be represented as

$$\delta_{\text{L}} = \frac{G_{\text{pc}}}{s} (P_{\text{g}}^{\star} - P_{\text{g}}), \tag{3}$$

where $G_{\text{pc}} = \frac{1}{sM + K_{\text{D}}}$. Using the expressions (1)–(3), and linearizing the system model around its steady-state, an approximate relationship between the reference and the actual active power injected from the converter is given by

$$\Delta P_{\text{g}} = \frac{\frac{K_{\text{s,conv}}}{M}}{s^2 + \frac{K_{\text{D}}}{M} s + \frac{K_{\text{s,conv}}}{M}} \Delta P_{\text{g}}^{\star}, \tag{4}$$

where $K_{\text{s,conv}}$ is the equivalent synchronizing-torque coefficient, given by $K_{\text{s,conv}} = \frac{E_{\text{conv}} E_{\text{g}} \cos(\delta_{\text{L}})}{X_{\text{f}}}$. In order to derive the control parameters, the closed-loop transfer function in (4) is shaped as a second-order low-pass filter, thus

$$\Delta P_{\text{g}} = \frac{\omega_{\text{n}}^2}{s^2 + 2\zeta\omega_{\text{n}}s + \omega_{\text{n}}^2} \Delta P_{\text{g}}^{\star}, \tag{5}$$

where ζ and ω_{n} are the damping ratio and natural frequency of the system, respectively. By comparing (4) and (5), it is possible to obtain

$$\zeta\omega_{\text{n}} = \frac{K_{\text{D}}}{2M}, \quad \omega_{\text{n}}^2 = \frac{K_{\text{s,conv}}}{M}. \tag{6}$$

It can be seen from (6) that for a constant $K_{\text{s,conv}}$, the virtual inertia is inversely proportional to the natural frequency of the power-control loop. This is similar to the operation of a synchronous generator, where larger inertia means slower power response of the machine. Once the value of the inertia, and hence ω_{n}, is chosen, K_{D} can be calculated based on the desired damping ratio, ζ. The damping ratio must be positive in order to obtain a damped response and is selected based on the desired location of the closed-loop poles. The rise-time of the controller is directly proportional to the damping ratio, and hence K_{D}.

As mentioned earlier, the PCC voltage (or alternatively the reactive power) controller generates the reference for the converter-voltage magnitude, $E_{\text{conv}}^{\star}$. The selection of this controller depends upon the function to be implemented by the converter, as well as on the strength of the grid at the connection point. A relatively strong grid (short-circuit ratio of 5) is modeled for the case studies in this paper and hence a reactive power controller is used. However, the analysis is also valid for a PCC voltage controller. The reactive power controller used here has the following structure

$$E_{\text{conv}}^{\star} = G_{\text{qc}} (Q_{\text{g}}^{\star} - Q_{\text{g}}), \tag{7}$$

where G_{qc} represents the transfer function of an integral regulator. Neglecting the losses in the filter reactor, the steady-state reactive power injected from the converter, Q_{g}, can be expressed as

$$Q_{\text{g}} = \frac{E_{\text{g}} [E_{\text{conv}} \cos(\delta_{\text{L}}) - E_{\text{g}}]}{X_{\text{f}}}. \tag{8}$$

Using the expressions (7) and (8), and linearizing the system model around its steady-state, an approximate relationship between the reference and the actual reactive power injected from the converter is given by

$$\Delta Q_{\mathrm{g}} = \frac{1}{\frac{X_{\mathrm{f}}}{K_{\mathrm{i,qc}} E_{\mathrm{g}} \cos(\delta_{\mathrm{L}})} s + 1} \Delta Q_{\mathrm{g}}^{\star}. \tag{9}$$

Thus, the gain of the integral regulator can be selected as

$$K_{\mathrm{i,qc}} = \frac{\alpha_{\mathrm{qc}} X_{\mathrm{f}}}{E_{\mathrm{g}} \cos(\delta_{\mathrm{L}})}$$

where α_{qc} can be approximated as the controller's loop-bandwidth. As the closed-loop transfer function in (9) is shaped as a first-order low-pass filter, the rise-time for the reactive power controller is approximately given by: $\frac{\ln(9)}{\alpha_{\mathrm{qc}}}$.

Impact of grid-frequency deviation on conventional VSM control and the suggested control modifications

Using (1), the steady-state active power injected from the converter, P_{g}, is given by

$$P_{\mathrm{g}} = P_{\mathrm{g}}^{\star} + K_{\mathrm{D}}(\omega_1 - \omega_{\mathrm{conv}}). \tag{10}$$

In steady-state, the angular speed of the virtual machine, ω_{conv}, will be equal to the actual angular frequency of the grid. Thus, if the grid frequency is at its nominal value, ω_1, $P_{\mathrm{g}} = P_{\mathrm{g}}^{\star}$. However, if the angular frequency of the grid deviates from its nominal value, a steady-state error in the active power injected from the converter will be experienced. Denoted by, $P_{\mathrm{e}} = P_{\mathrm{g}} - P_{\mathrm{g}}^{\star}$, this error is equal to

$$P_{\mathrm{e}} = K_{\mathrm{D}}(\omega_{\mathrm{e}}), \tag{11}$$

where ω_{e} is the steady-state deviation in the angular synchronous frequency of the grid. As it can be seen from (11), the steady-state error in the injected active power from the converter is directly proportional to the damping constant, K_{D}. As an example, a steady-state grid-frequency deviation of 0.5 Hz, which is the maximum permissible steady-state frequency deviation for all synchronous areas in Great Britain [7], will result in a steady-state error of 0.7 pu in the active power injected from the converter for $\omega_{\mathrm{n}} = 0.2$ pu, and $\zeta = 1.0$. One simple solution to fix this steady-state error could be to change the value of K_{D} to zero during steady-state and restore the value during transients. However, this would require a fast frequency estimation loop and a triggering logic, which could introduce additional dynamics to the system. Hence, two alternative approaches to eliminate the steady-state error in the active power injected from the converter are described in the following.

Virtual synchronous machine control with integrator or droop

A first solution to overcome the problem of steady-state error in the active power injected from the converter is to add an integral regulator to the VSM control structure as shown in Fig. 2(b). In this case, expression (1) is modified as

$$\theta = \frac{1}{s} \omega_{\mathrm{conv}} = \frac{1}{s} \left[\omega_1 + \frac{1}{sM + K_{\mathrm{D}}} (P_{\mathrm{g}}^{\star} - P_{\mathrm{g}}) + \frac{K_{\mathrm{i,pc}}}{s} (P_{\mathrm{g}}^{\star} - P_{\mathrm{g}}) \right], \tag{12}$$

where $K_{\mathrm{i,pc}}$ is the gain of the integral regulator. In this case G_{pc} is modified to, $G_{\mathrm{pc}} = \frac{1}{sM + K_{\mathrm{D}}} + \frac{K_{\mathrm{i,pc}}}{s}$. Assuming $K_{\mathrm{D}} = 0$ in (12), it can be seen that the effective value of virtual inertia is given by $\frac{M}{1 + K_{\mathrm{i,pc}} M}$. Hence, to prevent reduction in the effective value of virtual inertia, the integral gain should be selected such that $K_{\mathrm{i,pc}} M \ll 1$, or in other words $K_{\mathrm{i,pc}} \ll \frac{1}{M}$.

In some applications it can be beneficial to allow a small error in the active-power injection from the converter; in such a case, a droop regulator can be added to the VSM control structure, instead of an

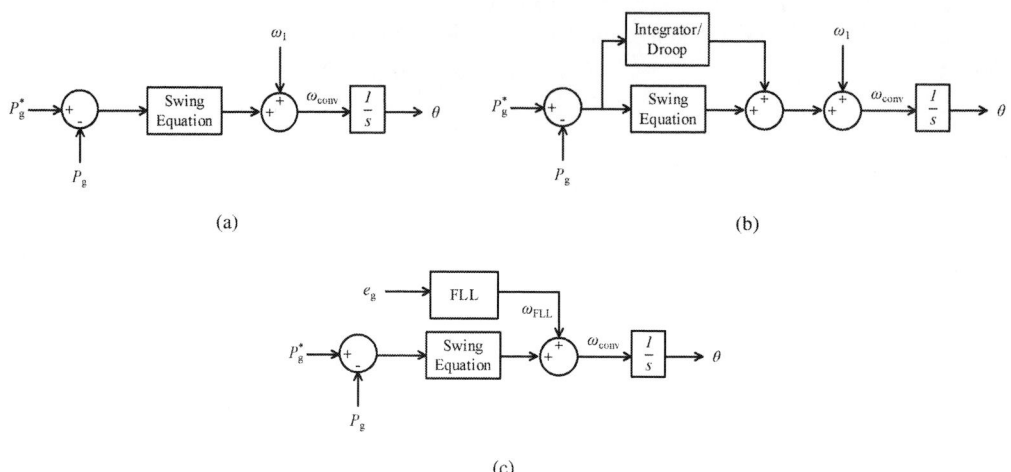

Fig. 2: Block diagram of (a) conventional VSM control structure; (b) VSM with integrator/droop; (d) VSM with frequency-locked loop.

integral regulator. In this case, expression (1) is modified as

$$\theta = \frac{1}{s}\omega_{\text{conv}} = \frac{1}{s}\left[\omega_1 + \frac{1}{sM+K_D}(P_g^\star - P_g) + \frac{K_{r,pc}}{1+s\tau_{r,pc}}(P_g^\star - P_g)\right],\tag{13}$$

where $K_{r,pc}$ is the droop gain and $\tau_{r,pc}$ is the time-constant of the droop regulator. In this case G_{pc} is modified to, $G_{pc} = \frac{1}{sM+K_D} + \frac{K_{r,pc}}{1+s\tau_{r,pc}}$. The droop gain is selected as, $K_{r,pc} = \frac{1}{D} - \frac{1}{K_D}$, where D is the desired droop. Following the previous approach and assuming $K_D = 0$ in (13), it can be seen that the effective value of virtual inertia is given by $\frac{M\tau_{r,pc}}{\tau_{r,pc}+K_{r,pc}M}$. Hence, to prevent any changes in the effective value of virtual inertia, $\tau_{r,pc}$ should be selected such that $\tau_{r,pc} \gg K_{r,pc}M$.

Virtual synchronous machine control with frequency-locked loop

An alternative strategy to remove the steady-state error in the active power injected from the converter, during steady-state frequency deviation, is to include a frequency-locked loop (FLL) in the VSM control structure. The FLL estimates the actual grid frequency, and feeds it into the active-power control loop as shown in Fig. 2(c), and hence modifying expression (1) to

$$\theta = \frac{1}{s}\omega_{\text{conv}} = \frac{1}{s}\left[\omega_{\text{FLL}} + \frac{1}{sM+K_D}(P_g^\star - P_g)\right],\tag{14}$$

where ω_{FLL} is the estimation of the actual grid frequency and is given as

$$\omega_{\text{FLL}} = \frac{K_{i,\text{FLL}}}{s}(e_g^{q,FLL}) + \omega_1; \quad s\hat{\theta}_g = \omega_{\text{FLL}} + K_{p,\text{FLL}}(e_g^{q,FLL}) = F_{\text{FLL}}(e_g^{q,FLL}) + \omega_1,\tag{15}$$

with $K_{p,\text{FLL}}$ and $K_{i,\text{FLL}}$ the proportional and the integral gains of the proportional-plus-integral (PI) regulator, respectively; $e_g^{q,FLL}$ representing the q-component of the PCC voltage in the rotating dq frame; and $F_{\text{FLL}} = \frac{K_{i,\text{FLL}}}{s} + K_{p,\text{FLL}}$, representing the transfer function of the frequency-locked loop. The gains of the PI regulator are selected as $K_{p,\text{FLL}} = 2\alpha_{\text{FLL}}/E_{g0}$ and $K_{i,\text{FLL}} = (\alpha_{\text{FLL}})^2/E_{g0}$, with α_{FLL} the loop-bandwidth of the frequency-locked loop [8]. In order to prevent reduction in the effective value of virtual inertia, α_{FLL} should be selected such that $\frac{K_{i,\text{FLL}}}{A_o} \ll \frac{1}{M}$, where $A_o = \frac{E_{g0}E_{s0}}{X_g}\cos(\hat{\delta}_{g0})$. E_s is the magnitude of the infinite-bus-voltage, X_g is the inductive reactance of the grid, and $\hat{\delta}_g$ is the estimated phase angle of the grid-voltage at steady-state with respect to the infinite-bus-voltage. It is of importance to consider that since the FLL here is neither used for fast frequency response nor used as a phase-locked loop, the loop-bandwidth of the FLL can be low, as the main objective of this control loop is to correct the steady-state error in the active power injected from the converter.

Admittance model of grid-connected converter

A generic frequency dependent ac-side input admittance of the grid-connected VSC, here denoted as Y_{conv}, is derived in this section. The corresponding expression for G_{pc} shall be used for the different control structures, as discussed in the previous section.

The space-vector of the converter-voltage reference in converter dq frame, which is defined by the angle θ, $\mathbf{e}_{\text{conv}}^\star$, is given by $\mathbf{e}_{\text{conv}}^\star = E_{\text{conv}}^\star$. In some cases, it might be convenient to include an additional damping term in the control loop, which acts directly on the converter-voltage reference [6]. In this case, the damping term consists of high-pass filtering of the current vector $\mathbf{i}_f$ multiplied by the active resistance R_a as

$$\mathbf{e}_{\text{conv}}^\star = E_{\text{conv}}^\star - H_a \mathbf{i}_f; \quad H_a = R_a \frac{s}{s + \omega_b}, \tag{16}$$

where, the filter bandwidth ω_b shall be selected smaller than ω_1, typically in the range of 0.1-0.2ω_1 [6].

With $\theta = \hat{\theta}_g + \delta_L$, the space-vector of the converter-voltage reference in stationary $\alpha\beta$-frame, $\mathbf{e}_{\text{conv}}^{s\star}$, is given as

$$\mathbf{e}_{\text{conv}}^{s\star} = \mathbf{e}_{\text{conv}}^\star e^{j\theta} = \mathbf{e}_{\text{conv}}^\star e^{j(\hat{\theta}_g + \delta_L)}. \tag{17}$$

The space-vector of the PCC voltage in the stationary $\alpha\beta$-frame, $\mathbf{e}_g^s$, is given by

$$\mathbf{e}_g^s = E_g e^{j\theta_g} = \mathbf{e}_g^{FLL} e^{j\hat{\theta}_g}, \tag{18}$$

where, $\theta_g = \omega_1 t + \theta_0$, and $\mathbf{e}_g^{FLL}$ represents the space-vector of the PCC voltage in the dq frame, which is defined by the angle, $\hat{\theta}_g$. From (18),

$$\mathbf{e}_g^{FLL} = E_g e^{-j(\hat{\theta}_g - \theta_g)} \approx E_{g0} + \Delta \mathbf{e}_g^{dq} - jE_{g0}(\hat{\theta}_g - \theta_g), \tag{19}$$

where the notation Δx represents a small-signal perturbation of a generic signal x around its steady-state value x_0, and $\mathbf{e}_g^{dq}$ represents the space-vectors of the PCC voltage in the dq frame defined by the angle, θ_g. Now using (19), $e_g^{q,FLL}$, can be written as

$$e_g^{q,FLL} = \Delta e_g^q - E_{g0}(\hat{\theta}_g - \theta_g). \tag{20}$$

Now using small-signal linearized model for (15) and (20) we get

$$\Delta \hat{\theta}_g = G_{\text{FLL}} \Delta e_g^q; \quad G_{\text{FLL}} = \frac{F_{\text{FLL}}}{s + E_{g0} F_{\text{FLL}}}. \tag{21}$$

G_{FLL} shall be selected as 0, if no FLL is used in the control structure. The space-vector of the converter-voltage reference in the grid dq frame that uses θ_g for coordinate transformation, $\mathbf{e}_{\text{conv}}^{dq\star}$, is given by

$$\mathbf{e}_{\text{conv}}^{dq\star} = \mathbf{e}_{\text{conv}}^\star e^{j(\delta_L + \hat{\theta}_g - \theta_g)}. \tag{22}$$

For the system shown in Fig. 1, the ac dynamics in the grid dq frame that uses the angle θ_g for coordinate transformation is given by

$$\mathbf{e}_{\text{conv}}^{dq} = \mathbf{e}_g^{dq} + j\omega_1 L_f \mathbf{i}_f^{dq} + R_f \mathbf{i}_f^{dq} + sL_f \mathbf{i}_f^{dq}, \tag{23}$$

where $\mathbf{i}_f^{dq}$ represents the space-vectors of the converters output current in the selected dq-reference frame. The VSM control is implemented in discrete-time-domain and the inherited time delays are represented through the transfer function $H_d = e^{-sT_d}$, where $T_d = 0.5T_s$, and T_s represents the sampling time. Correspondingly, the converter's output voltage, in the selected dq-reference frame, $\mathbf{e}_{\text{conv}}^{dq}$, is given by

$$\mathbf{e}_{\text{conv}}^{dq} = H_d \mathbf{e}_{\text{conv}}^{dq\star}. \tag{24}$$

From (22) the small-signal linearized model for the converter-voltage reference is given as

$$\Delta \mathbf{e}_{\text{conv}}^{dq\star} = e^{j\delta_{L0}} \Delta \mathbf{e}_{\text{conv}}^{\star} + j\mathbf{e}_{\text{conv},0}^{dq} \Delta\delta_L + j\mathbf{e}_{\text{conv},0}^{dq} \Delta\hat{\theta}_g. \tag{25}$$

Using small-signal linearized models for (3) and (16), the expressions for the active power injected into the grid $P_g = e_g^d i_f^d + e_g^q i_f^q$, the reactive power injected into the grid $Q_g = -e_g^d i_f^q + e_g^q i_f^d$, and (21), (25) can be represented in the scalar form as

$$G_{\text{PQv}} \begin{bmatrix} \Delta e_g^d \\ \Delta e_g^q \end{bmatrix} + G_{\text{PQc}} \begin{bmatrix} \Delta i_f^d \\ \Delta i_f^q \end{bmatrix} + F_{\text{PQ}} \begin{bmatrix} \Delta P_g^\star \\ \Delta Q_g^\star \end{bmatrix} = \begin{bmatrix} \Delta e_{\text{conv}}^{d\star} \\ \Delta e_{\text{conv}}^{q\star} \end{bmatrix}, \tag{26}$$

with

$$G_{\text{PQv}} = \begin{bmatrix} \dfrac{R_a i_{f0}^d i_{f0}^q + e_{\text{conv}0}^q i_{f0}^d}{s} & \dfrac{R_a i_{f0}^q i_{f0}^q + e_{\text{conv}0}^q i_{f0}^q}{s} - \dfrac{(e_{\text{conv}0}^q + R_a i_{f0}^q) G_{\text{FLL}}}{G_{\text{pc}}} \\ \dfrac{-R_a i_{f0}^d i_{f0}^d - e_{\text{conv}0}^d i_{f0}^d}{s} & \dfrac{-R_a i_{f0}^d i_{f0}^q - e_{\text{conv}0}^d i_{f0}^q}{s} + \dfrac{(e_{\text{conv}0}^d + R_a i_{f0}^d) G_{\text{FLL}}}{G_{\text{pc}}} \end{bmatrix} G_{\text{pc}}$$
$$+ \begin{bmatrix} \cos\delta_{L0} G_{\text{qc}} i_{f0}^q & -\cos\delta_{L0} G_{\text{qc}} i_{f0}^d \\ \sin\delta_{L0} G_{\text{qc}} i_{f0}^q & -\sin\delta_{L0} G_{\text{qc}} i_{f0}^d \end{bmatrix},$$

$$G_{\text{PQc}} = \begin{bmatrix} \dfrac{R_a i_{f0}^q E_{g0} G_{\text{pc}} + e_{\text{conv}0}^q E_{g0} G_{\text{pc}}}{s} - R_a & \cos(\delta_{L0}) G_{\text{qc}} E_{g0} \\ \dfrac{-R_a i_{f0}^d E_{g0} G_{\text{pc}} - e_{\text{conv}0}^d E_{g0} G_{\text{pc}}}{s} & -R_a + \sin(\delta_{L0}) G_{\text{qc}} E_{g0} \end{bmatrix},$$

$$F_{\text{PQ}} = \begin{bmatrix} \dfrac{-R_a i_{f0}^q G_{\text{pc}} - e_{\text{conv}0}^q G_{\text{pc}}}{s} & \cos\delta_{L0} G_{\text{qc}} \\ \dfrac{R_a i_{f0}^d G_{\text{pc}} + e_{\text{conv}0}^d G_{\text{pc}}}{s} & \sin\delta_{L0} G_{\text{qc}} \end{bmatrix}.$$

Now using (26), together with (23) and (24), the input admittance Y_{conv} of the VSC is derived as

$$\begin{bmatrix} \Delta i_f^d \\ \Delta i_f^q \end{bmatrix} = G_{\text{conv}} \begin{bmatrix} \Delta P_g^\star \\ \Delta Q_g^\star \end{bmatrix} - Y_{\text{conv}} \begin{bmatrix} \Delta e_g^d \\ \Delta e_g^q \end{bmatrix}, \tag{27}$$

where, the transfer matrices G_{conv}, and Y_{conv} are given by

$$G_{\text{conv}} = -[H_d G_{\text{PQc}} - Z_f]^{-1} H_d F_{\text{PQ}}; Y_{\text{conv}} = -[H_d G_{\text{PQc}} - Z_f]^{-1} [I - H_d G_{\text{PQv}}], \tag{28}$$

where, I represents an identity $[2 \times 2]$ matrix and Z_f is impedance matrix of the filter, given by

$$Z_f = \begin{bmatrix} sL_f + R_f & -\omega_1 L_f \\ \omega_1 L_f & sL_f + R_f \end{bmatrix}. \tag{29}$$

Table I: Test-System Data

System and Grid Parameters		VSM Control Parameters		VSM with Integrator Control Parameters		VSM with FLL Control Parameters	
L_g	0.2 pu	R_a	0.1 pu	R_a	0.1 pu	R_a	0.1 pu
R_g	0.02 pu	ω_n	0.2 pu	ω_n	0.2 pu	ω_n	0.2 pu
L_f	0.15 pu	ζ	1.0	ζ	1.0	ζ	1.0
R_f	0.015 pu	α_{qc}	0.1 pu	α_{qc}	0.1 pu	α_{qc}	0.1 pu
T_s	0.2 ms			$K_{i,pc}$	10.0 rad/s^2	α_{FLL}	0.01 pu

Impact of investigated control modifications on converter's passivity

In order to understand the contribution of each subsystem to the stability of the system, passivity-based approach is used. Following this approach, controlled power-electronic converters do not contribute negatively to the overall stability of the system if they act as a passive system, i.e., if the input admittance of the converter has positive real part for all frequencies. Being the input admittance, Y_{conv}, a $[2 \times 2]$ matrix, the converter is passive if and only if, Y_{conv} is stable (as it is the case for the system under consideration) and $[Y_{conv}(j\omega) + Y_{conv}^H(j\omega)] \geq 0, \forall \omega$, i.e., the matrix is positive semidefinite (the superscript ^{H} denotes the Hermitian conjugate). This is true only if the eigenvalues $\lambda_{1,2}[Y_{conv}(j\omega) + Y_{conv}^H(j\omega)]$ are non-negative [8]. For the frequency intervals where one of the two eigenvalues or both the eigenvalues are negative, there exists a risk of converter contributing negatively towards system stability. For the converter system under consideration, one of the two eigenvalues is always positive, and hence the minimum eigenvalue, λ_{min}, can be considered in the analysis.

In this section, the power-dissipation properties of the grid-connected VSC, controlled using the conventional VSM control, and the two modified VSM control structures discussed in the previous section, are compared. In order to make a fair comparison, the control parameters for the two control modifications are selected in a way that they give a similar rise-time for an active-power step. The system and the control parameters are set in accordance to Table I. The base values for the system power, voltage, and frequency are selected as 900 MVA, 230 kV and 50 Hz, respectively. In order to study the power-dissipation properties of the VSC, the operating point for the converter is selected as 0.8 pu active power output, while the reactive power output from the VSC is controlled to zero.

Figure 3(a) shows the frequency characteristic of λ_{min} in the low-frequency range. It can be seen from Fig. 3(a) that adding an integrator (or droop) or a frequency-locked loop to the conventional VSM control structure has only a slight negative impact on the power-dissipation properties of the converter. In both the cases, there is a slight increase in the negative peak value of λ_{min}, as compared to the conventional VSM control. In case of the VSM control with FLL, the frequency range where λ_{min} is negative slightly increases, however this is tightly linked to the selected loop-bandwidth of the FLL. In order to prevent large transients, a FLL (or a phase-locked loop) is normally required during converter energization, grid-synchronization for the first time and blocking stage of the converter, and hence VSM control with FLL can be recommended. Figure 3(b) shows the impact of α_{FLL} on the frequency characteristic of λ_{min}. It can be seen from Fig. 3(b) that an increase in the loop-bandwidth of FLL leads to a wider frequency range where the VSC exhibits a non-dissipative behavior (active region), thus indicating an increased risk of converter's negative contribution towards system stability. An increase in the bandwidth of FLL would also lead to a reduction in the effective value of virtual inertia as discussed in the previous section.

Simulation results

The system and the control parameters used to test the performance of the investigated control alternatives are specified in Table I. An average model of the VSC and discrete controllers are used for the simulation. For fairness of the comparison, the control parameters for the investigated control alternatives are selected in a way that they give a similar rise-time for an active power step as shown in Fig. 4(a) and Fig. 4(b). It can also be seen from Fig. 4(b) that adding an integrator (or droop) or a FLL to VSM control structure do not affect the effective value of virtual inertia from the converter. Figure 5 shows the dynamic and the

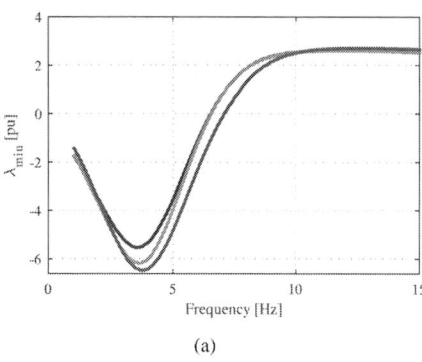

(a)

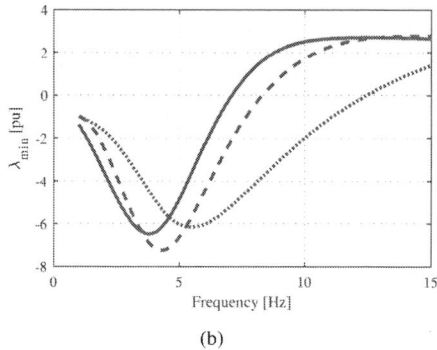

(b)

Fig. 3: (a) Trend of λ_{min} in the low-frequency range for conventional VSM control (Blue), VSM control with integrator (Red), VSM control with droop (Green); VSM control with FLL(Pink); (b) Impact of FLL bandwidth on λ_{min} in the low-frequency range, $\alpha_{FLL} = 0.01$ pu (solid), $\alpha_{FLL} = 0.04$ pu (dashed), $\alpha_{FLL} = 0.12$ pu (dotted).

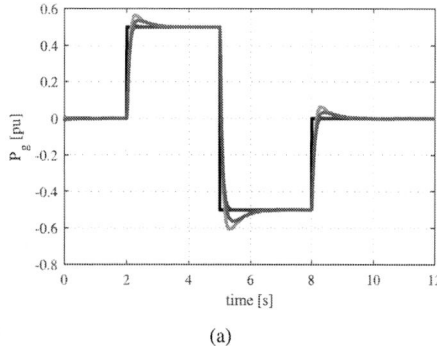

(a)

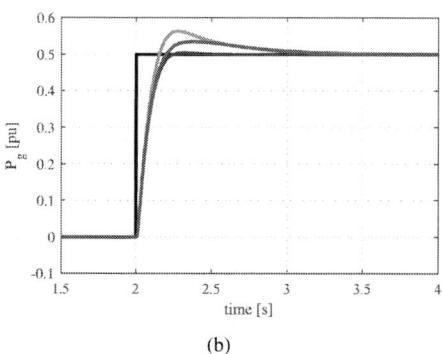

(b)

Fig. 4: (a) Active power step response for conventional VSM control (Dark blue), VSM control with integrator (Red), VSM control with droop (Green), VSM control with FLL(Pink); Power set point (Black); (b) Zoomed version of (a).

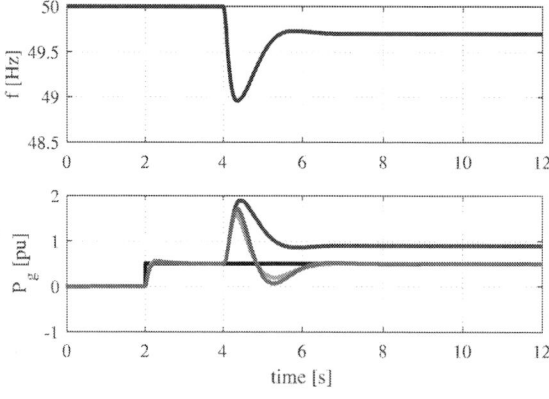

Fig. 5: Top: Grid-frequency variation over time; Bottom: Response of the VSC for conventional VSM control (Dark blue), VSM control with integrator (Red), VSM control with droop (Green), VSM control with FLL(Pink); Power set point (Black).

EPE'20 ECCE Europe

Assigned jointly to the European Power Electronics and Drives Association & the Institute of Electrical and Electronics Engineers (IEEE)

steady-state response of the VSC, for a variation in the grid-frequency, f. It can be seen from the Fig. 5 that, with the conventional VSM control, there is a steady-state error in the active power injected from the converter, if there is a deviation in the grid-frequency. This steady-state error matches the theoretical analysis and is governed by the expression (11). It can be seen clearly from Fig. 5 that there is no steady-state error in case of the VSM control with integrator and the VSM control with frequency-locked loop.

Conclusions

The impact of steady-state deviation in the grid-frequency on the performance of conventional VSM control structure for grid-connected converters is investigated in this paper. It is shown that, if the frequency of the grid deviates from its nominal value in the steady-state, the error in the active power injected from the converter can be very high and is dependent upon the tuning of the active power control loop, as well as the magnitude of grid-frequency deviation. Two control modifications, i.e., the VSM control with integrator (or droop) and the VSM control with frequency-locked loop, to address this issue with the conventional VSM structure are also suggested, and their impact on the active power response and the power dissipation properties of the converter are compared. The selection criteria for the integral gain, the parameters for droop controller, as well as the loop-bandwidth of the FLL, in order to prevent changes in the effective value of virtual inertia, is also presented. It is shown that both the control modifications result in similar power-dissipation properties of the converter. Finally, it is shown that during a steady-state deviation in the grid-frequency, the VSM control with droop has the ability to limit the steady-state error in the active power injected from the converter within the limits defined by the desired droop, D, whereas, the steady-state error in the active power injected from the converter can be reduced to zero in case of the VSM control with integrator and VSM control with FLL. A FLL (or a phase-locked loop) is normally required during converter energization, first-time grid-synchronization and blocking stage of the converter, and hence the VSM control with FLL can be recommended.

References

[1] H. Beck and R. Hesse, "Virtual synchronous machine," in *2007 9th International Conference on Electrical Power Quality and Utilisation*, Oct. 2007, pp. 1–6.

[2] L. Zhang, L. Harnefors, and H. Nee, "Power-synchronization control of grid-connected voltage-source converters," *IEEE Transactions on Power Systems*, vol. 25, no. 2, pp. 809–820, May 2010.

[3] Q. Zhong and G. Weiss, "Synchronverters: Inverters that mimic synchronous generators," *IEEE Transactions on Industrial Electronics*, vol. 58, no. 4, pp. 1259–1267, Apr. 2011.

[4] P. Rodriguez, I. Candela, C. Citro, J. Rocabert, and A. Luna, "Control of grid-connected power converters based on a virtual admittance control loop," in *2013 15th European Conference on Power Electronics and Applications (EPE)*, Sep. 2013, pp. 1–10.

[5] Q. Zhong, P. Nguyen, Z. Ma, and W. Sheng, "Self-synchronized synchronverters: Inverters without a dedicated synchronization unit," *IEEE Transactions on Power Electronics*, vol. 29, no. 2, pp. 617–630, Feb. 2014.

[6] L. Harnefors, M. Hinkkanen, U. Riaz, F. M. M. Rahman, and L. Zhang, "Robust analytic design of power-synchronization control," *IEEE Transactions on Industrial Electronics*, vol. 66, no. 8, pp. 5810–5819, Aug. 2019.

[7] "Network code on load-frequency control and reserves," *ENTSOE*, p. 22, Jun. 2013.

[8] M. Beza and M. Bongiorno, "Identification of resonance interactions in offshore-wind farms connected to the main grid by MMC-based HVDC system," *International Journal of Electrical Power and Energy Systems*, vol. 111, pp. 101–113, Oct. 2019.

A General Method to Damp Wind Turbine SSR with Different Transmission Systems

Ignacio Vieto and Jian Sun

Rensselaer Polytechnic Institute
110 8th St, Troy, NY, USA
Tel.: +(1) 518.276.8297
E-Mail: jsun@rpi.edu

Keywords

«Wind Energy», «Stability», «Damping», «SSR», «Impedance».

Abstract

This paper presents a general method to damp subsynchronous resonance (SSR) of type-III and type-IV wind turbines with different types of transmission systems. The method is based on reshaping the output impedance of wind turbines through a supplementary damping function added to existing turbine electrical control system. The types of transmission systems considered include long overhead transmission lines, overhead lines with series compensation, as well as high-voltage dc transmission.

Introduction

Type-III wind turbines are known to develop subsynchronous resonance (SSR) with series-compensated overhead transmission lines [1]-[2]. Recent work has also identified a similar SSR problem between type-III turbines and long overhead transmission lines without series compensation [3]. Resonance between type-IV turbines and weak or series-compensated grids has also been a subject of research in recent years [4]-[5]. In offshore wind farms with high-voltage dc (HVDC) transmission, the offshore HVDC converter serves as the "grid" for the turbines and typically exhibits capacitive impedance in certain frequency range below the fundamental, which may also lead to SSR with the wind farm.

A diverging or sustained resonance can be attributed to the negative or zero damping at the frequency of a system resonance. Negative damping in the subsynchronous frequency range is typically the results of control of converters, most noticeably the phase locked loop (PLL) and dc bus voltage control [6, 7]. In the case of type-III turbines, the so-called induction generator effects [6] also contribute to the negative damping in this frequency range. Recently developed analytical impedance models [6, 7] have made it possible to understand the root cause of system resonance and develop effective solutions.

Damping of SSR between type-III turbines and series-compensated transmission lines has been studied extensively [7]-[16] following a practical incident in 2009 [1]. The approaches proposed in the literature include controlling the series capacitor [8], using additional hardware such as FACTS devices [9], or installing passive filters [10]. While the use of system-level variables as control inputs has been proposed in some of these methods, recent works have focused on the use of local signals and embedding the damping function in turbine control [11-16], which is more desirable. Reference [14] examined 12 different points in a typical turbine control system where a damping function could be introduced and proposed the use of a washout filter with a constant gain to implement damping control. The design and analysis in [12] were performed using state-space modeling and eigenvalue analysis.

An impedance-based approach to SSR modeling and damping design for type-III turbines was presented in [13]. The damping function implemented was based on a washout filter with a high-frequency pole to attenuate switching ripple. The additional damping function was shown to introduce an equivalent resistance in series or parallel with the turbine impedance before the damping function was added. Because resonance and damping in electrical system are defined based on impedance, this impedance-based approach is more direct and insightful, and will be the basis for damping design in this work.

Compared to type-III turbines, type-IV turbines were considered less susceptible to SSR problems and the subject has not received as much attention. On the other hand, most factors that contribute to negative damping in type-III turbines, specially PLL and dc bus voltage control, are also present in type-IV turbines and can cause SSR with the transmission system. Indeed, a major incident involving type-IV turbines and overhead transmission lines that caused SSR protective shutdown of synchronous generators was described in [17]. Ref [18] provided more analysis of the root cause for such resonances.

This work presents a general method to damp SSR. The method can be applied to both type-III and type-IV turbines connected to different types of transmission systems, including long overhead transmission lines that resulting in very low short circuit ratio (SCR) for the wind turbines, overhead transmission lines with series compensation, as well as high-voltage dc transmission system. The damping function makes use of a narrowband band-pass filter [19] that allows better control of its effective frequency range. The design and performance analysis will be based on refined turbine impedance models that account for coupling effects ignored in earlier models and give more reliable prediction of system resonance close to the fundamental frequency.

Turbine Impedance Models and Characteristics

Accurate prediction of resonance at frequencies close to the fundamental requires impedance models that include dc bus dynamics and effects of coupling through the grid impedance [3]. These two dynamics are most significant at frequencies close to the fundamental and, in some cases, are the primary reason for system resonances. Small-signal models developed for the wind turbines for this purpose [3] will be briefly reviewed here. The key elements are two admittance functions that define the responses of a two-level voltage-source (VSC) converter to an ac terminal voltage perturbation: an admittance that defines the current at the perturbation frequency (s) and a transfer admittance that defines the current at the coupling frequency ($s - j2\omega_1$) [3]. When connected to a non-ideal grid, the transfer admittance creates an additional current at the perturbation frequency by coupling through the grid impedance and adds another admittance in parallel with the turbine.

Turbine Impedance Characteristics

To understand how SSR develops between a turbine and the grid, we will examine turbine impedance characteristics using the models presented in [3]. Fig. 1a) shows the impedance responses of a type-IV wind turbine in a 60 Hz grid. This wind turbine has standard vector control using PI regulators in dq-

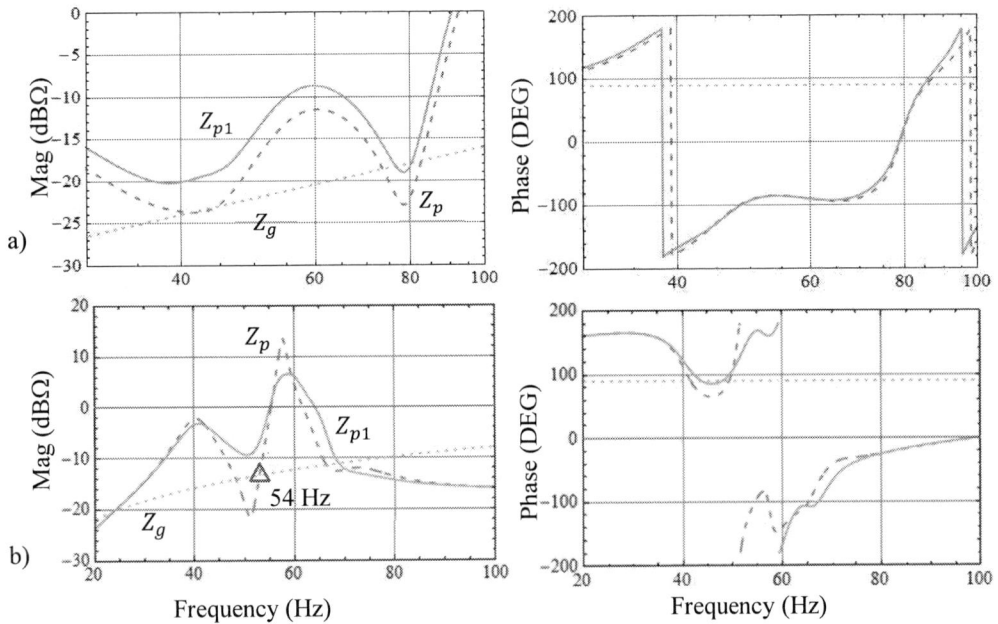

Fig. 1 a) Impedance of type-IV turbine considering coupling through the grid (blue-dashed), in comparison with the simplified model (green-solid) and impedance of the inductive grid with 1.33 SCR (orange-dots). b) Impedance of type-III wind turbine considering coupling through the grid (blue-dashed), in comparison with the simplified model (green-solid) and impedance of the grid with 1.33 SCR (orange-dots).

frame and the bandwidth of the control loops are adjusted within the normally expected ranges, i.e., 300 Hz current regulator, 30 Hz PLL and 20 Hz dc regulator. The bandwidth of the dc regulator correlates to the dips in the impedance magnitude at 40 and 80 Hz, meanwhile the bandwidth of the PLL correlates the phase jumps observed at 30 and 90 Hz. The effects of coupling can be seen in a few places. First, the overall impedance of the turbine is lower than its impedance when the coupling is ignored. This is due to the coupling acting as a second impedance connected in parallel with the turbine impedance. This effect is only significant below the second harmonic.

Similar responses are presented in Fig. 1b) for a type-III wind turbine. The turbine has standard controllers and regulator settings, as described in [7]. The impedance of the type-III turbine is different from the type-IV turbine because the effects of the induction machine are much more significant close to the fundamental than the effects of the line inductor used in the type-IV turbine. On the other hand, the induction machine does not directly contribute to the coupling, so the type-III turbine has its impedance altered in a less apparent manner. However, close to the fundamental, the impedance has a peak at 40 Hz and a valley at 50 Hz that are tied to the dc regulator and PLL dynamics, and these controllers contribute to the coupling admittance. In particular, the impedance magnitude at 50 Hz is more than 10 dB lower when the coupling is considered and we will show how this affects the stability analysis next. System resonance can be predicted by comparing the turbine and grid impedances, so the grid impedance is also included in Fig. 1. The grid impedance corresponds to an inductive grid with a SCR of 1.33.

Prediction of System Resonance

The results and plots discussed in the previous section will now be used to predict system resonance between the turbines and the grid. For both types of turbines, use of the simplified impedance models without the coupling effects indicates there is no stability problem because the phase different at each of the intersection point in the magnitude response is less than 180 degrees. However, use of the refined models show an intersection at 40 Hz with negative phase margin in the type-IV turbine case. This would predict an unstable resonance between the turbine and the grid. It should be noted that this intersection is mostly caused by the decrease in turbine impedance due to the coupling over the grid.

In the case of the type-III turbine, the refined models predict two intersections around 20 and 70 Hz, both with enough phase margin to guarantee stability. The dip in the turbine impedance at 50 Hz is lower in the refined model and this leads to two new intersections at 45 and 54 Hz. The intersection at 54 Hz has negative phase margin, which leads to instability. Notice that this additional crossover is not caused because the impedance is lowered over a broad frequency range as in the type-IV turbine, but because

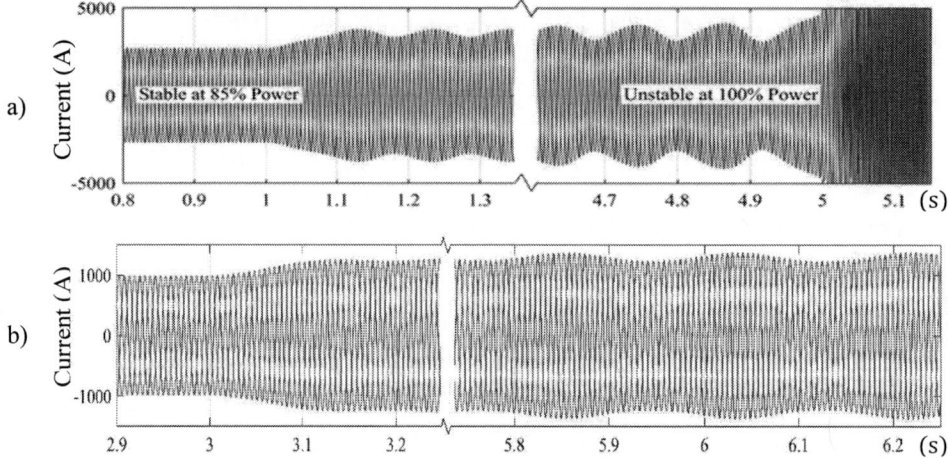

Fig. 2 a) Resonance in the ac current of a type-IV wind turbine connected to a weak inductive grid after power increases towards the rated power. b) Resonance in the ac current of a type-III wind turbine connected to a weak inductive grid when power increases towards the rated power.

the turbine impedance is greatly modified in a narrow frequency range where the control dynamics are most significant (in this example it is at the PLL bandwidth frequency).

These examples showed the importance to include coupling in the analysis of SSR with weak grid.

Time-Domain Responses

To validate the results of impedance analysis presented in the previous subsection, each of the case is simulated in Matlab/Simulink using a detailed EMT model of the turbine. The induction machine is simulated for the type-III turbine but the type-IV turbine is modeled by a current source and dc bus capacitor connected to the VSC. In each simulation, the turbine is started at a lower power level to avoid instability initially. Power is then ramped up towards the condition analyzed by the impedance method.

The results for the type-IV turbine are shown in Fig. 2a). The turbine operates stably up to 85% power. At $t = 1s$, the power is ramped up to 100% rated power and resonance starts. The amplitude of the resonance is initially low, but the impedance-based analysis predicted instability, so the simulation is left to keep running and at $t = 5s$ it becomes unstable. The frequency of the resonance is obtained through an FFT of the currents and seen to be 45 Hz and 75 Hz, which matches the impedance-based analysis. The current at 75 Hz is cause by the nonlinear response of the converter.

Simulation results for the type-III turbine case are presented in Fig. 2b). The turbine operates stably up to 75% power. At $t = 3s$ the power is increased towards 100% and the turbine enters an unstable resonance. Frequency analysis of the output current reveals two resonances at 65.5 and 54.5 Hz, which are mirrored around the fundamental. This validates the impedance-based stability analysis that predicted an unstable resonance at 54 Hz.

Active Damping of Wind Turbines

Different solutions to SSR have been proposed in the literature. In this work, we propose general solution that can be applied to both type-III and type-IV wind turbines working with different grids, including weak inductive grid, series compensated grid, or an HVDC converter. A sketch of the proposed damper is shown in Fig. 3, where $H_d(s)$ is a band-pass filter. The proposed approach allows us to reshape the turbine impedance within a narrow frequency range. The amount of damping can be designed by considering the effects of the damper on the turbine impedance. The damper can be viewed as a virtual impedance in series or parallel with the undamped turbine impedance. This concept has been applied to converters with LCL filters [20] and to type-III turbines with high-level of series compensation [13], and is generalized in this work.

Design Considerations

Existing methods to introduce damping to turbine impedance differ in the placement as well as design of the damping function. Some insertion points of the damping filter that have been proposed are the current signals [11], torque reference, modulating index directly [11], dc voltage error, etc. The damping filter can be a washout filter [11], a bi-quadratic filter [20], Kalman filter [12], band-pass filter [13], etc. The damping method proposed in this work employs the following band-pass filter:

$$H_d(s) = K_d(s) \frac{Q^{-1}\omega_0 s}{s^2 + Q^{-1}\omega_0 s + \omega_0^2}. \tag{9}$$

Key parameters used in (9) are as follows: $\omega_0 = \sqrt{\omega_h \omega_l}$ is the filter center frequency, ω_h is the upper filter corner frequency, ω_l is the lower corner frequency, $Q = \omega_0 \cdot (\omega_h - \omega_l)^{-1}$ is the filter quality factor, and $K_d(s)$ is the damper gain. This gain can be designed so that the damper acts as a resistor, a capacitor, or an inductor within its bandwidth as shown in Fig. 3b). However, the additional dynamics of the added virtual inductor and/or capacitor change the frequency response of the active filter in undesirable ways. The extra zero changes the high-frequency response from a band-pass filter to a high-pass filter. In the same way, the additional pole of the RC damper makes it a low-pass filter. The low-pass behavior cannot be implemented because it bypasses the regulator at the nominal frequency (dc for a current controller in dq-frame). Making the filter inductive can work if the high frequency gain of the damper is much lower than the gain of the PI regulator, effectively giving the normal controller path

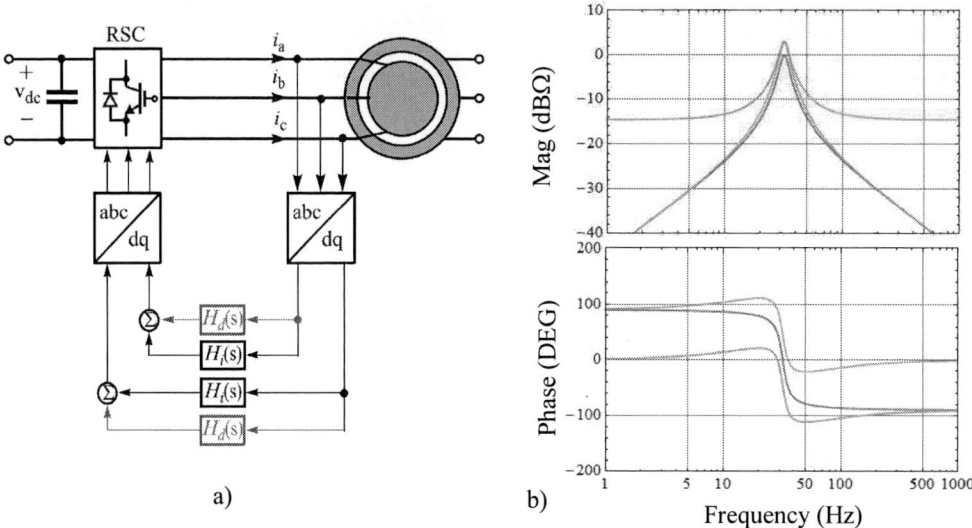

a)   b)

Fig. 3 a) Sketch of circuit and controller for the rotor-side converter with an active damper in parallel with the current regulator. b) Bode plot of bandpass filters considering a virtual resistor (blue), a virtual series RL circuit (orange) and a virtual series RC circuit (green). All three cases have a center frequency of 30 Hz and a quality factor of 5.34.

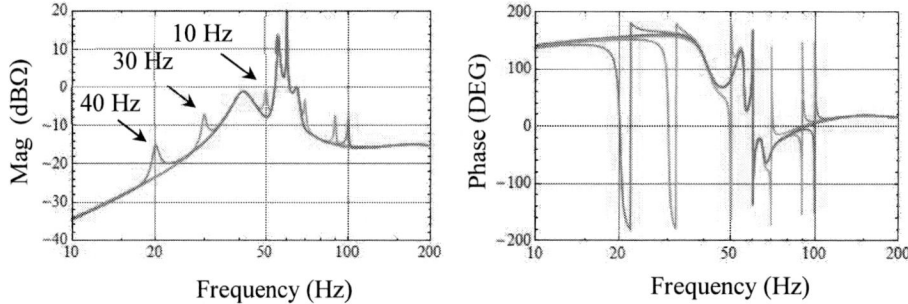

Fig. 4 Damped wind turbine impedance using the compensator from (9): input impedance without damping (blue). Damped turbine impedance with a filter with a quality factor of 10 and a variable center frequency at 10 Hz (orange), 30 Hz (green) and 40 Hz (red).

priority above its crossover frequency, but this implementation cannot be limited to a narrow frequency range, which is one of the stated goals of the proposed damper. Based on these considerations, we have chosen to design the damping gain $K_d(s)$ to emulate a virtual resistor within the bandwidth of the filter. For a desired virtual resistance R_d, the damper gain is $K_d = R_d V_{dc} Q \cdot \omega_0^{-1}$.

The chosen filter can be designed with a narrow bandwidth and high quality factor to target a specifically observed resonance with high resolution and without affecting the impedance at other frequencies. This method can also be used to design a wide bandwidth damper; however, this approach can lead to decreased stability and unintended consequences at high frequencies. The flexibility in the proposed damper design and its ability to target specific frequency range will be examined next based on the case involving the type-III turbine. Design for the type-IV turbine can follow the same principle and is simpler because there is no grid-connected machine.

A filter with a quality factor of 10 and 0.2 pu damping is chosen initially and the center frequency is varied between 10 Hz and 40 Hz. The results are shown in Fig. 4. A peak in the impedance can be seen at the chosen frequency, below and above the fundamental. This result also highlights why these active dampers might induce problems at unintended frequencies. The phase of the wind turbine impedance at frequency $\omega_1 - \omega_0$ is zero, but at frequency $\omega_1 + \omega_0$ it has considerable negative damping. Inverting the sign of the gain in the filter would reverse this behavior but it would not eliminate it. The effects at other frequencies are minimal except for at the mirrored harmonic frequency where the chosen filter adds negative damping to the wind turbine.

Comparison with Existing Method

The proposed damping scheme is now compared against three other dampers reviewed in the Introduction: the wideband filter method [13], the subsynchronous suppression filter [4], and stator virtual impedance method [21]. The wide-band filter is designed to have the same center frequency and quality factor as the one described in this work. The compensator gain is adjusted to provide the same amount of total compensation. The subsynchronous suppression filter is tuned to the same center frequency and bandwidth as the proposed filter. The gain of the suppression filter is adjusted to eliminate the effects of the current controller in a narrow frequency range. Finally, the virtual impedance design is tuned to compensate for the leakage inductance of the rotor windings.

The effects of the dampers on the overall turbine impedance are compared to each other and against the undamped impedance in Fig. 5. The proposed narrowband filter has its effects limited to around the desired damping frequency of 51 Hz, while the wideband filter affects the response in some way from around 40 Hz to around 80 Hz, except for a small frequency range close to the fundamental. The subsynchronous suppression filter acts in a very different manner than the virtual impedance damper - by minimizing the controller effect at the resonance frequency, the negative damping at the crossover frequency can be reduced. However, this approach leads to a wider and larger effect than the other dampers have. The effects of the virtual negative inductance are like the wide-band virtual resistor, but the virtual inductor method has a wider bandwidth. It can be concluded that the proposed damper has the highest frequency resolution, affects the overall impedance in the narrowest range, and has the least impact on the impedance outside the damping bandwidth.

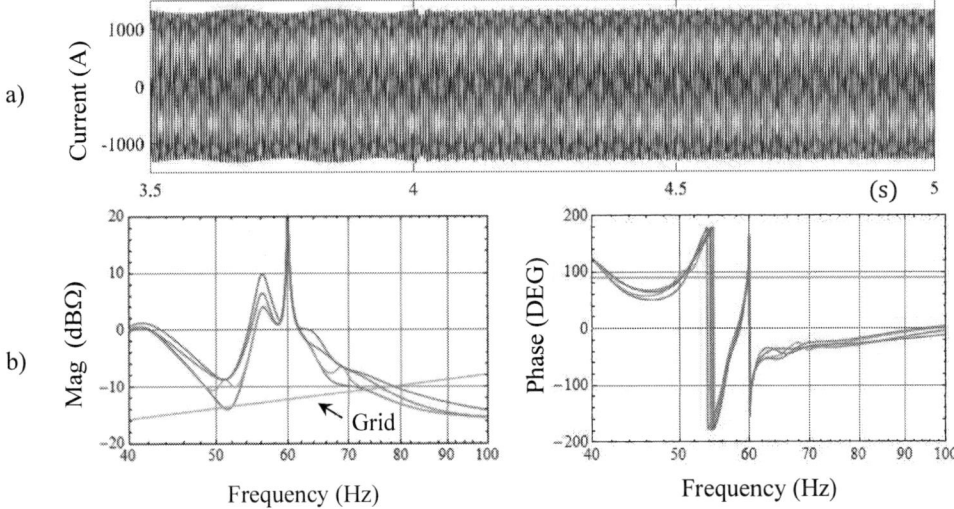

Fig. 5 a) Stable operation of the wind turbine with a weak inductive grid after the addition of damper controls. b) Effective wind turbine impedance without any damping in blue, with the damping proposed in this work in green, the wide-band damper in red, the subsynchronous suppression filter in purple and the virtual negative inductor damper in brown. The grid impedance is shown in orange.

Case Studies

Type-III Turbine with a Weak Inductive Grid

This case was already studied in Section II and the effects of the damper on the impedance system were examined in Section III. The tradeoff between the quality factor, high-cutoff frequency, and undesirable effects at the mirrored frequency must be examined when designing the filter. Designing the filter to have a smaller quality factor makes it effective over a wider range of frequencies but at the price of possible unintended effects on high-frequency impedance. This is a major issue because a positive damping at low frequencies is negative damping at high frequencies, making the system potentially unstable at supersynchronous frequencies. The filter cannot be designed with a very high quality factor either because the resonance frequency between the grid and the turbine is not fixed.

Based on these considerations, a low cutoff frequency of 1 Hz is chosen to block the dc component of the dq currents and a 10 Hz high cutoff is chosen to give the filter a wide enough bandwidth to damp the observed resonance. The corresponding filter are: $\omega_0 = 2\pi \times 3.16$, $Q = 0.35$, and $K_d = 0.00025$, which corresponds to an added damping equal to about 10% of the rotor resistance. The intersection that caused the 54 Hz resonance observed in Fig. 1 has disappeared.

To verify the design, the damper is included in the control loop and the resulting time-domain responses are presented in Fig. 5. The damper is initially disabled, and the resonance starts to increase in magnitude after the step change in power. The resonance quickly goes away when the damper is added at $t = 4$s.

Type-III Turbine with Series-Compensated Line or HVDC Converter

Next we test the method for type-III turbines connected to a series compensated grid and to a wind farm connected to an offshore HVDC converter. SSR in these two cases occurs at different frequencies and has different root causes. With the series compensated grid, the SSR is usually between 10-45 Hz for a 60 Hz fundamental, and the root cause is the capacitive impedance of the series compensated grid intersecting the inductive impedance of the wind turbine. The real part of turbine impedance is usually negative in this frequency range, as can be seen from Fig. 1b). For wind turbines connected to an HVDC converter, the mechanism to form an SSR is similar but the HVDC impedance can be capacitive with negative damping over a wider frequency range in which SSR may develop.

The dampers used in these two cases have the same architecture as used before but is designed with different gains. Even when the damper architecture remains the same, the resonance with the series compensated grid is solved by adding more phase margin at the intersection frequency while minimizing

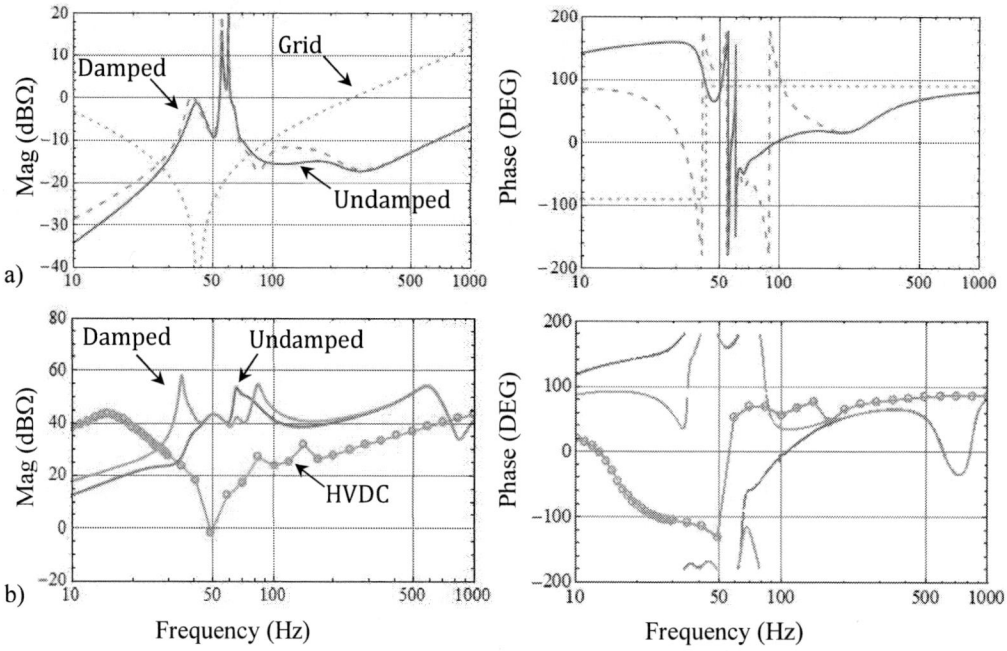

Fig. 6 a) Effective admittance of the type-III wind turbine without damping (blue), with damper controls included (green) and the series compensated grid (orange). b) Effective admittance of the type-III wind farm without damping (blue), with damping (green) and the HVDC offshore converter (orange).

its effect on the impedance magnitude. This is a fundamentally different approach than the one in Fig. 5 where the goal was to shape the magnitude to avoid the intersection.

To solve the resonance between the HVDC converter and the wind farm the design approach is a combination of the last two cases. The intersection cannot be avoided, but its frequency can be shifted towards a point where the HVDC converter has more damping and then add enough phase at the new intersection frequency to stabilize the system. In both cases, the damper has to add enough positive damping so that the phase margin at the crossover frequency is high enough (at least 30 degrees), but it must be limited so that its negative effects are minimized.

Based on these considerations, a low cutoff frequency of 20 Hz and a high cutoff of 45 Hz are chosen for the case with series compensated grid. For the case with an HVDC converter, the cutoffs are 13 Hz and 22 Hz. The resulting impedance responses are shown in Fig. 6 and compared with the corresponding grid impedance.

The time-domain results for are presented in Fig. 7. In the first case, the turbine is connected to a strong grid in parallel with a series compensated line. At 2.5 seconds the strong line is open such that the turbine is connected only to the compensated line. The resonance is immediate and unstable, and within 500 ms the system has currents that are over twice as high as the nominal system current. The proposed damper is added to the turbine and the same simulation is run, the damper is initially disconnected and then turned on 100 ms after the trip occurs. The stabilized results are also shown in Fig. 7, now the resonance starts to disappears at t=2.6s, and the turbine-grid system becomes stable.

The response of the turbine with a HVDC converter is slightly different. In this case, the wind farm cannot start because the 30 Hz is present regardless of operating power set-point. The unstable resonance is shown in Fig. 8a). The dampers are added to the turbines and the simulations run again, now the damper is added from the beginning of the simulation to help the system start up. Fig. 8b) shows the wind farm-HVDC system operating in a stable manner with the damper connected.

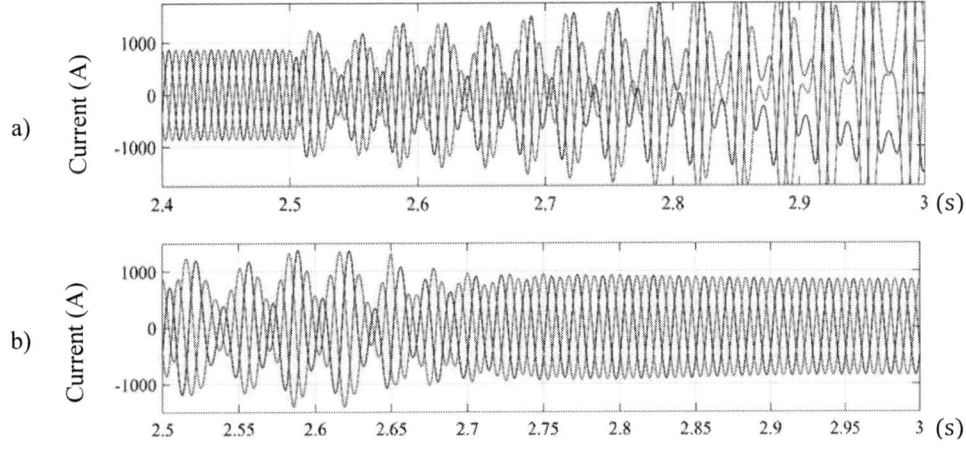

Fig. 7 a) Resonance in a type-III turbine connected to a series compensated transmission line. b) Stable operation of the wind turbine and the series compensated line after the addition of damper controls

Type-IV Turbine with Overhead Transmission Line

Finally, a case involving a type-IV wind turbine is presented. The undamped impedance and time-domain response are shown in Fig. 1. A similar damper to the one applied to the type-III turbine is designed and connected in parallel with the current regulator of the grid-connected converter. The damper is tuned to add damping between 13 and 22 Hz (in the controller reference frame) to mitigate the observed resonance around 70 Hz. The time-domain and impedance results are shown in Fig. 9, in which the damper is initially disconnected and turned on at $t = 3s$, the resonance disappears within 500 ms and the turbine operates in a stable manner.

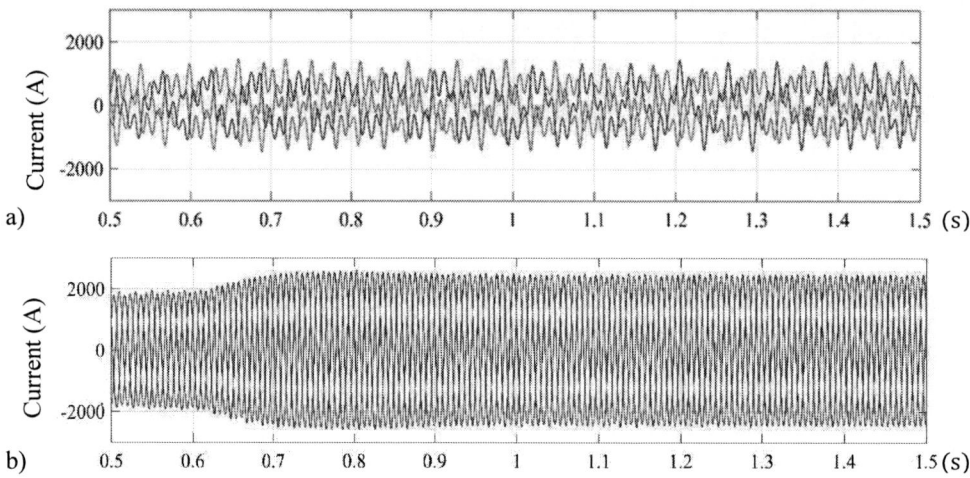

Fig. 8 a) A 30 Hz resonance between a wind farm and an MMC-based HVDC converter. b) Stabilization of the 30 Hz resonance between the wind farm and the HVDC station by damping of the turbine impedance.

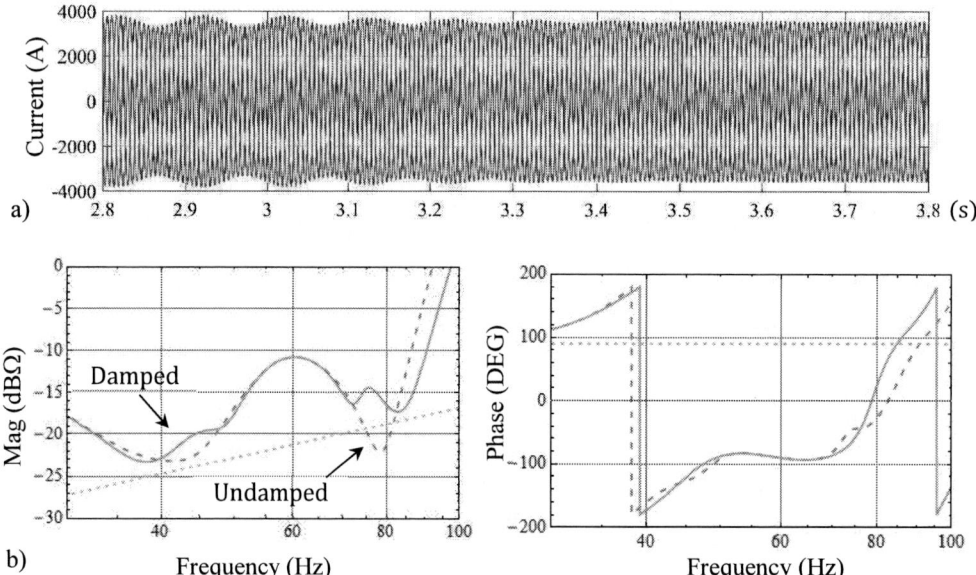

Fig. 9 a) Damped resonance in the ac current of a type-IV wind turbine with active damper. b) Effective impedance of the turbine with the damping (solid-green), without the damper (dashed-blue) and the grid impedance (dots-orange).

Summary

This work presented a new method to damp SSR between wind turbines and different transmission systems. The basis of the proposed method is a set of recently developed, small-signal impedance models that accurately model all dynamics close to the fundamental frequency that affect SSR. With these models, SSR can be more reliably predicted and damping filters can be better designed.

The proposed damping method is based on a narrowband band-pass filter that emulates a resistance in series with the wind turbine over a specific frequency range. It has several advantages over existing methods and can be applied to both type-III and type-IV turbines. The proposed design was evaluated for a wind turbine connected against a weak inductive grid and a series compensated grid as well as a wind farm connected to an HVDC converter. Detailed EMT simulation validated the effectiveness of the proposed design in each case.

References

[1] J. Adams, C. Carter, and S. H. Huang, "ERCOT experience with subsynchronous control interaction and proposed remediation," in *Proc. IEEE Transmission Distribution Conf. Expo.*, 2012, pp. 1–5.

[2] S. H. Huang and G. Yanfeng, South Texas SSR. ERCOT ROS Meeting, Available: http://www.ercot.com/content/wcm/key_documents_ lists/139265/10._South_Texas_SSR_ERCOT_ROS_May_2018_rev1.pdf.

[3] I. Vieto, G. Li and J. Sun, "Behavior, modeling and damping of a new type of resonance involving type-III wind turbines," in *Proc. 19th* IEEE *Workshop Control Modeling Power Electronics*, Padua, Italy, 2018, pp. 1-8.

[4] H. Liu, X. Xie, J. He, T. Xu, Z. Yu, C. Wang and C. Zhang, "Subsynchronous interaction between direct-drive PMSG based wind farms and weak AC networks," *IEEE Trans. Power Syst.*, vol. 32, no. 6, pp. 4708-4720, Mar. 2017.

[5] M. Beza and M. Bongiorno, "On the risk of subsynchronous control interaction in type 4 based wind farms," *IEEE Trans. Sustain. Energy*, vol. 10, no. 3, pp. 1410-1418, July 2019.

[6] I. Vieto and J. Sun, "Sequence impedance modeling and analysis of type-III wind turbines," *IEEE Trans. Energy Convers.*, vol. 33, no. 2, pp. 537-545, June 2018.

[7] I. Vieto and J. Sun, "Refined small-signal sequence impedance models of type-III wind turbines," in *Proc. 2018 IEEE Energy Conversion Cong. Expo.*, Portland, Or, 2018.

[8] H. Xie, B. Li, C. Heyman, M. de Oliveira and M. Monge," Subsynchronous resonance characteristics in presence of doubly-fed induction generator and series compensation and mitigation of subsynchronous resonance by proper control of series capacitor," *IET Renew. Power Gener.*, vol. 8, no. 4, pp. 411-421, May 2015.

[9] D. Suriyaarachchi, U. Annakkage, C. Karawita, D. Kell, R. Mendis, and R. Chopra, "Application of an SVC to damp sub-synchronous interaction between wind farms and series compensated transmission lines," in *Proc. IEEE Power Energy Soc. Gen. Meeting*, pp. 1-6, Jul. 2012.

[10] J. Sun, "Passive methods to damp AC power system resonance involving power electronics," in *Proc. 19th* IEEE *Workshop Control Modeling Power Electronics*, Padua, Italy, 2018, pp. 1-8.

[11] H. A. Mohammadpour, E. Santi, "SSR damping controller design and optimal placement in rotorside and grid-side converters of series compensated DFIG-based wind farm," *IEEE Trans. Sustain. Energy*, vol. 6, no. 2, pp. 388-399, Apr. 2015.

[12] A. E. Leon and J. A. Solsona, "Sub-synchronous interaction damping control for DFIG wind turbines," *IEEE Trans. Power Systems*, vol. 30, no. 1, pp. 419-428, Jan. 2015.

[13] I. Vieto and J. Sun, "Damping of subsynchronous resonance involving Type-III wind turbines," in *Proc. 16th IEEE Workshop Control Modeling Power Electronics*, Vancouver, Canada, 2015, pp. 1-8.

[14] P. H. Huang, M. El Moursi, W. Xiao, and J. Kirtley, "Subsynchronous resonance mitigation for series compensated DFIG-based wind farm by using two degree-of-freedom control strategy," *IEEE Trans. Power Syst.*, vol. 30, no. 3, pp. 1442-1454, May 2015.

[15] L. Fan, Z. Miao, "Mitigating SSR using DFIG based wind generation," *IEEE Trans. Sustain. Energy*, vol. 3, no. 3, pp. 349-358, Jul. 2012.

[16] H. A. Mohammadpour, A. Ghaderi, and E. Santi, "SSR damping in wind farms using observed-state feedback control of DFIG converters," *Elect. Power Syst. Res.*, vol. 123, pp. 57-66, 2015.

[17] J. Sun, M. Li, Z. Zhang, T. Xu, J. He, H. Wang and G. Li, "Renewable energy transmission by HVDC across the continent: System challenges and opportunities," *CSEE J. Power Energy Systems*, vol. 3, no. 4, pp. 353-364, Dec. 2017.

[18] I. Vieto, S. Rogalla and J. Sun, "On the potential of subsynchronous resonance of voltage-source converters with the grid," in *Proc. 2015 Wind Integration Workshop*, Brussels, October 2015.

[19] H. Zumbahlen, Linear Circuit Design Handbook, 1st ed. London, UK: Newnes/Elsevier, 2008.

[20] F. A. Gervasio, R. A. Mastromauro, D. Ricchiuto and M. Liserre, "Dynamic analysis of active damping methods for LCL-filter-based grid converters," in *Proc. IEEE 39th Annu. Conf. Industrial Electronics Society*, Vienna, Austria, 2013, pp. 671-676.

[21] Y. Song, X. Wang and F. Blaabjerg, "Doubly fed induction generator system resonance active damping through stator virtual impedance," *IEEE Trans. Industrial Electron.*, vol. 64, no. 1, pp. 125-137, Jan. 2017.

A Test Scheme for the Comprehensive Qualification of MMC Submodule Based on 10 kV SiC MOSFETs under High *dv/dt*

Xingxuan Huang[1], Shiqi Ji[1], Dingrui Li[1], Cheng Nie[1], William Giewont[1],
Leon M. Tolbert[1,2], and Fred Wang[1,2]
[1]Min H. Kao Department of Electrical Engineering and Computer Science,
The University of Tennessee
Knoxville, TN, USA
[2]Oak Ridge National Laboratory
Oak Ridge, TN, USA
E-Mail: xhuang36@vols.utk.edu

Acknowledgements

This work was primarily funded by PowerAmerica institute established by U.S. DOE through North Carolina State University. The authors thank Southern Company and Chattanooga Electric Power Board for their support. The authors also thank Wolfspeed for providing the 10 kV SiC MOSFETs. This work made use of the Engineering Research Center Shared Facilities supported by the Engineering Research Center Program of the National Science Foundation and U.S. DOE under NSF award number EEC-1041877 and the CURENT Industry Partnership Program.

Keywords

«Modular multilevel converter», «MMC submodule», «10 kV SiC MOSFET», «High dv/dt», «Qualification»

Abstract

A test scheme is designed to qualify MMC submodules based on 10 kV SiC MOSFETs comprehensively, including thermal design, insulation design, and operation under high *dv/dt*. In the test scheme, the essential step is the continuous test realized with the proposed ac-dc continuous test circuit with two MMC submodules in series. With the designed modulation scheme, two cascaded submodules are leveraged to generate high *dv/dt*. The submodule under test has to continuously withstand 2X normal *dv/dt* of 10 kV SiC MOSFETs, which could occur during the MMC converter operation. Higher *dv/dt* can also be generated to fully test the submodule. An open loop voltage balancing method is adopted to simplify the continuous test setup. Simulation and experimental results are provided to validate the proposed test scheme.

Introduction

Modular multilevel converter (MMC) has become increasingly more prevalent in various applications, thanks to its numerous benefits [1]-[3]. In medium voltage (MV) applications, especially, MMC has received growing attention, such as MVDC systems [1] and MV converters tied to the distribution grid [2]. The performance of MMC for MV applications can be further improved by the emerging 10 kV SiC MOSFETs. Compared to MMC submodules based on Si IGBTs, MMC submodules equipped with 10 kV SiC MOSFETs feature >10X switching frequency and higher submodule voltage, enabling a wide range of converter-level benefits such as simpler circuit, higher control bandwidth, power density, and efficiency [4]-[9].

The design and testing of MMC submodules are essential to the performance and operation of a MMC. The comprehensive testing and qualification of MMC submodules is necessary to ensure the robust operation of the MMC converter. Also, it is desirable to test submodules comprehensively so that problems can be found at the submodule level before assembling and testing the full converter, which is much more complex than one submodule. The qualification of MMC submodules based on 10 kV SiC MOSFETs is more crucial, considering the higher submodule voltage and much higher dv/dt and hence challenges in both insulation and noise immunity design [10]-[14].

The test scheme for submodules based on 10 kV SiC MOSFETs should be able to provide a condition similar to the real condition in a MMC, such as device current, dv/dt, and so on. Specifically, the test scheme should fully validate its thermal performance, insulation design, and the capability to withstand high dv/dt when the submodule operates as part of a MMC converter. For example, high dv/dt can distort PWM signals and falsely trigger protections, and such issues should be found during submodule testing, instead of during converter testing. Qualification of the gate driver for high voltage SiC devices is studied under high dv/dt conditions, however, only the gate driver is tested with a dc-dc continuous test setup [15]. Nevertheless, the testing and qualification of MMC submodules and MV converters based on high voltage SiC MOSFETs or IGBTs has not been covered in detail in previous literature, although experimental results are demonstrated in [5], [8].

This paper focuses on a simple test method to fully qualify a MMC submodule based on 10 kV SiC MOSFETs with half bridge topology. First, the proposed test scheme is introduced in detail, whose essential step is the ac-dc continuous test in which two MMC submodules are connected in series to resemble MMC operation with high dv/dt. MMC submodules which pass the qualification testing are able to operate as a robust building block of a MMC. Secondly, an open loop voltage balancing scheme is introduced to balance the voltage of two cascaded submodules. Detailed simulation and experimental results are provided to validate the test method for MMC submodules based on 10 kV SiC MOSFETs, followed by the conclusions of this paper.

Overview of proposed test scheme

The proposed test scheme for MMC submodules is a simple three-step scheme with the focus on the ac-dc continuous test. The first step is the component qualification and submodule assembly. Then, initial tests are conducted to check the gate loop and gate driver functions. In the first step, components should be tested and qualified individually before the submodule assembly, including gate driver, voltage sensor, MOSFETs, and busbar. After the submodule is assembled, the vital functions of the gate driver are examined, including rising/falling edge of gate-to-source voltage V_{gs}, feedback signal, short circuit protection as well as soft turn-off. The last step is the ac-dc continuous test which should be conducted carefully by starting from low dc-link voltage operation. Compared to the test scheme in [13], the proposed test scheme does not require double pulse test (DPT) and short circuit test for each MOSFET, and hence is much simpler, less time-consuming, and more efficient, which is important when many submodules and converters need to be tested. It is acceptable to skip DPT and short circuit test since the 10 kV/20 A SiC MOSFET and its package in the submodule have become more mature. Meanwhile, more comprehensive tests of the gate driver functions are required in the second step.

Overview of ac-dc continuous test

A proposed ac-dc continuous test circuit with two cascaded submodules is the core of the test scheme, as shown in Fig. 1(a). The proposed ac-dc continuous test circuit has a simple configuration, including a high voltage dc power supply, which is commercially available from various manufacturers, an input capacitor, two MMC submodules, and the load. Two MMC submodules should be connected in the same way as submodules are connected in the MMC. During the normal operation with balanced submodule voltage, the dc component of the submodule voltage is approximately equal to the input voltage V_g. Thereby, the high voltage dc power supply should be able to output the rated dc bus voltage of the MMC submodule. The load can be easily realized with arm inductors of the MMC and a resistive load. R_{load} can be adjusted to obtain the desired active power. Typically, the active power and the value of R_{load} are limited by the output current capability of the dc power supply. MMC submodules with different topologies can be tested with the proposed test circuit. This paper focuses on the testing of MMC submodules with a half bridge topology.

During the ac-dc continuous test, Submodule 1 in Fig. 1(a) is the submodule under test, with high dv/dt in its DC- terminal resulting from the switching actions in Submodule 2. Submodule 2 can be regarded as part of the ac-dc continuous test setup. The gate resistance of Submodule 2 can be reduced to further increase dv/dt in order to fully test the dv/dt immunity of the submodule under test. A fiber optic voltage probe which is able to withstand high common mode voltage can be used to monitor the gate signal, the output signal of the short circuit protection, and other signals, to evaluate the noise immunity under high dv/dt. In parallel with the ac-dc continuous test, the thermal design can be evaluated online by measuring device temperature with a fiber optic temperature sensor. Acoustic partial discharge detection method is effective in the online validation of the insulation design of a submodule under PWM voltage with high dv/dt [16].

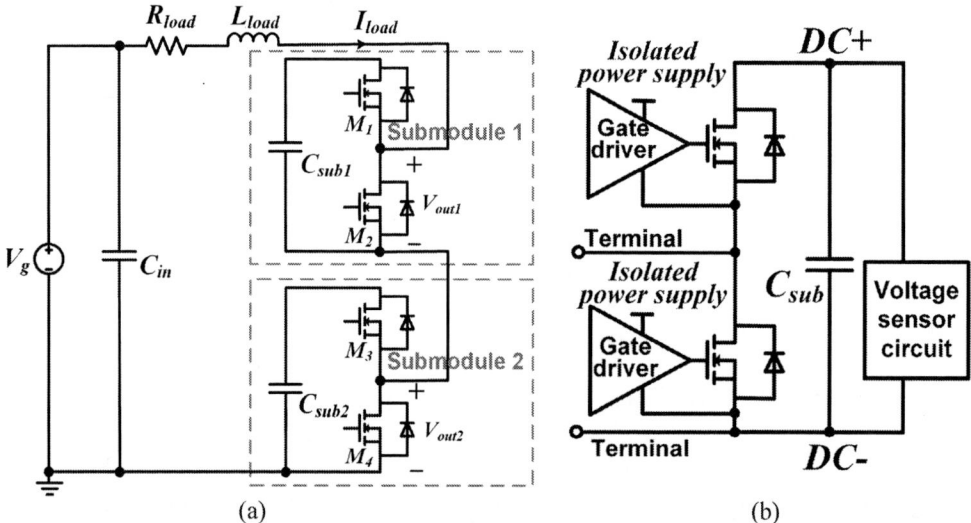

(a)   (b)

Fig. 1: (a) Circuit diagram of the proposed ac-dc continuous test circuit for the qualification of MMC submodules. (b) Overview of the MMC submodule with half bridge topology.

Modulation scheme

In the ac-dc continuous test, a modulation scheme is implemented to force the MMC submodule under test to undergo high dv/dt that may occur during the MMC converter operation. Cascaded submodules are leveraged to generate high dv/dt that is likely to occur in the converter. Bipolar SPWM modulation with the same modulation index is implemented in both Submodule 1 and 2. With zero phase shift between the PWM signals for Submodule 1 and 2, M_1 shares the same gate signal with M_3, and M_2 and M_4 receive the same gate signal.

If Submodule 2 and the submodule under test have the same switching speed and switching frequency, the source of M_1 in Submodule 1 will undergo the voltage rise from 0 to $2V_g$ with 2X normal turn-on dv/dt of M_1 as M_1 and M_3 turn on simultaneously. The source potential of M_1 drops from $2V_g$ to 0 with 2X normal turn-off dv/dt of M_1, as M_1 shuts off. Such switching actions with 2X normal dv/dt could occur in the MMC converter when two submodules in the same arm switch simultaneously. The voltage step change between 0 and $2V_g$ with 2X normal dv/dt results in 2X common mode (CM) current flowing through the gate driver and its isolated power supply. Also, the insulation capability of the isolated power supply for the gate driver can be fully tested in this case.

Reducing the gate resistance for MOSFETs in Submodule 2 will further increase the dv/dt for the submodule under test, >2X normal dv/dt and CM current can be realized. Meanwhile, the DC-terminal of the submodule under test undergoes dv/dt higher than the normal dv/dt of 10 kV SiC MOSFETs in a MMC. The dv/dt that the submodule under test normally experiences in operation can be controlled to be much higher for the purpose of evaluating the noise immunity margin. The noise immunity of the gate driver, especially the short circuit protection and the isolated power supply, can hence be fully tested. In fact, the high dv/dt can influence the normal circuit operation via CM current, radiation noise, and other mechanisms. The submodule under test can be tested and debugged to achieve the capability to handle high dv/dt. With such modulation scheme, the ac-dc continuous test setup is a suitable platform to test the submodule's capability to withstand high dv/dt during the MMC operation.

The ac-dc continuous test setup has the capability to validate the thermal and insulation design of the MMC submodule under test at the rated dc bus voltage and current. The rated dc bus voltage of the submodule can be realized by increasing V_g. The load voltage is a PWM-type voltage with step changes between V_g and $-V_g$. Normally the load current has a sinusoidal shape due to the high impedance of the load inductor at high frequency. The magnitude of the load current can be adjusted by changing the modulation index and the fundamental frequency. The peak value of the fundamental component of the load current $I_{fund,pk}$ can be estimated with the equation below.

$$I_{fund,pk} = \frac{mV_g}{2\pi f_{line}L_{load} + R_{load}} \tag{1}$$

In the equation, f_{line} is the fundamental frequency, and m is the modulation index, both of which are determined by the modulation signal. The dc component of the load current $I_{load,DC}$ is determined by the active power consumed in the continuous test setup, which can be calculated with the following equation.

$$I_{load,DC} = \frac{P_{loss} + R_{load}I_{RMS}^2}{V_g} \tag{2}$$

I_{RMS} is the RMS value of the load current, and P_{loss} is the power loss in the ac-dc continuous test setup. If the resistive load is not installed, the load current is almost pure AC current.

Voltage balancing of submodules

Submodule voltage balancing is essential to the ac-dc continuous test circuit with two cascaded MMC submodules. Closed loop submodule voltage balancing control is commonly used in the MMC converter [3], which can also be implemented in the ac-dc continuous test circuit. In order to further simplify the test setup and control, however, an open loop voltage balancing scheme is adopted.

The two cascaded submodules ideally have balanced submodule capacitor voltage, since the two capacitors always have the same current with the designed modulation scheme. In ideal conditions, M_1 and M_3 always conduct at the same time to achieve the natural submodule voltage balancing. In reality, the two submodules could have different dead times due to nonideal factors in the controller and the gate driver board. Also, there could be slight phase shift between V_{gs} of M_1 and M_3, or V_{gs} of M_2 and M_4. For example, the fiber optic transmitter or receiver with the same part number could have different propagation delay. Especially, longer dead time contributes to higher submodule capacitor voltage no matter what the direction of the load current is. Without any voltage balancing method, the voltage of the two submodules would diverge from each other, and the capacitor voltage of the MMC submodule with shorter dead time finally drops to zero. This can be explained by the difference between the average submodule capacitor current of the two submodules, which is named the offset current, I_{offset}. In the dc circuit model in Fig. 2(a), the voltage of the two submodules is marked as ΔV_{sub1} and ΔV_{sub2}, because the initial dc voltage is neglected. The dc voltage difference V_{offset}, defined as $\Delta V_{sub1} - \Delta V_{sub2}$, will continue to increase in magnitude as long as I_{offset} exists due to the different dead time, and the voltage of one submodule will eventually drop to zero.

The magnitude and polarity of I_{offset} is determined by the load current as well as the dead time difference and phase shift between V_{gs} for the MOSFETs of the two submodules. In fact, the magnitude and polarity of I_{offset} can be time-varying. For instance, assuming 20 ns phase shift between V_{gs} for M_1 and M_3 (V_{gs} for M_1 leading) and zero phase shift between V_{gs} for M_2 and M_4, the instantaneous I_{offset} is almost zero when the load current is positive, because M_1 and M_3 are synchronous devices whose body diodes conduct current during the dead time. If the load current is negative, the phase shift between V_{gs} for M_1 and M_3 will play a part, and the instantaneous $I_{offset}(t)$ can be described with the following equation.

$$I_{offset}(t) = \frac{1}{T_s}\left(\int_{t}^{t+DT_s} I_{load}(t)dt - \int_{t+\Delta t}^{t+\Delta t+DT_s} I_{load}(t)dt\right) \qquad (3)$$

In the equation, T_s is the switching period, D is the duty cycle, and Δt is the phase shift. The equation shows that the magnitude and polarity of I_{offset} depend on the instantaneous phase angle of the load current. In fact, it is likely that different dead time and slight phase shift between V_{gs} for M_1 and M_3 as well as between V_{gs} for M_1 and M_3 influence I_{offset} simultaneously. As a result, the average I_{offset} over a line cycle is not zero in steady state, and hence the voltage balancing cannot be achieved.

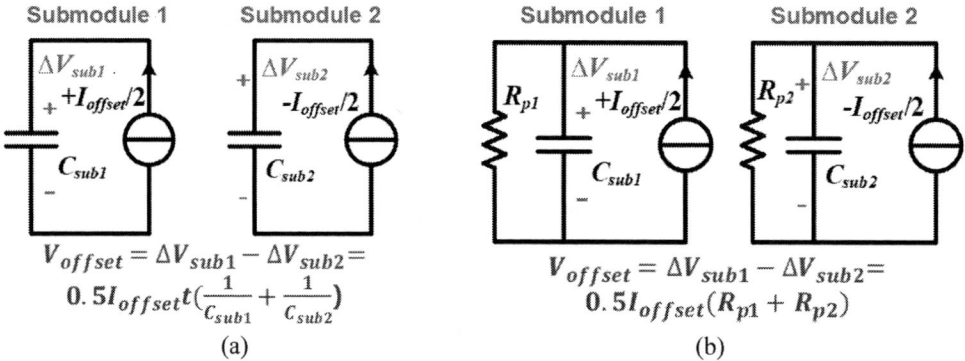

Fig. 2: (a) DC circuit model of the two MMC submodules used to study the submodule voltage difference V_{offset}. (b) DC circuit model of the two MMC submodules after adding a resistor in parallel with the capacitor.

An open loop method with an external parallel resistor is developed to suppress the impact of I_{offset} and achieve voltage balancing. As can be seen in Fig. 2(a), I_{offset} caused by numerous nonideal factors can only flow through the submodule capacitor, and hence keeps charging or discharging the submodule capacitor. After adding a resistor in parallel with the submodule capacitor, as shown in Fig. 2(b), all of the offset current will be absorbed by the resistor, and the submodule capacitor voltage will reach steady state. The dc voltage of the two submodules in steady state still has some difference V_{offest}, whose equation is shown in Fig. 2(b). V_{offest} is only determined by the added parallel resistor and the magnitude of I_{offset} and is independent of the capacitance and parasitics of the submodule capacitor.

It is difficult to calculate the average I_{offset} in a line cycle in an analytical way. However, the average I_{offset} can be obtained easily based on simulation results. The parallel resistor should be selected based on the trade-off between the V_{offest} and the power loss of the resistor. A small parallel resistance is attractive due to better voltage balancing results, yet bulky power resistors must be selected which may also require additional heatsinks or fans and lead to complicated experimental setup. A large parallel resistance results in simple test setup, but the large V_{offset} makes it difficult to control the voltage of the submodule under test in a convenient way, and increases the risk of device damage due to overvoltage.

Simulation results

The proposed ac-dc continuous test circuit with the open loop voltage balancing method is simulated in Matlab/Simulink. Parameters in the simulation are obtained from the built MMC submodules based on discrete 10 kV/20 A SiC MOSFETs from Wolfspeed [17] and the test setup which will be introduced in detail in next section. A 3D model of the half bridge submodule is displayed in Fig. 3, together with the 10 kV/20 A SiC MOSFET.

Fig. 4 shows simulation results of the case with zero phase shift between gate signals for the two submodules and 500 ns dead time for both submodules. The submodule voltage is hence perfectly balanced, and the PWM-type load voltage and the load current with sinusoidal shape can be seen in the simulation waveforms. The load current is regulated at ~6 A peak with almost zero DC component, since R_{load} is 1 Ω. In addition, Fig. 5 displays the simulated submodule voltage waveforms without and with the open loop voltage balancing method to prove its effectiveness. The case where two submodules have significantly different dead time is simulated since it provides a large I_{offset}. Without the added parallel resistor, the voltage of Submodule 2 with 100 ns shorter dead time keeps decreasing, and V_{offset} increases rapidly to ~800 V within 3 s. The calculated I_{offset} based on simulation results is 2.4 mA, and the calculated V_{offset} at t=3 s is 782 V based on the equation in Fig. 2(a). Then a relatively small parallel resistor is added in both submodules in order to achieve voltage balancing even faster and reduce computational burden in the simulation. After adding a 100 kΩ resistor, the voltage of two submodules is balanced, and reaches steady state with a constant V_{offset} of ~200 V, close to the calculated 240 V V_{offset}, based on the equation in Fig. 2(b). Like the real MMC operation, the submodule voltage has the line frequency ripple with a peak-to-peak value of ~300 V.

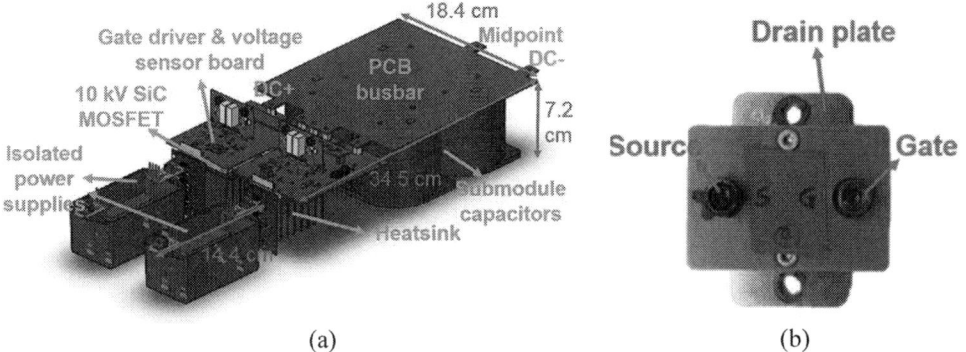

(a)   (b)

Fig. 3: (a) 3D model of the built MMC submodule based on 10 kV/20 A SiC MOSFETs. (b) Picture of the 10 kV/20 A SiC MOSFET from Wolfspeed.

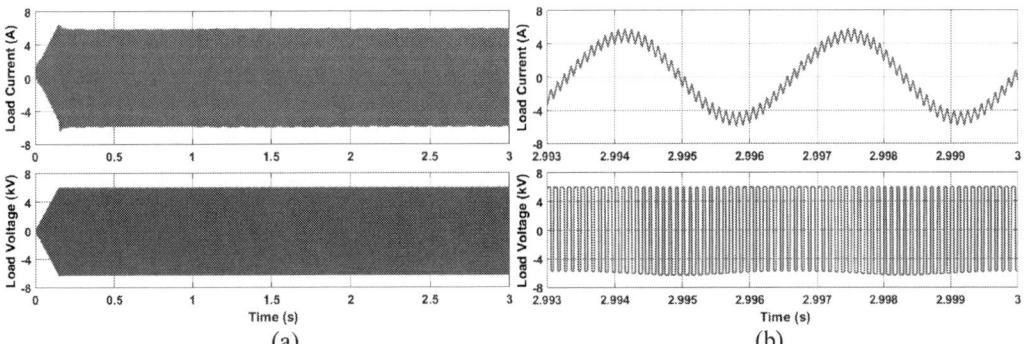

(a)   (b)

Fig. 4: Simulated waveforms of load voltage and load current at 6 kV dc-link voltage (modulation index m=0.25, R_{load} =1 Ω, L_{load} =175 mH): (a) Overview (V_g ramps up from 0 to 6 kV within 0.15 s); (b) Zoom-in waveforms.

Experimental setup and results

Experimental setup

The proposed ac-dc continuous test circuit is realized with the experimental setup shown in Fig. 6. The high voltage dc power supply from Spellman (Part Number: ST15P12) can output the dc voltage up to 15 kV with the maximum power of 12 kW. The input capacitor C_{in} is implemented with a capacitor bank whose equivalent capacitance is 46.7 μF. The load inductor is implemented with two high voltage inductors in series with a total inductance of 175 mH, which are also the arm inductors of the MMC. The resistive load is not installed due to the 0.8 A output current limit of the power supply. In fact, in order to reduce the output current ripple of the DC power supply, a 470 Ω resistor is inserted between the DC+ terminal of the power supply and C_{in}, otherwise an overcurrent fault will be reported from the power supply.

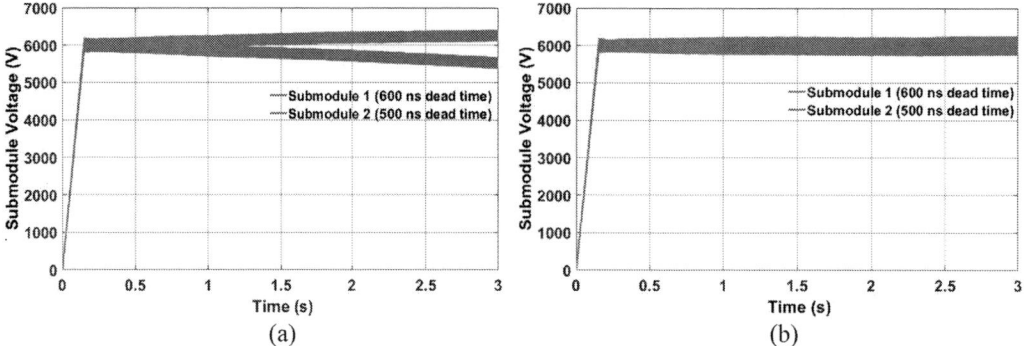

(a)   (b)

Fig. 5: Simulated submodule voltage waveforms (modulation index m=0.25, V_g= 6 kV, R_{load} =1 Ω, L_{load} =175 mH): (a) Without parallel resistor for voltage balancing; (b) With a 100 kΩ parallel resistor for voltage balancing.

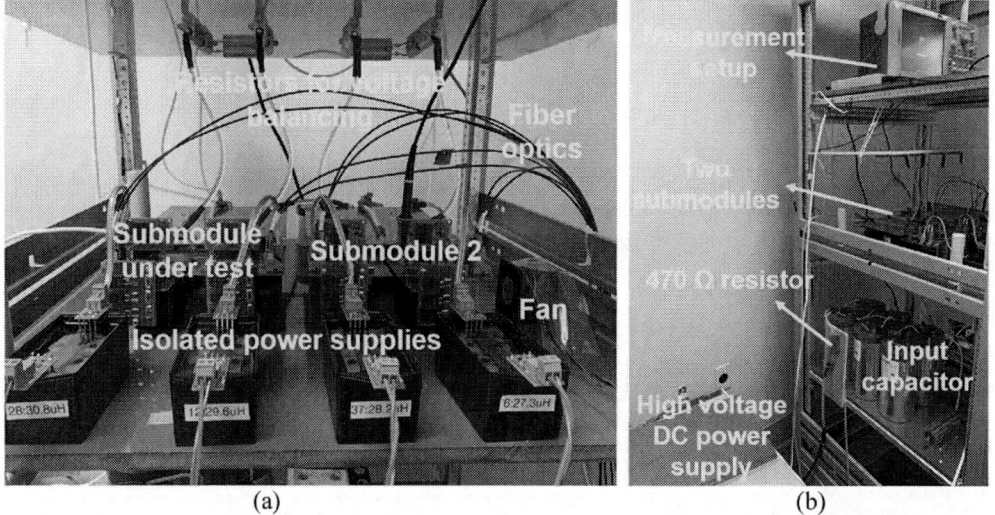

(a)   (b)

Fig. 6: Experimental setup of the ac-dc continuous test circuit: (a) Zoom-in view of the two cascaded MMC submodules; (b) Overview of the whole test setup (the load inductor is not visible).

Two cascaded MMC submodules are placed next to each other. Each MMC submodule has two 24 V isolated power supplies with 20 kV insulation capability to provide the auxiliary power [18], whose primary side is connected to a 110 V AC power adapter. Each submodule has a 500 kΩ resistor for voltage balancing which is composed of five 100 kΩ resistors (RH050100K0FE02 from Vishay) in series to support the continuous test up to 6 kV V_g. Without any additional heatsinks or fans, the 500 kΩ resistor has a power rating of 100 W, which is 39% higher than its expected power loss at 6 kV. The 500 kΩ resistor can support the test setup even when nonideal factors result in I_{offset} up to 4 mA and the voltage of one submodule up to 7 kV. As shown in Fig. 6, the resistors for voltage balancing are placed on a garolite board.

The cabinet is solidly grounded by connecting it with the grounded case of the high voltage dc power supply to achieve single-point grounding. The DSP controller and human-machine interface (HMI) are far away from the high voltage test setup and isolated via fiber optics. The selected fundamental

EPE'20 ECCE Europe

Assigned jointly to the European Power Electronics and Drives Association & the Institute of Electrical and Electronics Engineers (IEEE)

frequency is 300 Hz, which is higher than 60 Hz in order to limit the magnitude of the load current. The switching frequency of the 10 kV SiC MOSFETs is 10 kHz.

Experimental results

Experimental results of the ac-dc continuous test at 2.1 kV dc-link voltage are displayed in Fig. 7 and Fig. 8. Before the continuous test, both MMC submodules have passed the first two steps of the test scheme. Both submodules have the same gate resistance: 15 Ω for turn-on, and 3 Ω for turn-off. In terms of the measurement setup, a CWT Ultra Mini Rogowski coil from PEM is adopted for the load current measurement, and two differential voltage probes are used to measure the output voltage of the two submodules. V_{gs} of M_4 is monitored by a low voltage passive probe TPP1000 from Tektronix, whose source is solidly grounded. With a modulation index of 0.45, the sinusoidal load current with 10 kHz ripple has a peak value of ~3 A, which coincides with the estimation result and the simulation result. The low frequency ripple of the submodule voltage is also clearly indicated in Fig. 7.

The submodule voltage is well balanced with the help of the 500 kΩ parallel resistor. The voltage difference between the two submodules V_{offset} is less than 250 V with a dc-link voltage of 2.1 kV. In fact, only one pair of PWM signal is generated in the DSP controller, so the same gate signal is sent to the gate driver of M_1 and M_3 via fiber optics, as well as M_2 and M_4. The voltage difference is mainly attributed to the phase shift between the V_{gs} of M_1 and M_3, and M_2 and M_4, which is due to the propagation delay difference of the components in the gate driver board. According to the zoom-in waveforms in Fig. 8, the output voltage of the two submodules rises and drops almost simultaneously, with a phase shift of <20 ns. It is thereby proved that the designed modulation scheme forces the source of M_1 to undergo 2X dv/dt that M_2 and M_3 withstand. The dv/dt that M_1 and M_2 of the submodule under test experience continuously will be higher if Submodule 2 is modified to switch with higher dv/dt.

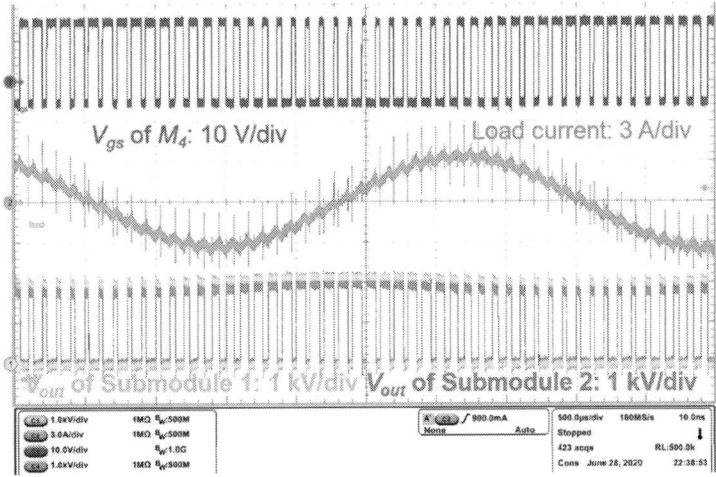

Fig. 7: Waveforms of the continuous test with the proposed ac-dc continuous test circuit at 2.1 kV.

The ac-dc continuous test is successfully conducted at 6 kV dc-link voltage, with the waveforms displayed in Fig. 9. Since the differential voltage probe is not capable of withstanding 6 kV common mode voltage with high dv/dt, the output voltage of the submodule under test (Submodule 1) cannot be measured any more. The output voltage of Submodule 2 is measured with a high voltage passive probe, P6015A from Tektronix. At 6 kV, the modulation index is reduced to 0.25 because the continuous operation of the load inductor only allows <9 A peak current. The maximum load current flowing through the inductor and MOSFETs is ~6 A. Because the source of M_1 experiences 2X normal dv/dt of the 10 kV SiC MOSFET, substantial displacement current flows through the effective parallel capacitance (EPC) of the load inductor, as clearly indicated in the measured load current waveform.

Measurement results of the output voltage of Submodule 2 show that the capacitor voltage of Submodule 2 varies from 5.1 kV to 5.4 kV. Therefore, it is estimated that the voltage of the submodule under test (Submodule 1) varies from 6.6 kV to 6.9 kV. The proposed open loop voltage balancing method effectively achieves voltage balancing in the ac-dc continuous test setup. With a larger I_{offset} due to the higher load current, the voltage difference V_{offset} is higher than 1 kV. In addition, more accurate submodule voltage measurement results and V_{offset} can be obtained with better high voltage differential voltage probes in the future.

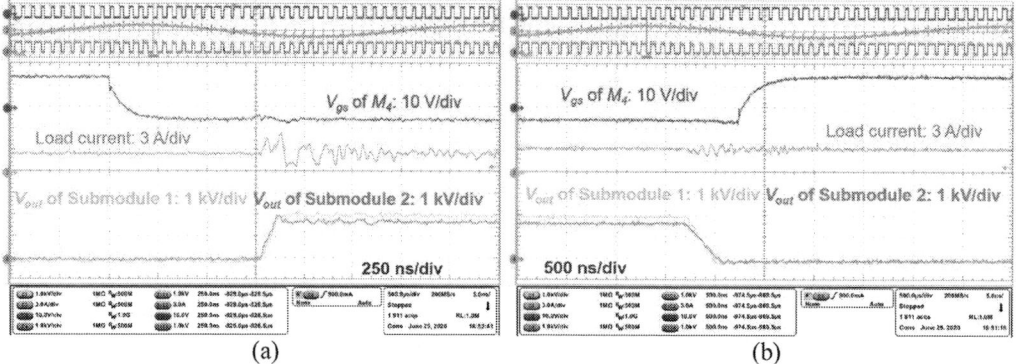

Fig. 8: Zoom-in waveforms of the continuous test with the proposed ac-dc continuous test circuit at 2.1 kV: (a) Submodule output voltage rises; (b) Submodule output voltage decreases.

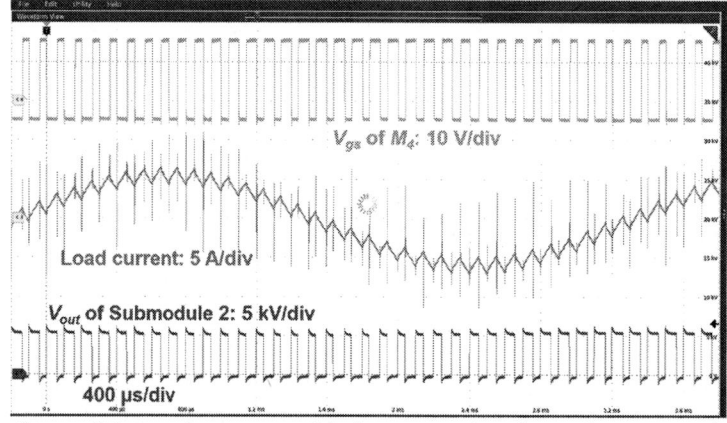

Fig. 9: Waveforms of the continuous test with the proposed ac-dc continuous test circuit at 6 kV.

Conclusions

This paper focuses on a simple test scheme to fully qualify MMC submodules based on 10 kV SiC MOSFETs. The test scheme provides a comprehensive and efficient qualification of the MMC submodule, including its thermal design, insulation design, and its capability to withstand high dv/dt. In the test scheme, an ac-dc continuous test circuit with two MMC submodules cascaded in series is developed to qualify the submodules as the final step. With the designed modulation scheme, the submodule under test needs to continuously withstand 2X normal dv/dt of 10 kV SiC MOSFETs, which could occur during the operation of a real MMC. In fact, the dv/dt that the submodule under test will undergo is controllable so that the capability of the submodule to operate normally under high dv/dt can be fully tested. An open loop voltage balancing method with the external parallel resistor is adopted to simplify the test setup. The developed continuous test circuit is fully validated with the built test setup and the ac-dc continuous test up to 6 kV.

References

[1] S. Jafarishiadeh, M. Farasat and A. K. Sadigh, "Medium-voltage DC grid connection using modular multilevel converter," *in Proc. of IEEE Energy Conversion Congress and Exposition (ECCE)*, Cincinnati, OH, 2017, pp. 2686-2691.

[2] H. Mohammadi and M. T. Bina, "A transformerless medium-voltage STATCOM topology based on extended modular multilevel converters," *IEEE Trans. Power Electron.*, vol. 26, no. 5, pp. 1534-1545, May 2011.

[3] S. Debnath, J. Qin, B. Bahrani, M. Saeedifard and P. Barbosa, "Operation, control, and applications of the modular multilevel converter: a review," *IEEE Trans. Power Electron.*, vol. 30, no. 1, pp. 37-53, Jan. 2015.

[4] D. Johannesson, M. Nawaz and K. Ilves, "Assessment of 10 kV, 100 A silicon carbide MOSFET power modules," *IEEE Trans. Power Electron.*, vol. 33, no. 6, pp. 5215-5225, June 2018.

[5] D. Rothmund, T. Guillod, D. Bortis and J. W. Kolar, "99.1 % efficient 10 kV SiC-based medium voltage ZVS bidirectional single-phase PFC AC/DC stage," *IEEE Journal of Emerging and Selected Topics in Power Electronics*, vol. 7, no. 2, pp. 779-797, June 2019.

[6] S. Ji, M. Laitinen, X. Huang, J. Sun, W. Giewont, F. Wang and L. M. Tolbert, "Short circuit characterization and protection of 10 kV SiC MOSFET," *IEEE Trans. Power Electron.*, vol. 34, no. 2, pp. 1755-1764, Feb. 2019.

[7] C. DiMarino, D. Boroyevich, R. Burgos, M. Johnson and G. Lu, "Design and development of a high-density, high-speed 10 kV SiC MOSFET module," in *Proc. of 19th European Conference on Power Electronics and Applications (EPE'17 ECCE Europe)*, Warsaw, 2017, pp. P.1-P.10.

[8] M. K. Das, C. Capell, D. Grider, S. Leslie, J. Ostop, R. Raju, M. Schutten, J. Nasadoski and A. Hefner, "10 kV, 120 A SiC half H-bridge power MOSFET modules suitable for high frequency, medium voltage applications," in *Proc. of IEEE Energy Conversion Congress and Exposition (ECCE)*, Phoenix, AZ, 2011, pp. 2689-2692.

[9] S. Madhusoodhanan *et al.*, "Solid-state transformer and MV grid tie applications enabled by 15 kV SiC IGBTs and 10 kV SiC MOSFETs based multilevel converters," *IEEE Trans. Ind. Appl.*, vol. 51, no. 4, pp. 3343-3360, July-Aug. 2015.

[10] B. Florkowska, M. Florkowski, J. Roehrich, and P. Zydron, "The influence of PWM stresses on degradation processes in electrical insulation systems," in *Proc. Annu. Rep. Conf. Electr. Insul. Dielectic Phenomena*, West Lafayette, IN, 2010, pp. 1-4.

[11] R. Färber, T. Guillod, F. Krismer, J. W. Kolar, and C. M. Franck, "Endurance of polymeric insulation foil exposed to dc-biased medium-frequency rectangular pulse voltage stress," *Energies*, vol. 13, no. 1, p. 13, Dec. 2019.

[12] D. N. Dalal et al., "Gate driver with high common mode rejection and self turn-on mitigation for a 10 kV SiC MOSFET enabled MV converter," in *Proc. of 19th European Conference on Power Electronics and Applications (EPE'17 ECCE Europe)*, Warsaw, 2017, pp. P.1-P.10.

[13] X. Huang, J. Palmer, S. Ji, L. Zhang, F. Wang, L. M. Tolbert, and W. Giewont, "Design and testing of a modular multilevel converter submodule based on 10 kV SiC MOSFETs," in *Proc. of IEEE Energy Conversion Congress and Exposition (ECCE)*, Baltimore, MD, 2019, pp. 1926-1933.

[14] D. Kranzer, J. Thoma, B. Volzer, D. Derix and A. Hensel, "Development of a 10 kV three-phase transformerless inverter with 15 kV silicon carbide MOSFETs for grid stabilization and active filtering of harmonics," in *Proc. of 19th European Conference on Power Electronics and Applications (EPE'17 ECCE Europe)*, Warsaw, 2017, pp. P.1-P.8

[15] A. Anurag, S. Acharya, G. Gohil and S. Bhattacharya, "Benchmarking and qualification of gate drivers for medium voltage (MV) operation using 10 kV silicon carbide (SiC) MOSFETs," in *Proc. of IEEE Applied Power Electronics Conference and Exposition (APEC)*, Anaheim, CA, USA, 2019, pp. 441-447.

[16] Y. Xu, C. Zhang, C. Gao, J. Wang, R. Burgos, D. Boroyevich, and M. Ren, "Insulation online monitoring for critical components inside SiC based medium voltage converter prototype," in *Proc. of IEEE Electric Ship Technologies Symposium (ESTS)*, Washington, DC, USA, 2019, pp. 484-491.

[17] X. Huang, S. Ji, J. Palmer, L. Zhang, D. Li, F. Wang, L. M. Tolbert, and W. Giewont, "A robust 10 kV SiC MOSFET gate driver with fast overcurrent protection demonstrated in a MMC submodule," in *Proc. of IEEE Applied Power Electronics Conference and Exposition (APEC)*, New Orleans, LA, USA, 2020, pp. 1813-1820.

[18] L. Zhang, S. Ji, S. Gu, X. Huang, J. Palmer, W. Giewont, F. Wang, and L. M. Tolbert, "Design considerations for high-voltage-insulated gate drive power supply for 10-kV SiC MOSFET applied in medium-voltage converter," *IEEE Trans. Ind. Electron.*, doi: 10.1109/TIE.2020.3000131.

PWM Gain Linearization Algorithm for Medium Voltage Source Inverter

Hamza El Jihad – Sami Siala – Elise Savarit
General Electric Power Conversion
18 Avenue du Quebec, 91140
Villebon-sur-Yvette, France
Tel.: +33 / (0) – 177 31 22 75.
E-Mail: hamza.eljihad@ge.com
sami.siala@ge.com
elise.savarit@ge.com
URL: http://www.gepowerconversion.com

Keywords

«Pulse Width Modulation (PWM)», «Converter control», «Optimal control», «IGBT», «Simulation».

Abstract

This paper deals with industrial Medium Voltage (MV) multilevel converters used in voltage source drives. The switching features (as minimum time of conduction) of the power electronic components induce some non-linearities on the output voltage, that increase the harmonics and limit the maximum power of the drive. To overcome these drawbacks, we will present an algorithm that keeps the linearity of the converter and by the way increases the drive power and mitigate some harmonics.

Introduction

A multilevel converter [1] delivers an output voltage versus the reference. The linearity of the output regarding the input is distorted (Fig 1) in respect to the intrinsic physical constraints (as minimum time of conduction) of the semiconductor components [2]. These constraints are reflected on the control and Pulse Width Modulation (PWM) algorithms: if one of the phase voltage references is in a non-linear area, it is deviated and clamped to either a superior or inferior threshold, which causes the non-linearity problems.

The idea of the algorithm presented in this paper is to use the benefit of the three phase system , and to apply the same deviation to all the phase voltage references every time that one (or more) of them is in a non-linear area (Fig 2).

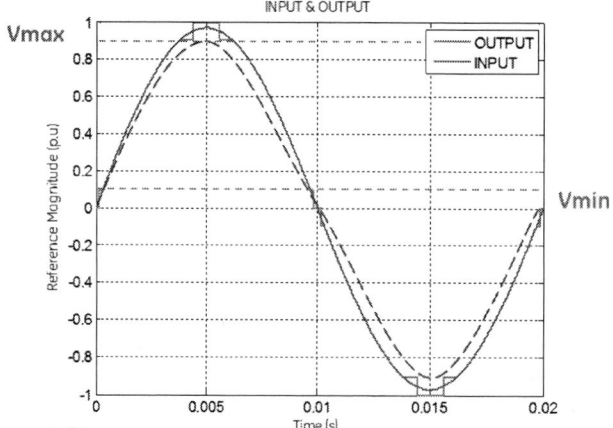

Fig 1: Typical response of the present clamping algorithm that induces the distortions

Operating principle of the algorithm

This new algorithm analyses the converter's voltage references of the three phases simultaneously and in real time in order to detect if one reference value or more are in a non-linear area. The algorithm clamps the reference(s) in question to the nearest threshold that respects the minimal time of conduction. The novelty of the algorithm is that it also reports the same deviation to the other references and checks if this operation does not put them in a non-linear area too. This way, the modification affects only the zero-sequence system and not the differential system.

In other words, this new proposed multilevel power converters voltage linearization approach is based on the algorithm presented in (Fig.3). As a first step, the voltage references of the three phases are read simultaneously in order to detect if at least one of the voltage references values is in an unauthorized area: ($|Vu,v,w \ ref| > Vmax$ or $|Vu,v,w \ ref| < Vmin$). Then the algorithm modifies the values of these voltage references if one of them is in an unauthorized area. This modification consists in "clamping" the voltage references to the nearest threshold respecting the minimal conduction time. Figure (2.a) shows an example of this modification in its classic method for a three levels power converter piloted using a Sinus PWM where the voltage references are sampled twice during a single PWM period (Asymmetric sampling). In this case, when the absolute value of the voltage reference of one phase is greater than the upper threshold $V_{max} = 1 - V_{min}$, it is the only one that is clamped to the nearest threshold (either +1 or $V_{max} = 1 - V_{min}$). The disadvantage of this method is that the phase to phase voltages (and so the differential mode and α,β components of the voltage vector) are not the same anymore, which induces the non-linearity of the output voltages regarding the voltage references. To overcome this problem, this new linearization approach also modifies in the same manner the other two voltage references that are not in the unauthorized area. This way, the differential mode is not changed, and the zero sequence is the only mode that is changed. (Fig 2.b) shows that when the blue voltage reference is this upper unauthorized area ($>V_{max}$), the other two voltage references are also changed. It is also important to check that this modification of the two other voltage references does not put one of them or both in the unauthorized area. If this occurs, the algorithm computes all the possible combinations that respect the minimal conduction time of the semi-conductor components associated the three phases of the power converter. The chosen combination is the one with the minimal quadratic error between the α,β components of the modified voltage references vectors and the targeted reference. This quadratic error is given at equation (1).

$$Error_i = \sqrt{\left(V_{\alpha ref} - V_{\beta ref}\right)^2 + (V_{\alpha i} - V_{\beta i})^2} \qquad (1)$$

With i : order of the ith combination

Each phase within the unauthorized range of modulation index has two clamp possibilities (Fig 2): upper or lower thresholds (going up or down) , reached by applying +deltaV1 and -delta V2 respectively. Table (1) shows the different thresholds regarding each non-linear area of the input reference voltage.

Table I: Thresholds regarding each non-linear area of the input reference voltage

Non-linear area	Upper threshold	Lower threshold
Around +1	1	Vmax = 1 − Vmin
Around 0	Vmin	-Vmin
Around -1	-Vmax = -1 + Vmin	-1

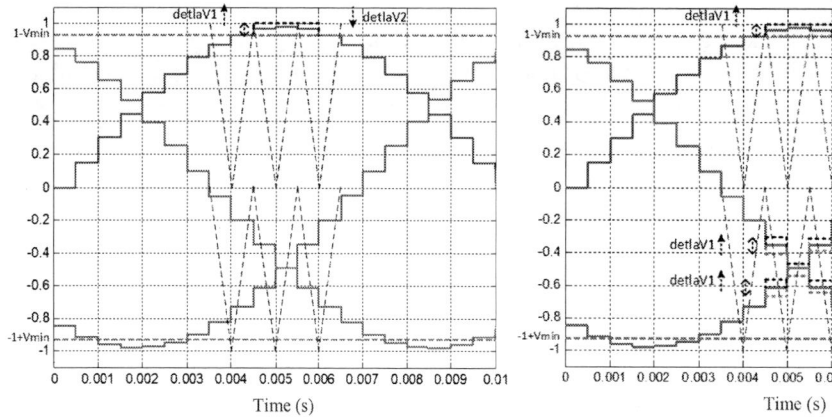

(a) **Classic** clamping method when one of the voltage references is greater than the upper threshold Vmax.

(b) **Proposed** clamping method when one of the voltage references is greater than the upper threshold Vmax.

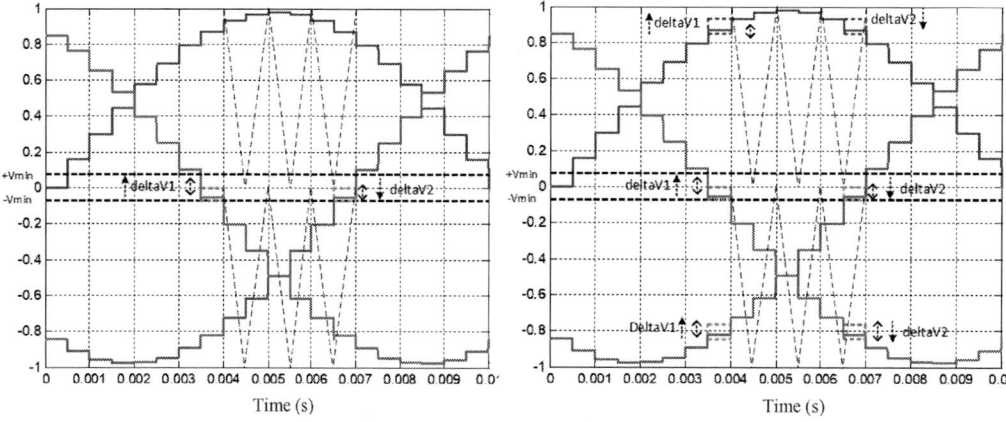

(c) **Classic** clamping method when one of the voltage references is zero crossing.

(d) **Proposed** clamping method when one of the voltage references is zero crossing.

Fig 2: Two possible ways of clamping the blue voltage signal in unauthorized area around +1: upper thresholds and 0: zero crossing

This leads to $2^3 = 8$ combinations possible. Being different from the targeted reference, the algorithm chooses the solution with the minimal quadratic error.

The distortions caused by the old algorithm tend to create a non-linearity of the modulation index that disturbs the control and regulation algorithms (Fig.4). Moreover, even if the semiconductor devices can stand a high switching frequency, the present clamping algorithms of the MV converter prevent from increasing the switching frequency because the non-linear area would start at a lower modulation index and we will not be able to go higher.

In case of using an output sinus filter, and as this distortion may generate a higher than acceptable harmonics [3], especially at the low frequencies. This leads to increase the size of the sinus filter [4, 5].

Flowchart of the new proposed linearization approach

(Fig.3) summarizes the different steps of this new linearization approach. The algorithm detects and identifies which one of the voltage references is within an unauthorized area. Then it sets two correction values and applies the lowest one to modify these voltage references so that it puts them back in the linear area. It also modifies the other voltage references in the same way, and then checks if all of them are in the linear area. If so, these new values are sent to the PWM generation bloc to be executed. If not, and if there is at least one correction value left, the algorithm applies it to the three phases at the same time and rechecks if all of them are in the linear area. If there is no correction value left, the algorithm applies the minimal correction value to the voltage reference phase which is within an unauthorized area. If two voltage references are within an unauthorized area, the algorithm computes eight combinations (Three phases with two possibilities) and then applies the one with the minimal quadratic error. At last, the new voltage references are sent to the PWM generation bloc to be executed.

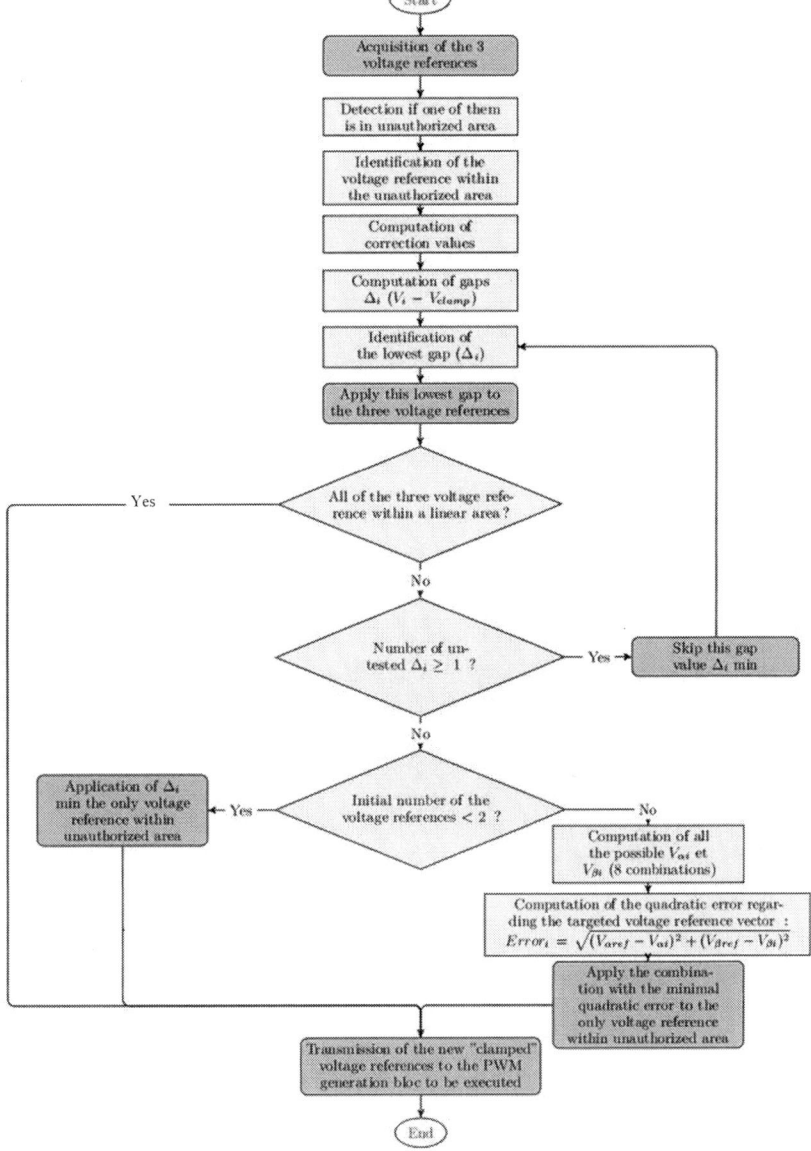

Fig 3: New output voltage linearization algorithm flowchart

Assessment of the proposed output voltage linearization approach

The aim of this section is to evaluate the performances of the proposed linearization method and to compare it regarding the classic method in terms of the converter's output voltage fundamental magnitude (Fig.4). To this end, multiple simulations are performed for a three levels PWM piloted converter.

Other aspects have also been considered for this evaluation, such as the performances of an Induction machine: Starting phase + steady state

Output voltage fundamental magnitude

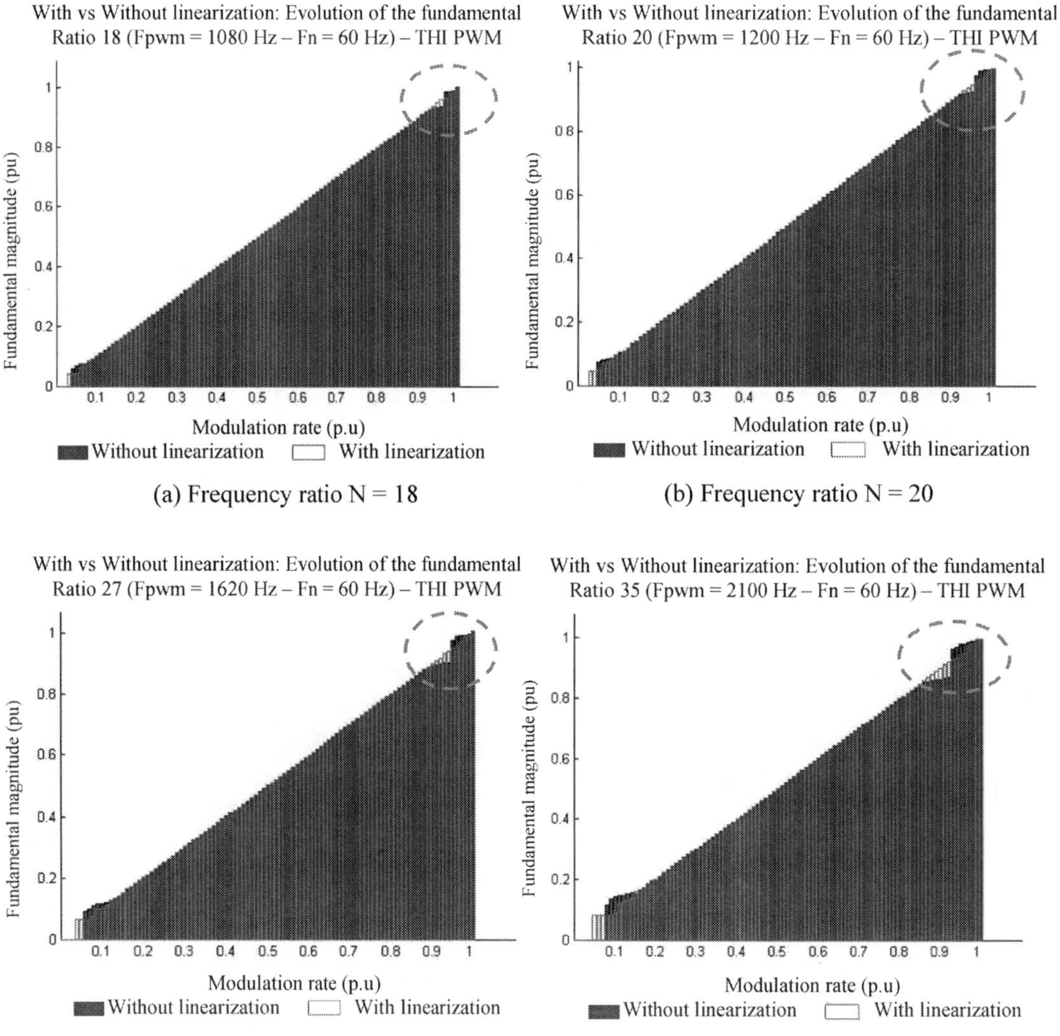

(a) Frequency ratio N = 18

(b) Frequency ratio N = 20

(c) Frequency ratio N = 27

(d) Frequency ratio N = 35

Fig 4: Evolution of the output voltage Fundamental: With / Without Linearization

(Fig.4) shows the evolution of the output voltage fundamental magnitude for different frequency ratios ($N = \frac{F_{PWM}}{F_{FUND}}$) with the two voltage clamping methods. The modulation rate is set in the range $[0,1]$ by 0.01 steps. The illustrated frequency ratios are $N = 18, 20, 27$ and 35. The fundamental frequency is set to 60 Hz.

The upper threshold V_{MAX} is given by the equation (2):

$$V_{MAX} = 1 - [F_{PWM} * \Delta T_{min}] \tag{2}$$

Where ΔT_{min} is the minimal conduction time of the semi-conductor components. In this case, $\Delta T_{min} = 70\ \mu s$.

Table II: Values of V_{MAX} for every frequency ratio

Frequency ratio N	18	20	27	35
V_{MAX}	0.924	0.916	0.886	0.853

The results presented in (Fig.4) confirm the efficiency of the proposed method that helps to attain a higher output voltage fundamental for modulation rates between $[0.8, 1]$, the classic method presents a clear distortion of the fundamental when the modulation rate is in the range $[0.8, 1]$.

Please note that only modulation rate around $[0.8, 1]$ are considered because the main concern of this new linearization approach is to achieve a higher voltage in the output of the converter and thus a higher power. The values of modulation rate around $[0, 0.1]$ are not considered because the starting phase is managed using a different PWM technique.

Performances in a complete power conversion chain

The objective of this paragraph is to evaluate the stability of a complete power conversion chain composed mainly of a three level PWM power converter and a squirrel cage induction machine, by implementing the proposed linearization approach in the pulse generation process.

The system under test has the following characteristics:

- Three-levels power converter:
  - DC link voltage: 5500 V
  - Line-to-line output voltage: 3300 V
  - Switching frequency: 1620 Hz
- Induction machine:
  - Stator resistance: 17 mΩ
  - Stator leakage inductance: 1.2 mH
  - Rotor resistance: 20 mΩ
  - Rotor leakage inductance: 0.6 mH
  - Magnetizing inductance: 30 mH
  - Mechanical inertia: 200 Nms²
  - Friction coefficient: 10 Nms
  - Pole pairs: 2

The system is tested in different phases using a scalar command (V/F constant):

- Phase 1: No-load starting
- Phase 2: Loading of the machine
- Phase 3: Increase of machine speed with constant load
- Phase 4: Unloading of the machine
- Phase 5: Heavy loading of the machine

In order to test the validity of the proposed linearization in the non-authorized areas (non-linear area), we will check the linearity of the converter's output voltage for modulation rates (m > 0.8). In the same way, we will check the behavior of machine torque, currents through their waveforms (Fig.5).

Please note that the following results are intended to study the new linearization algorithm in steady state. Transient effects on the currents and the torque and currents waveforms which are linked to the chosen regulation dynamic are not considered.

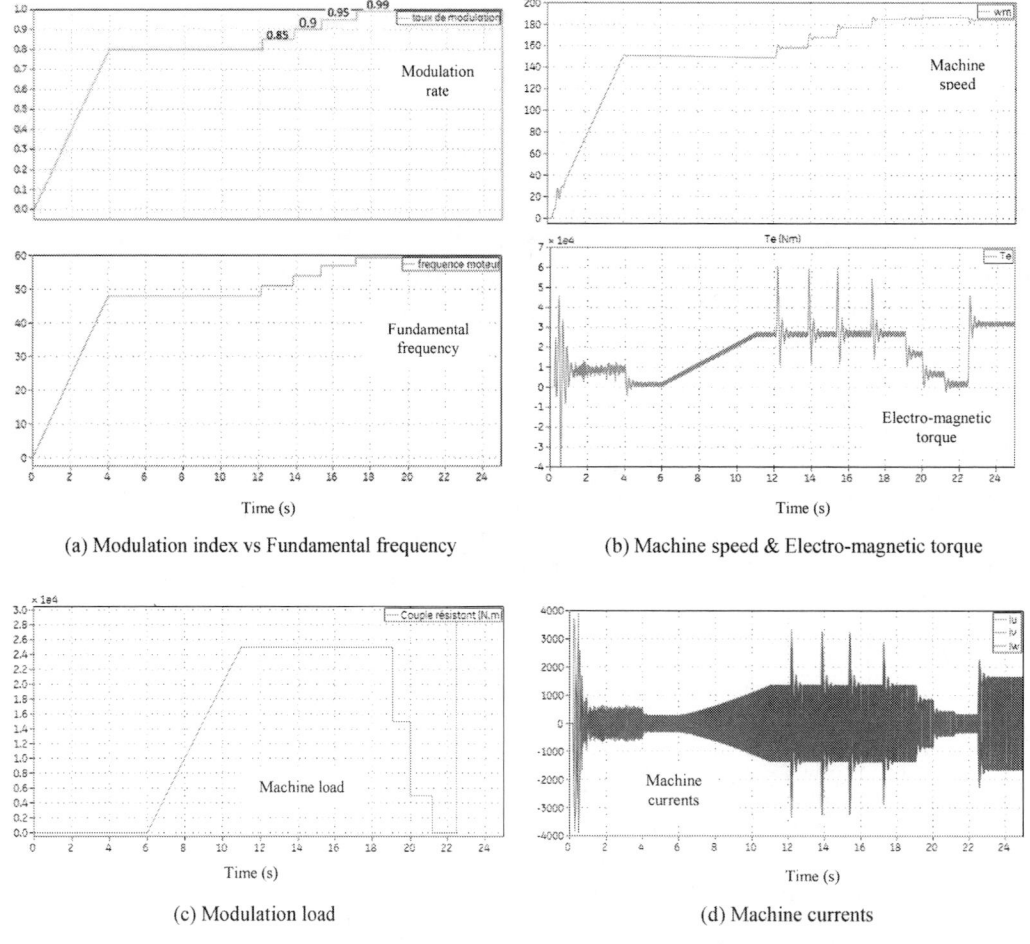

(a) Modulation index vs Fundamental frequency

(b) Machine speed & Electro-magnetic torque

(c) Modulation load

(d) Machine currents

Fig 5: Impact of the new voltage linearization method on the electrical machine quantities: Electro-magnetic torque and currents

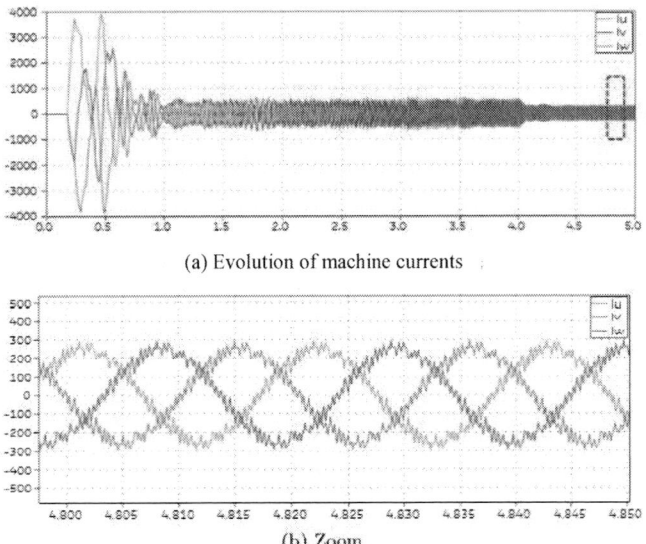

(a) Evolution of machine currents

(b) Zoom

Fig 6: Impact of the new voltage linearization method on the electrical machine quantities:
Machine currents (phase 1)

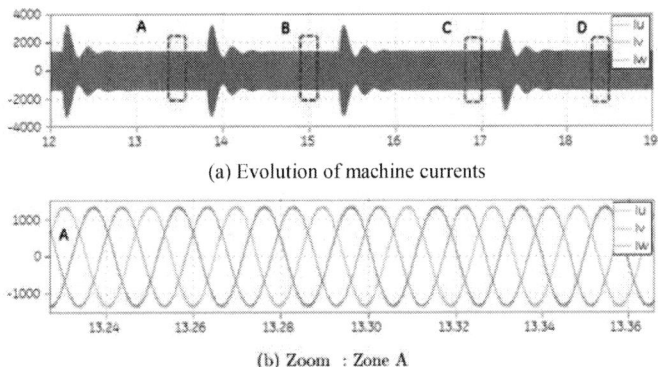

(a) Evolution of machine currents

(b) Zoom : Zone A

Fig 7: Impact of the new voltage linearization method on the electrical machine quantities:
Machine currents (phase 3)

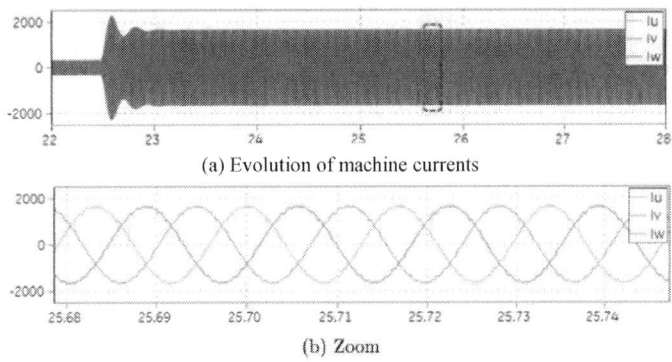

(a) Evolution of machine currents

(b) Zoom

Fig 8: Impact of the new voltage linearization method on the electrical machine quantities:
Machine currents (phase 5)

These tests (Fig.5, 6, 7 & 8) of a complete power conversion chain piloted using the new linearization method resulted in good performances over the whole speed range (modulation rate and output voltage range), and more importantly for modulation rates greater than 0.8 where the voltage references are near or within the un-authorized area. It can be pointed out that the application of the new linearization method has not caused any anomaly in the behavior of the fast dynamic quantities (stator currents and electromagnetic torque) and consequently in the behavior of the less dynamic quantities (such as speed) which are filtered by the inertia of the rotating parts. Rapid changes in motor speed and load references were performed, but this did not prevent the new linearization algorithm from performing well and delivering optimized voltage references for the power converter.

Conclusion

A novel algorithm allowing to linearize the power converters output voltage regarding their reference voltages has been proposed. Its integration in a pulse generation block of a 3-level converter as well as its compatibility with a complete power conversion chain (variable speed drive + Induction machine) have been simulated and verified. The overall operation presents very promising results, since the algorithm delivers reference voltages optimized for converter control. Note that the goal at the beginning of this study was to look for a new method for the reduction of the harmonic content of the output voltages of power converters. However, it turns out that the main advantage of this method lies in the linearization of the output voltage as a function of the input reference up to the theoretical maximum amplitude of modulating it (m = 100%).

The proposed linearization method has several advantages that can be split up in technical and commercial advantages:

A. Technical advantages
  ▪ The evolution of the fundamental (
  ▪
  ▪ Fig 4) of the converter's output voltage becomes linear regarding the modulation index. This leads to improve the control algorithms behavior because they will have the exact value of the fundamental instead of a clamped value
  ▪ Also, this new algorithm produces a significant gain in terms of maximal linear modulation index.

B. Commercial advantages
  ▪ For motor applications, the gain in modulation index means a gain of the applied voltage, and therefore, a gain of the total power
  ▪ For Active Front End applications, the gain in modulation index means a better behavior regarding High Voltage Ride Through
  ▪ In general, the gain on the modulation index is about 2/3 of the non-linear area (from the start of clamp to 1):
    o For low power / high switching frequency applications: 2000 Hz means that the non-linear area starts at about 86%. The modulation index is increased to about 95% (gain of + 9%)
    o For high power /low switching frequency applications: 750 Hz means that the non-linear area starts at about 94%. The modulation index is increased to about 98% (gain of + 4%).

Thanks to this algorithm, the converter voltage and power capabilities will increase by 4% to 9%.

References

[1] KOURO, Samir, MALINOWSKI, Mariusz, GOPAKUMAR, K., et al. Recent advances and industrial applications of multilevel converters. IEEE Transactions on industrial electronics, 2010, vol. 57, no 8, p. 2553-2580.

[2] BEN-BRAHIM, Lazhar. The analysis and compensation of dead time effects in three phase PWM inverters. In: Industrial Electronics Society, 1998. IECON'98. Proceedings of the 24th Annual Conference of the IEEE. IEEE 1998. P. 792-797.

[3] HOLMES, D. Grahame et LIPO, Thomas A. Pulse width modulation for power converters: principles and practice. John Wiley & Sons, 2003.

[4] EL JIHAD, Hamza, SIALA, Sami, SAVARIT, Elise, et al. Impact of The Carrier Phase Shift on the low Frequency Harmonics of Medium Voltage Drives. In: 2018 20th European Conference on Power Electronics and Applications (EPE'18 ECCE Europe). IEEE, 2018. p. P. 1-P. 9.

[5] EL JIHAD, Hamza, "Contribution to the study of multi-level medium voltage converters: Reduction of Low Frequency harmonics and linearization of their voltage", PhD thesis, University of Lorraine, France (2019).

Auto-commissioning of Acoustic Control of IM Drive using Bayesian Optimization

Michal Kroneisl, Václav Šmídl
Regional Innovation Centre for Electrical Engineering
University of West Bohemia
Pilsen, Czech Republic
Email: mikr@rice.zcu.cz

Acknowledgement

This research has been supported by the Ministry of Education, Youth and Sports of the Czech Republic under the project OP VVV Electrical Engineering Technologies with High-Level of Embedded Intelligence CZ.02.1.01/0.0/0.0/18_069/0009855.

Keywords

≪Induction motor≫, ≪Acoustic noise≫, ≪Predictive control≫, ≪Autotuning≫.

Abstract

Recently proposed model predictive acoustic control of an AC drive relies on availability of the transfer function from the drive current to acoustic pressure. Since such model is typically not available, we propose to consider its parameters to be unknown and use methods of Bayesian optimization to find them. Specifically, we define a acoustic control performance criterion and optimize it in an additional outer loop. A microphone is necessary in this operation to provide feedback for tuning of the parameters used in model predictive drive control algorithm. However, the microphone is removed after the tuning. Since single evaluation of the performance criteria is time consuming, we choose the Bayesian Optimization that is able to find optimal tuning with very few evaluations.

Introduction

Acoustic noise is a mostly negative phenomena associated with operation of electric drives. The noise of electric drives can be divided into 3 groups: mechanical noise (bearings, gears), aerodynamical noise (especially in air-cooled drives) and electromagnetic noise. The electromagnetic noise is caused by magnetic forces inside the motor. The electromagnetic noise depends on magnetic circuit design [1, 2] and feeding of the motor. The latter factor is noticeable when the motor is fed by an inverter in the case of adjustable speed drives. If the inverter is controlled by the Pulse Width Modulation (PWM), the voltage and current contains harmonics and the drive emits the well known "whistling" sound during acceleration or deceleration. Modulation techniques decreasing the noise tonality were developed, for example Random Pulse Width Modulation (RPWM) ([3], [4]) or PWM strategy injecting additional harmonics into the modulation signal [5]. Unfortunately these techniques cannot prevent mechanical resonances from occurring since they are not aware of their existence.

Recently proposed Acoustic MPC [6] is based on finite control set model predictive control (FCS-MPC) which provides the required switching combination directly to the inverter. Thus, it can influence even very high frequencies directly without the need for a modulator. To address the acoustic behavior of the drive, the controller needs to consider a transfer function from the drive stator currents to the acoustic pressure. Even if it is identified as proposed in [6], it still remains to fir additional model parameters and

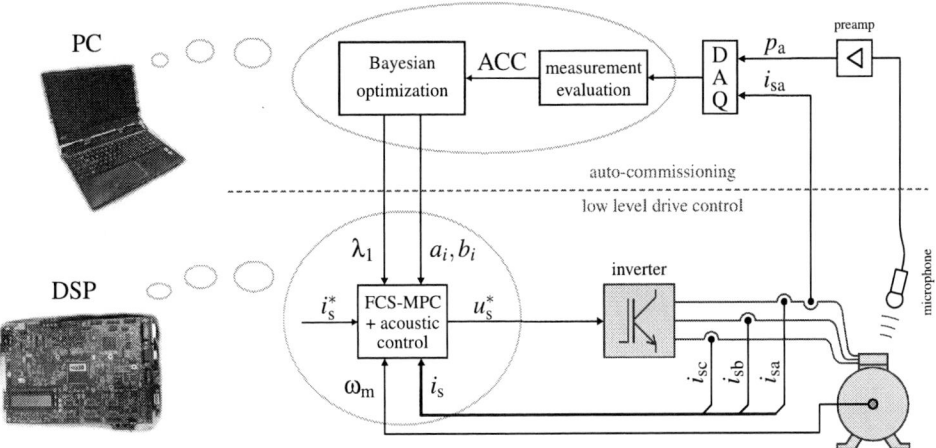

Fig. 1: Auto-commissioning of FCS-MPC acoustic control

design penalizations of the cost function. In this paper, we propose a two stage approach to model predictive control, where the upper supervisory loop (running in Matlab) provides parameters for the MPC, which runs in DSP and controls the real drive. In this paper, we demonstrate that both the penalizations and the model parameters can be tuned automatically.

We define the acoustic control cost that can be evaluated on the running drive and optimize this cost using a black-box optimization method. Since evaluation of the cost is time consuming, we seek a method that requires as few measurements as possible to find the optimum. This feature is characteristic for the Bayesian optimization (also known as Efficient Global Optimization) [7]. Other optimization techniques used in tuning of MPC controllers such as particle swarm optimization [8] typically require many more evaluations of the cost function.

The proposed approach is validated on a laboratory prototype of an induction machine drive of rated power 11 kW.

Acoustic Control of IM Drive

The acoustic drive control [6] is now briefly reviewed. We consider variable speed electric drive with an induction motor and a voltage source inverter. The drive is controlled by the FCS-MPC. The control algorithm uses mathematical model of the induction motor in rotating (dq) reference frame. State variables are composed of elements of the stator current vector $i_s = [i_{sd}, i_{sq}]$ and the rotor magnetic flux vector $\Psi_r = [\Psi_{rd}, \Psi_{rq}]$. The rotor magnetic flux is used to define the coordinate system, such that $\Psi_{rq} = 0$ and the state dynamics has only 3 equations:

$$\frac{di_{sd}}{dt} = -\alpha i_{sd} + \omega_{s\psi} i_{sq} + \beta \Psi_{rd} + \delta u_{sd}, \tag{1}$$

$$\frac{di_{sq}}{dt} = -\omega_{s\psi} i_{sd} - \mu i_{sd} - \eta \omega_{s\psi} \Psi_{rd} + \delta u_{sq}, \tag{2}$$

$$\frac{d\Psi_{rd}}{dt} = R_r \frac{L_m}{L_r} i_{sd} - \frac{R_r}{L_r} \Psi_{rd}. \tag{3}$$

Here, $\lambda = L_{s\sigma} + L_{r\sigma}\frac{L_m}{L_r}$, $\alpha = \frac{R_s + R_r\left(L_m^2/L_r^2\right)}{\lambda}$, $\beta = \frac{R_r\left(L_m/L_r^2\right)}{\lambda}$, $\delta = \frac{1}{\lambda}$, $\mu = \frac{R_s}{\lambda}$, $\eta = \frac{L_m/L_r}{\lambda}$, $L_r = L_m + L_{r\sigma}$ are constants calculated from the motor electric parameters. Angular speed $\omega_{s\psi}$ is speed of the rotor magnetic flux vector Ψ_r in stationary reference frame. For implementation in microprocessor the model is discretized using Euler approximation with sampling time T_s.

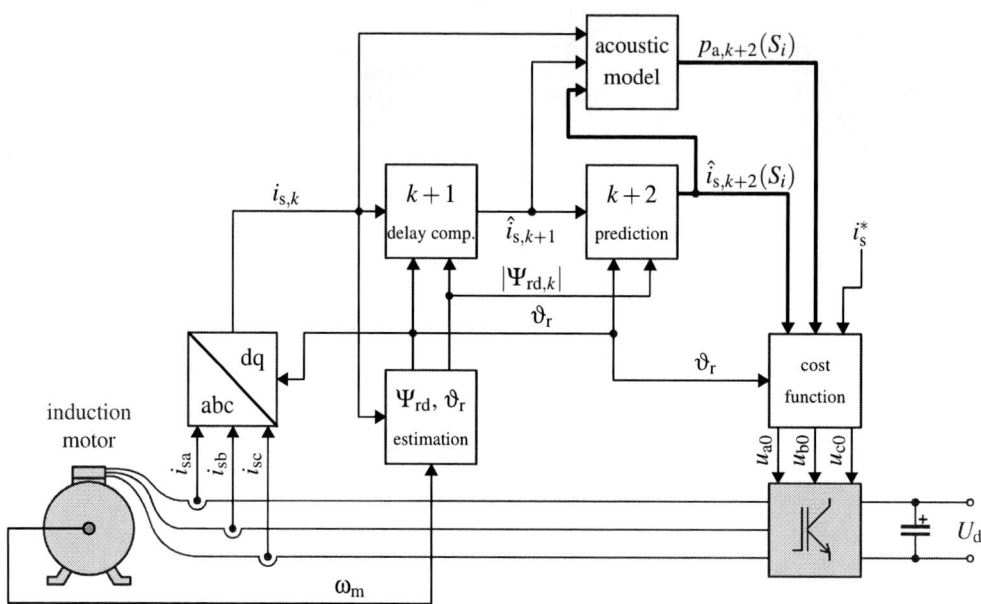

Fig. 2: Block diagram of the FCS-MPC acoustic control

Current tracking MPC

The elementary current tracking MPC is based on prediction of the stator currents in the next sampling period and selection of the switching combination that minimizes the distance of the predictions and the requested values:

$$g_{is}(x_{\text{pred}}, S_i) = \left| i_{\text{sd}}^* - \hat{i}_{\text{sd},k+2}(S_i) \right|^2 + \left| i_{\text{sq}}^* - \hat{i}_{\text{sq},k+2}(S_i) \right|^2 \tag{4}$$

where $S_i \in \{S_1, \ldots, S_7\}$ are switching combinations of the inverter. The predictions of the current require the knowledge of the rotor magnetic flux which is measurable. It is replaced by an estimate using equation (3). Equations (1) and (2) are used for prediction of stator current $i_{\text{sd}}, i_{\text{sq}}$ for all possible voltage vectors (switching combinations). The horizon of predictive control is 1 step long.

With this cost function, the control algorithm provides direct switching commands to the inverter which may remain unchanged for many time steps. Thus the algorithm does not have a fixed switching frequency and provides flat current spectrum as demonstrated in Figure 3. Since the mechanical and acoustic response of the motor depends on frequency, peaks in noise frequency spectrum can still appear, see Figure 3.

Acoustic drive control

For further improvement of the acoustic drive performance, we have to suppress the peaks in noise frequency spectrum by current spectrum shaping. Acoustic response around the peak is approximated by a simple linear transfer function

$$p_{\text{a}} = \frac{\sum_{i=0}^{n} b_i z^{-i}}{1 + \sum_{i=1}^{n} a_i z^{-i}} i_{\text{sd}}, \tag{5}$$

where coefficients $a_i, b_i, i = 1, \ldots, n$ are obtained by a filter design procedure. Specifically, we choose a band-pass Chebyshev Type I filter with center frequency 5300 Hz and passband f_b.

To minimize the acoustic pressure at this frequency, he cost function is designed as a compromise be-

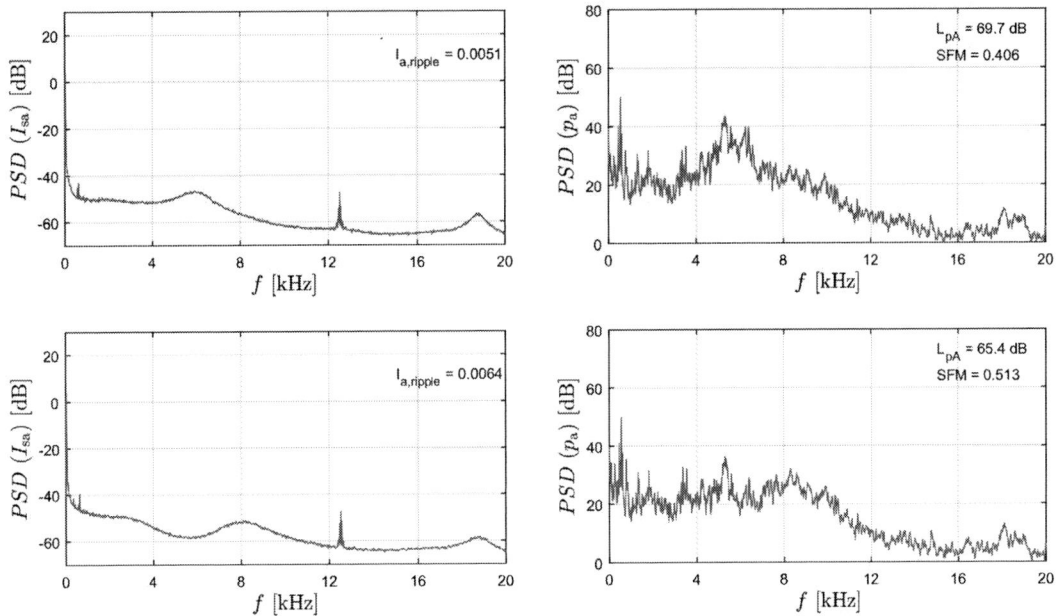

Fig. 3: Power Spectrum Density (PSD) of the stator current and the acoustic noise for basic current tracking FCS-MPC (top) and for FCS-MPC with tuned acoustic control (bottom). Measured during our experiment.

tween the current control and the acoustic noise:

$$g_{1\text{step}}(x_{\text{pred}}, S_i) = g_{is}(x_{k+2}, S_i) + \lambda_1 |p_{a,k+2}|^2 \tag{6}$$

where λ_1 is the weighting coefficient and $p_{a,k+2}$ is the output of the acoustic model (5). The complete control algorithm block diagram is shown in Figure 2.

Noise control algorithm tuning

The aim of FCS-MPC tuning is to find a balance between three contradictory criteria:

1. Low sound pressure level of the noise.

2. Flat frequency spectrum of the emitted acoustic noise (low tonality).

3. Low impact on distortion of the stator current.

For evaluation of our FCS-MPC algorithm in the auto-commissioning loop we define an Acoustic Control Cost function (ACC) that is a balanced compromise:

$$\text{ACC}(\lambda_1, f_b) = L_{\text{pA}} - 20\text{SFM} + 200I_{\text{a,ripple}}, \tag{7}$$

where L_{pA} is the A-weighted sound pressure level, SFM is the Spectral Flatness Measure and $I_{\text{a,ripple}}$ is a current distortion, which will be now defined.

The Spectral Flatness Measure express tonality of the sound. It is defined as a ratio of geometric mean and arithmetic mean of sound frequency spectrum [9]:

$$\text{SFM} = \frac{\sqrt[N]{\prod_{n=0}^{N-1} \text{FFT}(p_a, n)}}{\frac{1}{N}\sum_{n=0}^{N-1} \text{FFT}(p_a, n)}. \tag{8}$$

SFM is always between 0 and 1. Higher value of the spectral flatness means lower tonality of the sound and vice versa. We aim to make the noise of the drive less tonal (closer to white noise), so we need to maximize the SFM. The drive control cost function (6) shows that there is a trade-off between the current tracking ($g_{is}(...)$) and acoustic control. To prevent too severe degradation of the current, we evaluate the effective value of the current at frequencies higher than 70 Hz:

$$I_{a,ripple} = \sqrt{\sum_{n>70Hz} FFT^2(i_a, n)}, \tag{9}$$

We chose the 70 Hz cutoff frequency to prevent penalization of the fundamental frequency which is below 50 Hz in our system. When the current distortion is too severe due to high penalization of the acoustic response, the $I_{a,ripple}$ rises strongly. The proposed measure has similar meaning as the Total Harmonic Distortion with noise (THDn), but unlike THDn, it is not divided by the fundamental frequency magnitude which can vary according to mechanical load of the drive.

Bayesian Optimization

In each experimental measurement during auto-commissioning, we obtain information how the drive performs for a particular setting. Then we seek a new setting that has a chance to improve the performance. The Bayesian optimization approaches this task by constructing an approximation of the unknown function of performance by a non-parametric surrogate in the form of stochastic process, most often by a Gaussian process. The advantage of such choice is that it provides not only position of the most likely minima, but also uncertainty about our knowledge of the function value at any point. Thus, we can design a strategy of exploration of the unknown space. The strategy is expressed in the form of the acquisition function.

At each point, we assume that we have a set of previous results, i.e. measurements y at locations x. We seek a new point x^* where we expect maximum possible decrease of the unknown function. This is known as the expected improvement (EI) strategy. The improvement is distance at the new point x^* from the currently known minimum $z(x^*) = \min y - \mu(x^*)$,

$$EI(x^*) = z(x^*)\Phi(z(x^*), \sigma(x^*)) + \sigma(x^*)\phi(z(x^*), \sigma(x^*)) \tag{10}$$

where Φ and ϕ are the cumulative distribution and the pdf of Gaussian distribution, $\mu(x)$ and $\theta(x)$ are defined using conditional probability of a Gaussian distribution

$$\mu(x^*) = K(x^*, x)K(x, x)^{-1}y \tag{11}$$
$$\sigma(x^*) = K(x^*, x^*) - K(x^*, x)K(x, x)^{-1}K(x, x^*) \tag{12}$$

where y is a vector of available measurements. Matrices $K(x, x')$ are composed from substitutions of the new locations x and x^* into function $k(x, x')$ known as the kernel. In our work, we use Mattern 3/2 kernel

$$k(x, x') = \theta_0^2 e^{-\sqrt{3}r}(1 + \sqrt{3}r) \tag{13}$$
$$r^2 = (x, x')^T \Lambda (x - x') \tag{14}$$
$$\Lambda = diag([\theta_1, \theta_2]). \tag{15}$$

The kernel function has three hyper-parameters $\theta = [\theta_0, \theta_1, \theta_2]$ that are optimized using Type-II maximum likelihood estimation [10].

Autotuning algorithm

The autotuning algorithm is a Matlab script which runs in PC. It receives the measured data from the Data Acquisition Device (DAQ), computes the ACC index, and runs Bayesian optimization of the FCS-MPC parameters. The parameters are sent to Digital Signal Processor (DSP), which controls the real induction motor drive. The auto-tuning algorithm is now summarized in detail:

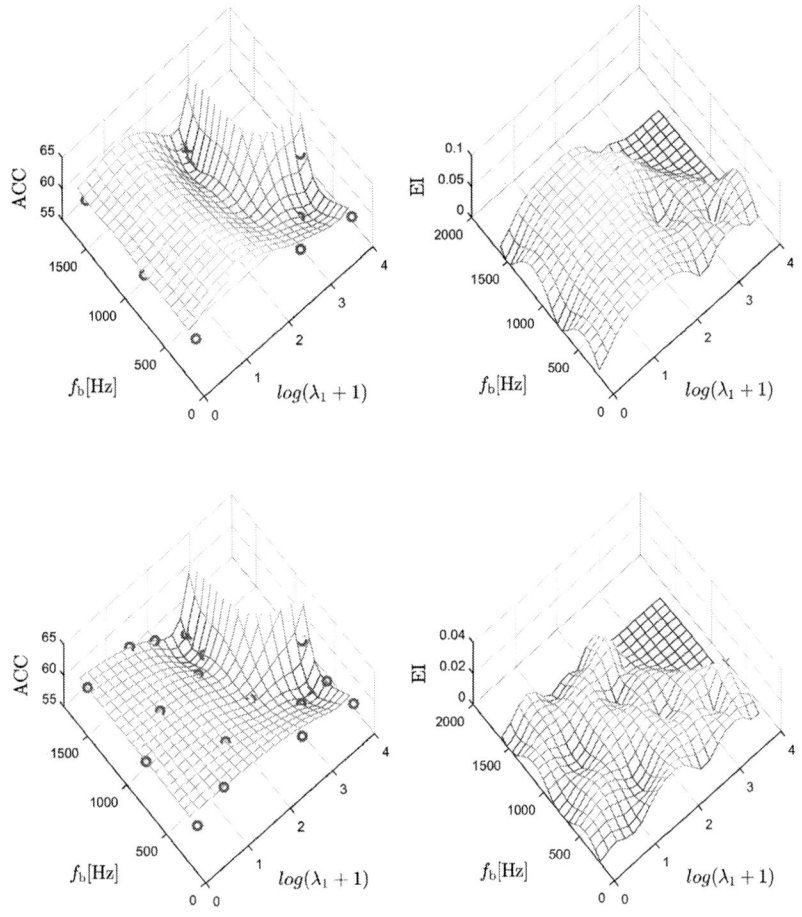

Fig. 4: Acoustic Control Cost (ACC) and Expected Improvement (EI) as functions of parameters. Iteration 11 (top) and iteration 19 (bottom).

Initialization: choose parameters and their possible ranges, choose a set of initial parameters (e.g. corners of the parameter ranges).

For a predefined number of requests do:

1. fit a Gaussian process with optimized hyper-parameters to existing results,

2. compute the expected improvement (10) on a grid of tuning parameters,

3. find tuning parameters with maximum expected improvement,

4. design filter for the selected parameters, upload filter coefficients to DSP,

5. run experiment on real IM drive with new coefficients,

6. collect experimental data from DAQ,

7. compute L_{pA}, SFM, $I_{a,ripple}$ and the Acoustic Control Cost (ACC),

8. add ACC and tuning parameters to the training set.

Experiment

The tuning parameters were chosen as the parameters of the acoustic transfer function in the form of IIR bandpass filter, and the penalization coefficient λ_1 of the MPC cost (6). The filter is Chebyshev Type I with center frequency fixed to 5300 Hz and variable passband $f_b \in \langle 200, 1600 \rangle$ Hz. The filter is designed by the Filter Design Toolbox from Matlab. The penalization coefficient λ_1 is expected to increase from zero, $\lambda_1 \in \langle 0, 5000 \rangle$. Note that the $\lambda_1 = 0$ corresponds to the conventional current tracking (4).

The parameters to be found by optimization are the passband of the acoustic filter and the penalization, $x = [f_b, \log(\lambda_1 + 1)]$. The logarithmic scale for the penalization coefficient was found to be more suitable for the task than the linear scale. The performance metric of Bayesian optimization is the Acoustic Control Cost (ACC). In every iteration the algorithm chooses new parameters with maximal Expected Improvement (EI) - see Figure 4. The first 9 iterations are made with fixed parameters grid, from iteration 10, the Bayesian optimization is used to get more precise results.

Conclusion

We proposed to tune parameters of the FCS-MPC acoustic control of IM drive using Bayesian optimization. The optimization runs on a PC in Matlab as a supervisory loop. We have demonstrated that it is able to tune the acoustic filter passband and the penalization coefficient with very few evaluations.

References

[1] D.-H. Cho and K.-J. Kim, "Modelling of electromagnetic excitation forces of small induction motor for vibration and noise analysis," *IEE Proceedings-Electric Power Applications*, vol. 145, no. 3, pp. 199–205, 1998.

[2] D.-J. Kim, J.-W. Jung, J.-P. Hong, K.-J. Kim, and C.-J. Park, "A study on the design process of noise reduction in induction motors," *IEEE Transactions on Magnetics*, vol. 48, no. 11, pp. 4638–4641, 2012.

[3] A. M. Trzynadlowski, F. Blaabjerg, J. K. Pedersen, R. L. Kirlin, and S. Legowski, "Random pulse width modulation techniques for converter-fed drive systems-a review," *IEEE Transactions on Industry Applications*, vol. 30, no. 5, pp. 1166–1175, 1994.

[4] A. M. Trzynadlowski, K. Borisov, Y. Li, L. Qin, and Z. Wang, "Mitigation of electromagnetic interference and acoustic noise in vehicular drives by random pulse width modulation," in *Power Electronics in Transportation, 2004*, pp. 67–71, IEEE, 2004.

[5] A. Ruiz-Gonzalez, M. Meco-Gutiérez, F. Perez-Hidalgo, F. Vargas-Merino, and J. R. Heredia-Larrubia, "Reducing acoustic noise radiated by inverter-fed induction motors controlled by a new pwm strategy," *IEEE transactions on industrial electronics (1982)*, vol. 22, no. 1, p. 228, 2010.

[6] M. Kroneisl, V. Šmídl, Z. Peroutka, and M. Janda, "Predictive control of im drive acoustic noise," *IEEE Transactions on Industrial Electronics*, 2019.

[7] B. Shahriari, K. Swersky, Z. Wang, R. P. Adams, and N. De Freitas, "Taking the human out of the loop: A review of bayesian optimization," *Proceedings of the IEEE*, vol. 104, no. 1, pp. 148–175, 2015.

[8] M. El Hachimi, A. Ballouk, and A. Baghdad, "Pso-mpc control of artificial pancreas," in *2018 4th International Conference on Optimization and Applications (ICOA)*, IEEE, 2018.

[9] A. Gray and J. Markel, "A spectral-flatness measure for studying the autocorrelation method of linear prediction of speech analysis," *IEEE Transactions on Acoustics, Speech, and Signal Processing*, vol. 22, no. 3, pp. 207–217, 1974.

[10] C. E. Rasmussen, "Gaussian processes in machine learning," in *Summer School on Machine Learning*, pp. 63–71, Springer, 2003.

Experimental EMI study of a 3-phase 100kW 1200V Dual Active Bridge Converter using SiC MOSFETs

Hadiseh GERAMIRAD[1,2], Florent MOREL[1], Piotr DWORAKOWSKI[1], Philippe CAMAIL[1], Bruno LEFEBVRE[1], Thomas LAGIER[1], Christian VOLLAIRE[1,2]

[1]SAS SuperGrid Institute, ITE
Villeurbanne, France.
E-Mail: Hadiseh.geramirad@supergrid-institute.com
URL: https://www.supergrid-institute.com

[2]Ecole centrale de Lyon, Ampère lab., CNRS 5005
Ecully, France.
E-Mail: Hadiseh.geramirad@ec-lyon.fr
URL: https://www.ec-lyon.fr

Acknowledgements

This work was supported by a grant overseen by the French National Research Agency (ANR) as part of the ''Investissements d'Avenir'' Program (ANE-ITE-002-01).

Keywords

«EMC/EMI», «Converter circuit», «dc/dc converters», «Silicon Carbide (SiC)», «Transformer».

Abstract

This paper presents a system-level EMI investigation in a new 3-phase 1.2kV 100kW insulated DC/DC converter. In this prototype, achieving full operation was prevented by false triggering of semiconductors due to large common-mode current through gate drivers. A passive filter embedded into the structure of the transformer to connect it to the ground is proposed as a corrective and cost effective solution to mitigate the electromagnetic interferences. New grounding layout enabled to reduce the common-mode current through gate drivers by up to 75% without decreasing the switching rate. This allowed to proceed to measurements on the prototype at nominal operation. The method is validated in simulation and extensive experimental results have been obtained from the considered prototype. This proved the feasibility of the proposed solution and showed the effect of the transformer design in EMI behavior of insulated DC/DC converters.

Introduction

Proper electromagnetic compatibility (EMC) design during development process of any type of electronic device is necessary to ensure the stable operation. The proper EMC design is a challenge when power semiconductors generate high-speed voltage and current transients. In switched mode power converters, the high switching speed of power semiconductors constitutes an important source of electro-magnetic interference (EMI). This means that the interference can disturb other devices in the close common electromagnetic environment. Therefore, electromagnetic compatibility (EMC) of the converter is a must to ensure not only its operation but also the proper function of other electronic equipment in its vicinity [1].

The common-mode (CM) current passes through the ground wire and return path which is caused usually by electrical capacitive coupling $\left(C \frac{dv}{dt} \right)$ [2,3]. Due to the complexity of the converter, the existence of many capacitive parasitic elements is unavoidable in the converter layout. These parasitic capacitances provide a low impedance path for high frequency CM current in the converter. These parasitic capacitances lead to circulating of the CM current in layout [8-10]. In switching power converters, CM current comes from the fact that switching operation involves charging and discharging of capacitances like semiconductor parasitic capacitances or capacitances in the layout of the converter. When compared to Si devices, wide band-gap semiconductors like GaN and SiC allow to operate at much higher switching frequency which leads to reduce the size of the passive components [4-6]. Alos higher switching speed enables converter designs with potentially higher efficiency. However, higher switching speed leads to fast variations of the voltages all over the layout of the converter which generates high CM current $\left(C \frac{dv}{dt} \right)$ [7]. Therefore, CM current mitigation is often a challenge in SiC or GaN based switching based converters.

Several options have been presented for mitigating EMC issues in DC-DC converters including reducing the EMI from the source and reducing the parasitic elements value [11-15]. There are plenty of researches discussing EMI issues in systems with WBG devices. Some of them deal with the package layout of the semiconductors in order to reduce the EMI [16]. Slowing down the switching speed in order to reduce the voltage variation in the layout of the converter is the most common solution but it increases losses. There are some studies focusing on the circulation of the CM current through the coupling paths [17,18]. The effect of coupling paths in redirecting and reducing the CM current from sensitive part of the circuit has been studied through the use of shielding, filtering, interconnection modification or the combination of all of them [17,18]. The coupling paths which correspond to physical layout must be taken into account seriously when it comes to the installation and the prototyping of the converter. The layout of the converter differs in each application which makes it difficult to have a general solution for different converters. Moreover, changing the parasitic elements of each components of the converter as well as the converter structure after prototype completion is time consuming and costly [19].

The CM current circulation is the major concern in this work. This current circulation was disturbing correct triggering of the semiconductors therefore it is considered as an obstacle for achieving the nominal operation of the converter. There are several studies illustrating the CM current circulation at simulation level and for small scale converters [20]. However experimental study of a high voltage/power converter has a great interest since estimating and simulating all the existing parasitic elements in this kind of converter is not an easy task. Hence this work gives an insight into the physical layout which contributes to CM current circulation in a 3-phases insulated converter. This article proposes a method which improves the operation of the converter. The method consists in adding a filter between the transformer star-point and ground. The benefits and limitations of the proposed method are presented.

This paper is organized as follows. The experimental platform is presented in the following section. Then the self-disturbance phenomenon which limited the operation of the converter is explained. After problem statement section, section introduces a solution to optimize CM current loops in the converter. The experimental results obtained from the proposed solution are depicted in this section. Finally, a conclusion is given. The article is finished by perspectives in the last section for improvements in future converter designs.

Experimental platform

The application under test in this work is a 3-phase Dual Active Bridge (DAB) converter (DAB3). In general, in DAB converters, two voltage-sources converters (VSC) are coupled through a medium-frequency transformer (MFT) (Figure 1). The leakage inductance of the medium frequency transformer is used to control power transfer. In a phase shift controlled DAB, the leakage of the transformer also allows soft switching operation with the help of output capacitance of the switches [21-22]. Besides transferring energy, MFT provides galvanic isolation (see Figure 1).

In the DAB3 prototype considered in this work, all legs inside both 3-phase VSCs are controlled to generate high-frequency square wave voltage (50% duty cycle). The voltages created by VSCs are phase shifted with respect to each other to control the power flow through the MFT. Thus the power can be transmitted from VSC1 to VSC2 (presented in Figure 1 as +P), or vice versa (shown in Figure 1 as -P). The novel 3-phase MFT prototype is presented in Figure 2. The transformer has been designed for 100kW power converter and 1.2kV [22-24]. The transformer ratio is set to 1:1 and the connection of windings is fixed to Yy (Figure 1). The 100kW SiC MOSFET based converter operates at 20kHz switching frequency.

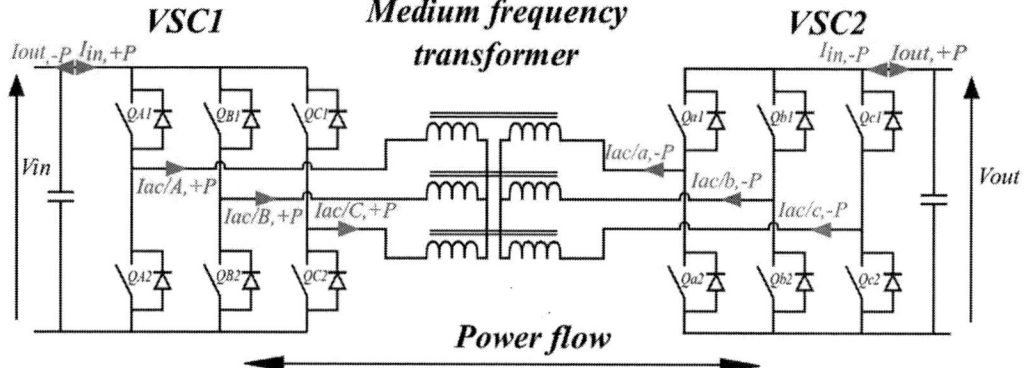

Fig. 1: 3-phase dual active bridge converter schematic

Fig. 2: 3-phase dual active bridge converter prototype [20]

Problem statement

As it has been explained in the introduction, there are many capacitive couplings in the converter layout. The schematic of the studied converter considering major capacitive couplings is shown in Figure 3. In practice, measuring the CM current passing through all the parasitic capacitances in the converter is difficult. In this paper, two groups of parasitic capacitances that contribute to the circulation of the CM current have been targeted. First, the parasitic capacitances in the structure of the transformer which allow the flow of CM current from one VSC to another one and also to the ground ($C_{interwinding}$ and C_{ng} in Figure 3). Secondly, there is a capacitive coupling in the isolated power supply of the gate drivers (C_{pg} in Figure 3) [25].

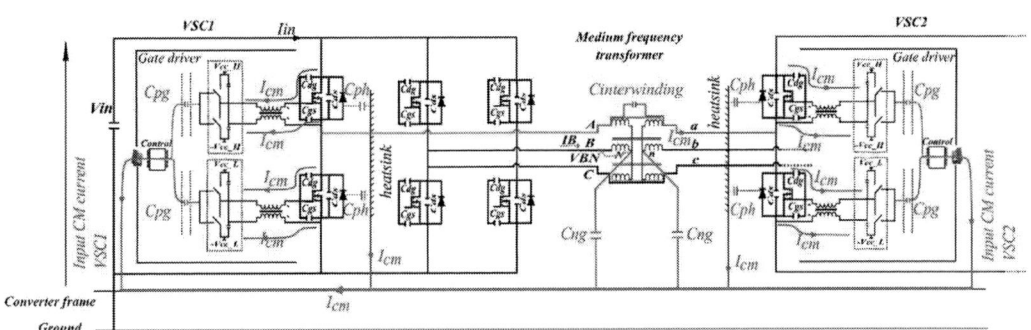

Fig. 3: Some parasitic capacitances and common-mode current paths in the converter.

The gate drivers provide an interface between the control circuit and the power semiconductors. Parasitic capacitances of the semiconductors (C_{dg} and C_{gs} in Figure 3) provide propagation paths to gate drivers for parasitic currents created by $\frac{dv}{dt}$ across the switches. With SiC MOSFETs, the high-speed high-voltage variation across switches induces a CM current which can disturb the gate bias and consequently jeopardize the correct triggering of the switch [26]. A significant amount of the CM current finds the gate drivers as a propagation path. This large CM current can continue flowing to the control circuit through the capacitive coupling in isolation stage of the gate driver. This may lead to disturbances of the control signals. This problem was preventing to reach the nominal operation of the considered converter. The altered control signal which leads to incorrect switching of SiC MOSFET is shown in Figure 4. It can be seen that the high side switch in phase B in secondary VSC at 1200V, 100kW, at the beginning of the period is not turning on which consequently stopped the operation of the converter.

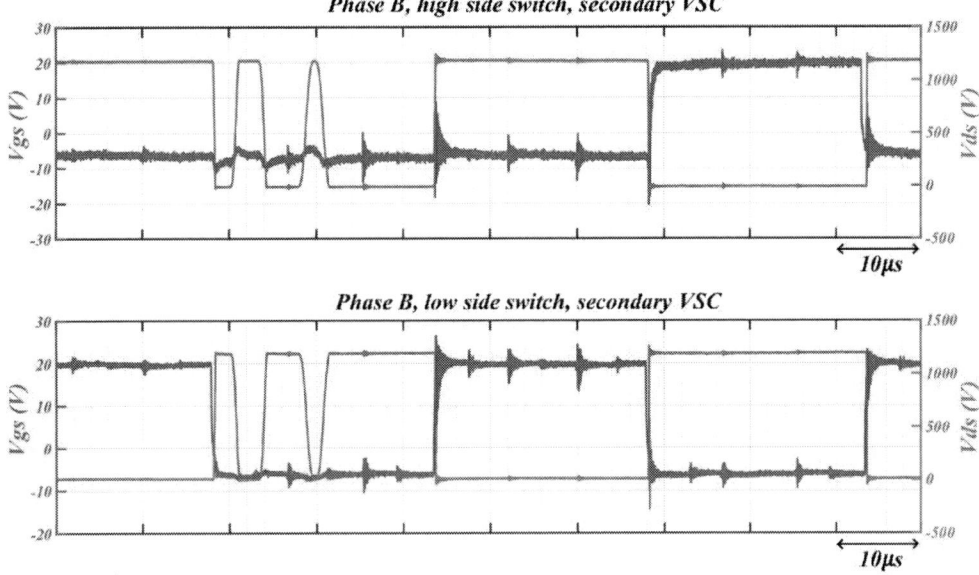

Fig. 4: Experimental result of false triggering of SiC MOSFET in secondary VSC, phase B, 1200V, 100kW.

The phenomenon occurred when the dc bus voltage was higher than 800V and when the transmitted power was reaching 100kW. The CM current of the gate driver (input CM current in Figure 3) of phase B has been measured (Figure 5).

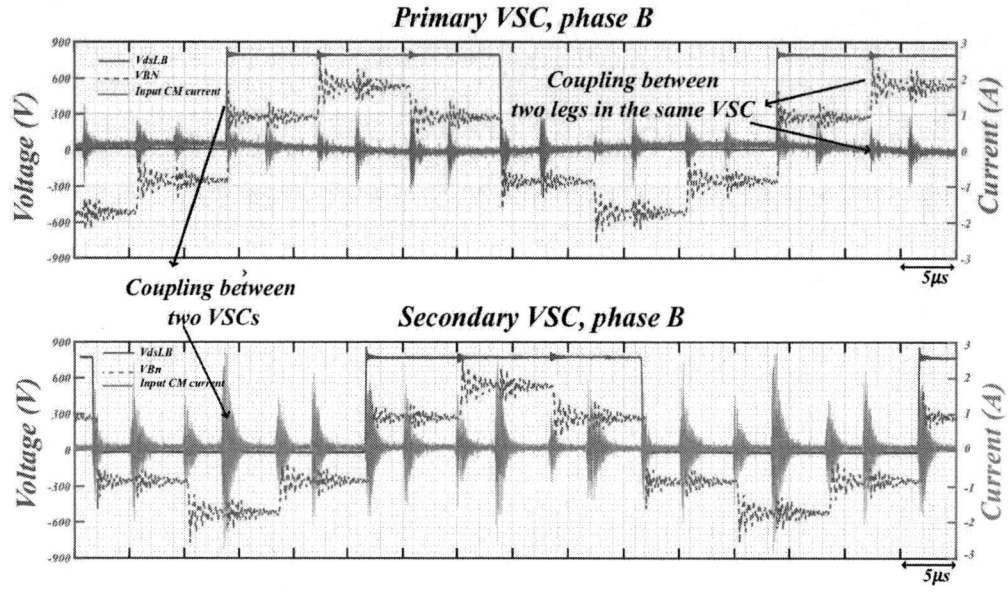

Fig. 5: Experimental results, measured CM current, 800V, 100kW.

The generated CM current pulse is repeated between the turn-on and turn-off transition of the corresponding switch. This shows that the CM current through a gate driver is not only due to the corresponding switch but also to the other switches in the same converter. Moreover, CM current pulses are observed in a gate driver power supply when a leg switches in the other VSC [27]. This shows that a current can pass from a VSC to the other. The presence of MFT parasitic capacitance ($C_{interwinding}$) explains this phenomena.

Based on above mentioned experiments, CM current through gate drivers has been suspected to be the cause of false triggering. The false triggering due to large amount of CM current is called here "self-disturbance" [26-27]. Therefore, in this work, "EMI performance" of the converter is assessed regarding to the magnitude of the input CM current of the gate drivers. In other words, solving the self-disturbance issue is addressed by reducing the magnitude of the CM current through the gate drivers.

Proposed solution and discussion

As it has been explained the parasitic elements in the layout provide paths for CM currents through gate drivers and solving EMI problems after prototyping is a challenge. In an optimized design, the solution has to be anticipated but in a completed prototype, the solution might be to redirect the CM from the victim of EMI to continue the operation of the converter. Adding a new path into the converter in order to divert the CM current from the victim of the EMI could be a better solution since the converter elements remain without changing. The influence of the parasitic elements can be reduced by creating a new circulation path for the CM current. Providing lower impedance path for CM current compared to isolation of the gate driver avoids gate drivers to experience large CM currents.

In order to limit the flow of CM current from the primary VSC to the secondary VSC and to reduce the input CM current of the secondary VSC, it is proposed to modify the electrical ground connection. In this way, the loops of the CM current are minimized. The proposed solution, is to connect the star-points (N in VSC1 and n in VSC2 in figure 3) of the MFT to the ground through a capacitor Cyg (see Figure 6). The main goal is to introduce another path to CM current in order to not pass through the gate drivers.

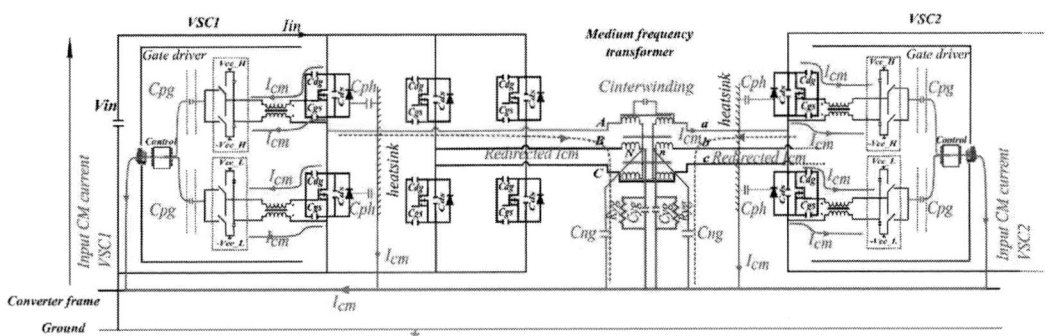

Fig. 6: Proposed solution for ground connection for CM current suppression of gate drivers (C_{yg} in green).

The MFT model parameters (see Figure 7) have been measured at 20kHz which is the switching frequency of the converter. The measurement has been done for each winding of the MFT with an impedance measurement device (see Table. I).

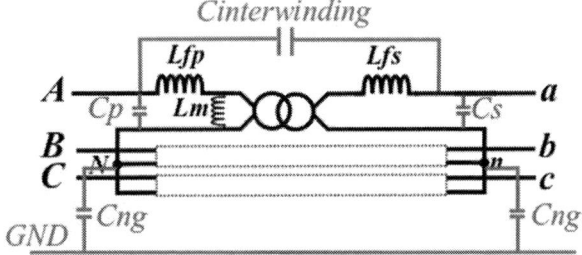

Fig. 7: MFT model [23]

Table I: MFT parameters based on impedance measurement in 20kHz [23]

Parameter	Unit	Value
L_{fp}	μH	17
L_{fs}	μH	17
L_m	mH	1
C_p	pF	30
C_s	pF	30
C_{ps}	pF	80
C_{ng}	pF	30

As it has been depicted in Table 1, the parasitic capacitance between the winding and ground C_{ng} is about 30 pF which is in the same order of magnitude as C_{pg} (20 pF) of the gate driver. The proposed additional star-point capacitor C_{yg} should have a capacitance high enough compared to the one in the galvanic isolation of the gate drivers. So as, it can provide a path of lower impedance for the CM current. Lower impedance in the ground connection makes the path preferable for the CM current to the ground and redirect the noise that was flowing through the transformer. The star-point connection prevents propagation of the CM current to the secondary VSC (see Figure 6, dashed line in green). The Cyg value of 100 pF was chosen. The DAB3 has been simulated in Matlab Simulink considering the model of the transformer (see Table. I) and including Cyg. The simulations have been carried out to evaluate the proposed solution in terms of voltage and current across the transformer. The simulation results are presented in Figure 8. The B phase voltage and current correspond to three cases: without Cyg, with Cyg and with Cyg and Ryg. It can be observed a significant voltage oscillation when the Cyg is added. However, adding the resistor Ryg damped the oscillation. The phase current is not influenced by Cyg or Ryg.

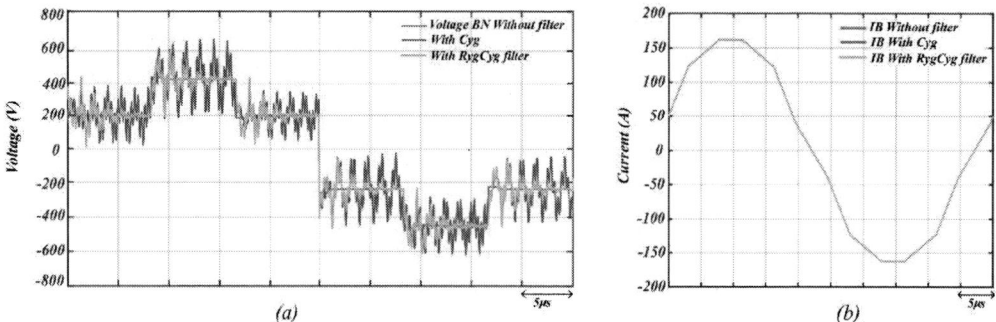

Fig. 8: Simulation results: Waveform with added filter at 1200V, 100kW (a) V_{BN} (b) I_B.

The proposed solution was applied to the prototype in order to verify its effectiveness and to measure the performance of the converter in a continuous test mode. Figures 9 illustrates the effect of the proposed connection in CM current of the gate drivers in VSC2. As it can be seen, the added capacitance shunts the CM current to the ground. Moreover, the magnitude of the transferred CM current from primary VSC to the secondary has been reduced. The peak of the CM current in secondary VSC is significantly reduced (by up to 75%) which allows converter operation above 800V. This confirms that self-disturbance was due to the level of CM current through the gate drivers.

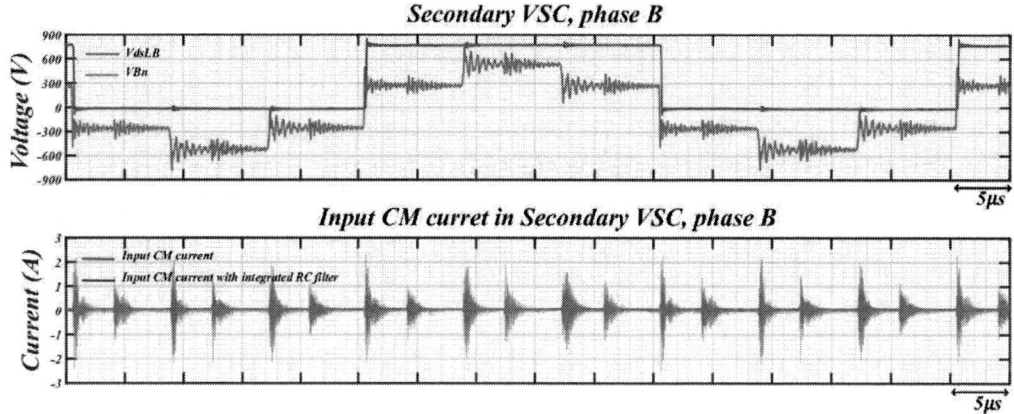

Fig. 9: Experimental results, measured CM current with integrated RC filter, 800V, 100kW.

The current and voltage in phase B have been measured and have been compared with the corresponding waveforms without added filter in order to show that they are not significantly changed due to the proposed solution. Figure 10 illustrates that the added capacitance has not changed the normal behavior of the converter. Moreover, thanks to the proposed solution it is possible to increase the voltage of the converter and reach the nominal operation of the converter (see Figure 11).

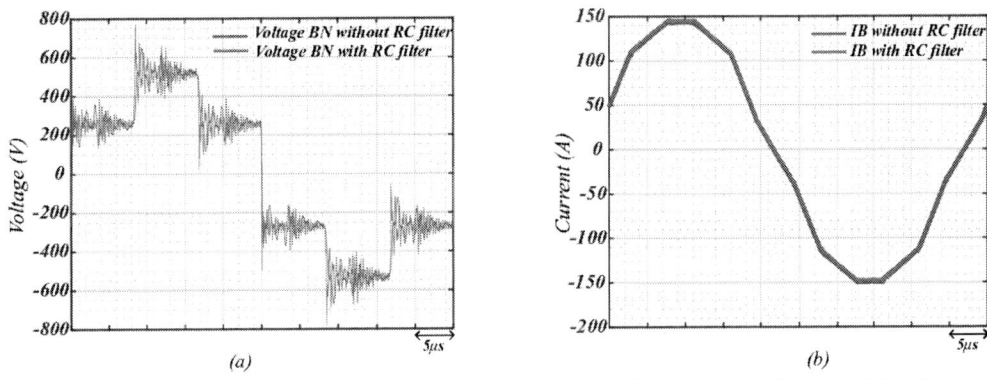

Fig. 10: Experimental results with and without RC filter at 800V, 100kW (a) Voltage in phase B (b) Current in phase B.

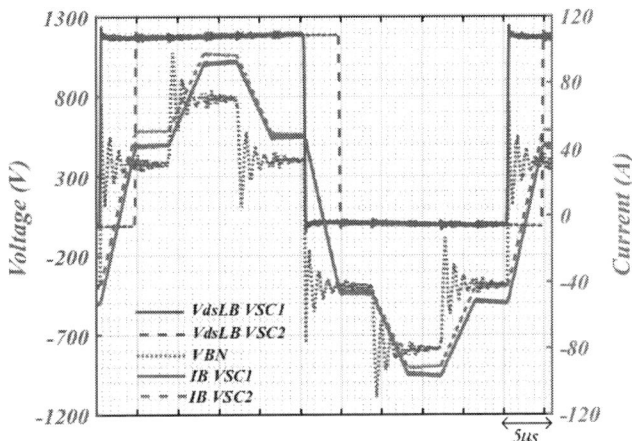

Fig.11: Experimental waveform of DAB3 with RC filter to star point of MFT, 1200V, 100kW.

Conclusion

This work gives an insight into the physical structure and the loops of the CM current in 3-phase dual active bridge prototype (DAB3). The experimental study in this work addressed the EMI problem of a 100kW, 1200V, SiC MOSFET based prototype. The big amount of the input CM current through gate drivers disturbed the control circuit of the converter. Indeed, the isolated supply of the gate drivers provides a propagation path for this current. Therefore, the amplitude of the CM current of the gate driver has been considered here as the indicator of EMI reduction for the prototype. As a corrective solution for a completed prototype, it has been proposed to connect the star-point of the transformer to the ground through a RC filter. This system-level strategy for ground connection of insulated converters allows to divert the CM current circulation from sensitive part of the system and reduce the magnitude of the CM current without decreasing switching speed. EMI behavior of the insulated DAB should be considered in an early stage of converter design, therefore, this work can help converter designer for further prototyping and installation.

The potential limitation of this study is the added resistance to the passive filter. In this specific design, implementing only the capacitance without the damping resistor adds voltage oscillations across the transformer terminals. The damping resistor reduces the voltage oscillations but it adds some power losses. Therefore, for further MFT designs the CM impedances should be considered as design constraints. At present, MFT design procedures rarely include this criteria in the optimization process. Also where the size specification allows to have a screen between primary and secondary windings, connecting the screen to the ground can minimize the CM current loop and provide a low impedance path for CM current.

References

[1] E. R. Ronan, S. D. Sudhoff, S. F. Glover and D. L. Galloway, "A power electronic-based distribution transformer," IEEE Power Engineering Review, vol. 22, pp. 61-61, 2002.

[2] M. Miloudi, A. Bendaoud, H. Miloudi, S. Nemmich and H. Slimani, "Analysis and reduction of common-mode and differential-mode EMI noise in a Flyback switch-mode power supply (SMPS)," in 2012 20th Telecommunications Forum (TELFOR), pp. 1080-1083, 2012.

[3] M. H. Nagrial and A. Hellany, "Radiated and conducted EMI emissions in switch mode power supplies (SMPS): sources, causes and predictions," in Proceedings. IEEE International Multi Topic Conference, 2001. IEEE INMIC 2001. Technology for the 21st Century., pp.54-61 2001.

[4] G. Aulagnier, M. Cousineau, T. Meynard, E. Rolland and K. Abouda, "High frequency EMC impact of switching to improve DC-DC converter performances," in 2013 15th European Conference on Power Electronics and Applications (EPE), pp. 1-9, 2013.

[5] M. R. Yazdani, H. Farzanehfard and J. Faiz, "Conducted EMI modeling and reduction in a flyback switched mode power supply," in 2011 2nd Power Electronics, Drive Systems and Technologies Conference, pp. 620-624, 2011.

[6] R. T. Naayagi, "Electromagnetic compatibility issues of dual active bridge DC-DC converter," in 2013 International Conference on Energy Efficient Technologies for Sustainability, pp. 699-703, 2013.

[7] J. L. Leslé, R. Caillaud, F. Morel, N. Degrenne, C. Buttay, R. Mrad, C. Vollaire and S. Mollov, "Optimum design of a single-phase Power Pulsating Buffer (PPB) with PCB-integrated inductor technologies," in 2018 IEEE International Conference on Industrial Technology (ICIT), pp. 782-787, 2018.

[8] J. L. Leslé, R. Caillaud, F. Morel, N. Degrenne, C. Buttay, R. Mrad, C. Vollaire and S. Mollov, "Multi-objective optimisation of a bidirectional single-phase grid connected AC/DC converter (PFC) with two different modulation principles," in 2017 IEEE Energy Conversion Congress and Exposition (ECCE), pp. 5298-5305, 2017.

[9] J. L. Lesle, R. Caillaud, F. Morel, N. Degrenne, C. Buttay, R. Mrad, C. Vollaire and S. Mollov, "Optimisation of an Integrated Bidirectional Interleaved Single-Phase Power Factor Corrector," in PCIM Europe 2018; International Exhibition and Conference for Power Electronics, Intelligent Motion, Renewable Energy and Energy Management, pp. 1-8, 2018.

[10] E. Rondon, F. Morel, C. Vollaire and J. Schanen, "Impact of SiC components on the EMC behaviour of a power electronics converter," in 2012 IEEE Energy Conversion Congress and Exposition (ECCE), pp. 4411-4417, 2012.

[11] M. R. Yazdani, H. Farzanehfard and J. Faiz, "Classification and comparison of EMI mitigation techniques in switching power converters-A review," Journal of Power Electronics, vol. 11, pp. 767-777, 2011.

[12] D. Hamza and P. K. Jain, "Conducted EMI noise mitigation in DC-DC converters using active filtering method," in 2008 IEEE Power Electronics Specialists Conference, pp. 188-194, 2008.

[13] M. Vilathgamuwa, J. Deng and K. J. Tseng, "EMI suppression with switching frequency modulated DC-DC converters," IEEE Industry Applications Magazine, vol. 5, pp. 27-33, 1999.

[14] Q. Wang, Z. An, Y. Zheng and Y. Yang, "Parameter extraction of conducted electromagnetic interference prediction model and optimisation design for a DC--DC converter system," IET Power Electronics, vol. 6, pp. 1449-1461, 2013.

[15] E. Rondon, F. Morel, C. Vollaire, M. Ferber and J. Schanen, "Conducted EMC prediction for a power converter with SiC components," in 2012 Asia-Pacific Symposium on Electromagnetic Compatibility pp. 1144-1150, 2012.

[16] B. Zhang and S. Wang, "An Overview of Wide Bandgap Power Semiconductor Device Packaging Techniques for EMI Reduction," in 2018 IEEE Symposium on Electromagnetic Compatibility, Signal Integrity and Power Integrity (EMC, SI PI), pp. 297-301, 2018.

[17] D. Fu, P. Kong, S. Wang, F. C. Lee and M. Xu, "Analysis and suppression of conducted EMI emissions for front-end LLC resonant DC/DC converters," in 2008 IEEE Power Electronics Specialists Conference, 2008.

[18] Y. H. Lee and A. Nasiri, "Analysis and modeling of conductive EMI noise of power electronics converters in electric and hybrid electric vehicles," in 2008 Twenty-Third Annual IEEE Applied Power Electronics Conference and Exposition, pp. 1952-1957, 2008.

[19] J. Liu, Y. Wang, D. Jiang and Q. Cao, "Fast prediction for conducted EMI in flyback converters," in 2015 IEEE International Conference on Computational Electromagnetics, pp. 247-249, 2015.

[20] Z. Quan and Y. Li, "Suppression of common mode circulating current for modular paralleled three-phase converters based on interleaved carrier phase-shift PWM," in 2016 IEEE Energy Conversion Congress and Exposition (ECCE), pp. 1-6, 2016.

[21] B.-R. Lin, K. Huang and D. Wang, "Analysis and implementation of full-bridge converter with current doubler rectifier," IEE Proceedings-Electric Power Applications, vol. 152, p. 1193–1202, 2005.

[22] T. Lagier and P. Ladoux, "Theoretical and experimental analysis of the soft switching process for SiC MOSFETs based Dual Active Bridge converters," in 2018 International Symposium on Power Electronics, Electrical Drives, Automation and Motion (SPEEDAM), pp. 262-267, 2018.

[23] T. Lagier, L. Chédot, F. W. L. Ghossein, B. Lefebvre, P. Dworakowski, M. Mermet-Guyennet and C. Buttay, "A 100 kW 1.2 kV 20 kHz DC-DC converter prototype based on the Dual Active Bridge topology," in 2018 IEEE International Conference on Industrial Technology (ICIT), pp. 559-564, 2018.

[24] P. Dworakowski, A. Wilk, M. Michna, B. Lefebvre, T. Lagier, "3-phase medium frequency transformer for a 100kW 1.2kV 20kHz Dual Active Bridge converter" 45th Annual Conference of the IEEE Industrial Electronics Society (IES), pp. 4071-4076, IECON 2019.

[25] L. Ghossein, F. Morel, H. Morel and P. Dworakowski, "State of the Art of Gate-Drive Power Supplies for Medium and High Voltage Applications," in PCIM Europe 2016; International Exhibition and Conference for Power Electronics, Intelligent Motion, Renewable Energy and Energy Management, pp. 1-6, 2016.

[26] H. Geramirad, F. Morel, B. Lefebvre, Ch. Vollaire and A. Breard, "Experiemental study of the self-disturbance phenomena in a half-bridge configuration of Si IGBT and SiC MOSFET switches," in PCIM Europe 2020; International Exhibition and Conference for Power Electronics, Intelligent Motion, Renewable Energy and Energy Management, 2020.

[27] H. Geramirad, F. Morel, P. Dworakowski, B. Lefebvre, T. Lagier, P. CAMAIL, C. Vollaire and A. Breard, "Conducted EMI reduction in a 100kW 1.2kV Dual Active Bridge converter," in PCIM Europe 2020; International Exhibition and Conference for Power Electronics, Intelligent Motion, Renewable Energy and Energy Management, 2020.

Modeling of a DAB under phase-shift modulation for design and DM input current filter optimization

Glauber de Freitas Lima, Yves Lembeye. Fabien Ndagijimana and Jean-Christophe Crebier
Univ. Grenoble Alpes, CNRS, Grenoble INP, G2Elab, F-38000 Grenoble, France
E-Mail: glauber.de-freitas-lima@g2elab.grenoble-inp.fr yves.lembeye@g2elab.grenoble-inp.fr fabien.ndagijimana@univ-grenoble-alpes.fr jean-christophe.crebier@g2elab.grenoble-inp.fr
URL: https://g2elab.grenoble-inp.fr/

Acknowledgements

This research work has been supported by an Auvergne Rhone Alpes FEDER funding for Mamaatec Project in partnership with MAATEL company, located in Moirans, France.

Keywords

«EMC/EMI», «Harmonics», «High frequency power converter», «Optimization», « Passive filter»

Abstract

This paper investigates the conducted Differential Mode Current (DMC) harmonic magnitudes and the Power Factor (PF) produced by a Dual Active Bridge (DAB) converter operated under single phase-shift modulation. Power factor and the first harmonic are modeled and analyzed though a dimensionless plane of static gain versus output parametrized current. It is shown that voltage and output current ranges have a significant impact on the magnitude of the reactive power and DMC filter needs, which can therefore compromise the performance of the DAB. Especially, the DMC should be considered in the DAB converter design optimization to avoid EMI filter oversizing or non-compliance with respect to EMC standards.

Introduction

Some of the future and trends of power electronic systems consists of multi-objective optimization designs such as power density, overall or peak efficiency, technology and economic viability [1]. Looking for best performance over large output voltage and/or current ranges while keeping high efficiency and power density requires a comprehensive mathematical modeling. For a power electronic converter, as presented in [2], assuring optimum power density while keeping high efficiency for the entire range of power is a challenging work as there is always a compromise between power density and efficiency. Therefore, analyzing the converter over a frame of operating points: static gain, input and output voltages and currents are fundamental for design optimization.

Particularly, it is known that the design of a DAB under phase-shift modulation requires a careful choice of the phase-shift, switching frequency and ac link inductance regarding reactive power [3], ZVS operation [4] and high-efficiency. Commonly, in the literature, these issues are usually treated with respect to output power [4], [5], making the analysis more specific to a single design. Besides output power information does not differentiate low voltage and high current from high voltage and low current applications. As it is demonstrated through the paper, for a DAB converter, even at same output power operating points, getting the input and output voltage and current information is essential to analyze its performance characteristics. Similar to [6], the plane of the static gain versus an output parametrized current will be used for the sake of a generic, dimensionless and still meaningful analysis. It is presented the modeling in the frequency domain of the input DMC normalized current of a lossless DAB, in such a way that the performance characteristic of power factor and DMC normalized harmonic impacts can be observed through this suitable plane. As it is presented a lossless theory, the power factor will represent the notion of efficiency, as the amount of reactive power impacts on both the efficiency [7] and the power density, whereas the first harmonic appearing in the lower limit of the frequency spectrum represent the effort for filtering to comply with regulations. Therefore, both analysis of power factor and DMC first harmonic represent how much the converter in a general way (filter, isolation, cooling, and etc.) should be oversized for a given output voltage and current operating ranges. Giving the desired input voltage, output voltage and current ranges, the designer will be able to find an optimum design with respect to different parameters as well as to identify the critical regions and limitations found in a DAB. To overcome these natural limitations or to enhance its performance and keeping same or similar converter topology, different strategies can be implemented, such as different dual phase-shift modulation [8], variable-tap transformer [9], frequency modulation [10] or even a combination of phase-shift and frequency modulation [11].

Besides proposing a guideline to design a DAB and to characterize its performance, the methodology proposed here can be used for other any converter in which analysis can be done using the same or similar output parametrized current [12], [13]. The dimensionless analysis therefore allows actual comparison among different power converters where tradeoffs become clearer to observe.

The assumptions made about the DAB presented throughout this paper are the following:
- The DAB is operated with classical single phase-shift modulation [14];
- The input voltage is assumed to be a constant voltage source;
- Only steady state operation is considered;
- Non-idealities such as propagation delays, dead-times [15], series parasitic resistances [16], magnetic and switching losses [17], [18] are not modeled;
- The ac link capacitance is considered to be infinite and its effect [19] is neglected;
- The magnetizing inductance is considered to be infinite and its effect [6] is also neglected
- Profile charge (light load and heavy load) has equal importance for all operating points;
- Even though bidirectional, only unidirectional power flow analysis are depicted;
- A simple lossless theoretical analysis is presented, and therefore, it does not allow finding separately the optimum switching frequency f_s and ac link inductance L_{ac}, but rather its product $f_s L_{ac}$.

Dual Active Bridge DMC Modeling in the Frequency Domain

The main waveforms of the DAB converter when operated under phase shift control are recalled. Fig. 1 (a) shows the topology of the DAB converter and the main voltage and current waveforms under phase-shift operation. Voltage waveforms correspond to AC squared voltages produced by the two H-bridges and applied on the input and output sides of the AC link. A generic input current of the DAB is described in Fig. 1 (b).

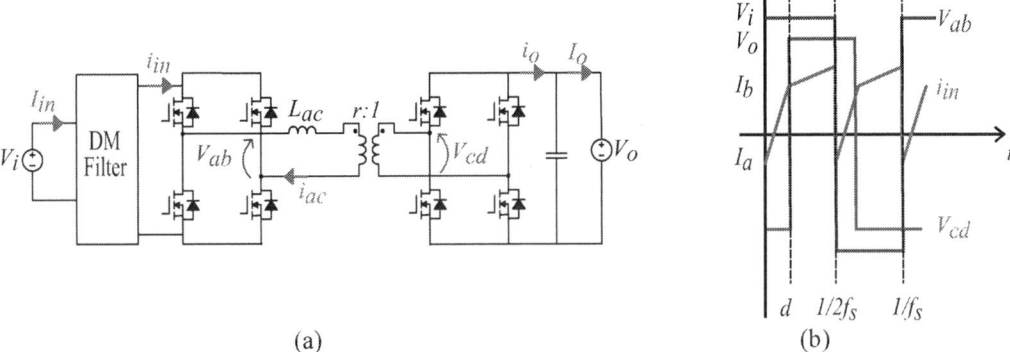

| (a) | (b) |

Fig. 1: (a) Ideal electrical equivalent circuit of a DAB; (b) Main time domain waveforms of a DAB.

In order to present an analysis that is generic, dimensionless and independent of the load resistance, a parametrization of the output current is presented in (1). The reconstruction of the input current in the time domain can then be normalized as presented in (4)-(7).

$$\gamma = \frac{2 f_s L_{AC} I_o}{r V_i} \tag{1}$$

$$M = \frac{r V_o}{V_i} \tag{2}$$

$$d = \frac{\alpha}{180^\circ} \tag{3}$$

Where:

- V_i is the input voltage (V);
- V_o is the output voltage (V);
- I_o is the average output current (A);
- f_s is the switching frequency (Hz);
- L_{ac} is the AC link inductance (H);
- α is the phase-shift in degrees from 0° to 180°;
- r is the transformer turns ratio.

$$\bar{I}_{in}(M,d,\gamma) = \frac{I_{in}(M,d,\gamma)}{I_o} = \frac{Md(1-d)}{\gamma r} \tag{4}$$

$$\bar{I}_a(M,d,\gamma) = \frac{I_a(M,d,\gamma)}{I_o} = \frac{M(1-2d)-1}{2\gamma r} \tag{5}$$

$$\bar{I}_b(M,d,\gamma) = \frac{I_b(M,d,\gamma)}{I_o} = \frac{M+2d-1}{2\gamma r} \tag{6}$$

$$\frac{i_{in}(M,d,\gamma,t)}{I_o} = \begin{vmatrix} \dfrac{i_1(M,d,\gamma)}{I_o} = \dfrac{2f_s(1+M)t}{\gamma r} + \bar{I}_a(M,d,\gamma) \quad 0 \le t \le \dfrac{d}{2f_s} \\[3mm] \dfrac{i_2(M,d,\gamma)}{I_o} = \dfrac{2f_s(1-M)(t-\frac{d}{2f_s})}{\gamma r} + \bar{I}_b(M,d,\gamma) \quad \dfrac{d}{2f_s} \le t \le \dfrac{1}{2f_s} \end{vmatrix} \tag{7}$$

Where:

- t is the time (s).

Applying Fourier series theory, the DM input current can be described in the frequency domain. The normalized coefficients are shown in (8), (9) and the reconstruction of the normalized input current, back to the time domain is then presented in (10). Notice that the frequency of the input current corresponds in theory to the double of the switching frequency as illustrated in Fig. 1 (b). Finally, it is presented the normalized values of the input harmonic current in (11). From it, the RMS value of the DMC and power factor are developed as presented in (12) and (13), respectively.

$$\bar{a}_n(M,d,\gamma,n) = \frac{a_n(M,d,\gamma)}{I_o} = -\frac{2\sin(\pi dn)^2 M}{\pi^2 n^2 \gamma r} \tag{8}$$

$$\bar{b}_n(M,d,\gamma,n) = \frac{b_n(M,d,\gamma)}{I_o} = \frac{M\sin(2\pi dn) + M\pi n - 2M\pi dn - \pi n}{\pi^2 n^2 \gamma r} \tag{9}$$

$$\frac{i_{in_{Fourier}}(M,d,n,f_s,\gamma,t)}{I_o} = \bar{I}_{in}(M,d,\gamma) + \sum_{n=1}^{\infty}\left(\bar{a}_n(M,d,\gamma,n)\cos(4\pi nf_s t) + \bar{b}_n(M,d,\gamma,n)\sin(4\pi nf_s t)\right) \tag{10}$$

Where:

- n is the harmonic order;

$$\bar{H}_{n_{in}}(M,d,\gamma,n) = \frac{H_{n_{in}}(M,d,\gamma,n)}{I_o} = \sqrt{\frac{\bar{a}_n(M,d,\gamma,n)^2 + \bar{b}_n(M,d,\gamma,n)^2}{2}} \tag{11}$$

$$\bar{I}_{in_{RMS}}(M,d,\gamma) = \frac{I_{in_{RMS}}(M,d,\gamma)}{I_o} = \sqrt{\bar{I}_{in}(M,d,\gamma)^2 + \sum_{n=1}^{\infty}\left(\bar{H}_{n_{in}}(M,d,\gamma)\right)^2} \tag{12}$$

$$\mathrm{Pf}(M,d) = \frac{\bar{I}_{in}(M,d)}{\bar{I}_{in_{RMS}}(M,d)} \tag{13}$$

The value of the phase-shift is not as significant, since in current controlled applications the feedback control imposes the desired phase-shift value. Therefore, knowing that the parametrized output current is given as in (14), the value of the phase-shift in equations (4) to (13) so far presented is substituted as a function of the parametrized current presented in (15). Besides, from now on the Power factor and DMC first harmonic analysis become dependent on only two variables: M and γ. Notice that two different values of the phase-shift are possible for the same output parametrized current. The domain of the phase-shit is comprehended from $0°$ to $180°$ for delivering power. However, it is known that there are no benefits in operating with phase-shit larger than $90°$ [3], that is d larger than 0.5.

$$\gamma = d(1-d) \tag{14}$$

$$d = \frac{1 \pm \sqrt{1-4\gamma}}{2} \Rightarrow \frac{1-\sqrt{1-4\gamma}}{2} \tag{15}$$

As d is a real number, it can be shown that the parametrized output current has a maximum value limited by 0.25, shown in (16). This shows that the possible values of the product $f_s L_{ac}$ are limited as presented in (17).

$$\gamma \leq \frac{1}{4} \tag{16}$$

$$f_s L_{ac} \leq \frac{rV_i}{8I_o} \tag{17}$$

As in [3], a necessary condition for achieving ZVS in the first and second H-bridges is that the currents $\overline{I_a}$ and $\overline{I_b}$ to be larger than zero, respectively. Considering (15), ZVS conditions in the first and second H-bridges follow as presented in (18) and (19), respectively.

$$M \leq \frac{1}{\sqrt{1-4\gamma}} \text{ if } M \geq 1 \tag{18}$$

$$M \geq \sqrt{1-4\gamma} \text{ if } M \leq 1 \tag{19}$$

Power factor and DMC first harmonic analysis

With (13), it is possible to evaluate the power factor of the DAB converter under single phase-shift modulation, as presented in Fig. 2 (a). Notice that even though a smaller γ (e.g. 0.01) presents better power factor, they become restricted to a unitary gain. Otherwise, the power factor is drastically degraded for a non-unitary voltage gain. For larger values of γ (e.g. 0.2) even though it allows larger voltage variation without highly compromising power factor, the unitary voltage gain in a given application would result in a power factor less than 0.8 usually. Therefore, it is advisable that the intervals in the γ axe [0 0.02] and [0.2 0.25] to be avoided if possible, while most optimum designs will be found in the interval [0.02 0.2]. This means that a maximum phase-shift equal to approximately $50°$ should be considered, while phase-shit as low as $3.7°$ will be very inefficient at non-unitary voltage gains, and therefore, another strategy such as, different modulations as aforementioned should be introduced when entering those regions. Considering this safe operation interval [0.02 0.2], it is already possible to discriminate the limitation of a DAB converter operating with ordinary single phase-shift modulation of being able to deliver safely, but not necessarily efficiently a minimum current of 10% of maximum current.

The plot of the normalized first harmonic distortion is presented in Fig. 2 (b). Notice that, as in the power factor results, there is a similar region in which γ and M should be prioritized; otherwise, the effort of the filter design starts to increase significantly. Another important conclusion to be pointed out is that, by inspection it is noticeable that the normalized values are quite symmetrical over the unitary gain.

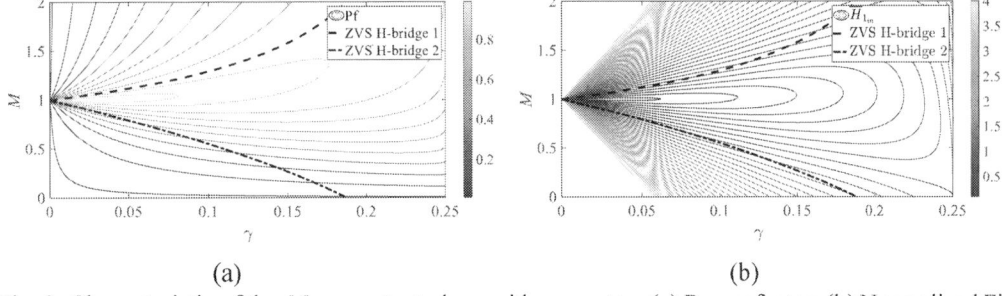

(a)   (b)

Fig. 2: Characteristic of the M vs γ output plane with respect to: (a) Power factor; (b) Normalized First harmonic. Including ZVS boundaries for primary and secondary H-Bridges.

Parametrized output load plane

The analysis of the power factor and first harmonic will be observed over the parametrized output plane shown in Fig. 2. On it, it will be placed a rectangular region formed by the intervals $[M_o \ M_f]$ (y axe) and $[\gamma_o \ \gamma_f]$ (x axe) as presented in Fig 3 (a). The domain in which the output parametrized current will operate is directly related to the percentage of minimum output current allowed, as presented in (20). Therefore, from (16), the maximum value of γ_o as a function of $I_{o\%}$ is presented in (21). For convenience, the parameter $M_{o\%}$ is defined in (22).

$$\gamma_o = I_{o\%}\gamma_f \tag{20}$$

$$\gamma_o \leq \frac{I_{o\%}}{4} \tag{21}$$

$$M_{o\%} = M_f - M_o \tag{22}$$

Where:

- M_o is the minimum static gain desired;
- M_f is the maximum static gain desired;
- γ_o is the initial parametrized output current;
- γ_f is the initial parametrized output current;
- $I_{o\%}$ is the minimum current desired in percentage of the rated current;
- $M_{o\%}$ is output voltage variation in percentage.

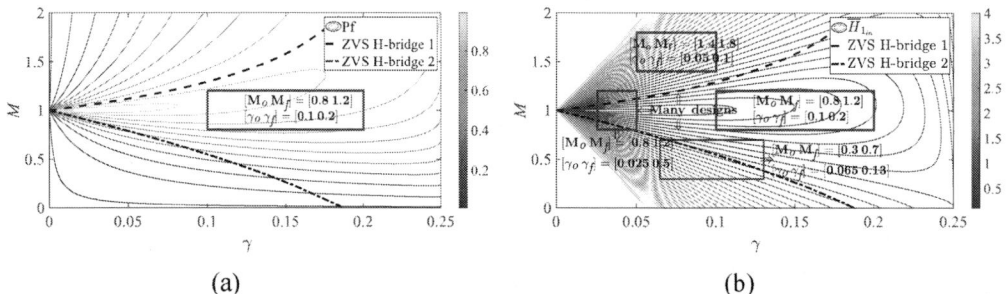

(a)	(b)

Fig. 3: Characteristic of the output plane M vs γ for $M_{o\%} = 0.4$ and $I_{o\%} = 0.5$ with respect to: (a) Power factor (one single design); (b) Normalized DMC first harmonic (different designs).

Then, the power factor and the first harmonic contained in this region will be analyzed and discussed further. It is important to know that, depending on the choice of the parameters, this rectangular region can be wider or not, and not symmetrical over the unitary gain. For example, γ varying [0.1 0.2] and M varying [0.8 1.2] allows minimum load of 50% of rated current and 40% of voltage variation, while γ varying [0.05 0.1] and M varying [1.4 1.8] allows the same 50% and 40% of variation, as presented in Fig. 2 (b), even though the area of the later region on the output plane is halved and shifted on the plane. A direct consequence from it is that a smaller γ_o allows designing a converter capable of delivering more range of output current variation more easily than larger ones. However, it should be noted that direct consequences with respect to ZVS region, power factor and first harmonic should be taken into account.

The parametrized plane underlines important observations in a straightforward way. For example:

- Since same output power can be expressed for different values of M, I_o and V_i, it is important to differentiate each scenario as they can result in completely different power factor or harmonic distortion;
- Besides, still considering same output power, high input voltage and low output current applications will necessarily require an increase of the product f_sL_{ac} to keep the same performance of low input voltage and high output current applications;
- To maintain the same power factor and DMC harmonics (i.e. to keep the same performance) insensitive to input voltage variations, a simple frequency modulation proportional to the input voltage variation appears to be very suitable.

Characterization and performance optimization of the DAB converter

As it can be seen for a given output current variation $I_{o\%}$ and voltage gain variation $M_{o\%}$ there is a set of $[\gamma_o \ \gamma_f]$ and $[M_o \ M_f]$, to be chosen that will result in an optimized design either with respect to power factor and/or first harmonic distortion. In this paper, a charge profile is related to this rectangular frame, but any other one, with different geometric formats, such as a battery charger could be analyzed and optimized through the presented methodology.

Power factor evaluation

To cover the entire rectangular region, a parameter Pf_{vol} is proposed as the integration in volume of the power factor divided by the volume of a cube that contains such region, as presented in (23) and depicted in Fig. 4 (a). By sweeping the values of M_o and γ_f for $I_{o\%} = 0.5$ and $M_{o\%} = 0.4$, it is possible to notice that there is a global maximum at $M_o = 0.922$ and $\gamma_f = 0.148$, resulting in $Pf_{vol} = 0.9$ as presented in Fig. 4 (b). However, considering that the gain is around the unitary gain, e.g. $M_o = 0.8$ and $M_o = 1.2$, equation (23) is adapted into (24), and it is observed a maximum point at $\gamma_f = 0.128$, resulting in $Pf_{vol} = 0.88$. An important conclusion is that these values mean that the turns ratio r can be used as an optimizing variable to produce such gains, and that symmetry over the unitary gain does not necessarily brings the best results with respect to this parameter Pf_{vol}.

$$Pf_{vol}\left(M_o,\gamma_f\right) = \frac{\int_{M_o}^{M_o+M_{o\%}} \int_{\gamma_f I_{o\%}}^{\gamma_f} Pf\left(M,\gamma\right) dM \, d\gamma}{M_{o\%}\gamma_f\left(1-I_{o\%}\right)} \quad \text{if } 0 \leq \gamma_f \leq 0.25 \tag{23}$$

$$P.f_{vol}\left(M_{o\%},\gamma_f\right) = \frac{\int_{1-\frac{M_{o\%}}{2}}^{1+\frac{M_{o\%}}{2}} \int_{\gamma_f I_{o\%}}^{\gamma_f} Pf\left(M,\gamma\right) dM \, d\gamma}{M_{o\%}\gamma_f\left(1-I_{o\%}\right)} \quad \text{if } 0 \leq \gamma_f \leq 0.25 \tag{24}$$

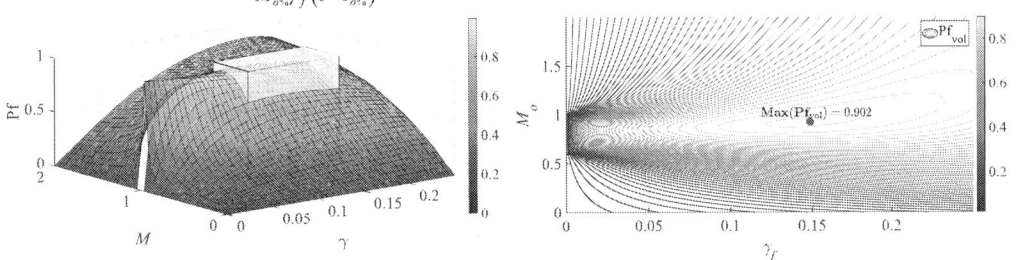

Fig. 4: (a) The volume of the power factor over a given range in the output plane; (b) The results of the parameter defined in (23) for $I_{o\%} = 0.5$ and $M_{o\%} = 0.4$.

Nevertheless, still considering this symmetry over the unitary static gain as in (24), for each $M_{o\%}$ varying from 0 to 1 and $I_{o\%}$ varying from 0 to 1, it is possible to take the optimum γ_f that will deliver the maximum value of Pf_{vol}, as presented in in (25) and depicted in Fig. 5 (a). For every optimum γ_f (25), and substituting its value in (24), it is possible to observe the results with respect to Pf_{vol} in which plot results is presented in Fig. 5 (b).

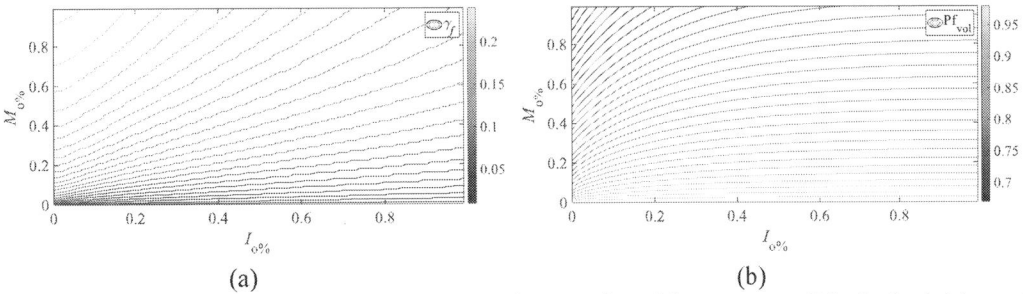

Fig. 5: (a) Values of γ_f (25) that results in the maximum value of the parameter Pf_{vol} for both $M_{o\%}$ and $I_{o\%}$ varying from 0 to 1; (b) Maximum value of Pf_{vol} considering (25).

$$\gamma_f\left(M_{o\%},I_{o\%}\right)=root\ \frac{d}{d\gamma_f}\left(\frac{\int_{1-\frac{M_{o\%}}{2}}^{1+\frac{M_{o\%}}{2}}\int_{\gamma_f I_{o\%}}^{\gamma_f}\mathrm{Pf}\left(M,\gamma\right)\mathrm{d}M\ \mathrm{d}\gamma}{M_{o\%}\gamma_f\left(1-I_{o\%}\right)}\right)\ \text{if } 0\le\gamma_f\le0.25 \tag{25}$$

Besides providing an optimization process, these plots show the natural limitation found in a DAB converter with simple phase-shit modulation regarding voltage and current range extension. Again, to overcome these issues, different techniques should be performed when entering in critical areas.

DMC first harmonic evaluation

From Fig. 6 (a), it is possible to realize that the worst-case normalized values are always located in the extreme coordinates (M_o,γ_o), (M_o,γ_f), (M_f,γ_o) and (M_f,γ_f). In order to identify more precisely, the first harmonic is verified for each voltage gain as presented in Fig 6 (a). Notice that, considering voltage variation over the unitary gain, the step-down mode results in equal or larger values than the step-up mode. For example, among the static gains varying from 0.8 to 1.2, the value of 0.8 is the one that results the highest value. Considering $M=0.8$, the value of γ that returns the least value for first harmonic is 0.138 resulting in $\overline{H}_{1_{in}}=0.45$. Once the value of γ_f has been chosen, the parameter in (26) allows to identify the consequences of decreasing current variation $I_{o\%}$, at a given voltage gain M. In Fig. 6 (b), it is presented the plot results of (26) for $M=0.8$ and a set of different γ_f.

$$\overline{\overline{H}}_{1_{in}}\left(M,\gamma_f,I_{o\%},1\right)=I_{o\%}\overline{H}_{1_{in}}\left(M,\gamma_f\times I_{o\%},1\right) \tag{26}$$

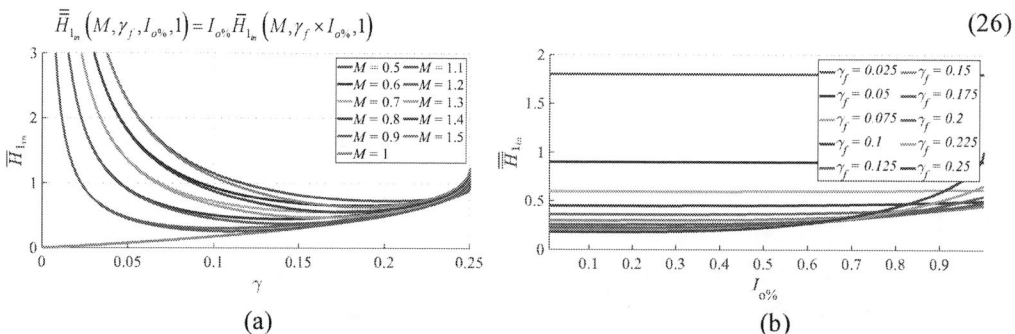

Fig. 6: (a) The value of the normalized DMC first harmonic for different static gains; (b) The effect of decreasing current for $M=0.8$ and different γ_f.

It is possible to demonstrate that lower currents will never affect more than the maximum designed current, and therefore, the filter design shall be designed considering the rated current as well as its worst voltage gain scenario, that is (M_o,γ_f). By inspection, notice that there is an interval where most γ_f should be designed contained in the interval from 0.1 to 0.2. Indeed, in this interval, the normalized DMC first harmonic is quite constant, especially for $M=0.8$.

DMC filter design

In this section, it is assumed that the frequency of the first harmonic will be larger than 150 kHz, that is, the lower limit of frequency band defined by CISPR class A and B. Therefore, if lower switching frequencies than 75 kHz are applied, it is necessary to consider the next highest harmonic order n larger than 150 kHz in the equations so far presented. Equation (11) will be brought to its absolute value and transformed into $dB\mu V$ in 50 Ω system for better visualization when designing the filter to meet CISPR emission limits, as presented in (27). The filter should be calculated for its worst-case condition, e.g. $M_o=0.8$. Then, it should be chosen the value of γ_f, e.g., 0.14 resulting in $\overline{H}_{1_{in}}=0.45$ as presented in previous section. Finally, the filter becomes a function of the attenuation required to comply with the EMC standards, approximately 60 $dB\mu V$, as presented in (28). As an example, $I_o=5$ A, $M_o=0.8$ and $\gamma_f=0.14$ results in an attenuation required of about 100 dB.

Then, as in [2], the attenuation value is used to calculate the filter that could be e.g., 1- or 2- LC filter structures, as depicted in Fig. 7 (a). The compromise of choosing each structure is related to the cutoff frequency and its required total realization effort, as shown in Fig. 7 (b). In general, higher filter orders enable designing a filter with a higher cutoff frequency, resulting in structures with smaller total volume. However, a careful study regarding parasitic resonances and loss effects in a filter should be taken into account for the sake of precisely optimization.

$$dB\mu V_{H_{1_{in}}}\left(M_o,\gamma,1\right)=20\log\left(\bar{H}_{1_{in}}\left(M,\gamma,1\right)\right)+20\log\left(I_o10^6\right)+20\log\left(50\right) \qquad (27)$$

$$Filter_{att}=dB\mu V_{H_{1_{in}}}\left(M_o,\gamma,1\right)-60dB\mu V \qquad (28)$$

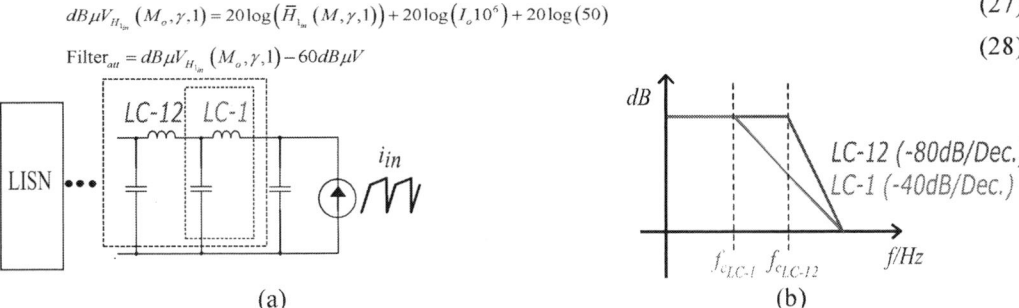

(a)   (b)

Fig. 7: (a) An example of filter design with 1 - or 2 - LC structures; (b) Filter attenuation per decade and the cutoff frequency for 1 - and 2 - LC structures.

Considering careful designs, the value of $\bar{H}_{1_{in}}$ can be roughly be approximate as worst cases equal to one, as in db such value does not impact more than 6 db if even deviated by twice as much the real value. This also means that the output current is the one that interferes the most in the filtering design.

Practical results

A DAB converter, designed for low voltage and power applications [20], [21], [22] (10 V to 20 V and 100 W) presented in Fig. 8 (a), was used to conduct practical results with respect to power factor. Considering its AC link inductor of 700 nH, an output current of 3 A and 1.5 A and input voltage of 10 V, the switching frequency were calculated as 330 kHz, in such a way that the parameter γ_f is set at 0.14 and γ_o is at 0.07. In Fig. 7 (b) it is presented the waveform of current measured in the AC link when I_o = 1.5 A and M_o=0.8 in which ideally presents the same RMS value of the input current. Once the average values of the input and output currents are measured, it is possible to derive the power factor and efficiency of the converter, as presented in Table I.

(a)   (b)

Fig. 8: (a) Low voltage and power DAB used for practical results; (b) Waveform of the current in the AC link inductor (light blue color) for I_o = 1.5 A and M_o=0.8.

Table I: Practical results

	I_o = 1.5 A M_o = 0.8		I_o = 1.5 A M_o = 1.2		I_o = 3.0 A M_o = 0.8		I_o = 3.0 A M_o = 1.2	
I_{in} (A)	1.38	1.2	2.15	1.8	2.91	2.4	4.4	3.6
I_{Lac_RMS} (A)	1.78	1.88	2.68	2.14	3.51	3.27	4.95	3.92
α (Degree°)	10.40	13.44	20.70	13.44	33.60	29.83	41.40	29.83

Pf	0.78	0.64	0.80	0.84	0.83	0.73	0.89	0.92
Efficiency (%)	85.8	100	84.7	100	83.1	100	81.1	100

At each operating point, it is presented both the practical and theoretical results of the lossless modeling here proposed for comparison purposed. The difference between theoretical and practical results is clearly related to AC series resistance, since voltage drop across this AC resistance sums up with the needed output voltage. This will cause a shifting in the static gain in such a way that step-down modes will approach the unitary gains improving its features, while step-up modes will move up further away, harming its characteristics. This means that, due to AC resistance and other non-idealities, the power factor and harmonics might present different results from the lossless theory.

Nonetheless, it is possible to notice that indeed the step-up mode presented better results than the step-down mode with respect to power factor, whereas with respect to efficiency, step-down modes presented better results. This shows that power factor and efficiency does not maintain a proportional or linear relation. Besides, regarding error between practical and theoretical results, non-idealities seems to effect more deeply low phase-shift intervals than high phase-shift ones. Therefore, a modeling accounting for at least AC series resistance is crucial for the performance characterization and design optimization of the DAB.

Conclusion

The methodology here proposed can be used to easily characterize in a generic way the performance of a DAB converter operating with sing phase-shift modulation; to optimize, identify or limit voltage and current ranges with respect to power factor parameters and DMC filter needs. Straightforward and meaningful converter design and optimization can be carried out with the parametrized plane, as the consequences of varying input voltage, static gain, output current, switching frequency and transformer turns ratio; and how those variable effects the performance of the DAB converter becomes well visible. Similar analysis could be performed in another converter topology. Besides, since dimensionless investigation is performed, actual comparison among different converter could be a possibility.

In this paper, the rectangular region is placed without weighting the importance of the current values. Therefore, it is suggested the developing of new parameters to consider such factor. As before mentioned, the charge profile, of a given application e.g., battery charger can be used over the output plane for optimization purposes.

As it could be noticed, a lossless theory does not allow identification of the parameter $f_s L_{ac}$ separately nor does the efficiency, while large errors of matching theoretical with practical results are susceptible. Therefore, future studies include the AC series resistive, parallel transformer and commutation losses as well as filter losses and parasitic, in such a way that the parameter of efficiency can be well analyzed in the same way as with power factor and harmonic distortion. At the end, for future studies, it is expected to have a methodology where any given optimum criteria can be evaluated: overall efficiency in a given output voltage and current ranges (profile charge), power density and EMC compliance.

References

[1] J. W. Kolar, J. Biela, S. Waffler, T. Friedli and U. Badstuebner.: Performance trends and limitations of power electronic systems, 2010 6th International Conference on Integrated Power Electronics Systems, Nuremberg, 2010, pp. 1-20.

[2] J. W. Kolar, F. Krismer, Y. Lobsiger, J. Muhlethaler, T. Nussbaumer and J. Minibock.: Extreme efficiency power electronics, 2012 7th International Conference on Integrated Power Electronics Systems (CIPS), Nuremberg, 2012, pp. 1-22.

[3] A. R. Rodríguez Alonso, J. Sebastian, D. G. Lamar, M. M. Hernando and A. Vazquez.: An overall study of a Dual Active Bridge for bidirectional DC/DC conversion, 2010 IEEE Energy Conversion Congress and Exposition, Atlanta, GA, 2010, pp. 1129-1135, doi: 10.1109/ECCE.2010.5617847.

[4] A. Rodríguez, A. Vázquez, D. G. Lamar, M. M. Hernando and J. Sebastián.: Different Purpose Design Strategies and Techniques to Improve the Performance of a Dual Active Bridge With Phase-Shift Control, in IEEE Transactions on Power Electronics, vol. 30, no. 2, pp. 790-804, Feb. 2015, doi: 10.1109/TPEL.2014.2309853.

[5] D. Costinett, R. Zane and D. Maksimovic.: Automatic voltage and dead time control for efficiency optimization in a Dual Active Bridge converter, 2012 Twenty-Seventh Annual IEEE Applied Power Electronics Conference and Exposition (APEC), Orlando, FL, 2012, pp. 1104-1111, doi: 10.1109/APEC.2012.6165956.

[6] M. N. Kheraluwala, R. W. Gascoigne, D. M. Divan and E. D. Baumann.: Performance characterization of a high-power dual active bridge DC-to-DC converter, in IEEE Transactions on Industry Applications, vol. 28, no. 6, pp. 1294-1301, Nov.-Dec. 1992, doi: 10.1109/28.175280.

[7] J. Wu, P. Wen, X. Sun and X. Yan.: Reactive Power Optimization Control for Bidirectional Dual-Tank Resonant DC–DC Converters for Fuel Cells Systems, in IEEE Transactions on Power Electronics, vol. 35, no. 9, pp. 9202-9214, Sept. 2020, doi: 10.1109/TPEL.2020.2971733.

[8] X. Liu et al..: Novel Dual-Phase-Shift Control With Bidirectional Inner Phase Shifts for a Dual-Active-Bridge Converter Having Low Surge Current and Stable Power Control, in IEEE Transactions on Power Electronics, vol. 32, no. 5, pp. 4095-4106, May 2017, doi: 10.1109/TPEL.2016.2593939.

[9] A. Jafari, M. S. Nikoo, F. Karakaya and E. Matioli.: Enhanced DAB for Efficiency Preservation Using Adjustable-Tap High-Frequency Transformer, in IEEE Transactions on Power Electronics, vol. 35, no. 7, pp. 6673-6677, July 2020, doi: 10.1109/TPEL.2019.2958632.

[10] J. Hiltunen, V. Väisänen, R. Juntunen and P. Silventoinen.: Variable-Frequency Phase Shift Modulation of a Dual Active Bridge Converter, in IEEE Transactions on Power Electronics, vol. 30, no. 12, pp. 7138-7148, Dec. 2015, doi: 10.1109/TPEL.2015.2390913.

[11] F. Jauch and J. Biela.: Combined Phase-Shift and Frequency Modulation of a Dual-Active-Bridge AC–DC Converter With PFC, in IEEE Transactions on Power Electronics, vol. 31, no. 12, pp. 8387-8397, Dec. 2016, doi: 10.1109/TPEL.2016.2515850.

[12] A. J. B. Bottion and I. Barbi.: A family of three-level DC-DC converters, 2013 Brazilian Power Electronics Conference, Gramado, 2013, pp. 115-122, doi: 10.1109/COBEP.2013.6785103.

[13] M. Gatti Bottarelli, I. Barbi, Y. Romulo De Novaes and A. Rufer.: Three-level quadratic non-insulated basic DC-DC converters, 2007 European Conference on Power Electronics and Applications, Aalborg, 2007, pp. 1-10, doi: 10.1109/EPE.2007.4417584.

[14] B. Zhao, Q. Song, W. Liu and Y. Sun, "Overview of Dual-Active-Bridge Isolated Bidirectional DC–DC Converter for High-Frequency-Link Power-Conversion System," in IEEE Transactions on Power Electronics, vol. 29, no. 8, pp. 4091-4106, Aug. 2014, doi: 10.1109/TPEL.2013.2289913.

[15] B. Zhao, Q. Song, W. Liu and Y. Sun.: Dead-Time Effect of the High-Frequency Isolated Bidirectional Full-Bridge DC–DC Converter: Comprehensive Theoretical Analysis and Experimental Verification, in IEEE Transactions on Power Electronics, vol. 29, no. 4, pp. 1667-1680, April 2014, doi: 10.1109/TPEL.2013.2271511.

[16] I. Aghabali, L. Dorn-Gomba, P. Malysz and A. Emadi.: Parasitic Resistance Effect on Dual Active Bridge Converter, IECON 2019 - 45th Annual Conference of the IEEE Industrial Electronics Society, Lisbon, Portugal, 2019, pp. 1932-1937, doi: 10.1109/IECON.2019.8926896.

[17] K. Zhang, Z. Shan and J. Jatskevich.: Estimating switching loss and core loss in dual active bridge DC-DC converters, 2015 IEEE 16th Workshop on Control and Modeling for Power Electronics (COMPEL), Vancouver, BC, 2015, pp. 1-6, doi: 10.1109/COMPEL.2015.7236501.

[18] F. Krismer and J. W. Kolar.: Accurate Power Loss Model Derivation of a High-Current Dual Active Bridge Converter for an Automotive Application, in IEEE Transactions on Industrial Electronics, vol. 57, no. 3, pp. 881-891, March 2010, doi: 10.1109/TIE.2009.2025284.

[19] R. Lenke, F. Mura and R. W. De Doncker.: Comparison of non-resonant and super-resonant dual-active ZVS-operated high-power DC-DC converters, 2009 13th European Conference on Power Electronics and Applications, Barcelona, 2009, pp. 1-10.

[20] Crebier, J.-C.; Phung, T.-H.; Nguyen, V.-S.; Lamorelle, T.; Andreta, A.; Kėachev, L.; Lembeye, Y.: DC-AC Isolated Power Converter Array. Focus on Differential Mode Conducted EMI, Electronics 2019, 8, 999.

[21] T. Lamorelle, Y. Lembeye and J. Crébier.: Handling Differential Mode Conducted EMC in Modular Converters, in IEEE Transactions on Power Electronics, vol. 35, no. 6, pp. 5812-5819, June 2020, doi: 10.1109/TPEL.2019.2947735.

[22] T. Lamorelle, S. Nguyen, J. C. Podvin, D. Rubio, Y. Lembeye and J. Crebier.: Multi-cell DC-DC converters - Input differential mode filtering generic design rules and implementation, PCIM Europe 2019; International Exhibition and Conference for Power Electronics, Intelligent Motion, Renewable Energy and Energy Management, Nuremberg, Germany, 2019, pp. 1-8.

Active Current and Energy Control for the Quasi-Three-Level Operation Mode of an Extended Modular Multilevel Converter Topology

Malte Lorenz, Jakub Kucka, and Axel Mertens
Institute for Drive Systems and Power Electronics
Leibniz University Hannover, Germany
Email: malte.lorenz@ial.uni-hannover.de

Acknowledgment

This work was supported by German Research Foundation (DFG) – Project 324166923.

Keywords

≪multilevel converters≫, ≪converter control≫, ≪high voltage power converters≫

Abstract

Recently, a Quasi-Three-Level Operation Mode for a new modular multilevel converter topology has been proposed. This paper presents a fundamental improvement of this operation mode which is achieved through a closed-loop branch current and energy control. The proposed approach uses the available degrees of freedom of the topology to actively balance the energies of the input capacitors and module capacitors. Moreover, the branch currents are controlled during both constant output voltage levels and transitions. In comparison to the operation mode from the literature, the proposed approach enables a stable operation with reduced input capacitances, module capacitances, peak branch currents, and converter losses. This is validated through simulations.

Introduction

A modular multilevel converter (MMC) can be easily scaled for applications requiring high voltages thanks to the series connection of modules. In comparison to the conventional two-level or three-level converters, the output voltage steps are reduced in height which lowers voltage stress on the isolation of electrical machines caused by cable reflection effects [1].

When applied to medium voltage drives, the conventional operation mode ensures low harmonic distortion at the output, but since the voltage fluctuation in the modules is inversely proportional to the output frequency, large module capacitors are required. The Low-Frequency Mode can improve this behaviour, but since additional current components are injected in the branches, the converter needs to be oversized when high torque is desired at low machine speeds [2].

To overcome these problems, the Quasi-Two-Level Operation Mode for MMCs in dc-ac configuration was proposed in [3]. The basic idea is that the MMC acts like a two-level converter and the output current flows through the branch in which no module capacitor is inserted. This way, the Quasi-Two-Level operation reduces the required module capacitance significantly [3, 4]. Thanks to the trapezoidal output voltage waveform, the reduction of cable reflection effects is still guaranteed [5, 4, 6]. Nevertheless, the harmonic distortion becomes similar to that of two-level converters.

To improve the harmonic distortion, the basic principle of the Quasi-Two-Level Operation Mode was extended to a Quasi-Three-Level Operation Mode in [7]. To enable this kind of operation, a new Quasi-Three-Level MMC topology was proposed which is described in the next section.

In comparison to [7], this paper presents a novel Quasi-Three-Level Operation Mode with closed-loop controllers. By controlling the energies of the capacitors within a modulation period, the input capacitance and the module capacitance can be further reduced. Moreover, the branch current controller reduces the peak branch currents and enables a stable operation despite a high efficiency of the converter.

Description of the Topology

The three-phase topology of the Quasi-Three-Level MMC (Q3L-MMC) is depicted in Fig. 1a. The dc input system and the ac output system are connected via three phase legs. Since every phase leg has an identical construction, details are shown only for the first phase leg.

Each phase leg consists of an upper branch (A) and a lower branch (B), comprising n_{bA} and n_{bB} half-bridge modules (HB), respectively. The additional clamping branch (C), that was introduced in this topology, contains half as many full-bridge modules (FB) $n_{bC} = \frac{1}{2}n_{bA} = \frac{1}{2}n_{bB}$. In series to the modules, the branch inductance L_b is connected in each branch. The branch losses are represented by the branch resistance R_b, connected in series in each branch (not depicted in Fig 1a).

The input system is connected to a constant voltage source V_i. Like in conventional NPC converters, the neutral point is provided by a centre-tapped capacitor, divided into C_{i1} and C_{i2}.

The MMC output voltage v_o is applied to an output system represented by the resistance R_o, the inductance L_o, and the equivalent internal voltage source $v_{o,s}$. The output frequency is assumed f_o. The common mode voltage between the input and the output system is v_{cm}.

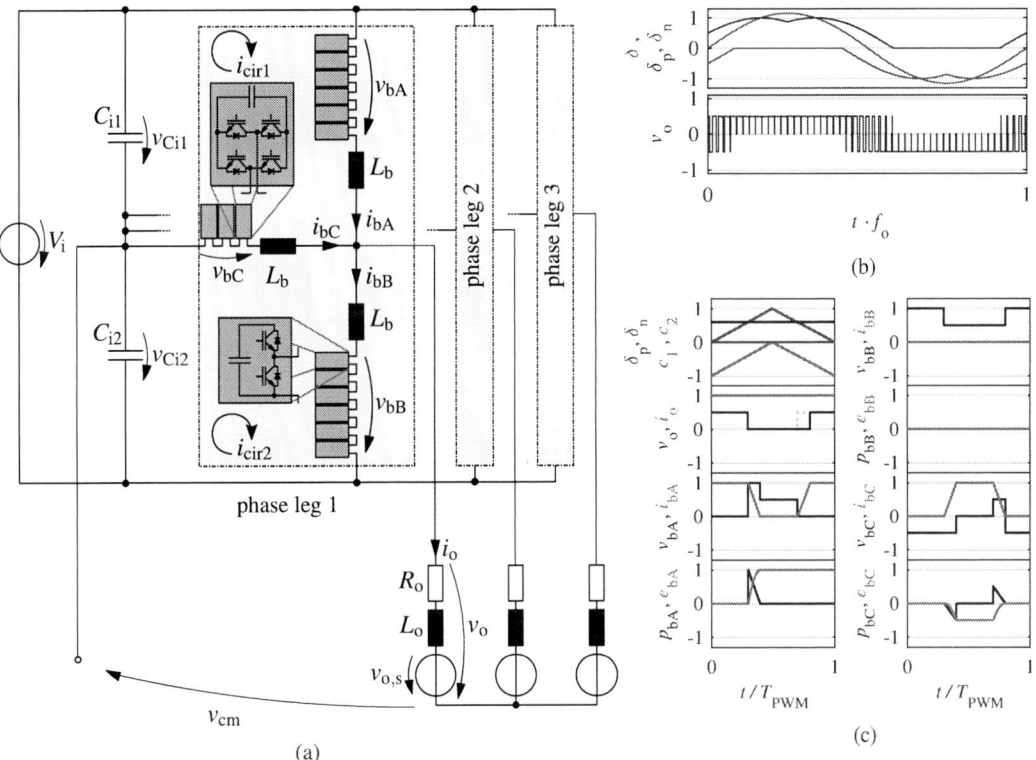

Fig. 1: (a) Three-phase topology of the Q3L-MMC with six half-bridge modules in branches A and B and three full-bridge modules in branch C, (b) Three-level output voltage modulation proposed in [8], (c) Waveforms showing the principle of the Quasi-Three-Level Operation Mode over a modulation period T_{PWM} (voltages are normalized to V_i, currents are normalized to i_o, powers and energies are normalized to their maxima)

Principle of the Quasi-Three-Level Operation Mode

The basic principle of the Quasi-Three-Level Operation Mode is shown in Fig 1. The MMC should behave like a conventional three-level converter, where the output voltage v_o is modulated with three voltage levels $\frac{1}{2}V_i$, $-\frac{1}{2}V_i$, and $0\,\text{V}$. For a stable operation, the voltages of the input capacitors $v_{C_{i1}}$ and $v_{C_{i2}}$ need to be balanced. Therefore, modulation methods can be used, which were introduced for conventional three-level converters with a centre-tapped capacitor. In this paper, the three-level modulation method that balances the centre-tapped capacitor from [8] is applied [see Fig. 1b]. For each sinusoidal duty cycle of the output voltages, two reference signals δ_p and δ_n are generated and then compared to two triangular carriers c_1 and c_2, leading to the branch voltages described in the left part of (1).

As it can be seen, in each case, one of the branch voltages is zero, signifying that no module capacitor is inserted in this branch. The branch that has currently zero voltage is referred to as passive branch. To minimize the energy fluctuation of the module capacitors, the output current flows through that passive branch. The currents of the two branches that are currently active (two from i_{bA}^*, i_{bB}^*, i_{bC}^*) are arbitrary and thus, a degree of freedom. The resulting branch currents are described in the right part of (1).

$$
\begin{pmatrix} v_{bA} \\ v_{bB} \\ v_{bC} \end{pmatrix} = \begin{cases} \begin{pmatrix} 0\,\text{V} \\ V_i \\ -\frac{1}{2}V_i \end{pmatrix} & \text{for } \delta_p > c_1 \\[1em] \begin{pmatrix} V_i \\ 0\,\text{V} \\ \frac{1}{2}V_i \end{pmatrix} & \text{for } \delta_n < c_2 \\[1em] \begin{pmatrix} \frac{1}{2}V_i \\ \frac{1}{2}V_i \\ 0\,\text{V} \end{pmatrix} & \text{else} \end{cases} \qquad \begin{pmatrix} i_{bA} \\ i_{bB} \\ i_{bC} \end{pmatrix} = \begin{cases} \begin{pmatrix} i_o + i_{bB}^* - i_{bC}^* \\ i_{bB}^* \\ i_{bC}^* \end{pmatrix} & \text{for } \delta_p > c_1 \\[1em] \begin{pmatrix} i_{bA}^* \\ -i_o + i_{bA}^* + i_{bC}^* \\ i_{bC}^* \end{pmatrix} & \text{for } \delta_n < c_2 \\[1em] \begin{pmatrix} i_{bA}^* \\ i_{bB}^* \\ i_o - i_{bA}^* + i_{bB}^* \end{pmatrix} & \text{else} \end{cases} \tag{1}
$$

At first consideration, it seems reasonable to set these arbitrary currents to zero, so that the inserted modules are not charged or discharged. However, if the passive branch changes and the output current needs to commutate between the branches, high voltages and currents can occur simultaneously in the branches during the transitions. This results in power peaks and branch energy shifts, that are visible in Fig. 1c.

In the proposed approach, the currents of the active branches are set to compensate for the energy shift caused by the transitions within each PWM period, leading to a low branch energy variation. These so-called compensating currents are low in comparison to the output current. Moreover, the branch currents will be controlled even during the transitions, facilitating a converter design with reduced input capacitances, module capacitances, converter losses, and peak branch currents.

Control System

The cascaded control structure for the three-phase Q3L-MMC is shown in Fig. 2. It consists of an input capacitor voltage balancing unit and three identical phase-leg control units.

Like in conventional three-level converters, the output currents are controlled with an external high-level control through the setpoints for the output voltages. The corresponding duty cycles δ_1, δ_2, and δ_3 are then forwarded to the input capacitor controller. As explained before, the input capacitor controller ensures the active balancing of the centre-tapped capacitor. The modulation method from [8] with the two reference signals δ_p and δ_n can further reduce the voltage fluctuation of the input capacitors C_{i1} and C_{i2} significantly.

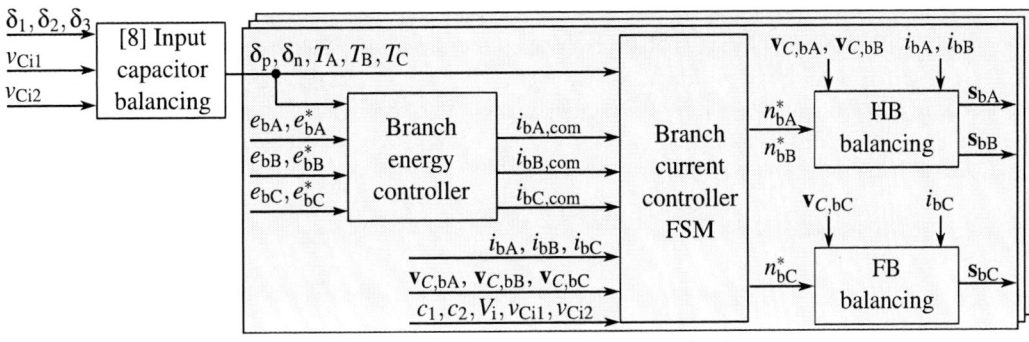

Fig. 2: Cascaded control structure for the Quasi-Three-Level Operation Mode

Phase-Leg Control Unit

The phase-leg control unit is based on the one proposed in [4]. The outer branch energy controller controls the energies stored in the module capacitors and forwards the desired compensating currents to the inner branch current controller. For a decoupled control of the output current, the branch currents are then transformed to circulating currents which are controlled in a finite state machine (FSM). To follow the setpoints of the branch voltages, the half-bridge and full-bridge balancing units insert or bypass the modules and ensure the voltage balancing of the module capacitors within the corresponding branch.

The energy controller determines the required branch powers p_{bA}^*, p_{bB}^*, and p_{bC}^* for the compensation of the branch energy disturbances within a PWM period T_{PWM}. For this task, a proportional controller with the gain G_p is employed (see Fig. 3). The calculation of the branch energies and their setpoints is shown in (2). For a simplified description, the variables $x \in (A, B, C)$ and $j \in (1, 2, ..., n_{bx})$ are introduced. The capacitance of the modules is C_{mod} and their voltages are $v_{C,bx}$.

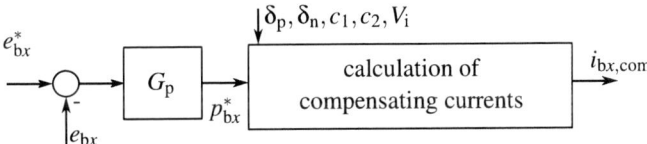

Fig. 3: Branch energy controller

$$e_{bx} = \frac{1}{2} C_{mod} \sum_{j=1}^{n_{bx}} v_{C,bx,j}^2 \qquad\qquad e_{bx}^* = n_{bx} \cdot \frac{1}{2} C_{mod} \cdot v_C^{*2} \qquad (2)$$

To generate the required branch powers, the arbitrary branch currents i_{bA}^*, i_{bB}^*, and i_{bC}^* are set to the compensating current values $i_{bA,com}$, $i_{bB,com}$, and $i_{bC,com}$, which are determined as described in the left part of (3). Therefore, the required branch powers are divided by the branch voltages according to (1) and multiplied with a relative factor, describing the duration of the corresponding branch having a non-zero voltage (active branches) within a PWM period. Here, T_A, T_B, and T_C are the durations of *State A* ($v_o = \frac{1}{2} V_i$), *State B* ($v_o = -\frac{1}{2} V_i$), and *State C* ($v_o = 0\,\mathrm{V}$), respectively.

Observing the left part of (3), it can be seen that the duty cycle of the Quasi-Three-Level Operation Mode should not exceed a maximum value of $\delta_{max} < 1.15$, because otherwise the duration of the active branches approaches zero and the compensating currents $i_{bA,com}$, $i_{bB,com}$ would reach very high values. Note that if no common mode voltage v_{cm} is injected, this maximum duty cycle is reduced even more to some values below one. The same applies to $i_{bC,com}$ for $\delta \to 0$, if *State A* and *State B* are not symmetrically divided over a PWM period at zero crossings, like it is achieved with the output voltage modulation method from [8] [see Fig. 1b].

The inner current controller is implemented as a finite state machine (FSM), controlling the branch currents during the *States A, B,* and *C* but also during the transitions (see Fig. 4). To decouple the controlled currents from the output current, the circulating currents i_{cir1} and i_{cir2} are introduced as control variables. The circulating currents flow as superposition currents through branches A and C, and B and C, respectively (see Fig. 1a). The setpoint values of the circulating currents, including the compensating currents, are given in the right part of (3).

$$
\begin{pmatrix} i_{bA,com} \\ i_{bB,com} \\ i_{bC,com} \end{pmatrix} = \begin{cases} \begin{pmatrix} 0 \\ \frac{p_{bB}^*}{V_i} \cdot \frac{T_{PWM}}{T_A+T_C} \\ \frac{p_{bC}^*}{-\frac{1}{2}V_i} \cdot \frac{T_{PWM}}{T_A+T_B} \end{pmatrix} & \text{for } \delta_p > c_1 \\[3em] \begin{pmatrix} \frac{p_{bA}^*}{V_i} \cdot \frac{T_{PWM}}{T_B+T_C} \\ 0 \\ \frac{p_{bC}^*}{\frac{1}{2}V_i} \cdot \frac{T_{PWM}}{T_A+T_B} \end{pmatrix} & \text{for } \delta_n < c_2 \\[3em] \begin{pmatrix} \frac{p_{bA}^*}{\frac{1}{2}V_i} \cdot \frac{T_{PWM}}{T_B+T_C} \\ \frac{p_{bB}^*}{\frac{1}{2}V_i} \cdot \frac{T_{PWM}}{T_A+T_C} \\ 0 \end{pmatrix} & \text{else} \end{cases} \qquad \begin{pmatrix} i_{cir1} \\ i_{cir2} \end{pmatrix} = \begin{cases} \begin{pmatrix} \frac{i_o}{2} + \frac{1}{2}i_{bB,com} - i_{bC,com} \\ \frac{1}{2}i_{bB,com} + \frac{1}{2}i_{bC,com} \end{pmatrix} & \text{for } \delta_p > c_1 \\[2em] \begin{pmatrix} \frac{1}{2}i_{bA,com} - \frac{1}{2}i_{bC,com} \\ \frac{-i_o}{2} + \frac{1}{2}i_{bA,com} + i_{bC,com} \end{pmatrix} & \text{for } \delta_n < c_2 \\[2em] \begin{pmatrix} \frac{-i_o}{2} + i_{bA,com} - \frac{1}{2}i_{bB,com} \\ \frac{i_o}{2} - \frac{1}{2}i_{bA,com} + i_{bB,com} \end{pmatrix} & \text{else} \end{cases}
$$

$$(3)$$

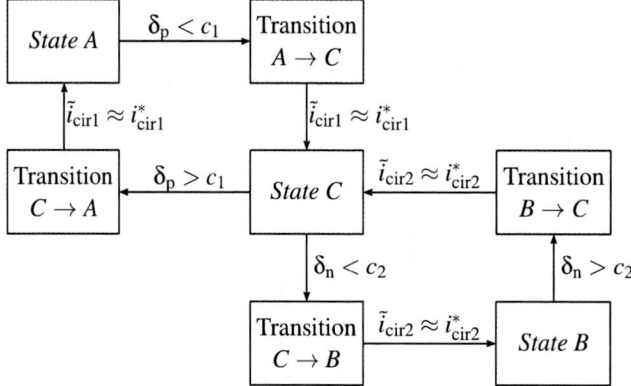

Fig. 4: Finite state machine of the branch current controller

Neglecting the compensating currents, it can be recognized that when a reference signal crosses a carrier, one circulating current has to change by $\pm i_o$ and the other one has to change by $\mp \frac{1}{2}i_o$ during the transitions. The dynamic of the circulating currents during the transitions can be described as in (4), assuming the branch resistances to be zero and the branch voltages to be constant.

$$
\begin{pmatrix} \Delta i_{cir1} \\ \Delta i_{cir2} \end{pmatrix} = \frac{1}{2L_b} \cdot \begin{pmatrix} -v_{bA} + v_{bC} + v_{Ci1} \\ -v_{bB} - v_{bC} + v_{Ci2} \end{pmatrix} \cdot t \qquad (4)
$$

For an accurate output voltage modulation and a minimal voltage fluctuation of the module capacitors, a short transition time is desired. For this purpose, the branch voltages are set during the transitions according to Table I.

Table I: Overview of Transitions

Transition	$i_o > 0$			$i_o < 0$		
	v_{bA}	v_{bB}	v_{bC}	v_{bA}	v_{bB}	v_{bC}
$A \to C$	$\approx V_i$	$\approx V_i/2$	$\approx -V_i/2$	$0\ V$	$\approx V_i/2$	$\approx V_i/2$
$C \to A$	$0\ V$	$\approx V_i/2$	$\approx V_i/2$	$\approx V_i$	$\approx V_i/2$	$\approx -V_i/2$
$B \to C$	$\approx V_i/2$	$0\ V$	$\approx -V_i/2$	$\approx V_i/2$	$\approx V_i$	$\approx V_i/2$
$C \to B$	$\approx V_i/2$	$\approx V_i$	$\approx V_i/2$	$\approx V_i/2$	$0\ V$	$\approx -V_i/2$

The FSM jumps into the next state (*State A, B,* or *C*) when the setpoints of the circulating currents of the following state are reached. Since the module balancing units insert and bypass the modules stepwise (explained below), the branch voltages do not change immediately to the next setpoint. Similar to [4], this voltage-time delay is considered in a predictive feedforward control for the circulating currents leading to the predicted circulating currents $\tilde{i}_{cir1}, \tilde{i}_{cir2}$.

During the *States A, B,* and *C*, the circulating currents are controlled by the active branches with a dead-beat controller. The dead-beat controller can be derived using (4), setting the voltage of the passive branch to zero and solving the equations for the voltages of active branches. The equation of the dead-beat controller with a period of T_{HF} is given in (5), where k indicates a discrete time step. The non-integer setpoint voltages of the active branches are then modulated with a high-frequency (HF) modulation in the same manner as in [4]. For a reasonable HF modulation frequency with a small branch current ripple, the branch inductance should not be too low.

$$\begin{pmatrix} v_{bA}(k+1) \\ v_{bB}(k+1) \\ v_{bC}(k+1) \end{pmatrix} = \begin{cases} \begin{pmatrix} 0 \\ -(i^*_{cir2}(k) - i_{cir2}(k))\frac{2L_b}{T_{HF}} + v_{Ci2}(k) - v_{bc}(k+1) \\ (i^*_{cir1}(k) - i_{cir1}(k))\frac{2L_b}{T_{HF}} - v_{Ci1}(k) \end{pmatrix} & \text{for } State\ A \\[2em] \begin{pmatrix} -(i^*_{cir1}(k) - i_{cir1}(k))\frac{2L_b}{T_{HF}} + v_{Ci1}(k) + v_{bC}(k+1) \\ 0 \\ -(i^*_{cir2}(k) - i_{cir2}(m))\frac{2L_b}{T_{HF}} + v_{Ci2}(k) \end{pmatrix} & \text{for } State\ B \\[2em] \begin{pmatrix} -(i^*_{cir1}(k) - i_{cir1}(k))\frac{2L_b}{T_{HF}} + v_{Ci1}(k) \\ -(i^*_{cir2}(k) - i_{cir2}(k))\frac{2L_b}{T_{HF}} + v_{Ci2}(k) \\ 0 \end{pmatrix} & \text{for } State\ C \end{cases} \qquad (5)$$

In each transition or *State A, B,* or *C*, the FSM calculates the number of modules n^*_{bA}, n^*_{bB}, and n^*_{bC}, which has to be inserted to the generate the desired branch voltages. These numbers are then forwarded to the module balancing units (see Fig. 2).

The module balancing units select the modules of the corresponding branch to be inserted or bypassed so as to guarantee balanced voltages of the module capacitors. The decision is based on the direction of the corresponding branch current, flowing through the currently inserted module or possibly flowing through the currently bypassed module, respectively. If a module needs to be inserted and the module current is positive, the module with the lowest voltage is inserted. If the module current is negative, the module with the highest voltage is inserted. If a module needs to be bypassed and the module current is positive, the module with the highest voltage is bypassed. If the module current is negative, the module with the lowest voltage is bypassed.

Furthermore, the module balancing units ensure a minimum delay time T_d between every switching instant. As a result, the branch voltage steps are small and thus the dv/dt of the output voltage is limited. Finally, the switching states s_{bA}, s_{bB}, and s_{bC} are routed to the half-bridge and full-bridge modules.

Simulation Results and Comparison

Before the comparisons are drawn, the simple complementary operation mode from [7], that neither controls the circulating currents nor the branch energies, is investigated. Whenever a reference signal crosses a carrier, the branch voltages are directly set to the next *State A, B* or *C* according to the left part of (1). For example, when the *State* changes from *A* to *C*, the required voltages of v_{bA} and v_{bC} are controlled and branch B passively blocks the resulting voltage (all IGBTs are in off-state). The module balancing units are implemented in the same way as described above, ensuring the balancing of the module capacitors and the small output voltage steps.

However, since there is no branch current control, the branch currents respond with a *RLC*-series-resonant behaviour, which is generally described by the resonant frequency f_r and the damping factor $\zeta = R \cdot \sqrt{\frac{C}{L}}$. This means that a reduction of the converter losses (proportional to the branch resistance), the module capacitance, or the input capacitance leads to higher peak branch currents, since the branch current overshoot depends on the damping factor. These challenging design aspects can also be found for MMCs with a non-controlled Quasi-Two-Level Operation Mode, investigated in detail in [9].

To demonstrate this effect, the approach from [7] is simulated with the original branch parameters (see Fig. 5a and Fig. 6a). Additionally, simulations are conducted with a ten times smaller module capacitance (see Fig. 5b and Fig. 6b) and a ten times smaller input capacitance and branch resistance (see Fig. 5c and Fig. 6c). All simulation parameters are listed in Table II.

Table II: Simulation Parameters

Parameter		Value		Unit
		Ref. [7]	Proposed method	
Input voltage	V_i	4	4	kV
Input capacitance	C_{i1}, C_{i2}	15 / 1.5	1.5	mF
Input resistance	R_i	10	10	$m\Omega$
Output inductance	L_o	15	15	mH
Output resistance	R_o	5	5	Ω
Branch inductance	L_b	2	200	μH
Branch resistance	R_b	50 / 5	0	$m\Omega$
Number of half-bridge modules branch A	n_{bA}	6	6	
Number of half-bridge modules branch B	n_{bB}	6	6	
Number of full-bridge modules branch C	n_{bC}	3	3	
Module capacitance	C_{mod}	2 / 0.2	0.2	mF
Setpoint value module capacitor voltage	v^*_{Cmod}	667	680	V
Output frequency	f_o	5	5	Hz
PWM frequency	f_{PWM}	1	1	kHz
Modulation index	M	0.8	0.8	
Delay between switching instants	T_d	1	1	μs
HF modulation frequency	f_{HF}	-	25	kHz
Branch energy controller gain	G_P	-	2000	1/s
Input capacitor controller gain acc. to [8]	G_{PI}	-4.2 / -0.417	-0.417	10^{-3}
Input capacitor controller time constant acc. to [8]	T_{PI}	31.83	31.83	ms

In Fig. 5a and Fig. 5b, the quasi-three-level output waveform and the sine output current with the RMS value of $I_o = 218\,\text{A}$ of the three-phase Q3L-MMC can be seen. The approach from [7] operates stably for the given branch parameters (indicated by the balanced module voltages $\mathbf{v}_{C,bA}$, $\mathbf{v}_{C,bB}$ and $\mathbf{v}_{C,bC}$). Of course, the voltage fluctuation is higher for the reduced module capacitance. As expected, the comparison of Fig. 6a to Fig. 6b shows that the branch current overshoot increases by the reduction of the module capacitance due to the lower damping factor. The same applies, if the input capacitance or the converter losses are reduced. This has to be considered in the design process, since the semiconductors have to be capable of conducting such currents. Moreover, a further reduction of the damping factor negatively impacts the stability of the operation, as it can be seen in Fig. 5c and Fig. 6c.

Active Current and Energy Control for the Quasi-Three-Level Operation Mode of an
Extended Modular Multilevel Converter Topology
LORENZ Malte

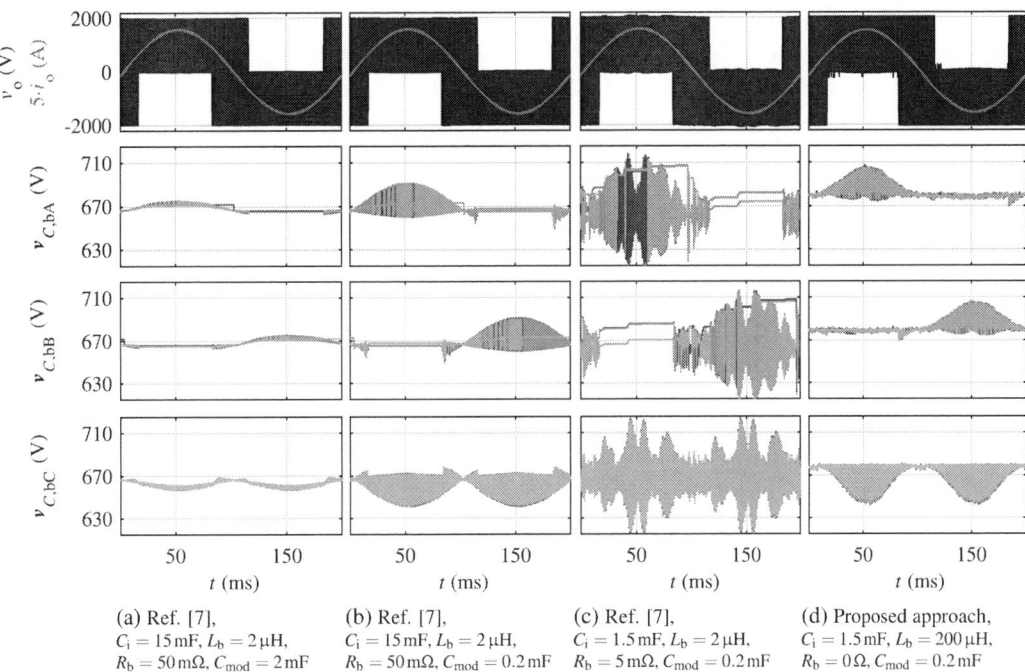

Fig. 5: Simulation results of the various Quasi-Three-Level Operation Modes: output voltage v_o, output current i_o and module capacitor voltages $\mathbf{v}_{C,\mathrm{bA}}$, $\mathbf{v}_{C,\mathrm{bB}}$, $\mathbf{v}_{C,\mathrm{bC}}$.

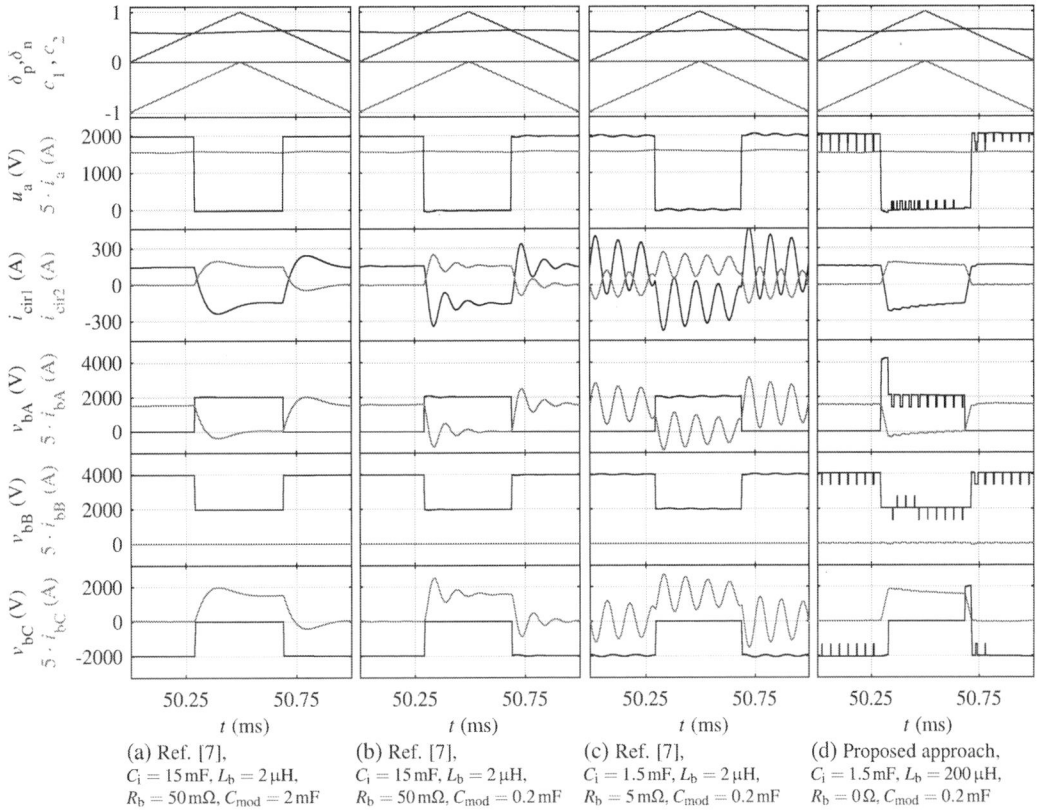

(a) Ref. [7],
$C_i = 15\,\mathrm{mF}$, $L_b = 2\,\mu\mathrm{H}$,
$R_b = 50\,\mathrm{m\Omega}$, $C_{\mathrm{mod}} = 2\,\mathrm{mF}$

(b) Ref. [7],
$C_i = 15\,\mathrm{mF}$, $L_b = 2\,\mu\mathrm{H}$,
$R_b = 50\,\mathrm{m\Omega}$, $C_{\mathrm{mod}} = 0.2\,\mathrm{mF}$

(c) Ref. [7],
$C_i = 1.5\,\mathrm{mF}$, $L_b = 2\,\mu\mathrm{H}$,
$R_b = 5\,\mathrm{m\Omega}$, $C_{\mathrm{mod}} = 0.2\,\mathrm{mF}$

(d) Proposed approach,
$C_i = 1.5\,\mathrm{mF}$, $L_b = 200\,\mu\mathrm{H}$,
$R_b = 0\,\Omega$, $C_{\mathrm{mod}} = 0.2\,\mathrm{mF}$

Fig. 6: Voltage and current waveforms of the Quasi-Three-Level Operation Modes over a single PWM period. The figure shows a detail of the simulation from Fig 5.

EPE'20 ECCE Europe

Assigned jointly to the European Power Electronics and Drives Association & the Institute of Electrical and Electronics Engineers (IEEE)

To demonstrate that the proposed approach works even without damping, the branch resistance is set to $0\,\Omega$. As mentioned, L_b is increased for a reasonable HF modulation frequency (see Table II).

In Fig. 5d, the quasi-three-level output waveform of the proposed approach can be seen. The balanced module capacitors indicate a stable operation. The voltage fluctuation of the module capacitors in Fig. 5b and Fig. 5d in branch C is almost equal. The voltage fluctuation of the module capacitors in branch A and B is slightly smaller for the prosed approach, because in contrast to the approach of [7], no modules or all modules are inserted in branch A or B during the transitions. Moreover, as it can be seen in the detailed view in Fig. 6b in comparison to Fig. 6d, the peak branch currents are not dependent on the damping factor in the proposed method and thus, the peak branch currents are lower. The HF modulation of the branch voltages can be seen for the proposed method in Fig. 6d. These additional switching instants are expected to increase the switching losses just slightly, since the branch currents of the active branches are low. Comparing Fig. 5c and Fig. 6c to Fig. 5d and Fig. 6d, it becomes apparent that since the proposed approach works also without damping, low module capacitances, input capacitances, and converter losses are generally feasible.

Conclusion

In this paper, a novel branch current and energy control for the Quasi-Three-Level Operation Mode is presented, that actively balances the input and module capacitors within a modulation period and controls the branch current during all output voltage levels and transitions. In comparison to the operation mode from [7], the proposed approach leads to lower peak branch currents and allows for a very low-damped branch design. Thus, the input capacitance and module capacitance can be further reduced while the converter efficiency can be increased. As a trade-off, the leg inductance has to be increased and the control becomes more complex. This trade-off seems acceptable, since the overall costs of the converter can be reduced.

In comparison to the Quasi-Two-Level Operation Mode, the topology requires additional semiconductors and capacitors. However, the Quasi-Three-Level Operation Mode facilitates a better THD of the output voltage and output current. Since the voltage fluctuation of the module capacitors is independent of the output frequency, the Quasi-Three-Level Operation Mode is suitable for variable-speed medium voltage drives, which require lower harmonic distortion.

References

[1] O. Drubel, *Converter applications and their influence on large electrical machines*, ser. Lecture notes in electrical engineering. Heidelberg: Springer, 2013, vol. 232.

[2] A. Antonopoulos, L. Angquist, S. Norrga, K. Ilves, L. Harnefors, and H.-P. Nee, "Modular multilevel converter ac motor drives with constant torque from zero to nominal speed," *IEEE Transactions on Industry Applications*, vol. 50, no. 3, pp. 1982–1993, 2014.

[3] A. Mertens and J. Kucka, "Quasi two-level PWM operation of an MMC phase leg with reduced module capacitance," *IEEE Transactions on Power Electronics*, vol. 31, no. 10, pp. 6765–6769, 2016.

[4] J. Kucka and A. Mertens, "Improved current control for a quasi-two-level PWM-operated modular multilevel converter," *IEEE Transactions on Power Electronics*, vol. 35, no. 7, pp. 6842–6853, 2020.

[5] J. Kucka, "Quasi-two-level PWM operation of modular multilevel converters : Implementation, analysis, and application to medium-voltage drives."

[6] F. Bertoldi, M. Pathmanathan, and R. S. Kanchan, "Quasi - two-level converter operation strategy for overvoltage mitigation in long cable applications," in *2019 IEEE International Electric Machines & Drives Conference (IEMDC)*. IEEE, 12.05.2019 - 15.05.2019, pp. 1621–1627.

[7] X. Wei, S. Lu, X. Deng, and S. Li, "A new quasi three-level hybrid modular multilevel converter," in *2018 IEEE International Power Electronics and Application Conference and Exposition (PEAC)*. IEEE, 04.11.2018 - 07.11.2018, pp. 1–6.

[8] R. Maheshwari, S. Munk-Nielsen, and S. Busquets-Monge, "Design of neutral-point voltage controller of a three-level NPC inverter with small dc-link capacitors," *IEEE Transactions on Industrial Electronics*, vol. 60, no. 5, pp. 1861–1871, 2013.

[9] J. Kucka and A. Mertens, "Designing a passively damped quasi-two-level-operated modular multilevel converter for drive applications," *IEEE Access*, vol. 8, pp. 80 218–80 232, 2020.

Torque ripple reduction technique for a switched reluctance motor

Krzysztof Jackiewicz, Arkadiusz Kaszewski, Andrzej Stras,
Bartlomiej Ufnalski, Tomasz Balkowiec
WARSAW UNIVERSITY OF TECHNOLOGY
Institute of Control and Industrial Electronics
75 Koszykowa St.
Warsaw, Poland
Tel.: +48 / (22) – 234-5120
E-Mail: krzysztof.jackiewicz@ee.pw.edu.pl
URL: http://www.isep.pw.edu.pl

Acknowledgements

The work has been financed by The National Centre for Research and Development for the project TECHMATSTRATEG1/346922/4/NCBR/2017 „Technologie materiałów półprzewodnikowych dla elektroniki dużych mocy i wysokich częstotliwości".

Keywords

«Digital control», «Variable speed drive», «Non-standard electrical machine», «Harmonics», «Switched reluctance drive»

Abstract

One of the main problems of switched reluctance motor are torque ripple, vibrations and high acoustic noise which decrease user comfort and cause faster machine and drive system wear out. In this paper a new approach for torque ripple minimization is presented, which is based on modified multi-oscillatory controller.

Introduction

Switched reluctance motor (SRM) is a type of synchronous electrical machine. The SRM rotor and stator are of salient type with concentrated windings placed on stator poles. The SRM has no coils nor permanent magnets on its rotor [1]. Usually, the rotor and stator are made from the same material i.e. transformer sheets. Constructional simplicity, absence of permanent magnets and high reliability make it an attractive choice for many applications like car industries, flywheel energy storage systems and variable speed wind generator [3][5]. On the other hand, indispensable double saliency of the motor structure is the main source of vibrations, high acoustic noise and torque ripple [2][3]. Torque ripple can be reduced by changing SRM geometry e.g. "introducing asymmetric inductance characteristic with non-uniform air-gap" [4] or by applying one of control techniques such as "current profiling algorithm based on semi-numerical model of switched reluctance machine (SRM)" [5] or "Torque Ripple Optimization for Acoustic Noise Reduction of SRM Drives via Fuzzy logic Control" [3]. However, if the motor is already built, the torque ripple reduction is only possible by applying an appropriate control strategy [3]. In this paper a new control method for torque ripple reduction in a 4-phase 8/6 SRM is presented. The control method is based on multi-oscillatory controller (MOSC), also known as multi-resonant controller [7][8][9]. MOSCs are used as a part of the speed control loop. Output signal of the speed controller is the reference signal for the current controllers. This approach allows to reduce torque ripple. Herein, a newly developed angle sampled MOSC (ASMOSC) is shown. ASMOSC is speed invariant and does not require resonant frequency adaptation. The proposed control system is verified by numerical simulations and experiments.

Principle of SRM operation

A typical 4-phase 8/6 SRM structure is presented in Fig. 1 [1]. The machine consists of 8 salient stator poles with windings and 6 salient rotor poles. Each stator pole is equipped with one coil. Two diametrically opposite coils are connected in series creating a phase winding. Each winding is galvanically separated from another one. The motor is operated by sequentially turning on and off consecutive phases along with rotor position [1]. The sequence of phases excitation is ABCD or DCBA for the reverse rotation direction. Magnetomotive force is produced if the stator pole is excited by a phase current. However, the torque is produced only when the rotor pole is out of alignment (Fig. 1a). Rotor tends to move towards minimum reluctance position, that is when two rotor poles and two excited stator poles are aligned. When this position is obtained another set of rotor poles with respect to consecutive stator poles is out of alignment. Consecutive phase is excited, and the movement of the rotor continues. Each phase is active within specific angle range θ_{on} and θ_{off} with respect to the magnetic axis of the active phase (Fig. 1b). The angles θ_{on} and θ_{off} are referred as driving angles and they depend on machine geometry, control strategy and operating point [1] [2]. θ_r is positive in clockwise direction and negative in counterclockwise direction. For motor operation θ_{on} and θ_{off} are negative [2].

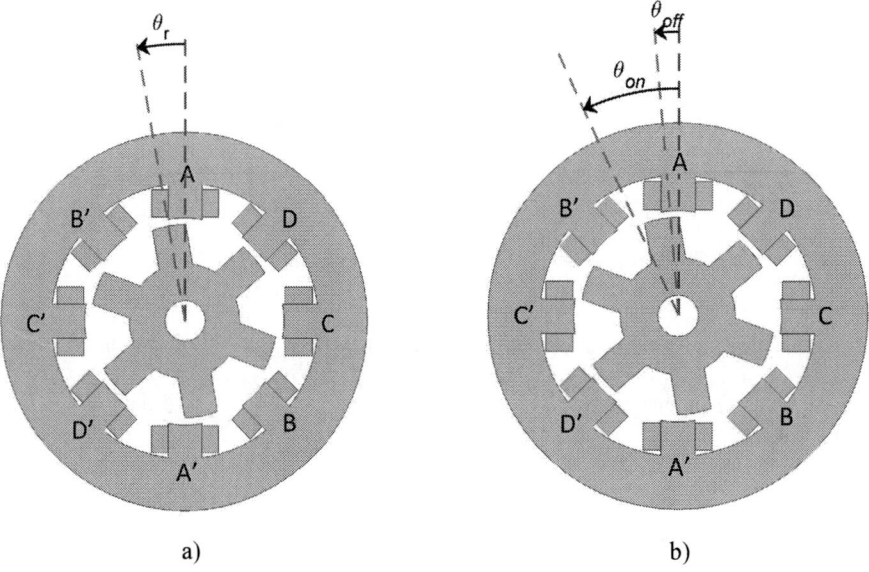

a)   b)

Fig. 1: Typical 4-phase SRM 8/6 structure with driving angles shown.

The torque produced by a single phase of SRM is a function of rotor position and phase current and can be expressed by simplified equation [1]:

$$T_{ePH} = \frac{\delta L_{PH}(\theta_{PH}, i_{PH})}{\delta \theta_{PH}} \frac{i_{PH}^2}{2} \quad , \tag{1}$$

where θ_{PH} is an angle between rotor pole and magnetic axis of the phase, L_{PH} is inductance of the phase and i_{PH} is the current of the phase. The reluctance of the magnetic circuit varies with rotation of the rotor. Because of that the value of the inductance L_{PH} changes alongside. This change is non-linear and so is the produced torque. Single phase torque vs current density and rotor angle characteristic is obtained from finite element analysis (FEA) and is presented in Fig. 2. As one can notice, positive driving torque is generated in a θ_{PH} ranging from -30° to 0°. Driving angles selection is based on torque vs rotor position and current characteristic the way high value of torque is generated for a given current.

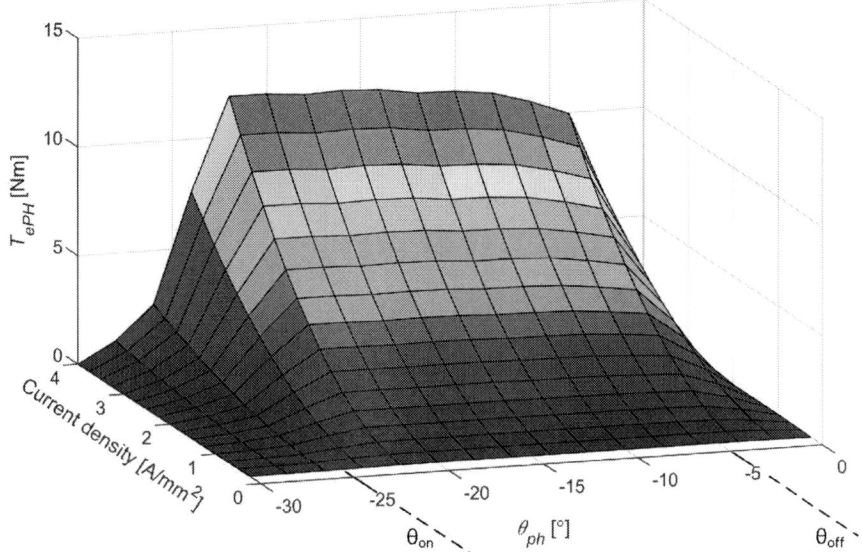

Fig. 2: Phase torque vs phase rotor position and current density.

Control system with PI controller

A typical control sequence for SRM operation is presented in [3]. Each phase is activated within selected rotor angle range (between θ_{on} and θ_{off}) and deactivated outside of that angle range. A phase activation sequence for $\theta_{on} = -30°$ and $\theta_{off} = 0°$ is presented in Fig. 3. Each phase current is regulated individually by a current controller (CC). A control system for SRM is shown in Fig. 4. Active phase module (APM) is responsible for selecting current reference of the phases based on rotor position and selected driving angles (Fig. 3). When phase is active, its current set point is set to reference current from speed controller. When phase is inactive, its current set point is set to 0. It should be noticed that the torque can be generated by more than one phase at the time depending on the defined θ_{on} and θ_{off} values (such a case is presented in Fig. 3)

Simulation of the typical control scheme were performed in PLECS software for SRM 8/6 running at 500 rpm and loaded with constant torque of 6 Nm. Phase currents and total torque of the SRM with typical control system are presented in Fig. 5.

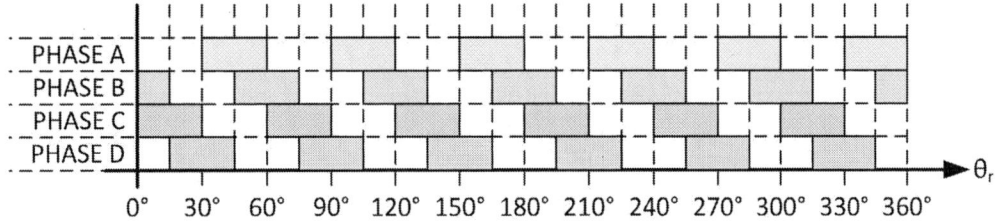

Fig. 3: Phase activation sequence for SRM.

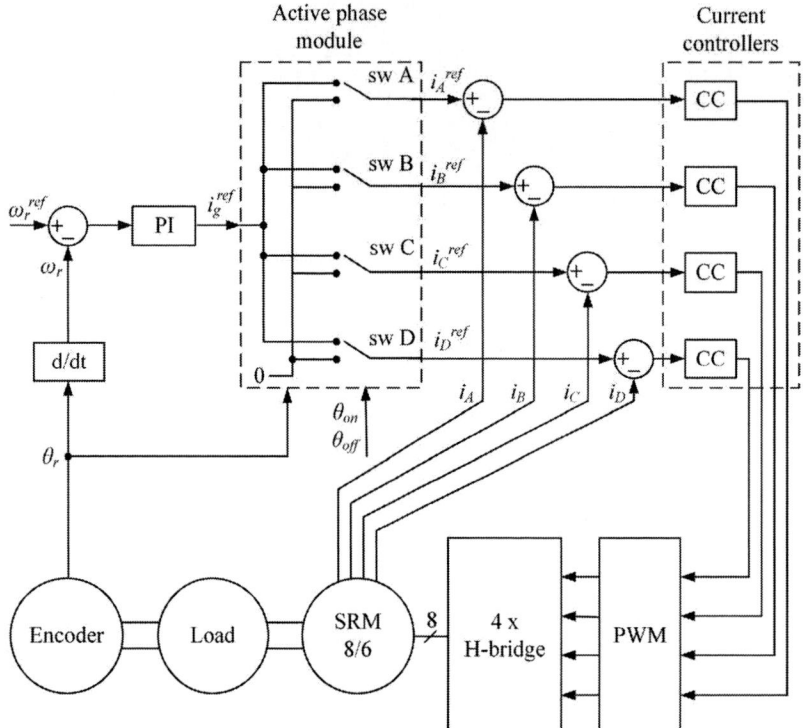

Fig. 4: Typical control system for SRM.

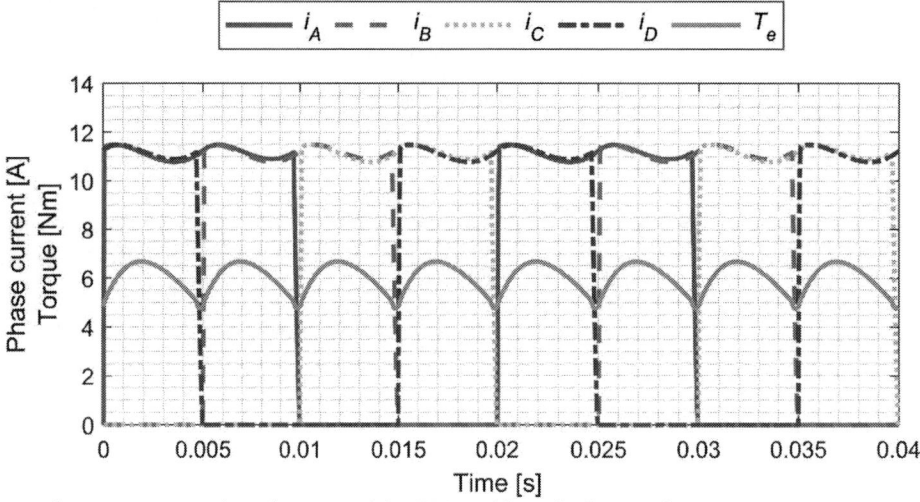

Fig. 5: Phase currents and total torque of the SRM with typical control system.

As seen in Fig. 5, the typical control system results in high torque ripple [3].

Control system with PI and ASMOSC - torque ripple reduction method

In SRM torque harmonics are periodic with respect to the angle position of the rotor and follow a specific frequency pattern of nk order, where $n = 1, 2, \ldots$ and k is a number of switches per revolution and can be calculated as:

$$k = \frac{360°}{\frac{360°}{N_r} - \frac{360°}{N_s}} = \frac{N_r N_s}{N_s - N_r} \; , \tag{2}$$

where N_s is a number of stator poles and N_r is a number of rotor poles. According to the internal model principle (IMP) [7] introducing MOSC to the control system enables disturbance rejection with zero steady state error. For this reason, the reduction of the torque ripples can be achieved by applying oscillatory terms in the closed loop system. The transfer function for conventional oscillatory term is:

$$G_{osc}(s) = k_n \frac{s}{s^2 + \omega_n{}^2} \; , \tag{3}$$

where k_n is the term gain and ω_n is the term resonant frequency. By matching ω_n with frequency of the torque oscillations it is possibble to reduce them. Total machine output torque is a sum of individual torques generated by machine phases:

$$T_e(t) = T_{ePH1}(t) + T_{ePH2}(t) + T_{ePH3}(t) + T_{ePH4}(t) \; , \tag{4}$$

Torque ripple will be zero if a sum of individual phase torques is constant. One approach is to introduce additional current component by providing another signal beside the one from PI speed controller. Finding this additional current component in general is not trivial and depends on many parameters like SRM construction, speed, driving angles etc. [5]. In the proposed approach, additional current component is generated by a MOSC [6] with speed error as input. In presented case, MOSC is equipped with 4 oscillatory terms. Resonant frequency of these controllers is set as follows:

$$f_{OSC1} = kn, \; f_{OSC2} = 2kn, \; f_{OSC3} = 3kn, \; f_{OSC4} = 4kn \; , \tag{6}$$

where n is rotor speed in revolutions per second. For variable speed drive each f_{osc} needs to be updated according to the actual motor speed. In a presented solution a discrete speed control loop consists of two independent regulators: PI speed controller and MOSC speed controller. PI speed controller works with constant sampling time while MOSC works with **constant angle step** which is time invariant. This means that the sample control error signal is periodic with respect to the rotational speed of the motor. The periodicity of control error is constant and independent of speed. In presented approach MOSC is sampled synchronously to constant rotor angle step in opposite to regular constant time-approach resulting in angle sampled MOSC (ASMOSC). Sampling pulses are generated by rotor angle change detector which creates a pulse every defined angle change (between 1/32 to 1/1 of encoder angle resolution). This feature enables ASMOSC to track and reduce torque ripple regardless of the SRM speed without the necessity to change ASMOSC resonant frequency. The SRM control scheme with ASMOSC for torque ripple reduction is presented in Fig. 7. Presented control method has been simulated with the same conditions as for typical control scheme. Phase currents and total torque produced by SRM with ASMOSC controller is presented in Fig. 8. Additional current component introduced by ASMOSC is visible in phase currents. Torque ripple generated by SRM are reduced in comparison to typical control system. Effects of ASMOSC being sampled synchronously with rotor position is clearly visible when speed of the SRM changes. Output of 4 oscillatory terms of ASMOSC and SRM speed is presented in Fig. 6.

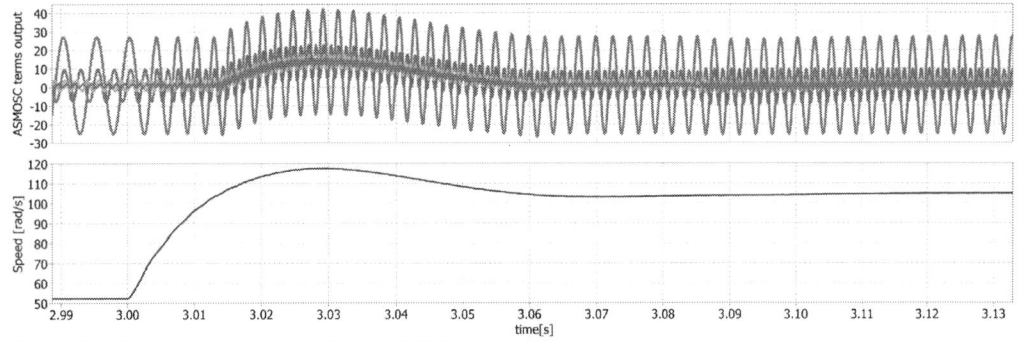

Fig. 6: Oscillatory terms output during SRM speed change.

In third second of simulation, a speed reference is changed from 500 rpm (52.3 rad/s) to 1000 rpm (104.7 rad/s). One can notice that frequencies of ASMOSC terms are directly related to SRM speed. Table I presents torque ripple comparison for typical and proposed control systems. Introduction of ASMOSC in control system reduced torque ripple from 41% to 10%.

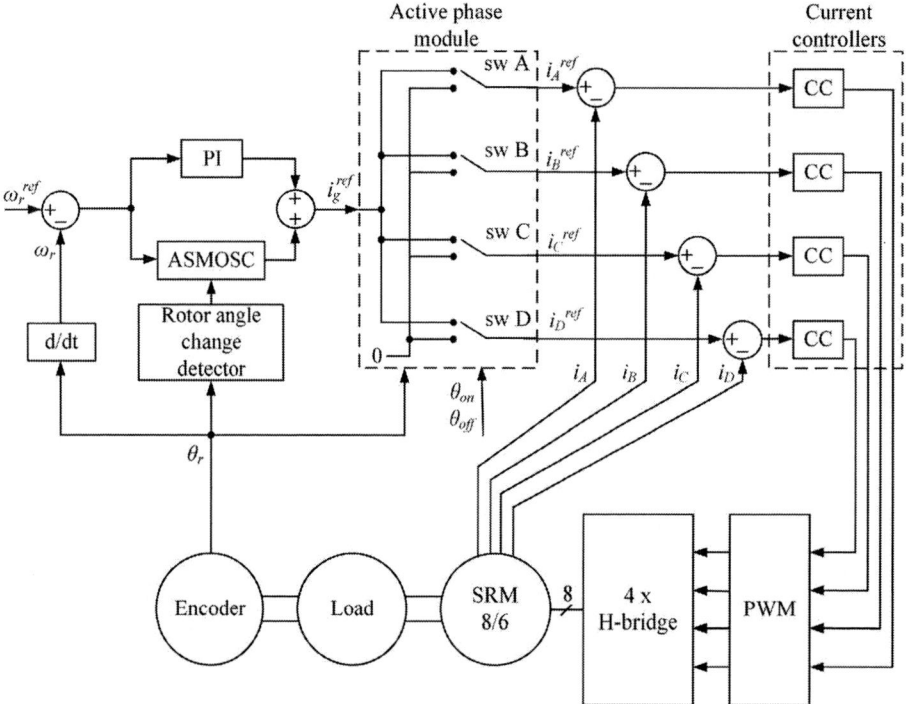

Fig. 7: Control system with ASMOSC for SRM.

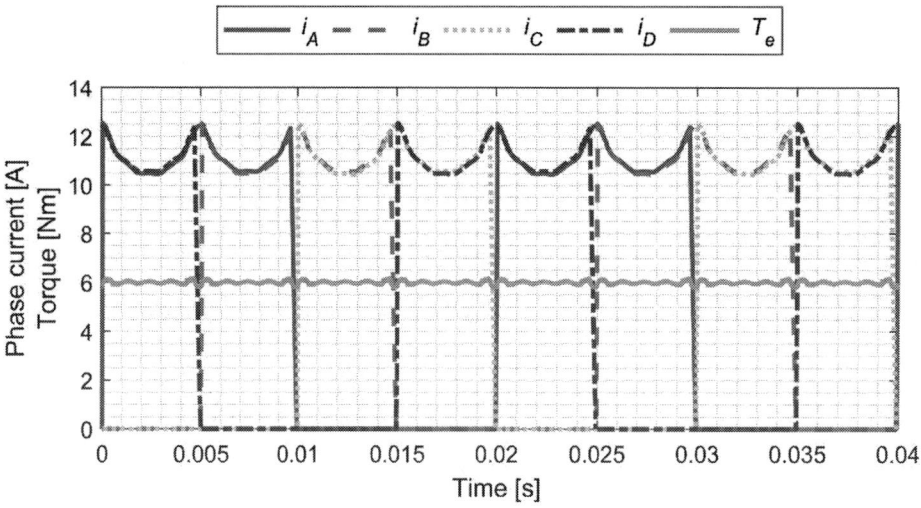

Fig. 8: Phase currents and total torque of SRM with ASMOSC.

Table I: Torque ripple comparison.

Control method	Torque min	Torque max	Torque ripple	Torque ripple (%)
Without ASMOSC	4.64 Nm	7.03 Nm	2.39 Nm	41
With ASMOSC	5.66 Nm	6.26 Nm	0.6 Nm	10

Laboratory setup

Power converter for a 4-phase 8/6 SRM is based on four H-bridges using 60 A, 650 V GaN transistors. Switching frequency is set to 250 kHz. Digital Signal Processor TMS320F28379D is used for controlling the converter. The SRM is equipped with position encoder generating 8192 pulses per revolution. A laboratory setup with power converter and SRM are presented in Fig. 9.

Fig. 9: Laboratory setup with power converter, control interface and SRM.

Experimental results

The proposed control method is verified on laboratory test bench (Fig. 9.). PMSM is used as a load for SRM. Speed reference is set to 300 rpm, which equals 5 rps. Load torque is 1 Nm. Torque ripple are measured indirectly by observing instantaneous angular rotor speed according to:

$$\frac{\delta\omega}{\delta t} = \frac{T_e - T_L}{J} \ , \tag{7}$$

where T_L is load torque and J is resulting drive system moment of inertia. Torque ripple reduction performance of the control system is analyzed with the help of root mean square value of the angular rotor speed AC component (ω_{AC}^{RMS}). Phase currents, reference current, rotor angular position and AC component of rotational speed are measured during experiment.

Initial FFT analysis of rotor angular speed shown the presence of additional harmonics that did not appear in the simulations, mainly: 1st, 6th, 12th and 18th. These harmonics are related to mechanical imperfections, magnetic anisotropy of SRM and cogging torque of PMSM. Moreover, in the case of tested SRM, levels of harmonics of order higher than 48th are low and can be omitted. For these reasons, in order to achieve high torque ripple reduction quality, control system is using ASMOSC with 6 resonant terms tuned for the following frequencies: $f_{OSC1} = n, f_{OSC2} = 6n, f_{OSC3} = 12n, f_{OSC4} = 18n, f_{OSC5} = 24n, f_{OSC6} = 48n$. Experimental results for control system without ASMOSC are presented in Fig. 10 and Fig. 11.

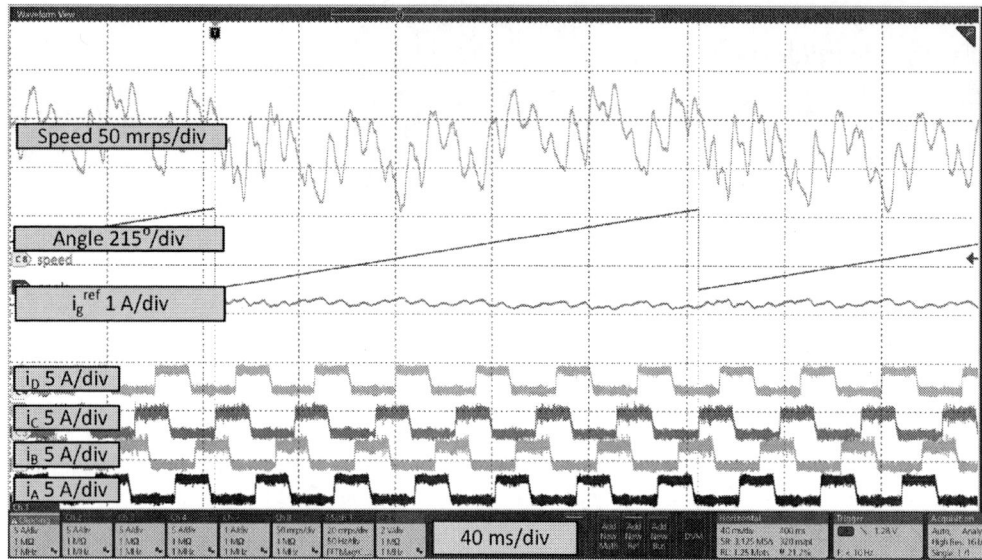

Fig. 10: Phase currents, reference current, rotor angular position and ac component of rotor angular speed for control system without ASMOSC.

It can be seen in Fig. 10 that the angular rotor speed in the drive system without ASMOSC is far from ideal constant shape. Fig. 11 shows the Fourier analysis of rotor angular speed. The angular speed contains mainly 1st harmonic (5 Hz), 6th harmonic (30 Hz), 12th harmonic (60 Hz), 18th harmonic (90 Hz), 24th harmonic (120 Hz) and 48th harmonic (240 Hz). For control system without ASMOSC $\omega_{AC}^{RMS} = 28.4\ mrps$.

Experimental results for control system with ASMOSC are presented in Fig. 12 and Fig. 13. Fig. 12 demonstrates that adding the ASMOSC to the control system introduces a specific AC component to the phase currents. This results in significantly lower ripple of the rotor angular speed. The spectrum of the angular rotor speed in the system with ASMOSC is shown in Fig. 13. It confirms that the content of harmonics in the speed is significantly reduced when ASMOSC is applied. For control system with ASMOSC $\omega_{AC}^{RMS} = 6.3\ mrps$.

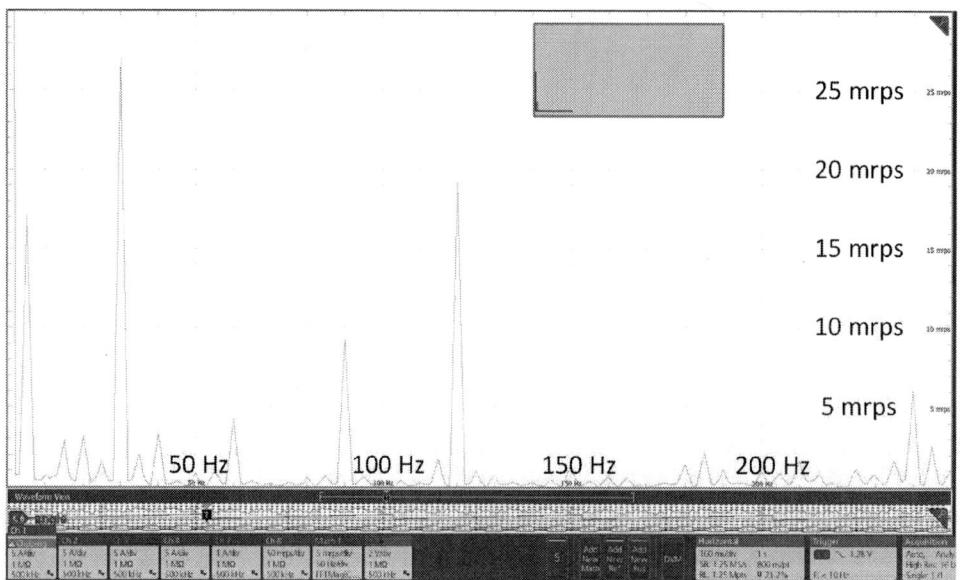

Fig. 11: Fourier analysis of angular rotor speed without ASMOSC.

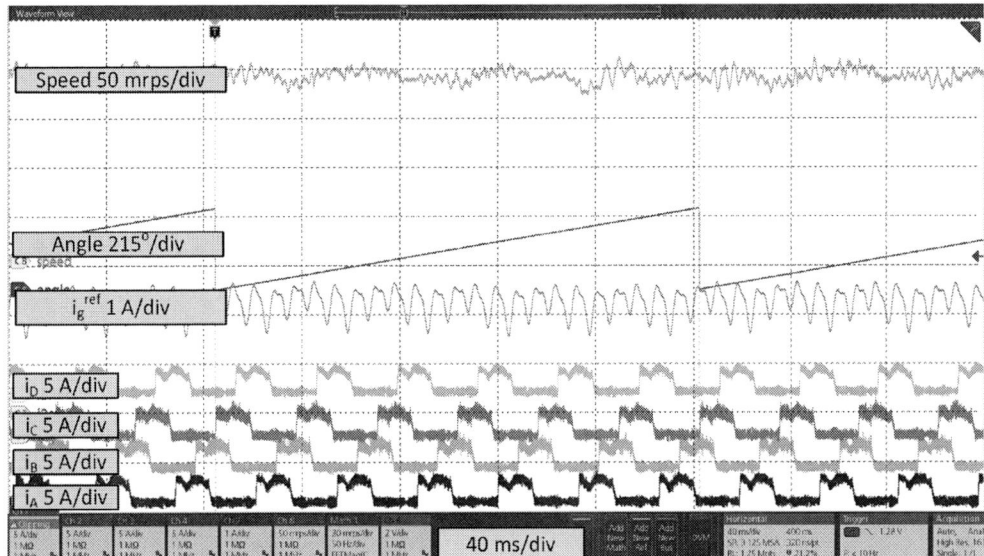

Fig. 12: Phase currents, reference current, rotor angular position and ac component of rotor angular speed for control system with ASMOSC.

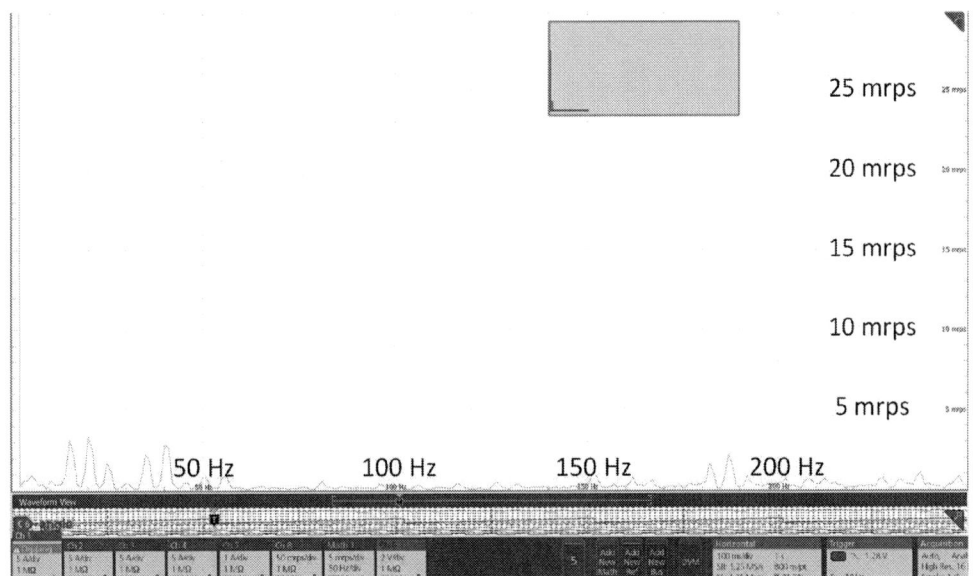

Fig. 13: Fourier analysis of angular rotor speed with ASMOSC.

Conclusion

The study shows that introducing novel Angle Sampled Multi-oscillatory controller in SRM allows to reduce torque ripple under variable speed conditions without the need of changing resonant-frequencies of oscillatory terms. Additional ASMOSC terms can be implemented to reduce other position-periodic torque ripples. An advantage of the presented method is that it does not require accurate parameter identification like resistance, inductance, or moment of inertia. Due to asynchronous nature of ASMOSC, the presented control system requires the usage of dual thread solution. It can be implemented with the usage of multicore digital signal controller. Patent is pending for presented control method for switched reluctance motor.

References

[1] Krishnan R.: Switched reluctance motor drives: modeling, simulation, analysis, design, and applications, Industrial Electronic Series, IBN 0-8493-0838-0, 2001

[2] Biernat A, Jackiewicz K. and Bienkowski K.: VIBRATION ANALYSIS OF SRM DESIGNED FOR MOTORING AND GENERATING OPERATION WITH SPREAD SPECTRUM CURRENT CONTROL, 19th European Conference on Power Electronics and Applications (EPE'17 ECCE Europe), 2017

[3] Divandari M. and Dadpour A.: Radial force and torque ripple optimization for acoustic noise reduction of SRM drives via fuzzy logic control, 2010 9th IEEE/IAS International Conference on Industry Applications - INDUSCON 2010, Sao Paulo, 2010

[4] Hieu P. T., Lee D., and Ahn J.: Design of 2-phase 4/2 SRM for torque ripple reduction, 2012 15th International Conference on Electrical Machines and Systems (ICEMS), Sapporo, 2012, pp. 1-6.

[5] Mehta S., Kabir M. A. and Husain I.: Extended Speed Current Profiling Algorithm for Low Torque Ripple SRM Using Model Predictive Control, 2018 IEEE Energy Conversion Congress and Exposition (ECCE), Portland, OR, 2018, pp. 4558-4563

[6] Ufnalski B., Galecki A., Kaszewski A. and Grzesiak L.: On the Similarity and Challenges of Multiresonant and Iterative Learning Current Controllers for Grid Converters Under Frequency Fluctuations and Load Transients, 2018 20th European Conference on Power Electronics and Applications (EPE'18 ECCE Europe), Riga, 2018, pp. P.1-P.10

[7] Jackiewicz K., Stras A., Ufnalski B. and Grzesiak L.: Comparative study of two repetitive process control techniques for a grid-tie converter under distorted grid voltage conditions, International Journal of Electrical Power and Energy System, Vol. 113, 2019

[8] Galecki A., Grzesiak L., Ufnalski B., Kaszewski A. and Michalczuk M.: Multi-oscillatory current control with anti-windup for grid-connected VSCs operated under distorted grid voltage conditions, 2017 19th European Conference on Power Electronics and Applications (EPE'17 ECCE Europe), Warsaw, 2017, pp. P.1-P.10

[9] Yepes A.: Digital resonant current controllers for voltage source converters Ph.D. thesis Spain: University of Vigo; 2011

Experimental Validation of the Performances of an Inverter Sized with Optimization Methods

Adrien Voldoire, Jean-Luc Schanen, Jean-Paul Ferrieux, Alexis Derbey
Univ. Grenoble Alpes, CNRS, Grenoble INP*, G2Elab
38000 Grenoble, France
E-Mail: jean-luc.schanen@g2elab.grenoble-inp.fr

Cyrille Gautier, Marwan Ali
SAFRAN TECH
78772 Magny-les-Hameaux, France

Keywords

«Voltage Source Inverters (VSI)», «Passive Component», «Design», «Test Bench», «Optimization»

Abstract

This work shows the experimental performances of a PWM inverter designed with a fast design automation tool. The chosen optimization method uses a deterministic algorithm with continuous equations. Analytical models are used to design active and passive components under loss, harmonic, thermal and efficiency constraints. Precise calorimetric test benches are used to measure the losses, and an original method is proposed to measure capacitor currents. Then, the results given by the optimization tool, presented as Pareto front, are discussed. A 10kW prototype was built to compare the constraints between the optimization results and the experimental values.

1. Introduction

The concept of More Electrical Aircraft consists in improving the energetic efficiency of aeronautic systems by increasing the electrical embedded vectors. It is now one of the key drivers of technological developments in aerospace. Power converters represent a significant part of the new technologies, and their power density still needs to be improved in order to save weight and therefore fuel consumption. Design automation using optimization method is an interesting way to reach this goal, defining the converter weight as the objective function, and considering the numerous aeronautical constraints. Sequential approach in the design may result in sub-optimal results, because several compromises are necessary during the choice of some parameters [1]. For example, the switching frequency has an opposite impact on the sizing of the semi-conductors and on the design of a low-frequency filter. Solving all tradeoffs simultaneously is the role of the optimization tool.

A deterministic algorithm, with a Sequential Quadratic Programming (SQP) method [2], is used to solve the global inverter design. This algorithm is very powerful to explore wide spaces of solutions with a large amount of constraints. However, it needs differentiable parameters and equations, which prevents from coupling the optimization tool with circuit or Finite Element Method (FEM) simulations. Since many parameters are discrete, and thus non-derivable, the solutions are found in a so-called "imaginary space", which is fully continuous [3]. Turn number of a winding is therefore not an integer, or all values of capacitance can be reached even if they are not available in a catalog. The optimization result is thus a kind of theoretical optimum, which is interesting in the pre-sizing step of a project [3], and allows fairly and quickly comparing two solutions. A possible drawback of gradient-based algorithms is to be stuck on local optima. If it is found that the result is sensitive to the initial conditions, a multi-start process is added to solve the issue.

The paper will briefly introduce all models used in the design automation. The main goal of this work is to prove the ability of the selected methodology to design accurately a power converter. Therefore, experimental validations of the optimization result will be presented, supported by an experimental setup including precise calorimetric test benches.

2. Analytical Models suitable for Deterministic Optimization

2.1. Introduction and specifications

This section summarizes the main models dedicated to a PWM inverter (Fig. 1), used in the optimization tool.

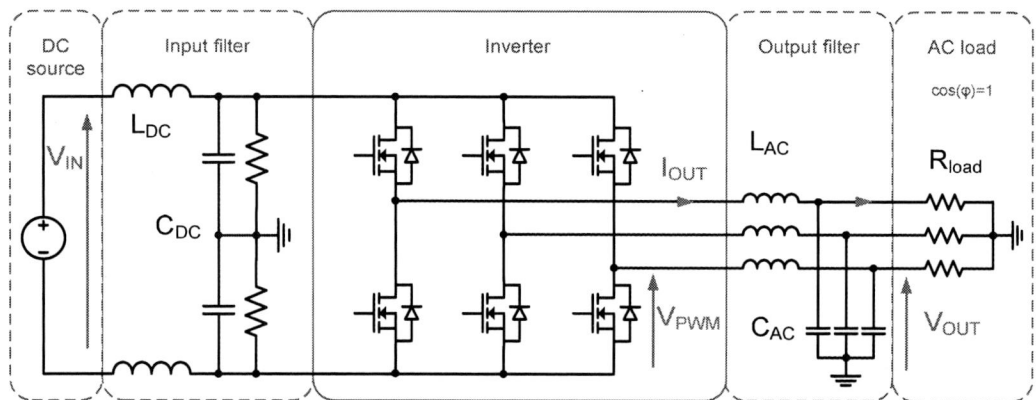

Fig. 1: Schematic of the PWM inverter

The specifications of the study and the main constraints driving the design of the converter are given in Table I.

Table I: Main specifications of the studied converter

Main input Data		
Load active power [W]	P_{load}	10 000
Load power factor	PF	1
Input Voltage [V]	V_{IN}	540
RMS AC Voltage (phase-neutral) [V]	V_{OUT}	115
Grid Frequency [Hz]	F_{GRID}	400
Switching Frequency [Hz]	F_{SW}	Variable
Constraints		
Junction temperature (MOSFET and diode) [°C]	T_J	< 150
Output voltage THD [%]	THD	< 3
Input bus voltage ripple [%]	ΔV_{DC}	< 1
Input source current ripple [%]	ΔI_{DC}	< 5
Volumetric losses of inductors [mW/cm³]	$P_{TOT\,VOL}$	< 500
Capacitor RMS current [A]	$I_{RMS\,C}$	< Inom capa AC
Global Efficency [%]	η	> variable

Analytical models are developed to be compliant with the optimization algorithm. A detailed survey about power electronics analytical models suitable for optimization is given in [4]. As the main interest of this paper is about validating the methodology rather than developing models, few equations will be given in this paragraph. A synoptic view of the problem is given in Fig. 2.

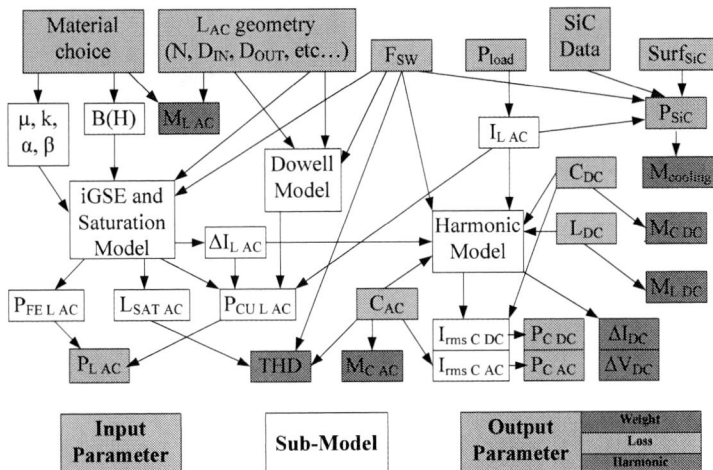

Fig. 2: Global synoptic view of the system

2.2. Signal analysis

On the AC side, the THD on the voltage and the RMS current in an AC filtering capacitor are calculated. The PWM spectrum is expressed with a double Fourier series leading to Bessel functions [5] (1).

$$V_{PWM}(k.m + p) = \frac{2.V_{DC}}{\pi.k} \sin\left(\pi \frac{k+p}{2}\right).J_p\left(\pi \frac{k.r}{2}\right) \tag{1}$$

with: k: switching frequency multiple variable, positive integer
p: sideband harmonic variable, relative integer
m: modulation index
h: harmonic index h=k.m+p
$J_p(k)$ p order, first type Bessel functions.

The transfer function (TF) of the filter is defined considering an L-C filter and a resistive load.

$$V_{out}(f) = TF(f, L_{AC}, C_{AC}, R_{load}).V_{PWM}(f) \tag{2}$$

The AC capacitor current is computed knowing the output voltage spectrum V_{out} at the terminals of the capacitor and its impedance.

The current spectrum in the DC side of the converter is useful to evaluate the DC voltage and current ripples, as well as the RMS DC capacitor current. It is worth noting that the contribution of the AC current ripple is considered in this analytical expression, developed in [6], what is important during optimization since large output AC ripples can be obtained during the exploration of the space of solution.

2.3. Component modeling

According to the aeronautical specifications given in Table I, this work considers silicon carbide (SiC) MOSFETs and diode with a voltage withstand of 1200V. To achieve an optimal result, the surface of semi-conductor chip is adapted to reach the best compromise between conduction and commutation losses. The computation of the commutation and conduction losses requires calculating the RMS and average current in the MOSFET and the diode [7]. A typical ratio of 1.5kg/kW is chosen to design the cooling system. However, the cooling system will not be taken into account in the experimental validation, since it is kept constant for practical reasons.

Toroidal cores are considered in the design of power inductors. Litz wire technology is used to decrease copper losses due to skin and proximity effects. Dowell method [8] enables calculating the variation of the resistance with the frequency. The chosen core material is iron-alloyed powder. In this case, the model takes into account a variable saturation state of the core. For example, the core can saturate during the current peak, what leads to a local increase of the current ripple. Considering saturation phenomenon is important when optimizing the weight, since inductors are contributing to a great part of the total converter weight. The full method is described in [9], consisting in calculating the induction ripple ΔB with Lenz law (3).

$$\Delta B_{HF}(t) = \frac{\frac{V_{in}}{2} - V_{out\,rms}\sqrt{2}.\sin(\omega.t)}{N.S.f_{sw}} \frac{1 + r.\sin(\omega.t)}{2} \tag{3}$$

This induction ripple is used in the analytical formulation of iGSE model [10] to compute iron losses (4).

$$P_{vol\,iGSE} = 2k_i \Delta B^{(\beta-\alpha)} \frac{1}{T} \sum_k \Delta B_{HF}(k)^\alpha.(t_{off}(k) - t_{on}(k))^{1-\alpha} \tag{4}$$

with: k_i: linear iGSE coefficient
α: Steinmetz coefficient for frequency
β: Steinmetz coefficient for induction
t_{off} and t_{on}: switching times

Finally, the current ripple is computed and is used for copper losses.

3. Test Benches

Among all the parameters that have to be checked, loss measurements of passive and active components are very challenging. Because of the switching waveforms, loss measurement by an electrical method (e.g. product of voltage and current) is not recommended. Calorimetric methods [11] are known to be more precise because it consists in a measure of the emitted heat, which gives directly the losses. This section introduces the two calorimetric test benches used for the measurement of semi-conductors losses, and AC inductor losses. In addition, the experimental setup to measure the current in the DC capacitor is indicated.

3.1. Semi-conductor loss measurement

The calorimetric measurement method for semi-conductors losses is set up in the case of water-cooling. Thus, the losses are proportional to the temperature rise and to the flow of water, as described by equation (5).

$$\Phi = Q.Cp.\Delta T \tag{5}$$

with ϕ the heat flow [W], Q the water flow [kg/s], Cp the water heat capacity [J/K/kg] and ΔT the temperature difference between hot and cold water [K].

The test bench based on this principle, introduced in Fig. 3, is used. With this setup, the maximum measurement error has been estimated below 2%.

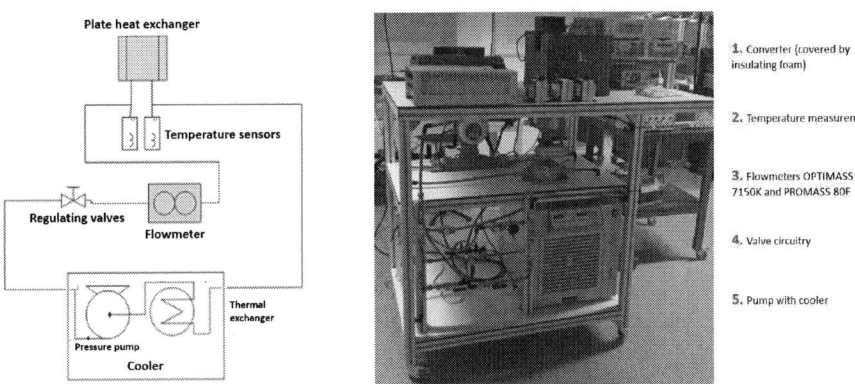

Fig. 3: Schematic and photo of the bench

Before validating the main performances of the inverter, this bench was used to verify the switching energy. Switching energies can be very different compared to the one given by the manufacturer, because the switching environment is different in the developed converter: gate driver and parasitic elements are strongly dependent on manufacturing and influence a lot switching energies.

In buck configuration, with experimental conditions given in Table II, many measurements were carried out with different switching frequencies and current.

Table II: Test conditions to measure switching energies

Config	Commutation	V_{IN} [V]	V_{OUT} [V]	L [mH]	ΔI [A]	T_J [°C]	Rg [Ω]
Buck	MOSFET-diode	400	200	1.0	< 10%	50	25

The low-side MOSFET of the switching leg is not controlled in this case (no active rectification). For one current, the slope of the loss-frequency curve gives the total switching energy, assuming that the Schottky diode has little recovery losses. The datasheet of the SiC power module used [12] gives the switching energies for a voltage of 600V. This is why a linear correction of 2/3 is applied on the measured energy to ensure a proper comparison.

Fig. 4 shows the measured losses for 3 output currents and many switching frequencies. With a linear interpolation, the switching energy is estimated and reported on the manufacturer datasheet. The results are close to the one given by the manufacturer, because the gate resistor was chosen equal to the value of the one used in the datasheet.

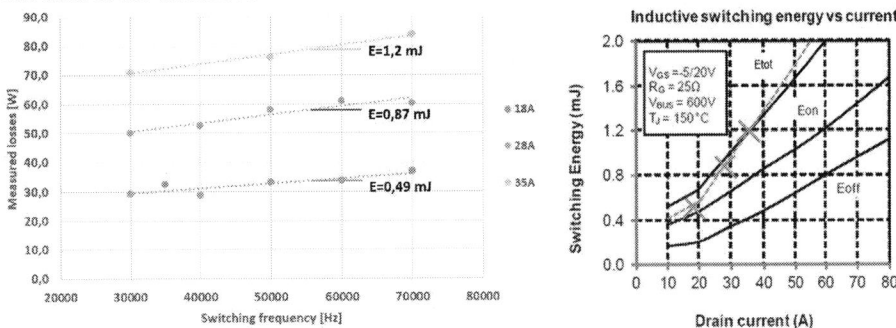

Fig. 4: Example of switching energy validation for 3 different currents

Other experiments are carried out with a gate resistor value divided by 3. In this case, the switching losses are half as much as before. This sensitivity, related to the converter manufacturing, makes this characterization mandatory for obtaining accurate prediction, even if this is time-consuming. A database of pre-existing measurement should be built in advance since it is not possible to conduct such measurement during the predesign step of a project.

3.2. Inductor loss measurement

The calorimetric method to measure the losses in an AC inductor is shown in Fig. 5. This test bench is introduced and used in [9]. Consequently, few details are given in this paper. The measurement consists in maintaining a constant temperature gradient between the oil and the environment. The environment temperature is kept constant by an oven (controller 2). The oil temperature is regulated by a PI controller, which controls a heating resistor. When the inductor heats, PI controller decreases the power injected in R1 in order to keep a steady temperature. The difference of injected power between the beginning and the end of the test is equal to the inductor losses. It was checked that the thermal conductivity of the system remains constant. The maximum measurement error is estimated below 1%.

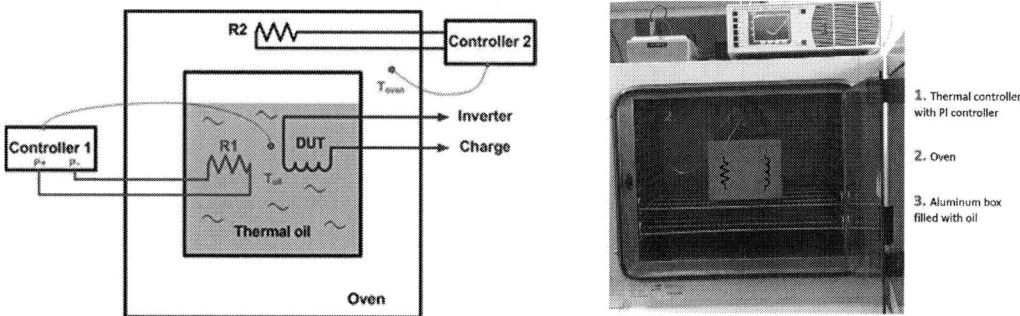

Fig. 5: Schematic of the calorimetric principle and photo of the bench

3.3. DC capacitor current measurement

Inserting a current sensor may modify the electrical path and change the measured current, especially in paralleled capacitors. It has been decided to measure a global current (Fig. 6) with a Rogowski coil embracing:

- the current between the MOSFETs and the DC capacitor
- the DC current coming from the source.

With a proper wire orientation in the coil, the two currents are subtracting, resulting in the DC capacitor current.

A "modular" converter core, shown in Fig. 7, is built with the possibility to measure these two currents. In fact, measuring the current between the MOSFETs and the DC capacitors is normally impossible because it flows in a PCB. To enable this measurement, the converter is made of two separated PCBs, one with the power modules, and one with the capacitors, linked by a metallic connector. This solution does not increase significantly the parasitic inductor, which is crucial to ensure a small voltage overshoot. Then, a Rogowski coil [13] is inserted to embrace the connector and the DC current.

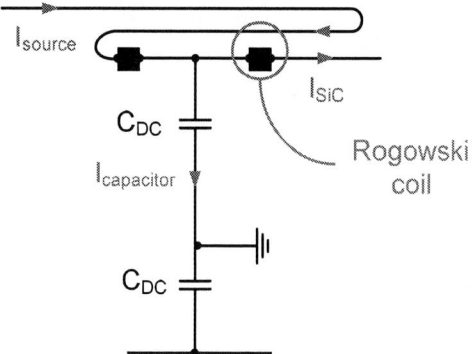

Fig. 6: Measurement method to get the DC capacitor current

Fig. 7: 3-Phase inverter developed (without filter)

Finally, all the experimental setup are used together to validate the converter performances. The overall experimental setup is given in Fig. 8. The converter design is presented in the next paragraph.

Fig. 8: Overall experimental setup

4. Comparison between Optimization Results and Experimental Results

4.1. From the "imaginary" world to the real converter

Optimization is carried out with different efficiency targets, always with the aim to minimize the weight. The Pareto front of Fig. 9 was obtained within 10 minutes by the algorithm. Among all of these optimal converters, it was chosen to build the inverter with 97.8% efficiency for validation.

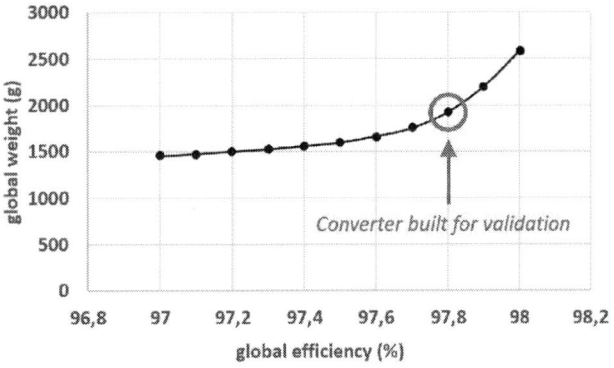

Fig. 9: Pareto Front weight-efficiency for a 10 kW inverter

The detailed design found by the algorithm is described in Fig. 10 with the optimal distribution of weight and losses. Half of the weight is in the AC inductors, which justify the modelling effort on this component to get an accurate result. Unsurprisingly, losses are distributed between the transistors and the AC inductors. Because of the impact of switching energies, semi-conductor losses are very sensitive to the switching frequency. Thus, global efficiency is directly linked to the switching frequency, placed at 40 kHz by the optimization tool for 97.8% efficiency.

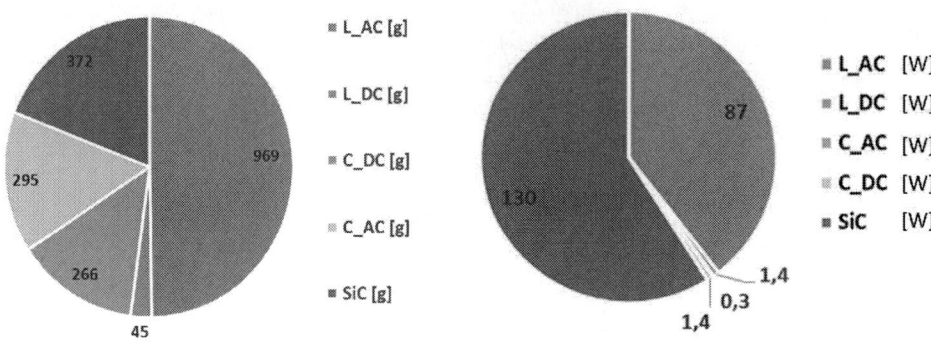

Fig. 10: Optimal weight and loss distribution given by the optimization tool at 97.8% of efficiency

The selected converter cannot be built right after the optimization. Using a deterministic approach involves working in an "imaginary" world, where every models and parameters are continuous and derivable. Therefore, it is necessary to get back to the "real" world to validate experimentally the design, using a dedicated procedure. The method to go back to the real world, described in [3], consists in assigning one by one the input variables to real values, which can be integers data or given by the manufacturer. Between every assignment, an optimization is carried to adapt the other input variables. The choice of assignment order of the variables depends on the importance of the variable and the gap between the two closest real values.

4.2. Global validation

Table III compares the theoretical performances of the converter given by design automation, and the performance of the developed converter.

Table III: Comparison between optimal results and experimental data

Losses	Unit	Optimization	Experiment	Error [%]
AC inductor losses	[W]	28.8	28.1	2.4
DC inductor losses	[W]	0.7	0.6	14
AC capacitor losses	[W]	0.1	0.1	0
DC capacitor losses	[W]	0.7	0.7	0
Semi-conductor losses	[W]	130	123	5.4
Global efficiency	[%]	97.8	97.88	0.08
Power quality				
RMS AC capacitor current	[A]	5.6	5.4	3.6
RMS DC capacitor current	[A]	21.9	22	0.5
Peak-peak DC source current ripple	[A]	0.9	0.6	33
Peak-peak DC bus voltage ripple	[V]	5.2	4	23
AC THD on voltage	[%]	2.1	1.8	14

The calculated losses are close to measured ones thanks to the detailed developed loss model and the precise calorimetric benches. Currents in the capacitors are also accurate and suffer of no doubt with the measurement setup described previously.

The measured voltage THD is lower than what the model expected. This is due to the saturation of the inductance along the low frequency period, which is considered in the worst case in the model. The measured DC voltage ripple is slightly different from the calculated one, because of the difficulty to measure an oscillation of few voltages on a 540 V_{DC} signal. Finally, a small absolute error appear on the DC source current ripple, because of the influence of the line parasitic inductor, which decrease this ripple.

To sum up, the difference between the theoretical values and the measured value is small, especially on crucial parameters. In addition, no abnormal phenomena's appeared during the multiple operation hours of the converter, except some low-frequency harmonics due to parasitic resonance, inducing nothing significant.

5. Conclusion

In this paper, analytical models for power electronics are selected and used to feed a fast optimization algorithm for the pre-design of inverters. Calorimetric test benches are presented to measure precisely active and passive component losses. They are used in the complete experimental validation of the characteristics a 10kW inverter sized with this design automation tool. Experimental results reveal the good accuracy that can be obtained with such a sizing tool.

This justifies the capability of a deterministic optimization tool to carry out an appropriate converter design. The fact of not using circuit simulation or Finite Element Simulation does not prevent from designing correctly a power converter. On the contrary, deterministic optimization makes it possible to find quickly the best design, which helps to take important decisions in the first step of a project. This ability is illustrated in [14], on the example of the selection of the best topology among multiple solutions.

References

[1] Y. Shen, S. Song, H. Wang, and F. Blaabjerg, 'Cost-Volume-Reliability Pareto Optimization of a Photovoltaic Microinverter', in *2019 IEEE Applied Power Electronics Conference and Exposition (APEC)*, Mar. 2019, pp. 139–146, doi: 10.1109/APEC.2019.8722043.

[2] P. T. Boggs and J. W. Tolle, 'Sequential Quadratic Programming', *Acta Numer.*, vol. 4, pp. 1–51, Jan. 1995, doi: 10.1017/S0962492900002518.

[3] M. Delhommais, J. Schanen, F. Wurtz, C. Rigaud, and S. Chardon, 'First order design by optimization method: Application to an interleaved buck converter and validation', in *2018 IEEE Applied Power Electronics Conference and Exposition (APEC)*, Mar. 2018, pp. 944–951, doi: 10.1109/APEC.2018.8341128.

[4] M. Mirjafari and R. S. Balog, 'Survey of modelling techniques used in optimisation of power electronic components', *IET Power Electron.*, vol. 7, no. 5, pp. 1192–1203, May 2014, doi: 10.1049/iet-pel.2013.0321.

[5] D. G. Holmes and T. A. Lipo, 'Modulation of ThreePhase Voltage Source Inverters', in *Pulse Width Modulation for Power Converters: Principles and Practice*, IEEE, 2003, pp. 215–258.

[6] A. Voldoire, J.-L. Schanen, J.-P. Ferrieux, C. Gautier, and C. Saber, 'Analytical Calculation of DC-Link Current for N-Interleaved 3-Phase PWM Inverters Considering AC Current Ripple', in *2019 21st European Conference on Power Electronics and Applications (EPE '19 ECCE Europe)*, Sep. 2019, p. P.1-P.10, doi: 10.23919/EPE.2019.8915183.

[7] A. Acquaviva and T. Thiringer, 'Energy efficiency of a SiC MOSFET propulsion inverter accounting for the MOSFET's reverse conduction and the blanking time', in *2017 19th European Conference on Power Electronics and Applications (EPE'17 ECCE Europe)*, Sep. 2017, p. P.1-P.9, doi: 10.23919/EPE17ECCEEurope.2017.8099052.

[8] P. L. Dowell, 'Effects of eddy currents in transformer windings', *Proc. Inst. Electr. Eng.*, vol. 113, no. 8, pp. 1387–1394, Aug. 1966, doi: 10.1049/piee.1966.0236.

[9] A. Voldoire, J. P. Ferrieux, J. Schanen, C. Rizet, C. Gautier, and C. Saber, 'Validation of Inductor Analytical Loss Models under Saturation Conditions for PWM Inverter', presented at the 2019 20th International Scientific Conference on Electric Power Engineering (EPE), 2019.

[10] K. Venkatachalam, C. R. Sullivan, T. Abdallah, and H. Tacca, 'Accurate prediction of ferrite core loss with nonsinusoidal waveforms using only Steinmetz parameters', in *2002 IEEE Workshop on Computers in Power Electronics, 2002. Proceedings.*, Jun. 2002, pp. 36–41, doi: 10.1109/CIPE.2002.1196712.

[11] D. Christen, U. Badstuebner, J. Biela, and J. W. Kolar, 'Calorimetric Power Loss Measurement for Highly Efficient Converters', in *The 2010 International Power Electronics Conference - ECCE ASIA -*, Jun. 2010, pp. 1438–1445, doi: 10.1109/IPEC.2010.5544503.

[12] Microsemi, 'APTMC120AM55CT1AG'. https://www.microsemi.com/existing-parts/parts/112756?catid=1349#resources (accessed Mar. 24, 2020).

[13] PEM Power Electronics Measurements, 'Sonde Rogowski CWT Ultra-mini'. Accessed: Apr. 16, 2020. [Online]. Available: https://www.es-france.com/index.php?controller=attachment&id_attachment=7132.

[14] A. Voldoire, J.-L. Schanen, J.-P. Ferrieux, B. Sarrazin, and C. Gautier, 'Experimental Comparison of the Performances of Interleaved and Coupled Inverters sized by Optimization', presented at the ECCE 2020, Detroit, Nov. 2020.

Influence of system parameters in variable speed AC-induction motor drives on parasitic electric bearing currents

Martin Weicker, Guilherme Bello, Dennis Kampen[1] and Andreas Binder
Institute of Electrical Energy Conversion
Darmstadt University of Technology
Landgraf-Georg-Straße 4,
64283 Darmstadt, Germany
Tel.: +49 6151/16-24191
E-Mail: mweicker@ew.tu-darmstadt.de
URL: https://www.ew.tu-darmstadt.de
[1] BLOCK Transformatoren-Elektronik GmbH
Max-Planck-Straße 36-46,
27283 Verden
Tel.: +49 4231 678-178
E-Mail: dennis.kampen@block.eu
URL: www.block.eu

Acknowledgements

We acknowledge the support of our industrial partners of the research group "bearing currents" especially *Block, Danfoss, Helukabel, Magnetec, SEW Eurodrive.*

Keywords

«Adjustable speed drive», «Asynchronous motor», «Breakdown», «Electrical machine», «IGBT», «Induction motor», «Measurement», «Passive filter», «Pulse Width Modulation (PWM)», «Test bench», «Three-phase system», «Variable speed drive», «Voltage Source Inverters (VSI)»

Abstract

High frequency bearing currents are a parasitic phenomenon that affects inverter-fed induction machines, causing damage and wear to the mechanical bearings, which might lead to unexpected failures, increase of the machine downtime and reduction of its lifespan. This paper describes how the bearing currents are affected by changes in the components of the electrical circuit that feeds the induction machine. The referred system consists of: Two-level voltage source inverter, different passive filters (a) common-mode ring filter, b) common-mode ring filter with small phase-to-ground capacitances, c) differential mode filter, d) dual-mode filter and e) du/dt-filter), different motor supply cables (shielded vs. unshielded cables with lengths of 50 m, 150 m or 300 m) and the two different induction motors (rated power of 11 kW and 15 kW, respectively). So, in total sixty system arrangements were tested. The highest reduction of bearing currents was measured with the dual-mode filter in all cable configurations.

1. Introduction

Bearing currents caused by the common-mode voltage of frequency inverters can damage the raceways of the bearings and lead to premature failure of the machine [1]. These currents depend on the parasitic capacitances of the motor, effective at high frequency, the ohmic/capacitive behavior of the bearing itself and the characteristics of the complete system (e.g. inverter, filter, motor supply cable) to which the machine is connected.

Four main types of inverter-induced bearing currents (IIBC) are well known in the literature: (i) circulating, (ii) electrical discharge machining (EDM), (iii) capacitive and (iV) rotor-to-ground bearing currents [2]. This paper focuses on high-frequency bearing currents (ii) associated with 4-pole cage induction machines below a rated power of 20 kW, which according to [3] and to the experimental observations in this work show almost only EDM bearing currents (ii) in case of non-grounded rotor.

1.1. EDM bearing currents occurrence mechanism

EDM bearing currents are produced by the fast switching behavior of the frequency inverter, that creates three phase PWM (Pulse Width Modulation) controlled phase-to-ground voltages u_{Ug}, u_{Vg}, u_{Wg}. The sum of the phase-to-ground voltages, differently than in a symmetric sinusoidal voltage system, is in general not zero at a given time instant. Thus, a non-zero common-mode voltage u_{com} occurs at the inverter output and consequently at the machine terminals (1).

$$u_{com} = \frac{u_{Ug} + u_{Vg} + u_{Wg}}{3} \tag{1}$$

This common-mode voltage u_{com} acts as the input of a voltage divider formed by the parasitic capacitances inside the induction machine, effective at high frequencies 10 kHz … 10 MHz (Figure 1), whose output is the voltage across the bearing u_b. These parasitic capacitances are the stator winding-to-rotor capacitance C_{sw-r}, the rotor-to-frame capacitance C_{r-f} and bearing capacitances $C_{b,DE}$ and $C_{b,NDE}$ of the drive end (DE) and non-drive end (NDE) side. The relation between the bearing voltage u_b and common-mode voltage u_{com} is called Bearing Voltage Ratio (*BVR*) and is calculated with (2).

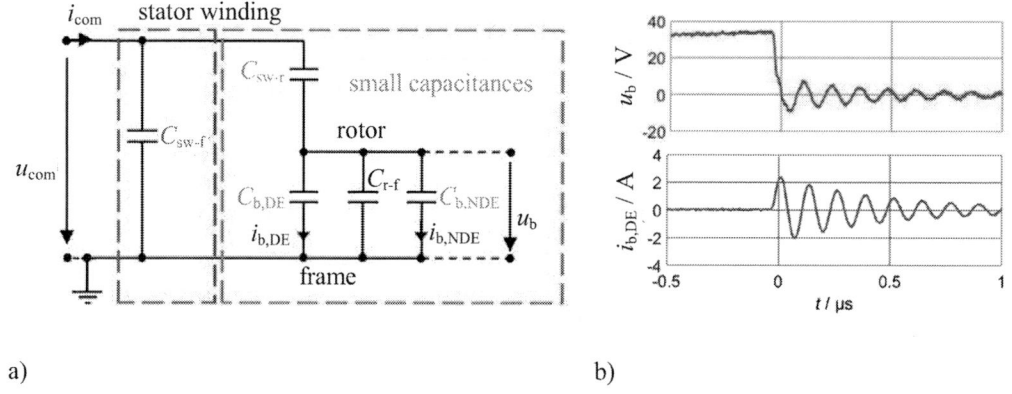

a) b)

Fig. 1: a) Equivalent circuit for calculation of the bearing voltage u_b from the common-mode voltage u_{com} via the HF machine capacitances of a cage induction machine, sw: stator winding, f: frame, r: rotor, b: bearing, DE: drive end, NDE: non-drive end,
b) Measured typical EDM bearing current at the tested 11 kW-induction machine, rotor speed 1000 rpm, switching frequency of the inverter $f_{IGBT} = 4$ kHz.

$$BVR = \frac{u_b}{u_{com}} = \frac{C_{sw-r}}{C_{sw-r} + C_{r-f} + C_{b,DE} + C_{b,NDE}} \tag{2}$$

If the maximum electrical field strength E at the lubricating oil film inside the bearing, which is given by the ratio between the maximum bearing voltage u_b and the minimum insulating oil film thickness h exceeds the breakdown threshold value $E = u_b/h > E_d$, then a voltage breakdown of the bearing voltage will occur, creating high frequency discharge current pulses called EDM bearing currents, depicted in Figure 1 b) [9], [12].

2. Measurement system and components

The used measurement system is shown in Figure 2. The two cage induction machines of 11 kW and 15 kW are fed each by the same IGBT-two-level-voltage source inverter with $U_d = 580$ V DC link voltage and a switching frequency of the inverter f_{IGBT} = 4 kHz, type FC-302 from Company *Danfoss*. The inverter is connected alternatively to the five passive inverter output filters (a) 1.1. common-mode ring filter, b) 1.2 common-mode ring filter with small phase-to-ground capacitances, c) 1.3 differential mode sine filter, d) 1.4 dual-mode sine filter, e) 1.5 du/dt-filter), manufactured by Company *Block*. Alternatively, an unshielded **NYY-J** or shielded **2YSLCYK-J** motor supply cable of 50 m, 150 m or 300 m length from Company *Helukabel* connects the feeding source to a 11 kW- or 15 kW-4-pole-cage induction machine.

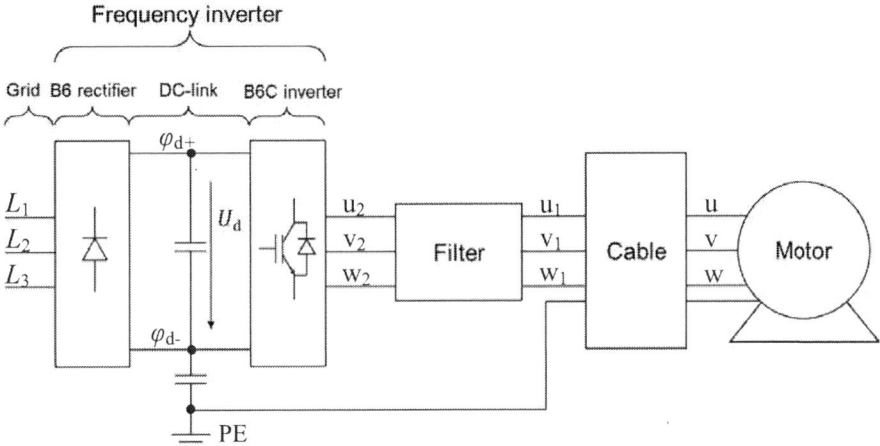

Fig. 2: Measurement set-up at the public 400 V AC grid L_1, L_2, L_3 with two-level-voltage source inverter, passive inverter output filter, motor supply cable and cage induction machine

For the measurement of bearing currents the bearing shields were re-equipped with an electrical insulation foil between bearing outer ring and bearing shield, depicted in Figure 3 a). A short copper loop bridging thus insulation is used to measure the bearing currents with a current probe type SS250 from Company *Iwatsu*. The bearing voltage at drive end (DE) is measured between the shaft, by contacting it by a carbon brush, and the machine housing, as depicted in Figure 3 b).

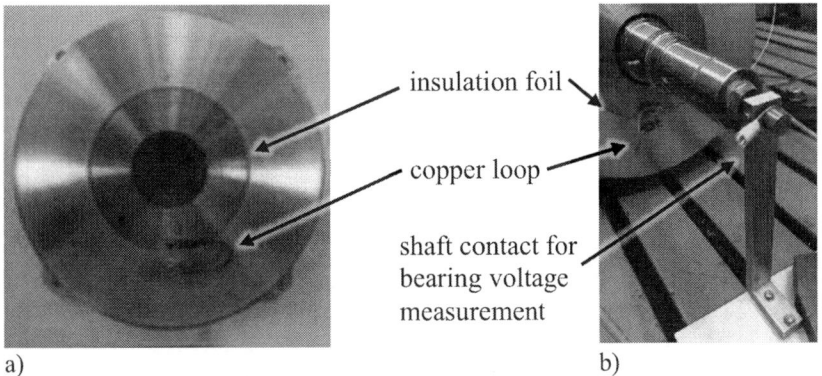

a)   b)

Fig. 3: a) Re-equipped bearing end shield with electrical foil insulation of the bearing seat and a short copper loop bridge to measure bearing currents with a current probe. b) Contacting the shaft at DE and at the machine housing for measuring in between the bearing voltage $u_{b,DE}$

2.1. Motor supply cables

The two alternative types of motor cables used in this work **NYY-J** and **2YSLCYK-J** share many similarities, such as similar external diameters (15 mm and 13 mm respectively) and same phase conductor cross section (4 mm^2). However, they differ by the fact that **NYY-J** is an unsymmetrical cable in relation to the position of the phase conductors to the ground conductor inside the cable (see Figure 4 a)), and it is an unshielded cable, while **2YSLCYK-J** has a symmetrical conductor arrangement and has an aluminum shield conductor layer (see Figure 4 b)). These differences are responsible for the different specific per phase and per meter inductance L'_{ph}, resistance R'_{ph} and capacitance C'_{ph} parameters, which lead to different wave impedance values, according to the approximation (3).

$$Z_0 = \sqrt{L'_{\text{ph}} / C'_{\text{ph}}} \tag{3}$$

The bibliographies [4] and [5] suggest that the motor supply cable characteristics are deeply related to voltage oscillations at the machine terminals caused by wave reflection effects, which may have an impact on the common-mode voltage u_{com} and consequently on the bearing currents.

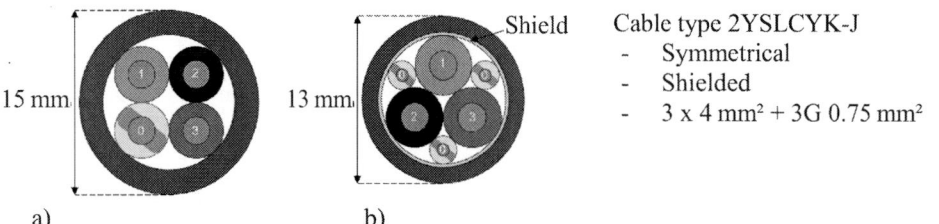

a)   b)

Fig. 4: Geometry of the used motor supply cables with a cross section per phase of 4 mm²: a) unshielded cable type NYY-J, b) shielded symmetrical cable type 2YSLCYK-J

In order to quantify their per phase characteristics, the cables were measured by an impedance analyzer type HP4192A LF from Company Hewlett-Packard, from which it was possible to derive the following values showed in Table I.

Table I: Measured per phase parameters for cable types NYY-J and 2YSLCYK-J at a frequency $f = 1$ kHz

Cable NYY-J			Cable 2YSLCYK-J		
Phase 1 and Phase 3			Phase 1, Phase 2 and Phase 3		
$R'_{\text{ph},1}$	$L'_{\text{ph},1}$	$C'_{\text{ph},1}$	$R'_{\text{ph},1}$	$L'_{\text{ph},1}$	$C'_{\text{ph},1}$
9.31 mΩ/m	0.633 µH/m	102.60 pF/m	14.01 mΩ/m	0.403 µH/m	116.70 pF/m
Phase 2					
$R'_{\text{ph},2}$	$L'_{\text{ph},2}$	$C'_{\text{ph},2}$			
9.44 mΩ/m	0.767 µH/m	82.54 pF/m			

2.2. Passive filters

Among the known mitigation techniques to reduce bearing currents, as discussed in literature [1], [7], [10], [11], one can highlight the passive filters at the inverter output, which do not require adjustments on the induction machine itself, because they eliminate the cause of the problem at its source. Thus, five types of filters were studied and compared to identify the most appropriate filter to reduce the EDM bearing currents. These filter numeration and type correspondence follow Table II. Their equivalent circuits are shown in Figure 5.

Table II: Filter numeration and its correspondence to the type and model

Numeration	Filter type	Model name
1.1	Common-mode ring filter (8 rings)	M-049
1.2	Common-mode ring filter (8 rings) with 5 nF Y-capacitors	M-049
1.3	Differential mode sine filter	SFB 400/16.5
1.4	Dual-mode sine filter	SFA 500/13
1.5	du/dt filter	B 1706040/17.5

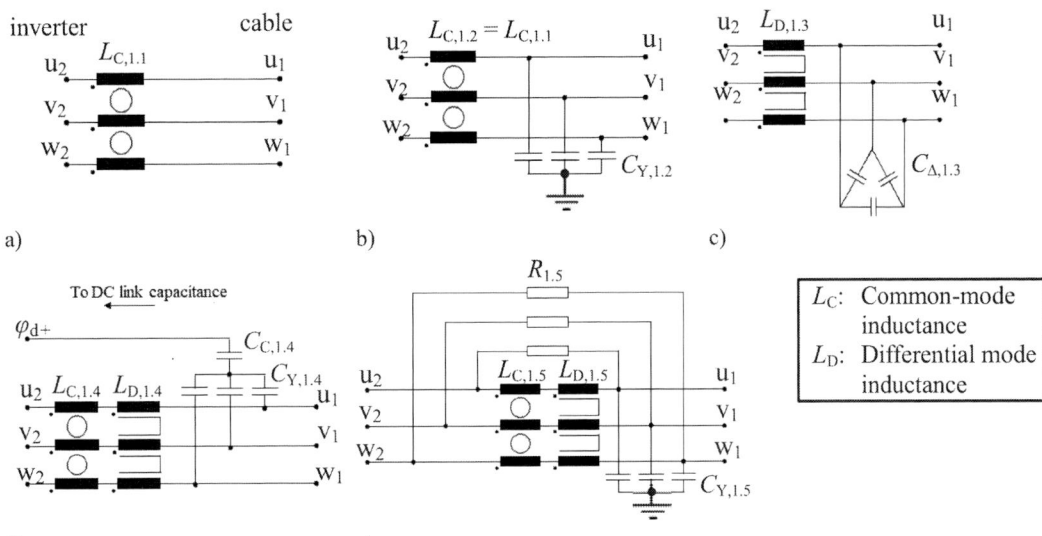

a)   b)   c)

d)   e)

Fig. 5: Equivalent circuits of the five inverter output filters: a) Common-mode ring filter (Filter 1.1); b) Common-mode ring filter with small phase-to-ground capacitors (Filter 1.2); c) differential mode sine filter (Filter 1.3); d) dual-mode sine filter (Filter 1.4); e) du/dt-filter (Filter 1.5).The terminals u_2, v_2, w_2 are connected to the inverter output terminals, u_1, v_1, w_1 to the input terminals of the 3-phase motor supply cable.

Both differential mode and dual-mode sine filters belong to the same class of passive filters, which are characterized for filtering the PWM switching frequencies and thus enabling a sinusoidal line-to-line voltage output, as it can be understood by their resonance frequencies: $f_{\mathrm{res}1.3} = 1.1$ kHz and $f_{\mathrm{res}1.4} = 1.3$ kHz, respectively.

However, they differ by the presence of a common-mode ring filter placed around the three phase conductors at the dual-mode sine filter terminals of Filter 1.4. Also, Filter 1.4 is characterized for having a physical connection to the positive DC-link voltage potential $\varphi_{\mathrm{d}+}$ of the frequency inverter. Thus, we can filter the common-mode voltage u_{com} and reduce the common-mode current i_{com} (Fig. 1a) by offering a path for the common-mode currents to flow and may be an effective mitigation technique for EDM bearing currents. The last filter tested during the experiments is a du/dt-filter, whose most relevant characteristic is to damp the voltage overshoots at the motor terminals.

2.3. Induction motors

The two tested squirrel cage induction machines of 11 kW and 15 kW rated power belong to the efficiency class IE3. Both machines have the same stator slot geometry and differ only by the iron length, which is $l_{\mathrm{Fe},11\mathrm{kW}} = 170$ mm and $l_{\mathrm{Fe},15\mathrm{kW}} = 220$ mm for the 11 kW- and 15 kW-machines, respectively. The main machine data is shown in Table III.

Table III: Main data of the investigated 11 kW- and 15 kW-induction machines

	11 kW-IM	**15 kW-IM**		**11 kW-IM**	**15 kW-IM**
P_{N}	11 kW	15 kW	f_{N}	50 Hz	
U_{N}	400/690 V Δ/Y		Thermal Class	155(F)	
I_{N}	21/12.2 A Δ/Y	29/16.7 A Δ/Y	m	117.7 kg	135.4 kg
$\cos\varphi$	0.81	0.80	η	91.4%	92.1%
n_{N}	1473 rpm	1474 rpm			

The EDM bearing currents effect on these motors is investigated by evaluating their *BVR*, which is done by measuring both the common-mode voltage u_{com} and bearing voltage u_b at the same time and use (2). This results in a *BVR* of 8.2% for the 11 kW-machine and 9.5% for the 15 kW-machine, as shown in Figure 6.

In comparison to other BVR-values found in literature for lower efficiency class cage induction machines (*BVR* = 2 ... 8%) [3] it is noticed that these two motors have a rather high *BVR* value.

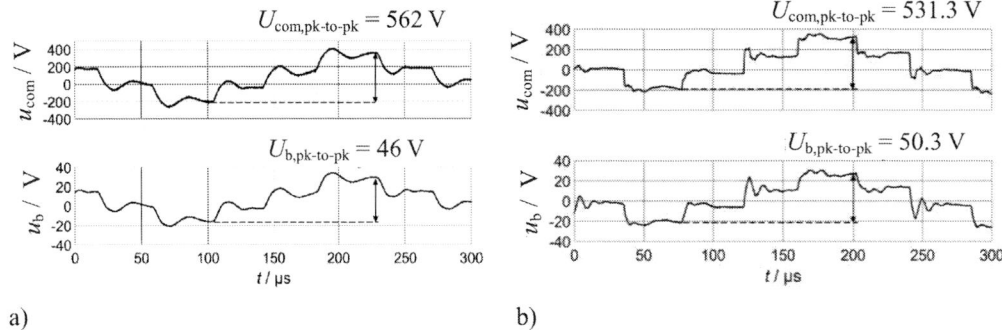

a)   b)

Fig. 6: Measured common-mode voltage u_{com} and bearing voltage u_b (NDE) for *BVR* calculation.

3. Common-mode voltage waveforms at motor terminals

As explained in Section 1.1, the formation of EDM bearing currents depends on the inverter output common-mode voltage value at the entrance of the parasitic capacitance circuit, shown in Figure 1. In the following, the bearing current amplitudes and the waveforms of the common-mode voltage u_{com} at the motor terminals for the different passive filters and cable length are examined.

Figure 7 shows a comparison of typical measured common-mode voltages at the 11 kW-induction motor terminals for an unshielded cable (NYY-J) with 50 m length. In literature [8] the influence of cable characteristics on motor terminal overvoltage due to voltage wave reflections is shown and will not be discussed here.

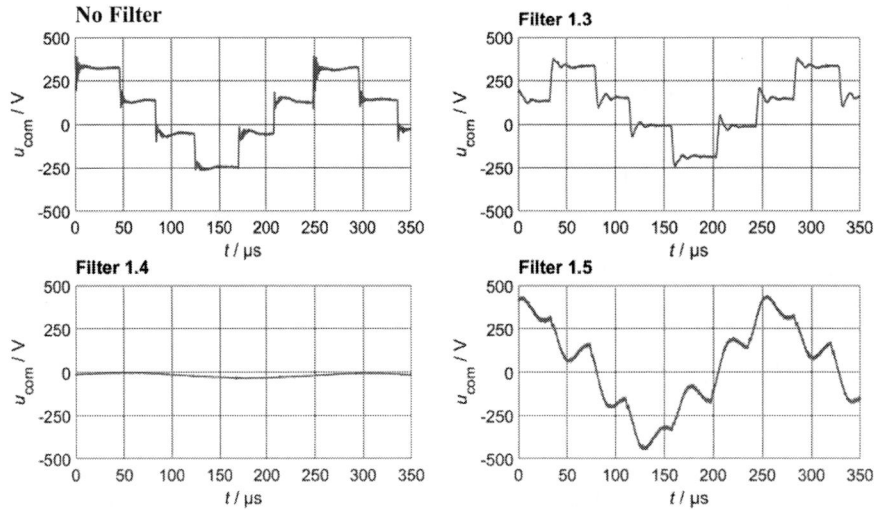

Fig. 7: Measured typical common-mode voltages at the 11 kW-induction motor terminals for: No Filter, Filter 1.3, Filter 1.4 and Filter 1.5, for an unshielded motor supply cable NYY-J with 50 m cable length

One observes in Figure 7 that all investigated filters reduce the voltage overshot HF oscillations caused by wave reflections at the motor terminals due to their low-pass filtering. Only Filter 1.4 successfully reduces the common-mode voltage nearly completely, which is due to the low low-pass resonance frequency of $f_{res,1.4} = 1.3$ kHz and the direct connection to the DC-link voltage. But this filter is the heaviest and more expansive one of all five investigated filters. The reduction of the common-mode voltage via (2) directly reduces the voltage u_b over the mechanical bearings of the machine and likewise reduces the EDM bearing currents.

4. Common-mode current measurements

The common-mode currents in the system Fig. 2 are the zero sequence current components, produced by the frequency inverter.

The amplitude of the common-mode currents is highly affected by the circuit capacitance values, which are dependent on the type of cable and the cable length. So experimental arrangements to study the influence of different system components on the common-mode current amplitudes were performed.

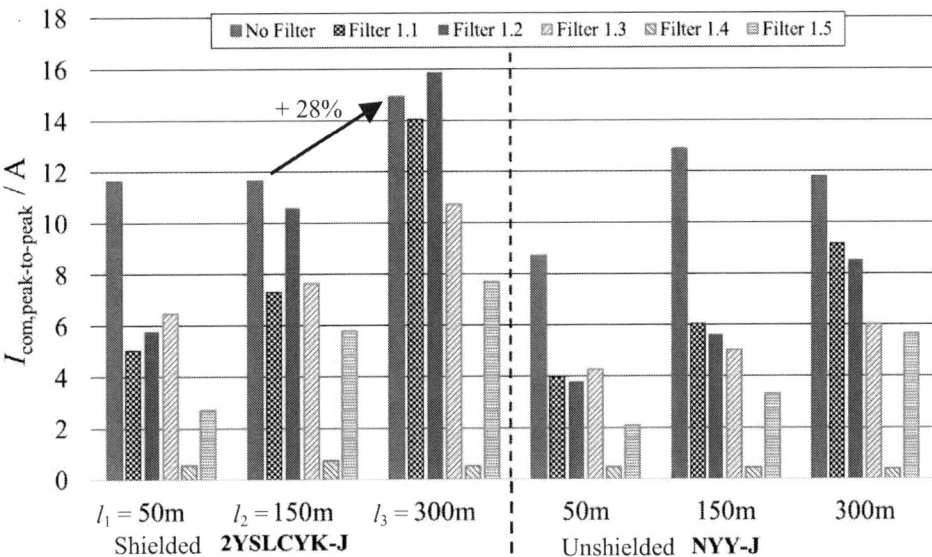

Fig. 8: Comparison of measured average peak-to-peak values of the common-mode current at the motor terminals for a motor rotation speed of 1500 rpm at the 15 kW-induction motor for: No Filter, Filter 1.1, Filter 1.2, Filter 1.3, Filter 1.4 and Filter 1.5, for two cable types: Shielded 2YSLCYK-J and unshielded NYY-J, three cable lengths: 50 m, 150 m and 300 m.

Figure 8 shows that longer cables tend to allow higher common-mode currents. For instance, an increase of 28% was observed when comparing cable lengths $l_2 = 150$ m and $l_3 = 300$ m without filter with the shielded cable 2YSLCYK-J. This is due to the higher phase capacitance values $C_3 = C'_{ph} \cdot l_3 > C_2 = C'_{ph} \cdot l_2$ in shielded cables, associated with higher common-mode currents i_{com}.

Additionally, Figure 8 shows that shielded cables, such as 2YSLCYK-J have in general higher common-mode currents values for the same cable length due to the higher per phase specific capacitance values (see Table I).

5. Measured bearing currents in dependence of rotor speed

The bearing currents measurements were performed for the speed range 0 ... 3000 rpm of the induction motors, varying from 150 rpm to 3000 rpm in steps of 150 rpm, which results in twenty speed points. Each point represented in Figure 9 corresponds to the average of 250 EDM bearing current peak-to-peak values, measured via an oscilloscope, controlled by a LABView-program.

The curve in Figure 9 compares the measurement results for four system setups (including the "no filter" case) for an unshielded cable NYY-J with a cable length of 50 m, and a motor with rated power of 11 kW. Each measurement was performed for a constant bearing temperature of $\vartheta_b = 50$ °C.

Testing at different speed values gives information how the bearings insulation film behaves over the speed. At rotor speeds below $n = 750$ rpm there is a significant reduction on the bearing current peak-to-peak values. This is due to thinner insulation film thickness h at lower speeds, which increases the electrical field strength E inside the bearings and triggers a bearing voltage breakdown at lower bearing voltages u_b, consequently reducing the EDM bearing current peak-to-peak values $I_{b,pk-to-pk}$ [6].

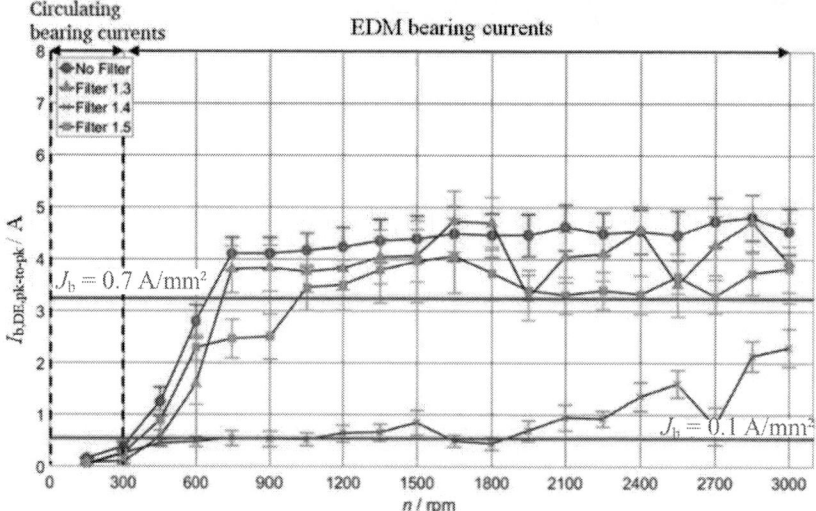

Fig. 9: Measured average peak-to-peak values of the EDM bearing current over the rotation speed at the DE-side of the 11 kW-induction motor for: No Filter, Filter 1.3, Filter 1.4 and Filter 1.5 for an unshielded motor supply cable NYY-J with 50 m cable length, bearing temperature $\vartheta_b = 50\ °C$.

Fig. 9 indicates as straight lines the two apparent bearing current density values (as bearing current peak vs. *Hertz*'ian area) that mark the limit below which no damage to the bearing will probably occur ($J_b < 0.1\ A/mm^2$) and above which it is almost certain that the bearing will be damaged ($J_b > 0.7\ A/mm^2$), according to [1]. The dual-mode sine filter (Filter 1.4) successfully reduces the bearing currents peak-to-peak values below its critical value. Although the EDM bearing currents are the focus of this paper, because they are the most observed type of IIBC on tested machines, they were not the only one. Also small circulating bearing currents were measured at low rotation speed ($n < 150\ rpm$) with partially metallic contact in the bearings. But the machine size is too small to build up a big shaft voltage, so no harmful circulating bearing current amplitudes occur [13].

Similar to Fig. 9, for the 15 kW-induction machine a reduced data set of measured speed points was used to measure EDM bearing currents. Results are depicted in Fig. 10 for all investigated Filters and 50 m unshielded cable type NYY-J.

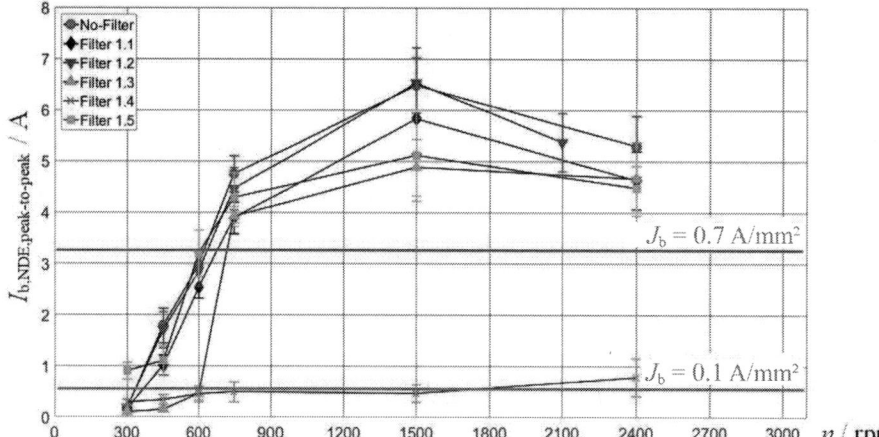

Fig. 10: Like Fig. 9, but with 15 kW-induction motor: No Filter, Filter 1.1, Filter 1.2, Filter 1.3, Filter 1.4 and Filter 1.5, for an unshielded cable NYY-J with 50 m cable length, bearing temperature $\vartheta_b = 50\ °C$.

6. Comparison of bearing currents amplitude between different system configurations

Summarizing the bearing current measurements for all the sixty different arrangements, it is possible to obtain the bar graphics in Figure 11, which give an overview of how each system configuration influences the EDM bearing current amplitudes at the same nominal rotation speed of $n = 1500$ rpm for a constant bearing temperature of $\vartheta_b = 50$ °C. Only the dual-mode filter 1.4 is capable of reducing strongly the EDM bearing currents as a common-mode effect.

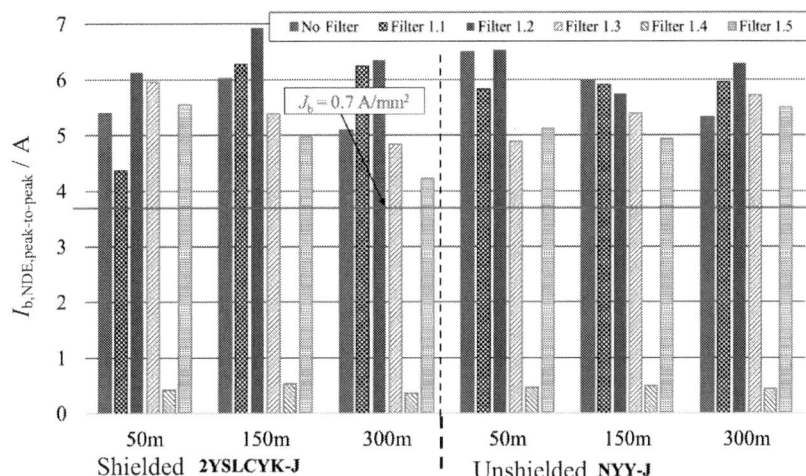

Fig. 11: Comparison of measured average peak-to-peak values of the bearing currents at NDE-side for nominal rotation speed of 1500 rpm in the 15 kW-induction motor for: No Filter, Filter 1.1, Filter 1.2, Filter 1.3, Filter 1.4 and Filter 1.5 for shielded cable type 2YSLCYK-J and unshielded cable type NYY-J, each cable with three different cable lengths: 50 m, 150 m and 300 m.

7. Conclusion

The dual-mode sine filter (Filter 1.4) with connection to the DC-link voltage is the most effective mitigation technique from all five tested inverter output filters. It enables a significant reduction on the EDM bearing currents peak-to-peak value by more than 90%. This is due to the fact that it filters the common-mode voltage u_{com} and therefore the common-mode current i_{com} by offering a path for these currents back to the voltage source. The other investigated filters (common-mode ring filters, differential mode sine filter and dv/dt-filter) cannot be considered as an effective mitigation method for suppressing EDM bearing currents.

Additionally, the experimental results also reveal that the shielding and the cable lengths have a minor influence on discharge bearing currents, especially when compared to the effect of the filters; but the cable type and cable lengths have influence on circulating and rotor-to-ground bearing currents, which are not investigated here.

Further, one can notice that higher EDM bearing currents are observed in the 15 kW-induction motor than in the 11 kW, which is related to the fact that the 15 kW-induction motor has a higher BVR (9.5%) than the 11 kW (8.2%).

References

[1] A. Muetze, A. Binder, H. Vogel, J. Hering, "Experimental evaluation of the endangerment of ball bearings due to inverter-induced bearing currents," *Proc. of the 2004 IEEE Industry Applications Conference*, Seattle, WA, USA, 03. - 07. October, 2004, vol. 3, pp. 1989 - 1995.

[2] A. Muetze, „Thousands of hits: on inverter-induced bearing currents, related work and the literature", *e&i – Elektrotechnik und Informationstechnik*, vol. 28, no. 11, pp. 382 - 388, 2011.

[3] A. Binder, A. Muetze, "Scaling Effects of Inverter-Induced Bearing Currents in AC Machines," *IEEE Transactions on Industry Applications*, vol. 44, no. 3, pp. 769 - 776, June 2008.

[4] G. Skibinski, R. Tallam, R. Reese, B. Buchholz, R. Lukaszewski, "Common Mode and Differential Mode Analysis of Three Phase Cables for PWM AC Drives," *Proc. of the 2006 IEEE Forty-First IAS Annual Industry Applications Conference*, Tampa, Florida, USA, 10. August, 2007, pp. 880 - 888.

[5] O. Magdun, A. Binder, C. Purcarea, A. Rocks, B. Funieru, "Modeling of asymmetrical cables for an accurate calculation of common mode ground currents," *Proc. of IEEE Energy Conversion Congress and Exposition*, 2009, San Jose, California, USA, 20. - 24. September, 2009, pp. 1075 - 1082.

[6] H. Tischmacher, I. P. Tsoumas and S. Gattermann, "Probability model for discharge activities in bearings of converter-fed electric motors," *Proc. of International Conference on Electrical Machines (ICEM),* Berlin, Germany, 02. - 05. September, 2014, pp. 1818 - 1824.

[7] H. Akagi, S. Tamura, "A Passive EMI Filter for Eliminating Both Bearing Current and Ground Leakage Current From an Inverter-Driven Motor," *IEEE Transactions on Power Electronics,* vol. 21, no. 5, pp. 1459 - 1469, 2006.

[8] G. Skibinski, D. Leggate and R. Kerkman, "Cable characteristics and their influence on motor over-voltages," *Proc. of APEC 97 - Applied Power Electronics Conference*, Atlanta, GA, USA, 23. - 27. February, 1997, pp. 114 - 121.

[9] E. Wittek, M. Kriese, H. Tischmacher, S. Gattermann, B. Ponick, G. Poll, "Capacitances and lubricant film thicknesses of motor bearings under different operating conditions," *Proc. of 2010 XIX International Conference on Electrical Machines*, 6 pages, CD ROM, Rome, Italy, 06. - 08. September, 2010.

[10] IEC/TS 60034-25:2014, Rotating electrical machines - Part 25: AC electrical machines used in power drive systems - Application guide (IEC/TS 60034-25:2014), Geneva, Switzerland.

[11] DIN EN 60034-24:2011, Rotating electrical machines – Part 24: Online detection and diagnosis of potential failures at the active parts of rotating electrical machines and of bearing currents – Application guide (IEC/TS 60034-24:2009), Geneva, Switzerland.

[12] Y. Gemeinder, M. Schuster, B. Radnai, B. Sauer, A. Binder, „Calculation and validation of a bearing impedance model for ball bearings and the influence on EDM-currents," *Proc. of the 21th International Conference on Electrical Machines (ICEM)*, Berlin, Germany, 02. – 05. September, 2014, pp. 1798 - 1804.

[13] M. Schuster, A. Binder, „Comparison of different inverter-fed AC motor types regarding common-mode bearing currents," *Proc. of the IEEE Energy Conversion Congress and Exposition (ECCE 2015)*, 21. - 24. September, 2015, Montreal, Canada, pp. 2762 - 2768.

[14] M. Burger, D. Kampen, "Drive system loss reduction by allpole sine filters", *Proc. of the International Exhibition and Conference for Power Electronics (PCIM) Europe*, 10. - 12. May, Nuremberg, 2016, pp. 144 - 147.

AUTHOR INDEX

Aarniovuori, Lassi 2829
Abbate, Carmine 2802
Abbosh, Amin..1006
Abdel-Rahim, Naser 352
Abdelhakim, Ahmed..................................2220
Abdelrahem, Mohamed 900
Abramson, Rose A. 1934
Abusara, Mohammad.................................471
Aganza-Torres, Alejandro..........................1813
Aguglia, Davide 3330
Ahmad, Bilal...3348
Ahmad, Faheem.......................................2987
Ahola, Jero...2753
Ait-Ahmed, Mourad 3289
Aizpuru, Iosu 251, 1205
Alam, M. M. 480, 1551
Alam, Muhammad Farhan 416
Alatise, Olayiwola 2241
Alawieh, Hadi 1685
Albach, Manfred 173, 193
Alexandre, Philippe 1905
Ali, Ahmed Ismail M. 1417
Ali, Marwan 1118, 2039
Ali, Mohammad 2743, 2763
Ali, Waqas..871
Alisar, Ibrahim 1205
Alishah, Rasoul Shalchi..............................460
Alkama, Kouceila 2564
Allard, Bruno 522, 829, 1470, 1874
Almaksour, Khaled 1700
Almeida, Bruno F. 3217
Alonso, C. 919
Alqatamin, Moath 65
Am, Sokchea...3172
Ammann, Ulrich 3137
Amrane, Fayssal 362
Anders, Erik..944
Andrade, Fabio 1400, 1841, 1850
Andresen, Jan...............................2303, 2545
Anzola, Jon 251
Aoustin, Yannick 1923
Arandia, Nerea 1524
Arazi, M...2881
Arrizabalaga, Antxon.................................1205
Arrozy, Juris 1067
Arruti, Asier 251
Artiglia, Melissa 2791
Aríztegui, Raquel González.........................1515

Asllani, Besar..1279
Avenas, Yvan 1685, 2564
Averbukh, Moshe 2439
Averous, Nurhan Rizqy...............................153
Azizian, Mohammadreza 2860
Baburajan, Silpa 2573
Bacha, S. 406
Bacha, Seddik..820
Baërd, H. 95
Bagaber, Bakr..2613
Bahman, Amir Sajjad 2704
Bahrani, Behrooz 787
Bai, Wenshuai 667
Baker, Erik 512
Bakran, Mark-M..............644, 686, 1252, 1533, 1831, 1885,
....2106
Bakri, R. 2554
Bakri, Reda 2078
Balkowiec, Tomasz 2029
Barazi, Yazan 1057
Barelli, Linda 292
Barg, Sobhi 416
Barwig, Markus 1460
Basic, D...37, 95
Basic, Duro 2938, 2957
Bauer, Pavol 1224, 1233, 1561, 3422
Bazin, Pascal 2467
Beczkowski, Szymon Michal 2987
Beerten, J. 1551
Belhaouane, Moez 1158, 1756
Bello, Guilherme 2049
Benchaib, A. 745
Benchaib, Abdelkrim 820, 1215
Bender, Vitor C. 3217
Bendfeld, Christian 163
Benjamin, Sébastien 3156
Benkhoris, Mohamed Fouad 3205
Bensebaa, S. 1363
Bentivegna, N..2293
Benzagmout, Abdelhadi 1905
Beranger, Bruno 2467
Berkani, M. 1363
Bernet, Steffen 124, 1569
Bertele, Felix 3137
Bertilsson, Kent 416, 460
Betto, Kento 3071
Betz, Robert Eric 927
Bevilacqua, Pascal....................................1279

Beza, Mebtu.................................969, 1952
Bhajana, V. V. Subrahmanya Kumar.................................2068
Bidini, Gianni 292
Biela, Juergen 2331, 2583, 2791
Biela, Jürgen 2230, 2409, 2446, 2475, 2513, 2524,
.................................2684, 2712, 2780, 2946
Bier, Anthony 2638
Bikinga, Wendpanga Fadel.................................2564
Binder, Andreas 2049
Birou, Camille 332
Bissal, Ara 871
Blaabjerg, F. 3237
Blaabjerg, Frede.................................1, 460, 810, 927, 2088, 2135,
.................................2220, 2393, 2573, 2888, 2898, 2928, 3119
Blanco, Marcos.................................1076
Blanquez, Francisco R. 1336
Blaquiere, Jean-Marc 1057
Blinov, Andrei 2996
Blume, Sebastian 2684
Böcker, Jan 2341
Böcker, Joachim 1638, 3024
Boersma, S..................................745
Bohlen, Oliver 707
Bohnke, M. 1613
Boige, François.................................1057
Boisaubert, Emile 3172
Bolzan, Thais E..................................3217
Bolzoni, A..................................1306
Bombois, X..................................745
Bongiorno, Massimo.................................969, 1952
Borcherding, Holger 608
Boulaud, Etienne.................................3172
Bourennane, Abdelhakim 1096
Bourguet, Salvy 2907
Bouscayrol, Alain 3330
Boutleux, Emmanuel 433
Boutry, Arthur 2366
Bozorg, Mokhtar.................................3247
Boškovic, N. 2812, 2820
Branca, Xavier 522
Briff, Pablo 9
Bringezu, Thilo.................................2780
Brockhage, Torben 1766
Brooks, Michael.................................2773
Bruyere, Antoine.................................1756
Brückner, Thomas.................................2938, 2957
Bründlinger, Roland 2723
Büdel, Johannes.................................1718
Bucher, A..................................3403
Bucher, Alexander 203
Budo, Kohei.................................1450
Buigues, Garikoitz 85

Burgos, Rolando.................................2366
Burgos-Mellado, Claudio.................................1354
Busatto, G. 3210
Busatto, Giovanni.................................2802
Buttay, Cyril.................................1106, 2265, 2366
Cacciato, M. 909
Cai, Pei 2135
Camail, Philippe.................................2000, 2265
Camara, M. B. 2881
Camurca, L..................................3305
Cardelli, Ermanno 292
Cárdenas, Roberto.................................1354
Carnielutti, Fernanda.................................2851
Caron, Hervé 1700
Carpita, Mauro 3247
Carpiuc, Sabin 962
Cascino, S. 2293
Castellazzi, Alberto 2210
Castellini, Simone 292
Castelltort, Arnaud 1747
Castiglia, V. 3237
Catellani, Stéphane.................................2638
Cavallaro, D. 2293
Chaiba, Azeddine 362
Chakraborty, Sajib.................................2320, 3111
Cheaito, Hassan.................................829
Chen, Linglin 1224, 1233
Chen, Qing 542
Chen, Yu 804
Chen, Zhengxin 637
Cheshire, Christoph 3137
Chevalier, Florian 1895
Chillón-Antón, Cristian.................................853, 1542
Chiumeo, Riccardo 424
Chraye, Hélène 3427
Chrin, Phok 3172, 3376
Chrzan, Piotr J..................................3054
Chédot, L. 183
Chédot, Laurent.................................433, 1106
Ciupageanu, Dana-Alexandra 292
Clerc, Guy.................................433, 829
Clerici, Alessio.................................424
Cochelin, Anne-Sophie 3426
Coelho-Medeiros, Rafael 1479
Colak, Ilknur 871
Colas, F. 1579, 2554
Colas, Frédéric 1158
Colmenero, Manuel.................................1336
Connaughton, Alexander.................................1866
Cordier, Julien.................................3272
Corentin, Darbas 1803
Cornea, Octavian.................................2192

Costa, François 503
Coujard, Clementine 1270
Cravero, Jean-Marc 3330
Crebier, Jean-Christophe 2010
Da Cunha, Julian 2312
Dabbabi, Asma 2907
Dahmen, Christopher 843
Dai, Jing 820, 1215, 1479
Dakyo, B. 2881
Dang, Ziyue 804
Danzer, Michael A. 707
Darivianakis, Georgios 998
Davari, Pooya 726, 1006, 2573, 3119, 3295
David, Romain 522
Davidson, Colin 2366, 3425
Davoodi, Amirali 2393
De Doncker, Rik W. 153, 163, 1627
De Jaeger, Jean-Claude 1895
De Jódar, Esther 1658
De La Grandiere, Hubert 3424
De Lauretis, Maria 512
De Mora, Pablo Rodriguez 1533
De Oliveira, Eduardo F. 2535
De, Dipankar 2210
De-Preville, Guillaume 9
Defrance, Nicolas 1895
Degrenne, N. 2865
Delamea, R. 2865
Delarue, Philippe 1158, 3330
Delette, G. 1613
Delhommais, Mylène 1737
Delpech, F. 2554
Demidov, Iurii 2419
Denis, Guillaume 1270
Dennetière, S. 2967
Derbey, Alexis 2039
Derkacz, Pawel B. 3054
Despesse, Ghislain 503
Despouys, Olivier 1373
Dessante, Philippe 1858
Devos, Guillaume 1858
Di Gregorio, Francesco 1747
Dieckerhoff, Sibylle 2274, 2341, 2603
Dierks, Rebecca 3091
Dietz, Armin 1316
Dincan, Catalin 2873
Dinkel, Daniel 1297
Dinulovic, Dragan 2773
Djerioui, Ali 3205
Dong, Dong 2366
Doppelbauer, Martin 1589
Douine, Bruno 1373

Doumiati, Moustapha 2251
Drabek, Pavel 2068
Dragicevic, Tomislav 2393, 2898
Driesen, J. 480
Driesen, Johan 27
Drofenik, Uwe 627
Duarte, J. 2812, 2820
Duarte, Jorge L. 1067, 2656
Duarte, Renan R. 3217
Duerbaum, Thomas 203
Dujic, Drazen 1776, 2486
Dürbaum, Thomas 173, 193, 1460
Dworakowski, Piotr 406, 1106, 2000, 2265, 3006
Džonlaga, Bogdan 1479
Ebersberger, Janine 3340
Ebrahimi, Reza 2860
Eckel, Hans-Günter 532, 1666, 2059, 2126, 3282
Eckerle, Richard 765
Ecrabey, Jacques 2467
Egrot, Philippe 1479
Eguia, Pablo 85
Ehlich, Martin 608
El Baghdadi, Mohamed 2320
El Jihad, Hamza 1982
Elizondo, Laura Ramirez 1561
Ellul, Racquel 1025
Elsabrouty, Ibrahim 871
Elsied, Moataz 3156
Elthokaby, Youssuf 352
Endisch, C. 551
Enjeti, Prasad 1390
Erenler, Yeliz 571
Errigo, F. 183
Escobar, Gerardo 1390
Escofficr, Réne 1470
Esfetanaj, Naser Nourani 2860, 3295
Eskandari, Bahman 863
Eslamian, Morteza 3064
Eslampanah, Vahid 2860
Espina, Enrique 1354
Etoz, Burhan 2241
Fadel, Maurice 3101
Fauth, Leon 3081
Fazli, Nastaran 2126
Fehr, Hendrik 1030, 2116
Ferreira, Jan Abraham 892
Ferrieux, Jean-Paul 2039
Finkenzeller, Michael 835
Fischer, Manuel 1168, 1605, 1709
Fogsgaard, Martin Bendix 2258
Foray, Etienne 1874
Forsyth, A. J. 1306

Fort, Jiri 1086
Founier, Etienne 3101
Francois, Bruno 362, 1700, 1756
Frédèric, Poitiers 1803
Frey, D. 406
Frey, David 2695
Freytes, Julián 9
Friebe, Jens 2743, 2763, 3081
Fromme, Christopher 2603
Frost, Damien 223
Fruchier, Olivier 1905
Fu, Siqi 302
Fuchs, Simon 2409, 2513
Fürst, Markus 1885
Galeshi, Soleiman 2695
Gamatié, Abdoulaye 1747
Gandolfi, Chiara 424
Ganjavi, Amir 726, 1006
Gao, Fei 1224, 1233
Gao, Jianbo 1262
Garate, José Ignacio 1524
Garbuio, Lauric 1685
García-Torres, Felix 134
Garnier, Laurent 2467
Gaubert, Jean-Paul 396, 2284
Gauthier, Jean-Yves 590, 736
Gautier, Cyrille 1118, 1858, 2039
Gautier, Maxime 1923
Geng, Zeyang 3198
Gensior, Albrecht 581, 1030, 2116
Gentejohann, Marius 2274
Georges, Didier 820
Gerada, Chris 1944
Gerada, David 1944
Geramirad, Hadiseh 2000
Gerstner, Michael 1316
Geske, Martin 2938, 2957
Geury, Thomas 2320
Geyer, Tobias 2145
Ghamrawi, Ahmad 2284
Ghanes, Malek 3205
Gholami-Khesht, Hosein 3119
Giacomazzo, M. 490
Gierschner, Sidney 2059, 2126
Giewont, William 1016, 1972
Giotakos, Panagiotis I. 2172
Girbau-Llistuella, Francesc 1542
Gireada, Mihaita 2192
Glac, Antonín 3257
Gladen, Marcel 1148
Glasberger, Tomas 3166, 3314
Gleissner, Michael 644, 686

Glushakov, Vasiliy V. 47, 56
Gnärig, Jan Lasse 1030
Golluccio, G. 3210
Golsorkhi, Mohammad S. 2733, 2898
Gomez, Juan S. 1354
Gomis-Bellmunt, Oriol 1542, 2977
Gonzalez, Jose Ortiz 2241
Gonzalez-Torres, J-C. 745
González-Fontderubinat, Paula 1542
Gosses, Kilian 143
Gou, Wanchao 153
Govaerts, G. 480
Gradinger, Thomas B. 627
Grainger, Brandon M. 65
Grbovic, Petar J. 2723
Grecki, Filip 627
Green, Tim C. 276
Green, Tim 2977
Griepentrog, Gerd 1675
Gruson, F. 1579
Gruson, Francois 1158
Gu, Chunyang 1944
Guerrero, Josep. M 3289
Gui, Qiuye 1030
Guichon, Jean Michel 2564
Guillaud, X. 1579
Guillaud, Xavier 1158, 1756, 1952
Guo, Mingzhu 718
Guo, Xuan 317
Gutierrez, A. 919
Gutierrez, Sebastian 598
Gärtner, M. 3403
Götting, Gunther 1289
Hackl, Christoph 900
Hage-Hassan, Maya 1858
Hai, Jie 231
Hallemans, L. 480
Hamid, Muhammad 1289
Hammerer, Horst 2182
Hammes, David 2059
Han, Hua 260, 285, 302, 326
Han, Lubin 370
Han, Weiji 765
Haq, Omer Ikram Ul 3393
Häring, Johannes 644, 686
Harnefors, Lennart 1952
Hartmann, Michael 2258
Harzig, Thibaut 65
Hase, Genki 467, 498
Hasenohr, C. 3403
Hasenohr, Christian 203
Hatori, Kenji 1489

Haug, Martin	2773
He, Maojun	804
He, Yuying	1410, 1435
Hegazy, Omar	2320, 3111
Hein, Lukas	3413
Helle, Lars	2873
Heller, M.	3403
Hénaux, Carole	3101
Henninger, Stefan	3179
Heredero-Peris, Daniel	853, 1542
Herkommer, Christian	1718
Hernandez, Fernando Davalos	598
Herwig, Daniel	1766
Heucke, Sören	2341
Heydari, Rasool	2733, 2898
Hideaki, Yano	1442
Higashihata, Takeshi	1489
Hijazi, A.	183
Hiller, Marc	3366
Hillermeier, Claus	1297
Himker, Niklas	979
Hinkkanen, Marko	2495
Hiraki, Eiji	386
Hirayama, Hiroshi	2376
Hiwatari, Daichi	2376
Hofer, Matthias	2403
Hoffmann, Felix	954
Hofmann, Harald	203
Homann, Michael	307
Hong, Yang	231
Horrein, Ludovic	3330
Horvatic, Iréna	3156
Houari, Azeddine	3205, 3289
Hu, Anliang	2946
Hu, Rui	2594
Huang, Han	1128
Huang, Pin-Yu	2666
Huang, Xingxuan	1972
Huisman, Henk	1067, 1186
Hulea, Dan	2192
Hussain, E. K.	471
Iannuzzo, Francesco	2258, 2704
Ibanez, Federico	598, 3034
Idarreta, Aitor	1205
Idir, Nadir	1793, 1895, 2078
Iman-Eini, Hossein	3305
Ingman, Jonny	563
Inoue, Sadayuki	19
Iraola, Unai	1205
Isaksson, Dan	1016
Ishihara, Hiroki	19
Isobe, Takanori	3358

Itoh, Jun-Ichi	1380, 2200
Jackiewicz, Krzysztof	2029
Jaeger, Johann	3179
Jakob, Roland	2957
Jaritz, Michael	2684
Jasim, Omar	9
Jean-Christophe, Olivier	1803
Jeannin, Pierre-Olivier	3054
Jehle, Andreas	2475
Jelena, Popovic	892
Jeong, Min	2409, 2513
Ji, Shiqi	1972
Jia, Ming	1627
Jiang, Jinhai	637
Jiaqi, Diao	231
Joebges, Philipp	1627
Jonokuchi, Hideki	2376
Jorge, Tenorio	1400
Joryo, Satoshi	3071
Jotwani, Ankit	1138
Joubert, Charles	522
Judge, Paul D.	276
Jun-Ping, He	1599
Junge, Patrick	2613
Juntunen, Raimo	3227
Junyent-Ferré, Adrià	2977
Justin, Elissa Cresenta Anak	2265
Jäppinen, Janne	563
Järvisalo, Heikki	1016
Jørgensen, Asger Bjørn	2987
Kado, Yuichi	2666
Kadri, R.	2554
Kahl, Tino	2603
Kahle, Karsten	1336
Kaiser, Ingmar	1666
Kallfass, Ingmar	1915
Kaminski, Nando	954
Kampen, Dennis	2049, 2153
Kanchan, R. S.	3393
Kang, Yong	797, 804
Karaventzas, Vasilios	2583
Karlsson, Martin	512
Kaszewski, Arkadiusz	2029
Kawabata, Yoshitaka	1177, 1243
Keel, Oliver	2791
Kefer, K.	3403
Kehl, Zdenek	3166
Keller, Christian	2938, 2957
Kennel, Ralph	542, 900, 1262, 2386, 3272
Kersten, Anton	765
Kesbia, Nasreddine	1685
Kestelyn, X.	1579

Ketchedjian, Vasken	1605
Keysan, Ozan	1823
Khanzadeh, Babak	1040
Kharezy, Mohammad	3064
Kikuchi, Naoto	2200
Killeen, Peter	777
Kim, Bunthern	3172, 3376
Kimura, Norihito	105
Kimura, Shota	377
Kindl, Vladimir	1086
Kirchenberger, U.	3403
Kirowitz, Thomas	2403
Kitagawa, Wataru	1425
Kitamura, Taishi	1177
Kiviniemi, Mika	563
Kjaer, Martin Vang	2888
Kjær, Philip	2873
Klass, Stefan	3272
Klier, Samantha	707
Koch, Dominik	880, 1915
Kohlhepp, Benedikt	266, 1460
Kojabadi, Hossein Madadi	2860
Komma, Thomas	835
Komrska, Tomáš	3257
Kondo, Keiichiro	2675
Kone, Lamine	1793
Kopacz, Rafal	2457
Korhonen, Juhamatti	1016
Kosan, Tomas	3166
Kouchaki, Alireza	3015, 3129
Koutroulis, Eftychios	3146
Krall, Felix	1196
Krim, Youssef	1700
Krischan, Klaus	1866
Kroneisl, Michal	1992
Krug, Dietmar	2059
Kucka, Jakub	2020, 3091
Kuder, Manuel	765
Kuebrich, Daniel	203
Kuhlmann, Kai	1718
Kukkola, Jarno	2495
Kumar, Dinesh	726, 1006, 2573
Kuring, Carsten	266
Kusaka, Keisuke	1380, 2200
Kuwana, Kazuki	1425
Kuwata, Akiko	19
Kwasinski, Alexis	2594
Kyyrä, Jorma	3348
Kärkkäinen, Hannu	2829
Kärkkäinen, Tommi J.	563
Kübrich, Daniel	266
Küster, Pierre	571

Labiano, Daniel	1205
Labouré, Eric	1858
Lacarnoy, Alain	654
Lacressonnière, Fabien	332
Ladoux, Philippe	1106, 2265
Lafon, Frederic	1793
Lafoz, Marcos	1076
Lagier, Thomas	1106, 2000, 2265
Lana, Andrey	2419
Langbauer, Thomas	1866
Langmaack, N.	241
Langwasser, M.	3305
Lapassat, N.	37
Larruskain, Marene	85
Lautner, Frank	1831
Lazaroiu, Gheorghe	292
Le Moigne, Philippe	1158, 2078
Le Métayer, Pierre	3006
Le, Hoai Nam	2200
Lee, Seong-Yong	2486
Leedham, Rob	871
Lefebvre, Bruno	1106, 2000, 2366
Lefebvre, S.	1363
Lefebvre, Stéphane	1118
Lehmann, Franziska	944
Lehn, Peter	223
Lembeye, Yves	2010, 2695
Lemmen, Erik	2350, 2358
Leo, Jacopo	75
Leppänen, Joonas	563
Leterme, Willem	276
Letrouvé, Tony	1700
Lexow, Daniel	3282
Li, Boyang	1289
Li, Chi	317
Li, Dingrui	1972
Li, Hui	2704
Li, Jiaqi	9
Li, Jing	1944
Li, Lang	285
Li, Qi	1262
Li, Tao	3044
Li, Weilin	2135
Li, Yongdong	317, 1944
Li, Yu	2386
Li, Zheming	2106
Liang, Chaohui	3044
Liang, Lin	370, 443
Liao, Jianquan	3385
Liao, Yuefeng	1498
Licari, John	1025
Lima, Glauber De Freitas	2010

Lin, Lei 937
Lin-Shi, Xuefang 590, 736
Liserre, Marco 3305
Liu, Cuicui 2505
Liu, Fuxin 1410, 1435
Liu, Libo 1289
Liu, Xudan 804
Liu, Yao 260
Liu, Zhangjie 260, 302, 326
Liukkonen, O. 2630
Llanos, Jacqueline 1354
Llonch-Masachs, Marc 1542
Locment, Fabrice 667, 677
Loisel, Rodica 2907
Loiselay, Florent 1648
Lomonova, E. A. 2812
Lorenz, Malte 2020
Ludois, Daniel C. 777
Lunz, B. 3403
Luo, Fang 370
Luo, Xian 153
Lutz, Josef 3413
Lutze, Marcel 1675
López-Alcolea, Fco. Javier 134
Ma, Yixiao 1435
Mabe, Jon 1524
Machmoum, Mohamed 2907, 3205, 3289
Maekawa, Sari 2838, 2844
Maerz, Martin 1316
Magambo, Jean Sylvio Ngoua Teu 2078
Maharana, Manoj Kumar 2068
Maharana, Suman 2210
Mahr, Florian 3179
Maier, Robert W. 1252, 2106
Mäkelä, Juha 3348
Mallwitz, R. 241, 490
Mallwitz, Regine 213, 1515
Maneiro, Jose 1106, 3006
Mannen, Tomoyuki 3358
Mannerhagen, Felix 3198
Mantellini, Mattia 1076
Mantzanas, Panagiotis 203
Mao, Saijun 892
Marcault, E. 919
Marchesoni, Mario 3321
Marciano, D. 3210
Marciano, Daniele 2802
Margueron, Xavier 2078
Marmolejo, Narciso G. 1589
Marquardt, Rainer 843, 1297
Martin, Christian 1874
Martin, Jérémy 2638

Martinez, Wilmar 598, 3034
Martire, Thierry 1905
Martínez-Gómez, Manuel 1354
März, Martin 143
Mateos, Felix Rodriguez 2583
Mattar, Rita 1118
Mattsson, Aleksi 1016
Maussion, Pascal 3376
Mayorga, John Paul 1658
Mazuela, Mikel 1205
McIntyre, Michael 65
Mehdi, Driss 2284
Meißner, Markus 124, 1569
Mendoza-Araya, Patricio 2917
Meneses, Javiera 2917
Meng, Qingchao 2446
Mercier, Adrien 1858
Mercier, Sylvain 2467
Mertens, Axel 979, 1766, 2020, 2303, 2545, 2613,
                            2743, 2763, 3091, 3340
Mesbahi, Tedjani 3205
Meynard, Thierry A. 654
Mezrag, Bachir 2564
Mezzetti, Margarita 944
Micallef, Alexander 1025
Miceli, R. 3237
Milas, Nikolaos T. 2172
Miletic, Zoran 2723
Millinger, Jonas 512
Minami, Masataka 467, 498
Mirtchev, Alex V. 2429
Mitani, Kohei 1425
Miyauchi, Tsutomu 377
Mizutani, Hiroto 386
Mohamed, Abdalla Hussein 115
Mohamed, Islam 352
Molina-Martínez, Emilio J. 134
Mollov, S. 2865
Molnar, Jan 3166
Monmasson, Eric 1118, 1823
Montero, E. Rodriguez 2098
Montesinos-Miracle, Daniel 853, 1542
Moraes, Tiago José Dos Santos 1693
Morel, Cristina 2251
Morel, F. 183, 406
Morel, Florent 2000, 2366
Morel, Hervè 1279, 1648
Mori, Osamu 386
Morici, Riccardo 1076
Morizane, Toshimitsu 1047, 3071
Mortimer, Benedict 163
Motegi, Shin-Ichi 1786

Mourouvin, Rayane	820
Mourtzis, Dimitris A.	2172
Muehlbauer, Markus	707
Muetze, Annette	1196
Müller, Jan-Kaspar	2763, 3340
Mumtaz, Muhammad Adnan	1326
Munk-Nielsen, Stig	2987
Muñoz, Fredy	3189
Muñoz, Javier	3189
Muntean, Nicolae	2192
Murata, Ryo	386
Musznicki, Piotr	3054
Nabatirad, Mohammadreza	787
Nada, Kaho	19
Nadh, Greeshma	1619
Nair, Durga S.	1619
Najera, Jorge	1076
Najjar, Mohammad	3015, 3129
Nakagaki, Akito	498
Nakamura, Keiichi	1489
Nakashima, Osamu	2376
Nakatani, Shota	1047
Nakazawa, Yosuke	2675
Narula, Anant	969, 1952
Natori, Kenji	2675
Navarro, Gustavo	1076
Navas, Alex F.	1354
Ndagijimana, Fabien	2010
Nee, Hans-Peter	2220
Neumann, Jessica	3101
Nevaranta, N.	2630
Ngoua-Teu, J-S	1613
Nguyen, Ngac Ky	1693
Nguyen, Van Sang	2638
Nicolas, Ginot	1803
Nie, Cheng	1972
Nie, Qingqing	797
Niemelä, M.	2630
Niemelä, Markku	563, 2829
Nikowitz, Mario	2403
Nisch, A.	3403
Nishida, Yasuyuki	1786
Nitzsche, Maximilian	880, 1605, 1709, 1915
Niu, Liyong	342
Norambuena, Margarita	2851
Nymand, Morten	3015, 3129
Obernolte, Urs	608
Oberschelp, Wolfgang	1727
Odriozola, Kepa	654
Oguma, Kenji	377
Ohta, Takahiro	2666
Okamori, Daichi	1047

Okazaki, Akihiro	2838
Okazaki, Yuhei	1040
Oliveira, Joao	1648
Olivier, Jean-Christophe	2251
Omori, Hideki	1047
Omrane, A.	2554
Ordoño, Ander	1524
Orfanoudakis, Georgios I.	3146
Ortega-Perez, Carmen	3330
Ota, Kenji	1489
Ottaviano, Andrea	292
Oumaziz, Amirouche	1096
Pace, Loris	1895
Páez, Juan	1106
Paez, J. D.	406
Palensky, Peter	810, 1561
Pallier, Joris	829
Palm, Herbert	707
Pan, Xuejiao	1128
Passalacqua, Massimiliano	3321
Patarroyo-Montenegro, Juan F.	1841
Patti, Davide	3210
Paulus, Sebastian	2303, 2545
Pavlicek, Vladimir	1086
Pawellek, A.	3403
Pawellek, Alexander	203
Payman, A.	2881
Pei, Xiaoze	342
Peller, Stefan	266
Pelosi, Dario	292
Peltoniemi, Pasi	3227
Peña, R. A.	183
Pendharkar, Ishan	1346
Peng, Han	797, 804
Peng, Hao	804
Peng, Tao	1498
Penin, Carolina	1905
Peralta, Patricio	75
Perenyi, Christian	65
Peretti, Luca	3393
Pérez-Molina, María José	85
Peric, V.	745
Peroutka, Zdenek	3257, 3314
Perriard, Yves	75
Petit, M.	1363
Petit, Mickael	1118, 3054
Petkovic, Marko	1776
Peyghami, Saeed	810, 2573
Pfeifer, Markus	1675
Pfeiffer, Jonas	571, 2535
Phulpin, Tanguy	618
Pidancier, Thomas	3247

Pietrzak-David, Maria	3376
Pilawa-Podgurski, Robert C. N.	1934
Pinheiro, Humberto	2851
Pinomaa, Antti	2419
Pinto, Rafael A.	3217
Pirsto, Ville	2495
Pitel, Ira	1390
Planson, Dominique	1279, 1648
Plaza, Jesus D. Vasquez	1841
Plissonnier, Marc	1470
Poebl, Monika	835
Polacek, Libor	1086
Pollet, Benjamin	503
Pommier-Petit, Pascal	829
Pool-Mazun, Erick I.	1390
Popuri, Madhuchandra	2068
Pouresmaeil, Edris	863
Pöyhönen, Santeri	2753
Prevost, Thibault	1270
Prieto, Dany	3101
Prieto-Araujo, Eduardo	2977
Pronin, Mikhail V.	47, 56
Puls, Simon	608
Pulvirenti, M.	909, 2293
Pursiainen, Jooa	3227
Putkonen, A.	2630
Pyrhönen, Juha	2829
Pyrhönen, O.	2630
Pyrhönen, Olli	2419
Qashlan, Ziyad H. S.	571
Qiang, Jin	163
Qoria, T.	1579
Qoria, Taoufik	1756
Queval, Loic	1473, 1479
Rabba, Heiko	307
Rabkowski, Jacek	2457, 2996
Radet, Hugo	332
Rahmoun, Yasser	2182
Rahul, Arun S.	1619
Rajput, Shailendra	2439
Ramm, Hannes	307
Ramírez-Scarpetta, J. M.	1850
Rasmussen, Tonny Wederberg	1138
Rathnayake, Hansika	726, 1006
Rault, P.	2967
Raute, R.	989, 1506
Rautio, Juuso	563
Ravyts, S.	480
Ravyts, Simon	27
Rayati, Mohammad	3247
Razzaghi, Reza	787
Rehlaender, Philipp	1638
Reigosa, Paula Diaz	1346, 2258
Reißenweber, Lukas	3263
Rekola, Jenni	3227
Ren, Chunpin	718
Restrepo, Jose Alex	1850
Retianza, Darian V.	1067
Retianza, Darian Verdy	1186
Rezaee, Ali Yahya	460
Richardeau, Frédéric	1057, 1096
Rietmann, Stefan	2712
Rigot, Valentin	618
Risch, Raffael	2230
Riu, Delphine	3156
Roa, Claudio	3189
Robet, Pierre-Philippe	1923
Roboam, Xavier	332
Robyns, Benoit	1700
Rodriguez, Jose	900, 2851
Rodriguez, José Luis	1205
Rokrok, Ebrahim	1756
Roncero-Sánchez, Pedro	134
Roose, T.	1551
Röser, Tobias	3137
Roth-Stielow, Jörg	880, 1168, 1605, 1709
Rouger, Nicolas	1057
Routimo, Mikko	2495
Rouzbehi, Kumars	863
Ru, Yang	231
Ruan, Xinbo	1410, 1435
Rubenbauer, Hubert	3179
Rufer, Alfred	696
Rute, Erwin	1354
Ruthardt, Johannes	1168, 1605, 1709
Saad, H.	2967
Saad, Yamen	3295
Saber, Christelle	1118
Sácz, Doris	1354
Sadarnac, Daniel	618
Saeedian, Meysam	863
Saggio, M.	2293
Sah, Gyanendra Kumar	532
Sahin, Ilker	1823
Saim, Abdelhakim	3289
Sakai, Kazuto	1442
Sakai, Norikazu	1489
Sakai, Yasuhiro	1489
Sakaria, Omar Ahmed	3295
Sakly, Jihen	618
Sakurazawa, Yoshiki	2675
Saleh, Bassem	2648
Salem, Qusay	1289
Sallot, P.	1613

Salvo, L. 909
Sánchez-Sánchez, Enric 2977
Sandelic, Monika 2088
Sandik, Diane -Perle 512
Sangwongwanich, Ariya 1, 2088
Sanseverino, A. 3210
Sanseverino, Annunziata 2802
Santos, Miguel 1076
Sari, A. 183
Sarraute, Emmanuel 1096
Sassatelli, Gilles 1747
Sathik, Mohd. Ali Jagabar 460
Saudemont, Christophe 1700
Savaghebi, Mehdi 2733, 2898
Savarit, Elise 1982
Sawada, Takashi 2623
Sayed, Mahmoud A. 1417
Scarpetta, Jose Miguel Ramirez 1400
Scelba, G. 909
Schafmeister, Frank 1638, 3024
Schanen, Jean-Luc 1685, 2039, 3054
Schiesser, Matthias 962
Schleippmann, Nico 143
Schlesinger, Richard 2524
Schlüter, Michael 2274
Schmidt, Dimitri 1168
Schmitt, Alexander 2182
Schmitt, N. 1363
Schmitt, Stefan 954
Schmitz, Jan 124, 1569
Schobre, Thorben 213, 1515
Schröder, Günter 1727
Schrödl, Manfred 2403
Schulte, Hendrik 1709
Schulz, Matthias 143
Schulz, Nicola 1346
Schulze, Torben A. 307
Schwabe, Christian 3413
Schütt, Michael 532
Sciacca, A. G. 909
Scicluna, K. 989, 1506
Sechilariu, Manuela 667, 677
Segur, R. 745
Seiler, Pascal 2331
Semail, Eric 1693
Sergeant, Peter 27, 115
Shah, Chirag 153
Sharkh, S. M. 471
Sharkh, Suleiman M. 3146
Shi, Guangze 326
Shinoda, Kosei 820, 1215
Shiozaki, Koji 2623

Shirakawa, Tomohide 386
Shousha, Mahmoud 2773
Si-Yuan, Cai 1599
Siala, S. 95
Siala, Sami 1982
Siebke, K. 490
Siemens, Ag 2603
Silventoinen, Pertti 563, 1016
Simola, Aleksi 2753
Singer, Arthur 765
Skala, Bohumil 1086, 3257
Smailus, Erik 1675
Šmídl, Václav 1992
Smit, A. 3403
Snook, Mark 871
Soeiro, Thiago Batista 1224, 1233, 1561
Sokur, Pavel V. 47
Soltau, Nils 1489
Song, Kai 637
Soumaoro, Ousmane 2251
Soupremanien, U. 1613
Sprunck, Sebastian 2535
Stadler, Alexander 3263
Staines, C. Spiteri 989, 1506
Stathis, Spyridon 2684
Staudt, Volker 1148
Stecca, Marco 1561
Steckler, Pierre-Baptiste 736
Stengl, Josef 1086
Stenglein, Erika 173, 193
Štepánek, Jan 3257
Stock, Alexander 1718
Stöckl, Johannes 2723
Stöttner, J. 551
Stotckaia, Anastasiia D. 47, 56
Stras, Andrzej 2029
Streit, Lubeš 3257
Strittmatter, Tobias 1346
Strunk, Robin 979
Su, Guoxing 1410
Su, Mei 260, 285, 302, 326, 1498
Suarez, Camilo 3034
Sugiyama, Kohei 1177
Sun, Jian 1962
Sun, Yao 260, 285, 302, 326, 1498
Svensson, Jan R. 1952
Tadano, Hiroshi 2623
Taheri, Shamsodin 863
Takahara, Takaaki 386
Takahashi, Hirotaka 377
Takahashi, Hiroyuki 1243
Takano, Tomihiro 19

Takeshita, Takaharu	1417, 1425, 1450
Takeuchi, Somi	1243
Talbert, Thierry	1905
Talla, Jakub	3257
Tan, Guoqiang	443
Tanaka, Ami	2844
Tanaka, Miwako	19
Tanaka, Nobuhiko	1489
Tang, Bojin	718
Tang, Houjun	1224, 1233
Tang, Xiaohu	1589
Tannhäuser, Marvin	2603
Tant, Jeroen	27
Tareilus, G.	241
Tarisciotti, Luca	1224, 1233
Tárraga, Sergio	1658
Tatakis, Emmanuel C.	755, 2172, 2429
Taul, Mads Graungaard	927
Tedesco, Davide	2802
Teigelkötter, Johannes	1718
Teirelbar, Ahmed	2648
Tenorio, Jorge	1850
Teramura, Keiko	377
Terbrack, C.	551
Thal, Eckhard	1489
Thiringer, Torbjörn	765, 1040, 3064, 3198
Tibola, Gabriel	2656
Tikhonov, Sergey	1638
Todd, R.	1306
Tolbert, Leon M.	1972
Torres, Alfonso Parreño	134
Torres, Esther	85
Torres, Fernando	3189
Torres, Jorge	1076
Torres, Jose Rueda	810
Touhami, Mustapha	503
Tran, Dai-Duong	2320, 3111
Traoré, Bakou	2251
Tremmel, Werner	2723
Tremouilles, D.	919
Trillaud, Frédéric	1373
Trochimiuk, Przemyslaw	2457
Tröster, Nathan	1168
Tsolaridis, Georgios	2331
Tsoumas, Ioannis	998, 2145, 2163
Turjanica, Pavel	1086
Turki, Faical	307
Twardon, N.	3403
Ufnalski, Bartlomiej	2029
Ulissi, Gabriele	2486
Umetani, Kazuhiro	386
Unruh, Roland	3024

Uwai, Shuto	1177
Vaccaro, Luis	3321
Valtee, Mikko	3227
Van Den Broeck, G.	480, 1551
Van Duivenbode, Jeroen	1186
Van Mierlo, Joeri	2320, 3111
Van Tichelen, P.	480
Vanfretti, L.	745
Vannier, Jean-Claude	1479
Vansompel, Hendrik	115
Vashishtha, Anushruti	1138
Vasquez-Plaza, Jesus D.	1850
Vázquez, Javier	134
Vecchia, Mauricio Dalla	27
Vechiu, Ionel	396
Velardi, F.	3210
Velardi, Francesco	2802
Velazco, Diego	433
Venet, P.	183
Verbelen, Florian	27
Vermeersch, Pierre	1158
Vermulst, Bas	2350, 2358
Viana, Caniggia	223
Videt, Arnaud	1793, 1895
Vienot, Stephane	1793
Vieto, Ignacio	1962
Villarejo, José	1658
Villegas, Carlos	962
Vip, Stephan	2303, 2545
Vitan, Danut	2192
Voborník, Ales	1086
Vogelsberger, M.	2098
Voigt, Matthias	944
Voldoire, Adrien	2039
Vollaire, Christian	2000
Vollmaier, Franz	1866
Vorontsov, Aleksey G.	47, 56
Votava, Martin	3314
Vu, Duc Tan	1693
Wada, Keiji	3358
Wallart, François	433, 736, 1106
Wang, Bo	450
Wang, Dian	677
Wang, Feng	2505
Wang, Fred	1972
Wang, Huai	1, 2888, 3295
Wang, Meiqi	1944
Wang, Qianggang	3385
Wang, Qiwu	1262
Wang, Tianqing	450
Wang, Xuehua	1410, 1435
Wang, Ziyue	443

Wankhede, Yugandhara H. 3081
Wasfi, Amr 2648
Watanabe, Hiroki 1380
Weicker, Martin 2049, 2153
Weimer, Julian 880, 1915
Weinert, Tristan 1727
Weiss, Sébastien 1793
Wernicke, Laurenz 2603
Weyh, Thomas 765
Wickramasinghc, Thilini 1470
Wijnands, Korneel 2350, 2358
Wilk, Andrzej 2265
Will, Frank 944
Winzer, Patrick 2182
Wolbank, T. 2098
Wondrak, W 3403
Wondrak, Wolfgang 644, 686
Wrona, Grzegorz 2457, 2996
Wu, Hailong 1693
Wu, Xiaohua 2135
Wunder, Bernd 143
Xi, Jiawen 342
Xia, Qingping 260
Xiang, Yusheng 1289
Xie, Jian 1289
Xin, Wei 370
Xiong, Weijing 1498
Xu, Chaoqun 718
Xu, Chen 937
Xu, Dianguo 450
Xu, Guo 1498
Xu, Junzhong 1224, 1233
Xu, Lie 1944
Xu, Xinwei 2656
Yakop, Netan 223
Yamada, Shota 1243
Yamashita, Daniela Yassuda 396
Yamazaki, Osamu 2675
Yamdeu, Mathias Tientcheu 3101
Yan, Xiaoxue 443
Yan, Zheng 1326
Yang, Bo 1944
Yang, Guang 637
Yang, Y 3237
Yang, Yongheng 2135, 2393, 2888
Yang, Zhiqing 153, 163
Yano, Takahiro 1177
Yao, Ran 2704
Yao, Wenli 2135
Ye, Shuaichen 2928
Ye, Zichao 1934
Yin, Tianxiang 937

Yu, Qihao 2350, 2358
Yu, Yong 450
Yu, Zhanqing 718
Yuasa, Hiroaki 105
Yüce, Firat 3366
Yuki, Kazuaki 2675
Yuratich, Michael A. 3146
Zacharias, Peter 571, 1813, 2535
Zafar, Talha 2773
Zanetti, E. 2293
Zaoskoufis, Konstantinos 755
Zare, Firuz 726, 1006
Zdanowski, Mariusz 2996
Zehelein, Matthias 880, 1915
Zeller, Valentin 1460
Zeng, Guang 3413
Zeng, Xianwu 342
Zhai, Dongling 718
Zhang, Haibo 1158, 1756
Zhang, He 1944
Zhang, Li 1128
Zhang, Peng 637
Zhang, Xiaokang 590
Zhang, Zhenbin 2386
Zhang, Ziqian 2505
Zhao, Biao 718
Zheng, Zedong 317
Zhou, Dao 2860, 2928
Zhou, Niancheng 3385
Zhu, Chunbo 637
Zhu, Q. 1306
Zhu, Yangming 450
Zhu, Yuanhao 326
Zhuo, Fang 2505
Zi-Fan, Li 1599
Ziegler, Philipp 1168, 1605, 1709
Zinchenko, Denys 2996
Zucuni, Jordan P. 2851
Zuolian, Liu 231

IEEE
445 Hoes Lane
Piscataway, NJ 08854-4141

ISBN 978-1-7281-9807-1